Vertebrate Endocrinology

Fifth Edition

science & technology books

Companion Web Site:

http://booksite.elsevier.com/9780123948151

Vertebrate Endocrinology, Fifth Edition

David O. Norris and James A. Carr

Resources for Professors:

- All figures from the book available as both Power Point slides and .jpeg files

ACADEMIC PRESS

Vertebrate Endocrinology

Fifth Edition

David O. Norris, Ph.D.
Professor Emeritus,
Department of Integrative Physiology,
University of Colorado at Boulder,
Colorado, USA

James A. Carr, Ph.D.
Professor,
Faculty Director, Joint Admission Medical Program,
Department of Biological Sciences,
Texas Tech University,
Lubbock, Texas, USA

AMSTERDAM • BOSTON • HEIDELBERG • LONDON • NEW YORK • OXFORD • PARIS
SAN DIEGO • SAN FRANCISCO • SINGAPORE • SYDNEY • TOKYO

Academic Press is an Imprint of Elsevier

Medical Illustrator: Wendy Beth Jackelow, MFA, CMI
Illustration preparation: Graphic World Illustration Studio
Acquiring Editor: Mara Conner
Development Editor: Megan Wickline
Project Managers: Karen East and Kirsty Halterman
Designer: Matthew Limbert

Photo Credits:
Frog: Male Glass Frog, *Hyalinobatrachium fleischmanni*, tending developing eggs. Photo courtesy of Jesse Delia
Copulating Bullsnakes, *Pituophis catenifer*. Photo courtesy of Alex Merrell
American Alligator Male, *Alligator mississippiensis*. Photo courtesy of Louis J. Guillette, Jr.
Tasmanian devil, *Sarcophilus harrisii*, a carnivorous marsupial mammal. Photo courtesy of Louis J. Guillette, Jr.
Male African Lion, *Panthera leo*. Photo courtesy of Louis J. Guillette, Jr.
Great White Shark, *Carcharodon carcharias*. Photo courtesy of Louis J. Guillette, Jr.
Laughing Kookaburra, *Dacelo novaeguineae*, of Australia. Photo courtesy of Louis J. Guillette, Jr.
Cedar waxwing, *Bombycilla cedrorum*. Photo courtesy of Louis J. Guillette, Jr.

Academic Press is an imprint of Elsevier
32 Jamestown Road, London NW1 7BY, UK
225 Wyman Street, Waltham, MA 02451, USA
525 B Street, Suite 1800, San Diego, CA 92101-4495, USA

Fifth edition 2013

British Library Cataloguing-in-Publication Data
A catalogue record for this book is available from the British Library

Library of Congress Cataloging-in-Publication Data
A catalog record for this book is available from the Library of Congress

ISBN: 978-0-12-394815-1

For information on all Academic Press publications
visit our website at elsevierdirect.com

Typeset by TNQ Books and Journals
www.tnq.co.in

Printed and bound by CPI Group (UK) Ltd, Croydon, CR0 4YY

We dedicate this volume to the many
Vertebrate Endocrinologists whose research efforts
are represented here as well as to the many students of endocrinology
that have helped us refine our understanding of this vast field and how
to communicate it effectively.
We wish to thank our editors, Megan Wickline and Mara Conner,
for their able assistance throughout this revision as well as the marvelous
illustrators who brought our ideas into colorful and informative visuals.
Additionally, we owe a great debt to our wives, Kay Norris and Deborah Carr,
for their patience and support during the preparation of this edition.

Contents

Preface to the Fifth Edition

Fifty years ago, I read two new books that shaped my professional life. The first was *A Textbook of Comparative Endocrinology* by Aubrey Gorbman and Howard A. Bern (1962) that contained virtually everything known about invertebrate and vertebrate endocrinology at that time. The second book was *Silent Spring* by Rachel Carson. Thus began my development as an "environmental endocrinologist" long before the term appeared in print. Since that time there has been exponential growth in our knowledge and it would take many volumes to adequately summarize it today. We thought the old lines drawn between the disciplines of endocrinology, neurobiology, and immunology were relatively clear 50 years ago but now they have become indistinct. All three fields employ chemical messengers affecting target cells through specific receptors, often involving the same messengers and receptors, and we now recognize that the nervous, endocrine, and immune systems each influence the activities of the others. Thus, a broader, integrated concept of chemical bioregulation for physiology and behavior has emerged that incorporates these three disciplines. Furthermore, a great variety of man-made chemicals related to products we use every day are appearing in aquatic and terrestrial environments at extremely low concentrations. These same chemicals are associated with endocrine, immunological, and neural disorders in laboratory rodents and very likely in humans. Similar impacts are being described for aquatic and terrestrial wildlife. Consequently, an understanding of endocrinology and other chemical regulatory mechanisms and their interactions is essential for our future on Earth.

Vertebrate or comparative "endocrinology" (i.e., bioregulation) has become too broad a field for any one person to keep abreast of, with hundreds of new publications appearing each year in the clinical and non-clinical literature. For example, in one month there may be (1) information about bioregulation in new species that influences our understanding of the evolution of endocrine systems, (2) discovery of new bioregulatory chemicals that contribute to coordination of activities within cells, between cells, and among organs and organ systems and even between organisms, (3) new understandings or treatments for a specific clinical disorder, and (4) new evidence of disruption of endocrine functions in humans and wildlife by chemicals previously believed to be safe.

In an attempt to make the content of *Vertebrate Endocrinology* more representative of this vast and rapidly changing field, I have added a co-author, Dr. James A. Carr, Professor of Biological Sciences at Texas Tech University to help with this task. Additionally, the illustrations (mostly in full color) have been completely redone to help students better conceptualize the major concepts. Rather than produce a compendium, we have attempted to provide a current overview and interpretation of chemical bioregulation of vertebrates with a basic framework for helping students access and understand the primary bioregulation literature. *Vertebrate Endocrinology* is organized to challenge advanced undergraduates and graduate students interested not only in mammalian or human endocrinology but those wishing to examine comparative vertebrate endocrinology as well.

David O. Norris

Professor Emeritus, Department of Integrative Physiology, University of Colorado at Boulder, Colorado, USA

James A. Carr

Professor, Faculty Director, Joint Admission Medical Program, Department of Biological Sciences, Texas Tech University, Lubbock, Texas, USA

Chapter 1

An Overview of Chemical Bioregulation in Vertebrates

Endocrinology as a scientific subdiscipline within physiology began a little over 100 years ago as the study of certain glands called **endocrine glands**, or glands of "internal secretion" that secreted their products into the blood. These secretions were called **hormones** (*hormon*, to stimulate or excite) because of their effects on distant **target cells**. Each hormone binds to a specific **receptor** molecule located in or on a target cell, and the resultant **hormone–receptor complex** causes a measurable change in the target cell. Many mechanisms employed in the vertebrate endocrine system have their counterparts among invertebrate animals as well as in microbial and botanical organisms. Originally a traditional field of specialization that focused on the endocrine glands and their secretions, endocrinology has expanded as a specialty within physiology and now deals with chemical regulation of virtually all biological phenomena in animals at the molecular, cellular, organism, and population levels of organization. The study of chemical regulation or **bioregulation** can be defined to include secretions of the endocrine system, nervous system, the immune system, and virtually all cells in the body that use chemical messengers to communicate with one another (Figure 1-1). The many secretions involved as internal chemical messengers can be called **bioregulators**.

Learning about the intricacies of how the activities of animals are regulated and coordinated by bioregulators is one of the most fascinating and complicated endeavors in biology. Every act that an animal performs is initiated, modulated, or blocked by bioregulators. Understanding the endocrine systems of invertebrates and vertebrate animals

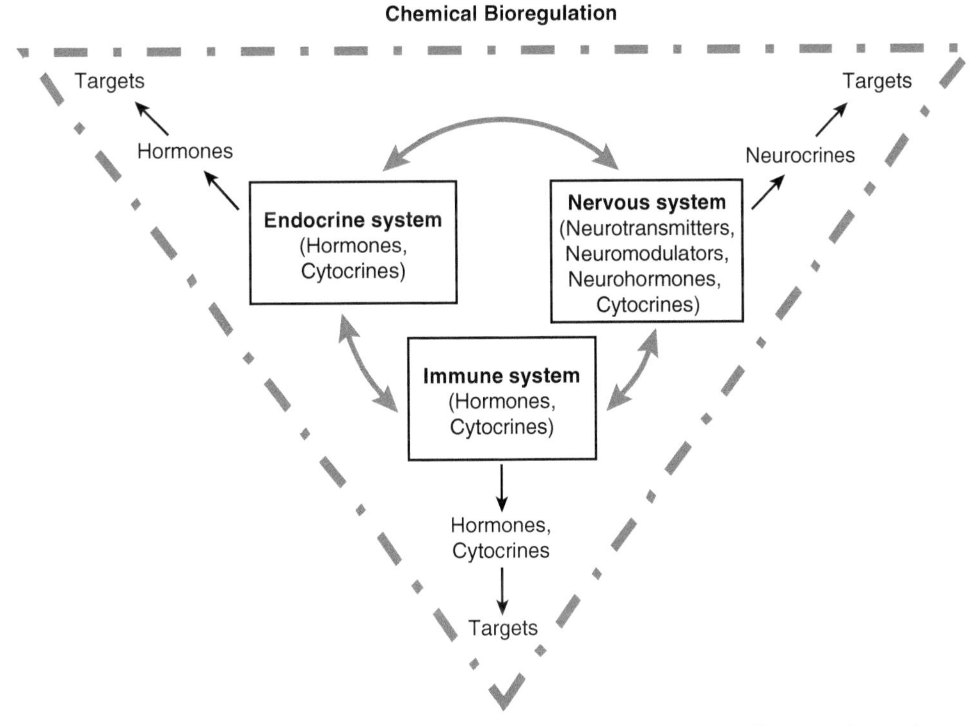

FIGURE 1-1 **Chemical bioregulation.** The endocrine system, nervous system, and immune system each secretes its own bioregulators: hormones, neurocrines, and cytocrines, respectively. However, all of these systems influence each other, and from a homeostatic viewpoint we can assume they function as one great bioregulatory system.

Vertebrate Endocrinology. http://dx.doi.org/10.1016/B978-0-12-394815-1.00001-X

is essential if we are ever to understand how bioregulatory mechanisms and systems evolved in animals and how they operate to maintain the vast array of living animal species. Furthermore, the continued health of each ecosystem depends on continued reproductive success of its component animal species.

Because the nervous system and endocrine system act together to integrate environmental information with bioregulation of physiology and behavior, a subdiscipline called **environmental endocrinology** has emerged within the more traditional approaches to endocrinology. This area of research not only is focused on natural environmental factors such as pheromones, behavior, light (e.g., photoperiod), and temperature but has also expanded to include effects of pesticides, heavy metals, and all manner of organic compounds added to the environment by human activities. These chemicals can alter normal bioregulatory mechanisms by mimicking or inhibiting the work of natural bioregulators. This interference of endocrine bioregulation by environmental pollutants and some natural chemicals is called **endocrine disruption**, and these chemicals are called **endocrine-disrupting chemicals (EDCs)** or **endocrine-active chemicals (EACs)**.

The application of the estrogenic mimic **diethylstilbestrol (DES)** to cattle and to pregnant women was recognized many years ago by John McLachlin and others as causing a disruption of normal endocrine function that has serious consequences for exposed animals and their offspring. Recent evidence suggests permanent effects of DES exposure that can be transmitted to subsequent generations in both laboratory mice and humans. The use of the pesticide **DDT** and **polychlorinated biphenyls (PCBs)** was banned in the United States when it was recognized that they were accumulating in the environment and were affecting the health of both wildlife and humans. However, it is now recognized that the potentially harmful nature of extremely low concentrations of estrogenic (feminizing and demasculinizing) pollutants in the environment represents a greater threat than ever before. Louis P. Guillette, Jr., in Florida, was the first to observe that a natural population of alligators exhibited abnormal sexual development after exposure in nature to low concentrations of DDT and its metabolite **DDE**. Soon, through the efforts of John Sumpter and numerous colleagues in the United Kingdom and Europe, evidence of reproductive disturbances surfaced that were a result of the exposure of fishes to incredibly low concentrations of estrogenic compounds present in wastewater effluents. We have now observed similar effects of wastewater effluents in the United States, as well. Knowledge of the intricate workings of the endocrine system is essential for understanding how such effects happen and to comprehend the enormous implications of these observations (see ahead).

I. THE COMPARATIVE VERTEBRATE APPROACH

Comparative vertebrate endocrinology emphasizes the evolution of bioregulatory systems and the discrete structures and bioregulators that constitute the vertebrate neuroendocrine and endocrine systems. Also of interest is the evolution of vertebrate endocrine glands and secretions from invertebrates. Vertebrate endocrine systems may be studied in a variety of ways. Some comparative endocrinologists may have a basic interest in fishes or reptiles and study their endocrine systems to better understand their ecology and evolution. Others may be interested in a specific phenomenon, such as aging, learning, or reproduction, and employ non-mammalian vertebrates as model systems in which to unravel basic mechanisms or evolutionary relationships. Certain procedures may be more readily performed on non-mammals, where processes are spread out stepwise over time, whereas in the mammal these steps all occur simultaneously. Comparative approaches may have direct applications for aiding our understanding of these same phenomena in mammals including humans. There are numerous examples of basic research in non-mammalian vertebrates that have had direct applications to human biology. For example, Spiedel made the first observation of neurosecretory neurons in the posterior spinal cord of fishes. Later, Ernst and Berta Scharrer extended this observation to the description of the hypothalamus–pituitary neuroendocrine system. Genetically controlled platyfish strains have provided a system for studying the bioregulation of aging and the reproductive system in relationship to genetic factors. Many species of fishes are used extensively as models for stress, growth, carcinogenesis, aging, and behavioral studies.

The toad urinary bladder was an excellent *in vitro* model for initial studies of the mechanism of action for the mineralocorticoid hormone aldosterone. Similarly, the amphibian ovarian follicle has provided an *in vitro* system for studying the bioregulation of oocyte maturation and the process of ovulation. Studies of non-mammals have been crucial for understanding mechanisms of tissue induction and chemical regulation of gene activity during embryonic development and differentiation. Numerous important hypotheses about development were first tested in amphibians. The discovery by Alberto Houssay that removal of the pituitary gland greatly reduced the severity of the removal of the pancreas in a toad became a model for studying diabetes mellitus using dogs. The origin of the gonadotropin-releasing hormone (GnRH) in neural ectoderm and the subsequent migration of GnRH neurons from the olfactory placode to the hypothalamus were first reported by Linda Muske and Frank Moore in an amphibian. Additional confirmation of the origin of the pituitary was performed using transplantation of normally pigmented toad embryonic cells into albino toad embryos and following their pigmented descendents.

Furthermore, ingenious experiments using chicken–quail chimaeras with cytologically distinct cell markers verified the neural origin of a number of endocrine cells. The lizard *Anolis carolinensis* was used by Richard Jones to study hormonal and neural control of the alternating pattern of ovulation by the paired ovaries, a phenomenon that also occurs in humans where it is more difficult to study. Avian systems have been used extensively for studies of development, neurobiology, immunology, cancer, and molecular genetics. F. Anne McNabb has used the quail as a model for the actions of a common pollutant, ammonium perchlorate, on the inhibition of thyroid function.

The discovery of pheromones among invertebrates and later studies in non-mammalian vertebrates as well as in mice and voles has led to new understandings in the roles of such secretions controlling mammalian reproductive and maternal behavior. Each of these systems has wide applicability to other vertebrates as well as to the mechanisms of hormone actions.

II. THE ORIGINS OF BIOREGULATION

Regulatory chemicals were probably essential for the survival of the first living cells both for coordination of internal events and for cell-to-cell interactions (Figure 1-2). Secretions that favored survival of the secreting cell no doubt led to further evolution of new chemical bioregulators. Thus, bioregulation probably had its origin in local secretions that affected nearby cells as well as bioregulators that affected internal cellular processes. The evolution of multicellular aggregates and eventually of multicellular organisms allowed further cell-to-cell types of bioregulation but more importantly also allowed for the eventual evolution of endocrine and neural bioregulation. The earliest appearance of neurons is noted in the most primitive of multicellular animals, the cnidarian invertebrates (Cnidaria), and these neurons secrete both peptides and non-peptides that function as typical neural and local bioregulators. The endocrine glands that secrete hormones did not appear until the type of internal transport mechanisms we call blood vascular systems developed. These bioregulatory systems have been termed **neuroendocrine systems** because they involve both neural and endocrine components. Complex neuroendocrine systems have evolved in annelids (Annelida), mollusks (Mollusca), insects, arachnids, and crustaceans (Arthropoda), as well as in the chordates (including the vertebrates).

III. CATEGORIES OF BIOREGULATORS

In addition to traditional endocrine regulation, we now recognize several other patterns of chemical bioregulation, as illustrated in Figure 1-3. The first of these involves the nervous system. Neurons produce **neurocrine** bioregulators called **neurotransmitters** or **neuromodulators** that are secreted into the synapses formed where they make direct connections with their target cells (typically other neurons, muscle cells, or gland cells). Once a neurocrine is bound to its receptor molecule, it brings about distinct changes in the postsynaptic cell. The parallelisms between endocrine cell and neuron, hormone and neurocrine, and target cell and postsynaptic cell are obvious. Only the location and chemical composition of the medium through which the bioregulator travels to reach its target cell separate hormones from neurocrines.

In the 1950s, Berta and Ernst Scharrer recognized that some neural regulators were released into the blood like hormones. These neural hormones were named **neurosecretions** or **neurohormones** to distinguish them from neurotransmitters, neuromodulators, and the traditional hormones. It seems that it was easier to formulate more definitions than to acknowledge that certain brain regions were also endocrine glands. Those neurons that secrete neurohormones are sometimes called **neurosecretory neurons** to distinguish them from the others. Although all neurocrines are actually neurosecretions, that latter term unaccountably has been reserved for the neurohormones.

Later, the separation of neural and endocrine systems became even more blurred for us when it was learned that some established hormones also are produced within the nervous system, where they function as neurotransmitters or neuromodulators (see Table 1-1). It soon became common knowledge that parts of the nervous system have control over certain portions of the endocrine system through direct innervation or via neurohormones. Likewise, hormones were seen to influence markedly not only the development of neural systems but also their activity. Hence, our concept of a neuroendocrine division within the endocrine system was established in physiology. Discovery of specific chemicals produced by diverse cellular types and released into extracellular fluids including blood, lymph, cerebral spinal fluid, and interstitial fluid has broadened the concept of chemical bioregulation still further to include cell-to-cell chemical communication that is not mediated via transport in the blood or at a synapse. Additional sources of hormones were discovered to come from traditionally non-endocrine tissues or organs such as the heart, the liver, adipose tissue, and skeletal muscle. The organization of the vertebrate endocrine system is outlined in Figure 1-4.

Chemical bioregulators were also discovered that are used for cell-to-cell communication within tissues. Such bioregulators that are secreted into extracellular fluid rather than the bloodstream are named "local hormones" or **cytocrines**. This category includes locally acting growth factors, mitogenic regulators, embryonic tissue-inducing substances, secretogogues (secretion-enhancing factors),

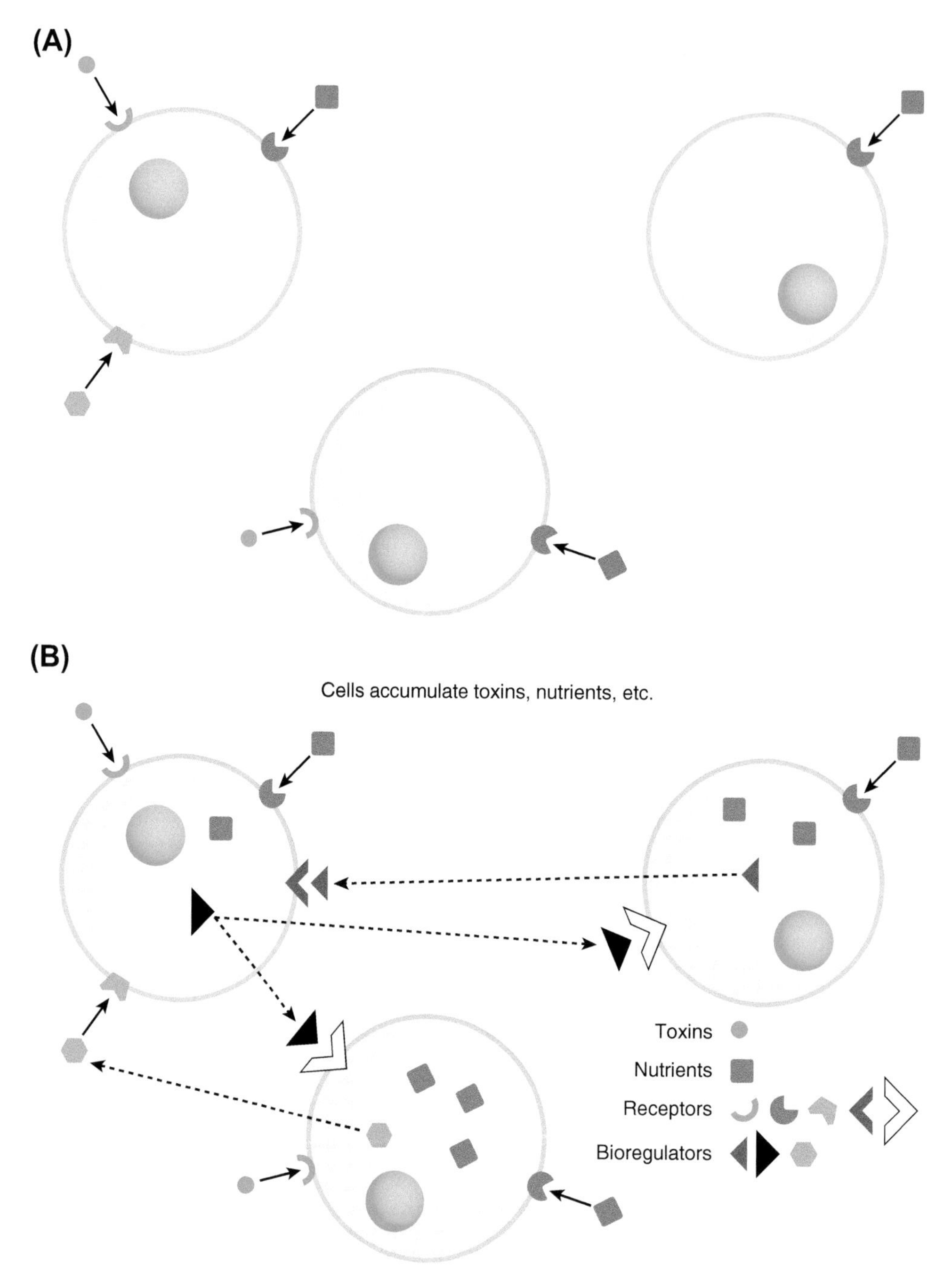

FIGURE 1-2 **Origins of chemical communication.** (A) Early cells living in the primordial seas developed "receptors" (here shown only on the cell membrane) for recognition of water-soluble toxins (blue circles) and nutrients (red squares) as well as internal "receptors" (not shown) for lipids that could readily pass through the membrane. Some of these "receptors" transferred these molecules into the cell for metabolism or detoxification. (B) In addition to accumulating molecules intracellularly, early cells also released special molecules into the environment that were detected via receptors on other cells and served as a mechanism for cell-to-cell communication. Various features of these ancient mechanisms for accumulation, detoxification, metabolism, and chemical communication have persisted in one form or another in all living cells to this day.

inhibitors, and immune regulators. If cytocrines also affect the emitting cell, they are sometimes termed **autocrines**. When they affect other cell types, they are called **paracrines**. However, both types of local cytocrine secretion into the general extracellular fluids are sometimes loosely referred to as **paracrine secretion**.

Chemicals released from neurons into the cerebrospinal fluid (CSF) do not quite fit the definition of neurohormone, as they are released into a filtrate of blood, but they do not quite fit the definition of paracrine secretion, either. For our purposes, we will refer to these bioregulators as neurohormones.

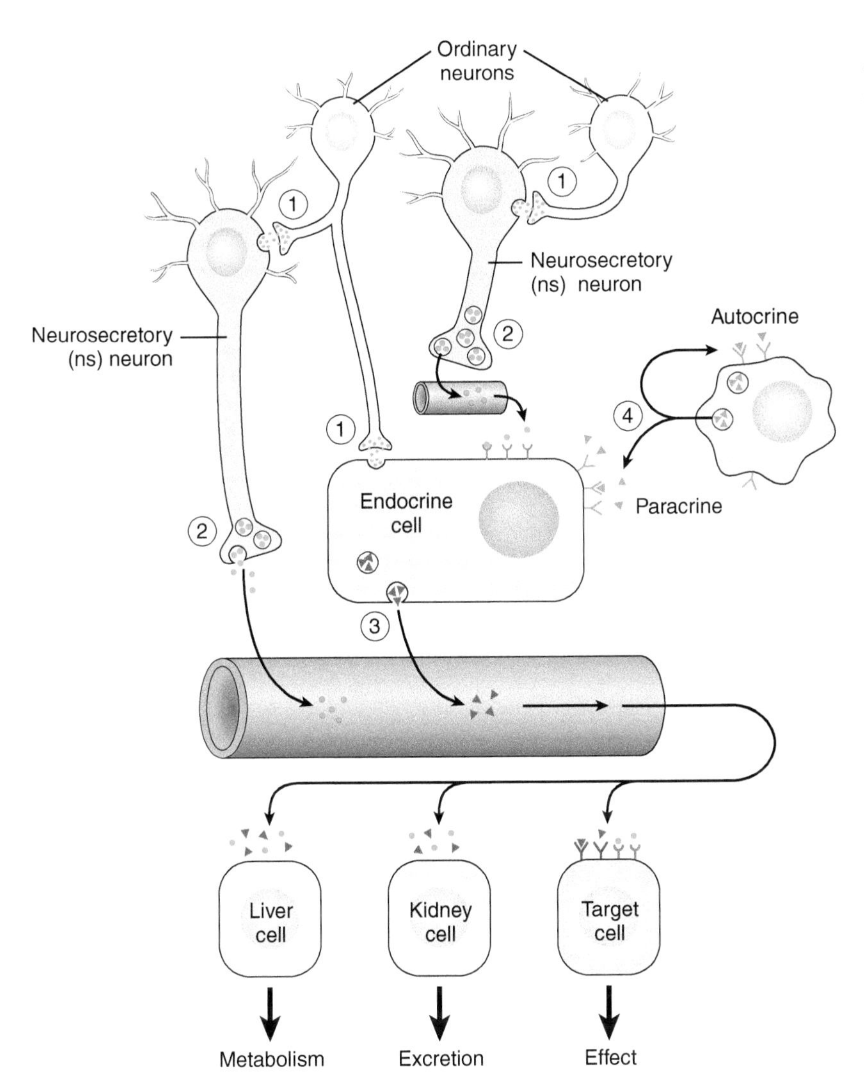

FIGURE 1-3 **Bioregulator organization.** Chemical communication involves neurocrines, including neurotransmitters or neuromodulators (1) and neurohormones (2), as well as hormones (3) and autocrine/paracrine regulators (4). The liver and kidney serve as major sites for the metabolism and excretion of bioregulators.

Intracellular chemical messengers that govern intracellular events have been called **intracrines**. These intracrine bioregulators would include chemicals such as the second messengers and transcription factors that are discussed in Chapter 3.

In its broadest sense, the study of bioregulation may include specific chemical messengers released by one organism into its environment that may affect the physiology or behavior of other individuals of that species or even of another species. A good name for these substances might have been "exocrines," but that term had already been assigned to the products of **exocrine glands** (e.g., salivary glands, sweat glands, mammary glands, and portions of the liver and pancreas) that secrete their products into ducts through which the secretions are conveyed to their sites of action on the free surface of the epithelium in such places as the digestive tract or the skin. These externally secreted bioregulators are called "ectohormones" or **semiochemicals** (*semio*, signal). Three subclasses of semiochemicals have been identified on functional bases. **Pheromones** are semiochemicals that act only on other members of the same species. **Primer pheromones** usually initiate a series of physiological events such as gonadal maturation. Signal or **releaser pheromones** trigger immediate behavioral responses such as sexual attraction or copulation. **Allelomones** are interspecific semiochemicals and are further separated into two types. If only the emitter of the semiochemical benefits from the effect on the other species, the allelomone is called an **allomone**. The well-known odor released by skunks is a dramatic example of an allomone that protects the skunk from would-be predators. When only the recipient species benefits, the allelomone may be termed a **kairomone**. The release of the simple metabolite l-lactate in the sweat of humans attracts female mosquitoes, which obtain the blood meal necessary for their reproduction. l-Lactate, then, could be classified as a kairomone, for there is no obvious benefit to the emitter and in fact it may harm the emitter.

In summary, intraorganismal bioregulation can be classified as endocrine (hormones), neurocrine (neurotransmitters, neuromodulators, neurohormones), paracrine (cytocrines, autocrines), or intracrine (intracellular regulatory messengers). Semiochemicals (pheromones and allelomones) are specialized for interorganismal communication. A listing of these types of bioregulators and their definitions are provided in Table 1-2.

Most bioregulators are peptides, proteins, or derivatives of amino acids. Some are lipids (e.g., steroids) and still others are nucleotides or nucleotide derivatives. A discussion of the chemical nature of regulators, how they are synthesized, how they produce their effects on targets, and how they are metabolized is the subject of Chapter 3. However, before examining these bioregulators more closely, we must consider some more general features of bioregulatory systems.

TABLE 1-1 Some Mammalian Neurocrine Bioregulators*

Class of Regulator	Example
Nonpeptides	Acetylcholine (ACh)
	Carbon monoxide (CO)
	Dopamine (DA)
	Epinephrine (E)
	γ-Aminobutyric acid (GABA)
	Glutamate
	Nitric oxide (NO)
	Norepinephrine (NE)
	Serotonin (5-HT)
Hypothalamic-releasing neuropeptides	Corticotropin-releasing hormone (CRH)
	Gonadotropin-releasing hormone (GnRH)
	Prolactin-releasing hormone (PRH)
	Somatostatin (SS or GHRIH)
	Thyrotropin-releasing hormone (TRH)
Other neuropeptides	Angiotensin II (Ang-II)
	Arginine vasopressin (AVP)
	Atrial natriuretic peptide (ANP)
	Brain natiuretic peptide (BNP)
	Cholecystokinin (CCK_8)
	Ghrelin
	Insulin
	Kisspeptin (Kp)
	Leptin
	Neuropeptide Y (NPY)
	Neuropeptide YY (PYY)
	peptide histidine isoleucine (PHI)
	Substance P (SP)
	Vasoactive intestinal peptide (VIP)

**Some of these molecules may function only as a neurotransmitter, neuromodulator, neurohormone, or paracrine regulator whereas others may perform multiple roles.*

A bioregulator has a distinct life history analogous to that of an organism, as illustrated in Figure 1-5 for a typical hormone. A bioregulator is born (synthesis), may exist first as an immature stage (inactive precursor molecule) that later is transformed to a mature form (metabolized to an active molecule), has a life (binds to receptors and produces an effect), and dies (is metabolized and/or excreted). It only lacks the ability to reproduce itself.

IV. GENERAL ORGANIZATION OF BIOREGULATORY SYSTEMS

As stated above, the endocrine system and the nervous system are the sources for most of the chemical messengers we have defined as bioregulators. Traditionally, the vertebrate neuroendocrine system includes the **brain** and the **pituitary gland** plus the classical endocrine glands they control: the **thyroid gland**, the paired **adrenal glands** and **gonads** (testes and ovaries), and the **liver**. In addition, there are **independent endocrine bioregulators**—that is, those not directly under the influence of the brain and/or pituitary. This would include the **adipose tissue, parathyroid glands**, **thymus**, heart, **endocrine pancreas,** organs of the **gastrointestinal tract, pineal gland,** and **kidney**. The major focus of this textbook is on the neuroendocrine systems of mammals and non-mammalian vertebrates. The independent endocrine glands are discussed for some special cases (for example, regulation of calcium homeostasis, which is discussed in Chapter 14) and when they interact with the bioregulators of the neuroendocrine system such as in the bioregulation of metabolism (Chapters 12 and 13).

The vertebrate neuroendocrine bioregulatory system (Figure 1-4) involves a major portion of the brain called the **hypothalamus**, that portion of the brain located directly above and anterior to the **pituitary gland**. The detailed organization and operation of this system are described in Chapters 4 and 5. Special groups of neurosecretory neurons in the hypothalamus produce a variety of neurohormones. Some of these neurohormones (i.e., releasing and release-inhibiting hormones) control the secretion by the pituitary gland of peptide and protein hormones called **tropic hormones** (tropic is derived from the Greek word *trophe*, referring to nutrition). These tropic hormones regulate the endocrine activities of the thyroid gland (secretes thyroid hormones; see Chapters 6 and 7), the adrenal cortex (secretes corticosteroids; Chapters 8 and 9), the gonads (secrete reproductive steroids; Chapters 10 and 11) and the liver (secretes an essential factor for growth; Chapter 4). Furthermore, some tropic hormones influence more general aspects of growth, metabolism, and reproduction and affect many non-endocrine target tissues. Additional neurohormones (**vasopressin** and **oxytocin**) are stored in part of the

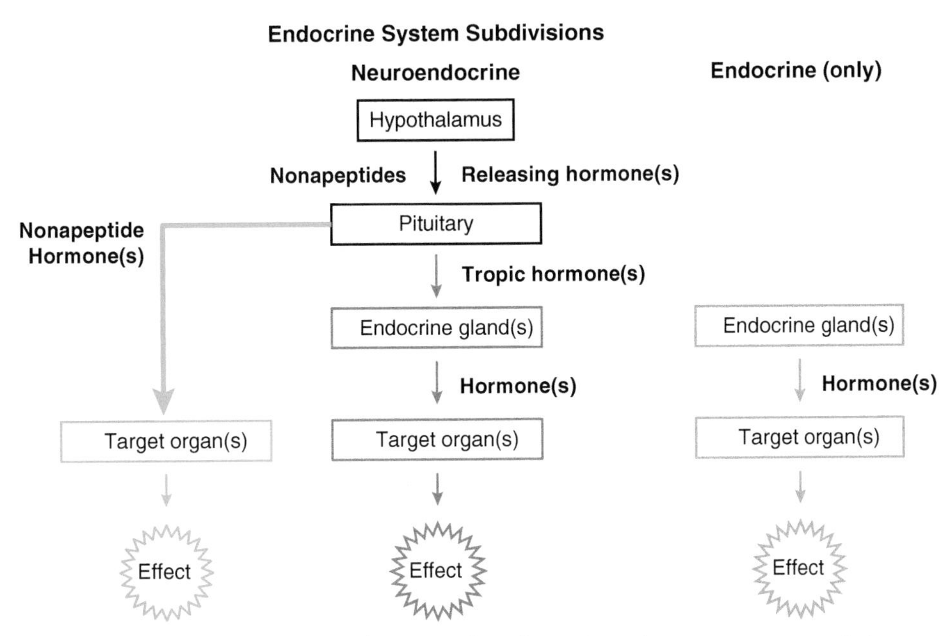

FIGURE 1-4 **Functional conceptualization of the endocrine system.** Input from other endogenous or exogenous factors can affect every level of regulation. The endocrine (only) glands typically respond to levels of chemicals in the blood, and, although innervated, their secretion is not directly controlled by the nervous system. Nonapeptide targets include the kidney and mammary gland as well as reproductive and vascular smooth muscle. Endocrine glands controlled by tropic hormones from the pituitary include the gonads, thyroid, adrenal cortex, and liver. Endocrine-only glands include the parathyroids, kidneys, heart, adipose tissue, and others.

pituitary until they are needed (Chapters 4 and 5). Vasopressin influences kidney function, blood pressure, and reproductive behavior, and oxytocin plays many reproductive roles related to both physiology and behavior.

Table 1-3 is a partial listing of bioregulators that are discussed in this book. In addition to their names and abbreviations, their sources, targets, and general effects on the targets are provided.

TABLE 1-2 **Types of Bioregulators**

Agent	Description	Examples
Neurotransmitter	Secreted by neurons into synaptic space	Acetylcholine, dopamine, substance P, GABA
Neuromodulator	Secreted by neurons into synaptic space; modulates sensitivity of postsynaptic cell to other neurotransmitters	Endorphins and various other neuropeptides
Neurohormone	Secreted by neurons into the blood or CSF; may be stored in neurohemal organ prior to release	TRH, CRH, oxytocin, dopamine
Hormone	Secreted by specialized nonneural cells into the blood	Thyroxine, GH, insulin
Cytocrine	Secreted by cells into the surrounding extracellular fluid; these local regulators typically travel short distances to nearby target cells	Somatostatin, norepinephrine
Paracrine	Secreted by cells and affect other cell types	Embryonic inducers, somatostatin, interleukins
Autocrine	Secreted by cells and affect emitting cell/type	Mitogenic agents, interleukins
Intracrine	Intracellular messengers; typically mediators of other regulators that bind to membrane receptors	cAMP, DAG, IP_3, cGMP, calmodulin, calcium ions
Semiochemical	Secreted into environment	Pheromones, allelomones

See Appendix A for explanation of abbreviations.

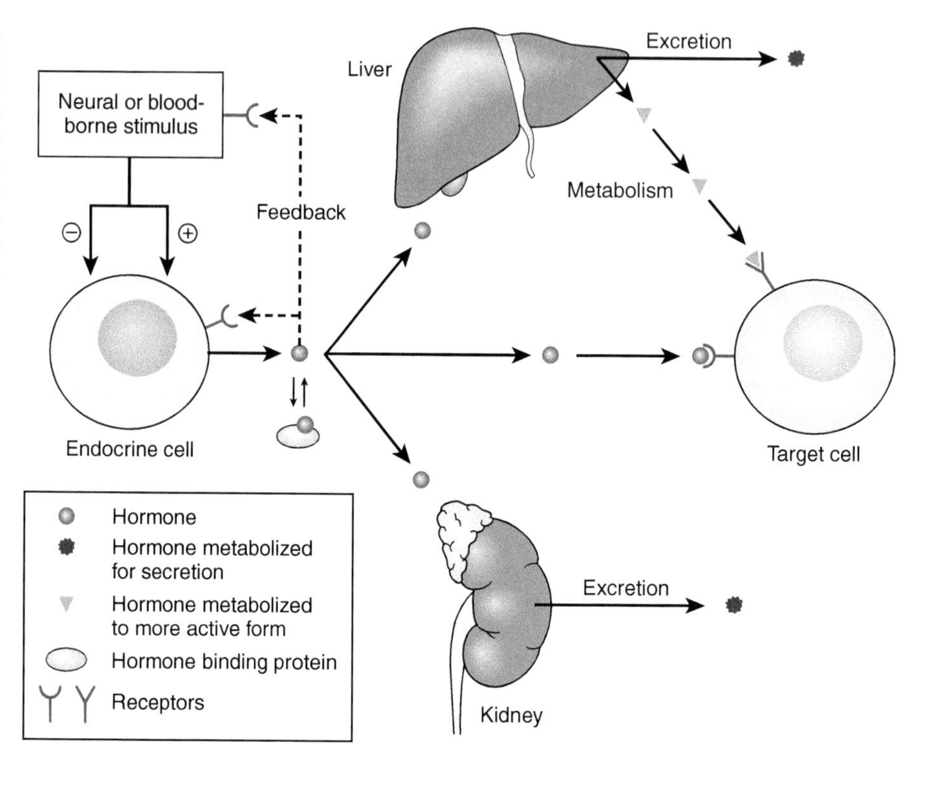

FIGURE 1-5 **Life history of a hormone.** A hormone is "born" in an endocrine cell and spends its short life "free" in the blood or bound to binding proteins. It may be metabolized and/or excreted ("die") before or after it binds to a target cell where it causes changes that result in its characteristic effect. In some cases, the hormone is secreted in an inactive form and must be metabolized to an active form before it can bind to its receptor and produce an effect.

V. CELL AND TISSUE ORGANIZATION OF BIOREGULATORY SYSTEMS

Endocrine and neuroendocrine cells can be identified easily on the basis of their cytological features as specialized secretory cells (Figure 1-6). Peptide-secreting cells have well-developed rough endoplasmic reticula and typically contain many protein-filled storage granules or vesicles (see Appendix F for a brief description of cellular structures). Mitochondria of peptide-secreting cells have flat, platelike cristae. The morphology and content of the protein storage granules may be used to differentiate specific types of endocrine cells. For example, the various tropic hormone-secreting cells of the anterior pituitary can be partially identified by the differential sizes of their storage granules or by special immunochemical methods (see Chapters 2 and 4). In contrast, steroid-secreting cells have well-developed smooth endoplasmic reticula, and their mitochondria have tubular cristae (compare steroid- and peptide-secreting cells as shown in Figure 1-6). Steroids are usually not stored in their cells of origin but lipid droplets containing cholesterol, the precursor steroid for their synthesis, are commonly observed.

Not only are neurosecretory neurons and regular neurons specialized elongated cells that are readily identifiable but they also contain discrete **synaptic vesicles** containing neurocrine products in their axonal tips that characterize them as secretory cells. Neurosecretory neurons tend to be larger than ordinary neurons. Neurons secreting peptides contain larger, dense granules than those secreting non-peptides such as catecholamines or acetylcholine. For example, the synaptic vesicles for acetylcholine are 30 to 45 nm in diameter and those for norepinephrine are about 70 nm, but peptide-containing vesicles are 100 to 300 nm.

In the central nervous system, the cell bodies of neurons are localized in groupings called nuclei, and their axons often form specific tracts connecting to other nuclei, blood vessels, or the cerebrospinal fluid, or they may exit from the central nervous system as nerves. In fact, neurosecretory neurons were first characterized as unique because their cell bodies in **neurosecretory nuclei** and axons in **neurosecretory tracts** contain materials that stain with unique dyes, distinguishing these cells from ordinary neurons. However, these general methods usually did not distinguish between different kinds of neurosecretory neurons. Modern immunological techniques now allow us to identify each type of neurosecretory or ordinary neuron with respect to its particular secretions (see Chapter 2).

Another factor that helped in the early identification of endocrine cells was their anatomical relationship to one another in forming discrete tissues (see Figure 1-7). Many endocrine cells are specialized epithelial cells that tend to be clumped in groups that are organized in one of the following ways (see Appendix F for descriptions of epithelia and other tissue types). The most common

TABLE 1-3 **The Mammalian Endocrine System: Major Secretions[a] and Actions**

Source[b] and Secretions	Target	Action
Anterior pituitary: produces tropic hormones		
Glycoprotein tropic hormones		
Thyrotropin (thyroid-stimulating hormone; TSH)	Thyroid gland	Synthesis and release of thyroid hormones
Luteinizing hormone (lutropin, LH)	Gonads	Androgen synthesis; progesterone synthesis; gamete release
Follicle-stimulating hormone (follitropin, FSH)	Gonads	Gamete formation; estrogen synthesis
Nonglycoprotein tropic hormones		
Growth hormone (somatotropin, GH)	Liver, cartilage, bone, adipose, muscle	Synthesis of IGF, proteins
Prolactin (mammotropin, PRL)	Mammary glands, epididymus	Synthesis of proteins
Corticotropin (adrenocorticostimulating hormone, ACTH)	Adrenal cortex	Synthesis of corticosteroids
Melanotropin (melanocyte- or melanophore-stimulating hormone, MSH)	Melanin-producing cells	Synthesis of melanin
Others products		
Tuberalin	Prolactin cells	Release of prolactin
Pituitary adenylate cyclase activating peptide (PACAP)	Numerous pituitary cells	Activates cAMP and enhances response to releasing hormones
Hypothalamus: neurosecretory nuclei produce neurohormones		
Hypothalamic-releasing hormones		
Thyrotropin-releasing hormone (TRH)	Anterior pituitary	Releases TSH
Gonadotropin-releasing hormone (GnRH)	Anterior pituitary	Releases LH/FSH
Corticotropin-releasing hormone (CRH)	Anterior pituitary	Releases ACTH
Somatostatin (GH-RIH or SST)	Anterior pituitary	Inhibits GH release
Growth hormone-releasing hormone (somatocrinin, GHRH)	Anterior pituitary	Releases GH
Prolactin release-inhibiting hormone (PRIH)	Anterior pituitary	Inhibits PRL release
Dopamine (PRL release-inhibiting hormone, PRIH)	Anterior pituitary	Releases PRL
Dopamine (MSH release-inhibiting hormone, MRIH)	Anterior pituitary	Inhibits MSH release
Other neurohormones		
Arginine vasopressin[c] (antidiuretic hormone, AVP)	Kidney	Water reabsorption
Oxytocin (OXY)	Uterus, vas deferens	Smooth muscle contraction
Endorphins/enkephalins	Pain neurons	Desensitizes pain neurons
Thyroid gland		
Thyroid hormones		
Thyroxine (T_4) and triiodothyronine (T_3)	Most tissues	Increase metabolic rate, control development and differentiation
Calcitonin (thyrocalcitonin, CT)[e]	Bone	Prevents resorption caused by parathyroid hormone

(Continued)

TABLE 1-3 The Mammalian Endocrine System: Major Secretions[a] and Actions—cont'd

Source[b] and Secretions	Target	Action
Gonads		
Ovary		
Estrogens (e.g., estradiol)	Primary and secondary sexual structures	Stimulate development
	Brain	Reproductive behavior
Progesterone	Uterus	Stimulates secretion by uterine glands
Inhibin	Anterior pituitary	Blocks FSH release
Testis		
Androgens (e.g., testosterone)	Primary and secondary sexual structures	Stimulates development and secretion
	Brain	Reproductive behavior[d]
Inhibin	Anterior pituitary	Blocks FSH release
Adrenal gland		
Adrenal cortex		
Aldosterone (A)	Kidney	Sodium reabsorption; potassium secretion into urine
Corticosterone (B)/Cortisol (F)	Liver, muscle	Conversion of protein into carbohydrates
Adrenal medulla		
Epinephrine/norepinephrine	Liver, muscle	Glycogen breakdown to glucose
Parathyroid gland		
Parathyroid hormone (parathormone, PTH)	Bone	Bone resorption or growth
	Kidney	Calcium reabsorption and phosphate secretion into urine
Endocrine pancreas		Increased glycogen storage
Insulin	Liver, Muscle	Increased glucose and amino acid uptake
	Adipose tissue	Inhibits fat hydrolysis
Glucagon	Liver, adipose tissue	Antiinsulin actions
Pancreatic polypeptide	Colon	Enhances muscle contraction
Somatostatin (paracrine substance)	Endocrine pancreas	Blocks release of pancreatic hormones
Gastrointestinal system		
Stomach		
Gastrin	Gastric glands of stomach	Stimulates acid secretion
Ghrelin	Brain	Stimulates feeding
Small intestine		
Secretin	Exocrine pancreas	Release of basic juice into duodenum
Cholecystokinin (CCK; same as pancreozymin-cholecystockinin, PZCCK)	Exocrine pancreas	Release of enzymes into duodenum
	Gallbladder	Contraction to eject bile into duodenum

TABLE 1-3 The Mammalian Endocrine System: Major Secretions[a] and Actions—cont'd

Source[b] and Secretions	Target	Action
Gastrin-releasing peptide	Stomach gastrin cells	Release of gastrin
Glucose-dependent insulinotropic peptide (gastric inhibitory peptide, GIP)	Endocrine pancreas	Release of insulin
Motilin	Stomach	Stimulates pepsinogen secretion and gastric motility
Somatostatin (paracrine action)	Small intestine	Inhibits release of other regulators
Vasoactive intestinal peptide (VIP)	Visceral blood vessels	Increases blood flow to
Liver		
Insulin-like growth factors (IGF-I, IGF-II)	Many tissues	Mitogenic effects
Adipose tissue		
Leptin	Brain	Inhibits feeding
Kidney		
Erythropoietin	Bone marrow	Stimulates RBC formation
Renin	Renin substrate in blood	Produces angiotensin
1,25-dihydroxycholecalciferol (1,25-DHC)[f]	Small intestine	Stimulates calcium absorption
Pineal gland		
Melatonin (neurohormone)	Brain	Controls thyroid, adrenal, and reproductive events
Immune system		
Thymus		
Thymosins	Lymphocyte-producing tissue	Production of lymphocytes
Macrophages/lymphocytes		
Interleukin 1 (autocrine/cytocrine)	Helper T cell	Activation
Interleukin 2 (autocrine/cytocrine)	Cytotoxic T cell	Activation

[a]In some cases, closely related molecular forms may be present and will be discussed at the appropriate time.
[b]Alternate names are given in parentheses, along with the most common abbreviation.
[c]Some mammals may rely on a different nonapeptide (e.g., lysine vasopressin or phenypressin; see Chapter 4).
[d]Secretory cells derived from ultimobranchial gland in mammals become incorporated into the thyroid (see Chapter 14).
[e]May require conversion into estrogens within certain brain target cells before effect is observed.
[f]Cholecalciferol is made in skin and converted in liver to precursor kidney uses to make 1,25-DHC (see Chapter 14).

orientation of secretory cells is to form folded sheets or **cords** of cells as seen in the pituitary gland or the adrenal cortex. In a few cases, the cells may form a spherical mass of one cell layer surrounding a fluid-filled space or lumen. This arrangement is termed a **follicle** and occurs in the thyroid gland of all vertebrates and in the pituitaries of some vertebrates. The lumen provides a unique storage site for secretions of the follicular cells. Sometimes, the endocrine cells will be separated into scattered clumps or **islets** of a few cells. Mixed islets containing several secretory cell types are best known in the pancreas, where they are called the **islets of Langerhans**.

Many cells that secrete bioregulators are not histologically distinct (i.e., do not form discrete tissues) and were not identified until precise immunochemical techniques were developed. One of the reasons why it took so long to identify the sources of gastrointestinal hormones was the

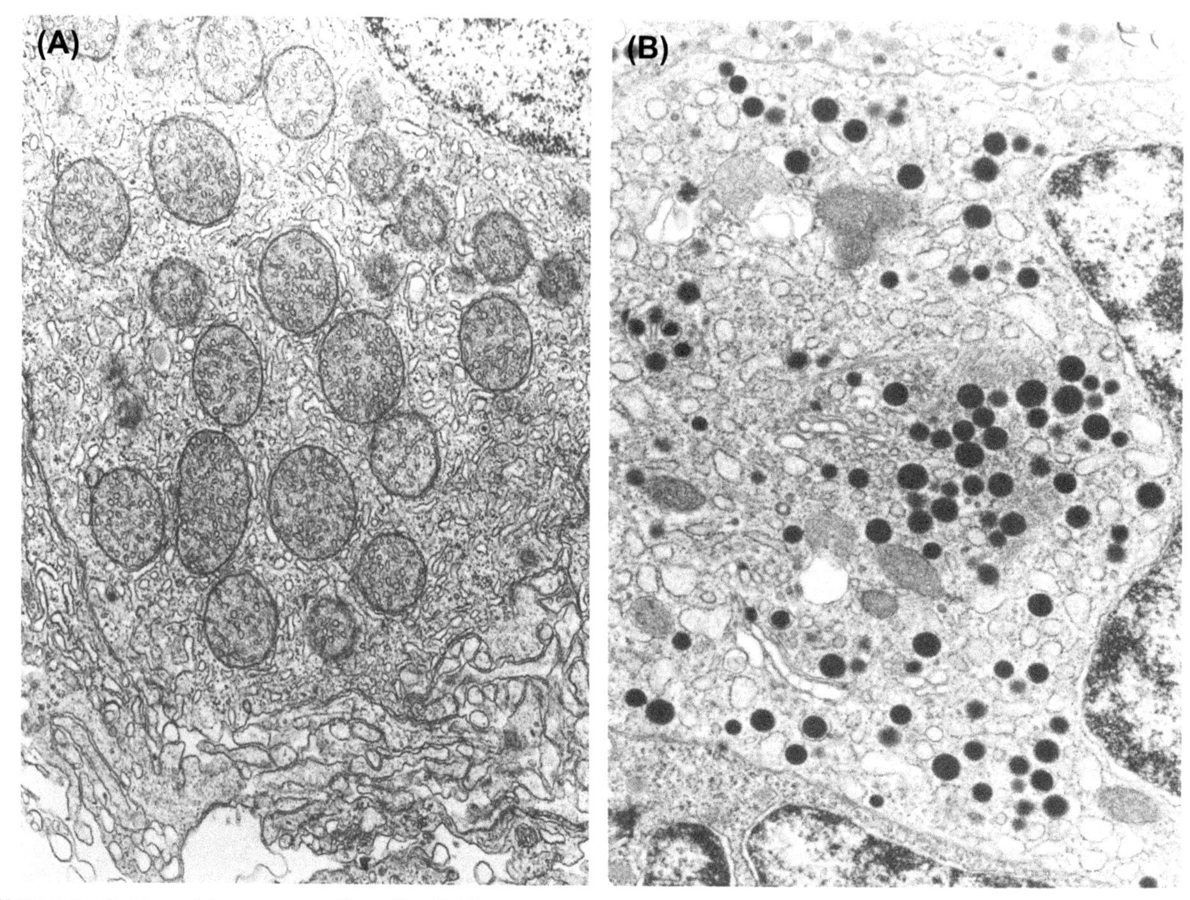

FIGURE 1-6 Cytology of hormone-secreting cells. (A) Microscopic appearance of a steroid-secreting cell. These adrenocortical cells, from juvenile salmon, secrete the steroid cortisol exhibit mitochondria with tubular cristae and an abundance of smooth endoplasmic reticulum. (B) A growth-hormone-secreting cell from the coho salmon (*Oncorhynchus kisutch*) showing dense secretory granules, well-developed Golgi apparatus, and mitochondria with plate-like cristae. *(Courtesy of Howard A. Bern and Richard Nishioka, University of California, Berkeley.)*

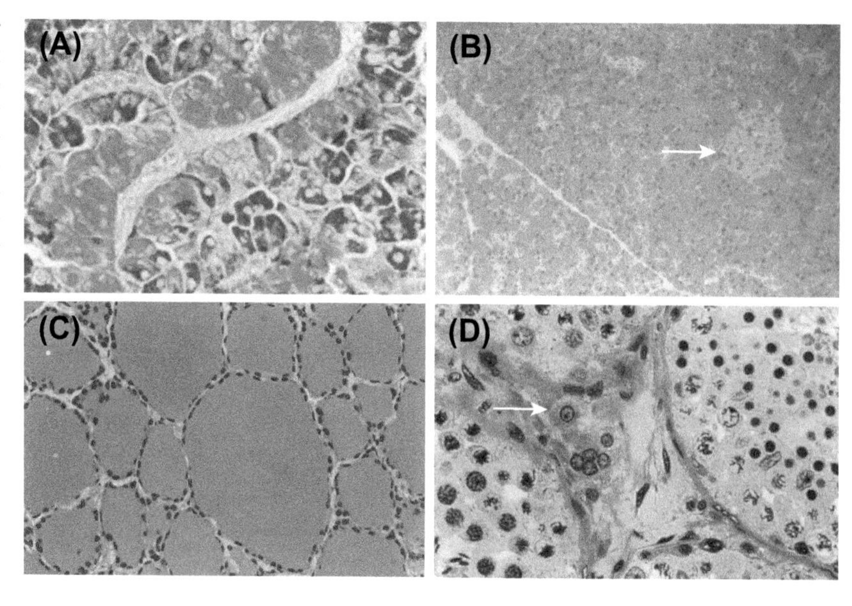

FIGURE 1-7 Hormone-secreting cells appear in many formations. (A) Cords of cells secreting growth hormone (orange) and gonadotropins (blue) in a pituitary gland. (B) Islet of insulin-secreting cells (arrow) embedded within the darker stained exocrine pancreas. (C) A collection of follicles (consisting of an epithelium surrounding a fluid-filled lumen) from a thyroid gland showing a thin epithelium and pink colloid (a protein suspension) filling the lumen of the follicle. (D) Isolated clusters of testosterone-secreting interstitial cells (arrow) located between seminiferous tubules in a testis.

TABLE 1-4 **Cellular Patterns of Secretion**

Secretory Pattern	Description	Example
Endocrine	Product secreted into the blood for transport internally to target tissues	Hormones
Exocrine	Product secreted into a duct that opens onto an external or internal surface	Sweat
Exocytosis	Product released from secretory cell via a process essentially the reverse of endocytosis	Peptide hormone release
Merocrine	Product secreted without visible damage to the secretory cell (involves exocytosis)	Thyroxine secretion
Apocrine	Product released by sloughing of "outer" or apical portion of secretory cell	Mammary gland milk
Holocrine	Product released through cell death and lysis	Sebaceous gland secretion
Cytogenous	Release of whole, viable cells	Spermatozoa

tendency for these secretory types to occur as **isolated endocrine cells** mixed in with many other cell types in the stomach and intestinal walls (see Chapter 12).

Another critical feature of neuroendocrine organization is the presence of an extensive vascular supply for endocrine cells and neurosecretory neurons. Endocrine glands typically are highly vascularized such that no secretory cell is far from a blood vessel. The axonal endings of many neurosecretory neurons terminate collectively in masses of capillaries to form what is called a **neurohemal organ**. These neurosecretory neurons release their neurohormones into the blood that flows through the neurohemal organ. Some neurosecretory neurons that release their products into the cerebrospinal fluid do not form axonal aggregates at common release sites.

Most regulatory cells employ **merocrine secretion**, where secretory products are released by exocytosis with no damage to the cell. In **apocrine secretion**, the apical portion or tip of the cell is sloughed along with stored secretions, whereas **holocrine secretion** involves lysis and death of the secretory cell. These latter two patterns are more characteristic of exocrine glands such as the mammary gland (apocrine) or the sebaceous glands of the skin (holocrine). **Cytogenous secretion** is the release of entire cells, such as sperm released from testes or ova released from ovaries. These secretory patterns are summarized in Table 1-4.

VI. HOMEOSTASIS

Bioregulatory mechanisms are the bases for controlling all physiology and behavior. It is through these mechanisms that homeostatic balance and survival in a harsh and dangerous environment are possible. Although Claude Bernard formulated the concept of homeostasis in the 19th century, it was the American physiologist Walter B. Cannon who in 1929 coined the term **homeostasis** to describe balanced physiological systems operating in the organism to maintain a dynamic equilibrium—that is, a relatively constant steady state maintained within certain tolerable limits. In Cannon's words:[1]

When we consider the extreme instability of our bodily structure, its readiness for disturbance by the slightest application of external forces and the rapid onset of its decomposition as soon as favoring circumstances are withdrawn, its persistence through many decades seems almost miraculous. The wonder increases when we realize that the system is open, engaging in free exchange with the outer world, and that the structure itself is not permanent but is being continuously broken down by the wear and tear of action, and is continuously built up again by processes of repair…

The constant conditions which are maintained in the body might be termed equilibria. That word, however, has come to have fairly exact meaning as applied to relatively simple physico-chemical states, in closed systems, where known forces are balanced. The coordinated physiological processes which maintain most of the steady states in the organism are so complex and so peculiar to living beings—involving, as they may, the brain and nerves, the heart, lungs, kidneys and spleen, all working cooperatively—that I have suggested a special designation for these states, homeostasis. The word does not imply something immobile, a stagnation. It means a condition—a condition which may vary, but which is relatively constant.

Cannon's original formulation of the homeostatic mechanism emphasized the maintenance of blood parameters such as osmotic pressure, volume, hydrostatic pressure, and levels of various simple chemicals such as calcium,

1. Cannon, W.B., *The Wisdom of the Body*, W.W. Norton, New York, 1932, pp. 20, 24.

sodium, and glucose. Cannon's viewpoint can be expanded to include all manner of physiological bioregulation at the level of the organism as well as at the molecular and cellular level. The concept of **allostasis** was developed to explain the complex role of bioregulators in integrating the response of many organ systems to changes in the environment (see Chapter 8).

When attempting to comprehend physiological systems, it is helpful to employ simplified models that simulate the various components of the system in a way that is easy to grasp and at the same time provide insights into how the system works as well as predictions on how it will respond to disturbances. In the following paragraphs, we will consider a very simple model of a basic bioregulatory mechanism operating for all physiological systems and provide some insight on how to use this model to understand complicated, integrated endocrine systems such as those discussed in later chapters.

A. A Homeostatic Reflex Model

In this model, **information (I)** is any stimulus that can provide quantitative or qualitative cues detectable in some way by the system. The basic model is depicted in Figure 1-8. The information is detected by a **receptor (R)** or transducer of some sort that translates (transduces) this information into the language of the bioregulatory system. For example, the opening of voltage-gated sodium channels may allow sodium ions to enter a neuron, depolarizing the cell membrane and inducing an action potential that in turn

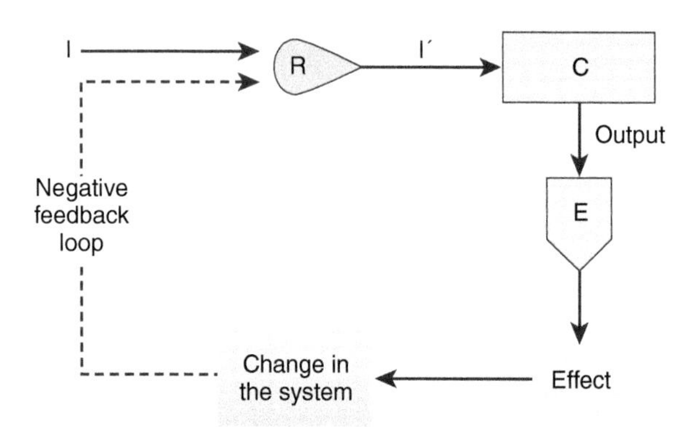

FIGURE 1-8 **A homeostatic model.** This simple homeostatic mechanism could represent a single cell as well as an endocrine or neurocrine unit. The mechanism involves the detection of information (I) by a receptor (R) that converts the information into biologically relevant cues or input (I′) and transmits this to the controller (C). The controller compares the input to a programmed set point and makes physiological adjustments as needed by producing output. This output travels by intracellular pathways (intracrines), neural axons (neurocrines), blood (hormones), or even extracellular fluid (autocrines/paracrines) to effectors (E) that in turn cause a change in the system that also feeds back via the same or different receptors to alert the controller that a change has occurred. Because this type of feedback drives the system toward the set point, it is called *negative feedback*.

causes a nerve impulse to be generated. The transduced information is now called **input (I′)** and is translocated to an integrating center called the **controller (C)**. The controller uses a preprogrammed set of instructions to compare the input with a **set point** and determines whether any adjustments are warranted. If the controller ascertains adjustments are needed to maintain or regain homeostatic balance, it will direct a message called **output (O)** (e.g., the nerve impulse and release of a neurotransmitter) to one or more **effectors (E)** (e.g., a postsynaptic muscle cell) which will perform some specific action (**effect, E_f**) which, in turn, will bring about corrective changes in the system (contraction of the muscle cell). The responsiveness of the controller may be influenced by input received from other homeostatic bioregulators. These additional inputs may enhance or reduce the output of the controller through altering its sensitivity to other input or by adjusting the set point.

Corrective changes signaled in response to output from the controller will alter the nature of the information originally perceived by the receptors. If the response has been sufficient and appropriate, the controller will be notified immediately by new information detected by receptors through a pathway called **feedback**. If the response was insufficient, the controller may increase its output to elevate the effector response. If an overcorrection or **overshoot** has occurred, the controller will alter its output accordingly and may even generate new output to other effectors to bring the system into line with the preprogrammed set point. Therefore, the set point represents the optimal physiological condition. The feedback loop that drives a physiological system toward the preprogrammed set point is called **negative feedback**. Any disturbance regardless of the direction of the disturbance (i.e., "up" or "down") causes a homeostatic reflex to be activated to maintain homeostasis. For example, when the blood level of a hormone decreases below its set point, it stimulates mechanisms that will increase its blood level back to the set point. Similarly, if the level of the hormone is greater than the set point, the mechanisms for its secretion will be decreased or inhibited by the controller until the blood level falls back to the set point. In either case, this would be considered negative feedback because it drives the system toward the set point from either above or below.

Consider a controlled-temperature room as a physical model of a simple control system similar to what has been described for a biological system. The programmable thermostat represents both the receptor and controller components; an air conditioner and a heater are the effectors. Mechanical deformations produced in a bimetal strip (receptor) exposed to the air temperature (information) of the room are transduced into electrical current (input). The controller compares this input with the preprogrammed temperature (set point) and electrically turns on or off (output) the appropriate effector to maintain a constant temperature (homeostasis). The new air temperature will be detected by the same receptor and new input will be sent to the controller that will continue to make adjustments *up or down* as needed to drive the system toward the set point (negative feedback).

Under some conditions, feedback may drive a system away from the preprogrammed condition, usually to a higher level but sometimes to a lower level. Such a feedback loop is called **positive feedback**, and it drives a system to a different level of activity. Positive feedback is invoked where a rapid change is required, may be associated with an emergency type response, or might be responsible for short-term adaptations to complete a series of changes. The rapid influx of sodium during generation of an action potential, the physiological stress response (Chapter 8), certain events during ovulation, and the induction of labor causing birth in mammals (Chapter 10) are all examples of events that employ positive feedback. In general, positive feedback is important for short-term events but is detrimental over longer time periods and can lead to disease or even the death of the animal if it persists. In contrast, long-term negative feedback generally is advantageous to long-term survival as it helps maintain homeostasis in the face of environmental or internal changes. Negative feedback is the most common type of feedback in physiological systems.

In certain instances, changes in a regulated variable are anticipated through **feedforward regulation**, which accelerates homeostatic responses and minimizes fluctuations in the regulated variable. For example, the regulation of internal body temperature involves a classical negative-feedback loop based on the temperature of the blood flowing to the brain. However, changes in body surface temperature before internal temperatures are affected can send additional neural input to the brain that begins making appropriate adjustments in temperature production and conservation or dissipation of heat to ward off changes in internal body temperature predicted by the input from the skin. Secretions invoked by the digestive system following appearance of glucose in the small intestine result in secretion of insulin from the endocrine pancreas even before blood sugar has become elevated, the normal homeostatic stimulus causing insulin release. Thus, increased levels of insulin appear ready to direct glucose entering the blood from the intestine into cells for use or storage. These are examples of feedforward regulation.

To apply this homeostatic reflex model to any bioregulatory mechanism, one must ask a series of simple questions. First of all, on what information does the system cue? What are the receptors and where are they located? What sort of input is generated by this stimulus and how is it conducted to the controller? What is the controller and where is it located? What is the set point? What sort of output is generated? What are the effectors and what effects are produced? What changes in the system result and how does feedback occur? Is there negative, positive, or feed-forward regulation involved?

When examining chemically regulated systems, you will soon discover that complicated responses involve the integration of many different simple homeostatic reflexes. For example, the "controller" for one system may actually be the "effector" of another homeostatic reflex, and there may be pathways that modulate the responsiveness of a controller or effector to other stimuli. An excellent example of the integration of individual reflexes is the **neuroendocrine reflex**.

A typical vertebrate neuroendocrine reflex is represented by the adrenal endocrine axis in Figure 1-9. The liver cell can be considered an effector (target cell) whose activity is regulated by a steroid hormone (e.g., cortisol) from the adrenal cortex. This would make the adrenal cortex the controller, but the adrenal cortex is also an effector for the anterior pituitary that controls its secretion of cortisol through output of a tropic hormone, corticotropin or ACTH. In turn, the anterior pituitary is also an effector for a neurohormone (corticotropin-releasing hormone, CRH) from the hypothalamus (arbitrarily labeled the main controller in this diagram). Only the major or primary feedback loop of cortisol is shown, but other feedback loops may be operating in this system (see Chapters 4 and 8). In this neuroendocrine reflex, we have indicated that the hypothalamic controller receives information via the blood but it also can be affected by information accumulated through a variety of receptors associated with other reflexes or from the environment. Two neural reflex loops are shown, each with its own receptor and controller that might send modulating input to the hypothalamic controller, altering the set point or telling the hypothalamus to ignore the set point altogether. In spite of the apparent complexity of this neuroendocrine reflex as compared to our basic model, one simply asks the same series of questions given above for each level in the system until an understanding is achieved of the integrated whole.

B. Endocrine Disruption of Homeostasis

Traditionally, clinical endocrinology has dealt with disorders of the endocrine system that involve a disruption of homeostasis. Because of the broad actions of hormonal chemical regulators, homeostasis can be profoundly affected by endocrine imbalances. A listing of some well-characterized endocrine disorders is provided in Table 1-5. Many disorders are discussed in later chapters after the normal endocrine physiology is described.

A recent focus for endocrinologists is the potential for disruption of endocrine functions in natural ecosystems

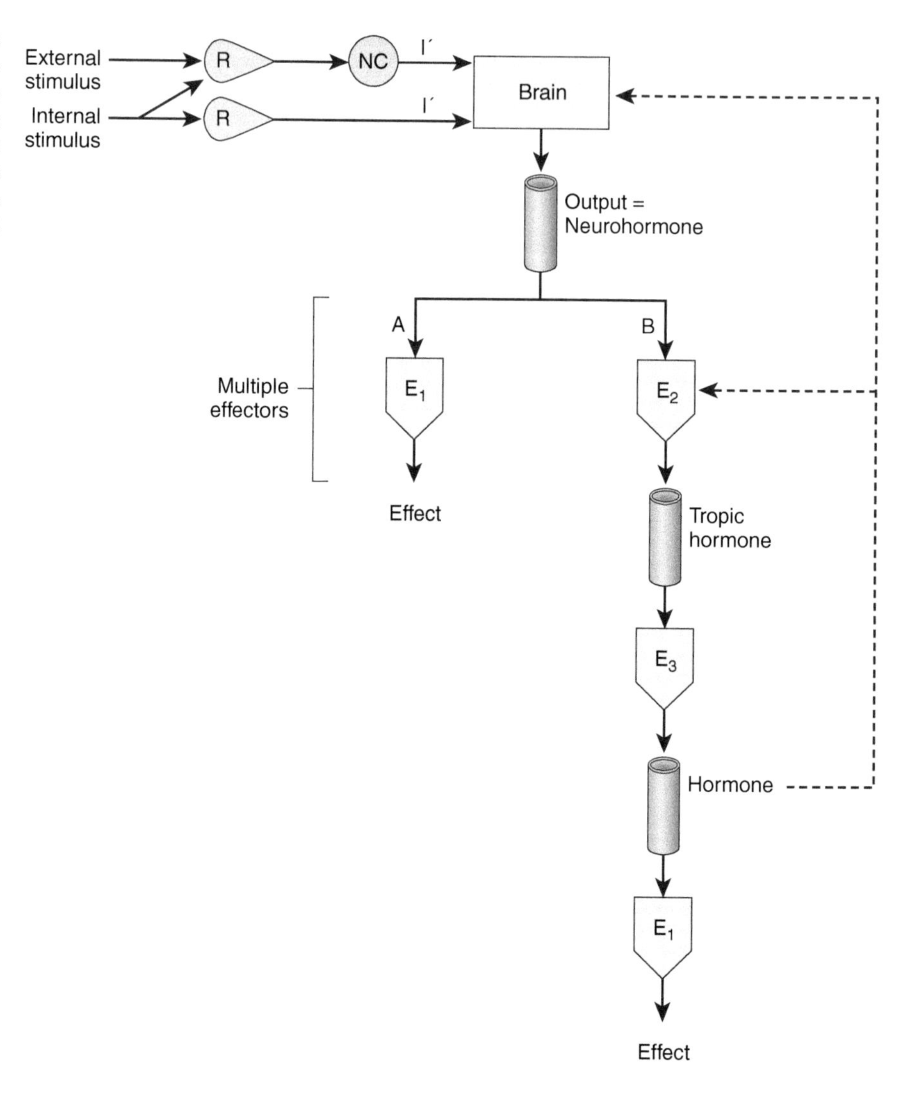

FIGURE 1-9 **A complex neuroendocrine homeostatic system.** This mechanism involves the interaction of neurohormones from the brain (controller), tropic hormones from the pituitary (E_2), and hormones from a variety of endocrine glands to control more complicated events. Note that multiple receptors, multiple effectors (targets), and multiple feedback loops may occur.

caused by the presence of discarded chemicals in our surroundings, exposures at work, or daily contact such as through the use of pharmaceuticals, plastics, certain detergents, personal care products, and food. Initially, they were called EDCs because we became aware of them through their disruptive effects on vertebrate reproduction and thyroid function. Some prefer more general names such as endocrine-active chemicals (EACs) because they may act as disrupters or simple mimics of hormones. Still others prefer to designate them as part of a variety of chemicals called "emerging contaminants" because we now recognize a much broader range of effects in animals beyond the endocrine system. We use the acronym EDC because our focus deals primarily with the endocrine system. These EDCs are produced by human activities (i.e., they are anthropogenic) and can mimic natural bioregulators; they can prevent their synthesis, transport, or normal actions or alter their rates of metabolism and/or excretion. EDCs represent a wide variety of compounds of diverse origins (Table 1-6). For example, some of them are pesticides, including insecticides (e.g., DDT and its metabolite DDE), herbicides (e.g., atrazine, glyophosate), and fungicides (e.g., vinclozolin). Others are industrial or mining byproducts such as heavy metals, **dioxins**, and PCBs. Estrogenic or anti-androgenic (feminizing) chemicals such as **phthalates**, **nonylphenols**, and **bisphenyl A (BPA)**, leach from a variety of plastics used in food and beverage packaging, dental sealants, aerosol sprays, latex paints, cosmetics and personal care products, and many other products we use everyday. **Phytoactive chemicals** are endocrine-disrupting chemicals produced in plants that find their way into animal systems either through diet or as a result of industrial activities (e.g., pulp and paper mill effluents dumped into rivers). Humans excrete natural hormones as well as prescription and non-prescription pharmaceuticals taken for therapeutic purposes which pass to wastewater treatment plants and can concentrate in aquatic systems, affecting wildlife as well as the drinking water of downstream human communities.

TABLE 1-5 Clinically Relevant Endocrine Disorders

Disease	Description
Acromegaly	Inappropriate and continued secretion of growth hormone by a tumor of pituitary cells which leads to soft tissue swelling and hypertrophy of the skeletal extremities (usually in the third or fourth decade).
Addison's disease	Adrenocortical insufficiency resulting from a deficient production of glucocorticoids and/or mineralocorticoids due to a destruction of the adrenal cortex.
Cretinism	Characterized by a permanent neurological and skeletal retardation and results from an inadequate output of thyroid hormone during uterine and neonatal life; may be caused by iodine deficiency, thyroid hypoplasia, genetic enzyme defects, or excessive maternal intake of goitrogens.
Cushing's disease	Hypercortisolism resulting from the presence of small pituitary tumors which secrete ACTH leading to excess production of cortisol by the adrenals.
Cushing's syndrome	The circumstance of glucocorticoid excess without specification of the specific etiology; it may result from endogenous causes but is more commonly iatrogenic (physician-induced).
Diabetes insipidus	A deficient secretion of vasopressin which is manifested clinically as *diabetes insipidus*; it is a disorder characterized by the excretion of an increased volume of dilute urine.
Diabetes mellitus	A disease characterized by a chronic disorder of intermediary metabolism due to a relative lack of insulin which is characterized by hyperglycemia in both the postprandial and fasting state (see also types I and II *diabetes mellitus*).
Diabetes mellitus Type I (insulin-dependent)	The form of diabetes that usually appears in the second and third decades of life and is characterized by a destruction of the pancreas B cells; this form of the disease is normally treated with daily administration of insulin.
Diabetes mellitus Type II (insulin-independent)	The form of diabetes often arising after the fourth decade, usually in obese individuals; this form of the disease does not normally require treatment with insulin.
Feminization	Feminization of males, usually as manifested by enlargement of the breasts (gynecomastia) which can be attributed to an increase in estrogen levels relative to the prevailing androgen levels.
Froehlich's syndrome	A condition usually caused by craniopharyngioma (a tumor of the hypothalamus) which results in a combination of obesity and hypogonadism; sometimes termed adiposogenital dystrophy
Galactorrhea	The persistent discharge from the breast of a fluid that resembles milk and that occurs in the absence of parturition or else persists postpartum (4–6 months) after the cessation of nursing.
Gigantism	This condition appears in the first year of life and is characterized by a rapid weight and height gain; affected children usually have a large head and mental retardation; to date no specific endocrine abnormalites have been detected.
Goiter	Goiter may be defined as a thyroid gland that is twice its normal size; endemic goiter is the major thyroid disease throughout the world. Goiter is frequently associated with a dietary iodine deficiency; in instances of sporadic goiter it may occur as a consequence of a congenital defect in thyroid hormone synthesis.
Grave's disease	An autoimmune disease characterized by the presence in the serum of long-acting thyroid stimulator (LATS) that is an antibody that activates the receptor for TSH. Grave's disease is the most common cause of thyrotoxicosis.
Gynecomastia	Abnormal breast enlargement which may occur in males during puberty.
Hermaphroditism	True hermaphroditism is defined as the presences of both testicular and ovarian tissue in the same individuals; pseudohermaphroditism is a discrepancy between gonadal and somatic sex.
Hirsutism	An increase in facial hair in women which is beyond that cosmetically acceptable; this condition may be associated with a number of masculinizing disorders including Cushing's syndrome, congenital adrenal hyperplasia, and polycystic ovary syndrome.
Hyperaldosteronism	An inappropriate secretion of aldosterone. It can occur as a primary adrenal problem (e.g., adrendal tumor) or can be secondary to other metabolic derangements that stimulate its release; it is often characterized by inappropriately high levels of plasma renin.

(Continued)

TABLE 1-5 Clinically Relevant Endocrine Disorders—cont'd

Disease	Description
Hyperparathyroidism	Inappropriately high secretion of PTH leading to hypercalcemia. Frequently associated with the hyperparathyroidism is a metabolic bone disease characterized by excessive bone calcium reabsorption; frequently attributable to an adenoma of the parathyroid gland.
Hypoparathyroidism	Inappropriately low secretion of PTH, leading to hypocalcemia; the disease is either idiopathic or iatrogenically induced.
Klinefelter's syndrome	Typically characterized by male hypogonadism; the presence of extra X chromosomes is likely the fundamental underlying etiological factor. It is characterized by varying degrees of decreased Leydig cell function and seminiferous tubule failure.
Myxedema	Hypothyroidism clinically manifested by the presence of a mucinous edema; the disease may appear at any time throughout life and is attributable to disorders of the thyroid gland or to pituitary insufficiency.
Osteomalacia	A bone disease in adults characterized by a failure of the skeletal osteoid to calcify; it is usually caused by an absence of adequate access to vitamin D.
Polycystic ovary syndrome (PCOS)	A complex of varying symptoms ranging from amenorrhea to anovulatory bleeding, often associated with obesity and hirsutism. The term denotes an absence of ovulation in association with continuous stimulation of the ovary by disproportionately high levels of LH.
Premature ovarian failure or insufficiency	Patient typically presents with high FSH levels, low estrogen levels, and amenorrhea. Many potential causes include genetic disorders, autoimmune disease, smoking, and ovarian resistance. Secondary ovarian failure is caused by the failure of the pituitary gland to secrete FSH.
Pseudohypoparathyroidism	A familial disorder characterized by hypocalcemia, increased circulating levels of PTH, and a peripheral unresponsiveness to the hormone; afflicted individuals frequently are of short stature, with mental retardation and short metacarpals and/or metatarsals.
Rickets	A failure in the child of the skeletal osteoid to calcify; it is usually caused by an absence of adequate amounts of vitamin D; it is characterized by a bowing of the femur, tibia, and fibulas
Turner's syndrome	A condition present in females with a 45, XO chromosome pattern (i.e., complete absence of the X chromosome). The XO individual is typically short with a thick neck and trunk and no obvious secondary sex characteristics.
Zollinger-Ellison syndrome	Tumors of the pancreas which result in excessive secretion of gastrin; the afflicted subject has recurrent duodenal ulcers and diarrhea caused by hypersecretion of gastric acid.

Pharmaceutical compounds with hormonal activity, such as the birth control steroid **ethynylestradiol (EE_2)**, or neuroactive chemicals, such as fluoxetine, are designed to resist degradation and hence may be very persistent in the environment.

Conventional sewage treatment systems are not designed to remove all of these EDCs (and in some cases may increase their potency), and even small concentrations (nanograms/liter) may produce effects in wildlife. Many of these compounds have been shown in clinical, laboratory, and field studies to be estrogenic (e.g., DDT, 4-nonylphenol, BPA, EE_2), androgenic (e.g., trenbolone), antiandrogenic (e.g., vinclozolin, phthalates), anti-thyroid (e.g., PCBs, perchlorate), or anti-adrenal (e.g., cadmium). Although most of these compounds end up in biosolid wastes, sufficient endocrine and neuroactive compounds, for example, may pass with the processed sewage effluent into freshwater and estuarine environments, where effects on reproduction and behavior of fishes and other wildlife have been described. The nanogram or microgram/liter levels of EDCs in wastewater effluents are well below the standardized "safe" toxicity levels determined by traditional toxicology. Unfortunately, biological receptors can detect natural bioregulators and bioregulator mimics at much greater dilutions than traditional chemistry.

Increased incidences of breast, prostate, and testicular cancer and reduced sperm counts have been reported in human populations in developed countries. Decreases in the proportion of live male to live female births have occurred since the 1990s, although the prior ratio was consistent for the previous 500 years. Puberty has been occurring at progressively earlier ages, especially in the past two decades. These are, of course, correlative changes only; the causes of these changes are not established and could be very complex in nature. However, increased incidences of abnormal reproductive development, especially

TABLE 1-6 Classification of Some Known Endocrine-Disrupting Chemicals (EDCs)

By usage
- Propellants/surfactants (e.g., nonylphenols)
- Plant substances (e.g., genistein)
- Personal care products (e.g., phthalates, parabens)
- Pesticides (e.g., DDT, methoxychlor, atrazine, Roundup)
- Pharmaceuticals (e.g., diethylstilbestrol, ethinylestradiol, antidepressants)
- Plasticizers (e.g., phthalates, bisphenol A)
- Phytoestrogens (e.g., genistein)

By chemistry
- Alkylphenol ethoxylates and derivatives (includes nonylphenols)
- Dioxins
- Heavy metals
- Polychlorinated biphenyls (PCBs)
- Perchlorate
- Phthalates
- Steroids

By endocrine-disrupting actions
- Androgenic
- Anti-adrenal
- Anti-androgenic
- Anti-estrogenic
- Anti-thyroid
- Estrogenic
- Obesogenic

in newborn males, are correlated directly with exposures to estrogenic compounds as are some cases of precocial puberty in girls. The discovery of feminized fishes in wastewater-effluent-dominated streams and estuaries in Europe, Asia, and North America has been attributed directly to the presence of mixtures of estrogenic substances of human origin (see Figure 1-10). Because hormone mimics add to and hormone inhibitors detract from natural bioregulators already present, there is theoretically no level for any EDC that may not have consequences in natural populations. This concept of "no safe level" is in marked contrast to traditional toxicology, which could readily define a "safe" level when using death or disease induction as evidence of toxicity.

Endocrine-disrupting chemicals may be additive in producing effects in animals when they all operate through the same mechanism (such as by binding to the same receptor). For example, BPA, nonylphenol, EE_2, and the natural hormone estradiol all bind to the estrogen receptor and produce estrogenic effects if present in sufficient amounts. However, it has been well documented that mixtures of these chemicals are estrogenic even when each is present at a dose that by itself will not produce an estrogenic effect if the total amount of estrogen is sufficient to stimulate enough receptors (see Figure 1-11). Even mixtures of chemicals working through different pathways (e.g., PCBs and dioxins) can produce additive effects on reproduction.

Timing of the exposure to EDCs may be critical. The effects of a given EDC might appear during gamete preparation, embryonic or postembryonic development, sexual maturation, or breeding. It might cause a change of sex, reduce survival, reduce fertility, prevent normal reproduction, or accelerate senescence. The presence of "normal" levels of estrogenic phthalates in the urine of pregnant women as well as the consumption of vegetarian diets during pregnancy are correlated with increased incidence of abnormal genitalia in males at birth. Exposure of

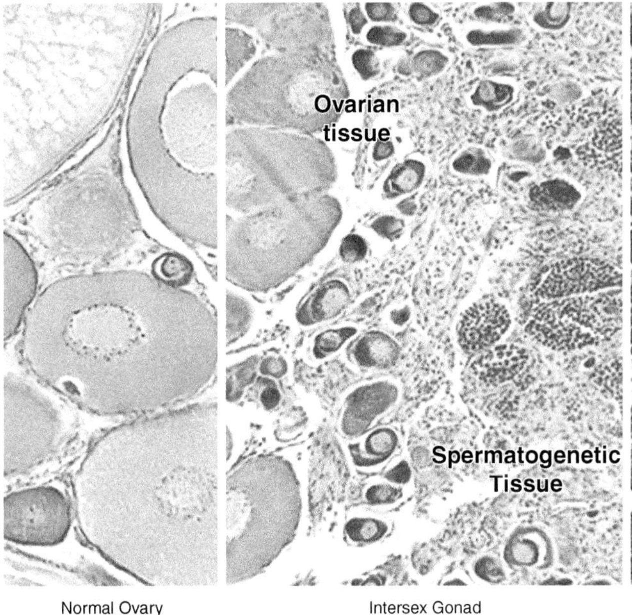

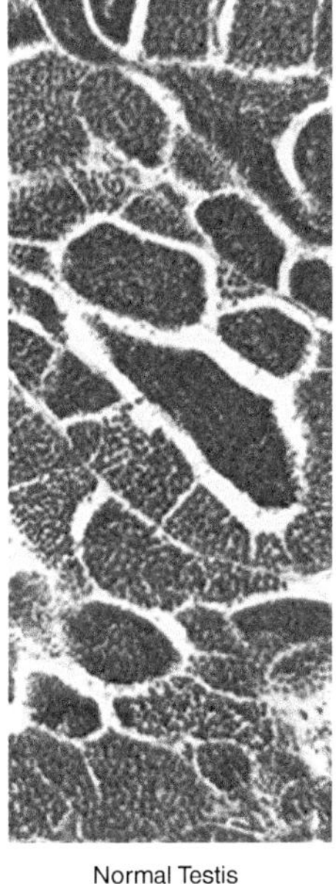

FIGURE 1-10 **Intersex gonad from white sucker (*Catostomus commersoni*).** The left panel illustrates the normal ovary from fish at a reference site above a wastewater treatment plant (WWTP). The middle panel shows ovarian tissue to the left and spermatogenetic tissue to the right in an intersex gonad of a fish collected downstream of the discharge from a WWTP. To the right is a section through a normal testis from the reference site. Intersex fish also produce the female estrogen-dependent protein vitellogenin. *(Courtesy of Alan Vajda, University of Colorado, Denver.)*

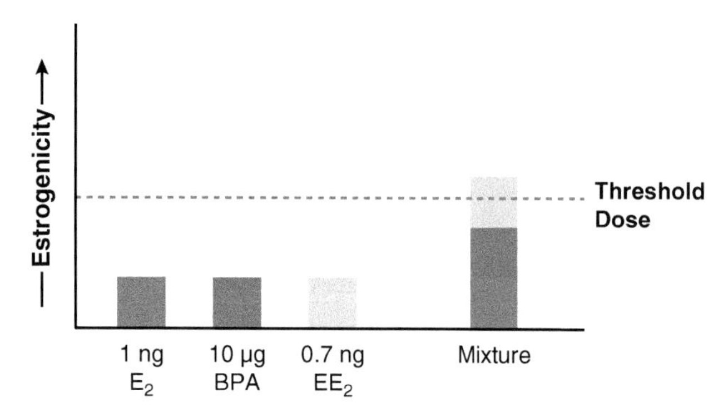

FIGURE 1-11 **Additive effect.** The additive effect of a mixture of estrogenic chemicals working through the estrogen receptor but with very different affinities for the receptor. Each chemical by itself at the dose indicated has some stimulatory ability but not enough to reach the threshold dose necessary to get an overt effect. However, when all three are present in a mixture, their effects are additive and an estrogenic effect occurs.

pregnant women to thyroid inhibitors such as PCBs or perchlorate could alter development of the nervous system and induce mental deficiencies (see Chapter 6). Studies on achievement of children exposed to significant levels of PCBs during development have demonstrated that children of mothers exposed to PCBs through their diet while pregnant performed poorly on intelligence tests as compared to control populations.

1. Epigenetic Effects During Development and Transgenerational Epigenetic Effects of EDCs

Embryonic exposure to certain EDCs may produce epigenetic effects in an individual that do not appear until years or decades later. **Epigenesis** is the progressive change in gene transcription that makes differentiation of an early embryo into multiple cellular types and organs possible. Gene transcription is controlled by patterns of methylation of the DNA or the chemical alteration of histone proteins in chromatin that prevent genes from being expressed. Demethylation and the addition of acetyl groups (acetylation) to histone proteins via the enzyme **histone acetyltransferase (HAT)** allows genes to be transcribed. Thus, the pattern of DNA methylation, for example, becomes established in a certain somatic cell type during development and remains the same in that cell type for the life of the animal. In the somatic cells of the body, some of these genes may be active all of the time. Some genes may be active early whereas others are not activated until later in life. The basic methylation and histone acetylation patterns of the zygote are retained in the resulting germ cells that reside in the gonads. When eggs or sperm are produced, that pattern is retained to direct early development of a new individual following fertilization of an egg. The ability of a cell or an organism to inherit different phenotypes associated with DNA methylation and histone acetylation, without any change in the sequence of the DNA in the gene being transcribed, is referred to as **epigenetics**.

During development, there are distinct periods when a general demethylation of the DNA occurs followed by remethylation (Figure 1-12). If the pattern of methylation is changed accidentally or by exposure to environmental chemicals during these periods, abnormal patterns of gene activation or repression may result in all descendants of that cell type and manifest later in the life of that individual as

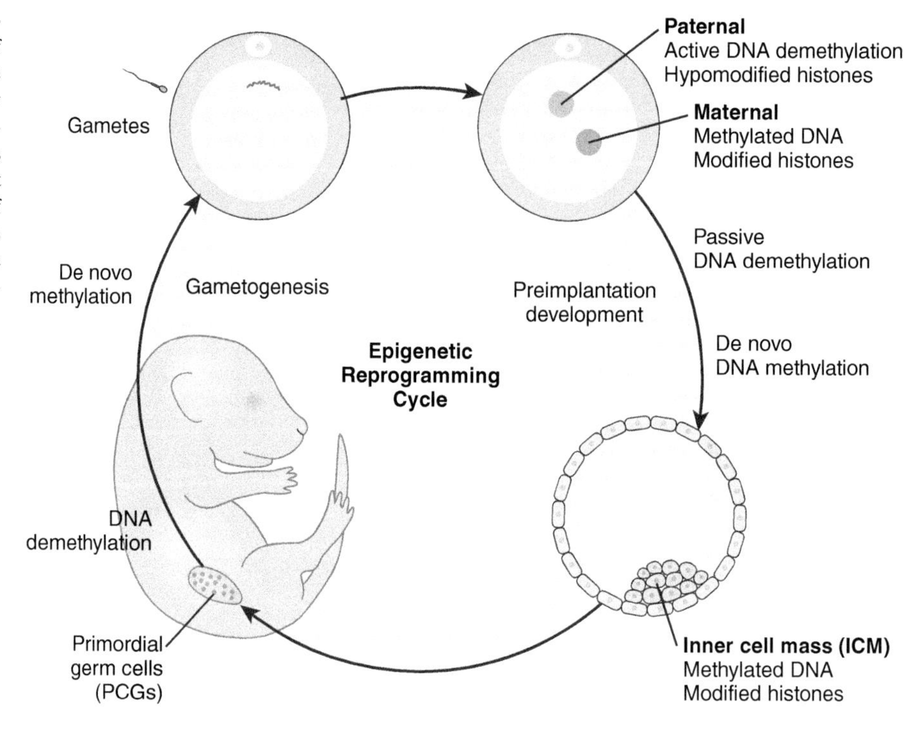

FIGURE 1-12 **Epigenetic programming.** In this example, a general period of DNA demethylation takes place in the testes when sperm are made followed by remethylation. After fertilization, demethylation allows for functioning of genes necessary for very early development following fertilization. A later period of methylation and histone modification is shown. *(Adapted with permission from Morgan, H.D. et al.,* Human and Molecular Genetics, ***14****, R47–R58, 2005.)*

various kinds of disorders including cancer. The exposure of mammalian embryos to certain estrogenic (feminizing) and anti-androgenic (demasculinizing) chemicals has influenced the incidence of reproductive cancers when those offspring become adults and age. Studies of laboratory rodents exposed while in the uterus demonstrate that these effects are permanent and are passed on to their offspring over several generations through epigenetic mechanisms.

BOX 1A Histone Acetylation and Hormone-Sensitive Cancer

Acetylation or deacetylation of lysine residues on histone proteins is generally thought to increase or decrease gene transcription, respectively, and histone acetylation/deacetylation is often an important mechanism in hormone action, especially for steroid hormones. The enzymes responsible for acetylating and deacetylating histones—histone acetyltransferases and histone deacetylases (HDAC)—are becoming increasingly important targets for therapies to treat hormone-sensitive tumors. In particular, HDAC inhibitors have shown promise in causing suppression of tumor growth in prostate, ovarian, and breast cancer. Although the precise mechanisms underlying HDAC inhibitor suppression of tumor growth remain an active area of investigation, HDACs are overexpressed in some hormone-sensitive tumors and may be involved in the invasion of blood vessels (angiogenesis) that accompanies the growth of solid tumors found in breast and prostate cancer patients.

2. Broader Implications of EDCs

Most studies of EDCs have been focused on vertebrates, but many invertebrate species are sensitive to many of the same EDCs because they exhibit many of the same biochemical pathways and mechanisms. Chemical bioregulatory and communication mechanisms are determined by similar genes, and these ancient mechanisms have been subjected to thousands of manmade chemicals within just a few decades. Thus, systems that evolved over millions of years are being subjected to significant levels of chemicals that can interfere with or mimic critical pathways and processes necessary for reproduction and survival. And this is happening in a very short time. There are surprisingly many parallels in bioregulation with respect to the specific compounds involved and their receptors between the millions of known invertebrates and the relatively small number of vertebrate animals. Already, numerous examples of endocrine disruption have been described in annelids (e.g., earthworms), mollusks (e.g., snails), echinoderms (e.g., sea stars), and arthropods (e.g., insects and crustaceans). No one knows yet the actual impact of EDCs on natural ecosystems, but the potential for devastating effects is there. Recognition of the real and potential consequences of this unique pollution problem has led to recent worldwide attention and considerable controversy. The economic importance of these compounds and the potential financial consequences of these disruptive observations versus the health of wildlife and human populations provide a focus for the controversy. Certainly, endocrine disruption and its consequences for the future of life on Earth should be a major focus of comparative and clinical endocrinologists both now and in the future.

VII. ORGANIZATION AND GOALS FOR THIS TEXTBOOK

This book is organized along rather traditional lines with an emphasis on the hypothalamus–pituitary axes followed by considerations of the regulation of metabolism and calcium physiology. Chapter 2 describes the tools used by endocrinologists of the past and present to provide the reader with a background in how we have learned what we know and hope to learn from future studies. Chapter 3 introduces most of the bioregulators discussed in this book, their chemistry, their mode of synthesis, their mechanism of action, and how they are metabolized or excreted. Organization and functioning of the mammalian systems are discussed in Chapter 4 (Hypothalamus and Pituitary), Chapter 6 (Thyroid), Chapter 8 (Adrenal), Chapter 10 (Reproduction), and Chapter 12 (metabolism). Because all of the terminology is primarily based on the mammalian systems, the endocrine systems of the rest of the vertebrates and discussion of their evolution follow each mammalian chapter to provide a complete treatment of the vertebrates (Chapters 5, 7, 9, 11, and 13). A discussion of calcium regulation in all vertebrates can be found in Chapter 14.

As each endocrine gland or endocrine axis is discussed, the reader will find not only a discussion of the anatomy and detailed chemical events that shape our physiology and behavior but also descriptions of some of the classical endocrine disorders related to normal endocrine events. Additionally, current information on EDCs that target these same systems in wildlife and humans will be provided. Each chapter concludes with suggested readings providing the documentation behind information found in the chapters as well as in-depth treatments beyond the scope of this book for those who might want to pursue topics further.

STUDY QUESTIONS

1. How did you define *endocrinology* and *endocrine system* prior to reading this chapter? Has your view changed since reading this chapter? In what way(s)?
2. Name and define the general categories of bioregulators.
3. How is the endocrine system organized?

4. Define *homeostasis* and describe how a homeostatic system operates.
5. How can an increase or a decrease in a particular hormone still be termed negative feedback on a regulating cell?
6. What is meant by *endocrine disruption* and why should we be concerned about it?
7. What is meant by *epigenetic change*, and what are the implications of transgenerational epigenetic change?

SUGGESTED READING

Endocrinology

Bentley, P.J., 1998. Comparative Vertebrate Endocrinology, third ed. Cambridge University Press, Cambridge, U.K.

Dawson, A., Sharp, P.J., 2005. Functional Avian Endocrinology. Narosa Publishing House, New Delhi.

Goodman, H.M., 2008. Basic Medical Endocrinology, fourth ed. Elsevier, Burlington, MA.

Hadley, M.A., Levine, J.E., 2006. Endocrinology, sixth ed. Benjamin Cummings, San Francisco, CA.

Heatwole, H., 2005. Amphibian Biology. In: Endocrinology, vol. 7. Surrey Beatty & Sons, Chipping Norton, U.K.

Kovacs, W.J., Ojeda, S.R., 2011. Textbook of Endocrine Physiology. Oxford University Press, New York.

Lovejoy, D.A., 2005. Neuroendocrinology: An Integrated Approach. John Wiley & Sons, New York.

Melmed, S., Polonsky, K.S., Reed, P., Larson, M.D., Kronenberg, H.M., 2012. William's Textbook of Endocrinology, twelfth ed. W.B. Saunders, Philadelphia, PA.

Morgan, H.D., Santos, F., Green, K., Dean, W., Reik, W., 2005. Epigenetic programming in mammals. Human and Molecular Genetics 14, R47−R58.

Nelson, R.J., 2011. An Introduction to Behavioral Endocrinology, fourth ed. Sinauer, Sunderland, MA.

Papoutsoglou, S.E., 2012. Textbook of Fish Endocrinology. Nova Science Publishers, Hauppauge, NY.

Reinecke, M., Zaccone, G., Kapoor, B.G. (Eds.), 2006. Fish Endocrinology. Oxford and Science Publishers, Enfield, NH.

Endocrine Disruption

Colborn, T., Dumanoski, D., Myers, J.P., 1996. Our Stolen Future. Dutton, New York.

Diamanti-Kandarakis, E., Bourguignon, J.-P., Giudice, L.C., Hauser, R., Prins, G.S., Soto, A.M., Zoeller, R.T., Gore, A.C., 2009. Endocrine-disrupting chemicals: an Endocrine Society scientific statement. Endocrinology Reviews 30, 293−342.

Diamanti-Kandarakis, E., Gore, A.C., 2012. Endocrine Disruptors and Puberty. Springer, New York.

Gore, A.C., 2010. Endocrine-Disrupting Chemicals: From Research to Clinical Practice. Humana Press, Totowa, NJ.

Guillette, L., Crain, D.A., 2000. Environmental Endocrine Disrupters. Taylor & Francis, New York.

Kime, D.E., 1998. Endocrine Disruption in Fish. Springer, New York.

Langston, N., 2010. Toxic Bodies: Hormone Disruptors and the Legacy of DES. Yale University Press, New Haven, CT.

Norris, D.O., Carr, J.A., 2006. Endocrine Disruption: Biological Bases for Health Effects in Wildlife and Humans. Oxford University Press, New York.

Sparling, D.W., Linder, G., Bishop, C.A., 2000. Ecotoxicology of Amphibians and Reptiles. The Society for Environmental Chemistry and Toxicology (SETAC), Pensacola, FL.

Chapter 2

Methods to Study Bioregulation

The basic method of scientific investigation has changed little over the years, but dramatic advances in observational, manipulative, and analytical tools during the last 50 years have resulted in a virtual explosion in our knowledge of animal biology that parallels similar events in chemistry, physics, engineering, and other scientific disciplines. For example, our observational powers have been increased several orders of magnitude by advances in the field of microscopy. Miniature radiotransmitters have made it possible to tag secretive animals such as rattlesnakes and follow their natural migrations without disturbing them. Sensitive techniques in chemistry such as high-performance liquid chromatography and mass spectroscopy can be used to analyze volumes as small as a few microliters that may contain only a few nanograms or even picograms of an important molecule. From a tiny sample, one can determine the chemical structure of a molecule and, if it is a peptide, then construct a gene that will direct the synthesis of large amounts of the peptide. And, we can tell when a gene begins to turn on production of a peptide even before we can detect the final peptide product. This tremendous capability of probing the activities of a single cell has created an additional problem for the endocrinologist. How does one distinguish between random noise in the system and changes that have relevance to the organism at the physiological level? The endocrinologist must be able to bridge the gap between molecular information and physiological responses in organismal events such as metabolism or reproduction, or link molecular information to clinical disorders.

The advent of computers and associated technology has not only automated and accelerated many of our laboratory procedures but has also greatly augmented our ability to analyze complex sets of data. Computers also can be used to simulate natural conditions and to construct models of natural phenomena that we can manipulate and use to predict events in nature.

I. THE SCIENTIFIC METHOD

The process used by animal biologists to investigate the lives and activities of animals is the **scientific method**. We seek facts called *data* (singular, *datum*) and organize them into hypotheses, theories, or laws that give these data order and meaning. Many scientists are engaged primarily in the processes of gathering data. These data may be accumulated through careful observation or by use of planned experiments. Creative scientists also take data and try to organize them to form generalizations. From a series of observations, the scientist might formulate a **hypothesis** (Greek *hypothesis*, supposition) or a predictive statement, a statement of what may be the true explanation of certain phenomena. Until it is tested experimentally, it is only a working hypothesis that must be supported or rejected on the basis of experimental results. If the hypothesis does not hold up to the test, it must be rejected or possibly revised and retested. If supported, the scientist may choose to test it more vigorously or formulate additional hypotheses on the relationships observed.

Hypotheses must be testable through observation or experimentation. The proposal that life on Earth originated from outer space would be difficult to test, as would the notion that dinosaurs became extinct because other animals ate their eggs. All scientific generalizations and hypotheses must be tested no matter how "self-evident" or "unlikely" they might appear. Hypotheses or theories that are contradicted by data obtained from valid testing procedures must be modified or discarded. Hence, the scientist must distinguish between what we *believe* to be true, which may be true or false, and what we *know* to be true based on the results of tested hypotheses. In the final analysis, however, science often relies on the human interpretations of known facts, and scientists must constantly evaluate what they know and what they believe.

Testing of any hypothesis involves rigorous attention to details. Unless the test is reliable, the data obtained can neither contradict nor support the hypothesis or theory. Although it is possible to disprove hypotheses, it usually is not possible to prove one with a single observation or experiment. The data obtained may support the hypothesis, but because there are frequently many other ways one might test the hypothesis, it is not yet proven beyond all doubt. Hence, the experimental design, analytical tools employed, and methods of data analysis are critical to testing hypotheses.

The more ways scientists test a hypothesis, the more confident they become of its validity. When a preponderance of new data supports a generalization, it becomes a **theory** (Greek *theoria*, speculation). Continued testing of the theory never stops. Every theory must be re-examined and modified if necessary as new data are accumulated; however, be aware that a scientific theory is not just an

Vertebrate Endocrinology. http://dx.doi.org/10.1016/B978-0-12-394815-1.00002-1

educated guess, as "theory" is often used by the nonscientist or as is suggested by the ancient origin of the word. A scientific theory is an established concept based on accumulated data. Once a theory is accepted with certainty due to the weight of supporting data, it becomes a **law** or **principle**. Some people refer to it as **dogma**. However, even principles are still subjected to retesting and revision of their parts when new data are not explained by the reigning theory or dogma.

A. Controlled Experimental Testing

Science has accepted ways in which hypotheses or theories may be tested through observation or experimentation. There are techniques for making observations or for designing experiments and evaluating the result so that prejudice (bias) on the part of either investigator or subject is eliminated. Procedures are rigidly followed to control for inadvertent biases produced by the methodologies used, and an attempt is made to control other factors that might influence the results. This must be done so that scientists can be confident that the outcome of an experiment is a consequence of certain manipulations only. A **variable** is an event or condition that is subject to change. The scientist may allow or cause one or more variables to change while keeping all others constant. The changed or manipulated variable is the **independent variable** and those variables that are altered as a result of manipulating the independent variable are called **dependent variables**. One group of organisms serves as the **experimental group** and is subjected to the changed independent variable. The independent variable is unchanged for a second group known as the **control group**. In the simplest situations, there is only one independent variable but multiple variables are commonly examined in endocrine research.

One type of experimental control is a complete match to the manipulated animal minus only the factor being tested. If one group of rats received injections of a drug or hormone dissolved in a solvent (called a **vehicle**), the appropriate control would be a second group of similar rats (matched for age, body weight, sex, history, etc.) receiving injections of an equivalent amount of vehicle without the drug or hormone (to control for effects of administering the injection). If a researcher wishes to investigate the effects of removal of some body organ, the appropriate control would not be an unoperated animal, because the anesthesia or the surgical procedure could influence the result. A more appropriate control would be a **sham-operated** animal, one that was also anesthetized and surgically disturbed but in which the gland in question was left intact. Looking at drug treatments of surgically altered animals requires a more complicated design (see Table 2-1).

One of the consistent flaws in experimental studies is the omission of **initial controls**. Frequently, scientists will collect or purchase animals and set up a controlled laboratory experiment with appropriate experimental and control (e.g., sham-operated, vehicle-injected) groups. At the conclusion of the experiment, they measure a dependent variable and determine whether it is different in the two test groups; however, because they did not know where the two groups of animals were at the beginning of the experiment with respect to the dependent variable, the interpretation of the results may be limited (see Figure 2-1). For example, if animals in experimental group A have significantly larger

TABLE 2-1 Comparison of Adequate and Inadequate Controls[a]

Treatment 1	Treatment 2 Adequate Control	Inadequate Control	Treatment 3	Treatment 4
One Independent Variable: Surgery				
Surgically altered animals	Sham-operated animals	Unaltered animals	—	—
One Independent Variable: Injection				
Animal injected with chemical regulator in vehicle (a vehicle is some medium)	Animal injected with vehicle only	Uninjected animal	—	—
Two Independent Variables: Surgery and Injection				
Surgery plus chemical regulator in vehicle	Sham operated plus chemical in vehicle	—	Surgery plus vehicle only	Sham-operated plus vehicle only

[a]*A simple experiment involving only one independent variable (either surgery or injection) and a more complicated experiment involving both independent variables. Why are some attempts to establish control groups labeled as inadequate? Note that although two treatment groups are all that is necessary when manipulating a single variable, four are needed for two variables. How many experimental groups would be needed if three independent variables were examined simultaneously* (hint: $2^1 = 2$)?

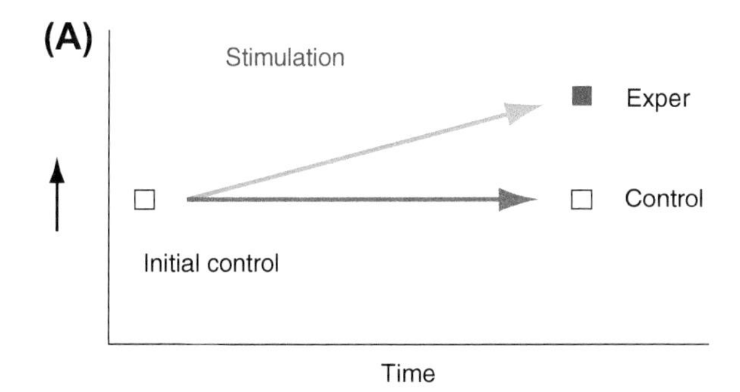

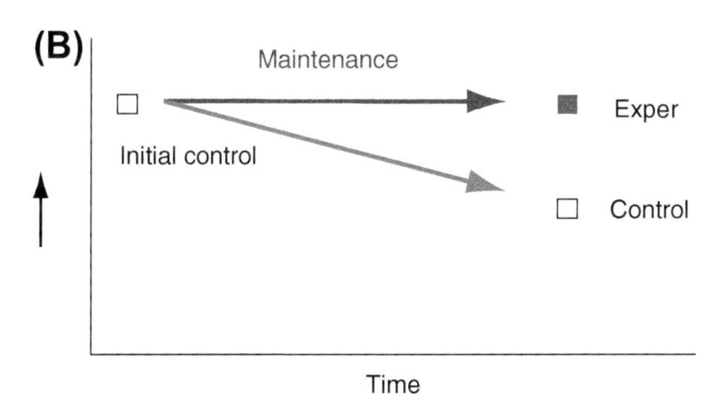

FIGURE 2-1 **Importance of initial controls.** Comparison of data for animals taken at the beginning of the experiment (initial controls) allows one to determine whether the experimental treatment stimulated the animals (A) or simply maintained the initial conditions (B) while the "controls" declined.

"grumbacks" after two weeks in the laboratory than those of the control group B, are they larger (1) because the treatment stimulated "grumback" development in Group A, or (2) because the treatment prevented regression of the "grumback" in Group A but allowed it to regress in Group B? If we knew the "grumback" size at the start of the experiment, we could easily decide between these alternatives.

Appropriate controls are often more difficult to establish in human subjects either because of moral and ethical issues or because of complicating psychological factors. It is well known, for example, that subjects expecting certain responses to treatments may respond differently from other subjects told to anticipate different responses. Consequently, in studies of the effects of a drug on the performance of long-distance runners, some subjects receive the real drug and the others are given an inactive substitute or a **placebo** (Latin, "I will please") administered in the same manner to control for psychological factors. In a **double-blind study**, neither the subjects nor the people administering the drug or placebo know which subjects are receiving which treatment. Another type of control is to switch treatments after a suitable period of observation, without telling the subjects, and then observing the subjects for a second period. These approaches also can be useful when working with nonhumans.

It is much easier to control experimental variables under laboratory conditions, especially in cellular or tissue culture, than in nature. **Control sites** for field research actually do not exist because of the myriad of environmental factors that the investigator cannot control at two locations and which will not be the same. For example, close examination of a site contaminated with cadmium and comparing it to a site without measurable cadmium may reveal not only subtle differences in natural variables (such as photoperiod, temperature, food items) but also the presence of unique chemicals at the "control site" not present at the experimental site. Consequently, we call the comparison site a **reference site** rather than a control site, acknowledging our awareness that they are not identical.

Finally, scientists must be careful not to bring their own biases when either designing an experiment or interpreting the results of observations or experimentation. For example, a scientist who has an intellectual stake in supporting an ingenious hypothesis she or he developed must guard against **confirmation bias**, the tendency to confirm his or her hypothesis regardless of what the results may be indicating. As humans, it often is difficult for us to ignore things we have been taught to believe and acknowledge that the data are not supporting what we already believed was true.

B. Representative Sampling

An important method used by biologists of all types is the examination of a **representative sample**. It is rarely feasible to test all potential subjects in a population or all the individuals of one kind of animal, so the scientist must establish a representative sample to test. Again, care must be exercised to ensure that the subjects have not been identified with any biases that might influence the results. For example, human subjects with robust thyroid function may be inappropriate subjects in which to test the vulnerability of the thyroid to a particular chemical, as they are the least likely subjects to show a response and hence do not represent the "at risk" portion of the human population.

The size of the representative sample is equally important and may be influenced by the availability of animals or appropriate human subjects, expected variability in factors being measured, statistical test criteria, previous experience of the investigator, and so forth.

Once scientists determine a representative sample, they need a reliable method of analyzing experimental results and comparing appropriate control groups to see if their results are valid and representative of the population in question. At the end of an experiment, changes in dependent variables of the experimental group are compared to the same variables in the control group. The branch of mathematics called **statistics** is the scientific tabulation and treatment of data. Statistical treatments provide tests for determining if the results are highly reproducible (that is, the probability that the differences observed in a dependent variable between experimental and control group are significant).

Reports of experimental data must be examined carefully not only to see if statistical analyses were performed but also to determine whether the appropriate statistical test was performed. Thus, it is essential that all researchers have a background in statistical methods for analyzing data and designing experiments. The design of a study must reflect knowledge of what appropriate statistical tests are to be applied when the experiment is concluded. Many unpublished and some published studies are difficult to interpret because of errors in experimental design or in the application of inappropriate statistical tests.

C. The Dose–Response Relationship

Biological systems are very sensitive to the quantity of available chemical bioregulators, and small changes in concentration may have profound effects on physiology and behavior. This influence of concentration on physiology and behavior is summarized in the **dose–response relationship**. We can use a graph to illustrate this by comparing the physiological or behavioral response (dependent variable) on the *y*-axis with the dose of bioregulator or drug (independent variable) on the *x*-axis (Figure 2-2). Typically, there is a minimally effective dose necessary to produce any response. This is often called the **threshold dose**. Smaller doses producing no response are called **subthreshold doses**. The highest subthreshold dose is often termed the **no observed effect dose.** Some test systems show an **all-or-none response**, meaning that the effect, if produced at all, occurs at its maximal intensity and adding more regulator will not give a greater response (Figure 2-2A).

Increasing the dose progressively, however, may produce a greater response up to some maximal level (sometimes termed the optimal level) beyond which further increases in regulator actually produce a reduction in the response (Figure 2-2B). This reduction may be due to interfering effects of the regulator on other systems or may be simply a toxic action that decreases the ability of the system to respond, or possibly due to feedback (see ahead). Generally speaking, the natural physiological range for the normal regulator in the animal is between the threshold dose and the maximal dose. Notice that if you ran an experiment with only one dose of a hormone and measured a certain response in this type of system, you could not distinguish between producing a physiological response or a response to a supramaximal dose. When designing any experiment that involves treatment with a natural

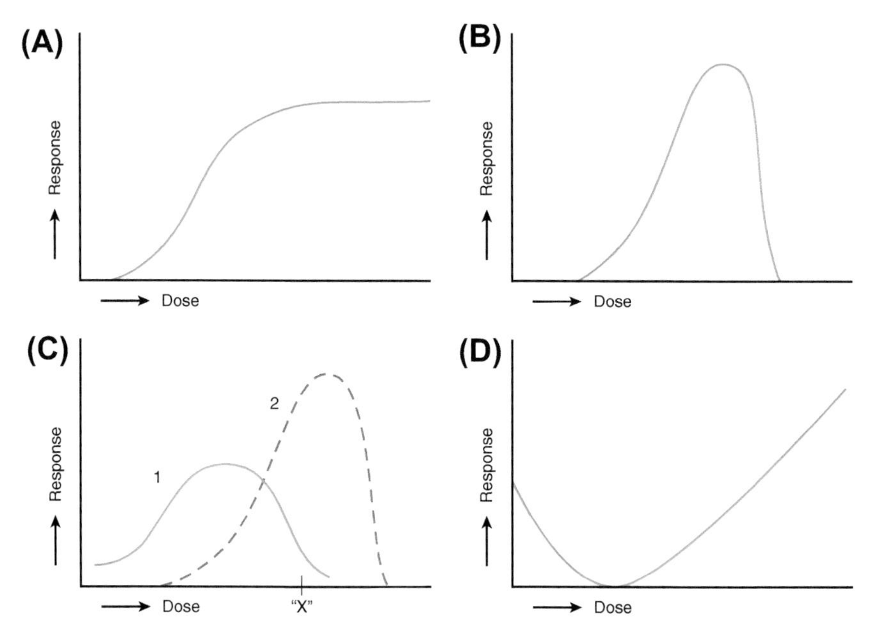

FIGURE 2-2 **Dose–response relationships.** (A) A system may respond to a threshold dose and then continue to show increased response with increasing dose until the process reaches a maximum. Additional bioregulator then produces no increased response over a wide range. (B) Relationship of optimal response to an "optimal" dose. A dose too low to produce a response is considered subthreshold. Doses above the "optimal" dose produce diminished responses and may indicate a toxicity to the tissue or animal. Alternatively, the bioregulator may reach a threshold that triggers accelerated removal of the bioregulator and hence a reduction in response at higher doses. Such doses often are considered to be "pharmacological," although any level of a regulator that exceeds natural levels also could be termed "pharmacological." Notice that when observing the response of a single dose you might not know whether you were above the optimal dose or below it. Consequently, when the data are not readily available, multiple doses should be employed to determine a useful dose–response relationship. (C) Species-specific or tissue-specific dose–responses. An "optimal" dose for one tissue or one animal might be either subthreshold or pharmacological for another. (D) J-shaped or U-shaped dose–response curves are common in biological systems when dealing with very low doses.

bioregulator or other chemical such as antagonistic or agonistic drugs, it is essential to ascertain either from the literature or experimentally what the appropriate physiological dose should be. Excessive doses may produce effects via different pathways that might influence your interpretation of the data.

When testing actions of a chemical substance on physiology and behavior, it is often difficult to separate direct actions from unanticipated actions on other tissues that might influence the results. Some drugs useful for studying one system in fact may produce toxic side effects of a **paradoxical nature** on other tissues. For example, the drug thiourea is an effective inhibitor of thyroid gland function. However, the response of the liver to thiourea produces effects opposite to those caused by surgical thyroidectomy, making interpretation of the effects of thiourea on the whole animal open to question.

The graphic illustration of the relationship between the concentration of a chemical substance and its action is called a **dose–response curve**. It is important to realize that comparison of the dose–response relationship in different tissues, in different individuals, or in different species may yield very different results (Figure 2-2C). Consequently, one has to be very careful when using data based on one species for designing an experiment with another species or when extrapolating from doses observed in a chemical solution, a cell culture, or a tissue culture to the intact animal.

Dose–response data for biological systems often exhibit a J-shaped pattern (Figure 2-2D). In this case, doses below "threshold" or the no observed effect dose may be stimulatory. Whereas simple in vitro systems may not show this phenomenon, intact organisms commonly do. The explanation lies in the complex interactions between different thresholds for different tissues of an intact organisms to respond to a bioregulator as well as changes in the ability of the animal to metabolize and/or excrete the bioregulator. This phenomenon dose–response pattern is called **hormesis** (see readings at end of this chapter). Hormesis may have serious implications for clinical situations, experimental studies, and in environmentally induced endocrine disruption.

D. Occam's Razor and Morgan's Canon

There is an old principle of logic called **Occam's razor** that should be invoked when interpreting data. Occam's razor states that if several explanations are compatible with the evidence, the simplest one should be considered the most probable one. Experience tells us that complicated explanations may not be necessary to explain a phenomenon when a simple one will do. A modification of Occam's razor called **Morgan's canon** was formulated in the late 1800s with respect to interpreting animal behavior. Morgan's canon says that, when considering at what level a behavior might be controlled by hormones and/or by the nervous system, we should invoke the lowest level (that is, the least complicated) that will work. These are important principles to keep in mind as you continue your studies of physiology. This is not meant to imply that more complicated explanations are always incorrect but to emphasize that we do not have to make them more complicated than necessary.

E. Biological Rhythms

The release of hormones and some other chemical bioregulators often occurs in bursts (**phasic secretion** or pulses of hormonal release) rather than at a continuous and constant rate (**tonic secretion**). Many of these bursts exhibit distinct, predictable hourly, diurnal (daily), monthly, or seasonally cyclic patterns of secretion that are termed **biological rhythms** (Figure 2-3). For example, maximal levels of testosterone in the blood of human males normally occur between 2 and 6 a.m. due to increases in the frequency and/or amplitude of the pulsatile release.

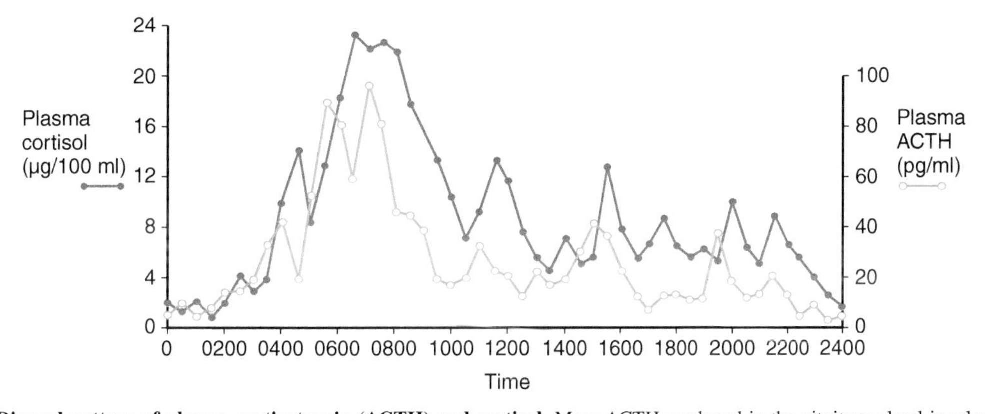

FIGURE 2-3 **Diurnal pattern of plasma corticotropin (ACTH) and cortisol.** More ACTH produced in the pituitary gland is released in the early morning than at other times of day. Because ACTH stimulates the adrenal cortex to secrete cortisol, the latter shows a matching rhythm in blood concentration trailing that of ACTH. *(Adapted with permission from Vinson, G.P., "The Adrenal Cortex," Prentice Hall, Upper Saddle River, NJ, 1993.)*

Similarly, monthly fluctuations for estradiol and progesterone blood plasma levels in human females recur with precise regularity. A predictable increase of androgens in the blood of the Italian frog *Rana esculenta* occurs every spring, signaling the onset of the spring breeding season. As one might expect, not only can hormone levels vary predictably, but the sensitivity of target cells may also show diurnal or seasonal fluctuations. For example, the responsiveness of the pigeon crop to the stimulatory actions of the tropic hormone prolactin varies with time of day. Furthermore, the metabolic responses of killifish to a given dosage of prolactin varies with the season of the year in which the fish are treated.

The existence of biological rhythms suggests another caution when interpreting data. Failure of a bioregulator to produce an effect when applied at 11 a.m. on September 15 may not mean that this bioregulator has no action in this animal but that the animal is not responsive to this dose at this time. If we had tried a different time of day or another time of the month or season of the year, we might have observed different results. If the bioregulator had produced an effect, we still could not conclude that we would always see that effect. For example, animals of different ages may vary considerably in their sensitivity to that bioregulator regardless of the time of administration.

The discovery of biological rhythms has opened many new areas of research and has opened old data to new interpretations. Today, experimental designs more frequently consider the times for treatment and the repeating of experiments with diurnal or seasonal treatments. Increasingly, experiments are conducted to understand what internal and external factors are responsible for regulating these rhythms.

II. METHODS OF ENDOCRINE ANALYSIS

In recent years, the development and application of new techniques have caused a revolution in endocrinological studies resulting in many new and exciting discoveries. Some of these developments as well as older techniques are discussed briefly here.

A. Extirpation–Observation and Replacement–Observation

Early studies of hormone actions were limited to gross manipulations and observations because there were no precise ways to estimate levels of chemical regulators. Putative hormones were identified by first removing a gland or other organ (**extirpation**) and observing the effects of removal on the organism. Then an attempt was made to replace the lost tissue through transplants or grafts or by administering extracts of the tissue (**replacement**). If continued observations verified that the symptoms caused by extirpation were relieved by the replacement therapy, evidence for the existence of a chemical bioregulator was established.

A related early technique that evolved from the classical approaches involved the use of a **bioassay system**, a system employing animals or animal parts that could provide a quantitative or at least a qualitative estimate of the presence of a bioregulator in a tissue extract. Such bioassays made it possible to determine if a chemical fraction of a gland had the biological activity that represented the sought-after bioregulator, and the quantitative assays could provide an estimate of how much of the bioregulator was present. Through their use, bioassays made it possible to obtain seasonal data on fluctuations in bioregulator levels. Furthermore, they could be used to isolate bioregulators and eventually to obtain highly purified preparations. Bioassays made it possible to identify and test suspected agonists and antagonists, too, and bioassays were used to validate the modern methods now used to estimate bioregulator levels in tissues and body fluids.

Hormones and other bioregulators operate at minute concentrations, typically in the range of micrograms, nanograms, or picograms per milliliter of blood or extracellular fluids. This means 10^{-6}, 10^{-9}, or 10^{-12} g/mL, or 1 millionth, 1 billionth, or 1 trillionth of a gram. These natural levels are often referred to as *physiological* and should be considered when designing experiments to observe the actions of a bioregulator. Treatment levels that produce circulating levels in excess of natural levels are termed **pharmacological doses** (commonly in the milligram range). Interpretations of such pharmacological studies may be complicated with respect to understanding natural functions of bioregulators. Large doses also greatly increase the probability of paradoxical or toxic side effects.

Until the advent of sensitive biochemical procedures that allowed us to measure actual hormone levels, endocrinology was limited to extirpation/replacement ("inject 'em and inspect 'em") or crude estimates based on bioassays. Many of these studies emphasized pharmacological doses that produced excessive and possibly unnatural responses. Numerous molecular approaches have led to a recent revolution in our understanding of the nature of chemical bioregulation and have produced an unprecedented explosion in data generation. Some of these techniques are described below.

B. Imaging

Visualization has been an integral part of endocrine assessment since the beginning. Initial observations of entire organisms soon led to examination of tissues through **light microscopy** and eventually of cells and subcellular structure by **transmission electron microscopy (TEM)**. **Fluorescence light microscopy** is used for localization within cells or tissue of naturally fluorescing compounds or compounds labeled by addition of a molecule that fluoresces when illuminated by light of the appropriate wavelength. Fluorescence light microscopy may be performed using epifluorescence or by using laser scanning microscopy. Many recent advances in laser scanning microscopy have provided endocrinologists with sophisticated tools to optically section cells to reveal and reconstruct cellular detail. Both **confocal laser microscopy** and **multiphoton laser scanning microscopy** use beams of laser light to excite a labeled probe within a cell. The fluorescence emission can then be captured and reconstructed three dimensionally by computer software. In this way an entire cell can be optically sectioned to provide exquisite detail about the precise location or activity of the fluorescent probe (Figure 2-4). The lasers used in multiphoton laser scanning microscopy produce longer wavelengths than lasers used in confocal laser microscopy, thereby causing less damage to living cells and allowing endocrinologists to view and reconstruct changes in fluorescence emitted from living cells (Figure 2-5).

Whole body scanning techniques, or **tomography**, produce images of entire organs or organisms that can be viewed with the aid of a computer. Specific whole-body scanning techniques include **computer axial tomography (CAT scans)**, **magnetic resonance tomography (MRT)**, and **positron emission tomography (PET scans)**. CAT scans rely on x-rays to construct two-dimensional slices that can be converted to a three-dimensional image by a computer. MRT was originally called *nuclear magnetic*

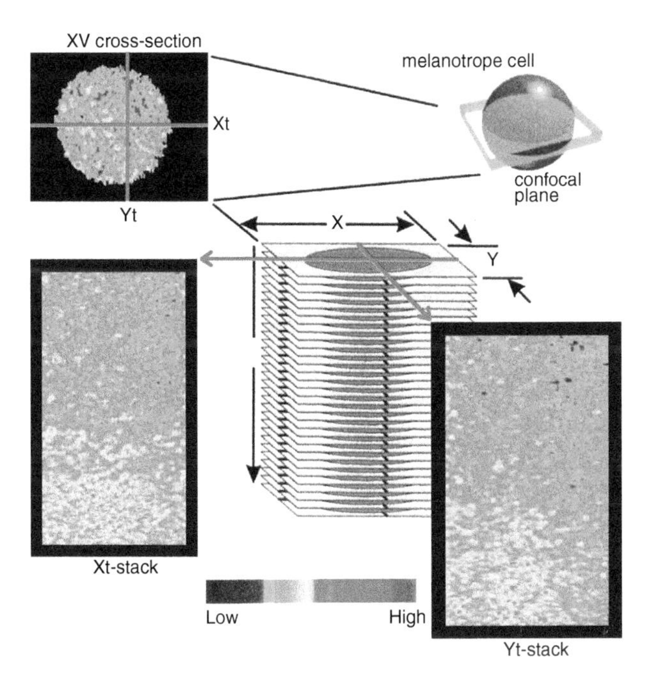

FIGURE 2-4 **Confocal laser microscopy.** Application of confocal laser microscopy to the visualization of calcium movement in a single pituitary melanotrope cell. By vertically stacking successive images through the cell, a 3D image is created. Changes in fluorescence caused by calcium movement (red, high; blue, low) can be detected in multiple planes (X, Y, t) sliced through the cell. *(Courtesy of Dr. Bruce Jenks, Department of Cellular Animal Physiology, Radboud University Nijmegen. From Koopman, W.J. et al.,* Biophysics Journal, ***81**, 57–65, 2001.)*

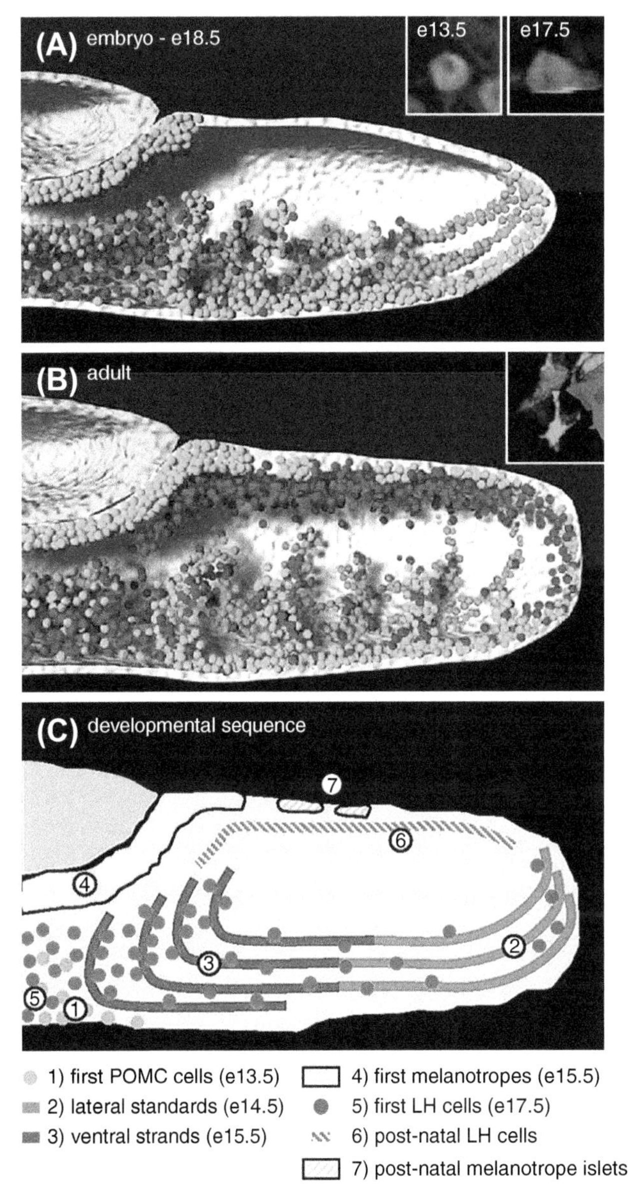

FIGURE 2-5 **Reconstruction from multiphoton and confocal microscopy.** Elucidating the pattern and timing of pituitary LH (purple) and POMC (green) cells in the developing (A) and adult (B) mouse pituitary as reconstructed from multiphoton and confocal microscopy images. Panel C outlines the developmental sequence for cell appearance based upon the microscopic analysis. POMC cells first appear in the ventral anterior lobe (1) and then extend laterally and ventrally into the AL (2, 3). The first POMC cells in the intermediate lobe (melanotropes) appear next (4). LH cells then appear in the AL, first in the ventral AL (5) and then in the dorsal AL (6). Islands of melanotropes are found in the dorsal AL postnatally. *(Reprinted with permission from Budry, L. et al.,* Proceedings of the National Academy of Sciences USA, ***108**, 12515–12520, 2011.)*

resonance imaging (NMRI), but "nuclear" was dropped to avoid an association by patients with radiation, which is not involved in this procedure. This form of tomography relies on the bipolar nature of water molecules. Because one end of the water molecule is positively charged with respect to the opposite end that is negatively charged, water molecules can be aligned in a magnetic field. Although the precise nature of this process requires a discussion of quantum mechanics, suffice it to say that this behavior by the water molecules can be used by a computer to provide a three-dimensional image. PET scans do employ the use of common metabolic molecules that are labeled with short-lived radioisotopes (e.g., ^{11}C, ^{13}N, ^{15}O, ^{18}F) that produce minimal radiation exposure to the subject. With a PET scan, it is possible to measure the uptake of labeled glucose or other molecules by brain cells, thus allowing the investigator or clinician to determine regions of increased or decreased activity. PET scans can be useful in locating a rapidly developing tumor, for example, that might exhibit heightened metabolic activity.

C. Radioimmunoassay

Once purified molecules became available for experimentation, a number of techniques developed that allowed endocrinologists to routinely measure bioregulator levels in body fluids, localize specific molecules in a particular cell or part of a cell, and monitor these parameters under a multitude of experimental conditions. The first major technical breakthrough was the development of the **radioimmunoassay**, or **RIA**, by Rosalind Yalow and Solomon Berson. While studying the occurrence of insulin resistance in patients with diabetes mellitus, these investigators noted that these people had antibodies to insulin in their blood. In part because they had difficulty in convincing the scientific community of the validity of their observations, Yalow and Berson went ahead and showed that the addition of excess insulin could displace radioactively labeled insulin from these antibodies. This demonstration, together with their fully developed mathematical treatment of these interactions, provided the foundations for RIA, a technique that quickly revolutionized the entire field of endocrinology. Today, many RIAs are used routinely to measure all sorts of bioregulators in nanogram and even picogram quantities, a feat never possible with bioassay systems. Because of his early death at age 54, Berson did not share the Nobel Prize awarded to Yalow in 1977 in recognition of the importance of their contribution to the entire field of physiology and medicine.

Development of any RIA relies on the availability of a pure source of bioregulator that can be used to induce formation of highly specific antibodies against it. Second, one must be able to radioactively tag or label a quantity of the bioregulator. It is assumed that the antibody cannot distinguish between an unlabeled or "cold" molecule of the bioregulator and a labeled or "hot" one and will bind each ligand with equal affinity. A scientist simply sets up a balanced and carefully controlled competition between cold and hot bioregulators for the binding sites on antibody molecules. The competition is designed with a constant amount of antibody and constant amount of hot ligand in the reaction mixture.

When no cold ligand is present, the amount of hot ligand bound to the antibody at the end of a prescribed period of time is defined as 100% binding. Then, additional mixtures are prepared with increasing quantities of cold ligand. As the quantity of cold ligand increases, the competition for antibody binding sites increases in favor of the cold ligand so that less and less hot ligand is bound; consequently, the percentage of hot ligand bound decreases with the addition of increasing amounts of cold ligand. Then, the antibody–ligand complexes are precipitated, leaving the unbound hot and cold ligands in the supernatant. By measuring the radioactivity of either the precipitate or the supernatant, the percentage of bound (or unbound; also called *free*) hot ligand (dependent variable) can be plotted against the known concentrations of cold ligand added (independent variable) (Figure 2-6). Radiation-detection equipment (scintillation counter or gamma counter) is used to measure the amount of free or bound radioactivity. The plot of percent bound in Figure 2-6 constitutes a standard curve from which the quantity of cold ligand in a sample of unknown quantity (the unknown) can be estimated. A similar competition is set up with the same quantity of antibody and hot ligand plus a sample of the unknown solution in which we wish to determine the level of cold ligand. By determining the percent bound for the unknown sample (y-axis), the quantity of cold ligand can be estimated by extrapolating to the x-axis.

At first, this technique proved useful only for peptides, but soon it became possible to trick antibody-synthesizing cells to make specific antibodies against all manner of chemical substances. Hence, we can measure thyroid hormones and steroids with the same ease that we measure insulin or thyrotropin (TSH).

D. High Performance Liquid Chromatography/Spectroscopy

Chemists, while searching for techniques to improve the ability to separate closely related molecules from one another, developed a sophisticated separation system known as **high-performance liquid chromatography**, or **HPLC**. This technique is a modification of column chromatography that operates under high pressures and relies on differential solubility of molecules in the solvents employed to wash the columns and on the affinities of these

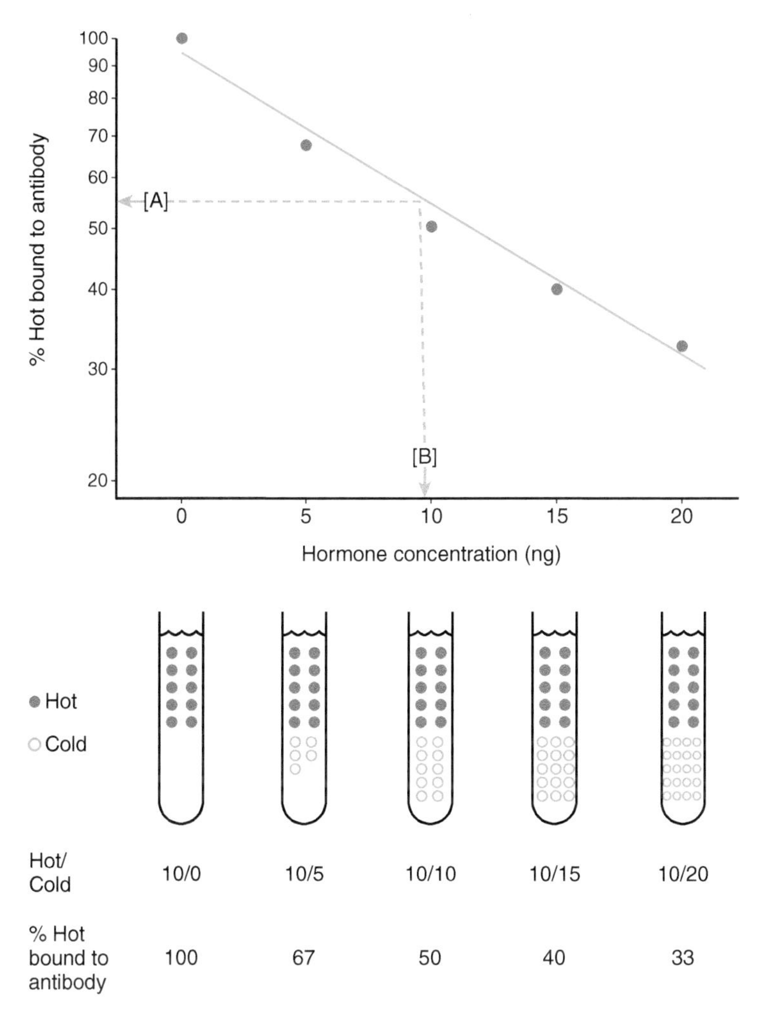

FIGURE 2-6 **Standard curve for radioimmunoassay (RIA).** A standard curve is prepared by placing the same amount of antibody and radiolabeled (hot) hormone into every test tube and adding a different known amount of unlabeled (cold) hormone to each tube so that cold and hot hormone will compete for the same binding sites on the antibody. The greater the amount of cold hormone added, the lower will be the amount of radioactivity bound to the antibody. The antibody with its bound load of hot and cold hormone is precipitated from the solution and either the radioactivity of the supernatant or of the precipitate is counted. Thus, one can plot the percentage of hot hormone bound to the antibody against the amount of cold hormone in the solution to produce a standard relationship or what is called a *standard curve*. If an additional tube is prepared with the same amounts of antibody and hot hormone but with an unknown amount of cold hormone (such as might be present in a blood sample), one can determine the percentage of hot hormone bound (A) and extrapolate from the standard curve to estimate how much cold hormone (B) was present in the blood sample.

same molecules for the substances of which the columns are made. A mixture of molecules in a particular solvent is applied to the column and the solvent is collected in aliquots as it comes off at the bottom. If properly designed, all of one type of molecule will come off in the same aliquot. What aliquot contains a given type of molecule depends on the affinity of each type for the column versus the solvent. Special spectroscopic detectors allow investigators to identify individual molecules and quantify them in various mixtures. Such an approach allows investigators to quantitatively determine all of the steroids and steroid metabolites or all of the biogenic amine neurotransmitters and their metabolites in a given sample simultaneously (see Figure 2-7). In contrast, the use of RIA requires a separate analysis for each molecule you expect to find.

Mass spectroscopy techniques (see later discussion of proteomics, ahead) following chromatographic separation have further refined analytical chemistry such that it is possible to measure multiple chemicals very accurately in a single sample when present at extremely low concentrations (e.g., ng/L or parts per trillion). These approaches are now being applied to measure bioregulators as well as endocrine-disrupting chemicals (EDCs) in blood and other body fluids as well as in environmental aquatic samples. Thus a complete steroid profile can be obtained from a single blood sample or all of the chemical impurities of interest can be detected at extreme dilution in a water sample.

E. Immunohistochemistry

Another analytical use for antibodies is to localize particular bioregulators, synthesizing enzymes, or degrading enzymes in tissues, cells, or parts of cells. This approach, called **immunohistochemistry** (or immunocytochemistry), employs an antibody made in one species (say, a mouse) against the specific molecule (antigen) you wish to localize in the brain of a song sparrow or other animal. There are several variations of this technique based on a simple procedure (see Figure 2-8). First, make the mouse antibody that would be a mouse gamma globulin protein and apply that to sections of song sparrow tissue placed on

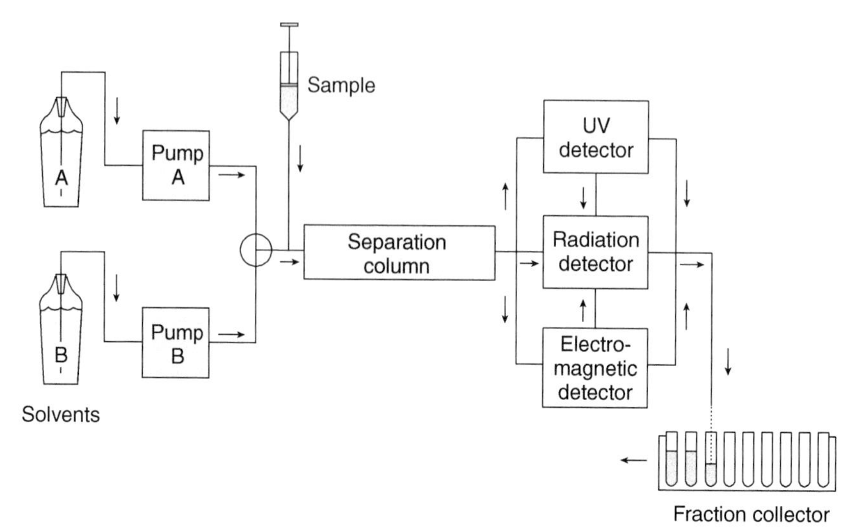

FIGURE 2-7 **High-performance liquid chromatography (HPLC).** A sample is applied to a separation column in a particular solvent. Additional solvents can be mixed and pumped through the separation column and, depending on the affinity of the sample components for the column and the solvents, the sample components will leave the column at different times and pass through some kind of detector. Some compounds, such as steroids, are best detected and quantified by their absorbance of ultraviolet light (UV), whereas others (e.g., catecholamines) can be quantified using electrochemical detectors. HPLC can be coupled to radiation detectors for identifying a labeled molecule or specific metabolic products of a radioactive precursor, and the radiation detector may be used in tandem with either UV or electrochemical detectors. Various fractions from the column may be recovered by using a fraction collecting device.

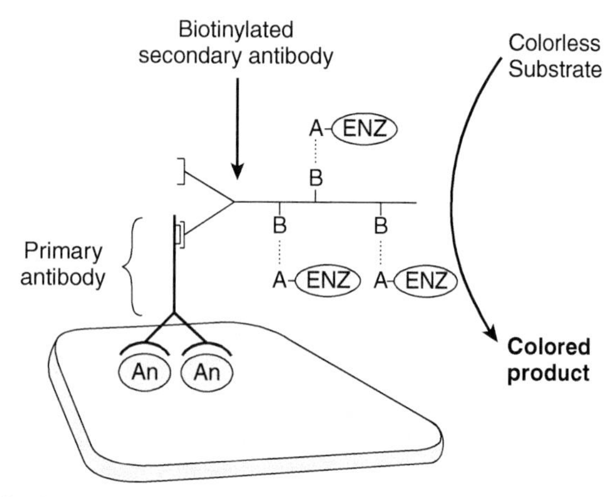

FIGURE 2-8 **Immunocytochemistry.** This method requires a pure source of antigen to make a specific antibody (primary antibody) in a mouse (or rabbit, etc.). Although the primary antibody usually is applied to sections of cells or tissues, this method also can be used on whole blocks of tissue. A secondary antibody is made against mouse immunoglobulins (for example, in a goat) to amplify the location of the primary antibody that is bound *in situ* to the antigen. The secondary antibody can be complexed to an enzyme, a fluorescing compound, or a radioactive marker for detecting the exact location of the antigen in a cell or tissue. This approach can be used for detecting the presence or absence of an antigen or to determine the number of reactive cells, etc. It also can be coupled with other techniques to estimate the intensity of the reaction and relate that to the quantity of antigen present.

a microscope slide. Theoretically, the antibody will bind only to cells that contain the antigen. In another mammalian species, make an antibody against mouse gamma globulin and conjugate that with an enzyme known as a peroxidase. This is also applied to the tissue sections. This second antibody will bind to the gamma globulin that has bound previously to the song sparrow antigen. Next, add a substrate for the peroxidase enzyme that results in a colored product that will be localized in the cell containing the antigen—gamma globulin—anti-gamma globulin—peroxidase complex. This complex can then be viewed with the microscope (Figure 2-9).

There are other variations on this basic technique using different enzyme-substrate markers that allow one to examine more than one antigen in a single section. **Immunofluorescence** is a modification that attaches a compound to the anti-gamma globulin that will fluoresce under certain wavelengths of light (usually ultraviolet) instead of using the peroxidase enzyme.

Another related approach is the **enzyme-linked immunoabsorbant assay (ELISA)**, which is commonly used to detect the presence of a specific molecule in blood plasma. Molecules of a specific bioregulator are coated onto walls of special microtiter plates (commonly used in immunological studies) and are used to compete with free molecules in plasma or a tissue extract for a specific antibody. The peroxidase—antiperoxidase method is then used to reveal the immobilized bioregulator—antibody complex.

F. Techniques for Determining the Number and Characteristics of Receptors

For a protein to be a receptor for a particular bioregulator, it must have certain properties. Because of the small number of bioregulator molecules usually present, a receptor must have a high **affinity** for the bioregulator; that is, a strong

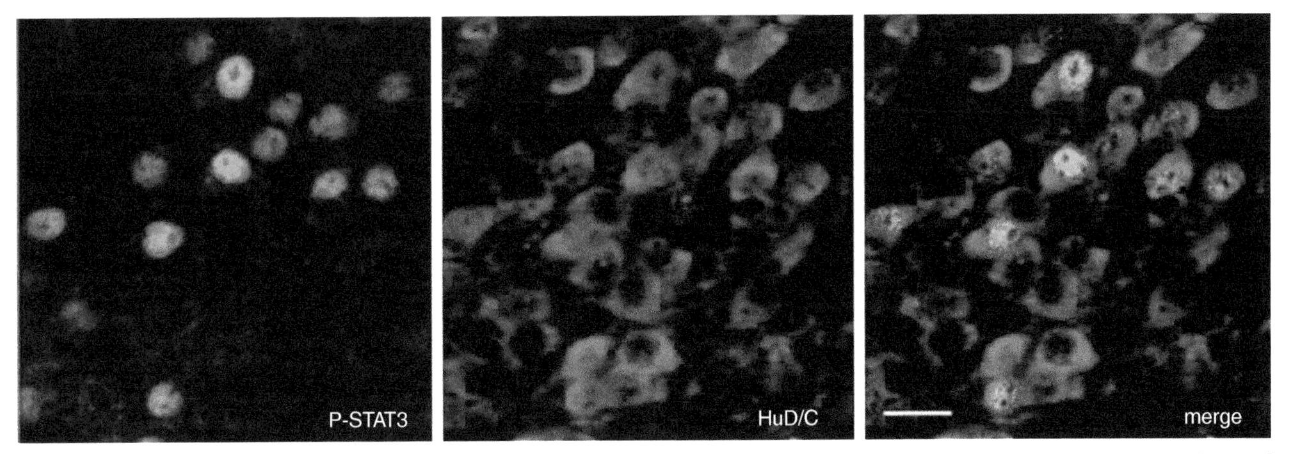

FIGURE 2-9 **Confocal laser microscopy identifies multiple antigens in a single cell.** Phosphorylated STAT3 (signal transducer and activator of transcription 3) is upregulated in cells of the mouse hypothalamus expressing the neuronal marker HuD/C after intraperitoneal injection of the appetite-regulating hormone leptin. *(From Frontini, A. et al.,* Brain Research, ***1215****, 105–115, 2008.)*

tendency to bind the bioregulator. Second, the receptor should have high **specificity** for the bioregulator and little tendency to bind other molecules. A third feature of receptors is their low **capacity**. This means that all available receptor sites are occupied at relatively low concentrations of bioregulator; that is, the receptor is said to be **saturated**. Finally, the distribution of a putative receptor should correspond to the known target tissues for the bioregulator and should be correlated with some biological effect.

Many bioregulators may "stick" to proteins other than their specific receptors, especially in cell or tissue homogenates where the process of disrupting the normal cell architecture may unmask potential binding sites normally not available to the bioregulator. Thus, it is important to distinguish **specific binding** (high-affinity, low-capacity proteins) from **nonspecific binding** (low-affinity, high-capacity binding proteins). These nonspecific binding sites do not saturate unless huge doses of bioregulator are applied. To distinguish between specific and nonspecific binding, we take advantage of the availability of radioactively labeled bioregulators to determine total binding capacity of a cell or tissue homogenate. Then, by adding an excess of unlabeled bioregulator to another sample also containing the labeled bioregulator, a competition is set up for the small number of high-affinity, low-capacity sites (the true receptors) causing labeled bioregulator to be displaced from only the specific binding sites. Because the nonspecific binding sites have such a high capacity, there is little competition occurring there. This competition is done using a range of unlabeled concentrations, and the difference between total binding (homogenate sample without unlabeled excess) and the nonspecific binding (homogenate with unlabeled excess) provides an estimate of specific binding (Figure 2-10). However, since specific binding is a function of both the affinity of the receptors as well as

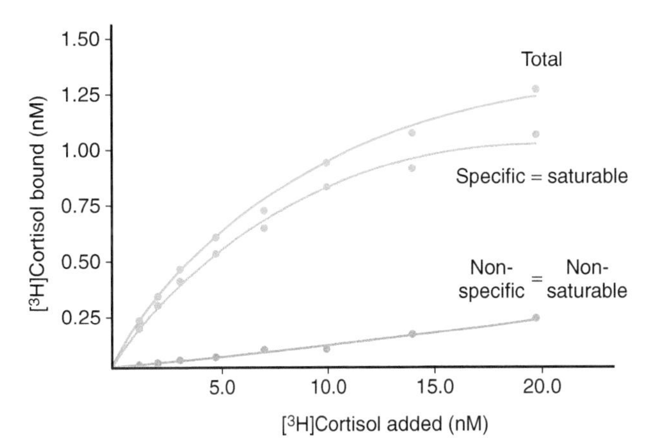

FIGURE 2-10 **"Specific" and "nonspecific" binding by receptors.** Determination of saturable (receptor or "specific") and nonsaturable (nonreceptor or "nonspecific") binding of cortisol to a liver cytosolic preparation using ^{3}H-cortisol as the ligand. Nonsaturable binding is not affected much by the amount of ligand because it has a high capacity. This is determined by adding a large excess of unlabeled cortisol in an attempt to displace the saturable receptors (specific binding) while not affecting the nonsaturable (nonspecific) binding. The difference between the binding observed with and without the excess of unlabeled ligand is defined as the saturable or specific binding. *(Adapted from Chakraborti, P.K. and Weisbart, M.,* Canadian Journal of Zoology, ***65****, 2498–2503, 1987.)*

the number of receptors, simply demonstrating specific binding cannot be used for comparative purposes. Therefore, investigators compare the kinetics of the bioregulator–receptor binding, much like you would do for studies of substrate–enzyme interactions, to provide an estimate of both the affinity of a receptor for the specific bioregulator in question as well as an estimate of the number of receptors present. The most commonly used procedure for kinetic studies is the **Scatchard analysis**, a relationship first employed by George Scatchard in 1949 (Figure 2-11). The analysis of the experimentally derived data is similar to that done for enzymatic kinetics. Although

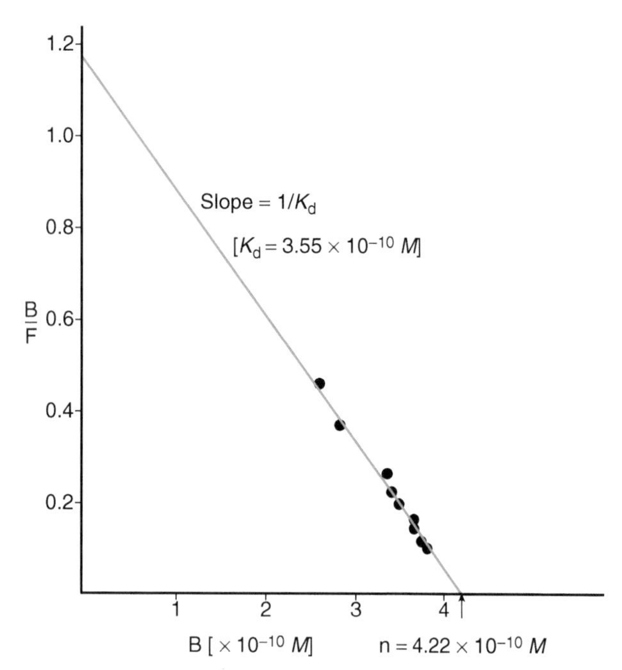

FIGURE 2-11 **Scatchard plotting method for receptor determination.** The Scatchard plot is described in the text. *(From Bolander, F.F., "Molecular Endocrinology," Academic Press, San Diego, CA, 1989.)*

this method involves certain assumptions and necessitates careful, repeatable laboratory procedures, it provides information on affinity (slope of the line), receptor number (y-axis intercept), and purity of the preparation (straight line versus curvilinear plot).

A bioregulator bound by a specific receptor can generally be termed a **ligand**. The chemical kinetics for reversible enzyme–substrate interactions you likely have studied previously are appropriate for describing ligand–receptor interactions. Normally, a ligand [L] binds reversibly with its receptor [R] to form a **ligand–receptor complex** [LR]:

$$\underset{\text{free ligand}}{[\mathrm{L}]} + \underset{\text{unoccupied receptor}}{[\mathrm{R}]} \rightleftharpoons \underset{\text{bound ligand = occupied receptor}}{[\mathrm{LR}]}$$

The affinity or association of the ligand for the receptor is described by an **association rate constant, k_1**, and the tendency for LR to dissociate to L + R is described by the **dissociation rate constant, k_2**. At equilibrium,

$$k_1[\mathrm{L}][\mathrm{R}] = k_2[\mathrm{LR}]$$

This can be expressed as

$$\frac{[\mathrm{L}][\mathrm{R}]}{[\mathrm{LR}]} = \frac{k_2}{k_1} = K_d$$

where K_d is the dissociation constant. The reciprocal of the dissociation constant is the association constant, K_a, and

$$k_1 = \frac{K_d}{K_a}$$

Therefore

$$K_a = \frac{[\mathrm{LR}]}{[\mathrm{L}][\mathrm{R}]}$$

From this equation, you can see that the K_a is equal to the ratio of bound to free ligand, or [LR]/[L].

If a fixed number of target cells (in other words, a fixed number of receptors) is incubated in replicates with increasing concentrations of ligand in each replicate, the number of occupied receptors also will increase until all of the available receptors are complexed to ligand (saturation or 100% bound). At saturation, the number of bound ligand molecules equals the number of available receptors while the ratio of bound ligand to free ligand approaches zero. If we plot the bound/free ratio for each replicate incubation against the bound concentration of ligand, the resulting straight line will intercept the y-axis of our graph at a point that defines the total number of receptors or the receptor capacity. This graph (Figure 2-11) is called a **Scatchard plot**. The slope of the line on the Scatchard plot is equal to the negative value of the K_a (or $-1/K_d$). Thus, determining only the bound ligand and the ratio of bound ligand to free ligand where the number of receptors is kept constant and only the concentration of free ligand varies allows one to extrapolate the number of receptors and the association constant for that particular receptor.

Most Scatchard plots, however, do not yield a straight line but rather a downward curving line. The most common explanation for repeated observations of this sort is that one is dealing with a heterogeneous mixture of high-affinity and low-affinity sites. Special mathematical methods are available to separate these binding sites and allow differentiation of the numbers of high- and low-affinity sites.

III. MOLECULAR BIOLOGY AND BIOREGULATION

In the past few years, the field of genomics has elucidated the entire genomes of certain species from humans to zebrafish to invertebrate chordates and has contributed immensely to the arsenal of molecular techniques available to the clinical and experimental endocrinologist for investigating bioregulation. More recently, efforts focused on the study of proteins, proteomics, are providing new tools to evaluate cellular events. Whereas the human genome, for example, contains only about 22,000 different genes, there are about 400,000 translated proteins and peptides that result from alternative splicing of RNAs and post-translational processing of the resulting proteins (see Chapter 3), indicating that knowledge based on genomics alone cannot explain the complexity of organisms. Only about 2% of genomes are involved directly in the synthesis of proteins. Nevertheless, genomic and proteomic tools are revolutionizing our understanding of bioregulation as well as how we approach the study of bioregulation.

A. Genetic and Genomic Approaches in Endocrinology

Use of the **polymerase chain reaction (PCR)** is another amplifying technique that allows the investigator to increase a small quantity of DNA to amounts that can be analyzed readily. PCR can be used to create cDNA probes, to detect mRNA synthesis even before new peptides or proteins are measurable, to screen for mutations of specific genes, and to amplify a specific mRNA using reverse transcription (making DNA from RNA instead of making RNA from DNA) coupled with PCR to amplify the product. cDNA probes can be prepared that bind to specific mRNAs and ascertain when a certain gene is turned on in a given cell. (cDNA is "copy" DNA made from RNA using a reverse transcription process.) The application of PCR technology to amplification of mRNAs provides a new way to measure even weakly stimulated gene activity following application of a bioregulator. **Real-time PCR**, also called *quantitative real-time PCR*, is a modification of the basic PCR technique that allows endocrinologists to simultaneously amplify and quantify mRNAs of interest. Microarray hybridization techniques using **DNA chips** (or gene chips) (Figure 2-12), sometimes encoding up to 44,000 DNA sequences, allow endocrinologists to measure changes in the expression of thousands of genes at one time.

Sites on DNA that interact with bioregulator receptors (such as steroid or thyroid hormone receptors, for example), or that bind to other transcription factors or proteins, can be identified using a technique called the **chromatin immunoprecipitation** or **ChIP assay** (Figure 2-13). In this technique, proteins and DNA within a cell lysate are allowed to bind together and are then precipitated with the use of a specific antibody. The bound DNA can then be sequenced. ChIP assays can also be used to identify histone alterations within chromatin (see discussion of epigenetics and transgenerational effects of EDCs in Chapter 1).

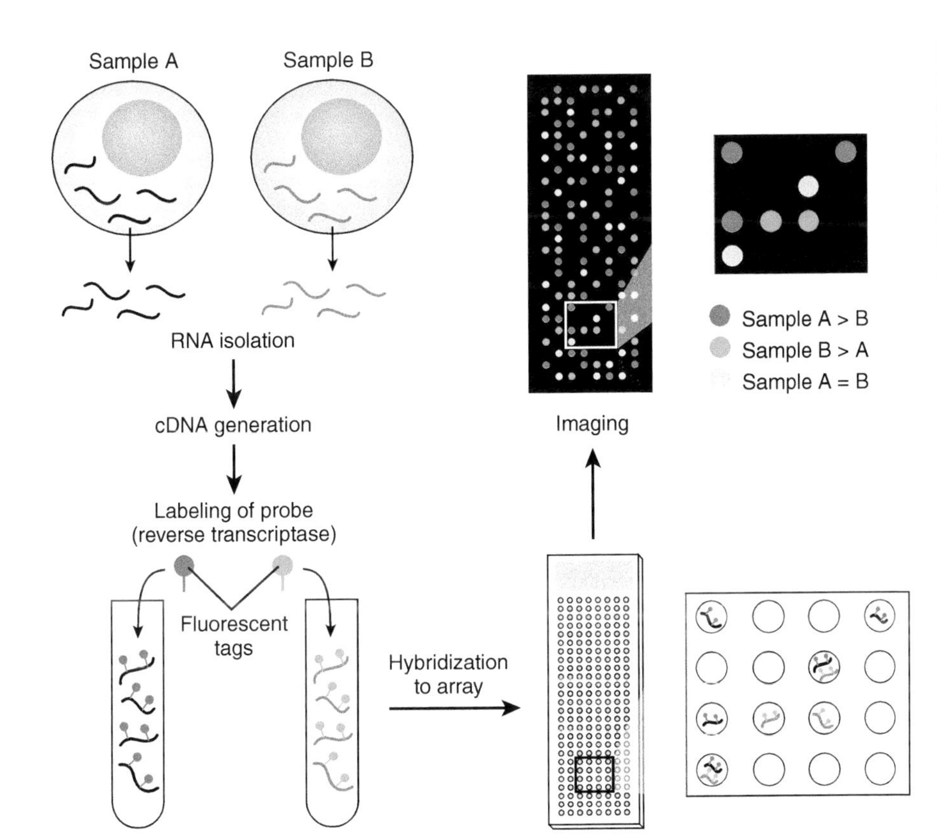

FIGURE 2-12 **Microarray hybridization.** Basic protocol for determining large scale changes in gene expression using microarray hybridization. RNA is isolated from cells or tissues and reverse transcribed to cDNA. cDNA is then used to produce fluorescently labeled copies of cRNA that are hybridized on microarrays containing as many as 44,000 gene targets.

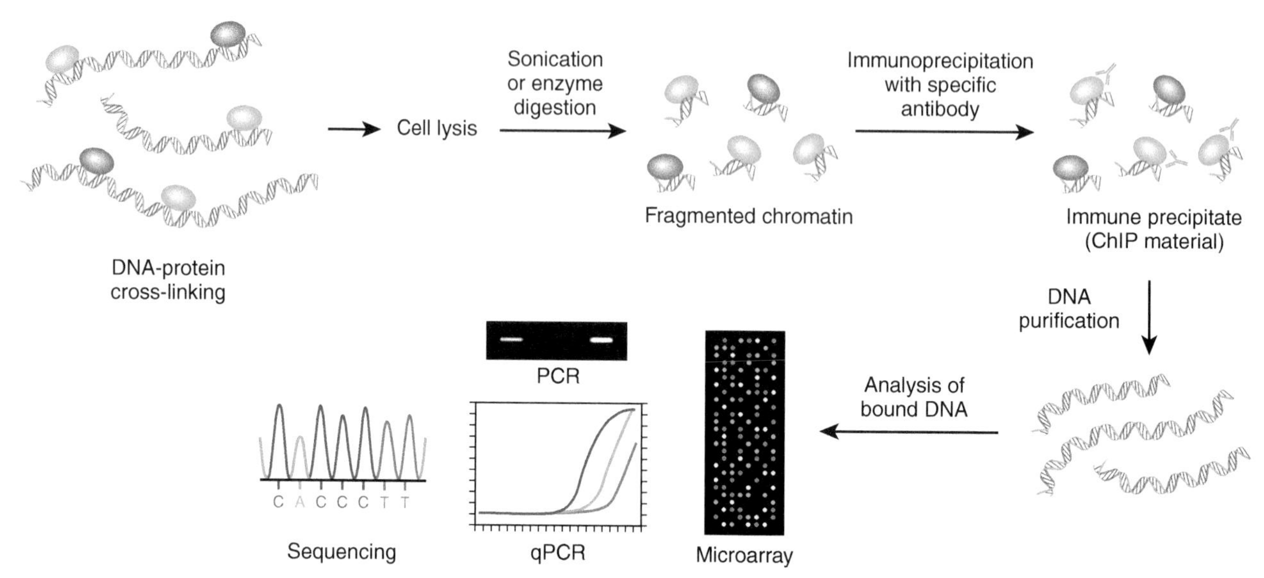

FIGURE 2-13 **Chromatin immunoprecipitation (ChIP) assay.** See text for explanation. *(Adapted with permission from Collas, P. and Dahl, J.A.,* Frontiers in Bioscience, *13, 929–943, 2008.)*

The presence of computerized banks of gene sequences for some or all of the genes in a given genome has proven extremely useful in identifying genes of interest to endocrinologists in other organisms. These approaches have helped discover related genes in other species as well as related gene products. Genomics has provided the identity of many receptor molecules (initially called *orphan receptors*) that eventually led to the discovery of their natural ligands and new insights into bioregulation. Being able to analyze entire gene sequences rather than just the amino acid sequence of the active peptide that represents only a fragment of the entire gene has enabled comparative endocrinologists to construct much more meaningful phylogenetic trees and thus gain a better grasp on evolutionary relationships.

Transgenic animals are lines of animals produced by inserting multiple copies of a specific gene or genes into a fertilized egg or early embryo. These genes may be responsible for producing a specific hormone such as growth hormone (GH). Insertion of GH genes into mice results in production of an exceptionally large phenotype, and this trait is passed on to their offspring. In other cases, investigators may insert genes complexed to "reporter" genes that code for some very unique protein and are linked to a known hormone response element where the hormone–receptor complex binds to the DNA (see Box 2A and Chapter 3). This reporter gene will only function once the appropriate transcription factor is present. Insertion of a gene for making the enzyme luciferase is a popular reporter gene as is the gene for making **green fluorescent protein (GFP)**. Over the last decade endocrinologists also have taken advantage of an elegant transgenic mouse model that adds molecular "switches" that tell the gene when to turn on or turn off and in which cells (in essence creating a "double" transgenic mouse). The Cre/Lox system, originally developed by Brian Sauer to engineer microbes, has been widely adapted to mammalian cells and mice for the

BOX 2A Gene Reporter Assays

Endocrinologists have increasingly turned to cell culture systems that express a single hormone receptor for rapidly testing a wide variety of compounds for the ability to interact and potentially disrupt estrogen, androgen, and thyroid hormone receptor signaling. In cell-culture-based gene reporter assays such as the yeast estrogen screen, or YES assay, cells are transgenically modified to overexpress a single hormone receptor—for example, the estrogen receptor in the case of the YES assay. The hormone receptor gene is attached to a "reporter" molecule that can be easily visualized or otherwise detected. Green fluorescent protein (GFP), luciferase, or the *lacz* gene in bacteria are common reporter molecules. Some reporter assays use yeast cells so that the receptor protein can be synthesized by pathways that are similar to higher eukaryotic cells but which do not normally possess hormone receptor to reduce background interference. A major advantage for cell-culture-based reporter assays is that they can be used for so-called high-throughput testing, in order to test thousands of chemicals in a relatively short period of time. In 1998, Congress authorized the U.S. Environmental Protection Agency to develop an Endocrine Disruptor Screening Program to test the more than 85,000 chemicals released into the environment for various commercial purposes. Gene reporter assays will obviously play a critical role in such testing because of their simplicity and the sheer number of chemicals that are required to be tested for endocrine-disrupting effects.

study of cell-specific gene expression. Cre (short for cyclization recombination) is a natural DNA recombinase enzyme that recognizes the *loxP* site on DNA to catalyze recombination between two consecutive *loxP* sites in the DNA. The DNA that lies between the *loxP* sites is thus excised. In order to generate mice with cell-specific expression of the gene of interest (Figure 2-14A), one must genetically engineer a mouse to contain a *lox* STOP cassette between the gene and promoter, thus inhibiting transcription of the gene of interest. When mated with a mouse that expresses Cre recombinase expressed in a particular cell phenotype, the *lox* STOP cassette is excised and the gene of interest is expressed only in that subset of cells that express Cre. Alternatively, by flanking a gene of interest with *loxP* sites, the gene can be excised when combined with Cre (Figure 2-14B).

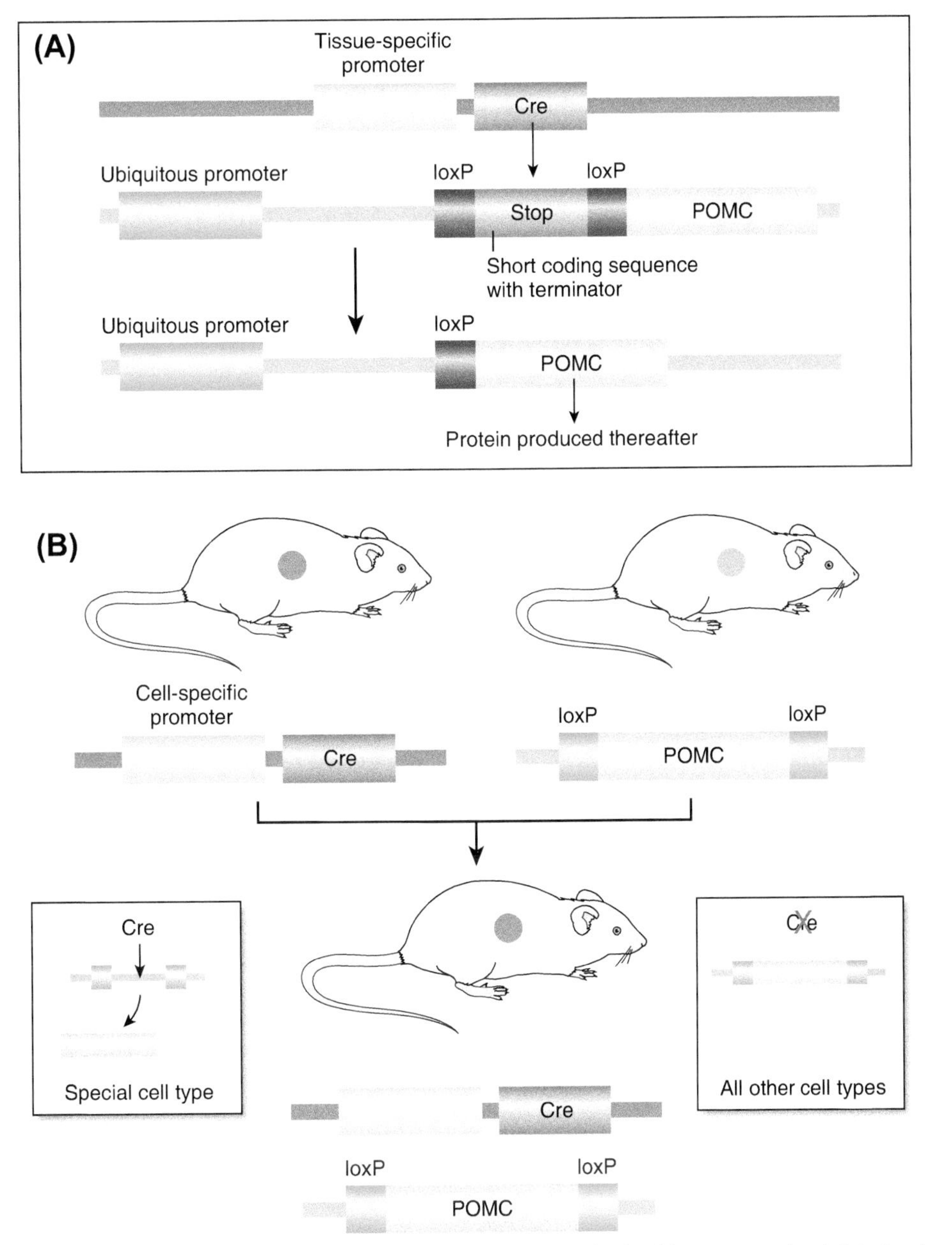

FIGURE 2-14 **Transgenic animals.** Use of Cre/loxP system for producing transgenic animals with gene expression deficits in selected target cells. (A) The pro-opiomelanocortin (POMC) gene cannot be expressed when loxP (locus of X-over P1) sites are present surrounding the stop sequence, despite the ubiquitous promoter that is active in many different cell types. When recombined with the gene encoding the DNA recombinase Cre, which is expressed only in special cell types because of the tissue-specific promoter, the loxP sites are deleted and the POMC sequence can be transcribed. (B) Transgenic mice expressing Cre in special cell types are mated with mice containing a construct in which the target gene is surrounded by loxP sites. Offspring will lack target gene expression in only those cells expressing Cre.

Genetic engineering techniques have allowed us to develop sensitive assays for the determination of minute quantities of hormones. For example, the use of yeast cells transfected with genes for the human estrogen receptor (called the **YES assay**) has provided a sensitive assay for detecting estrogenic compounds in solutions such as wastewater effluents or river water. The YES assay can identify the presence of estrogenic chemicals even when conventional chemical methods such as RIAs were unable to detect them.

Knockout (KO) animals in which a nonfunctional gene replaces the normal gene have proven very useful. Not only do they demonstrate the roles such genes normally play, but they can also provide model animals for studying disease states or for other molecular studies. Hence, βERKO mice are mice homozygous for a gene defective at producing the β-form of the estrogen receptor (ER). Endocrinologists also are increasingly using **RNA interference (RNAi)** techniques to silence genes of interest. RNA interference encompasses a broad set of tools that use modified RNA sequences to bind to mRNA and thereby reduce or eliminate translation of a protein. **Morpholinos** are small nucleotide sequences in an antisense orientation that also can be used to suppress, or "knock down," the expression of genes of interest.

B. Proteomics

Proteomics includes a variety of approaches ranging from protein separation, identification, and quantification to sequence and structural analyses of these molecules. Furthermore, proteomics is concerned with protein-to-protein interactions at the molecular and cellular levels as well as with posttranslational processing of these essential gene products. The workhorse of modern proteomics is a technique called **two-dimensional (2D) gel electrophoresis** (Figure 2-15). In the first step, proteins are separated based upon their isoelectric point (pI) using isoelectric focusing in a tube- or strip-shaped gel. In the second step, proteins are further separated using standard SDS-polyacrylamide gel electrophoresis (SDS-PAGE). Because it is unlikely that any two proteins will have identical migration patterns in both gels, many proteins can be separated with high resolution in one sample. Individual proteins can then by isolated from the SDS-PAGE gel and analyzed by liquid chromatography combined with mass spectroscopy (LC-MS or HPLC-MS) to identify their amino acid structure (see earlier discussion on high-performance liquid chromatography/ spectroscopy). Additional analytical approaches employed in proteomics research are x-ray crystallography and tomography, MRT, and a myriad of new approaches to studying protein–DNA interactions (see ChIP assay, above).

IV. ANIMAL MODELS

One of the most useful approaches in physiology research at all levels has been the development of specific species for use by many investigators studying a common problem.

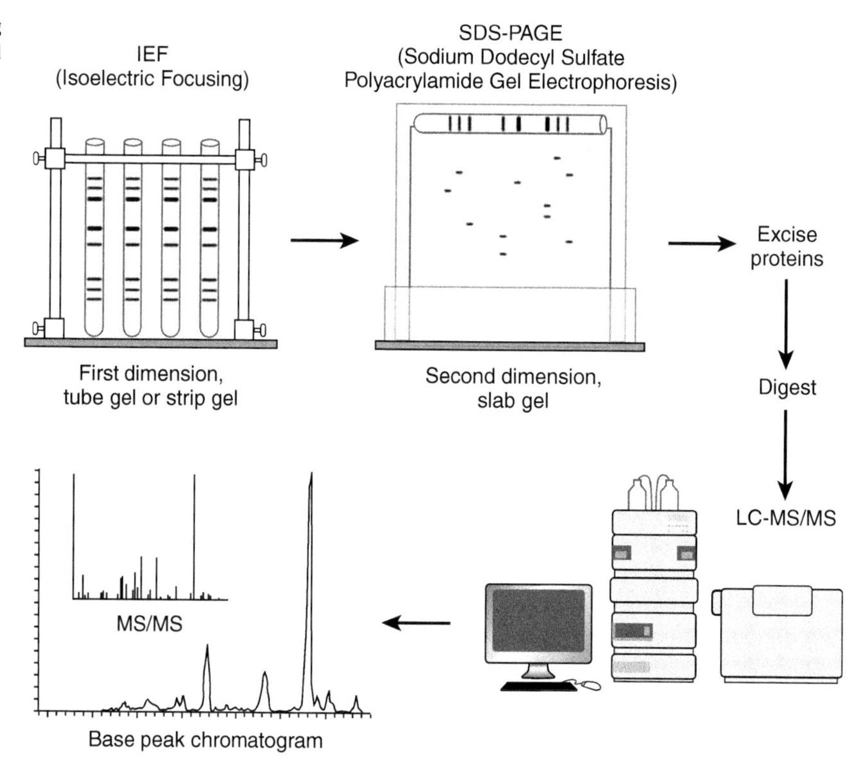

FIGURE 2-15 **Methodology for identifying proteins using 2D gel electrophoresis followed by LC-MS/MS.** See text for explanation.

Some of these species have been bred artificially for generations specifically for research purposes (for example, laboratory mice and rats) and still other organisms have been genetically engineered to express or not express particular traits. Natural and genetically modified animal models for studying clinical disorders have provided considerable insight for understanding of the onset and progression of endocrine-related disorders such as diabetes, obesity, and Alzheimer's disease. Often, non-mammalian species also have proven to be useful models for experimental research. Examples include the use of amphibian metamorphosis for the study of thyroid hormone and receptor interaction during development, the frog amphibian bladder for molecular investigations of the role of aldosterone in regulating sodium transport, and the nematode worm *C. elegans* in aging studies, as well as the development of specific bioassays for hormones employing specific species, such as the pigeon crop sac assay for prolactin or the fathead minnow, *Pimephales promelas*, serving as an indicator for estrogen in wastewater treatment plant effluents and polluted rivers.

A. Statistics

Regardless of the approaches used, proper statistical processing of experimental data is critical for establishing a database upon which to draw scientific conclusions and formulate new, testable hypotheses. Understanding which statistical test is appropriate for a given experimental design and what is required for its application should be an integral part of every experiment. Thus, knowledge of basic statistical procedures is essential before attempting to design experiments to test any hypothesis. Data collected may consist of a series of precise measurements or numerical scales based on subjective evaluations of such things as color reactions, behaviors, physical appearance, etc., determined under experimental and control conditions. The nature of these data and the number of variables examined dictate what statistical tests can be used. Use of initial controls, positive or negative controls, etc., as discussed previously will also influence the choice of analysis for the data. A working knowledge of statistics is essential for proper conduction of all endocrine research.

STUDY QUESTIONS

1. Identify some common biases that may interfere with a scientist's ability to interpret data.
2. How does a scientific theory differ from a scientific hypothesis?
3. Can you think of an example of a *falsifiable* hypothesis?
4. When testing the effects of a bioregulator, why is it important to also test the effects of the vehicle solution that the bioregulator is delivered in? What other types of controls might be important to include in designing an experiment to test the effects of a bioregulator?
5. What is meant by *confirmation bias*?
6. Identify factors that may influence the shape of a dose–response curve. What is the significance of a J-shaped dose–response curve?
7. Describe how a radioimmunoassay works and why this method relies on competition between labeled and unlabeled forms of a bioregulator. What are the advantages and disadvantages of radioimmunoassay methods over bioassay or ELISA approaches?
8. Imagine that you have identified a gene that encodes a novel peptide that may act as a bioregulator. You chemically synthesize the peptide.
   (a) Describe what approach you would take to *visualize* whether the peptide is expressed within individual cells within the pituitary gland.
   (b) How would you determine whether the gene encoding the peptide is differentially expressed under different physiological conditions?
   (c) Describe the approach that you would take to measure concentrations of the peptide in blood plasma.
9. Distinguish between *multiphoton* and *confocal laser microscopy*.
10. How do endocrinologists determine the sequence of DNA that may bind to a transcription factor or intracellular hormone receptor?
11. Why is it important to distinguish between specific and nonspecific binding when examining the ligand binding characteristics of a receptor?
12. Describe the types of information that can be determined from analysis of a Scatchard plot.
13. You have isolated one pituitary cell phenotype and are maintaining the cells in culture (*in vitro*). You are interested in determining which genes may be differentially transcribed when the cells are exposed to various bioregulators or drugs. What approach might you take to look at differential expression of thousands of genes at one time? Similarly, what approach might you take to broadly screen for changes in the proteins that are synthesized?
14. Describe how the Cre/loxP approach can allow an investigator to manipulate tissue specific gene expression.

SUGGESTED READING

Books

Bolander, F.F., 1989. Molecular Endocrinology. Academic Press, San Diego, CA.
Burry, R.W., 2010. Immunocytochemistry: A Practical Guide for Biomedical Research. Springer, New York.

Diaspro, A. (Ed.), 2002. Confocal and Two-Photon Microscopy: Foundations, Applications and Advances. Wiley-Liss, New York.

Handwerger, S., Aronow, B.J. (Eds.), 2008. Genomics in Endocrinology: DNA Microarray Analysis in Endocrine Health and Disease. Humana Press, Totowa, NJ.

Articles

Brabant, G., Cain, J., Jackson, A., Kreitschmann-Andermahr, I., 2011. Visualizing hormone actions in the brain. Trends in Endocrinology & Metabolism 22, 153–163.

Budry, L., Lafont, C., El Yandouzi, T., Chauvet, N., Conéjero, G., Drouin, J., Mollard, P., 2011. Related pituitary cell lineages develop into interdigitated 3D cell networks. Proceedings of the National Academy of Sciences USA 108, 12515–12520.

Calabrese, E.J., Baldwin, L.A., 2003. Toxicology rethinks its central belief. Nature 421, 691–692.

Carter, D., 2006. Cellular transcriptomics—the next phase of endocrine expression profiling. Trends in Endocrinology & Metabolism 17, 192–198.

Chakraborti, P.K., Weisbart, M., 1987. High affinity cortisol receptor in the liver of the brook trout, *Salvelinus fontinalis*. Canadian Journal of Zoology 65, 2498–2503.

Collas, P., Dahl, J.A., 2008. Chop it, ChIP it, check it: the current status of chromatin immunoprecipitation. Frontiers in Bioscience 13, 929–943.

Clynen, E., De Loof, A., Schoofs, L., 2003. The use of peptidomics in endocrine research. General and Comparative Endocrinology 132, 1–9.

Frontini, A., Bertolotti, P., Tonello, C., Valerio, A., Nisoli, E., Cinti, S., Giordano, A., 2008. Leptin-dependent STAT3 phosphorylation in postnatal mouse hypothalamus. Brain Research 1215, 105–115.

Koopman, W.J., Scheenen, W.J., Errington, R.J., Willems, P.H., Bindels, R.J., Roubos, E.W., Jenks, B.G., 2001. Membrane-initiated Ca(2+) signals are reshaped during propagation to subcellular regions. Biophysics Journal 81, 57–65.

Piret, S.E., Thakker, R.V., 2011. Mouse models for inherited endocrine and metabolic disorders. Journal of Endocrinology 211, 211–230.

Sauer, B., 1998. Inducible gene targeting in mice using the Cre/lox system. Methods 14, 381–392.

Walker, D.M., Gore, A.C., 2011. Transgenerational neuroendocrine disruption of reproduction. Nature Reviews Endocrinology 7, 197–207.

Chapter 3

Synthesis, Metabolism, and Actions of Bioregulators

The chemical properties of each bioregulatory molecule are keys to understanding much of its physiology. They prescribe not only how a bioregulator is synthesized but also how it is secreted and transported to its target site, where its receptors are located, how it produces its effects on a specific target, and how it is metabolized or inactivated. Most bioregulators, such as monoamines, small peptides, polypeptides, or proteins, are at home in aqueous media. In sharp contrast, steroids, thyroid hormones, and eicosanoids have low solubility in aqueous media, and, unlike peptides, more readily pass through cell membranes. These hydrophobic bioregulators have markedly different synthetic pathways and processes of secretion, transport, and action on target cells. Because the features of each group of bioregulators are uniquely tied to their chemical composition, each major chemical type is discussed separately: (1) amino acids, amines, peptides, and proteins; (2) steroids; (3) thyroid hormones; and (4) eicosanoids.

I. AMINO ACIDS, AMINES, PEPTIDES, AND PROTEINS

Most neurotransmitters, neuromodulators, neurohormones, and classical hormones as well as many bioregulators common to interstitial fluids are composed of linear sequences of amino acids linked together by peptide bonds (see Table 3-1). These peptides vary in length from a tripeptide known as *thyrotropin-releasing hormone* (TRH) to large molecules of 200 or more amino acids (e.g., growth hormone, GH). Some common bioregulators are single amino acids (e.g., the neurotransmitter glutamate) or modified amino acids such as the catecholamines, indoleamines, and thyroid hormones.

A. Catecholamine Bioregulators

A **catechol** is an unsaturated six-carbon ring (phenolic group) with two hydroxyl groups attached to adjacent carbons (dihydroxyphenol) (see Figure 3-1). Attachment of the catechol ring to a side chain with an amine group characterizes a **catecholamine**. Catecholamines are synthesized in the cytosol from the amino acid tyrosine (also called *hydroxyphenylalanine*) (see Figure 3-1). Addition of one more hydroxyl (OH) group to the phenolic ring of tyrosine is followed by removal of the carboxyl group from the alanine portion to yield a catecholamine.

Three important catecholamines are synthesized from tyrosine by neurons in the brain and ganglia as well as in cells of the adrenal medulla. The addition of an OH group to tyrosine by the rate-limiting enzyme **tyrosine hydroxylase** yields dihydroxyphenylalanine, or **DOPA**. Next, the carboxyl group is removed by the enzyme **DOPA decarboxylase** to form the catecholamine **dopamine**, a common neurotransmitter in the central nervous system. In some neurons, an additional enzyme, **dopamine β-hydroxylase**, converts DA to the catecholamine neurotransmitter **norepinephrine** by addition of an OH group to the former alanine side chain. In still other neurons, norepinephrine is further altered by addition of a methyl group to the amine group to make the

TABLE 3-1 Amino Acid Composition of Some Peptide Bioregulators

Name	Bioregulator Type	Structure
Thyrotropin-releasing hormone[a]	Neurohormone	Pyro-E-H-P-$CONH_2$
Substance P[b]	Neuromodulator	NH_2-R-P-K-P-Q-Q-F-F-G-L-M-$CONH_2$
Glucagon	Hormone	NH_2-H-S-Q-G-T-F-T-S-D-Y-S-K-Y-L-D-S-R-R-A-Q-D-F-V-Q-W-L-M-N-T-COOH

See Appendic C for explanation of amino acid designations.
[a]*Amino (pyro) and carboxyl(amide) ends are modified.*
[b]*Substance P is amidated at the carboxy end.*

Vertebrate Endocrinology. http://dx.doi.org/10.1016/B978-0-12-394815-1.00003-3

FIGURE 3-1 Synthesis of catecholamines. Catecholamines may be synthesized from either of the amino acids phenylalanine or tyrosine. The rate-limiting enzyme for this pathway is tyrosine hydroxylase. Depending upon which enzymes are active in a cell, the final secretory product may be dopamine, norepinephrine, or epinephrine. A catechol group consists of a benzene ring with two adjacent hydroxyl groups attached (highlighted).

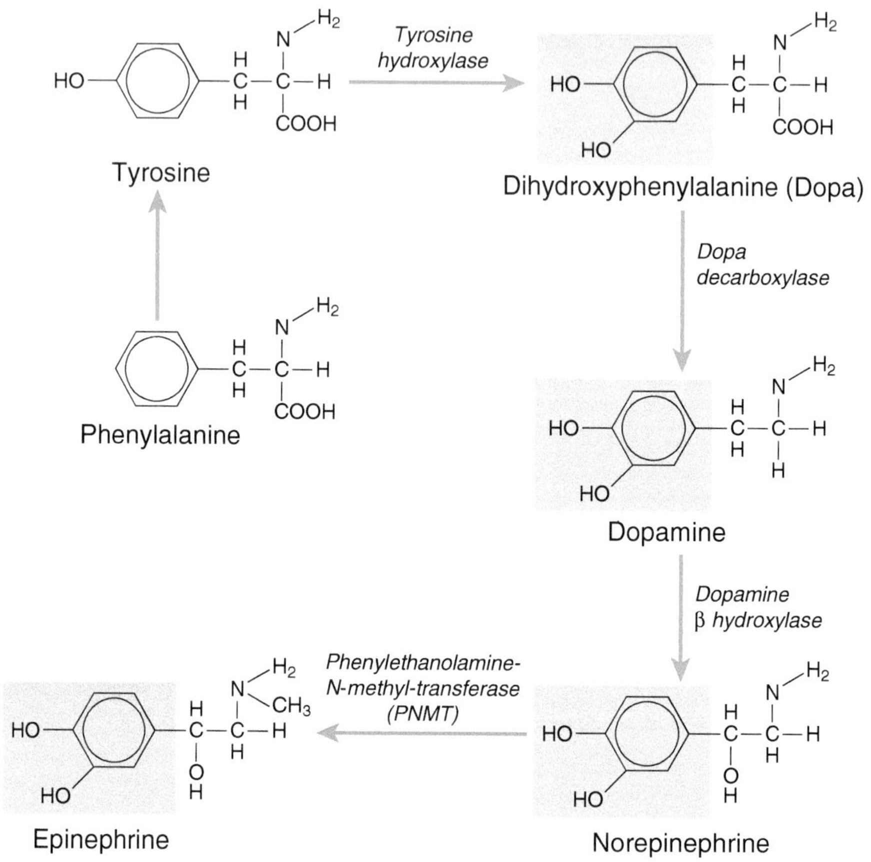

Tyrosine

Tyrosine hydroxylase

Dihydroxyphenylalanine (Dopa)

Phenylalanine

Dopa decarboxylase

Dopamine

Dopamine β hydroxylase

Norepinephrine

Phenylethanolamine-N-methyl-transferase (PNMT)

Epinephrine

catecholamine neurotransmitter **epinephrine**. This last conversion is catalyzed by the enzyme **phenylethanolamine *N*-methyltransferase (PNMT)**. Tyrosine hydroxylase immunoreactivity is often used as a cytochemical marker to locate catecholamine-secreting cells, although it does not indicate whether the final secretory product of that cell is dopamine, norepinephrine, or epinephrine.

Catecholamines are more than just neurotransmitters. Dopamine, norepinephrine, and epinephrine can all be released into the circulation where they exhibit endocrine functions; for example, the hypothalamus releases dopamine that acts as a neurohormone to inhibit release of prolactin from the pituitary gland (see Chapter 4). The adrenal medulla, a modified sympathetic ganglion, secretes norepinephrine and epinephrine into the blood in response to neural signals directed via sympathetic nerve pathways coming from the hypothalamus. Although traditionally recognized as hormones, these adrenal medullary secretions also could be called neurohormones. Finally, in the central nervous system, there is evidence that norepinephrine is acting as a paracrine bioregulator, too.

Release of catecholamine neurotransmitters is followed by their partial reuptake and recycling and/or degradation by the secreting neuron to free the postsynaptic receptors and turn off the postsynaptic cell. The intraneuronal enzyme complex called **monoamine oxidase (MAO)** is responsible for degrading catecholamines. Local neuroglial cells also may participate in the degradation of these neurotransmitters. A second enzyme, **catechol-*O*-methyl transferase (COMT)** in the nervous system, is important in the degrading of catecholamines by nearby neuroglial cells. Norepinephrine and epinephrine in the peripheral circulation are metabolized primarily by liver MAO and aldehyde oxidase to produce slightly different metabolites than those found in the brain.

Numerous agonists (mimics) and antagonists (inhibitors of receptor binding) for catecholamines have been developed as well as inhibitors of MAO and COMT. Other drugs have been developed that block release of the catecholamines. A partial listing of these compounds is provided in Table 3-2. The development of specific agonists and antagonists of neurotransmitters has allowed the investigation of the roles of catecholamine neurotransmitters in the regulation of endocrine function by the brain.

B. Indoleamine Bioregulators

Synthesis of the indoleamines is outlined in Figure 3-2. The amino acid tryptophan is hydroxylated by the enzyme tryptophan hydroxylase to yield 5-hydroxytryptophan. This

TABLE 3-2 Some Pharmacological Compounds That Alter Catecholamine Activity

Compound	Action
Amphetamine	Potent agonist of α-and β-adrenergic receptors that binds especially well to central nervous system receptors but poorly to peripheral receptors
Ephedrine	An agonist that binds to α- and β-adrenergic receptors
Haloperidol	Antagonist that binds competitively to dopaminergic receptors and blocks effects of dopamine on its target cells
Harmaline	Inhibitor of monoamine oxidase (MAO), an important enzyme for degradation of catecholamines
Isoproterenol	β_1-and β_2-adrenergic receptor agonist; mimics both epinephrine and norepinephrine
Phentolamine	An α-antagonist that competively blocks norepinephrine action
Propranolol	Antagonist that binds competitively to β_1-and β_2-adrenergic receptors and prevents actions of epinephrine or norepinephrine; consequently, propranolol is called a β-blocker
Pyrogallol	Blocks the enzyme catechol *O*-methyltransferase (COMT), an enzyme necessary for metabolism of catecholamines
Reserpine	Blocks reuptake of catecholamines by presynaptic neurons

Sertraline—Inhibitor of dopamine reuptake by presynaptic cells; also inhibits serotonin reuptake
Clonidine—α_2-adrenergic receptor agonist
Phenylephrine—α_1-adrenergic receptor agonist
Apomorphine—Dopamine receptor agonist

metabolite is converted by L-aromatic amino acid decarboxylase to the neurotransmitter 5-hydroxytryptamine, or **serotonin**. A form of MAO degrades serotonin to inactive **5-hydroxyindoleacetic acid (5-HIAA)**. In the pineal gland, the enzyme ***N*-acetyltransferase (NAT)** alters serotonin to an *N*-acetylated form that in turn is altered by **hydroxyindole-*O*-methyltransferase (HIOMT)** to the pineal neurohormone, **melatonin**, or *N*-acetyl-5-methoxytryptamine (see Chapter 4). NAT is considered to be the rate-limiting enzyme for melatonin synthesis. P450 cytochrome enzymes primarily in the liver metabolize melatonin to 6-hydroxymelatonin, which is then conjugated with sulfate and excreted by the kidneys.

Melatonin is secreted primarily during the dark (**scotophase**) and appears to be important in regulating cyclical functions as well as having negative influences on thyroid and reproductive functions. Daytime levels are very low. Cessation of melatonin secretion in children can lead to precocial sexual maturation (see Chapter 10). Some agonists and antagonists of serotonin and melatonin functions are provided in Table 3-3.

C. Peptide and Protein Bioregulators

Similar to most peptides and proteins destined for export from the cell, peptide and protein bioregulators are synthesized at the ribosome-studded **rough endoplasmic reticulum (RER)** in the cytoplasm according to directions encoded in nuclear genes composed of deoxyribonucleic acid (DNA). Genes are responsible for the production of structural (e.g., microtubule) and functional (e.g., enzymes, antibodies) proteins. A molecule of ribonucleic acid (RNA) is transcribed from the molecule of DNA by an enzyme called **RNA-polymerase**. This enzyme attaches to the promoter region of the DNA. Following transcription of the RNA, intervening sequences corresponding to the nucleotides making up the **introns** of the parent DNA are excised from the RNA, and the remaining pieces (representing **exons** in the DNA) are spliced together to form the messenger RNA (mRNA) that leaves the nucleus and travels to the ribosome of the RER for translation into an amino acid sequence (i.e, a polypeptide). Alternative processing of the original RNA transcript sometimes results in **splice variants** of mRNA that can result in translation of different forms of the peptide originally coded in the nuclear DNA.

RNA polymerase does not always have ready access to promoter sites on the DNA, and one or more **transcription factors** may be necessary to expose the promoter to RNA polymerase. Transcription factors are cytoplasmic proteins that typically migrate into the nucleus, bind to the bioregulatory sites on a gene, and enhance transcription of that gene.

1. Translation and Posttranslational Events

The nucleotide sequence of mRNA is translated into a linear sequence of amino acids at ribosomes. For products destined for export from the cell, this sequence of amino acids is called a **prepropeptide** (Figure 3-3). A special sequence of amino acids located at the amino terminal end of the prepropeptides is termed the **signal peptide**. The newly synthesized signal peptide sequence is synthesized first and recognized quickly by a special particle called a **signal recognition particle**. Attachment of the signal recognition particle to the translation complex halts further translation. The signal recognition particle also recognizes a specific **docking protein**, a special receptor embedded in the RER membrane. Thus, the prepropeptide is attached to the RER membrane, the

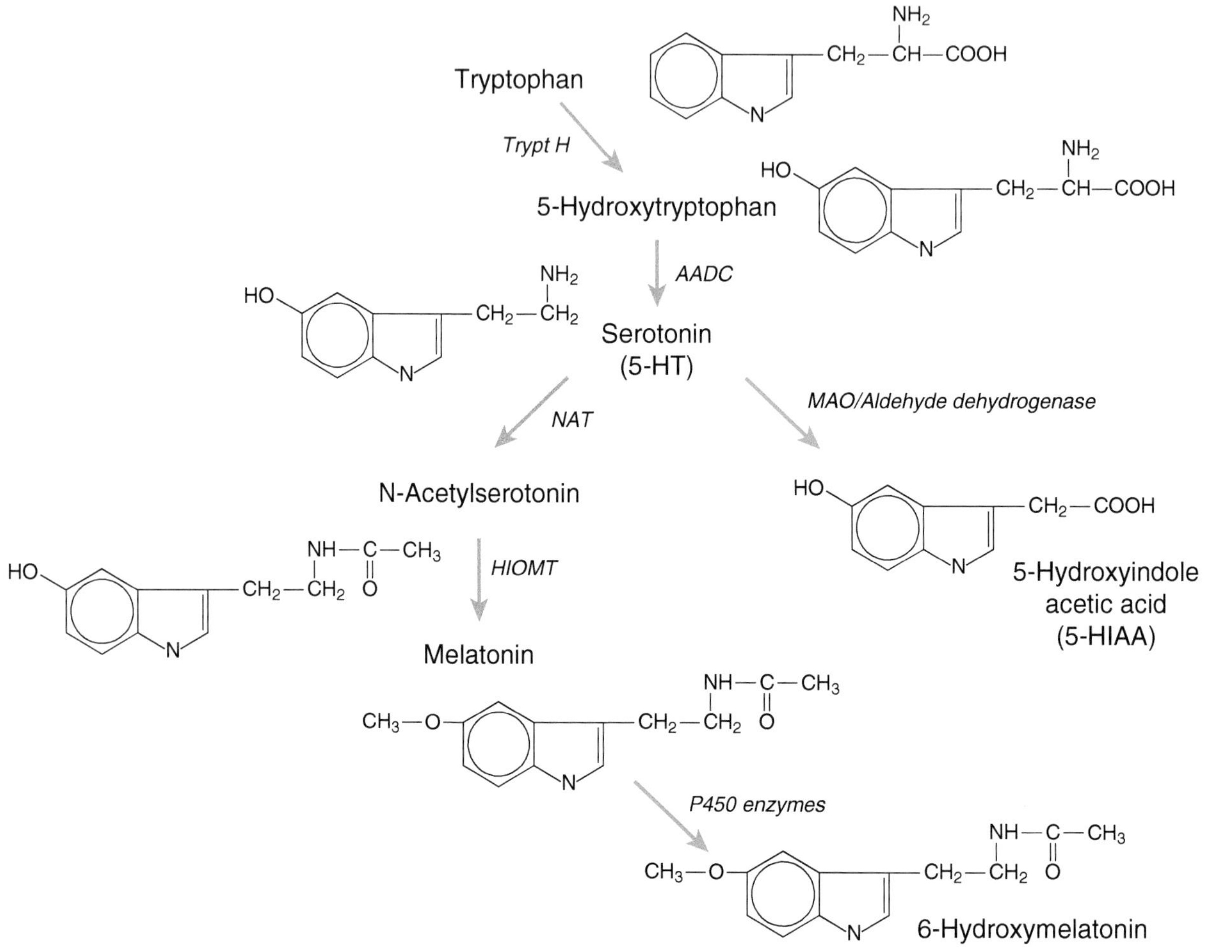

FIGURE 3-2 **Synthesis of indolamines: serotonin and melatonin.** 5-HIAA is a principal metabolite of serotonin and 6-hydroxymelatonin is the principal metabolite of melatonin. The rate-limiting enzyme for melatonin synthesis is *N*-acetyltransferase (NAT). Abbreviations: AADC, aromatic-L-amino acid decarboxylase; HIOMT, hydroxyindole-*O*-methyltransferase; MAO, monoamine oxidase; TryptH, tryptophan hydroxylase.

signal recognition particle detaches, and translation is resumed on membrane-bound ribosomes of the RER. While the hydrophobic signal peptide is firmly attached to the membrane, the remainder of the prepropeptide, called the **propeptide**, is synthesized and intruded through the membrane into the cisterna of the RER. Once in the cisterna, the propeptide can migrate to vesicle-forming regions of the RER and be packaged into vesicles for translocation to the Golgi apparatus. The signal peptide does not enter the RER.

Additional posttranslational processing of the propeptide may occur within the RER, in the Golgi apparatus, or possibly in the storage granules prior to release from the cell. Typically, enzymes will remove a portion or portions of the propeptide to produce the final peptide or peptides destined for export. Endocrinologists refer to the precursor forms of a peptide hormone as a **preprohormone** (prepropeptide) and a **prohormone** (propeptide). For example, the prohormone for the pancreatic hormone insulin consists of a long polypeptide folded through the formation of

TABLE 3-3 **Some Pharmacological Compounds That Alter Indolamine Activity**

Compound	Action
Fenfluramine	Blocks reuptake of serotonin by presynaptic neurons
Fluoxetine	Selective serotonin reuptake inhibitor (SSRI)
Harmaline	Inhibitor of monoamine oxidase (MAO), an important enzyme for degradation of indoleamines
Lysergic acid diethylamide	Antagonist of serotonin at receptor sites
Methysergide	Antagonist of serotonin at receptor sites

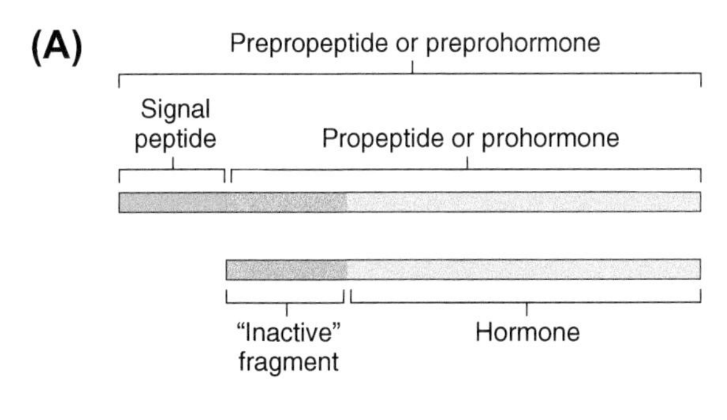

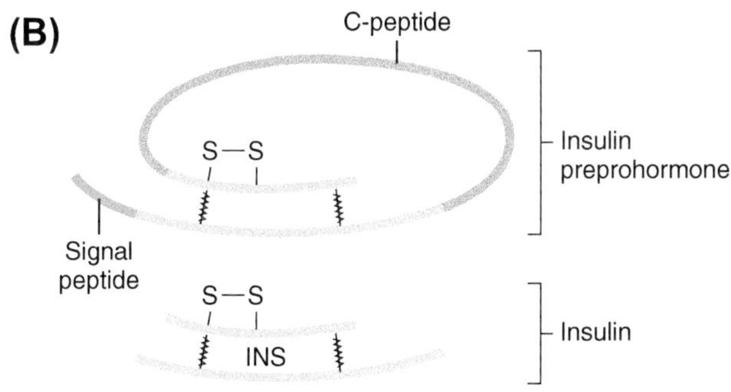

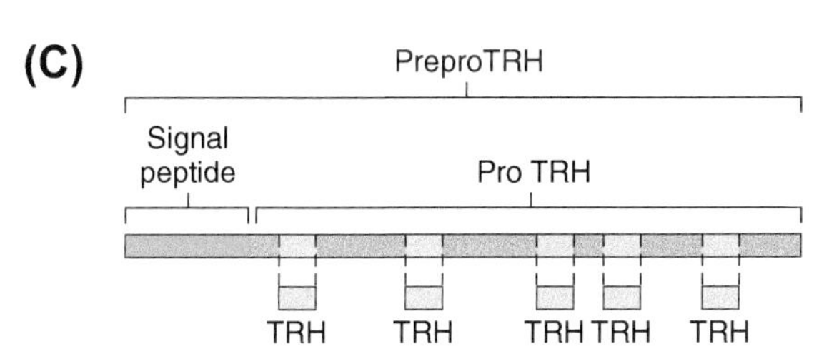

FIGURE 3-3 **Synthesis of export peptides.** (A) The product of mRNA produced at the ribosome is the preprohormone. The signal peptide is necessary to connect the prohormone to the endoplasmic reticulum and is cut off from the prohormone, which then enters the cisternae of the endoplasmic reticulum. The prohormone is later cleaved to produce an inactive fragment and the definitive hormone. Typically, both the inactive fragment and the hormone will be released from the cell. Sometimes the entire prohormone may be released, as well. (B) The hormone insulin is synthesized from the preprohormone by first removing the signal peptide, folding the single peptide chain of the prohormone and cleaving it in two places to yield a connecting C-peptide fragment and the hormone insulin that now appears to be made of two separate polypeptide chains. Some proinsulin is secreted along the the C-peptide and insulin. (C) Five copies of the TRH tripeptide are produced by multiple cleavages of each prohormone.

disulfide bonds between cysteine residues located in different parts of the chain. A special enzyme, a prohormone convertase, cleaves off a connecting sequence known as the **C-peptide**, leaving what appears to be two separate peptides (A-peptide and B-peptide) joined together by disulfide bonds. This resulting molecule is known as insulin (see Figure 3-3). Both the C-peptide and insulin are released from the cell although no peripheral function is known for the C-peptide. Enzymatic cleavage may occur at several sites along the prohormone as is the case for the tripeptide (TRH) (see Figure 3-3). This results in production of five identical TRH tripeptides from each prohormone molecule, thus amplifying the amount of neurohormone synthesized.

For certain peptide products, other compounds such as carbohydrates or lipids may be added as well. Posttranslational processing may involve additions to the basic prohormone as well as deletions. For example, some peptides are amidated or acetylated. Such changes increase their resistance to degradation and improve their binding to receptors. In the synthesis of **glycoprotein hormones (GpHs)** in the anterior pituitary, carbohydrates are complexed to two separate propeptides which are then joined together to yield the biologically active hormones. In other cases, there is no posttranslational processing, and the translated peptide enters the RER ready for export.

Another variation in posttranslational processing can produce more than one biologically active species from the same preprohormone depending on what processing enzymes are involved in different cell types. For example, the prohormone **proopiomelanocortin (POMC)** is cleaved to produce the pituitary tropic hormone **corticotropin (ACTH)** in a particular cell type in the pars distalis region of the pituitary, whereas different processing enzymes in pars intermedia cells of the pituitary release **melanotropin (α-MSH)** from proopiomelanocortin (see Chapter 4 for details about proopiomelanocortin and other enzymatic products).

2. Homologies in Peptide and Protein Structure

Analysis of bioregulatory peptides and proteins shows that it is possible to group molecules with common amino acid sequences into families of related peptides. In many cases, the genes responsible for directing the synthesis of structurally related or homologous peptides are also similar structurally with respect to their nucleotide sequences that ultimately code for these peptides and proteins. Some examples of homologous families are provided in Table 3-4, Table 3-5, and Figure 3-4. Many argue that the peptides or proteins that represent a single family had, in an evolutionary sense, a common ancestral gene that, through duplications and subsequent independent, random mutations, evolved via natural selection into a family of closely related genes. In the case of the glycoprotein pituitary hormones that are each composed of two peptide subunits (α and β), there has been little change in the gene coding for the α-subunit, but the β-subunit shows considerable variation related to the different functions that have evolved for these hormones (see Chapter 4 for more details).

Smaller variations may exist between particular peptides or proteins isolated from different species. For example, the primary structure of pituitary ACTH isolated from humans differs by only one amino acid from ACTH produced in sheep and pigs. Because ACTH is cleaved from a much larger prohormone, one should expect many more differences between the genes that produce ACTH in these species.

TABLE 3-4 Families of Peptide Bioregulators

Family Member	No. of AA Residues	No. AA Common to Molecule 1 or Other Homologous Feature
Neurohypophysial nonapeptides		
1. Arginine vasopressin (AVP)	9	9
2. Lysine vasopressin (LVP)	9	8
3. Arginine Vasotocin (AVT)	9	8
4. Mesotocin (MST)	9	7
5. Oxytocin (OXY)	9	7
6. Phenypressin (PVP)	9	8
Glucagon-secretin family		
1. PACAP-27	27	27
2. Secretin	27	11
3. PHI/PHAM	27	14
4. VIP	28	19
5. Glucagon	29	10
6. GLP-1	30	
7. GLP-2	35	
8. GIP	42	6
9. GHRH (GRH,GRF)	44	8
10. PACAP-related peptide (PRP)	48	
Insulin-like peptides		
1. Insulin ($\alpha + \beta$ chain)	51	51
2. Relaxin ($\alpha + \beta$ chain)	54	53
3. IGF-1 (single chain)	70	24
4. IGF-2 (single chain)	67	29
Endothelins		
1. Endothelin 1	21	21
2. Endothelin 2	21	18
3. Endothelin 3	21	15
4. Sarafotoxin S6b	21	14
CRH-like peptides		
1. Human CRH	41	41
2. Sheep CRH	41	34
3. Sauvagine	39	18
4. Urotensin 1 (carp)	41	22
5. Urocortin-1	40	

TABLE 3-4 Families of Peptide Bioregulators—cont'd

Family Member	No. of AA Residues	No. AA Common to Molecule 1 or Other Homologous Feature
POMC-related peptides		
1. ACTH 39	39	39
2. α-MSH	13	1–13 of ACTH
4. β-endorphin	31	
5. DynorphinA	16	1–4 like β-endorphin
6. met-Enkephalin	5	50–63 of β-endorphin & 1–5 of dynorphin

See text or Appendix A for explanation of abbreviations.

TABLE 3-5 Families of Human Protein Bioregulators

Family Name	Members	Molecular Weight[a]	Common Features
Pituitary/placental glycoproteins	1. TSH	32,000	α Subunit of 89 AA, β Subunit, 112 AA
	2. FSH	32,000	α Subunit of 89 AA, β subunit,115 AA
	3. LH	32,000	α Subunit of 89 AA, β subunit,115 AA
	4. CG	38,000	Like LH in structure and action
Nonglycoprotein pituitary/placental hormones	1. GH	22,500	191 AA
	2. Prolactin	23,000	198 AA
	3. CS	22,000	191 AA
TGF-β superfamily	1. TGF-β1		Composed of α and β subunits
	2. Bone morphogenic proteins (BMPs)	12–30,000	Composed of α and β subunits
	3. Anti-müllerian Hormone (AMH)	140,000	Composed of α and β subunits
	4. Inhibins	29–32,000	Composed of α and β subunits
	5. Activins	28,000	Composed of α and β subunits

[a]*Approximate molecular weights.*
See Appendix A for explanation of abbreviations.

The sequence of amino acids gives rise to the primary structure of a peptide or protein. The secondary and tertiary structures that are based largely on the primary structure also may be similar, resulting in considerable overlap in their three-dimensional structures. This may cause overlap with respect to the receptors that bind these similar molecules. Such overlap in binding may yield confusing experimental results, especially when large doses of a bioregulator are applied. In such cases, the applied bioregulator normally may not compete effectively for binding to receptors of a chemically related bioregulator. However, by adding excessive amounts, sufficient binding to a related receptor may produce actions that are not physiologically associated with that particular bioregulator.

3. Transport of Peptide Bioregulators to Target Cells

Completed peptide and protein bioregulators along with peptide fragments separated during posttranslational processing are packaged into secretory vesicles or granules at the Golgi apparatus and are retained or stored in the cytoplasm until released from the cell through exocytosis into the interstitial fluids or blood. Bioregulators released into interstitial fluids (i.e., cytocrines, autocrines,

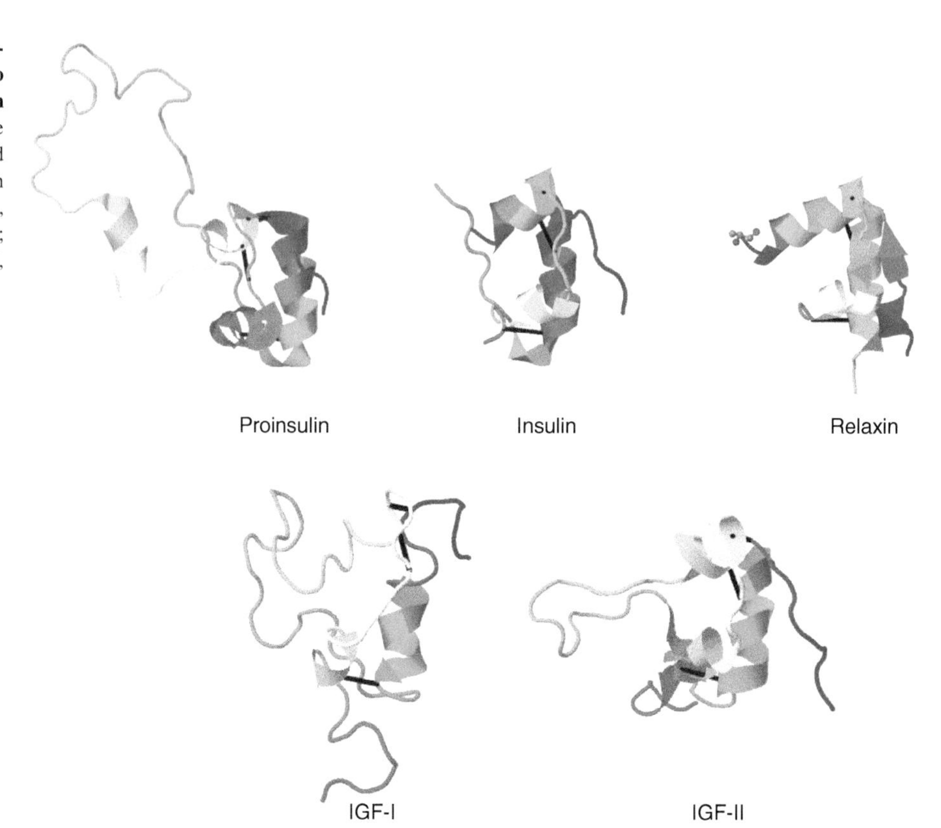

FIGURE 3-4 **Colorized representations of the insulin peptides (amino terminus in blue, carboxy terminus in red).** Disulfide bonds are indicated by the thick black lines. Peptides are modeled after protein data base structures (human pro-insulin, code 2KQP; human T insulin, code 1MSO; human relaxin, code 6RLX; human IGF-I, code 1BQT; human IGF-II, 1IGL).

neurotransmitters, neuromodulators) rely on diffusion and any movement of the interstitial fluid that might occur to allow them to reach their target sites more rapidly. In the blood, bioregulatory peptides may complex to specific plasma proteins that enhance their transportability in the blood, slow their rate of metabolism by blood peptidases, and may even facilitate interactions with target cells.

The time for removal or rate of disappearance of any bioregulator from the blood through degradation, excretion, or removal by target cells is termed its **biological half-life**. This usually is determined by following the rate of removal of radioactively labeled molecules from the blood. The time required to remove or clear half of the labeled dose from the blood is the half-life. In general terms, a larger peptide will have a longer biological half-life than a smaller peptide (see Table 3-6). Notable exceptions are the smallest peptides such as TRH that are modified during post-translational processing in a manner that slows their enzymatic degradation while in the blood. A number of hormones possess special modifications or activities that prolong their presence in the blood (see below).

D. Receptors for Amine, Peptide and Protein Bioregulators

Although amino acids, amines, peptides, and proteins are soluble in blood plasma and interstitial fluids, they cannot readily pass through cell membranes that are composed largely of phospholipids and cholesterol. Instead, they bind to **receptor proteins** embedded in the target cell membrane. A **target cell** for a peptide bioregulator is a cell possessing receptor proteins in its external membrane that specifically bind the bioregulator. A molecule that binds to a binding site on a receptor is termed a **ligand**. Receptor binding of a ligand employs a specific lock-and-key mechanism like that of an enzyme and its substrate. Binding of the ligand is followed by observable changes in the target cell that are characteristic for each ligand–receptor complex. These changes could involve (1) opening or closing of ion channels affecting membrane potentials and/or secretion, (2) activating or inactivating of an enzyme, (3) initiating a series of reactions or cascading events, and/or (4) activating transcription factors that alter gene activity. Changes caused by such mechanisms could affect development and differentiation, cellular biochemistry, and morphology, as well as general aspects of physiology and behavior.

A receptor molecule consists of several functional regions or **domains**. One of these domains has a three-dimensional **binding site** for the bioregulator that recognizes the shape of this bioregulator and will not bind any other molecule unless it closely or exactly conforms to that shape (Figure 3-5). This relationship is like the lock-and-key fit so often described for an enzyme (the receptor, in this case) and its substrate (the bioregulator). Once a ligand is

TABLE 3-6 Biological Half-Lives of Some Mammalian Hormones

Hormone	Approximate Mol. Wt.	Species	Approximate half-life (min)
TRH	300	Mice	2
OXY	900	Rat	2
		Human	5
AVP	900	Rat	3–4
α-MSH	1300	Dog	2
GnRH	1000	Pig	12
		Dog	5
		Human	2–5
Gastrin	1700	Human	7–8
ACTH	3900	Rat	1–4
		Pig	5–7
		Human	5–29
Glucagon	2900	Human	5–10
β-Endorphin	3100	Human	15
Calcitonin	3200	Rat	2
		Human	3
		Pig	2–5
CCK	3300	Human	2
Insulin	5100	Pig	6
		Human	3–4
Proinsulin	8200	Pig	20
		Human	18–25
GH	20,000	Human	20–30
PRL	19,800	Rabbit	16
		Cow	29
LH	32,000[b]	Human	136
FSH	32,000[b]	Human	220
Cortisol	300[b]	Human	90
Aldosterone	300[b]	Human	35
T_4	800	Human	7 days
T_3	700	Human	24 hr

[a]*Number of amino acids or*
[b]*molecular weight; except for glycoproteins, the number of amino acids × 100 gives an approximate estimate of molecular weight of any peptide (120 × is more accurate but not as easy to calculate).*
See Appendix A for explanation of abbreviations.

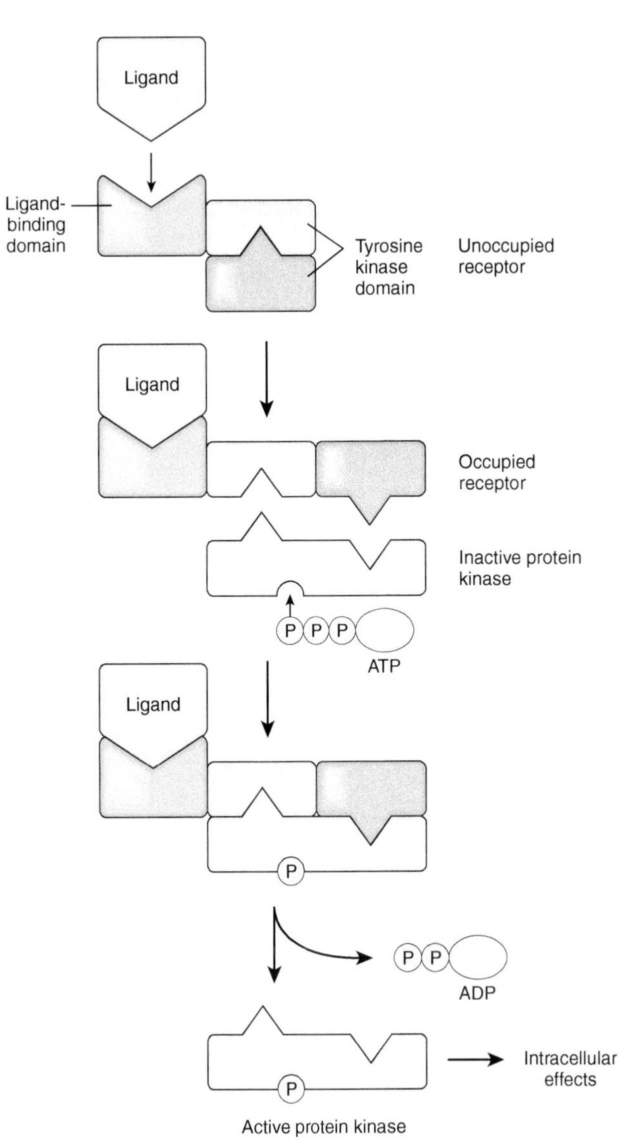

FIGURE 3-5 **Ligand/receptor fit.** As occurs for enzymes and substrates, ligands bind to particular domains on receptor molecules and typically cause conformational changes in the receptor that are important for initiating a response in a target cell. This hypothetical illustration imagines a ligand that binds to a receptor with tyrosine kinase activity. Once occupied, the receptor changes shape and now interacts with ATP and a protein kinase. The protein kinase is activated by phosphorylation and can now produce other effects in the cell. The receptor releases its ligand for degradation and returns to its unoccupied state. Alternatively, the ligand–receptor complex may be internalized prior to release of ligand and/or degrading of the receptor.

bound to a receptor, the receptor changes from **unoccupied receptor** to **occupied receptor**. A responsive target cell typically has thousands of receptors for a specific ligand. Many non-target cells may express a low number of receptors for the same ligand but are unresponsive for reasons discussed below.

There are several classes of membrane receptor types, each of which is specific for different ligands or groups of closely related ligands. Because a portion of these receptors passes through the membrane, they can be described as **transmembrane receptors** (Figure 3-6). Some receptors have innate enzymatic activity following binding of a ligand, whereas others may produce their effects secondarily through other kinds of proteins.

Receptors are often characterized by the types of cellular activity they produce. Insulin, **growth hormone (GH)**, and **insulin-like growth factor-I (IGF-I)**, as well as **epidermal growth factor (EGF)**, **colony-stimulating factor (CSF)**, and **platelet-derived growth factor (PDGF)**, have receptors possessing tyrosine kinase enzymatic activity. **Tyrosine kinases** are a general class of enzymes that phosphorylate tyrosine residues in proteins. A number of receptors have been characterized as having innate protein kinase activity with the ability to phosphorylate tyrosine in other proteins. There are also kinases that phosphorylate other amino acids, such as **serine** or **threonine kinases**.

Receptor tyrosine kinases have several molecular domains (Figure 3-6). The **extracellular domain** is responsible for binding a specific ligand and is represented by the α-subunit. This domain is unique for each type of receptor and the ligand it binds preferentially. The **transmembrane domain** crosses the plasma membrane once to connect the extracellular domain to the β-subunit that houses the **intracellular** or **tyrosine kinase domain**. The receptor also contains one or more **regulatory domains**, and there is also a **juxtamembrane region** that separates the intracellular domain from the plasma membrane and the transmembrane domain.

Receptor tyrosine kinases for growth factors form dimers after binding their ligand, and experimental studies show that formation of dimers is necessary to activate fully the tyrosine kinase domain. The receptor tyrosine kinases for insulin and insulin-like growth factors apparently are effectively dimerized prior to addition of ligand. Typically, both tyrosine kinase receptors bind a ligand before forming dimers. In the case of GH, one molecule of GH binds to two receptors that activate two kinases. This complex is called a **Janus kinase**, named for the two-faced Roman god who guarded gates.

Numerous studies have shown that occupied membrane receptors eventually are internalized through formation of endosomes (see ahead). The activity of these receptors on the membranes of endosomes may be important in maintaining sustained effects of a ligand following internalization of the ligand–receptor complex. However, studies employing mutant forms of insulin and EGF receptors that fail to undergo internalization when occupied show that internalization is not a requisite for transduction of the external signal into the cell. Once occupied, these mutant receptors exhibit tyrosine kinase activity followed by the typical intracellular events normally associated with that ligand.

The receptor tyrosine kinases undergo rapid autophosphorylation that makes them capable of interacting with other proteins. Phosphorylation of tyrosine residues facilitates binding of other proteins to the receptor tyrosine kinase. There is evidence for interactions of receptor tyrosine kinase with intracellular serine and threonine protein kinases that may carry the message to other intracellular sites.

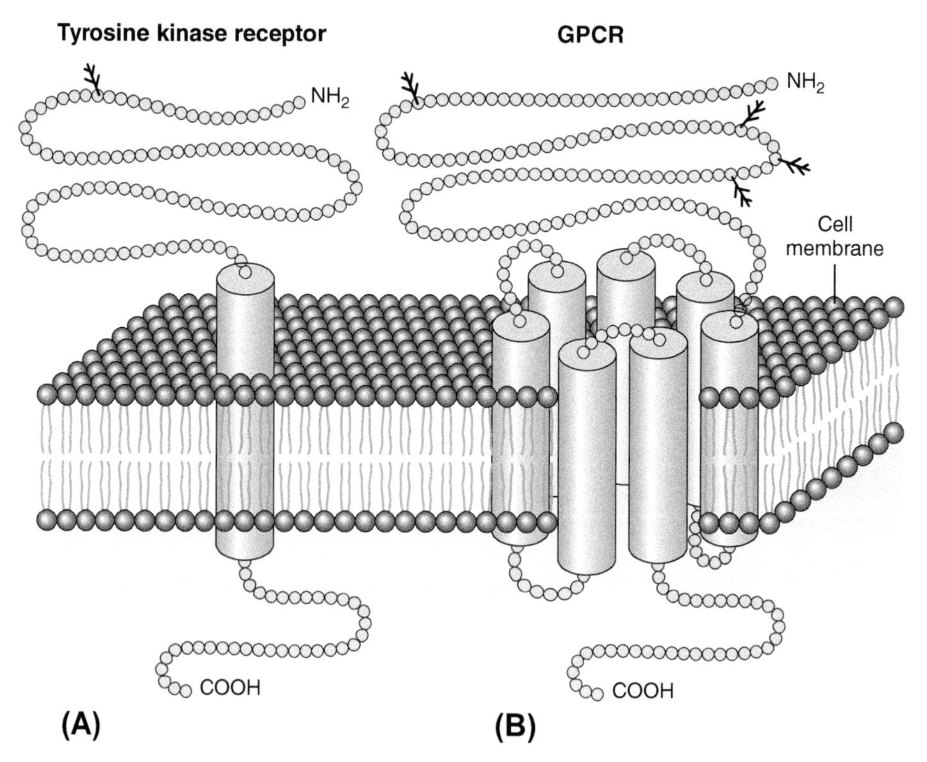

FIGURE 3-6 **Transmembrane receptors.** (A) A tyrosine kinase receptor has a ligand-binding extracellular domain, a single transmembrane domain, and an intracellular domain that acts as a tyrosine kinase. (B) A G-protein-coupled receptor (GPCR) also has three domains. The extracellular domain is responsible for binding the specific ligand. The transmembrane domain traverses the membrane seven times before ending in the cytoplasmic domain that is coupled with a G-protein. The type of G-protein is dependent upon the type of receptor. Occupied receptors often form dimers prior to activation of intracellular events.

Molecules in the secretin–glucagon family of peptides, which includes **vasoactive intestinal peptide (VIP)**, **pituitary adenylate cyclase-activating polypeptide (PACAP)**, **glucagon**, and **secretin**, all have similar sequences of amino acids and often bind to the same receptor proteins. Growth hormone and **prolactin (PRL)** from the pituitary have similar molecular structure as do their receptors. Similarly, there may be more than one kind of receptor for a given ligand, of which each differs slightly in its composition and binding characteristics. Thus, different tissues may contain different receptor forms for the same ligand (Table 3-7).

An important group of receptors are the **G-protein-coupled receptors (GPCRs)**. These G-proteins are embedded within the cell membrane and were so named because they can bind and hydrolyze the nucleotide **guanosine triphosphate (GTP)**. The transmembrane domain of these GPCRs traverses the cell membrane seven times (see Figure 3-6). Because of this snake-like traversing of the membrane, GPCRs are sometimes referred to as **serpentine receptors**. G-proteins are critical for the action of bioregulators employing this mechanism that generate formation of "second messengers" in the cytosol of the target cells that mediate the ligand's actions (see ahead).

TABLE 3-7 Cell Membrane Receptor Types

Cell Membrane Receptor Types
Enzyme-linked receptors (one transmembrane unit)
Tyrosine kinase activity
EGF family
Insulin family
PDGF family
NGF family
Serine kinase activity
TGFβ
Activin
AMH (MIS)
Guanylate cyclases
cGMP generating
GPCR (7 transmembrane units; G-protein linked) receptors
β-Type receptors(increase Ca^{2+}; increase cAMP)
β-Adrenergic (βAR)
α-Adrenergic receptors (αAR)
Dopamine receptor
Serotonin receptors
Muscarinic type (increase Ca^{2+}; decrease cAMP)
Neurokinin type
Non-neurokinin type
Calcitonin receptor
Fibronectin-like receptors
Class 1: Cytokine family
Erythropoietin receptor
Interleukin-1 receptor
Class2: TNF/IFN family
TNF receptor
Interferon (IFN) receptor
Ubiquitin receptor

See Appendix A for explanation of abbreviations.

1. Multiple Receptor Subtypes

Each bioregulator must have a specific receptor associated with its target cell in order to produce a localized and specific effect; however, there may be more than one kind of receptor for a given bioregulator resulting in different tissue-specific responses to the same bioregulator. Receptor variants or subtypes may be the product of gene duplications or may be due to splice variants formed during their synthesis. Cell membrane receptors are often complexes of several proteins, each the product of different genes. As we learn more about bioregulator actions, we note that there are small differences in cell membrane receptors for a given bioregulator in different tissues that influence binding and may even be connected to different intracellular mechanisms. Typically, because of variances in the molecular composition of receptors, there is differential sensitivity of receptors to agonists or antagonists (see Box 3A). This differential sensitivity often becomes incorporated into the name of the receptor type; for example, there are two types of **cholinergic receptors** for the neurotransmitter **acetylcholine** in heart muscle and skeletal muscle. The cholinergic receptors of heart muscle cells are stimulated by **muscarinic acid**, a substance obtained from certain mushrooms, but those of skeletal muscle are not. In turn, skeletal muscle cholinergic receptors are activated by **nicotine** whereas heart muscle receptors are not. Consequently, we refer to the heart receptors as **muscarinic cholinergic receptors** and those of skeletal muscle as **nicotinic cholinergic receptors**. **V2 receptors** for arginine vasopressin are linked to a different cellular mechanisms of action than are **V1 receptors**, and arginine vasopressin produces markedly different intracellular events in different tissues depending on the type of receptor present.

Receptors for the catecholamine neurotransmitters are designated as **α-receptors** or **β-receptors**. These receptors involve several proteins each and exist in a variety of subforms (e.g., $\beta 1$, $\beta 2$). Norepinephrine has a higher

BOX 3A Agonists, Antagonists, and Inverse Agonists

The development of agonists and antagonists for various bioregulatory ligands has been instrumental in experimental studies of the mechanism of action of ligands as well as for clinical studies and development of drug therapies for various endocrine disorders. Agonists bind to the natural receptor for a particular ligand and mimic its action. A strong agonist may bind even better than the natural ligand (i.e., it may have a higher affinity for the receptor) as opposed to a weak agonist. Similarly, an antagonist also binds to the same receptor but does not activate it. Strong antagonists may bind so tightly to the receptor that they prevent binding of the natural ligand. The discovery that some agonists or antagonists were more effective in certain target tissues than in others has led to the discovery of multiple receptor types.

One of the first ligands to be recognized as having multiple receptor types was epinephrine. First, we discovered two receptor types, the α-adrenergic receptor and the β-adrenergic receptor. Soon, subtypes of each receptor were discovered (α_1, α_2, β_1, etc). The binding of epinephrine and its agonists to different adrenergic receptor (AR) subtypes in an array is similar to the keying of locks in a building used by multiple groups of people (Box Table 3A-1). Like a master key, epinephrine binds to and activates all adrenergic receptor subtypes. Phenylephrine is like a submaster key that opens all of the α-sublocks but none of the β-sublocks.

Various endocrine-disrupting chemicals (EDCs) present as environmental pollutants may produce their effects through different receptor subtypes. Thus, knowledge of receptor specificity is important when evaluating their effects. For example, phytoestrogens bind principally to estrogen receptor β (ERβ), whereas estrogenic phthalates bind to ERα. Because these receptor subtypes appear in different tissues in the body, it would influence where you should look for possible effects.

Inverse agonists may bind to the same receptor as an agonist but typically have the opposite effect on the target cell. Many hormone receptors exist in a partial state of activity even in the absence of a ligand bound to the receptor. Some chemicals thought previously to be antagonists, with no ability to activate the receptor on their own, inhibit this spontaneous receptor activity and thus are called *inverse agonists.* Many hormone receptors exhibit spontaneous agonist-independent activity and can, in theory, be targeted by bioregulators or drugs that act as inverse agonists.

BOX TABLE 3A-1 Relationship Between a Ligand and Its Receptor Subtypes

Keys	Locks	Ligands	Receptor Subtypes
Master key	Opens all locks	Epinephrine	Binds to all ARs
Submaster key α	Opens all α locks	Phenylephrine	Binds to all αARs
Submaster key β	Opens all β locks	Isoproteronol	Binds to all βARs
Key α 1	Opens α1 locks	Clonidine	Binds to α_1ARs
Key α 2	Opens α2 locks	Prazosin*	Binds to α_2ARs
Key β 1	Opens β1 locks	ICI 89,406*	Binds to β_1ARs
Key β 2	Opens β2 locks	Salbutamol	Binds to β_2ARs

**Antagonist (related key) that fits into a specific AR subtype binding site (lock) and blocks access to any agonist but doesn't activate receptor (doesn't open the lock).*

affinity for the α-receptors, whereas epinephrine binds well to either α- or β-receptors. Biochemists have synthesized specific agonists (mimics) and antagonists (blockers) for each adrenergic receptor and its subtypes. Thus, there can be multiple ligands (keys) with differing specificities for receptors (locks) as illustrated in Box Table 3A-1. Dopamine **D1-receptors** are distinct from **D2-receptors** and also are characterized by binding different synthetic agonists and antagonists.

The mechanisms of action for these multiple receptor types may be very different, with one type operating through one mechanism and another type through a different mechanism of action. However, the biological meaning of variance in receptor types is not always so obvious. In some cases, we may be looking at evolving receptors moving toward greater selectivity and tissue specificity for ligands. Consequently, we may not be able to assign an "adaptive value" at this time, leading to confusion about what their roles might be. This diversity of receptor types has allowed for development of useful pharmacological tools for research and clinical drug therapies.

E. The Second-Messenger Concept

Following binding of the bioregulator to its receptor, a new set of events are initiated in the target cell that are specific to that bioregulator and its receptor. How is this orchestrated? Generally speaking, the bioregulator can be considered the **first messenger** carrying a signal from the secreting cell to a target cell. The first messenger binds to its receptor and may initiate the synthesis of cytosolic **second-messenger** molecules that carry the bioregulatory signal into the interior of the cell. Because the production of a second messenger occurs through enzymatic activity, the message of relatively few molecules of first messenger is thereby amplified as many second-messenger molecules are produced for each first messenger bound to a receptor and for as long as the receptor is occupied by the ligand. The need for this amplification at least in part relates to the short half-life of second messengers. It is critical that there be enough membrane receptors and second messengers available to amplify the message sufficiently to bring about specific internal changes in the target cell.

In the 1960s, Earl Sutherland and his coworkers discovered the first second messenger**, cyclic adenosine monophosphate (cAMP)**, while attempting to elucidate the mechanism of action for epinephrine on glycogen breakdown in skeletal muscle and liver cells. Epinephrine has its actions on muscle and liver cells by first binding to what we know now as a cell membrane-bound GPCR. Sutherland was awarded a Nobel Prize for this pioneering work in 1972. Much later, the discovery of the membrane-bound intermediate G-protein operating between the first and second messengers led to a Nobel Prize in 1994 for Alfred Gilman and Martin Robdell. Although the G-protein logically could be called a second messenger, this latter term has been retained for the type of cytosolic messenger originally described by Sutherland's group. Several types of second-messenger systems have been discovered that involve the interaction of an occupied receptor with a specific G-protein and results in the production of second messengers.

Activated G-proteins may interact directly with a **signal-generating enzyme** adjacent to or located in the membrane but on the cytosolic side. When a signal-generating enzyme is activated, it catalyzes the synthesis of a second messenger from a specific substrate. The second messenger then initiates intracellular events normally associated with the actions of the first messenger on that cell. Thus, through the second-messenger system, a small amount of bioregulator can be amplified to hundreds of active molecules that bring about specific changes in a target cell. Because the ligand–receptor complex is not located at a fixed site in the membrane, it can migrate within the membrane and contact more than one G-protein, thus possibly amplifying its effect even further.

G-proteins may also interact with **ion channels** in the cell membrane. Opening of a calcium channel can influence the entrance or exit of calcium ions (Ca^{2+}); when a sufficient number of channels are affected, intracellular or membrane events are altered. Hence, an ion such as Ca^{2+} that enters the cell and causes specific changes, such as activation of an enzyme, following binding of a bioregulator to its receptor also can be called a second messenger. Other types of receptors can influence Ca^{2+} influx by mechanisms that do not involve G-proteins.

Second messengers also may alter cytoplasmic phosphorylating systems. This may result in additional effects including activation of transcription factors resulting in delayed effects on protein synthesis, altered functioning of enzymes, etc. Such generalized effects may even mimic some events activated by other bioregulators that are capable of altering phosphorylating systems (see discussion of kinase cascade ahead).

1. The cAMP Second-Messenger System

Seconds after epinephrine binds to its receptor on the target cell, a unique compound is synthesized following activation of an enzyme called **adenylyl cyclase** that synthesizes **cAMP** following hydrolysis of the energy rich molecule, adenosine triphosphate (ATP) (Figure 3-7). In this novel form of AMP, the phosphate forms two bonds instead of one to the adenine base.

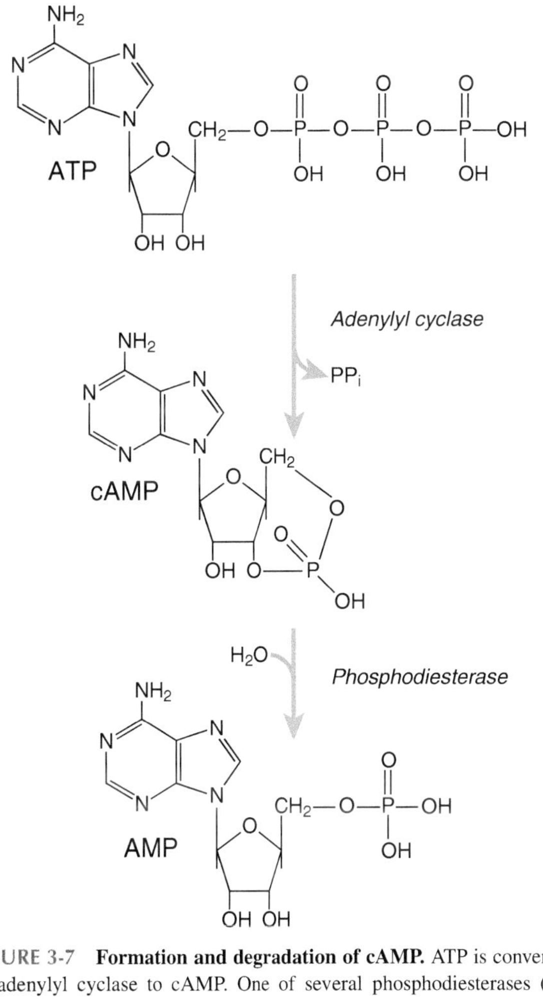

FIGURE 3-7 **Formation and degradation of cAMP.** ATP is converted by adenylyl cyclase to cAMP. One of several phosphodiesterases (see Table 3-8) inactivates cAMP by converting it to ordinary AMP.

The sequence of events following binding of epinephrine to its target cell are outlined in Figure 3-8. In order to activate adenylyl cyclase, the epinephrine-occupied receptor first interacts with a particular G-protein called a **G_s-protein**, which in turn interacts with adenylyl cyclase and generates many molecules of cAMP (amplification). Additional studies show that administration of cAMP can mimic all the actions of epinephrine, supporting its role as a true intracellular messenger.

When the receptor is unoccupied, the adenylyl cyclase is inactive and GDP is bound to the G_s-protein. The G_s-protein consists of three peptide subunits: α-, β-, and γ-subunits (Figure 3-9). The occupied receptor interacts with the β- and γ-subunits of the G_s-protein, allowing the dissociation of GDP and the binding of GTP to the α-subunit of the G_s-protein. The α-subunit then dissociates from the β- and γ-units to interact with the catalytic portion of the adenylyl cyclase and generate cAMP. Intrinsic GTPase-degrading activity of the α-subunit converts the GTP back to GDP, reducing its stimulatory effect on the catalytic domain of adenylyl cyclase. The α-subunit then returns to its association with the other G_s-subunits.

The specific cellular response to cAMP as a second messenger depends on the cell type involved (Figure 3-10). Once generated by epinephrine in a liver or skeletal muscle cell, cAMP can repeatedly combine with **protein kinase A (PKA)**, a cytosolic enzyme that phosphorylates the inactive enzyme **phosphorylase kinase**, which in turn converts **phosphorylase-b** into its active form, **phosphorylase-a**. Active phosphorylase-a triggers the enzymatic breakdown of glycogen to provide glucose–phosphate for energy production in muscle cells and glucose–phosphate and free glucose in liver cells (see Chapter 12 for information on metabolism). In an adipose or fat cell, epinephrine produces cAMP, which phosphorylates PKA, which in turn activates a **hormone-sensitive lipase (HSL)**, resulting in hydrolysis of fats. cAMP also may produce effects via other cellular mechanisms associated with the same or different first messengers. For example, in cardiac muscle cells, PKA activates calcium pumps that release Ca^{2+} into the cytosol and shorten the intervals between contractions.

In addition to the cytoplasmic bioregulatory elements associated with cAMP, a **cAMP regulatory element binding protein (CREB)** has been isolated from the nucleus and identified as a transcription factor that can regulate gene transcription. This could explain some of the delayed effects of peptide hormones on protein synthesis observed following initial enzymatic activation through cAMP. Bioregulatory elements like CREB as well as other transcription factors could be considered "third messengers" that mediate nuclear events.

Cell membranes contain another G-protein called a **G_i-protein** that can inhibit the cAMP mechanism in the following manner. Apparently, certain bioregulators can

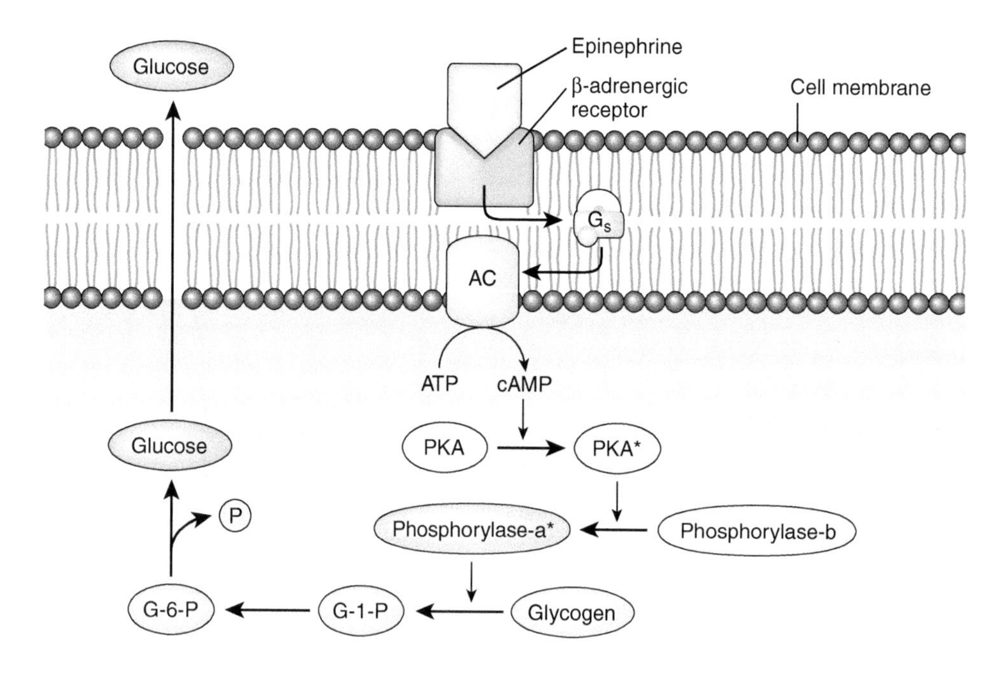

FIGURE 3-8 **Actions of cAMP within target cell.** After binding to its cell membrane β-adrenergic receptor, the ligand, epinephrine, produces different effects in target cells by first stimulating production of cAMP, which converts inactive protein kinase A (PKA) to active PKA*. In liver and skeletal muscle, PKA phosphorylates the enzyme glycogen synthetase, converting it from an active to an inactive form and thus reducing glycogen synthesis (not shown). PKA* converts inactive enzyme phosphorylase *b* through an additional phosphorylation to its active form (Phosphorylase a*) and causes hydrolysis of glycogen to release glucose-1-phosphate (G1P), which in liver can be converted to free glucose and free phosphate (P). Free glucose leaves the cell via mediated transport. AC, adenylyl cyclase. A comparision of epinephrine's actions through PKA in other tissues is provided in Figure 3-12.

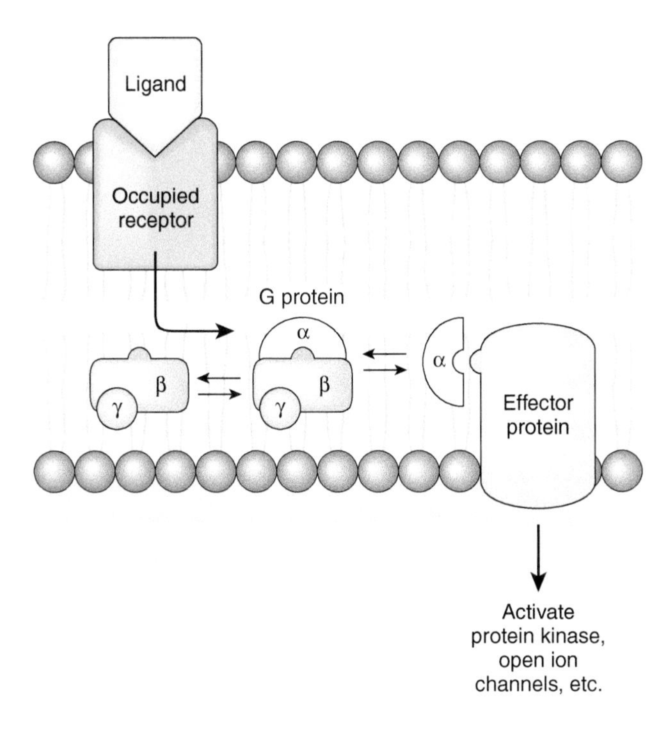

FIGURE 3-9 **G-proteins consist of three subunits.** The α-subunit that has innate GTP-binding and hydrolyzing capacity can separate from the other subunits following interaction with an appropriate, occupied receptor. The free α-subunit interacts with a membrane channel protein or an enzyme that generates a second messenger. Once the GTP has been hydrolyzed, the subunits recombine.

bind to a receptor that employs the G_i-protein which, like the G_s-protein, consists of three subunits. The G_i-protein also has a unique α-subunit but its β- and γ-subunits are very similar to those of the G_s-protein. The G_i-protein α-subunit interacts with the G_s protein—adenylyl cyclase complex and prevents cAMP generation. Several bioregulators that antagonize the actions of cAMP-dependent peptide bioregulators are known to operate through a G_i-protein (see Figure 3-11).

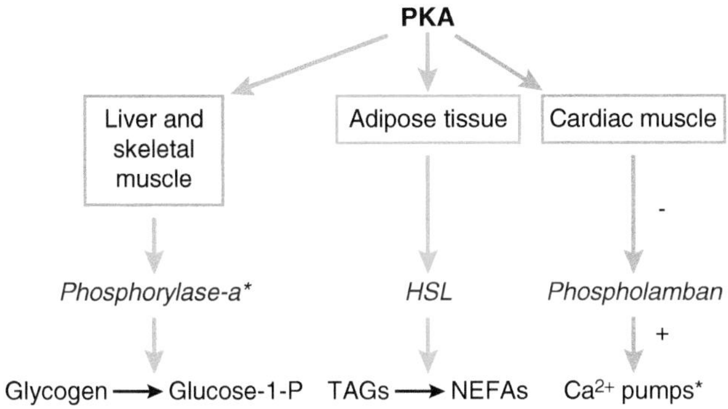

FIGURE 3-10 **Subsequent actions of PKA following epinephrine activation in different tissues.** Abbreviations: HSL, hormone-sensitive lipase; NEFAs, non-esterified fatty acids; TAGs, triacylglycerides or fats; phospholamban is an inhibitor of calcium pumps.

Several drugs have proven very useful for studying cAMP second-messenger mechanisms. Adenylyl cyclase can be activated directly by a diterpene called **forskolin**, obtained from an Indian medicinal herb, *Coleus forskohli*. This plant had been used with some success for centuries in India to treat heart disease, respiratory ailments, convulsions, and insomnia. **Cholera toxin** obtained from pathogenic bacteria blocks the GTPase activity of the α-subunit of the G_s-protein. It is this action in the digestive tract that leads to continuous diarrhea, resulting in excessive water losses and eventually death due to dehydration of the person infected with cholera bacteria. Another bacterial product, **pertussis toxin**, causes ADP-ribosylation of a cysteine on the α-subunit of the inhibitory G_i-protein, reducing its affinity for GTP. This prevents the G_i α-subunit from interacting normally, causing a slight rise in cAMP levels and preventing the action of bioregulators that normally work through the G_i-protein.

2. Inositol Trisphosphate and Diacylglycerol as Second Messengers

Some GPCRs interact with yet another G-protein, the **G_q-protein**, which activates an additional signal-generating enzyme complex called **phospholipase C (PLC)** (see Figure 3-12). The substrate for PLC is **phosphatidylinositol bisphosphate (PIP_2)**, which is cleaved to produce *two* second messengers: **inositol trisphosphate (IP_3)** and **diacylglycerol (DAG)**, each of which has independent actions. The peptide hormone arginine vasopressin produces its effects through a G_q-protein in some of its target cells, although in other target cells it activates cAMP. IP_3 interacts with IP_3-receptors on the external membrane of the endoplasmic reticulum to release stored Ca^{2+} into the cytosolic compartment of the cell. In muscle cells, these ions bind to special proteins that initiate contraction. In other cells, the resulting increase in intracellular Ca^{2+} usually involves a cytoplasmic calcium-binding peptide called **calmodulin**. Calmodulin can produce a variety of cellular effects depending on the cell type involved. For example, calmodulin can combine with and activate the major enzyme necessary for cAMP degradation (cAMP-specific phosphoidiesterase), thus diminishing the effectiveness of cAMP as a second messenger (see ahead). DAG and Ca^{2+} activate the enzyme **protein kinase C (PKC)**, which in turn may activate other cytoplasmic enzymes. Furthermore, DAG may serve as a substrate for production of arachidonic acid, a precursor for the synthesis of eicosanoids, a unique collection of lipid bioregulators (see ahead).

3. cGMP as a Second Messenger

The role of **cyclic guanosine monophosphate (cGMP)** as a second messenger has not been studied to the same extent as the cAMP system but is similar to the cAMP

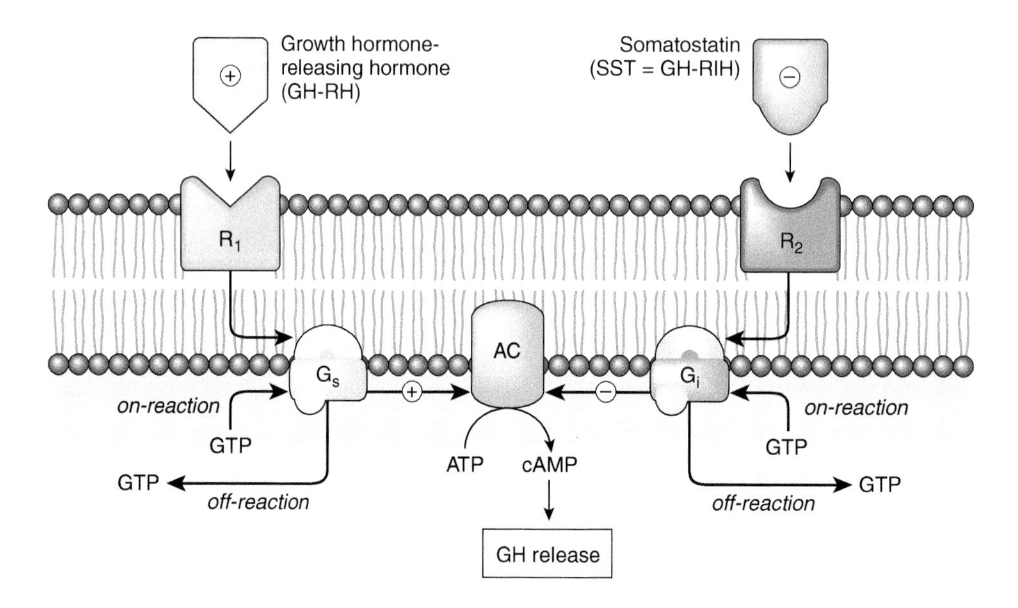

FIGURE 3-11 **G-protein interactions and inhibition of cellular reactions.** Growth hormone (GH)-releasing hormone binds to its receptor (R_1) and activates the G_s-protein that turns on adenylyl cyclase (AC) to synthesize cAMP from ATP. cAMP acts as a second messenger to mediate release of GH. Somatostatin (SST), after binding to the R_2 receptor, works through an inhibitory G_i-protein to prevent the activation of AC. Thus, in the presence of SST, it is difficult to stimulate GH release except through the addition of exogenous cAMP. A similar mechanism operates in the antagonism of norepinephrine by acetylcholine in cardiac muscle. *(Adapted with permission from Frohman, L.A. and Jansson, J.O.,* Endocrine Reviews, *7, 223–253, 1986. © The Endocrine Society.)*

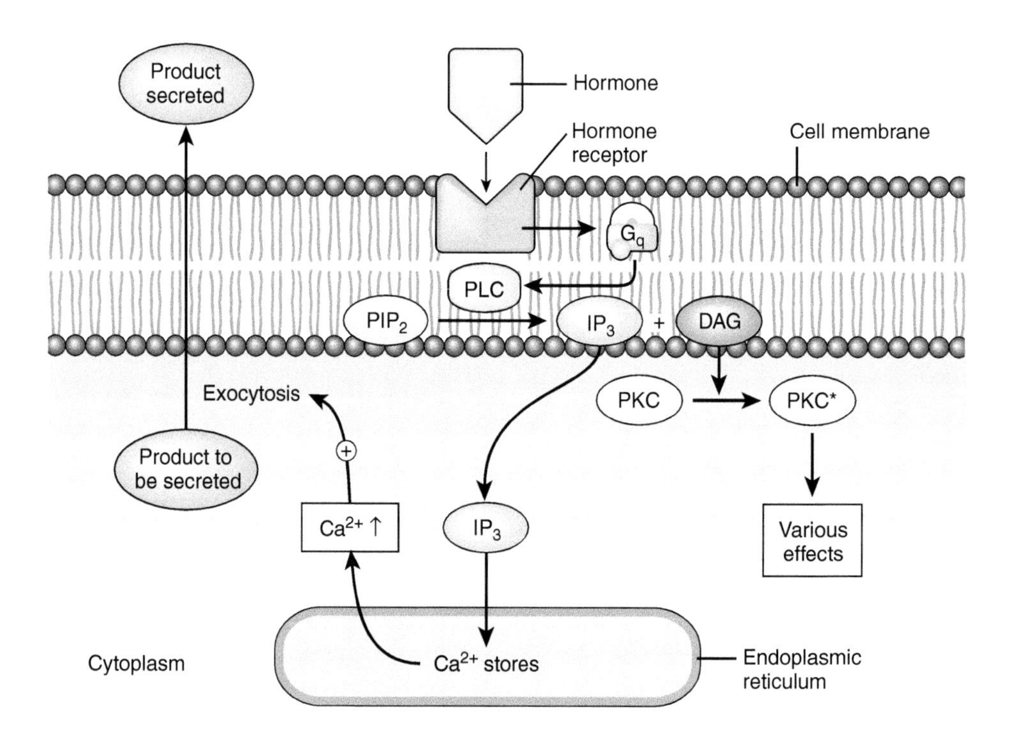

FIGURE 3-12 **IP_3 and DAG as second messengers.** Schematic representation of the action of a chemical regulator working through a G_q-protein to activate the enzyme phopholipase C (PLC) and generating the second messengers inositol trisphosphate (IP_3) and diacylglycerol (DAG) from phosphatidylinositol bisphosphate (PIP_2). IP_3 releases intracellular Ca^{2+}, which may interact with secretory vesicles and induce exocytosis of some product (e.g., hormone, secretory protein). DAG may activate phosphokinase C (PKC*) and produce additional phosphorylations and various effects.

mechanisms in many ways. The signal-generating enzyme **guanylyl cyclase** is activated by a small family of peptide bioregulators that includes **atrial natriuretic peptide (ANP)**. Guanylyl cyclase appears to be an integral part of the receptors for these peptides and, following binding of the appropriate ligand, produces cGMP from GTP. One of the actions of testosterone on the penis is activation of an enzyme, **nitric oxide synthetase (NOS)**. This enzyme is responsible for production of **nitric oxide (NO)**, which in turn elevates cGMP levels and contributes to induction of the erectile response by the penis. Oxytocin also elevates cGMP through activation of NO production.

4. Calcium Flux as an Intracellular Messenger

Recent studies of hormone action have discovered a common theme involving Ca^{2+} cycling across the cell membrane as a potential mechanism for sustaining a cellular response even hours after the initial binding of the hormone. This has led to development of a model system by Howard Rasmussen and his colleagues using mammalian adrenal cortical cells to study the involvement of Ca^{2+} in the action of certain bioregulators. Initial observations of the action of the octapeptide **angiotensin-II (Ang-II)** on the sustained release of the mineralocorticoid hormone **aldosterone** found only a transient rise in cytosolic Ca^{2+} following binding of Ang-II to cells of the adrenal cortex, yet a prolonged release of aldosterone was observed (see Chapter 8). However, it was noted that there was an increase in Ca^{2+} influx as well as efflux. They discovered that this Ca^{2+} cycling was caused by a **calcium-calmodulin-dependent protein kinase**. Furthermore, not only does this enzyme phosphorylate and activate the Ca^{2+} pump, but it also behaves as a plasma membrane-associated transducer that converts the Ca^{2+} cycling message into a sustained cellular event: release of aldosterone.

When a ligand–receptor complex interacts with the enzyme PLC, the substrate PIP_2 is converted to IP_3 and DAG. The IP_3 causes Ca^{2+} release from the endoplasmic reticulum, thereby producing a transient increase in intracellular Ca^{2+}. These Ca^{2+} ions bind to calmodulin, resulting in the activation of PKC that in the presence of DAG becomes associated with the plasma membrane and stimulates its Ca^{2+} pump. As long as DAG remains in the plasma membrane, the PKC remains active. All of these events lead to a submembrane increase in Ca^{2+} that is believed to facilitate the phosphorylating activities of PKC and to bring about a prolonged response.

Aldosterone is responsible for directing reabsorption of Na^+ from the urine by kidney cells and returning Na^+ to the blood (see Chapter 8). An appropriate ratio of Na^+ to K^+ in blood and extracellular fluids is essential for maintaining normal plasma membrane potentials on cells. Inappropriate release of aldosterone could lead to death caused by neural and muscular dysfunction. This is prevented by the sensitivity of the Ca^{2+} channels in adrenocortical cells to circulating levels of K^+.

F. Turning Off the Response to Bioregulators

Once a bioregulator binds to its receptors, the target cell is stimulated to produce second messengers until the occupied receptor becomes unoccupied again. In order to terminate the action on the target cell, the bioregulator must be separated from its receptor and the second messengers also must be inactivated. Several mechanisms have evolved in cells to turn off the response of the target cell once it has been stimulated. Different mechanisms turn off the response at the receptor and at the second-messenger level.

1. Fate of Membrane-Bound Ligands

The earliest clues to the fate of first-messenger ligands came from studies of the action of the neurotransmitters acetylcholine and norepinephrine on their target cells. In the case of acetylcholine, an enzyme called **acetylcholinesterase (AChE)** located in the postsynaptic cell membrane was found to degrade acetylcholine to acetic acid and choline, freeing the receptors from ligand until more acetylcholine was released from the presynaptic cell. Other studies showed that the norepinephrine was either retaken up by the presynaptic neuron, where it was recycled, or it was taken up by nearby glial cells that degraded it intracellularly. However, the fate of peptide ligands soon was shown to follow a unique pathway.

The first reports of finding peptide bioregulators inside target cells were assumed to be an artifact of the methods employed to demonstrate their presence, as it was believed then that they did not enter cells but bound only to the cell membrane. We now know that subsequent to occupation and activation of receptors, occupied receptors with their ligands migrate along the surface of the cell membrane to specialized regions or pits that are involved in formation of internalized vesicles called **endosomes** (or phagosomes). These pits are membrane regions where specific proteins (e.g., **clathrin, caveolin**, or **dynamin**) have accumulated (Figure 3-13). Endosome formation results from endocytosis, and each endosome consists of a small sphere of cell membrane, with occupied receptors having the ligand-binding side extending into its lumen. A variety of internal **Rab proteins** are involved in distribution of endosomes within the cell depending on their eventual fate. They may contact lysosomes containing hydrolytic enzymes that fuse with the endosomes to form **endolysosomes** (or phagolysosomes). The hydrolytic enzymes may degrade both the receptors and the ligands. In other cases, endosomes or portions of an endosome may be directed back toward the

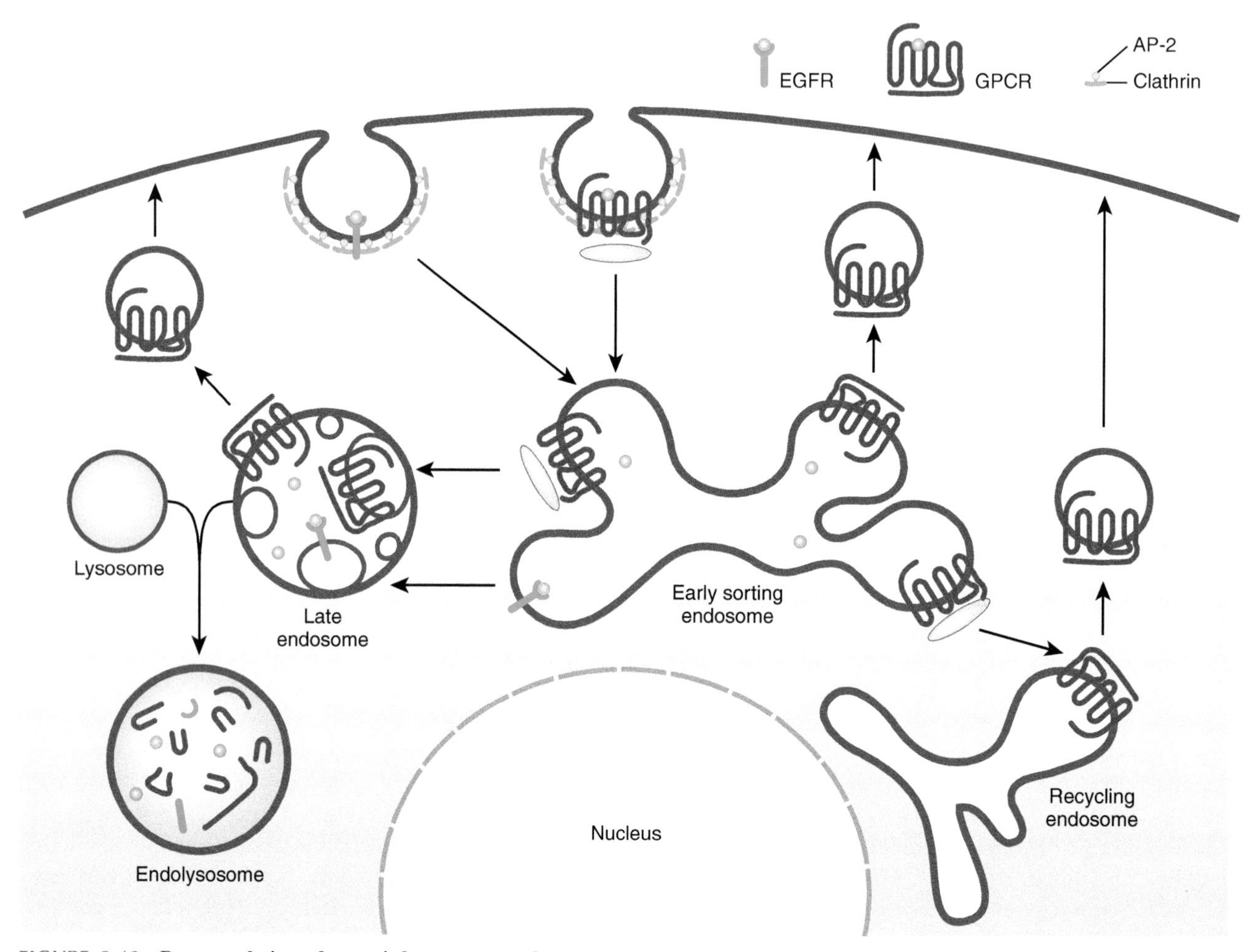

FIGURE 3-13 **Downregulation of occupied receptors and receptor recycling.** Occupied epidermal growth factor receptors (EGFRs) and G-protein-coupled receptors migrate along the cell membrane to locations where endosomes form (via endocytosis). These sites may be associated with the special proteins such as clathrin. The early sorting endosomes direct the fates of the internalized receptors, with some directed to late endosomes, which fuse with lysosomes to form endolysosome, usually resulting in degradation of both ligand and most or all of the receptors. Some receptors may be directed to recycling endosomes, and the receptors are recycled directly to the cell surface.

cell membrane, and the associated receptors without their ligands may be recycled back to the cell surface.

The discovery of this internalization of occupied receptors partially explains the phenomenon of **downregulation** of receptors observed in most bioregulatory systems following arrival of the ligand at its target. Downregulation refers to the observed reduction in receptor number that typically occurs as evidenced by reduced sensitivity of the target cell following the initial stimulation by the bioregulator. A later increase in receptors or an **upregulation** of receptors may follow ligand-stimulated protein synthesis. Thus, up- and downregulation can be considered mechanisms for altering the dynamic range of a hormone-sensitive system. In many developmental systems, the first contact of a ligand with the target cell may result in upregulation of receptors and accelerated binding and subsequent actions of the ligand on the target cell. Once the target cell has responded, downregulation occurs frequently so that the cell cannot be restimulated until after a period of recovery and reestablishment of a sufficient receptor population in the cell membrane.

2. Receptor and G-Protein Interaction

Phosphorylation of the β-adrenergic receptor prevents it from interacting with the G_s-protein. Once activated, the β-subunit and γ-subunit complex of the G_s-protein can activate a protein kinase called **β-adrenergic receptor kinase (βark)**, which then allows a cytosolic protein, **β-arrestin**, to bind to the phosphorylated receptor. In this state, the occupied receptor is unable to continue its interaction with the G_s-protein.

3. Fate of Second Messengers

Just as it is important to remove the first messengers to allow a cell to stop responding to a bioregulator, so second messengers also must be degraded. Specific enzymes are present in the cytosol to degrade second messengers. The best known of these enzymes are the **phosphodiesterases** that degrade cyclic nucleotides (see Table 3-8). One such phosphodiesterase has a high affinity for cAMP and rapidly destroys it. Binding of cAMP to PKA and activation of a specific metabolic pathway in a target cell are thus dynamic events governed by the generation of cAMP and its rate of degradation by a particular phosphodiesterase. Certain drugs called **methylxanthines** are potent inhibitors of phosphodiesterase and can be used to potentiate the actions of a bioregulator by prolonging the half-life of cAMP. One of these drugs, **caffeine**, is used commonly in experimental studies. Another drug, **theophylline**, is administered to augment the natural action of epinephrine in order to alleviate asthmatic symptoms in humans. IP_3 is reduced progressively to inositol by **phosphomonoesterases**. These esterases are inhibited by lithium ions. The successful treatment of certain forms of mental depression with lithium may be related to its enhancing actions on IP_3 effects.

4. Inactivation in the Blood and/or Excretion of Bioregulators

In addition to the inactivation of peptide bioregulators by their target cells, many of the smaller peptides are degraded rapidly by peptidase enzymes that are present in the blood. These peptidases attack one end of the peptide and remove amino acids one-at-a-time until the ability of the peptide to be recognized by its receptor is diminished, usually progressively, and then lost. Eventually, the entire peptide is demolished. Even larger peptides may be partially degraded while in the blood. Consequently, many peptides of differing numbers of amino acids derived from a secreted bioregulator may be present in the circulation at the same time, all or only some of which may be fully active. For example, the first 34 amino acids of parathyroid hormone have full biological activity as do larger fragments up to the parent bioregulator of 84 amino acids whereas fragments smaller than the first 34 have little or no activity.

G. Effects of Membrane-Bound Bioregulators on Nuclear Transcription

In addition to the early actions on target cells by bioregulators that employ membrane-bound receptors, delayed effects on protein synthesis were well known long before we could explain how this effect might be mediated. The physical distance separating the cell membrane from the nucleus is very small by our standards, only about 20 to 30 μm, but this is a huge separation on a molecular scale.

Studies of protein kinase actions have led to the recent development of one hypothesis to link the cell membrane to nuclear events. It is called the **kinase cascade hypothesis** and has evolved from studies of cell division regulation. Cell biologists in many laboratories working on the actions of **mitogens** (factors that stimulate cell division) or growth factors provided the first connections. Mitogens, such as EGF and many hormones, can cause hyperplasia of target cells (e.g., the effect of IGFs on cartilage and the effect of TSH on goiter formation in the thyroid). New evidence suggests that mitogens induce a cascade of protein-kinase-dependent intracellular events not unlike those described for second-messenger systems (Figure 3-14). The EGF receptor is one of a family of membrane receptors that when occupied form a dimer with tyrosine kinase activity. The occupied receptor dimer phosphorylates itself, allowing it to interact with an intracellular heterodimer consisting of two component proteins: **growth factor receptor-binding protein (GRB2)** and the product of

TABLE 3-8 Classification of Mammalian Phosphodiesterases (PDEs)[a]

Class	Unique Features of the PDE	Factors that Increase PDE Activity
1. Calcium-calmodulin-activated PDE (5 forms described)	Higher affinity for cGMP	GnRH, muscarinic cholinergic agonists
2. cGMP-activated PDE	Have allosteric cGMP-binding site	ANP
3. cGMP-inhibited PDE (3 forms described)	cGMP increases can inhibit cAMP action	Insulin, glucagon, dexamethasone[b]
4. cAMP-specific PDE	cAMP is the intracellular moderator	FSH, PGE, TSH, β-adrenergic agonists, Calmodulin
5. cGMP-specific PDE (3 forms described)	Found in retina of eye; employs special G-protein called transducin	Light

[a]*See Appendix A for explanation of abbreviations.*
[b]*Dexamethasone is a very potent synthetic glucocorticoid.*

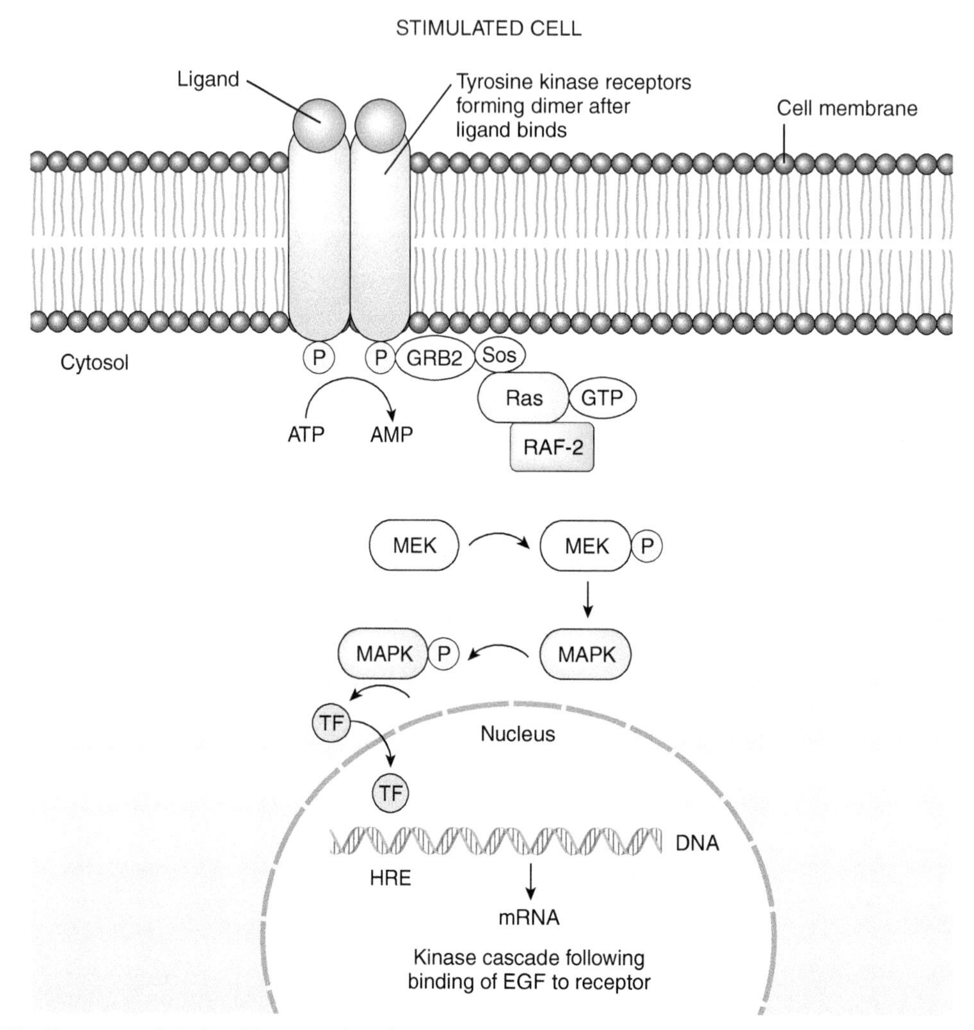

FIGURE 3-14 **The kinase cascade induced by a tyrosine kinase receptor.** Occupied receptors interact with a series of protein kinases resulting in production of transcription factors (TFs), which as "third messengers" enter the nucleus and alter transcription. See text for an explanation of the abbreviations. The cascade can also be activated by cross-talk (see Figure 3-17).

a gene first named the **son of sevenless (SOS)** for its role in regulating eye development in fruit flies. Once the heterodimer is activated by occupied EGF receptor, the SOS component interacts with a specific GTP-binding protein (G-protein) called **Ras**, previously identified as a common relay protein for cell growth factors. Ras then exchanges its GDP for a GTP and now binds to a cytoplasmic protein kinase called **Raf-1 protein**, which in turn phosphorylates another kinase termed **MEK**. Once activated, MEK phosphorylates a cytoplasmic complex of enzymes known as **mitogen-activated protein kinases (MAPKs).** Among the many proteins phosphorylated by these MAPKs are certain **transcription factors (TFs)** that are translocated to the nucleus, where they bind to DNA and ultimately stimulate cell division.

The kinase cascade hypothesis as outlined here is oversimplified. Evidence suggests there may be additional branches that allow alternative routes for the final activation of the MAPKs. For example, occupied insulin receptors can activate this mitogenic pathway but apparently employ two unique intracellular proteins that interface with the kinase cascade. Nevertheless, the kinase cascade hypothesis provided the first comprehensive explanation of how interactions with a cell membrane receptor can result in nuclear stimulation.

Additional evidence suggests roles for bioregulators that activate PKC and stimulate transcription by employing members of the **Jun–Fos** family of transcription factors. PKA activated by cAMP causes the production of CREB, which in turn binds to cAMP response elements (CREs) on DNA to increase or decrease the transcription of other genes. These observations imply a general role for protein kinases in all the actions of membrane-bound occupied receptors.

1. Cross-Talk

No bioregulator works in a physiological vacuum but rather does its thing amid a complex landscape of simultaneous events. Some bioregulators are known to act cooperatively or antagonistically through effects at their release sites or through interactions at the cell membrane of the target cell. The discovery that many different bioregulators produce their effects through common molecular pathways such as the kinase cascade or via similar second messengers or transcription factors has led to a concept called **cross-talk** that provides another explanation for how different bioregulators can influence cellular processes cooperatively or antagonistically. The possibility of cross-talk emphasizes the importance of not assuming that all observations of a bioregulator's action are due only to that bioregulator acting through its receptor (Figure 3-15). Another level of cross-talk may involve the interactions of different hormone systems, where one system, such as the thyroid system, affects the activity of the reproductive system (see Figure 3-16).

II. STEROID BIOREGULATORS

The chemical term **steroid** refers to a variety of lipoidal compounds, all of which possess the basic structure of four carbon rings known as the cyclopentanoperhydrophenanthrene ring or **steroid nucleus** (Figure 3-17). There are many naturally occurring steroids, including cholesterol, 1,25-dihydroxycholecalciferol, the bile salt cholic acid, the adrenocortical steroids or corticosteroids, and the gonadal sex steroids: androgens, estrogens, and progestogens. The most commonly occurring steroids of these categories are listed in Table 3-9.

1,25-Dihydroxycholecalciferol (1,25-DHC) regulates calcium absorption by cells of the intestinal epithelium. It is

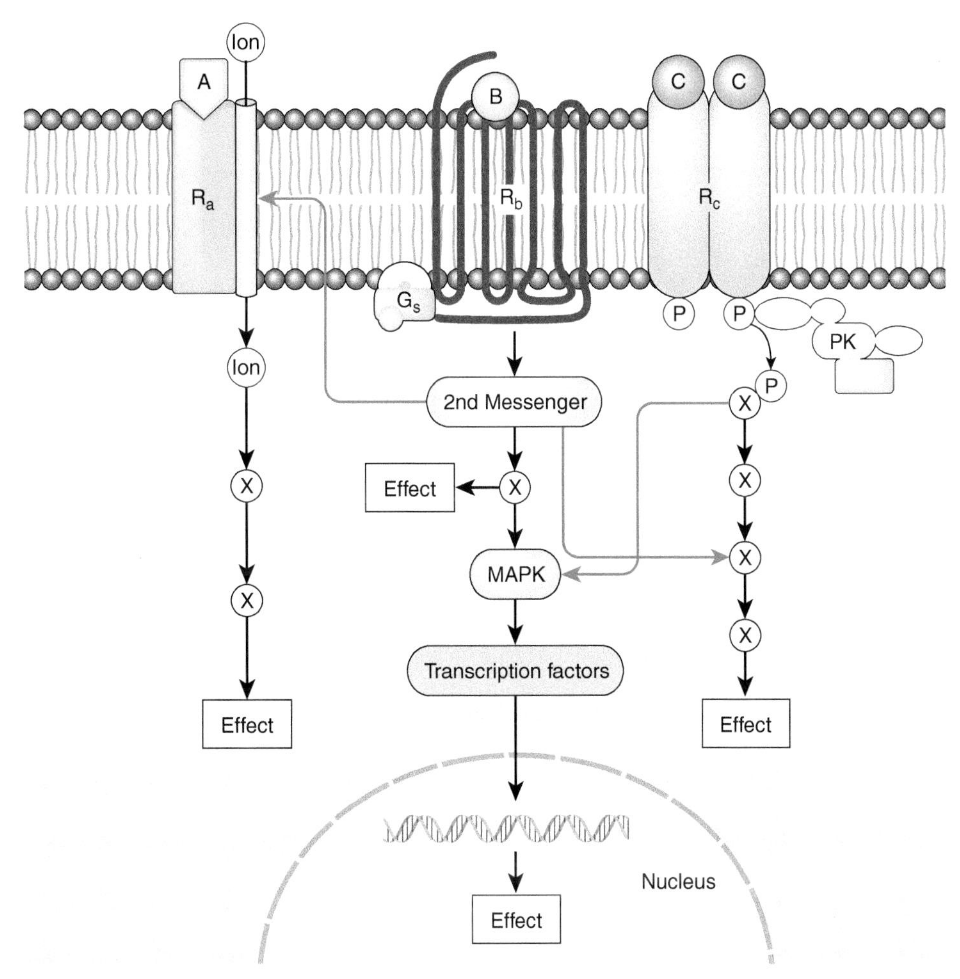

FIGURE 3-15 **Cross-talk between intracellular hormone actions.** Hormone A binds to an ion channel receptor that regulates ion influx. Hormone B binds to a GPCR that operates through a G_s second-messenger system. Hormone C binds to a tyrosine kinase receptor that activates a protein kinase complex (PK) that initiates a kinase cascade ending with MAPK activation of a transcription factor. Red arrows represent each hormone's mechanism of action. Black arrows represent possible cross-talk effects on the mechanisms of the other hormones affecting this cell. Such interactions could by stimulatory or inhibitory in nature. "X" represents unidentified intermediate steps.

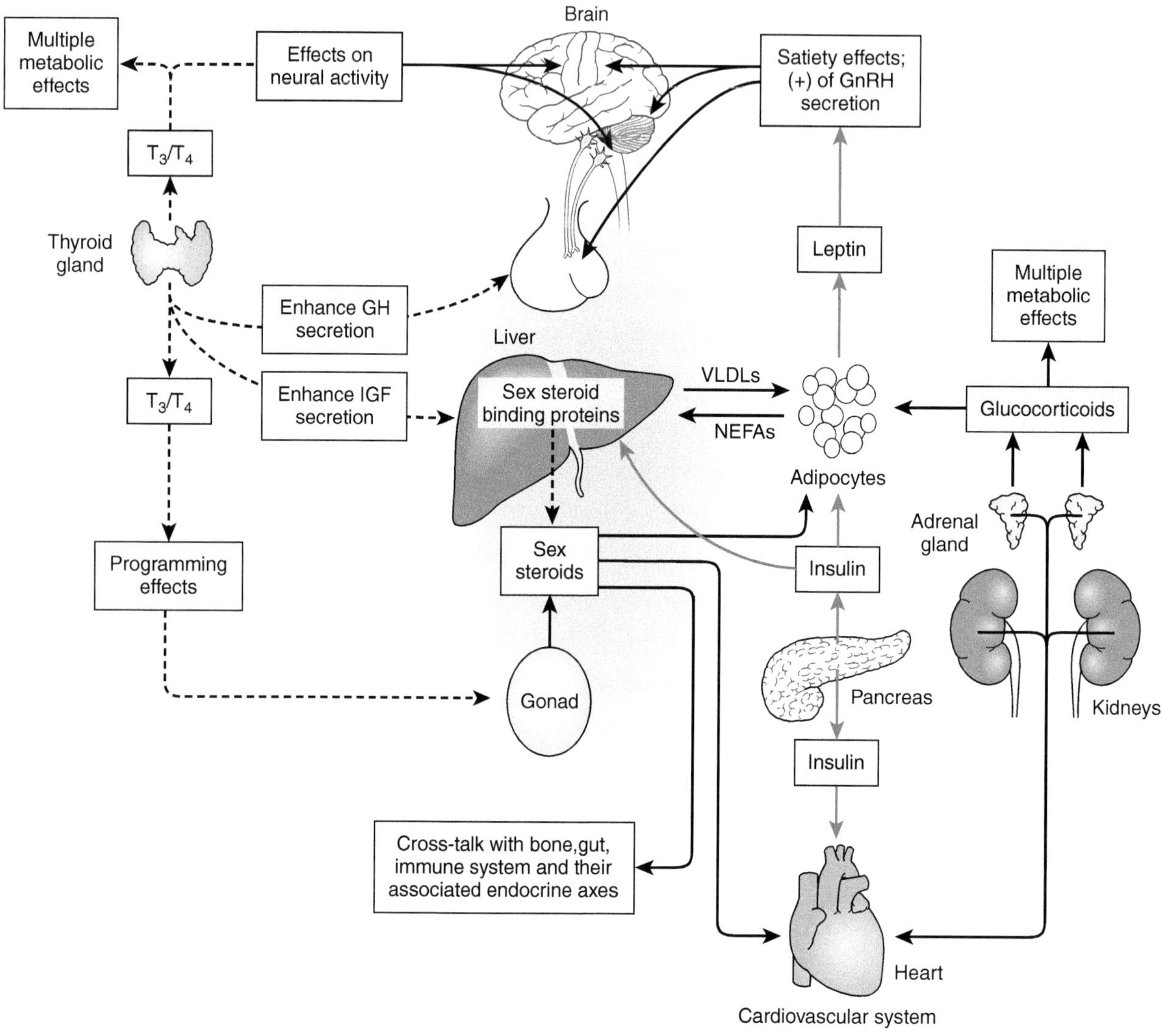

FIGURE 3-16 **Another kind of cross-talk.** This diagram illustrates some of the ways in which bioregulators communicate with other pathways such that one bioregulator system may influence the effectiveness of another. See Appendix A for abbreviations. *(Adapted from Damstra, T. et al., "Global Assessment of the State-of-the-Science of Endocrine Disruptors," World Health Organization, Geneva, Switzerland, 2002, p. 20.)*

synthesized through the sequential cooperation of skin, liver, and kidney. 1,25-DHC is especially important for growing children and pregnant women to increase uptake of adequate Ca^{2+} for normal bone development and growth. The initial step in its synthesis depends on the actions of sunlight on skin. An intermediate of 1,25-DHC, **vitamin D** (the "sunshine vitamin"), is commonly added to milk in many north temperate countries to ensure an adequate supply of 1,25-DHC for absorption of the Ca^{2+} provided in milk. Cod liver oil is another rich source of vitamin D.

Androgens (Figure 3-18) are defined as compounds that stimulate development of male characteristics; that is, they are masculinizing agents. In males, the primary source for circulating androgens, such as **testosterone**, is the testis, where **luteinizing hormone (LH)** from the pituitary gland stimulates their synthesis and release into the blood. Other important androgens are **dihydrotestosterone (DHT)**, **androstenedione**, and **dehydroepiandrosterone (DHEA)**. In females, the adrenal cortex and the ovaries also synthesize androgens that play important roles (see Chapter 10). During pregnancy, the fetal adrenal becomes an important source of androgens for the placenta (see Chapters 8 and 10).

Estrogens (Figure 3-18) are compounds capable of stimulating cornification in epidermal cells lining the vagina of castrate female rats. Cornification involves production of the structural fibrous protein keratin, a major intermediate filament that helps form the cytoskeleton of cells in skin and hair. A more common test used today for

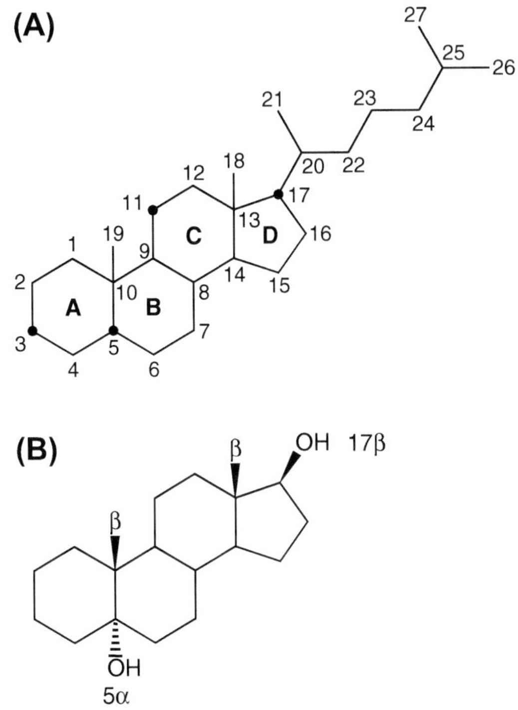

FIGURE 3-17 **The steroid nucleus.** (A) There are 17 carbons in the nucleus with two additional carbons (18, 19) attached to carbons 13 and 10, respectively. The four rings of the nucleus are labeled A, B, C, and D. The side chain of carbons 20 to 27 is attached to the steroid nucleus at carbon 17 in the β-configuration and is indicated as a solid line. Some of the asymmetric carbons of the nucleus are designated as enlarged dots where the lines representing the covalent bonds intersect. (B) Atoms attached to an asymmetric carbon in the α-configuration are designated with a dashed line as indicated for 5α OH. Those attached in the β-configuration (including carbons 18 and 19) are indicated with a solid wedge.

demonstrating estrogen activity involves the use of yeast cells that have been transfected with the estrogen receptor gene. Although traditionally labeled as "female hormones," estrogens are synthesized in males and play important roles just as androgens do in females (see Chapter 10). Estrogens, such as **estradiol**, also stimulate proliferation and vascularization of the uterine mucosa or endometrium. They are converted from androgens by the enzyme **aromatase**. Aromatase synthesis is stimulated by FSH from the pituitary. Additional, important estrogens are **estrone** and **estriol**.

A **progestogen** (Figure 3-19) is a compound that maintains pregnancy (i.e., they are progestational) or the secretory condition of the uterine endometrium during the luteal phase of the ovarian cycle. Progestogens (= progestins) such as **pregnenolone**, **progesterone**, **17α-hydroxypregnenolone**, and **17α-hydroxyprogesterone**, are produced by all steroidogenic tissues as intermediates in the synthesis of most of the other steroid hormones. The role of progesterone as a mammalian reproductive hormone, however, is also well established (see Chapter 10).

There are two subcategories of **corticosteroids**: glucocorticoids and mineralocorticoids (Figure 3-19). **Glucocorticoids**, such as **cortisol** and **corticosterone**, can influence protein and carbohydrate metabolism under certain conditions, and their major physiological role appears to be an influence on peripheral utilization of glucose (see Chapters 8 and 12). **Mineralocorticoids**, such as **aldosterone**, affect sodium and potassium transport mechanisms in nephrons of the kidney (see Chapter 8). It must be emphasized that these two terms become rather confusing when applied to non-mammalian vertebrates, because a given molecule that possesses glucocorticoid activity in mammals may have mineralocorticoid activity in non-mammals. Cortisol, for example, is a glucocorticoid in humans, but it also is a mineralocorticoid in teleost fishes (see Chapter 9). Corticoids are synthesized by the outer portions of the adrenal gland (the adrenal cortex) or its homologue in non-mammals (often called **interrenal tissue**) following stimulation by pituitary ACTH (see Chapters 8 and 9). It should be obvious that these definitions for the various categories of steroid hormones are functionally derived.

Although naturally occurring steroids that exhibit a particular hormonal activity in vertebrates (estrogenic, androgenic, etc.) are structurally similar to one another, many structurally unrelated compounds may have similar hormonal activity when tested and could be classified accordingly (see Figure 3-20). For example, **diethylstilbestrol (DES)** is a powerful synthetic estrogen that readily binds to estrogen receptors, but it is not a steroid. Additional plant sterols (phytoestrogens), components of detergents (alkylphenols and their derivatives such as nonylphenols), compounds leached from plastics (e.g., bisphenol A, nonylphenol, phthalates), and certain pesticides (e.g., DDT), as well as a number of polychlorinated chemicals from industrial production (dioxins, PCBs), are known to bind to estrogen receptors and mimic the naturally occurring estrogens (see discussion of endocrine-disrupting chemicals in Chapters 1 and 10).

Although we have long known that the brain is a target for steroids and that some of these steroids may be converted by neural enyzmes to more active forms for binding to receptors, we must now include the brain as a steroid-synthesizing tissue. **Neurosteroids** are so named to emphasize the fact that they are synthesized *de novo* in neural tissue and to distinguish them from steroids modified by neural cells after their accumulation from the blood or cerebrospinal fluid. In recent years, it has been demonstrated that steroids, especially pregnenolone and DHEA, are synthesized within neuroglial cells of the central and peripheral nervous systems and play important roles, such as in the myelination process. Numerous steroidogeneic enzymes have been identified in neural cells except for those involved in corticosteroid synthesis. Neurosteroids

TABLE 3-9 Some Common Steroids

Class	Example	Number of Carbons	Major Site(s) of Synthesis
Progestogens	Progesterone	21	Adrenal cortex Male: testis Female: ovary, adipose tissue
Glucocorticoids	Cortisol	21	Adrenal cortex: zona fasciculata
	Corticosterone	21	Adrenal cortex: zona fasciculata
Mineralocorticoids	Aldosterone	21	Adrenal cortex: zona glomerulosa
Androgens	Testosterone	19	Male: testis Female: ovary, adrenal cortex
	Dihydroepiandrosterone (DHEA)	19	Adrenal cortex: zona reticularis
Estrogens	Estradiol	18	Ovary, testis
Vitamin D	Cholecalciferol	25	Skin (in presence of UV light)
	25-Hydroxycholecalciferol	25	Liver (made from cholecalciferol)
	1,25-Dihydroxycholecalciferol	25	Kidney (made from 25-hydroxycholecalciferol)
Cholesterol	Cholesterol	27	Liver, gonads, adrenal cortex

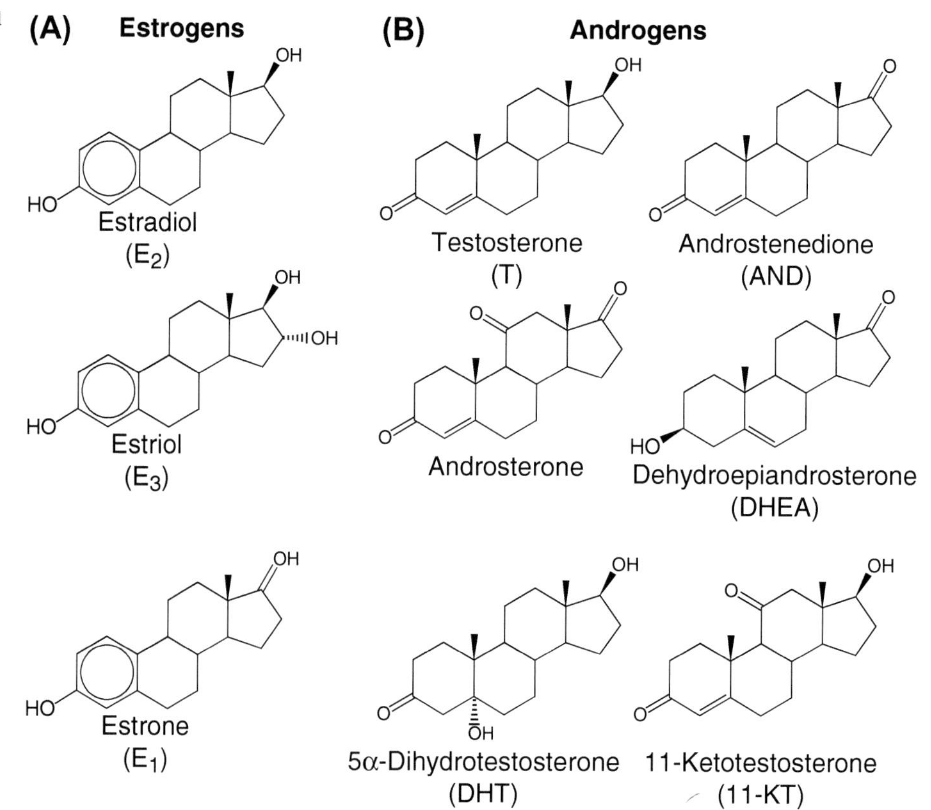

FIGURE 3-18 (A) C_{18} estrogens and (B) C_{19} androgens.

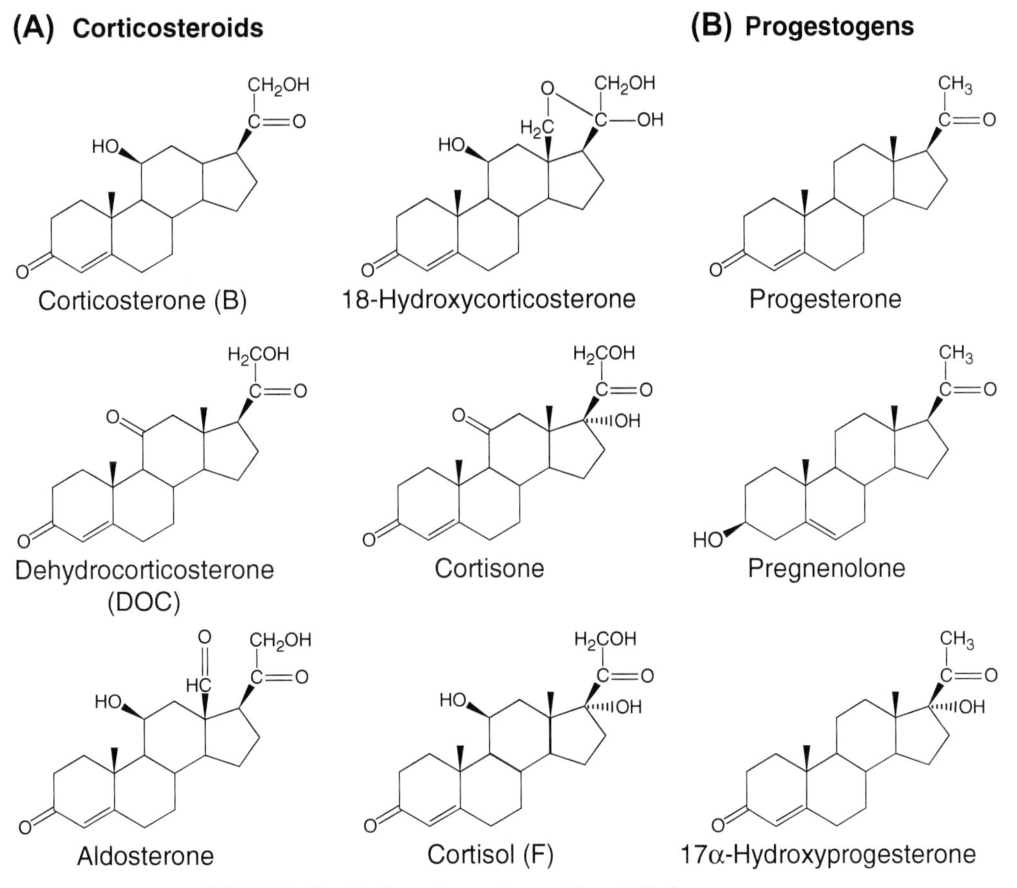

FIGURE 3-19 (A) Some C_{21} corticosteroids and (B) C_{21} progestogens.

are believed to function locally (paracrines) and probably do not appear in the general circulation.

A. Steroid Structure and Nomenclature

The literature on steroid hormones is very confusing to the uninitiated in part because of the inclusion of chemical designations in their names and also by the multiplicity of trivial and chemical names in the literature for a single molecule. Cortisol, for example, in addition to being designated by two different imposing chemical names (11β,17α,21-trihydroxypregn-4-ene-3,20-dione; 11β17α,-21-trihydroxy-Δ^4-pregnene-3,20-dione) has three trivial or common names (cortisol, 17-hydroxycorticosterone, and hydrocortisone). Furthermore, cortisol was known as Reichstein's compound M and Kendall's compound F before its chemical structure was elucidated. The common abbreviation for cortisol has become "F" for Kendall's unknown fraction.

In order to use the established literature, it is necessary to become familiar with the various trivial names as well as to understand the bases for the chemical names. A working knowledge of the chemical nomenclature for steroids is not really necessary for the treatments provided in this textbook, although those students interested in details of synthesis and metabolism of steroids will find knowledge of the chemical nomenclature invaluable. Furthermore, a glance at the chemical structure of a steroid also tells you a great deal about its biological role. Table 3-10 summarizes special designations used in the nomenclature of steroids.

The steroid nucleus contains 17 carbons, each designated by a number and arranged into four rings (Figure 3-17). Steroids are named chemically by relating each steroid to a saturated hypothetical parent hydrocarbon compound (i.e., one that has no double bonds present) and modifying this name with one or more prefixes and no more than one suffix to designate the specific compound (Figure 3-21). These hypothetical parent compounds are **estrane** (18 carbons, or C_{18}), **androstane** (C_{19}), **pregnane** (C_{21}), and **cholestrane** (C_{27}). Naturally occurring estrogens (all C_{18} compounds) and androgens (C_{19} compounds) possess the basic carbon skeleton of estrane and androstane, respectively, whereas the names for corticoids and progestogens (both C_{21} compounds) are related to pregnane for naming purposes. Cholestrane is employed for naming cholesterol and cholesterol derivatives including bile salts and the vitamin D compounds.

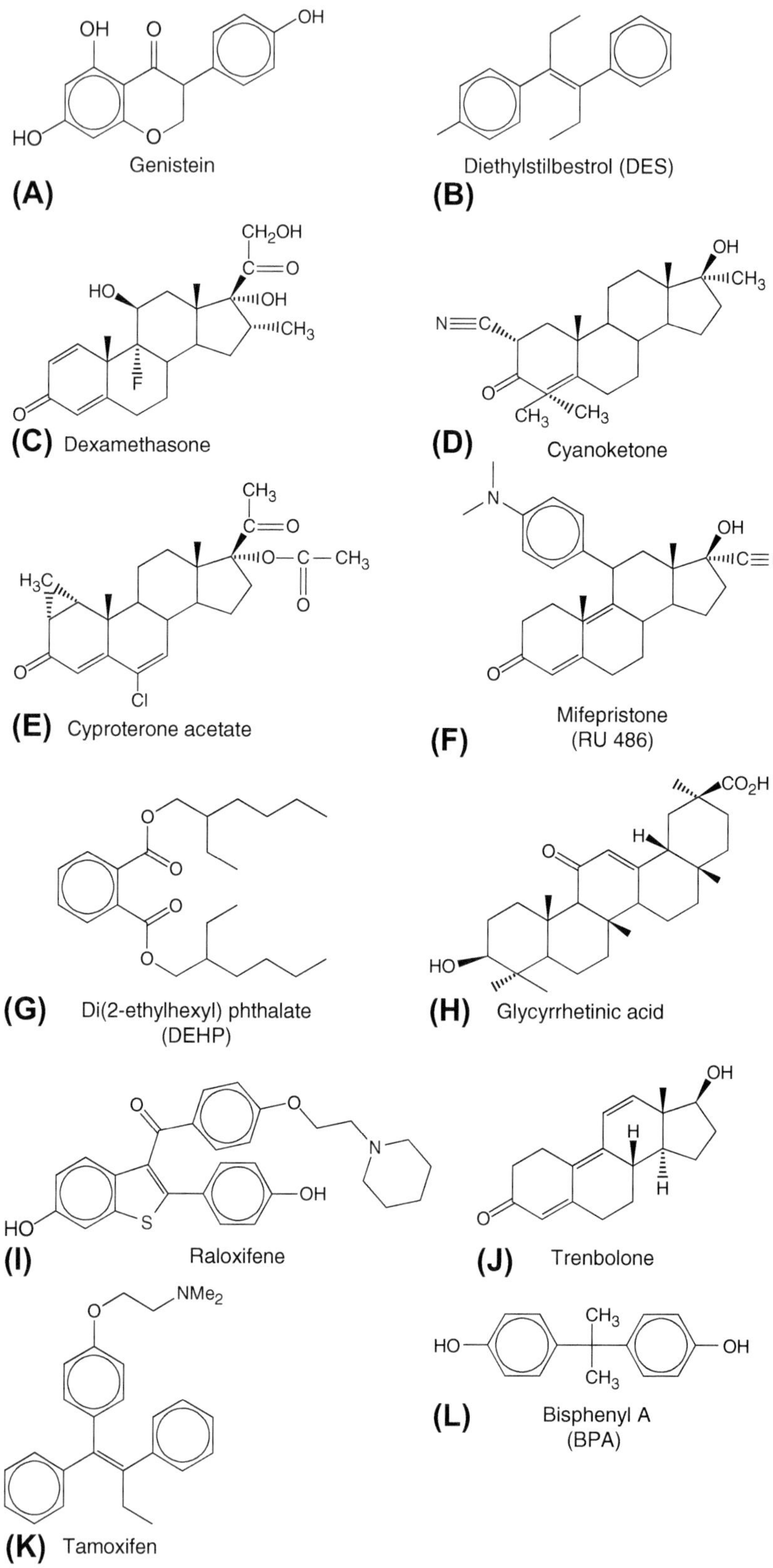

FIGURE 3-20 **Some synthetic steroids and nonsteroids with steroid-like activity.** (A) Genistein is a phytoestrogen found in clover and other plants. (B) Diethylstilbestrol is a potent estrogen. (C) Dexamethasone is a synthetic glucocorticoid that contains fluorine and is more potent than any of the naturally occurring ones. (D) Cyanoketone is a steroid that inhibits the enzyme that normally converts the steroid pregnenolone to progesterone. (E) Cyproterone acetate is an antiandrogen and blocks androgen binding to receptors. (F) Mifepristone, or RU 486, is an antiprogesterone and an antiglucocorticoid. (G) Diethyl-hexylphthalate (DEHP) has effects on several HP axes. (H) Glycyrrhetinic acid is found in licorice and has weak corticosteroid activity. (I, K) Selective estrogen receptor modulators (SERMs) roloxifene and tamoxifen. (J) Trenbolone, a potent synthetic androgen. (L) Bisphenyl A (BPA), an estrogenic chemical.

TABLE 3-10 Summary of Special Designations in Steroid Nomenclature

Designation	Explanation
Δ	Location of double bond
-ene	One double bond in steroid nucleus
-diene	Two double bonds in steroid nucleus
-triene	Three double bonds in steroid nucleus
hydroxyl-	Hydroxyl (OH) substituted for hydrogen on nucleus
-ol	Hydroxyl (OH) substituted for hydrogen on nucleus
oxa-	Ketone (=O) substituted for hydrogen on nucleus
keto-	Ketone (=O) substituted for hydrogen on nucleus
-one	Ketone (=O) substituted for hydrogen on nucleus
α	Atom or atoms attached to a given carbon of the steroid nucleus projects away from viewer
β	Atom or atoms attached to a given carbon of the steroid nucleus projects toward viewer
Arabic number	Indicates location of substitution or double bond

1. Presence of Double Bonds in the Steroid Nucleus

The position of double bonds between carbon atoms within the steroid nucleus was formerly designated by the Greek letter delta (symbol, **Δ**) followed by the superscripted number of the lowest-numbered steroid carbon with which the bond is involved. Thus, Δ^4 would indicate a double bond between carbons 4 and 5 in the steroid nucleus. Currently, a separate scheme employs the term **ene** to refer to one double bond, **diene** for two double bonds, and **triene** for three double bonds in the nucleus, and the location of each bond is indicated by the number of the carbon preceding the location of the double bond; for example, Δ^4-pregnene refers to a double bond in pregnane between carbons 4 and 5. Nevertheless, much of our steroid-related terminology still employs the older Δ terminology.

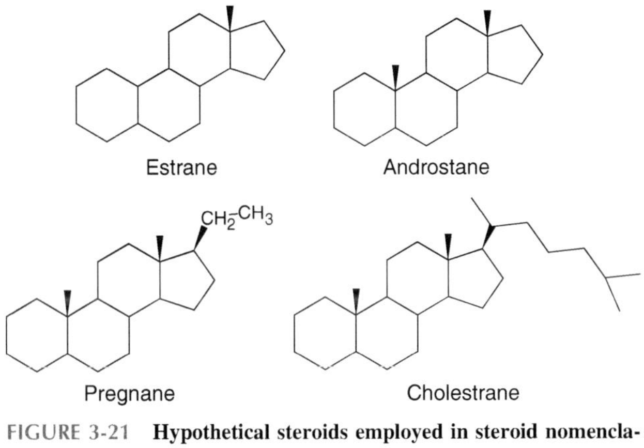

FIGURE 3-21 Hypothetical steroids employed in steroid nomenclature. These compounds do not exist and are used only for the purposes of constructing the chemical names for the four major groups of steroid hormones.

2. Common Substitutions to the Steroid Nucleus

Substituted groups applied to the steroid skeleton are designated according to the number of the carbon atom in the steroid skeleton to which they are attached. For example, **17-hydroxy** refers to a hydroxyl group (–OH) attached to carbon number 17. Hydroxyl groups are usually indicated as prefixes on the chemical name unless they are the only substitution on the steroid nucleus, in which case the hydroxyl group is designated by the suffix **-ol**. Ketone groups (=O) substituted on the steroid nucleus are indicated by the prefix **oxo-** or **keto-** or by the suffix **-one**; thus, **-3,20-dione** would refer to two ketone groups attached at positions 3 and 20, respectively, as would **-3,20-dioxo-**. In naming steroids with both hydroxyl and keto groups, the latter takes priority as the suffix (-one) over the hydroxyl (-ol).

3. Stereoisomerism

One carbon atom is capable of forming four covalent bonds with other atoms. Only two of these bonds are used when a given carbon is incorporated into the steroid nucleus and bound to two other carbon atoms. Because of the necessary bonding angles dictated by the tetrahedral shape of the carbon atom, a steroid does not exist in a flat plane as most diagrams of their chemical structures would suggest. The remaining two sites for each carbon atom incorporated into the steroid nucleus can form covalent bonds with hydrogen, oxygen, or other carbon atoms. Depending on how and where the carbon atom appears in the nucleus, it may be able to bind hydrogen or oxygen to either side. One of these bonds will project toward the viewer when examining the steroid, as depicted in Figure 3-17, and the other will project away from the viewer—that is, into the page. Such carbon atoms are referred to as **asymmetric carbons**. There are asymmetric carbons at positions 3, 5, 11, 16, and 17 of the steroid nucleus as well as for carbon 20 in the side chain. These numbers appear in their chemical names and sometimes carry over to the trivial names because they form important bonds with ketone or hydroxyl groups.

Two three-dimensional isomers (**stereoisomers**) can be formed by substituting one hydrogen on a given asymmetric carbon with, for example, a hydroxyl group. The chemical formulas would be the same (i.e., the same number of C, H, and O atoms), but the three-dimensional structure of each stereoisomer would differ according to whether the added −OH group projected out from or into the page. A spatial designation of α is used if the substituted group projects away from the viewer, and β is used if the group projects toward the viewer. In estrogens, which have an aromatic A-ring in the steroid nucleus, the carbon at position 3 can be neither α or β. In C_{21} steroids, carbon 20 is connected in the β-configuration to carbon 17. Any other attachment to carbon 17 of a C_{21} steroid must be in the α-position, and it is not necessary to indicate this in the chemical name.

In two-dimensional diagrams, the α-position is indicated by a slashed or dotted line, whereas the β-position is designated with a solid line connecting the substituted group to its carbon (Figure 3-17). Hydrogen atoms usually are not designated, but their presence is implied. Estradiol occurs in two isomeric forms, **17β-estradiol** and **17α-estradiol**. Notice how the trivial name "estradiol" has been embellished with details derived from its chemical name (diol refers to 2 OH groups, one of which is on carbon 17, an asymmetric carbon, and in either the α- or β-conformation). The complete chemical names for these two estradiol molecules are 1,3,5-estratriene-3,17α-diol and 1,3,5-estratriene-3,17β-diol. It was generally accepted that 17β-estradiol was the important estrogen of vertebrates and that 17α-estradiol had little or no estrogenic activity. However, studies in both fishes and mammals have demonstrated the effectiveness of 17α-estradiol in binding to and activating estrogen receptors even though they do not bind to plasma-binding proteins. Furthermore, studies in mammals have verified the synthesis of 17α-estradiol in the brain of castrate and adenalectomized rats. Much remains to be learned about the possible roles of 17α-estradiol, however, and we will continue to consider 17β-estradiol to be the major estrogen of vertebrates. Henceforth, we shall use "estradiol" alone to designate 17β-estradiol.

The chemical structures of several steroids are diagramed in Figures 3-18 and 3-19, and the chemical and trivial names for some of these steroids are indicated in Table 3-11. Can you determine the basis for the chemical name assigned to each steroid?

B. Steroid Synthesis

All vertebrate steroid bioregulators are synthesized from **cholesterol**, a C_{27} steroid (Figure 3-22). Cholesterol is synthesized from acetate (acetyl-coenzyme A) produced via glycolysis or via fatty acid oxidation. The synthesis of the steroid nucleus from acetate units is termed **steroidogenesis**, although often this term is used by endocrinologists for only the synthesis of steroid hormones from cholesterol. Steroidogenesis begins with an involved series of enzymatically catalyzed reactions that converts acetate into cholesterol. Following synthesis of a relatively long hydrocarbon chain, a complex cyclization step results in closure of the carbon skeleton into the steroid nucleus. Cholesterol synthesized in this manner may be used directly in the biosynthesis of the various steroid hormones but cannot be degraded back toward acetate. Excess cholesterol is converted by the liver into bile salts for excretion.

Most cholesterol is synthesized in the liver and is released into the blood as lipid droplets coated with protein (for details, see Chapter 12). Adrenal cortex, ovaries, and testes also can synthesize cholesterol but more commonly utilize plasma lipoprotein complexes absorbed from the gut or synthesized in the liver as sources of cholesterol (see Chapter 12).

Cholesterol, bile salts, and the steroid hormones are synthesized primarily in the liver (cholesterol, bile salts), the gonads (estrogens, androgens, progesterone), the placenta (estrogens, progesterone), the adrenal cortex (corticosteroids, androgens), and in the brain (neurosteroids). In addition, stromal cells present in adipose tissue may be important in estrogen synthesis, especially in postmenopausal women and in aging men. The pathways for the synthesis of these steroids are provided in Figures 3-22, 3-23, 3-24, and 3-25. Vitamin D compounds made from cholesterol are made by the cooperative activities of skin, liver, and kidney. The pathway for forming 1,25-DHC differs markedly from that for the gonadal and adrenal steroids and will be described in Chapter 14 with the discussion of Ca^{2+} regulation.

1. Key Enzymes in Gonadal and Adrenal Steroid Biosynthesis

Identification of certain key synthetic enzymes and quantification of their activity levels are often used as indicators of the biosynthesis of steroid hormones. More than one scheme has been proposed for naming these enzymes (see Table 3-12), and some enzymes may have several different names. The newest methods involve applying a specific name for the gene that is written in italics (e.g., *CYP19*). Each gene is responsible for an enzyme's production, and either a name related to the gene name but without italics (e.g., CYP19) or a more descriptive name (e.g., $P450_{aro}$ or aromatase) is used for the enzyme. The more descriptive names will be used here to simplify connecting each enzyme to its action and location within the cell. Many of these enzymes are members of a class of membrane-bound enzymes known as the **P450 cytochromes**, familiar to physiologists for their role in oxidative phosphorylation. Others are cytosolic proteins.

TABLE 3-11 Some Vertebrate Steroid Hormones

Category	Trivial name	Chemical name[a]
Androgens	Testosterone	17β-Hydroxy-4-androsten-3-one
	Androstenedione	4-Androstene-3,17-dione
	Dehydropiandrosterone	3β-Hydroxy-5-androsten-17-one
Corticoids	Aldosterone	11β,21-Dihydroxy-3,20-dioxo-4-pregnen-18-ol
	Cortisol	11β,17,21-Trihydroxy-4-pregnene-3,20-dione
	Corticosterone	11 β,21-Dihydroxy-4-pregnene-3,20-dione
	11-Deoxycorticosterone	21-Hydroxy-4-pregnene-3,20-dione
Estrogens	Estradiol-17β	1,3,5(10)-Estratriene-3,17β-diol
	Estrone	3-Hydroxy-1,3,5(10)-estratriene-17-one
	Estriol	1,3,5(10)-estratriene-3,16α,17β-triol
Progestogens	Pregnenolone	3β-Hydroxy-5-pregnen-20-one
	Progesterone	4-Pregnene-3,20-dione

[a] *There are variations in how these molecules may be named.*

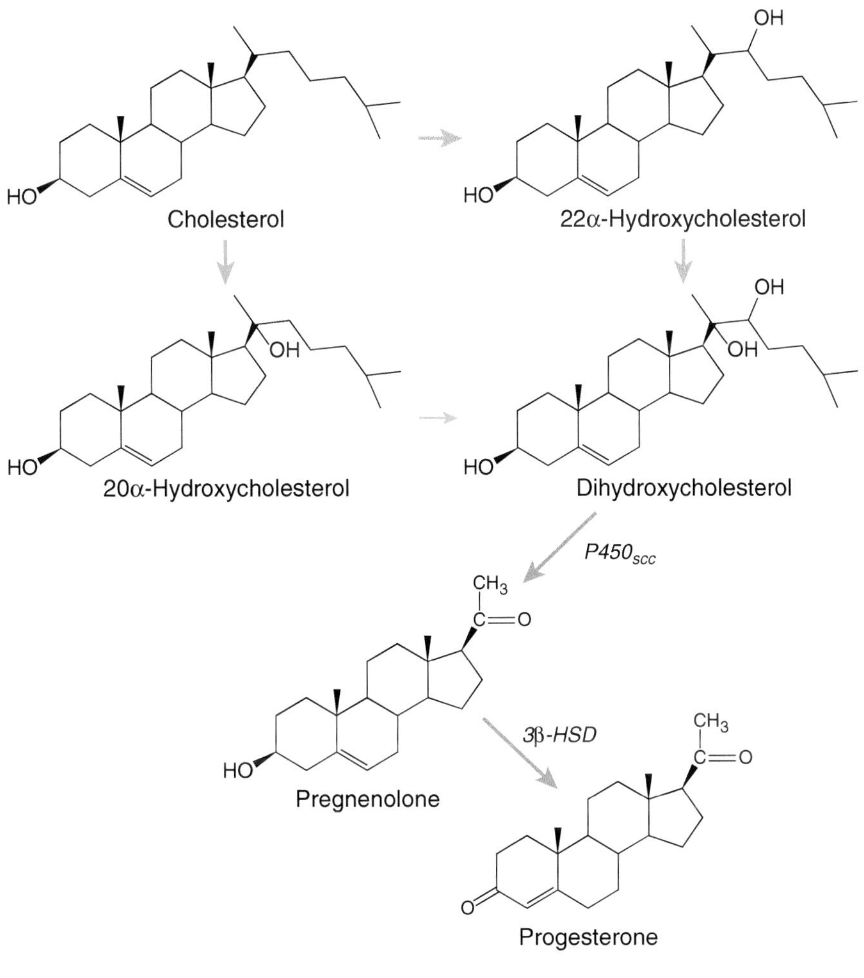

FIGURE 3-22 Synthesis pathway for progesterone.

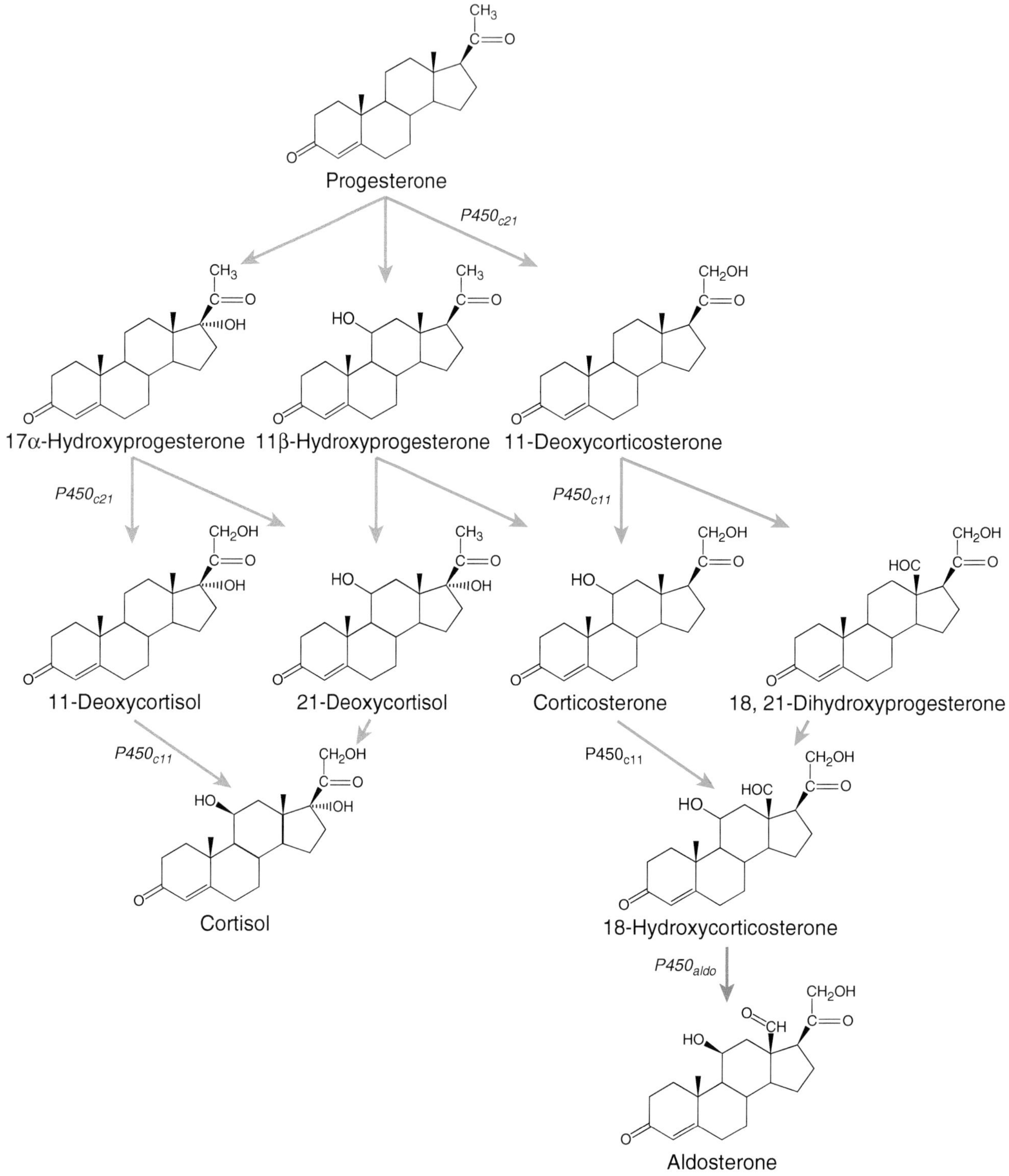

FIGURE 3-23 **Synthesis of corticosteroids from progesterone.**

The first step in the biosynthesis of these steroid hormones involves **side-chain hydrolysis**, which removes most of the hydrocarbon side chain of cholesterol to yield a C_{21} intermediate, **pregnenolone**. Removal of the cholesterol side chain is accomplished by the **side-chain cleaving enzyme ($P450_{scc}$)** (CYP11A). A **steroidogenic acute regulating protein (StAR)** discovered by Doug Stocco's group at Texas Tech University facilitates the transfer of cholesterol from the outer mitochondrial membrane to the inner membrane where $P450_{scc}$ is located.

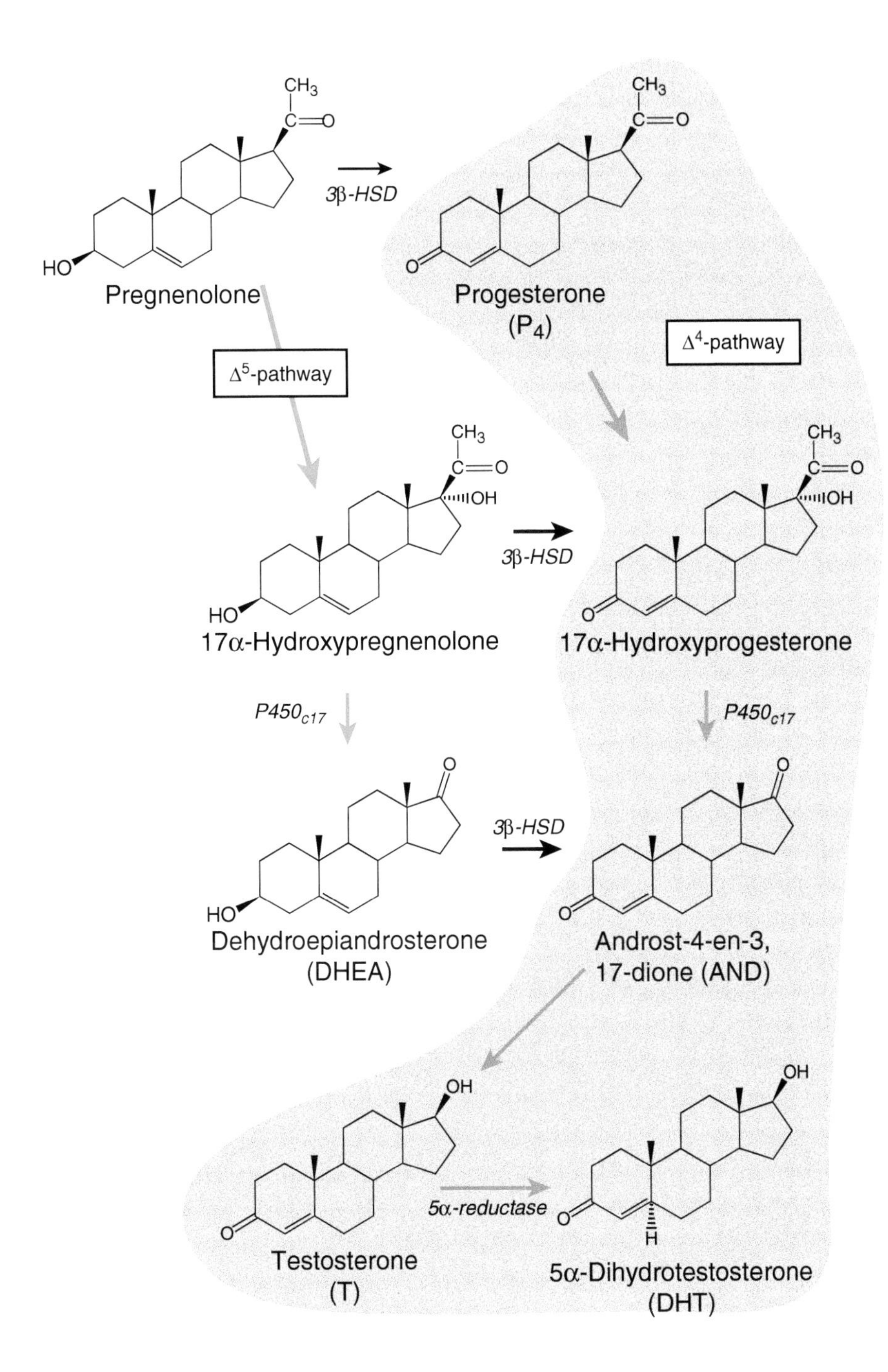

FIGURE 3-24 **Δ^4- and Δ^5-pathways for androgen synthesis.** The Δ^5-pathway typically occurs in the adrenal cortex and usually stops with the production of DHEA or DHEAS. In ovaries of some species, this pathway may lead to testosterone and eventually to estrogen synthesis. Testes employ only the Δ^4-pathway (highlighted). Note that the enzyme 3β-hydroxysteroid dehydrogenase (3β-HSD) can convert several Δ^5- steroids into Δ^4-steroids.

Patients unable to convert cholesterol to corticosteroids but who have demonstrable $P450_{scc}$ provided the clue to the existence of the StAR protein.

A key step following side-chain hydrolysis is the conversion of pregnenolone to progesterone, which involves moving the double bond from the B ring to the A ring (change from Δ^5 to Δ^4) and converting the β-OH group on carbon 3 to a ketone (=O; see Figure 3-22). This step is accomplished by the cytosolic enzyme, **Δ^5,3β-hydroxysteroid dehydrogenase (3β-HSD)**. The drug **cyanoketone** is known to competitively inhibit the activity of 3β-HSD because of the structural similarity of cyanoketone (Figure 3-20) to pregnenolone (Figure 3-22) and its ability to block access of pregnenolone to the enzyme's catalytic site. Cyanoketone effectively blocks gonadal steroid biosynthesis and reduces circulating levels of all adrenal and gonadal steroids. It also prevents synthesis of corticosteroids by adrenal cells.

In the adrenal cortex, progesterone is metabolized through several pathways to produce the glucocorticoids

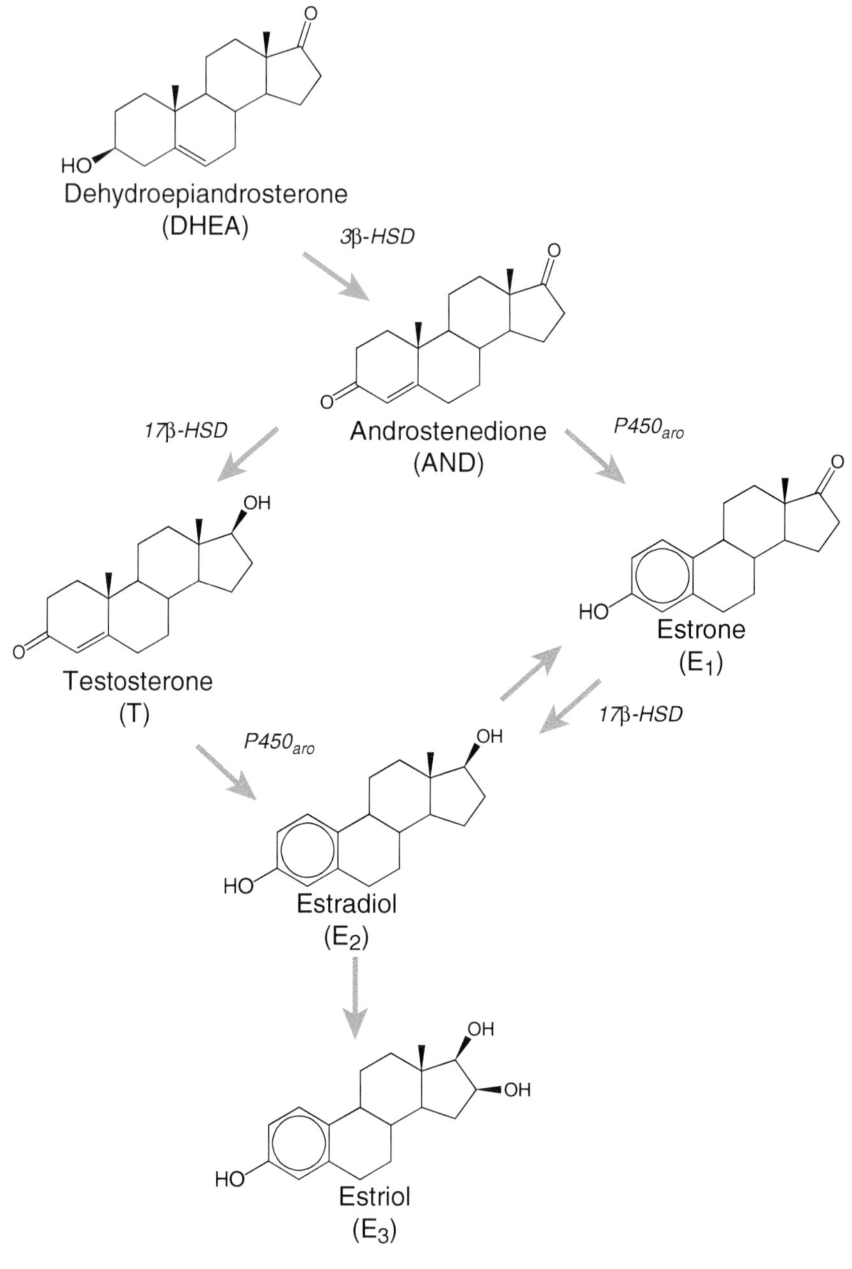

FIGURE 3-25 **Synthesis of estrogens from androgens.** Estrogen synthesis requires either prior synthesis of an androgen or an external source of androgen.

cortisol and corticosterone and the mineralocorticoid aldosterone (Figure 3-23). Two enzymes regulate the production of the glucocorticoids, a C21-hydroxylase ($P450_{c21}$; CYP21) and an 11β-hydroxylase ($P450_{11\beta}$; CYP11B1). An additional enzyme, aldosterone synthetase ($P450_{aldo}$; CYP11B2) converts corticosterone to aldosterone. The drug **metyrapone** selectively inhibits 11β-hydroxylase, thus blocking corticoid synthesis in adrenocortical cells.

In the gonads, progesterone is converted to androstenedione by the enzyme **17β-hydroxysteroid dehydrogenase (17β-HSD)** to testosterone, the principle androgen of all male vertebrates. Additionally, the enzyme **5α-reductase** converts testosterone into DHT, an important androgen for development of the penis and scrotum in males (Figure 3-24). Furthermore, testosterone may be converted by the enzyme aromatase (**$P450_{aro}$**; CYP19), which results in loss of one carbon and aromatization of the A ring to yield estradiol, a C_{18} estrogen. This same enzyme can convert androstenedione into estrone, which can be converted enzymatically into estradiol, as well (Figure 3-25). The presence of aromatase activity in a cell or tissue is an indicator of the ability to transform androgens into estrogens. $P450_{aro}$ plays an important role in fetal development, as levels of this enzyme are high in

TABLE 3-12 Names for Key Steroidogenic Enzymes

Gene	Enzyme Name	P450 Abbreviation
CYP21	C_{21}-Hydroxylase	$P450_{c21}$
CYP11A	Cholesterol side-chain cleavage, 20-22 Desmolase	$P450_{SCC}$
CYP17	17α-Hydroxylase, 17,20-Lyase	$P450_{C17}$
CYP19	Aromatase	$P450_{ARO}$
CYP11B1	11β-Hydroxylase	$P450_{C11}$
CYP11B2	Aldosterone synthetase	$P450_{ALDO}$
CYP1A1	Aryl hydrocarbon hydroxylase	$P450_{1A1}$
HSD3B2	3β-hydroxysteroid dehydrogenase, $\Delta^{5\text{-}4}$-isomerase	—
SRD5A2	5α-reductase	—

the fetal liver of rats and humans. Aromatase activity occurs in adult vertebrate brains, adipose tissue, and certain bone cells, where it converts androgens into estrogens. Testosterone, like estradiol, occurs in both α and β forms. α-Testosterone can be converted to 17α-estradiol by $P450_{aro}$. Furthermore, 17α-estradiol can be formed from estrone.

Although DHT is considered to be a non-aromatizable androgen, it can be modified, at least in some mammals, by enzymes to **5α-androstane-3β,17β-diol (3β-diol)**, which binds and activates one form of the estrogen receptor found in brain and prostate gland (see ERβ, ahead). 3β-diol is chemically an androgen, but it can produce distinct effects via an estrogen receptor in these tissues.

The enzyme 3β-HSD can work on a variety of steroid substrates. If 3β-HSD acts early in the sequence by converting Δ^5-pregnenolone to Δ^4-progesterone, the subsequent enzymatic transformations of progesterone are referred to as the **Δ^4-pathway**. When pregnenolone is not converted to Δ^4-progesterone, it is metabolized along the **Δ^5-pathway**. 3β-HSD may act at other points in the Δ^5-pathway and convert other intermediates into components of the Δ^4-pathway—for example, DHEA to androstenedione (see Figure 3-24). Δ^4 androgens are more active in binding and activating androgen receptors than are Δ^5 androgens. Both of these molecules are C_{19} compounds. The significance of the Δ^4- and Δ^5-pathways in adrenal, testicular, and ovarian steroid syntheses is discussed in Chapters 8 and 10. Pregnenolone and progesterone can be converted to DHEA and androstenedione, respectively, by the enzyme **17α-hydroxylase** (**P450c17**; CYP17; = 17,20-lyase).

2. Cytological Aspects of Gonadal and Adrenal Steroid Biosynthesis

Biosynthesis of steroid hormones is, cytologically speaking, a very complex affair involving cytosolic, mitochondrial, and smooth endoplasmic reticulum (SER) enzymes. The importance of the latter is demonstrated by observations that a major cytological feature of steroidogenic cells is the abundance of SER.

C. Transport of Steroid Hormones in Blood

Steroids are nonpolar compounds and consequently are not very soluble in aqueous solutions such as blood. Furthermore, free steroids readily diffuse or are transported through cellular membranes and rapidly disappear from the blood as a result of the activities of the liver and kidneys. The association of circulating steroids with plasma proteins results in their being retained at higher concentrations for longer times in the circulation. These **plasma binding proteins** reduce the removal of active steroid hormones by the liver or kidney and their excretion via the urine. Progesterone and glucocortioids are transported in plasma by a protein called **corticosteroid binding globulin (CBG)**, which is a member of family of serine protease inhibitors (**serpins**). CBG, also known as serpinA6, does not bind estrogens or androgens. Testosterone and estradiol are transported by **sex hormone binding globulin (SHBG)**. Thus, relatively high titers of steroid hormones can be maintained in the circulation, providing maximal local titers of dissociated free steroid and increasing the probability of their entering appropriate target tissues. As mentioned earlier, plasma-binding proteins may facilitate the entrance of steroids into their target cells.

D. Mechanisms of Steroid Action

The classical mechanism of action for steroids involves activation of receptors and the induction of gene transcription—that is, a **genomic action** where occupied receptors act as transcription factors. More recently, the involvement of steroids with membrane receptors has been recognized as a mechanism for producing rapid responses in target cells such as those described for peptide and protein hormones. These actions do not involve gene transcription and are often termed **non-genomic actions**.

1. A General Model of Genomic Steroid Action

The original genomic model for steroid hormone action was proposed for estrogens by Elwood Jensen in the early 1970s (see Figure 3-26). Although the model has undergone considerable revision since that time, such as the location of unoccupied receptors, the basic features are still

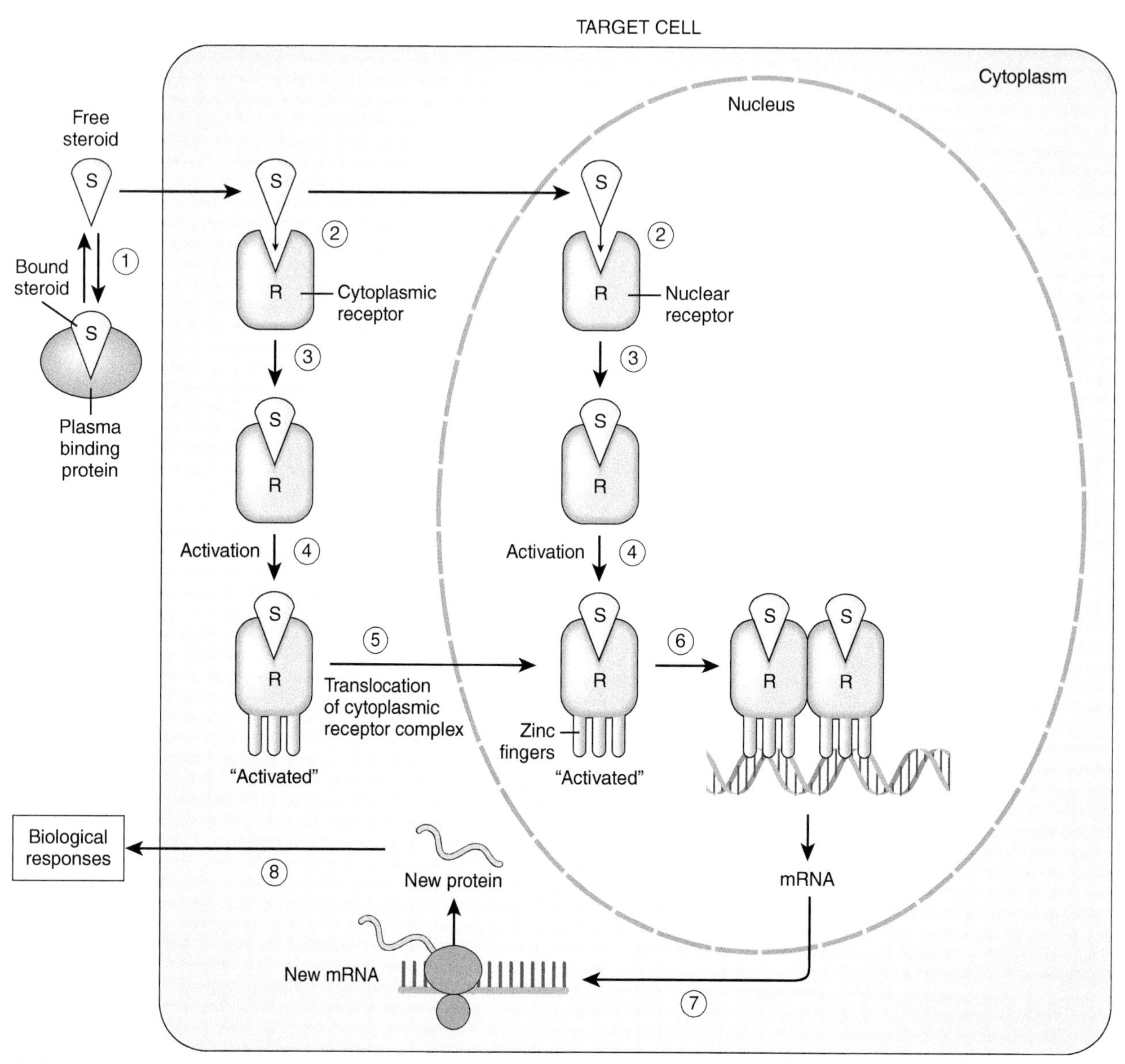

FIGURE 3-26 **Genomic mechanism of activation by steroids.** The interaction of steroids with different genes in a target cell may direct the synthesis of structural proteins such as cytoskeletal elements or receptors as well as enzymes. These enzymes may produce a variety of effects within the cell. The genomic mechanism of action for thyroid hormones is very similar, with the emphasis more on nuclear location of unoccupied receptors. Additionally, occupied thyroid hormone receptors form heterodimers with RXR receptors (see Figure 3-33).

valid. Once these steroids enter their target cells, they diffuse into the cytosol or nucleus, where they bind to specific protein receptors. Because the unoccupied receptors for estrogens originally were found in the nucleus of target cells, these **steroid receptors** and related receptor molecules have been termed **nuclear receptors**. Although additional research has shown that the location of estrogen, progesterone, and androgen receptors may be either cytosolic or nuclear depending on the target cell, all of these receptors are generally termed "nuclear receptors." In contrast to the other steroid receptors, corticosteroid receptors appear to be exclusively cytoplasmic and are translocated to the nucleus following binding of the appropriate ligand. Once a steroid hormone has bound to its receptor, it dimerizes with another occupied receptor, and the complex becomes a **ligand-activated transcription factor** that can now interact with DNA to initiate transcription.

Regardless of their location, all of these unoccupied steroid receptor proteins are complexed with a number of **chaperone proteins** (often called **heat-shock proteins**) that are involved with maintaining the three-dimensional shape of the receptor prior to binding with its ligand (see Figure 3-27). Once a steroid binds to its receptor, some of the charparone proteins are released. Part of the steroid receptor protein is complexed to zinc ions (Zn^{2+}). The interaction of

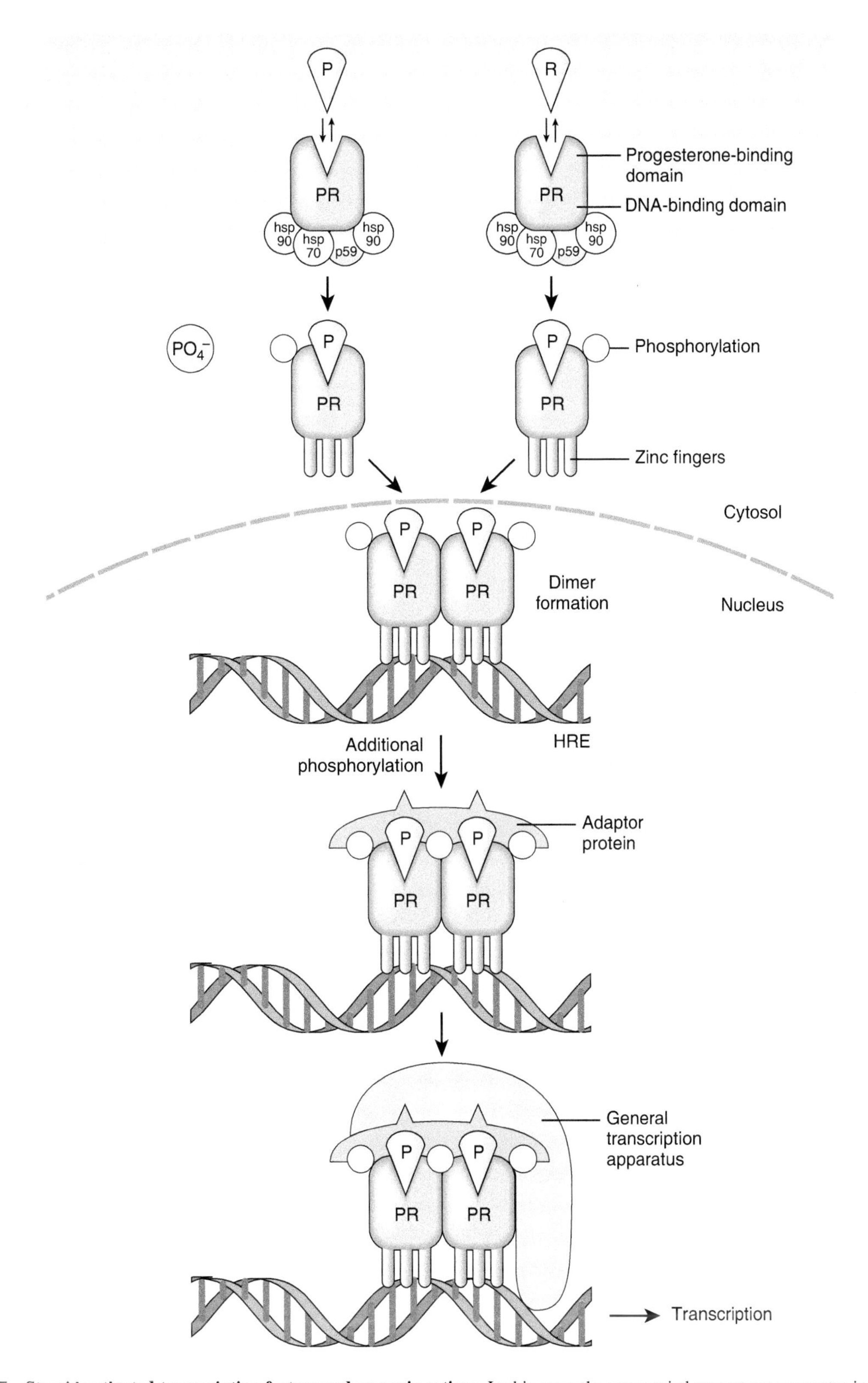

FIGURE 3-27 **Steroid-activated transcription factors and genomic actions.** In this example, unoccupied progesterone receptor is associated with several molecular chaperones including several heatshock proteins. Once occupied, the receptors are phosphorylated, lose some of their heatshock proteins, translocate to the nucleus, and form homodimers. Following binding of the receptor dimer to the HRE (PRE) site on nuclear DNA, a second phosphorylation occurs and an adapter protein complex is recruited that facilitates interaction with the general transcription apparatus. RNA polymerase activity and hence transcription are thereby modulated. Other steroids work in a similar manner. Thyroid hormone also operates this way; however, it forms heterodimers in the nucleus prior to binding to the TRE on DNA (see Figure 3-33). *(Adapted with permission from McDonnell, D.P.,* Trends in Endocrinology and Metabolism, *6, 133–138, 1995.)*

certain amino acid residues with Zn^{2+} causes the peptide chain to develop special loops that have been called **zinc fingers** (Figure 3-28). Binding of the steroid and loss of chaperone proteins exposes the zinc fingers. Interaction of the exposed zinc fingers with the major grooves of DNA allows transcription of mRNA to occur. Steroid and **thyroid receptors (TRs)** that directly affect gene transcription have zinc fingers that are highly specific for binding only to certain gene promoters called **hormone response elements (HREs)** that occur only on certain genes. For example, the estrogen receptor has a sequence of 80 amino acids that forms two zinc fingers for binding to a particular DNA sequence, allowing only a specific promoter site to be exposed for transcription. HREs are named according to the steroid—receptor complex that binds to it; for example, the HRE for estrogens is termed ERE, for glucocorticoids it is termed GRE, and for androgens it is called an ARE.

Once a specific ligand-activated transcription factor has bound to its HRE, additional proteins may complex with the ligand-activated transcription factor. These **nuclear adapter proteins** aid binding of occupied receptor to HREs, influence interactions with RNA polymerase, and facilitate RNA transcription. A different suite of nuclear adapter proteins may be associated with different steroid receptors although some are found associated with a variety of HREs. These nuclear adapter proteins also may vary in different tissues for the same steroid, resulting in activation of HREs on different genes. Furthermore, occupied receptors of a specific ligand may have to compete with other species of occupied receptors for the same nuclear adapter proteins. Once a gene is activated in the target cell, it may direct the synthesis of structural proteins or functional proteins such as peptide bioregulators or enzymes.

It generally is accepted that once a steroid receptor is occupied, it is phosphorylated and forms a dimer and that

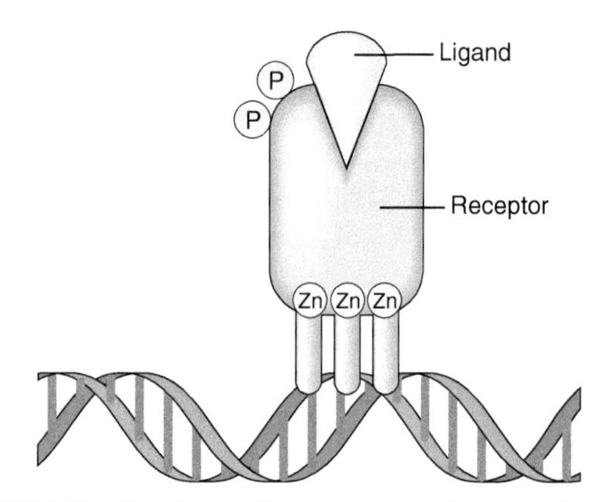

FIGURE 3-28 **Zinc fingers.** Certain sequences of amino acids can fold around zinc (Zn) ions to form projections called *zinc fingers*. These zinc fingers are associated with the DNA-binding domains of steroid receptors and facilitate binding to HREs on the DNA. Only a monomer is depicted here. P, phosphate.

a second phosphorylation occurs after the dimer binds to the HRE. Estrogens, androgens, progestogens, and corticosteroids always form homodimers, whereas vitamin D_3 (1,25-DHC) and bile acids form heterodimers with other receptors (Figure 3-29). The resultant transcription and synthesis of new proteins by the target cells bring about the events classically associated with the actions of these hormones.

2. Membrane Receptors for Non-Genomic Steroid Actions

Typically, there is a considerable delay (hours) between contact of the steroid bioregulator with its target cell and manifestation of its genomic-based effects, rather than almost immediate changes as described earlier for membrane receptors acting through second-messenger systems. Most of these actions of steroids occur on the time scale of the delayed genomic effects seen with second-messenger systems; however, numerous steroid actions are very rapid, such as the effects of corticosteroids on behavior or of progesterone on oocyte maturation. Evidence of membrane receptors initially was discovered for corticosteroids in the amphibian brain and for progesterone in the amphibian oocyte. This was followed by the discovery that a large proportion of the estrogen receptors located in the rat brain operate through membrane receptors, too. These membrane steroid receptors may operate through second messenger systems or via other mechanisms to produce rapid effects in their target cells. At least one of these receptors is a GPCR (GPCR 30) that binds estradiol and links to the IP_3 second-messenger pathway.

3. Metabolism of Steroids Prior to Receptor Interactions in Target Cells

In some androgen-specific target cells, testosterone is first altered chemically in the cytoplasm before it migrates to the nucleus and binds to a receptor. Testosterone may be converted by 5α-reductase to DHT in certain target cells (e.g., in brain, prostate) or by aromatase to estradiol (e.g., in brain). Thus, unoccupied nuclear receptors in these cells are not specific for testosterone but for one of its metabolites. In contrast, cortisol, which can bind to either cortisol or aldosterone receptors, is prevented from binding to mineralocorticoid receptors because it is enzymatically altered to the metabolite cortisone, which is unable to bind to these receptors (see Chapter 8).

4. Steroid Receptors and the Nuclear Receptor Superfamily

Regardless of whether unoccupied intracellular steroid receptors are present in the cytosol or the nucleus, all belong to a superfamily of nuclear receptors. This **nuclear receptor**

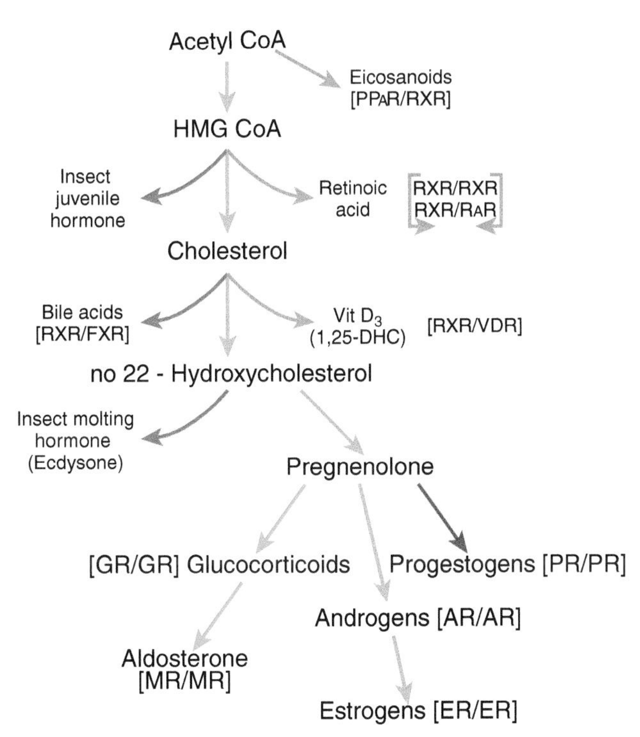

FIGURE 3-29 **Steroids and synthetically related lipid bioregulators.** Formation of dimer ligand–receptor complexes involves heterodimer formation with the exception of the vertebrate steroids that form only homodimers prior to activating gene response elements. Dimers are indicated in brackets. See Appendix A or text for explanation of abbreviations.

superfamily includes more than 300 receptors, most of which have been identified from gene sequences. Many of these receptors have no known natural ligand, and these molecules initially were termed **orphan receptors**. Some authorities subdivide the nuclear receptors into as many as six subfamilies based on common structural features (Table 3-13). One subfamily includes all of the steroid receptors except those for the C_{27} steroids (hydroxycholesterol, bile acids, and vitamin D). The **steroid nuclear receptor subfamily** can be separated into two subgroupings: the glucocorticoid-like receptors (includes receptors for androgens, progestogens, and mineralocorticoids) and the estrogen-like receptors (estrogen receptors as well as estrogen-related receptors). Some researchers include a third grouping here that includes the vitamin D receptor as well as some non-steroid receptors. Some nuclear receptors form heterodimers with other receptor types when occupied, whereas the occupied androgen, estrogen, progesterone, and corticosteroid receptors all form homodimers.

The ancestral vertebrate steroid receptor is believed to have been similar to the **estrogen-related receptor (ERR)** present in the primitive lamprey as well as in other vertebrates. The ERR gene is considered to be the source from which the **estrogen receptor (ER)** gene evolved after a genome duplication event. Following a second genome duplication, one copy of the ER gene presumably gave rise to the gene that produces the **progesterone receptor (PR)**. Following a proposed duplication of the PR gene, it gave rise to the **androgen receptor (AR)** gene and later to the **glucocorticoid receptor (GR)** and **mineralocorticoid receptor (MR)** genes.

Meanwhile, the ER gene duplicated again and diverged into two distinct forms we find in mammals, ERα and ERβ. These receptors exhibit somewhat different tissue distributions and have different affinities for various estrogenic molecules. Receptor subtypes have been identified for other steroid receptors as well. Of the reproductive steroid receptors, ERs have been studied most extensively. ERα was discovered first and is the most widely distributed ER in the body. ERβ is absent from liver and is the only ER in the gastrointestinal tract. Furthermore, ERα is less selective in its acceptance of ligands and is the target for most estrogenic endocrine-disrupting chemicals such as nonylphenols, phthalates, and bisphenol A, as well as for estrogenic pharmaceuticals such as ethinylestradiol (EE_2), DES, and the antiestrogen **tamoxifen**. The antiestrogen **raloxifene** antagonizes ERβ. Various drugs that affect estrogen receptors are termed **selective estrogen receptor modulators (SERMs)**.

5. Corticosteroid Action

Two classes of corticosteroid receptors have been described that are located in the cytosol of target cells and form

TABLE 3-13 Some Nuclear Receptor Superfamily Members

Endocrine Receptors	Abbreviation(s)	Endocrine Ligands
Mineralocorticoid	MR (GR type 1)	Mineralocorticoids
Glucocorticoid	GR (GR type 2)	Glucocorticoids
Progesterone	PR	Progesterone
Androgen	AR	Androgens
Estrogen	ERα,β	Estrogens
Retinoic acid	RARα,β,γ	Retinoic acids (RAs)
Thyroid	TRα,β	Thyroid hormones (T_3 & T_4)
Vitamin D	VDR	Vit D_3, LCA
Adopted Orphans		**Endogenous & Exogenous Ligands**
Retinoid X	RXRα,β,γ	9-cis-RA, DHA
Peroxisomal	PPARα,β,γ	Fatty acid
Proliferator-activated		
Liver X	LXR	Oxysterols
Farnesoid X	FXR	Bile acids
Pregnane X	PXR	Xenobiotics
Orphans[1]		**Endogenous Ligands Uncertain**
Estrogen-related receptor	ERRα,β,γ	Synthetic steroids
	HNF-4α,γ	Fatty acids?
	RORα,β,γ	Fatty acids, sterols?
Steroidogenic factor 1	SF-1	Phospholipids?
	LRH-1	Phospholipids?
	GCNF	?
	PNR	?
	TLX	?
	TR2,4	?
	NGFI-Bα,β,γ	?
	COUP-TFα,β,γ?	?
	RVRα,β	?
	DAX-1	?
	SHP	?

[1] *Many of these may function as simple transcription factors without ligands.*

homodimers when occupied. The MR binds aldosterone and glucocorticoids and the GR binds only cortisol or corticosterone. The MR binds both aldosterone and the two glucocorticoids, so an alternative scheme exists for naming these receptors. Because both receptor types bind glucocorticoids, they also have been called **type 1 glucocorticoid receptor** (GR_1 = MR) and **type 2 glucocorticoid receptor** (GR_2 = GR). The first scheme (MR, GR) is used here for simplicity, but keep in mind that the MR also can bind glucocorticoids as this is of clinical importance with respect to glucocorticoid therapies. In target cells that are involved in the regulation of mineral balance, a special form of the enzyme **11β-hydroxysteroid dehydrogenase (HSD11B2)** converts cortisol that enters the cytosol into cortisone which does not bind effectively to the MR. Thus, these cells will respond only to aldosterone even when plasma cortisol levels are elevated during normal physiological conditions.

E. Metabolism and Excretion of Steroid Hormones

Steroid hormones are dissociated from their receptors and metabolized by the target cell or the liver that possesses enzymes capable of altering the specific steroids and rendering them biologically inactive and water soluble. The liver performs the major task of steroid inactivation by removing free steroids from the circulation. Metabolism of steroids typically involves reduction or removal of side chains or attached groups, or both, as well as combining with other molecules (conjugation) such as a **gluconate** (formed from glucose) to produce a glucuronide. Steroids also may be conjugated with **sulfate**. The relative emphasis on sulfate or glucuronide varies depending on the steroid and/or the species involved; for example, in humans, estradiol is excreted primarily as the glucuronide derivative. The conjugates are water soluble and when released into the blood will no longer bind effectively to serum proteins or enter cells and bind to receptors. Consequently, they are filtered from the blood by the kidney and are added to the urine. Some of these steroids are metabolized, added to the bile, and excreted via the intestinal route.

Steroids also are metabolized via oxidative pathways as well as by reductions. These oxidized metabolites originally were not identified because they occurred in a fraction of urine usually discarded when researchers were isolating urinary steroids for analysis. Oxidized steroid metabolites may represent a substantial portion of the total metabolites for a given steroid. For example, it is estimated that from 12 to 36% of circulating cortisol in humans is converted to oxidized **corotic acids** (cortolic and cortolonic acids) (Figure 3-30).

Androgens of both adrenal and gonadal origin are found in the urine primarily as sulfates and can be

BOX 3B An Ancient "Orphan" Receptor

The gene responsible for the synthesis of the cytoplasmic aryl hydrocarbon receptor (ahR) is one of the oldest genes known in organisms. It is a member of the basic helix–loop–helix PAS gene family. The ahR gene has been identified in a nematode worm, indicating that it existed at least 500 MYBP. The ahR gene contains a region known as the Per–ARNT–Sim region and exhibits sequence homologies with Per and Sim (genes identified in the fruit fly *Drosophila*) as well as ARNT, the aryl hydrocarbon nuclear translocator protein.

The ahR is an orphan receptor, having no known endogenous ligand; however, it can bind a number of polyaromatic hydrocarbons. It has the highest affinity for a dioxin known as 2,3,7,8-tetrachlorodibenzo-*p*-dioxin (TCDD). Dioxins are toxic products of combustion that are naturally produced by forest fires and volcano activity, but they also are produced by human activities through many industrial practices involving combustion of organic materials. Once the ahR has bound a ligand, it sheds heat-shock chaperone protein (hsp90), and the occupied ahR enters the nucleus, where it forms a heterodimer with ARNT protein. The ahR–ARNT complex binds to specific response elements on a number of CYP genes such as *CYP1A1*. These genes direct the synthesis of enzymes that aids cells to detoxify dioxins as well as other contaminants, including pesticides and drugs. Studies conducted in both mammals and fishes demonstrate that glucocorticoid levels often observed in stressed animals are sufficient to potentiate the actions of the ahR. Furthermore, numerous laboratory studies have linked dioxin binding by the ahR to alterations in steroid action and/or metabolism, probably through cross-talk as well as through effects on the production of metabolizing enzymes.

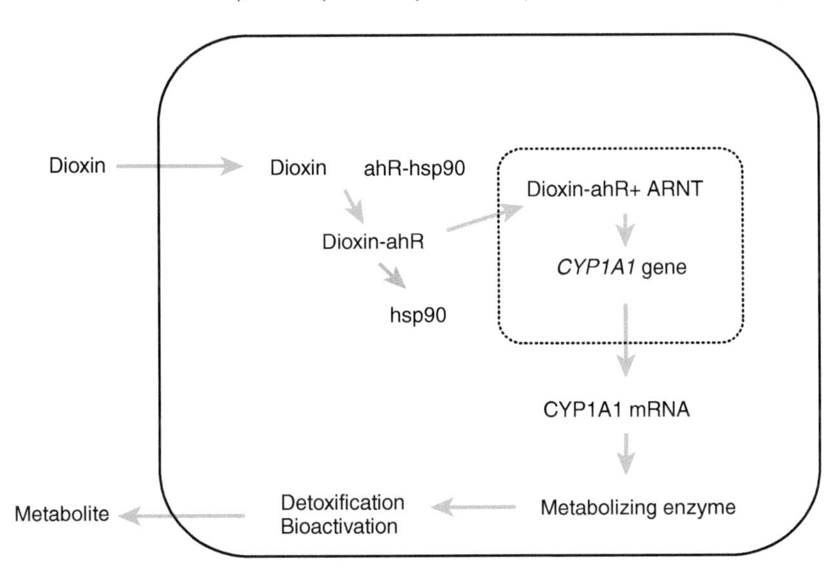

BOX FIGURE 3B-1 **Dioxin actions.** Dioxins bind to the arylhydrocarbon receptor (ahR). The ligand–receptor complex enters the nucleus and dimerizes with the aryl hydrocarbon nuclear translocator protein (ARNT) and activates the *CYP1A1* gene. This gene produces a metabolizing enzyme that destroys a variety of potential toxicants including dioxins.

measured chemically as **17-ketosteroids**. In addition, some androgens are found as conjugates with glucuronic acid. Estrogens may be excreted as either glucuronides or sulfates, such as estriol glucosiduronate or estrone sulfate. **Catechol estrogens** are also common excretory products (Figure 3-30; compare to catechol amines). Higher production of one catechol estrogen is associated with breast cancer induction. Progestogens are excreted mainly as glucuronides of pregnanediol or pregnanetriol; however, pregnanediol may be formed from corticosteroid metabolism (that is, from deoxycorticosterone) as well as from progesterone. Pregnanetriol is formed principally from 17α-hydroxyprogesterone, but small amounts also may be produced from 11-deoxycortisol and 17α-hydroxypregnenolone. The corticosteroids are primarily excreted as C_{21} **17-hydroxycorticosteroids** after conjugation with glucuronic acid.

Under some conditions, membrane-bound **sulfatase** enzymes can remove the sulfate and liberate free steroids, which can then enter the cell. Sulfatase activity is very high in the placenta, where the sulfated androgen **DHEA sulfate (DHEA-S)**, produced in the fetal adrenal, is desulfated and converted to estrogen. A female fetus is thus protected from the potentially masculinizing actions of this fetal androgen by sulfation. There also are cases known where sulfated steroids can bind to cellular membrane proteins; for example, pregnenolone sulfate produced by certain brain cells is known to bind to and suppress the GABA receptor on nearby neurons. The ability of bacteria associated with wastewater treatment plants (WWTPs) to remove both sulfates and gluconates from steroid metabolites results in the reactivation of excreted steroids that may then find their way into aquatic systems via WWTP effluents. Estrogenic compounds in WWTP effluents have been associated with

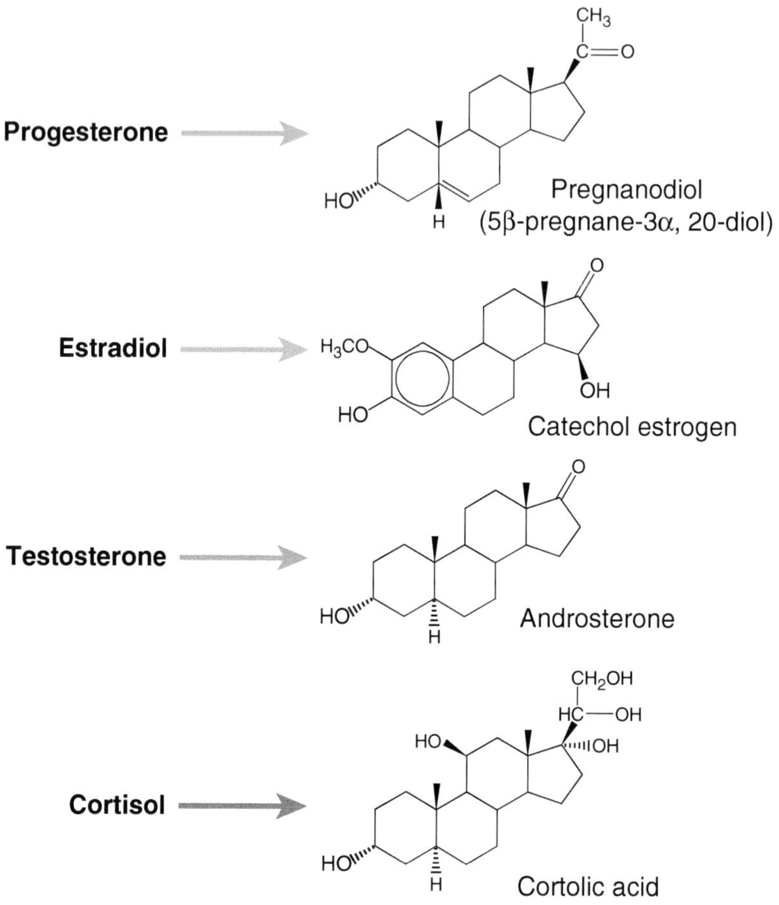

FIGURE 3-30 **Some nonconjugated steroid metabolites.** *(Adapted with permission from Norman, A. and Litwack, G., "Hormones," 2nd ed., Academic Press, San Diego, CA, 1997, p. 84.)*

the feminization of male fishes in rivers throughout the world.

A number of chemical methods formerly were employed to identify and quantify various steroid metabolites in urine, but these approaches have given way to radioimmunoassay techniques, ELISAs, and sophisticated chromatographic separation and identification by mass spectroscopy techniques (see Chapter 2).

III. Thyroid Hormones

Thyroid hormones are important developmental and metabolic bioregulators synthesized in the follicular cells of the thyroid gland under the influence of TSH. Details of this process in association with the structure of thyroid follicular cells are provided in Chapter 6 and only an overview of the chemical events is given here. The original molecule with thyroid hormone activity was isolated from the thyroid gland by Edward C. Kendall in 1919 (see Box 3D) and was named **thyroxine**. It was known to contain four iodide atoms attached to a unique compound called a **thyronine** and was thought to be the thyroid hormone. The chemical name for thyroxine is **3,5,3′,5′-tetraiodothyronine** (Figure 3-31). Because of the presence of four iodide atoms per molecule (tetraiodo), the abbreviation for thyroxine became $\mathbf{T_4}$. In 1952, J. Gross and Rosalind Pitt-Rivers discovered a second form of thyroid hormone lacking one of the iodides found in T_4. This molecule was named **3,5,3′-triiodothyronine** or **triiodothyronine** and the abbreviation became $\mathbf{T_3}$. Many endocrinologists now consider T_3 to be the biologically active form of thyroid hormone and T_4 to be primarily a precursor. Although T_4 is not a propeptide, it may be considered as a prohormone for T_3.

A. Structure and Synthesis of Thyroid Hormones

T_4 is derived from two molecules of iodinated tyrosine (Figure 3-32). Tyrosine is first incorporated into a large globular protein called **thyroglobulin (Tgb)** and then iodinated by the enzyme **thyroid peroxidase (TPO)**. Inorganic iodide accumulated by thyroid cells from the blood is converted following the hydrolysis of hydrogen peroxide (H_2O_2) by TPO to "active iodide" that can than attach to tyrosine. An iodinated tyrosine has two iodides

BOX 3C Steroid-Metabolizing Enzymes

Steroids are metabolized by a group of generalized enzymes that can operate on a wide variety of substrates including drugs and contaminants from the environment. Many of these enzymes can be induced by the presence of a suitable substrate (Box Table 3C-1). These enzymes are classified as phase I (activational) enzymes or as phase II (conjugation) enzymes. Activation usually involves addition of a hydroxyl group that may result in a more reactive molecule. Conjugation occurs by addition of gluconate, glutathione, or sulfate and typically makes the molecule more soluble in water and allows it to be excreted more readily. Phase II reactions may directly follow a phase I reaction for a given substrate.

Phase I reactions are carried out primarily by cytochrome P450 monooxygenases although numerous other enzymes may participate in phase I reactions. These monooxygenases are produced by the CYP gene family that generally are membrane-bound hemoproteins associated with the smooth endoplasmic reticulum. In addition to their involvement in metabolizing steroid hormones prior to their excretion, these same enzymes can metabolize a wide variety of drugs and pollutants including caffeine, pesticides, PCBs, and phthalates.

Phase II reactions are catalyzed by a great variety of enzymes, including glucuronosyl transferase, glutathione transferases, and sulfotransferases, transaminases, and esterases, among others. These enzymes participate with phase I enzymes to help cells eliminate the products of phase I activations.

BOX TABLE 3C-1 Induction of CYP Enzymes

Some Substrates Capable of Inducing CYP Gene Expression in Mammals

Gene Family	Mammalian Subfamily	Substrate Examples
CYP1	1A, 1B	PAHs,[1] pesticides, natural estrogens, arachidonic acid, caffeine, indoles, flavenoids
CYP2	2A-2G, 2J	Various drugs, steroids, pesticides, caffeine, alcohol
CYP3	3A	Various drugs, steroids, PAHs, pesticides, caffeine
CYP4	4A, 4B, 4F	PCBs,[2] phthalates, arachidonic acid, lauric acid

Phase II reactions are catalyzed by a great variety of enzymes including glucuronosyl transferease, glutathione transferases, and sulfotransferases, transaminases, esterases, etc. These enzymes participate with Phase I enzymes to help cells eliminate the products of Phase I activations.
[1]*PAH = Polynuclear aromatic hydrocarbons.*
[2]*PCB = polychlorinated biphenyls; includes dioxins; aroclors are commercial mixtures of PCBs.*
Adapted with permission from Celander, M. 1999. Impact of stress in animal toxicology. In P. H. M. Balm (ed.) "Stress Physiology in Animals." CRC Press, Boca Raton, FL, pp. 246–278.

attached at positions 3 and 5 of the phenolic ring, respectively, and the iodinated tyrosines are retained as part of the polypeptide backbone of the Tgb molecule. Attachment of one iodide at position 3 yields **3-monoiodotyrosine (MIT)**, and addition of a second iodide produces **3,5-diiodotyrosine (DIT)**. Two DITs are joined by TPO through removal of the phenolic ring of one DIT and the attachment of it to the hydroxyl group extending from the phenolic group of the other DIT to yield a unique amino acid called a **thyronine**, which has four iodides attached. The remaining fragment of DIT after removal of the iodinated ring is called **dehydroalanine** (alanine missing one hydrogen). The carbons of the recently added phenolic ring are designated by a prime (′) symbol, which appears in the chemical name of the resultant **3,5,3′5′-tetraiodothyronine**, or T_4. When thyroglobulin containing T_4 molecules is hydrolyzed in the thyroid cell, T_4 is released and then enters the blood. In most cases, T_4 is the major circulating thyronine and is converted peripherally by a **type 1 deiodinase (D1)** enzyme in the liver or in target cells (see Table 3-14) to T_3 by the removal of one iodide from the outer ring. Most circulating T_3 is formed from T_4 by the liver, but a small proportion of the circulating T_3 is a result of deiodination of T_4 by a thyroid D1 deiodinase prior to release from the thyroid gland. Deiodinases are unique **selenoproteins** that contain a selenocysteine selonol group (compared to a cysteine thiol group) in the active site, giving them unusual catalytical power.

Thyroid hormones are hydrophobic at normal blood pH and hence poorly soluble in plasma. However, target cells possess **organic anion transporter polypeptides (oatps)** that can transport thyroid hormones inward. The major thyroid transport appears to be a specific transporter called **monocarboxylate transporter 8 (MCT8)**.

B. Transport of Thyroid Hormones in the Blood

Most of the circulating T_4 (about 99%) and T_3 (about 90%) are bound reversibly to plasma proteins and only a small

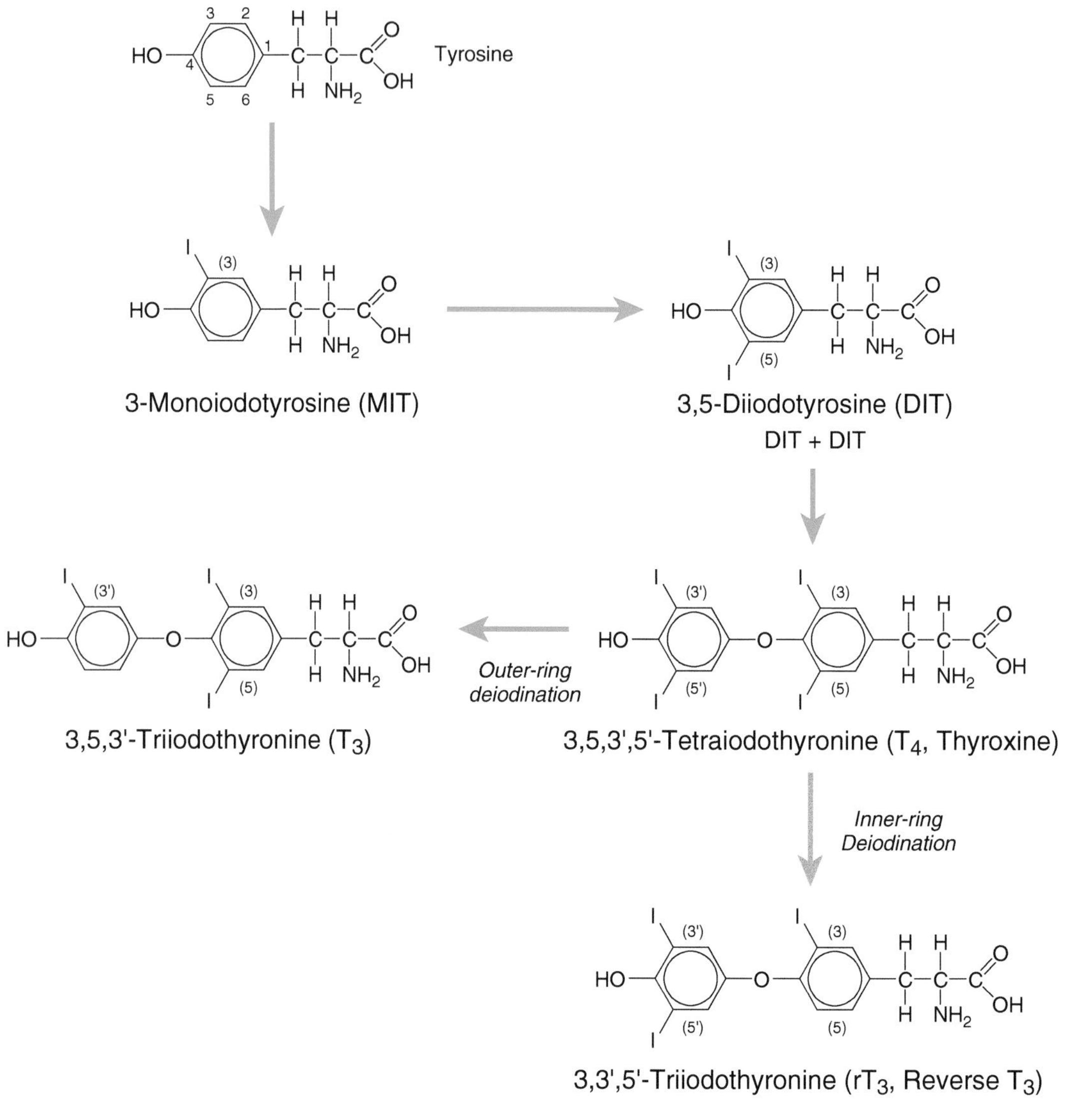

FIGURE 3-31 **Thyroxine, precursors, and some deiodinated metabolites.** Synthesis of MIT, DIT, T_4, and conversion of T_4 to T_3 and rT_3 by thyroid deiodinase enzymes (see text and Figure 3-32 for details).

percentage is free in the blood. Binding to plasma protein by thyroid hormones is considered helpful for their transport in the blood, because T_3 and T_4 are somewhat hydrophobic and are not highly soluble in blood. Binding to plasma proteins also slows their elimination by the kidneys. When free thyroid hormones leave the blood and enter the liver or a target cell, additional hormones dissociate from the transport proteins, instantly replenishing the pool of free hormones. Thus, the binding proteins provide a ready reservoir of thyroid hormones in the circulation that are ready for immediate usage.

Several different serum proteins are capable of binding and transporting thyroid hormones. In humans, for example, about 75% of the bound hormones are linked to the α_2-globulins called **thyroid-binding globulin (TBG)**. Similar to CBG, TBGs are a member of the serpin protein family (serpinA7). An additional 15% and 10%, respectively, are bound to **prealbumin (TBPA**; also called **transthyretin, TTR**) and **albumin (TBA)**.

C. Mechanism of Thyroid Hormone Action on Target Cells

Thyroid receptors belong to the same superfamily of nuclear receptor types that includes the steroid receptors and retinoic acid receptors (see Table 3-13). The molecular mechanism of action for thyroid hormones appears to be similar to that described previously for steroids (see Figure 3-33). Thyroid hormones enter target cells where they migrate to the nucleus and bind to specific nuclear receptor proteins. Following binding to nuclear receptors, occupied thyroid hormone receptors form dimers and initiate nuclear gene

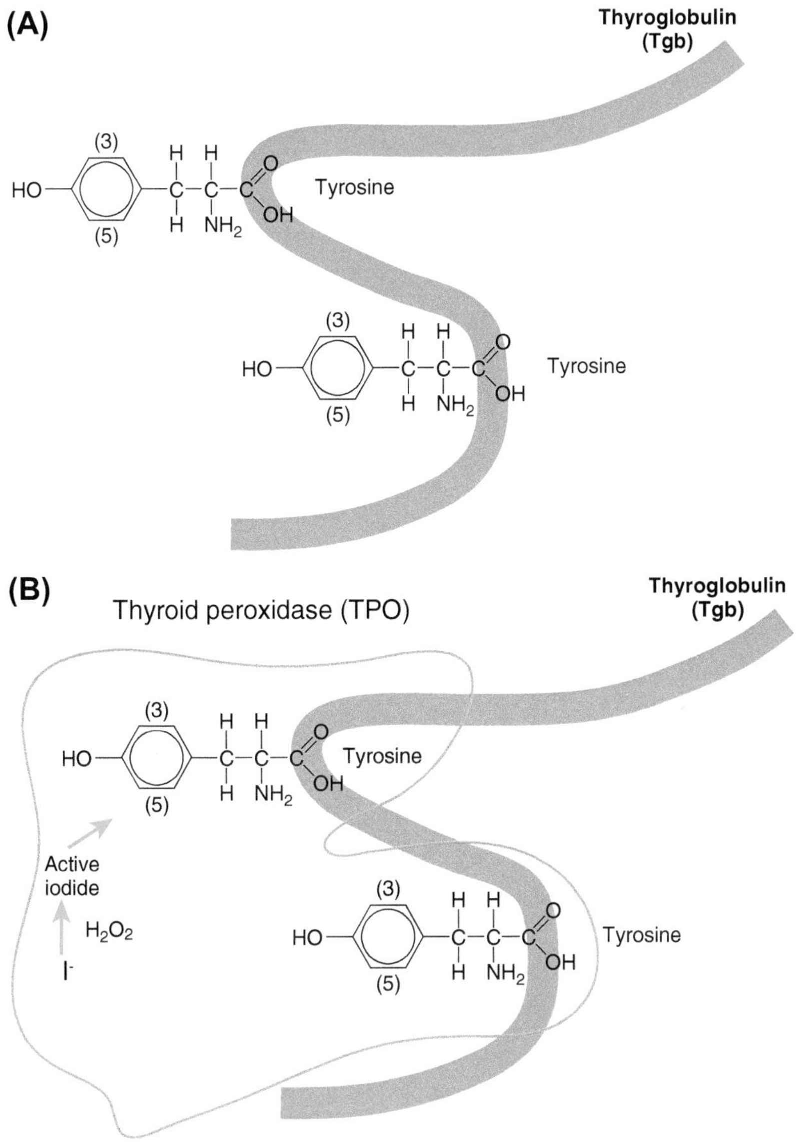

FIGURE 3-32 **Roles of thyroid peroxidase (TPO) in thyroxine (T_4) synthesis.** (A) Tyrosine molecules are incorporated into the polypeptide backbone of thyroglobulin. (B) The enzyme TPO converts iodide to "active iodide." (C) TPO then attaches the active iodides to the phenolic ring of the tyrosines to form diiodothyronines (DITs). (D) TPO removes the hydroxyphenyl group from one DIT to another to form a thyronine (3,3′,5,5′-tetaiodothyronine, T_4) leaving behind a modified alanine called *dehyroalanine*.

transcription, which results in the synthesis of new proteins. Although the nucleus is the major site identified for location of unoccupied thyroid hormone receptors, cytoplasmic and membrane receptors are also known.

Nuclear receptors for thyroid hormones have been isolated and characterized. Typically, they have greater affinity for T_3 than for T_4, supporting the hypothesis that conversion of T_4 to T_3 is requisite for thyroid hormone action. Target cells for thyroid hormones are equipped with a **type 2 deiodinase (D2)** enzyme to accomplish this conversion. Mitochondrial receptor proteins also have been demonstrated, and these mitochondrial receptors are associated with observed effects of thyroid hormones on mitochondrial protein synthesis and oxidative metabolism in target cells (see Chapter 6).

The genomic actions of thyroid hormones on nuclear DNA are similar to what we described earlier for steroids (Figure 3-33). At least two major isoforms of nuclear TRs have been demonstrated in humans, mice, and rats, suggesting there may be a separate functional role for each type. In the rat, two forms, TRα-1 and TRβ-1, are found in all target tissues, whereas another form, TRβ-2, is limited to the pituitary, hypothalamus, and other brain areas.

Occupied TRs bind to **thyroid response elements (TREs)** in or near specific genes in target cell nuclei but may do so in a variety of ways. Monomers of occupied receptor may bind directly to TREs but they do not produce much transcription by the thyroid-responsive gene. Similarly, occupied TR homodimers bind for such a short time that they also produce little transcription. However,

FIGURE 3-32 (*continued*).

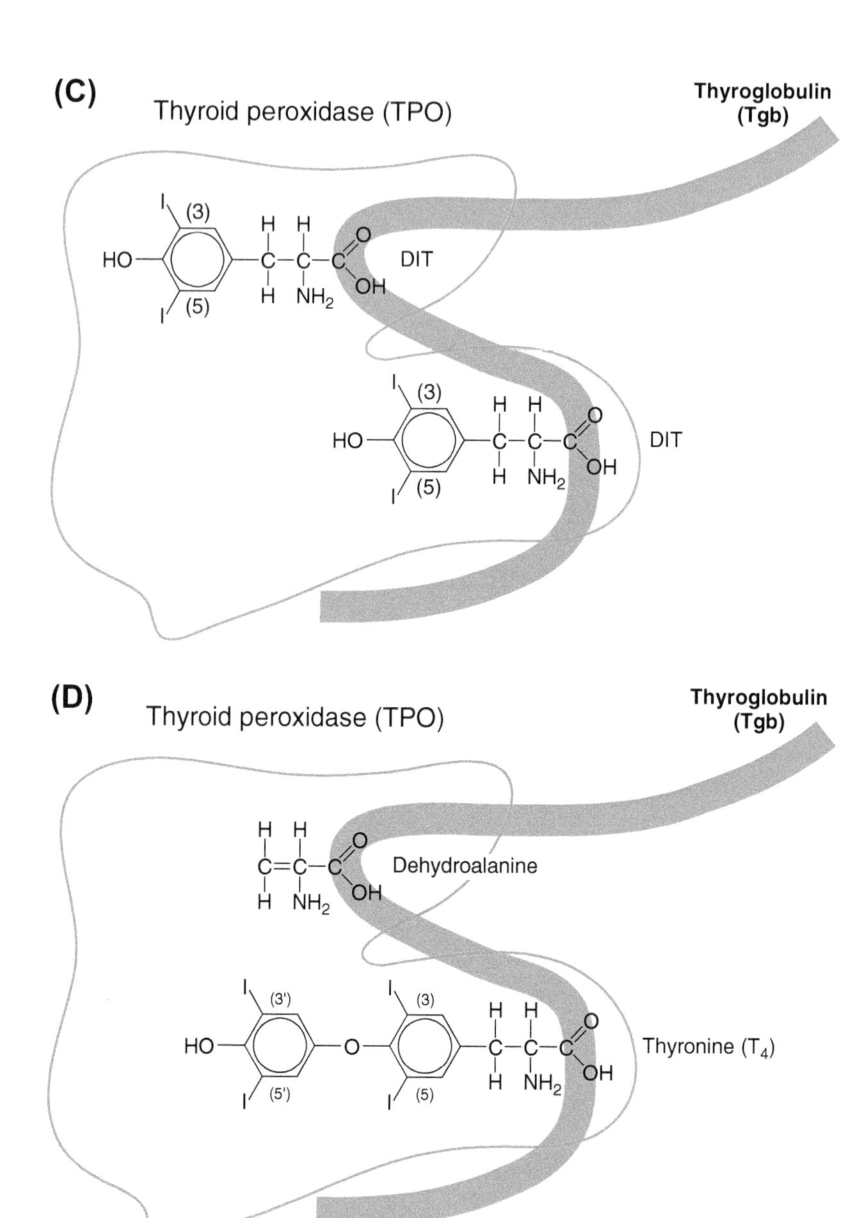

formation of a unique heterodimer provides maximal stimulation of transcription. An occupied TR monomer may form a heterodimer with the **retinoic acid receptor (RAR)** or more commonly with the orphan **retinoid X receptor (RXR)**. Additionally, occupied TR may form a heterodimer with **T_3 receptor auxiliary protein (TRAP)** that also aids in TRE activation.

D. Metabolism of Thyroid Hormones

Deiodination is not only a mechanism for activation T_4 by conversion to T_3, but it is also an important mechanism for degrading both T_4 and T_3 to inactive compounds (see Table 3-14). Thyroxine has several metabolic fates after being released from the thyroid gland. Approximately 33 to 40% is converted to T_3 by removal of 5′-iodide in the liver. This is accomplished by a D1 deiodinase and represents the source for most of the circulating T_3 in many species. About 15 to 20% of the circulating T_4 is deaminated and decarboxylated in the liver to form **tetraiodothyroacetate (TETRAC)**, which is very soluble in water, is an antagonist of the integrin αvβ3 receptor (see Chapter 6), and is excreted rapidly in urine or bile (Figure 3-34). T_3 also may be converted similarly to form **triiodothyroacetate (TRIAC)** or the latter can be formed by D1 deiodination of TETRAC. Some glucuronides of T_4 and to a lesser extent of T_3 also may be formed. All of these metabolized molecules are more soluble in water, cannot

TABLE 3-14 Types of Deiodinase Activity

Characteristic	Type of Deiodinase		
	I (D1)	II (D2)	III (D3)
Location	Liver, kidney, thyroid	Brain, pituitary, placenta, brown adipose tissue (BAT)	Brain, skin, placenta
Substrate preference	Inner or outer ring	Outer ring	Inner ring
Effect of PTU[a]	Inhibition	No effect	No effect
Reactions catalyzed	$T_4 \rightarrow T_3$	$T_4 \rightarrow T_3$	$T_3 \rightarrow T_2$
	$T_4 \rightarrow rT_3$	$rT_3 \rightarrow T_2$	$T_4 \rightarrow rT_3$
	$rT_3 \rightarrow T_2$		
	$T_3 \rightarrow T_2$		

[a]*PTU, propylthiouracil; rT_3, reverse T_3.*

bind effectively to plasma binding proteins or receptors, and are quickly eliminated via the bile or urine.

In recent years, it was discovered that approximately one-half of the circulating T_4 is eventually converted by deiodination to a unique form of T_3: 3,3′,5′-triiodothyronine, or **reverse T_3 (rT_3)** (Figure 3-31). This conversion is accomplished by deiodination by D1 or by a D3 deiodinase that removes an iodide from the internal phenolic ring of a thyronine. Reverse T_3 is degraded or excreted more rapidly than normal T_3. Reverse T_3 may act as an endogenous ligand on a truncated form of the nuclear TR that is located in the cytoplasm (see Chapter 6). As a result of its rapid clearance, rT_3 levels in the blood are rather low and hence escaped detection for many years; however, increases or decreases in circulating T_3 levels are always accompanied by reciprocal changes in rT_3 levels. The liver may continue to deiodinate T_3 and rT_3, removing the remaining iodides to produce **3,3′-diiodothyronine (T_2)** and then **monoiodothyronines.**

IV. Eicosanoids

The **eicosanoids** are small lipids derived from a common precursor, **arachidonic acid**. In the early 1930s, Maurice Goldblatt in England and U.S. von Euler in Sweden independently discovered the first eicosanoids, the **prostaglandins** (**PGs**). Elucidation of the structure and functions of these important bioregulatory compounds and documentation of their synthetic pathways resulted in the awarding of the 1982 Nobel Prize in Physiology or Medicine to three prostaglandin researchers: Sune Bergstrom, Bengt Samuellson, and John Vane.

Prostaglandins were named on the belief that their source in men was the prostate gland. Since then, we have learned that they are secreted by many tissues of both men and women. The PGs have many diverse actions, including stimulation of smooth muscle contraction in the intestine and uterus, vasodilation (but may cause vasoconstriction

BOX 3D The Isolation of Hormones from the Thyroid Gland and Adrenal Cortex by Edward C. Kendall

Edward C. Kendall was the biochemist credited with first isolating thyroxine from the thyroid gland and isolating the major steroid hormones from the adrenal cortex. In the early 1900s, it was known that the thyroid gland contained a chemical capable of mitigating hypothyroidism. Kendall developed a bioassay that he ultimately used to isolate thyroxine from more than 6,500 pounds of pig thyroid glands. Although the chemical structure of thyroxine was later determined in 1926 by C.R. Harrington and colleagues, it was Kendall's original work in isolating thyroxine that led to this discovery. Kendall went on to spend about 20 years isolating the various steroid hormones synthesized by the adrenal cortex. Interestingly, this effort appears to have been accelerated during World War II because of rumors that German scientists were actively purifying adrenal gland extracts for compounds that could be used by Luftwaffe pilots to combat high altitude sickness. Kendall's group at the Mayo Clinic (along with the laboratory of Tadeus Reichstein at the Pharmaceutical Institute in Basel, Switzerland) eventually isolated and determined the structures of the major steroids produced by the adrenal cortex. Before the chemical structures of the adrenal steroids were known, the compounds were identified by letters based on the migration patterns on paper or alumina chromatography columns: compound A for 11-dehydrocorticosterone, compound B for corticosterone, compound E for cortisone (or 11-dehydrocortisol, Reichstein's compound S), and compound F (Reichstein's substance M) for cortisol. In 1950, Kendall and Reichstein, along with Philip Hench, were awarded the Nobel Prize for Physiology and Medicine for these discoveries.

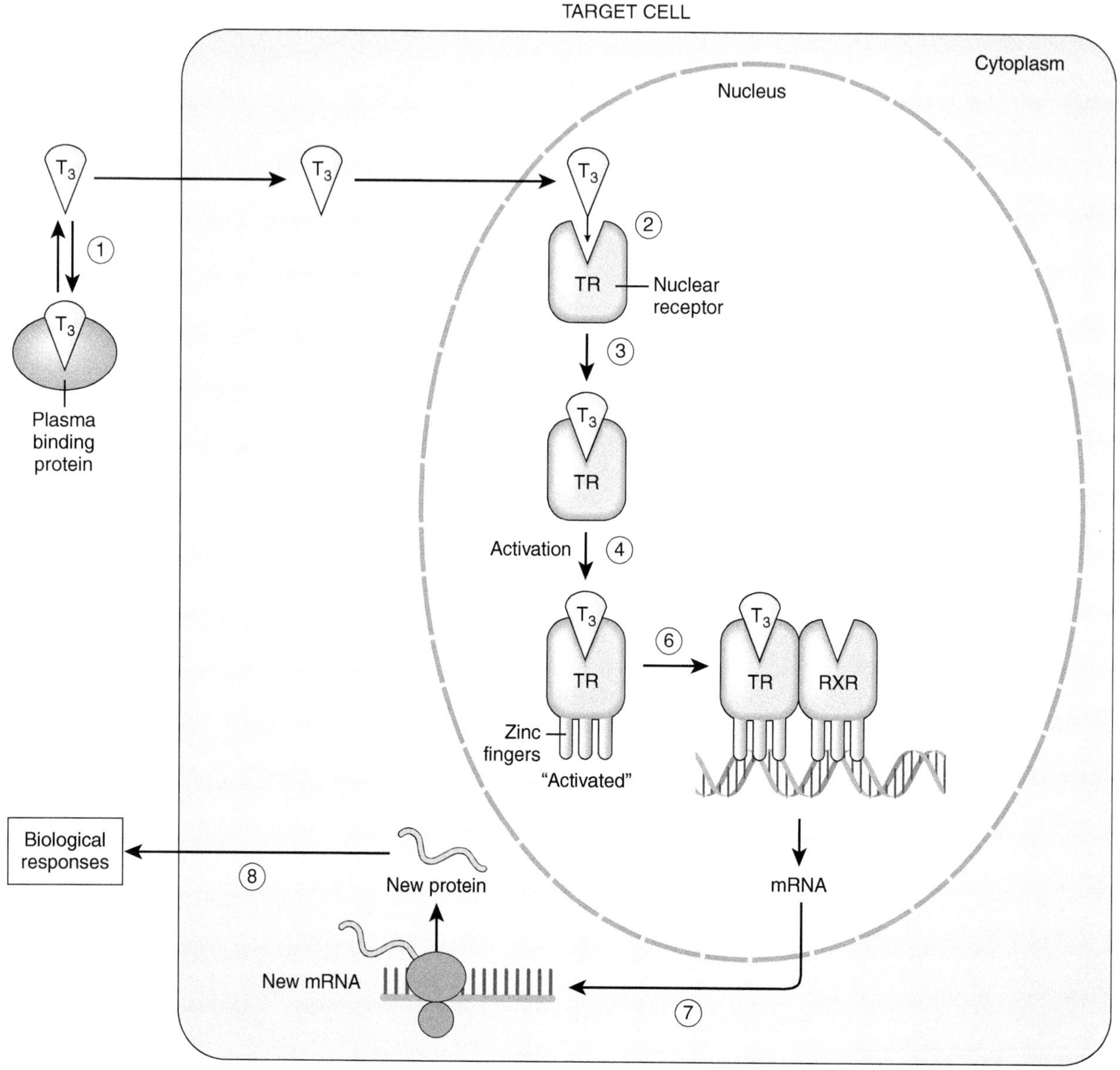

FIGURE 3-33 **Genomic mechanism of action for triiodothyronine (T_3).** Typically, unoccupied thyroid hormone receptors (TRs) are localized in the nucleus and once occupied form heterodimers with RXRs prior to binding to the thyroid response element (TRE) in the target gene. Thyroxine (T_4) typically is converted to T_3 in the cytosol prior to binding with the TR.

in certain vessels), and modulation of central nervous system function. They also stimulate synthesis of corticosteroids, testosterone, and a variety of specific enzymes. One PG ($PGF_{2\alpha}$) is believed to be the uterine luteolytic substance in certain mammalian species (see Chapter 10). Prostaglandins also reduce progesterone synthesis by the corpus luteum, induce ovulation and lactation in rodents, and may be involved in the induction of labor. Furthermore, they may induce inflammation and fever. The anti-inflammatory and antipyretic (fever-decreasing) action of aspirin is due to its inhibition of PG synthesis. **Prostacyclin (PGI_2)**, another form of prostaglandin, is a potent inhibitor of blood platelet aggregation and inhibits blood clotting.

Researchers discovered the **thromboxanes** during studies on prostaglandin metabolism and action. Thromboxane A_2 released from platelets causes translocation of free calcium ions to bring about changes associated with the shape of blood platelets. It is this change in platelet shape that allows platelets to aggregate and facilitate clotting. Other thromboxanes may be released from the platelet, causing local constriction of vascular smooth muscle. This paracrine effect might enhance clotting by reducing the diameter of the arterioles and slowing blood flow through the capillary beds of damaged tissues.

Leukotrienes are synthesized and released by mast cells and white blood cells, often in response to injury. They contribute to inflammatory or allergic responses

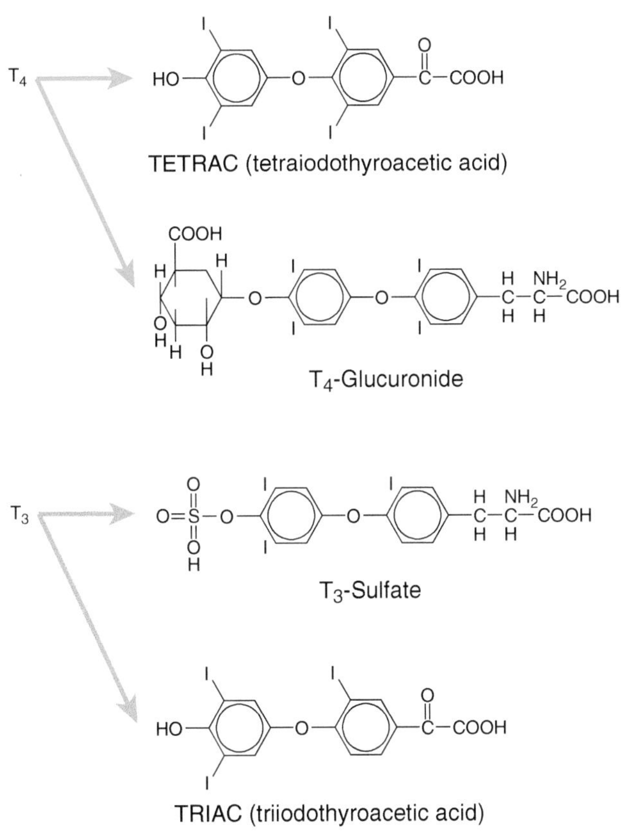

FIGURE 3-34 **Additional thyroid hormone metabolites.** See text for explanation.

by causing contraction of vascular smooth muscle and by increasing vascular permeability. Increased levels of leukotrienes have been associated with allergic reactions, asthma, cystic fibrosis, septic shock, and a number of other disorders.

A. Chemical Structure of Eicosanoids

Prostaglandins are all related structurally to **prostanoic acid**, and they can be separated into four classes (E, F, A, and B) on the basis of structural differences (Figure 3-35). Bergstrom succeeded in elucidating the structures for 16 PGs in 1956. The most commonly occurring PGs are PGE_1, PGE_2, and $PGF_{2\alpha}$. Prostacyclin (PGI_2) represents a unique class of prostaglandins called PGI. The leukotrienes consist of at least 15 related compounds occurring in five structural classes: A, B, C, D, and E (Figure 3-36). Thromboxanes are very short-lived molecules, some of which are depicted in Figure 3-36.

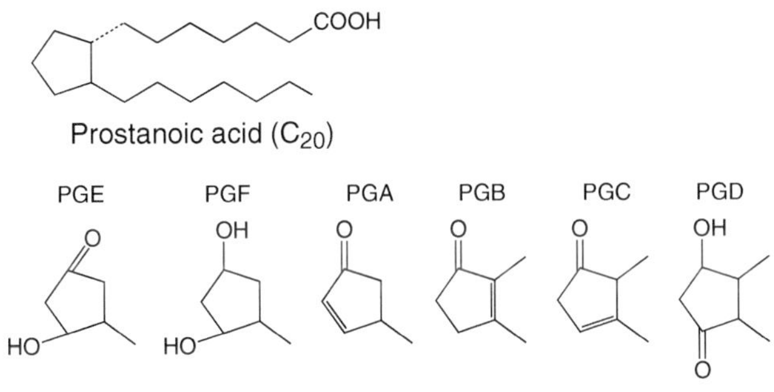

FIGURE 3-35 **Prostaglandin structures.** The basic chemical formula for the prostaglandins is that of prostanoic acid. These C20-lipids are divided into classes (A, B, F, etc.) based on substitutions to the five-membered carbon ring. Modifications of the side chains result in different forms within a class, each designated by a subscript (e.g., $PGF_{2\alpha}$, PGE_1).

B. Biosynthesis and Actions of Eicosanoids

Arachidonic acid is synthesized from α-linolenic acid derived from linoleic acid, an essential fatty acid, by the enzyme Δ^6-desaturase. Once formed, arachidonic acid can be converted to any of the eicosanoids (Figure 3-36). Cyclooxygenase is an enzyme that transforms arachidonic acid into endoperoxides which are used to synthesize prostaglandins, prostacyclin, or thromboxanes. Drugs such as aspirin and indomethacin inhibit cyclooxygenase and block the synthesis of prostaglandins and thromboxanes. Certain flavenoids obtained from plants also block cyclooxygenase. A separate enzyme, 5-lipoxygenase, forms the leucotrienes from arachidonic acid. This enzyme is not inhibited by aspirin or indomethacin but can be inhibited by specific lipoxygenase inhibitors as well as by plant flavenoids.

Eicosanoids produce their effects by binding to the **peroxisome proliferator-activated (PPAR)** receptor. PPARs are also members of the same superfamily of nuclear receptors that includes steroids and TRs and are in the same subgrouping as the TRs. Occupied PPARs, like occupied TRs, form heterodimers with RXRs prior to binding to gene response elements.

V. Summary

Neurotransmitters, neurohormones, and hormones occur as peptides, proteins, modified amino acids (indoleamines, catecholamines, thyroid hormones), or lipids (steroids, eicosanoids). Peptides and proteins neurohormones and hormones are synthesized as preprohormones and are modified posttranslationally to prohormones and then cleaved again to the final product. These processes usually occur intracellularly. Additional posttranslational modifications may occur as well (e.g., addition of carbohydrates, dimer formation, amidation, acetylation). Peptides are secreted from endocrine epithelial cells by exocytosis and travel in the plasma unattached or bound to plasma proteins. Most of the steroids (including androgens, estrogens, progesterone, corticosteroids, and 1,25-DHC) and thyroid hormones, as well as many amines, are transported primarily bound to specific plasma proteins in equilibrium with a small quantity of free hormones.

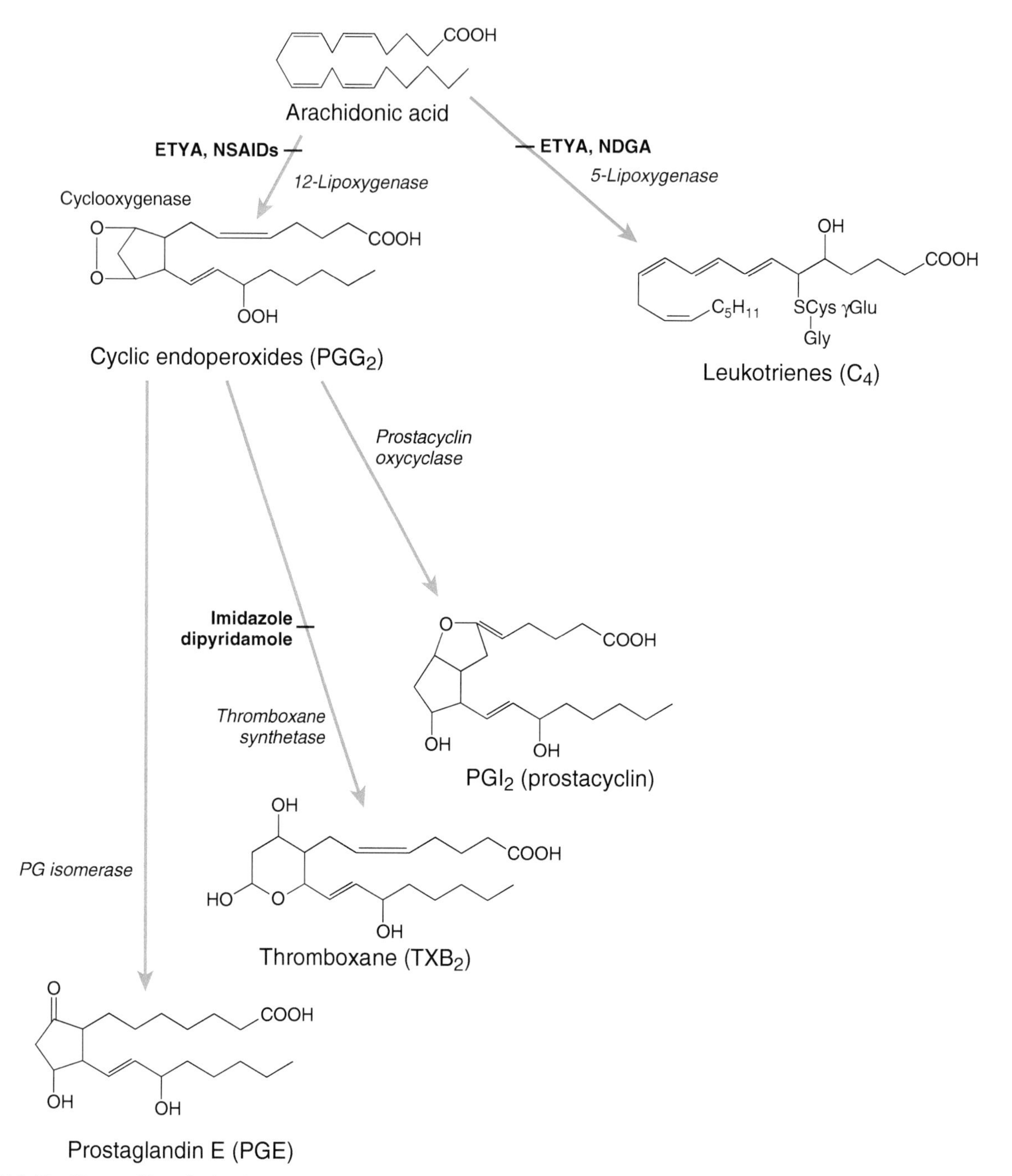

FIGURE 3-36 **Eicosanoid synthesis.** The precursor for all eicosanoids, arachidonic acid, can be synthesized from diacylglycerol (DAG) which has dual roles as second messenger and eicosanoid precursor. NSAIDs (nonsteroidal antiinflammatory drugs) and ETYA (eicosatetraynoic acid) inhibit the enzyme cyclooxygenase and block prostaglandin and thromboxane synthesis. ETYA and NDGA (nordihydroguaiaretic acid) block the enzyme 5-lipoxygenase and prevent leukotriene synthesis. EYTA is a modified form of arachidonic acid that competes for any enzyme that normally uses arachidonic acid as its substrate. *(Adapted with permission from Bolander, F.F., "Molecular Endocrinology," Academic Press, San Diego, CA, 1989.)*

Peptides, proteins, indoleamines, catecholamines, thyroid hormones, and many other bioregulators are formed from one or more amino acids. The amino acid arginine is also the substrate for synthesis of the gaseous bioregulator, nitric oxide. Steroid hormones are all derived from cholesterol, and the eicosanoids have a common precursor, arachidonic acid.

Peptide, protein, and amine bioregulators bind to receptors located on the cell surface and typically produce their actions through production of second messengers. Some of these hormones work through mechanisms that employ a variety of G-proteins that influence levels of second messengers such as cAMP, IP_3, and DAG, which in turn may affect the activity of specific protein kinases or permeability of the cell membrane. Other receptor complexes may exhibit protein kinase activities themselves and produce their effects through phosphorylation of specific proteins or by activation of phosphorylation cascades. Initial effects of these bioregulators usually alter membrane transport in (uptake) or out (secretion) of the cell or activate intracellular enzyme systems. Many of these membrane ligands also produce a delayed effect on protein synthesis through activation of cytosolic transcription factors. Cell surface receptors may cause a flux of Ca^{2+}, which also functions as a second messenger.

Steroids and thyroid hormones enter the cytoplasm where they may bind to what are called nuclear receptors although they are sometimes located in the cytosol. Occupied receptors for these hormones typically form homo- or heterodimers and act as ligand-activated transcription factors, binding to HREs and affecting nuclear mRNA production (genomic actions) and subsequent protein synthesis. In a few cases, certain steroids may be ligands for receptors located in the cell membrane. Some steroids and T_4 may be modified by intracellular enzymes prior to binding to receptors.

Bioregulators are metabolized in a variety of ways to inactivate them so that activated cells can recover. Metabolism may involve inactivation by enzymes associated with the cell surface or by intracellular enzymes in the target cells or nearby cells. Reuptake of amines by presynaptic neurons or adjacent cells is another method for inactivating the target cell. Steroids and thyroid hormones are often metabolized by addition of special components (e.g., sulfate, glucuronides) that increase their solubility in water and accelerate their excretion via the urine. Sensitivity of target cells also can be influenced by increases (upregulation) or decreases (downregulation) in receptor number.

Study Questions

Synthesis and Metabolism of Bioregulators

1. What are the major chemical groups that constitute most of the endocrine bioregulators?
2. Describe the synthetic pathway for catecholamines, naming the critical enzymes and intermediates in the pathway.
3. Describe the synthetic pathway for indoleamines, naming the critical enzymes and intermediates in the pathway.
4. How are catecholamines and indoleamines metabolized?
5. Describe the process for synthesizing peptides and proteins for export from a cell.
6. How does the synthesis process differ for peptides and proteins that are exported and those that are used within the cell?
7. What is a splice variant and what is its significance?
8. What mechanisms are employed to prolong the life of a bioregulator in the circulation?
9. How are bioregulators inactivated?
10. What are the consequences of receptor downregulation and upregulation for a cell?
11. In what ways do glycoprotein bioregulators differ from peptide bioregulators?
12. What are the important chemical differences among the various steroid bioregulators?
13. Describe the pathway for the synthesis of steroid bioregulators from cholesterol. Name all key intermediates and key enzymes involved.
14. What is the difference between the Δ^4 and Δ^5 pathways?
15. What happens if a particular enzyme in a synthetic pathway for a catecholamine, indolamine, steroid, or thyroid hormone is missing? How might the consequences differ for steroids where the pathways are branched versus catecholamines where the pathway is a single sequence?
16. What is *cross-talk* and why is it important in understanding how bioregulators work?
17. Describe the life history for each chemical category of hormone covered in this chapter (e.g., steroid, peptide).

Mechanisms of Bioregulator Actions

1. List and characterize the different types of receptors that peptides, amines, and proteins utilize. How do these differ from steroid and thyroid hormone receptors and how are they similar?
2. What are receptor domains and how do they differ for membrane and nuclear receptors?
3. What is a *second messenger*? What are the major second messenger systems and how do they produce effects within a target cell?
4. Compare and contrast mechanisms of action that work via membrane receptors and either cytoplasmic or nuclear receptors.
5. What is the significance of having multiple receptor types for one bioregulator?
6. Name a specific hormone that operates via each mechanism of action described in this chapter.

SUGGESTED READING

General

Baker, M.E., 2003. Evolution of adrenal and sex steroid action in vertebrates: a ligand-based mechanism for complexity. Bioessays 25, 396–400.

Bolander, F.F., 1989. Molecular Endocrinology. Academic Press, San Diego, CA.

Damstra, T., Barlow, S., Bergman, A., Kavlkock, R., Van der Kraak, G., 2002. Global Assessment of the State-of-the-Science of Endocrine Disruptors. World Health Organization, Geneva, Switzerland.

Escriva, H., Delaunay, F., Laudet, V., 2000. Ligand binding and nuclear receptor evolution. Bioessays 22, 717–727.
Guengerich, F.P., 1993. Cytochrome P450 enzymes. American Scientist 81, 440–447.
Thornton, J.W., 2001. Evolution of vertebrate steroid receptors from an ancestral estrogen receptor by ligand exploitation and serial genome expansion. Proceedings of the National Academy of Sciences USA 98, 5671–5676.

Catecholamines and Indolamines

Ma, X., Idle, J.R., Krausz, K.W., Gonzalez, F.J., 2005. Metabolism of melatonin by human cytochromes P450. Drug Metabolism and Disposition 33, 489–494.

Protein, Peptide, and Amine Mechanisms of Action

Conti, M., 2000. Phosphodiesterases and cyclic nucleotide signaling in endocrine cells. Molecular Endocrinology 14, 1317–1327.
Dupre, D.J., Robataille, M., Rebois, R.V., Hebert, T.E., 2009. The role of Gβγ subunits in the organization, assembly and function of GPCR signaling complexes. Annual Review of Pharmacology and Toxicology 49, 31–56.
Frohman, L.A., Jansson, J.O., 1986. Growth hormone-releasing hormone. Endocrine Reviews 7, 223–253.
Gaborik, Z., Hunyady, L., 2004. Intracellular trafficking of hormone receptors. Trends in Endocrinology and Metabolism 15, 286–293.
Gavi, S., Shumany, E., Wang, H.-Y., Malbon, L.C., 2006. G-protein coupled receptors and tyrosine kinases: crossroads in cell signaling and regulation. Trends in Endocrinology and Metabolism 17, 46–52.
Kelly, P.A., Djiane, J., Edery, M., 1992. Different forms of the prolactin receptor: insights into the mechanism of prolactin action. Trends in Endocrinology and Metabolism 3, 54–59.
Naor, Z., Benard, O., Seger, R., 2000. Activation of MAPK cascades by G-protein-coupled receptors: the case of gonadotropin-releasing hormone receptor. Trends in Endocrinology and Metabolism 11, 91–99.
Re, R.N., 2003. The intracrine hypothesis and intracellular peptide hormone action. Bioessays 25, 401–409.
Turgeon, J.L., Waring, D.W., 1992. Functional cross-talk between receptors for peptide and steroid hormones. Trends in Endocrinology and Metabolism 3, 360–365.
Vassart, G., Parmentier, M., Libert, F., Dumont, J., 1991. Molecular genetics of the thyrotropin receptor. Trends in Endocrinology and Metabolism 2, 151–156.
Vilardaga, J.-P., Bünemann, M., Feinstein, T.N., Lambert., N., Nikolaev, V., Engelhardt, S., Lohse, M.J., Hoffmann, C., 2009. Minireview: GPCR and G-proteins: drug efficacy and activation in live cells. Molecular Endocrinology 23, 590–599.
Werry, T.D., Sexton, P.M., Christopoulos, A., 2005. 'Ins and outs' of seven-transmembrane receptor signaling to ERK. Trends in Endocrinology and Metabolism 16, 26–33.

Steroid Hormones

Charalampopoulos, I., Remboutsika, E., Margioris, A.N., 2008. Neurosteroids as modulators of neurogenesis and neuronal survival. Trends in Endocrinology and Metabolism 19, 301–307.
Cheung, J., Smith, D.F., 2000. Molecular chaperone interactions with steroid receptors: an update. Molecular Endocrinology 14, 939–946.
Compagnone, N.A., Mellon, S.H., 2000. Neurosteroids: biosynthesis and function of these novel neuromodulators. Frontiers in Neuroendocrinology 21, 1–56.
Hammond, G.L., 1990. Molecular properties of corticosteroid binding globulin and the sex-steroid binding proteins. Endocrine Reviews 11, 65–79.
Maggi, A., Ciana, P., Belcredito, S., Vegeto, E., 2004. Estrogens in the nervous system: mechanisms and nonreproductive functions. Annual Review of Physiology 66, 291–313.
Mellon, S.H., Griffin, L.D., 2002. Neurosteroids: biochemistry and clinical significance. Trends in Endocrinology and Metabolism 13, 35–43.
Miller, W.L., 1988. Molecular biology of steroid hormone synthesis. Endocrine Reviews 9, 295–318.
Miller, W.L., 2007. StAR search: what we know about how the steroidogenic acute regulatory protein mediates mitochondrial cholesterol transport. Molecular Endocrinology 21, 589–601.
Santen, R.J., Brodie, H., Simpson, E.R., Siiteri, P.K., Brodie, A., 2009. History of aromatase: saga of an important biological mediator and therapeutic agent. Endocrine Reviews 30, 343–375.
Walker, E.A., Stewart, P.M., 2003. 11β-hydroxysteroid dehydrogenase: unexpected connections. Trends in Endocrinology and Metabolism 14, 334–339.

Steroid Mechanisms of Action

Edwards, D.P., 2005. Regulation of signal transduction pathways by estrogen and progesterone. Annual Review of Physiology 67, 335–376.
Foradori, C.D., Weiser, M.J., Handa, R.J., 2008. Non-genomic actions of androgens. Frontiers in Neuroendocrinology 29, 169–181.
Giguere, V., 2002. To ERR in the estrogen pathway. Trends in Endocrinology and Metabolism 13, 220–225.
Hermanson, O., Glass, C.K., Rosenfeld, M.G., 2002. Nuclear receptor coregulators: multiple modes of modification. Trends in Endocrinology and Metabolism 13, 55–60.
Kalaitzidis, D., Gilmore, J.D., 2005. Transcription factor cross-talk: the estrogen receptor and NF-κB. Trends in Endocrinology and Metabolism 16, 46–52.
Levin, E.R., 2009. Plasma membrane estrogen receptors. Trends in Endocrinology and Metabolism 20, 477–482.
McDonnell, D.P., 1995. Unraveling the human progesterone receptor signal transduction pathway: insights into antiprogestin action. Trends in Endocrinology and Metabolism 6, 133–138.
Nettles, K.W., Greene, G.L., 2005. Ligand control of coregulator recruitment to nuclear receptors. Annual Review of Physiology 67, 309–333.
Norman, A., Litwack, G., 1997. Hormones, second ed. Academic Press, San Diego, CA.
Picard, D., 2006. Chaperoning steroid hormone action. Trends in Endocrinology and Metabolism 17, 229–235.
Privalsky, M.L., 2004. The role of corepressors in transcriptional regulation by nuclear hormone receptors. Annual Review of Physiology 66, 315–360.
Rahman, F., Christian, H.C., 2008. Non-classical actions of testosterone: an update. Trends in Endocrinology and Metabolism 18, 371–378.
Saunders, P.T.K., 2005. Does estrogen receptor β play a significant role in human reproduction? Trends in Endocrinology and Metabolism 16, 222–227.

Sugiyama, N., Barros, R.P.A., Warner, M., Gustafsson, J.-A., 2010. ERβ: recent understanding of estrogen signaling. Trends in Endocrinology and Metabolism 21, 545–552.

Wang, H., Eriksson, H., Sahlin, L., 2000. Estrogen receptors α and β in the female reproductive tract of the rat during the estrous cycle. Biology of Reproduction 63, 1331–1340.

Thyroid Hormones

Sutija, M., Joss, J.M.P., 2006. Thyroid hormone deiodinases revisited: insights from lungfish: a review. Journal of Comparative Physiology B 176, 87–92.

Thyroid Hormone Actions

Lefebvre, P., Benomar, Y., Staels, B., 2010. Retinoid X receptors: common heterodimerization partners with distinct functions. Trends in Endocrinology and Metabolism 21, 676–683.

Williams, G.R., Brent, G.A., 1995. Thyroid hormone response elements. In: Weintraub, B.D. (Ed.), Molecular Endocrinology: Basic Concepts and Clinical Implications. Raven, New York, pp. 217–239.

Yen, P.M., Chin, W.W., 1994. New advances in understanding the molecular mechanisms of thyroid hormone action. Trends in Endocrinology and Metabolism 5, 65–72.

Chapter 4

Organization of the Mammalian Hypothalamus–Pituitary Axes

The **pituitary gland** or **hypophysis** was called the "master gland" because its hormones were shown to control many diverse systems that are essential for survival and reproduction. The discovery of the complex control of the pituitary by the **hypothalamus** of the brain transferred our attention to the brain as being the master gland, and the pituitary became just another component of this major system for homeostatic regulation (see Figure 4-1). An overview of this system was provided in Chapter 1. Basically, neurosecretory (NS) neurons in the hypothalamus secrete **hypothalamic-releasing hormones** (Table 4-1) that travel to the pituitary gland where they regulate secretion of the pituitary **tropic hormones** (Table 4-2). These tropic hormones target certain peripheral endocrine glands (e.g., the thyroid gland) that in turn release their hormones (e.g., thyroid hormones) into the blood. These target gland hormones affect specific non-endocrine targets of their own and feedback on the hypothalamus and/or the pituitary.

The embryonic vertebrate brain develops as four major regions known as the **telencephalon** (most anterior), **diencephalon**, **mesencephalon**, and **rhombencephalon** (most posterior). The telecephalon and diencephalon comprise the **forebrain**, whereas the mesencephalon and rhombencephalon become the **midbrain** and **hindbrain**, respectively. During subsequent development, each of these regions differentiates into major components of the adult brain. The telencephalon becomes the **olfactory bulbs**, the **olfactory lobes**, and the **cerebral hemispheres**. The diencephalon differentiates into three regions: the dorsal **epithalamus**, the central **thalamus**, and the ventral **hypothalamus**. Most of the diencephalon becomes the thalamus, a major relay

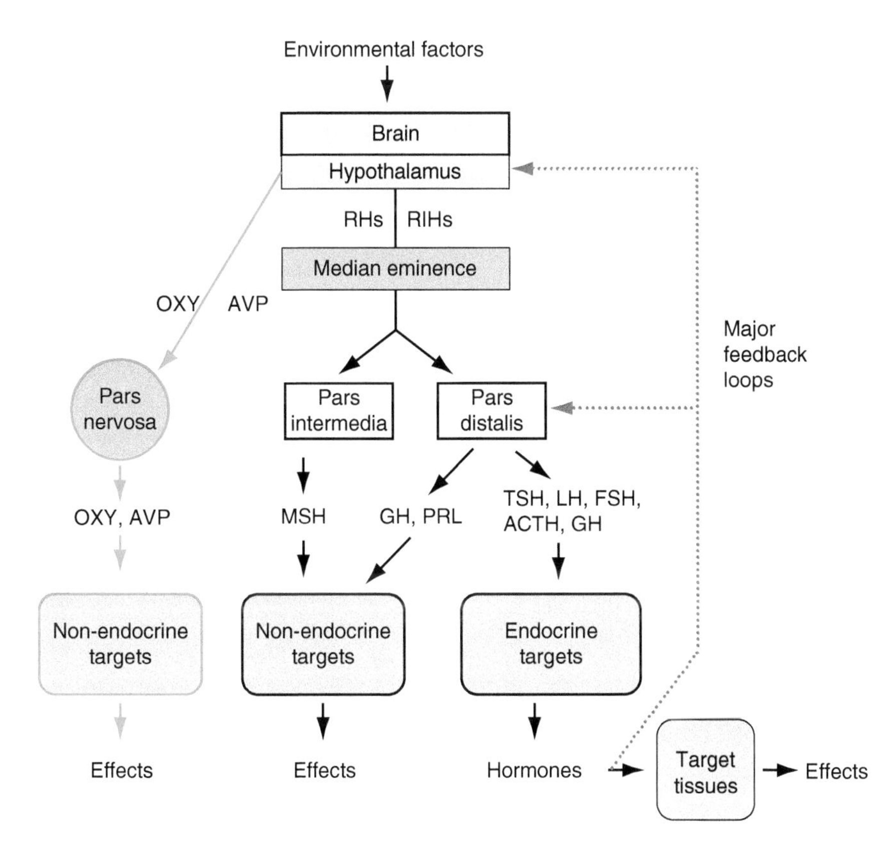

FIGURE 4-1 **The vertebrate neuroendocrine system.** The hypothalamus secretes RHs and RIHs, stores them in the median eminence, and releases them to the pars distalis and pars intermedia where they regulate release of tropic hormones. The activity of the hypothalamus is influenced by a variety of exogenous environmental factors via the central nervous system. The tropic hormones affect some non-endocrine targets, but most have endocrine targets that in turn release hormones that have specific effects on target cells and feedback at the hypothalamus or adenophypophysis (pars distalis). The hypothalamus also secretes the nonapeptides oxytocin (OXY) and vasopressin (AVP), which are stored in the pars nervosa and released into the blood through which they travel to non-endocrine targets. Only one type of long feedback is shown, but short (tropic hormone feedback) and ultrashort feedback (neuropeptides) may also influence hypothalamic function.

Vertebrate Endocrinology. http://dx.doi.org/10.1016/B978-0-12-394815-1.00004-5

TABLE 4-1 Major (Primary) and Secondary Factors Affecting Tropic Hormone Release

Tropic Hormone	Primary Hypothalamic Hormone	Source in Hypothalamus	Secondary Hypothalamic Factors	Source in Hypothalamus
TSH	(+) TRH	PVN	(+) VIP (+) CRH	PVN PVN/ARC
FSH	(+) GnRH	POA/ARC	(+) NPY (−) GnIH (−) CRH	ARC ? DMN
LH	(+) GnRH	POA/ARC	(+) Galanin* (+) NPY (−) GnIH (−) CRH	POA/ARC ARC ? DMN
ACTH	(+) CRH	PVN	AVP enhances CRIH ?	PVN**
PRL	(−) DA	PERIV/ARC	(+) VIP (+) TRH ?	POA/SON PVN?
GH	(−) SST (GHRIH, GRIH); (+) GHRH (in absence of SST)	AMG/PERIV ARC	(+) GRH = GHRH (+) Ghrelin	
MSH	(−) DA (MRIH)	ARC		

*Colocalized with GnRH.
**Colocalized with CRH.
(−) = inhibits tropic hormone release; (+) = stimulates tropic hormone release.
See text or Appendix A for explanation of abbreviations.

station between higher and lower portions of the brain. The floor or ventral portion of the diencephalon becomes the hypothalamus, containing various NS nuclei, which are sources for the neurohormones involved with regulation of pituitary function. The epithalamus is derived from the roof of the diencephalon and gives rise to the endocrine **epiphysial complex**, which includes the **pineal gland**. The mesencephalon gives rise primarily to the **optic tectum**, and the rhombencephalon differentiates into the **cerebellum** and **medulla**.

Several distinct axes are involved in the hypothalamus, a portion of the pituitary gland and a specific target endocrine gland. Each axis represents a neuroendocrine link between the nervous system and the endocrine system and is innervated by non-neurosecretory or ordinary neurons that produce classical neurotransmitters and are usually identified by their neurotransmitter phenotype. These classical neurotransmitters include acetylcholine (cholinergic neurons), epinephrine (adrenergic neurons), gamma aminobutyric acid (GABA, so-called gabaergic neurons), glutamate (glutaminergic neurons), and serotonin (serotonergic neurons). Classical neurotransmitter pathways originate from many different brain areas to innervate hypothalamic neurosecretory neurons (Figure 4-2). Many peptides also have been identified in non-neurosecretory neurons and may act as neuromodulators and possibly neurotransmitters. Both these ordinary neurons and the NS neurons are responsive to effects of hormones present in the blood or cerebrospinal fluid (CSF). Each **hypothalamus−pituitary (HP) axis** consists of the neurosecretory neurons within NS nuclei of the hypothalamic region of the brain, specific cell types in the pituitary that secrete tropic hormones, and the endocrine glands directly controlled by these tropic hormones. In this chapter, we focus on four of these axes: the **hypothalamus-pituitary-thyroid (HPT) axis**, **hypothalamus-pituitary-gonad (HPG) axis**, **hypothalamus-pituitary-adrenal (HPA) axis**, and **hypothalamus-pituitary-hepatic (HPH)** or **liver axis.** In addition to these four major axes, we also will consider several pituitary hormones as well as some hypothalamic neurohormones released directly into the general circulation to non-endocrine targets.

Only the mammalian HP axes will be discussed here, and the homologous axes in non-mammalian vertebrates will be discussed in Chapter 5. Although in an evolutionary sense such a discussion should begin with agnathan fishes and proceed through the other fishes to tetrapods, the mammalian system is better understood and provides the nomenclature with respect to structures, hormones, and functions that later were applied to non-mammalian systems. In fact, the entire field of endocrinology has developed in similar fashion from mammalian

TABLE 4-2 Synonyms, Abbreviations, Cellular Source, Targets, and Actions for Mammalian Tropic Hormones

Name	Abbreviation[a]	Synonyms	Other Abbreviations	Cellular Source	One Target	One Action on Target
Prolactin	PRL	Mammotropin, luteotropin, luteotropic hormone	LTH	Lactotrope	Mammary gland	Stimulates milk synthesis
Growth hormone	GH	Somatotropin, somatotropic hormone	STH	Somatotrope	Muscle	Stimulates incorporation of amino acids into protein
Corticotropin	ACTH	Adrenocorticotropic hormone, adrenocorticotropin		Corticotrope	Adrenal cortex	Stimulates synthesis and secretion of corticosteroids
Melanotropin	MSH	Intermedin, melanocyte- or melanophore-stimulating hormone		Pars intermedia (Melanotrope)	Melanocyte, etc.	Stimulates synthesis of melanin pigment
Thyrotropin	TSH	Thyroid-stimulating hormone		Thyrotrope	Thyroid gland	Stimulates synthesis of thyroid hormone
Follicle-stimulating hormone	FSH	Follitropin		Gonadotrope	Gonad	Stimulates follicular development in females and spermatogenesis in males; estrogen secretion in females
Luteinizing hormone	LH	Interstitial cell-stimulating hormone, lutropin	ICSH	Gonadotrope	Gonad	Stimulates androgen and progesterone synthesis in females and androgen secretion in males

[a] *The names in the first column and the abbreviations in the second column are used throughout the text.*

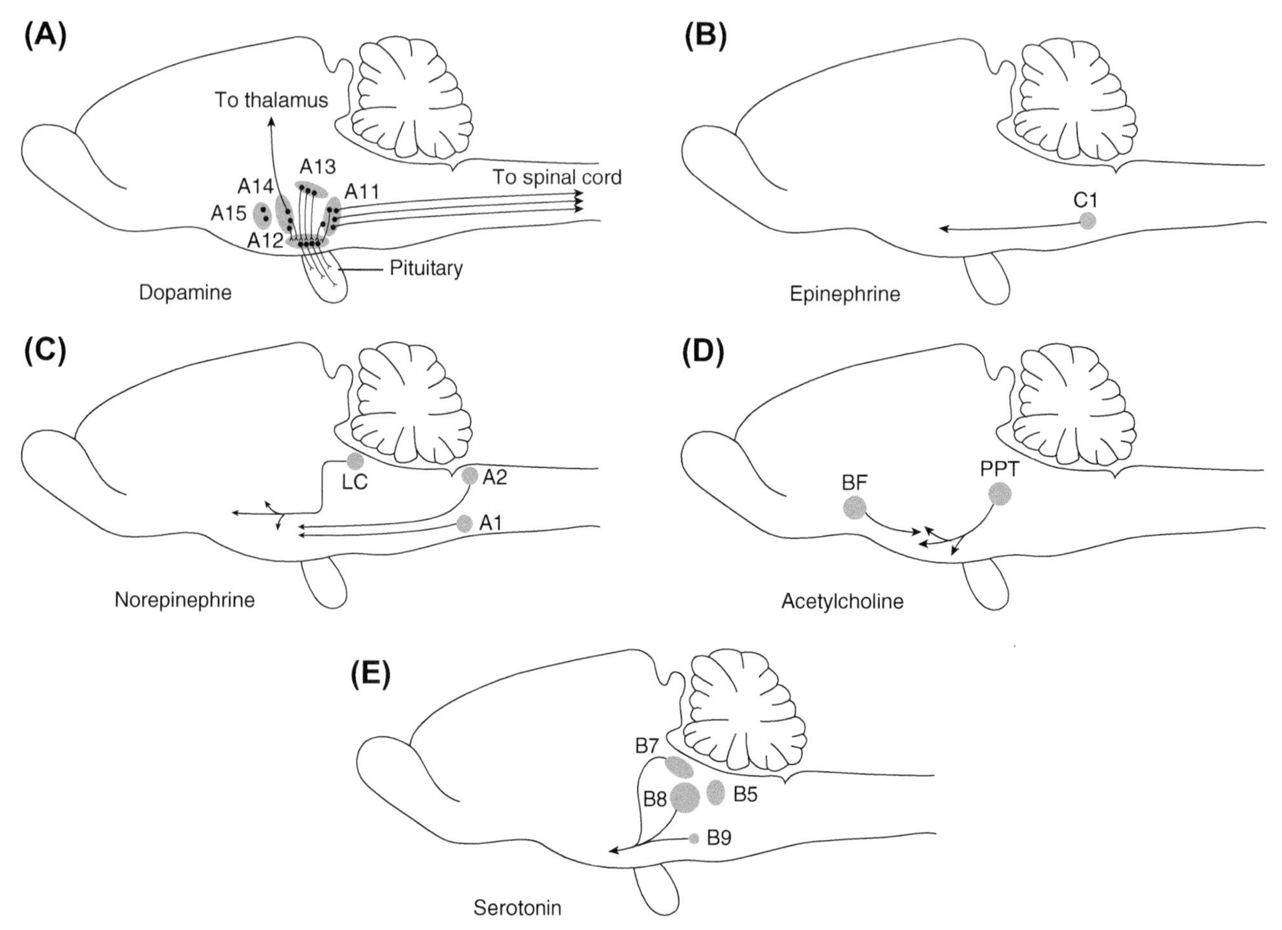

FIGURE 4-2 **Pathways for classical neurotransmitter innervation of the hypothalamus.** Nomenclature for (A) dopamine, (B), epinephrine, (C) norepinephrine, (D) acetylcholine, and (E) serotonin pathways is based upon Dahlstrom and Fuxe (1964). Abbreviations: BF, basal forebrain; LC, locus coeruleus; PPT, pedunculopontine tegmental nucleus. *(Adapted with permission from Cooper, J.R. et al., "The Biochemical Basis of Neuropharmacology," Oxford University Press, New York, 2002.)*

investigations, largely of a clinical orientation and motivation, and then applied to wild mammals as well as to nonmammalian vertebrates.

I. THE MAMMALIAN PITUITARY

The pituitary or hypophysis of adult mammals is located ventral to the brain just posterior to the optic chiasm, and it remains attached to the hypothalamus by a stalk-like connection (Figure 4-3). Endocrinologists today generally refer to this structure as the *pituitary gland* rather than the *hypophysis*, although both terms are technically correct. The term "pituitary gland" is derived from the Latin word for phlegm (*pituita*), because it was once thought to be the source of this important humour. The term "hypophysis" is derived from the Greek words for growth (*physis*) and below (*hypo*) (i.e., below the brain). The pituitary gland is separable into two regions, the **adenohypophysis** and the neurohemal **neurohypophysis**. The anatomical terminology used here for all vertebrate pituitaries is based on Green (1951) and Purves (1961).

The pituitary gland is located directly beneath the third ventricle of the brain. The third ventricle is a cavity continuous with the other ventricles of the brain and the central canal of the spinal cord. It is filled with cerebrospinal fluid (CSF). Studies in the early 1800s determined that the pituitary developed through an apparent fusion of a ventral growth or evagination from the diencephalon, the **infundibulum**, with an ectodermal sac known as **Rathke's pouch** (Figure 4-4). The latter developed as an inward pocketing or invagination off the anterior roof of the oral cavity called the *stomodeum*; hence, it was concluded that cells forming the adenohypophysis arose from non-neural ectoderm. However, numerous studies primarily employing amphibian and bird embryos have shown that the secretory cells of the adenohypophysis and the neurosecretory neurons of the hypothalamus have a common origin from the neural ridge of the embryo (neuroectoderm) (Figure 4-5) and migrate during development to the tissues

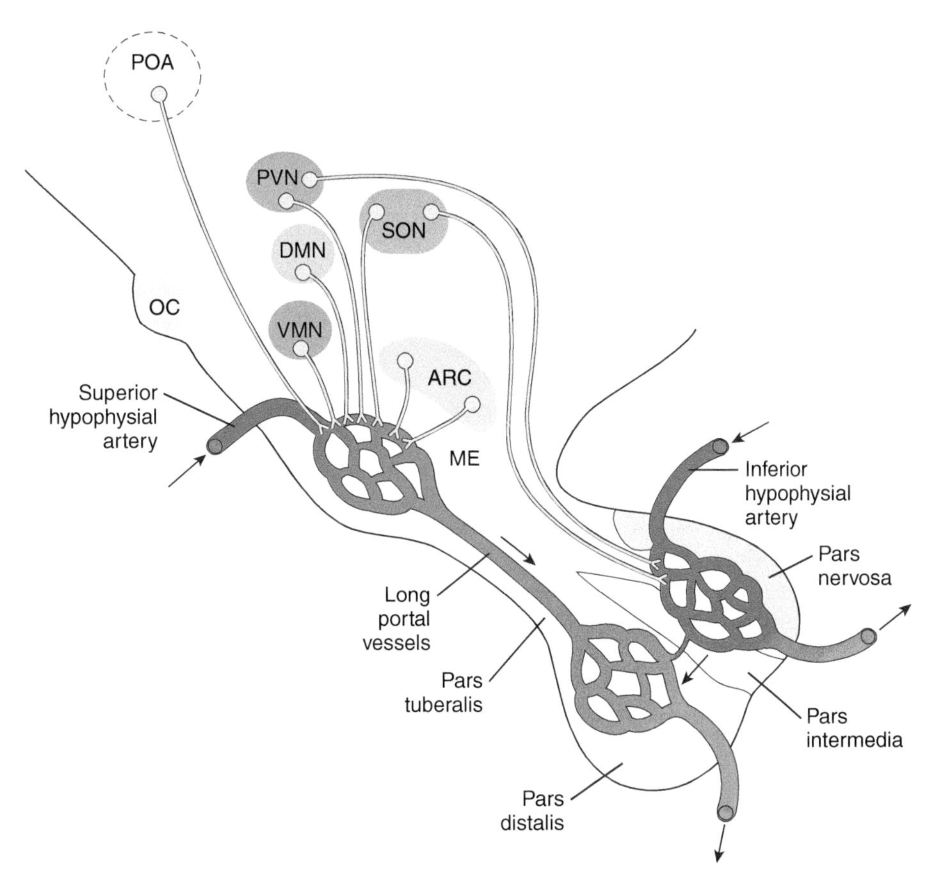

FIGURE 4-3 **A generalized mammalian hypothalamus–pituitary axis.** The hypothalamus contains several neurosecretory centers including the arcuate (ARC) nucleus, dorsomedial nucleus (DMN), paraventricular nucleus (PVN), preoptic area (POA), supraoptic nucleus (SON), and the ventromedial nucleus (VMN). The median eminence (ME) and pars nervosa are separate neurohemal areas. The superior hypophysial artery supplies blood to the capillaries in the ME. The ME is connected by hypothalamic portal blood vessels passing through the pars tuberalis to the pars distalis where releasing hormones from the hypothalamus stimulate tropic hormone release. The pars nervosa stores nonapeptides and has a separate blood supply. OC, optic chiasm.

that ultimately form the hypophysis and hypothalamus. Evidence also suggests that the hormone-secreting cells of the pituitary are of neural origin (see Box 4A). The neurosecretory cells that control reproduction have their origins in a portion of the neural ridge that gives rise to the nasal placodes (olfactory system), and these neurons migrate along the olfactory nerve eventually to take up residence in the hypothalamus (see ahead).

The adenohypophysis is an epithelial glandular structure (*adeno*, gland) and can be subdivided into three anatomical regions or pars (= bodies): the **pars anterior** or **pars distalis**, the **pars tuberalis**, and the **pars intermedia**. Pars

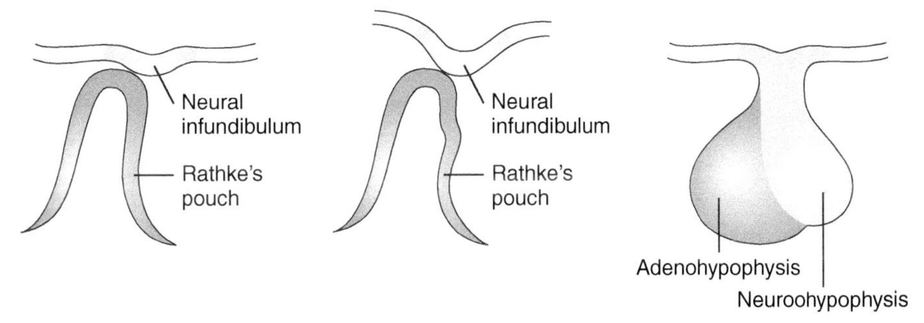

FIGURE 4-4 **Pituitary gland development.** Early understanding of development of the stomatodeal epidermis into the adenohypophysis from Rathke's pouch and the neurohypophysis from the neural infundibulum. The pars intermedia developed at the point of contact between the neuroinfundibulum and Rathke's pouch. *(Adapted from Dubois, P.M. and El Amraouci, A.,* Trends in Endocrinology and Metabolism, *6, 1–7, 1995. © Elsevier Science, Inc.)*

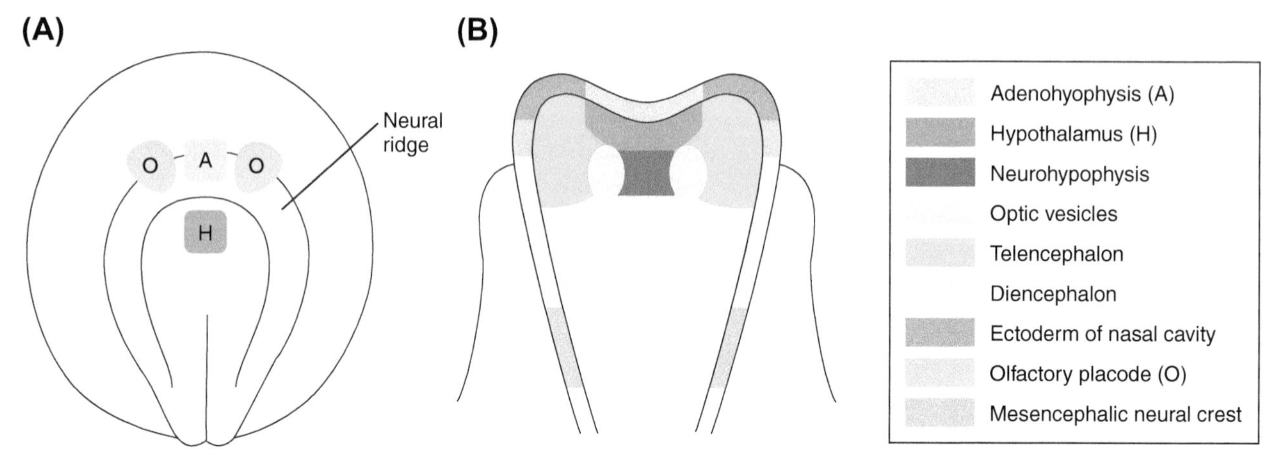

FIGURE 4-5 **Anterior neural ridge origin of the hypothalamic and pituitary endocrine cells.** (A) Dorsal view of neurula stage of anuran embryo showing close proximity for origins of hypothalamic neurons (H), pituitary endocrine cells (A), and the olfactory placodes (O). GnRH cells of the hypothalamus originate from the olfactory placodes. (B) Similar origins are shown for an avian embryo. *(Redrawn from several sources.)*

distalis is used more commonly than pars anterior. Each region of the adenohypophysis is distinguished by cytological features as well as by anatomical relationships to the neurohypophysis. Two subregions can be identified in the neurohypophysis: the more anterior **pars eminens** or **median eminence** and the **pars nervosa**. (Note: The term "pars eminens" is rarely used today.) These neurohemal regions consist mostly of aminergic and peptidergic axonal endings mixed with blood capillaries and what are probably modified neuroglial cells called **pituicytes**. An extensive venous portal system, the **hypothalamus–hypophysial portal system**, develops between capillary beds in the median eminence of the neurohypophysis and capillary beds in the pars distalis of the adenohypophysis (Figure 4-3). This portal system carries blood from the median eminence directly to the epithelial cells of the pars distalis. The capillaries of the pars nervosa connect directly with the general venous drainage system, and there are no associated portal vessels. Although there are other portal systems in the body, endocrinologists often use "portal system" to refer only to this one.

The portal system forms a neurovascular link between the hypothalamus and the pituitary gland as described in the pioneering anatomical studies of Wislocki. Blood containing hypothalamic-releasing hormones flows from the median eminence to the pars anterior, and the venous drainage from the latter carries pituitary tropic hormones into the general circulation. Additional data suggest that, due to the low blood pressure in this system, there may be significant blood flow from the adenohypophysis to the hypothalamus as well. This **retrograde flow** may prove to be important for feedback actions of pituitary hormones on the hypothalamus.

Sympathetic nerve fibers innervate pituitary blood vessels and influence blood flow through the portal system. In addition, peptidergic fibers containing a variety of neuropeptides have been demonstrated to innervate these vessels although their significance is unknown.

A. Subdivisions of the Adenohypophysis

The adenohypophysis of most mammals consists of three subdivisions: pars tuberalis, pars intermedia, and pars distalis. The **pars tuberalis** consists of a thin layer of cells projecting rostrally (anteriorly and dorsally) from the adenohypophysis. It is in contact with the median eminence of the neurohypophysis, and the portal vessels of the portal system pass near or through the pars tuberalis *en route* to the capillary beds of the pars distalis.

The portion of the adenohypophysis that makes contact with the pars nervosa of the neurohypophysis is defined as the pars intermedia. Indeed, differentiation of the pars intermedia occurs only if physical contact takes place between the developing adenohypophysis and the infundibulum that will become the pars nervosa.

The major portion of the adenohypophysis originally was designated as the **pars anterior**, and a variety of cellular types were identified there by selective staining procedures. In these animals lacking a true pars intermedia, the pars anterior was called the pars distalis. However, this latter term is now applied routinely as if it were synonymous with the pars anterior, especially since pars distalis was the appropriate term for adult humans and was so named in most textbooks. Because of extensive use of pars distalis in the literature for both anatomical designations, we use the term pars distalis exclusively in this textbook. Likewise, the term **anterior lobe** is applied loosely to mean either the pars distalis or the pars anterior plus the pars intermedia. **Posterior lobe** often is used in place of neurointermediate lobe, but at other times this term indicates only the pars nervosa in animals lacking a pars intermedia. Because of their

BOX 4A APUD Cells and Paraneurons

Among the vertebrates, extensive analyses of the embryonic origins of endocrine glands have provided a partial understanding of the evolution of vertebrate neuroendocrine systems. It was once thought that neurosecretory cells and the neurohemal portion of the pituitary were derived from embryonic neural ectoderm, whereas the portion of the pituitary that secretes tropic hormones was derived from non-neural ectoderm. The liver, thyroid, parathyroids, thymus, and gastrointestinal endocrine cells were believed to come from embryonic endoderm. The gonads and kidneys clearly develop from embryonic mesoderm (see Appendix F for a brief description of embryonic tissues). Endocrinologists believed that the fundamental differences among these groups of glands reflected their embryological origins until sophisticated biochemical techniques coupled with careful developmental studies demonstrated a common origin for most cells responsible for production of bioregulators.

Many of these cells, previously thought to be ectodermally and endodermally derived, possessed a property previously ascribed only to amine-secreting neurons. Cells with the ability to accumulate amino acid precursors and convert these precursors into biologically active amine neurotransmitters by removing their carboxyl acidic group were described as **APUD** cells: **amine content and amine precursor uptake and decarboxylation** cells. This property also was shared by melanin-producing skin cells that, similar to the amine-secreting neurons of sympathetic ganglia, were derived from special embryonic neural ectoderm cells called **neural crest** cells. During early development, the neural crests develop as paired masses of cells at intervals along the embryonic nerve cord, migrate to other locations, and give rise to the sympathetic ganglia, melanin-producing skin cells, secretory cells of the adrenal medulla, and components of the skull and branchial skeletal elements. The discovery of this APUD property in many peptide-secreting endocrine and neuroendocrine cells prompted Pearse and Takor (1976) to propose that all cells with APUD characteristics have the same neural origins, including parathyroid cells, calcitonin-secreting cells of the ultimobranchial body, gastrointestinal endocrine cells, and even some pituitary cells. Later, elegant experiments with bird embryos by LeDouarin and LeLievre verified that the calcitonin-secreting cells of the avian ultimobranchial glands were of neural crest origin, although the matrix of the gland itself developed as an outpocketing of gut endoderm. They transplanted quail neural crest cells with a cytological marker into chicken embryos and observed them migrating to and finally residing in the ultimobranchial glands. Similar evidence was obtained for the parathyroid gland secretory cells in the frog *Rana temporaria*. Earlier efforts that showed a neural origin for the tropic hormone-secreting cells of the pituitary as well as the neurosecretory neurons of the hypothalamus apparently have been confirmed in studies by Kikuyama and his associates who transplanted pigmented neural ridge tissue into albino Japanese toads, *Bufo japonicus*, and were able to trace their migration into the neuroendocrine system. Thus, we now recognize that most endocrine and neuroendocrine cells are of neural origin with the exception of the steroid-secreting cells of the gonads and adrenal cortex (mesodermal), endocrine cells of the kidney (mesodermal), and possibly the thyroid follicular cells.

While this shared property of endocrine and neuroendocrine cells suggests a common neural origin, the physiological significance of the APUD characteristics has remained unclear until recently. In 2001, Blackmore and colleagues showed that at least one role of the amine transporter lies within the secretory vesicles of peptide producing endocrine and neuroendocrine cells. The amine transporter located in the secretory granule membrane alters peptide processing (cleaving of amino acids from the parent peptide) by causing changes in pH within the secretory granules.

In addition to the recognition of APUD characteristics between neural cells and established endocrine cells, it was noted that neuroendocrine-like cells appeared not only diffusely distributed within the gastrointestinal tract but also in gills of fishes, lungs of amphibians and mammals (so called "clear" cells), and in skin. These neural-like cells were termed **paraneurons** by Fujita and were found to synthesize many of the same bioregulators found in the nervous system. Some paraneurons have APUD characteristics and some are even innervated. A few function as chemosensory receptors. Some paraneurons have been demonstrated to have paracrine-like actions locally, and this may be a general role for these paraneurons found in epithelia.

variable meanings, the terms anterior lobe and posterior lobe are not used here.

1. The Pars tuberalis

The pars tuberalis is characteristic of all tetrapod vertebrates, but understanding of its physiological significance is limited. The presence of cells containing certain tropic hormones has led to the suggestion that the pars tuberalis is only an extension of the pars distalis related primarily to reproduction. However, recent studies have demonstrated that the pars tuberalis is an important endocrine link between the pineal gland and the prolactin-secreting cells of the pars distalis (see pineal gland discussion ahead).

Structurally the cells of the pars tuberalis are connected to the cerebrospinal fluid of the third ventricle in the brain through cellular processes originating in modified **ependymal cells** known as **tanycytes**. Ependymal cells are epithelial cells that line the ventricles of the brain and form a protective layer that surrounds the nervous system. It has been suggested that tanycytes may selectively remove molecules, including various types of regulators, from cerebrospinal fluid and transfer them to cells of the pars tuberalis, causing the latter to release their stored products.

Although this is a highly speculative idea, such an interesting anatomical relationship demands some imaginative research to provide a better understanding of both tanycytes and the cells of the pars tuberalis.

2. Pars Intermedia

Only one glandular cell type appears in the mammalian pars intermedia, and it is responsible for secretion of the peptide hormone melanocyte-stimulating hormone or **melanotropin** (**α-MSH**). An alternative name for α-MSH is *melanophore-stimulating hormone*, based on its effects in a unique pigment cell not found in mammals, the melanophore (see Chapter 5). In mammals, α-MSH stimulates skin cells, known as **melanocytes**, to synthesize a brown pigment, **melanin**, which causes increased deposition of pigment in the skin or hair. The term **neurointermediate lobe** designates both the pars intermedia and the pars nervosa as an anatomical unit although they are not functionally related. In some species, the pars intermedia is separated from the remainder of the adenohypophysis by a cavity or cleft. Some mammals, such as whales, manatees, elephants, armadillo, pangolin, beaver, and adult humans, lack a pars intermedia. Most of those mammals lacking a pars intermedia lack hair and/or have few if any integumentary melanocytes.

3. Pars Distalis

Five cellular types are present in the pars distalis and are responsible for secretion of six pituitary tropic hormones: **corticotropin**, or adrenocorticotropic hormone (**ACTH**); **thyrotropin**, or thyroid-stimulating hormone (**TSH**); **growth hormone** (**GH**) or somatotropin; **prolactin** (**PRL**); and two **gonadotropins**, or gonadotropic hormones (**GTHs**). The two GTHs are **follicle-stimulating hormone** or follitropin (**FSH**) and **luteinizing hormone** or lutropin (**LH**), both of which are named for their effects in female mammals but have equally important and similar roles in males. All of these hormones are polypeptides or proteins. In addition, α-MSH, a peptide secreted from the pars intermedia, often is included as one of the tropic hormones. The pituitary tropic hormones, alternative names for them, their targets, and their general physiological roles are summarized in Table 4-2. The names most commonly found in the modern literature on vertebrate endocrinology are emphasized in this textbook.

In addition, peptides known as **lipotropins (LPH)** and endorphins (e.g., **β-endorphin)** may be released from the adenohypophysis. Although not listed here as tropic hormones, they may perform endocrine functions. β-endorphin is one of several peptides known as **endogenous opioid peptides (EOPs)** that bind to the same receptors as the drug morphine, an exogenous opioid derived from the plant product opium. The term "endorphin" as a contraction of "endogenous morphine" was suggested for the natural internal substance that bound to what had been called **opioid receptors** in the central nervous system (CNS).

B. Cellular Types of the Adenohypophysis

Differentiation of the five pars distalis cell phenotypes requires the interaction of a number of transcription factors (e.g., **pituitary-specific transcription factor, Pit-1**), differentiation or growth factors, and **steroidogenic factor 1 (SF1)**, as outlined in Table 4-3. The cellular phenotypes in the pars distalis and the pars intermedia first were distinguished by utilizing special dyes in particular staining combinations.

TABLE 4-3 Transcription Factors Involved in Pituitary Tropic Cell Development

Transcription Factor	Abbreviation	Cell Type Affected	Action
Corticotrope upstream transcription factor	CUTE	Corticotrope	Designates POMC gene expression specificity
Leukemia inhibitory factor	LIF	Corticotrope	Designates POMC gene expression specificity
Lim/homeobox gene-3	LHX3	All except corticotropes	Necessary for differentiation
Paired-like homeodomain transcription factors 1 and 2	Pitx 1 and 2	All	Early pituitary differentiation
Pituitary-specific transcription factor	Pit-1	Thyrotropes, lactotropes, somatotropes	Necessary for differentiation
Rathke's pouch homeobox	Rpx	All	Earliest marker for pituitary primordium; unknown target genes
Steroidogenic factor 1	SF-1	Gonadotropes	Makes cells responsive to GnRH
Thyrotrope embryonic factor	TEF-1	Thyrotropes	Necessary for differentiation

The electron microscope also has been used to characterize cellular types of the pars distalis on the basis of general cellular morphology and the size and shape of electron-dense cytoplasmic storage granules containing tropic hormones. A combination of ultrastructural, tinctorial (staining with dyes), and immunocytochemical techniques leaves little doubt as to the cellular origins of the tropic hormones. Ultrastructural features of some pars distalis cells can be seen in Figure 4-6. More recently immunocytochemical methods (Chapter 2) have been used to determine cellular phenotypes in the pars distalis.

1. Cytology of the Pars Distalis

Originally using differential staining combined with experimental manipulations, it was determined that there was one cellular type responsible for synthesis and release of each tropic hormone with one exception. These initial observations were later supported and refined by immunocytochemical studies using antibodies prepared against each pituitary hormone. In the following descriptions, the traditional mammalian designation for each of these five cellular types is given. The cells responsible for secreting tropic hormones are designated with the suffix -**trope**. Because tropic hormones were formerly called "trophic hormones," some authors still refer to the cell types with the suffix "-troph."

The **thyrotrope** is the least common of the secretory cell types in the pars distalis. Thyrotropes have long cytoplasmic processes and contain spherical secretory granules. They occur primarily in the anterior–medial portion of the pars distalis and show little variations with sex or age.

Despite the chemical similarity of GTHs and TSH (see ahead), the **gonadotrope** has been readily distinguishable from the thyrotrope by cytological and immunological techniques. Gonadotropes represent about 15 to 20% of the pars distalis cells and are distributed throughout the pars distalis. Two populations of spherical or slightly irregular secretory granules can be distinguished on the basis of size. At least three gonadotropic subtypes were identified by differences in immunoreactivity. One subtype contains only FSH, one contains only LH, and the third contains both LH and FSH.

Two separate cell types in the pars distalis are considered to be sources for GH and PRL, respectively (Figure 4-7). The **somatotrope** is the most abundant cell type in the pars distalis, representing about 50% of the cells, and is found mostly along the lateral margins of the pars distalis. The second acidophil, called a **lactotrope**, secretes PRL. The lactotropes account for between 10 and 25% of the cells in the pars distalis, with the lower figure being common in men and nulliparous women (never having borne children). There are relatively few PRL-secreting cells in children. Lactotropes are distributed throughout the pars distalis, often found associated with gonadotropes. At least two lactotropes have been identified using ultrastructural criteria. One is very common, a sparsely granular cell with smaller spherical, oval, or irregular granules. The second type is uncommon, is densely granular, and typically occurs adjacent to capillaries. A third lactotrope has been described, the **mammosomatotrope**, which secretes both GH and PRL, especially during pregnancy.

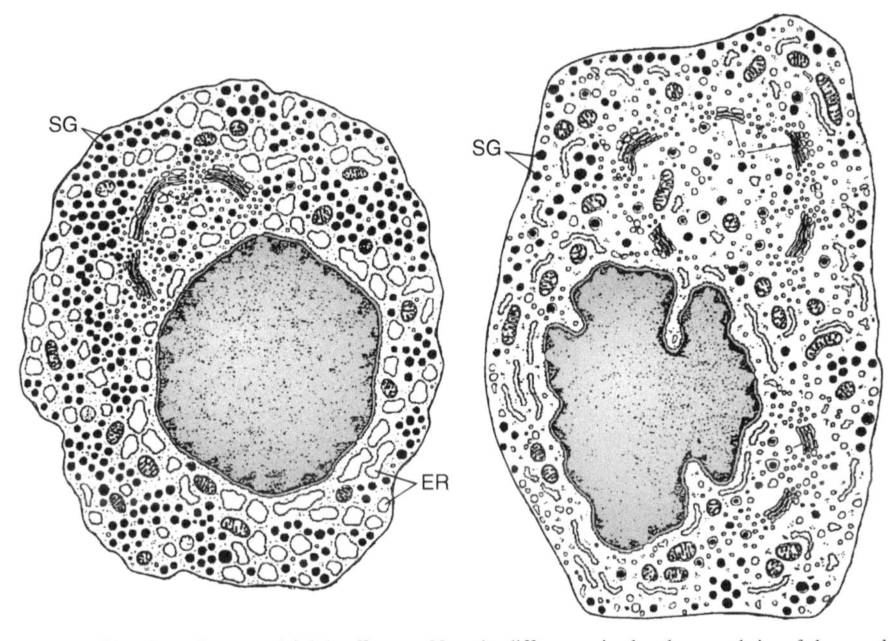

FIGURE 4-6 **Gonadotrope (left) and corticotrope (right) cell types.** Note the differences in abundance and size of electron dense granules. Compare to the lactotrope and somatotrope in Figure 4-7. *(Adapted with permission from Norman, A.W. and Litwack, G., "Hormones," 2nd ed., Academic Press, San Diego, CA, 1997.)*

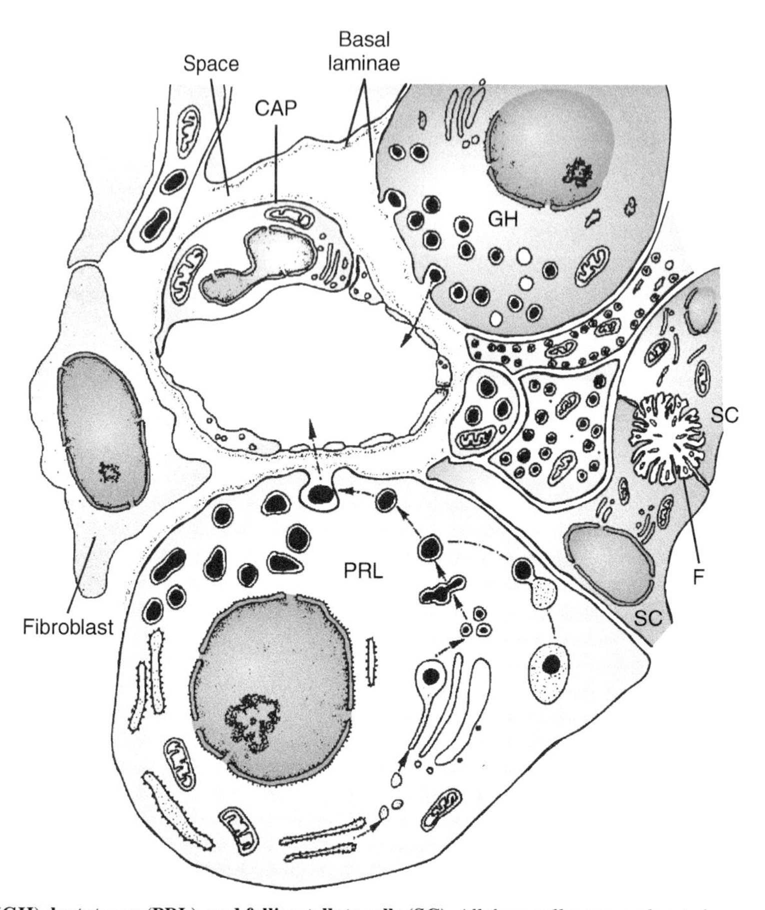

FIGURE 4-7 **Somatotrope (GH), lactotrope (PRL), and follicostellate cells (SC).** All three cell types are located near a capillary (CAP). The GH and PRL cells are distinguished by the different size and relative abundance of secretion granules, whereas the SC are not granulated and show a tendency to form a follicular structure (F) where they contact one another. Compare to gonadotropes and corticotropes in Figure 4-6. *(Adapted from Baker, B.L., Yu, Y. U. Cell and Tissue Research 156, 443–449, 1975.)*

Corticotropes producing ACTH are located in a central wedge within the pars distalis and represent 10 to 15% of the total cells. They contain a variety of granules that are somewhat larger than those of thyrotropes and are immunoreactive for ACTH, LPH, and β-endorphin.

Corticotropes also are immunoreactive for the protein **cytokeratin** that occurs in the perinuclear area and typically is not found in other tropic cells. The number of corticotropes does not vary with age or sex but may vary markedly in a number of pathological states (see Chapter 8).

Careful studies of hormone distribution, receptors present in pituitary tropic cells, and the diversity of hormone genes expressed by each tropic cell have painted a much more dynamic picture of the pars distalis than previously thought. Not only do we have multiple populations of GTH-secreting cells and PRL-secreting cells, but we are also finding some cells that are making a greater variety of tropic hormones than previously suspected. For example, not only do some gonadotropes produce GTHs but they may also produce GH. Furthermore, these GTH–GH cells possess receptors for both hypothalamic-releasing hormones, suggesting that these cells are secreting both hormones. Corticotropes have been observed to vary with respect to the type of receptors they express for **corticotropin-releasing hormone (CRH)** and in their ability to also bind the neuropeptide vasopressin. Using many of the molecular techniques described in Chapter 2, investigators have discovered the presence of mRNAs for more than one tropic hormone in pituitary tropic cells implying that, if the appropriate signals are received by these cells, they could begin secretion of alternative hormones. These new developments suggest a much more dynamic pituitary cytology than previously believed.

A sixth cellular type found in the mammalian pars distalis is the nongranulated cell that is not distinguished by selective staining techniques. Nongranulated cells may represent inactive, depleted, or undifferentiated cells, and some of the latter may differentiate into hormone-secreting cells, depending upon the stage of development or physiological conditions or in response to experimental manipulations. One type of nongranulated cell is the **null cell** that has no special histologic, immunoreactive, or ultrastructural features other than the presence of a few small cytoplasmic granules. Null cells are thought to be the source of certain pituitary adenomas. A special type of nongranulated cell, the **follicostellate cell**, has been observed in all vertebrates with the aid of the electron microscope. Follicostellate cells exhibit **S-100 protein**, a characteristic of neuroglial cells in the brain, and the S-100 protein is not found in any other cells of the adenohypophysis. The cytoplasmic processes of these glial-related stellate (star-shaped) cells are very long and form a sort of network or reticulum between capillaries throughout the pars distalis. They are referred to as *follicular* because of the way their stellate processes will sometimes surround or enclose tiny spaces. Each of these follicles consists of an extracellular space surrounded completely by processes of the follicostellate cells and are filled with fluid (Figure 4-7). Microvilli and sometimes cilia project into the follicular lumina of these follicles. The follicostellate cells may perform a supportive or nutritional function and are known to act as phagocytic scavenger cells. They are probably not the source of any tropic hormones, but they do produce paracrine secretions including **interleukin (IL-6), basic fibroblastic growth factor (bFGF)**, and **vascular endothelial cell growth factor (VEGF)**. Secretions from follicostellate cells in culture attenuate release of GH, PRL, and LH following administration of substances that normally evoke their release. Finally, recent evidence suggests that follicostellate cells have properties of stem cells and may differentiate into other cell types.

2. Cytology of the Pars Intermedia

The **melanotrope** is the only glandular epithelial cell in the mammalian pars intermedia. The pars intermedia also contains nongranulated stellate cells of unknown function that are interspersed among the α-MSH-secreting cells. In most mammals, the cleft that separates the pars intermedia from the pars distalis is lined by ependymal-like cells called **epithelial cleft cells**. The cleft cells often are ciliated, and, while they resemble the ciliated ependymal cells that line the brain ventricles, their functional role has not been worked out.

3. Cytology of the Pars Tuberalis

Several cell types have been reported in the pars tuberalis of mammals. One cell type reacts specifically with antibody to pituitary LH, and a second type specifically binds antibody to TSH. Occasionally, one or two rare cells are observed in primates that bind antibody to ACTH and GH. Another special cell type is the tuberalin cell that secretes **tuberalin**, a peptide that stimulates PRL secretion by lactotropes in the pars distalis (see ahead). The majority of pars tuberalis cells are chromophobic in most mammals, but all are stainable types in humans. The pars tuberalis may represent a "fragment" of the pars distalis, and it may function as an additional, but limited, source of tropic hormones.

C. Subdivisions of the Neurohypophysis

The mammalian neurohypophysis consists of two distinct neurohemal components, the median eminence and the pars nervosa. The median eminence is defined as the more anterior portion of the neurohypophysis that has a blood supply in common with the adenohypophysis—specifically, the portal system. (Note that in some terminologies the median eminence is considered to be a subdivision of the hypothalamus but is considered by Green as a neurohemal subdivision of the neurohyphophysis.) An abundant but separate blood supply characterizes the pars nervosa (Figure 4-3), which is that posterior portion of the neurohypophysis in contact with the pars intermedia. In species lacking a pars intermedia, the pars nervosa is defined on the basis of what neuropeptides are present (see ahead). Both the median eminence and the pars nervosa are composed of capillaries, pituicytes, and axonal tips of NS neurons originating in hypothalamic NS nuclei. Pituicytes probably are derived from ependymal or neuroglial cells, and they could play a role as supportive elements or may be involved actively in storage and release of neurohormones from the neurohypophysis, similar to the role of stellate cells of the adenohypophysis.

II. THE MAMMALIAN HYPOTHALAMUS

The neurosecretory nuclei of the hypothalamus and preoptic area produce neurohormones that are stored in the neurohypophysis. Axons from these nuclei travel to either the median eminence (Figures 4-8 and 4-9) or the pars nervosa (Figure 4-10). The neurohormones associated with

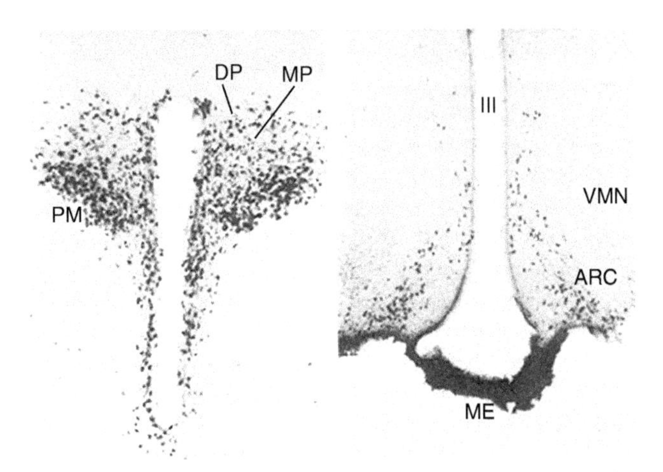

FIGURE 4-8 **Projections from the hypothalamus to the median eminence.** Cells in the paraventricular nucleus (left) and arcuate nucleus (ARC) (right) are stained with wheat germ agglutinin after injection of the tracer into the median eminence (ME). Abbreviations: III, third ventricle; DP, dorsal parvocellular subdivision of the paraventricular nucleus; MP, medial dorsal parvocellular subdivision of the paraventricular nucleus; PM, magnocellular paraventricular nucleus; VMN, ventromedial nucleus. *(Reprinted with permission from Lechan, R.M. et al.,* Brain Research, ***245****, 1–15, 1982.)*

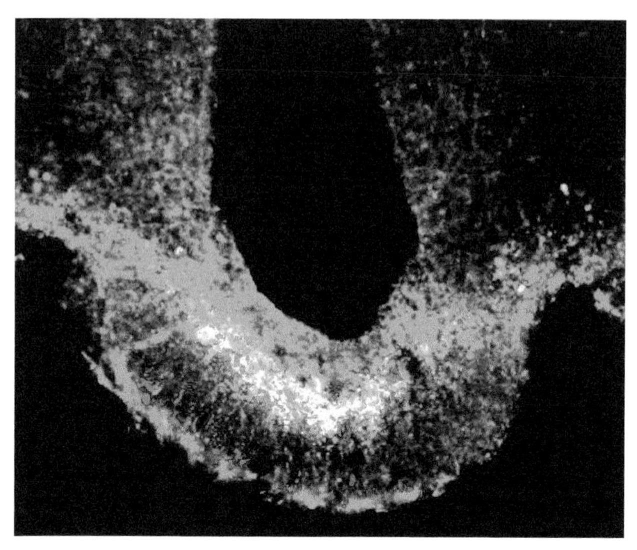

FIGURE 4-9 **The rat median eminence.** In this photograph, immunoreactive cytokine (stromal cell-derived factor 1, in green) and AVP (in red) are co-localized (yellow) in nerve fibers innervating the internal layer of the rat median eminence. *(Reprinted with permission from Callewaere, C. et al.,* Journal of Molecular Endocrinology, *38, 355–363, 2007.)*

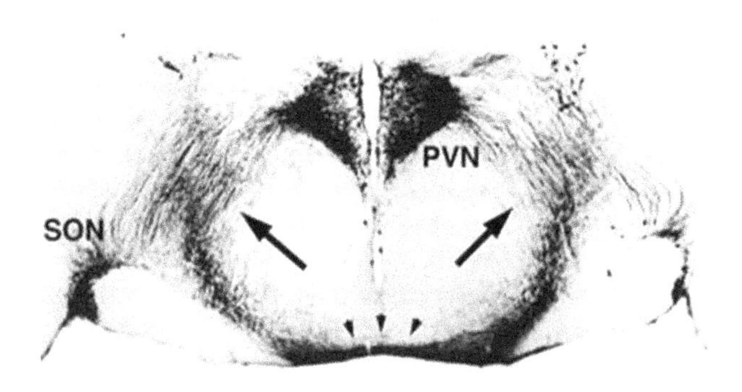

FIGURE 4-10 **Projections from the hypothalamus to the pars nervosa.** Magnocellular neurons in the paraventricular nucleus (PVN) project (arrows) ventrally to the supraoptic nucleus (SON) where they join (arrowheads at bottom of figure) fibers innervating the pars nervosa. *(Reprinted with permission from Lechan, R.M. and Toni, R., Functional anatomy of the hypothalamus and pituitary, Chapter 3B, www.endotext.org, 2012.)*

the median eminence and adenohypophysis are the hypothalamic-releasing hormones and can be identified as either **releasing hormones (RHs)** or **release-inhibiting hormones (RIHs)**, depending on whether they stimulate or inhibit tropic hormone release from the adenohypophysis (Table 4-1). Initially, these neurohormones were referred to as *factors* until their specific chemical structures were known. Some researchers still refer to some or all of them as releasing factors or release-inhibiting factors rather than hormones. These regulating hormones are mostly small peptides composed of as few as three to as many as 44 amino acids. One is simply a catecholamine (dopamine) derived from a single amino acid, tyrosine. The neurohormones associated with the pars nervosa are all very similar in structure (Table 4-4). Each is a peptide consisting of nine amino acid residues. The name *octapeptide* was used originally for these neurohormones because the two cysteine residues at positions 1 and 6 form a disulfide bridge to become cystine. Formation of this disulfide bond results in conversion of six amino acid residues into the characteristic five amino acid-ring structure with a side chain of three amino acids, hence the original name of octapeptide. However, because there are nine residues numbered 1 to 9, these peptides now are referred to as the **nonapeptide neurohormones** of the pars nervosa.

Five nonapeptides are stored in nerve terminals innervating the adult mammalian pars nervosa and synthesized in cell bodies of the hypothalamus although not all occur in the same species (see Table 4-4). These nonapeptides include the neutral nonapeptide **oxytocin (OXY)**, as well as the basic nonapeptides **arginine vasopressin (AVP)**, **lysine vasopressin (LVP)** in suiform mammals, and **phenypressin (PVP)** in macropodid marsupials (kangaroos and wallabies). In addition, **mesotocin (MST)**, which is characteristic of non-mammalian tetrapods (see Chapter 5), is found in some species as well as OXY and in others instead of OXY.

The region of the mammalian brain that controls pars distalis function consists primarily of bilateral (paired) neurosecretory nuclei, including the **suprachiasmatic nucleus (SCN)**, **ventromedial nucleus (VMN)**, **dorsomedial nucleus (DMN)**, **supraoptic nucleus (SON)**, **paraventricular nucleus (PVN)**, **periventricular nucleus (PERIV)**, and **arcuate nucleus (ARC)** (Figure 4-11). Although technically the paired nuclei of the preoptic area (POA) are not anatomically part of the hypothalamus, they usually are included in discussions of hypothalamic control because they function as part of the HPG axis. Collectively, these bilateral nuclei are responsible for producing the hypothalamic-releasing neurohormones that regulate release of tropic hormones from the hypophysis and for production of the nonapeptide neurohormones of the pars

TABLE 4-4 Vasopressin-Like Peptides

Name	Amino Acid Sequence
Vasopressin-Like Peptides	
Arginine vasopressin	C-Y-F-Q-N-C-P-R-G-NH_2
Lysine vasopressin	C-Y-F-Q-N-C-P-*K*-G-NH_2
Phenypressin	C-*F*-F-Q-N-C-P-R-G-NH_2
Arginine Vasotocin	C-Y-*I*-Q-N-C-P-R-G-NH_2
Oxytocin-Like Peptides	
Oxytocin	C-Y-I-Q-N-C-P-L G-NH_2
Mesotocin	C-Y-I-Q-N-C-P-*I*-G-NH_2

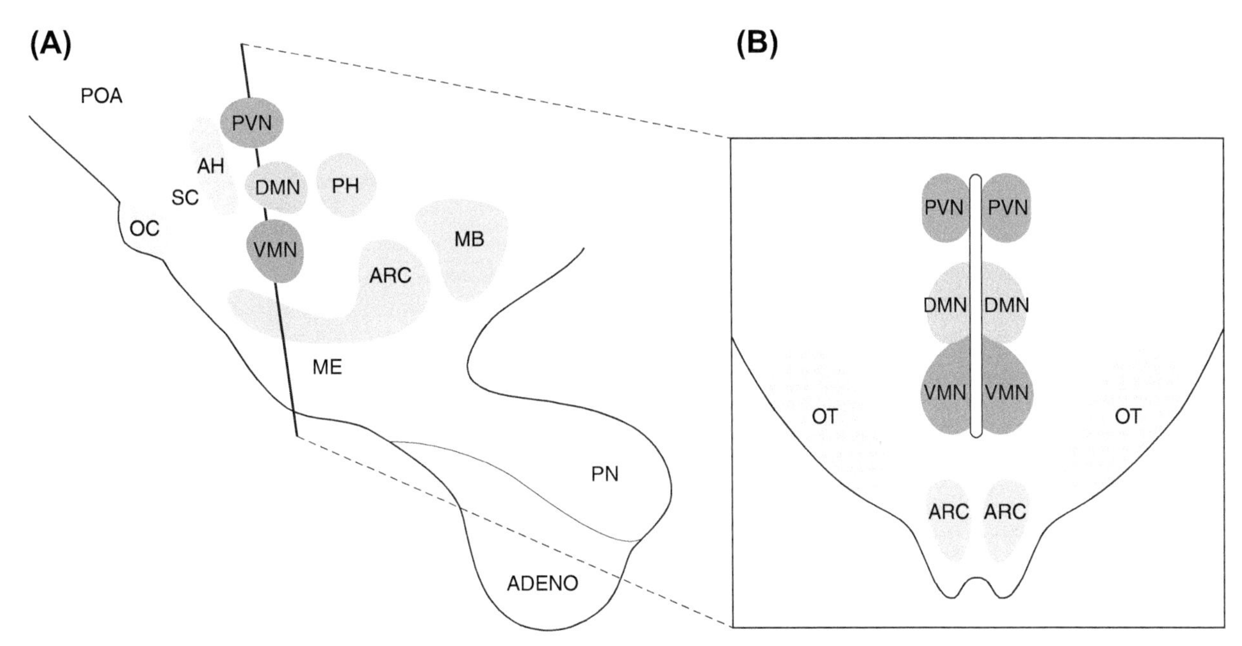

FIGURE 4-11 **Hypothalamic neurosecretory centers.** (A) Sagittal section of brain showing nuclei on one side. (B) Cross-section of brain showing paired nature of nuclei. The nuclei depicted here are as follows: AH, anterior hypothalamic; ARC, arcuate; DMN, dorsomedial; PH, posterior hypothalamic, POA, preoptic area; paraventricular, PVN; suprachiasmatic, SC; ventromedial, VMN. Abbreviations: ADENO, adenohypophysis; MB, mammillary body; ME, median eminence; OC, optic chiasm; OT, optic tract; PN, pars nervosa.

nervosa. In the HPG axis, it was noted in both male and female rats that the releasing hormone contents of the VMN, DMN, and PVN on the right side were greater than on the left side of the brain. Other studies support a functional dominance of nuclei on the right side of the brain in control of gonad function as well as for thyroid and adrenal cortex (see references provided at the end of this chapter). Although the possibilities of such functional asymmetries as shown in the rat will not be a focal point in future discussions, the student should never assume that paired structures, whether they occur in the brain or as paired structures elsewhere in the body, are always the same on both sides of the body.

Much of our initial knowledge about the functional roles for these hypothalamic NS centers was accumulated from observing the effects on pituitary tropic hormone secretion of electrical or chemical lesions of discrete hypothalamic nuclei or of localized electrical stimulation in the hypothalamus and POA as well as from studies involving implants of crystalline hormones into these regions. Some cautions for interpretation of data obtained from such studies are in order, however. The use of disruptive lesions, for example, requires careful bilateral placement of comparable lesions and leaves some uncertainty as to exactly what was destroyed by the lesions. Alterations in pituitary function following placement of lesions might involve destruction of the neurosecretory neurons that elaborate a given hypothalamic releasing hormone or may only disrupt a NS tract. Furthermore, the lesion might have damaged non-neurosecretory neurons that would normally modulate the activity of certain neurosecretory neurons. Damage to vascular elements of the median eminence might also alter tropic hormone release patterns. Ideally, secretion of all tropic hormones should be monitored following placement of a particular lesion; yet, for practical reasons, this is rarely done. Usually only one (e.g., TSH) or at most two tropic hormone systems are examined, whereas others (e.g., PRL, LH, FSH, ACTH, GH in this example) often are ignored. Finally, it is difficult to establish suitable controls for some of these procedures. In spite of such drawbacks, the use of these approaches in combination with immunocytochemical techniques has helped to establish the location of neurosecretory centers responsible for secreting each hypothalamic neuropeptide.

A. Sexual Differences in the Hypothalamus

A variety of sexual dimorphisms have been described in the mammalian brain, especially in the hypothalamus. These dimorphisms are related to differing reproductive functions by the hypopthalamic centers of males and females, and they are important in puberty and in the regulation of reproductive cycles and reproductive behavior (see Chapter 10). Androgens secreted early in development are generally responsible for establishing these dimorphisms although estrogens may play a role in females. Extensive studies in rodents have established that conversion of androgens to estrogens by the enzyme **P450 aromatase** ($P450_{aro}$; see Chapter 3) is involved in establishing the masculinization of the hypothalamus, but it is not clear whether this is the mechanism of the androgenic effect in other mammals including primates. In the **medial preoptic area (mPOA)** of the rat, there is a striking difference in the density of neurons and size of the nucleus that is larger in males (2.5 to 5× that of the female). This area is known as the **sexually dimorphic nucleus (SDN)**. Although similar differences have been observed in other mammals (e.g., guinea pigs, gerbils, ferrets, monkeys, and humans), mice do not exhibit this SDN. Studies have shown that male rats switched from a diet containing soy phytoestrogens (e.g., genistein) to a phytoestrogen-free diet exhibited a significantly reduced SDN, suggesting that exogenous estrogen exposure might be responsible for this dimorphism. A marked sexual dimorphism also has been described in the mammalian **vomeronasal organ (VNO)** and the associated accessory olfactory bulb, some regions of the bed nucleus of the stria terminalis, and the medial amygdala, with these regions being more developed in males than in females. These regions all exhibit steroid receptors and are especially responsive to pheromones in rodents.

The presence of sexual dimorphisms in humans and their associations with androgens are controversial. Anatomical studies that have demonstrated differences in the human brain are not always repeatable, and studies using brain imaging techniques of living brains are needed. No differences in the region referred to as the SDN in humans is present in children, and there is no experimental evidence demonstrating that androgens affect sexual differentiation of this region.

B. Hypothalamic-Releasing Hormones

Numerous studies have confirmed that the hypothalamus exerts a direct influence over functioning of the adenohypophysis. Microscopic observations indicate the absence of neural connections between the mammalian hypothalamus and the adenohypophysis such as seen in teleost fishes. The discovery of the portal system led to the establishment of what is now termed the **neurovascular hypothesis:** Hypothalamic neurohormones released into the portal circulation are responsible for controlling tropic hormone release from the adenohypophysis. Severing the portal connections or transplanting the pituitary to some avascular site elsewhere in the body causes marked changes in the secretory pattern of the tropic hormones. These operations usually are followed by a marked reduction in circulating levels of TSH, FSH, LH, GH, and ACTH, whereas PRL and α-MSH levels increase. These observations led to the initial interpretation that release of TSH, GH, ACTH, and the GTHs primarily is under stimulatory control (via releasing hormones) and that release of α-MSH and PRL normally is under inhibitory control (via release-inhibiting hormones). This initial interpretation soon was modified as we learned more about the many factors that influence tropic hormone release (see ahead). If the severed blood vessels of the portal system are allowed to regenerate so that blood may again flow from the median eminence to the adenohypophysis, the normal secretory patterns for the tropic hormones resume. These latter observations support strongly the neurovascular hypothesis of hypothalamic control over tropic hormone release in mammals.

In recent years, many hypothalamic regulatory hormones have been proposed and their chemical identities established. Andrew Schally and Roger Guillemin shared a Nobel Prize for the initial isolation and characterization of **thyrotropin-releasing hormone (TRH)** and **gonadotropin-releasing hormone (GnRH)**, respectively. Soon, with the advent of vastly improved biochemical techniques, other regulatory neurohormones were identified.

Each hypothalamic regulatory hormone is named for the tropic hormone it was first shown to influence and is designated according to whether it causes release (RH) or is release inhibiting (RIH). The relative importance of RHs or RIHs for the various tropic hormones differs for each tropic hormone but is consistent for different mammalian species as well as for many non-mammalian vertebrates (see Chapter 5).

C. Control of Hypothalamic Hormone Release

Release of hypothalamic regulatory hormones is influenced primarily by neural activity and negative or positive feedback mechanisms of certain hormones. The following accounts represent a generalized pattern of control in mammals, but individual species may vary significantly from this pattern.

The predominant feedback loop involves production of bloodborne hormones or metabolites resulting from the actions of tropic hormones on specific peripheral target cells (Figure 4-1). This feedback may affect neurosecretory neurons in the hypothalamus directly. Feedback may also alter the sensitivity of pituitary cells to hypothalamic RHs and RIHs or affect other neurons that innervate the hypothalamic neurosecretory neurons. Feedback effects by tropic hormones and even by RHs and RIHs have been documented, but, for simplicity, we will focus our attention on the predominant feedback mechanism described above. You will recall from Chapter 1 that most feedback is of the negative type but that enhancement of responsiveness or even positive feedback may occur, such as the estrogen induction of the midcycle GTH surge characteristic of female mammals (see Chapter 10).

BOX 4B Hypothalamic-Releasing Factors

The concept that hypothalamic factors regulate anterior pituitary hormone secretion has its beginnings in conversations that took place between Dr. Geoffrey Harris, widely considered the father of modern neuroendocrinology, and a reproductive endocrinologist named Francis Marshall when the two scientists were at Cambridge in the mid-1930s. Marshall's idea was that sensory pathways act via neuronal reflexes to regulate anterior pituitary hormone secretion—for example, timing the secretion of reproductive hormones to environmental signals of light and temperature. Starting as a medical student in the late 1930s, Harris went on to demonstrate that electrical signals precisely applied to discrete areas of the hypothalamus could elicit the secretion of luteinizing hormone, corticotropin, thyrotropin, oxytocin, and vasopressin, but electrical pulses applied directly to the anterior pituitary had no effect. Together with an anatomy colleague, John Green, Harris in 1947 proposed that hypothalamus-releasing factors reach the anterior pituitary via a "humoral relay through the hypophysial portal vessels."

In 1954, the race to discover the hypothalamic-releasing factors was begun in earnest, with the search for a corticotropin-releasing factor by Andrew Schally and Roger Guillemin (Box Figure 4B-1). This work involved purifying hundreds of thousands of sheep hypothalamic extracts on enormous separation columns 2 m in length, with one separation run sometimes using several thousand sheep hypothalami in a single run. The structure of the first hypothalamic-releasing factor, TRH, was reported in 1969. The discovery of GnRH followed soon after. The sheer cost and labor force required to conduct this work eliminated most of the competing labs in the area and Guillemin and Schally shared the Nobel prize in 1977 along with Rosalyn Yalow for the discovery of radioimmunoassay.

Roger Guillemin Roger Burgus
first structure of TRF from mass spectrometry 9 June 1969

BOX FIGURE 4B-1 **Discovery of the chemical structure of TRH.** Roger Guillemin (left) and Roger Burgus examining the structure of thyrotropin-releasing factor ([pyro]Glu-His-Pro-NH_2) as deduced by mass spectrometry, June 9, 1969, in their laboratory at the Baylor College of Medicine in Houston, Texas. *(Reprinted with permission from Guillemin, R.J.,* Journal of Endocrinology, ***184****, 11–28, 2005.)*

Ironically, it was not until after the Nobel Prize was awarded that an former student of Guillemin's, Wylie Vale, isolated corticotropin-releasing factor in 1981, nearly 30 years after work to identify this peptide had begun. The day-to-day laboratory work involved in discovering these novel peptides was documented in a 1979 book entitled "Laboratory Life: The Social Construction of Scientific Facts," by sociologists of science Bruno Latour and Steve Woolgar.

1. Hypothalamic Hormones (Factors)

Regulation of tropic hormone release is accomplished directly by hypothalamic neurohormones as illustrated in Figure 4-12. In addition, neuronal input to these NS neurons can modulate their activity as can direct feedback from circulating hormones. Most of the RHs and RIHs are peptides as are some of the neuronal neurotransmitters that influence their release. In addition, a variety of non-peptide neurotransmitters (e.g., acetylcholine, dopamine, GABA, glutamate, norepinephrine, serotonin) (see Figure 4-2) affect tropic hormone release through their effects at the hypothalamic level. A specific neurotransmitter may stimulate release of one tropic hormone and inhibit release of another depending on the receptor type present and the morphology of their synaptic connections. Conversely, two or more different molecules might justifiably be termed an RH or RIH for a particular tropic hormone.

Many pharmacological studies of nervous regulation of hypothalamic/POA neurosecretory centers have been conducted, employing neurotransmitters or drugs that either mimic (agonists) or block (antagonists) the activity of various known neurotransmitters (see Chapter 3). Studies of this type have led to identification of neurons that regulate release of individual hypothalamic releasing hormones. For example, application of dopamine to cultured pituitary cells with and without cocultured hypothalamic/POA tissue has made it possible to distinguish between the indirect stimulatory activity of dopamine on LH release via a neurotransmitter role in the hypothalamus/POA and its direct inhibitory action as a neurohormone on PRL release. Utilization of catecholamines and related drugs *in vivo* and *in vitro* also has contributed much to our understanding of neuronal regulation of hormone release. It is important to point out, however, that while pharmacological identification of neurotransmitters regulating hypothalamic/POA neurosecretory centers is important and informative, knowledge of the *neural circuits* controlling neurosecretory centers is ultimately what must be determined to understand how various

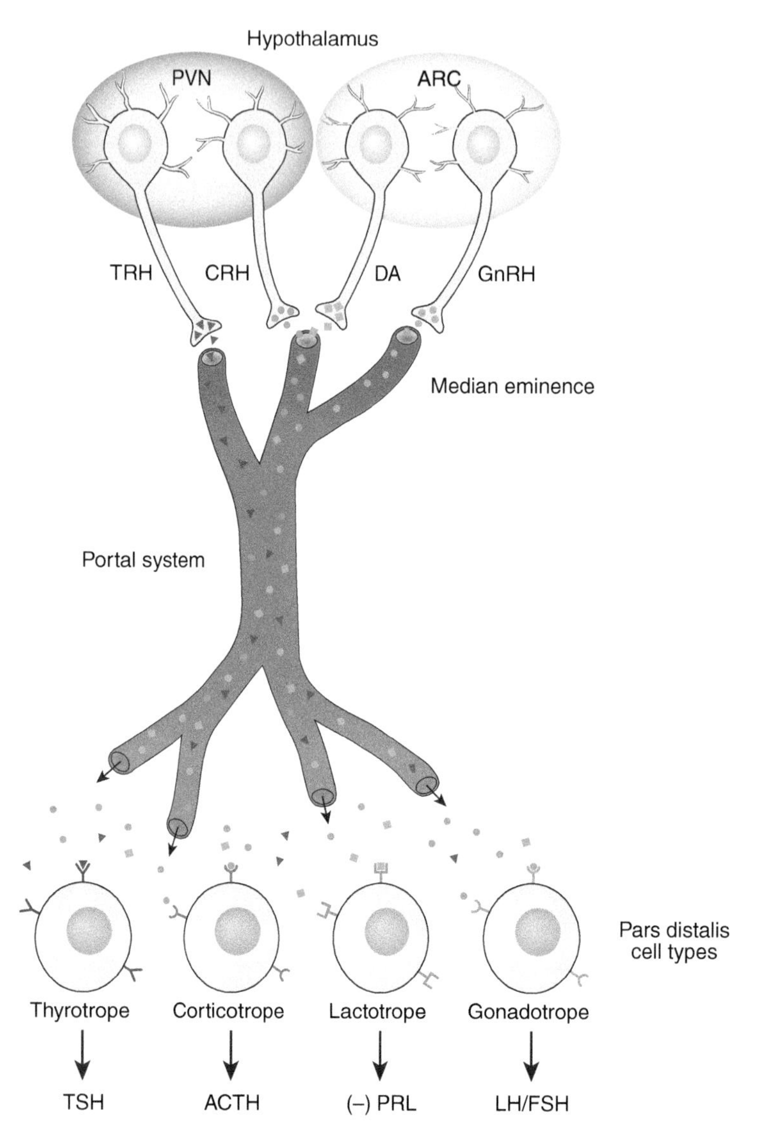

FIGURE 4-12 **Origin and targets for some hypothalamic-releasing and release-inhibiting hormones.** Note that each hypothalamic hormone travels through the portal system and binds to receptors on different pars distalis cell types, evoking tropic hormone release. Abbreviations: ACTH, corticotropin; ARC, arcuate nucleus; CRH, corticotropin-releasing hormone; DA, dopamine (prolactin release-inhibiting hormone); FSH, follicle-stimulating hormone; GnRH, gonadotropin-releasing hormone; LH, luteinizing hormone; PRL, prolactin; PVN, paraventricular nucleus; TSH, thyrotropin; TRH, thyrotropin-releasing hormone.

sensory signals relate changes in the environment to the hypothalamus–pituitary axes.

The response of neurosecretory neurons to specific neurotransmitters and neuromodulators or of pituitary cells to neurohormones is determined by the presence of specific receptors for these substances on the cell membranes of the neurosecretory neurons. For example, the inhibitory action of dopamine on PRL release mentioned above is accomplished through the binding of dopamine to receptors in the plasma membrane of the PRL-secreting cells. The ergot alkaloids, such as **ergocornine** and **ergocryptine**, can mimic the action of dopamine by binding to another receptor on the PRL cell membrane termed an α-receptor. Stimulation of this α-receptor by a PRL–RH might ordinarily evoke PRL release, but the use of these α-receptor blocking drugs (α-blockers) can completely inhibit hormone release even in the absence of the PRL–RIH.

D. Paracrine Factors in the Adenohypophysis

The discovery of many known regulatory peptides in cells of the adenohypophysis first suggested possible paracrine roles for tropic cell secretions as well as possible paracrine secretions from some of the non-granulated cell types.

Follicostellate cells, for example, have been proposed to form a chemical communication network throughout the adenohypophysis. Most of the supporting data for paracrine and sometimes autocrine functions for various peptides come from culture systems in which the density and physical relationship of cell types may be very different from conditions within the pituitary gland. For example, renin, renin substrate, and **angiotensin II (Ang-II)** have been reported from gonadotropes in the rat and from lactotropes in humans. Release of PRL is stimulated in cultures of both rat and human cells by Ang-II; however, the importance of these factors *in vivo* is still unclear.

An important paracrine regulator in the adenohypophysis is the polypeptide known as **pituitary adenylate cyclase activating peptide (PACAP)**. PACAP is a member of the secretin-glucagon family of peptides (see Table 3-4 and Figure 4-13). After translation, the PACAP prohormone is processed into two forms that functionally are indistinguishable (27 and 38 amino acids). PACAP enhances the release of certain pars distalis tropic hormones when their normal releasing hormones work through a GPCP receptor that activates cAMP production. It appears that PACAP is secreted by a variety of cell types in the pituitary in addition to the stellate cells.

A **natriuretic peptide** (see Chapter 8) known as **CNP** may be an important paracrine factor in the hypothalamus and the pituitary. CNP is related to **atrial natriuretic peptide (ANP)**, which is important in the body's effort to combat chronic hypertension. Specific receptors for CNP are found on GnRH neurons in the ARC nucleus and on gonadotropes in the pars distalis. The exact role of CNP in GTH release has not been established.

III. TROPIC HORMONES OF THE ADENOHYPOPHYSIS

Numerous bioassays have been developed for quantitatively measuring tropic hormone activity, although current techniques for measuring gene action and immunological identification procedures generally have superseded the use of bioassays. A brief survey of bioassays for tropic hormones can be found in Appendix D.

The tropic hormones are separable into three distinct chemical categories (Table 4-5). The hormones within each category exhibit considerable overlap in chemical structures (that is, amino acid sequences) and in some cases overlap in biological activities as well, especially when administered in pharmacological doses. Category I includes the glycoprotein hormones (GpHs): TSH, FSH, and LH. Each of these hormones is composed of two polypeptide subunits, each containing specific carbohydrate moieties. GH and PRL constitute the category 2 tropic hormones. Both PRL and GH are fairly large, folded polypeptide chains, and they exhibit considerable structural and some functional overlap. Category 3 includes smaller peptides: ACTH, α-MSH, LPH, and endorphins. These category 3 molecules have a common prohormone, have overlapping amino acid sequences, and exhibit some overlap in their biological actions.

In addition to pituitary tropic hormones, certain tropic hormones of similar chemical structure and biological activity are produced in the placental mammals. As many as five tropic-like hormones have been isolated from the chorionic (fetal) portion of the placenta (see Chapter 10), including **chorionic gonadotropin (CG)**, which is primarily LH-like in both structure and function, and **chorionic somatomammotropin (CS)**, which has some GH but

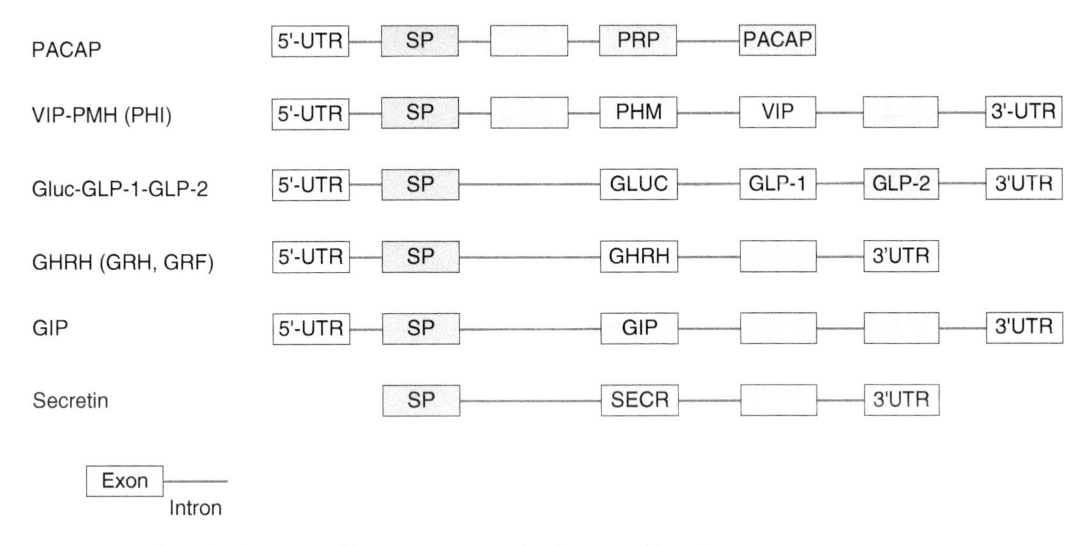

FIGURE 4-13 **Human genes for the secretin-glucagon family of peptides.** SP, signal peptide; see Appendix A or text for other abbreviations. *(Adapted with permission from Sherwood, N.M., Krueckl, S.L., McRort, J.E. Endocrine Reviews 21, 619–670, 2000.)*

TABLE 4-5 Chemical Categories of Some Placental Peptide/Protein Hormones

Category	Name	Site of Synthesis
I	Thyrotropin (TSH)	Adenohypophysis: pars distalis
	Luteinizing hormone (LH)	Adenohypophysis: pars distalis
	Follicle-stimulating hormone (FSH)	Adenohypophysis: pars distalis
	Chorionic gonadotropin (CG)	Placenta
	Chorionic thyrotropin (CTSH)	Placenta
	Menopausal gonadotropin (MG)	Adenohypophysis: pars distalis
II	Growth hormone (GH)	Adenohypophysis: pars distalis; placenta
	Prolactin (PRL)	Adenohypophysis: pars distalis; placenta
	Chorionic somatomammotropin (CS)	Placenta
III	Corticotropin (ACTH)	Adenohypophysis: pars distalis
	α-Melanotropin (α-MSH)	Adenohypophysis: pars intermedia
	β-Endorphin	Adenohypophysis: pars distalis and pars intermedia
	Chorionic corticotropin (CC)	Placenta

mostly PRL-like activity. Both a **chorionic thyrotropin (CT)** and a **chorionic corticotropin (CC)** have been isolated from human placentas and may be secreted in other mammals. Pregnant mares produce large quantities of a placental GTH that has both strong FSH-like and LH-like properties. This glycoprotein hormone is termed **pregnant mare serum gonadotropin (PMSG)**. The importance of these placental hormones in pregnancy is discussed in Chapter 10. A variant pituitary GTH that occurs in postmenopausal women is called **menopausal gonadotropin (MG)**. Human MG is basically FSH-like and is produced by the postmenopausal adenohypophysis. Large amounts are secreted in response to failure of the ovaries to produce adequate levels of estrogens.

Much of our initial knowledge concerning the structure and function of tropic hormones initially came about as a result of the availability of pituitary glands and placentas from slaughtered domestic livestock. Huge quantities of starting tissue were needed in those days to yield 1 mg of pure hormone. Modern molecular and biochemical techniques have reduced drastically the amount of tissue needed and increased the efficiency of extraction and consequently the availability of purified tropic hormones from many animal sources. Large quantities of peptide or protein hormone can now be synthesized through recombinant DNA methods or *in vitro* translation methods using cell free extracts, as long as one knows the sequence of the gene or mRNA of interest. Techniques of genetic engineering have made it possible to develop culture systems to synthesize large quantities of a specific tropic hormone for research and clinical uses.

Because there are many structural variations in these polypeptides when isolated from different animals, and some corresponding differences in biological activity, it is important to designate the source of the hormone used in experimental studies. This is especially true when using mammalian hormones in non-mammals where a molecule that performs a particular function in mammals may provide different results in a non-mammal. Investigators who study mammalian tropic hormones usually designate the source of the hormone such as bovine (cattle), ovine (sheep), porcine (pig), cervine (deer), equine (horse), caprine (goat), or murine (rodent). An additional lowercase letter preceding the abbreviation of a tropic hormone usually designates the species source. For example, bovine GH is designated bGH, GH prepared from human genes is hGH, and bullfrog GH is bfGH. This method of designating the source may cause some confusion as purified hormones from a greater number of species become available, but it is a useful shorthand for indicating the source.

The activities of the various tropic hormones first were determined by bioassays (see Chapter 3). The classical bioassays for each tropic hormone can be found in Appendix D. These biological approaches still may be used in the biochemical isolation and characterization of tropic hormones, especially in non-mammals and in studies where purified hormones are not available. Once highly purified hormones became available, radioimmunoassays (RIA) and ELISAs (see Chapter 3) were developed for the mammalian tropic hormones and now are routinely employed to measure circulating levels.

A. Category I Tropic Hormones

All of the mammalian GpH tropic hormones examined to date are composed of two peptide subunits (Figure 4-14) with an assortment of carbohydrate moieties attached. Molecular weights for these glycoproteins are about 32 kDa.

BOX 4C Antibody Specificity and Measurement of Tropic Hormones

There are a number of drawbacks to widely employing RIAs for measurement of circulating peptide/protein hormone levels. Production of antibodies against hormones purified from pituitary glands may result in an antibody that reacts against some portion of a preprohormone or prohormone that is released from the cell rather than against the circulating biologically active form. Use of such antibodies that interact with differing proportions of active and inactive hormones might yield results that do not correlate with biological bioassay data, as preprohormones and prohormones may react with the antibody but not bind to the receptors on target cells. Some peptide hormones may have different biological potencies with respect to receptor binding and activation but may bind equally well to the antibody used in the RIA or ELISA. Furthermore, the close similarities in structure among the various tropic hormones of a given category (e.g., PRL, GH, CS) may result in cross-reactivities to the antibody produced against only one hormone because all in this category have similarities in key stretches of amino acids (**epitopes**) that bind to the antibody. The extremely similar structures of GHs and PRLs within and among species make measurements by RIAs for either PRL or GH difficult, and extreme caution should be used when interpreting RIA data, especially the use of antibodies to mammalian PRL to measure PRL levels in non-mammals. There is a common heptapeptide core (seven amino acids) in α-MSH, ACTH, and LPH, and this similarity in structure is reflected clearly by overlapping biological activity. Corticotropin has considerable α-MSH activity, whereas MSH has weak ACTH activity. These overlaps in function affect interpretation of bioassayable data on α-MSH activity; that is, which peptide is actually being measured by the bioassay? Similar questions could be raised with respect to the actual specificity of α-MSH or ACTH antibodies for RIAs or ELISAs, although these days it is relatively easy to generate an antibody specifically designed to recognize a part of the MSH peptide that includes the amidated C-terminal amino acid (the amino acid in position 13 of ACTH is not amidated) or amino acids in the C-terminus of ACTH that are not present in MSH.

The specificity of the antibody for the structure of the purified hormone antigen makes it difficult to use one antibody prepared against oFSH to estimate circulating levels of FSH in another species where differences in amino acid sequences of this species' tropic hormones might result in reduced cross-reactivity to the biologically active hormone or the cross-reactivity to another glycoprotein (LH or TSH). These problems of variability and structural similarities make it absolutely essential that any RIA be validated in several ways (including by bioassay), especially if the species used for the antibody preparation is not phylogenetically close to the species in which it is being used to measure hormone levels. The development of the immunoradiometric assays of recent years has improved the selectivity of identifying specific molecules in plasma samples with less interference from closely related molecules or fragments. For example, antibodies made against the unique β-subunits of the GpH hormones (LH, FSH, TSH) avoids the complications of their all possessing the identical α-subunit (see text for details).

The biological half-lives for TSH and LH in mammals are about 60 minutes, whereas that of FSH is about 3× longer. The longer half-life for FSH is attributed at least in part to differences in its unique carbohydrate components.

Each glycoprotein tropic hormone consists of two subunits, an **α-subunit (GpHα)** and a **β-subunit (GpHβ)**. The GpHα-subunit is identical in all three adenohypophysial GpHs as well as in CG. The GpHβ-subunit is specific to each hormone and is responsible for its unique biological activity. Hence, RIA procedures that employ antibodies made against GpHβ are more accurate and show less cross-reactivity with other glycoprotein hormones. Likewise, RT-PCR or qRT-PCR methods to study changes in gene expression would employ upstream and downstream primers targeted against mRNA encoding the specific GpHβ.

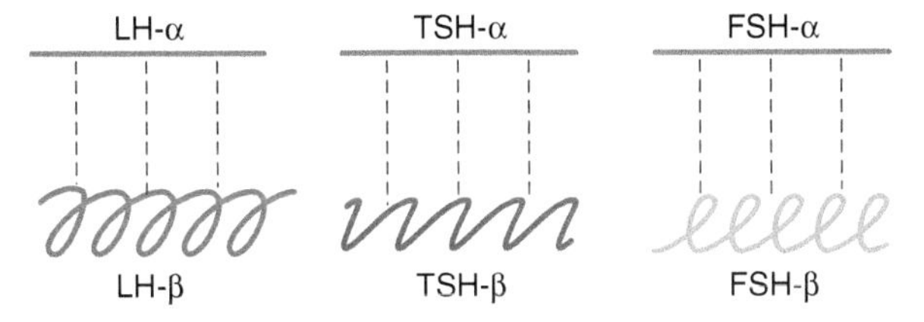

FIGURE 4-14 **Generalized structures of pituitary glycoprotein hormones.** The α-subunit is common to all three hormones but the β-subunit coded for by a different gene is unique to each hormone and is responsible for the type of biological activity shown by the mature heterodimer.

Glycoprotein subunits are each synthesized as a separate prosubunit. Each prosubunit is coded by a different gene and modified posttranslationally (including the addition of carbohydrates), and then GpHα and the appropriate GpHβ are coupled to form a heterodimer. Although the GpHα-subunits are nearly identical in amino acid composition, there is considerable variation among the GpHβ of LH, FSH, and TSH. The GpHα- and LHβ-subunit genes for hLH are located on different chromosomes (6 and 19, respectively). Both h*LHβ* and h*CGβ* genes occur on chromosome 19, suggesting that the h*CGβ* gene arose by a relatively recent duplication of the *LHβ* gene that occurred about the time mammals evolved from reptiles. Considerable overlap occurs between GpHβ-subunits of hLH and hCG, which have very similar biological activities.

The carbohydrate components represent 15 to 30% of the molecular weight of the glycoprotein subunits and the resulting heterodimer hormones. Glycoprotein hormones also show considerable specificity in their carbohydrate composition. For example, FSHs contain larger quantities of sialic acid than do the others, and the sialic acid is largely associated with the FSHβ-subunit. Sialic acid protects FSH from rapid degradation by the liver. Treatment of FSH with the enzyme neuraminidase selectively removes sialic acid, reduces the biological activity of FSH, and allows it to be degraded more rapidly.

It is relatively easy with chemical procedures to dissociate these GpHs into their respective subunits. These separated subunits have little if any biological activity when administered to animals. It is possible to recombine the dissociated subunits and restore full biological activity. A GpHα can be combined with any GpHβ, resulting in a fully active hormone characteristic of the source of the GpHβ. Thus, when a GpHα isolated from a TSH molecule is combined with FSHβ, a GpH with FSH activity results. Thyrotropes and gonadotropes produce excessive amounts of GpHα, indicating that regulation is directed at the unique β-subunit genes.

1. LH Actions

Synthesis of androgens in both males and females is caused by LH action on the testes and ovaries. It also can be caused by CGs. LH acts through a membrane-bound G-protein-based receptor (GPCR) connected to a cAMP second-messenger system (see Chapter 3). Gamete release (sperm release in males and ovulation in females) also is under the control of LH. In females, LH causes formation of the corpus luteum from the ruptured ovarian follicles remaining after ovulation and also may stimulate the corpus luteum of the ovary to secrete progesterone (see Chapter 10).

Because three pituitary tropic hormones were named for their actions in females (PRL, FSH, and LH), an effort was mounted some years ago to rename LH for its action on the androgen-producing cells that occur between the seminiferous tubules of the testis rather than for inducing corpus luteum formation (luteinization) in the female. Hence, it was suggested that LH be renamed the **interstitial cell-stimulating hormone (ICSH)** for its action on the steroidogenic interstitial cell of the testis (also called the Leydig cell) and the interstitial cell of the ovary (see Chapter 10). Although the use of ICSH occurs sporadically in the literature, LH has prevailed and is used today by most endocrinologists.

2. FSH Actions

Like LH, FSH binds to its GPCR and stimulates cAMP production as a second messenger. Whereas the major actions for LH are stimulation of androgen synthesis and gamete release in both sexes, FSH is primarily involved with gamete preparation—that is, ovarian follicle development in females and spermatogenesis in males. In females and to a lesser extent in males, FSH also stimulates the conversion of androgens into estrogens through the induction of the enzyme $P450_{aro}$ (see Chapter 3). This enzyme also is very important in converting androgens to estrogens in the male brain.

3. TSH Actions

TSH also operates via a GPCR connected to a cAMP-dependent mechanism to increase synthesis of thyroid hormones, cause release of stored thyroid hormones, and secondarily increase iodide uptake by cells of the thyroid. However, these measurements provide no consistent information concerning rates of thyroid hormone synthesis and release (see Chapters 3 and 6 for more details of TSH-stimulated thyroid events).

Humans may produce variant TSHs, one of which is associated with a pathological condition known as Graves' disease. Normal hTSH has a biological half-life of about 0.25 hours. The so-called **long-acting thyroid stimulator (LATS)** in Graves' disease has a biological half-life of 7.5 hours. Furthermore in rats, TSH causes maximal radioiodide uptake in 4 hours whereas LATS continues to produce elevated uptake 12 hours after administration. LATS is not a product of the pituitary but is an aberrant immunoglobulin that is not influenced by negative feedback of elevated thyroid hormones (see Chapter 6 for more information on Graves' disease).

B. Category 2 Tropic Hormones

Two pituitary tropic hormones, GH and PRL, plus the placental tropic hormone, CS, comprise category 2. Multiple copies of the genes for hGH and hCS are found in humans on chromosome 17, whereas multiple copies of the hPRL gene occur on chromosome 6. Hence, multiple forms may occur in the plasma of one individual. PRL and GH are large, single polypeptide hormones of similar structure (Figure 4-15) and molecular weight (about 22 to 23 kDa). Human CS is similar to both hGH and hPRL, although in other mammals CS may be structurally more like PRL than GH. There is an 85% homology between hGH and hCS as well as considerable overlap with hPRL, hence the name "somatomammotropin." CS also is known as **placental lactogen**, but this older name, though still in use, does not reflect its GH-like actions. It is estimated that duplication of the GH gene and evolution of the CS gene occurred between 85 and 100 MYBP. This is a relatively recent event compared to the separation of the GH and PRL genes, estimated to have occurred about 400 MYBP in non-mammals.

The human placenta also produces the pituitary forms of GH and PRL. Placental PRL accumulates in amniotic fluid. A smaller (16 kDa) variant of pituitary PRL also has been isolated from the rat placenta. A 20-kDa variant of

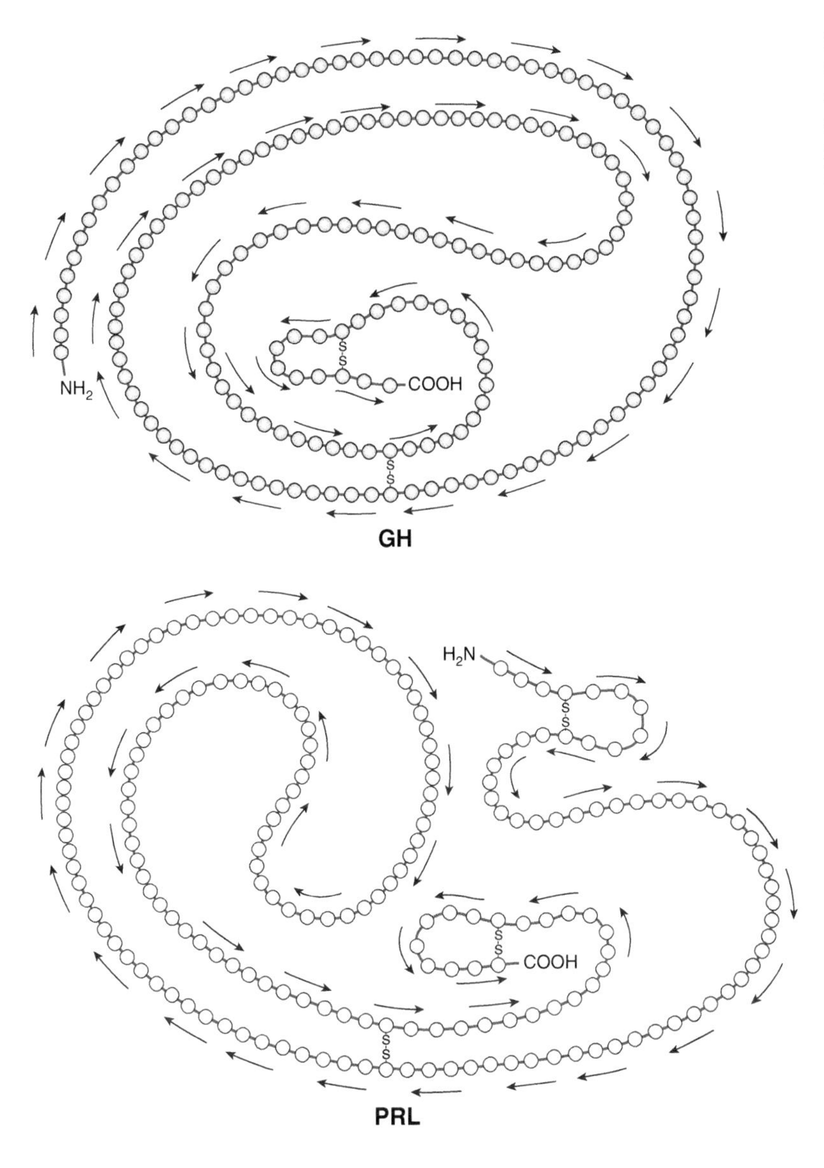

FIGURE 4-15 **Comparison of growth hormone (GH) and prolactin (PRL).** Both hormones are of comparable size, exhibit considerable overlap in amino acid sequence, and may have similar actions in some systems. GH typically has two disulfide bonds, whereas PRL typically has three.

pituitary hGH is screted during the second half of pregnancy.

Both GH and PRL appear in the circulation as monomers as well as dimers (e.g., "big" GH) or oligomeres (e.g., "big-big" GH). These multiple forms are measureable by RIA but the monomers have greater biological activity. In addition, GH, at least in humans and rabbits, is known to interact with a plasma **GH-binding protein**, complicating further the picture of circulating levels. Because of the heterogeneity of category 2 hormones, all references to GH or PRL are to the normal monomeric forms unless indicated.

Because of their structural similarity, GH and PRL produce a number of common effects on osmoregulation (renal function, intestinal fluid absorption), selective tissue growth (prostate gland, sebaceous gland), lactation, and other processes. However, PRL and hCS, unlike GH, only have weak effects on body growth and metabolism.

Membrane receptors for GH and PRL are monomeric proteins that span the cell membrane only once, yet they lack innate enzymatic activity (see Chapter 3). Neither PRL nor GH receptors are activate adenylyl cyclase. Both GH and PRL receptors undergo dimerization and then interact with a **Janus kinase (JAK)** and a transcription factor called **signal transducer and activator of transcription (STAT)** protein. The JAK is an intracellular tyrosine kinase that phosphorylates STAT, allowing the latter to enter the nucleus and alter gene transcription. Other proteins, such as **mitogen-activated protein kinase (MAPK)**, may be phosphorylated and cause other effects.

1. GH Actions

Growth hormone is often described as a protein anabolic hormone because it stimulates incorporation of amino acids into proteins and has a negative effect on nitrogen excretion. Growth hormone represents about one-half of the total hormone content of the human adenohypophysis, which emphasizes its importance in adults as well as during the years of maximal growth. It has been characterized chemically as a protein composed of 191 amino acids (MW = 21.5 kDa) having a biological half-life in blood of 20 to 40 minutes. hCS also consists of 191 amino acids, of which 161 are identical to those in hGH; yet, as mentioned earlier, hCS has rather low GH activity. Although GH for clinical purposes is produced using recombinant DNA technology today, crude preparations of GH from human cadavers were used to treat patients as recently as the mid-1980s and were linked to isolated cases of Creutzfeldt–Jacob disease due to contamination with brain tissue and disease-causing agents called *prions*.

Crude GH preparations consist of a collection of protein isohormones produced by multiple genes. Each form is thought to have its own actions, and collectively they produce all the effects normally attributed to pituitary GH activity. The gene responsible for synthesis of the 21,500-dalton form of GH has been cloned and inserted successfully into the genome of mice. Growth rates of mice with the inserted genes and growth rates of their offspring are about twice that of normal mice. This approach to GH therapy has important implications for future treatment of GH-based growth deficiencies in humans.

Growth hormone stimulates transport of amino acids into cells and stimulates protein synthesis, especially by skeletal muscle cells. It cooperates with insulin to channel amino acids, fatty acids, and carbohydrates into storage following a meal. Furthermore, GH becomes an important regulator of blood glucose and amino acid utilization in the absence of insulin during short-term and long-term starvation.

Circulating levels of hGH are highest during the period of maximal growth (ages 2 to 17 years). A daily secretory rhythm becomes established at about 4 years of age and continues throughout adult life. This pattern of GH secretion is both irregular and spontaneous, depending upon the physiological state of the individual, but episodes of GH release are frequently correlated with the onset of deep sleep.

Optimal growth-promoting actions of GH are obtained in hypophysectomized animals only when thyroid hormones are administered together with GH. This relationship between thyroid hormones and GH has been described as a synergism; that is, the growth response elicited by combined therapy with thyroid hormones and GH in hypophysectomized animals is greater than predicted by adding together the responses obtained with each hormone administered alone. Either thyroid hormones or GH will reinitiate some growth in hypophysectomized animals, but complete resumption of normal growth requires combined therapy. Furthermore, intact animals that exhibit thyroid deficiencies grow slowly and abnormally (see Chapter 6).

Thyroid hormones may influence synthesis of GH in intact rats but act peripherally to enhance GH potency in hypophysectomized animals. Thyroid hormones maintain a "responsive state" in target cells so they are more sensitive to GH and other regulators (see Chapter 6 for more details of this "permissive" effect).

The effects of steroid hormones on growth are complex. Androgens and estrogens can increase the responsiveness of human tissues to hGH but to a lesser extent than do thyroid hormones. The mechanism of this steroid effect is not understood. Steroids, especially androgenic ones, have important effects on amino acid and carbohydrate metabolism unrelated to the roles of GH (see Chapter 12). Androgens are known to stimulate protein synthesis and hypertrophy of skeletal muscle, and estrogens selectively increase protein synthesis in the uterus. Conversely, the increase in androgens and estrogens associated with the onset of puberty causes cessation in proliferation of the epiphyseal plates at the ends of long bones of the appendicular skeleton and render these tissues unresponsive to GH. This results in a permanent cessation of growth in stature.

Direct metabolic actions of GH on protein synthesis, amino acid transport, and lipolysis have been reported in several tissues. However, these growth effects of GH are mediated indirectly by the GH-stimulated production of two peptide regulators in the liver or, in some cases, directly in target tissues. These peptides were first called **sulfation factors** because of effects on incorporation of sulfate into cartilage during GH-stimulated cartilage growth, a phenomenon that could not be invoked by direct application of GH to cartilage cells *in vitro*. Later, they became known as **somatomedins**, as they mediated the actions of the somatotropic hormone, GH. We now know that these peptide growth stimulators are structurally related to insulin and have some insulin-like activity, in addition to their growth-promoting actions, due to their ability to bind to the insulin receptor. Later, they acquired the names of **insulin-like growth factors (IGF-I, IGF-II)**. IGF-II is secreted primarily during fetal growth, and IGF-I is secreted primarily in children and adults. The IGF-I receptor is a tyrosine kinase membrane receptor and acts by phosphorylating a variety of proteins in different target cells. The liver, then, can be considered a target endocrine gland for GH because it synthesizes and releases IGFs into the circulation, thus constituting the HPH axis. IGFs are transported in the blood while complexed to specific plasma **IGF-binding proteins**. In other target tissues, IGFs synthesized by GH target cells may be necessary for producing GH-linked effects in these targets (e.g., in cartilage and bone). Levels of both IGF-I and IGF-binding proteins are depressed in hypothyroid patients and elevated in hyperthyroid patients, indicating that thyroid hormones can affect GH actions through their actions at the liver as well as at other GH target cells.

In adult mammals, IGF-I binds to a receptor that is similar to the receptor for insulin and that can even bind insulin weakly. In contrast, IGF-II and its receptor are primarily a fetal growth factor with a unique receptor, and it may be secreted under the influence of hCS in humans. IGF-II has only weak insulin-like activity and does not bind well to adult insulin or IGF receptors. However, both IGFs are powerful mitogens in their appropriate target cells (i.e., they stimulate mitosis). Circulating IGF-I in humans increases at about 6 to 8 years of age and peaks during puberty. Lower, relatively constant levels of IGF-I are characteristic of adults.

2. *Prolactin Actions*

Prolactin consists of a single chain of 199 amino acids (23 kDa) and, like GH, occurs as multiple isohormones. It produces a variety of distinctive actions in animals, including effects associated with reproduction, growth, osmoregulation, and the integument (Table 4-6). Furthermore, PRL may produce synergistic actions with ovarian, testicular, thyroid, and adrenal hormones. The best-known action for PRL is the lactogenic effect on the mammary gland of females for which the hormone was named. PRL stimulates DNA synthesis, cellular proliferation, and the synthesis of milk proteins (casein and lactalbumin), free fatty acids, and lactose by the glandular epithelium of the mammary gland. hCS from the placenta produces a similar effect.

In some species (e.g., rat, sheep), PRL may influence the synthesis of progesterone by the corpus luteum of the postovulatory ovary. This action was responsible for the older name for PRL, **luteotropic hormone (LTH)**. There also is evidence in male mammals for effects of PRL on certain sex accessory structures. These reproductive actions of PRL are discussed in Chapter 10 with respect to the overall regulation of reproduction in mammals.

Like the situation for GH, PRL actions on the mammary gland and possibly on other targets involve an interaction with additional hormones. Estrogens favor cell proliferation and growth of the mammary gland, making the mammary more responsive to PRL. Glucocorticoids also potentiate the actions of PRL in all species examined. Progesterone inhibits PRL actions on the mammary gland and can block lactogenesis. One hypothesis suggests that progesterone competes for glucocorticoid binding and/or blocks gene activation by glucocorticoids. The stimulatory actions of insulin on the mammary gland may be related to its IGF-like activity that mimics an action of PRL.

C. Category 3 Tropic Hormones

This category comprises several hormones derived from the same precursor, a prohormone known as **proopiomelanocortin (POMC)** and includes ACTH, α-MSH, LPH, and the EOP β-endorphin. POMC-related peptides are found in cells of the brain, the pars distalis, and the pars intermedia

TABLE 4-6 Prolactin Action in Mammals

Actions related to reproduction
- Mammary development and lactation
- Preputial gland size and activity
- Synergism with androgen on male sex accessory glands
- Luteotropic in rodents
- Fertility in dwarf mice
- Increased androgen binding in human prostate
- Stimulation of glucuronidase activity in rodent testis
- Parental behavior
- Decreased copulatory activity in male rabbits
- Advanced puberty in rats
- Vaginal mucification in rats
- Antiovulatory and antiluteinizing actions in rats
- Relaxation of uterine cervix in rats
- Reduced catabolism of progesterone by rat uterus
- Inhibition of myometrial contractions
- Increased estradiol binding by rat uterus
- Deceased GHT release

Actions related to growth and development
- Mammary development
- Sebaceous and preputial gland growth
- Hair growth
- Erythropoietic actions
- Renotropic actions
- Spermatogenetic actions
- Male sex accessory development

Actions related to water and electrolyte balance
- Lactation
- Inctreased Na^+ retention at renal level
- Corticotropic

Actions on integumentary structures
- Mammary development and lactation
- Sebaceous and preputial gland size and activity
- Hair maturation

Actions on steroid-dependent targets or synergisms with steroid
- Mammary growth (ovarian steroids)
- Milk secretion (corticosteroids)
- Sebaceous and preputial gland secretion (gonadal and cortical steroids)
- Growth and secretion of male sex accessory glands (androgens)
- Luteotropic action (estrogens?)
- Renal Na^+reabsorption (aldosterone?) and renotropic action (androgens)
- Spermatogenesis (androgens)
- Advanced puberty (gonadal steroids)
- Hair growth (androgens corticosteroids)
- Vaginal mucification in rats (estrogen and progesterone)

(when this pituitary lobe is present). All of the category 3 molecules produced in the pituitary are a result of variations in posttranslational processing of POMC, and some products exhibit considerable overlap in their biological activities due to possession of similar amino acid sequences (see Figure 4-16).

The prohormone POMC is differentially cleaved into different end products through the action of two different prohormone convertase (PC) enzymes, PC1 and PC2, which are differentially expressed in corticotropes of the pars distalis and melanotropes of the pars intermedia (Figure 4-16). In corticotropes, POMC is cleaved to produce ACTH (39 amino acids), an *N*-terminal **16K-fragment** (MW = 16,000) with no known biological activity, and a large form (91 amino acids) of LPH known as β-lipotropin, or **β-LPH,** representing the C-terminal portion of POMC. β-LPH may be cleaved again to form a 58-amino-acid fragment called **γ-LPH** (consisting of residues 1 to 58 of β-LPH) and **β-endorphin** (residues 61 to 91 of β-LPH). In melanotropes of the pars intermedia, ACTH is cleaved further, acetylated on the *N*-terminus, and amidated on the *C*-terminus to yield **α-MSH** (*N*-acetyl ACTH 1-13 amide) and a **corticotropin-like peptide** called **CLIP** (residues 18 to 39 of ACTH). In the brain, **proenkephalin** (enkephalin, meaning "in the head") is the precursor for either the opioid pentapeptide **met-enkephalin** or the opioid pentapeptide **leu-enkephalin,** which differ only at position 5. **Dynorphin** is a larger opioid peptide occurring in two forms (13 or 16 amino acids) derived from a POMC-like prohormone called **prodynorphin** (also known as proenkephalin B). There are two smaller, partly homologous peptide versions of the dynorphins called **α-** and **β-neoendorphins.** The enkephalins, dynorphins, and endorphins all have the same four or five *N*-terminal amino acids that are important for binding to opiate receptors.

In the pars intermedia melanotropes, additional post-translational modification of POMC end products occurs after prohormone convertase action on POMC. The *C*-terminus of α-MSH and β-endorphin are amidated by the enzyme **peptidylglycine alpha-amidating monooxygenase (PAM)**. PAM and its isoforms also play an important role in amidating neuropeptides in hormone- and neurohormone-producing cells through the body. The *N*-termini of α-MSH and β-endorphin are acetylated by an interaction with *N*-acetyltransferase and acetyl coenzyme A. Virtually all of the α-MSH and β-endorphin secreted by melanotropes is acetylated prior to release. Interestingly, *N*-acetylation of β-endorphin eliminates the ability of this peptide to bind to mu (morphine) opioid receptors, thus eliminating its potency as an analgesic peptide.

1. Melanotropin (α-MSH)

In mammals, the epidermal melanin-producing cell is the melanocyte that synthesizes melanin under the influence of α-MSH but extrudes it into the extracellular compartment

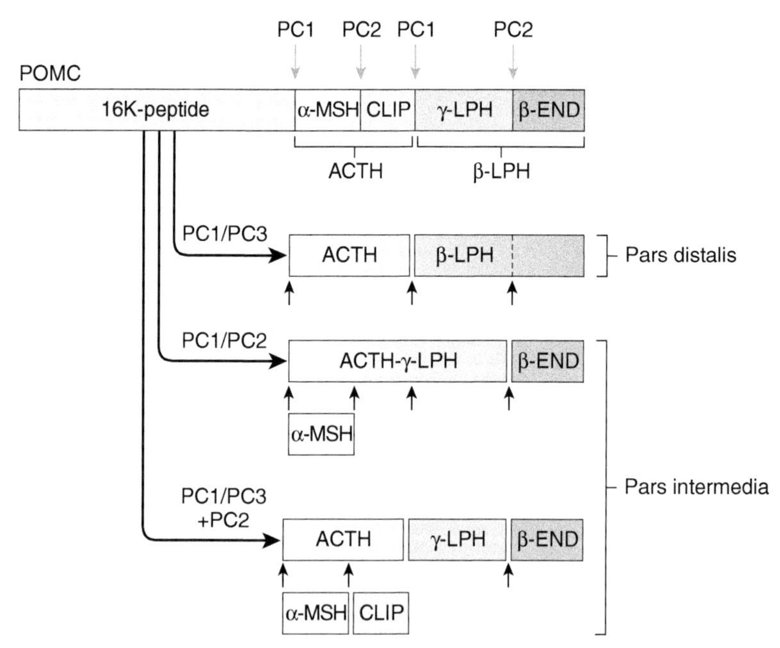

FIGURE 4-16 **Fates of proopiomelanocortin (POMC) in pituitary cells.** The convertase enzyme PC1 (=PC3, PC1/PC3) hydrolyzes POMC in corticotropes of the pars distalis to yield mainly corticotropin (ACTH) and β-lipotropin (β-LPH). Corticotropes may also release β-endorphin (β-END) and γ-LPH from the β-LPH fraction, perhaps through action of PC1 or presence of PC2 in some cells. In some melanotropes within the pars intermedia, PC2 separates the major fragment of ACTH into α-MSH (melanotropin), a major inactive fragment consisting of part of ACTH and γ-LPH, and releases β-END. If PC1 and PC2 are both active in the melanotrope, the result is α-melanotropin (α-MSH), corticotropin-like peptide (CLIP), γ-LPH, and β-END. The 16K-peptide is inactive but in some groups contains the sequence known as γ-MSH. *(Adapted with permission from Tanaka, S.,* Zoological Science*, **20**, 1183–1198, 2003.)*

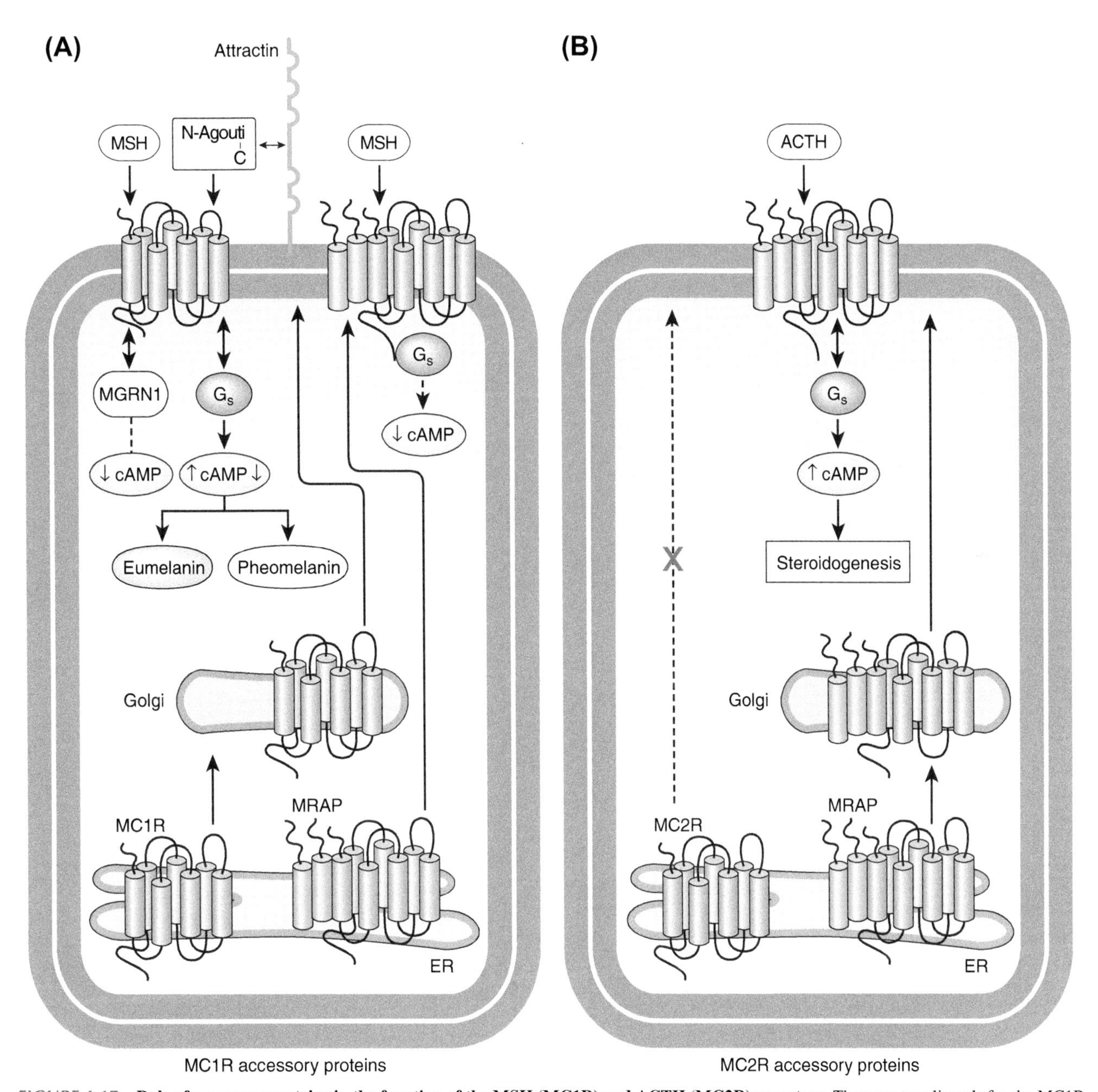

FIGURE 4-17 **Role of accessory proteins in the function of the MSH (MC1R) and ACTH (MC2R) receptors.** There are two ligands for the MC1R (A): MSH which promotes dark (eumelanin) synthesis and agouti signaling protein (Agouti), which promotes light (pheomelanin) production. Agouti protein interaction with the MC1R requires a separate accessory protein, attractin. MC1R signaling is inhibited by the cytosolic protein mahogunin ring finger-1 (MGRN1). MSH receptors can be sent to the plasma membrane without the accessory protein (melanocortin receptor accessory protein, MRAP), but when MRAP is coupled with the MSH receptor the potency of MSH is actually reduced. In contrast, MRAP is required for transfer of the ACTH receptor to the plasma membrane (B). *(Adapted with permission from Cooray, S.N. and Clark, A.J.,* Molecular and Cellular Endocrinology, ***331****, 215–221, 2011.)*

where it is accumulated in **keratinocytes** (keratin-containing epithelial cells). α-MSH acts on the **melanocortin-1 receptor (MC1R)** to stimulate cAMP and melanin synthesis in melanocytes (Figure 4-17). Animals that change from a white winter coat to a brown summer coat employ the services of α-MSH to stimulate melanin production for the summer coat. Hypophysectomy of the short-tailed weasel during the winters causes the summer coat to be white like the winter coat. Treatment of hypophysectomized weasels with either α-MSH or ACTH (which has some inherent α-MSH-like action) is sufficient to cause regrowth of the normal brown summer coat. Interestingly, coat or pelage color can be dramatically affected by the interaction of an additional circulating protein called the **agouti signaling peptide**. This 131 amino acid protein was discovered in 1994 and is found circulating in the blood where it acts as an inverse agonist (Chapter 3) on the MC1R to inhibit melanin synthesis, instead favoring the production of the yellow pigment pheomelanin (Figure 4-17). Animals over-expressing the *agouti* gene and subsequently overproducing agouti signaling peptide tend to have yellow fur color and also tend to be obese, as a related form of the agouti protein blocks the appetite-suppressing effects of α-MSH in the hypothalamus (see Chapter 12).

The adenohypophyses of all vertebrates tested by bioassay (see Appendix D) possess α-MSH activity, including some mammals and all birds that lack a pars intermedia. This activity may reside in ACTH or possibly in LPH that also contains some MSH-like sequences (see ahead). As mentioned above, α-MSH and CLIP are released following the hydrolysis of ACTH in melanotropes. α-MSH usually is acetylated prior to release, slowing its degradation and increasing its biological activity. It is not possible to distinguish MSH-like activity caused by α-MSH or by ACTH or ACTH fragments that also bind readily to α-MSH receptors and activate them. The physiological roles for α-MSH in pigmentation are not known in birds or in most mammals, but a role in feeding behavior has been described (see Chapter 12).

Two additional forms of MSH have been isolated from the mammalian pars intermedia. One of these forms, **β-MSH** is found within the LPH sequence and if separated from the rest of the peptide will exhibit MSH activity. Similarly, a **γ-MSH** occurs as part of the LPH sequence of amino acids. It is not clear that either of these melanotropins is of any physiological relevance because under normal circumstances they do not appear to be released as free peptides into the circulation. Thus, α-MSH probably is the true melanotropin. Consequently, only α-MSH is discussed below, but the reader should be aware that the other peptide forms bind to the same receptors as α-MSH, although with lower affinity.

2. Corticotropin (ACTH)

Corticotropin acts on a G-protein-coupled receptor called the **melanocortin-2 receptor (MC2R)** and stimulates the adrenal cortex (see Chapter 8) to secrete glucocorticoids (cortisol and/or corticosterone), hormones that alter protein and carbohydrate metabolism (see Chapters 8 and 12). ACTH purified from several mammalian sources (e.g., bovine, porcine, ovine, human) consists of 39 amino acids in a single peptide chain with a molecular weight of about 4500. Amino acids 1 to 23 of ACTH have full biological activity at the MC2R, 1 to 19 have 80% of full activity, but the fragment 1 to 16 has very little ACTH biological activity. Amino acids 24 to 39 are obviously outside of that region of the molecule responsible for its biological activity. All ACTH fragments containing residues 1 to 13 also have α-MSH activity because of the 'message' sequence His-Phe-Arg-Trp. Hence, although CLIP has considerable amino acid homology to part of the intact ACTH molecule, it has no ACTH-like or α-MSH-like biological activity because it lacks the essential first 13 amino acids. The presence of multiple-sized fragments of $ACTH_{39}$ in the circulation has implications for the efficacy of RIA procedures for both ACTH and α-MSH with antibodies that may or may not have overlapping affinities (see also Box 4C). The amino acid sequences of several mammalian ACTHs are compared in Table 4-7.

3. Lipotropins (LPHs)

In addition to its role as a precursor for endorphins, LPHs have been proposed as hormones that stimulate lipolysis in adipose tissue (that is, hydrolysis of fats to free fatty acids and glycerol). Lipotropins, presumably of pituitary origin, have been identified in the systemic circulation, but levels of circulating LPHs have not been linked to observed changes in lipid metabolism, leaving open the question of any physiological role for LPHs. Other lipolytic hormones appear to be much more potent than the LPH peptides, further questioning their importance as lipolytic factors *in vivo* (see Chapter 12). LPHs may have some importance as sources for the production of endorphins (see ahead).

4. The Endorphins and Enkephalins

Morphine is an opiate analgesic (pain-killing) drug that binds to mu opiate receptors in the central nervous system. Scientists postulated that there also would be endogenous compounds that produce analgesic opiate-like (morphine)

TABLE 4-7 Structural Variation of Mammalian ACTHs

Source of ACTH	1	....	25	26	27	28	29	30	31	32	33	....	39
Porcine	Ser	....	Asn	Gly	Ala	Glu	Asp	Glu	Leu-	Ala	Glu	....	Phe
Ovine	Ser	....	Asp	Gly	Ala	Glu	Asp	Glu	Ser	Ala	Gln	....	Phe
Bovine	Ser	....	Asn	Gly	Ala	Glu	Asp	Glu	Ser	Ala	Gln	....	Phe
Human	Ser	....	Asn	Gly	Ala	Glu	Asp	Glu	Ser	Ala	Glu	....	Phe

effects on the central nervous system. A search for endogenous analgesics has resulted in identification and chemical characterization of two groups of EOPs. The larger EOPs include the dynorphins, β-endorphin, and some *C*-terminal hydrolysis products of β-endorphin that are acetylated at the *N*-terminal end. However, as mentioned previously, these additional alterations to β-endorphin markedly reduce its analgesic properties. The pentapeptide enkephalins also bind to opioid receptors. The distribution of endorphins in the pituitary and central nervous system parallels that observed for ACTH and LPHs, indicating that they are products of POMC hydrolysis. The enkephalins and dynorphins are produced from different prohormones by separate sets of neurons. Painful stimuli elevate levels of endorphins and enkephalins in the CSF, and they appear to exhibit the features required for endogenous opiate-like agents. The endorphins function as neuromodulators or neurotransmitters within the central nervous system through their morphine-like actions. The action of morphine, a non-peptide, is blocked by closely related pharmaceuticals such as **naloxone** (Figure 4-18). The effects of endorphins also are blocked by naloxone, implying closeness in mechanisms of action for morphine and the endorphins that bind to similar receptors. Three types of opioid receptors have been identified with differing affinities for the various opioids (see Table 4-8).

In addition to their involvement with pain perception, endorphins influence release of neurotransmitters affecting tropic hormone release and can inhibit OXY release (see Figure 4-19). When endorphins bind to mu opiate receptors they cause efflux of K^+, thereby hyperpolarizing the neuron, so in almost all cases they inhibit neuronal activity. The possible roles for endorphins and enkephalins as regulators of behavior, especially as related to painful stimuli, and their possible endocrine implications represent an exciting area of neuroendocrinology.

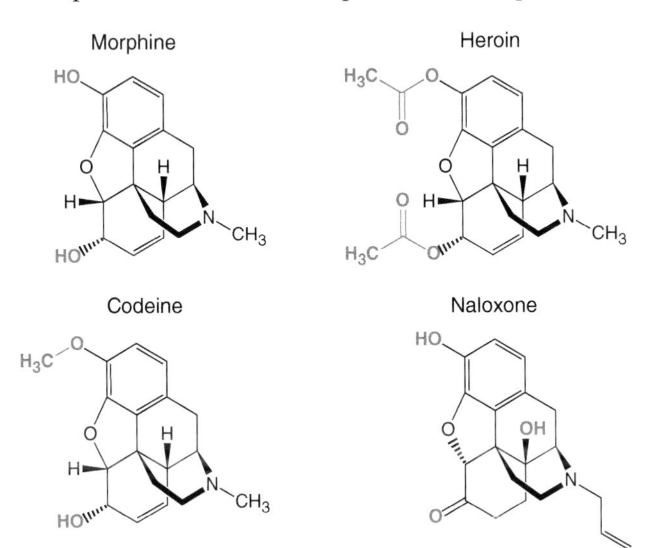

FIGURE 4-18 **Chemical structures of opiate and opiate receptor antagonist (naloxone) drugs.** Three common opiates (morphine, heroin, codeine) and the opiate receptor antagonist naloxone differ according to the groups (red) attached to the carbon rings.

TABLE 4-8 **Affinity of Opioids for Major Opioid Receptor Types**

	Receptor Types		
Opioid	**μ Receptor (mu)**	**δ Receptor (delta)**	**κ Receptor (kappa)**
Morphine	High	Low	None
Naloxone (antagonist)	High	Low	None
β-Endorphin	High	Low	None
Enkephalins	Low	High	None
Dynorphin	None	Low	High

IV. REGULATION OF TROPIC HORMONE SECRETION IN MAMMALS

The nature and actions of the various RHs and RIHs involved, what the major feedback loops are, and how other neural factors may influence secretion of a tropic hormone are discussed ahead for each of the tropic hormones. In addition, an overall scheme that summarizes or models each tropic hormone regulatory system is provided (see Figure 4-20).

A. Regulation of Thyrotropin Secretion

Release of TSH is under stimulatory control as evidenced by the lack of TSH release following disruption of the portal vessels or explanting of the adenohypophysis. The major factor controlling release is the hypothalamic tripeptide known as **TRH** or **thyroliberin**. The TRH prohormone may have from four to seven copies of the TRH peptide embedded within itself. There is no confirmed thyroid release-inhibiting hormone, but a number of other hypothalamic factors may be involved in regulating TSH release. The major factors controlling TSH release are summarized in Table 4-1.

1. Thyrotropin-Releasing Hormone (TRH)

The tripeptide TRH was the first hypothalamic regulatory hormones to be identified chemically (see Box-4B). It is produced in the neurosecretory neurons of many nuclei but in highest concentration in the PVN that sends TRH-immunoreactive fibers to the median eminence. TRH appears in the portal system blood following electrical stimulation of the appropriate regions of hypothalamus, and causes release of

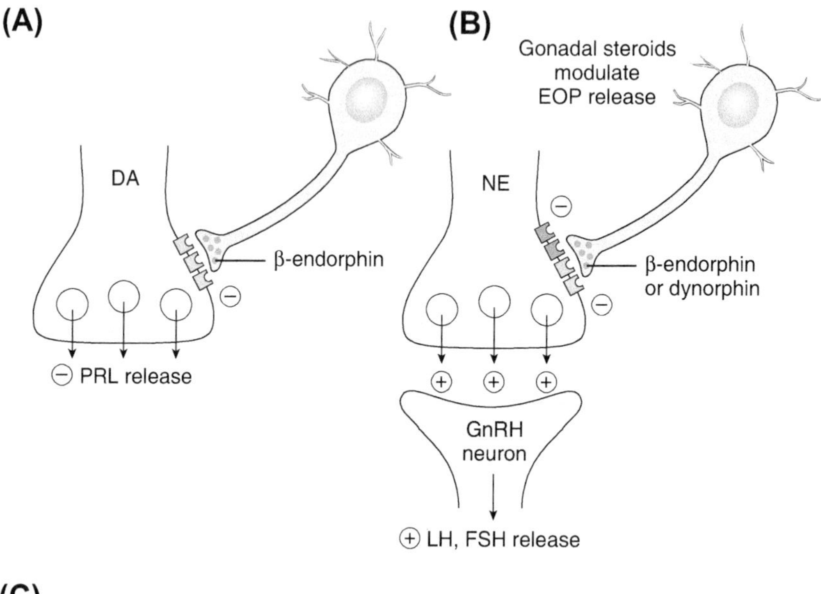

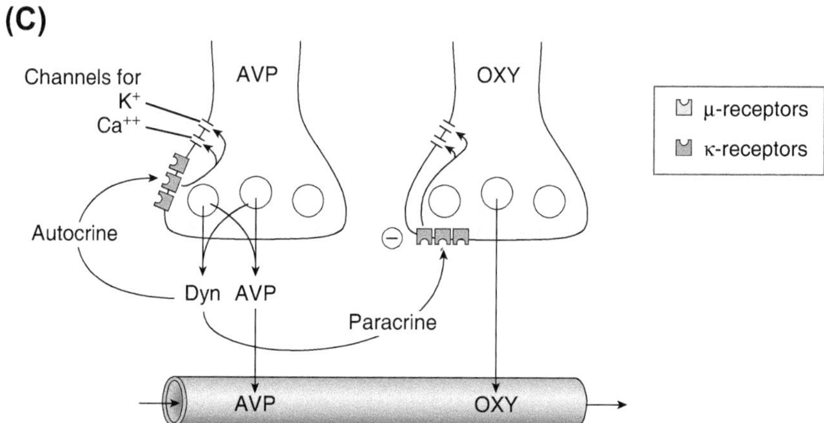

FIGURE 4-19 **Endogenous opioid peptides (EOPs) as neuromodulators.** EOPs affect hormone release in at least three ways as exemplified with specific examples here. They can (A) increase PRL release by blocking the release of dopamine (DA) which normally blocks PRL release; (B) inhibit norepinephrine (NE) stimulation of GnRH release; or (C) prevent oxytocin (OXY) release when arginine vasopressin (AVP) is being released from the pars nervosa. Dyn, dynorphin. *(Adapted from Brown R.E. "An Introduction to Neuroendocrinology," Cambridge University Press, 1994.)*

TSH *in vivo* and *in vitro* from the adenohypophysis. TRH binds to a GPCR that works through inositol trisphosphate (IP_3) to activate phosphokinase C (PKC) (see Chapter 3). In turn, PKC phosphorylates transcriptions factors, such as Pit-1, that can activate transcription of the *TSHβ* gene. IP_3 also elevates intracellular Ca^{2+} and facilitates release of TSH from the thyrotrope. A second neurohormone, **vasoactive intestinal peptide (VIP)**, stimulates the cAMP second-messenger system and also activates TSH secretion.

Extrahypothalamic TRH also is present in other brain regions, the spinal cord, the pineal gland, and the neurohypophysis as well as in some other tissues. The common occurrence of TRH outside the hypothalamus and its presence in extrahypothalamic regions of the nervous system of mammals, non-mammalian vertebrates, and even invertebrates has led to the suggestion that TRH may also function as a neuromodulator or neurotransmitter. Administration of synthetic TRH causes depression of firing in certain brain neurons, and pituitary-like TRH receptors have been demonstrated in many brain areas.

The biological half-life for TRH in peripheral blood is very short (e.g., 2 minutes in mice) apparently because peptidases in the blood rapidly inactivate TRH. Were it not for posttranslational modifications of both the *C*- and *N*-terminal amino acids that slow peptidase degradation of TRH, this tripeptide would be destroyed even more rapidly (a pyroglutamate on the *N*-terminal end and an amidated *C*-terminal end).

2. Other Neural Factors Affecting TSH Secretion

Both stimulatory and inhibitory neural control of TRH release occurs in mammals. Experimental studies in rodents have demonstrated that TRH neurons in the PVN are innervated by dopaminergic and norepinephrine-secreting neurons. Stimulation of norepinephrine-secreting neurons can evoke release of TRH, whereas dopamine is an inhibitor of TRH release. The stimulatory role of norepinephrine in humans is supported but not confirmed. Serotoninergic fibers have been shown to inhibit TRH release although some studies indicate a stimulatory role for

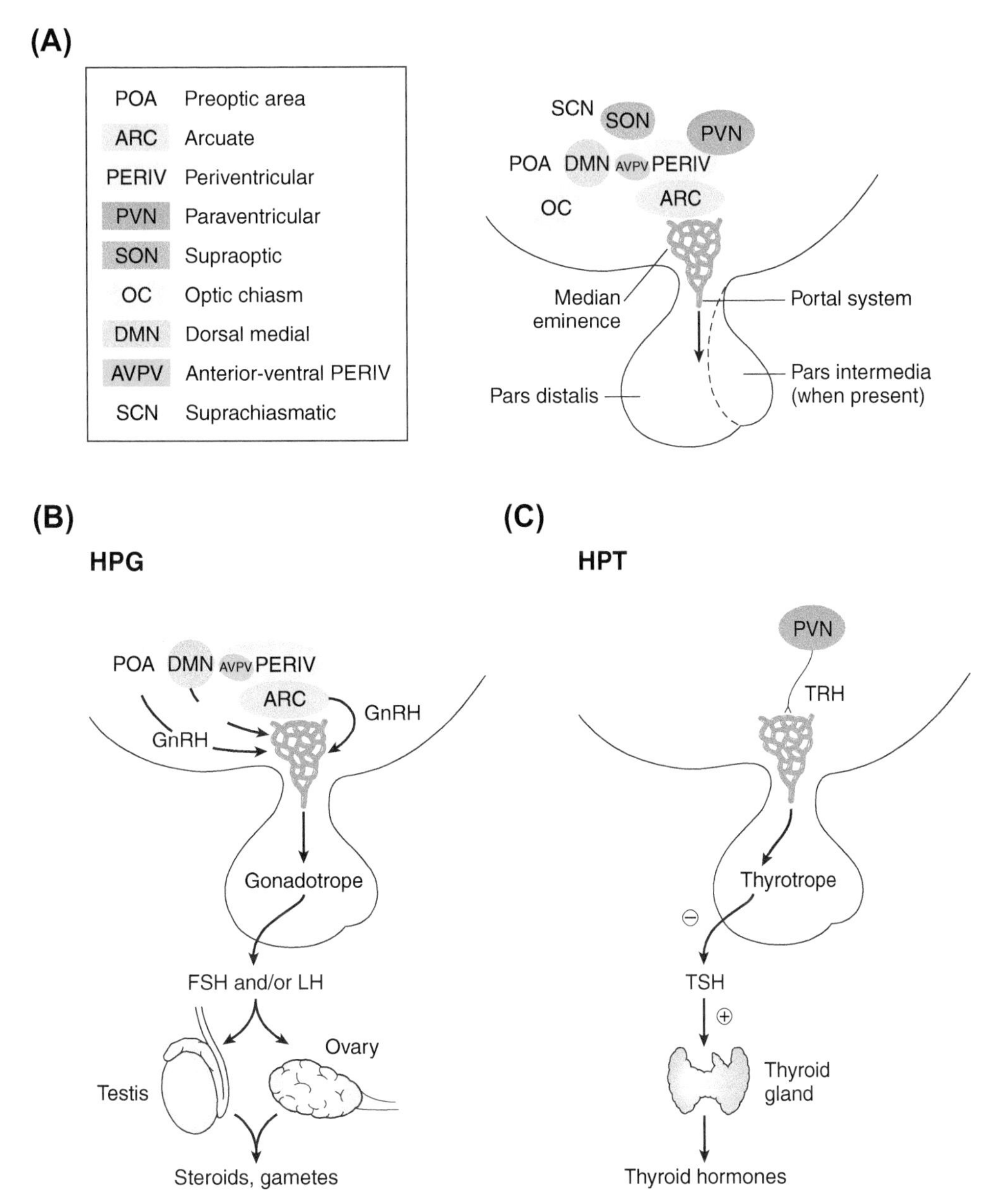

FIGURE 4-20 **Hypothalamic factors affecting tropic hormone release.** (A) Identification of principle sources for hypothalamic hormones. (B) Control of LH and FSH release. (C) Control of TSH release. (D) Control of ACTH release. (E) Control of GH release. (F) Control of PRL release. (G) Control of α-MSH release. See Appendix A or text for explanation of abbreviations and text for descriptions of regulation.

serotonin. Effects of serotonin in humans are based largely on use of pharmacological agents known to influence serotonergic neurons or serotonin receptors. These studies are somewhat controversial in interpretation, and no clear role has been established.

The peptide **leptin** stimulates TRH mRNA production and releases TRH but only from hypothalamic neurons of the PVN. It is suggested that peptides implicated in appetite control, such as leptin, α-MSH, **neuropeptide Y (NPY)**, or **galanin**, can influence TRH release in mammals and are related to thyroid hormone involvement in metabolism (see Chapters 6 and 12).

Somatostatin, another hypothalamic neurohormone associated with GH secretion (see ahead), can function as a TSH-RIH, but its physiological role has not been confirmed. Release of somatostatin can be inhibited by adrenergic α_2-agonists which is in keeping with the known stimulatory action of norepinephrine on TSH release.

3. Feedback Effects on TSH Secretion

Feedback of thyroid hormones occurs primarily at the thyrotropic cells in the pars distalis and reduces their sensitivity to TRH. The plasma concentration of T_4 is considered to be the most important plasma cue, and

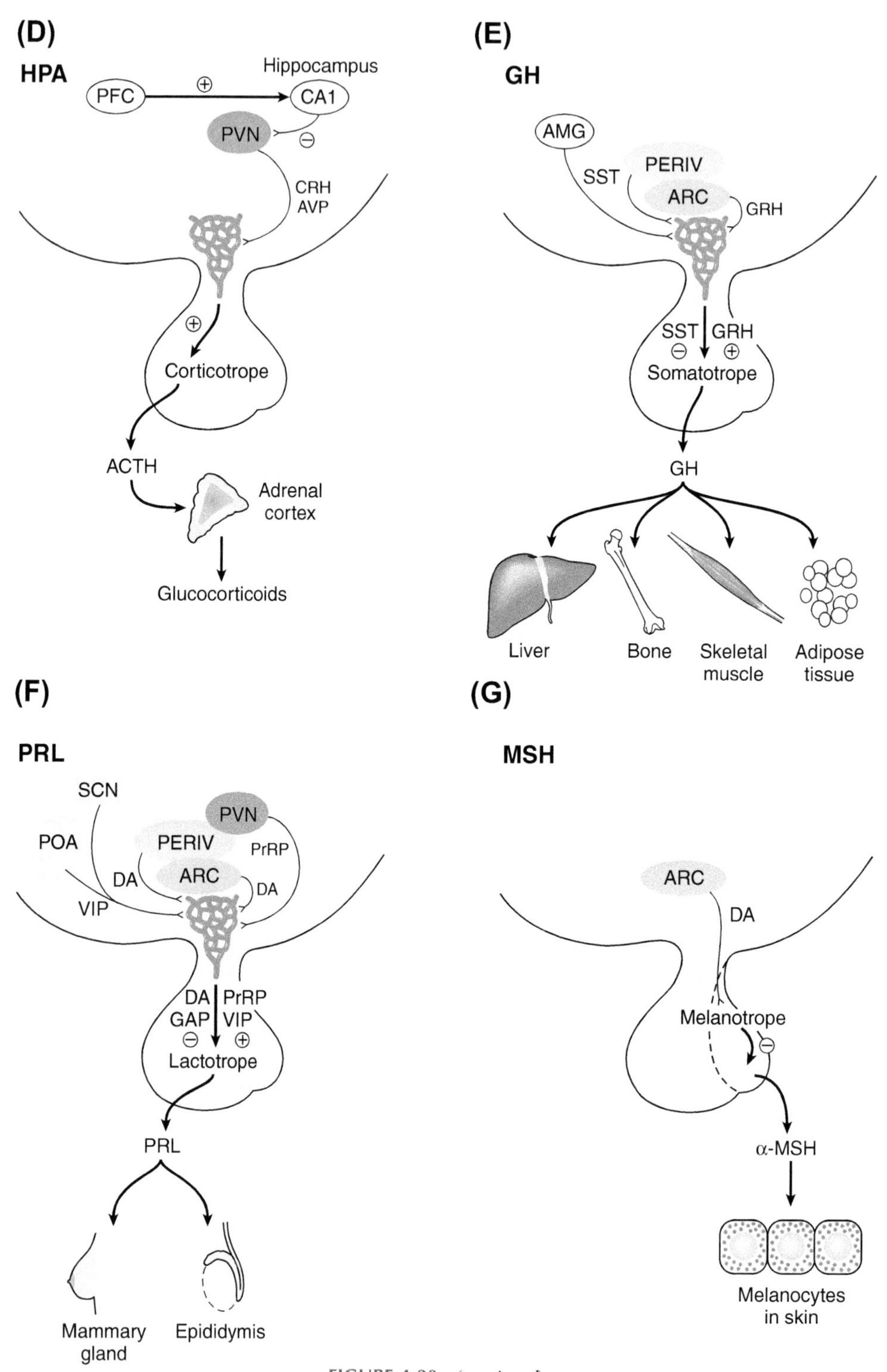

FIGURE 4-20 *(continued)*.

pituitary thyrotropes contain a deiodinase enzyme that converts T_4 to T_3 prior to binding to nuclear thyroid receptors (TRs). Occupied TRs interfere with the activation of other transcription factors such as Pit-1 that normally would release TSH. Additional evidence suggests that inhibition of the hypothalamic dopaminergic and serotonergic neurons by thyroid hormones is also important in feedback regulation of the HPT axis. Studies with other neurotransmitters (e.g., acetylcholine, histamine, GABA) and related pharmacological agents exhibit diverse effects but have provided little insight into the control of TSH release. They do suggest there can be many subtle influences on the HPT axis, and this may be important in clinical studies as well as in implications of environmental pharmaceuticals on possible endocrine disruption of the HPT axis.

B. Regulation of GTH Secretion by GnRH

Release of the GTHs is largely under stimulatory control similar to that described for TSH. A **gonadotropin inhibitory hormone (GnIH)** has been isolated from avian brains (see Chapter 5), and these observations were extended to three rodent species in 2006 (see ahead). Release of both LH and FSH is caused by the decapeptide (10 amino acids) (Table 4-1) gonadotropin-releasing hormone (GnRH). Because earlier studies had focused on the actions of this peptide on release of LH, it was first named **luteinizing hormone-releasing hormone (LHRH)**. However, its dual action causing release of both LH and FSH implies that GnRH is a more appropriate name, thus it is commonly used today. Existence of a specific **FSH–RH** has long been speculated on physiological evidence, but the actual molecule responsible has remained elusive. Major neural and neurohormonal factors affecting GTH secretion are summarized in Table 4-1 and Figure 4-20B.

Since mammalian GnRH was first synthesized, hundreds of synthetic analogues have been created, including analogues that are even more potent stimulatory agents than native GnRH, as well as inhibitors that can block the action of endogenous GnRH. One super-releaser has approximately 150 times the potency of native GnRH. All of the releasing analogues liberate both FSH and LH under most conditions. Under certain circumstances, exogenous GnRH can inhibit gonadotropin secretion, too (see ahead).

It was soon discovered that a second form of GnRH is present in the mammalian midbrain brain but was not associated with the HPG axis. This second GnRH in mammals is identical to a form first found in the midbrain of chickens (called **chicken II**) as well as in other vertebrates (except the laboratory mouse). Soon other molecular forms of this decapeptide were isolated from various species and in different brain regions, and each was named for the species source. Several nomenclature schemes have been used for naming GnRH molecules according to what species in which they were first discovered, where they are found within the brain (Figure 4-21), or what gene is responsible for their production (see also Chapter 5). The GnRH form in the hypothalamus that is responsible for GTH release in mammals first was designated as **mGnRH** just as the midbrain form was called cGnRH-II (the hypothalamic form in the chicken was designated as cGnRH-I). The second scheme identified **GnRH-1** as the form produced in the hypothalamus and stored in the

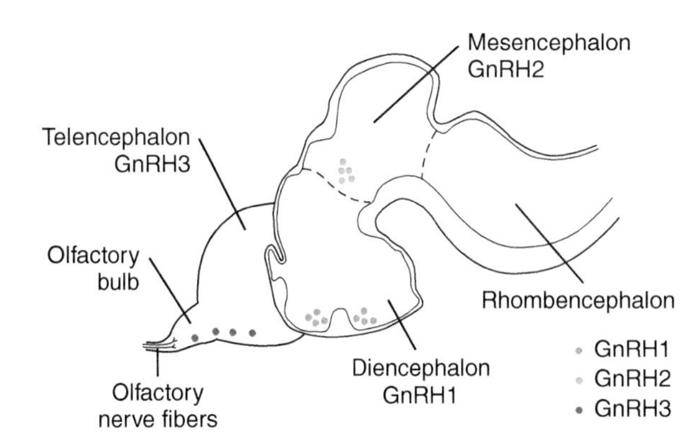

FIGURE 4-21 **Distribution of the forms of GnRH in the mouse brain.** See text for explanation of abbreviations. *(Adapted with permission from Whitlock, K.E.,* Trends in Endocrinology and Metabolism, ***16****, 145–151, 2005. © Elsevier Science, Inc.)*

median eminence. The midbrain form was called GnRH-2 and the telencephalic form was called GnRH-3. These latter two forms are not involved in regulating pituitary function. Most recently, genome analysis in a variety of vertebrate species has identified the evolution of three unique gene clusters associated with GnRH genes, and they have been termed GnRH-1, GnRH-2, and GnRH-3. Unfortunately, this latter terminology overlaps with the names for the anatomical designations but not necessarily with the physiological roles for the many different vertebrate forms of GnRH (see Chapter 5). However, in mammals, GnRH-1 is the hypothalamic form that releases LH and FSH from the pituitary and is part of the GnRH-1 gene cluster.

Endogenous GnRH-1 originates in neurosecretory neurons located in the POA and/or the medial basal hypothalamus, depending on the species. In rats and mice, most of the GnRH-1 immunoreactive cells sending fibers to the median eminence are located in the POA, anterior to the optic chiasm. In the guinea pig and human, the medial basal hypothalamus (in or near the ARC nucleus) is the major residence for GnRH-1 neurons.

GnRH-1 cells also contain other regulatory molecules including the **delta sleep-inducing peptide (DSIP)** and the peptide galanin. Nothing is known about the importance of DSIP in GnRH neurons with respect to reproduction, but galanin may be released along with GnRH and is itself a releaser of LH. Galanin has no effect on FSH release, however. Furthermore, galanin content varies in GnRH neurons of females during the reproductive cycle. Synthesis of galanin in these neurons peaks just prior to ovulation and may participate in the generation of the midcycle surge in LH that brings about ovulation and subsequent corpus luteum formation in the ovary (see Chapter 10).

Neurons that secrete GnRH-1 do not form a discrete, compact nucleus but instead develop an interconnected network that produces an endogenous, synchronized, pulsatile pattern of GnRH-1 release. This network is called the **GnRH pulse generator**. Special GnRH neurons called **GT-1 neurons** (a special cell line derived from transgenic mice) exhibit an oscillatory pattern of GnRH release *in vitro* and, together with some related cell lines, have provided a model system for studying the GnRH pulse generator.

GT-1 neurons synthesize and exhibit pulsatile release of GnRH but do not carry any biochemical glial cell markers. The absence of glial markers supports the contention that these cultures consist only of neurons. The pulsatile nature of their secretion implies an endogenous oscillator controlling this behavior, and experiments suggest an autocrine role for GnRH-1 to control its own release. From these experiments, researchers have concluded that operation of the endogenous oscillator depends on Ca^{2+} influx through **voltage-sensitive calcium channels (VSCCs)** and an autocrine positive feedback of released GnRH that causes additional GnRH release. The pulsatile nature of GnRH-1 secretion is essential to stimulate FSH and LH secretion from the pituitary. Hence administration of chronic, non-pulsatile doses of GnRH or GnRH analogues may prevent rather than stimulate gonadotropin release, possibly due to GnRH-induced down regulation of GnRH receptors.

Ultrashort negative feedback has been reported for GnRH-1 on normal GnRH-secreting neurons of the rat hypothalamus both *in vivo* and *in vitro*. The mechanism for this autofeedback is not clear, but it could occur through recurrent collateral fibers of GnRH-1 neurons that synapse on their own cell body or dendrites. GT-1 neurons do have GnRH receptors, and binding of GnRH-1 to these receptors is associated with a rapid, dose-dependent increase in intracellular Ca^{2+}. This results in a two-phase response involving an initial phospholipase C-mediated, IP_3-dependent mobilization of Ca^{2+} (see Chapter 3) and a sustained entrance of Ca^{2+} through VSCCs. The phospholipase C system activates other internal mechanisms involving phospholipase D and diacyl glycerol (DAG) that sustain activation of protein kinase C.

Initial binding of GnRH-1 activates additional GnRH release (positive feedback) but soon is followed by inhibition and loss of spontaneous pulsatility (negative feedback). These dual autocrine actions of GnRH-1 result in regular pulsatile episodes of GnRH release into the medium. Additional mechanisms for activating GnRH-1 release in GT-1 cells have been demonstrated, and these may influence the *in vivo* responses of GnRH-secreting neurons. Among the agents that can affect Ca^{2+} mobilization in GnRH cells are the peptides called **endothelins** (Figure 4-22) and catecholamines (dopamine, norepinephrine, and epinephrine). Receptor channels for glutamate and GABA are also present on GnRH-1 neurons and may be important modulators of the hypothesized endogenous oscillator controlling pulsatile GnRH-1 release.

1. GnRH-1 Action

GnRH-1 operating through a cAMP-dependent mechanism causes release of both FSH and LH from gonadotropes in the adenohypophysis, although a greater amount of LH release is always observed following administration of synthetic GnRHs. Prolonged treatment with most GnRH agonistic analogues results in downregulation of GnRH receptors, however, and chronic administration causes a reduction in GTH release. Pulsatile administration of GnRH-1 or GnRH analogues is necessary to stimulate natural patterns of GTH secretion. The half-life of GnRH-1 is very short, for it is rapidly degraded in peripheral plasma. The short-term success of several potent synthetic analogues of GnRH-1 appears to be related to their relative resistance to degradation by blood peptidases. GnRH-2 (= cGnRH-II) has been identified immunologically in the midbrain as well as in the pineal gland of some species. Like TRH, GnRH-2 can cause depression of neural function in the central nervous system; thus, GnRH-2 probably plays a physiological role as a neuromodulator or neurotransmitter.

2. Origin of GnRH-1 cells

Studies of early amphibian development showed that the precursors of adenohypophysial cells as well as certain hypothalamic neurons have their origins in portions of the embryonic neural ridge (Figure 4-5). From these studies, we have learned in both amphibians and mammals that one of these neural ridge derivatives, the **nasal placode**, is the origin of GnRH-1-secreting neurons of the hypothalamus as well as olfactory neurons that eventually make up the olfactory, terminal, and vomeronasal nerves. Nasal placodes are bilateral structures responsible for development of both left and right olfactory tracts. The olfactory neurons migrate first from the nasal region and contact the forebrain, inducing formation of the olfactory bulbs (Figure 4-23). The neurosecretory neurons follow an extracellular matrix laid down by olfactory neurons along their migration route and are identifiable by their secretion of GnRH-1. The degree of migration of these neurosecretory progenitor cells in different species probably explains the variability in GnRH-1 distribution within the brain. Ablation (surgical destruction or extirpation) of both nasal placodes in amphibians results in the absence of GnRH-1 cells in the hypothalamus. Unilateral ablation results in absence of GnRH cells only on the side of the ablation. Similarly, a genetically based abnormality known as **Kallmann's syndrome** in humans results in a failure of nasal placode cells to migrate and produces patients with anosmia and hypogonadism (see Chapter 10).

3. Other Neural Factors Affecting Gonadotropin Secretion

Neurons in discrete areas of the hypothalamus express a gene called ***kiss1*** that encodes a prohormone that is

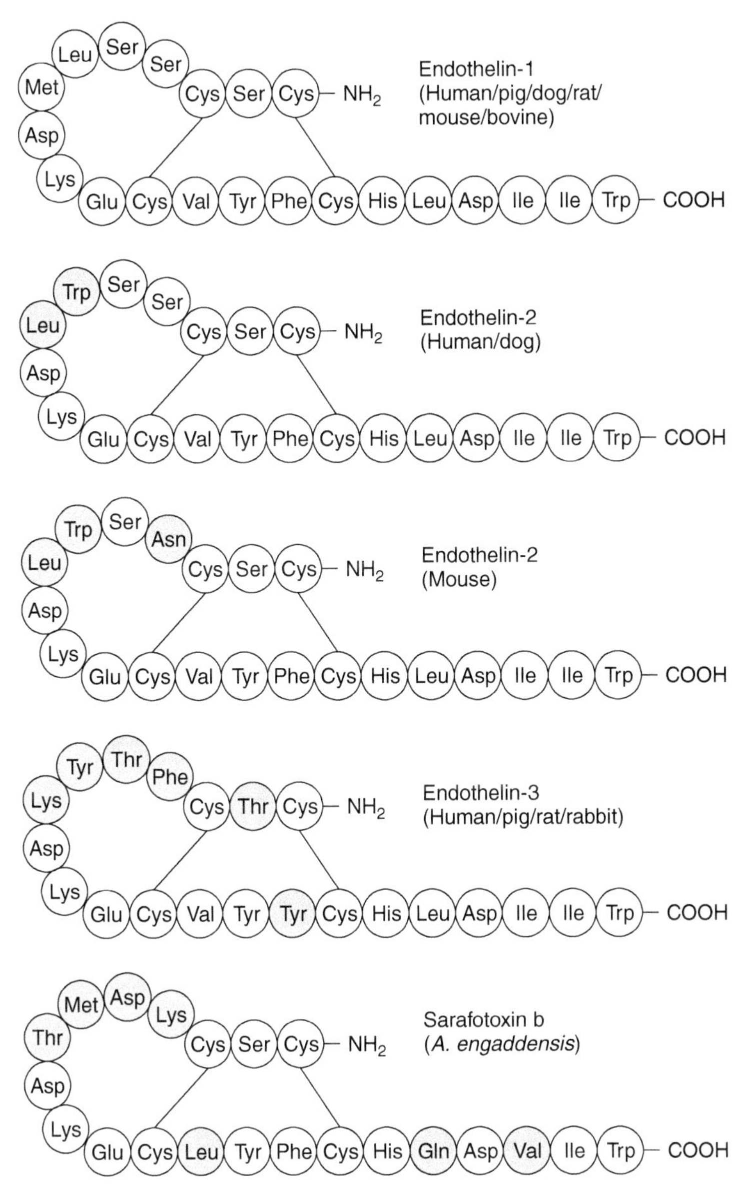

FIGURE 4-22 **Structure of endothelins and the related sarafotoxin from snake venom.** The colored circles represent amino acids not present in the sequence of endothelin-1. Amino acid abbreviations are explained in Appendix C. *(Adapted with permission from Masaki, T.,* Endocrine Reviews, ***14****, 256–268, 1993. © The Endocrine Society.)*

subsequently processed to form a group of peptides called **kisspeptins (Kp).** The 145-amino-acid Kp prohormone is cleaved to produce a number of Kp products that vary in size from 54 to 10 amino acids, although all active Kp peptides possess a 10-amino-acid peptide in the C-terminus that is amidated and is critical for binding to the Kp receptor (kisspeptin 1R, **kissR1**). In rodents there are two main groups of neurosecretory neurons in the ARC and the **anteroventral periventricular area (AVPV)** that express the *kiss1* gene and produce Kps (Figure 4-24). Kiss1 neurons are key upstream stimulators of GnRH-1 neurons (Figure 4-25). Estradiol acting on estrogen receptors (ERs) in AVPV kiss1 neurons causes Kp secretion from kiss1 neurons, innervating GnRH-1 neurons and enhancing GnRH-1 secretion into the pituitary portal circulation. Kps act on a putative kissR1 receptor also called **G-protein-coupled receptor 54 (GPR54)** (Figure 4-26). Kp binding to its receptor hypopolarizes GnRH neurons by blocking potassium efflux and opening calcium channels in the plasma membrane allowing for calcium influx and exocytosis of GnRH-containing secretory vesicles (Figure 4-26).

The 10 amino acid peptide GnIH is produced within neurosecretory neurons in the paraventricular nucleus of the hypothalamus that project to the median eminence and release GnIH into the pituitary portal circulation (Figure 4-27). GnIH acts on a membrane receptor located on gonadotropes to directly inhibit FSH and LH secretion. Melatonin produced by the pineal gland stimulates GnIH synthesis and secretion (Figure 4-27).

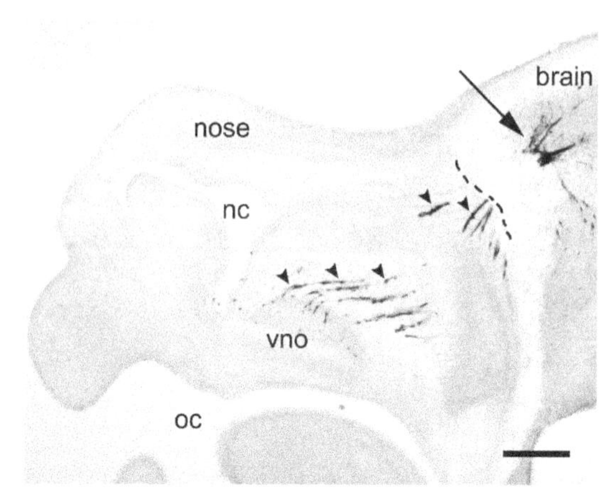

FIGURE 4-23 **Gonadotropin-releasing hormone (GnRH) migration in the mouse embryo.** Immunostaining of GnRH sagittal section of the head on embryo day 13.5. Arrow indicates GnRH neurons already located in the brain, whereas the arrow heads designate migrating GnRH neurons still in the nasal compartment. Abbreviations: nc, nasal cavity; oc, oral cavity; vno, vomeronasal organ. The dotted line designates the cribiform plate separating the nasal region from the brain. *(Courtesy of Drs. John Gill and Pei-San Tsai, University of Colorado.)*

Secretion of GnRH also is under the stimulatory control of noradrenergic neurons in both rats and primates. Activity of these noradrenergic neurons as well as the GABA-secreting neurons can be depressed by neurons secreting β-endorphin. The only other neurotransmitter with distinct stimulatory actions on gonadotropin release is **glutamate**, an excitatory amino acid neurotransmitter. The actions of cholinergic, serotonergic, and dopaminergic, factors on GnRH-1 and GTH release are based largely on pharmacological studies and are considered controversial due to the variety of reported effects.

4. Feedback Effects on Gonadotropin Secretion

Gonadal steroids (estrogens and androgens) generally have negative feedback effects on GnRH-1 release at the level of the hypothalamus and at the gonadotropes, reducing their sensitivity to GnRH. Positive feedback by estrogens also occurs in the gonadal axis of female mammals and is responsible for the midcycle surge of LH that stimulates ovulation. Although estrogens inhibit Kp release from kiss1 neurons in the ARC, estrogens stimulate Kp release from the Kp neurons in the AVPV. As estrogen levels gradually increase the positive effect predominates, leading to the LH surge. GABA-secreting neurons that innervate GnRH-1 neurons also can stimulate GnRH-1 release. These GABA neurons also possess ERs and may be responsible for the GnRH-1 surge. Part of the inhibitory action of endorphins on GnRH-1 release is through inhibition of these GABA neurons. Rupture of the ovarian follicle at ovulation causes a drastic reduction in estrogen synthesis and a marked increase in secretion of progesterone by the corpus luteum in the ovary, reinstating a predominantly negative feedback on hypothalamic secretion of GnRH-1 (see Chapter 10 for more details).

Selective negative feedback on FSH secretion occurs at the gonadotrope, reducing FSH output but not LH output in response to GnRH-1. This feedback is accomplished through FSH-dependent production by the gonads of a protein known as **inhibin**. The predominance of LH in the midcycle surge of GTH at ovulation in females is due in part to this selective action of inhibin. Galanin released at this time from the hypothalamus enhances release of LH, thus contributing to the LH surge that precedes ovulation. Further discussion of inhibin and other peptides involved in ovulation can be found in Chapter 10.

C. Regulation of Growth Hormone (GH) Secretion

Although initial studies involving transection of the portal system or explant of the pituitary implied that release of

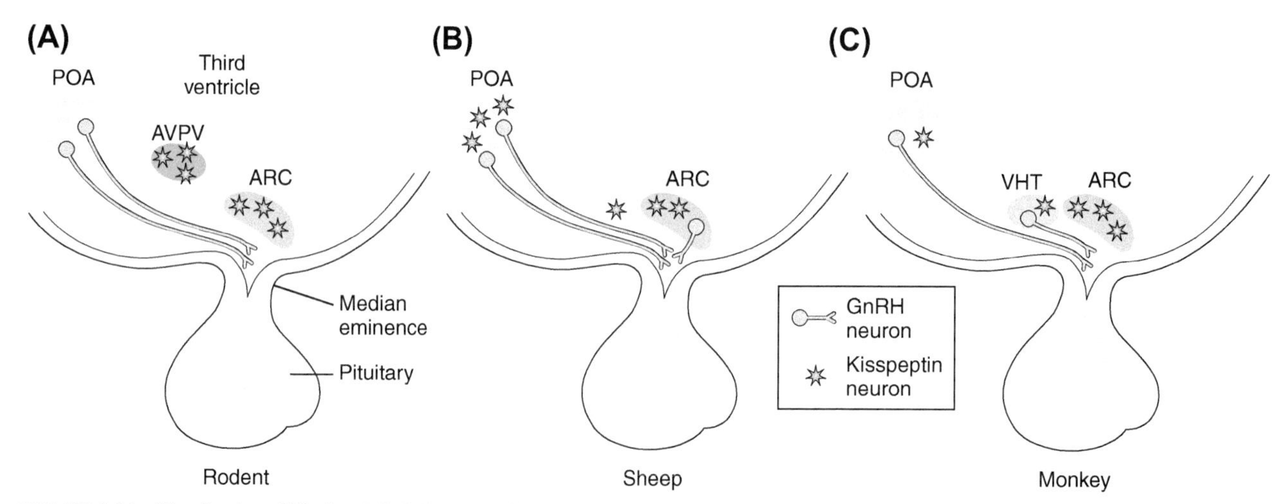

FIGURE 4-24 **Distribution of kiss1 and GnRH neurons in mammal brains.** *(Adapted with permission from Colledge, W.H.,* Trends in Endocrinology and Metabolism, ***20****, 115–121, 2009.)*

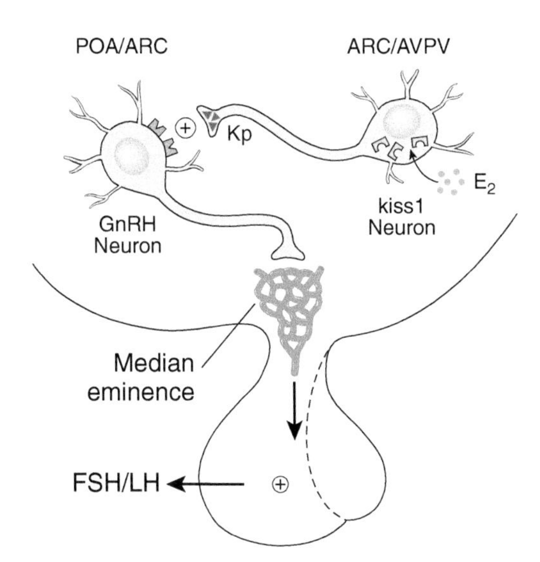

FIGURE 4-25 **Kiss1 and control of gonadotropin secretion.** E_2 binds to ERs (yellow) in kiss1 neurons, resulting in activation (positive feedback) in the AVPV and inactivation (negative feedback) in the arcuate nucleus. These kiss1 neurons innervate and may activate secretion of GnRH from GnRH neurons in the POA and/or ARC via kisspeptin (Kp) acting through the GPR54 receptor (purple), thus increasing gonadotropin secretion (FSH/LH) from the gonadotropes in the pars distalis of the pituitary. *(Adapted with permission from Smith, J.T. et al.,* Reproduction, *131, 623–630, 2006.)*

GH was under stimulatory hypothalamic control, we now know it is under both inhibitory and stimulatory control by neurohormones. Secretion of GH is episodic; that is, it occurs in bursts separated by longer intervals of lowered release; for example, in rats, GH is secreted at intervals of about 3 hours. In humans, maximal episodes of GH release occur during sleep (Figure 4-28). The major neural and neurohormonal factors influencing GH secretion are summarized in Table 4-1 and Figure 4-20E.

1. GH Release-Inhibiting Hormone (Somatostatin)

Release of GH *in vivo* is controlled primarily by a GH release-inhibiting peptide (GHRIH), more commonly called **somatostatin (SST)**. However, even in the absence of SST, a releasing hormone known variously as somatocrinin, somatoliberin, or **growth hormone-releasing hormone (GHRH = GRH = GRF)** is necessary to stimulate GH release. The episodic bursts of GH release occur during periods of lowered SST levels and elevated GHRH.

The PERIV nucleus of the anterior hypothalamus and the amygdala of the limbic system are the sources for most of the SST-containing nerve terminals in the median eminence. This tetradecapeptide (14 AA; Table 4-1) is a strong inhibitor of GH release. SST_{14} appears to work through a GPCR linked to a GTP-dependent G_i-protein that prevents the elevation of cAMP necessary to provoke GH release (see Chapter 3, Figure 3-11). SST also occurs in extrahypothalamic nervous tissue (both brain and spinal cord). Multiple SST-receptors have been identified and their differential distribution in the CNS and peripheral tissues described. A number of possible neuromodulator/neurotransmitter functions in the central nervous system have been suggested, similar to the case for the other major releasing hormones. The mucosa of the stomach also secretes SST where it locally inhibits release of the gastric hormone, gastrin. In the pancreatic islets, SST inhibits both glucagon and insulin release (see Chapter 12). A larger form of SST (**SST_{28}**) was extracted first from the gut and pancreatic tissues and later was shown to exist in nerve endings of the median eminence along with the tetradecapeptide. Studies have since shown that SST_{28} is released in equal or greater amounts than SST_{14} into portal blood. Furthermore, somatotrope receptors show stronger binding affinity for the larger peptide. Both forms of SST apparently are produced from the same prohormone.

2. Growth Hormone-Releasing Hormone (GHRH)

A potent peptide releaser of GH was first isolated and characterized from a human pancreatic tumor associated with an unusual case of excessive GH secretion resulting in a clinical condition called *acromegaly* (see clinical discussion ahead). The same gene was later found to be active in the hypothalamus. In human, porcine, bovine, ovine, and caprine mammals, GHRH isolated from the ARC nucleus is composed of 44 amino acid residues and is related chemically to the secretin-glucagon family of peptides (see Chapter 3, Table 3-4). In the rat, GHRH consists of only 43 amino acids and differs considerably in amino acid composition from the other mammalian GHRHs. The significance of this deviation in rat GHRH structure is unknown.

GHRH-immunoreactive fibers extend from the ARC nucleus to the median eminence. Some neurons in the ARC nucleus contain dopamine and neurotensin as well as GHRH. Others exhibit GHRH colocalized with dopamine and galanin. A potent GHRH-releasing peptide was discovered and later identified as the peptide **ghrelin,** previously found in the gastrointestinal tract. Ghrelin, like GHRH, is found in neurons of the ARC nucleus and probably plays an important role in GHRH secretion and hence in GH release (Figure 4-29). It also has important effects on appetite (see Chapter 12).

3. Other Neural Factors Affecting GH Secretion

There is no shortage of investigations that have reported the effects of pharmacological and/or disruptive techniques on GH secretion; however, there are considerable conflicting data, and it is difficult to determine whether effects observed are direct or indirect. Furthermore, it is often difficult to ascertain whether an effect is mediated through control of SST or GHRH release. For example, an increase

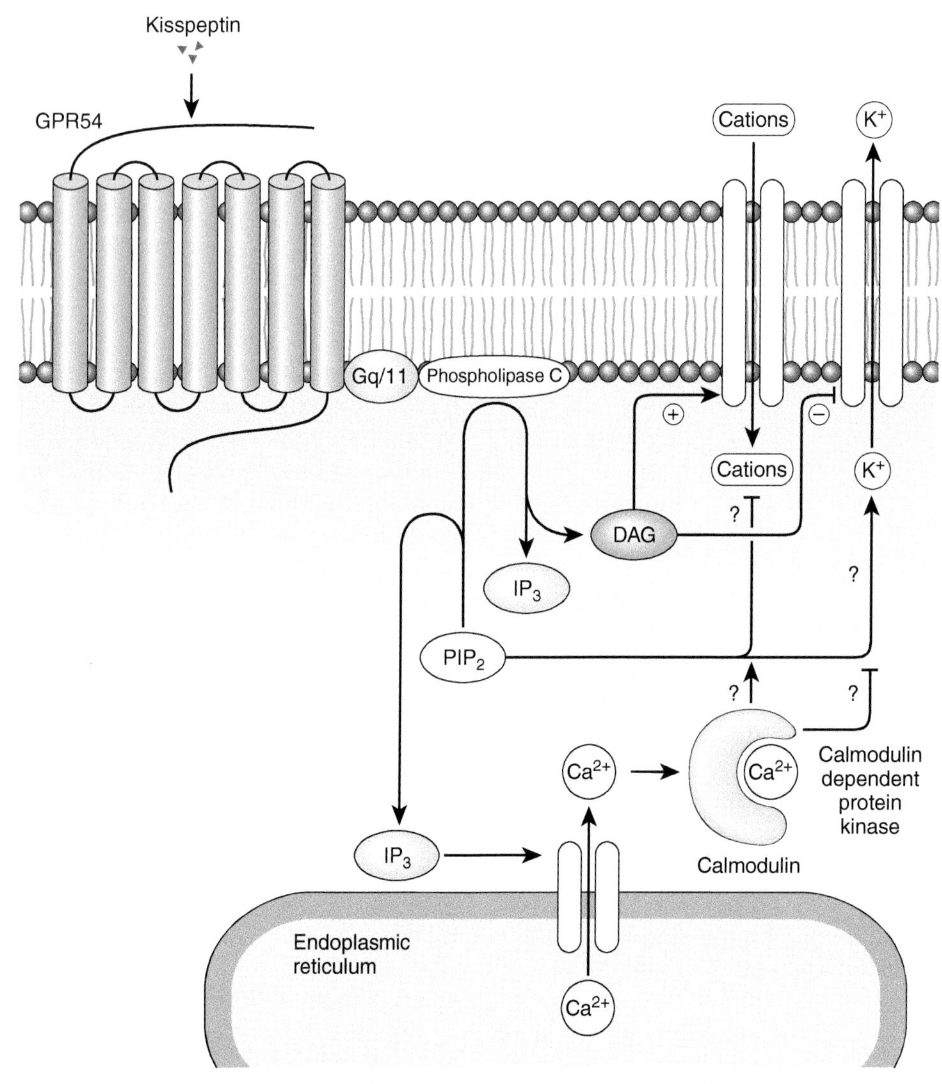

FIGURE 4-26 **Action of Kisspeptin on GnRH release via GPR54 receptors.** Kisspeptin enhances GnRH secretion by binding to the GPR54 receptor and increasing the inositol triphosphate (IP_3) second-messenger pathway. As a result, diacylglycerol activates (+) non-selective cation channels and inactivates (−) membrane potassium channels, leading to depolarization of the GnRH neuron. These channels also may be modulated directly by phosphatidylinositol bisphosphate (PIP_2) or calmodulin-dependent protein kinases. *(Adapted with permission from Colledge, W.H.,* Trends in Endocrinology and Metabolism, ***20****, 115–121, 2009.)*

in GH secretion could represent either a decrease in SST release or an increase in GHRH release, or both.

Factors that have been shown to elicit GH release under certain conditions include neurotransmitters (norepinephrine, dopamine, serotonin, GABA, NPY), opiates, VIP, thyroid hormones, glucocorticoids, and estrogens. Inhibition of GH secretion has been reported following treatment with a variety of neural factors (GABA, serotonin, substance P, neurotensin), large quantities of glucocorticoids, and androgens. From this brief summary, it seems clear that catecholamines are stimulatory. Serotonin is usually stimulatory, probably acting through effects on norepinephrine-secreting neurons. The dichotomous actions of GABA could be due to its ability to inhibit release of SST and GHRH under different conditions. Substance P and neurotensin increase SST release, explaining their prevention of GH secretion.

Some of the confusion as to roles of various regulators on GH release stems from species differences. For example, stress, which is usually associated with elevated glucocorticoids, generally increases GH secretion in primates, inhibits GH secretion in rodents, and has no effect in domestic ungulates. Moderate exercise causes GH release in humans. Total food deprivation increases GH secretion in primates as well as in domestic ungulates.

4. Feedback Effects on GH Secretion

Many of the actions of GH on body growth are mediated by IGF-I and IGF-II produced in the liver and secreted into the general circulation under the direction of GH rather than by

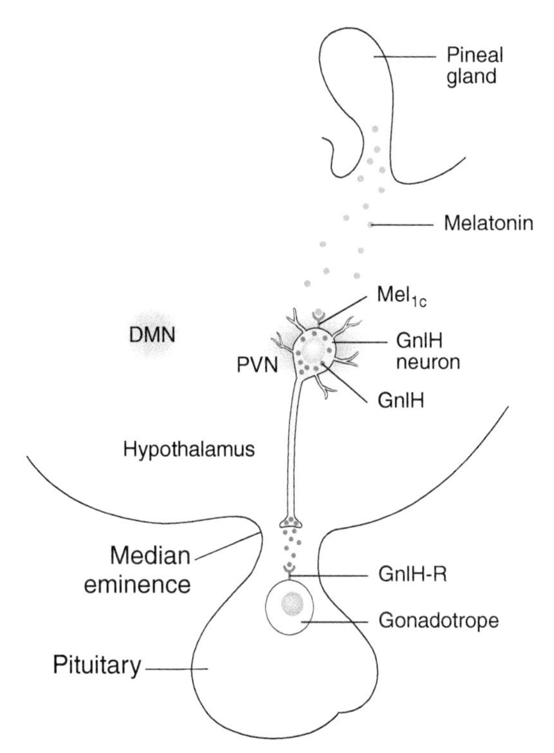

FIGURE 4-27 **Melatonin released from the pineal gland stimulates GnIH secretion from neurons in the PVN.** See text for explanation. *(Adapted with permission from Tsutsui, K. et al.,* General and Comparative Endocrinology, *153, 365–370, 2007.)*

GH itself. Plasma IGF-I reduces release of GH through negative feedback at the hypothalamus, where it stimulates SST release and inhibits secretion of GHRH. IGF-I also reduces the sensitivity of somatotropes in the pituitary to GHRH.

GH produces metabolic effects including elevation of glucose and free fatty acid levels as well as decreases in

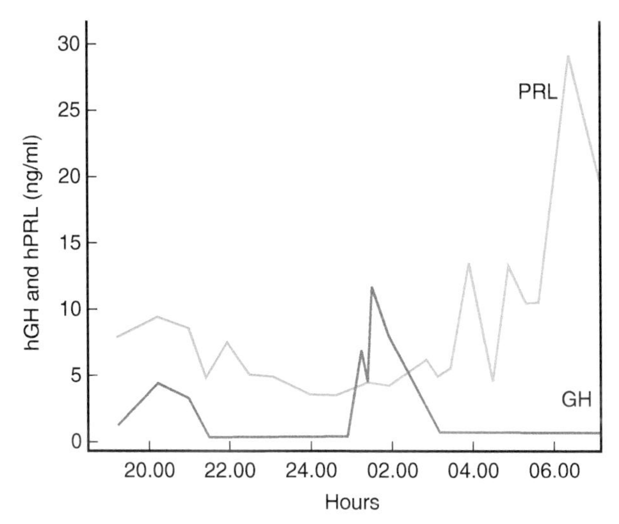

FIGURE 4-28 **Diurnal patterns of human GH and PRL release.** *(Adapted with permission from Baulieu, E. and Kelly, P., "Hormones: From Molecules to Disease," Chapman & Hall, London, 1990, p. 206.)*

plasma amino acids. These metabolic products can affect GH release, but it appears they play no major physiological role. Elevated glucose can decrease GH output, and insulin-induced hypoglycemia elevates GH release. High levels of free fatty acids also depress GH levels, and, conversely, depressed fatty acid levels elevate GH release. Both glucose and free fatty acid feedback effects are consistent with the reduction in both following insulin injection.

GH release is especially sensitive to one amino acid, arginine. Infusion of arginine causes a marked increase in GH release that is not blocked by IGF-I. However, other amino acids are not very effective stimulators of GH release, and no elevation in circulating GH is observed following consumption of a high-protein meal. Hence, the physiological importance of amino acid feedback is unclear.

D. Regulation of Prolactin (PRL) Secretion

The primary control over PRL release, as mentioned earlier, is inhibitory in mammals, and the pituitary spontaneously releases PRL when freed either surgically or chemically from hypothalamic control. This is in marked contrast to the situation with GH where both the absence of SST and the presence of GHRH are necessary to elicit secretion; consequently, ectopic or explanted pituitaries spontaneously release large quantities of PRL. Nevertheless, there is considerable evidence for stimulatory hypothalamic control of PRL secretion during lactation.

In lactating mammals, suckling at the teat sends neural input to the hypothalamus and evokes release of the nonapeptide oxytocin into the circulation, causing contraction of myoepiothelial cells lining the ducts of the mammary glands resulting in milk-ejection. This **neuroendocrine reflex** also stimulates a delayed release of PRL that stimulates the mammary gland to replace lost milk. Major neural and neurohormonal factors affecting PRL release are summarized in Table 4-1 and Figure 4-20F.

1. Prolactin Release-Inhibiting Hormones (PRIH)

The consensus is that mammalian **prolactin release-inhibiting hormone (PRIH)** is the catecholamine neurotransmitter, dopamine, which is a potent inhibitor of PRL release *in vivo* or *in vitro* (Figure 4-30). Dopamine is released from hypothalamic neurons in the ARC and PERIV nuclei of the hypothalamus as a neurohormone, and dopamine is elevated in the portal system under physiological conditions where PRL release is inhibited. Two additional PRIH candidates have been proposed. The first is GABA, which blocks PRL release *in vivo* and *in vitro* by direct action on lactotropes. However, large amounts of GABA are required to inhibit PRL release, and such levels are not found in either portal or peripheral blood samples,

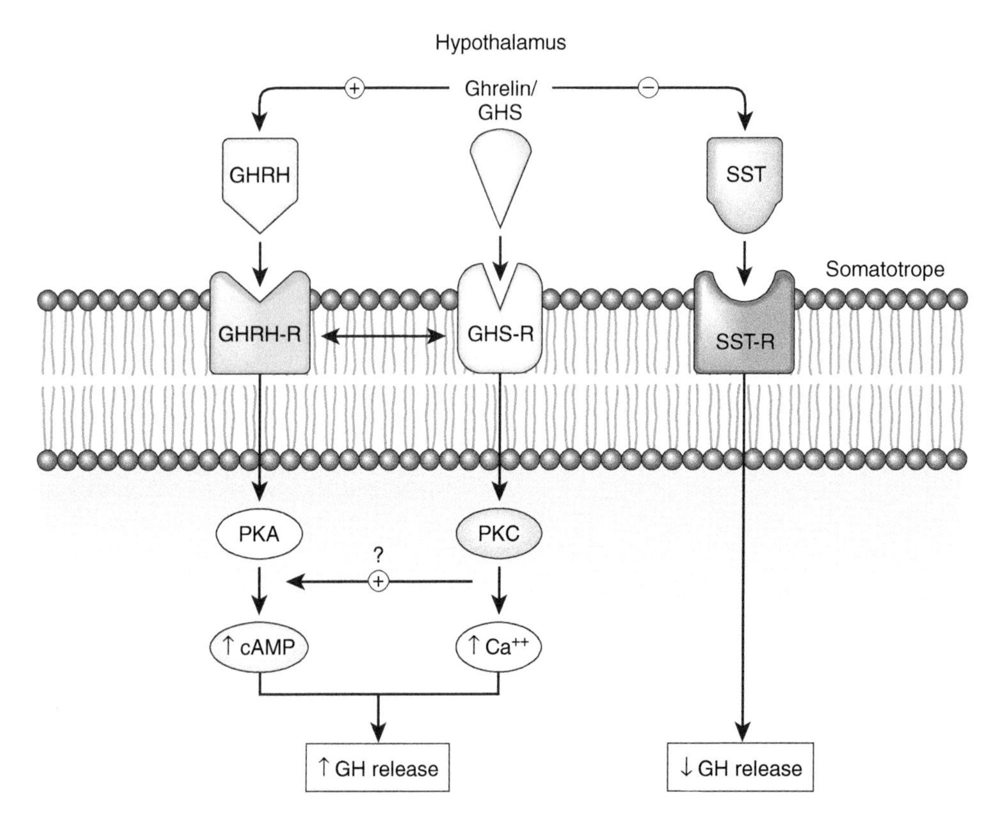

FIGURE 4-29 **Multireceptor regulation of GH secretion.** GHRH, ghrelin, and somatostatin (SST). GHRH and ghrelin each act independently to increase GH secretion through the GHRH receptor and growth hormone secretagogue receptor (GHS), respectively. In addition, there are synergistic effects on GH release when both GHRH and ghrelin activate their receptors. Ghrelin also acts within the hypothalamus to stimulate GHRH secretion and inhibit SST secretion. SST also may act within the somatotrope to blunt GHRH and ghrelin action through G_i activation (see Chapter 3). *(Adapted with permission from Lengyel, A.M.J.,* Arquivos Brasileiros de Endocrinologia & Metabologia, ***50****, 17–24, 2006.)*

indicating that GABA probably is not a physiological PRIH. There is also support for a peptide PRIH that can inhibit PRL release. This peptide PRIH is the 56-amino acid C-terminal fragment of the GnRH prohormone. Inhibition of PRL release often accompanies GTH secretion; however, it is not confirmed that this **GnRH-associated peptide (GAP)** is a physiological regulator.

2. Prolactin-Releasing Hormones

Like the situation for GH, there is evidence for an endogenous factor or factors that stimulate PRL release. This **PRL-releasing hormone (PRH)** at first was thought to be the tripeptide TRH because administration of synthetic TRH stimulates PRL release as well as TSH release, at least under some conditions. However, the neuropeptide VIP stimulates PRL release during suckling when TRH is ineffective. Administration of TRH to suckling mice causes elevated TSH release but no increase in PRL release such as caused by VIP. Furthermore, TSH release is not observed during normal suckling, arguing against TRH as being responsible for PRL release at that time. A peptide first isolated as an orphan receptor was found to be a potent releaser of PRL both *in vitro* and *in vivo*. This **prolactin-releasing peptide (PrRP)** releases PRL when administered in low doses but is not as potent as TRH. PrRP apparently is widely distributed in the brain and can modulate many unrelated physiological events including lactation, stress, body weight homeostasis, feeding behavior, and gastric motility. It also is a potent releaser of PRL in some non-mammals (see Chapter 5). Additional studies have shown that PrRP affects appetite (see Chapter 12). The peptide **tuberalin**, produced in the pars tuberalis, can stimulate release of PRL from lactotropes and is believed to be an important regulator related to photoperiodic effects on PRL release (see pineal gland discussion ahead).

3. Other Neural Factors Affecting Prolactin Secretion

Several neuropeptides evoke PRL release under experimental conditions, including neurotensin, epidermal growth factor (EGF), and EOPs. The opioids apparently block the activity of dopamine-secreting neurons that normally prevent PRL release. SST also can inhibit PRL release. GABA, which can directly inhibit PRL release at the lactotrope, can also stimulate PRL release through actions on hypothalamic neurons. The physiological significance of these actions by SST and GABA is not clear.

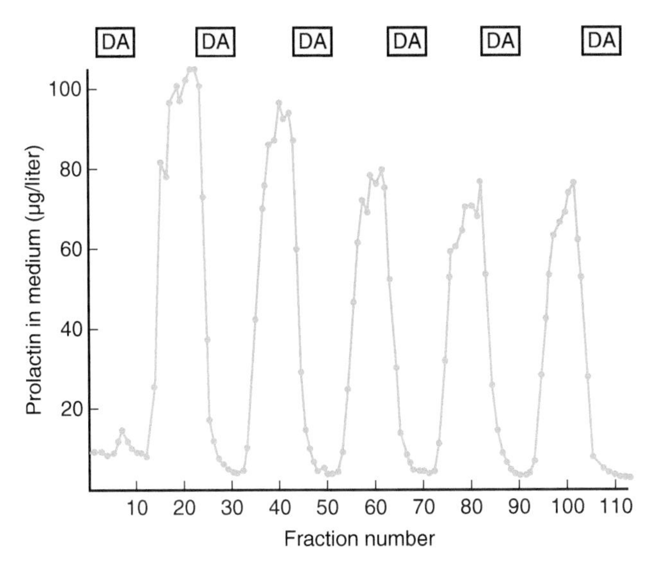

FIGURE 4-30 **Response of perifused rat pituitary lactotropes to dopamine (DA).** Additions of DA immediately depresses the spontaneous secretion of PRL normally seen *in vitro*. *(Adapted with permission from Martini, L. and Besser, G.M., "Clinical Neuroendocrinology," Academic Press, New York, 1977.)*

4. Feedback and Other Chemical Effects on Prolactin Secretion

The major feedback loop established in rats involves PRL feeding back directly on hypothalamic dopaminergic neurons. PRL also increases GABA synthesis and release into the portal circulation, supporting its inhibitory action on the lactotrope as part of the feedback mechanism rather than as a true PRIH. Estrogens modulate PRL release in rats and to a lesser extent in sheep by enhancing the sensitivity of lactotropes to hypothalamic PRHs. This effect may involve an estrogen-induced reduction in their sensitivity to dopamine by altering receptor levels and/or intracellular second-messenger systems. Some GABA-secreting neurons exhibiting ERs are inhibited by estrogens and may release some PRH-secreting neurons from tonic GABA inhibition. Gonadotropes release the octapeptide Ang-II in response to GnRH-1. Ang-II then binds to specific receptors on nearby lactotropes. This paracrine action on the lactotrope may have some importance for PRL release.

E. Regulation of Corticotropin (ACTH) Secretion

Corticotropin secretion by corticotropes of the pars distalis is primarily under stimulatory regulation by the neuropeptide CRH. Physiological evidence has been presented that suggests the existence of an inhibitory hypothalamic factor as well. Although ACTH and α-MSH are synthesized by hydrolysis of the same preprohormone in the melanotrope and the corticotrope (see below), there are many differences in the regulation of their release. Secretion of ACTH stimulates synthesis and release of cortisol (e.g., humans) or corticosterone (e.g., rats) or both (e.g., deer) from the adrenal cortex, and the circadian rhythm of ACTH secretion is reflected in a similar rhythm for corticosteroid release (see Chapter 8). In humans, there is a gradual decrease in ACTH plasma levels with the onset of sleep to a minimum at about midnight (see Chapter 2, Figure 2-3). Then, a series of bursts of ACTH release occurs, culminating in maximal plasma levels of ACTH and cortisol at about 6 a.m. Following these peaks, there is a gradual decline in both hormones during the daylight hours with occasional bursts of secretion. Release of ACTH may occur following a midday meal but not an evening meal. In nocturnally active rats, the pattern of corticosterone secretion is the reverse of that seen in diurnal humans. Major neural and neurohormonal factors affecting ACTH release are summarized in Table 4-1 and Figure 4-20D.

1. Corticotropin-Releasing Hormone (CRH) or Factor (CRF)

CRH consists of 41 amino acid residues, and it releases ACTH from the corticotrope *in vivo* and *in vitro*. The highest hypothalamic levels of CRH in rats and humans are in the PVN and ARC nucleus. In rats and humans, AVP and CRH have been colocalized in the same hypothalamic neurons. CRH in rats occurs in the highest concentration within neuronal endings located in the median eminence where it exceeds concentrations in other brain regions by 10 to 100 times. CRH occurs in other forebrain nuclei (e.g., central nucleus of the amygdala) as well as in other regions of the brain (e.g., substantia nigra of the midbrain). In addition to its role in ACTH secretion from the pituitary, CRH probably functions as a neuromodulator in other brain regions including the olfactory bulb, cerebral cortex, cerebellum, hippocampus, and amygdala.

Human and rat CRHs are identical peptides but oCRH differs from these at seven residues. There is considerable homology of CRH (at 20 positions) to **sauvagine**, a 40-residue peptide isolated from skin of the frog *Phyllomedusa sauvagii*, as well as overlap with the structure of a hypotensive agent **urotensin I**, isolated from the spinal cord and urophysis of teleostean fishes (see Chapter 5).

Additional members of the CRH-like family of peptides in mammals are called **urocortins** (Table 4-9). Three urocortins have been identified: **urocortin I (Ucn-I)**, **urocortin II (Ucn-II), and urocortin III (Ucn-III)**. Ucn-I is present in neurons of the midbrain Edinger–Westphal nucleus as well as in cells of the GI tract, testes, cardiac muscle, thymus, spleen, and skin, suggesting diverse roles. Ucn-II and III are found primarily in the hypothalamus. Two receptors that bind CRH and the urocortins also have been described and usually are called **CRH-R1** and **CRH-R2** (Figure 4-31). CRH and Ucn-I both bind to **CRH-R1**,

TABLE 4-9 Distribution of Basic and Neutral Nonapeptides in Mammals

Mammals		Basic	Neutral
Prototheria	Echidna, platypus	AVP	OXY
Metatheria	Common opossum, four-eyed possum	LVP	OXY
	American opossum, bandicoot	LVP	MST
	Dasyuroids	AVP	MST
	Macropodids (kangaroos wallabies)	LVP, PVP	OXY, MST
Eutheria	Suiformes (hippo, wild pigs)	LVP, AVP	OXY
	Suiformes (domestic pig)	LVP	OXY
	All others	AVP	OXY

See text or Appendix A for explanation of abbreviations.

which is the more prevalent receptor type in the brain and pituitary. All three urocortins bind to CRH-R2, which has been demonstrated in the hypothalamus as well as in peripheral sites including the digestive tract, arteries, lung, and skeletal muscle. CRH binds strongly to CRH-R1 but binds much less strongly to CRH-R2 than do the urocortins. When administered directly into the brain, CRH and the urocortin peptides stimulate many physiological and behavioral responses associated with stress (Chapter 8) including elevated blood pressure, gastric emptying, reduced food intake, and stimulation of E and NE secretion from the adrenal medulla. When injected into the peripheral blood stream urocortins as well as injected CRH inhibit gastric secretion, decrease blood pressure, and reduce inflammation. These differences are likely due to differences in the number and location of CRH-R1 and -R2 receptors. A splice variant of CRH-R2 (βCRH-R2) with affinity for Ucn-II and Ucn-III occurs in heart muscle and is involved with regulation of cardiac output.

ACTH release also is stimulated by the nonapeptide AVP, and there is good evidence to support a physiological role for AVP in enhancing the responsiveness of corticotropes to CRH. AVP is co-localized with CRH in a subpopulation of the same hypothalamic neurons. During stress (see Chapter 8), the ratio of secreted AVP to CRH increases from 2:1 to as high as 9:1. Apparently, AVP released from the median eminence acts directly upon corticotropes to markedly increase their sensitivity to CRH and prevents shutdown of ACTH secretion via corticosteroid feedback on CRH release (see ahead). CRH-like bioactivity in the SON is due primarily to the

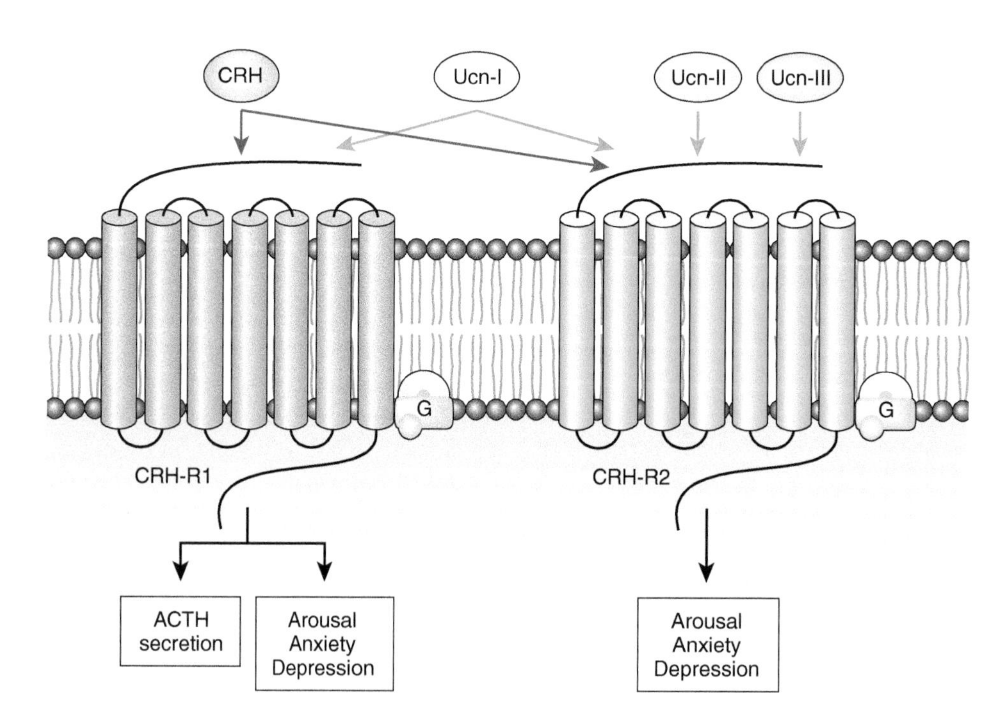

FIGURE 4-31 **Actions and interactions of corticotropin-releasing hormone (CRH) and urocortins (Ucn).** Two CRH receptors have been identified, CRH-R1 and CRH-R2. CRH has a higher affinity for the CRH-R1 receptor and greater affect on ACTH release than does Ucn-I. Similarly, Ucns all have a much greater affinity for the CRH-R2 receptor.

presence of AVP and not CRH whereas the major CRH-like bioactivity in the PVN is due to CRH.

Although the Brattleboro strain of rats lacks the ability to synthesize AVP, these rats still show a stress response. However, extracts from Brattleboro hypothalami exhibit only 20% of the CRH activity demonstrated in normal rats. Addition of AVP restores ACTH-releasing activity to 100% in these rats.

CRH is also a potential bioregulator of LH and GH release at the level of the hypothalamus. These effects do not involve opioid neurons but the exact mechanism is unknown. CRH can stimulate GHRH release and inhibit SST release, which might explain its effect on GH release. The reduction in LH release when CRH is elevated may provide a mechanism whereby stress shuts down the reproductive axis.

2. Other Neural Factors Affecting ACTH Secretion

In addition to CRH and AVP, release of ACTH occurs following administration of several classical neurotransmitters (norepinephrine, epinephrine, serotonin, acetylcholine) and a variety of neuropeptides (OXY, VIP, PHI, Ang-II). Although all of these compounds have been reported within nerve fibers of the median eminence, none of them appears to be of major importance in influencing ACTH release. GABA, however, has direct inhibitory actions on paraventricular CRH neurons as well as on corticotropes in rats. In humans, hypothalamic GABA stimulates release of ACTH but GABA of extrahypothalamic origin inhibits ACTH secretion.

3. Feedback Effects on ACTH Secretion

Negative feedback occurs through the actions of glucocorticoids both on corticotropes of the pars distalis as well as directly on CRH neurons in the hypothalamus and indirectly on non-CRH neurons in brain areas outside the hypothalamus (see Chapter 8). Glucocorticoids are known to feed back via mineralocorticoid receptors (MR = GR-1; see Chapter 3) on CA1 neurons in the hippocampus that in turn innervate CRH neurons in the hypothalamus, but bind to glucocorticoid receptors (GR = GR-2) in the corticotropes. During chronic stress as mentioned previously, the output of AVP, which is not subject to glucocorticoid feedback, increases. As a result, the sensitivity of corticotropes to CRH is enhanced. This action probably helps to sustain prolonged elevation of glucocorticoids in the face of negative feedback on CRH release during chronic stress.

F. Regulation of Melanotropin (α-MSH) Secretion

Unlike primates, which have a functioning pars intermedia only during the fetal and neonatal periods, adult rodents have a functioning pars intermedia that when transplanted to an ectopic site or placed *in vitro* releases more α-MSH, indicating that chronic inhibition of α-MSH release is the normal condition. Although evidence exists for stimulatory factors regulating α-MSH release, it seems that removal of inhibitory controls is most important for eliciting α-MSH release. A summary of major neural and neurohormonal factors affecting α-MSH release is provided in Figure 4-20G.

1. Melanotropin-Releasing (MRH) and Release-Inhibiting Hormones (MRIH)

It is not always clear whether or α-MSH release is controlled by a neurohormone or by direct innervation from hypothalamic neurons. Dopamine appears to be the physiological **melanotropin release-inhibiting hormone (MRIH)** in numerous species. Several peptides possess intrinsic MRIH-like activity, but they may not be physiological relevant. Extrahypothalamic MRIH activity has been demonstrated, and at least one of the synthetic releasers of α-MSH is an antidepressant when administered to humans.

2. Feedback Effects on α-MSH Secretion

There does not appear to be any direct feedback mechanism operating on α-MSH release. Stimulation of melanocytes by α-MSH does not release anything into the circulation that would be a feedback candidate. Perhaps the strong negative control of α-MSH release precludes the requirement for negative feedback.

V. THE NONAPEPTIDE HORMONES

The hypothalamic nonapeptide neurohormones are synthesized in the SON and the PVN. Most of these neurons project their axons to the pars nervosa, although some neurons connect to the median eminence. Nonapeptides stored in the pars nervosa can be released from neurosecretory neuronal endings directly into the general circulation in response to neural stimulation. The targets for these hormones are located at considerable distances from the pars nervosa (e.g., kidney, mammary gland, uterus).

Two neurohypophysial nonapeptide hormones (usually AVP and OXY) are present in the pars nervosa of most adult mammals. LVP is a variant of AVP produced by members of the mammalian order Suina, which includes peccaries, domestic pigs, and the hippopotamus. Most of the species in this group secrete both AVP and LVP in addition to OXY, but the domestic pig secretes only LVP. PVP, a unique vasopressin-like molecule, replaces vasopressin in the pars nervosa of marsupials.

Fetal mammals secrete a molecule that at first appears to be a hybrid of AVP and OXY, being composed of the side chain of AVP with the ring structure of OXY. This nonapeptide is known as **arginine vasotocin (AVT)** and is characteristic of adult non-mammalian vertebrates (see Chapter 5). The pineal gland of at least some adult mammals also contains AVT.

The vasopressins (AVP, LVP) and PVP function as antidiuretic agents, increasing the ability of the kidneys to reabsorb water from the glomerular filtrate, reducing urine volume (antidiuresis). At higher doses, vasopressins cause vasoconstriction and can elevate blood pressure (pressor effect). This action may increase glomerular filtration and water excretion (diuresis). AVT produces similar actions in non-mammals and may play an osmoregulatory role in fetal mammals.

As mentioned earlier, OXY stimulates contraction of myoepithelial cells lining the ducts of the mammary glands and causes ejection of milk. The stimulatory action of OXY on smooth muscle of the uterus is related to the induction of labor and the birth process. OXY also stimulates contractions in oviducts as well as in the vas deferens (sperm duct) of males.

Nonapeptide hormones, of course, are synthesized as part of larger propeptides. Early studies established the presence of additional peptides in pars nervosa extracts that lacked biological activity in the nonapeptide bioassays. These peptides first were thought to play some carrier function to bring the nonapeptides to the secretory granules and were named **neurophysins**. The first stains specific for staining the secretions of the pars nervosa that allowed us to trace the neurosecretory neurons to their source in the hypothalamus were actually staining these neurophysins. We now recognize two distinct propeptides, **prooxyphysin** (also called prooxytocin) and **propressophysin** (provasopressin) as the prohormones for OXY and the vasopressins, respectively (Figure 4-32). When prooxyphysin is hydrolyzed, it yields OXY plus **neurophysin I**, a peptide of 92 amino acids. Hydrolysis of propressophysin yields vasopressin, **neurophysin II** (93 amino acids), and a short, unnamed glycopeptide (39 amino acids).

A. The Biological Actions of Vasopressins

As we have already discussed, the main physiological role for vasopressins appears to be an antidiuretic action on the

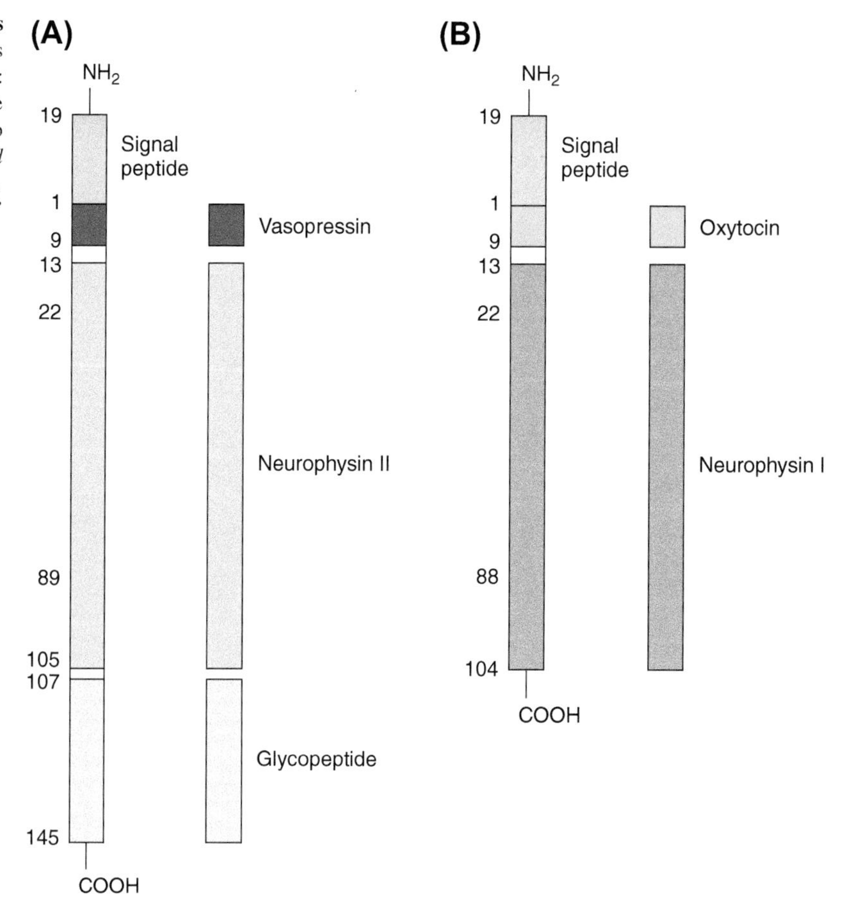

FIGURE 4-32 **Nonapeptide preprohormones and products.** (A) Prepropressophysin undergoes posttranslational processing to yield three peptides: vasopressin, neurophysin II, and a glycopeptide called copeptin. (B) Preprooxyphysin gives rise to two peptides: oxytocin and neurophysin I. *(Adapted with permission from Baulieu, E. and Kelly, P., "Hormones: From Molecules to Disease," Chapman & Hall, London, 1990, p. 206.)*

kidney with a secondary role in elevating blood pressure through effects on vascular smooth muscle. OXY causes contraction of reproductive tract smooth muscles in both males and females and contraction of myoepithelial cells lining the ducts of the mammary gland in females.

A brief examination of the amino acid composition of these nonapeptides reveals the bases for their biological activities. Antidiuretic and pressor activities apparently require a basic amino acid at position 8, and these actions are enhanced by phenylalanine at position 3. Peptides with neutral amino acids at position 8 exhibit predominately OXY-like actions, and the appearance of isoleucine at position 3 enhances that effect (see Table 4-4).

As already mentioned, vasopressins and OXY may act as neurotransmitters or neuromodulators in the central nervous system, and AVP enhances release of ACTH from the pituitary. Furthermore, a number of metabolic actions for both neurohormones have been reported, but it is not clear how important these putative roles are in relation to the dominant metabolic hormones such as GH, glucocorticoids, and insulin (see Chapter 12).

1. Antidiuresis and Blood Vascular Effects: Vasopressins

In the kidney, blood pressure determines the **glomerular filtration rate** at which water and dissolved solutes are filtered through capillary tufts known as glomeruli and enter the nephrons as the **glomerular filtrate**. Normally, the filtrate lacks plasma proteins and the cellular components of the blood plasma, and, initially, the concentrations of solutes such as Na^+ and glucose in the filtrate are identical to blood plasma. Numerous mechanisms operate to return solutes and most of the water back to the blood vascular system. Interference with the reabsorption process causes production of a larger than normal volume of urine (i.e., diuresis). Should reabsorption increase, more fluid is reabsorbed and antidiuresis may result in less urine production. Regulation of water reabsorption is critical to maintaining normal blood volume and blood pressure. Likewise, any marked changes in blood volume and concommitant changes in blood pressure will have effects on glomerular filtration rate and may lead to antidiuresis (e.g., by decreased blood pressure) or diuresis (e.g., due to increased blood pressure).

Vasopressins act on the cells lining the collecting ducts by binding to V_2 receptors (GPCR type) on the serosal surface of the cells of the collecting duct in the kidney (Figure 4-33). This V_2 receptor activates cAMP-based intracellular mechanisms that produce two effects. Phosphorylation of special cytoplasmic proteins called **aquaporins** causes them to interact with the luminal cell surface and allow water to enter the collecting duct cell. In addition, genes are activated to synthesize more aquaporin molecules. Water that had diffused into the cell then diffuses out through additional aquaporin molecules on the basal surface (see Figure 4-33). From here, water moves osmotically into the blood, resulting in an increase in total blood volume and an increase in blood pressure. This reuptake of water from the glomerular filtrate

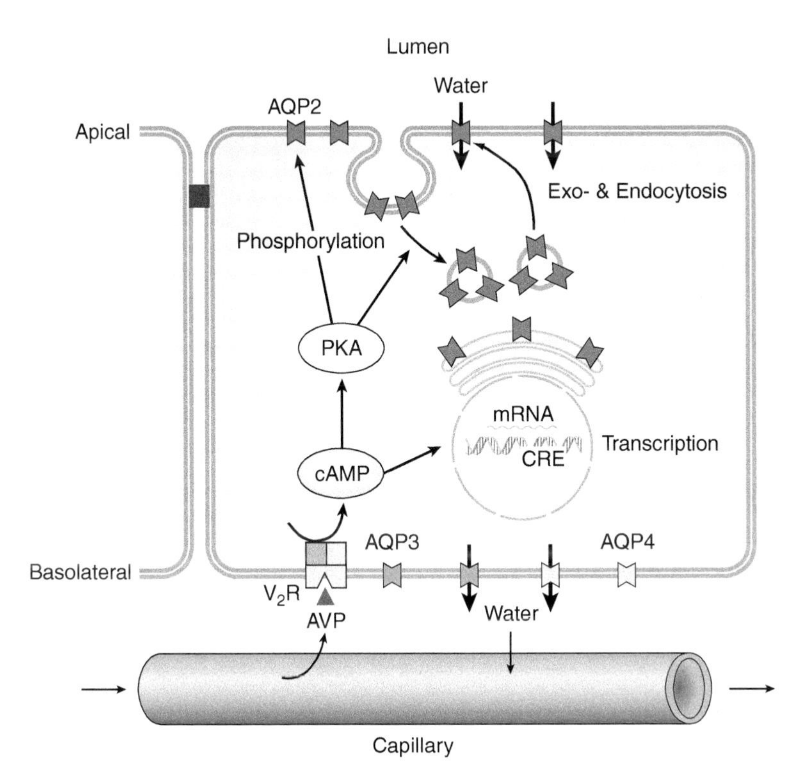

FIGURE 4-33 **Vasopressin actions on aquaporins and water transport.** Vasopressin binds to the V_2-receptor and via a cAMP second-messenger system stimulates synthesis of aquaporin-2. cAMP also activates protein kinase A that through phosphorylation directs the movements of newly synthesized aquaporin-2 molecules to the cell surface where they can participate in water transport. *(Adapted with permission from Sasaki, S. et al.,* Annual Review of Physiology, ***60****, 199–220, 1998.)*

reduces the volume of urine formed and elevates its solute concentration.

An increase in blood osmotic concentration and/or a decrease in blood pressure trigger the release of vasopressin which causes increased water reabsorption, antidiuresis, and a homeostatic decrease in blood osmotic concentration and/or elevation of blood pressure. Similarly, an increase in blood pressure and/or a decrease in blood osmotic concentration repress vasopressin secretion, causing diuresis with a corresponding drop in blood volume and pressure. Thus, vasopressins are important in the minute-to-minute regulation of blood volume and pressure as water uptake occurs in the intestine and water is lost from the body via, sweat, respiration, feces, and urine. A secondary action of vasopressins occurs in the brain where they stimulate thirst. Consumption of water will also add fluid to the blood vascular system and will increase blood pressure.

High doses of vasopressins cause contraction of arteriole smooth muscle and elevate blood pressure. This in turn increases the glomerular filtration rate sufficiently to override the normal antidiuretic action of the hormone and produces a net diuresis. Secretion of sufficient vasopressin to bring about arteriole constriction and attendant diuresis probably occurs following severe drops in blood pressure such as following hemorrhage. Neural regulation as well as other endocrine regulators of blood pressure, such as the renin–angiotensin system (see Chapter 8), may be more important regulators of blood pressure than vasopressins, however, and the vasopressins are primarily antidiuretic hormones.

2. Natriuretic Peptides (NPs)

Chronic high blood pressure can cause the release of **atrial natriuretic peptide (ANP)** from the heart into the general circulation (Figure 4-34). Initially named for its discovery in the artrium, ANP also is synthesized in the ventricles and is produced in the brain as well. The brain produces a second but less potent NP called **BNP**, which is also produced by the heart. Studies show that the major circulating forms of NPs are from the heart and not from the brain. A third NP, **CNP**, also has been described. NPs primarily produce their effects at the level of the kidney and at the adrenal cortex but also have effects in the central nervous system, accelerating sodium loss and hence promoting diuresis and lowering blood pressure. In addition, ANP inhibits release of vasopressin and further accelerates diuresis in an attempt to compensate homeostatically for the high blood pressure. NPs are released under conditions of chronic hypertension and have been shown to block secretion of aldosterone and vasopressins (see Chapter 8).

3. Other Factors Affecting Vasopressin Secretion

Ethyl alcohol has an inhibitory effect on release of vasopressins; hence, consumption of alcoholic beverages produces a diuresis by increasing blood volume, blood pressure, and glomerular filtration rate as well as through reduction in the efficiency of the reabsorption of water in the kidney. This inhibition of vasopressin release by alcohol may be prolonged even after blood pressure returns to normal so that excessive dehydration occurs, contributing to production of the severe headaches often associated with hangovers.

4. AVP and Corticotropin Release

The hypothalamus also is the source of AVP that potentiates ACTH release from the adenohypophysis, but, in this case, AVP is released directly into the portal system and does not come from the pars nervosa. This effect on ACTH release apparently is very important in prolonging the stress response to a chronic stressor (see Chapter 8).

B. Uterotonic and Milk Ejection Activities: Oxytocin

The contraction of uterine smooth muscle caused by OXY was almost immediately recognized as a methodology for artificially inducing labor in women, and it has been employed extensively at pharmacological levels for this purpose as well as to stimulate clamping down of the

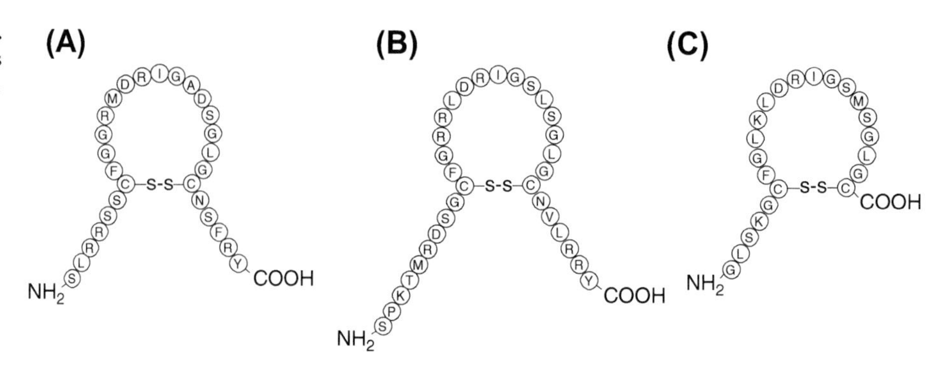

FIGURE 4-34 **Natriuretic peptides.** *(Adapted from Samson, W.K.,* Trends in Endocrinology and Metabolism, *3, 86–90, 1992. © Elsevier Science, Inc.)*

uterine muscles after birth to reduce postpartum bleeding. It was many years, however, before it was proven that OXY actually participates in the natural induction of labor in humans and other mammals.

OXY not only produces rhythmic contractions in the female reproductive tract but is also responsible for rhythmic contractions of the vas deferens and epididymis during ejaculation of sperm by males. The pleasurable sensation of orgasm involves rhythmic contractions of reproductive smooth muscle in both men and women and is induced by OXY as well.

In women, cows, and other female mammals, release of milk from the mammary glands is also induced by OXY. Exposure of mammary glands to estrogens stimulates development of the glands and the myoepithelial cells that line the ducts. Prolactin causes the epithelial cells of the mammary glands to synthesize milk, and OXY causes contraction of the myoepithelial cells lining the ducts to cause milk ejection.

Release of OXY is brought about through a neuroendocrine reflex. Suckling of the newborn on the nipple sends neural impulses to the brain that reach the hypothalamus and direct the release of OXY from the pars nervosa into the general circulation. Sufficient OXY in the circulation also can stimulate rhythmic oviduct contractions, leading, in the mother, to a pleasurable sensation during suckling.

Like AVP, administration of OXY has effects on tropic hormone release; however, no distinct endogenous role for OXY as a releasing hormone has been verified.

C. Nonapeptide Neurohormones and Behavior

Parental behavior in mammals is affected by nonapeptides. Among a group of rodents known as voles, some species such as the prairie vole (*Microtus ochrogaster*) are monogamous, whereas other species such as the meadow vole (*Microtus pennsylvanicus*) are non-monogomous. Both males and females of the monogomous prairie voles spend considerable time tending to pups in their nest (60 to 70% of their time), whereas male meadow voles spend only about 15% of their time tending the nest and even females spend only 35%. The distribution of OXY neurons differs in the brains of these two species, and corticosterone and AVP levels are high in prairie voles during periods of parental behavior. Maternal behavior in mice is regulated by nine genes, of which only one gene produces the prohormone for OXY. AVP not only induces parental behavior in males but also increases aggression toward strange males. In rats, OXY induces lordosis posture in females, a position of submission and acceptance to mounting and copulation by an amorous male. OXY also stimulates female sheep (ewes) to nurture their young.

Peptides closely related to OXY and vasopressins are localized in neurons of a number of invertebrates (see Chapter 5, Table 5-12) and have been shown to have behavioral roles. One of these peptides, **conopressin**, regulates ejaculatory behavior in male snails and regulates egg laying and reproductive behavior in female snails. **Annetocin** in earthworms is expressed in the subesophageal ganglion and regulates egg laying and other reproductive movements.

Administration of OXY via a nasal spray is claimed to increase the sense of trust between people when compared to others receiving a placebo nasal spray. Researchers suggest that this "trust" role for OXY may be important in the formation of bonds between offspring and their mother. The possibility that there may be roles for neural nonapeptides in human behavior may have important implications for parental social behavior.

VI. THE PINEAL GLAND

The human pineal gland was described by Galen during the second century as a structural (supportive) element within the brain. Much later, in 1646, Rene Descartes reasoned that the pineal was the location of the soul because it was an unpaired structure within an otherwise bilateral brain. It was, however, not until three centuries later that scientists determined what the soul was doing through the pineal gland. McCord and Allen in the early 1900s observed that pineal extracts caused blanching (lightening of the skin) of amphibian larvae by causing a concentration of melanin within the melanophores. Later, the pineal was discovered to play an important role in the control of seasonal reproduction.

The epithalamus represents the roof of the diencephalic portion of the brain that has differentiated into a variety of secretory structures in vertebrates as well as into photoreceptors (Figure 4-35). The **pineal gland** or **epiphysis** and the nearby **parapineal** are known as the **epiphysial complex**. Two additional prominent dorsal evaginations of the brain occur in the epithalamus: the **paraphysis** and the **dorsal sac**.

The pineal complex is connected to an adjacent ependymal structure, the **subcommissural organ of Dendy (SCO)**. The ependymal cells of the SCO produce a secretion rich in disulfide bonds and cysteine. This secretion is similar to that observed in the pineal ependyma. The major secretory product of the SCO is a non-cellular fiber that in some species extends into the central canal of the spinal cord for its entire length. This structure is known as **Reissner's fiber**. Its significance is not clear. The SCO and its Reissner's fiber have been described in vertebrates from cyclostomes to mammals. Once it was supposed that Reissner's fiber was involved in regulating posture through tension produced in it by flexion of the body. This tension presumably operated through influences of Reissner's fiber on pressure-sensitive

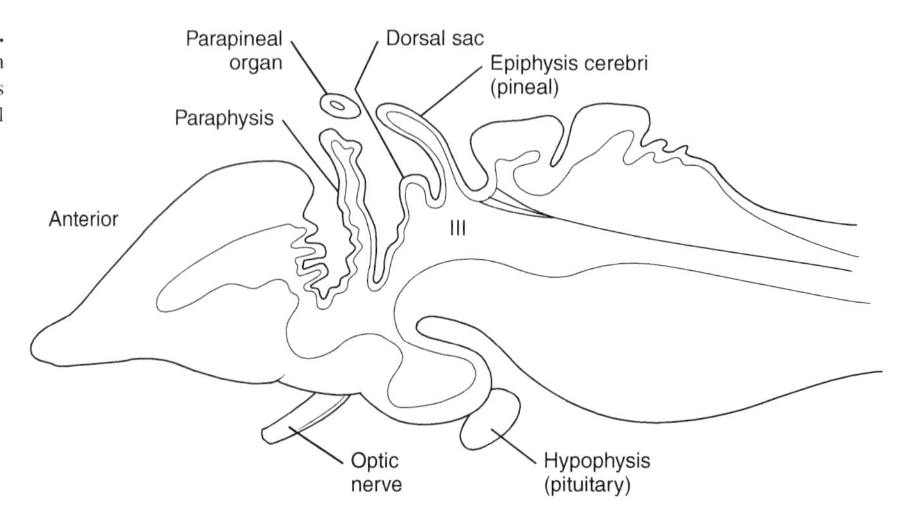

FIGURE 4-35 **Epithalamic structures.** Among the evaginations that develop from the roof of the epithalamus are the epiphysis cerebri or pineal, the parietal or parapineal organ, the dorsal sac, and the paraphysis.

neurons. A more plausible suggestion is the possibility that Reissner's fiber contributes to formation of cerebrospinal fluid. Formation and dissolution of the fiber into the CSF has been documented as a temperature-dependent process in the frog *Rana esculenta*. Reissner's fiber also binds biogenic amines (epinephrine, norepinephrine) present in the CSF in both *R. esculenta* and mammals (cow, cat). Studies with mammals and reptiles suggest a functional relationship among the pineal complex, the SCO, and the adrenal cortex, but the nature of that relationship remains somewhat obscure. Cytological activation of the SCO in the lizard *Lacerta s. sicula* is correlated positively with seasonal activities of adrenal cortical cells and of testicular steroidogenic cells. The actual role or roles for the SCO and its secretory products must await further research, but preliminary data would suggest it is somehow related to activity of the pineal complex.

1. The Pineal Gland and Melatonin

Many years after the observation that pineal extracts caused the skin of frogs to lighten, Lerner and his coworkers succeeded in isolating and characterizing the active skin-lightening agent, 5-methoxyl-*N*-acetyltryptamine or **melatonin**. Subsequently, a number of biologically active indoleamines and related compounds have been isolated from pineal tissue including melatonin and **serotonin (5-HT)**. Both melatonin and serotonin are secreted by the pineal into the blood and each has its unique actions. The initial substrate for synthesis of the indoleamines is the amino acid tryptophan that is converted to serotonin and then to melatonin (see Chapter 3, Figure 3-2).

B. Functions of the Pineal Gland in Mammals

A central action for the pineal gland is the regulation of endogenous rhythms, a role probably related to the primitive role of the epiphysial complex as a photoreceptive organ. The mammalian pineal gland also is implicated as an inhibitor of reproductive and thyroid activity. Recently, the pineal has come under scrutiny as a potential regulator of aging and of the immune system. Some confirmed actions of melatonin are summarized in Table 4-10.

1. The Pineal and Endogenous Rhythms

Melatonin in the blood exhibits a distinct diurnal rhythm. Levels are greater at night than during the day in almost every species examined, and the circadian rhythm usually persists under constant dark conditions. Plasma melatonin rhythm is a consequence of a circadian rhythm in activity of the rate-limiting enzyme, **N-acetyltransferase (NAT)** (Figure 4-36). This enzymatic rhythm is controlled by neural signals from the SCN of the hypothalamus that probably controls a number of circadian rhythms in mammals. The SCN is predominately composed of inhibitory GABA neurons that communicate with many other brain areas and apprise them of the photoperiod pattern. In mammals, information on photoperiod detected by the retina is responsible for entraining the SCN to light/dark cycles. Blood levels of melatonin are greatest during the dark or the **scotophase** of the day—night cycle and lowest during the **photophase** (daylight portion). Nighttime melatonin secretion occurs in three basic patterns (with numerous variations) showing differences in latency of the response after light disappears or in the temporal relationship to when light returns (Figure 4-37).

Most studies of pineal activity and light have been done in rodent species that show greater sensitivity to light than do large mammals such as humans and sheep. Nevertheless, similar mechanisms appear to be operating in most species. Light stimulates the retina of the eye that generates neural impulses via two pathways to alter pineal secretion

TABLE 4-10 Summary of Nonreproductive Actions of Melatonin and the Pineal in Mammals

Target	Description of Action
Melanophores (melanocytes)	Melatonin implants in weasels, *Mustela erminea*, cause them to grow white coats (typical of winter) in the spring instead of brown coats
Connective tissue	Pinealectomy reduces permeability of subcutaneous connective tissue
Parathyroid	Pinealectomy of rat caused hypertrophy of parathyroids, which was reduced by administration of pineal extract or melatonin
Cardiovascular system	Vasopressor activity reported for pineal extracts, probably due to presence of AVT
Immune response	Chronic administration of pineal extracts caused leukocytosis, lymph node hypertrophy, and an increase in mitotic activity in the spleen; probably it was a simple immunological response to antigens in the extract
Thyroid	Melatonin or pineal extracts inhibit thyroid function, possibly through regulation of TSH release from the adenohypophysis

BOX 4D Melanotropin and Retinal Regulation of Circadian Rhythms

A special pigment, melanopsin, was first discovered in the light-sensitive melanophores of larval amphibian skin. Melanopsin later was found in a small percentage of ganglion cells in the retinas of mice and humans. Melanopsin is sensitive to blue light wavelengths, and melanopsin-containing ganglion cells are responsible for pupillary reactions to light. Additionally, these cells regulate circadian rhythms in some blind animals and humans. Some of these special ganglion cells innervate neurons of the SCN and provide the SCN with information about the day–night cycle.

These discoveries not only explain the retention of circadian rhythm control in blind animals that have intact retinas but also explain why the use of blue wavelengths has been successful for treatment of various disorders ranging from circadian dysrhythmia (cause of "jet lag"), health and performance problems in night-shift workers, and **Seasonal Affective Disorder (SAD)** common to people living near the poles.

(Figure 4-38). The **retinohypothalamic pathway** innervates the SCN that in turn operates through the brain stem and intermediolateral cell column of the spinal cord to reduce the activity of sympathetic fibers traveling from the **superior cervical ganglion (SCG)** to the pinealocytes of the pineal gland. Normally, these postganglionic fibers release norepiniephrine, which increases cAMP in the pinealocytes, causing secretion of melatonin. These norepinephrine-secreting neurons also release the peptide NPY that modulates the responsiveness of pineal cells to norepinephrine (Figure 4-39). Elevated cAMP is associated with increased activity of NAT and subsequent melatonin synthesis. Light shining on the retina reduces NE input to the pineal, reducing cAMP levels, NAT activity, and melatonin synthesis via this pathway. Lesions in the retinohypothalamic pathway or the SCN do not necessarily abolish pineal secretory rhythms and led to the discovery of a second pathway between the retina and the brain stem that travels via the inferior accessory optic tract (Figure 4-38).

2. The Pineal Gland and Embryonic Development

During pregnancy, the fetus must rely on the mother not only for nutrients and removal of metabolic wastes but also information about the environment. Evidence supports a role for melatonin from the mother in providing the fetus with information on the day–night cycle. Thus, melatonin acts like a pheromone to the fetus to appraise it of environmental conditions. Even after birth, melatonin passes from the mother to the newborn through milk until the offspring's pineal begins its own rhythmic secretion.

3. Pineal Secretions and Reproduction

Major effects of the pineal gland in mammals are related to reproduction and are most pronounced in species that breed only during spring or fall. In 1941, Fiske reported that keeping rats under conditions of constant light increased the frequency of estrus (a time of enhanced female receptivity to the male; see Chapter 10). Several years later, Wurtman discovered that pinealectomy also increased the frequency of estrus in rats maintained under normal photoperiods, and a surge of investigation was launched into possible roles of photoperiod, the pineal gland, and melatonin in controlling sexual maturity and reproductive cycles in mammals. Animals whose annual reproductive cycles are controlled by photoperiod (length of photophase and/or scotophase) are termed *photoperiodic*. A mass of data soon appeared to suggest that melatonin released from the pineal gland acted through either the blood or cerebrospinal fluid on the hypothalamus or directly on the pituitary to lower circulating LH levels in photoperiodic animals. As mentioned earlier, light inhibits sympathetic input to the pineal, resulting in decreased melatonin synthesis followed by increased levels of LH leading to estrus.

FIGURE 4-36 **Serotonin and melatonin levels related to photoperiod and activity.** Both the quantity of melatonin and activity of the rate-limiting enzyme *N*-acetyl transferase (NAT) in the pineal gland of chickens increases during the photophase. Similar observations have been made in mammals. HIOMT, hydroxyindole-*M*-transferase. *(Adapted with permission from Binkley, S.A., "The Clockwork Sparrow: Time, Clocks, and Calendars in Biological Organisms," Prentice-Hall, Englewood Cliffs, NJ, 1990.)*

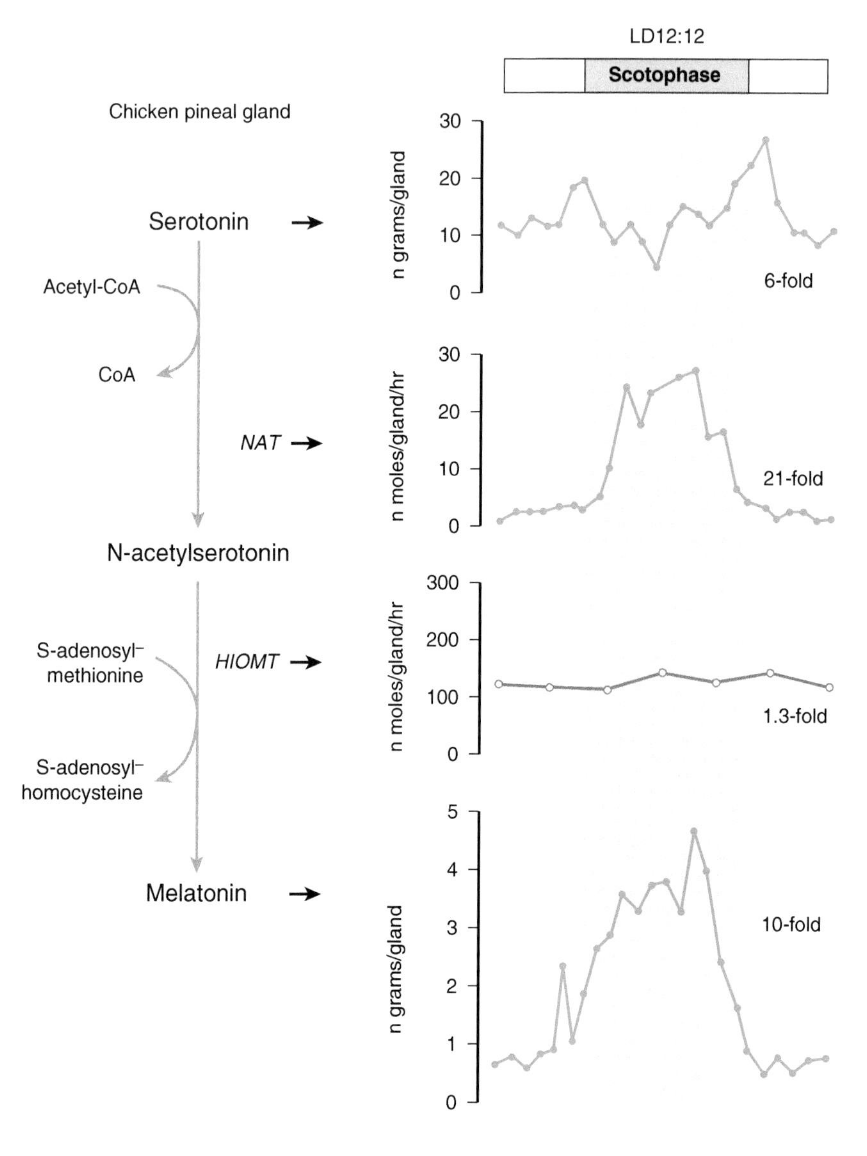

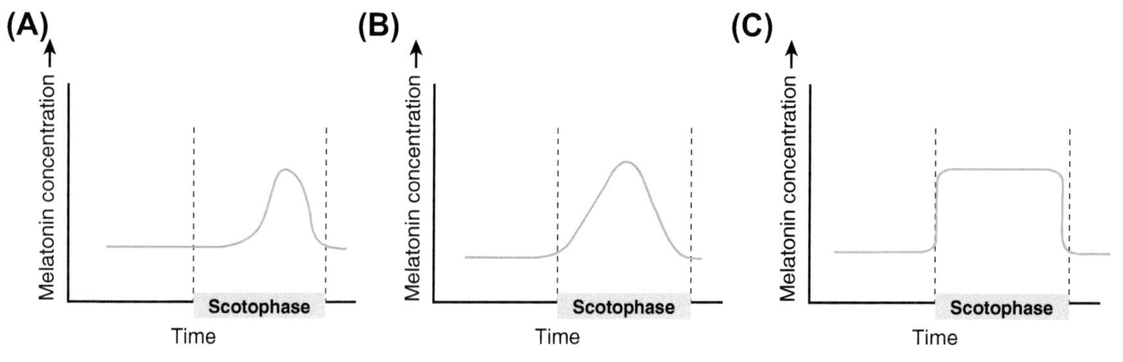

FIGURE 4-37 **Patterns of melatonin secretion.** Three distinct nocturnal secretory patterns are shown. (A) Increased secretion observed only during the second half of the photoperiod (house mouse, Syrian hamster). (B) Most common pattern where secretion begins soon after darkness, peaks at mid-photophase, and decreases prior to onset of photophase. (C) Maximal secretion reached immediately as soon as scotophase begins and continues to secrete at a more or less constant rate until the lights go on (Siberian hamster, domestic sheep). *(Adapted with permission from Reiter, R.J.,* Endocrine Reviews, ***12****, 151–180, 1991. © The Endocrine Society.)*

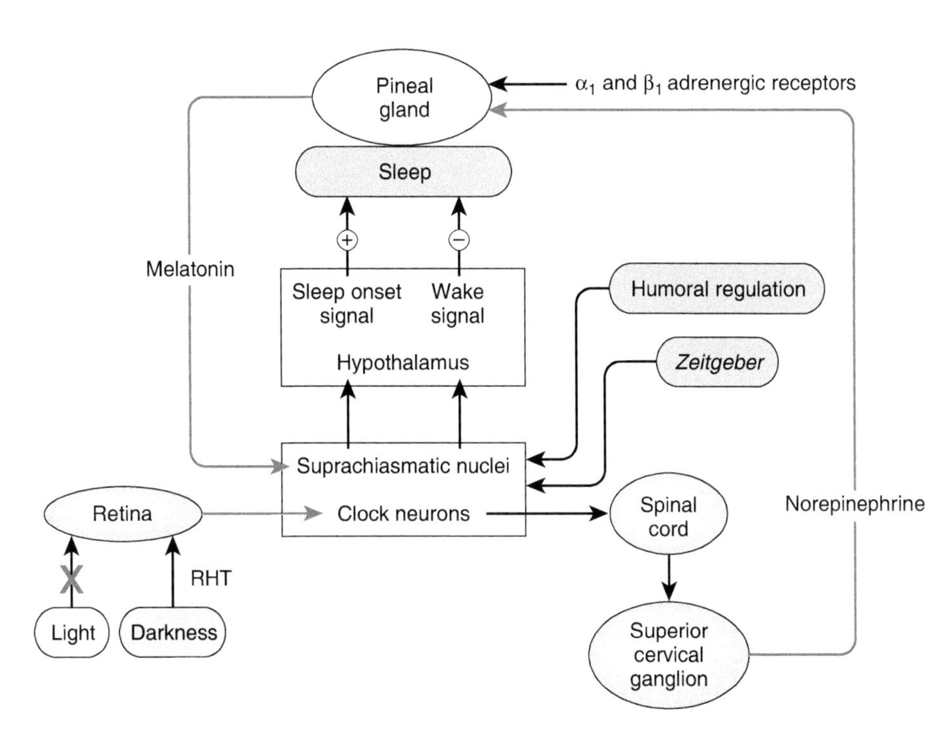

FIGURE 4-38 **Light and melatonin secretion.** Melatonin secretion is regulated by ambient light and the sympathetic nervous system (both inhibitory) as well as hormonal signals from the periphery. RHT, retino-hypothalamic tract.

Considerable evidence has accumulated from clinical studies that implicate the pineal gland in controlling the onset of puberty in humans, although little experimental work on humans is available. Circulating melatonin decreases by 75% between the ages of 7 to 12 years when LH levels are observed to rise. Furthermore, many cases of precocial puberty, especially in males, are associated with non-secreting pineal tumors that reduce melatonin output and allow for an early release of LH and consequent gonadal stimulation (for additional information on puberty and reproduction, see Chapter 10).

Many studies of the relationship between pineal and gonadal function have been conducted with the golden hamster (*Mesocricetus auratus*). This animal exhibits marked gonadal collapse when subjected to short photoperiods (less than 12 hours of light per day) when the daily period of elevated plasma melatonin is longest. Pinealectomized golden hamsters do not exhibit gonadal collapse when subjected to short photoperiods, and subdermal melatonin implants (administered in silastic capsules) cause testicular atrophy in hamsters maintained on long photoperiods.

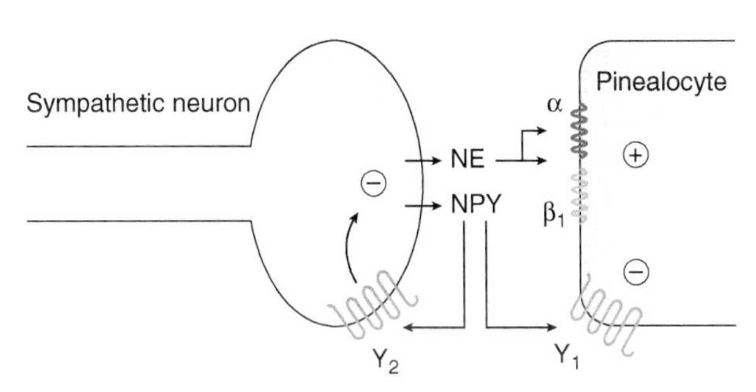

FIGURE 4-39 **Neuropeptide Y (NPY) and its interactions with norepinephrine (NE).** Activation of the release of melatonin is accomplished by NE secreted by sympathetic postgangionic neurons from the superior cervical ganglion. NPY acts as a local inhibitor via Y1 receptors to shut off the response to NE and via Y2 receptors to prevent additional NE release.

Some studies have provided a different explanation for the effects of light on estrus. Pinealectomy or injection of massive doses of melatonin never produces marked effects on rat reproduction, and some workers have not found any effect of melatonin on rat reproduction. One study, in fact, reported stimulation of rat gonads by melatonin treatments. If rats are made anosmic (olfaction blocked either mechanically or surgically) and are blinded, more marked gonadal atrophy occurs than was seen following pinealectomy alone.

For many mammals, longer photoperiods stimulate gonadal development only at certain times and at other times the animal does not respond; that is, it is **photorefractory**. Many photoperiodic animals exhibit photorefractoriness after breeding and cannot be induced to reenter the breeding condition. In the ferret, melatonin may be responsible for bringing the animal out of photorefractoriness, thus supporting a stimulatory gonadal role for melatonin. Treatment with melatonin restores the gonadal growth response to long photoperiod in photorefractory (postbreeding) ferrets maintained on artificially long photoperiods. In Djungarian hamsters, photorefractoriness appears in a different form. Short photoperiods cause gonadal collapse due to inhibitory effects of elevated melatonin secretion on the reproductive endocrine axis; however, after a period of continued short-day exposure, the animals develop photorefractoriness to the short photoperiod, and the gonads undergo recrudescence (= regrowth).

Ependymal cells of fetal human and rat pineals synthesize AVT. If AVT is administered to neonatal mice during the period when the brain is undergoing sexual differentiation, increased growth of reproductive organs is observed upon entering adulthood. In contrast, if AVT is administered after the brain has undergone sexual differentiation, the growth of accessory organs and in some cases growth of the gonads themselves is inhibited.

The hypertrophy of the remaining ovary after unilateral ovariectomy of mice is a response to increased GTH levels caused by an effective reduction in circulating estrogens (a feedback effect). AVT administered intraperitoneally or directly into the third ventricle of the brain prevents this **compensatory ovarian hypertrophy (COH)**. Much less AVT is required if it is administered through the third ventricle than if it is given intraperitoneally. Several related basic nonapeptides, including AVP and LVP, also inhibit COH, but OXY is ineffective. All of the nonapeptides active at inhibiting COH have an identical ring structure and a basic amino acid at position 8. Treatment of these active molecules with the reducing agent mercaptoethanol disrupts the disulfide bridges necessary for maintenance of the ring structure. Such reduced nonapeptides no longer prevent COH; in fact, they enhance it.

Several hypothalamic peptides, including AVT, OXY, AVP, TRH, and SST, are present in human pineals. In addition to sympathetic innervation coming indirectly from the SCN through the SCG, numerous peptidergic fibers are found in the pineal gland that originate in other brain regions (e.g., the PVN and habenular nucleus). Immunoreactive VIP, NPY, AVP, and OXY have been demonstrated in pineal nerve endings. This peptidergic innervation probably modulates pineal function; for example, receptors for VIP are present on pinealocytes and binding of VIP to these receptors activates a cAMP-mediated increase in NAT function. This mechanism enhances the stimulation of cAMP and NAT by norepinephrine.

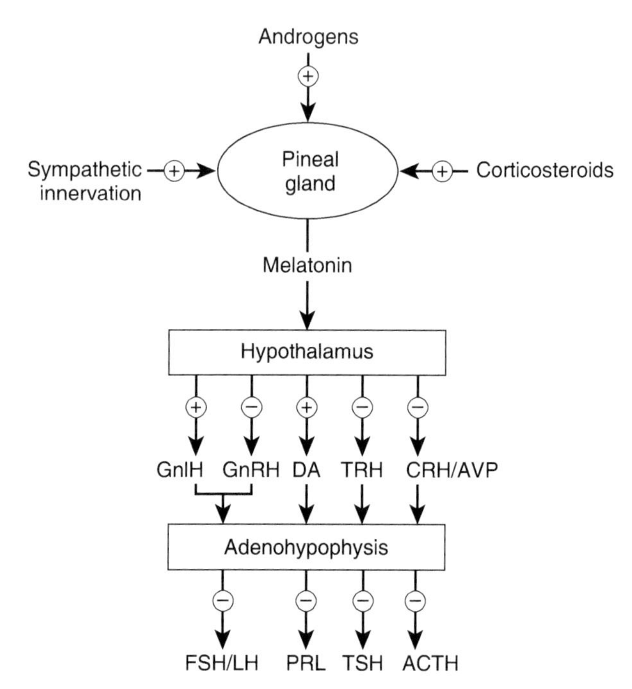

FIGURE 4-40 **Melatonin regulates hypothalamic functions.** The pineal receives input from hormones as well as sympathetic neural innervation that affects melatonin secretion. Melatonin in turn can block secretion of hypothalamic-releasing hormones (GnRH, TRH, CRH/AVP) as well as stimulate DA release, the PRL release-inhibiting hormone. Gonadal, thyroid, and adrenal hormones may provide negative feedback input through the pineal.

4. The Pineal Gland and Tropic Hormones

In addition to the inhibition of GTH secretion, melatonin may have important influences on secretion of other tropic hormones as well (Figure 4-40). Thyroid function in at least some mammals is strongly affected by photoperiod that appears to be acting through the control of melatonin secretion. Melatonin treatment reduces thyroid function presumably by limiting hypothalamic secretion of TRH and not by a direct action at the thyrotropic cells of the pars distalis.

The rise in PRL release observed in rats at the onset of the photophase has been linked to a reduction in melatonin release. Long photoperiods are correlated with increased PRL secretion in ruminant ungulates (sheep, cattle, goats), and melatonin treatment decreases PRL secretions in both sheep and goats although not in cattle. Pinealectomy or denervation of the pineal sometimes produces increased PRL secretion in goats and sheep but has no effect in cattle. Similarly, pineal activity is associated with seasonal breeding in sheep and goats but is not associated with reproduction in cattle. This difference among domesticated ungulates may be due to artificial selection as a result of selective breeding of cattle, sheep, and goats over many centuries, but information is lacking for wild ungulates. The explanation for this pineal effect apparently lies with the pars tuberalis. Cells in the pars tuberalis that produce tuberalin possess melatonin receptors, and their secretion is inhibited by melatonin. Thus, pinealectomy removes melatonin inhibition and can result in secretion of PRL.

Evidence for an effect of melatonin on ACTH release relies mainly on a few observations. Exogenous corticosteroids stimulate melatonin release that in turn might be involved in reducing ACTH release as part of a negative feedback loop. Furthermore, whereas the HPA axis reduces immune functions, melatonin, possibly acting at the hypothalamus and reducing ACTH release, can enhance certain immune functions. Melatonin also may enhance directly the functioning of the immune response system, and it may increase immune surveillance and decrease the incidence of cancer. Although supported by experiments in rodents, clinical studies have not substantiated such claims for humans.

5. The Pineal Gland and Aging

Biological aging is a complex phenomenon that is complicated often by attendant pathologies. There are many theories as to the causes and progression of aging and no single explanation seems adequate. One popular theory of aging involves the formation and accumulation of **free radicals**, compounds that can interact with and damage particular proteins, phospholipids, nucleic acids, and sugars. One of the most dangerous free radicals is produced during the breakdown of hydrogen peroxide. The presence of melatonin in certain *in vitro* systems reduces free radicals, whereas 5-HT, the predominant pineal secretion during daylight hours, elevates free radicals. In a number of pathological disorders, free radicals are known to be responsible for cell damage related to protein aggregations (e.g., Parkinson's disease, atherosclerosis, muscular dystrophy, multiple sclerosis, and rheumatoid arthritis). As a mammal ages, the SCN of the hypothalamus, which can send important regulatory messages to the pineal as well as to cells controlling pituitary function, becomes dysfunctional, and the pineal gland reduces production of melatonin and elevates production of 5-HT.

Another influence of the pineal on aging may be related to effects of melatonin on the immune response system. In contrast, circulating 5-HT of pineal origin may impair immune functions. Because decreased immune functions in general are associated with aging, melatonin may have a dual retarding effect on the aging process (e.g., by reducing free radicals and enhancing immune surveillance). Clinical studies do not support the use of melatonin as an anti-aging drug, however, and because of the many potential effects of melatonin on other systems The Endocrine Society recommends extreme caution in its use for any purpose.

F. Other Factors Affecting the Pineal

In addition to the well-known actions of light on the pineal gland, hypophysectomy, stress, and gonadal steroids all influence pineal function. Some of these effects suggest that the pineal may participate in feedback mechanisms of steroid hormones on tropic hormone secretion (Figure 4-40). For example, androgens (e.g., testosterone, dihydrotestosterone) inhibit the **monoamine oxidase (MAO)** enzymatic activity in the pineal that normally degrades the melatonin precursor 5-HT, which allows for increased melatonin secretion. Increased melatonin can reduce LH release and in turn decrease androgen levels. In contrast, estrogens can increase MAO activity and decrease pineal secretion of melatonin, bringing about enhancement of GTH release. Hypophysectomy or administration of histamine also reduces melatonin synthesis by reducing activity of the necessary enzyme **hydroxyindole-M-tranferase (HIOMT)**. Acute stress stimulates pineal activity, presumably through enhanced sympathetic stimulation. Chronic stress (for example, starvation) is associated with elevated corticosteroids that reduce pineal MAO activity and allow for increased melatonin production. Thus, stress working through these two pathways may depress reproductive function by reducing activity of the pineal gland.

G. Extrapineal Sources of Melatonin

The **harderian gland**, or Harder's lacrimal gland, was described by Harder in 1694 in the red deer. It is located directly behind and around the eye in all vertebrates that possess nictitating membranes (reptiles, birds, and most mammals). In humans, the harderian gland is poorly developed (rudimentary). Reddish porphyrin pigments present in the harderian gland undergo fluctuations correlated with lighting conditions. Prior to 12 days of age, the rat harderian gland contains little porphyrin pigment. Blinded 12-day-old rats exhibit an increase in pineal 5-HT as well as HIOMT activity during the scotophase, but this rhythm is abolished if the harderian glands also are removed. Melatonin synthesis has been demonstrated in the rat harderian gland, and continuous illumination causes enlargement of the rat harderian gland and an increase in HIOMT activity. Harderian HIOMT differs somewhat from the HIOMT found in the pineal gland and from that found in the retina of the eye. In contrast, continuous illumination decreases pineal weight and pineal HIOMT activity. The importance of these observations to overall involvement of pineal indoles or harderian indoles to observations on reproduction or other pineal-influenced processes remains to be determined.

The retina of the eye may be another viable source for melatonin, but the nocturnal increase in circulating melatonin is clearly of pineal origin. Retinal melatonin appears to regulate melanin distribution within the retina. Pinealectomy does not eliminate melatonin from the circulation but typically reduces it to daytime levels. Circulating melatonin during the photophase appears to be due to retinal and/or harderian gland production.

VII. CLINICAL ASPECTS OF THE NEUROENDOCRINE SYSTEM

Only disorders associated with the pars nervosa nonapeptides, GH secretion, and general pituitary dysfunction are considered here. Discussions of clinical problems that involve the HPT, HPG, and HPA axes are covered after specific discussions of these axes in later chapters (see Chapters 6, 8, 10, 12, and 14).

A. Disorders of the Hypothalamus

A rare clinical disorder of the hypothalamus is neurogenic **diabetes insipidus**. The patient produces an abnormally large volume (3 to 30 liters/day) of dilute urine. The term *diabetes* comes from the Greek meaning "to flow through" and *insipidis* comes from the French word for "tasteless" (compared to sweet urine of a person with diabetes mellitus that contains glucose). The resulting diuresis is most commonly due to the absence of AVP, which normally controls water reabsorption in the kidney. Forty to 50% of such patients are **idiopathic** (i.e., denoting a disease of unknown cause) and exhibit no other evidence of neuroendocrine dysfunction. Some of these patients have been identified at autopsy as having degeneration of the SON and the PVN. About 15% of the cases are related to the presence of tumors within the brain, producing pressure on the SON and PVN and indirectly reducing secretion of AVP. Physical damage (such as a brain lesion) or consequences of certain infections (e.g., encephalitis) account for the remainder.

Three other forms of diabetes insipidis exist. **Nephrogenic diabetes insipidus** is the result of a failure of the kidney tubules to respond to normal or above normal levels of AVP. This could be due to abnormal V2 receptors or to an inadequate numbers of receptors. A defect in the aquaporins also could be responsible. Damage to the thirst center in the hypothalamus can result in an abnormal increase of water intake, causing increased blood pressure that inhibits vasopressin secretion, contributing to excessive diuresis. This condition is called **dipsogenic diabetes insipidis**. During pregnancy, **gestational diabetes insipidis** may occur due to excessive production of an enzyme, placental vasopressinase, that degrades AVP. Additionally, several complications of pregnancy can induce this condition.

The **syndrome of inappropriate antidiuresis (SIAD)** is caused by excessive AVP release. High levels of AVP result in excessive water retention, elevated blood pressure, and reduced urine production. Drugs such as demeclocycline block the action of AVP on the kidney and are used to treat this condition. However, long-term use of such drugs induces nephrogenic diabetes insipidus. Medications that control AVP release from the pars nervosa often are not uniformly effective and can influence other hypothalamic functions.

B. Disorders Associated with the Pineal Gland

Tumors within the central nervous system are responsible for a number of disorders including **sexual precocity** (accelerated sexual maturation; see Chapter 10 for a technical description of precocity). Precocity is much more common in males. About one-fourth of precocity cases are correlated with the presence of a pineal tumor, and 95% of these occur in males. A number of other cases of sexual precocity are associated with hypothalamic tumors, most of which also are found in males. One type of tumor associated with precocity, the **hamartoma**, occurs in the posterior hypothalamus. It consists of masses of partially disoriented glial and ganglion cells or of normal cells located in abnormal sites. Some hamartomas secrete GnRH, which could explain their effects on early sexual maturation. Other causes for precocity are discussed in Chapter 10.

Melatonin is thought to play a critical role in **circadian dysrhythmia**: the disruption of normal body rhythms due to air travel. Crossing one or more time zones in a 24-hour period results in a significant change in the photoperiod relationship and causes biological rhythms to get out of phase with the day–night cycle, hence dysrhythmia. Crossing numerous time zones within 24 hours magnifies this disruptive effect. People often feel out of sorts when these rhythms are disrupted. Oral administration of melatonin is claimed to help reset the pineal and restore normal rhythms so that travelers feel in tune with their environment. Similarly, melatonin has been claimed to have sleep-promoting effects that would help one adjust to new time zones. Many elderly people claim that self-administered melatonin helps them sleep as well, although clinical studies are still controversial on this point.

Melatonin has been implicated in a variety of psychiatric disorders. Alterations in melatonin production are known to be associated with SAD, bipolar disorder, unipolar depression, bulimia, anorexia, schizophrenia, panic disorder, and obsessive–compulsive disorder. The manic high in bipolar disorder is associated with elevated melatonin secretion, whereas the depression phase is accompanied by reduced melatonin secretion. High and low melatonin are also correlated with low and high glucocorticoid levels, respectively, in bipolar disorder. It is not clear whether the pineal is involved in the onset of these disorders or is a consequence of other causative disturbances for these disorders.

C. Disorders of the Adenohypophysis

Pituitary chromophobe adenomas are the most common source of pituitary-related problems. An adenoma is a benign or noncarcinogenic glandular tumor. They rarely secrete any hormones (occasionally GH and rarely PRL, TSH, or ACTH), and their effects usually are due to pressure on the brain or optic chiasm caused by growth of the tumor. Most patients experience severe headaches and visual disturbances (even blindness can result). Sometimes the production of one or more pituitary hormones may be reduced. Other adenomas may secrete excessive amounts of PRL (prolactinomas). Rarely, TSH, ACTH, or GTHs may be products of pituitary adenomas, and their clinical impacts are discussed in Chapters 6, 8, and 10, respectively. Surgical removal of the adenoma is the most common treatment and is highly successful.

Acromegaly is a spectacular, although rather rare, disorder of GH regulation that has been publicized heavily. It affects from 3 to 40 individuals per million people in the United States each year. When it occurs in children, it can lead to **gigantism**. It is a well-known disorder because acromegaly was the first disorder of the pituitary gland to be recognized and because it can produce "giant" individuals. Acromegaly is caused by overproduction of GH due either to the absence of adequate SST to suppress GH release or by the absence of negative feedback to suppress release. Growth hormone-secreting tumors are unaffected by normal control mechanisms and release GH autonomously. Approximately half of acromegalic patients are deficient in one or more additional pituitary hormones, usually the GTHs, due to the growth of the GH tumor that compresses other cell types. Not only do such patients exhibit excessive growth, but body proportions become distorted. When acromegaly develops in adults, only the body proportions become distorted, as growth in stature has ceased and cannot be reinitiated with any quantity of GH. In adults, cartilage in joints proliferates under the influence of GH-dependent IGFs, resulting in abnormally proportioned hands and elongate jaws. The nose and ears, whose supporting tissue is cartilage, enlarge markedly, causing a distortion in appearance. There also are marked effects on other systems; for example, excessive sweating and secretion of sebum by the skin, enlargement of the heart, and hypertension may develop. Life expectancy is shortened considerably for these victims. Use of excessive amounts of GH by people to augment muscle growth in order to improve physical performance can produce effects similar to those of acromegaly. Excessive use of exogenous GH also has been associated with increased incidence of colon cancer.

Removal of the pituitary GH tumor has been the classical method for treating acromegaly although this can also produce deficiencies of other pituitary hormones. Use of a synthetic GH-receptor antagonist that blocks formation of IGFs can bring growth into normal limits although it does not address the initial cause of the disorder.

In contrast, lack of sufficient GH during early life can cause short stature; however, other explanations for subnormal growth are known. Furthermore, it is important to distinguish between **short stature** with normal body proportions and **dwarfism**, where the individual exhibits distorted features. **Laron-type dwarfism** occurs in humans with normal GH levels but very low IGF blood levels. Due to possession of a mutant allele, the liver of these patients lacks GH receptors and does not secrete IGFs. GH-binding proteins are absent from the blood as well. In contrast, African pygmies lack the binding proteins for GH and show the normal pattern for growth proportions but exhibit short stature. The inability to maintain normal circulating levels of GH causes total growth to be reduced.

Partial or total **hypopituitarism** refers to selective or total absence of pituitary hormones. These defects may reside in the adenohypophysis itself (primary disorder) or be due to hypothalamic dysfunction (secondary disorder).

There has been considerable interest in the role of EOPs on mental disturbances such as depression and schizophrenia. Clinical studies, however, have yielded mixed results, and it is not clear whether mental illness is associated with EOPs. CRH and TRH deficiencies or excesses also have been correlated with certain psychological disturbances.

VIII. SUMMARY

The adenohypophysis produces tropic hormones and consists of a pars distalis, a pars intermedia, and a pars tuberalis. The embryonic origin of the secretory cells in the adenohypophysis appears to be neural in origin similar to the neurohypophysis and hypothalamus. The pars distalis produces GH, PRL, GTHs, (FSH, LH), ACTH, TSH, and endorphins (EOPs). Four of the tropic hormones and their endocrine target glands constitute the major hypothalamus–pituitary (HP) axes of the neuroendocrine system: HPG (gonad), HPT (thyroid), HPA (adrenal), and HPH (hepatic or liver). The posterior portion of the adenohypophysis is the pars intermedia, which is responsible for synthesis of α-MSH and endorphins. A number of mammals lack a pars intermedia. The pars tuberalis contains some stainable cell types and secretes tuberalin, which stimulates PRL release.

The neurohypophysis forms from the infundibulum and consists of an anterior neurohemal area, the median eminence, and a more posterior neurohemal structure, the pars nervosa. The median eminence stores the hypothalamic-releasing hormones (RHs and RIHs) produced in hypothalamic nuclei that regulate tropic hormone release from the adenohypophysis. The portal system connects the median eminence to the adenohypophysis and brings RHs and RIHs to the tropic-hormone-producing cells. The pars nervosa has no common blood supply with the adenohypophysis. It is responsible for storage of nonapeptide neurohormones (usually AVP and OXY) produced in the hypothalamus until they are released into the general circulation.

There are three categories of tropic hormones based on chemical structure. Category 1 includes the glycoproteins LH, FSH, and TSH. Category 2 includes the large peptides GH and PRL. Category 3 includes the very similar smaller peptides ACTH, α-MSH, and LPH. Structural similarities suggest that LPH may be related to the endorphins, which are linked to opiate actions on the mammalian central nervous system. Placentas of mammals produce up to four tropic-like hormones, including CG (LH-like), CS (PRL-like and GH-like to some degree), CT (TSH-like), and CC (ACTH-like), as well as PRL and GH. Pregnant mare serum gonadotropin is a chorionic-type GTH with both FSH- and LH-like properties. Menopausal gonadotropin is FSH-like and is produced by the pituitary of postmenopausal women.

The neurosecretory nuclei controlling tropic hormone release are generally located in the preoptic area or the ventral region of the hypothalamus. Aminergic and peptidergic neurons terminate in the median eminence where they regulate release of RHs and RIHs into the portal circulation. These neurosecretory neurons are innervated by regular neurons (e.g., aminergic, cholinergic, serotonergic, and/or peptidergic) that influence their release. These regulating neurons may be responsive to hormones or other factors as well.

TSH release is stimulated by TRH secreted primarily by neurons in the PVN. TRH release can be increased by VIP, norepinephrine, and epinephrine and inhibited by serotonin and dopamine. Thyroid hormones feedback primarily on the thyrotropes with only minor effects in the hypothalamus.

Gonadotropin release (LH and FSH) is induced by pulsatile secretion of GnRH-1 from the ARC that in turn is stimulated by norepinephrine but inhibited by opioid-secreting neurons. Negative feedback occurs generally through actions of gonadal steroids at the hypothalamus neurons or on gonadotropes. Inhibin secreted by the gonads feeds back specifically to limit FSH release from the gonadotrope, whereas galanin from the hypothalamus enhances LH release. Positive feedback by estrogens appears to operate primarily through Kp neurons in the ARC and AVPV that control GnRH-1 release and possibly through GABA neurons that innervate GnRH-1 neurons.

GH release is under strong inhibitory control by hypothalamic SST, but GHRH is necessary to evoke release in the absence of SST. SST comes primarily from the PERIV nucleus and the amygdala of the limbic system. GHRH is produced in the ARC nucleus. Many neural and neurosecretory factors (e.g., ghrelin) can affect GH release but the pathway and physiological importance of most observations are unclear. Catecholamines are always stimulatory. Many actions of GH are mediated by IGFs, and the latter form the primary negative feedback on GH release. Metabolites (glucose, free fatty acids, arginine) can affect GH release but their physiological role is uncertain.

PRL release is under strong inhibitory control by dopamine acting as a neurohormone from the ARC and PERIV nuclei. It also can be inhibited by GAP, although a physiological role for GAP is not clear. Unlike the situation for GH, no PRH seems necessary to get PRL release in the absence of DA inhibition. During suckling, however, hypothalamic VIP is an important stimulant. Estrogens enhance the sensitivity of lactotropes to VIP. Tuberalin from the pars tuberalis stimulates PRL release. Although TRH has been shown to induce PRL release, it may not be a physiological releaser. Negative feedback may occur through direct actions of PRL on hypothalamic neurons. The possible role of PrRP in stimulating PRL release is controversial.

ACTH secretion is stimulated by CRH, and AVP enhances the sensitivity of corticotropes to CRH. Corticotropes also produce β-LPH from which they release β-endorphin. β-LPH, β-endorphin, and ACTH are produced from the same preprohormone, POMC. CRH is produced primarily in the PVN and ARC nucleus, and greatest levels are observed in the median eminence. AVP in the median eminence comes primarily from CRH neurons in the PVN. Glucocorticoids provide negative feedback directly on corticotropes and on hypothalamic CRH neurons as well as indirectly via several brain areas outside the hypothalamus (see Chapter 8).

GABA neurons play an important inhibitory role over CRH release. In chronic stress, glucocorticoid feedback is in part overcome by elevated AVP.

Release of melanotropin (α-MSH), like PRL, is inhibited by dopamine, but the dopamine is secreted from nerve terminals directly innervating the pars intermedia. Melanotropes first produce ACTH from POMC and then hydrolyze it further to α-MSH and CLIP prior to release. Like corticotropes, they also release β-endorphin into the general circulation. There is little evidence for negative feedback on α-MSH release.

Neurosecretory neurons in the SON and PVN produce the nonapeptides AVP and OXY, which are stored in the pars nervosa. A few mammals make LVP as well as AVP, and marsupials produce PVP instead of AVP. Some marsupials secrete MST. Fetal mammals produce AVT instead of AVP and OXY.

The major secretory product of the pineal gland or epiphysis is melatonin, although peptides such as AVT may be secreted as well. The pineal receives innervation from several sources. In mammals, the most important pathway is the sympathetic innervation of pinealocytes which is controlled by two retinal pathways both passing through the superior cervical ganglion. One of these pathways involves the hypothalamic SCN, a major determiner of biological rhythms in mammals. Pineal secretion of melatonin is inhibited by light, and the pineal gland plays a major role in mediating seasonal and daily endocrine activity primarily through effects on the hypothalamus. Melatonin generally acts as an anti-gonadal, anti-adrenal, or anti-thyroid factor. It is also an anti-PRL factor through its inhibition of tuberalin secretion. The pineal may have a role in delaying aging and enhancing immune surveillance.

STUDY QUESTIONS

1. Describe the major regions of the pituitary gland and their embryonic origins.
2. Identify the derivation of the various names used to identify the pituitary gland and its components: (a) hypophysis; (b) pituitary; (c) adenohypophysis; (d) neurohypophysis; (e) median eminence; (f) pars distalis; (g) pars intermedia; (h) pars nervosa.
3. Imagine that you are a molecule of CRH that has just been secreted from a nerve ending in the median eminence and answer the following questions:
   (a) In what part of the brain or pituitary were you synthesized?
   (b) Were you synthesized by an epithelial cell, a nerve cell, or a glial cell?
   (c) What biosynthetic building blocks were used in your synthesis? Carbohydrates? Fatty acids? Amino acids?
   (d) In your journey through the pituitary gland, which cell type will you stimulate? How will you recognize the appropriate cell and stimulate it?
4. We sometimes describe the stimulus for secretion of hormones from the adrenal glands, gonads, and thyroid glands as beginning with releasing factors in the hypothalamus. In your opinion is this an accurate representation of how an endocrine axis responds to predictable and unpredictable changes in the environment? Explain your answer.
5. Identify examples within the major classes of neurotransmitters that innervate the hypothalamus and regulate releasing hormone secretion.
6. Explain how dopamine can be both a hormone and a neurotransmitter in regulating endocrine function.
7. Describe the significance of APUD cells.
8. Identify five hormone-producing cell types in the pars distalis. How many hormone-producing cell types are in the pars intermedia?
9. In some mammals (such as primates) the pars intermedia is entirely degenerated in adults, whereas other mammalian species (rodents, for example) maintain an active pars intermedia throughout most of their lifetime. What does this tell you about the physiological significance of MSH released from the pars intermedia in mammals?
10. Corticotropes in the pars distalis and melanotropes in the pars intermedia both express the POMC gene but secrete entirely different peptides. Describe how posttranslational processing of POMC differs in these two pituitary lobes.
11. Describe the role of accessory proteins in melanocortin receptor function.
12. How do GnRH, GnIH, and kisspeptin neurons interact to regulate gonadotropin secretion?
13. How many types of GnRH are present in the mammalian brain? Do all GnRH neurons in the mammalian brain regulate gonadotropin secretion? Explain your answer.
14. How many genes must be transcribed to produce one active molecule of TSH?
15. Which pituitary hormones can be referred to as *heterodimeric glycoproteins*? How do the subunits for these hormone influence their biological activity?
16. What is the target tissue for growth hormone in regulating bone growth?
17. Morphine is an alkaloid extracted from poppy plants that binds to mu opioid receptors in the mammalian brain. How does the structure of morphine differ from opioids that are encoded by genes expressed within neurons in the brain? Is there a morphine gene expressed in the human brain? If not, does this mean that morphine cannot be produced in the mammalian brain? Explain your answer.
18. Describe how vasopressin is produced from prepropressophysin.
19. Describe the role of vasopressin in regulating water movement in the collecting duct of the nephron. What cellular events are initiated in collecting duct cells after vasopressin binds to its receptor?
20. Is there a melatonin gene expressed by cells in the pineal gland? If not, how is melatonin synthesized?
21. Is the pineal gland innervated by the retina? If not, describe how signals from the retina ultimately direct melatonin secretion from the pineal gland.

SUGGESTED READINGS

Books

Bell, W.B., 2012. The Pituitary. Nabu Press, Charleston, SC.

Binkley, S.A., 1990. The Clockwork Sparrow: Time, Clocks, and Calendars in Biological Organisms. Prentice-Hall, Englewood Cliffs, NJ.

Cooper, J.R., Bloom, F.E., Roth, R.H., 2002. The Biochemical Basis of Neuropharmacology, eighth ed. Oxford University Press, New York.

Conn, P.M., Freeman, M.E., 2010. Neuroendocrinology in Physiology and Medicine. Humana Press, Totowa, NJ.

Fink, G., Pfaff, D.W., Levine, J., 2011. Handbook of Neuroendocrinology. Academic Press, Boston, MA.

Imura, H., 1994. The Pituitary Gland. In: Comprehensive Endocrinology Revised Series, second ed. Raven Press, San Diego, CA.

Lovejoy, D.A., 2005. Neuroendocrinology: An Integrated Approach. John Wiley & Sons, New York.

Motta, M., 1991. "Brain Endocrinology. In: Comprehensive Endocrinology Revised Series, second ed. Raven Press, San Diego, CA.

North, W.G., 1993. The Neurohypophysis: A Window on Brain Function, vol. 689. New York Academy of Sciences, New York.

Swearingen, B., Biller, B.M., 2010. Diagnosis and Management of Pituitary Disorders. Humana Press, Totowa, NJ.

Articles

General/Miscellaneous

Blackmore, C.G., Varro, A., Dimaline, R., Bishop, L., Gallacher, D.V., Dockray, G.J., 2001. Measurement of secretory vesicle pH reveals intravesicular alkalinization by vesicular monoamine transporter type 2 resulting in inhibition of prohormone cleavage. Journal of Physiology 531, 605–617.

Brabant, G., Cain, J., Jackson, A., Kreitschmann-Andermahr, I., 2011. Visualizing hormone actions in the brain. Trends in Endocrinology and Metabolism 22, 153–163.

Brann, D.W., Mahesh, V.B., 1992. Excitatory amino acid neurotransmission: evidence for a role in neuroendocrine regulation. Trends in Endocrinology and Metabolism 3, 122–126.

Boyd, C.A., 2001. Amine uptake and peptide hormone secretion: APUD cells in a new landscape. Journal of Physiology 531, 581.

Callewaere, C., Banisadr, G., Rostène, W., Parsadaniantz, S.M., 2007. Chemokines and chemokine receptors in the brain: implication in neuroendocrine regulation. Journal of Molecular Endocrinology 38, 355–363.

Dahlström, A., Fuxe, K., 1964. Evidence for the existence of monoamine-containing neurons in the central nervous system. I. Demonstration of monoamines in cell bodies of brainstem neurons. Acta Physiologica Scandinavica 232 (Suppl), 1–55.

Dubois, P.M., El Amraouci, A., 1995. Embryology of the pituitary. Trends in Endocrinology and Metabolism 6, 1–7.

Evans, J.J., 2002. The anterior pituitary gland is mysterious, alluring and useful. Archives of Physiology and Biochemistry 110, 3–8.

Fannon, S.A., Vidaver, R.M., Marts, S.A., 2002. Sex, cells and signals in the developing brain. Trends in Neuroscience 25, 334–335.

Fauquier, T., Lacampagne, A., Travo, P., Bauer, K., Mollard, P., 2002. Hidden face of the anterior pituitary. Trends in Endocrinology and Metabolism 13, 304–309.

Gershon, M.D., 1993. Development of the neural crest. Journal of Neurobiology 24, 141–145.

Guillemin, R., 2005. Hypothalamic hormones a.k.a. hypothalamic releasing factors. Journal of Endocrinology 184, 11–28.

Hammer, G.D., Ingraham, H.A., 1999. Steroidogenic factor-1: its role in endocrine organ development and differentiation. Frontiers in Neuroendocrinology 20, 199–223.

Kobayashi, H., Yamaguichi, Y., Uemura, H., 1999. The median eminence: a mediator in the regulation of the pituitary by the brain. In: Rao, D.S.P., Peter, R.E. (Eds.), "Neural Regulation in the Vertebrate Endocrine System". Kluwer Academic/Plenum, New York, pp. 1–22.

Lechan, R.M., Toni, R., 2012. Functional anatomy of the hypothalamus and pituitary Chapter 3B. www.endotext.org.

Moore, R.Y., 1996. Neural control of the pineal gland. Behavior and Brain Research 73, 125–130.

Pearse, A.G.E., Takor, T.T., 1976. Neuroendocrine embryology and the APUD concept. Clinical Endocrinology 5 (Suppl), 229s–244s.

Raisman, G., 1997. An urge to explain the incomprehensible: Geoffrey Harris and the discovery of the neural control of the pituitary gland. Annual Review of Neuroscience 20, 533–566.

Reiter, R.J., 1991. Pineal melatonin: cell biology of its synthesis and its physiological interactions. Endocrine Reviews 12, 151–180.

Schwartz, J., Cherny, R., 1992. Intercellular communication within the anterior pituitary influencing the secretion of hypophysial hormones. Endocrine Reviews 13, 453–475.

Stojilkovic, S.S., 2001. A novel view of the function of pituitary folliculo-stellate cell network. Trends in Endocrinology and Metabolism 12, 378–380.

Toni, R., Malaguti, A., Benfenati, F., Martini, L., 2004. The human hypothalamus: a morphofunctional perspective. Journal of Endocrinology Investigation 27 (Suppl. 6), 73–94.

Zhu, X., Gleiberman, A.S., Rosenfeld, M.G., 2007. Molecular physiology of pituitary development: signaling and transcriptional networks. Physiological Reviews 87, 933–963.

Zhu, X., Wang, J., Ju, B.-G., Rosenfeld, M.G., 2007. Signalling and epigenetic regulation of pituitary development. Current Opinion in Cell Biology 19, 605–611.

Hypothalamic RHs and RIHs

Bakker, J., Balm, M.J., 2000. Neuroendocrine regulation of GnRH release in induced ovulators. Frontiers in Neuroendocrinology 21, 220–262.

Ben-Shlomo, A., Melmed, S., 2010. Pituitary somatostatin receptor signaling. Trends in Endocrinology and Metabolism 21, 123–133.

Colledge, W.H., 2009. Kisspeptins and GnRH neuronal signaling. Trends in Endocrinology and Metabolism 20, 115–121.

Fekete, C., Lechan, R.M., 2007. Negative feedback regulation of hypophysiotropic thyrotropin-releasing hormone (TRH) synthesizing neurons: role of neuronal afferents and type 2 deiodinase. Frontiers in Neuroendocrinology 28, 97–114.

Fukusumi, S., Fujii, R., Hinuma, S., 2006. Recent advances in mammalian RFamide peptides: the discovery and functional analyses of PrRP, RFRPs and QRFP. Peptides 27, 1073–1086.

Hauger, R.L., Grigoriadis, D.E., Dallman, M.F., Plotsky, P.M., Vale, W.W., Dautzenberg, F.M., 2003. International Union of Pharmacology. XXXVI. Current status of the nomenclature for receptors for corticotropin-releasing factor and their ligands. Pharmacological Reviews 55, 21–26.

Kiaris, H., Chatzistamou, I., Papavassiliou, A.G., Schally, A.V., 2011. Growth hormone-releasing hormone: not only a neurohormone. Trends in Endocrinology and Metabolism 22, 311–317.

Kriegsfeld, L.J., Mei, D.F., Bentley, G.E., Ubuka, T., Mason, A.O., Inoue, K., Ukena, K., Tsutsui, K., Silver, R., 2006. Identification of a gonadotropin-inhibitory system in the brains of mammals. Proceedings of the National Academy of Sciences USA 103, 2410–2415.

Krsmanovic, L.Z., Hu, L., Leung, P.K., Feng, H., Catt, K.J., 2010. Pulsatile GnRH secretion: roles of G protein-coupled receptors, second messengers and ion channels. Molecular and Cellular Endocrinology 314, 158–163.

Lengyel, A.M.J., 2006. From growth hormone-releasing peptides to ghrelin: discovery of new modulators of GH secretion. Arquivos Brasileiros de Endocrinologia & Metabologia 50, 17–24.

Neill, J.D., 2002. Minireview: GnRH and GnRH receptor genes in the human genome. Endocrinology 143, 737–743.

Roa, J., Navarro, V.M., Tena-Sempere, M., 2011. Kisspeptins in reproductive biology: consensus knowledge and recent developments. Biology of Reproduction 85, 650–660.

Schwanzel-Fukuda, M., Jorgenson, K.L., Bergen, H.T., Weesner, G.D., Pfaff, D.W., 1992. Biology of normal luteinizing hormone-releasing hormone neurons during and after their migration from olfactory placode. Endocrine Reviews 13, 623–634.

Smith, J.T., Clarke, I.J., 2010. Gonadotropin inhibitory hormone function in mammals. Trends in Endocrinology and Metabolism 21, 255–260.

Smith, J.T., Clifton, D.K., Steiner, R.A., 2006. Regulation of the neuroendocrine reproductive axis by kisspeptin-GPR54 signaling. Reproduction 131, 623–630.

Sun, B., Fujiwara, K., Adachi, S., Inoue, K., 2005. Physiological roles of prolactin-releasing peptide. Regulatory Peptides 126, 27–33.

Tsutsui, K., Bentley, G.E., Ubuka, T., Saigoh, E., Yin, H., Osugi, T., Inoue, K., Chowdhury, V.S., Ukena, K., Ciccone, N., Sharp, P.J., Wingfield, J.C., 2007. The general and comparative biology of gonadotropin-inhibitory hormone (GnIH). General and Comparative Endocrinology 153, 365–370.

Tsutsui, K., Ubuka, T., Bentley, G.E., Kriegsfeld, L.J., 2012. Gonadotropin-inhibitory hormone (GnIH): discovery, progress and prospect. General and Comparative Endocrinology 177, 305–314.

Whitlock, K.E., 2005. Origin and development of GnRH neurons. Trends in Endocrinology and Metabolism 16, 145–151.

Wierman, M.E., Pawlowski, J.E., Allen, M.P., Xu, M., Linseman, D.A., Nielson-Preiss, S., 2004. Molecular mechanisms of gonadotropin-releasing hormone neuronal migration. Trends in Endocrinology and Metabolism 15, 96–102.

Gonadotropins

Bernard, D.J., Fortin, J., Wang, Y., Lamba, P., 2010. Mechanisms of FSH synthesis: what we know, what we don't, and why you should care. Fertility and Sterility 93, 2465–2485.

Dode, C., Hardelin, J.-P., 2004. Kallmann syndrome: fibroblast growth factor signaling insufficiency? Journal of Molecular Medicine 82, 725–734.

TSH

Magner, J.A., 1990. Thyroid-stimulating hormone: biosynthesis, cell biology, and bioactivity. Endocrine Reviews 11, 354–385.

Szkudlinski, M.W., Grossmann, M., Weintraub, B.D., 1996. Structure-function studies of human TSH: new advances in design of glycoprotein hormone analogs. Trends in Endocrinology and Metabolism 7, 277–286.

Tixier-Vidal, A., Faivre-Baumann, A., 1992. Ontogeny of thyrotropin-releasing hormone biosynthesis and release of hypothalamic neurons. Trends in Endocrinology and Metabolism 3, 59–64.

GH and PRL

Amselem, S., Duquesnoy, P., Goossens, M., 1991. Molecular basis of Laron dwarfism. Trends in Endocrinology and Metabolism 2, 35–40.

Asa, S.L., Tannenbaum, G.S., 2006. Cell–cell communication in the pituitary: orchestrator of pulsatile growth hormone secretion? Trends in Endocrinology and Metabolism 17, 299–300.

Barkum, A.L., 1992. Acromegaly. Trends in Endocrinology and Metabolism 3, 205–210.

Baumann, G., 1991. Growth hormone heterogeneity: isohormones, variants, and binding proteins. Endocrine Reviews 12, 424–449.

Baxter, R.C., 1993. Circulating binding proteins for the insulinlike growth factors. Trends in Endocrinology and Metabolism 4, 91–96.

Ben-Jonathan, N., Lie, J.-W., 1992. Pituitary lactotrophs: endocrine, pararcrine, juxtacrine, and autocrine interactions. Trends in Endocrinology and Metabolism 3, 254–258.

Ben-Jonathan, N., Hnasko, R., 2001. Dopamine as a prolactin (PRL) inhibitor. Trends in Endocrinology and Metabolism 22, 724–763.

Corpas, E., Harman, S.M., Blackman, M.R., 1993. Human growth hormone and human aging. Trends in Endocrinology and Metabolism 14, 20–39.

Dieguez, C., Casanueva, F.F., 1995. Influence of metabolic substrates and obesity on growth hormone secretion. Trends in Endocrinology and Metabolism 6, 55–59.

Freeman, M.E., Kanyicska, B., Lerant, A., Nagy, G., 2000. Prolactin: structure, function, and regulation of secretion. Physiological Reviews 80, 1523–1631.

Goffin, V., Binart, N., Touraine, P., Keey, P.A., 2002. Prolactin: the new biology of an old hormone. Annual Review of Physiology 64, 47–67.

Müller, E.E., Locatelli, V., Cocchi, D., 1999. Neuroendocrine control of growth hormone secretion. Physiological Reviews 79, 511–607.

Samson, W.K., 1992. Natriuretic peptides: a family of hormones. Trends in Endocrinology and Metabolism 3, 86–90.

Soares, M.J., Konno, T., Alam, S.M., 2007. The prolactin family: effectors of pregnancy-dependent adaptations. Trends in Endocrinology and Metabolism 18, 114–121.

White, M.M., Samson, W.K., 2006. Closing the link in short-loop prolactin feedback. Trends in Endocrinology and Metabolism 17, 217–218.

POMC Derivatives

Bicknell, A.B., 2008. The tissue-specific processing of pro-opiomelanocortin. Journal of Neuroendocrinology 20, 692–699.

Cooray, S.N., Clark, A.J., 2011. Melanocortin receptors and their accessory proteins. Molecular and Cellular Endocrinology 331, 215–221.

Lightman, S.L., Windle, R.J., Ma, X.-A., Harbuz, M.S., Shanks, N.M., Julian, M.D., Wood, S.A., Kershaw, Y.M., Ingram, C.D., 2002. Hypothalamic–pituitary–adrenal function. Archives of Physiology and Biochemistry 110, 90–93.

Snyder, S.H., Pasternak, G.W., 2003. Historical review: opioid receptors. Trends in Pharmacological Sciences 24, 198–205.

Tanaka, S., 2003. Comparative aspects of intracellular proteolytic processing of peptide hormone precursors: studies of proopiomelanocortin processing. Zoological Science 20, 1183–1198.

Vuong, C., Van Uum, S.H., O'Dell, L.E., Lutfy, K., Friedman, T.C., 2010. The effects of opioids and opioid analogs on animal and human endocrine systems. Endocrine Reviews 31, 98–132.

Endothelins

Battistini, B., D'Orléans-Juste, P., Sirois, P., 1993. Endothelins: circulating plasma levels and presence in other biologic fluids. Laboratory Investigation 68, 600–628.

Macrae, A.D., Bloom, S.R., 1992. Endothelin: an endocrine role. Trends in Endocrinology and Metabolism 3, 153–157.

Masaki, T., 1993. Endothelins: homeostatic and compensatory actions in the circulatory and endocrine systems. Endocrine Reviews 14, 256–268.

Stojilkovic, S.S., Catt, K.J., 1992. Neuroendocrine actions of endothelins. Trends in Pharmacological Sciences 13, 385–391.

Nonapeptides

Juul, K.V., 2012. The evolutionary origin of the vasopressin/V2-type receptor/aquaporin axis and the urine-concentrating mechanism. Endocrine 42 (1), 63–68.

Murphy, D., Konopacka, A., Hindmarch, C., Paton, J.F., Sweedler, J.V., Gillette, M.U., Ueta, Y., Grinevich, V., Lozic, M., Japundzic-Zigon, N., 2012. The hypothalamic-neurohypophyseal system: from genome to physiology. Journal of Neuroendocrinology 24, 539–553.

Sasaki, S., Ishibashi, K., Marumo, F., 1998. Aquaporin-2 and -3: representatives of two subgroups of the aquaporin family colocalized in the kidney collecting duct. Annual Review of Physiology 60, 199–220.

Urano, A., Ando, H., 2011. Diversity of the hypothalamo-neurohypophysial system and its hormonal genes. General and Comparative Endocrinology 170, 41–56.

Zingg, H.H., Laporte, S.A., 2003. The oxytocin receptor. Trends in Endocrinology and Metabolism 14, 222–227.

Chapter 5

The Hypothalamus–Pituitary System in Non-Mammalian Vertebrates

Here we describe the major features of the hypothalamus–pituitary (HP) axes of non-mammalian vertebrates and their associated releasing hormones, tropic hormones, and nonapeptide hormones as contrasted to those previously discussed for mammals in Chapter 4. All of the axes described for mammals appeared early in the evolutionary history of vertebrates and we can examine the roles of the **hypothalamus–pituitary–gonad (HPG), hypothalamus–pituitary–thyroid (HPT),** and **hypothalamus–pituitary–adrenal (HPA)** axes in the jawed fishes and non-mammalian tetrapods. Hormones and anatomical features described in Chapter 4 that also appear in non-mammals are not redefined although their full names appear with abbreviations the first time they are used. If the reader is not familiar with non-mammalian vertebrates, a brief discussion of the evolutionary relationships and importance of various vertebrate taxonomic groups listed here can be found in Appendix B. The Phylum Chordata includes two protochordate invertebrate groups, the cephalochordates and the tunicates or urochordates, as well as the vertebrates. The evolutionary relationships among these groups are under current debate, but they are believed to have diverged about 500 MYBP from a common ancestor. We will refer to the protochordate groups at times when discussing the evolution of various elements of the HP axes.

The HP system is unique to the vertebrate chordates, and no invertebrate chordate, or for that matter any other invertebrate phylum, exhibits an endocrine gland so closely integrated with the brain as the vertebrate pituitary. However, there is evidence of a possibly homologous system in a group of primitive cephalochordates commonly known as lancelets or amphioxus (*Asymmetron* spp., *Branchiostoma* spp., *Epigonichthys* spp.). A shallow epithelial groove located in the oral cavity of *Branchiostoma*, the most studied genus, appears in close proximity to the simple dorsal nervous system of this animal (Figure 5-1). It is known as **Hatschek's pit** and has been found to react positively to antibodies against **substance P, met-enkephalin, cholecystokinin (CCK)**, mammalian **luteinizing hormone (LH)**, and the enzyme aromatase ($P450_{aro}$), which converts certain androgens into estrogens (see Chapter 3). Hatschek's pit also expresses **Pit-1**, a transcription factor unique to the developing vertebrate pituitary. Genomic analyses of protochordates have demonstrated genes related to important bioregulators of the vertebrate HP axes, including **kisspeptin (Kp), Kp receptor (GPR54 or Kiss1R), gonadotropin-releasing hormone (GnRH)** receptors, and the ancient glycoprotein **thyrostimulin** (see ahead). Several genes for steroidogenic enzymes (see Chapter 3), including $P450_{c11}$, $P450_{c17}$, 17β-hydroxysteroid dehydrogenase (17β-HSD), 3β-HSD, and 5α-reductase, are present as well as genes for the **estrogen receptor (ER)** and the **steroid acute regulatory protein (StAR)**. Additionally, thyroid axis genes including thyroid hormone receptor, **Na-I symporter (NIS)**, and thyroid peroxidase have been demonstrated, although interestingly a thyroglobulin gene appears to be absent. Curiously, genes for **prolactin (PRL), growth hormone (GH), corticotropin (ACTH), corticotropin-releasing hormones (CRH)**, and **thyrotropin-releasing hormone (TRH)** appear to be absent.

In the group of invertebrate chordates, the tunicates, we find a neural gland complex consisting of a cerebral ganglion and the dorsal strand (Figure 5-2) that also is associated with the oral cavity. A number of neuropeptides of endocrine significance may be present in tunicates, including ACTH, PRL, GnRH, and nonapeptides. In addition, the **proopiomelanocortin (POMC)** processing enzymes, PC1 and PC2 (described in Chapter 4), have been reported. However, examination of the genome for *Ciona*, a sessile tunicate, indicates PRL, GH, ACTH, and CRH genes are absent, although Kp, GnRH receptors, and several steroidogenic enzymes are present. However, a CRH-related family of peptides known as the **teneurin C-terminal-associated peptides (TCAPs)** are present in tunicates as well as all other metazoan phyla. These TCAPs are products of the teneurin gene that produces a transmembrane protein found primarily in the nervous systems of all metazoans. In spite of the inconsistencies, the association of all three structures (pituitary, Hatschek's pit, neural gland complex) with the nasal placodes together with the presence of similar bioregulators and enzymes are strongly suggestive of homology.

With few exceptions, the brain structures responsible for secretion of neuropeptides in vertebrates are paired areas, although contributions of right and left nuclei may not always be equal, as discussed in Chapter 4. However, in discussing the presence of these secretory areas or nuclei and

Vertebrate Endocrinology. http://dx.doi.org/10.1016/B978-0-12-394815-1.00005-7

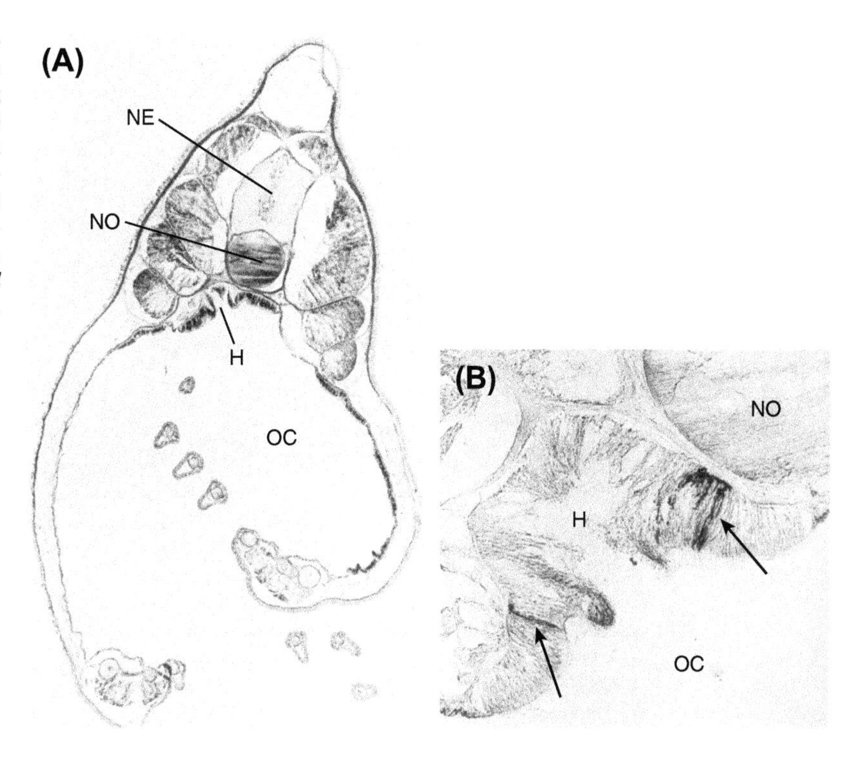

FIGURE 5-1 **Hatschek's pit in the cephalochordate amphioxus.** (A) Cross-section through the oral cavity (OC) of an adult animal showing two basic chordate features: dorsal nerve cord (NE) and notochord (NO). Hatschek's pit (H) appears on the dorsal pharyngeal surface similar to the location of Rathke's pouch in vertebrate embryos. (B) Enlargement of Hatschek's pit showing immunoreactive metenkephalin-like material (arrows). *(Reprinted with permission from Nozaki, M. and Gorbman, A.,* Zoological Science, ***9****, 387–395, 1992.)*

their related neuropeptides, it is often cumbersome to refer always to their paired nature. In the following accounts, the reader should assume that all nuclei are paired or bilateral even though they may be referred to in the singular.

A major developmental difference in formation of the brains of fishes and tetrapods has made it more difficult to identify homologous brain centers. In development of the teleost brain, for example, the neural tube turns itself outward, whereas in the tetrapod brain, the growth occurs inward. Thus, the hippocampus in mammals is located deep within the brain, whereas in teleosts the putative hippocampus occurs on the lateral dorsal surface of the brain.

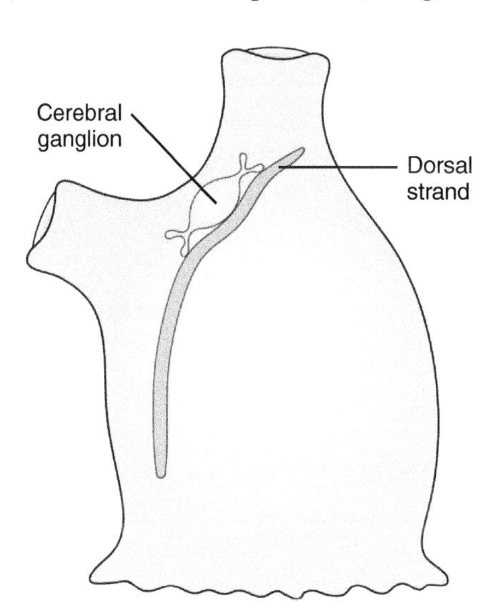

FIGURE 5-2 **The neural complex of a tunicate.** *(From Kawamura, K., Kouki, T., Kawahara, G., Kikuyama, S. 2002. Hypophyseal development in vertebrates from amphibians to mammals.* General and Comparative Endocrinology ***126****, 130–135.)*

I. THE PITUITARY GLAND OF FISHES: ANATOMICAL CONSIDERATIONS

The piscine HP system is separable into the same major divisions as that of mammals: hypothalamus, **neurohypophysis**, and **adenohypophysis** (Figure 5-3). Much of the following account is based on pioneering anatomical studies of the HP axis in primitive fishes done by Michael

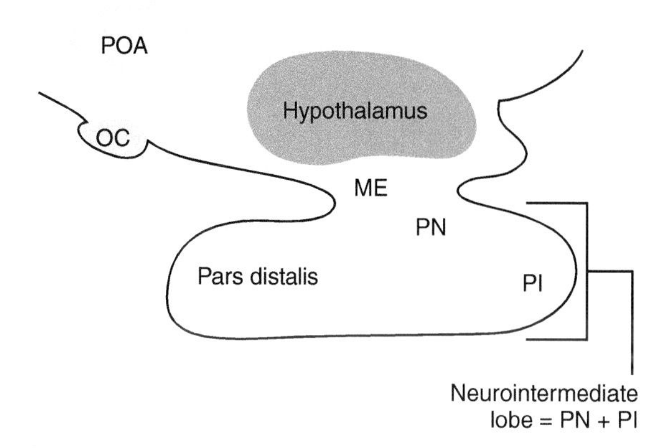

FIGURE 5-3 **Generalized hypothalamo–hypophysial axis in fishes.** Note the lack of a Pars tuberalis in fishes. Abbreviations, ME, median eminence; OC, optic chiasm; PI, pars intermedia; PN, pars nervosa; POA, preoptic area.

Lagios. These and other studies are summarized in "The Pituitary Gland: A Comparative Account" by Holmes and Ball (see reading list at the end of the chapter). Some marked differences from mammals occur in fishes. The fish adenohypophysis consists of a **pars distalis** and a **pars intermedia** but lacks a **pars tuberalis**, although a possibly homologous structure, the **pars ventralis**, occurs in elasmobranchs (chondrichthyeans). The neurohypophysis of lampreys, sharks, and non-teleost bony fish groups is separable into a **median eminence** and a **pars nervosa**. The median eminence is connected to the adenohypophysis by a portal system; however, there is no median eminence and no portal system in hagfishes and teleosts. Hagfishes apparently rely on diffusion of factors from the hypothalamus, whereas the axons of the hypothalamic neurons make direct contact with pituitary cells in teleosts.

In most fishes, the pars distalis of the adenohypophysis is differentiated into two subregions or zones, each with its special cell types. Two different terminologies have been proposed for the subregions of the piscine adenohypophysis. The nomenclature proposed by Green is used here in favor of the alternative system proposed by Pickford and Atz because the Green system is similar to mammalian terminologies (Figure 5-4) and is used most commonly in the literature. During this discussion, each term is defined initially, followed by the italicized terms proposed by Pickford and Atz, in parentheses. There are three distinct zones recognized in the fish adenohypophysis by both schemes. The most anterior and rostral (dorsal) portion of the piscine adenohypophysis often consists of follicles of cells and is termed the **rostral pars distalis** (*proadenohypophysis*). The remainder of the pars distalis comprises the **proximal pars distalis** (*mesoadenohypophysis*). The third region is the **pars intermedia** (*meta-adenohypophysis*). Each of these regions of the adenohypophysis is readily distinguished cytologically, and generally each contains different cell types that produce different tropic hormones. The cell types found in each region and the hormones they are thought to produce are summarized in Appendix G. The general anatomy and the distribution of cell types in various fish pituitaries are illustrated in Figure 5-5. Although alternative schemes have been proposed, the classification scheme for pituitary cell types used here is the same one described for mammals in Chapter 4 (acidophils, basophils, chromophobes; gonadotropes, thyrotropes, etc.).

The pars intermedia of most fishes is intimately interdigitated with the pars nervosa of the neurohypophysis to form a **neurointermediate lobe (NIL)** (Figure 5-3).

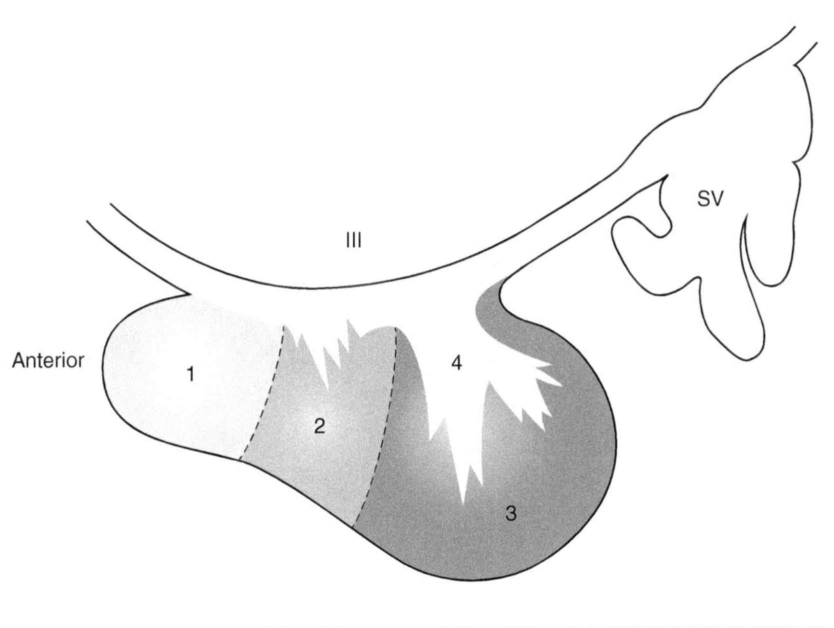

	Zone	Green (1951)	Pickford and Atz (1957)
Adenohypophysis	1	Rostral pars distalis	Proadenohypophysis
	2	Proximal pars distalis	Mesoadenohypophysis
	3	Pars intermedia	Meta-adenohypophysis
Neurohypophysis	4	Pars nervosa	Pars nervosa

FIGURE 5-4 **Regions of the teleost adenohypophysis.** The terminologies of Green (1951) and Pickford and Atz (1957) are compared for the three histologically distinct regions (1, 2, and 3). The saccus vasculosis (SV) is not part of the pituitary but is a prominent nearby structure. Only the Green nomenclature is used in the text.

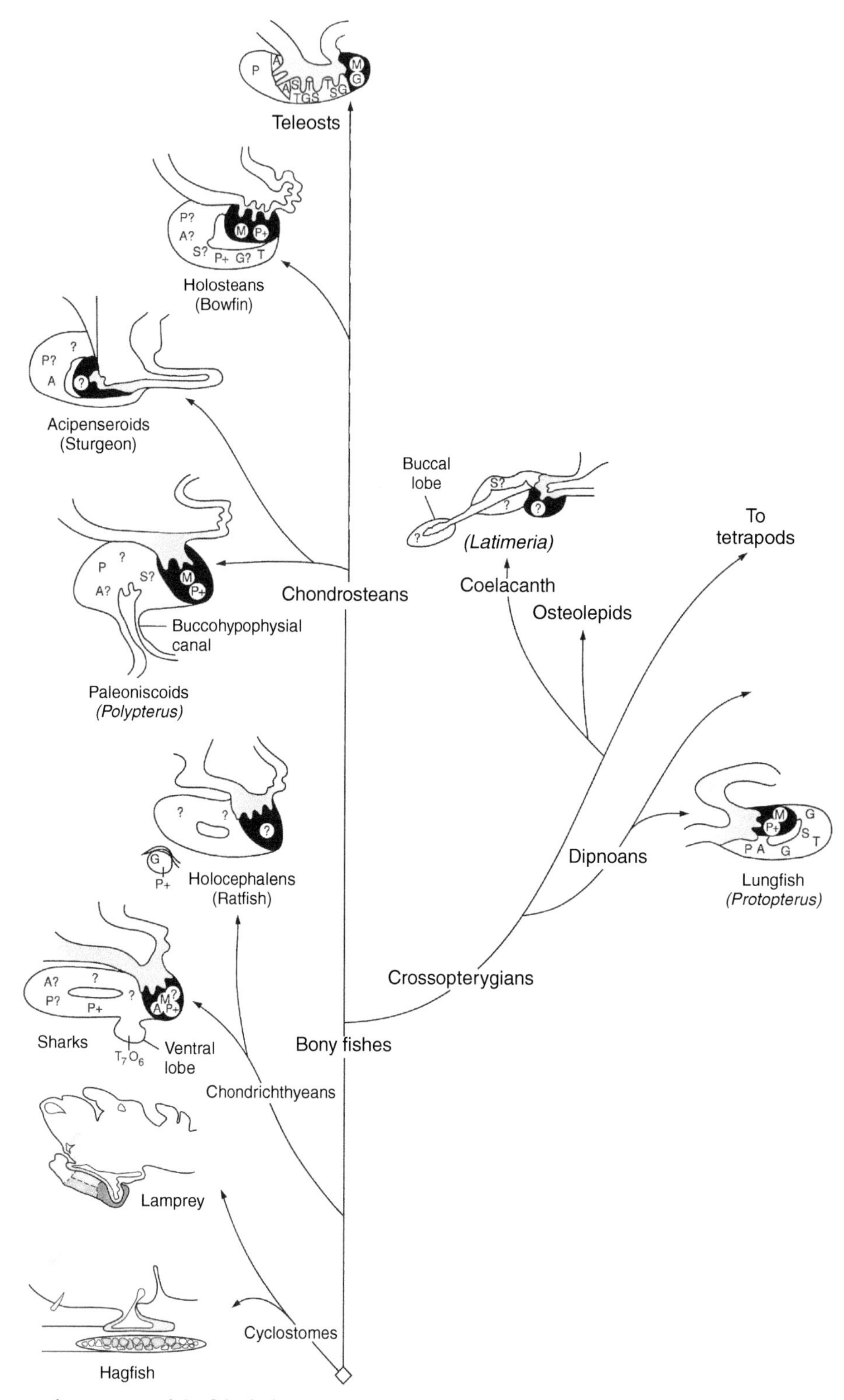

FIGURE 5-5 **Comparative anatomy of the fish pituitary.** The neurohypophysis is indicated by the darker color and the pars intermedia is shown in black. Approximate distributions of cell types secreting tropic hormones are indicated by letters. A, corticotropin; G, gonadotropins; M, melanotropin; P, prolactin; P+, somatolactin; S, growth hormone or somatotropin; T, thyrotropin, ?, unidentified. *(Adapted with permission from Schreibman, M. P. (1986). Pituitary Gland. In "Vertebrate Endocrinology: Fundamentals and Biomedical Implications. Volume 1, Morphological Considerations" (P.K.T. Pang and M.P. Schreibman, eds), pp 11–56. Academic Press, Orlando, Florida.)*

Posterior to the neurointermediate lobe in cartilaginous fishes and in most bony fishes is a unique structure formed from the floor of the diencephalon called the **saccus vasculosus (SV)** (Figure 5.4). This structure is probably derived from ciliated glial cells called *ependymal cells*. It is especially prominent in some groups, but its function is unknown. The SV is a prominent feature with possible endocrine function that evolved among early jawed fishes (see ahead). For our discussions of hypothalamic control of hypophysial functions, we shall focus on two regions in which the majority of neurosecretory (NS) neurons reside. The first area is located just anterior and dorsal to the optic chiasm and is called the **preoptic area (POA)**. This region marks the telencephalic and diencephalic boundary and usually is considered part of the telencephalon. Endocrinologists often include it when they are discussing the hypothalamic control of pituitary function as though it were part of the hypothalamus, a habit continued here. The second region includes the hypothalamus proper. With the exception of the **preoptic nucleus (PON)** of hagfishes, the NS nuclei of the HP system are paired structures. Other telencephalic and mesencephalic structures may produce similar peptides or may be targets for fibers secreting peptides usually associated with the HP system, but they are not thought to be central players in the regulation of endocrine function.

A. Agnathan (Jawless) Fishes

Our knowledge of the neuroendocrine system of agnathan fishes is largely due to the early studies by Aubrey Gorbman that have been continued largely through the efforts of Stacia Sower and her colleagues at the University of New Hampshire as well as by Japanese scientists, most notably Hiroshi Kawauchi.

1. The Myxinoids (Hagfishes)

The Atlantic and Pacific hagfishes possess the most primitive HP system among the chordates (Figure 5-6). The hagfish system lacks many of the features that characterize other piscine and tetrapod groups. Furthermore, hagfishes are much more primitive in many respects than even their closest living agnathan relatives, the lampreys (Petromyzontidae).

The origin of the adenohypophysis of hagfishes appears to be from endoderm rather than from ectoderm, presenting an additional puzzle with respect to the origin of the pituitary. Furthermore, it raises the question of possible homology of the hagfish adenohypophysis to that of other vertebrates and supports the viewpoint that hagfishes are aberrant vertebrates, are not on the mainline evolutionary pathway, and cannot readily be compared to vertebrates.

The hagfish adenohypophysis is not differentiated into subregions; that is, there is no discernable pars distalis or pars intermedia. The adenohypophysis is composed primarily of nonstainable cells with rare PAS(+) basophils or an occasional acidophil. Electron micrographs of the hagfish adenohypophysis show rare granular cells with cytoplasmic granules of 100 to 200 nm in diameter. These granular cells are believed to represent the two rare

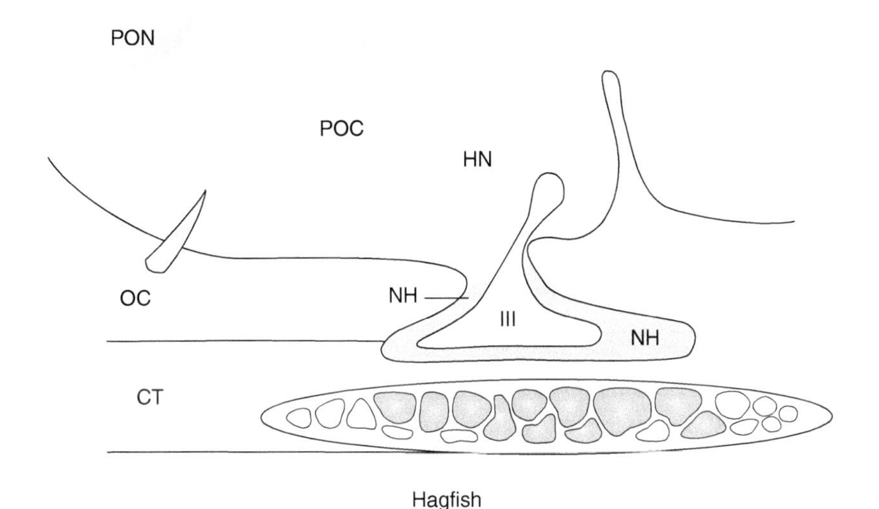

FIGURE 5-6 **Neuroendocrine systems of Atlantic hagfish (*Myxine glutinosa*).** Colored cells in adenohypophysis represent gonadotropes. Abbreviations: AH, adenohypophysis; CT, connective tissue; HN, hypothalamic nucleus; NH, neurohypophysis; POC, postoptic commissure; PON, preoptic nucleus; III, third ventricle. *(Adapted with permission from Sower, S.A. and Kawauchi, H., in "Hormones and Reproduction of Vertebrates. Vol. 1. Fishes" (D.O. Norris and K.H. Lopez, Eds.), Academic Press, San Diego, CA, 2011, pp. 193–208.)*

stainable cell types identifiable with the light microscope. When hagfish adenohypophysial tissue is cultured *in vivo*, no observable changes take place in either granular or nongranular cells.

Certain cells of the myxinoid adenohypophysis exhibit cytological modifications where they make contact with the neurohypophysis. These altered cells collectively are termed **modified adenohypophysial tissue**, and it has been proposed that this apparent induction by neurohypophysial tissue represents phylogenetically the origin of the pars intermedia. In other vertebrates, the pars intermedia develops following contact of the presumptive adenohypophysis with the neurohypophysis.

2. Lampreys (Petromyzontids)

In the lampreys (Figure 5-7), immunocytochemical staining has identified corticotropes mainly in the rostral pars distalis. An acidophilic cell containing GH/PRL-like immunoreactivity occupies the dorsal portion of the proximal pars distalis. The ventral portion of the proximal pars distalis consists of a basophilic cell that produces a **glycoprotein hormone (GpH)**. Additionally, a melanotrope has been identified in the pars intermedia.

The distinct pars nervosa and the pars intermedia form a well-developed neurointermediate lobe. Peptidergic neurons terminate in the pars nervosa where the nonapeptide neurohormone **arginine vasotocin (AVT)** is stored. A second anterior neurohemal region is associated with the pars distalis and has been referred to as a *median eminence*; however, this structure is devoid of portal blood vessels so it is unlikely that it functions as a median eminence. Studies do show that materials can diffuse readily from the brain into the adenohypophysis via this putative median eminence.

B. Chondrichthyean Fishes

The chondrichthyean fishes represent a side road off the main line of vertebrate evolution. Once a dominant marine group, they are represented today by selachians or elasmobranchs (the sharks, rays, and skates) and a small unusual group of fishes known as the holocephalans (chimaeras or ratfishes). However, they do retain numerous features of the HP system that probably evolved in ancient fishes and gave rise to this group as well as to the bony fishes.

1. Sharks, Rays, and Skates (Elasmobranchs)

The elasmobranch HP system possesses two anatomical features not found in agnathan fishes. The pars ventralis represents a fourth subdivision of the adenohypophysis unique to elasmobranchs. It is located ventral to the proximal pars distalis, to which it is connected by a stalk (Figure 5-5). Localization of GTH and TSH activity in the pars ventralis has prompted some investigators to suggest that the pars ventralis is homologous to the **pars tuberalis** of the tetrapod adenohypophysis but this seems unlikely (see Chapter 4). The terms *rostral* and *proximal* may be somewhat misleading when applied to the elasmobranch pars distalis because of the presence of the ventral lobe, which is probably homologous

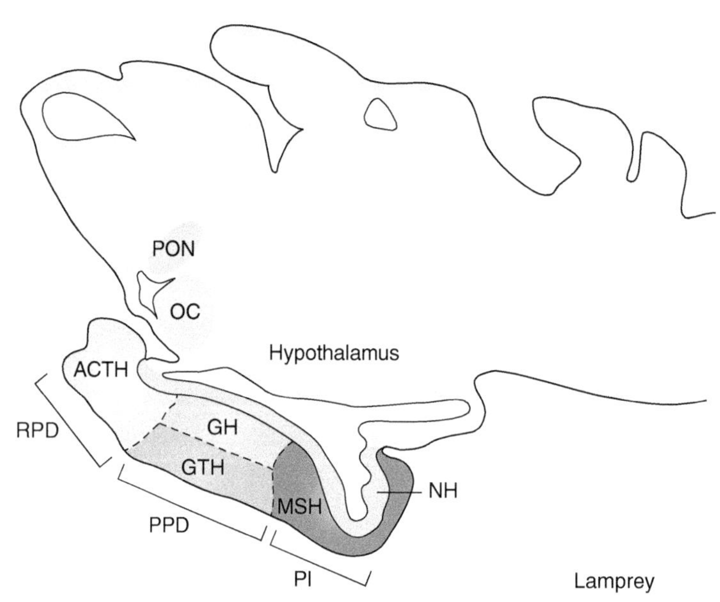

FIGURE 5-7 **Neuroendocrine systems of sea lamprey (*Petromyzon marinus*).** Abbreviations: NH, neurohypophysis; PI, pars intermedia; PON, preoptic nucleus; PPD, proximal pars distalis; RPD, rostral pars distalis. See Appendix A for other abbreviations. *(Adapted with permission from Kawauchi, H. and Sower, S.A.,* General and Comparative Endocrinology, ***148****, 3–19, 2006. Sower, S.A. and Kawauchi, H., Reproduction in agnathan fishes: lampreys and hagfishes. In "Hormones and Reproduction of Vertebrates. Vol. 1. Fishes" (D.O. Norris and K.H. Lopez, Eds.), Academic Press, San Diego, CA, 2011, pp. 193–208.)*

to part of the proximal pars distalis of teleost fishes. These regions are sometimes designated as the rostral, median, and ventral lobes of the pars distalis. The cell types in the elasmobranch pituitary are described in Appendix G. The second feature appearing for the first time in elasmobranchs is the saccus vasculosus derived from the ependymal cells lining the third ventricle and located immediately posterior to the neurointermediate lobe. It is not as well-developed as its homologue in bony fishes, but it does possess the unique **coronet** cell type characteristic of the saccus vasculosus of bony fishes (see Figure 5-8). A variety of neurons have been described in the elasmobranch saccus as well as cells immunoreactive for **melanin-concentrating hormone (MCH)** colocalized with **melanotropin (α-MSH)** in the saccus vasculosus of the shark (*Scyliorhinus* sp.). The function of the saccus vasculosus is unknown.

The neurointermediate lobe shows extensive interdigitation of neural and endocrine cells. The pars nervosa stores some unique nonapeptides in addition to AVT (see ahead). These hormones are presumably produced in the hypothalamus and are stored in the neurohemal portion of neurointermediate lobe. The pars intermedia contains only one cell type associated with α-MSH activity. A well-developed median eminence with a portal system connects it to the pars distalis. The median eminence consists of anterior and posterior neurohemal areas. The posterior region receives both peptidergic and aminergic NS axons and appears to be linked by portal vessels to the proximal pars distalis but not to the ventral lobe. There is considerably less NS material in the anterior neurohemal area that appears to be connected by capillaries to the rostral pars distalis. It is tempting to suggest that the median eminence has differentiated in these fishes to increase the efficiency of delivering hypothalamic neurohormones to specific adenohypophysial cells.

2. *Ratfishes (Holocephalans)*

The pituitary of ratfishes is readily subdivided cytologically into rostral pars distalis, proximal pars distalis, and pars intermedia. Although several cell types have been demonstrated with selective staining procedures, no experimental studies have verified which cells produce which tropic hormones. Like elasmobranchs, holocephalans possess a unique region associated with the adenohypophysis, called the **pharyngeal lobe**. This structure is located in the roof of the mouth outside the cranium and is not connected to the adenohypophysis proper. The pharyngeal lobe consists of follicles and may be homologous to the follicular rostral pars distalis of bony fishes. On the other hand, it may be homologous to the ventral lobe of the elasmobranch and the buccal lobe of the coelacanth pituitary (see ahead). The ratfish neurohypophysis includes a prominent median eminence connected to the rostral pars distalis and the proximal pars distalis by a portal system. The pars nervosa is mingled with the pars intermedia of the adenohypophysis to form a typical neurointermediate lobe. AVT and **oxytocin (OXY)** are present in the ratfish pars nervosa and represent a unique occurrence of OXY among chondrichthyean fishes. A well-developed saccus vasculosus is also present.

C. Bony Ray-Finned Fishes (Actinopterygians)

Four distinct groups of ray-finned fishes are recognized traditionally by most authorities, and their more common names are used here (the scientific taxonomic names of these groups may vary with different authorities). Three of these groups (polypterans, chondrosteans, and holosteans) are considered to represent the more primitive bony fish groups, whereas the fourth group (teleosts) is recognized as being the most evolutionary advanced, most abundant, and most diverse of these fish groups.

1. *Polypteran Fishes (Polypterus and Calamoichthyes)*

The ancient African polypterid fishes possess all the typical piscine pituitary features including rostral and proximal divisions of the pars distalis, a neurointermediate lobe, and

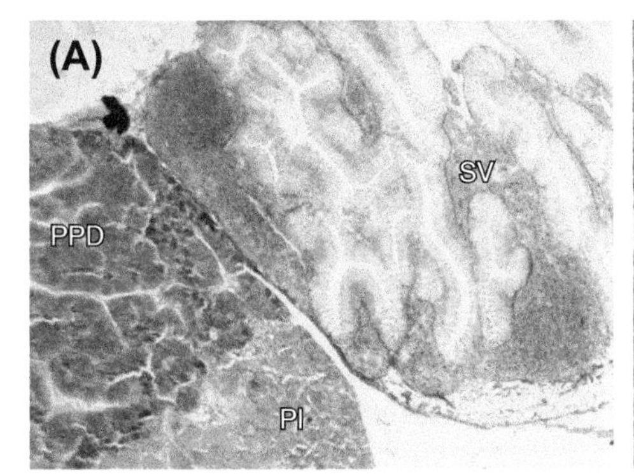

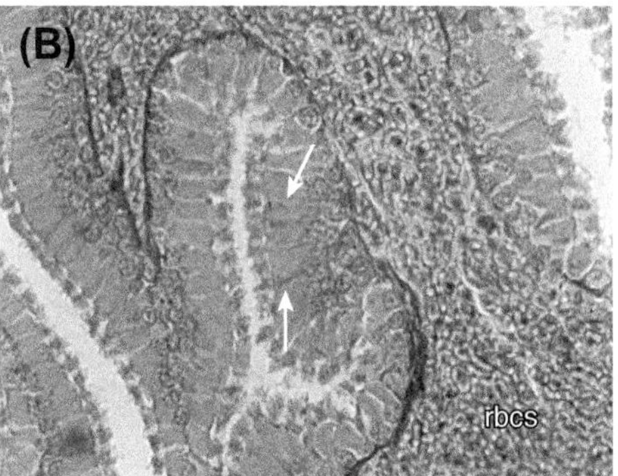

FIGURE 5-8 **The saccus vasculosus of rainbow trout (*Oncorhynchus mykiss*).** (A) Note the closeness of the saccus vasculosus (SV) and adenohypophysis (PPD, PI). (B) Higher magnification of SV. Arrows indicate coronet cells. Note accumulation of secretory material along the SV lumina. rbcs, red blood cells.

a saccus vasculosus. As adults, they retain a connection between the hypophysis and the mouth cavity called the **buccohypophysial canal** (*bucco*, mouth). This canal is believed to be a remnant originally connecting the pituitary to the oral cavity (Figure 5-5). The buccohypophysial canal or duct is lined with a weakly staining cell type that does not appear to be associated with production of any tropic hormones. The polypteran pars nervosa contains AVT and an OXY-like nonapeptide secreted by the preoptic nucleus that is known as **isotocin (IST)**. These two neurohormones are characteristic of all ray-finned fishes. A portal system with a defined median eminence is well developed. Peptidergic fiber tracts travel from the PON via the median eminence to terminate in the pars nervosa.

2. Chondrostean Fishes (Sturgeons, Paddlefishes)

There is less information concerning the HP system of chondrostean fishes than for any other ray-finned bony fishes; however, chondrostean fishes are very large, long-lived species with very limited distributions and hence are difficult to study. There is no buccohypophysial canal in chondrosteans, but a hypophysial cavity is present. This cavity separates the pars distalis and pars intermedia and may be homologous to the buccohypophysial canal and/or the hypophysial cleft described in some mammals. The pars distalis consists of a rostral zone and a proximal zone (rostral and proximal pars distalis). Numerous follicles occur throughout the pars distalis, and their lumina, which are filled with a basophilic colloidal material, are considered by some to be remnants of the hypophysial cavity. The entire pars distalis may be homologous to the proximal pars distalis of teleostean fishes, and the terms *rostral pars distalis* and *proximal pars distalis* may not be applicable to the two apparent zones of the chondrostean pars distalis. The pars intermedia of sturgeons is large and closely associated with the pars nervosa, forming a typical neurointermediate lobe. The pars nervosa is basically hollow and is similar in appearance to the saccus vasculosus. Both peptidergic and aminergic fibers have been reported in the pars nervosa.

3. Holostean Fishes (Gars and Bowfin)

Adult holostean fishes do not exhibit a buccohypophysial duct nor is there any hypophysial cleft, although a transient hypophysial cleft does appear during development of the pituitary. The adenohypophysis consists of a follicular rostral pars distalis, a proximal pars distalis, and a pars intermedia that interdigitates with the pars nervosa to form a neurointermediate lobe. The median eminence is connected to the pars distalis by a well-developed portal system. In addition, a limited number of peptidergic and aminergic axons penetrate the pars distalis; however, only aminergic fibers are associated with the median eminence of the bowfin, *Amia calva*. Here, in these near-relatives of the teleost fishes, is the modest beginning of a shift from neurovascular control to neuroglandular control of the adenohypophysis directly by neurons of the hypothalamus as is seen among the teleosts.

4. Teleost Fishes

The HP systems of numerous teleost species have been studied in detail. In view of the vast adaptive radiation that teleosts have undergone, it is not surprising to find considerable variability in this system. Although only a relatively few species have been examined, they have been examined in greater detail than for other piscine groups, and the various cell types have been identified and their secretions characterized chemically. There is a strong tendency for localization of cell types in particular regions of the adenohypophysis (Figures 5-9 and 5-10). This anatomical arrangement has aided scientists in determining what hormone each cell type produces. Unlike the situation with mammals, two kinds of GTH-secreting cells have been demonstrated in the teleost proximal pars distalis.

Teleosts exhibit extensive innervation of the pars distalis by preoptic and hypothalamic neurons. There is no distinct portal system and hence no true median eminence in most species examined, and the control of tropic hormone release is **neuroglandular** by direct peptidergic or aminergic innervation rather than **neurovascular** as found in teleost ancestral groups and in the tetrapods. Some neurons make direct synaptic connections to the pituitary cells, whereas the axons of others end on the pituitary basement membrane, and their secretions must diffuse through the basement membrane to reach receptors on the pituitary cells. Examples of some peptidergic neurons in the sea bass pituitary are provided in Table 5-1.

The pars intermedia is intimately associated with the pars nervosa of the neurohypophysis to form an elaborate neurointermediate lobe as described for the elasmobranchs and more ancient bony fishes. Direct innervation is responsible for controlling α-MSH release from the pars intermedia. In the early 1990s, a new teleost hormone was discovered associated with pars intermedia of Japanese flounder (*Paralichthys olivaeus*) by M. Ono and associates and from the Atlantic cod (*Gadus morhua*) by Marilyn Rand-Weaver and coworkers. This new hormone was named **somatolactin (SL)** because it was similar in structure to both GH and PRL. Thus, in addition to α-MSH-secreting cells, a unique cell type, the **PIPAS cell**, is present in the pars intermedia of most teleosts and secretes **SL** (see ahead). A similarly staining cell also occurs in the pars intermedia of some more ancient bony fishes (i.e., *Amia* and *Polypterus*), and SL has been characterized from sturgeon and lungfish pituitaries.

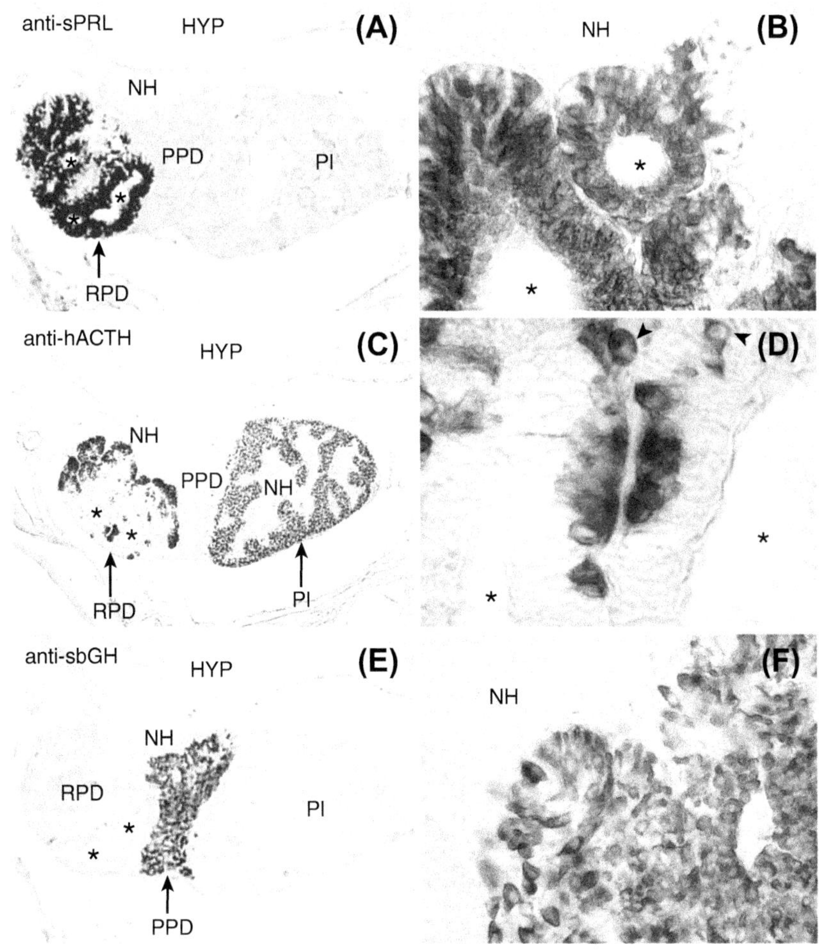

FIGURE 5-9 **Prolactin (PRL), corticotropin (ACTH), and growth hormone (GH) in the teleost pituitary.** Immunoreactive demonstration of (A,B) lactotropes, (C,D) corticotropes, and (E,F) somatotropes in the pituitary of *Alosa spidissima*. Asterisks indicate cavities. Abbreviations: HYP, hypothalamus; NH, neurohypophysis, PI, pars intermedia, PPD, proximal pars distalis; RPD, rostral pars distalis. *(Reprinted with permission from Laiz-Carrion, R. et al.,* General and Comparative Endocrinology, ***132****, 454–464, 2003.)*

5. The Lobe-Finned Fishes (Sarcopterygians)

Although they once thrived in the ancient oceans and radiated in many directions, only two lobe-finned groups have living representatives. The dipnoans or lungfishes are represented by three genera comprising seven species. The two species of living crossopterygians are thought to be very close to the long extinct crossopterygian fishes that gave rise to amphibians. The HP system of the Australian lungfish (*Neoceratodus forsteri*) is similar to that of primitive bony fishes, including the presence of a neurointermediate lobe. In contrast, the pituitary systems of the African and South American lungfishes structurally are more similar to those of tetrapods, especially amphibians (see Figures 5-5 and 5-11). For example, there is less regionalization of cell types in the adenohypophysis and no neurointermediate lobe; however, this similarity with amphibians could be the result of parallel or convergent evolution. In general, hypothalamic regulation in lungfishes appears to be neurovascular as in primitive actinopterygians rather than neuroglandular as in teleosts. The pars tuberalis, a landmark tetrapod feature, however, is missing in all living lungfishes as well as in the coelacanths.

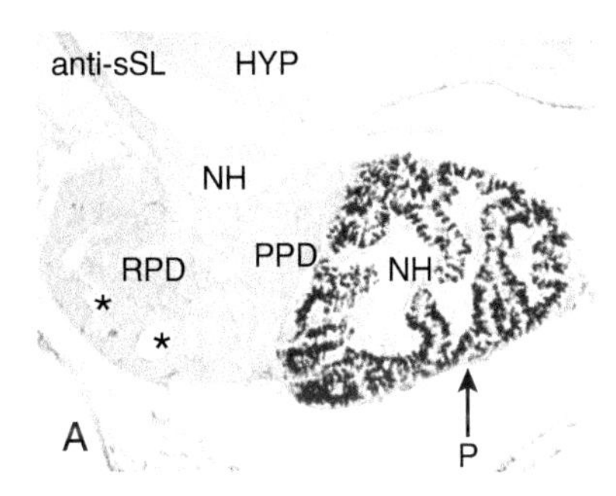

FIGURE 5-10 **Somatolactin (SL) cells in teleost pars intermedia.** Immunoreactive SL cells in the pars intermedia (P) of the neurointermediate lobe of the teleost *Alosa sapidissima* demonstrated with use of an antibody to salmon SL (anti-aSL). See Figure 5-9 for an explanation of abbreviations. *(Reprinted with permission from Laiz-Carrion, R. et al.,* General and Comparative Endocrinology, ***132****, 454–464, 2003.)*

TABLE 5-1 Presence of Neuropeptides in Hypothalamic Neurons of the Sea Bass (*Dicentrarchus Labrax*)

Region	Secretory Cell Type	Peptides Present
Rod	ACTH-cells	SS, GHRH, CRH, AVT, IST, SP, NT, GAL
	PRL-cells	GAL
Pod	GH-cells	SS, GHRH, AVT, IST, CCK, SP, NPY, GAL
	GTH-cells	GnRH, IST, GHRH, AVT
	TSH-cells	GHRH, CRH, AVT, IST, SP, GAL
PI	MSH cells	GHRH, CRH, MCH, AVT, IST, SP
	PAS(+)-cells	GHRH, CRH, MCH, AVT, IST, SP

Note: See text or Appendix A for explanation of abbreviations for the peptides listed.
Adapted with permission from Moons, L., et al., 1989. General and Comparative Endocrinology, 73, 270.

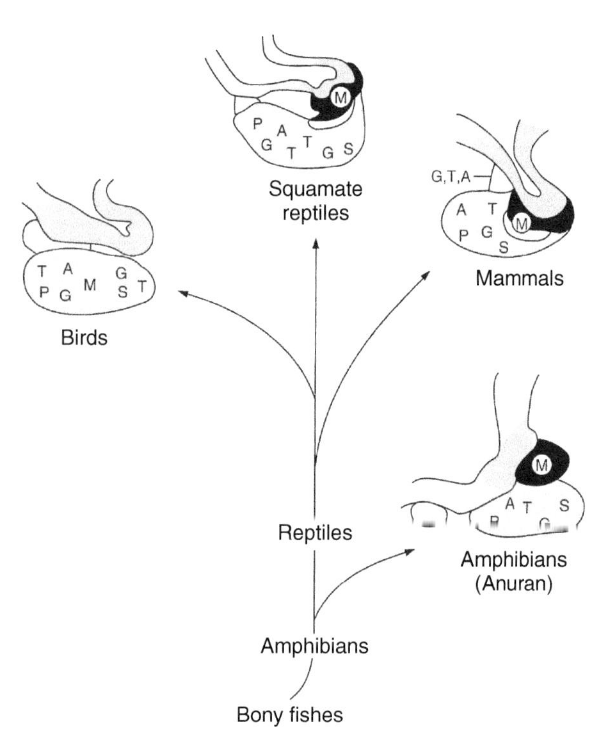

FIGURE 5-11 **Comparative anatomy of the tetrapod pituitary.** See Figure 5-4 for an explanation of symbols. *(Adapted with permission from Schreibman, M.P., in "Vertebrate Endocrinology: Fundamentals and Biomedical Implications. Vol. 1. Morphological Considerations" (P.K.T. Pang and M. Schreibman, Eds.), Academic Press, Orlando, FL, 1986, pp. 11–56.)*

In contrast to lungfishes, the HP system of the coelacanths is much more piscine-like than it is tetrapod-like. The general anatomy of the adenohypophysis is similar to that of the elasmobranchs (Figure 5-5). There is a **buccal lobe** in the pars distalis that, based on the cell types present there, may be homologous to the ventral lobe of elasmobranchs. The pars distalis may be subdivided into rostral and proximal portions with stainable cell types appearing that are distributed similarly to those of teleosts. The pars intermedia forms a typical neurointermediate lobe with the pars nervosa as in other bony fishes. Unlike teleosts, there is a distinct median eminence connected to the adenohypophysis by a well-developed portal system, particularly in the proximal region. However, some NS axons appear to penetrate the proximal pars distalis similar to the situation described for the holostean fish *Amia calva*. Although the sarcopterygian fishes represent descendents of fishes basal to the ancestral line of tetrapods, extant members of this group provide no transitional stages with respect to anatomical features of the HP system.

II. THE PITUITARY OF TETRAPOD VERTEBRATES: ANATOMICAL CONSIDERATIONS

The neuroendocrine systems of tetrapods exhibit most of the characteristics described for mammals but lack some major features that characterize the fishes (Figure 5-11). The median eminence and pars nervosa are well-developed, distinct neurohemal structures. In tetrapods, there is no remnant of a saccus vasculosus. The amphibian pars distalis shows a tendency for cell regionalization, and the pars distalis of reptiles and birds consists of two somewhat distinct zones; however, these zones are not anatomically separated as are the rostral and proximal zones in fishes, although the distribution of cell types is similar from anterior to posterior. The pars tuberalis is a consistent tetrapod feature and appears to contain some secretory cells. A putative role for the pars tuberalis in controlling PRL release from the pars distalis has been proposed for mammals (see Chapter 4), but no role has been established for any non-mammal.

A. Amphibians

The amphibian HP axis has most of the features characteristic of the tetrapod system, including the presence of a pars tuberalis (Figure 5-11). The pars nervosa of the amphibian neurohypophysis receives peptidergic fibers originating in the magnocellular neurons of the PON and is the storage site for at least two nonapeptide hormones (AVT and the oxytocin-like **mesotocin, MST**). There is no tendency for development of a neurointermediate lobe in amphibians. There is less interdigitation between the pars nervosa and pars intermedia compared to fishes, and the term *neurointermediate lobe* generally is not used in tetrapods. The amphibian adenohypophysis consists of pars tuberalis, pars intermedia, and pars distalis. Based upon ultrastructural comparison of cytoplasmic granules and other features, there appear to be two separate granular cell types in the anuran pars tuberalis similar to those described in mammals (Figure 5-12). There also appears to be a neural pathway extending from the ependymal lining of the third ventricle to the pars tuberalis, but no functional correlation has been reported.

The amphibian pars intermedia has virtually no vascular supply, but it is directly innervated by peptidergic and aminergic neurons that originate in various nuclei of the hypothalamus. These characteristics of the pars intermedia are shared by all tetrapod groups. The pars distalis is not separable into discrete regions although there is a tendency for some regionalization of cell types (Figure 5-13). Caudate amphibians exhibit greater regionalization than do anuran species. Careful cytological studies have been performed on many amphibian species, often coupled with experimental manipulations, bioassays, or correlation with specific life history events. The cytology of the pituitary of *Rana temporaria* can serve as an example of the basic amphibian condition (see Appendix G).

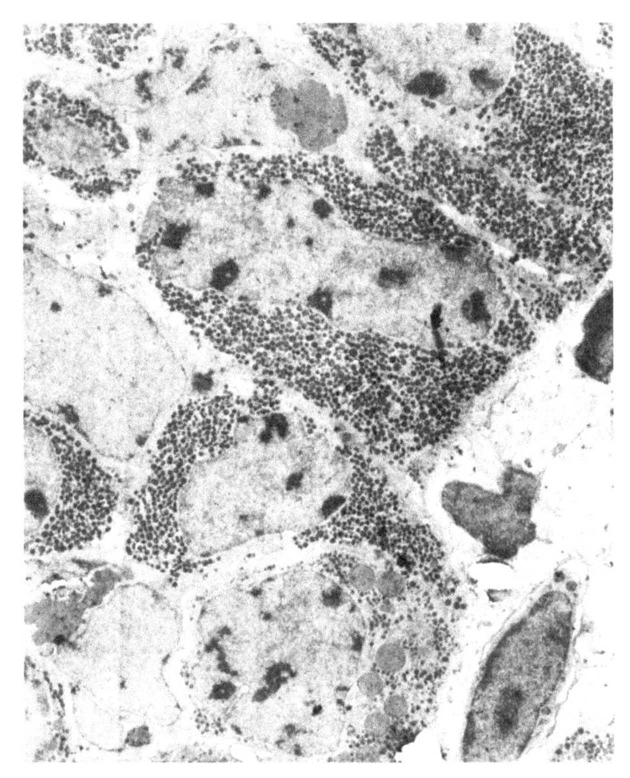

FIGURE 5-12 **Amphibian pars tuberalis.** Transmission electron micrograph of granular cells in the pars tuberalis of the frog, *Rana pipiens*. Two cell types can be identified on the basis of granule size. *(Photograph courtesy of Dr. Kevin T. Fitzgerald.)*

B. Reptiles

The adenohypophysis is well developed in reptiles, and it consists of a pars distalis, a pars tuberalis, and a pars intermedia (Figure 5-11). The Rhynchocephalia (*Sphenodon punctatus*), Chelonia (turtles), and Crocodilia (crocodiles, alligators, etc.) are thought to represent the basal reptilian condition. The pars distalis appears as two distinct regions reminiscent of the condition in bony fishes. The pars distalis is separable into a rostral or **cephalic lobe** and a **caudal lobe**, and the distribution of cell types in these lobes is similar to that described for fishes (see Appendix G). The pars tuberalis is well-developed in the Rhynchocephalia, Chelonia, and Crocodilia, but it is greatly reduced and sometimes absent in lizards and all adult snakes. There is no explanation for the absence of the pars tuberalis in squamate reptiles. Although the intermediate lobes of many lizards and burrowing snakes are markedly reduced or absent, some reptiles have well-developed intermediate lobes. It is especially well developed in certain chelonians, crocodilians, most snakes, and in the anoline lizards.

C. Birds

The avian HP system differs from that of other tetrapods (and fishes) in that the pars intermedia is absent in adults of all species, a condition seen in many reptiles as well as some mammalian groups, including primates. A well-developed pars tuberalis is present, and the pars distalis consists of cephalic and caudal lobes homologous to those described for reptiles (Figure 5-11). Unlike mammals, the primary capillaries of the portal system in birds lie superficially or in grooves on the surface rather than penetrating the pars tuberalis and do not form the complex vascular bed seen in mammals. In some birds (i.e., pigeon, Japanese quail, and white-crowned sparrow), the median eminence is separable into an anterior neurohemal area and a more posterior neurohemal area similar to that described for elasmobranchs. Each of these areas has its own portal connection to the pars distalis, which consists of a cephalic and a caudal lobe. It is not certain how widespread this phenomenon of two media eminentia is among other avian species. However, such potential regionalization of both the median eminence and the pars distalis could represent a mechanism to increase the efficiency of delivery of hypothalamic neurohormones to cell types regionalized in various parts of the pars distalis. This neurovascular specialization is analogous to the system of direct innervation of pars distalis cells that was observed in teleosts.

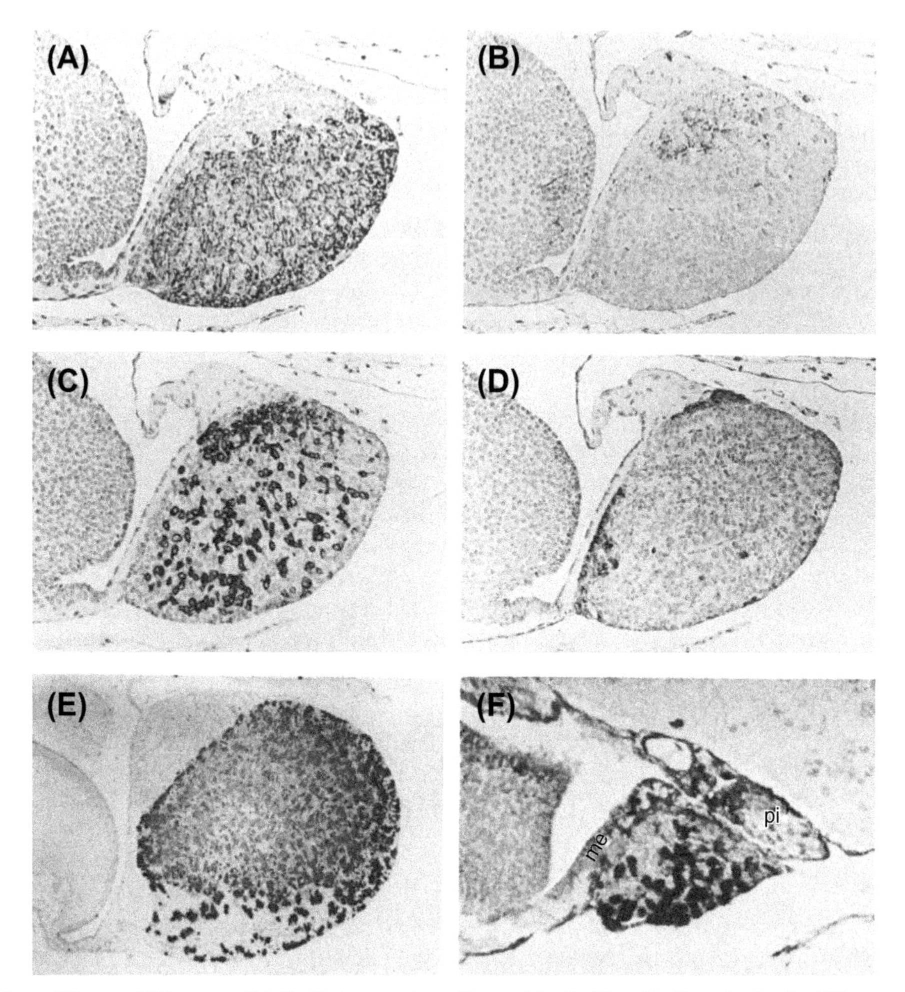

FIGURE 5-13 **Cytology of the amphibian pars distalis.** Dark-appearing cells are stained with antibodies selective for different pituitary hormones. (A) Immunoreactive prolactin-secreting cells. (B) Immunoreactive growth hormone-secreting cells. (C) Immunoreactive FSH-secreting cells. (D) Immunoreactive LH-secreting cells. (E) Immunoreactive corticotropin-secreting cells. (F) Immunoreactive thyrotropin-secreting cells. (A–D) Japanese newt (*Cynops pyrrhogaster*); (E) adult bullfrog (*Rana catesbeiana*); (F) postmetamorphic toad (*Bufo calamita*). *(Parts A to D are courtesy of Shigeyasu Tanaka and Sakae Kikuyama, Waseda University, Tokyo. Part E is reprinted with permission from Tanaka, S. et al.,* General and Comparative Endocrinology, *77, 88–97, 1990. Part F is reprinted with permission from Garcia-Navarro, S. et al.,* General and Comparative Endocrinology, *71, 116–123, 1988.)*

III. THE HYPOTHALAMUS OF NON-MAMMALIAN VERTEBRATES

The general organization of the hypothalamus within vertebrates is relatively simple in phylogenetically ancient groups, with neurons in the tetrapod hypothalamus migrating farther away from the third ventricle and forming more distinct nuclei from amphibians to reptiles and then to birds and mammals. It is possible to follow homologies of certain nuclei through the vertebrates. Many of the same hypothalamic regulators are present throughout the vertebrates. Interspersion of a vascular bed in the median eminence for transporting releasing hormones from the hypothalamus to the pituitary is termed a neurovascular connection. Neurovascular control of adenohypophysial hormone secretion by hypothalamic neurohormones is the basic pattern found in vertebrates with two notable exceptions. There are no vascular or neural connections between the hagfish brain and the adenohypophysis, and any control by the brain requires diffusion of neural factors to the pituitary. Among teleosts, there is no vascular connection between the hypothalamus and the adenohypophysis; instead, the axons of hypothalamic neurons travel directly to the pituitary. This neuroglandular pattern of control involves direct innervation of the adenohypophysis.

A. The Hypothalmaus of Agnathan Fishes

The hagfish brain has several NS regions (see Figure 5-6). Anterior and dorsal to the optic chiasm is the unpaired PON, which appears to produce NS products that are stored in the neurohypophysis, a neurohemal area comparable to the mammalian pars nervosa. The PON of hagfishes, unlike in other fishes, is an unpaired structure. There is, however, no neurohemal region in the Atlantic hagfish comparable to the median eminence of most other vertebrates, although a very primitive anterior neurohemal area has been described for the Pacific hagfish (*Eptatretus stoutii*) and has been termed a *median eminence*. The presence of GnRH has been demonstrated in the Atlantic hagfish (*Myxine glutinosa*), and it has been shown to vary seasonally with reproductive state. There are more NS nuclei identifiable in the lamprey brain than have been demonstrated in their relatives, the hagfishes (Figure 5-14). Two major NS centers occur in the lamprey brain anterior and dorsal to the optic chiasm: the **medial** and **lateral POAs** as well as the PON. Two additional NS areas are found in the hypothalamus: the **anterior hypothalamic area (AHA)** and the **dorsal** and **ventral periventricular arcuate nuclei.** Both GnRHs and TRH have been reported in the lamprey brain.

B. The Hypothalamus of Chondrostean and Holostean Fishes

The hypothalamus of chondrostean fishes contains a well-developed PON that provides peptidergic fibers to the pars nervosa. The **lateral tuberal nucleus (LTN, nucleus lateralis tuberis)** appears for the first time in bony fishes and consists of peptidergic and aminergic neurons connected to a well-developed median eminence. The median eminence is separated from the pars distalis by a connective tissue sheath so that no neurons penetrate the pars distalis. A portal system is present and probably conducts neurohormones from the median eminence to the pars distalis. The PON of holostean fishes has separated into two distinct portions: the dorsal **pars magnocellularis** made up of larger NS neurons and the ventral **pars parvocellularis** of smaller cells. Peptidergic fibers from the PON pass to the pars nervosa where nonapeptide hormones (AVT and IST) are stored. Aminergic fibers also appear in the pars nervosa and are believed to come at least in part from the LTN.

C. The Hypothalamus of Teleost Fishes

The diversity of teleosts makes generalizations about the hypothalamic distribution of neuropeptides very difficult. Furthermore, development and extensive application of immunohistochemistry to the teleost brain has resulted, as it did for mammals, in the discovery that neuropeptides may appear in many neurons that are not associated directly with the activity of the HP system. For example, in addition to the hypothalamus, immunoreactive GnRH neurons may be found in the most anterior olfactory regions of the brain as well as in the midbrain. Similarly, the amount of TRH present in the telencephalon typically is much greater than in the hypothalamus. Also, CRH neurons send processes to other brain regions. The anatomy of the brain of the Japanese eel (*Anguilla japonica*) serves as a representative but not necessarily typical teleost (Figure 5-15). There are numerous neurosecretory centers in the teleost brain and considerable variation in the names applied to them. Hence, only general descriptions are provided here, and the

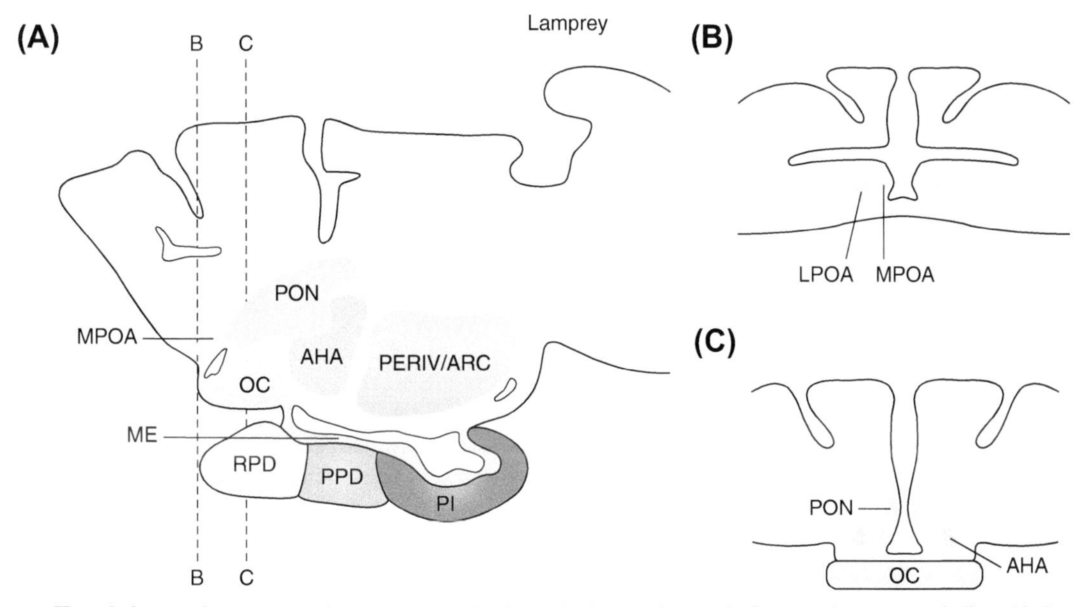

FIGURE 5-14 **Hypothalamus of sea lamprey (*Petromyzon marinus*).** (A) Sagittal section. (B,C) Cross-sections at levels indicated in Part A. See text or Appendix A for an explanation of abbreviations. *(Adapted with permission from Matsumoto, A. and Ishii, S., "Atlas of Endocrine Organs: Vertebrates and Invertebrates," Springer-Verlag, Berlin, 1989.)*

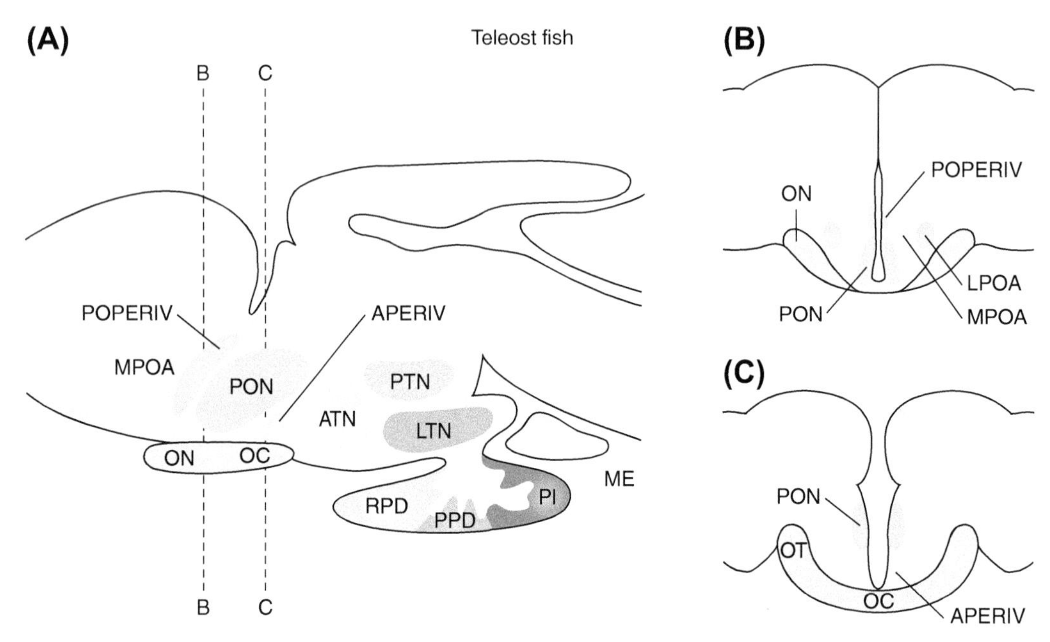

FIGURE 5-15 **Hypothalamus of a representative teleost, the Atlantic eel (*Anguilla anguilla*).** (A) Sagittal section. (B,C) Cross-sections at levels indicated in Part A. APERIV, anterior periventricular nucleus; ATN, anterior tuberal nucleus; ON, optic nerve; OT, optic tract; POPERIV, periventricular preoptic nucleus; PTN, posterior tuberal nucleus. See text or Appendix A for other abbreviations. *(Adapted with permission from Matsumoto, A. and Ishii, S., "Atlas of Endocrine Organs: Vertebrates and Invertebrates," Springer-Verlag, Berlin, 1989.)*

interested reader should consult the references at the end of this chapter for specific sources. Axons of neurosecretory neurons follow well-defined tracts to the pars distalis or the neurointermediate lobe of the pituitary, whereas others may connect with a variety of other brain structures. The POA consists of several specialized areas in the eel. In addition to the PON containing parvocellular and magnocellular neurons, there also are medial and lateral POAs as well as the anterior periventricular nucleus of the anterior hypothalamus which is a major source of dopaminergic axons ending in the pars distalis. The **suprachiasmatic nucleus (SCN)** also sends axons to the pituitary as well as to other brain areas. The teleost LTN may be separated into anterior, posterior, ventral, and lateral parts that are principally responsible for secretion of several releasing hormones.

D. The Hypothalamus of Amphibians

The prominent POA (Figure 5-16) contains several NS centers including the lateral and medial POAs and the PON we saw in the fishes. In addition, there is a distinct magnocellular region of the POA that is continuous with a well developed SCN. **Dorsal hypothalamic (DH)** and **ventral hypothalamic (VH)** nuclei as well as an **infundibular nucleus (IFN)** are present, and the IFN is homologous to at least part of the major hypophysiotropic region of the arcuate nucleus in the mammalian hypothalamus. The exact relationship between the IFN of amphibians and the LTN of fishes is uncertain. The distribution of various neuropeptide-producing neurons in the hypothalamus resembles the pattern seen in mammals. While there is no distinct PVN, CRH and AVT neurons are located in close proximity to each other in the POA and magnocellular nuclei as well as in the IFN. Dopamine-producing neurons are found in various hypothalamic areas including the SCN, and these neurons provide the major inhibitory control of α-MSH secretion from the pars intermedia in amphibians. TRH neurons are widely spread throughout the hypothalamus in the POA, SCN, and IFN. Mammalian GnRH (GnRH I) neurons are located in the POA. Sexual dimorphism has been described in the amphibian brain. The number of neurons in the anterior POA of leopard frogs (*Rana pipiens*) is dependent on androgens, whereas the number of neurons in the posterior tuberculum is estrogen dependent. Furthermore, the distribution of AVT also is sexually dimorphic in the amphibian brain.

E. The Hypothalamus of Reptiles

The reptilian hypothalamus is depicted in Figure 5-17. In reptiles, there is no longer a PON but the medial and lateral POAs are present. The PON of anamniotes has separated into the **supraoptic nucleus (SON)** and the **paraventricular nucleus (PVN)** that are characteristic of all amniote vertebrates. These nuclei consist of peptidergic NS neurons that produce the octapeptide neurohormones AVT and MST. Their axons terminate in the pars nervosa. The hypothalamus is separated into **anterior**, **dorsal**, and **posterior hypothalamic areas**, **paraventricular organs**,

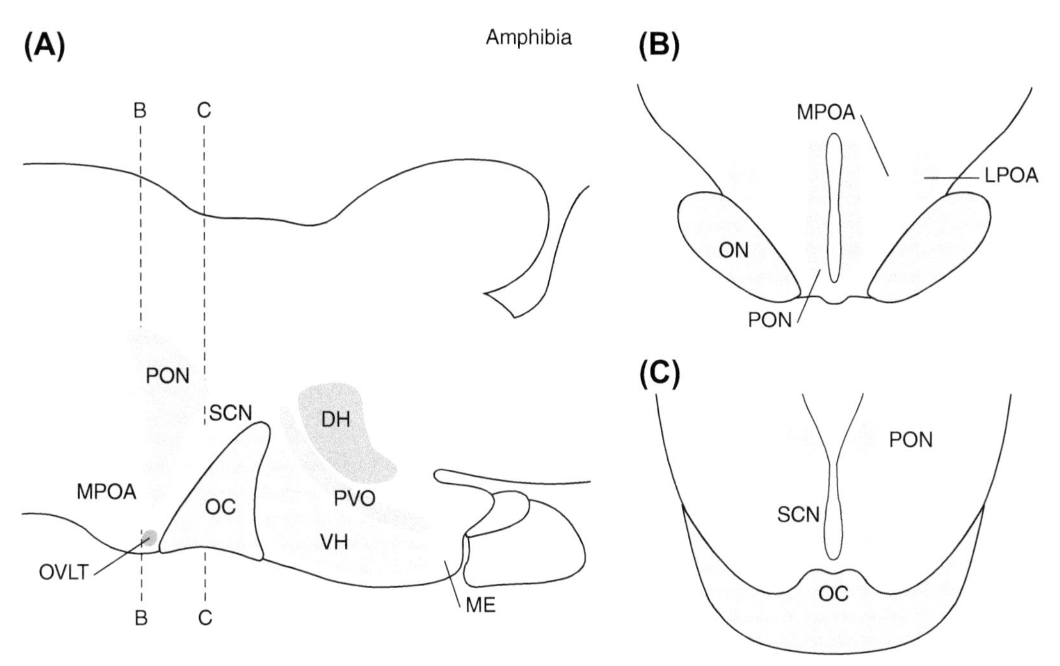

FIGURE 5-16 **Hypothalmaus of an amphibian, the American bullfrog (*Rana catesbeiana*).** (A) Sagittal section. (B,C) Cross-sections at levels indicated in Part A. PVO, paraventricular organ. See text or Appendix A for other abbreviations. *(Adapted with permission from Matsumoto, A. and Ishii, S., "Atlas of Endocrine Organs: Vertebrates and Invertebrates," Springer-Verlag, Berlin, 1989.)*

periventricular nucleus (PERIV), and both **ventral (VMN)** and **dorsal (DMN) medial nuclei**. The IFN, also called the **arcuate (ARC) nucleus**, is considered to be homologous to the VH of the amphibian IFN and the ARC nucleus (= IFN) of birds as well as to the major hypophysiotropic area of the mammalian hypothalamus. Aminergic and peptidergic NS fibers originate in this nucleus and terminate in the median eminence.

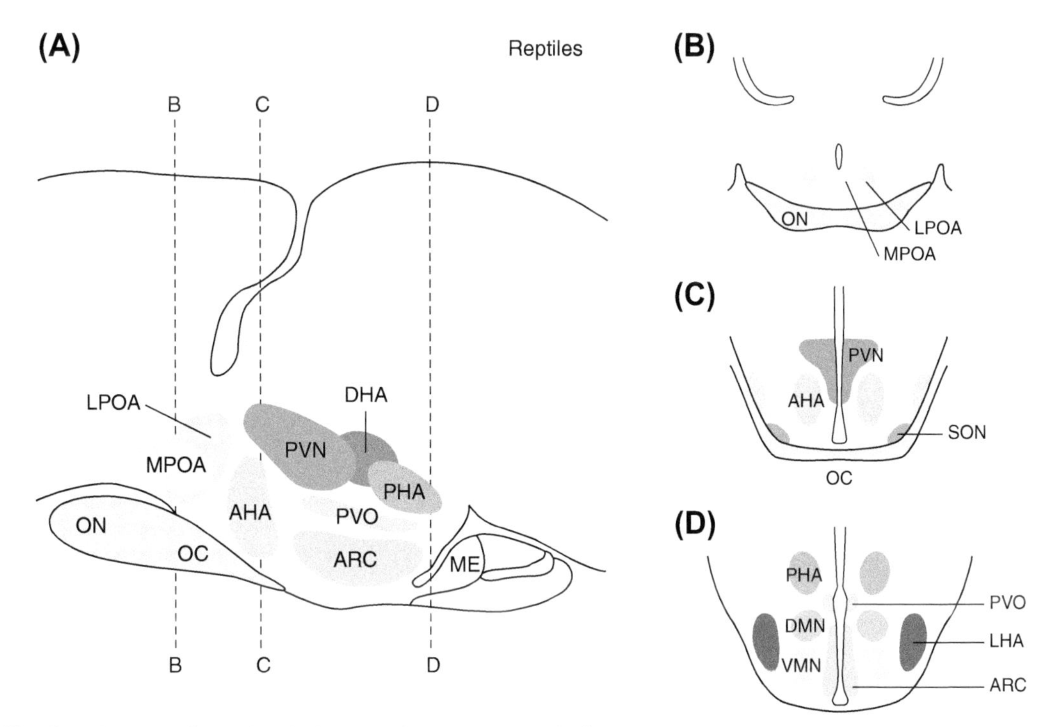

FIGURE 5-17 **Hypothalamus of a reptile, the Japanese forest ratsnake (*Elaphe conspicillata*).** (A) Sagittal section. (B,C,D) Cross-sections at levels indicated in Part A. AHA, DHA, LHA, PHA = anterior, dorsal, lateral, posterior hypothalamic area, respectively; PVO, paraventrocular organ. See text or Appendix A for other abbreviations. *(Adapted with permission from Matsumoto, A. and Ishii, S., "Atlas of Endocrine Organs: Vertebrates and Invertebrates," Springer-Verlag, Berlin, 1989.)*

F. The Hypothalamus of Birds

The hypophysiotropic region of the hypothalamus in birds supplies aminergic and peptidergic fibers to the median eminence. The region anterior and dorsal to the optic chiasm is occupied by the medial anterior POA, a dorsal PON, SCN, PVN, and SON (Figure 5-18). As in reptiles and mammals, the SON and the PVN are primarily responsible for secretion of the nonapeptide neurohormones. The hypothalamus itself includes the anterior and lateral hypothalamic areas, **paraventricular organ**, and an IFN (= ARC). Two sexually dimorphic brain regions are known for birds. In quail, males exhibit a larger medial PON (mPON) with greater levels of aromatase activity. This nucleus controls male copulatory behavior and is innervated by both AVT and catecholaminergic neurons. Testosterone (T) blood levels are responsible for activating this center but T must be converted by aromatase to estradiol (E_2) to have any effect (see Chapter 3). Administration of either E_2 or T, then, can induce male copulatory behavior. Any reduction in circulating T levels such as would occur following castration or placement of the birds on a non-stimulatory photoperiod (short days) results in regression of the mPON. The second sexually dimorphic region is the song control region of passerine (song) birds which is larger in males that do the majority of singing. In wrens that sing duets, there is no marked difference in this region whereas there is a marked difference in one species of wren in which only the male sings. Although clearly a sex-related phenomenon, it is not clear that this dimorphism has a hormonal basis.

IV. TROPIC HORMONES OF NON-MAMMALIAN VERTEBRATES

The same categories of tropic hormones are found in non-mammals as were discussed for mammals in Chapter 4. Each category represents a gene family. Category 1 includes the **glycoprotein hormones (GpH)** LH, FSH, and TSH. Growth hormone (GH) and PRL as well as SL in teleosts constitute hormones of Category 2, and the derivatives of POMC represent Category 3: ACTH, α-MSH, **corticotropinlike peptide (CLIP)**, and **β-endorphin**.

A. Category 1: Gonadotropins (GTHs) and Thyrotropin (TSH)

All vertebrate GpHs, as described for mammals, are heterodimers composed of an α-subunit and a β-subunit (see Chapter 4, Figure 4-14). **Follicle-stimulating hormone (FSH),** LH, and **thyrotropin (TSH)** share a common α-subunit (GpHα) and each has a unique β-subunit (three distinct GpHβs). An additional GpH named **thyrostimulin** was discovered in mammals consisting of two previously unknown subunits named **GPA2** (corresponding to GpHα) and **GPB5** (corresponding to GpHβ) (see Figure 5-19). These new subunits appeared early in

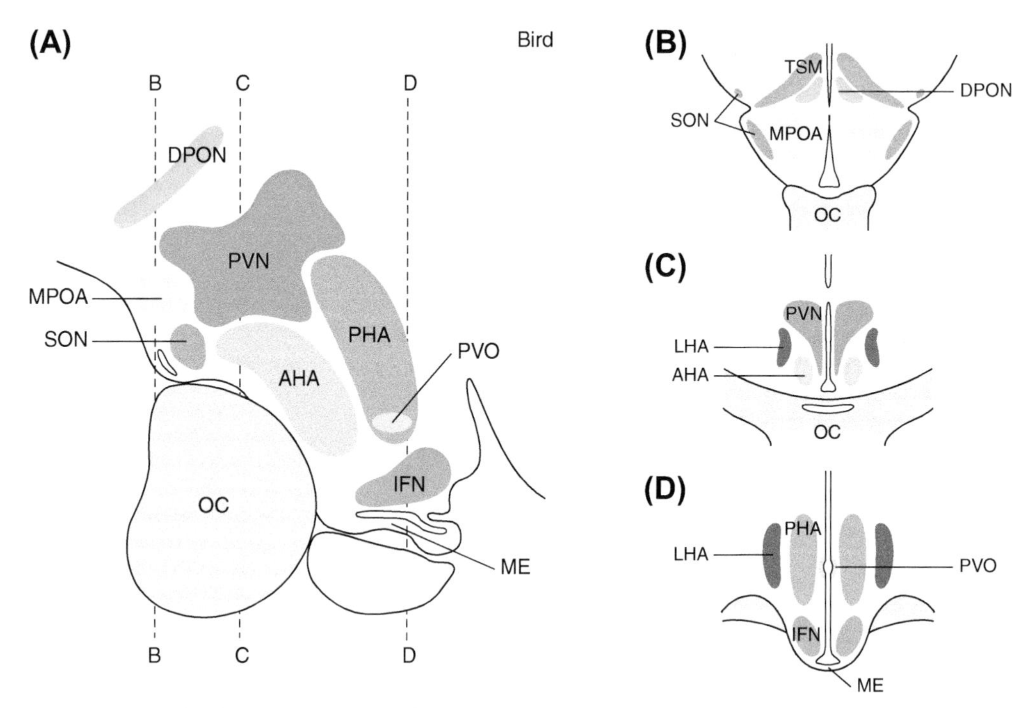

FIGURE 5-18 **Hypothalamus of a bird, the Japanese quail (*Coturnix coturnix japonicus*).** (A) Sagittal section. (B,C,D) Cross-sections at levels indicated in Part A. TSM, septomesencephalic tract. See Figure 5-17 or Appendix A for other abbreviations. *(Adapted with permission from Matsumoto, A. and Ishii, S., "Atlas of Endocrine Organs: Vertebrates and Invertebrates," Springer-Verlag, Berlin, 1989.)*

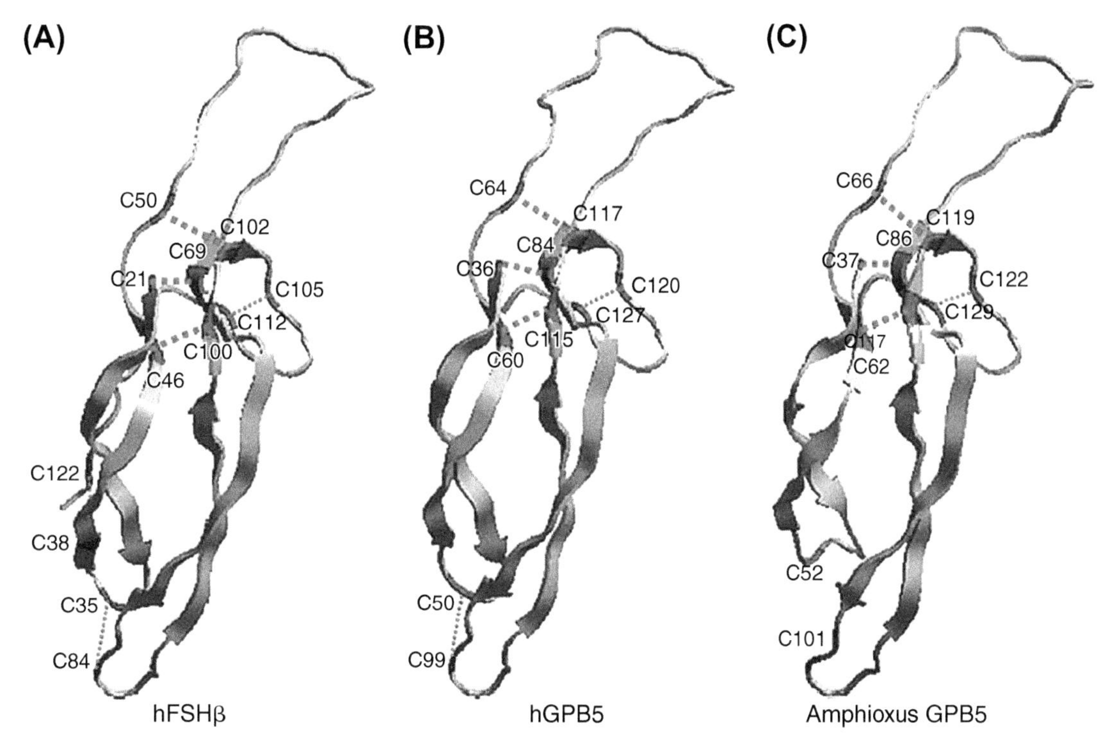

FIGURE 5-19 **A comparison of human FSH and GPB5 with GPB5 from amphioxus, a protochordate.** *(Reprinted with permission from Tando, Y. and Kubokawa, K.,* General and Comparative Endocrinology, ***162****, 328–339, 2009.)*

animal evolution and are retained in many invertebrate and vertebrate animals. Generally, these subunits are found close to one another in the genomes on the same chromosome of most bilateral animals with the exception of the tetrapod vertebrates. Although thyrostimulin was named because of its ability to activate the TSH receptor, the functions of these subunits in animals appear to be closely related to development.

Early experimental studies focused on the actions of mammalian GTHs in lower vertebrates. Eventually, GTHs were purified from non-mammalian species and their roles clarified. Generally, the hormonal control of reproduction in non-mammals involves two GTHs with LH-like and FSH-like activities. The apparent failure to find two GTHs in squamate reptiles may be due to derived secondary reliance on only one GTH, with synthesis of the other repressed. Non-mammalian GTHs can be expected to differ markedly in structure and function when compared to their mammalian counterparts, as well as among different non-mammalian groups (e.g., teleosts vs. reptiles) and occasionally even among closely related species. Caution must be applied before assuming that an LH of species X will be more LH-like in its action when tested in species Y or that a molecule labeled as an LH when applied to another species may induce FSH-like effects. This diversity in actions of GTHs may be due in part to variability in receptor structures that might recognize different regions of these molecules or respond to subtle changes in shape of the same region in a GTH molecule due to variations in amino acid sequences and/or carbohydrate additions.

It was proposed a number of years ago that the most primitive glycoprotein hormone was an LH-like molecule. Later, the LH-like gene duplicated, and one copy became modified to produce a TSH-like hormone. In this scenario, FSH diverged even later from TSH. Support for this hypothesis was based in part on observations of inherent thyroid-stimulating activity of mammalian LHs tested in more primitive vertebrates. However, category 1 tropic hormones have now been isolated from an array of jawed fish groups and all of the tetrapod groups, and many of the β-subunits have been examined for homologies. TSHβ, LHβ, and FSHβ subunits cluster as separate groupings regardless of the species from which they are derived (Figure 5-20). Furthermore, LHβ and FSHβ appear to be more closely related to one another whereas TSHβ subunits form a separate cluster. This relationship is in keeping with the hypothesis that the ancestral GpH gene underwent duplication, and one gene diverged into a TSHβ (*tshb*) gene and a GTHβ (*gthb*) gene. Following a second genome duplication, the primitive *gthb* gene diverged into an LHβ gene (*lhb*) and an FSHβ (*fshb*) gene. This later creation of the *fshb* gene would have had less time to diverge from the *lhb* gene than the *tshb* gene has had. The single exception seems to be the LHβ subunit isolated from tilapia (teleost) that is more like the TSHβ subunits of other vertebrate groups than its is like other vertebrate LHβ subunits.

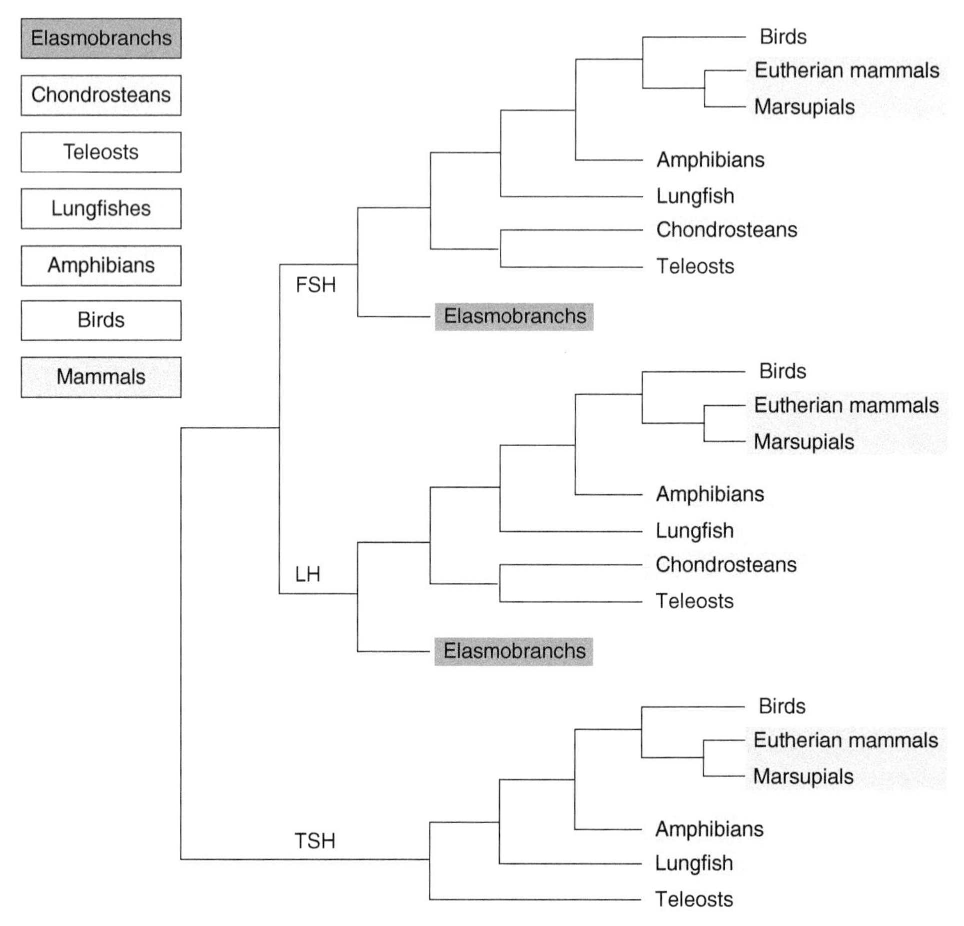

FIGURE 5-20 **Phylogeny of β-subunit genes for the glycoprotein tropic hormones.** Note that the genes for each tropic hormone form a distinct cluster for all species examined. Although not shown, FSHβ and LHβ genes from a reptile (Reeves's turtle) cluster near the respective bird gonadotropin. *(Based on studies by Oba, Y. et al.,* Comparative Biochemistry and Physiology B, ***129****, 441–448, 2001; Querat, B. et al.,* Biology of Reproduction*,* ***70****, 356–363, 2004.)*

1. Gonadotropins (GTHs) in Agnathan Fishes

One GTH is found in the proximal pars distalis of the lamprey, and the use of cDNA cloning techniques has identified a unique GTHβ subunit whose expression is stimulated by lamprey GnRH. This may be closer to the first *gthb* gene. No GTH data are available for hagfishes.

2. Gonadotropins (GTHs) in Chondrichthyean Fishes

Gonadotropin activity is present in both the proximal pars distalis and the ventral lobe of elasmobranchs. Antibody to mammalian GTHs binds to cells of the ventral lobe. Bioassay of the elasmobranch proximal pars distalis reveals the presence of an LH-like GTH that stimulates oocyte maturation in the clawed frog *Xenopus laevis*. Both LHβ and FSHβ subunits have been sequenced from a shark (*Scyliorhinus canicula*) and shown to be unique from those of all other gnathostomes (see Figure 5-20).

3. Gonadotropins (GTHs) in Bony Fishes

Two glycoprotein GTHs consisting of α- and β-subunits like the GTHs of tetrapods and their specific genes have been characterized from more than 50 species of teleost fishes. Initially, they were termed **GTH-1** and **GTH-2**, but now are referred to as FSH and LH, respectively. Gonadal maturation (spermatogenesis and oocyte development) is controlled by the *fshb* gene responsible for FSH secretion, whereas gamete release (spermiation and final oocyte maturation) is controlled by the *lhb* gene responsible for LH secretion. Although both hormones show similar potencies in stimulating steroidogenesis, LH apparently is released primarily during the final step of gonadal maturation when it stimulates synthesis of the steroid **17,20β-dihydroxy-4-pregnen-3-one (17,20β-DHP)**. Initially, 17,20β-DHP was named the **oocyte maturational steroid hormone** or **maturation-inducing hormone (MIH)** but later it was shown to cause spermiation in males, too. Immunocytochemical studies have

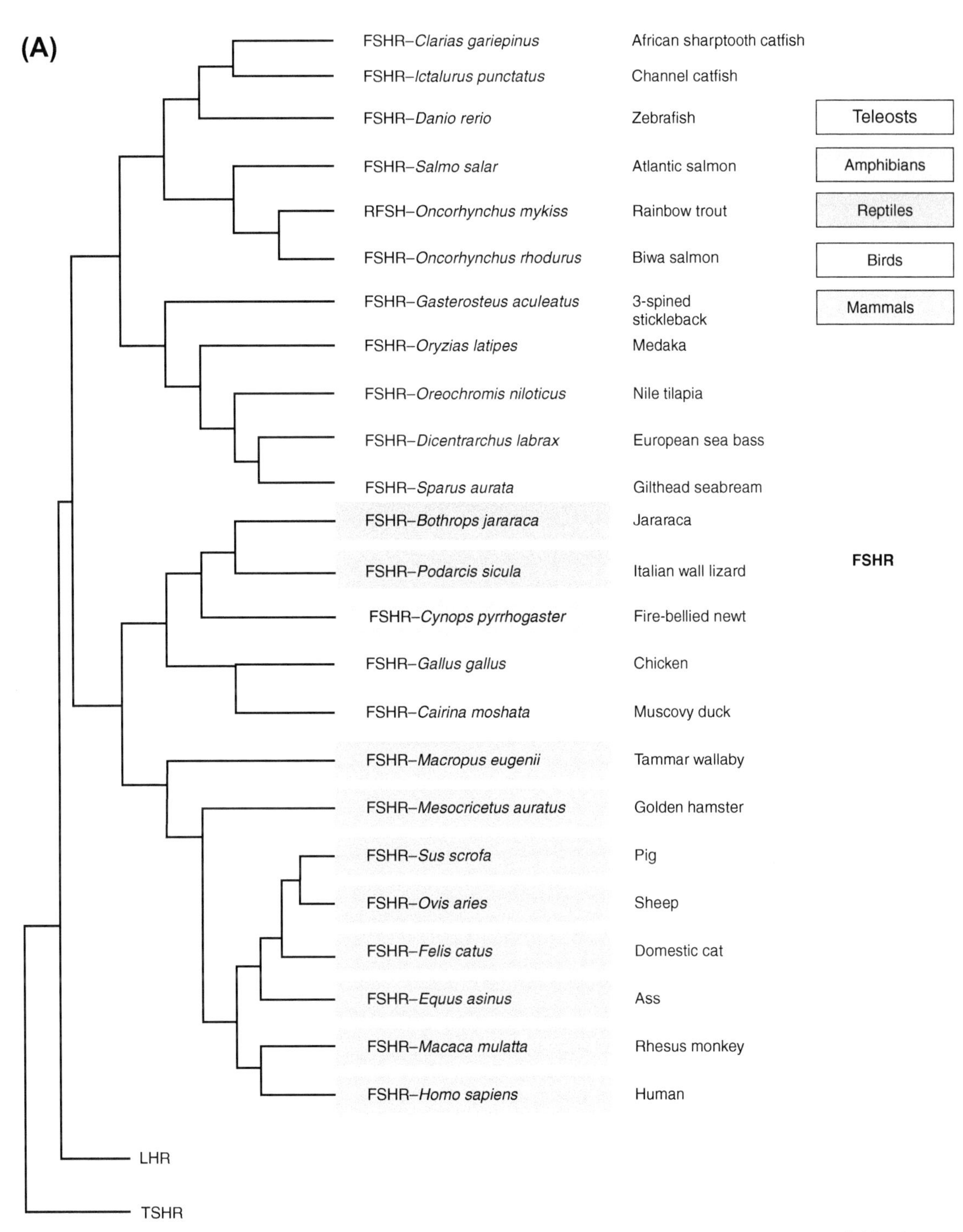

FIGURE 5-21 **GpH Receptor evolution within vertebrates.** (A) Follicle-stimulating hormone receptor (FSHR). (B) Luteinizing hormone receptor (LHR). (C) Thyrotropin receptor (TSHR). Note that the teleosts can be separated into two groups indicating gene differences coding for somewhat different receptors. *(Adapted with permission from Levavi-Sivan, B. et al.,* General and Comparative Endocrinology*, **165**, 412–437, 2010.)*

demonstrated that FSH and LH are produced in separate pituitary cell types, unlike the situation in tetrapods where a single cell type produces both GTHs.

Both LH receptors (LHRs) and FSH receptors (FSHRs) occur in teleost gonads. Both are **G-protein-coupled receptors (GPCRs)**, and analysis of the the genes responsible for their production shows they are related to the respective mammalian receptors and those of other tetrapods (that is, all LHRs are alike, whereas all FSHRs are alike but differ from the LHR gene cluster (Figure 5-21). The extracellular domains of the teleost receptors differ considerably from those of the cognate mammalian receptors, however.

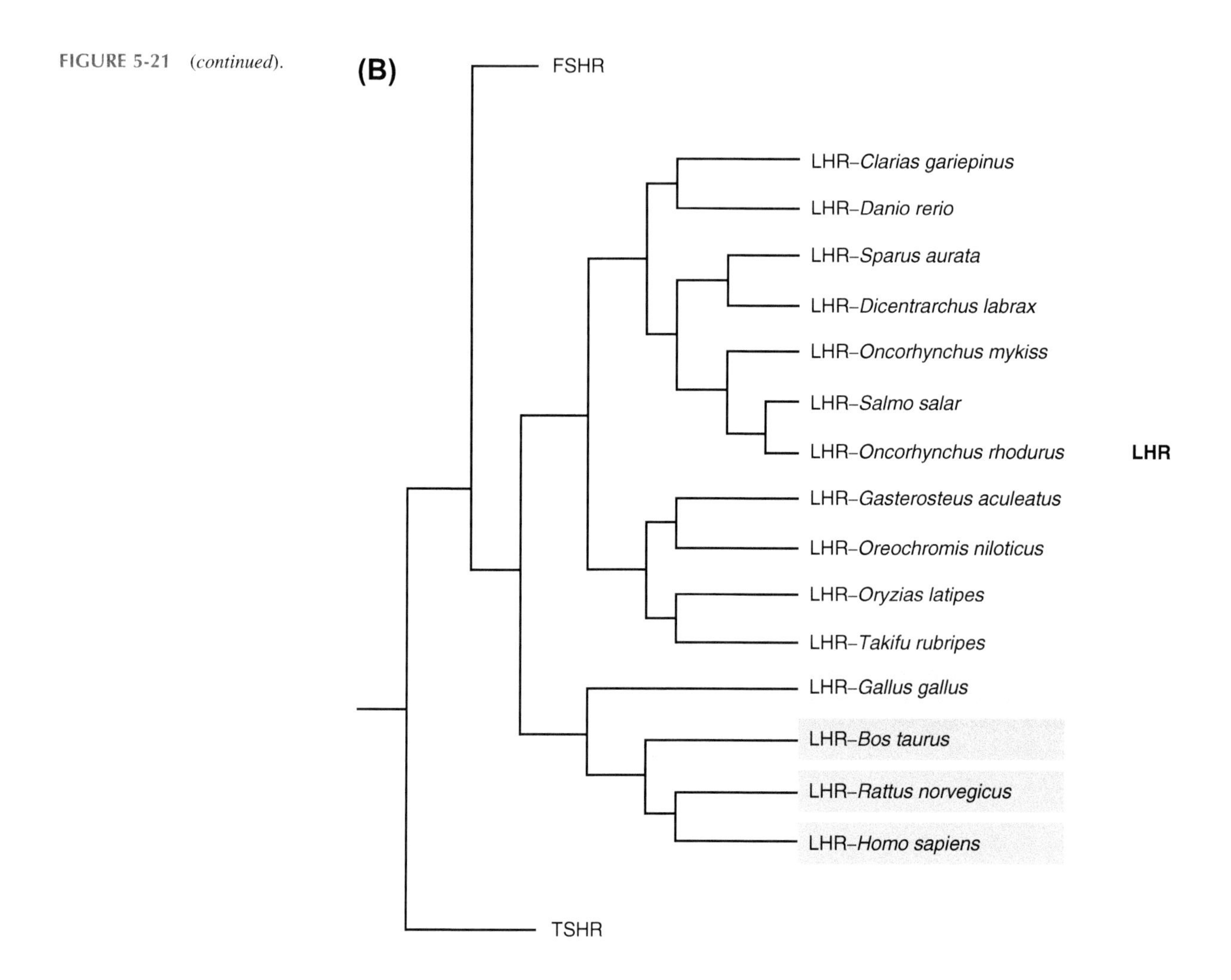

FIGURE 5-21 *(continued)*.

4. Gonadotropins (GTHs) in Amphibians

Two distinct GTHs have been isolated and purified from several urodeles and anurans that are similar to GTHs of other tetrapods (Figure 5-21). In male anuran and urodele amphibians, FSH stimulates spermatogenesis, whereas LH stimulates androgen synthesis and spermiation. Follicle development and estrogen synthesis in females are regulated by FSH, whereas androgen and **progesterone (P_4)** secretion is regulated by LH. Because of its effect on P_4 synthesis, LH is also associated with final oocyte maturation (see Chapter 11). These observations are similar to what is known for other tetrapods. Amphibian LH is responsible for stimulating androgen synthesis as well as for spermiation and ovulation. The secretion of P_4 by the short-lived corpora lutea appears to be dependent on LH.

Some species specificity is associated with these GTHs that may be due to differences in the structure of the GTH protein, the carbohydrates attached to the protein, or species differences in the binding domain of the receptor. For example, salamander (*Ambystoma*) LH is not as effective as bullfrog LH in raising plasma androgen levels in bullfrogs. Conversely, *Ambystoma* LH is more effective than bullfrog LH in *Ambystoma*, although the latter was considerably more effective at raising androgen levels than *Ambystoma* FSH. Similarly, bullfrog LH is more effective than ovine LH in stimulating P_4 synthesis in amphibians. Bullfrog LH also stimulates reptilian and avian thyroids, reflecting a structural closeness between LH and TSH molecules that is probably the basis for these observations.

An *fshr* gene has been characterized for the newt *Cynops pyrrhogaster* that is similar to FSHRs of other tetrapods (Figure 5-21). Both FSHRs and LHRs have been demonstrated in the gonads of anuran species.

5. Gonadotropins (GTHs) in Reptiles

Studies employing injections of mammalian hormones into squamate reptiles (snakes and lizards) have emphasized

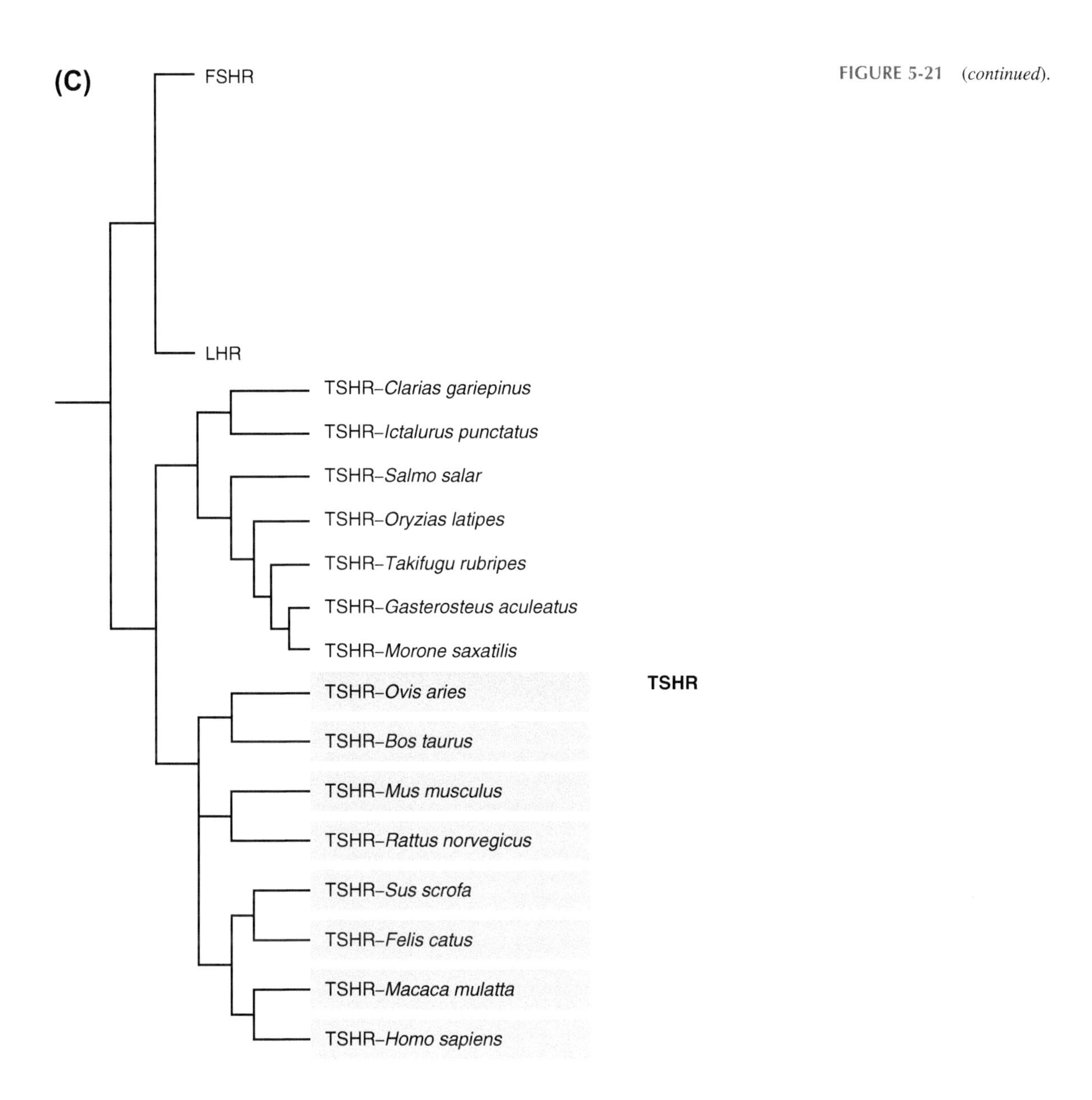

FIGURE 5-21 *(continued)*.

a role for FSH-like GTHs but suggest no specific role for LH-like hormones. Earlier data might be interpreted as a consequence of a very short biological half-life for injected mammalian LH or that mammalian FSH might be similar enough to both a squamate FSH and LH to possess both activities. Purification of gonadotropic activity from squamate reptiles, however, has verified that only one GTH that is chemically similar to FSHs is present, although both FSH-like and LH-like GTHs have been isolated from other reptiles (i.e., chelonians and crocodilians). Presence of only one molecular species of GTH in squamates does not rule out the possibility that this molecule has intrinsic FSH-like and LH-like activities. Because both FSH- and LH-like GTHs are found in other reptiles, the squamate condition is probably secondarily derived.

There is great variability in the responses of reptilian tissues to GTHs purified from different vertebrate groups. For example, the testicular weight of hypophysectomized *Anolis carolinensis* is readily maintained by FSH-like hormones purified from mammals, birds, reptiles, or amphibians, but this testicular parameter is very insensitive to LH-like gonadotropins from any source. Curiously, plasma androgen levels in hypophysectomized *A. carolinensis* may be elevated by either FSH-like or LH-like GTHs from almost any source. Similar non-specific effects of GTHs on plasma levels of androgens or on androgen synthesis by minced testes *in vitro* have been observed for crocodilians. The situation in chelonians is not clear, as testes of some species do not respond to LH-like hormones *in vitro*, whereas at least one species of turtle (*Chrysemys picta*) does respond to *in vivo* injections of mammalian LH but not as well as it responds to mammalian FSH. The reptilian ovary, like the testis, exhibits broad sensitivity to mammalian and other FSHs and LHs. Non-reptilian FSHs are generally the most effective GTHs for treating reptiles, although there are some conflicting reports regarding this point.

6. Gonadotropins (GTHs) in Birds

Birds conform to the typical mammalian pattern of two GTHs, LH and FSH, although avian species are rather insensitive to mammalian FSH and LH preparations. Two separate GTHs have been extracted and partially purified from domestic galliform birds; however, studies of the duck support the existence of only one GTH. Chicken LH stimulates interstitial cells in the testis of chickens or Japanese quail, whereas chicken FSH stimulates the seminiferous tubules. Mammalian LH and avian LH are reported to be more effective than FSHs in stimulating androgen production by chicken interstitial cells *in vitro*. Some conflicting reports suggest that avian FSH and LH are more nearly equal in their actions on minced pigeon or chicken testes, however. Reptilian LH is more effective than reptilian FSH in stimulating androgen production by the avian testis. Purified amphibian GTHs (both LH-like and FSH-like) are equally effective, although their overall activity in birds is low.

7. Thyrotropin (TSH) in Vertebrates

Thyrotropins have been isolated from all gnathostome vertebrate groups. The presence of thyrostimulin subunits as well as the presence of a thyroid glycoprotein receptor in lampreys suggests the formation of the HPT axis occurred in the earliest vertebrates. Mammalian TSHs stimulate thyroid function in representatives of all vertebrate groups, although they are ineffective in hagfishes. Similarly, pituitary extracts from most non-mammals exhibit TSH-like activity when tested in mammals. The specific actions of TSH on non-mammalian thyroid glands and the functions of thyroid hormones in non-mammals are similar to those for mammals (see Chapters 4, 6, and 7). Although non-mammalian TSHs have not been studied to the extent that non-mammalian GTHs have, it is apparent that genes encoding β-subunits of teleosts, amphibians, and birds cluster with mammalian TSHβ subunits and are distinctly separate from vertebrate GTHβ subunits (Figure 5-20). Interestingly, the TSHβ subunit from sturgeon is more like those of tetrapods than like several species of teleosts. Furthermore, TSH receptors genes isolated from amago salmon cluster with tetrapod TSH receptor genes but not with fish or tetrapod GTH receptor genes. Purified bullfrog TSH is thyrotropic in both anurans and caudate amphibians but is ineffective on thyroids of reptiles and birds, indicating considerable variability in these gene sequences through time.

B. Category II Tropic Hormones: Prolactin (PRL), Somatolactin (SL), and Growth Hormone (GH) Family

As discussed in Chapter 4, PRL and GH are chemically very similar proteins. This similarity in structure has resulted in considerable confusion about the actions of these hormones in non-mammals, especially when based on the observations of treating non-mammals with mammalian PRL and GH. Among the teleost fishes, the GH gene has given rise to the related SL gene. Phylogenetic relationships among GHs, PRLs, and SL genes can be seen in Figures 5-22, 5-23, and 5-24. A similar phylogenetic scheme is found for GH, PRL, and SL receptors (Figure 5-25).

Early studies of purified PRL produced more than 300 different effects in vertebrates, although many of these effects may have little physiological relevance. Most of the reported actions of PRL (Table 5-2) may be grouped into six general categories: (1) water and electrolyte balance, (2) growth and development, (3) metabolism, (4) behavior, (5) reproduction, and (6) immunoprotection.

The most primitive role for PRL is the regulation of osmotic–ionic balance, with the other actions being acquired later among vertebrates during evolution. Certainly, the major role for PRL in teleosts is related to osmotic regulation in freshwater. Even in humans, PRL has retained some osmoregulatory actions during development, and release of PRL can be induced following alterations in blood osmotic pressure. Regardless of the multiplicity of

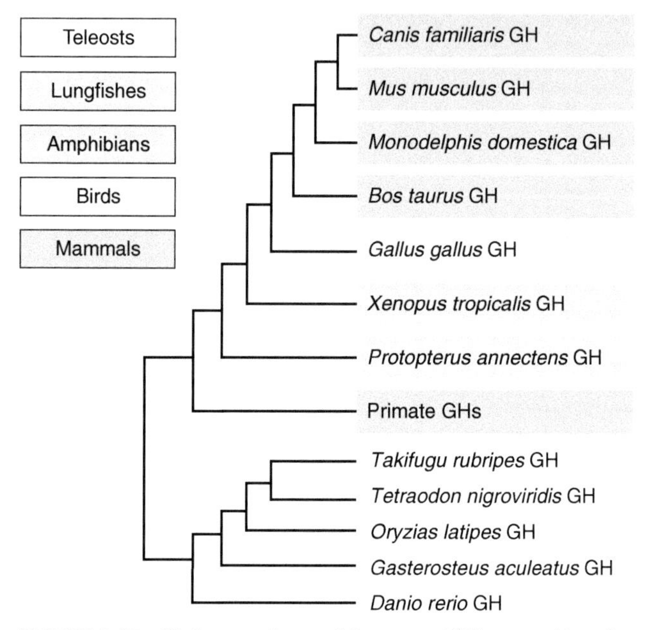

FIGURE 5-22 **Phylogeny of growth hormone (GH) genes.** Note that primate GH is distinctly different from other mammals. *(Adapted with permission from Daza, D.O. et al.,* Annals of the New York Academy of Sciences, ***1163****, 491–493, 2009.)*

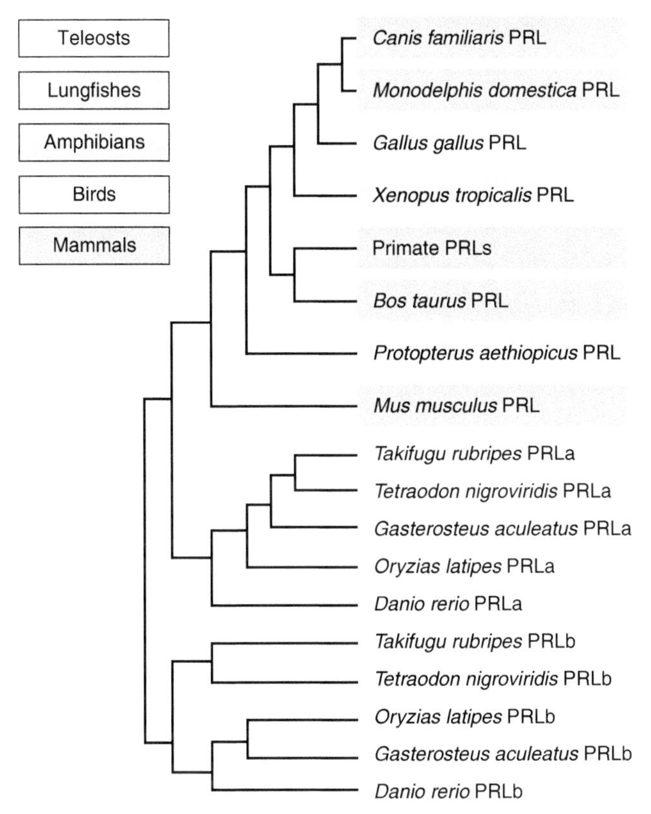

FIGURE 5-23 **Phylogeny of Prolactin (PRL) genes.** Two forms of PRL (PRLa, PRLb) are present in all teleosts tested as a result of an additional gene duplication in teleosts. *(Adapted with permission from Daza, D.O. et al.,* Annals of the New York Academy of Sciences, ***1163****, 491–493, 2009.)*

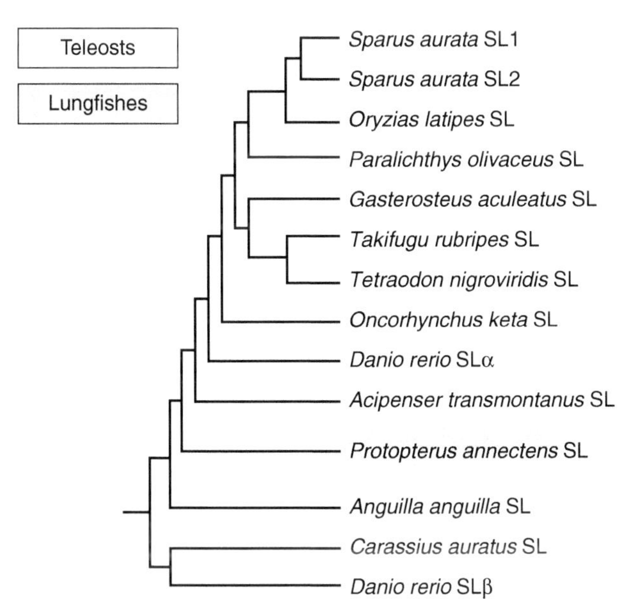

FIGURE 5-24 **Phylogeny of somatolactin (SL) genes within fishes.** Note that two distinct forms (α and β) are found in some species, including zebrafish and rainbow trout (not shown here). Note that the African lungfish SL is very much like teleost SL. *(Adapted with permission from Daza, D.O. et al.,* Annals of the New York Academy of Sciences, ***1163****, 491–493, 2009.)*

roles or even the establishment that it may be primarily an osmoregulatory hormone, the name *prolactin* that was established in mammals for this hormone is here to stay.

PRL provides an excellent example for demonstrating the evolution of endocrine systems. There have been evolutionary changes in the structure of the PRL molecule as evidenced by the failure of fish PRLs to work in avian and mammalian bioassays, although mammalian PRLs retain activity when tested in fishes. The antigenic portion of fish PRLs is similar to mammals as evidenced by the binding of mammalian PRL antibodies to the lactotropes of the fish rostral pars distalis. Furthermore, new PRL target tissues have evolved (for example, the crop sac of birds and the mammary glands of mammals) that possess receptors specific for the "newer" portions of the molecules. This is evidenced by the failure of the fish PRLs to activate responses in avian and mammalian tissues. It is unfortunate that there are no bioassays employing reptilian tissues and that purified reptilian PRLs are not available to fill in this part of the evolutionary changes that appear to be supported by studies with the bioassays of the other major groups.

Mammalian GHs are effective in most non-mammals, and most non-mammalian preparations exhibit GH activity in mammals. The differences among vertebrate GHs are exemplified by their varied effectiveness when tested in a single system such as the rat tibia bioassay (see Appendix D). Comparative functional studies are hampered further by the structural similarity of GH and PRL. For example, PRL has clear growth-promoting actions in larval amphibians.

Structure analyses of GH cDNA from fishes and tetrapods suggests a molecular phylogeny that recognizes a distinct dichotomy with most teleost genes diverging from other bony fish and tetrapods genes (Figure 5-22). Interestingly, lungfish GH and amphibian GH form a separate cluster. GH has been characterized from adult sea lampreys and shown to stimulate **insulin-like growth factor (IGF)** mRNA production in the lamprey liver as it does in mammals (see Chapter 4). The confirmation of a GH gene in lampreys, together with the absence of assayable PRL, suggests that GH is the most primitive member of category 2 tropic hormones and that PRL appeared subsequent to a GH gene duplication. IGF genes have been cloned from a hagfish, the cephalochordate amphioxus, and a tunicate, although GH has been identified only in the hagfish. The crocodilians and birds form a subcluster distinct from other tetrapods. Primate GHs are distinctly mammalian and do not appear to be closely related structurally to non-mammalian hormones.

The third member of this polypeptide family is the hormone SL, which is unique to the teleost fishes and lungfishes. The SL gene appears to have evolved from the GH gene in teleosts, although the SL protein differs markedly from both GH and PRL in having six or seven

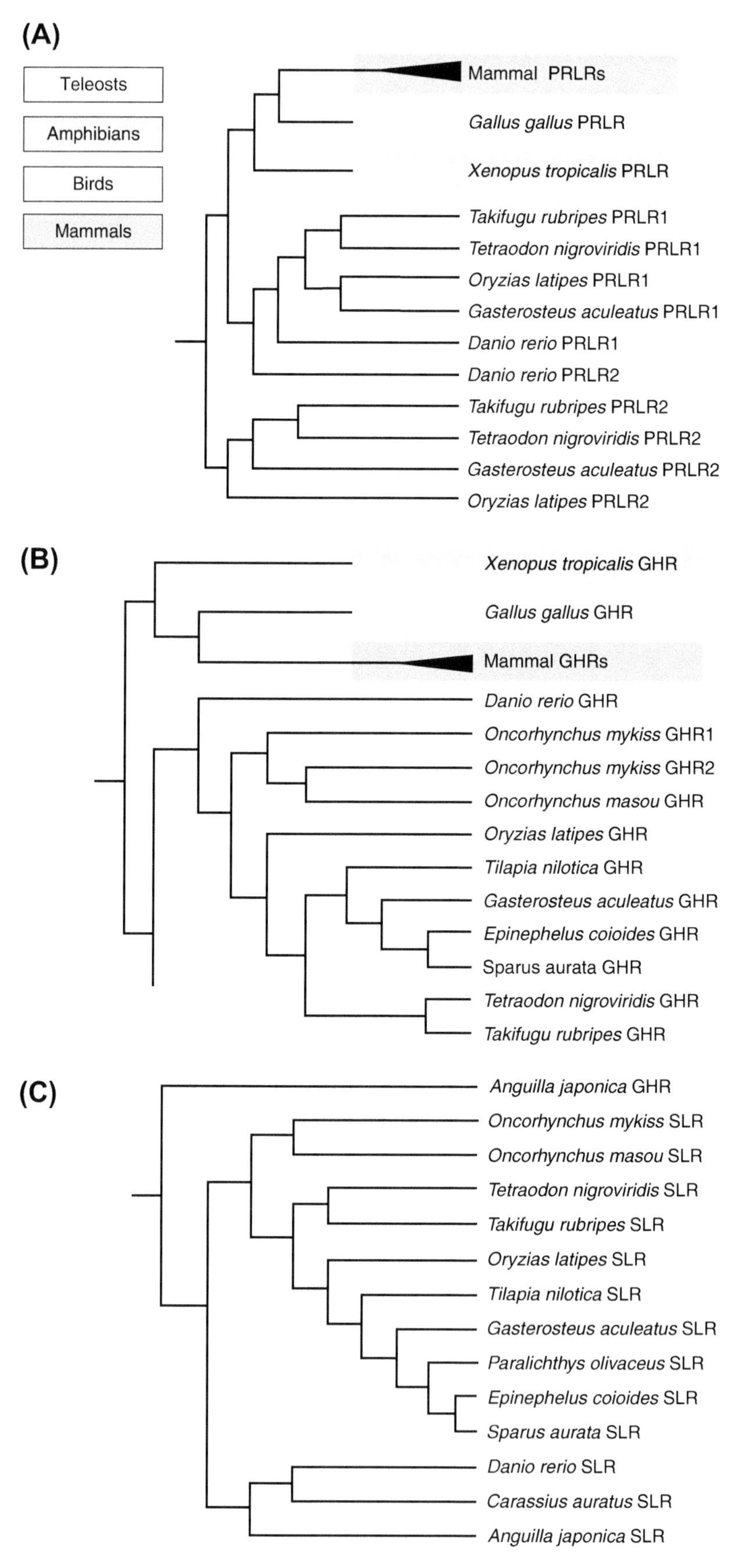

FIGURE 5-25 **Phylogeny receptor genes for prolactin (A), growth hormone (GH), and somatolactin (SL) receptors.** Note the anomalous position of the GHR for the Japanese eel (*Anguilla japonica*) that is clustered with SLRs. *(Adapted with permission from Daza, D.O. et al.,* Annals of the New York Academy of Sciences, ***1163****, 491–493, 2009.)*

TABLE 5-2 General Effects of Prolactin in Non-Mammalian Vertebrates

	Teleosts	Amphibians	Reptiles	Birds
Reproduction	X	X	X	X
Osmoregulation	X	X	X	X
Growth and Development	X	X	X	X
Skin Pigmentation	X	X		
Molting			X	X[1]

[1]*Stimulates feather growth only.*

disulfide bonds instead of the two or three disulfide bonds characteristic of GHs and PRLs.

1. Category 2 Hormones (PRL and GH) in Agnathan Fishes

Because of their importance as the representatives of the most primitive vertebrate group, lampreys have been studied extensively with respect to the origins of GH and PRL. Chondrichthyean species, on the other hand, have been largely ignored. Although lamprey pituitaries contain one cell type that binds both GH and PRL antibodies prepared from teleost hormones, it appears that they only produce GH. Lamprey pituitaries contain bioassayable PRL activity, based on responses observed in the xanthophore-expanding *Gillichthyes* bioassay (see Appendix D), but this may simply be due to structural similarity of the lamprey GH and fish PRL. Hagfish pituitaries do not contain bioassayable PRL activity although they do possess a single GH/PRL immunoreactive cell type. A single GH gene has been identified in the sea lamprey, and GH stimulates IGF production in lamprey liver.

2. Category 2 Hormones (PRL and GH) in Chondrichthyean Fishes

Growth hormone has been isolated from the blue shark, but information on GH from other chondrichthyean species is lacking. Shark GH is structurally unique from lungfish GH and the GHs of several teleosts examined. Prolactin has not been characterized from any species in this group, although testing of pituitary extracts from elasmobranchs (sharks and rays) in the *Gillichthyes* bioassay indicates PRL activity in the rostral pars distalis. No bioassays of holocephalan (ratfishes) pituitaries have been reported. IGF activity has been demonstrated for two sharks, *Squalus acanthias* and *Mustelus canis*, although no definite link to GH has been established.

3. Category 2 Hormones (PRL, GH, and SL) in Bony Fishes

Not only do bony fishes produce GH and PRL from the pars distalis, but they also secrete the related hormone SL from PIPAS cells of the pars intermedia. Madeleine Olivereau in France correlated the cytological appearance of PIPAS cells in the pars intermedia of teleosts with low-calcium environments. Curiously, in salmonid fishes (such as rainbow trout) that lack PIPAS cells in the pars intermedia, SL is produced in the pars distalis. These SL-immunoreactive cells are stimulated cytologically by placing rainbow trout in low-calcium environments and are cytologically repressed when the fish are placed in high-calcium environments.

Among the many actions of PRL in fishes, it plays an important role in osmotic regulation in freshwater teleosts (Table 5-2). Ion and water movements in gill, kidney, intestine, urinary bladder, and skin are altered by PRL. Prolactin also stimulates mucus secretion by gill, intestine, and skin. Mucus secretion and parental care by both male and female blue discus fish (*Symphysodon aequifasciata*) are stimulated by PRL. This mucus is used by their offspring as a food source, foreshadowing the stimulation of milk secretion in mammals by PRL and analogous to the effects of PRL on the bird crop (see ahead). Additionally, PRL has been implicated in the expression of male parental behavior by bluegill sunfish (*Lepomis machrochirus*) and three-spined sticklebacks (*Gasterosteus aculeatus*).

Several teleost PRLs and lungfish PRL have been purified. Teleost PRLs have little biological activity in amniote bioassays, but they work well in amphibians and teleost bioassays (see Appendix D). Purified PRLs prepared from teleosts are all missing the *N*-terminal disulfide bond that characterizes PRLs from sturgeon, lungfishes, and all tetrapods. Hence, teleost GHs structurally resemble mammalian GH more closely than do mammalian PRLs (see fish GH and PRL molecules in Figure 5-26 and compare to mammalian GH and PRL hormones in Chapter 4, Figure 4-15).

Mammalian GHs are very active in promoting the growth of fishes, and polypterid, chondrostean, holostean, and teleost pituitaries exhibit GH activity in the mammalian tibia bioassay (see Appendix D). Additionally, there is a marked influence of thyroid hormones on growth in teleosts. Thyroid hormones accelerate growth in fishes, and thyroid hormone treatment restores growth rates to normal in radiothyroidectomized trout (*Oncorhynchus mykiss*) (see Chapter 7). Although it has not been demonstrated conclusively, it is reasonable to assume a synergistic relationship between thyroid hormones and GHs similar to that reported for mammals (see Chapter 4).

Secretion of GH correlates directly with induction of **IGF-I** mRNA in livers of growing coho salmon and

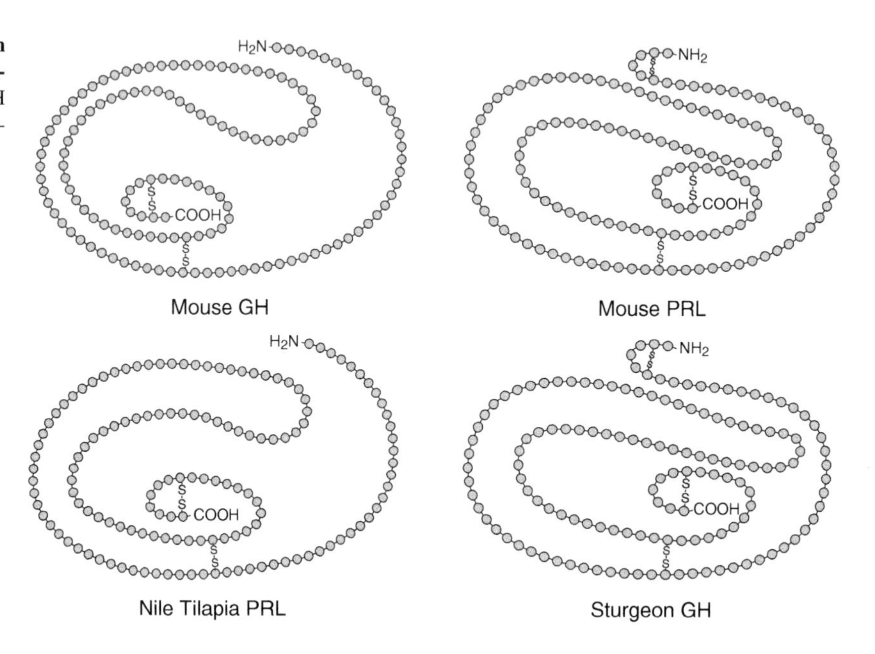

FIGURE 5-26 **Comparison of teleost prolactin (PRL) and growth hormone (GH) to mammalian forms.** Note that tilapia PRL is like mouse GH with respect to the disulfide bonds, whereas sturgeon GH structure is more like mouse PRL.

similar relationships between GH and IGF-I levels have been reported for rainbow trout, long-jawed mudsucker (*Gillichthyes*), and Japanese eels (*Anguilla*). Transfer of juvenile coho salmon prematurely into seawater decreases IGF-I levels and growth, although GH levels increase. In coho salmon, GH appears to be important for seawater adaptation, although it does not seem to be important in all teleosts examined. IGF-binding proteins are described for several species, and GH-binding proteins have been characterized from rainbow trout. These observations suggest that involvement of GH and IGF-I in the growth of teleosts is similar to mammals although considerable work remains to be done.

Like PRL, SL in bony fishes is connected with multiple physiological actions. For example, SL produces effects on pigment cells, Ca^{2+} balance, reproduction, and metabolism. Rising plasma levels of SL also are associated with smolting of seaward-migrating salmonids and with gonadal maturation associated with their return to freshwater for spawning. Plasma SL peaks at spawning in Pacific salmon (*Oncorhynchus keta*, *O. kisutch*, *O. nerka*) and in *Mugil cephalus*, although it is not known if these correlations are related to metabolism or to reproductive functions. Furthermore, SL stimulates steroidogenesis in ovarian or testicular fragments isolated from coho salmon, although SL is less potent in this regard than is FSH. This possible steroidogenic role for SL is similar to effects reported for exogenous GH and PRL in several teleosts and suggests the overlapping actions of these three hormones may be a reflection of their chemical similarities.

Isolation and characterization of SLs from diverse teleost species have revealed two subfamilies of SLs (Figure 5-24). Some species exhibit only one type, whereas others have both an SL and an SLα (e.g., zebrafish, rainbow trout). The discovery of these dissimilar SL genes may explain in part the diversity of functions so far ascribed to SLs. Genes similar to SLα have been cloned from sturgeon (a chondrostean fish) and the African lungfish, and there is evidence for SL in a shark as well. The binding of SLs to **GH receptors (GHR1** and **GHR2)** is reported but it is not clear which receptor binds SL under physiological conditions.

4. Category II Hormones (PRL and GH) in the Amphibians

PRLs from most amphibians cross-react with antibody to rat GH, and amphibian PRL is thought by some to be the larval GH of amphibians; however, PRL from tiger salamanders (*Ambystoma tigrinum*) does not cross-react with rat GH antibody. In addition to possible influences on larval growth when administered to larvae, PRL is antimetamorphic in both anurans and caudate amphibians; that is, it blocks metamorphosis from the aquatic larva to the semiterrestrial or terrestrial juvenile form. However, overexpression of the PRL gene in transgenic frogs does not inhibit metamorphosis but simply prevents tail resorption and even promotes tail growth. Nevertheless, some PRL also is essential for normal metamorphosis (see Chapter 7 for a discussion of the endocrine control of amphibian metamorphosis). PRL induces water-drive behavior in newts and salamanderistics (see ahead) and influences secondary sexual characteristics associated with breeding. Integumentary effects of PRL related to water balance have been observed in newts and salamanders indicating that PRL affects water balance in anurans and caudate

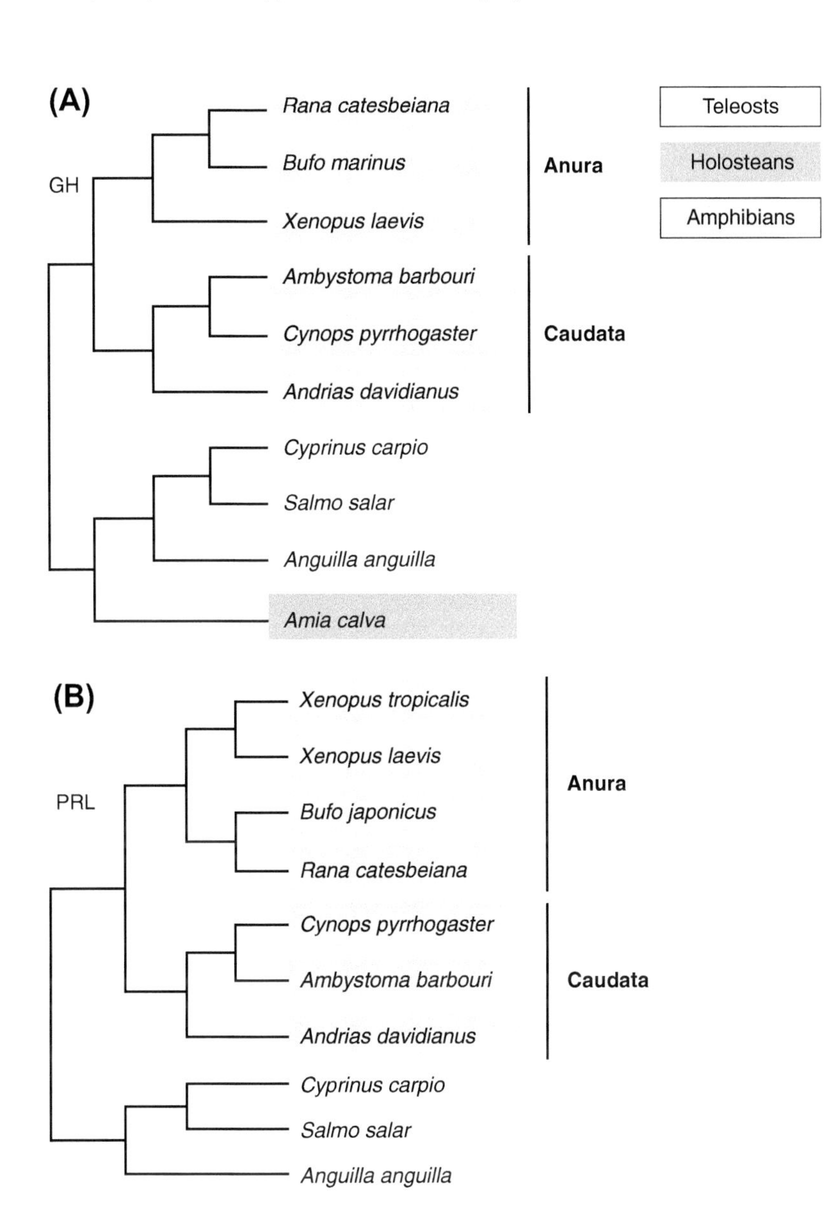

FIGURE 5-27 **Phylogeny of Growth Hormone (GH) and Prolactin (PRL) in Amphibia.** (A) GH. (B) PRL. Some fishes (teleosts and holosteans) are included for comparison. *(Adapted with permission from Yang, L. et al.,* General and Comparative Endocrinology, ***165****, 177–180, 2010.)*

amphibians as it does in fishes. Japanese workers have isolated amphibian PRLs and developed specific radioimmunoassays for studying possible functions of PRL in activities such as breeding migrations.

Both GH and PRL have been isolated from several frogs, salamanders, and newts (see Figure 5-27). The large number of similarities in the physical properties of amphibian and mammalian GHs and the similarities in amino acid composition suggest that there has been considerable conservatism expressed in the evolution of tetrapod GH genes. Purified frog GHs are not as effective as bovine GH in the rat tibia bioassay, however (see Appendix D for GH bioassays). Overexpression of the GH gene facilitlates growth in larval frogs and accelerates sexual maturity, suggesting some differences in GH and PRL action in larval amphibians. IGF activity has been demonstrated in *Bufo woodhousei* and *Xenopus laevis*, but no link to either PRL or GH has been established. This is an important area that demands more attention in all non-mammalian groups.

5. Category II Hormones (PRL & GH) in the Reptiles

The reptiles are the only major vertebrate group in which a PRL bioassay has not been developed, and little work has been done with PRL in reptiles. Mammalian PRL stimulates growth in juvenile snapping turtles (*Chelyldra serpentina*) and lizards (*Lacerta s. sicula*). Appetite is also stimulated in *Lacerta* by PRL. A possible effect on water balance has been reported in turtles in which PRL influenced glomerular filtration. Purified reptilian PRLs give positive responses in other vertebrate bioassays (Appendix D), and this PRL activity appears to be confined to the cephalic (rostral) lobe of the pars distalis. Mammalian GH, like PRL, stimulates growth in juvenile snapping turtles and in the lizard *Lacerta*, but the sites of action for GH and PRL in the lizard appear to be different. Growth hormone and PRL both stimulate appetite in *Lacerta*. A marked increase in growth of the digestive tract (**splanchnomegaly**) is caused by GH. Purified GH has been prepared from the caudal lobe of adult snapping turtle pituitaries and sea turtles. Many characteristics of these GHs are similar to those of other tetrapods. Turtle GHs are very effective in the rat tibia bioassay (Appendix D). Crocodilian GH is structurally closer to GH isolated from chickens that other reptiles, an observation that fits with the known phylogenetic closeness of these reptiles to birds.

6. *Category II Hormones (PRL and GH) in Birds*

PRL and GH have been isolated from domesticated birds (e.g., chicken, duck, turkey, Japanese quail). The amino acid composition and electrophoretic properties of chicken PRL are similar to mammalian PRLs, and chicken PRL is structurally and functionally distinct from chicken GH. GH activity has been demonstrated in pituitaries of chickens and turkeys by means of the mouse tibia bioassay (Appendix D). However, anti-bovine GH antibody apparently does not cross-react with chicken pituitary extracts nor does purified GH from duck pituitaries cross-react with rat GH antibody. These observations suggest unique differences in bird GHs that are probably more similar to crocodilian GHs than to mammalian GHs.

PRL plays several essential roles in avian reproduction. The first established role for PRL in birds was the stimulation of a nutritious, cytogenous secretion by the pigeon crop sac called **crop milk**, which is coughed up and fed to young birds. This led to the development of the first non-mammalian bioassay for PRL, the **crop sac bioassay** (Appendix D). A second reproductive role for PRL in some birds is the development of a **brood patch**, a ventral portion of the body that becomes defeathered and highly vascularized in some species during egg incubation. This brood patch allows the transfer of warmth to the eggs during incubation. Estrogens synergize with PRL in formation of the brood patch. Elevated PRL is associated with males in a number of bird species where these males exhibit considerable parental care. A third role for PRL is to stimulate premigratory fattening and induction of migratory restlessness in birds. This restless behavior, or **Zugunruhe,** appears just prior to the actual migration of many north temperate bird species. PRL secretion also has been implicated in the induction of photorefractoriness in migratory birds (see Chapter 11).

C. Category III Tropic Hormones: The POMC Group

As in mammals, ACTH-, α-MSH-, CLIP-, and endorphin-like peptides are all derived from the same precursor, POMC, as discussed in Chapter 4. In corticotropes of non-mammals (Figure 5-28), POMC is hydroylzed to yield ACTH and β-endorphin, whereas in the melanotrope the end products are α-MSH (*N*-acetyl $ACTH_{1-13}$ amide), CLIP, β-endorphin, and *N*-acetylated and *C*-terminus shortened forms of β-endorphin that do not bind to the mu opioid receptor and are biologically inactive. The presence of additional peptide sequences with melanotropin activity (i.e., β-MSH, γ-MSH, and δ-MSH) have proven useful in examining phylogenetic relationships, although, as discussed in Chapter 4, only α-MSH probably has bioregulatory roles.

Five melanocortin receptors (MC1R, MC2R, etc.) have been identified, with MC2R having the highest affinity for ACTH and the others binding α-MSH (see Chapter 4). Examination of the ACTH sequence (which includes the α-MSH sequence as the first 13 amino acids) indicates an MCR binding sequence in amino acid residues 6 to 9 as well as a strong affinity sequence (15 to 19) that specifically binds to MC2R, explaining why ACTH binds to both α-MSH and ACTH target cells but α-MSH cannot bind to adrenal cells having only the MC2R receptor (Figure 5-29).

Although there is considerable variation in the composition of the POMC proprotein sequence within each major group of gnathostome vertebrates (Figure 5-30), the ancestral gnathostome POMC gene form is found in at least some members of each group. The POMC complex of tetrapods has not been so closely examined as in the fishes with respect to the evolution of POMC. Robert Dores, Hiroshi Kawauchi, Stacia Sower, and others have focused on the evolution of this complex group of bioregulatory peptides in fishes. Analysis of POMC proprotein sequences identifies three major groupings among the fishes: chondrichthyeans, lungfishes and coelacanth, and ray-finned bony fishes (Figure 5-30). In this latter grouping, chondrostean (sturgeons, paddlefishes) and holostean (gar) POMCs exhibit the basic gnathostome pattern that is distinct from teleost POMCs. Teleosts differ from other vertebrates in the absence of the γ-MSH sequence. Elasmobranchs are unique in their possession of yet another MSH-like sequence named δ-MSH. The lungfishes and coelacanth have many POMC features that are more like tetrapods than like other fishes. Curiously, the lamprey has been found to have two separate POMC-like genes—**proopiocortin (POC)** and **proopiomelanotropin (POM)**—that result in production of ACTH and β-endorphin and two MSHs plus a somewhat different β-endorphin, respectively (see ahead).

The roles of ACTH in regulation of adrenal functions in non-mammalian vertebrates are discussed in Chapter 9, and

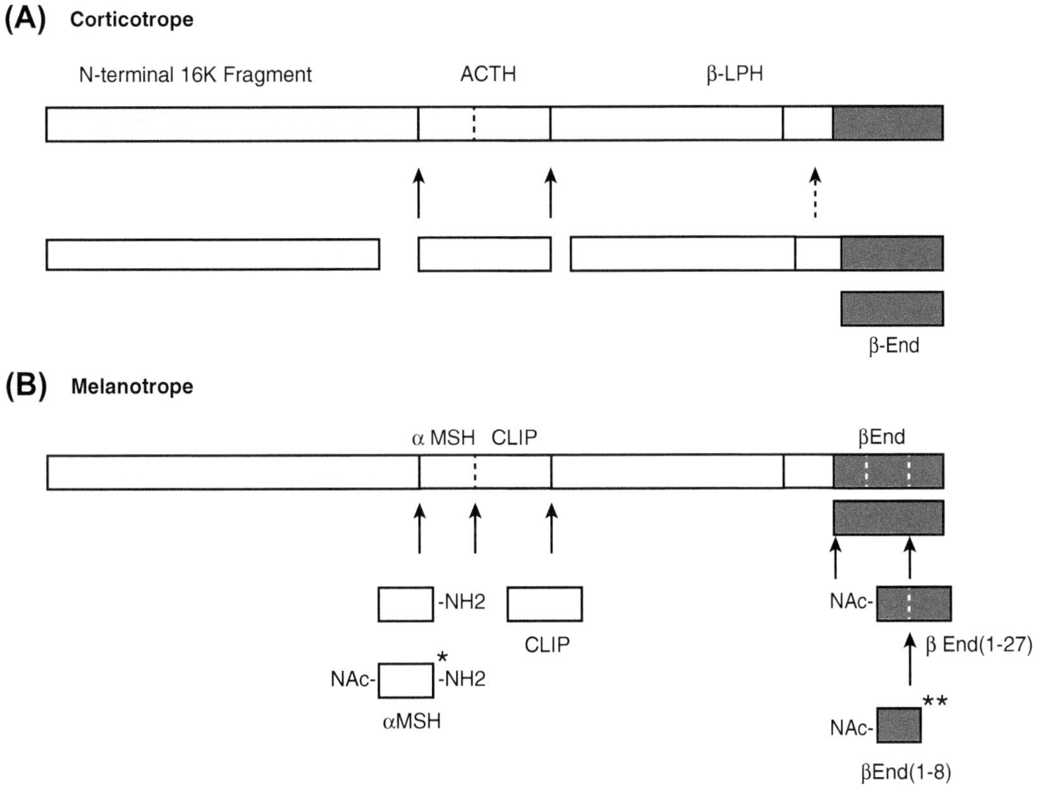

FIGURE 5-28 **Hormonal content within POMC.** (A) Modifications that take place in a corticotrope. (B) Modifications that take place in a melanotrope. *(Adapted with permission from Dores, R.M. and Baron, A.J.,* Annals of the New York Academy of Sciences, ***1220****, 34–48, 2011.)*

only some general statements will be made here. The actions of pitutiary derived α-MSH on pigmentation are discussed in this chapter, whereas the role of neuronally derived α-MSH in the metabolism of non-mammalian vertebrates is discussed in Chapter 13.

1. ACTH Actions in the Fishes

Little is known about ACTH actions in the more primitive fish groups, and not much more is known about teleosts. Treatment of lampreys with mammalian ACTH causes cytological changes in the interrenal cells (interrenal cells are homologous to the adrenocortical cells of mammals; see Chapter 9) that are suggestive of stimulation. Injection of CRH elevates plasma levels of 11-deoxycortisol, the putative lamprey glucocorticoid, presumably by elevating pituitary ACTH secretion. ACTH has been purified from lamprey and shark pituitaries, but its effects on interrenal tissue have not been studied. ACTH in teleosts stimulates secretion of cortisol.

2. ACTH Actions in Non-Mammalian Tetrapods

Activity of the steroidogenic enzyme **Δ^5,3β-hydroxysteroid dehydrogenase (3β-HSD)** is enhanced in the bullfrog interrenal following treatment with ACTH. Blood levels of

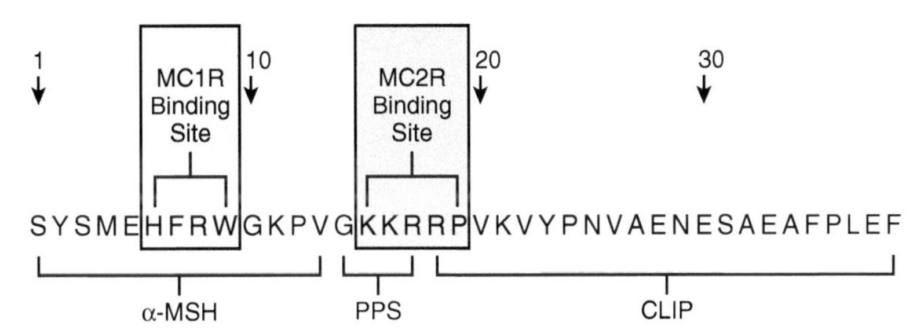

FIGURE 5-29 **Binding domains of the α-MSH/ACTH peptide for melanocortin receptors.** The motif HFRW binds to MC1R, MC3R, MC4R, and MC5R whereas the KKRRP motif is believed to bind to the MC2R. PPS is the posttranslational processing site removed when ACTH is cleaved in melanotropes. Removal of PSS results in lost of binding ability to MC2R by the products (α-MSH or CLIP). *(Adapted with permission from Dores, R.M. and Baron, A.J.,* Annals of the New York Academy of Sciences, ***1220****, 34–48, 2011.)*

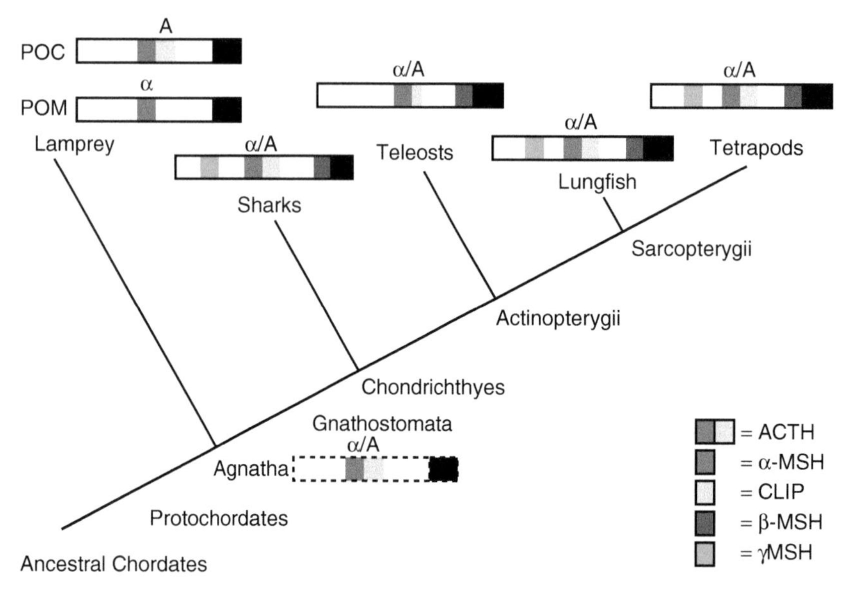

FIGURE 5-30 **Evolution of POMC.** The appearance of two separate POMC-like molecules in corticotropes (POC) and melanotropes (POM) of lampreys may indicate a derived condition that appeared in the modern lampreys after their separation from the ancestral vertebrate condition (as suggested within the dashed box). *(Adapted with permission from Dores, R.M. and Baron, A.J.,* Annals of the New York Academy of Sciences, ***1220**, 34–48, 2011.)*

corticosterone are elevated by ACTH treatment and lowered by hypophysectomy in amphibians, lizards, turtles, and birds. In addition, bird adrenals also respond to PRL and GH as well as to ACTH. In all tetrapods examined, ACTH seems to work through a cAMP-dependent mechanism as it does in mammals.

3. α-MSH Actions in Non-Mammals

Many vertebrates show changes in pigmentation or pigmentary patterns that are correlated with environmental factors or particular events in their life histories. Rapid pigmentary responses are under direct neural control, and the slower responses are generally the result of endocrine control or a combination of neural and endocrine control. Vertebrates possess a variety of specialized, pigmented cells referred to as **chromatophores** (see Table 5-3). Chromatophores singly or in special combinations are the bases for color patterns associated with the integument. **Physiological color changes** involving displacement of pigment granules within a pigment cell are characteristic of many fishes, amphibians, and some reptiles. However, a physiological role in pigment changes has not been established for α-MSH in fishes where most of their pigment changes apparently are under direct, dual sympathetic–parasympathetic innervation, allowing for very rapid responses. In amphibians and some reptiles, α-MSH as well as melatonin influence physiological color changes. All vertebrates exhibit increases in the number of chromatophores and/or increases in the amount of pigment contained within or in the vicinity of the chromatophores in response to α-MSH. This type of color change is termed **morphological color change**.

TABLE 5-3 **Vertebrate Chromatophores**

Chromatophore	Organelle	Pigment	Color
Melanophore	Melanosomes	Melanins	Brown, black (yellow, red)
Iridophore (guanophore, leukophore)	Reflecting platelets	Guanine, adenine, hypoxanthine, uric acid	–
Xanthophore	Pterinosomes	Pteridines	Yellow, orange
	Carotenoid vesicles	Carotenoids	Yellow, orange, red
Erythrophore	Pterinosomes	Pteridines	Red, orange
	Carotenoid vesicles	Carotenoids	Yellow, orange, red

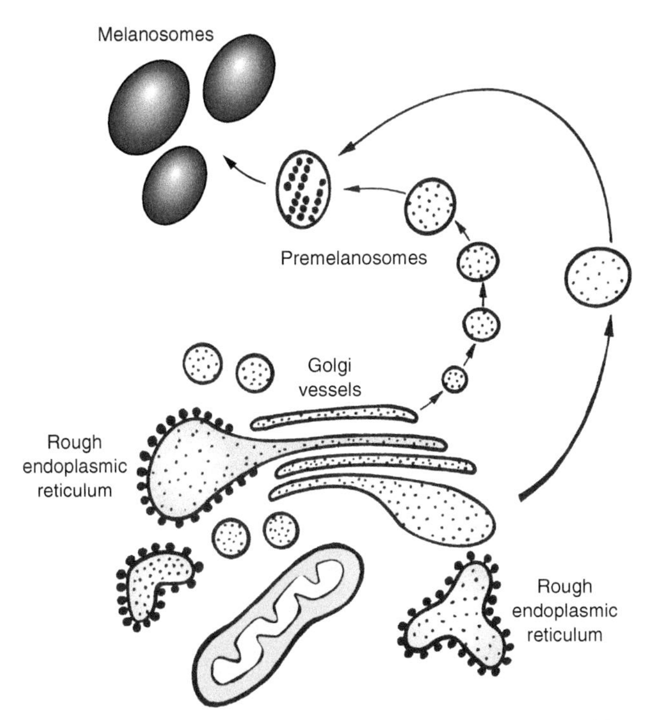

FIGURE 5-31 **Production of melanosomes.** Premelanosomes differentiate from Golgi vesicles. They exhibit tyrosinase activity and synthesize melanin by polymerizing tyrosine. Once melanization of these organelles is complete, they are called *melanosomes*.

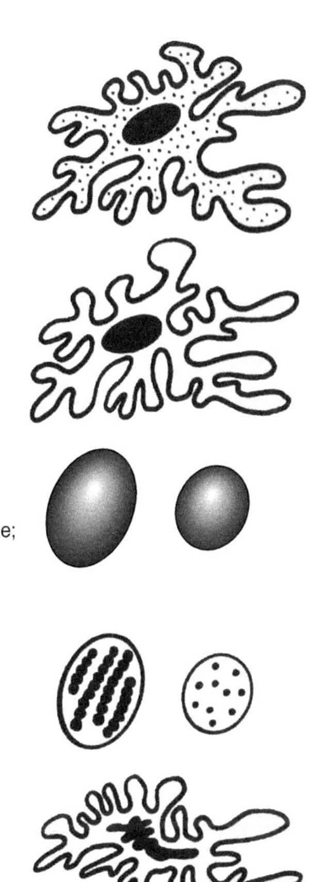

FIGURE 5-32 **Melanophores and melanocytes.** This terminology is based on the Sixth International Pigment Cell Conference held in 1966. Melanophores can disperse their melanosomes uniformly in the cytoplasm or concentrate them to varying degrees around the nucleus.

Among the most important chromatophores of amphibians and some fishes are the dermal **melanophores,** which contain melanin granules concentrated in special organelles called **melanosomes** (Figure 5-31). Melanophores differ from **melanocytes,** which deposit their melanin products extracellularly (Figure 5-32). When melanosomes are concentrated around the nucleus of the melanophore, the skin appears lighter than when they are dispersed throughout the cytoplasm. The degree of concentration of melanosomes is inversely related to the darkness of the skin (see Appendix D). Coloration patterns in some vertebrates may be determined by the distribution of different types of chromatophores and by the relationship of dermal melanophores to other chromatophores. For example, the dark spots on the skin of the leopard frog (*Rana pipiens*) are due to concentrations of melanocytes and extracellular deposition of melanin, whereas adjacent regions of the skin that may vary from light green to black are occupied exclusively by dermal melanophores and other chromatophores. The green color is produced by interactions of other chromatophores overlying the melanophores that are masked when the melanosomes are dispersed throughout the cytoplasm of the melanophore.

The dermal melanophore is the major target for α-MSH, and, in a few cases, other chromatophores may be affected. The actual mechanism of how melanosomes migrate in and out of the stellate processes of the melanophore is not completely understood. It appears that microtubules are essential for aggregation of the melanosomes, and these may be activated by α-MSH through the second-messenger cAMP. Any experimental treatment known to interfere with microtubule formation, such as application of the drug **colchicine**, blocks melanosome aggregation. Ultrastructural observations, however, in some amphibians melanophores do not support the presence of microtubules oriented properly to cause melanosomes to disperse, and the role of microtubules needs further clarification.

3a. α-MSH Activity in Agnathan Fishes

In sea lampreys (*Petromyzon marinus*), immunoreactive α-MSH is present in the pars intermedia (Figure 5-33), and hypophysectomy causes paling of these animals. Furthermore, treatment with α-MSH causes melanosome

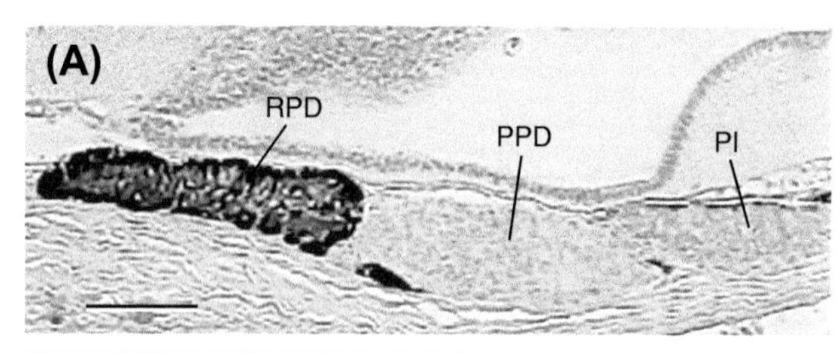

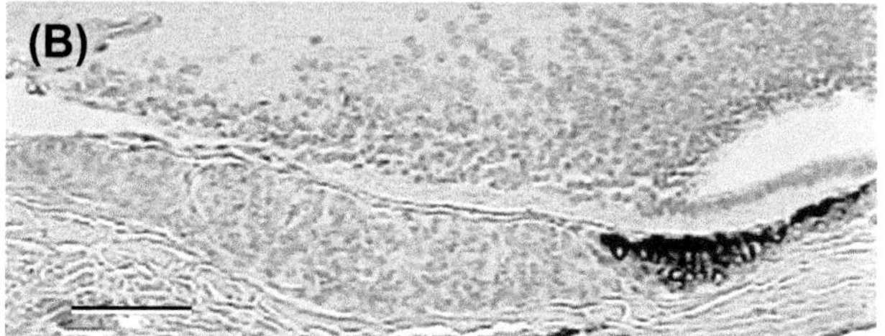

FIGURE 5-33 **Immunoreactive ACTH in lamprey pituitary gland.** (A) Dark brown staining for immunoreactivity to sea lamprey ACTH in the rostral pars distalis (RPD) of *Mordacia*. (B) Immunoreactive MSH in the pars intermedia (PI). *(Reprinted with permission from Takahashi, A. et al.,* General and Comparative Endocrinology, ***148**, 72–78, 2006.)*

TABLE 5-4 **Hormones and Drugs Affecting Melanosome Distribution in Melanophores of a Teleost Fish**

Chemical Factor	Aggregation (Lightening of skin)	Dispersion (Darkening of skin)
Serotonin antagonists		+
Adrenergic antagonists		+
Depressants		+
Cholinergic activators		+
Catecholamines	+	
Melatonin	+	
Serotonin	+	
Adrenergic agonists	+	

Note: (+), effective.

dispersion in lampreys. Interestingly, lamprey POMC contains two unique MSH sequences not found in other vertebrates (MSH A that is similar to β-MSH and MSH B). Hagfishes (Myxinoidea) do not adapt their body color when placed on different backgrounds suggesting no pigmentation role for α-MSH.

3b. α-MSH Activity in Chondrichthyean Fishes

Sharks adapt to background changes very slowly, requiring up to 100 hours to achieve maximal adaptation. Hypophysectomy abolishes this ability to "background adapt," whereas ectopic transplants of the neurointermediate lobe cause permanent darkening regardless of the background upon which the shark is placed. Melanophore responses to MSHs are paralleled by similar responses in xanthophores. In addition to α-MSH, the elasmobranch POMC contains β-MSH and γ-MSH as well as a unique sequence among vertebrates named δ-MSH. As in other vertebrates, α-MSH is probably the only important MSH sequence.

3c. α-MSH Activity in Bony Fishes: Teleosts

The pigmentary responses of most fishes are under the control of direct aminergic innervation of melanophores and other chromatophores (Table 5-4) and α-MSH does not play a major role in physiological color changes. The melanophores of most species do not respond to either hypophysectomy or exogenous α-MSH, whereas some show a limited response to α-MSH following denervation of the melanophores. Dispersion of pigments in melanophores of the goldfish *Carassius auratus* and erythrophores of the European minnow *Phoxinus phoxinus* occurs following treatment with α-MSH. In contrast, melanophore dispersion can be induced only in denervated melanophores of the killfish (*Fundulus heteroclitus*) because of the presence of overriding neural stimuli in intact fish. α-MSH may play some role in morphological color changes in teleosts. Treatment of xanthic goldfish (goldfish normally lacking any melanophores) with α-MSH stimulates melanophore differentiation and melanogenesis. Hypophysectomy of *F. heteroclitus* causes a reduction in the number of melanophores, and treatment with α-MSH results in an increase in the number of melanophores. The peptide MCH was isolated first from teleost pituitaries before it was demonstrated in mammals. It is a potent cyclic heptadecapeptide (17 amino acids) that causes contraction of teleost melanophores. MCH also blocks both ACTH and α-MSH release from teleost pituitaries.

3d. α-MSH Activity in Amphibians

Direct control of amphibian melanophores is under endocrine regulation, and there is evidence for direct innervation of amphibian melanophores by sympathetic neurons in some species, including the African clawed frog, *X. laevis* (now called *Silurana laevis*). Pigment organelles of another pigment cell, the **xanthophore**, are normally expanded in most amphibians. In the canyon tree frog (*Hyla arenicolor*), however, the xanthophores are normally in an aggregated condition, and they may be dispersed by the application of α-MSH. An endogenous role for α-MSH on xanthophore pigment organelles in tree frogs has not been confirmed nor is it known how widespread this phenomenon might be. The aggregation of reflecting platelets in the **iridophore**, another special chromatophore, is stimulated by either cAMP or α-MSH. These iridophores possess α- and β-adrenergic receptors, and stimulation of the α-receptors produces dispersion of the reflecting platelets of the iridophore. These data suggest a possible role for catecholamines in iridophore regulation. Although MCH is present in the anuran brain, no melanin-concentrating action has been demonstrated. In fact, MCH can cause melanophore expansion. Additional work is needed to elucidate the physiological importance of MCH in amphibians. Generally, the concentration of melanosomes in dermal melanophores of amphibians associated with lightening of the animals is caused by melatonin from the pineal gland (see ahead).

3e. α-MSH Activity in Reptiles

Reptiles exhibit a variety of mechanisms for regulating dermal pigment cells, including neural and endocrine mechanisms. Unlike the well-established pattern of dopaminergic innervation in most amphibians, no innervation of the pars intermedia has been observed at either the light or electron microscopic level in several lizard species. The primitive tuatara of New Zealand, *Sphenodon punctatus*, also exhibits no innervation of the pars intermedia, implying this may be a basic pattern that was established in the earliest of reptiles or possibly in their amphibian progenitors. Innervation of pigment cells does occur in some reptiles, and pigmentary control of dermal melanophores may be by neural or endocrine mechanisms or both.

Color changes in the true chameleon *Chameleo pumilis* are under direct neural control, and α-MSH has no effect on the skin chromatophores. Melanophores of the closely related species *Chameleo jacksoni*, however, respond to both α-MSH and ACTH. The opposite extreme is found in the American chameleon *Anolis carolinensis*, in which there is no general neural control over melanophore responses, and melanosome dispersion can be readily induced *in vitro* by application of α-MSH or cAMP to isolated pieces of *Anolis* skin. However, there is a patch of skin near the eye that darkens only in response to catecholamines. Pretreatment with α-adrenergic-blocking agents obliterates the response of *Anolis* melanophores to α-MSH. Horned lizards, in the genus *Phrysonoma*, also exhibit both neural and hormonal regulation of melanophores. Neither xanthophores nor iridophores of *A. carolinensis* show any response to α-MSH. It is not known whether any non-melanin-containing chromatophores of other reptiles show any regulatory control but, in fact, α-MSH produces weak melanosome dispersion in amphibian and reptilian melanophores.

3f. α-MSH Activity in Birds

Birds lack a pars intermedia and hence the normal source of α-MSH; hence, feather pigments (including melanin) are under the control of gonadal, thyroidal and gonadotropic hormones in birds. The only reported action for α-MSH in birds possibly is related to a developmental action. Embryonic implants of chicken pituitaries cause formation of black feathers where normally only white feathers would develop. This effect can be mimicked by treatment with either α-MSH or ACTH, but this only suggests the presence of MCRs and does not mean there is a physiological role for MSH in birds.

4. Endorphins in Non-Mammals

β-endorphin and related molecules are produced by hydrolysis of POMC in the pituitaries of all vertebrates (or from POC and POM in lampreys). Endorphins are usually *N*-acetylated except in the cartilaginous dogfishes although it is *N*-acetylated in the ratfish. No physiological role has been established for the pituitary endorphins of non-mammals, although neuronally produced endorphins are assumed to have an analgesic action.

V. COMPARATIVE ASPECTS OF HYPOTHALAMIC CONTROL OF PITUITARY FUNCTION IN NON-MAMMALS

Numerous studies have documented the distribution of mammalian-like neuropeptides in the brains of non-mammals, but relatively little is known about the functional roles for most of these neuropeptides in non-mammals. In general, there is more plasticity in non-mammals with respect to the actions of neuropeptides that bear specific names related to their roles in mammals. When making comparisons, it is important to bear in mind that hypothalamic control in teleosts is primarily neuroglandular rather than neurovascular as in other fish groups and in tetrapods. Hence, teleost hypothalamic control involves many of the same chemicals but they perform as neurotransmitters rather than neurohormones.

A. Category I: Regulation of Glycoprotein Hormones (GpHs)

The focus in studies of GpH regulation has been on the GTHs rather than on TSH. In fact, it has been and continues to be a problem to determine what neuropeptides are the endogenous bioregulators of TSH secretion in many cases. The story for GTH release is similar to the pattern described for mammals.

1. Control of GTH Release

The embryonic origin of GnRH-secreting cells in non-mammals is similar to that described for mammals in

TABLE 5-5 GnRH Amino Acid Sequences

	1	–	–	2	3	4	5	6	7	8	9	10	
Vertebrates													
Mammals	pQ	-	-	H	W	S	Y	G	L	R	P	G	amide
Guinea pig	pQ	-	-	Y	W	S	Y	G	V	R	P	G	amide
Chicken-I	pQ	-	-	H	W	S	Y	G	L	Q	P	G	amide
Frog	pQ	-	-	H	W	S	Y	G	L	W	P	G	amide
Sea bream	pQ	-	-	H	W	S	Y	G	L	S	P	G	amide
Pejerrey/Medaka	pQ	-	-	H	W	S	F	G	L	S	P	G	amide
Herring	pQ	-	-	H	W	S	H	G	L	S	P	G	amide
Catfish	pQ	-	-	H	W	S	H	G	L	N	P	G	amide
Whitefish	pQ	-	-	H	W	S	Y	G	M	N	P	G	amide
Chicken-II	pQ	-	-	H	W	S	H	G	W	Y	P	G	amide
Dogfish	pQ	-	-	H	W	S	H	G	W	L	P	G	amide
Lamprey-II	pQ	-	-	H	W	S	H	G	W	F	P	G	amide
Salmon	pQ	-	-	H	W	S	Y	G	W	L	P	G	amide
Lamprey-III	pQ	-	-	H	W	S	H	D	W	K	P	G	amide
Lamprey-I	pQ	-	-	H	Y	S	L	E	W	K	P	G	amide
Protochordates													
Tunicate-1	pQ	-	-	H	W	S	D	Y	F	K	P	G	amide
Tunicate-2	pQ	-	-	H	W	S	L	C	H	A	P	G	amide
Tunicate-3	pQ	-	-	H	W	S	Y	E	F	M	P	G	amide
Tunicate-4	pQ	-	-	H	W	S	N	Q	L	T	P	G	amide
Tunicate-5	pQ	-	-	H	W	S	Y	E	Y	M	P	G	amide
Tunicate-6	pQ	-	-	H	W	S	K	G	Y	S	P	G	amide
Tunicate-7	pQ	-	-	H	W	S	Y	A	L	S	P	G	amide
Tunicate-8	pQ	-	-	H	W	S	L	A	L	S	P	G	amide
Tunicate-9	pQ	-	-	H	W	S	N	K	L	A	P	G	amide
Other Invertebrates													
Sea urchin	pQ	V	H	H	R	F	S	G	W	R	P	G	amide
Octopus	pQ	N	Y	H	F	S	N	G	W	H	P	G	amide
Aplysia	pQ	N	Y	H	F	S	N	G	W	Y	A	-	amide
Limpet	pQ	H	Y	H	F	S	N	G	W	K	S	-	amide
Marine worm	pQ	A	Y	H	F	S	H	G	W	F	P	-	amide
Leech	pQ	S	I	H	F	S	R	S	W	Q	P	-	amide

The chordate GnRHs are all decapeptides. Invertebrate GnRH-like peptides consist of 11 or 12 amino acids. (*Adapted with permission from Roch, G.J., Busby, E.R., Sherwood, N.M. Evolution of GnRH: Diving deeper. General and Comparative Endocrinology* ***171****, 1–16., 2011*).

Chapter 4 (see also Figure 4-23). A considerable body of research has accumulated on the characterization of GnRHs in non-mammals and on their evolutionary relationships. Release of GnRH is controlled by hypothalamic Kp similar to that described for mammals in Chapter 4. In addition, a **gonadotropin-inhibitory hormone (GnIH)** was discovered in birds and later in mammals. Chemically, GnIH is an RF-amide dodecapeptide; the *C*-terminal end of the peptide is capped by arginine (R) and phenylalanine (F), thus RF. This discovery may prove to have a considerable impact on our understanding of reproductive bioregulation in all vertebrates.

1a. Gonadotropin-Releasing Hormone (GnRH)

Numerous molecular variants of GnRH have been described in non-mammals, including several forms in the tunicates (Table 5-5). There are two ways to describe these diverse molecules that have GTH-releasing activity in vertebrates. One system involves simply naming them for the first species in which that molecular variant was discovered. More recently, investigators are using a method based on the origins of the cells producing the GnRHs. In this scheme, GnRH-1 refers to GnRH-secreting cells that migrate from the olfactory epithelium to the preoptic–hypothalamus area and extend their axons to the median eminence or directly to the pituitary in the case of teleosts. GnRH-2 refers to GnRH cells derived from the midbrain. GnRH-3 cells migrate from the olfactory placode but cease their movement and remain in the telencephalon. One of the confusing aspects of this nomenclature scheme is that the same GnRH molecular form may be classified differently in diverse species depending on its distribution. A comparison of the two schemes is illustrated in Chapter 4, Figure 4-21. In some cases, GnRH-III (or GnRH-3) may refer to the third GnRH discovered in that group of fishes. We use a combination of both schemes here; for example, salmon GnRH (sGnRH) from the hypothalamus that regulates pituitary function would be designated as a GnRH-1.

In early studies of GnRH, mammalian GnRH (mGnRH-1) was isolated from sheep, pig, rat, mouse, and human as well as from amphibians. Two GnRH variants were found in chickens and were arbitrarily named **chicken-I (cGnRH-I)** and **chicken-II (cGnRH-II)**. These two forms are found in both birds and reptiles, with chicken-I being a GnRH-1 in reptiles and birds and chicken-II being the common GnRH-2 in all vertebrate groups.

Several different molecular forms of GnRH-1 are found among the fishes. For example, salmon were first found to possess a decapeptide (sGnRH) that is found in several other teleost species as well. Other unique GnRH-1 types occur in a catfish (**cfGnRH**) and the dogfish shark (**dfGnRH**). Lampreys produce three unique GnRHs (lGnRH-I, lGnRH-II, lGnRH-III). Two additional teleost forms found in sea bream (sb-GnRH) and medaka (md-GnRH) are classified as GnRH-1 molecules. Frequently, more than one GnRH form may release GTH when applied exogenously (Table 5-6), but typically only one is distributed appropriately between brain and pituitary to release GTHs (i.e., acts as a GnRH-1).

TABLE 5-6 Factors Affecting Gonadotropin (GTH) Release in Non-Mammalian Vertebrates

Factors	Teleosts	Amphibians	Reptiles	Birds
GnRH (mammal)	+	+	+	+
GnRH (salmon)	+	+	+	
GnRH (lamprey)	+	0	0	
GnRH (chicken I)	+	+	+/0	+
GnRH (chicken II)	+		+	
Dopamine	–	–		
VIP			0	+
NYP	+			
Serotonin	+			

Note: (+), stimulatory; (–), inhibitory; 0, no effect; (+/–), (+/0), (–/0), varies by species.
See text for explanation of abbreviations.

The work of many investigators led to a scheme for the possible evolution of vertebrate GnRH forms involving an early gene duplication and subsequent variation of one resultant gene and relative conservatism of the other (Figure 5-34). This scheme is based in part on the observation that cGnRH-II is found in the midbrain of all gnanthostome vertebrates and in part on the structures of the various decapeptides that are found with cGnRH-II and hence is hypothesized to be closest to the ancestral form. Later, following a gene duplication, one form of cGnRH-II gave rise to a second GnRH that became associated with release of pituitary GTHs. This second gene further mutated among teleosts into several forms and later gave

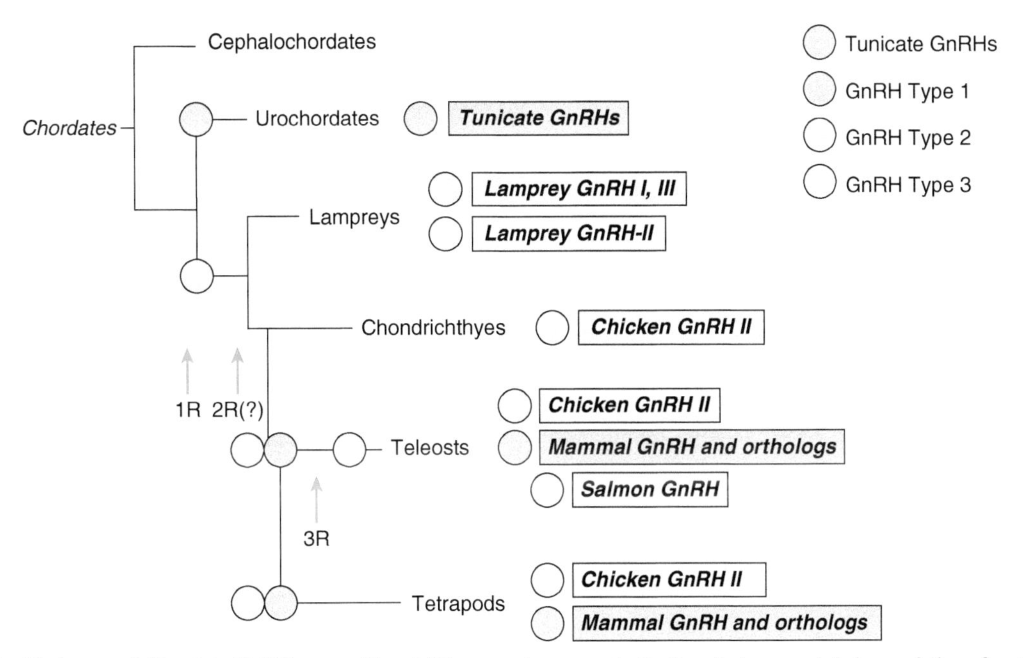

FIGURE 5-34 **Phylogeny of Chordate GnRH genes.** 1R and 3R represent genome duplications that occurred during evolution of vertebrates. 2R is a proposed additional partial or complete genome duplication. *(Adapted with permission from Sower, S.A. et al.,* General and Comparative Endocrinology, ***161****, 20–29, 2009.)*

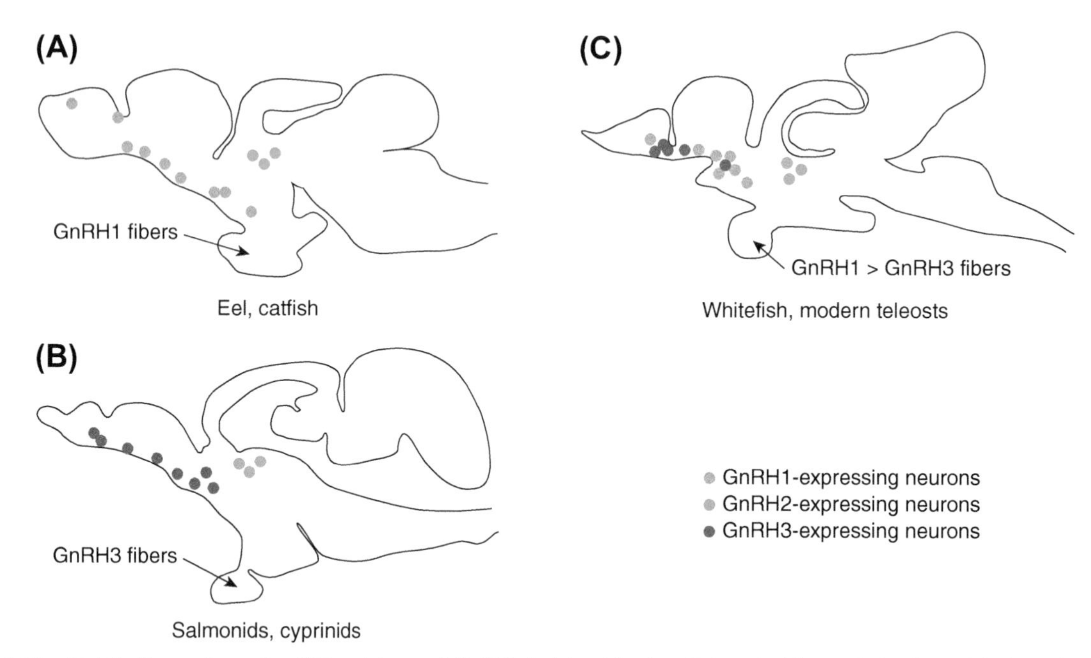

FIGURE 5-35 **Distribution patterns for different types of GnRHs in teleost brains.** *(Adapted with permission from Kah, O. and Dufour, S., in "Hormones and Reproduction of Vertebrates. Vol 1. Fishes" (D.O. Norris and K.H. Lopez, Eds.), Academic Press, San Diego, CA, 2011, pp. 15–42.)*

rise to the form found in living reptiles and birds. Lampreys and hagfishes may have diverged from the ancestral pattern as evidenced by the presence of three unique GnRHs that should not be considered ancestral forms for gnathostomes. Caution must be applied whenever phylogenetic schemes are based upon only a short fragment of a gene product (e.g., ten amino acids), and such analyses should include the analysis of the entire GnRH gene.

The discovery of GnRH-like peptides among a variety of invertebrate species and their association with reproduction certainly has important implications for understanding the evolution of these peptides. Nine GnRH-like molecules have been isolated from tunicates (see Table 5-5), and both chicken and mammalian GnRH-1 stimulate gamete release in the tunicate *Ciona intestinalis*. GnRH-like immunoreactive neurons occur in mollusks such as *Aplysia* and in a possible chordate ancestor, the hemichordate *Saccoglossus bromophenolosus*. Exposure of a chitin (*Mopalia*) to either lamprey GnRH-I or tunicate tGnRH-2 dissolved in seawater stimulates gamete release. Molluscan GnRH-like molecules are slightly larger peptides than the traditional decapeptide of chordate GnRH molecules, although there is considerable amino acid homology (Table 5-5).

Teleosts as a group show considerable variation in the molecular forms of GnRH molecules and in the distribution of GnRH gene types within the brain (Figure 5-35). Although the distribution of type 2 GnRH is consistent, pituitary regulation of of gonadotropin secretion may be regulated by either a type 1 GnRH or type 3 GnRH, or a mixture of both, depending on the species.

Whereas the number of GnRH forms present is fewer as we progress from tunicates to mammals, the same reduction seems to be true of GnRH receptors (Figure 5-36). Receptors for GnRH (GnRH-Rs) are found in several brain regions as well as in the pituitary gland. At least three types (and as many as five in some species) of GnRH-R have been described in teleosts although their distribution and relationship to the different forms of GnRH genes are not clear. Amphibians have as many as three genes coding for GnRH-Rs. Mammals have two GnRH-Rs but a number of mammals have lost one of the genes or it has been inactivated (e.g., humans, chimpanzees, sheep, cattle, some rodents).

1b. Kisspeptin (Kp)

Small peptides called **FR amides** were so named because their amine terminals ended with the amino acids arginine (R) and phenyalanine (F). The first RFamide peptide discovered was the cardioexcitatory **FMRFamide** (Phe–Met–Arg–Phe–NH_2) neuropeptide isolated from the ganglia of a mollusk. Kisspeptins represent another unique group of RF amides implicated in the regulation of GnRH release from the hypothalamus in mammals (see Chapter 4). Secretion of GnRH in non-mammals appears to be regulated by Kps as well. As in mammals, Kps bind to a specific receptor (GPR54 or kiss1r) located on GnRH neurons.

In teleosts, there are two Kp genes (*kiss1* and *kiss2*) (Figure 5-37) whose roles vary in different species. A scheme outlining regulation of GnRH and LH secretion in a generalized teleost by Kp neurons and DA neurons is provided in Figure 5-38. In medaka (*Oryzias latipes*), *kiss1*

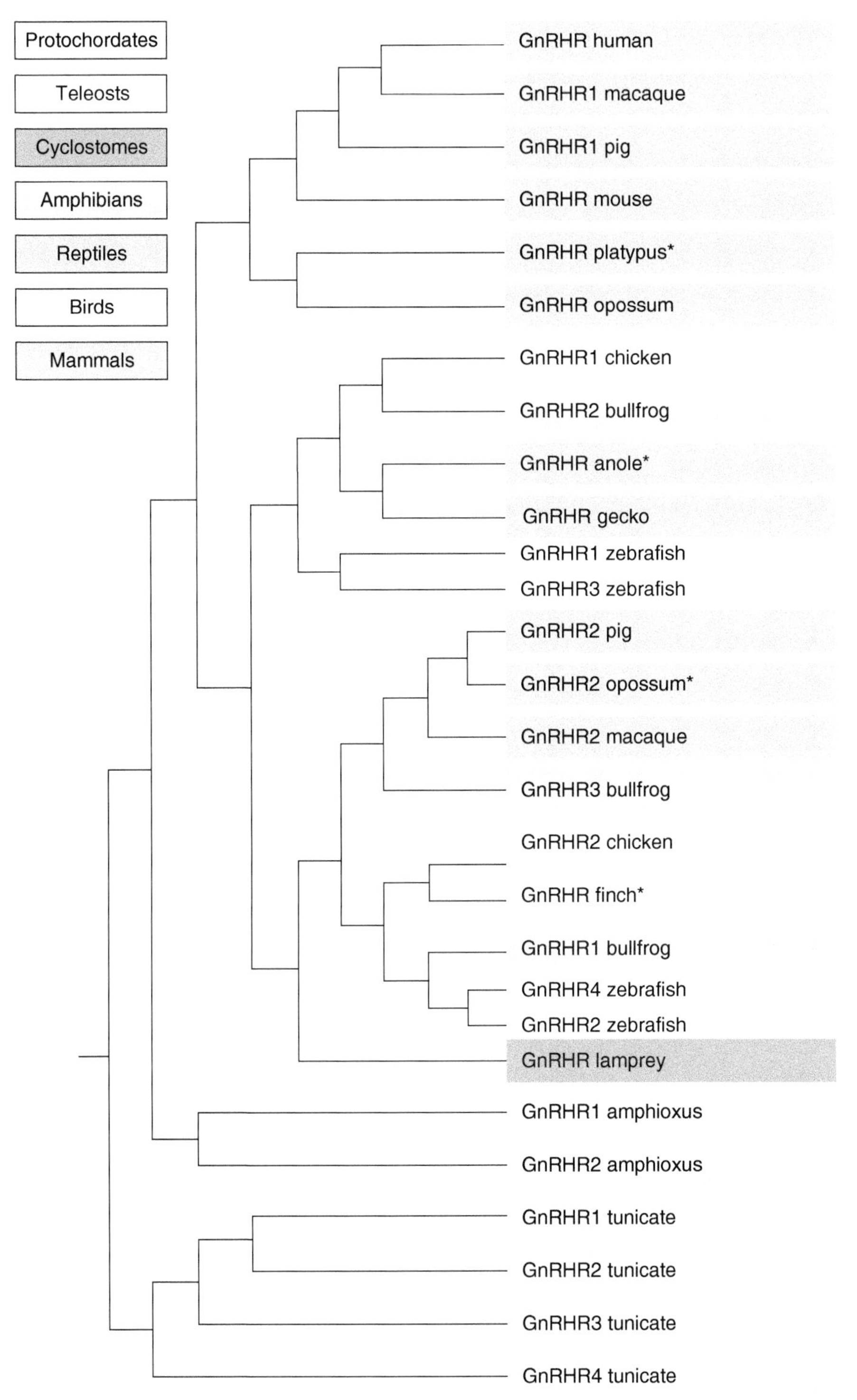

FIGURE 5-36 **GnRH-R gene evolution.** Note the inclusion of tunicate and amphioxus receptors in this phylogeny. *Indicates new data provided by this source. *(Adapted with permission from Roch, G.J. et al.,* General and Comparative Endocrinology, ***171****, 1–16, 2011.)*

from the **ventral tuberal nucleus (VTN)** appears to be the active Kp, and *kiss1* mRNA levels are elevated by exposure to long photoperiods or to E_2, whereas ovariectomy results in a reduction of *kiss1* mRNA in these neurons. *Kiss1* also causes LH secretion in goldfish but *kiss2* is ineffective. In contrast, it is *kiss2* in zebrafish that stimulates synthesis of β-FSH and β-LH subunits, and *kiss1* is inactive. The distribution of GnRH and *kiss1* neurons in the zebrafish brain is shown in Figure 5-39.

Limited studies of amphibians (*X. laevis* and *X. tropicalis*) and some lizards indicate the presence of *kiss2*-type Kps. *Kiss1* also has been reported in *X. laevis*.

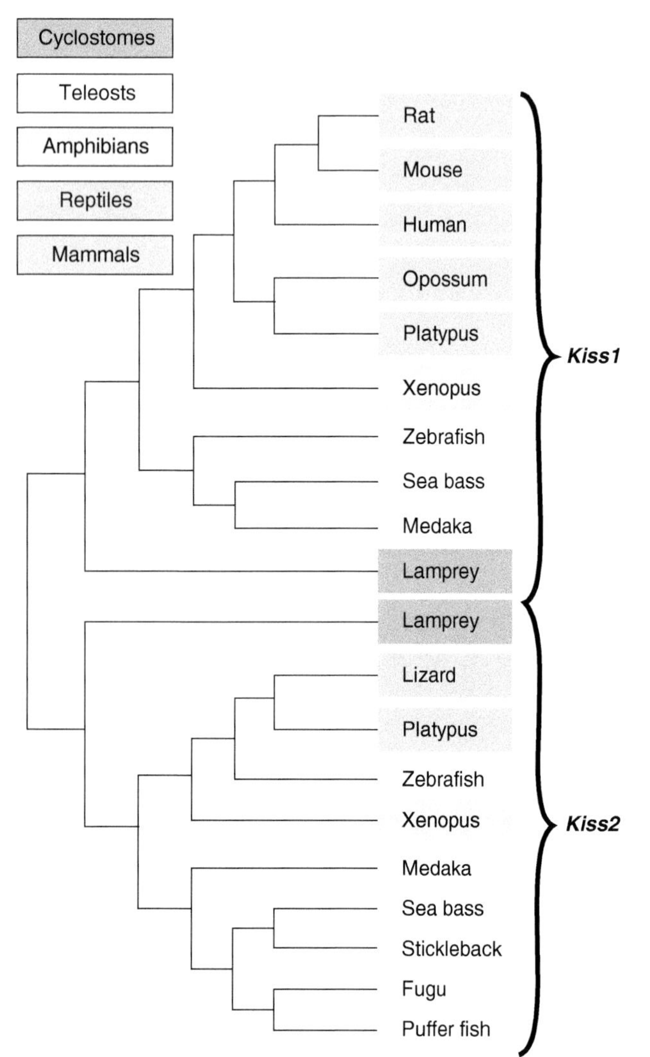

FIGURE 5-37 **Phylogeny of *kiss1* & *kiss2* based on cDNAs.** *Adapted with permission from Zohar, Y., Muñoz-Cueto, J.A., Elizur, A., Kah, O. (2010). Neuroendocrinology of reproduction in teleost fish.* Gen. Comp. Endocrinol. *165, 438-455.*

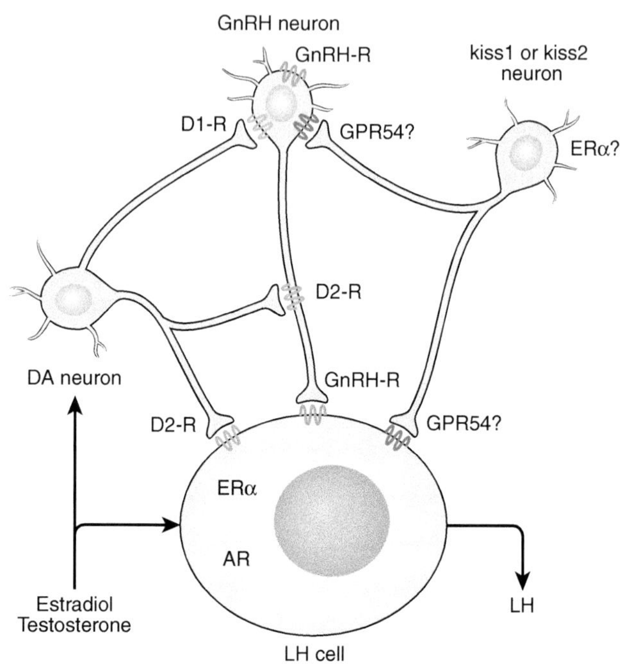

FIGURE 5-38 **Kisspeptin (kiss1 or kiss2) and dopamine (DA) regulation of gonadotropin secretion in teleosts.** Inhibitory DA neurons innervate both GnRH neurons and luteinizing hormone-secreting gonadotropes (LH cell). Kiss1 or kiss2 neurons stimulate GnRH release and possibly have direct effects on LH release operating through GPR54 (kiss1r) receptors. Gonadal steroids (estradiol, testosterone) feedback on the LH cell and the DA neuron (and possibly on the kiss1 or kiss2 neurons). *(Adapted with permission from Zohar, Y., et al., Neuroendocrinology of reproduction in teleost fish. General and Comparative Endocrinology **165**, 438-455, 2010.*

1c. Gonadotropin Release-Inhibiting Hormone (GnIH)

While searching for RFamide types of neuropeptides in the avian brain, K. Tsutsui and associates discovered a novel RF amide neuropeptide in the quail brain that inhibits GTH release in a dose-dependent manner. This GnIH was found to

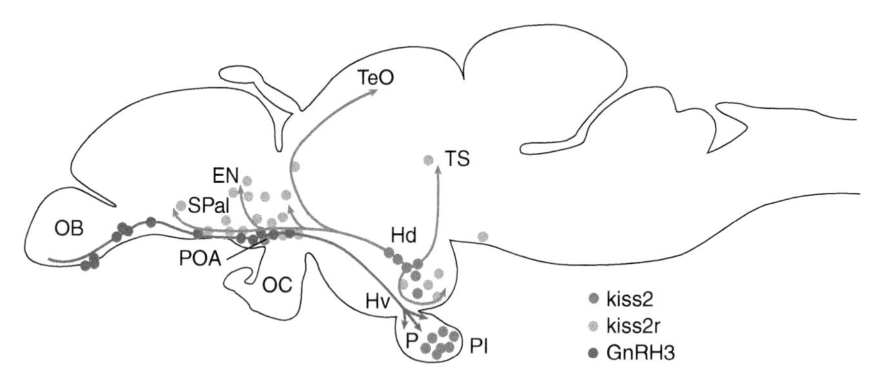

FIGURE 5-39 **The Kiss2–GnRH system in zebrafish brain.** The zebrafish (*Danio rerio*) produces both kiss1 and kiss2, but only the latter has been linked to the regulation of GnRH release in the zebrafish. Abbreviations: EN, enteropeduncular nucleus; Hd, dorsal hypothalamus; Hv, ventral hypothalamus; P, pituitary; PI, pars intermedia; OB, olfactory bulb; POA, preoptic area; SR, superior raphe; TS, torus semicularis. *(Adapted with permission from Servili, A. et al.,* Endocrinology*, 152, 1527–1540, 2011.)*

TABLE 5-7 Factors Affecting Thyrotropin (TSH) Release in Non-Mammalian Vertebrates

			Amphibia			
Factor	**Chondrichthyes**	**Teleosts**	**Urodele**	**Anuran**	**Reptiles**	**Birds**
TRH			$+^{a}$	$+^{a}$	+	+
CRH		+	+	+	+	+
VIP					+	+
SST		−				
Sauvagine					$+^{b}$	
Urotensin I					$+^{b}$	

Note: (+), stimulatory; (−), inhibitory; 0, no effect; (+/−), (+/0), (−/0), varies by species. See text for explanation of abbreviations.
[a]*Adults only.*
[b]*Possibly due to similarity is structure to CRH.*

reside principally in the PVN and GnIH-immunoreactive (ir) axons extended from the PVN to the median eminence. This ir-GnIH peptide also is present in the PVN of at least seven songbird species. A gene has been cloned that produces a prohormone of 173 amino acids that includes the sequence of GnIH plus two similar related peptide sequences (**GnIH-RP1** and **GnIH-RP2**). This gene also has been characterized in the white-crowned sparrow, where the deduced sequence of the GnIH peptide is similar to the quail neuropeptide. Administration of purified GnIH to white crown sparrows decreases GTH levels in a dose-dependent manner. Additionally, pinealectomy reduces the activity of the GnIH gene and decreases GnIH content of the PVN and the median eminence, whereas melatonin restores GnIH levels in pinealectomized birds in a dose-dependent manner.

1d. Other Factors Affecting Gonadotropin Release

Numerous neuropeptides, amine neurotransmitters, and steroids alter GTH release in vertebrates (see Table 5-6). These amines may operate at the level of the GnRH neurons in the brain or at the pituitary. For example, dopamine directly inhibits release of GTHs from the teleost and amphibian pituitaries. Combinations of GnRH receptor agonists and dopamine receptor antagonists, such as the AMPHIPLEX formulation developed by Vance Trudeau and colleagues, are particularly effective in stimulating reproduction in a number of fish and amphibian species. Feedback regulation of GTH release has been demonstrated in non-mammals and resembles that described for mammals. In teleosts, there is evidence for both positive feedback effects and negative feedback effects on GTH secretion depending on the stage of the reproductive cycle. Furthermore, exposure at very low levels (e.g., 0.1 to 1 ng/L) to exogenous natural and synthetic estrogenic compounds, including estradiol and the pharmaceutical **ethynylestradiol (EE_2)** commonly found in wastewater treatment plant effluents, can inhibit reproductive development in adult teleosts in both laboratory and field experiments, presumably by blocking GTH secretion. Similar observations of steroids on the inhibition of GTH secretion are known for other tetrapods as well.

2. Control of Thyrotropin (TSH) Release

Although mTRH is an effective releaser of TSH in mammals and is present in the brain of all non-mammalian vertebrates examined so far, it is relatively ineffective in releasing TSH but has been found to release a host of other tropic hormones (see Table 5-7). Release of TSH in amphibians, reptiles, and birds is more responsive to mCRHs than TRH, and several studies suggest CRH is the "TRH" of larval amphibians, too. **CRH receptors (CRH-R2)** present in bullfrog and chicken pituitaries apparently are involved in the stimulation of TSH release, but it is not known if this release is due to binding of CRH to CRH-R2 in the thyrotrope cell membrane or if TSH release is a result of a paracrine action following activation of another cell type. In mammals, CRH has greatest affinity for CRH-R1 and CRH-like peptides (e.g., urotensins and urocortins) show greater affinities for CRH-R2. Thus, the molecule we call "TRH" in mammals is probably not the "TRH" molecule of non-mammals.

Experimental evidence suggests that the control of TSH release is less specific in non-mammals. Thyrotropes of non-mammalian tetrapods may release TSH not only in response to TRH and CRH but also to exogenous mGnRH, mGHRH, **vasoactive intestinal peptide (VIP)**, **pituitary adenylate cyclase activating polypeptide (PACAP)**, and the CRH-like peptides sauvagine and urotensin I (see Table 5-7). **Somatostatin (SST)** blocks TSH release in amphibians. In chickens, TRH can stimulate TSH as well as GH release and the effect of TRH on both tropic hormones is inhibited by SST.

Teleosts represent an even greater anomaly in the control of TSH release. Not only are they unique among the vertebrates in having neuroglandular control rather than neurovascular control over TSH release, but TSH release is also under inhibitory control.

B. Category II: Growth Hormone (GH), Prolactin (PRL), and Somatolactin (SL)

Growth hormone is a major metabolic hormone in non-mammals that regulates early growth and adult metabolism as described for mammals (see Chapter 4). Prolactin appears to be primarily an osmoregulatory hormone in teleost fishes and has larval growth effects and anti-metamorphic roles (see Chapter 7) in amphibians. In birds, PRL plays important reproductive roles (see Chapter 11). Somatolactins have been implicated in many actions within teleosts including metabolism, reproduction, pigmentation, and stress. Neither GH nor PRL has been studied adequately in reptiles. Release of both category II tropic hormones are very similar to what has been observed in mammals. Notable exceptions are the neural control of PRL release exhibited in teleosts and the fact that PRL release in birds is primarily under stimulatory control. The ancestral member of this group appears to be GH and is thought to have given rise to the PRL and SL genes following gene duplications (see Figure 5-36). Only GH is found in lampreys, supporting this viewpoint.

1. Control of GH Release

Patterns of GH secretion are species-specific and often sexually dimorphic. Furthermore, secretion patterns can vary with age. Multiple daily peaks of GH have been reported in teleosts, with higher peaks occurring during the scotophase. Release of GH is altered predictably by mammalian hypothalamic neuropeptides (Table 5-8). Treatments of teleosts, amphibians, reptiles, and birds with mGHRH, PACAP, or ghrelin also are effective at causing GH release. Release of GH is also caused by the tripeptide TRH in at least some teleosts, amphibians, reptiles, and birds.

Somatostatin inhibits GH release in teleosts and birds but appears to be ineffective in amphibians and reptiles. Several molecular forms of SST, including SST-14 and SST-28, have been reported for various non-mammals. Where studied, IGF-1 appears to produce a negative feedback effect on GH secretion, implicating its role in the actions of GH as well as in normal feedback regulation of GH secretion. Treatment with some other factors affect GH release, but it is not clear that they all play a major role.

2. Control of PRL Release

Observations on the effects of neural factors on PRL release in non-mammals are summarized in Table 5-9. As in mammals, dopamine exhibits the major inhibitory control over PRL release in teleosts and amphibians but is ineffective in birds. Regulation of PRL release in reptiles is not known. Secondary stimulation of PRL in teleosts,

TABLE 5-8 Factors Affecting Growth Hormone (GH) Release in Non-Mammalian Vertebrates

Hormone	Teleosts	Amphibians	Reptiles	Birds	Mammals
GHRH	+	+	+	+	+
Ghrelin	+	+		+	+
SST	−	0[a]	0	−	−[b]
PACAP	+	+	+	+	+/−
IGF-1[c]	−			−	−
TRH	+/−	+	+	+/−	+
CRH	+		+		
GnRH	+/−				
NE	−				−
5-HT	−			0	+
NPY	+				+

Note: (+), stimulatory; (−), inhibitory; 0, no effect; (+/−), (+/0), (−/0), varies by species. See text for explanation of abbreviations.
[a]*No effect although preprosomatostatin (PPSS) is present.*
[b]*In pigs, low SST (10^{-15}M) is (+); high SST (10^{-7}M) is (−).*
[c]*IGF-1 action is due to negative feedback on somatotropes.*

TABLE 5-9 Factors Affecting Prolactin (PRL) Release in Non-Mammalian Vertebrates

Factor	Teleosts	Amphibia	Reptilia	Aves
Dopamine	–	–		–
SST	–			
GAP	–	–		
VIP				+
TRH	+	+	+	+
Copeptin[a]	+			
PrRP	+	+		+

Note: (+), stimulatory; (–), inhibitory; 0, no effect; (+/–), (+/0), (–/0), varies by species. See text for explanation of abbreviations.
[a]*Released from IST prohormone.*

amphibians, and birds is caused by administration of TRH or **prolactin-releasing peptide (PrRP)** as seen in some mammals (see Chapter 4). In a variety of teleosts, PrRP is effective at causing PRL release *in vivo* and *in vitro*, and PrRP-positive neurons innervate cells of the rostral pars distalis. In some of these teleosts, PrRP activates the PRL gene in addition to indicating that PrRP is a physiological regulator of PRL secretion.

In mammals, VIP is an active releaser of PRL (see Chapter 4), and the major control of PRL release in birds is via stimulatory control by VIP. However, in teleosts and amphibians, VIP inhibits PRL secretion. This inhibitory action of VIP is somewhat unexpected in light of its stimulatory role in birds and mammals. **Peptide histidine isoleucine (PHI)** is a VIP-like peptide that also inhibits PRL release in teleosts.

C. Category III: The POMC Group

Although there is considerable structural overlap between the peptides resulting from hydrolysis of the common precursor POMC in non-mammals and mammals, there are some marked differences in their regulation by hypothalamic factors.

1. Control of ACTH Release

An ancient CRH-like family of peptides developed very early in metazoan evolutionary history although its evolution in the chordates went in a different direction than among the invertebrate phyla. Several gene expansion events resulted in gene duplications among the chordates responsible for CRHs and several unique CRH-like peptides, including urotensin 1, urocortin-1, urocortin-2, and urocortin-3 (see Figure 5-40).

Mammalian CRHs appear to be effective releasers of ACTH in teleosts, amphibians, and birds as evidenced by their ability to increase corticosteroid secretion (Table 5-10). Release of TSH is also stimulated by CRH, and this could be the main mechanism for controlling TSH release in non-mammals as mentioned earlier. Hence, CRH appears to have dual roles (Figure 5-41). The distribution of CRH immunoreactivity in reptiles is supportive of a similar role. In teleosts, CRH-like peptides (i.e., sauvagine and urotensin-I) also can evoke ACTH release, but these CRH-like peptides so far have proven ineffective in amphibians. As described for mammals (see Chapter 4), either AVT or **arginine vasopressin (AVP)** can stimulate ACTH release in amphibians or birds, but the details of this effect have not been elucidated. The appearance of immunoreactive AVT-like material in the pars nervosa, together with CRH in the median eminence of bullfrog larvae at metamorphosis, supports an attendant releasing role for AVT. CRH levels in the POA are correlated with stress as well as foraging behavior in juvenile frogs. CRH and AVT are colocalized in neurons of the PVN in the snake *Natrix maura*.

It is important to recognize that CRH has other functions in vertebrates besides releasing ACTH from corticotropes. For example, CRH increases locomotory activity and sympathetic nervous system activity in vertebrates. Furthermore, the CRH molecule also is a potent releaser of TSH in amphibians as mentioned earlier.

2. Control of α-MSH Release

Dopamine has proven to be the most potent hypothalamic inhibitor of α-MSH release, and several investigators have proposed that dopamine is the hypothalamic **α-MSH release-inhibiting hormone (MRIH)** in all non-mammals as well as in mammals (see Chapter 4). Both aminergic and peptidergic fibers penetrate into the dogfish pars intermedia. It has been hypothesized that the peptidergic fibers control synthesis of α-MSH and that aminergic fibers are responsible for controlling α-MSH release.

Neurotransmitter and neuromodulator regulation of α-MSH secretion from the anuran pars intermedia has served as a model system for understanding the complex interaction of factors (Table 5-11) that regulate hormone secretion from pituitary cells. Amphibian melanotropes are innervated by a wide variety of neurotransmitter and neuropeptide-producing neurons including serotonin neurons, dopamine and neuropeptide Y neurons originating in the SCN, norepinephrine-producing neurons originating from the brain stem, and CRH neurons from the hypothalamus. The fact that such a diverse array of brain areas innervate anuran melanotropes speaks to the variety of sensory modes that may impact α-MSH secretion in addition to retinal inputs innervating the SCN. Many of these same pathways appear to play a role in the regulation of the teleost pars intermedia (Table 5-11).

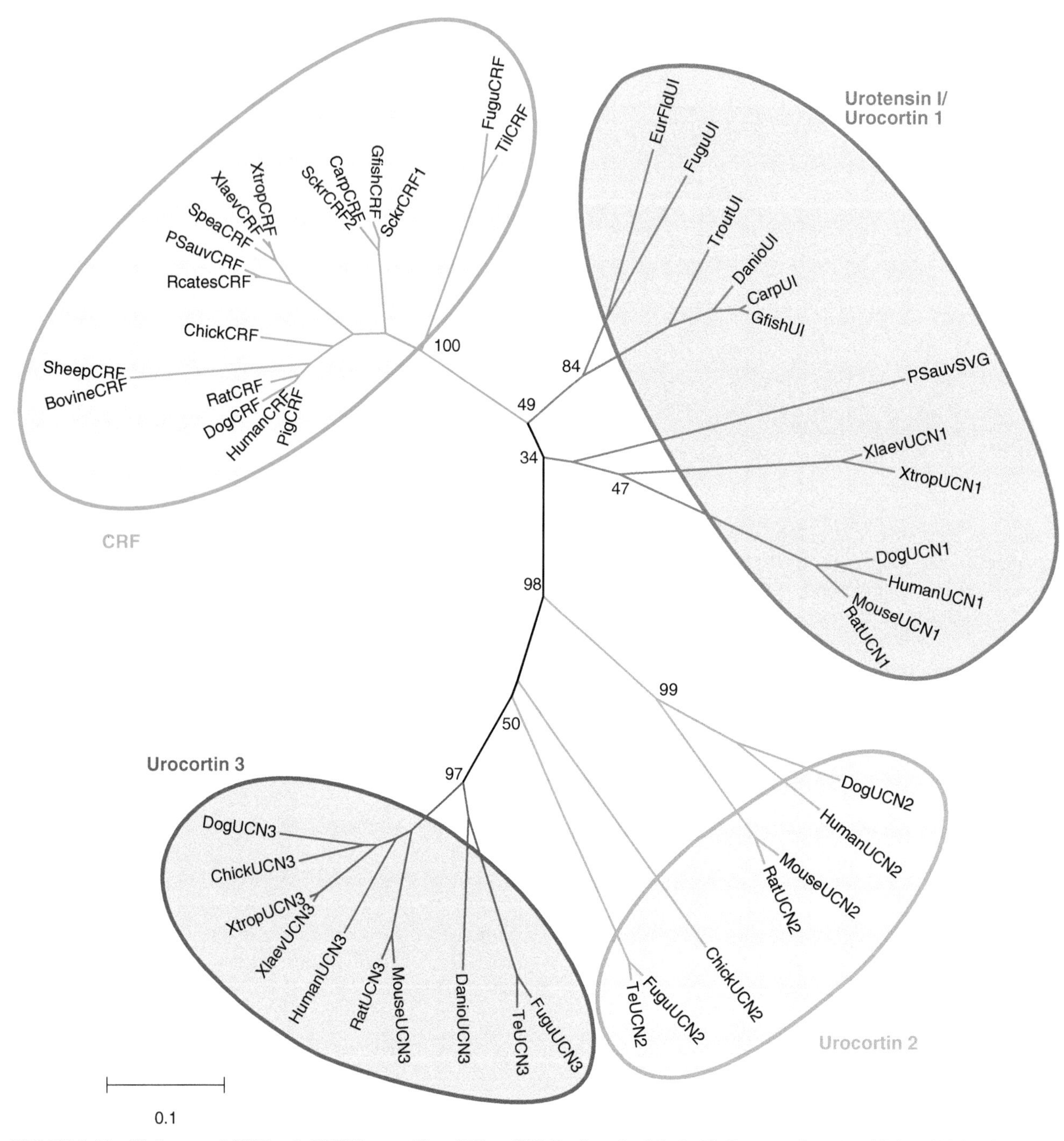

FIGURE 5-40 **Phylogeny of CRH and CRH-like peptides.** CRF = CRH; Danio, zebrafish; EurFl, European flounder; Gfish, goldfish; PSauv, *Phylomedusa sauvageii;* Rcates, bullfrog; Sckr, white sucker; Te, pufferfish; Til, tilapia; Xlaev, X. laevis; Xtrop, X. tropicalis. *(Adapted with permission from Boorse, G.C. et al.,* Endocrinology, ***146****, 4851–4860, 2005.)*

VI. NONAPEPTIDE NEUROHORMONES OF THE NON-MAMMALIAN PARS NERVOSA

The discovery among invertebrates (mollusks, annelids, protochordates) of genes producing prohormones that are similar to the vertebrate prohormones of the nonapeptides of the pars nervosa suggests that these invertebrate peptides are ancient bioregulators that have acquired new structures and functional roles in vertebrates. These prohormones in non-mammals are similar to the prohormones of mammalian nonapeptides (Figure 5-42). With the exception of the agnathan fishes, non-mammals typically produce two nonapeptides as described for mammals: one basic nonapeptide and one neutral nonapeptide. In virtually every case, the basic peptide is AVT, and this nonapeptide is the only one

TABLE 5-10 Factors Affecting Corticosteroid Release Via Corticotropin (ACTH) Release in Non-Mammalian Vertebrates

Factors	Teleosts	Amphibians	Reptiles	Birds
CRH (ovine)	+	+		+
Urotensin-I	+	0		
Sauvagine	+			
AVT/AVP		+		+

Note: (+), stimulatory; (−), inhibitory; 0, no effect; (+/−), (+/0), (−/0), varies by species.

found in the agnathans. AVT structurally appears to be a hybrid between mammalian AVP and OXY but actually evolved earlier than the mammalian nonapeptides. There is considerable variation in amino acid sequences among the neutral nonapeptides of non-mammalian vertebrates. Most of the vertebrate nonapeptides and their structures are provided in Table 5-12 as is a comparison with some invertebrate peptides. The phylogeny of the vertebrate nonapeptide prohormones is provided in Figure 5-43.

Agnathan fishes produce only one nonapeptide, AVT, whereas other non-mammalian vertebrates produce two: one member of a variety of neutral nonapeptides plus the same basic peptide, AVT. In most bony fishes the neutral nonapeptide is IST, whereas in most amphibians, reptiles, and birds the neutral nonapeptide is **mesotocin (MST)**. Elasmobranchs produce some unique neutral nonapeptides (see Table 5-12). For example, the spotted dogfish shark produces **phasvatocin (PhaT)** and **asvatocin (AsvT)** whereas the spiny dogfish produces a different pair of neutral nonapeptides, **aspargtocin (AspT)** and **valitocin (ValT)**. Rays and skates, however produce only one neutral nonapeptide, **glumitocin (GLT)**. In contrast, the holocephalan ratfish actually produces OXY, seemingly a unique appearance for this nonapeptide outside of mammals. Although most amphibians secrete MST, two additional neutral peptides have been found in isolated species: **seritocin** (Ser_5Ileu_8–oxytocin) (*Bufo regularis*) and **Val_4-MST** (*Plethodon shermani*). Furthermore, anuran amphibians produce

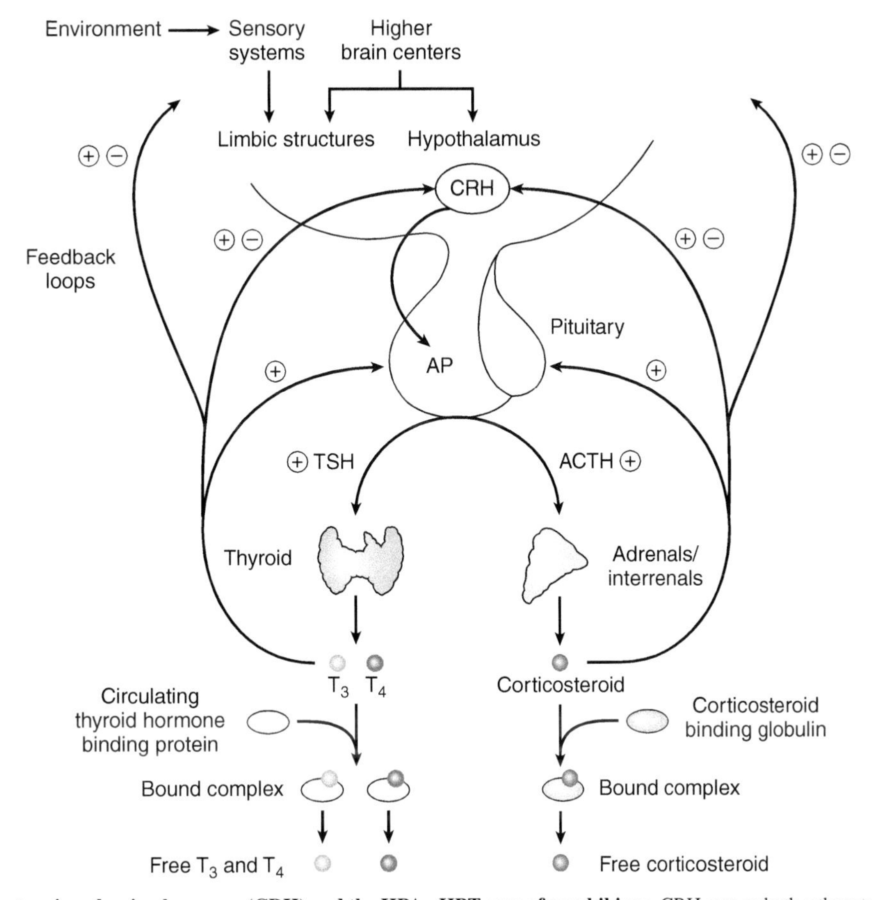

FIGURE 5-41 **Corticotropin-releasing hormone (CRH) and the HPA–HPT axes of amphibians.** CRH acts as both a thyrotropin-releasing hormone and a corticotropin-releasing hormone in amphibians. The system is influenced by environmental factors acting through various neural centers that all feed information to the hypothalamus. *(Adapted with permission from Denver, R.J.,* General and Comparative Endocrinology, *164, 20–31, 2009.)*

TABLE 5-11 Factors Affecting Melanotropin (α-MSH) Release in Non-Mammalian Vertebrates

Factors	Teleosts	Amphibians	Reptiles
TRH	+	+	+/0
CRH (ovine)	+	0	
Urotensin-I	+	0	
Sauvagine	+		
MCH	−		
Dopamine	−	−	

Note: (+), stimulatory; (−), inhibitory; 0, no effect; (+/−), (+/0), (−/0), varies by species. See text for explanation of abbreviations.

peptides called **hydrins** from the AVT peptide. Like AVT, these peptides also influence water balance.

The neurohypophysial nonapeptides have been identified in only a relatively small number of species with absolute certainty. The proliferation of forms in the chondrichthyean fishes, amphibians, and mammals (see Chapter 4) suggests that we may eventually find a greater variety of these nonapeptide molecules in much larger taxa such as the teleost fishes or even in the birds and reptiles than have been identified to date.

A. Nonapeptide functions in Non-Mammals

In general, studies of nonapeptide functions in non-mammals have focused on reproductive roles rather than osmoregulatory roles. In part, this was a consequence of early bioassays of AVT showing that it has considerable OXY-like activity on reproductive smooth muscle (see Appendix D). Spawning reflexes in fishes as well as sperm release and/or egg laying in amphibians, reptiles, and birds are stimulated by AVT. Furthermore, the sensitivity of oviducts to AVT is dependent on previous exposure to steroid hormones. Two AVT receptors have been isolated from anuran amphibians: **VT1aR** and **VT2R**. They are all GPCR types. The major AVT receptor seems to be the VT2R that works through cAMP. The VT1aR works through phosphokinase C. A unique **VT1bR** has been isolated from the newt *Cynops pyrrhogaster* that is similar to the mammalian VT1bR. The shell gland of Japanese quail (*Coturnix coturnix japonica*) exhibits an **oxytocin gene-related receptor (VT3R)** that binds AVT released from the pars nervosa during oviposition.

Kidney function is affected by AVT in non-mammals where it generally acts to conserve water. The actions of AVT on the kidney of non-mammals parallel those described for AVP in mammals. The skin and urinary bladder are also iono-osmoregulatory AVT targets. In contrast, a diuretic response to MST has been reported in anuran amphibians.

The physiological roles for the neutral nonapeptides are poorly studied, and we know little of the biological roles for MST in lungfishes and tetrapods (including the marsupials), IST in bony fishes, or the variety of neutral nonapeptides found in chondrichthyeans and amphibians. AVT and not IST or MST seems to be the more potent nonapeptide in non-mammals at stimulating reproductive smooth muscle. Amphibian GPCRs for MST (MTRs) are similar to mammalian OXY receptors and have been associated with Na^+ and water uptake from the urinary bladder.

A limited number of studies have investigated the roles of nonapeptides on non-mammalian behaviors For example, AVT is known to influence calling behavior in frogs as well as in some birds. The response of female frogs to calling males is increased by AVT, and clasping behavior associated with reproduction in male rough-skinned newts (*Taricha granulosa*) is induced by AVT (see Chapter 11). The molecular counterparts of vertebrate nonapeptides, **conopressin** in mollusks and **annectocin** in annelids, also have been linked to reproductive behaviors, suggesting an ancient behavioral role for these peptides.

B. Evolution of Nonapeptide Hormones

Numerous attempts to develop meaningful phylogenetic trees based on the distribution of the nonapeptides met with

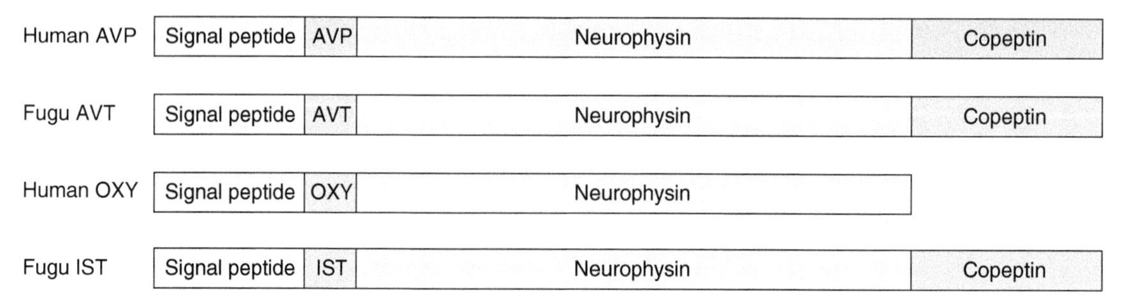

FIGURE 5-42 **Nonapeptide preprohormones.** Comparison of human AVP and OXY preprohormones with those of AVT and IST, respectively, from Japanese fugu, a teleost fish. See text for explanation of abbreviations. *(Adapted with permission from Gwee, P.-C. et al.,* BMC Evolutionary Biology, *8, 93, 2008.)*

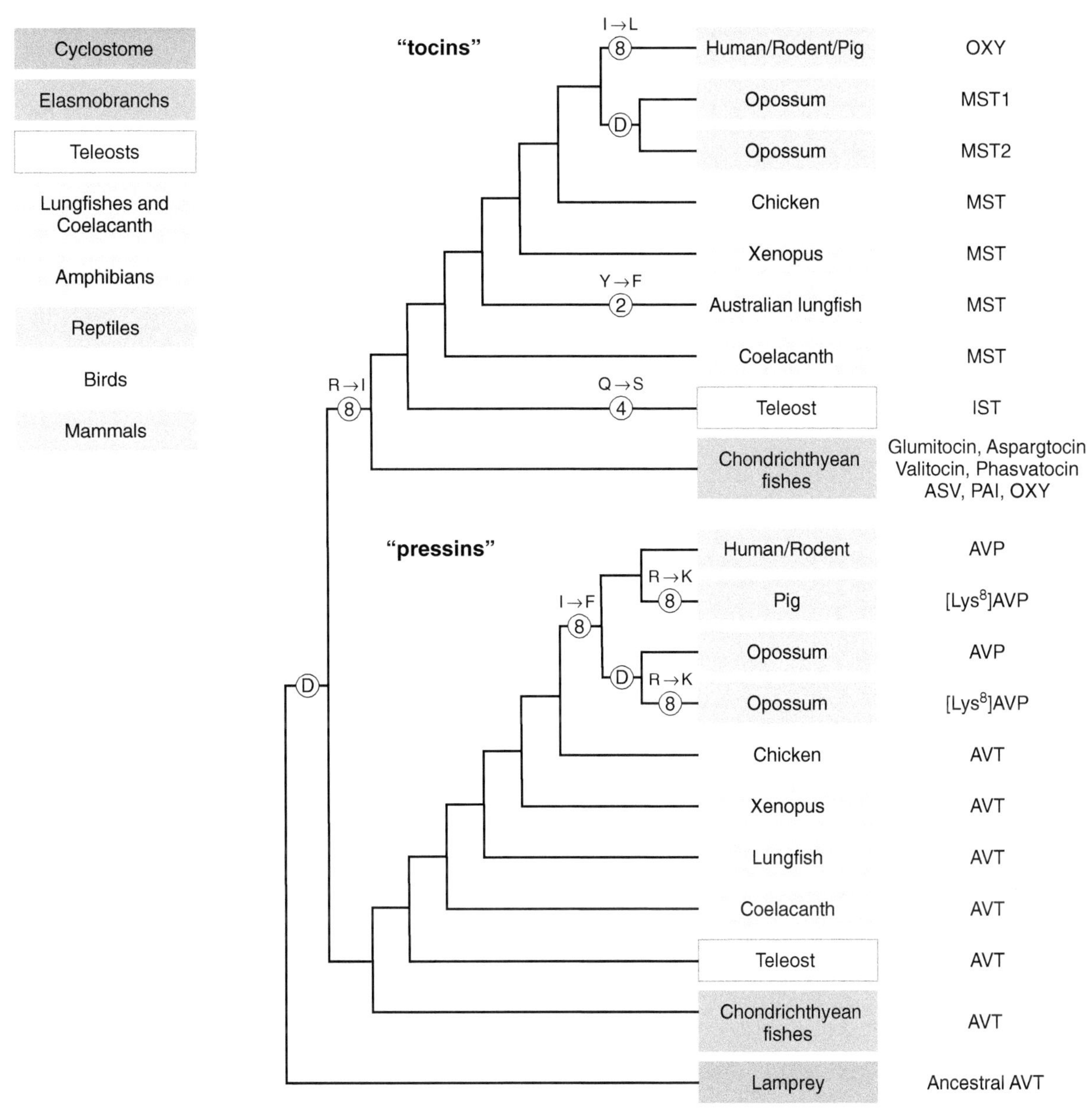

FIGURE 5-43 **Evolution of neurohypophysial nonapeptides genes.** The letter "D" within the circle represents a gene duplication event. The numbers within the circles denote the position of the amino acid that has been substituted (shown above the circle). Note that the presence of MST and PYP in some marsupials has not been included. See text for an explanation of abbreviations. *(Adapted with permission from Gwee, P.-C. et al.,* BMC Evolutionary Biology, *8, 93, 2008.)*

limited success. One reason lies in the early attempts to construct phylogeny based on only a small fraction of the prohormone; consequently, much of the genetic information was overlooked in these earlier schemes. However, recent phylogenies based on genetic analysis and comparison of prohormones support the general pattern described using only the nonapeptide amino acid sequences.

It is very likely that AVT represents the most primitive (and possibly ancestral) form of nonapeptide in vertebrates from which the others evolved following one or two duplications of the AVT gene. This gene evolution has involved the development of the MST gene following duplication of the AVT gene in sarcopterygian fishes and early amphibians, with the MST gene later giving rise to the OXY gene following another duplication. The AVT gene later evolved into the AVP gene of mammals. Much later than the emergence of MST, duplication of the AVT gene within the teleosts resulted in evolution of the IST gene. The prohormones for AVT, AVP, and IST all include the **copeptin** sequence as well as a neurophysin sequence. The copeptin sequence is absent from MST and OXY prohormones.

TABLE 5-12 Amino Acid Sequences of Vertebrate Nonapeptides

Vasopressin homologs		
Arginine vasopressin	C-Y-F-Q-N-C-P-R-G-(NH_2)	Mammals
[Lys[8]]vasopressin	C-Y-F-Q-N-C-P-K-G-(NH_2)	Pigs, hippopotamus, some marsupials
Phenypressin	C-F-F-Q-N-C-P-R-G-(NH_2)	Some marsupials
Arginine vasotocin	C-Y-I-Q-N-C-P-R-G-(NH_2)	Non-mammals
Oxytocin homologs		
Oxytocin	C-Y-I-Q-N-C-P-L-G-(NH_2)	Mammals, some marsupials, platypus, ratfish (*H.colliei*)
Mesotocin	C-Y-I-Q-N-C-P-I-G-(NH_2)	Some marsupials, non-eutherian tetrapods, some lungfishes
[Phe[2]]mesotocin	C-F-I-Q-N-C-P-I-G-(NH_2)	Australian lungfish
Isotocin	C-Y-I-S-N-C-P-I-G-(NH_2)	Ray-finned fishes
Glumitocin	C-Y-I-S-N-C-P-Q-G-(NH_2)	Skates
Valitocin	C-Y-I-Q-N-C-P-V-G-(NH_2)	Sharks (*S. acanthias*)
Aspargtocin	C-Y-I-N-N-C-P-L-G-(NH_2)	Sharks (*S. acanthias*)
Asvatocin	C-Y-I-N-N-C-P-V-G-(NH_2)	Sharks (*S. canicula; T. scyllium*)
Phasitocin	C-Y-F-N-N-C-P-I-G-(NH_2)	Sharks (*T. scyllium*)
Phasvatocin	C-Y-F-N-N-C-P-V-G-(NH_2)	Sharks (*S. canicula*)

See Appendix C for explanation of amino acid symbols.
Adapted from Gwee, P.-C., et al., 2008. BMC Evolutionary Biology, 8, 93.

VII. THE EPIPHYSIAL COMPLEX OF NON-MAMMALS

Almost all vertebrates exhibit one or two epithalamic structures that constitute the epiphysial complex (Figure 5-44). The components of this complex are the pineal organ and a more anterior projection, the parapineal organ. The pineal gland secretes the indoleamines **melatonin** and serotonin and possibly some small peptides, including AVT, into the blood and cerebrospinal fluid. As seen for mammals (Chapter 3, Figure 3-2), the enzymes **hydroxylindole-O-methyl transferase (HIOMT)** and **N-acetyl transferase (NAT)** are essential for conversion of tryptophan to serotonin and then to melatonin.

Numerous studies have documented actions of melatonin on the behavior and physiology of fishes,

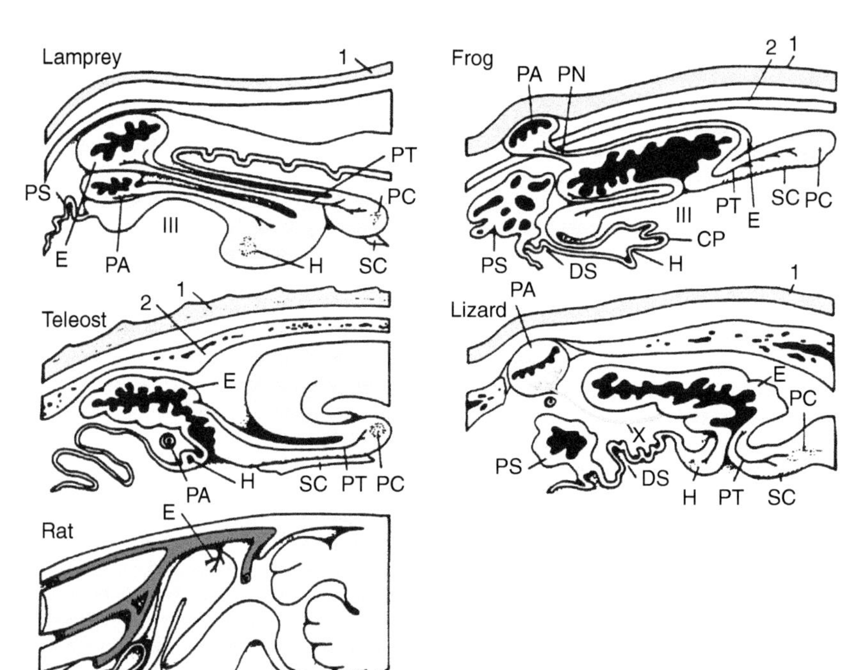

FIGURE 5-44 **The epiphysial complex.** The pineal gland and associated structures are shown for a generalized teleost, frog, lizard, and mammal (rat). 1, skin; 2, skull; III, third ventricle. Abbreviations: CP, choroids plexus; DS, dorsal sac; E, epiphysis (pineal), H, habenular commissure; PA, parietal (parapineal) organ; PC, posterior commissure; PN, pineal nerve; PS, paraphysis; PT, pineal tract; SC, subcommissural organ; X, parietal nerve. *(Adapted with permission from Bentley, P.J. "Comparative Vertebrate Endocrinology", Cambridge University Press, 1976.)*

amphibians, reptiles, birds, and mammals. Many of these actions are related to the mediation of photoperiods on reproduction via pineal secretions. Pineal actions are mediated via effects on various targets including the hypothalamus and pituitary as well as on peripheral targets such as gonads.

In fishes, amphibians, and some reptiles (lizards), the organs of the epiphyseal complex are basically sac-like diverticula that are more or less open to the third ventricle of the brain. They consist of a basal portion composed of sensory and ependymal (supportive) cells and may have an attached stalk with a distal end vesicle that contacts the dorsal brain case. These structures probably arose in primitive fishes as a pair of diverticula that for some reason later changed positions relative to one another. A well-developed parapineal organ has been retained only in cyclostomes and lizards, whereas the pineal organ is found in all vertebrate groups with the exception of crocodilians. In anamniotes, as well as in lizards, the pineal organ has retained both its sensory and endocrine functions; however, in other reptiles as well as birds and mammals, the pineal appears to be only an endocrine structure and often is termed simply the pineal gland.

The anteriormost evagination of the epiphysial complex that develops from the telencephalon is known as the **paraphysis**. The paraphysis is best seen in amphibians as a highly vascularized, sac-like diverticulum (Figure 5-44); it may function similarly to the choroid plexus in producing cerebrospinal fluid. The dorsal sac arises in most vertebrates as an epithalamic evagination just posterior to the paraphysis but anterior to the epiphysial complex. It is especially prominent in the ganoid fishes (chondrosteans, holosteans) and becomes less conspicuous in teleost fishes. In most vertebrates the dorsal sac contributes to formation of the choroid plexus, and these structures may be indistinguishable in the adult.

The pineal complex is connected to an adjacent ependymal structure, the **subcommissural organ (SCO)** (Figure 5-44). The ependymal cells of the SCO produce an aldehyde fuchsin-positive secretion rich in disulfide bonds and cysteine. This secretion is similar to that observed in the pineal ependyma. The major secretory product of the SCO is a noncellular fiber that in some species extends into the central canal of the spinal cord for its entire length. This structure is known as **Reissner's fiber**. Its significance is not clear (see Chapter 4). Cytological activation of the SCO in the lizard *Lacerta s. sicula* has been correlated positively with seasonal activities of adrenal cortical cells and of testicular steroidogenic cells. The actual role or roles for the SCO and its secretory products must await further research, but preliminary data would suggest it is somehow related to activity of the pineal complex.

The retina has become the transducer for photoperiodic control of pineal function, operating through retinal cells that project to the the SCN (see Chapter 4). Evidence indicates that the retina of non-mammals produces melatonin and may also function in photoperiodic regulation, possibly interacting with or independent from the pineal (Figure 5-45).

A. Roles for the Pineal in Non-Mammalian Vertebrates

The first role discovered for melatonin was the regulation of melanophores in the skin of amphibians with respect to exposure to light. Pineal function has been studied most extensively in fishes and non-mammalian tetrapods, especially with respect to photoperiodic stimuli and daily and seasonal biological rhythms. Although there is controversy concerning the control of seasonal reproduction by melatonin, numerous reproductive roles have been demonstrated in non-mammals (see ahead). Throughout vertebrate evolution, there has been a progressive decrease in the role of the pineal as a photoreceptor and an increasing dependence upon the lateral eyes and the SCN-pineal pathway to control photoperiodically linked events such as seasonal reproduction and the secretion of melatonin. A summary of some of melatonin's actions in non-mammalian vertebrates is provided in Table 5-13.

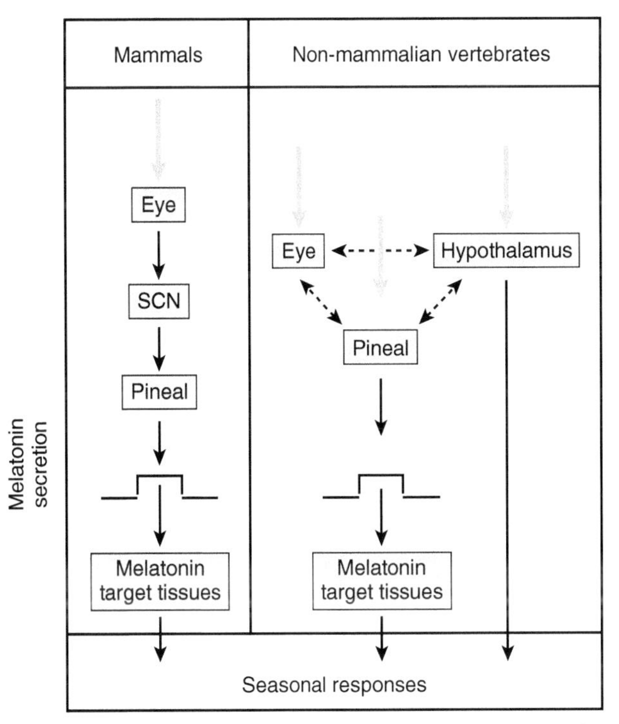

FIGURE 5-45 **Comparison of mammalian and non-mammalian melatonin systems.** SCN, suprachiasmatic nucleus. See text for comparison. *(Adapted with permission from Hazlerigg, D.G. and Wagner, G.C.,* Trends in Endocrinology and Metabolism, *17, 83–91, 2006.)*

TABLE 5-13 Melatonin Effects on Non-Mammalian Vertebrates

	Vertebrate Group				
Effect	Agnathans	Teleosts	Amphibians	Reptiles	Birds
Concentration of melanosomes or inhibition of melanin synthesis	Yes	Yes	Yes	Yes	
Preferred temperature		Increase	Decrease	Decrease	
Thermogenic effect on body temperature				Decrease	Decrease
Gonad function					
Adult		Stimulate/inhibit	Inhibit	Inhibit	Inhibit
Juvenile					Stimulate
Thyroid function	Stimulate?		Inhibit		

1. Agnathan Fishes: Cyclostomes

The epiphysial complex of cyclostomes consists of a pineal organ and a parapineal organ, although the latter may be lacking in some species. Both organs possess variably shaped end vesicles that project dorsally against the roof of the brain case. The end vesicles contain sensory and ependymal cells structurally organized to suggest that they are photoreceptors. Efferent neural fibers from the pineal organ end at the posterior commissure, whereas those from the parapineal terminate at the habenular commissure. These organs probably relay photoperiodic information to other regions of the brain. Nocturnal blanching has been observed in the ammocetes larvae of some lampreys, and epiphysial levels of HIOMT are correlated with nocturnal lightening of the skin in *Geotria australis*. Removal of the epiphysial complex causes persistent expansion of melanophores in *Lampetra planeri* and *G. australis* but not in *Mordax mordacia*, which lacks a parapineal organ and does not exhibit nocturnal lightening. Thus, melanophore expansion may be mediated via the parapineal. Hypophysectomy of cyclostomes also results in blanching due to the absence of α-MSH that normally stimulates melanophore expansion. The role of the parapineal may be to inhibit release of α-MSH, but melatonin could be the factor responsible for the observed melanophore contraction. There is evidence for a possible thyroid effect of the pineal organ of lampreys. Pinealectomy inhibits metamorphosis of the ammoecete to the adult body that normally is prevented by elevated thyroid hormones.

2. Chondrichthyean Fishes

The shark pineal organ contains both photosensitive and supportive cells. No experimental data, however, have been reported, and the functional importance of the pineal is not known for these fishes. A parapineal organ has not been described for any species in this group. This vertebrate taxon demands further investigation.

3. Bony Fishes: Teleosts

The epiphysial complex of teleosts consists of a pineal organ (Figure 5-44) that is extremely variable in both size and degree of development. There is usually a prominent lumen, which in some cases is open to the third ventricle. Photosensory cells and ependymal cells are present in the pineals of several species. A reduced parapineal has been described in some teleosts.

The effects of the pineal and melatonin in teleosts vary greatly, and it is difficult to generalize their roles for the group as a whole. Even the pattern of nocturnal secretion of melatonin seen in the tetrapods is not consistent in teleosts, with the pineal of some species secreting only during the night like tetrapods (Figure 5-46), some exhibiting daytime peaks only, and others exhibiting no rhythmic pattern of secretion. A generalized summary for photoperiodic regulation of reproduction in teleosts exhibiting nocturnal patterns of pineal secretion is provided in Figure 5-47.

Pigmentation, responses to light, and changes in thyroid activity are influenced by the pineal organ in some teleosts. Melanophore changes have been reported in several species following injection of pharmacological amounts of melatonin. Some of these species also respond in a similar manner to epinephrine treatment. Other species respond only to epinephrine and exhibit no reaction to melatonin. Circulatory melatonin levels exhibit no correlations to adaptation by rainbow trout to different backgrounds,

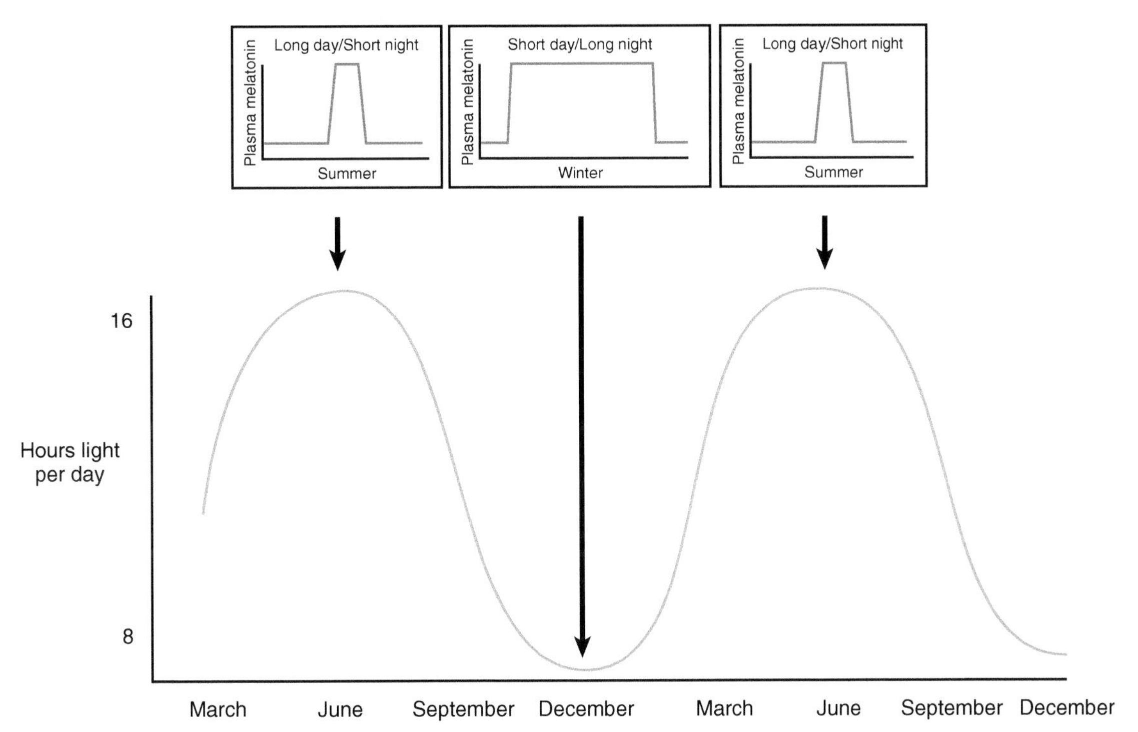

FIGURE 5-46 **Changes in melatonin secretion patterns with seasonal changes in day length (hours of light) in a generalized teleost fish.** Such changes have been associated with initiation of reproductive development in some species. *(Adapted with permission from Bromage, N. et al.,* Aquaculture, ***197****, 63–98, 2001.)*

indicating that melatonin may not be a factor influencing normal pigmentary responses in this species.

Control of pineal secretion appears to be determined by the presence or absence of light in certain teleosts and not by some endogenous mechanism or rhythm. For example, rainbow trout pineal glands *in vitro* exhibit no endogenous secretory rhythm in melatonin secretion and release melatonin only when in darkness. Similar observations have been reported for goldfish pineals. Melatonin also is produced in the retina of several teleost species. Levels of melatonin in the rainbow trout retina may exceed levels reported for the pineal. It is proposed that melatonin may cause concentration of pigment in retinal melanophores to increase sensitivity of retinal cells in dim light. However, neural connections to the hypothalamus and the SCN have been described in some species, suggesting dual pathways through which photoperiod could produce effects on teleost physiology, whereas mammals exhibit only the retinal

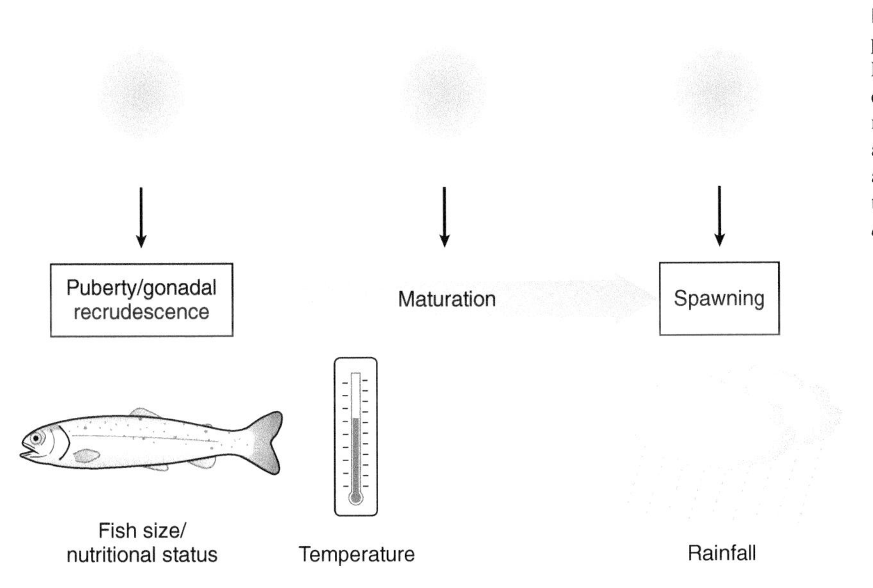

FIGURE 5-47 **Photoperiodic control of puberty/recrudescence and spawning in fishes.** Photoperiod effects vary by season with respect to different reproductive events. These effects are modulated by permissive factors such as fish size at that time, nutritional status, water temperature, and rainfall (which can affect salinity, temperature, etc.). *(Adapted with permission from Bromage, N. et al.,* Aquaculture, ***197****, 63–98, 2001.)*

pathway (see Figure 5-42). However, the pineal gland of some teleosts may be the only site responding directly to light and may not be influenced through the lateral eyes and the SCN pathway.

The pineal of the white sucker (*Catostomus commersoni*) responds to light intensity and/or thermal gradients. Shielding the pineal from light causes these fish to choose the warmer portion of horizontal temperature gradients or the better illuminated portions of chambers maintained at a constant temperature. When the shield is removed, the fish return to their original preference.

Structural correlations have been described between the pineal organ and phototactic responses by fishes. Species with a translucent covering over the pineal organ (a definitive pineal spot) exhibit predominantly positive phototaxis, whereas species with a pigmented, opaque skeletal covering do not show phototaxis. Species that have pigment cells located so that dispersal and concentration of pigment granules could regulate the intensity of light reaching the pineal organ exhibit responses varying from positive phototaxis to no response. The pineal organ might be involved in some other functions for the species showing no phototaxis. The presence of a pineal spot is more common in deep-sea fishes than in freshwater or shallow-water marine species and may relate to the influence of light on vertical migrations performed by deep-sea fishes.

Pinealectomy of *Poecilia reticulata* (guppy) causes pituitary enlargement and hyperplasia of the thyroid. This effect also occurs in the killifish (*Fundulus heteroclitus*) if pinealectomy is performed during the winter months (December through March). Pinealectomy between February and June produces no effect on the thyroid, however. Pituitary and thyroid of the characin, *Astyanax mexicanus*, are not affected by pinealectomy, but both stimulation and inhibition of the goldfish thyroid have been described.

4. The Pineal in Amphibians

The epiphysial complex of amphibians consists of a pineal organ, dorsal sac, choroid plexus, SCO, paraphysis, and often a parapineal organ (Figure 5-44). The proximal (basal) portion of the amphibian epiphysial complex contains HIOMT activity and melatonin. The retina of the eye has the same features. Pinealectomy reduces circulating melatonin to daytime levels, suggesting that the retina may be responsible for basal levels of melatonin in the blood; however, a retinal–hypothalamic pathway has not been described in amphibians.

The role of melatonin on melanophores in larval amphibians was first suggested by the observations of McCord and Allen almost a century ago. Since that time, it has been shown that melatonin is the pineal agent responsible for the blanching of tadpoles or larval salamanders when held in the dark. As little as 0.1 ng/mL medium causes aggregation of melanin granules in melanophores of *X. laevis* tadpoles. Plasma melatonin exhibits a distinct daily rhythm determined by photoperiod that is unaffected by temperature, suggesting that the pineal could provide a cue to regulate seasonal events, as well.

Attempts to relate pineal function or melatonin with thyroid function have been somewhat equivocal in amphibians. Pinealectomy of tadpoles of the midwife toad (*Alytes obstetricans*) accelerates metamorphosis, but a similar operation in larvae of the newt *Taricha torosa* is without effect. Earlier observations in *Bufo americanus*, however, indicate that feeding mammalian pineal to tadpoles accelerated metamorphosis. Pinealectomized larval tiger salamanders (*Ambystoma tigrinum*) exhibit decreased thyroidal uptake of injected radioiodide, but neither purified melatonin nor commercial bovine pineal powder influences iodide uptake of intact larvae.

Reproduction may be under inhibitory influence of the pineal organ, at least in anurans. Accelerated gonadal development follows pinealectomy of *A. obstetricans* and the tree frog *Hyla cinerea*. Gonadotropin-induced ovulation from *Rana pipiens* ovaries *in vitro* is inhibited by addition of melatonin to the culture medium. Bovine pineal extract similarly inhibited human chorionic gonadotropin-induced spermiation in male *Rana esculenta*, but purified melatonin had no effect. This dichotomy of melatonin's influence in male and female anurans warrants further investigation. The possible influence of pineal principles on reproduction in urodeles has not been studied.

Unlike the other jawed vertebrates, some anurans have retained a well-developed parapineal end vesicle known also as the **frontal organ** or *stirnorgan* (Figure 5-44). Because of the presence of photosensory cells, the parapineal is often referred to as the **parietal eye** in these species. Photic information collected by the parietal eye is conducted directly to the pineal, and the parietal eye functionally may be just an extension of the pineal.

5. The Pineal in Reptiles

Reptiles can be separated into several groups on the basis of the anatomy of the epiphysial complex. Melatonin has been localized in the blood, pineal gland, and retinas of snakes, lizards, and turtles. Modern crocodilians have no pineal, parapineal, or parietal structures. Although a pineal is absent in alligators, melatonin, presumably of retinal origin, is present in the blood. Lizards possess an elaborate sac-like, pigmented pineal organ containing both photosensory and ependymal cells (Figure 5-44). The lumen of the lizard pineal lies close to the third ventricle but does not join with it. A parapineal organ penetrates the skull, forming a parietal spot on the surface. As in amphibians, the parapineal is often termed the parietal eye. Turtles and snakes have retained only the basal (glandular) portion of the pineal organ and have lost the end vesicle and stalk of the pineal organ as well as the complete parapineal organ. Nevertheless, the turtle pineal organ is the largest and the best-developed epiphysial structure among reptiles. Perhaps photic stimuli are conveyed to the pineal via the retinal–hypothalamic pathway in these animals.

The parietal eye of lizards has been examined with respect to thyroid function, thermoregulation, and reproduction. The lizard parietal eye contains HIOMT activity, suggesting that it also synthesizes melatonin, and removal of the parietal eye stimulates thyroid hyperplasia and oxygen consumption (a thyroid-sensitive process; see Chapters 6 and 7). These data suggest that melatonin or some other principle inhibits some aspects of pituitary function, although direct effects on the thyroid are not ruled out. It is suggested that perhaps the reptilian parietal eye is a photo/thermal radiation dosimeter that monitors solar radiation and, in turn, regulates daily activity patterns of lizards. Indeed, excision of the pineal or parietal eye alters thermal responses of numerous lizard species, and this may well be a major functional role for the epiphysial complex in lizards.

Definitive effects of the epiphysial complex and melatonin inhibition on reproduction also have been reported in lizards. For example, excision of the parietal eye of the lizard *A. carolinensis* stimulates ovarian development in reproductively quiescent animals. This effect is blocked by administration of melatonin. In this lizard, long photoperiod and warm temperatures induce the onset of gonadal recrudescence. The parietal eye thus appears to be the transducer through which photoperiod influences reproduction in these lizards. Treatment of male red-sided garter snakes with melatonin inhibits courtship behavior in males recently emerged from winter dens, suggesting that long photoperiod (short scotophase) is a permissive factor allowing for expression of breeding behavior that is prevented by elevated melatonin.

6. The Pineal in Birds

As seen in mammals, the pineal organ of birds has been reduced to the glandular basal portion, or pineal gland. No parapineal or remnant thereof is present, and distinct photoreceptors are absent in the pineal. Structurally, the avian pineal exhibits considerable diversity, and it is composed of several cell types. The avian pineal is innervated by sympathetic fibers as reported for mammals.

Avian pineals are biochemically like their mammalian counterpart, with similar secretion patterns. Variations have been reported in HIOMT and NAT activities with respect to lighting conditions, but the most dramatic effects involve NAT. Activity of this enzyme exhibits a marked increase with onset of the scotophase, a peak at about the middle of the scotophase, and a decrease rapidly following the onset of the photophase (see Chapter 4, Figure 4-37) . Brief exposure to light at the peak of NAT activity causes a rapid reduction in photophase levels. Melatonin rhythms that correlate with rhythms in pineal NAT activity have been reported for brain, pineal, retina, and serum of birds. The brain and especially the hypothalamus in birds may be the primary site of action for melatonin and may explain the effects of melatonin on gonadal function, thermoregulation, and locomotor activity.

A progonadal role for the pineal has been suggested by some studies. For example, pinealectomy inhibits androgen synthesis, whereas administration of melatonin stimulates androgen synthesis, presumably by altering gonadotropin release from the adenohypophysis. Pinealectomy of quail delays ovarian development, an observation that also supports a progonadal role. However, melatonin injections can decrease gonadal weight, suggesting that the progonadal agent is not melatonin but might be a peptide that normally overpowers effects of endogenous melatonin on reproduction. Marked species differences may occur, however, especially between migratory and non-migratory species. For example, no effects of melatonin treatment on

BOX 5A Evolution of Melatonin's Functions

An intriguing hypothesis, proposed by William Gern, for the original function of melatonin and the evolution of other functions is based on the presence of melatonin-synthesizing systems in retinas, parietal eyes, and pineals and on the observations that pineals of more primitive vertebrate groups also are photoreceptive. Evidence suggests that both retinal and pineal melatonin exhibit nighttime (scotophasic) peaks of synthesis. This hypothesis proposes that melatonin was initially a local hormone for regulating the distribution of melanosomes in the retina. During the day, melanosomes are dispersed in retinal pigment cells that protect the photoreceptors from intense light. At night, elevated melatonin causes concentration of melanosomes and allows dim light to maximally stimulate the photoreceptors. This sensitivity of melanosomes to melatonin is retained in the melanophores of skin and brain ependyma of both modern fishes and amphibians. Similar mechanisms presumably would operate in the photoreceptive outer portion of the pineal, in the parietal eye, and in the amphibian frontal organ. The increase in melatonin synthesis during the scotophase causes a greater proportion of melatonin to appear in the blood. Consequently, the scotophasic elevation in melatonin is a reliable internal cue for obtaining information about seasonal photoperiods. Information concerning the length of the scotophase is reflected in circulating melatonin levels. Thus, according to Gern's hypothesis, the diurnal rhythm in melatonin has been co-opted as the bloodborne signal entraining a number of other internal events during the evolution of vertebrates.

parameters of gonadal development and their relationship to photoperiod could be demonstrated in either white-throated sparrows or border canaries.

Neural control of melatonin secretion in some birds also differs markedly in contrast to mammals where NE stimulates pineal secretion. Increased activity of sympathetic fibers from the superior cervical ganglion decreases melatonin secretion in birds, and NE turnover is greatest during the photophase. This inhibitory action of NE appears to involve G_i-proteins in pinealocytes and subsequent reduction in cAMP formation, NAT activity, and melatonin synthesis and release.

Pinealectomy abolishes endogenous body temperature rhythms as well as free-running locomotor activity rhythms in house sparrows (*Passer domesticus*). Effects on locomotor activity appear to involve two pathways, one of which can bypass the pineal organ. Pinealectomized birds that have lost free-running locomotor activity in the dark still exhibit entrainment to light–dark cycles, supporting the presence of a bypass system, presumably a retinal–hypothalamic pathway.

VIII. SUMMARY

The progressive evolution of hypothalamic centers in vertebrates is evident from examination of different vertebrates. The pattern of regulation has branched within the fishes, with the most recent fishes (teleosts) exhibiting predominantly neuroglandular control of tropic hormone release. Primitive fishes, including the chondrichthyeans, and tetrapods have specialized in neurovascular control with the development of a distinct median eminence and a hypophysial portal blood system. The distinct regionalization of tropic cell types in the fish adenohypophysis is less marked in tetrapods. The adenohypophysis of fishes is usually separable into a rostral pars distalis, a proximal pars distalis, and a pars intermedia. The pars tuberalis of tetrapods is absent in fishes but a unique ependymal structure is evident, the saccus vasculosus. Lungfishes lack both a pars tuberalis and the saccus vasculosus.

All of the mammalian tropic hormones have a molecular counterpart among the non-mammals. Many non-mammalian functions can be stimulated by mammalian tropic hormones. The established occurrence of at least one hormone from each category in all gnathostomes (jawed vertebrates) suggests that early in vertebrate evolution three cell types differentiated, and each began elaboration of one of three types of molecules: glycoproteins (TSH, LH, FSH), large peptides (GH, PRL, SL), and POMC-related peptides (ACTH, α-MSH, β-endorphin), respectively). These three primitive molecular types attained functional significance as tropic hormones and gave rise via amino acid substitutions, modified cleavage of prohormones, or both, to the additional hormones that characterize each category.

Non-mammals generally produce the same tropic hormones as do mammals, and there is considerable homology among them. Nevertheless, there is clear evidence of changes in the functions of these molecules related both to structural changes in the tropic hormones themselves and alterations in receptors such that new functions have evolved in some cases.

Regulation of the release of tropic hormones in non-mammals follows the mammalian pattern for the most part, although there seems to be less specificity among receptors. Many of the same neuropeptides are present in the brains of non-mammals as described for mammals but they show greater variations in their effects, often causing release of different tropic hormones than implied by their mammalian names.

The most primitive nonapeptide is AVT, which, unlike other basic peptides, has both pressor and uterotonic activity. Jawless fishes (agnathans) have only AVT, but all other vertebrates studied exhibit at least one basic nonapeptide (AVT in non-mammals) and one neutral nonapeptide (IST in most bony fishes; MST in most amphibians, reptiles, birds, and marsupial mammals; oxytocin in placental mammals). The variability of neutral peptides in elasmobranchs and amphibians suggests that there were numerous forms of nonapeptides in early gnathostomes, most of which were lost in more advanced groups, or that these many variations were peculiar events to the chondrichthyean fishes. Three GPCRs for AVT (VT1a, VT1b, VT2) and one GPCR for MST (MTR) have been described in amphibians.

The major secretory product of the pineal gland by the actions of HIOMT and NAT is the indoleamine melatonin although peptides such as AVT may play roles as well. Pineal secretion is inhibited by light, and the pineal gland plays a major role in mediating seasonal and daily endocrine activity primarily through effects on the hypothalamus. Evidence exists for retinal–hypothalamic pathways regulating pineal function as in mammals, and dual control mechanisms may exists in fishes, amphibians, and reptiles. Melatonin generally acts as an antigonadal (except possibly in some birds) and antithyroid factor. In teleosts and amphibians, melatonin also has effects on body pigmentation. There is evidence for melatonin controlling thermoregulation by teleosts and reptiles.

STUDY QUESTIONS

1. Compare and contrast the anatomical organization of the HP axes among the major vertebrate groups. What major changes have occurred in this system?

2. Discuss the evolution of the tropic hormones for each of the three tropic hormone categories: (a) GpHs, (b) GH-PRL-SL, and (c) POMC.
3. Describe the major actions of each tropic hormone in each vertebrate group. What functions occur uniformly among all groups?
4. How might the evolution of tropic hormones and their receptors be important?
5. Compare and contrast the hypothalamic control over release of each tropic hormone across the major vertebrate groups.
6. Discuss the evolution of the nonapeptide hormones with respect to their structures and their functions.
7. Compare the retinal and pineal patterns of photoperiodic control of mammalian physiology (Chapter 4) with that of non-mammals.

SUGGESTED READING

Books

Bagnara, J.T., Hadley, M.E., 1973. Chromatophores and Color Change: The Comparative Physiology of Animal Pigmentation. Prentice-Hall, Englewood Cliffs, NJ.

Binkley, S.A., 1990. The Clockwork Sparrow: Time, Clocks, and Calendars in Biological Organisms. Prentice Hall, Englewood Cliffs, NJ.

Dawson, A.D., Sharp, P.J., 2005. Functional Avian Endocrinology. Narosa Publishing House, New Dehli.

Holmes, R.L., Ball, J.N., 1974. The Pituitary Gland: A Comparative Account. Cambridge University Press, Cambridge, U.K.

Matsumoto, A., Ishii, S., 1989. Atlas of Endocrine Organs: Vertebrates and Invertebrates. Springer-Verlag, Berlin.

Pickford, G.E., Atz, J.W., 1957. The Physiology of the Pituitary Gland of Fishes. New York Zoological Society, NY.

Articles

General

Green, J.D., 1951. The comparative anatomy of the hypophysis with special reference to its blood supply and innervation. American Journal of Anatomy 88, 225–311.

Gwee, P.-C., Amemiya, C.T., Brenner, B., Venkatesh, B., 2008. Sequence and organization of coelacanth neurohypophysial hormone genes: evolutionary history of the vertebrate neurohypophysial hormone gene locus. BMC Evolutionary Biology 8, 93. http://dx.doi.org/10.1186/1271-2148-8-93.

Insel, T.R., Young, L.J., 2000. Neuropeptides and the evolution of social behavior. Current Opinion in Neurobiology 10, 784–789.

Kawamura, K., Kouki, T., Kawahara, G., Kikuyama, S., 2002. Hypophyseal development in vertebrates from amphibians to mammals. General and Comparative Endocrinology 126, 130–135.

Kawauchi, H., Sower, S.A., 2006. The dawn and evolution of hormones in the adenohypophysis. General and Comparative Endocrinology 148, 3–19.

Nozaki, M., Gorbman, A., 1992. The question of functional homology of Hatschek's pit of amphioxus (*Branchiostoma belcheri*) and the vertebrate adenohypophysis. Zoological Science 9, 387–395.

Nozaki, M., Oshima, Y., Miki, M., Shimotani, T., Kawauchi, H., Sower, S.A., 2005. Distribution of immunoreactive adenohypophysial cell types in the pituitaries of the Atlantic and Pacific hagfish, *Myxine glutinosa and Eptatretus burger*. General and Comparative Endocrinology 143, 142–150.

Oba, Y., Hirai, T., Yoshiura, Y., Kobayashi, T., Nagahama, Y., 2001. Fish gonadotropin and thyrotropin receptors: the evolution of glycoprotein hormone receptors in vertebrates. Comparative Biochemistry and Physiology B 129, 441–448.

Ozaki, M., Ominato, K., Takahashi, A., Kawauchi, H., Sower, S.A., 2001. Adenohypophysial cell types in the lamprey pituitary: current state of the art. Comparative Biochemistry and Physiology B 129, 303–309.

Querat, B., Arai, Y., Henry, A., Akama, Y., Longhurst, T.J., Joss, J.M.P., 2004. Pituitary glycoprotein hormone subunits in the Australian lungfish and estimation of the relative evolution rate of these subunits within vertebrates. Biology of Reproduction 70, 356–363.

Santos, S.D., Bardet, C., Bertrand, S., Escriva, H., Habert, D., Querat, B., 2009. Distinct expression patterns of glycoprotein hormone-2 and -5 in a basal chordate suggest independent developmental functions. Endocrinology 150, 3815–3822.

Schreibman, M.P., 1986. Pituitary gland. In: Pang, P.K.T., Schreibman, M. (Eds.), Vertebrate Endocrinology: Fundamentals and Biomedical Implications. Morphological Considerations, Vol. 1. Academic Press, Orlando, FL, pp. 11–56.

Sower, S.A., Fremat, M., Kavanaugh, S.I., 2009. The origins of the vertebrate hypothalamic–pituitary–gonadal (HPG) and hypothalamic–pituitary–thyroid (HPT) endocrine systems: new insights from lampreys. General and Comparative Endocrinology 161, 20–29.

Sower, S.A., Kawauchi, H., 2001. Update: Brain and pituitary hormones of lampreys. Comparative Biochemistry and Physiology B 129, 291–302.

Sower, S.A., Kawauchi, H., 2011. Reproduction in agnathan fishes: lampreys and hagfishes. In: Norris, D.O., Lopez, K.H. (Eds.), Hormones and Reproduction of Vertebrates. Fishes, Vol. 1. Academic Press, San Diego, CA, pp. 193–208.

Sueiro, C., Carrera, I., Ferreiro, S., Molist, P., Adrio, F., Anadon, R., Rodriguez-Moldes, I., 2007. New insights on saccus vasculosus evolution: a developmental and immunohistochemical study in elasmobranchs. Brain, Behavior and Evolution 70, 178–204.

Tando, Y., Kubokawa, K., 2009. Expression of the gene for ancestral glycoprotein hormone β subunit in the nerve cord of amphioxus. General and Comparative Endocrinology 162, 328–339.

Gonadotropins

Kanda, S., Okubo, K., Oka, Y., 2011. Differential regulation of the luteinizing hormone genes in teleosts and tetrapods due to their distinct genomic environments: insights into gonadotropin beta subunit evolution. General and Comparative Endocrinology 173, 253–258.

Levavi-Sivan, B., Bogerd, J., Mañanos, E.L., Gomez, A., Lareyre, J.J., 2010. Perspectives on fish gonadotropins and their receptors. General and Comparative Endocrinology 165, 412–437.

Zhang, L., Kessler, A.E., Tsai, P.-S., 2007. Characterization and steroidal regulation of gonadotropin beta subunits in the male leopard frog. *Rana pipiens*. General and Comparative Endocrinology 150, 66–74.

Thyrotropin

Han, Y.-S., Liao, I.-C., Tzeng, W.-N., Yu, J.Y.-L., 2004. Cloning of the DNA for thyroid stimulating hormone β-subunit and changes in activity of the pituitary–thyroid axis during silvering of the Japanese eel, *Anguilla japonica*. Journal of Molecular Endocrinology 32, 179–194.

Okada, R., Kobayashi, T., Yamamoto, K., Nakakura, T., Tanaka, S., Vaudry, H., Kikuyama, S., 2009. Neuroendocrine regulation of thyroid-stimulating hormone secretion in amphibians. Annals of the New York Academy of Sciences 1163, 262–270.

Growth Hormone and Prolactin

Daza, D.O., Sundström, G., Larsson, T.A., Larhammar, D., 2009. Evolution of the growth hormone–prolactin–somatostatin system in relation to vertebrate tetraploidizations. Annals of the New York Academy of Sciences 1163, 491–493.

Gahete, M.D., Duran-Prado, M., Luque, R.M., Martinez-Fuentes, A.J., Quintero, A., Gutierrez-Pascual, E., Cordoba-Chacon, J., Malagon, M.M., Garcia-Navarro, F., Castaño, J.P., 2009. Understanding the multifactorial control of growth hormone release by somatotropes: lessons from comparative endocrinology. Annals of the New York Academy of Sciences 1163, 137–153.

Laiz-Carrion, R., del Mar Segura-Noguera, M., del Pilar Martin del Rio, M., Mancera, J.M., 2003. Ontogeny of adenohypophyseal cells in the pituitary of the American shad (*Alosa sapidissima*). General and Comparative Endocrinology 132, 454–464.

Manzon, L.A., 2002. The role of prolactin in fish osmoregulation: a review. General and Comparative Endocrinology 125, 291–310.

Wan, G., Chan, K.M., 2010. A study of somatolactin actions by ectopic expression in transgenic zebrafish larvae. Journal of Molecular Endocrinology 45, 301–315.

Yang, L., Meng, Z., Liu, Y., Zhang, Y., Liu, X., Lu, D., 2010. Growth hormone and prolactin in *Andrias davidianus*: cDNA cloning, tissue distribution and phylogenetic analysis. General and Comparative Endocrinology 165, 177–180.

Zhu, Y., Stiller, J.W., Shaner, M.P., Baldini, A., Scemama, J.-L., Capehart, A.A., 2004. Cloning of somatolactin and cDNAs in zebrafish and phylogenetic analysis of two distinct somatolactin subtypes in fish. Journal of Endocrinology 182, 509–518.

POMC Peptides

Dores, R.M., Baron, A.J., 2011. Evolution of POMC: origin, phylogeny, posttranslational processing, and the melanocortins. Annals of the New York Academy of Sciences 1220, 34–48.

Dores, R.M., Qais Majeed, Q., Komorowski, L., 2011. Observations on the radiation of lobe-finned fishes, ray-finned fishes, and cartilaginous fishes: phylogeny of the opioid/orphanin gene family and the 2R hypothesis. General and Comparative Endocrinology 170, 253–264.

Jenks, B.G., Galas, L., Kuribara, M., Desrues, L., Kidane, A.H., Vaudry, H., Scheenen, W.J., Roubos, E.W., Tonon, M.C., 2011. Analysis of the melanotrope cell neuroendocrine interface in two amphibian species, *Rana ridibunda* and *Xenopus laevis*: a celebration of 35 years of collaborative research. General and Comparative Endocrinology 170, 57–67.

Klovens, J., Schiöth, H.B., 2005. Agouti-related proteins (AGRPs) and Agouti-signaling peptide (ASIP) in fish and chicken. Annals of the New York Academy of Sciences 1040, 363–367.

Takahashi, A., Kawauchi, H., 2006. Evolution of melanocortin systems in fish. General and Comparative Endocrinology 148, 185–194.

Takahashi, A., Nakata, O., Moriyama, S., Nozaki, M., Joss, J.M.P., Sower, S.A., Kawauchi, H., 2006. Occurrence of two functionally distinct proopiomelanocortin genes in all modern lampreys. General and Comparative Endocrinology 148, 72–78.

TRH

Galas, L., Raoult, E., Tonon, M.-C., Okada, R., Jenks, B.G., Castaño, J.P., Kikuyama, S., Malagon, M., Roubos, E.W., Vaudry, H., 2009. TRH acts as a multifunctional hypophysiotropic factor in vertebrates. General and Comparative Endocrinology 164, 40–50.

GnRH, GnIH, and Kisspeptins

Bentley, G.E., Jensen, J.P., Kaur, G.J., Wacker, D.W., Tsutsui, K., Wingfield, J.C., 2006. Rapid inhibition of female sexual behavior by gonadotropin-inhibitory hormone (GnIH). Hormones and Behavior 49, 550–555.

Biran, J., Ben-Dor, S., Levavi-Sivan, B., 2008. Molecular identification and functional characterization of the kisspeptin/kisspeptin receptor system in lower vertebrates. Biology of Reproduction 79, 776–786.

Dufour, S., Weltzien, F.A., Seibert, M.E., Le Belle, N., Vidal, B., Vernier, P., Pasqualini, C., 2005. Dopaminergic inhibition of reproduction in teleost fishes: ecophysiological and evolutionary implications. Annals of the New York Academy of Sciences 1040, 9–21.

Fernald, R.D., White, R.B., 1999. Gonadotropin-releasing hormone genes: phylogeny, structure, and functions. Frontiers in Neuroendocrinology 20, 224–240.

Filby, A.L., van Aerle, R., Duitman, J., Tyler, C.R., 2008. The kisspeptin/gonadotropin-releasing hormone pathway and molecular signaling of puberty in fish. Biology of Reproduction 78, 278–289.

Gorbman, A., Sower, S.A., 2003. Evolution of the role of GnRH in animal (metazoan) biology. General and Comparative Endocrinology 134, 207–213.

Kah, O., Dufour, S., 2011. Conserved and divergent features of reproductive neuroendocrinology in teleost fishes. In: Norris, D.O., Lopez, K.H. (Eds.), Hormones and Reproduction of Vertebrates. Fishes, Vol 1. Academic Press, San Diego, CA, pp. 15–42.

Kim, D.-K., Cho, E.B., Moon, M.J., Sumi Park, S., Hwang, J.-I., Kah, O., Sower, S.A., Vaudry, H., Young Seong, J.Y., 2011. Revisiting the evolution of gonadotropin-releasing hormones and their receptors in vertebrates: secrets hidden in genomes. General and Comparative Endocrinology 170, 68–78.

Millar, R.P., 2005. GnRHs and GnRH receptors. Animal Reproduction Science 88, 5–28.

Northcutt, R.G., Muske, L.E., 1994. Multiple embryonic origins of gonadotropin-releasing hormone (GnRH) immunoreactive neurons. Developmental Brain Research 78, 279–290.

Osugi, T., Daukss, D., Gazda, K., Ubuka, T., Kosugi, T., Nozaki, M., Sower, S.A., Tsutsui, K.L., 2012. Evolutionary origin of the structure and function of gonadotropin-inhibitory hormone: insights from lampreys. Endocrinology 153 (5), 2362–2374.

Pawson, A.J., Morgan, K., Maudsley, S.R., Millar, R.P., 2003. Type II gonadotropin-releasing hormone (GnRH-II) in reproductive biology. Reproduction 126, 271–278.

Roch, G.J., Busby, E.R., Sherwood, N.M., 2011. Evolution of GnRH: diving deeper. General and Comparative Endocrinology 171, 1–16.

Servili, A., et al., 2011. Organization of two independent kisspeptin systems derived from evolutionary-ancient kiss genes in the brain of zebrafish. Endocrinology 152, 1527–1540.

Somoza, G.M., Miranda, L.A., Strobl-Mazzulla, P., Guilgur, L.G., 2002. Gonadotropin-releasing hormone (GnRH): from fish to mammalian brains. Cellular and Molecular Neurobiology 22, 589–609.

Terasawa, E., Kurian, J.R., Guierriero, K.A., Kenealy, B.P., Hutz, E.D., Keen, K.L., 2010. Recent discoveries on the control of gonadotropin-releasing hormone neurons in nonhuman primates. Journal of Neuroendocrinology 22, 630–638.

Whitlock, K.E., 2005. Origin and development of GnRH neurons. Trends in Endocrinology and Metabolism 16, 145–151.

Yaron, Z., Gur, G., Matemed, P., Rosenfeld, H., Elizur, A., Levavi-Sivan, B., 2003. Regulation of fish gonadotropin. International Review of Cytology 225, 131–185.

Zohar, Y., Muñoz-Cueto, J.A., Elizur, A., Kah, O., 2010. Neuroendocrinology of reproduction in teleost fish. General and Comparative Endocrinology 165, 438–455.

CRH and Urocortins

Boorse, G.C., Crespi, E.J., Dautzenberg, F.M., Denver, R.J., 2005. Urocortins of the South African clawed frog, *Xenopus laevis*: conservation of structure and function in tetrapod evolution. Endocrinology 146, 4851–4860.

Balment, R.J., Song, W., Ashton, N., 2005. Urotensin II. Ancient hormone with new functions in vertebrate body fluid regulation. Annals of the New York Academy of Sciences 1040, 66–73.

De Groef, B., Van der Geyten, S., Darras, V.M., Kühn, E.R., 2006. Role of corticotropin-releasing hormone as a thyrotropin-releasing factor in non-mammalian vertebrates. General and Comparative Endocrinology 146, 62–68.

Denver, R.J., 2009. Stress hormones mediate environment–genotype interactions during amphibian development. General and Comparative Endocrinology 164, 20–31.

Lovejoy, D.A., Jahan, S., 2006. Phylogeny of the corticotropin-releasing factor family of peptides in the metazoa. General and Comparative Endocrinology 146, 1–8.

Lovejoy, D.A., Rotzinger, S., Barsyte-Lovejoy, D., 2009. Teneurin C-terminal-associated peptides and corticotropin-releasing factor superfamilies. Annals of the New York Academy of Sciences 1163, 215–220.

Lowry, C.A., Moore, F.L., 2006. Regulation of behavioral responses by corticotropin-releasing factor. General and Comparative Endocrinology 146, 19–27.

Okada, R., Miller, M.F., Yamamoto, K., De Groef, B., Denver, R.J., Kikuyama, S., 2007. Involvement of the corticotropin-releasing factor (CRF) type 2 receptor in CRF-induced thyrotropin release by the amphibian pituitary gland. General and Comparative Endocrinology 150, 437–444.

Neurohypophysial Peptides

Acher, R., Chauvet, J., Rouille, Y., 1997. Adaptive evolution of water homeostasis regulation in amphibians: vasotocin and hydrins. Biology of the Cell 89, 283–291.

Boyd, S.K., 2013. Amphibian neurohypophysial peptides. In: Kastin, A. (Ed.), The Handbook of Biologically Active Peptides, second ed. Elsevier, Burlington, MA.

Boyd, S.K., 2013. Vasotocin modulation of social behaviors in amphibians. In: Choleris, E., Pfaff, D., Kavaliers, M. (Eds.), Oxytocin, Vasopressin and Related Peptides in the Regulation of Behavior. Cambridge University Press, Cambridge, U.K.

Gwee, P.-C., Amemiyat, C.T., Brenner, S., Venkatsch, B., 2008. Sequence and organization of coelacanth neurohypophysial hormone genes: evolutionary history of the vertebrate neurohypophysial gene locus. BMC Evolutionary Biology 8, 93. http://dx.doi.org/10.1186/1471-2148-8-93.

Srivastava, R., Cornett, L.E., Chaturvedi, C.M., 2010. Age-dependent expression of AVT and its oxytocic-like receptor VT3 in the shell gland of Japanese quail. Corturnix coturnix japonica. General and Comparative Endocrinology 165, 47–52.

Urano, A., Ando, H., 2011. Diversity of the hypothalamo-neurohypophysial system and its hormonal genes. General and Comparative Endocrinology 170, 41–56.

Pineal

Bromage, N., Porter, M., Randall, C., 2001. Seasonal variations in the pattern of melatonin secretion melatonin cycle: the environmental regulation of maturation in farmed finfish with special reference to the role of photoperiod and melatonin. Aquaculture 197, 63–98.

Falcon, J., Miguad, H., Muñoz-Cueto, J.A., Carillo, M., 2010. Current knowledge on the melatonin system in teleost fish. General and Comparative Endocrinology 165, 469–482.

Hazlerigg, D.G., Wagner, G.C., 2006. Seasonal photoperiodism in vertebrates: from coincidence to amplitude. Trends in Endocrinology and Metabolism 17, 83–91.

Maronde, E., Stehle, J.H., 2007. The mammalian pineal gland: known facts, unknown facets. Trends in Endocrinology and Metabolism 18, 142–149.

Chapter 6

The Hypothalamus–Pituitary–Thyroid (HPT) Axis of Mammals

The thyroid gland (Figure 6-1) is unique among vertebrate endocrine glands in that it stores its secretory products (thyroid hormones) extracellularly. It is possibly the most highly vascularized endocrine gland in mammals and appears to be one of the oldest vertebrate endocrine glands phylogenetically (see Chapter 7). Thyroid function is regulated by the **hypothalamus–pituitary–thyroid (HPT)** axis (see Chapter 4). **Thyrotropin (TSH)** from the pituitary stimulates the synthesis of **tetraiodothyronine** or **thyroxine (T_4).** To synthesize T_4, the amino acid tyrosine is incorporated into a glycoprotein called **thyroglobulin (Tgb)**. **Tyrosine** is iodinated and then two iodinated tyrosines are enzymatically linked together to form T_4. Some T_4 is partially deiodinated to form the more active thyroid hormone **triiodothyronine (T_3)** prior to release from the thyroid gland. The structure of these compounds and the sequence of their synthesis are provided in Chapter 3 (Figure 3-32).

Thyroid hormones influence many aspects of reproduction, growth, differentiation, and metabolism. Many of these actions occur cooperatively with other hormones, and the thyroid hormones enhance their effectiveness. This cooperative role for thyroid hormones is referred to as a **permissive action** whereby thyroid hormones produce changes in target tissues that "allow" these tissues to be more responsive to another hormone, to neural stimulation, or possibly to certain environmental stimuli such as light. Thyroid hormones may maintain maximal sensitivity to other regulating agents in many types of tissues. The importance of thyroid hormones is reflected in the observation that the incidence of thyroid disease in humans is exceeded only by the incidence of metabolic syndrome and diabetes mellitus (see Chapter 12 for a discussion of these latter disorders).

Although rarely lethal, thyroid disorders have widespread effects in humans due to their many actions with other hormones. Of the more than 13 million thyroid disorders diagnosed in the United States, 11 million occur in women. This difference in incidence between males and females is not understood, although many cases are linked to pregnancy. Thyroid deficiencies that occur during pregnancy can have devastating effects on the development of the nervous system of the fetus, leading to extensive mental deficiencies. Thyroid deficiencies develop so gradually that they often are not recognizable, even by physicians.

I. SOME HISTORICAL ASPECTS OF THYROID PHYSIOLOGY

Either deficient or excessive production of thyroid hormones may lead to serious pathological states with overt symptoms (Table 6-1). The first description of thyroid disease was of abnormal enlargement of the thyroid in humans recognized by Chinese physicians about 3000 BC. As a remedy, they recommended ingestion of seaweed and burned sponge or desiccated deer thyroids. The first two substances contained therapeutic quantities of iodide and the last sufficient thyroid hormones to alleviate the pathological symptoms in most cases. Hypothyroid deficiencies of this sort were not recognized in Western culture as clinical disorders until many centuries later. The **cretinism** syndrome was described clinically in Europe in 1526. Cretinism is manifest very early in life as a consequence of severe thyroid deficiency. This syndrome is characterized by dwarfism and a number of other physical abnormalities in addition to severe mental retardation, slow mental and physical activity, bradycardia (slowing of heart beat), and hypothermia (decreased body temperature). From 1880 to 1890, another clinical disorder in adults, **myxedema**, was linked to hypothyroid function. Myxedematous symptoms in adults are related to abnormal accumulation of water and protein throughout the body as well as to other disturbances in general metabolism. Accumulations of protein and fluid in the skin alter facial features, causing the patient to appear expressionless. In later stages of the disorder, the sufferer becomes less interested in both self and environment and, if untreated, would eventually enter a coma and die. **Juvenile myxedema** is similar to cretinism except that early growth and development are normal but severe retardation develops in later childhood. All of these clinical syndromes have the same basic cause: hypofunctioning of the thyroid gland.

Bauman discovered in 1896 that an organic iodine-containing compound could be extracted from thyroid glands. Subsequently, it was demonstrated that this "thyroidin" substance could reverse the adverse effects of

Vertebrate Endocrinology. http://dx.doi.org/10.1016/B978-0-12-394815-1.00006-9

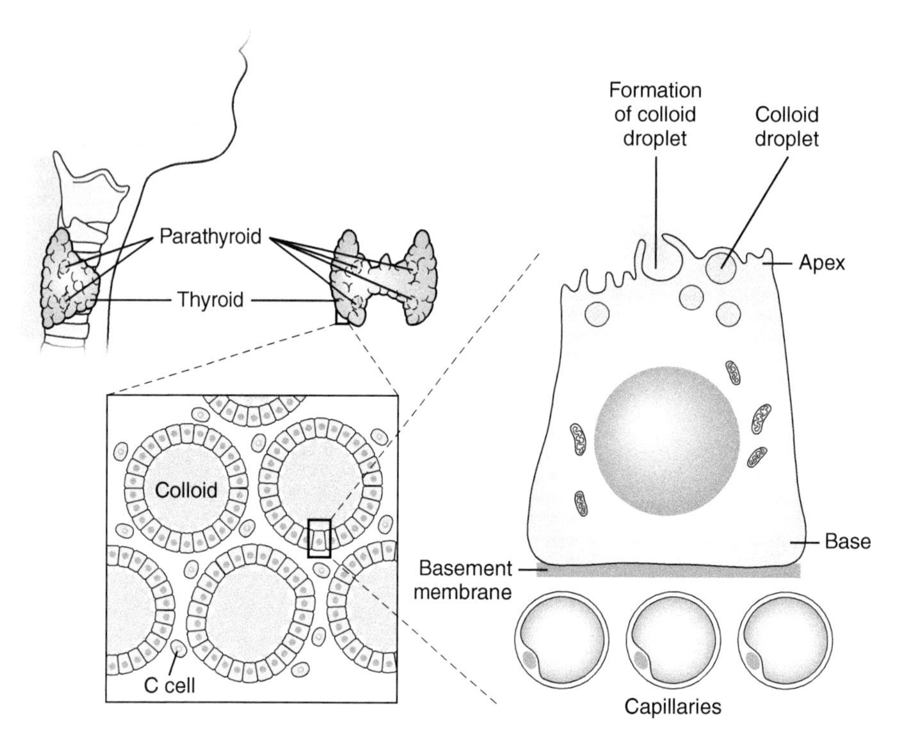

FIGURE 6-1 **The mammalian thyroid.** The thyroid gland is located in the neck region. It consists of many hollow follicles, each of which is filled with a proteinaceous fluid called colloid which is secreted by the follicle cells. Thyroxine synthesized by the follicle cells is stored in the colloid. The C-cells or parafollicular cells are of ultimobranchial origin and secrete the calcium-regulating hormone, calcitonin (see Chapter 14). *(Adapted with permission from McNabb, F.M.A., "Thyroid Hormones," Prentice Hall, Upper Saddle River, NJ, 1993.)*

TABLE 6-1 Symptoms of Thyroid Deficiency and Hyperactivity in Humans

Type of Symptom	Hypothyroid	Hyperthyroid
Appearance	Myxedema; deficient growth	Exophthalmos
Behavior	Mental retardation; mentally and physically sluggish; somnolent; sensitive to cold	Often quick mentally; restless, irritable, anxious, hyperkinetic; wakeful; sensitive to heat
Metabolism	Hypophagia; low basal metabolic rate; reduced QO_2[a] of liver, kidney, and muscle *in vitro*; decrease in oxidative enzymes; constipation	Hyperphagia; high basal metabolic rate; increased QO_2 of liver, kidney, and muscle *in vitro*; increased oxidative enzymes; diarrhea
Muscle function	Weakness; hypotonia	Weakness; fibrillary twitchings, tremors

[a] *QO_2, respiratory quotient.*

iodide deficiency. In the early 1900s, the thyroid gland and its hormones were implicated in elevating basal metabolic rate, primarily through effects on certain tissues (e.g., liver, kidney, muscle). This observation has strongly influenced the direction of thyroid research in mammals as well as in many non-mammalian vertebrates. The action of thyroid hormones on metabolism is reflected in clinical thyroid states (Table 6-1).

The iodine-containing hormone T_4 was isolated, crystallized, and reported by Edward C. Kendall in 1915. This event marked a significant milestone, not only in thyroid research but also in endocrinology as a whole, for T_4 was the first hormone to be isolated in pure form. It was not until 1952, however, that the second thyroid hormone, T_3, was identified by J. Gross and R. Pitt-Rivers. This second thyroid hormone was soon found to be more potent than T_4

as a thyroid hormone and is considered by many to be the active form of thyroid hormone (see ahead).

It was the discovery of **anti-thyroid drugs** in the early 1940s, as well as the ready availability of radioactive isotopes of iodide (**radioiodide**) following developments in nuclear physics, that provided diagnostic tools for assessing thyroid function (radioiodide) and chemicals (anti-thyroid drugs) suitable for blocking thyroid function. Radioiodide proved useful for clinical and laboratory work and provided a label for thyroid molecules that made it possible to elucidate the details of thyroid hormone synthesis, metabolism, and mechanisms of action.

II. DEVELOPMENT AND ORGANIZATION OF THE MAMMALIAN THYROID GLAND

The mammalian thyroid gland consists of many epithelial follicles encapsulated by a connective tissue sheath. The gland is highly vascularized, with a dense capillary network surrounding each follicle. The thyroid vasculature receives cholinergic innervation, and the follicle cells receive adrenergic (norepinephrine and dopamine) and peptidergic (e.g., **neuropeptide Y, NPY; vasoactive inhibitory peptide, VIP**) innervation. Direct innervation of the follicle cells may play an important role in overall thyroid regulation (see ahead).

Development of the thyroid gland begins by formation of a ventral bud in the floor of the embryonic pharynx (endoderm) between the first and second pharyngeal pouches. The gland initially differentiates as cellular cords that later separate into clusters of cells destined to become thyroid follicles. The cells of a cluster secrete a protein-rich fluid termed **colloid** that accumulates extracellularly in the center of the cluster. This secretory activity eventually leads to formation of a colloid-filled space, the **lumen** of the follicle, surrounded by a single layer of epithelial cells, the **thyroid epithelium** (Figure 6-1). The portion of the follicular cell that borders on the lumen of the follicle is known as the apex or **apical portion**. The apical portion of each thyroid follicle cell is linked to the adjacent cell by occluding or tight junctions, thus forming an impenetrable barrier between the cells. The nucleus is generally found in the **basal portion** of the cell that is farthest from the lumen and closest to the capillaries that surround each follicle.

In addition to capillaries and follicles, **parafollicular** or **C cells** occur in the regions between or adjacent to the follicles. Parafollicular cells may occur within the follicular epithelium or may even form separate follicular structures in some species. These cells are derived from another pharyngeal derivative, the **ultimobranchial body**, and secrete a hypocalcemic hormone, **calcitonin**, that influences calcium metabolism (see Chapter 14). A comparison of parafollicular cells and follicular cells (Table 6-2) emphasizes their different structural and functional features. In some mammals, the parathyroid glands also may be embedded within the mass of the thyroid (Figures 6-1 and 6-2). The parathyroids, like the parafollicular cells, have their origin nearby from the embryonic pharynx and in some species become embedded in the mass of thyroid follicles during development. The parathyroid glands secrete the hypercalcemic hormone **parathyroid hormone (PTH)** and are also discussed in Chapter 14. Unlike the thyroid follicular cells that appear to be of endodermal origin, the parafollicular cells and the secretory cells of the parathyroid glands actually are of neural crest origin and migrate into the embryonic glands (see Chapter 4, Box 4A).

TABLE 6-2 Comparison of Characteristics of Thyroid Follicular and Parafollicular Cells

Thyroid Follicular Cell	Thyroid Parafollicular Cell
Absence of secretion granules	Large number of eosinophilic granules, 0.2-μm diameter; stain with silver nitrate
Endoplasmic reticulum cisternae of larger diameter, containing flocculent precipitate like that found in albumin-secreting cells	Many mitochondria and high level of the mitochondrial enzyme α-glycerophosphate dehydrogenase
Carbohydrate added at Golgi apparatus, which is rather inconspicuous in these cells	No lumenal surface present
	Nucleus more irregular in outline than those of follicular cells
Enlargement of Golgi apparatus from TSH treatment	Golgi apparatus prominent
Binds antibody to thyroglobulin but not to calcitonin	Binds antibody to calcitonin
Cytology not altered by high blood calcium level	Degranulation due to high blood calcium level
Readily accumulates iodide	

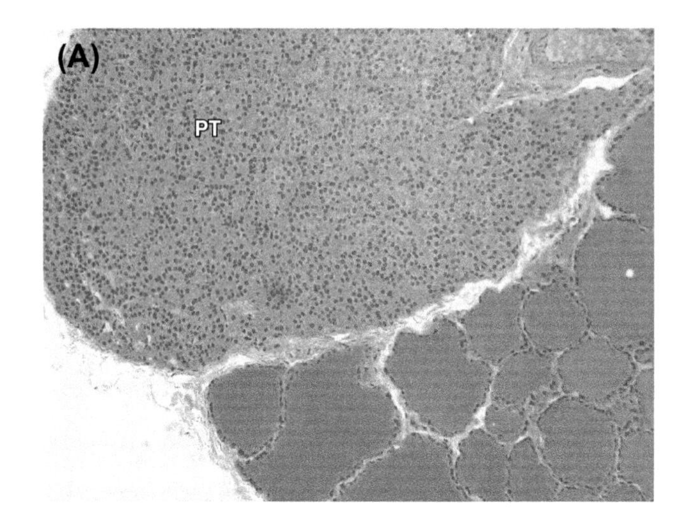

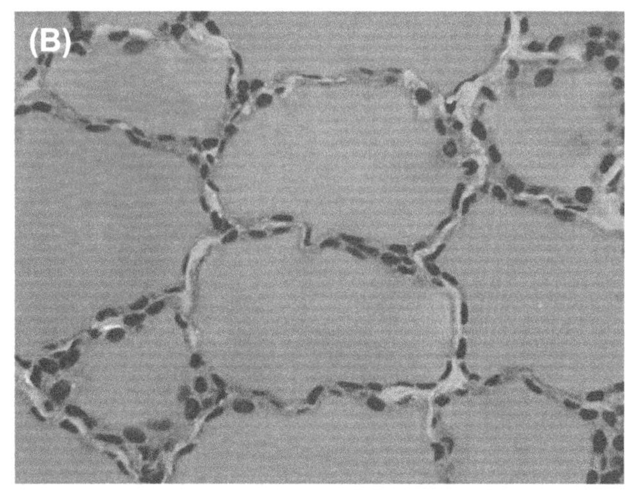

FIGURE 6-2 **Thyroid and parathyroid glands.** (A) Low magnification of compact parathyroid gland (above) embedded in the thyroid gland consisting of colloid-filled follicles (below). (B) High magnification of thyroid follicles with squamous epithelium surrounding colloid.

III. SYNTHESIS, SECRETION, ACTION, AND METABOLISM OF THYROID HORMONES

The events related to the ability of thyroid follicles to synthesize and release thyroid hormones are discussed separately for simplicity, but it is important to keep in mind that many of these events may be occurring simultaneously (see Figure 6-3). Details of the synthesis may be found in Chapter 3 and only a brief summary is provided here in relation to cytological events. The processes discussed in this section include:

1. Accumulation of inorganic iodide by follicular cells
2. Synthesis of Tgb, a glycoprotein that contains numerous tyrosine residues for hormone synthesis
3. Binding of inorganic iodide to tyrosine residues in Tgb
4. Synthesis of T_4 from iodinated tyrosines
5. Storage of Tgb containing T_4 in the lumen of the follicle
6. Engulfing of colloid by follicular cells and hydrolysis of Tgb to release T_4 and subsequent conversion of some T_4 to T_3
7. Entry of T_3 and T_4 into the general circulation and their transport to targets

A. Dietary Iodide and Iodide Uptake

The principal source for inorganic iodide is dietary. In certain portions of the world, environmental iodide, normally obtained through consumption of plants, is in short supply in the soil and hence in plants grown there—for example, the Great Lakes and Rocky Mountain regions of the United States and much of northeastern Europe, as well as regions of Australia, Africa, and Asia (Figure 6-4). Consequently, in these regions, human diets are low in naturally occurring iodide, and hypothyroid states commonly are encountered unless an iodide supplement is used. Iodide deficiency prevents normal synthesis of thyroid hormones, and feedback mechanisms cause excessive stimulation of thyroid glands by TSH. At one time, hypothyroid **goiters** or enlarged thyroids were common in people who inhabited these low-iodide regions, or "goiter belts," but the addition of iodized salt to the diet restores thyroid hormone synthesis to normal and has almost eliminated this condition in developed countries. (The term "goiter" or "goitre" originally meant any tumor or abnormal glandular enlargement in the neck but has come to mean an enlarged thyroid.) Unfortunately, in spite of efforts to provide supplemental iodide to human diets, hypothyroidism due to low iodide still is rampant in many Asian and African countries where it has devastating effects on the well-being of millions of people. Tendencies to use sea salt and other types of salts naturally low in iodide instead of common iodized salt in some countries (e.g., United States, Australia) also places some at risk for hypothyroidism. Although not specifically iodized, processed cows milk now contains less iodide because milk processing facilities have stopped using iodide-based antibacterial cleaning agents in favor of non-iodide products.

Once ingested, inorganic iodide is readily absorbed from the intestine into the blood and is selectively accumulated by thyroid follicular cells. Energy-dependent, active transport mechanisms in both the follicular basal and apical cell membranes are specific for iodide. Iodide is co-transported with Na^+ (**Na-I symporter**, **NIS**) at the basal membrane (see Box 6A) and transported across the apical membrane into the colloid by an apical porter called

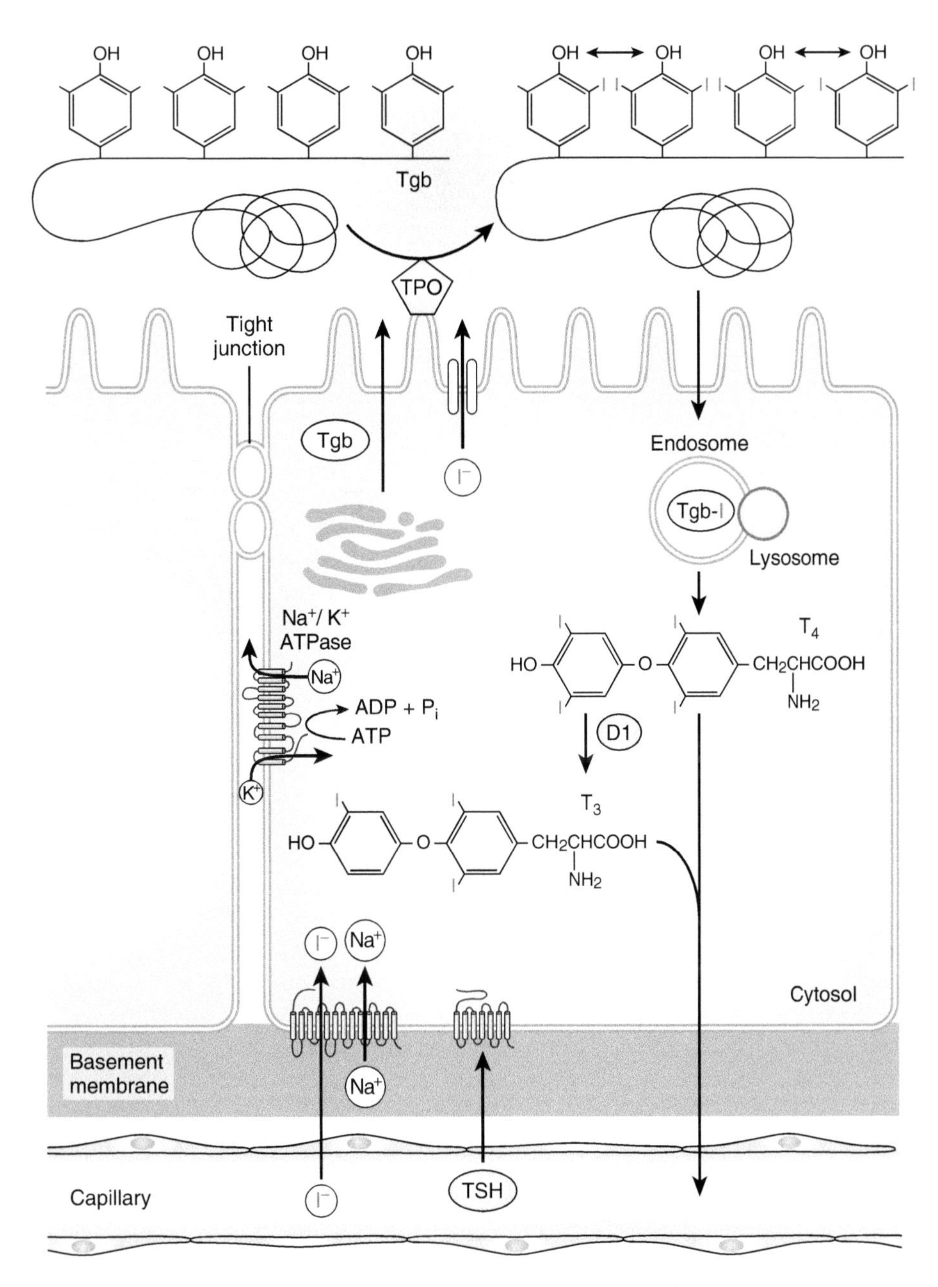

FIGURE 6-3 **Thyroid hormone biosynthesis.** The sodium iodide symporter (NIS) transports Na^+ and I^- across the basolateral plasma membrane of a follicular cell. The Na^+/K^+ ATPase maintains the sodium diffusion gradient required for operation of the NIS. The enzyme thyroid peroxidase (TPO) located at the apical surface is responsible for activating I^-, for iodinating thyroglobulin (Tgb), and for coupling iodinated tyrosines to form T_4. Release of thyroid hormones requires engulfing colloid (endocytosis) to form intracellular endosomes that merge with lysosomes to form an endolysosome. This results in degradation of Tgb and liberation of T_4 into the cytosol, where a type-1 deiodinase (D1) converts some of it to T_3 and rT_3 (not shown). These products then pass from the basal surface of the cell into the blood.

pendrin. Increased amounts of Tgb in the follicular lumen increases production of pendrin. Thus, uptake of iodide from the blood is enhanced by translocation of inorganic iodide at the apical surface and its conversion to organically bound forms (that is, iodinated tyrosines). The resulting concentration of iodide in the thyroid gland normally exceeds plasma levels by 20 to 40 times and under some conditions may be even greater. Thyroidal iodide accumulation is not affected by other halide anions including Cl^-, Br^-, and Fl^-, but the iodide pump can be inhibited by excessive amounts of iodide. Patients with **pendrin syndrome** have a defective pendrin gene; they are unable to retain iodide in the thyroid gland and hence are thyroid hormone deficient.

BOX 6A Iodide Transport Defects

Iodide transport defects (ITDs) are disorders caused by problems with NIS function. The NIS protein is a large 13-transmembrane protein encoded by a gene (*SLC5A5*) within a large family of sodium dependent transporters. Any mutation that alters NIS glycosylation sites (Asn sites in extracellular loops 4 and 7), I^- or Na^+ transport sites, or sites required for translocation to the plasma membrane could in theory impair normal iodide transport into thyroid follicle cells.

Patients with ITDs exhibit virtually no iodide uptake by the thyroid gland and display hypothyroidism and possibly goiter. ITDs may be caused by a variety of factors, including inability to translate NIS mRNA, problems in the co-transport of Na^+ by the NIS that is required for I^- transport, problems with translocation of the NIS to the basolateral plasma membrane of thyroid follicle cells, or truncation of parts of the NIS protein rendering it unable to transport either Na^+ or I^-. Many ITDs are inherited. Box Table 6A-1 shows some of the known mutations in the human NIS and the aberrant phenotypes that arise from these mutations. Box Figure 6A-1 shows the locations of these mutations within the secondary structure of the NIS protein. One mutation that has gained recent interest is G93R, which causes a switch from glycine to arginine at position 93. G93R-mutated NIS proteins travel normally to the plasma membrane but are entirely inactive in transporting iodide. The G93 position appears to play a critical role in the conformational change in the NIS that takes place as Na^+ and I^- move from the outside to the inside of the thyrocyte.

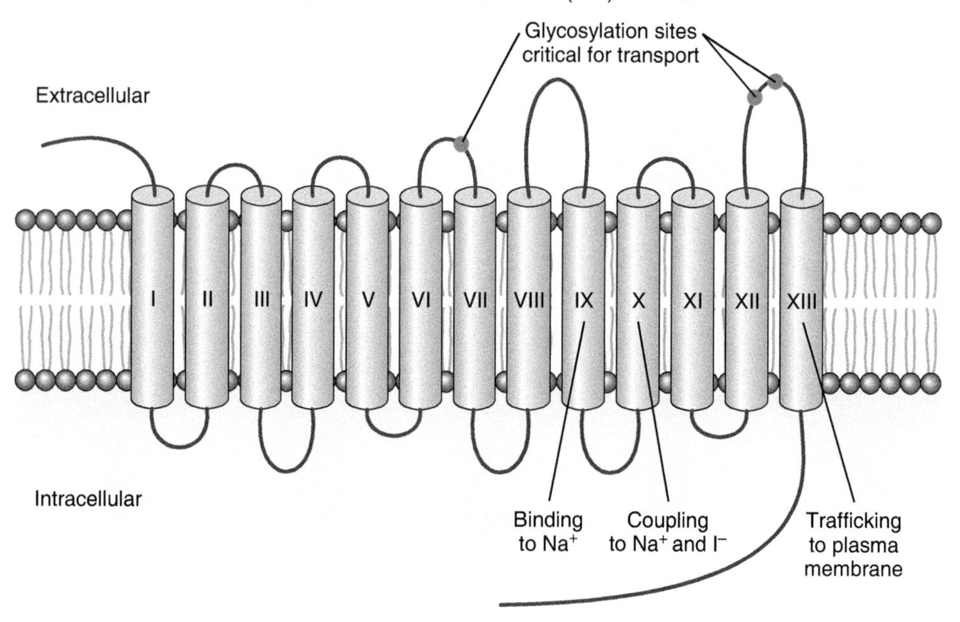

BOX FIGURE 6A-1 **Secondary structure of the human sodium iodide symporter (NIS) protein.** Each of the 13 transmembrane domains is labeled by a roman numeral. Mutations in the NIS known to cause iodide transport defects (ITDs) have led to a better understanding of the functional regions of the NIS protein as indicated. *Adapted with permission from Spitzweg, C., Morris, J.C., 2010. Genetics and phenomics of hypothyroidism and goiter due to NIS mutations. Molecular and Cellular Endocrinology 322., 56-63.*

BOX TABLE 6A-1 Some Known Mutations in Human NIS Leading to Iodide Transport Defects (ITDs)

Mutation in hNIS	Mutation Phenotype	Patient's Country of Origin
V59E	ITD; reduced I^- uptake, exhibits altered intramembrane helix-helix interaction	Japan
G93R	ITD; reduced I^- uptake, NIS unable to undergo conformational change during Na^+/I^- exchange	Japan
R124H	ITD; no perchlorate sensitive I^- uptake, cause unknown	Cameroon
Q267E	ITD; reduced NIS turnover number	Mexico
C272X	ITD; produces truncated protein devoid of I^- transport	Brazil

BOX TABLE 6A-1 Some Known Mutations in Human NIS Leading to Iodide Transport Defects (ITDs)—cont'd

Mutation in hNIS	Mutation Phenotype	Patient's Country of Origin
T354P	ITD; correct targeting to plasma membrane but cannot transport I^- due to lack of Na^+ binding	Japan
G395R	ITD; Substitution of uncharged amino acid with small side chain at position 395 with larger amino acid results in lower Vmax but no change in Km for either I^- and Na^+ transport, coupling of Na^+/I^- transport is presumably affected	Canada
G543E	ITD; trafficking defect, NIS not targeted to plasma membrane	Japan

Mutations are read from left to right. For example, V59E means a change from valine to glutamic acid at amino acid position 59. TMS, Transmembrane segment.

1. Measurement of Iodide Dynamics

The events of iodide uptake, accumulation, and binding to tyrosine initially were elucidated using radioiodide. Uptake of radioiodide, like that of the normal isotope (^{127}I), is stimulated by TSH from the adenohypophysis, and there is no discrimination among the various radioiodide isotopes in the formation of organically bound iodide associated with thyroid hormone synthesis. For many years, the radioisotope employed most frequently for iodide uptake studies was ^{131}I, a strong beta- and gamma-emitting isotope with a short radiation half-life (about 8 days). This particular isotope could be detected in blood or tissues with relatively unsophisticated detection equipment because of the high-energy gamma radiation it emits. Other isotopes, ^{125}I and ^{123}I, now are employed in most thyroid studies, making use of the lower energy radiation they produce.

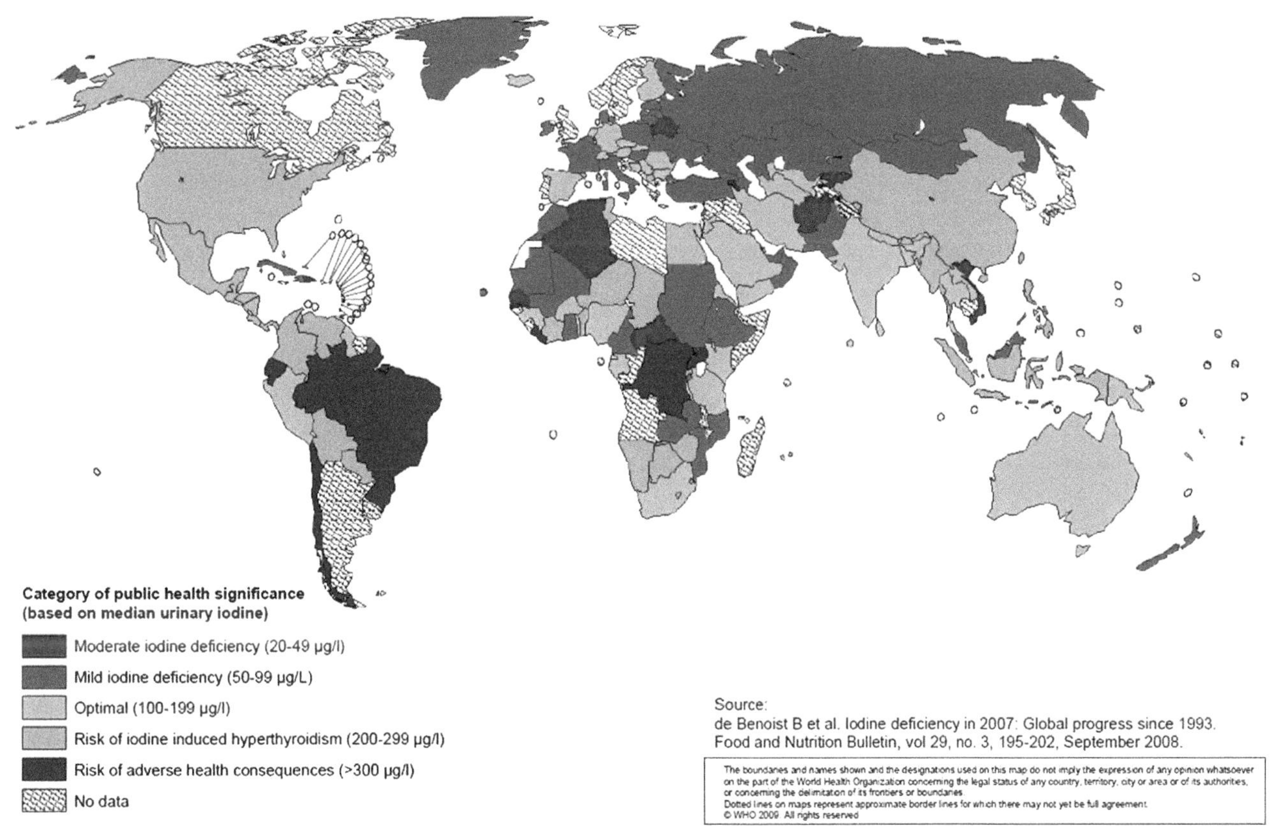

FIGURE 6-4 **Worldwide location of iodide-poor regions.** The shaded portions indicate the iodide-poor regions.

TABLE 6-3 Comparison of Radioisotopes to Pertechnetate

Isotope	Half-life	Relative Damage to Tissues (rad[a]/μCi[b])	Common Usage
^{131}I	8 days	1000	Radiothyroidectomy
^{125}I	60 days	650	Radioimmunoassay, synthesis & imaging studies
^{123}I	13.3 hr	2	Uptake & imaging
TcO_4^-		0.2	Uptake & imaging

[a]A rad (radiation absorption dose) is a unit of damage to tissue caused by exposure to radiation. A rad is proportional to the energy level of the gamma ray emitted by the isotope.
[b]A curie (Ci) is unit of radiation emission concerned with the number of atomic nuclei that disintegrate per second and release gamma rays.

These isotopes emit beta and low-energy gamma radiation and are suitable for high-resolution autoradiography at both the level of light and electron microscopes. They also are more suited for metabolic studies and are safer to use than ^{131}I. However, ^{131}I is still the isotope of choice when attempting to selectively destroy hyperactive follicles (radiothyroidectomy).

Today, radioisotope uptake for clinical imaging and assessment of thyroid gland function in human subjects is often done using **pertechnetate ($^{99}TcO_4^-$)**, a compound containing the radioisotope **technicium, ^{99}Tc**. Pertechnetate is absorbed from the blood by thyroid cells similar to radioiodide but is not incorporated into thyroid hormones and eventually is excreted from the body. It has very little radioactivity compared to radioidide isotopes (Table 6-3) and is much safer to use.

Calculation of a rate for radioisotope accumulation following administration of a given dose provides a quantitative estimate of the degree of TSH stimulation and a reflection of pituitary TSH release. Hence, measurement of radioiodide or pertechnetate uptake and accumulation provides a simple and rapid method for estimating endogenous activities of the HPT axis as well as a means to assess responsiveness of thyroid follicular cells to exogenous TSH. Usually, radioisotope uptake is expressed as percent uptake of the injected dose at some predetermined time following administration of the radioisotope (e.g., 24 hours) (Table 6-4).

It is important to understand that the ability to accumulate and bind iodide organically is not a feature unique to thyroid follicular cells, and many cell types will accumulate some iodide. Pigment cells such as melanophores and melanocytes, pigmented retinal cells, and the epithelial cells found in sweat glands, salivary glands, lactating mammary glands, and kidney tubules readily accumulate radioiodide following injection and may add radioiodide to their secretions. Furthermore, oocytes of many oviparous (egg-laying) vertebrates readily accumulate large amounts of radioiodide. This ovarian accumulation is associated with the normal process of ensuring a source of iodide in the egg that can be used by the young animal for early synthesis of thyroid hormones until an adequate dietary source becomes available. In addition, thyroid hormones, presumably of maternal origin, have been found in the oocytes of several teleosts and at least one amphibian. Marsupials, placental mammals, and other live-bearing vertebrates are known to transfer iodide and thyroid hormones to the developing young from the maternal blood or via the milk to suckling newborns.

Thyroid hormones may not be released into the circulation in proportion to the uptake and binding of radioiodide, however. Uptake, binding, and release of thyroid hormones are separate events independently influenced by a variety of

TABLE 6-4 Comparison of Thyroid Function in a Monotreme (Echidna), Marsupial (Bandicoot), and Placental Mammal (Rabbit)[a]

	Euthyroid Parameters			Effect of Thyroidectomy	
	Iodide Uptake (% Injected Dose)	Plasma T_4 (nmol/liter)	Plasma T_3 (nmol/liter)	BMR[b]	Body Temperature
Echidna, *Tachyglossus aculeatus*	6.4	15.7	0.7	No effect	No effect
Bandicoot, *Perameles nasuta*	13.7	22.0	1.5	Decrease	No effect
Rabbit, *Oryctolagus cuniculus*	22.9	57.9	6.9	Decrease	No effect

[a]*Data from Hurlbert and Augee,* Physiology and Zoology, ***55****, 220–228, 1982.*
[b]*BMR, basal metabolic rate.*

BOX 6B Iodine and Radiation Exposure from Nuclear Plant Disasters

^{125}I and ^{131}I are produced by nuclear reactors as a byproduct of uranium fission, although ^{131}I is by far a more common end-product of nuclear fission. While both isotopes can be used safely in small amounts in clinical and research settings, dangerous amounts of radioiodine isotopes can be released into the atmosphere during a nuclear plant disaster. Such a disaster occurred in April 1986 at the Chernobyl nuclear power plant in the Ukraine and more recently at the Fukushima Daiichi nuclear power plant in March 2011 as a result of the Tohoku earthquake and tsunami. Radioactive iodine could be measured in drinking water and dairy products in Tokyo through May of 2011 and could be measured as far away as California, where it was concentrated into leafy structures of coastal vegetation such as kelp.

The major threat posed by radioiodine released into the atmosphere is thyroid cancer. After the Chernobyl nuclear disaster, thyroid cancer rates skyrocketed in individuals who were older than 15 to 25 at the time of the incident (Box Figure 6B-1). Based upon a report by the United Nations Scientific Committee on the Effects of Atomic Radiation, as of 2005, 6,000 Ukrainian, Russian, and Belarusian residents had been diagnosed with thyroid cancer, a number three times greater than normal rates for this cancer. Because the half-life of ^{131}I is 8 days, it is generally believed that the use of countermeasures during the period of greatest exposure can be effective in mitigating the impact of radioiodine on thyroid gland function. After the Fukushima disaster, the Japanese government issued more than 230,000 potassium iodide tablets. Potassium iodide tablets are believed to mitigate the effects of radioiodine by saturating iodide transport into the thyroid gland, thereby preventing ^{131}I from entering the thyroid gland. Unfortunately, the efficacy of this treatment in preventing thyroid cancers in residents of Fukushima prefecture will not be known for years.

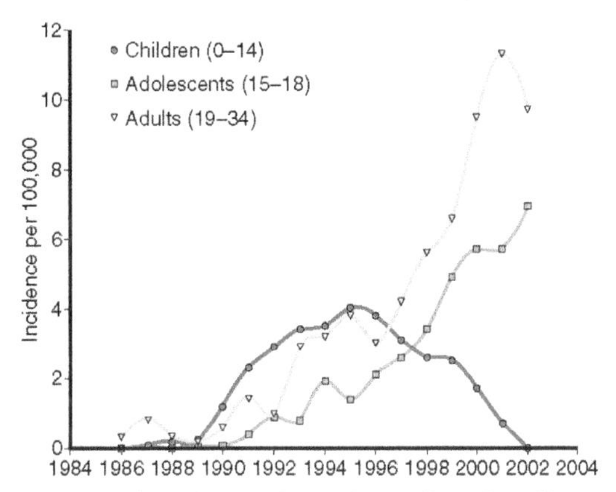

BOX FIGURE 6B-1 **Incidence of thyroid cancer following the Chernobyl nuclear plant disaster of 1986.** (*Adapted with permission from Demidchik, Y.E. et al.*, International Congress Series, ***1299****, 32–38, 2007* and *Cardis, E. et al.*, Journal of Radiation Protection ***26****, 127–140, 2006.*)

factors, as evidenced in the following discussions. Nevertheless, measurement of the uptake of radioisotopes is a rapid and convenient method to assay for TSH activity with respect to endogenous levels or exogenous treatments, and it is widely employed in thyroid diagnosis and research in animals.

B. Biosynthesis and Release of Thyroid Hormones

As described in Chapter 3, the synthesis of thyroid hormones in the follicular cells involves the synthesis of Tgb at the rough endoplasmic reticulum; it is packaged into membrane-bound secretion granules in the Golgi apparatus. Tgb-containing tyrosine is released by exocytosis into the colloid where the formation of active iodide and the binding of active iodide to some of these tyrosines forms **diiodotyrosine (DIT)**. This step is accomplished in the colloid by the enzyme **thyroid peroxidase (TPO)**, which is embedded in the apical membrane of the follicular cell. The final step, the linking together (coupling) of two DITs contained within Tgb to form the iodinated thyronine, T_4, is also performed by TPO.

1. Hormone Release: Hydrolysis of Thyroglobulin

Release of thyroid hormones following administration of TSH is not linked directly to iodide uptake and iodothyronine synthesis. Thyrotropin independently stimulates engulfment of colloid by the follicular cell and its intracellular hydrolysis (Figure 6-3). Autoradiographic studies indicate that the first event observed following TSH administration is the engulfment of colloid droplets through a process of endocytosis to form **endosomes** that migrate from the apical portion of the cells toward the basal portion. Endosomes fuse with lysosomes that contain a number of hydrolytic enzymes to form "fusion droplets" or **endolysosomes**. As the endolysosomes continue migrating toward the basal portion of the cell, they become progressively degranulated, because of hydrolysis of Tgb and diffusion of the hydrolysis products into the cytosol. Hydrolysis of Tgb within the endolysosome releases DIT, T_4, dehydroalanine, and other amino acids.

A **type I deiodinase (D1)** present in the cytosol of thyroid follicular cells plays an important role in iodide metabolism (see Chapter 3 for a description of the types of deiodinase enzymes). This enzyme can deiodinate both the inner and outer ring of a thyronine and can deiodinate iodotyrosines as well. Most of the T_4 released from Tgb rapidly exits from the follicular cell, but a small number of T_4 molecules may be partially deiodinated to form T_3 or **reverse T_3 (rT_3)**. The amount of deiodination depends on the species. DIT is also deiodinated, leading progressively to formation of MIT and tyrosine plus free iodide. Thus, deiodination is in part a conservation mechanism to reuse inorganic iodide and tyrosine. Approximately 85 to 90% of the iodide released through deiodination of MIT, DIT, and T_4 enters a second iodide pool within the follicular cell and is then available for diffusion into the colloid and reuse for iodination of tyrosines in newly synthesized Tgb.

BOX 6C Iodolipids

T_3 and T_4 are not the only chemicals produced by the thyroid gland that are covalently bound to iodine. The thyroid gland produces a group of chemicals called **iodolipids**, so named because they are synthesized from lipid precursors. Two major types of iodolipid are synthesized by thyroid follicle cells: 6-iodo-5-hydroxy-8,11,14-eicosatrienoic acid (6-iodolactone) from arachidonic acid (AA) and 2-iodohexadecanal (2-IHDA) (see Box Figure 6C-1) an iodoaldehyde that is synthesized from the plasmalogen plasmenylethanolamine. 6-Iodolactone is formed from AA by peroxidase enzymes in the presence of hydrogen peroxide and I^-. 2-IHDA is formed by poration of reactive oxidized iodine (I+) into the vinyl ether group (−CH2 −O−CH=CH−) of plasmenylethanolamine, a plasmalogen (Box Figure 6C-1). Formation of iodolactones and iodoaldehydes is dependent upon NIS activity as evidenced by the inhibitory action of perchlorate (a selective competitor of the NIS) on incorporation of I^-. Iodolipids can be synthesized by peroxidase enzymes, and in thyroid tissue this is accomplished by thyroid peroxidase.

Iodolactones and iodoaldehydes have interesting effects on thyroid cell function. Iodolactone mediates the inhibitory effect of I^- on thyroid cell proliferation. 2-IHDA is believed to be responsible for mediating the Wolff−Chaikoff effect, as this iodolipid inhibits NADPH-dependent H_2O_2 production in thyroid cells by directly inhibiting NADPH oxidase and by inhibiting thyroid peroxidase activity.

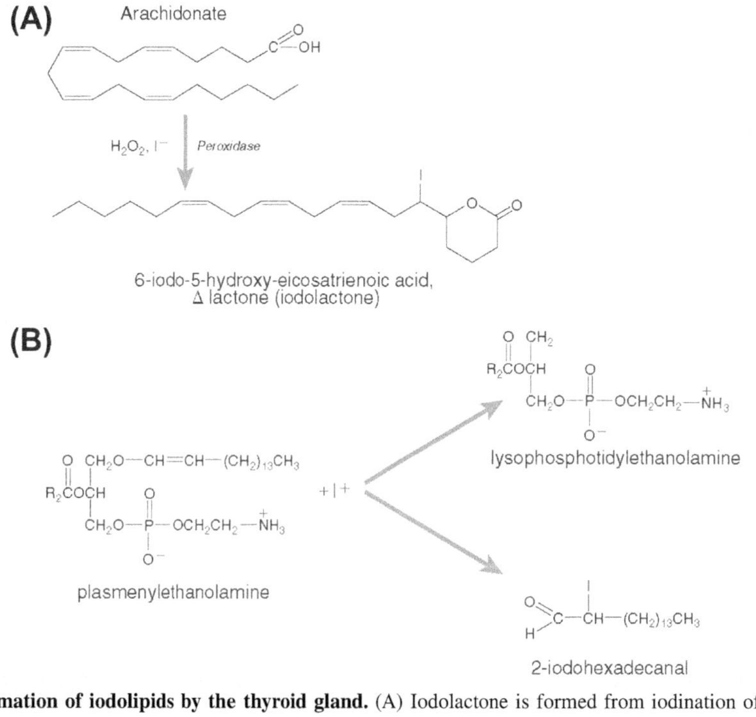

(A) Arachidonate

H_2O_2, I^- Peroxidase

6-iodo-5-hydroxy-eicosatrienoic acid, Δ lactone (iodolactone)

(B)

plasmenylethanolamine

+I+

lysophosphotidylethanolamine

2-iodohexadecanal

BOX FIGURE 6C-1 Formation of iodolipids by the thyroid gland. (A) Iodolactone is formed from iodination of arachidonate. (B) 2-Iodohexadecanal is synthesized by iodination of the phospholipid plasmenylethanolamine. Synthesis of both iodolipids by the thyroid gland requires

2. *Peripheral Deiodination and Metabolism of Thyroid Hormones*

Only about one-seventh to one-half of the circulating T_3 is produced by T_4 deiodination within the thyroid follicular cell (depending on the species and/or physiological parameters), and the remainder is produced through peripheral deiodination of T_4 by the liver. Details of the deiodination of thyroid hormones are provided in Chapter 3.

D. Mechanism of Action of Thyroid Hormones

Once T_4 enters a target cell (see Chapter 3), it is converted rapidly to T_3 by a cytoplasmic type III deiodinase (D3). T_3 diffuses into the nucleus and typically binds to a nuclear **thyroid hormone receptor (TR)**, although cytosolic and membrane receptors also have been described on target cells. The most effective nuclear receptor complex for interacting with a **thyroid hormone response element (TRE)** involves a heterodimer of occupied TR plus a **retinoid X receptor protein (RXR)**. Homodimers of TR are less effective in binding to and activating the TRE. Occupied receptors influence the synthesis of new proteins by regulating gene transcription. Nuclear TRs have greater affinity for T_3 than for T_4, supporting the hypothesis that T_3 is the active form of thyroid hormone. Mitochondrial receptor proteins for thyroid hormones also have been demonstrated, and these mitochondrial receptors may be associated with observed effects of thyroid hormones on mitochondrial protein synthesis and oxidative metabolism (see ahead).

Some permissive actions of thyroid hormones may be a consequence of effects of thyroid hormones on nuclear-directed synthesis of adenylyl cyclase or on the availability of ATP through their actions on mitochondria or on both. The levels of adenylyl cyclase and ATP would influence the effects of hormones that normally produce their actions through some cAMP-dependent mechanism (see Chapter 3). Another mechanism for their permissive actions could be related to thyroid-hormone-induced synthesis of receptor proteins for other regulators.

Evidence is accumulating for the importance of a membrane thyroid hormone receptor. Thyroid hormones bind to the extracellular region of the **integrin receptor**, a receptor that is critical for attachment of many cell types to the **extracellular matrix (ECM)**. Interestingly, T_4 binds with greater affinity than T_3 to this membrane receptor. Given the central importance of integrin receptors in cell attachment, inflammation, angiogenesis, and cell migration, this receptor may mediate many different aspects of thyroid hormone effects on cellular interaction with the ECM.

IV. FACTORS THAT INFLUENCE THYROID FUNCTION IN MAMMALS

The hypothalamus and pituitary exert direct control over thyroid gland functions but in turn are influenced by environmental factors working through the nervous system. Other hormones (e.g., melatonin) may alter thyroid function. Many additional factors (e.g., diet) influence thyroid state in mammals and hence may influence processes controlled by other hormones by altering permissive roles.

A. Endocrine Factors Affecting Thyroid Gland Function

The hypothalamus exerts regulatory control over the release of TSH from the pars distalis via secretion of TRH (see Chapter 4). Thyroid hormones themselves play important roles through negative feedback on the thyroid axis. In addition, the pineal gland may have a negative influence on thyroid function under certain conditions.

1. *TRH and TSH*

The hypothalamus is the source of TRH that is released from the median eminence and travels to the pituitary thyrotrope via the hypothalamo–hypophysial portal system. Thyrotropin release is stimulated by TRH, which in turn produces an increase in circulating thyroid hormones. Both synthesis and release of TSH are regulated through an IP_3 second-messenger system (see Chapters 3 and 4). Calcium ions also are involved in release of TSH from the thyrotrope. Sensitivity of the thyrotrope to TRH may be affected by paracrines secreted from other pituitary cells.

TSH enhances the expression of the NIS gene and uptake of iodide as well as the synthesis of thyroglobulin and thyroid hormones. In addition, TSH induces endocytosis and subsequent hydrolysis of Tgb, causing thyroid hormones to be released into the blood. Continued stimulation by TSH causes structural changes in the follicular cells that are related to thyroid hormone synthesis and release. The follicular cells in an inactive or unstimulated follicle are usually flat or squamous cells. Thyrotropin can cause such flat cells to assume a cuboidal or even columnar shape, resulting in visible thickening of the follicular epithelium and even increasing the size of the entire gland. Much of this enlargement of the follicular cells is due to an increase in rough endoplasmic reticulum and Golgi apparatus for Tgb synthesis. This increase in cellular size due to increased cellular growth is referred to as **hypertrophy**. Chronically stimulated thyroid glands may exhibit hypertrophy as well as **hyperplasia** (an increase in cellular numbers due to mitotic divisions by the stimulated cells).

Hypertrophy or hyperplasia or both can lead to formation of a goiter.

The increase in the cellular portion of the follicle due to hypertrophy and the concomitant reduction in colloid are reflected in a change in the diameter of the follicle with respect to the thickness of the epithelium and/or to the volume of the lumen. The ratio of follicle diameter to thickness of the epithelium or diameter of the follicular lumen changes predictably with TSH levels and frequently has been used as a measure of the degree of stimulation by TSH. Generally, a "stimulated" histology is indicative of thyroid hormone deficiencies and enhanced TSH secretion to compensate for these deficiencies. Other factors, such as cold stress, may be operating at the hypothalamus, however, to elevate TSH secretion above that normally maintained through negative feedback by the thyroid hormones. This effect also may cause follicular cell hypertrophy.

One of the first cellular events that occurs in follicular cells following administration of TSH is activation of adenylyl cyclase and resultant increase in the intracellular levels of cAMP. Coincident with increased iodide uptake, formation of organic iodide, and endocytosis of colloid is an increase in glucose oxidation that may be caused by cAMP. Glucose oxidation is thought to be the "driving force" for endocytosis and iodination, the latter involving oxidation of pyridine nucleotides and formation of H_2O_2. By controlling reactions such as glucose oxidation, cAMP could mediate several different cellular events associated with the action of TSH on the follicular cells.

2. T_3 and T_4 Feedback Effects

The release of TSH is regulated by negative feedback produced by thyroid hormones, and the administration of exogenous thyroid hormones decreases circulating TSH and associated thyroid gland activities. The major site for negative feedback is on the thyrotropic cells directly and not the hypothalamic thyrotropic center responsible for TRH production. Thyrotropes contain receptors that bind T_3 more effectively than T_4. Occupied thyroid hormone receptors interfere with the cAMP-dependent releasing mechanism such as that stimulated by VIP. Type II deiodinase present in thyrotropes converts most of the T_4 that enters the thyrotrope to T_3, resulting in enhanced action with receptors leading to negative feedback. Thyrotropin levels seem to be maintained by direct negative feedback, and the role of TRH may be to override the system during times of increased demand for thyroid hormones. In other words, thyroid hormones determine the level of TSH secretion that regulates daily thyroid gland activities, whereas the hypothalamus adjusts that level through TRH secretion as dictated by other neural factors and/or by environmental cues.

3. The Epiphysial Complex

The pineal gland of the epiphysial complex in mammals (see Chapter 4) has been implicated as an inhibitory factor regulating thyroid function. This action on thyroid function is probably mediated through inhibitory effects of melatonin on hypothalamic TRH release. Photoperiodic influences on thyroid activity also may be mediated through inhibition or stimulation of the pineal gland.

B. Non-Endocrine Factors Affecting Thyroid Gland Function

Thyroid activity is affected by a number of chemicals in the diet that block one or more biochemical steps in thyroid hormone synthesis. In addition, a number of environmental factors, such as photoperiod and temperature, can influence the activity of this system through the brain, pineal gland, and pituitary.

1. Direct Neural Control of the Thyroid Gland

Although it has been known for quite some time that the thyroid gland is innervated by nerve endings, the origin of this innervation and its biological significance have remained a mystery. Recent evidence indicates a direct innervation of the thyroid gland by neurons of the suprachiasmatic nucleus (see Chapter 4) and **intermediolateral cell column of the spinal cord (IML)**, the location of preganglionic sympathetic nervous system neurons. This suggests direct communication between the brain and the thyroid gland, which may coordinate regulation of metabolism and heat production (see below) and circadian changes in thyroid activity without the involvement of TSH.

2. Diet

Low iodide availability reduces the synthesis of thyroid hormones and leads to development of hypothyroid conditions. Most seriously affected by the reduction in iodide is the synthesis of DIT, which in turn reduces the amount of T_4 that can be synthesized. An excessive level of blood iodide may inhibit uptake and accumulation of iodide by the follicular cells, a phenomenon known as the Wolff–Chaikoff effect. This effect is transitory and, even in the presence of elevated blood iodide, the thyroid gland will begin making thyroid hormone again due to reduced expression of the NIS gene. The cellular basis for the Wolff–Chaikoff effect is not entirely known but may partly involve local feedback by iodolipids (see above).

In nature, it would be most unusual for a mammal to be subjected to an excess of iodide, but several instances of excessive iodide are known for humans. Ingestion by Japanese fisherman of large quantities of seaweed, which is naturally rich in iodide, induces a hypothyroid state. In the early 1980s, occurrence of high levels of iodide in milk and fast foods due to artificial additives or contaminants raised medical concerns of potential induction of hypothyroidism, especially in children whose nervous systems would be especially sensitive to insufficient thyroid hormone. However, the low levels of iodide added to commercial salt preparations do not pose a threat with respect to normal salt intake and are an important deterrent to the development of hypothyroidism due to insufficient iodide availability in both children and adults. As mentioned previously, the iodide content of milk has decreased markedly in recent years.

Reduced caloric intake or fasting depresses circulating T_3 with (in rats) or without (in humans) depression of T_4 levels. This is an adaptive response that limits growth or metabolic rate when energy sources are low. Studies in rats suggest that overfeeding (especially of carbohydrates) elevates T_3 levels and similar carbohydrate intake might be related to hyperactivity in children.

3. Chemical Inhibitors of Iodide Uptake

Certain anions are effective in blocking accumulation of iodide by follicular cells through competitive inhibition of iodide transport. **Thiocyanate (SCN^-)**, **perchlorate ions (ClO_4^-)**, and pertechnetate (TcO_4^-) are particularly effective at blocking iodide uptake. Because of its ability to compete with iodide for uptake by thyroid cells, small doses of radioactive pertechnetate are used commonly in place of radioiodide for determining thyroid activity. High levels of pertechnetate may be useful as a blocking agent, as it is less toxic than some other agents, especially SCN^-. Experimentally, all of these agents can be used to block thyroid function and particularly to block iodide uptake mechanisms, but some are found naturally in food and water; for example, the cassava root (Figure 6-5), an important dietary source of carbohydrate in tropical countries, is loaded with thiocyanate. Some vascular plants such as cauliflower contain a glycoside of thiocyanate that can be converted to free thiocyanate in the body; however, one would have to ingest about 10 kg of cauliflower per day to produce any serious effects on thyroid function unless dietary iodide was extremely low. Perchlorate anions can be found naturally in the soil in dry and arid areas (including the surface of Mars!) but are a major source of contamination in groundwater and surface waters near military and aerospace facilities that use perchlorate as an explosive. The Colorado River accumulates considerable amounts of perchlorate as it flows from Colorado through the southwestern United States.

FIGURE 6-5 The cassava root is an important dietary staple in tropical countries but is rich in thiocyanate which blocks iodide uptake by the thyroid gland.

4. Chemical Inhibitors of Iodination

Compounds that interfere with thyroid hormone synthesis by inhibiting iodination of tyrosines are called **goitrogens**. The resultant reduction in circulating hormones causes increased TSH secretion as a consequence of reduced negative feedback. Continuous stimulation of the thyroid gland by TSH results in enlargement of the thyroid and production of a goiter. Typically, goitrogens block formation of active iodide and hence prevent iodination of tyrosines. Intracellular accummulation of iodide secondarily inhibits iodide uptake. Because agents that selectively inhibit iodide uptake secondarily block thyroid hormone synthesis and also can lead to goiter formation, the inhibitory anions, such as perchlorate, are often called goitrogens.

Several anti-thyroid drugs (collectively called **thionamides**) are used clinically to block formation of active iodide, including PTU (propylthiouracil), as well as newer drugs such as **methimazole** (Figure 6-6). These drugs are not as toxic as some of the anions described earlier. They interfere with the peroxidase system responsible for generation of H_2O_2. Treatment with such drugs can be used experimentally to chemically thyroidectomize an animal reversibly. Certain reduced compounds such as ascorbic acid, reduced glutathione, and reduced pyrimidines also remove H_2O_2 from the system and can block formation of active iodide.

Many flowering plants of the family Brassicae (e.g., cabbage, Brussels sprouts, rutabaga, and turnips) naturally contain a compound known as **progoitrin** that can be converted by a specific enzyme to a goitrogenic compound called **goitrin** (Figure 6-6). If sufficient quantities of goitrin are absorbed from the intestine into the general circulation, the synthesis of thyroid hormones is impaired and a hypothyroid state ensues. People or animals that consume large quantities of these plants are at risk for hypothyroidism that may be accentuated if coupled with an iodide-poor diet. Cooking these plants normally destroys the enzyme that converts progoitrin to goitrin, but progoitrin

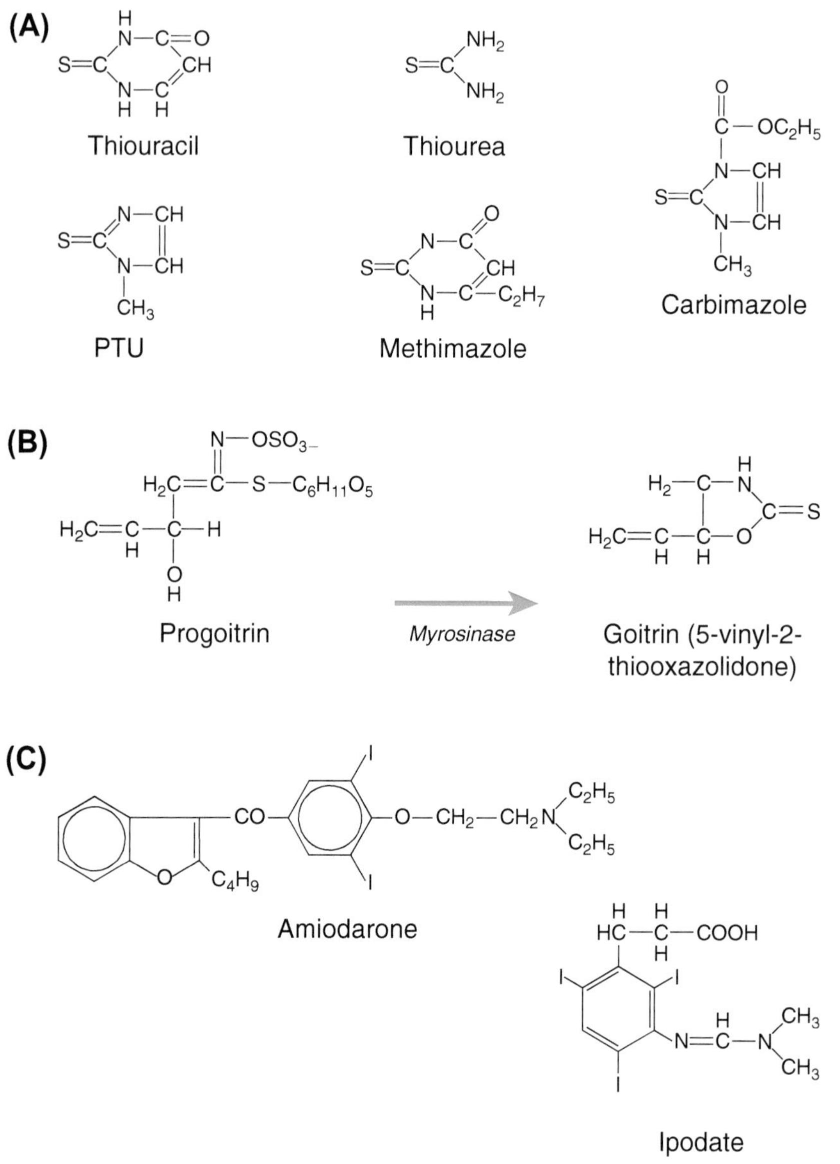

FIGURE 6-6 **Some thyroid inhibitors.** (A) Thiourea, thiouracil, propylthiouracil (PTU), carbimazole, and methimazole are all goitrogens that block iodide uptake and/or the iodination and coupling reactions. (B) Goitrin is a naturally occurring goitrogen that is made from the precursor progoitrin by the enzyme myrosinase. (C) Ipodate and amiodarone block liver deiodinases. PTU also blocks type-1 deiodinase activity.

itself is not affected by the quantity of heat applied in cooking the vegetables. However, certain bacteria that are part of the human intestinal flora are capable of converting ingested progoitrin into goitrin, thus reversing at least some if not all of the protective effect of cooking.

5. Specific Inhibitors of Deiodination and of Receptor Binding

Deiodination of T_4 to T_3 can be prevented by treatment with drugs such as **iopanoate** or **ipodate** (Figure 6-6). These drugs were injected initially as radio-opaque agents that were removed from the blood by the liver and secreted into bile where they aided visual examination of the gallbladder in which they accumulated. They later were discovered to be very potent blockers of deiodinating enzymes. It is also possible to block thyroid hormone action with the drug **amiodarone**, which binds to thyroid receptors and blocks binding of thyroid hormones. This drug was first used to treat cardiac arrhythmia before its anti-thyroid role was discovered. We now know that amiodarone also is an inhibitor of deiodinase activity, giving it another mechanism for reducing thyroid functioning. As previously mentioned, the type I deiodinase (D1) found in the thyroid gland and in the liver is inhibited by PTU, although D2 and D3 deiodinases are unaffected by this drug. Hence, PTU not only blocks T_4 synthesis but also blocks conversion of T_4 to T_3.

6. Natural and Anthropomorphic Environmental Factors

Environmental factors such as photoperiod and temperature may affect thyroid hormone secretion rates through neuroendocrine mechanisms. Such factors may influence synthesis and release of hypothalamic and hypophysial hormones or may alter thyroid function by affecting neurons that innervate the thyroid gland. Internal biological clocks may be related to the actions of environmental factors in regulating thyroid cycles. Cyclical variations have been reported for thyroid hormones on both a diurnal and seasonal basis and probably involve pineal melatonin. Internal secretory rhythms of hypothalamic regulators might be influenced by environmental factors or might regulate the sensitivity of other effectors to external factors. Feedback by thyroid hormones or actions of other hormones in the internal environment may alter sensitivity as well.

Polychlorinated biphenyls (PCBs) (Figure 6-7) that have accumulated in the environment are endocrine-disrupting chemicals (EDCs) in that they inhibit thyroid function. Fishes in the Great Lakes area of the United States have accumulated sufficient PCBs to affect thyroid function in humans consuming them on a regular basis. Exposure of pregnant women to high dietary levels of PCBs results in children with reduced learning capabilities, presumably due to permanent effects of a hypothyroid condition during development of the nervous system.

Polybrominated diphenyl ethers (PBDEs) arise mainly from fire-retardant materials. PBDEs are widespread in the environment, and it is generally assumed that nearly all people have been exposed to low levels of PBDEs at some point in their life. Concentrations of PBDEs in some human populations have increased over the last two decades. PBDEs are structurally similar to thyroid hormones (Figure 6-7) and, while there is general agreement that PBDE exposure alters thyroid function, the mechanism of PBDE action on the thyroid axis is still being studied.

Contamination of water supplies by **perchlorate** poses another threat to animals and humans due to its accumulation in drinking water. The most common sources of ammonium perchlorate are from military activities and automobiles, and contamination of the environment with perchlorate ions is most prevalent around military installations. Areas of heaviest contamination in the United States include southern California and western Texas. Perchlorate ions, as indicated earlier, have a goitrogenic action and block thyroid function. The increasing concentration of perchlorate ions in drinking water are of special concern, especially for pregnant women due to the possibility of thyroid deficiency affecting fetal nervous system development.

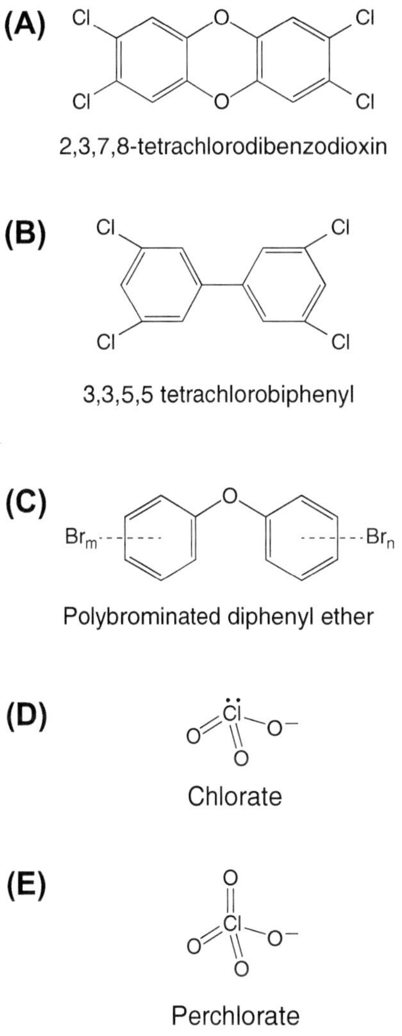

(A) 2,3,7,8-tetrachlorodibenzodioxin

(B) 3,3,5,5 tetrachlorobiphenyl

(C) Polybrominated diphenyl ether

(D) Chlorate

(E) Perchlorate

FIGURE 6-7 **Some environmental thyroid disrupting contaminants.** Dioxins (A) and polychlorinated biphenyls (PCBs) (B) arise from a number of diverse industrial sources and can alter normal thyroid hormone metabolism and action. Polybrominated diphenyl ethers (PBDEs) (C) are ubiquitous contaminants in the environment that arise mainly from fire-retardant materials. PBDEs have a striking structural resemblance to T_4 and T_3 and have been reported to alter plasma thyroid hormone levels and thyroid hormone metabolism. Chlorate (D) and perchlorate (E) anions potently block thyroid iodide transport and therefore inhibit thyroid hormone synthesis. They are found in the environment as a result of aerospace and military waste and agricultural use (chlorate is applied as a defoliant).

V. BIOLOGICAL ACTIONS OF THYROID HORMONES IN MAMMALS

Thyroid hormones affect many diverse tissues and influence major processes such as metabolism, growth, differentiation, and reproduction. They are responsible for maintaining a general state of well-being for many cells so that they are capable of maximal responses to other stimuli.

A. Metabolic Actions

The effects produced by thyroid hormones on mammalian metabolism include a calorigenic or **thermogenic action** (heat-generating) as well as specific effects related to carbohydrate, lipid, and protein metabolism. In general, thyroid activity in mammals is greater during prolonged periods of cold stress (winter) than during warmer periods. Thermogenic actions of thyroid hormones become more meaningful when considered together with the actions of other hormones on metabolism (see Chapter 16). Many of these metabolic actions are possibly permissive actions occurring in cooperation with other hormones such as epinephrine and growth hormone (see Chapters 4 and 12).

Thermogenic actions of thyroid hormones are restricted to certain tissues and are involved in physiological responses to cold stress. They can accelerate the rate at which glucose is oxidized in these tissues and thus increase the amount of metabolic heat produced in a given time. This elevated heat production can be used to warm the body. Accelerated glucose oxidation is reflected in an increased **basal metabolic rate (BMR)** as measured by an increased in rate of oxygen consumption. In contrast, decreased nutrient intake operates through neural mechanisms that reduce thyroid hormone secretion and lower metabolic rate. As mentioned earlier, there is evidence that thyroid hormone receptors exist in mitochondria, and thyroid hormones induce increased synthesis of several mitochondrial respiratory proteins, especially cytochrome C, cytochrome oxidase, and succinoxidase. In brown adipose tissue (a tissue especially important for thermogenesis), T_3 but not T_4 stimulates production of a unique mitochondrial protein known as **uncoupling protein 1 (UCP-1)** in a dose-dependent manner. The high vascularity of brown adipose tissue in rodents, for example, is visibly different from yellow adipose tissue, which primarily stores lipids. Although humans are thought to lack brown adipose tissue, apparently some of the yellow adipose tissue functions as though it were brown.

UCP-1 is one of a group of mitochondrial proteins that are upregulated by thyroid hormones in several tissues, but of these proteins only UCP-1 has been linked to uncoupling oxidative phosphorylation and heat production. This mitochondrial action to augment oxidative metabolism would be advantageous in adapting to chronic cold stress. Long before the discovery of UCP-1, thyroid hormones were postulated to "uncouple" oxidative phosphorylation, which would decrease the efficiency of ATP synthesis in the mitochondria and increase the quantity of heat released per mole of glucose oxidized. It is not clear if the ability of thyroid hormones to increase the total rate of glucose oxidation, or the postulated uncoupling, is more important in heat production, but both would definitely contribute to chronic cold stress adaptation. Rapid cold responses are mediated primarily by epinephrine from the adrenal medulla (Chapter 8) rather than by thyroid hormones. TSH receptors also have been identified in rat brown adipose cells, and treatment of warm acclimated rats with TSH results in an upregulation of mRNA levels for both D2 deiodinase and UCP-1. Although acute cold exposure also caused upregulation of D2 deiodinase and UCP-1, it caused downregulation of TSH receptors, supporting uninvolvement of the HPT axis in the regulation of acute cold stress homeostasis.

In addition to increasing glucose oxidation, thyroid hormones cause hyperglycemia and may secondarily stimulate lipolysis (hydrolysis of fats). These actions may in part be associated with permissive potentiation of the hyperglycemic and lipolytic actions of epinephrine and/or glucocorticoids (see Chapters 8 and 12). Thyroid hormones may alter nitrogen balance and can be either protein anabolic (through enhancement of GH actions) or catabolic, depending on the tissue being examined and under what experimental conditions it is examined.

In many non-hibernating mammals such as beaver and muskrat, thyroid activity is depressed during the winter months. Hypothyroidism has been described for hibernating ground squirrels and badgers, but there does not appear to be a causal relationship between reduced thyroid function and the onset of hibernation. Increased thyroid activity also is correlated with arousal. Additional field studies employing sophisticated methods for assessing thyroid functions are needed before the endocrine factors related to either onset or termination of hibernation can be established.

B. Growth and Differentiation

Thyroid hormones are essential for normal growth and differentiation in mammals as evidenced in cretinism and juvenile myxedema in humans (see ahead). These growth-promoting actions of thyroid hormones are closely related to the role of pituitary growth hormone (GH), and they probably represent a permissive action on GH secretion and on GH-sensitive target cells. Thyroid hormones also may stimulate **insulin-like growth factor (IGF-I)** production and hence augment the actions of GH on its target tissues (see mechanism of GH action in Chapter 4).

Nervous tissue development is markedly affected by reduced thyroid hormones during differentiation of the nervous system. T_3 is required for accelerating several aspects of neuron and glial cell function, including myelin formation, development of intracellular signaling pathways, synapse formation, the extension and growth of dendrites, and nerve cell migration. In brain tissue, T_4 is converted to T_3 in a type of glial cell called an **astrocyte** before entering neighboring nerve cells by a transporter

BOX 6D Thyronamines

As discussed in Chapter 4, the enzyme aromatic L-amino acid decarboxylase (AADC) is present in many amine precursor uptake and decarboxylase (APUD) cells where it can convert biologically inactive catecholamine (l-DOPA) and indoleamine (5-hydroxytryptophan) precursors to dopamine and serotonin, respectively. Recently it has been shown that decarboxylation of thyroid hormones results in biologically active metabolites called **thyronamines (TAM)**. Two forms of TAM have been detected so far in mammals, 3-iodothyronamine (3-T_1AM) and thyronamine (T_0AM) (Box Figure 6D-1). Initially, it was suspected that AADC may carry out the conversion of T_3 to 3-T_1AM, but neither T_3 or T_4 is a substrate for AADC, and the enzyme (so-called iodothyronine decarboxylase) responsible for decarboxylation of T_3 remains a mystery. Administration of exogenous TAMs causes effects that are opposite to those of T_3, including reduced metabolic rate and respiratory depression and hypothermia. The effects of TAMs are mediated through an interaction with a G-protein-coupled membrane receptor called the trace amine associated receptor 1 (TAAR1) (Box Figure 6D-2). The physiological role of TAM is unknown.

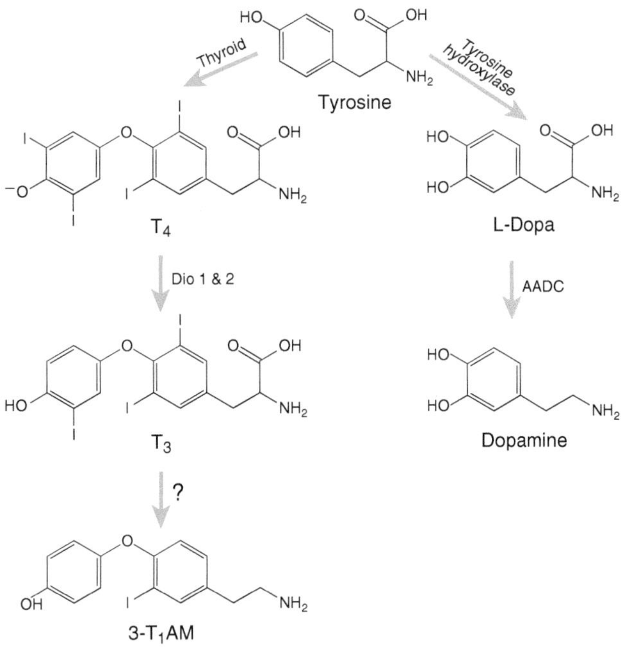

BOX FIGURE 6D-1 **Alternative routes for decarboxylation (removal of COOH) and biosynthesis of thyronamines and dopamine.** Recent studies indicate that an as of yet undiscovered "iodothyronine decarboxylase" enzyme (question mark) is responsible for the decarboxylation of T_3 to 3-T_1AM, the principle thyronamine and one of two thyronamines (TAMS) found *in vivo*. *(Adapted with permission from Hoefig, C.S. et al.,* Molecular and Cellular Endocrinology, ***349****, 195–201, 2012.)*

named MCT8 (monocarboxylate transporter 8; see Chapter 3) (Figure 6-8). Inherited mutations in MCT8 lead to severe mental retardation. Not surprisingly, hypothyroidism during gestation and early childhood in humans also seriously impairs differentiation and functioning of the nervous system. Similar observations have been made in other mammals, as well. Furthermore, a reduction in mental activity can occur in hypothyroid adults, too, emphasizing the continued importance of thyroid hormones in normal nervous system function. Exposure of pregnant women to endocrine-disrupting anti-thyroid chemicals such as PCBs and perchlorate can have devastating effects on the fetus that are carried into adulthood. The early fetus depends on maternal thyroid hormones as well as for iodide throughout development and is thus critically dependent on the mother's thyroid condition (Figure 6-9).

Thyroid hormones can affect differentiation processes in adults, too. For example, replacement of hair in adult mammals is stimulated by thyroid hormones. The postnuptial molt cycle in harbor seals (*Phoca vitulina*) involves thyroid hormones and cortisol from the adrenal cortex. Hair loss is correlated with low thyroid function and high cortisol levels, whereas resumption of hair growth is correlated with increased T_4 and return of cortisol to basal levels. Thyroid activity also is related to molting of hair in other mammals including red fox and mink. Changes in hair are seen in hypothyroid humans as well.

BOX 6D Thyronamines (*Continued*)

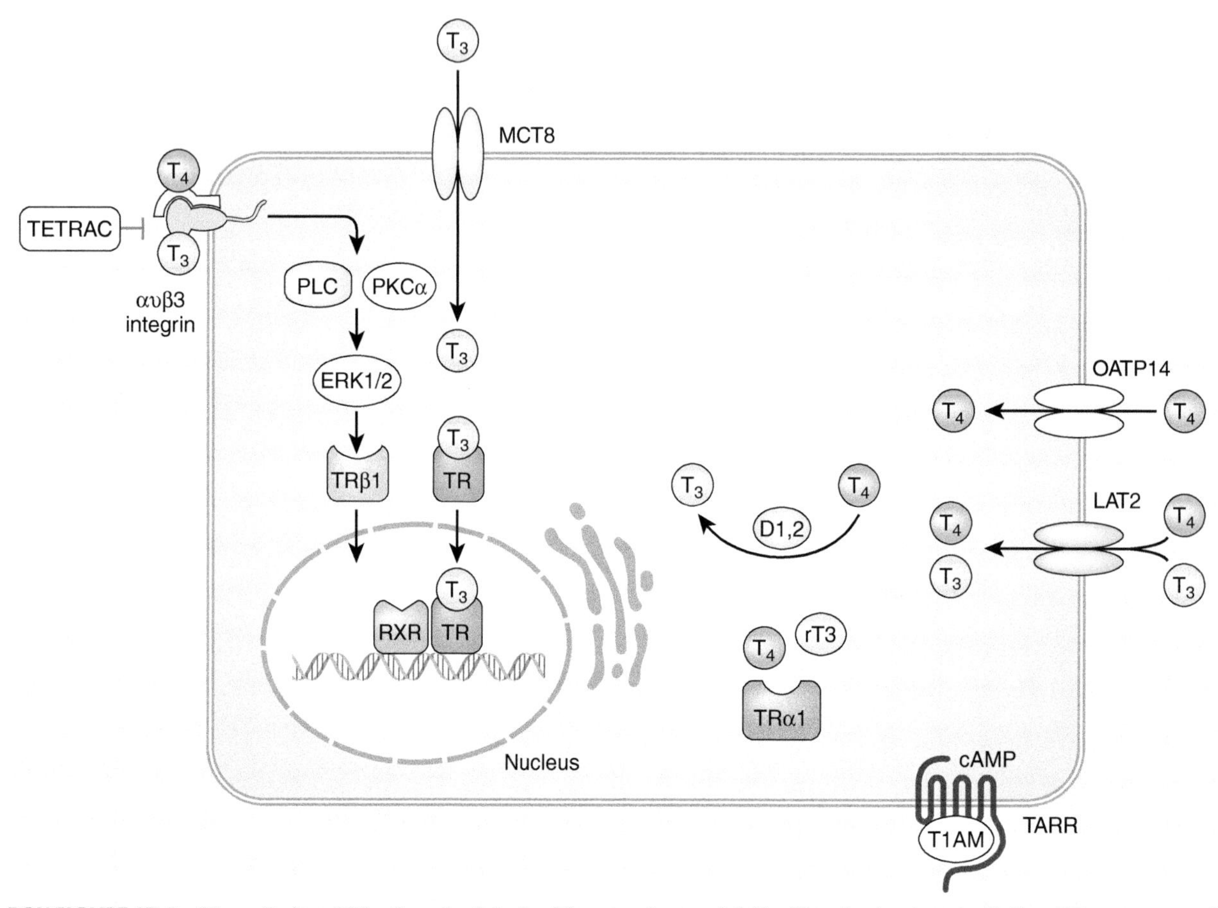

BOX FIGURE 6D-2 **Theoretical model for the potential role of thyronamines modulating T_3 action in a target cell.** T_3 and T_4 are transported by various proteins (LAT2, MCT8, OATP14) into the target cell where they can interact with the thyroid hormone receptor (TR) to alter transcription and protein synthesis. Alternatively, T_3 and T_4 can interact with a domain on the extracellular portion of the integrin receptor to elicit effects through the protein kinase C and phospholipase signaling pathways. These effects can be blocked by an antagonist of the thyroid hormone membrane receptor tetraiodothyroacetic acid (TETRAC). Thyronamines (TAMs) are ligands for the G-protein-coupled receptor TAAR. TAMs can modulate cellular activity by activating this receptor and elevating intracellular cyclic AMP levels. Abbreviations: αυβ3 integrin, vitronectin receptor; D1,2, type 1 and type 2 deiodinase; ERK1/2, extracellular-signal-regulated kinases; LAT2, L-type amino acid transporter 2; MCT8, monocarboxylate transporter 8; OATP14, organic anion transporter 14; PKC, protein kinase C; PLC, phospholipase C; RXR, retinoic acid X receptor; TAAR, trace amine associated receptor; TRα1 (TRΔα1), cytosolic variant of TR. *(Adapted with permission from Piehl, S. et al.,* Endocrine Reviews, ***32****, 64–80, 2011.)*

C. Reproduction

Another cooperative/permissive role for thyroid hormones occurs with respect to gonadal development and function (see Chapter 10). In general, sexual maturation is delayed in hypothyroid mammals. In hypothyroid males, spermatogenesis may occur, but androgen synthesis is low. Similarly, ovarian weight is reduced and ovarian cycles become irregular in hypothyroid females. These reproductive correlations to hypothyroidism have been attributed to reduced gonadotropin (GTH) levels and can be alleviated by treatment with GTHs or thyroid hormones. Experimental studies also support the notion that thyroid hormones influence GTH release through an effect at the level of the hypothalamus.

VI. CLINICAL ASPECTS OF THYROID FUNCTION

Although thyroid disorders are among the most common of human endocrine problems, hypothyroid conditions are most common and include inherited disorders as well as environmentally induced disorders (e.g., dietary deficiencies or EDCs such as PCBs and perchlorate). In general, thyroid function declines with advancing age, making the

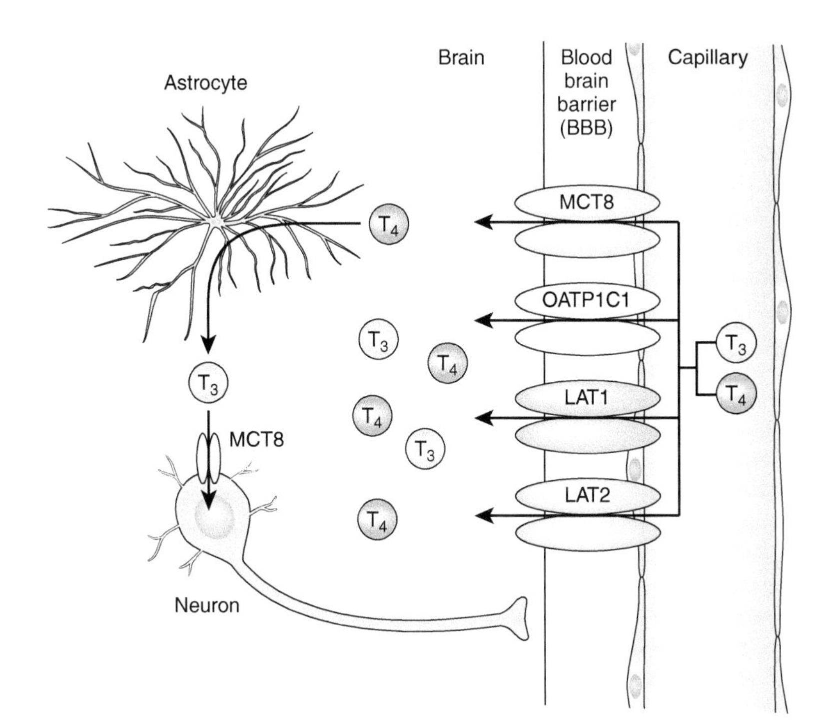

FIGURE 6-8 **Organic anion transport proteins that transport thyroid hormones across the blood–brain barrier (BBB) and into neurons.** Monocarboxylic acid transporter 8 (MCT8), organic anion transporting polypeptide 1C1 (OATP1C1), and large neutral amino acid transporters 1 and 2 (LAT 1 and 2) transport T_3 and T_4 across the blood–brain barrier. Once transported across the blood–brain barrier, T_4 is deiodinated to T_3 by neighboring astrocytes and then transported into neurons by MCT8. *(Adapted with permission from Kinne, A. et al.,* Thyroid Research, ***4****(Suppl. 1), 1–10, 2011.)*

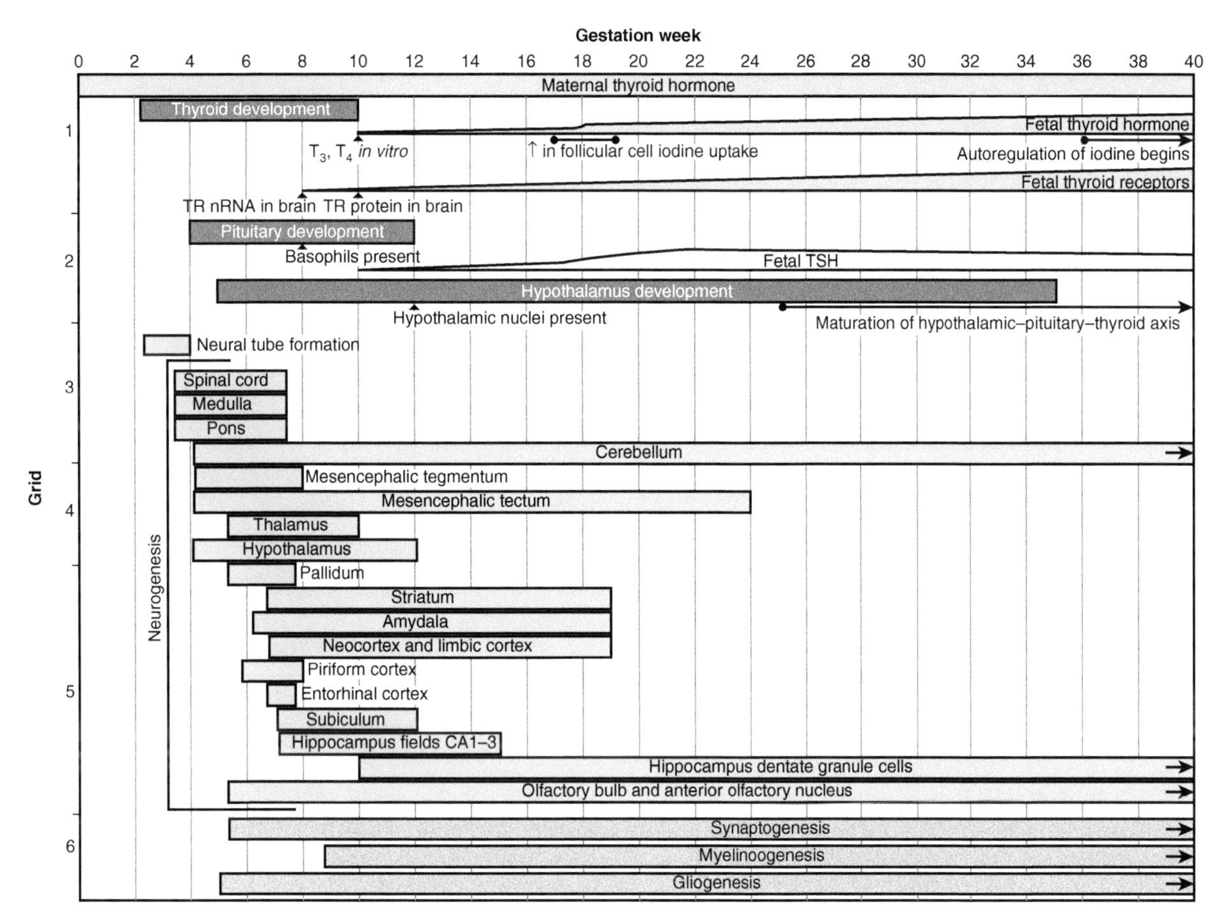

FIGURE 6-9 **Thyroid hormones and development of the nervous system in humans.** Note that many critical events in the nervous system are correlated with periods of thyroid hormones secretion. *(Adapted with permission from Howdeshell, K.L.,* Environmental Health Perspectives, ***110****(Suppl. 3), 337–348, 2002.)*

elderly more at risk for hypothyroidism. Hyperactive thyroid disorders are less common, and thyroid carcinomas are rare. Either hypothyroid or hyperthyroid conditions may be associated with goiter formation, although their occurrence is more common with hypothyroidism. Generalized symptoms of thyroid disorders are listed in Table 6-1. Although only human disorders are described here, similar conditions can develop in other mammals, including domestic dogs and cats.

A. Thyrotoxicosis and Hyperthyroidism

Thyrotoxicosis is a general term referring to an excess of thyroid hormone. If this condition results from thyroid hypersecretion, it is known as **hyperthyroidism**. Primary hyperthyroidism may be due to toxic multinodular goiters consisting of multiple aggregates of small, hyperactive follicles (**Marine–Lenhart syndrome**), or several large TSH-dependent hyperactive follicles (**Plummer's disease**). Follicular adenomas are sometimes autonomously hyperactive as well. Circulating TSH levels typically are low when autonomously hyperactive multinodular goiter or adenomas are present.

Hyperthyroidism may be of a secondary nature caused by a rare pituitary adenoma of TSH-secreting cells. Certain cancerous tumors such as choriocarcinomas of the placenta may elaborate TRH-like or TSH-like molecules that stimulate thyroid activity. Secretory tumors of these types typically are insensitive to any feedback by thyroid hormones.

Graves' disease is another form of secondary hyperthyroid state that occurs in less than 0.25% of the human population but is much more common in women than men (8 to 1). It appears to be mediated by an immunoglobulin known as **LATS (long-acting thyroid stimulator)**. Evidence from twin studies suggests that susceptibility to Graves' disease has a genetic basis. The symptoms of Graves' disease include protruding eyes (exophthalmos), weight loss, increased appetite, restlessness, heat intolerance, fatigue, muscle cramps, tremors, frequent bowel movements, menstrual irregularities, goiter, rapid heartbeat, changes in sex drive, heart palpitations, and blurred or double vision. The basis for production of LATS by the immune system is not understood, but it is an antibody that binds to TSH receptors in the thyroid cell membrane and activates them even in the absence of circulating TSH. These individuals present a hyperthyroid state with very low circulating TSH. The favored treatment, especially in children and adolescents, is to use anti-thyroid drugs such as PTU or methimazole. Drug treatment, however, does not alleviate the cause of the disorder. Furthermore, long-term remission occurs in only 20 to 30% of the patients, and chronic treatment with PTU can cause severe hepatitis. A partial reduction of thyroid tissue by treatment with large doses of radioiodide (^{131}I) is another approach for treating Graves' disease; it is much more effective than anti-thyroid drug therapy (80 to 90% long term remission). The resultant reduction in thyroid tissue reduces thyroid hormone production. Surgical removal of a portion of the thyroid is sometimes employed and is as effective as radioiodide treatment (90 to 95% remission). All of these approaches reduce thyroid hormone production and alleviate the symptoms of hyperthyroidism, but none addresses the cause of the disorder: LATS production by the immune system.

Rarely, hyperthyroidism is due to an ectopic source of thyroid hormones. For example, ovarian dermoid tumors can synthesize sufficient thyroid hormones to bring about hyperthyroidism. **Juvenile thyrotoxicosis** is characterized by elevated thyroid hormones in children, and these patients exhibit nervousness, tremor, accelerated heart rate, and goiter. This syndrome usually occurs in children beyond age 10 (80% of the cases). Exophthalmos is present in about half of these children, and normal weight gain usually is retarded, although growth in stature may be unaffected.

A somewhat rare but dramatic hyperthyroid condition is **thyrotoxic crisis** or **thyroid storm**. This disorder involves a sudden increase in thyroid secretion, severe hypermetabolism, fever, and some other, more variable symptoms. It only occurs in hyperthyroid patients and may be precipitated following incomplete thyroidectomy, interruption of anti-thyroid therapy, or even as a reaction to an infection or tooth extraction. It also may be induced by periods of excessive summer heat. The actual cause of thyroid storm is not certain and the condition could encompass a variety of different causes. Perhaps increased levels of UCP-1 contribute to this apparent uncoupling of oxidative processes.

Thyroid hormone resistance is caused by mutation of genes responsible for synthesis of nuclear thyroid receptors, usually the *TRβ* gene. Mutated receptors result in reduced tissue responses to thyroid hormones. Such patients may appear euthyroid (i.e., exhibit normal TSH and thyroid hormone levels) but show overt symptoms of hypothyroidism.

B. Myxedema and Hypothyroidism

Myxedema is the extreme clinical condition in adults where no thyroid hormones are secreted. In these patients, swelling of the skin and subcutaneous tissues is caused by the extracellular accumulation of a high-protein fluid. **Hypothyroidism** refers to any condition where levels of thyroid hormones are insufficient due to defects of primary (at the thyroid) or secondary (hypothalamus or pituitary) origin. It is especially serious in children because of marked effects on both general growth and neural development. The term **juvenile hypothyroidism** refers to cases of hypothyroidism in children that do not lead to severe retardation in somatic and intellectual development. When growth and development are markedly retarded, it is called **cretinism**. This syndrome is remarkably common: 1 in every 8500 births exhibits cretinism. However, if hypothyroidism in the neonate is detected at birth by using a simple thyroid test, it can be alleviated with thyroid hormone therapy so that growth and development from that time are normal. Because of poor medical care and widespread dietary iodide deficiency, however, cretinism is a serious, crippling disorder in many Asian and African countries.

Juvenile hypothyroidism does not refer to the fetus that is dependent upon the thyroid state of the mother. Hence, it is possible, as stated previously, that hypothyroidism during pregnancy can reduce neural development and that can be further affected by continued hypothyroidism in the neonate.

There are a number of symptoms characteristic of hypothyroidism, including rough and dry skin, yellow pallor, coarse scalp hair, hoarse voice, and slow thought and action (Table 6-1); however, sometimes the hypothyroid person exhibits none or only a few of these symptoms. Obesity is often listed as a characteristic of hypothyroidism but it is not always present and occasionally occurs in patients presenting hyperthyroidism. Conversely, persons suffering from secondary hypothyroidism are often thin as is frequently the case for hyperthyroid patients. Exophthalmos, usually correlated with hyperthyroid states, may occur occasionally in primary myxedema. These changes are progressive, and they can occur very gradually, which makes it difficult to diagnose early stages of hypothyroidism. However, reduced circulating thyroid hormones and/or TSH levels usually are indicative of hypothyroidism.

C. Goiters

Any enlarged thyroid is referred to as a goiter regardless of the cause or nature of the enlargement. Actually, there are four kinds of clinical goiters. The first and most common kind is a **hypothyroid goiter** caused by failing thyroid hormone production resulting in a shortage of T_3 and T_4. Circulating TSH levels are elevated because of reduced negative feedback, and this increased TSH causes enlargement of the thyroid gland and formation of a goiter.

The second type is the **hyperfunctioning goiter**, which is not as common as hypothyroid goiters. Typically, circulating thyroid hormones are high due to spontaneously active follicular cells, and TSH levels are low due to feedback. Two subtypes of hyperfunctional goiters are described. **Diffuse thyrotoxic goiter** is often termed Graves' disease, whereas **toxic nodular goiters** are associated with Plummer's or Marine–Lenhart syndromes (see above).

The third goiter type is the **hyperfunctioning goiter of pregnancy**. During normal pregnancy, there is an increase in thyroxine-binding globulins and a consequent decrease in free T_4 and T_3 in maternal plasma. This results in elevated TSH through decreased negative feedback and in a slight thyroid enlargement to maintain normal functional levels of thyroid hormones during pregnancy. This condition usually has no serious consequences during pregnancy for either mother or fetus, especially if iodide supplies are adequate, and typically returns to normal in women after birth.

The fourth kind of goiter might be termed **normal functioning goiters**, which develop in people with otherwise normal thyroid function. Such enlargements have many different causes including infiltration of the gland with tuberculosis bacteria, syphilitic bacteria, or parasites. The presence of adenomas or carcinomas that do not secrete thyroid hormones could produce an enlarged thyroid without affecting thyroid physiology. Inflammation due to autoimmune disease (i.e., thyroiditis) can sometimes cause enlargement of the thyroid.

D. Thyroiditis

Thyroiditis is a general term often applied to a collection of gene-linked autoimmune disorders involving production of antibodies that attack thyroid proteins, especially the peroxidase enzyme and sometimes Tgb. Thyroid cells also may be attacked by lymphocytes that are attracted by the actions of antibodies with thyroid antigens. As with all thyroid disorders, thyroiditis is more prevalent in women than in men. It is sometimes the cause of hypothyroidism that follows pregnancy due to a rebound in immune activity following suppression during pregnancy.

The most common types are **Hashimoto's thyroiditis** (struma lymphomatosa) and **Reidel's thyroiditis** (struma fibrosa). Unlike Graves' disease, a hyperthyroid autoimmune disease, thyroiditis disorders all are associated with hypothyroidism. It is one of the most common causes for goiter among adolescents without iodide deficiency and is characterized by the presence of numerous lymphoid follicles, diffuse infiltration of the thyroid by lymphocytes, extensive increase in the connective tissue components of the thyroid gland, some changes in follicular structure, and a marked reduction in production of thyroid hormones. In Reidel's thyroiditis, the gland is progressively replaced by fibrous connective tissue. In either type of thyroiditis, a gradual reduction in circulating thyroid hormones brings about an elevation of TSH secretion (negative feedback) that augments goiter formation. Antibodies produced against thyroid iodoproteins and peroxidase generally are demonstrable in the blood of these patients.

E. Inherited Thyroid Disorders

More than 20 heritable defects in thyroid gland metabolism are known and may be classified according to the type of causative metabolic defect. For example, there are several defects related to unresponsiveness of thyroid cells to TSH including insufficient numbers of receptors or abnormal receptors. There also have been numerous mutations discovered in the gene responsible for synthesis of one particular thyroid hormone receptor, TRβ-1. These mutations are linked to generalized resistance to thyroid hormones. In a number of these cases, resistance to thyroid hormones has been correlated with the presence of **attention-deficit hyperactivity disorder (ADHD)** in children.

Inherited defects also occur in iodide transport mechanisms (as discussed above in Box 6A) and in the iodination and coupling of iodinated tyrosines. Several defects have been identified in relation to synthesis of abnormal Tgb. Deficiencies in thyroid deiodinase (D1) and total body deiodinases as well as the presence of abnormal plasma transport proteins sometimes occur as heritable errors. Mutations in the MCT8 transporter gene are responsible for the X-chromosome linked disorder called **Allan–Herndon–Dudley syndrome**, one of the first heritable disorders linked with mental retardation. **Pendred syndrome** leads to deafness and goiter formation and is caused by a defect in the *SLC26A4* gene which encodes the pendrin transporter, a member of a large family of facilitated diffusion anion transporters.

F. Euthyroid Sick Syndrome

Many non-thyroid conditions can alter thyroid function, so that even though the person is actually euthyroid, he or she may appear to be hyper- or hypothyroid. For example, fasting, anorexia nervosa, protein/calorie malnutrition, and untreated diabetes mellitus are all metabolic disturbances that bring about marked decreases in circulating T_3 and increases in rT_3. Thyroid hormones also are altered in a variety of liver and renal diseases as a consequence of numerous infections and following myocardial infarctions. Some conditions may alter levels of T_4, TSH, T_3, and/or rT_3. These alterations in thyroid parameters by peripheral disorders constitute the so-called **euthyroid sick syndrome**, reflecting normal responses of the euthyroid person to the generalized disease state that presents superficially as abnormal thyroid function.

VII. SUMMARY

The functional unit of the vertebrate thyroid gland is the thyroid follicle that is unique for storing T_4 extracellularly in the colloid of the follicular lumen. Normal thyroid gland functioning depends upon a constant supply of iodide in the diet as well as upon regulatory stimuli from the hypothalamus (TRH) and the adenohypophysis (TSH). Iodide is accumulated by the follicular cells and at the apical cell surface is incorporated into tyrosine residues of large glycoproteins collectively referred to as Tgb. Iodide is added to tyrosine by TPO to form DIT. Two DITs within the Tgb are coupled by thyroid peroxidase to form T_4, and the iodinated Tgb is stored in the follicular lumen. Endocytosis of colloid droplets followed by intracellular hydrolysis of Tgb in endolysosomes frees T_4 to enter the circulation. Some T_4 is degraded to T_3 and rT_3 by a type I deiodinase present in thyroid cells, and these products also may enter the blood. The same deiodinase converts DIT back to tyrosine and inorganic iodide for new synthesis of Tgb and subsequent iodination of its tyrosine residues. Synthesis and release of thyroid hormones are stimulated by TSH. Anti-thyroid drugs (thionamides, such as PTU and methimazole) and certain anions (SCN^-, $HClO_3^-$, TcO_4^-) may be termed goitrogens and interfere with iodide uptake or the iodination process, or both, causing thyroid deficiencies that may lead to goiter formation.

Thyroid hormones are transported in the circulation bound to plasma proteins, and only a small proportion of free hormones is present. Metabolism of most of the T_4 to T_3 and rT_3 occurs in the liver by the action of D1 deiodinase, with some T_3 arising from the thyroid or in target tissues prior to binding to intracellular receptors. Mitochondrial and nuclear receptors also occur in target tissues. These receptors have greater affinity for T_3 than for T_4, supporting the hypothesis that deiodination of T_4 to T_3 is the important first step in its mechanism of action.

It is difficult to characterize the actions of thyroid hormones with respect to their functions. In addition to their direct actions on development and oxidative metabolism, thyroid hormones play numerous permissive roles. They maintain the "responsiveness" in many cells that allows these cells to become more sensitive to other endocrine or neural stimuli. Thyroid hormones also enhance the secretion and actions of other hormones. Processes that are affected by permissive or cooperative actions include growth, metabolism, and reproduction.

The action of thyroid hormones on peripheral tissues following specific binding of the hormone to nuclear receptors is followed by activation or inhibition of mRNA transcription and eventual enzyme syntheses. Occupied TRs commonly form heterodimers with RXRs prior to interacting with the TRE in the DNA. The permissive actions of thyroid hormones may be related to such events as stimulation of the synthesis of components of second-messenger systems, upregulation of receptors for another regulator, effects on structural components, etc. Binding of thyroid hormones to mitochondrial receptors or indirect actions on mitochondria, especially in liver and kidney, may be important in the thermogenic and oxidative actions of thyroid hormones on BMR and in chronic cold stress.

STUDY QUESTIONS

1. What is myxedema and how is this related to thyroid function?
2. What is cretinism and how is this related to altered thyroid gland function?
3. Does the vertebrate genome contain genes for T_3 and T_4? Explain.
4. Imagine that you are an iodide ion. Outline your journey from the blood, through the thyroid follicle cell, and into the colloid, identifying the relevant transport proteins and enzymes involved in each step of your voyage. Repeat this exercise for tyrosine. How does tyrosine enter the thyroid follicle cell and where is tyrosine added to the Tgb molecule?
5. Describe the role that endocytosis plays in thyroid hormone synthesis and secretion.
6. Describe the role that the sodium potassium ATPase protein plays in establishing the sodium gradient across the basolateral plasma membrane of the thyrocyte and why this is critical for iodide transport.
7. What role do apical tight junctions play in thyroid follicle cell function? How do tight junctions contribute to polarity or compartmentalization of thyroid cell function?
8. How does the role of the sodium iodide symporter differ from that of pendrin with respect to iodide transport in thyroid follicle cells?
9. What are iodolipids? How are they synthesized and how do they differ structurally from T_3 and T_4?
10. Why would nuclear power plant workers be instructed to take potassium iodide tablets in the event of a reactor meltdown?
11. Why are epidemiologists concerned about thyroid cancer rates in the wake of the Fukushima nuclear power plant that occurred in March of 2011?
12. Describe how mutations in discrete parts of the NIS gene can lead to problems in NIS posttranslational processing, trafficking to the plasma membrane, and binding to Na^+ or I^-.
13. How are changes in thyroid follicle cell size linked to blood concentrations of TSH?
14. Identify the products obtained when type 1, 2, or 3 deiodinases use T_4 as a substrate. What is reverse T_3 and how is it produced?
15. Describe the role of negative feedback in the regulation of the HPT axis.
16. How is dietary iodide linked to normal thyroid gland function? Can high dietary iodide intake adversely affect thyroid function?
17. Identify three environmental iodide uptake inhibitors.
18. What is a thyronamine? Describe how thyronamines are synthesized from thyroid hormone.
19. Identify three distinct mechanisms for thyroid hormone receptor interaction on target cells. Include mechanism(s) that may mediate thyronamine action. How do thyronamines differ from thionamides?
20. Identify examples of EDCs that are known to disrupt the HPT axis. Can you identify EDCs that interfere with TH synthesis, metabolism, and receptor interaction?
21. Outline the mechanisms underlying thyroid hormone action on metabolism and body temperature. Identify the roles of brown adipose tissue and uncoupling protein in mediating these effects.
22. What role do thyroid hormones play in normal brain development?
23. How do thyroid hormones gain access to nerve cells in the brain?
24. Identify three causes of thyrotoxicosis.
25. What evidence suggests that Graves' disease is an autoimmune disease?
26. Identify abnormalities in thyroid function that may lead to goiter formation.
27. Identify three heritable thyroid axis disorders.

SUGGESTED READING

Books

Braverman, L.E., Cooper, D. (Eds.), 2012. Werner and Inghar's The Thyroid: A Fundamental and Clinical Text, tenth ed. Lippincott Williams & Wilkins, New York.

Burrow, G.N., Oppenheimer, J.H., Volpe, R., 1989. Thyroid Function and Disease. W.B. Saunders, Philadelphia, PA.

Greer, M.A., 1990. The Thyroid Gland. In: Comprehensive Endocrinology Review Series. Raven Press, San Diego, CA.

McNabb, F.M.A., 1993. Thyroid Hormones. Prentice Hall, Upper Saddle River, NJ.

Melmed, S., Polonsky, K.S., Larsen, R.P., Kronenberg, H.M., 2011. Williams Textbook of Endocrinology: Expert Consult-Online and Print, twelfth ed. W.B. Saunders, Philadelphia, PA.

Articles

Bernal, J., 2002. Action of thyroid hormone in brain. Journal of Endocrinological Investigation 25, 268–288.

Bizhanova, A., Kopp, P., 2010. Genetics and phenomics of Pendred syndrome. Molecular and Cellular Endocrinology 322, 83–90.

Cheng, S.Y., Leonard, J.L., Davis, P.J., 2010. Molecular aspects of thyroid hormone actions. Endocrine Reviews 31, 139–170.

Christophe, D., Vassart, G., 1990. The thyroglobulin gene: evolutionary and regulatory issues. Trends in Endocrinology and Metabolism 1, 356–362.

Davis, P.J., Leonard, J.L., Davis, F.B., 2008. Mechanisms of nongenomic actions of thyroid hormone. Frontiers in Neuroendocrinology 29, 211–218.

De Groef, B., Decallonne, B.R., Van der Geyten, S., Darras, V.M., Bouillon, R., 2006. Perchlorate versus other environmental sodium/iodide symporter inhibitors: potential thyroid-related health effects. European Journal of Endocrinology 155, 17–25.

Dohán, O., Carrasco, N., 2003. Advances in Na(+)/I(−) symporter (NIS) research in the thyroid and beyond. Molecular and Cellular Endocrinology 213, 59–70.

Dumont, J.E., Maenhaut, C., Lamy, F., 1992. Control of thyroid cell proliferation and goitrogenesis. Trends in Endocrinology and Metabolism 3, 12–17.

Flamant, F., Gauthier, K., Samarut, J., 2007. Thyroid hormones signaling is getting more complex: STORMs are coming. Molecular Endocrinology 21, 321–333.

Fliers, E., Klieverik, L.P., Kalsbeek, A., 2010. Novel neural pathways for metabolic effects of thyroid hormone. Trends in Endocrinology and Metabolism 21, 230–236.

Hoefig, C.S., Renko, K., Piehl, S., Scanlan, T.S., Bertoldi, M., et al., 2012. Does the aromatic l-amino acid decarboxylase contribute to thyronamine biosynthesis? Molecular and Cellular Endocrinology 349, 195–201.

Howdeshell, K.L., 2002. A model of the development of the brain as a construct of the thyroid system. Environmental Health Perspectives 110 (Suppl. 3), 337–348.

Kinne, A., Schülein, R., Krause, G., 2011. Primary and secondary thyroid hormone transporters. Thyroid Research 4 (Suppl. 1), 1–10.

Kohn, L.D., Suzuki, K., Nakazato, M., Royaux, I., Green, E.D., 2001. Effects of thyroglobulin and pendrin on iodide flux through the thyrocyte. Trends in Endocrinology and Metabolism 12, 10–16.

König, S., Neto, V.M., 2002. Thyroid hormone actions on neural cells. Cellular and Molecular Neurobiology 22, 517–544.

Lanham, S.A., Fowden, A.L., Roberts, C., Cooper, C., Oreffo, R.O., Forhead, A.J., 2011. Effects of hypothyroidism on the structure and mechanical properties of bone in the ovine fetus. Journal of Endocrinology 210, 189–198.

Marsili, A., Zavacki, A.M., Harney, J.W., Larsen, P.R., 2011. Physiological role and regulation of iodothyronine deiodinases: a 2011 update. Journal of Endocrinology Investigation 34, 395–407.

Miller, M.D., Crofton, K.M., Rice, D.C., Zoeller, R.T., 2009. Thyroid-disrupting chemicals: interpreting upstream biomarkers of adverse outcomes. Environmental Health Perspectives 117, 1033–1041.

Piehl, S., Hoefig, C.S., Scanlan, T.S., Köhrle, J., 2011. Thyronamines: past, present, and future. Endocrine Reviews 32, 64–80.

St. Germain, D.L., 1994. Iodothyronine deiodinases. Trends in Endocrinology and Metabolism 5, 36–42.

Silva, J.E., 1993. Hormonal control of thermogenesis and energy dissipation. Trends in Endocrinology and Metabolism 4, 25–32.

Spitzweg, C., Morris, J.C., 2010. Genetics and phenomics of hypothyroidism and goiter due to NIS mutations. Molecular and Cellular Endocrinology 322, 56–63.

Studer, H., Gerber, H., 1991. Intrathyroidal iodine: heterogeneity of iodocompounds and kinetic compartmentalization. Trends in Endocrinology and Metabolism 2, 29–34.

Twyffels, L., Massart, C., Golstein, P.E., Raspe, E., Van Sande, J., Dumont, J.E., Beauwens, R., Kruys, V., 2011. Pendrin: the thyrocyte apical membrane iodide transporter? Cell Physiology and Biochemistry 28, 491–496.

Visser, W.E., Friesema, E.C., Jansen, J., Visser, T.J., 2008. Thyroid hormone transport in and out of cells. Trends in Endocrinology and Metabolism 19, 50–56.

Yen, P.M., 2001. Physiological and molecular basis of thyroid hormone action. Physiology Reviews 81, 1097–1142.

Zoeller, R.T., Dowling, A.L.S., Herzig, C.T.A., Iannacone, E.A., Gauger, K.J., Bansal, R., 2002. Thyroid hormone, brain development, and the environment. Environmental Health Perspectives 110 (Suppl. 3), 355–361.

Clinical Articles

Andersson, M., de Benoist, B., Rogers, L., 2010. Epidemiology of iodine deficiency: salt iodisation and iodine status. Best Practice & Research Clinical Endocrinology & Metabolism 24, 1–11.

Edwins, D.L., McGregor, A.M., 1990. Pregnancy and autoimmune thyroid disease. Trends in Endocrinology and Metabolism 1, 296–300.

Fisfalen, M.-E., DeGroot, L.J., 1995. Graves' disease and autoimmune thyroiditis. In: Weintraub, B.D. (Ed.), Molecular Endocrinology: Basic Concepts and Clinical, Implications. Raven Press, San Diego, CA, pp. 319–370.

Fisher, D.A., 2002. Fetal-perinatal thyroid physiology. In: Eugster, E.A., Pescovitz, O.H. (Eds.), Developmental Endocrinology: From Research to Clinical Practice. Humana Press, Totowa, NJ, pp. 135–149.

Hauser, P., Zametkin, A.J., Martinez, P., Vitiello, B., Matochik, J.A., Mixson, A.J., Weintraub, B.D., 1993. Attention-deficit-hyperactivity disorder in people with generalized resistance to thyroid hormone. New England Journal of Medicine 328, 997–1040.

Kopp, P., Pesce, L., Solis-S, J.C., 2008. Pendred syndrome and iodide transport in the thyroid. Trends in Endocrinology and Metabolism 19, 260–268.

Leung, A.M., Pearce, E.N., Braverman, L.E., 2011. Iodine nutrition in pregnancy and lactation. Endocrinology and Metabolism Clinics of North America 40, 765–777.

O'Connor, G., Davies, T.F., 1990. Human autoimmune thyroid disease: a mechanistic update. Trends in Endocrinology and Metabolism 1, 266–274.

Ron, E., 2007. Thyroid cancer incidence among people living in areas contaminated by radiation from the Chernobyl accident. Health Physics 93, 502–511.

Saranac, L., Zivanovic, S., Bjelakovic, B., Stamenkovic, H., Novak, M., Kamenov, B., 2011. Why is the thyroid so prone to autoimmune disease? Hormone Research in Paediatrics 75, 157–165.

Usala, S.J., Weintraub, B.D., 1991. Thyroid hormone resistance syndromes. Trends in Endocrinology and Metabolism 2, 140–144.

Wong, T.K., Hershman, J.M., 1992. Changes in thyroid function in nonthyroid illness. Trends in Endocrinology and Metabolism 3, 8–12.

Zimmermann, M.B., 2009. Iodine deficiency. Endocrine Reviews 30, 376–408.

Chapter 7

The Hypothalamus–Pituitary–Thyroid (HPT) Axis of Non-Mammalian Vertebrates

The thyroid hormones, **triiodothyronine (T_3) and thyroxine (T_4)**, are ancient molecules, and they appear in a number of invertebrate phyla where they affect development (Table 7-1). The vertebrate **hypothalamus–pituitary–thyroid (HPT) axis** can be traced to its beginnings among the primitive protochordates. All of the features of the mammalian HPT axis are present in vertebrates, and many are found in the protochordates, where thyroid hormones control developmental processes, including metamorphosis.

In all vertebrates, iodide is accumulated from the environment and sequestered within the thyroid cells in response to **thyrotropin (TSH)** secreted from the pituitary gland (Table 7-2). In response to TSH, these cells produce **thyroglobulin (Tgb)** and store it in the follicular lumen. **Thyroid peroxidase (TPO)** is responsible for iodinating tyrosines to form **diiodotyrosine (DIT)** in Tgb and coupling two DITs to produce T_4. TSH also stimulates hydrolysis of Tgb to release T_4, which is converted by **deiodinases** to T_3. Vertebrate thyroids are sensitive to the same **goitrogens** as described for mammals in Chapter 6 (Table 7-2). Circulating levels of T_3 and T_4 in some representative vertebrates are provided in Table 7-3. Chapters 3, 4, 5, and 6 provide the basic details and

TABLE 7-1 Occurrence of Thyroid Hormones and Their Precursors in Animals

	MIT	DIT	T_3	T_4
Phylum Porifera (sponges)	+	+		
Phylum Cnidaria (corals)	+	+		
Phylum Annelida (clam worm)	+	+	+	+
Phylum Arthropoda (cockroach, fly)	+	+		+
Phylum Chordata: Protochordates	+	+	+	+
Phylum Chordata: Vertebrates	+	+	+	+

TABLE 7-2 Some Effects of Thyrotropin and Thyroid Inhibitors on ^{131}I Uptake[a] by Thyroids of Larval Salamanders, *Ambystoma tigrinum*

N	Mean Body Weight (g ± SEM)	Daily Injections; Pretreatment for 7 Days	Mean Thyroid Uptake (% Injected Dose ± SEM)
6	11 ± 0.5	None (control)	4.6 ± 1.6
6	8 ± 0.6	0.25µg TSH[b]	12.4 ± 2.5
6	11 ± 0.6	2.5µg TSH	29.2 ± 4.7
6	17 ± 16	25µg TSH	38.3 ± 1.0
6	9 ± 0.6	2.5µg TSH + 10 µg PTU[c]	10.2 ± 2.9
6	7 ± 1.0	2.5µg TSH +0.5mg NaSCN[d]	3.9 ± 0.7

[a]*Radioiodide uptake was determined 24 hr after intraperitoneal injection of 5 µCi of ^{131} I.*
[b]*Ovine TSH (NIH-TSH-S6).*
[c]*PTU, propylthiouracil.*
[d]*Sodium thiocyanate.*

Vertebrate Endocrinology. http://dx.doi.org/10.1016/B978-0-12-394815-1.00007-0

TABLE 7-3 Blood Levels of Thyroid Hormones in Selected Vertebrate Species

Class/order	Species	Conditions	T_3 (ng/mL)	T_4 (ng/mL)
Agnatha/Cyclostomata	*Petromyzon marinus* (sea lamprey)	Mature females	0.3-1.2	54-108
Chondrichthyes	*Potamotrygon motoro*[a] (Freshwater stingray)	Females	0.2	5.43
		Males	0.2	1.85
	Paratrygon aiereba[a] (Freshwater stingray)	Females	0.1	5.8
	Sphyrna tiburo[a] (Bonnethead shark)	Preovulatory	0.04	2.11
		Pregnant	0.14	3.62
	Triaenodon obesus (White-tipped reef shark)	Males & Females Males & Females	0.67	4.9
	Dasyatis sabina (Atlantic skate)	Males & Females	2.17	17
Osteichthyes/Teleostei	*Gadus morhua* (cod)	Winter	18-27	8-12
		Summer	2-10	1-2
	Psuedopleuronectes americanus (winter flounder)		2-19	2-16
	Pleuronectes platessa (plaice)		3-9	3-13
Amphibia/Anura	*Pelophylax (Rana) ridibunda* (Eurasian marsh frog)		0.03	0.35
Amphibia/Caudate	*Ambystoma tigrinum* (tiger salamander)	Sexually mature larvae (neotene)	0.1	0.3-9
	Ambystoma mexicanum (Mexican axolotl)	Neotene	0.01	0.37
		Metamorph	0.06	2.12
Reptilia/Squamata	*Alligator mississippiensis*	Seasonal variations in juveniles		2-12
	Trachydosaurus ragostus (shingleback lizard)	At preferred temperature	0.18	2
	Calotes versicolor (garden lizard)	Prior to hibernation	1.2	4.3
		During hibernation	0.6	1
Reptilia/Crocodilia	*Crocodylus johnstoni* (freshwater crocodile)	At preferred temperature	0.33	2.5
Reptilia/Chelonia	*Chelodina longicollis* (long-necked tortoise)	At preferred temperature	0.18	0.42
Aves/Galliformes	*Aptendytes patagonicus* (King penguins)	Chicks, First summer	0.13	3.9
		Chicks, Second summer (moult)	1.6	35

[a]*Unpublished data courtesy of Dr. Mikki McComb-Kobza*

operation of the vertebrate HPT axis as well as details of thyroid hormone synthesis, actions, and metabolism that apply to all vertebrates.

I. EVOLUTION OF THE THYROID GLAND AND ITS FUNCTIONS

The fundamental structure of the thyroid gland structurally is conserved in vertebrates and has a direct connection to an invertebrate chordate structure. In vertebrates, the thyroid generally occurs as one or two masses of highly vascularized follicles surrounded by a connective tissue capsule or as scattered follicles throughout the pharyngeal region, as is the case for most fishes (Figures 7-1 and 7-2). Regardless of any gross morphological differences, follicular structure and function are mammalian-like in all the jawed vertebrates with respect to iodide metabolism and hormone synthesis.

Current hypotheses on the origin of the thyroid gland (see Figure 7-3) rely on the presence of thyroid hormone receptors (TRs) among invertebrates ancestral to the protostome–deuterostome divergence some 550 MYBP followed by the active synthesis of thyroid hormones and their association with development among invertebrate deuterostomes. This was followed by the appearance of discrete organs for synthesis of thyroid hormones among protochordates that gave rise to the thyroid gland in vertebrates.

A. Invertebrate Origins of Thyroid Hormones and Evolution of the Thyroid Gland

Iodine compounds such as iodate as well as iodide are found in all organisms. It has been proposed that iodinated compounds such as **3-monoiodotyrosine (MIT)** and **3,5-diiodotyrosine (DIT)** were important signaling chemicals in early multicellular organisms. The incorporation of iodide into tyrosine residues in a protein involving formation of MIT and DIT and iodinated thyronines (T_3 and T_4) occurs repeatedly among the invertebrate phyla (Table 7-1). Initially, the incorporation of iodide was a general surface phenomenon that became concentrated in the oral regions of many animals, and this eventually led to development of a distinct thyroid gland (see Figure 7-3). However, relatively few studies have examined these iodinated molecules and their functions among invertebrates.

One hypothesis for the origin of these primitive iodoproteins is that they first served some enzymatic or structural function and that the basic mechanism became modified into a hormonal synthetic pathway. Another hypothesis may be that these iodine-containing compounds were obtained originally from feeding activities by protochordates and later were utilized opportunistically as regulatory substances following digestion in the gut that liberated the thyronines. Absorption of T_4 liberated during digestion of these glycoproteins and deiodination of T_4 to T_3 might have been important prerequisites for development of a regulatory role. The retention of these glycoproteins, the ability to accumulate iodide, and the formation of iodinated thyronines were acquired by certain cells of the **endostyle** (Figure 7-4). These specialized cells later became the follicular cells of the thyroid gland as evidenced from studies in lampreys where the iodide-concentrating cells of the endostyle of the ammocetes larva of the lamprey are reformed into follicles of the thyroid gland during metamorphosis into the adult body form. The production of extracellular colloid and its resultant accumulation in the follicular lumen do not occur in modern cyclostomes. These abilities may have been acquired in other fishes after the cyclostomes branched from the mainline of vertebrate evolution.

The first evidence in chordates of formation of thyroid hormones comes from studies of the endostyle of protochordates such as the tunicates and especially the cephalochordates known as amphioxus. The endostyle is a mucus-secreting gland located in the pharyngeal region. This mucus, a glycoprotein mixture, is involved in feeding mechanisms that trap food particles brought into the pharynx by water currents created by movements of pharyngeal cilia. These cilia then move the mucus with trapped food into the digestive tract. Most of the elements of the thyroid axis have been described in amphioxus (Figure 7-5). The protochordate endostyle expresses **thyroid transcription factor 1 (ttf-1)**, a product of the *Ttf1* gene (also called *Nkx2.1*). In mammals, ttf-1 is necessary for development of the thyroid gland as well as of the lung, another derivative of the anterior digestive tract. Specialized cells near the opening of the endostyle of protochordates accumulate iodide from seawater and use it to iodinate tyrosines incorporated into the mucus glycoproteins secreted by these cells. The enzyme TPO is responsible for iodination of tyrosine and coupling them to form T_4. Although a vertebrate-like gene for Tgb has not been identified in the amphioxus genome, another glycoprotein is involved. Deiodinase is present in amphioxus and converts T_4 to T_3. There is some debate about the chemical structure of the most primitive bioactive thyroid hormone. Despite the fact that the cephalochordate amphioxus synthesizes both T_4 and T_3, T_3 binds poorly to the amphioxus **thyroid receptor** (TR) and is not very effective at inducing metamorphosis in these animals. Rather, it is a decarboxylated metabolite of T_3, 3,3′,5-triiodothyroacetic acid or TRIAC that is most effective in stimulating amphioxus metamorphosis. TRIAC is also produced in humans, accounting for roughly 14% of T_3 metabolism.

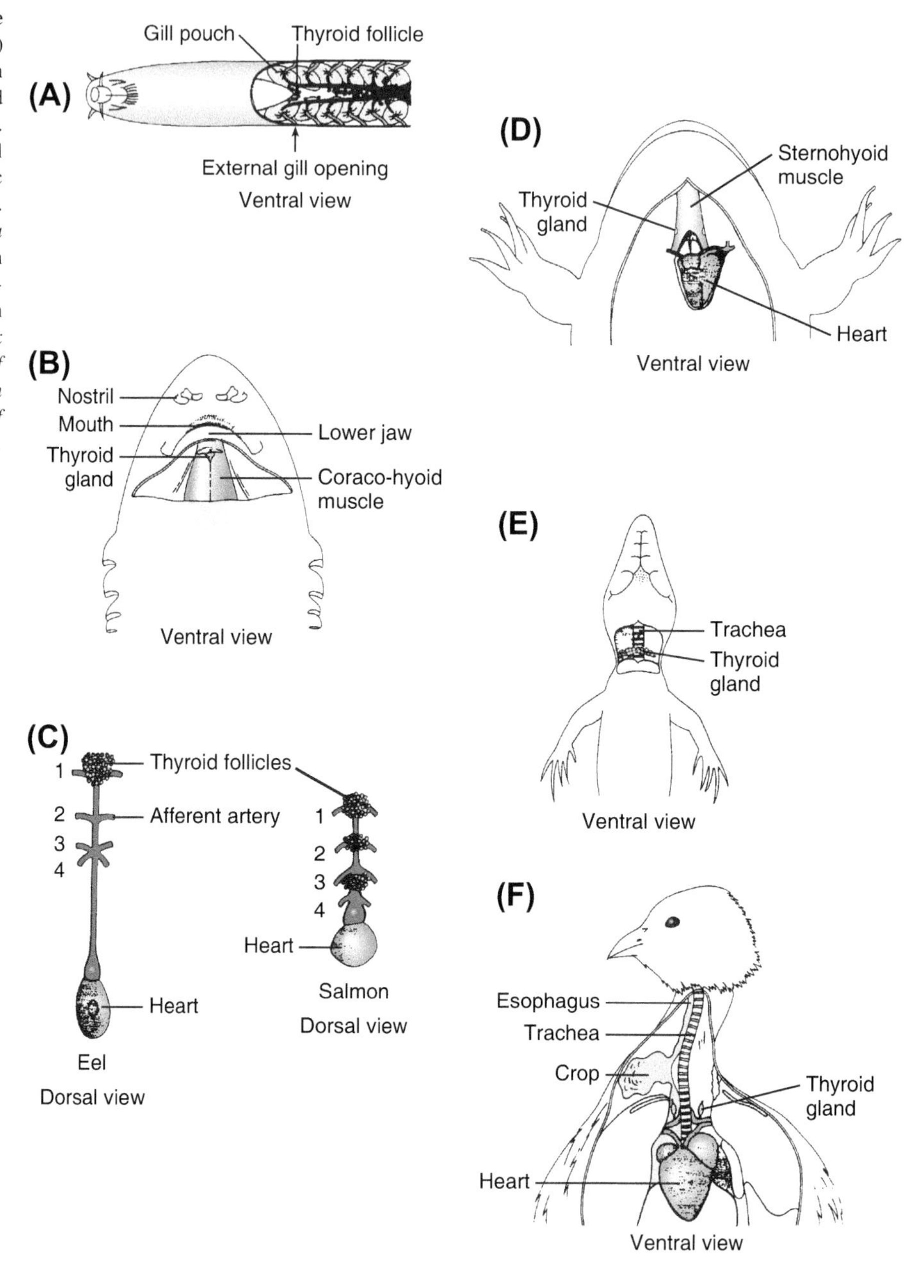

FIGURE 7-1 **Location of thyroid tissue in non-mammalian vertebrates.** (A) Scattered thyroid follicles in a hagfish (*Eptatretus burgeri*). (B) Discrete thyroid gland of the shark *Triakis scyllium*. (C) Diffuse thyroids of the Japanese eel (*Anguilla japonica*) (left) and the Pacific salmon (*Oncorhynchus masou*) (right). (D) Paired thyroids in the bullfrog (*Rana catesbeiana*). (E) Medial thyroid gland in neck of the lizard *Takydromas tachydromoides*. (F) Paired thyroid glands in a bird, the Japanese quail (*Coturnix coturnix japonicus*). *(Dissections of thyroid regions adapted with permission from Matsumoto, A. and Ishii, S., "Atlas of Endocrine Organs," Springer-Verlag, Berlin, 1989.)*

The most primitive thyroid condition in vertebrates is considered to be that found in the adult cyclostomes in which only scattered follicles may be present. However, extracellular storage of thyroid hormones does not occur in the cyclostome follicle, and iodoproteins are retained in the follicular cells of these primitive fishes. Most teleosts also have a diffuse thyroid consisting of follicles that are distributed throughout the pharyngeal region, but there is colloid in the follicles. In all other vertebrates, the follicles are organized into discrete glands and appear to function like those of mammals (see Chapter 6 for details of follicular activities).

B. Origins of Thyroid Functions

Acceptance of the simple notion that thyroid cells evolved from specialized cells of the endostyle of protochordates leaves some basic questions unanswered. For example, why were these iodinated compounds first synthesized? What was their primitive role? What selective forces were responsible for adoption of these iodinated compounds as metabolic regulators?

Perhaps an early functional role for thyroid hormones was regulation of development. Thyroid hormones are associated with a dramatic developmental change from

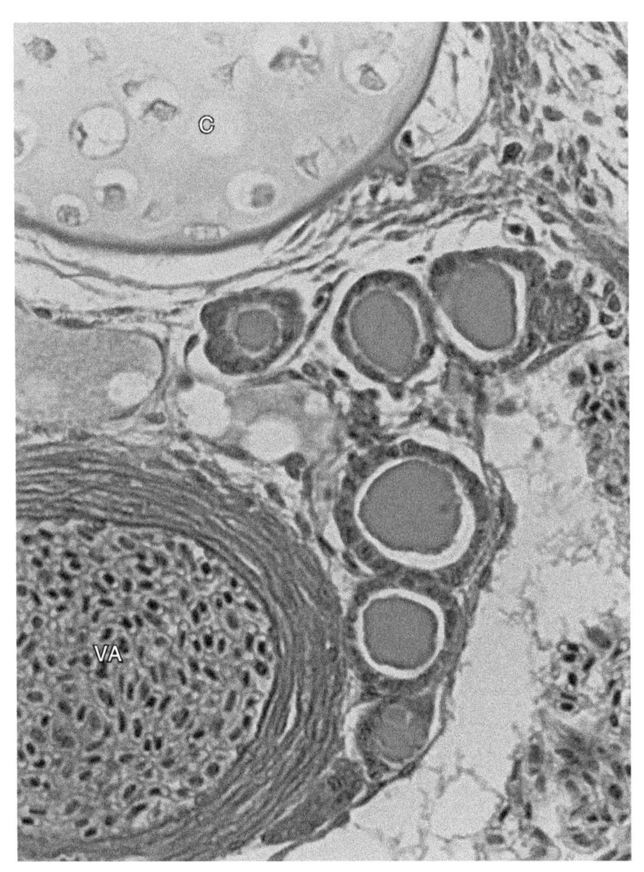

FIGURE 7-2 **Salmon thyroid follicles.** Cross-section through the lower jaw of a fingerling chinook salmon (*Oncorhynchus tshawytscha*) near the second aortic arch. Five thyroid follicles filled with colloid appear to the right of the ventral aorta (VA) and below the supporting cartilage (C).

a larval body form to an adult body form known as **metamorphosis** in a number of invertebrate phyla, including the protochordate amphioxus (Figure 7-6). The metamorphosis of larvae of the tunicate protochordate *Ciona intestinalis* is blocked by exposure to the potential goitrogen thiourea. Metamorphosis in fishes and amphibians includes specific gene activation by thyroid hormones, resulting in biochemical, physiological, structural, and behavioral changes as well as a major revision of the ecological role of the organism (see ahead). General morphological events as well as development of the central nervous system are seen as being thyroid hormone dependent from amphioxus to mammals. The accumulation of thyroid hormones and iodide in eggs of many species also relates to the importance of thyroid hormones in early development. Because of the early association of thyroid hormones with the digestive tract, it has been suggested that thyroid hormones might first have played a role in osmoregulation as well as in development.

An additional hypothesis states that thyroid function evolved hand in hand with endocrine control of reproduction and that the basic function for thyroid hormones is associated primitively with gonadal maturation (Table 7-4). This hypothesis is supported by a variety of observations, including the close parallel of thyroid activity and reproductive cycles in elasmobranchs and bony fishes and assumes that the roles in development, growth, etc., were acquired later.

C. Regulation of Vertebrate Thyroids

In Chapters 4 and 5, we described the regulation of thyroid function in vertebrates by the HPT axis and its hormones (**thyrotropin-releasing hormone, TRH**; **corticotropin-releasing hormone, CRH**; and TSH). The focus for feedback regulation in the HPT axis is at both the hypothalamus and pituitary. This system maintains a sufficient level of bound and free T_4 that can be converted peripherally by a D1 deiodinase to T_3 in liver and kidney or by a D2 deiodinase in target cells of most species. A D3 deiodinase is present in all vertebrates examined and deals primarily with the inactivation of T_4 and T_3 (see Chapter 3 for information about the types of deiodinases and their actions).

Pituitary thyrotropes and TSH are apparently absent in agnathans, and these cells may have evolved later from gonadotropes. This proposed origin for thyrotropes is supported by their cytological similarity to gonadotropes, their location in the adenohypophysis of elasmobranchs (pars ventralis) and teleosts (proximal pars distalis), and by the biochemical similarity of TSH-subunit genes to those for the gonadotropins (GTHs; see Chapters 4 and 5). Exogenous thyroid hormones and gonadal steroids both have inhibitory actions on thyrotropes as well as on gonadotropes in teleosts. Similar influences of thyroid hormones on gonadal function and, reciprocally, gonadal steroids on thyroid function have been reported in amphibians, reptiles, and birds. The observations that mammalian luteinizing hormone (LH) stimulates thyroid function in fishes may be interpreted as supportive of this relationship or it may be due simply to a chance similarity in the structures of mammalian LH and teleost TSH or absence of teleost TSH receptor specificity. These same GTHs are ineffective when tested on mammalian thyroids.

D. Prolactin and Its Interactions with the Thyroid Axis

A general antagonism has been reported between **prolactin (PRL)** and the thyroid axis of fishes and non-mammalian tetrapods. Studies have demonstrated what appears to be a goitrogenic action of PRL directly on thyroids of teleosts, amphibians, lizards, and birds, but the peripheral antagonism between T_4 and PRL in certain tissues of larval

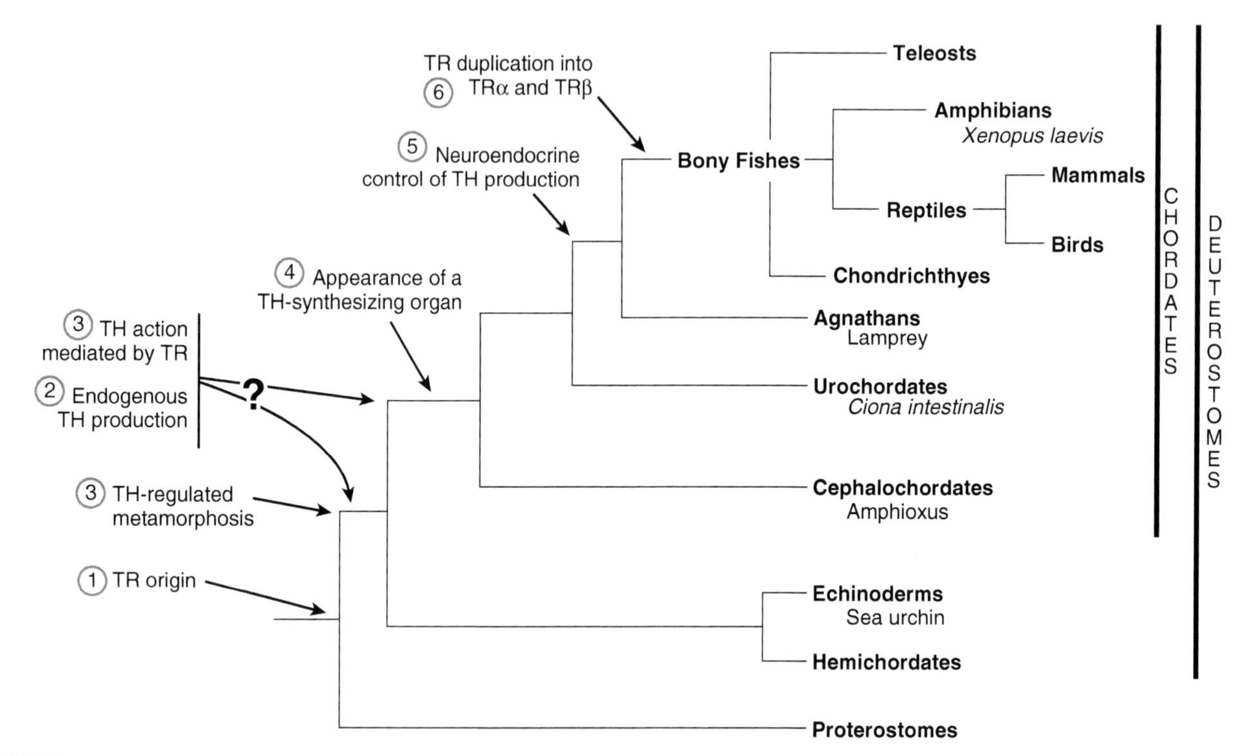

FIGURE 7-3 **Evolution of major events in thyroid physiology.** Thyroid receptors (TRs) appeared prior to the divergence of the proterostome and deuterostome invertebrates (1). Endogenous thyroid hormone production was next (2), and linkage to TRs and involvement with metamorphosis represent perhaps the earliest roles of thyroid hormone, appearing first among deuterostome invertebrates (3). A thyroid-synthesizing organ first appears among the protochordates (4), but neuroendocrine control probably originated among the first vertebrates as evidence from extant cyclostomes (5). *(Adapted with permission from Paris, M. et al.,* Integrative and Comparative Biology, *50, 63–74, 2008.)*

amphibians is best known. The significance of an antagonistic interaction between thyroid hormones and PRL in fishes, reptiles, and birds is not clear. In amphibians, the administration of anti-PRL agents (PRL antibodies or ergot derivatives) enhances responses of larval amphibians to endogenous and exogenous thyroid hormones. Early on these observations suggested an endogenous role for PRL in preventing premature metamorphosis (see ahead), although more recent studies in tadpoles overexpressing the PRL gene suggest a more limited role in antagonizing certain actions of T_4 in peripheral tissues. It is important to recognize that numerous investigations suggest that, although high levels of PRL are antimetamorphic, lower levels are essential for successful metamorphosis in amphibians.

Synthetic mammalian TRH causes PRL release from pituitaries of bullfrogs, turtles, birds, and mammals but not from the pituitaries of red-spotted newts. These observations question the biological significance of demonstrations that PRL produces anti-thyroid effects. They also suggest that the role of the TRH tripeptide, which is abundant in the amphibian brain, is that of a PRL-releasing agent, especially in light of many failed attempts to demonstrate a role for TRH in the activation of the amphibian thyroid axis (see Chapter 5).

E. Surgical, Chemical, and Radiothyroidectomy

A basic approach employed in thyroid studies involves hypophysectomy or thyroidectomy or a combination of the two followed by classical replacement therapy. Sometimes, however, it is desirable to make an animal only slightly hypothyroid or reversibly hypothyroid, or both. Chemical thyroidectomy is reversible and involves administration of a chemical goitrogen (see Chapter 6) such as **methimazole**, **propylthiouracil (PTU)**, or **thiourea (TU)** at a predetermined dose for a given period. Withdrawal of the goitrogen may then allow the animal to return to a euthyroid condition for comparisons. With this approach, changes in thyroid function before, during, and after treatment may be examined in each individual. Caution must be exercised in that some goitrogens (e.g., TU) have been shown to produce effects on other tissues (especially the liver) that do not appear following surgical thyroidectomy. Such "nonspecific" actions of chemical inhibitors are well known to investigators and must be considered when evaluating experimental results.

Large doses of radioiodide often are employed as therapeutic agents in humans to destroy excessive amounts of thyroid tissue in certain hyperthyroid conditions.

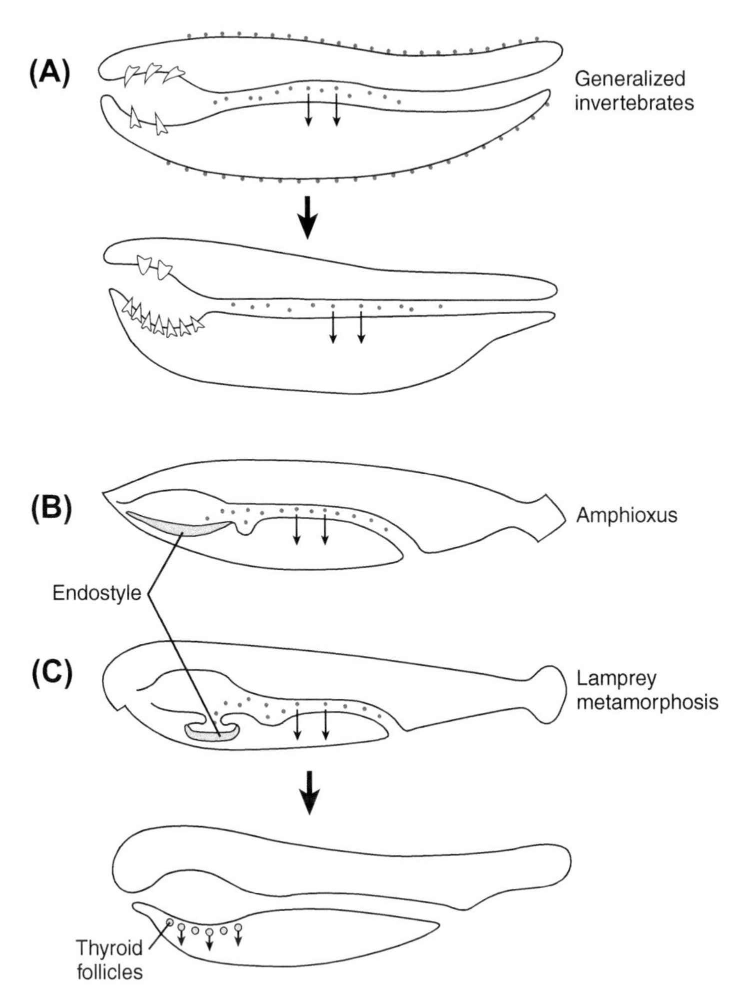

FIGURE 7-4 **Generalized pattern for evolution of the thyroid gland.** (A) Initially, iodinated mucoproteins were distributed over the body surface of marine invertebrates and in the anterior digestive tract and later became restricted more to the mouth region, where iodinated mucoproteins related to a ciliary–mucus feeding mode were released from the mouth region and entered the gut. Once in the gut, the iodinated mucoproteins were digested, liberating iodinated tyrosines and thyronines. (B) The cephalochordate amphioxus has iodinated protein production confined to the endostyle in the mouth and pharynx. (C) Metamorphosis of the larval endostyle to a thyroid gland occurs during transformation of the ammocete larvae to the adult lamprey.

Accumulated radioiodide destroys cells because of the destructive effects of radiation on the cell that has incorporated it. Relatively large doses can be used to completely destroy thyroid tissue. This approach has proven useful especially for some teleost fishes where it is impossible to surgically remove all of the thyroid tissue that occurs as scattered follicles. Radiothyroidectomy, like chemical thyroidectomy, must be interpreted with caution because radioiodide accumulation may occur in other tissues (see Table 7-5 and Figure 7-7) and result in destructive changes that may not be thyroid related. The general effects of whole-body radiation from this powerful gamma emitter could be a concern as well.

II. COMPARATIVE THYROID PHYSIOLOGY

The following accounts and Table 7-6 emphasize the biological actions of thyroid hormones on vertebrate target tissues. The secretion of TSH and the effects of various hypothalamic hormones on TSH secretion were discussed in Chapter 5, and the focus here is on the functions of T_3 and T_4.

A. Agnathan Fishes: Cyclostomes

As stated earlier, the endostyle of the larval lamprey becomes the thyroid gland during metamorphosis of the larval ammocoete into the adult lamprey. The transcription factor ttf-1 is also present in the lamprey endostyle, supporting a connection between protochordate endostyles and that of a larval vertebrate. No stimulatory role for thyroid hormones has been verified in cyclostome fishes, although in lampreys the binding of iodide by the larval endostyle increases following administration of T_4. Circulating levels of T_4 decrease markedly as free-living, filter-feeding ammocetes larvae of sea lampreys (*Petromyzon marinus*) undergo metamorphosis to the ectoparasitic adult lampreys.

Hormone synthesis in agnathans differs from other vertebrates in that organic binding of iodide and storage of the iodinated proteins occurs intracellularly. Hypophysectomy does not alter thyroid function in adult lampreys or in

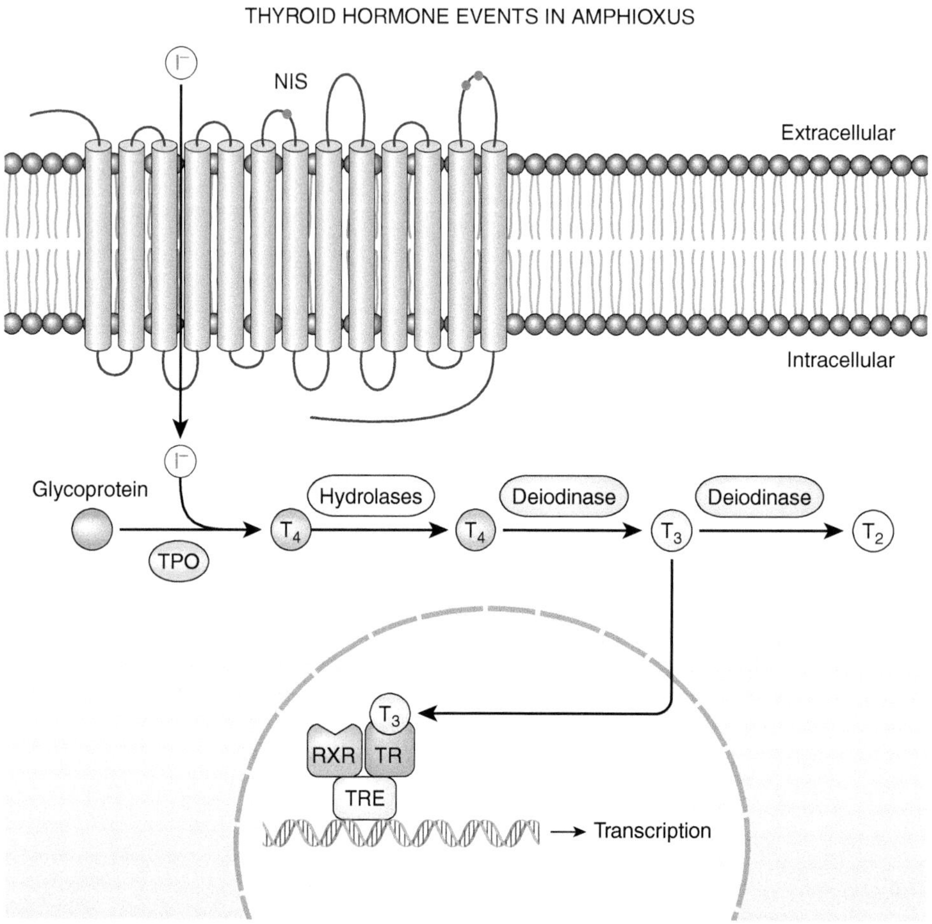

THYROID HORMONE EVENTS IN AMPHIOXUS

FIGURE 7-5 **Amphioxus thyroid hormone events.** Genomic evidence for the beginnings of the thyroid hormone synthesis and mode of action of thyroid hormones is present in this protochordate. Extracellular iodide is transported by a sodium-iodide pump symporter (NIS), incorporated into a glycoprotein forming T_4 through the actions of a thyroid peroxidase (TPO). The T_4-containing glycoprotein is hydrolyzed to release T_4, which is converted to triiodothyronine (T_3) by a deiodinase. T_3 binds to a thyroid receptor (TR) and forms a dimer with the RXR receptor prior to binding to a thyroid response element (TRE) and altering the target gene's activity (transcription). A gene coding for thyroglobulin apparently is absent in amphioxus.

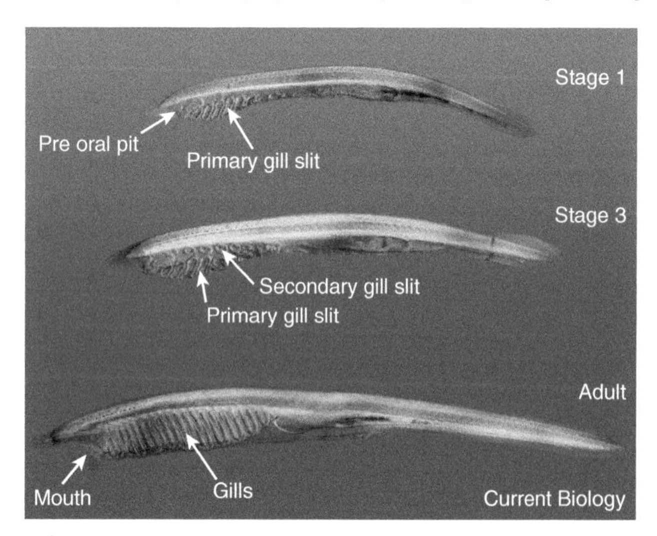

FIGURE 7-6 **Metamorphosis of the cephalochordate amphioxus.** This process is controlled by thyroid hormones synthesized in the endostyle. The adult amphioxus closely resembles the ammocoetes larva of the vertebrate lamprey (Agnatha, Cyclostomata). *(Photographs courtesy of Dr. Mathilde Paris.)*

hagfishes, and TSH has not been demonstrated in an agnathan pituitary. These observations suggest that evolution of thyrotropes and TSH either occurred first in jawed fishes, possibly in the ancient placoderms, or in another agnathan group that gave rise to jawed fishes and has persisted in all other extant vertebrates. T_4 is present in serum from mature female sea lampreys at levels comparable to those reported for ammocetes of this species. Much lower levels of T_4 are reported for mature male sea lampreys. T_3 occurs in the serum of both immature and mature sea lampreys, but the levels are higher in the mature individuals. These data suggest a relationship to reproductive maturation, but definitive data are needed before a firm relationship can be accepted.

B. Chondrichthyean Fishes

Only limited data are available with respect to thyroid function in sharks, in which the HPT axis appears to be well established. Most studies on sharks are related to effects of thyroid hormones on reproduction or oxygen consumption,

TABLE 7-4 Selected Experimental Results Relating Thyroid and Reproduction in Subavian Vertebrates

Vertebrate	Treatment/Observation	Effect on Reproduction
Agnathan fishes		
Sea lamprey	Elevated thyroid hormones	Occurs during spermiation and at ovulation
Jawed fishes		
Sharks	Thyroidectomy	Prevents annual gonadal recrudescence[a] by blocking vitellogenesis or incorporation of yolk into oocytes
	Stimulated thyroid histology	Coincides with sexual maturation
Teleosts	Elevated thyroid hormones in blood	During gonadal development
	Decreased thyroid hormones in blood	At spawning
	Goitrogen exposure	Inhibits gonadal recrudescence; can cause gonadal atrophy in both sexes
	Presence of goiters in salmon from Great Lakes	Associated with low egg production; low gonadal steroid levels in blood
	Thyroid hormone treatments	No effect or mild enhancement of gonadal maturation
Amphibians	Surgical or chemical thyroidectomy of frogs	Prevents ovulation
	Treatment with thyroxine of thyroidectomized frogs	Permits ovulation
	Treatment of anuran liver with thyroid hormones	Enhances action of estrogen on vitellogenesis
	Thyroid histology	Not correlated with reproductive activities
	Goitrogen treatment of tadpoles of some anurans	Prevents testicular development; may allow sex reversal to females
	Elevated thyroxine in the blood of salamanders	Corresponds to period of gonadal growth in males and females
Reptiles	Excess or insufficient thyroid hormones	Imparied ovarian growth; loss of steroid-secreting cells in testes
	Thyroxine treatment	Causes atrophy of steroid-secreting cells in testes
	Decreased thyroid hormones in blood and thyroid gland histology	Correlated with peak reproductive activity and high androgen levels
	Elevated thyroxine levels in blood	Decreased androgen levels in males

[a]*Normal regrowth of the gonad during the next breeding season.*

and the thyroid systems of other chondrichthyean fishes are virtually unknown.

1. Thyroid and Reproduction in Elasmobranchs

Thyroid cycles are positively correlated with reproductive cycles in sharks. However, increased thyroid activity observed in at least one species is correlated with migratory behavior related to reproduction rather than with reproduction or gonadal maturation *per se*. This area requires further investigation.

2. Thyroid and Oxygen Consumption in Elasmobranchs

Late embryos of *Squalus suckleyi* exhibit a transient increase in oxygen consumption following treatment with T_3 or T_4, but this response cannot be maintained by continued treatment with thyroid hormones. Treatment with PTU has no effect on oxygen consumption in this species. These observations do not provide strong support of a role for thyroid hormones in oxidative metabolism.

TABLE 7-5 Accumulation of Radioiodide by Thyroids and Gonads of Sexually Mature Vertebrates

Class	Species	Sex	Radioiodide Uptake gonad/thyroid ratio
Osteichthyes	*Micropterus dolomieu* (smallmouth bass)	M F	0.012 3.780
Amphibia	*Ambystoma tigrinum* (tiger salamander, neotene)	M F	0.045 0.512
Aves	*Coturnix coturnix japonica* (Japanese quail)	F	4 to 10

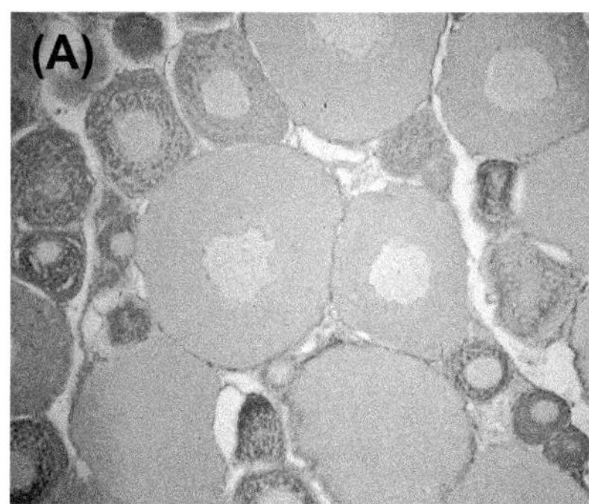

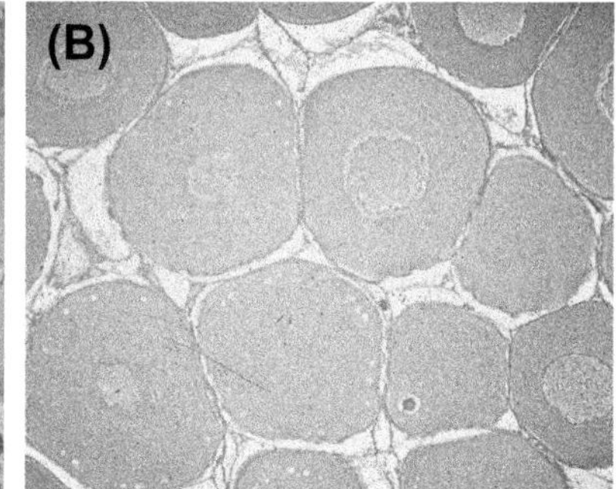

FIGURE 7-7 **Oocytes of control (A) and radioiodide-treated (B) yearling rainbow trout (*Oncorhynchus mykiss*).** The ovaries of fishes and other non-mammalian vertebrates readily concentrate iodide in the oocytes. The control ovary at the left exhibits a size range of developing oocytes. The ovary at the right is from a fish that accumulated radiodide (^{131}I) shortly after hatching. Although the radioactivity quickly decayed to undetectable levels, the ovary at the right contained only the largest class of oocytes when examined a year later.

TABLE 7-6 Some Major Actions of Thyroid Hormones in Vertebrates[a]

Vertebrate Group	Development	Molting	Growth	Metabolism	Reproduction	Neural Behavior	Neural Differentiation	Thermogenesis
Sharks					+	+	+	
Bony fishes	+		+		+	+		
Amphibians	+	+[b]	–[c]		-	+	+	
Reptiles		+	+	?	+			
Birds	+	+	+	+	+/–	+		+
Mammals	+	+	+	+	+	+	+	+

[a]+, stimulatory; –,inhibitory. The stimulatory action of thyroid hormones on metabolism seems to be associated with the evolution of homeothermy.
[b]Urodeles only. Molting in anurans is controlled by corticosteroids.
[c]Prevents growth in larval amphibians by inducing metamorphosis.

3. Thyroid and Neural Differentiation in Elasmobranchs

Differentiation of hypothalamic neurosecretory centers is accelerated in the embryos of the oviparous shark *Squalus suckleyi* following treatment with T_3 or T_4. This effect of thyroid hormones is manifest in both the preoptico-hypophysial fiber tracts and in the neurohypophysis. These data are suggestive of a role for thyroid hormones in nervous tissue differentiation and maturation of the HP system.

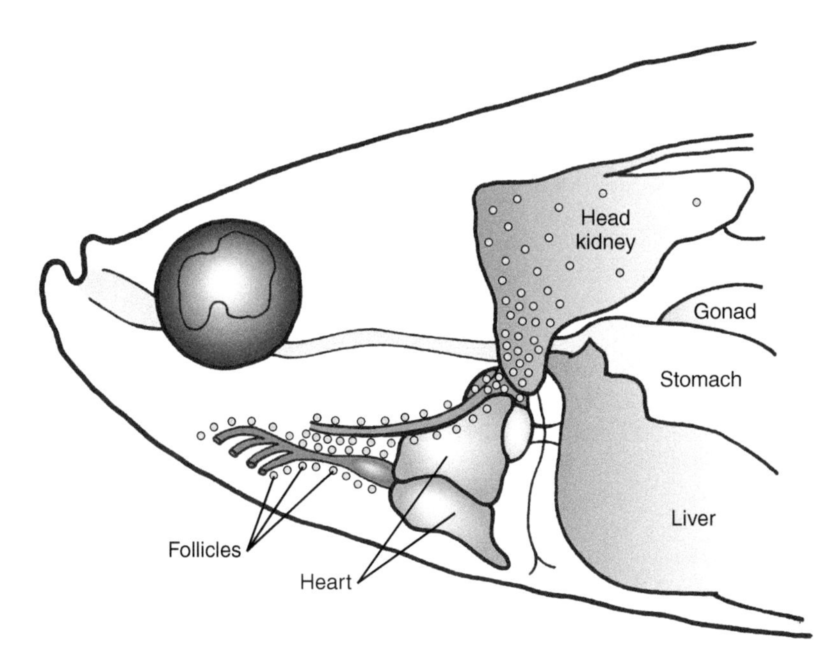

FIGURE 7-8 **Heterotopic thyroid tissue.** This teleostean fish *Xiphophorus maculatus* has thyroid follicles (yellow dots) located not only throughout the pharyngeal region but also in the heart, head kidney, and sometimes even in the gonads and liver. *(Adapted with permission from Baker-Cohen, K.F., in "Comparative Endocrinology" (A. Gorbman, Ed.), John Wiley & Sons, New York, 1959, pp. 283–319.)*

C. Bony Fishes: Chondrosteans

Some limited data for sturgeons indicate that peak thyroid activity coincides with spawning behavior. Although thyroid hormone levels were extremely low in cultured lake sturgeon (0.3 and 0.2 ng/mL for T_4 and T_3, respectively), wild spawning sturgeon had much higher levels (0.83 and 1.31, respectively). Furthermore, thyroid treatments can reverse the degenerative changes that occur in gonads of captive lake sturgeon, suggesting a direct relationship between thyroid hormones and reproduction. Curiously, lake sturgeon thyroid glands contain 10 times more T_3 than T_4, suggesting considerable thyroid deiodination compared to other vertebrates. Additional studies are needed in chondrostean fishes to examine other possible roles of thyroid hormones and to confirm this suggestive relationship with reproduction.

D. Bony Fishes: Teleosts

The teleost thyroid consists of scattered thyroid follicles throughout the pharyngeal region. The diffuse nature of the teleost thyroid gland, with only a few exceptions (for example, tuna and parrotfishes), makes assessment of thyroid gland function difficult and renders surgical thyroidectomy impossible. Most of the thyroid tissue is located in the pharyngeal area, where follicles are usually found between the second and fourth aortic arches (see Figures 7-2 and 7-8). Because of the absence of a covering connective tissue sheath that holds the follicles into one or more masses, thyroid follicles are frequently found outside this pharyngeal region. These extrapharyngeal thyroid follicles are termed accessory or **heterotopic thyroid** follicles because of their location outside the normal site. Heterotopic thyroid follicles occur commonly in species of some of the more recent teleost families (Figure 7-8). Relatively large numbers of thyroid follicles may be found embedded within the head kidney of some species and occasionally in other locations such as the pericardium or ovary.

Thyroid function has been studied intensively in teleost fishes, and considerable information is available with respect to iodide uptake, hormone synthesis, secretion rates, clearance rates, and metabolism of thyroid hormones. Several features of teleost thyroid function differ markedly from what we know about mammals. Probably the most striking difference is that neuroglandular hypothalamic control in all species studied is inhibitory, with dopamine, TRH, and **somatostatin (SST)** capable of preventing TSH release. However, both TRH and catecholamines are stimulatory in salmonids. Teleosts readily accumulate iodide from the surrounding water through their gills, and some species maintain high circulating levels of protein-bound inorganic iodide. Consequently, these fish rarely show iodide deficiencies in nature, and their thyroid iodide metabolism is resistant to excessive iodide inhibition. A notable exception is the appearance of goitrous thyroids in coho salmon (*Oncorhynchus kisutch*) transplanted to the Great Lakes of the United States, a historically iodide-deficient environment.

There is negligible enterohepatic circulation of thyroid hormones or their metabolites in teleosts, and thyroid hormones are readily lost in urine, too. The active form of thyroid hormone is T_3, and its production is controlled largely through the activity of deiodinases. Teleosts have the same deiodinases as found in mammals—D1, D2, and D3 (see Figure 7-9)—and they are produced by homologous genes: *doi1*, *doi2*, and *doi3*. Treatment of striped parrotfish (*Scarus iseri*) with T_3 causes activation of *doi3* genes in liver, brain, and testes, as well as *doi2* in both ovaries and testes. Curiously, administration of T_4 generally has no effect on circulating levels of T_3.

Thyroid hormones travel through the blood mostly bound to plasma proteins. Studies show binding to albumin as well as to a binding protein, **transthyretin (TTR)**. Unlike the situation in mammals, teleost TTR has a greater affinity for T_3 than for T_4. TTR occurs in many target tissues, including skeletal muscle, testes, gills, and the pituitary, and this may be the basis for accumulation of T_3 in skeletal muscle reported by some investigators.

Teleost target tissues may contain a variety of nuclear thyroid receptors similar to those described in mammals. However as many as four different TRs each produced by a separate gene occur in some species. The most universal teleost TRs are TRα1, TRα2, and TRβ (see Figure 7-10).

Virtually all aspects of fish physiology are influenced by thyroid hormones, and thyroid hormones are essential for all aspects of the lives of fishes. Because of the many permissive actions of thyroid hormones and their close ties to metabolism, it often is difficult to discern which effects are direct actions of thyroid hormones and which are indirect, as indicated in the following sections.

1. Thyroid and Development in Teleosts

Thyroid hormones are responsible for a posthatching metamorphosis in many species of teleosts. Probably the most dramatic metamorphosis is that of flounders and other flatfishes which involves migration of the eye and attendant neural structures from one side of the head to the other (Figure 7-11) while the mouth and associated structures migrate to the opposite side. The adult animal then behaves with the left or right side (depending on the species) acting as the ventral surface with the mouth located there for bottom feeding and the other side having both eyes and acting as the dorsal surface. Levels of plasma T_4 rise dramatically during flounder metamorphosis as does cortisol whereas T_3 levels remain low. Measurements of T_3 turnover are lacking, however.

Anadromous salmonid fishes have a complicated life cycle (Figure 7-12) beginning with eggs laid in freshwater where early development leads to the parr stage. The **parr–smolt transformation** or **smoltification** occurs prior to seaward migration in many salmonids, such as steelhead trout (*Oncorhynchus mykiss*) and coho salmon (*Oncorhynchus kisutch*). It involves a sort of metamorphosis from a sedentary, cryptically marked fish, the parr, with a freshwater physiology to an active, silvery smolt capable of osmoregulating in seawater. The process of smoltification is very complex and involves pulses of several hormones, including T_4, cortisol, and PRL (Figure 7-13). Treatment with T_4 can induce the deposition of guanine in the scales, giving the fish a silvery appearance of a smolt (Figure 7-14).

Following smoltification in the spring, the smolt then migrates downstream to the ocean where it will remain for two or more years until it returns to spawn. Thyroid follicles of adult salmonids exhibit marked hypertrophy and hyperplasia (Figure 7-15) during their return migration to freshwater for spawning. It is not known if this is indicative of increased thyroid function related to metabolism or some other aspect of the migration or if it is the result of a hypothyroid condition.

2. Thyroid and Growth in Teleosts

Growth in fishes appears to be influenced by thyroid state as described for mammals (see Chapter 6). Growth is accelerated by thyroid hormones, and retardation of growth is observed in salmonids following treatment with goitrogens or radiothyroidectomy. Normal growth is resumed in radiothyroidectomized salmonids following administration of T_4 (Figure 7-16), suggesting an interaction with GH. Radiothyroidectomy also results in numerous skeletal abnormalities in these fish. The stimulatory effects of thyroid hormones on growth may be largely due to permissive actions (see Chapter 6).

3. Thyroid and Reproduction in Teleosts

There is a strong positive correlation in many teleosts between thyroid state and reproductive cycles. Thyroid hormones consistently stimulate precocial gonadal maturation, whereas hypothyroid conditions inhibit or retard gonadal development. In numerous species, thyroid activity is greatest at spawning. Increased thyroid activity as well as glucocorticoid secretion is coincident with migrations and spawning in Pacific salmon, and many of the effects on reproduction attributed to thyroid hormones may be a consequence of their metabolic actions related to the energy demands of reproductive and migratory behaviors. Although there have been many studies conducted to verify a direct action of thyroid hormones on reproduction, we are left with a definite strong association and strong involvement in reproduction.

4. Thyroid and Oxygen Consumption in Teleosts

Although the role of thyroid hormones in oxidative metabolism is clearly established for mammals, the situation for fishes is very different. Observations have yielded opposing results with respect to the importance of thyroid

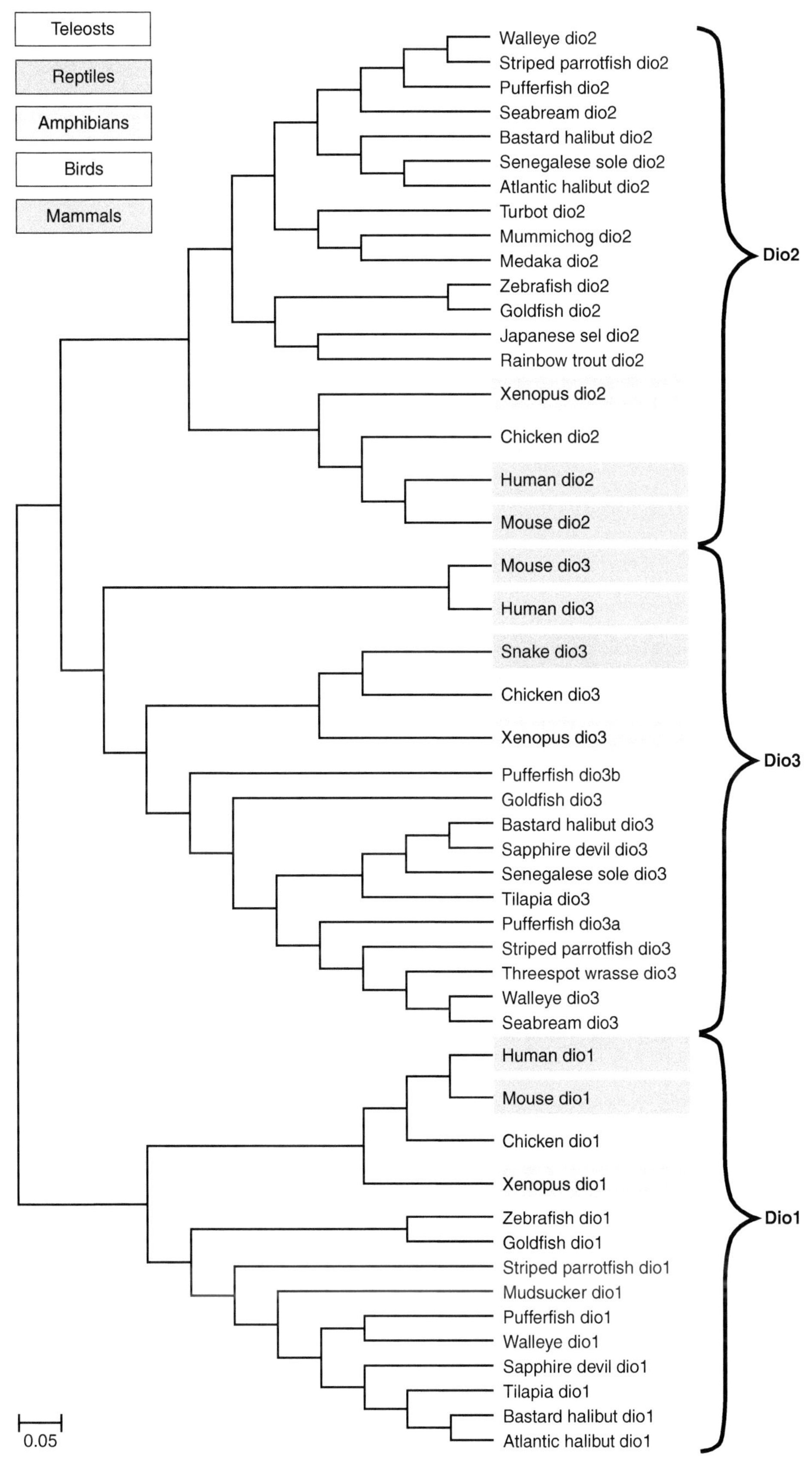

FIGURE 7-9 **Phylogeny of deiodinase genes in vertebrates.** Some species are omitted for whom there is evidence for deiodinases because the responsible genes have not been analyzed. Note that *Xenopus* has all three genes although evidence for other amphibians does not support the presence of a D1 deiodinase. *(Adapted with permission from Johnson, K.M. and Lema, S.C.,* General and Comparative Endocrinology, *172, 505–517, 2011.)*

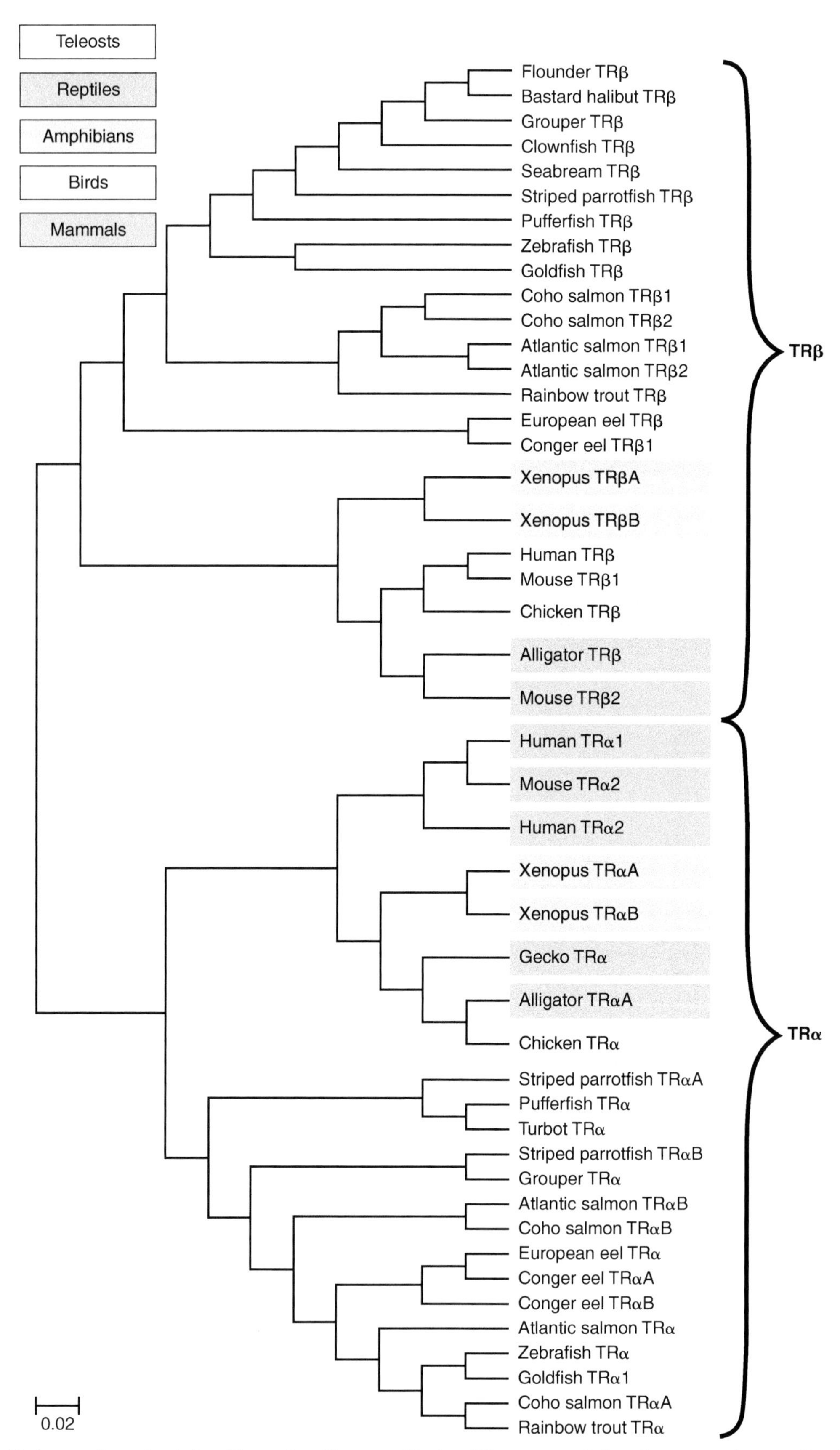

FIGURE 7-10 **Phylogeny of vertebrate thyroid receptors.** Tissue-specific thyroid hormone regulation of gene transcripts encoding thyroid hormone receptors in striped parrotfish (*Scarus iseri*). *(Adapted with permission from Johnson, K.M. and Lema, S.C.,* General and Comparative Endocrinology, *172, 505–517, 2011.)*

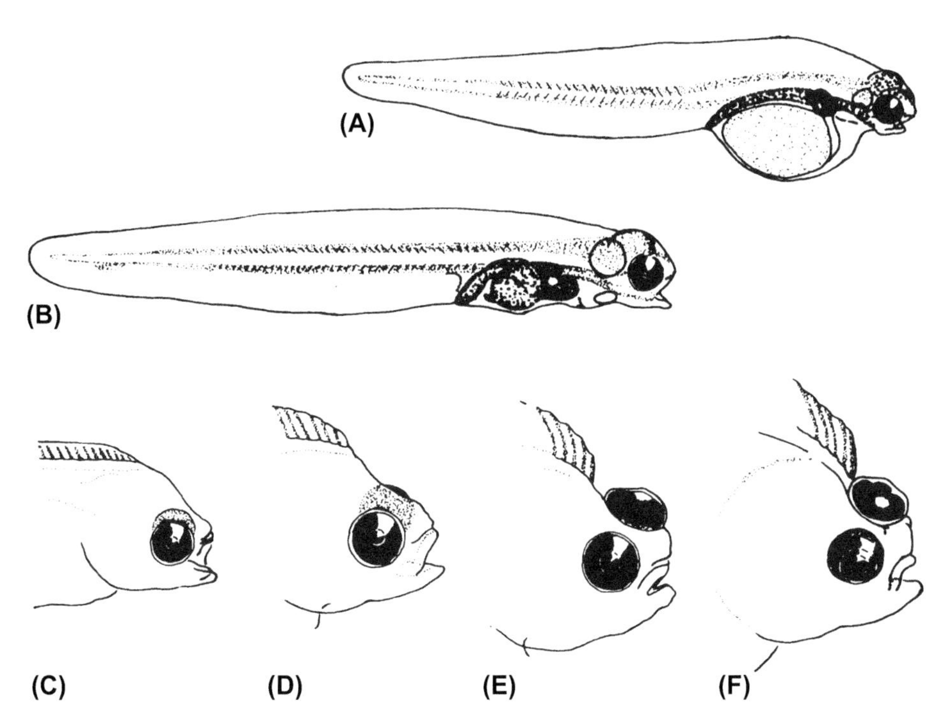

FIGURE 7-11 **Metamorphosis of the flounder *Pleuronectes platessa*.** During metamorphosis (C-F), which is controlled by thyroid hormones, one eye migrates from one side of the body to the other. *(Adapted with permission from Blaxter, J.H.S., in "Fish Physiology" (W.S. Hoar and D.J. Randall, Eds.), Academic Press, San Diego, CA, 1988, pp. 11A, 1–58.)*

hormones in oxygen consumption by teleost fishes. Most studies suggest that neither thyroid hormones nor goitrogens have any effects on oxygen consumption. A positive correlation has been reported for cyclic seasonal variations in thyroid activity and oxygen consumption by the Asian stinging catfish (*Heteropneustes fossilis*). Studies of oxygen consumption are difficult to interpret because temperature, changes in behavior, and potential side effects of treatments used to render the fish hypothyroid may have direct effects on metabolism that indirectly alter oxygen consumption.

5. Thyroid and Osmoregulation in Teleosts

The endocrine regulation of osmoregulation is primarily under the control of prolactin and cortisol with respect to maintaining Na^+ balance (see Chapter 9), and **arginine vasotocin (AVT)** is responsible for water balance (see Chapter 5). Calcium regulation is probably controlled by **stanniocalcin-1** from the **corpuscles of Stannius** embedded in the kidney, **calcitonin** from the ultimobranchial body found in the pharyngeal region (see Chapter 14), and **somatolactin** from the pars intermedia (see Chapter 5). Less direct roles in ionic and osmotic regulation are attributable to catecholamines, thyroid hormones, and factors from the pineal gland (see Chapter 5) as well as the caudal neurosecretory system (see Chapter 9). Thyroid hormones may enhance seawater adaptation and may influence migratory behavior, especially in species that migrate between salt and fresh water. These proposed actions may be related to a permissive type of action rather than to a causative role for thyroid hormones or may be only correlated events and not causally related.

6. Behavioral Actions of Thyroid Hormones in Teleosts

Thyroid hormones are elevated in relation to spawning, premigratory, and migratory behaviors. Increased thyroid function has been reported for a variety of migrating fishes, and premigratory restless behavior has been correlated with increased thyroid function. Thyroid activity is greater in young salmon smolts migrating to the ocean and in adults during their upstream return to freshwater for spawning. Again, the action of thyroid hormones may only enhance metabolic adjustments and/or the sensitivities of neural components to environmental stimuli.

E. Bony Fishes: Sarcopterygians

The function of thyroid hormones in lungfishes has received only a cursory examination. Thyroid hormones have been linked to a particularly fascinating aspect of the life history of African lungfishes: the ability to survive periods of drought while encased in a "cocoon." Awakening of the estivating lungfish *Protopterus annectens* from the

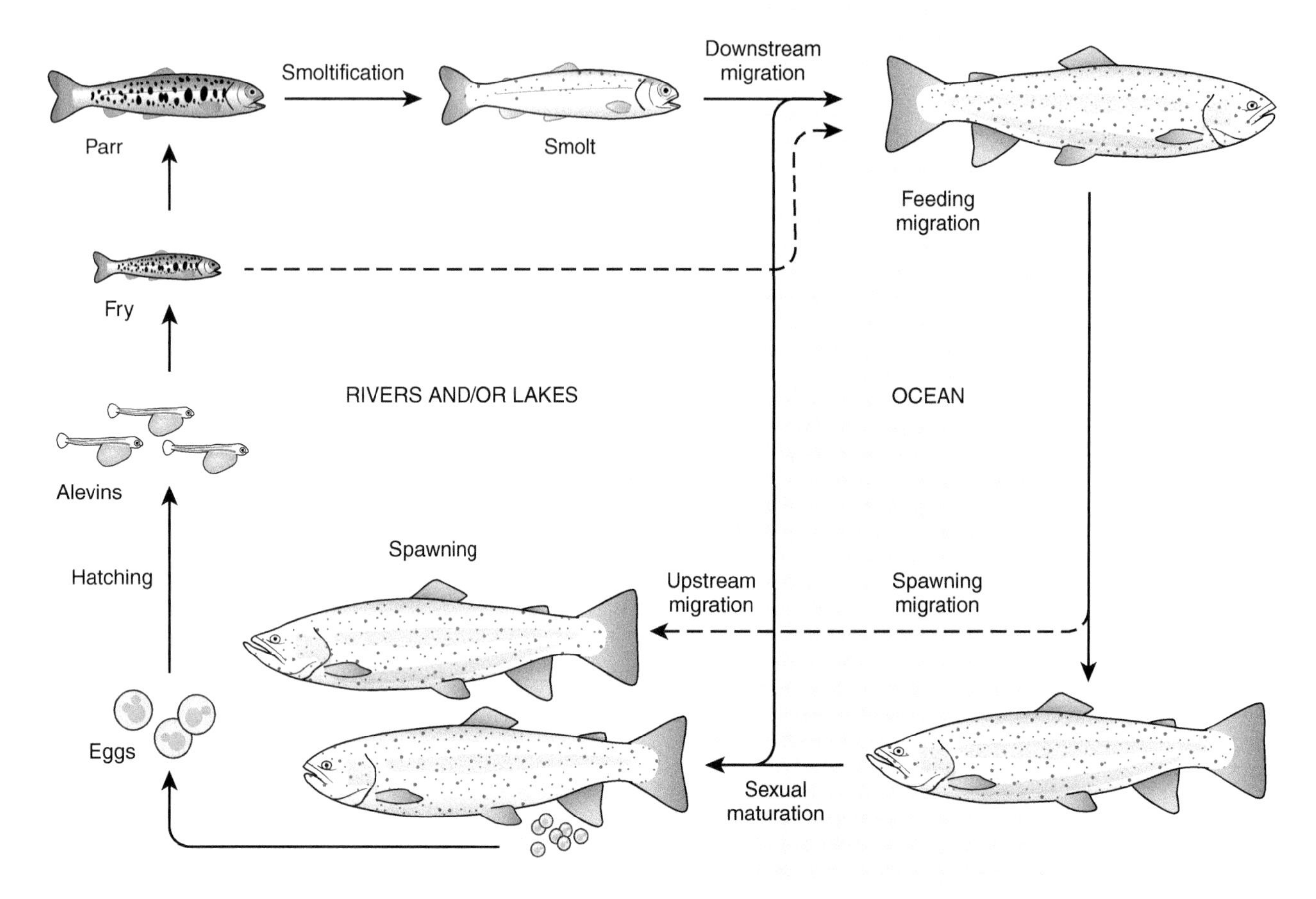

FIGURE 7-12 **Life cycle of Pacific salmon and steelhead (*Oncorhynchus* spp.).** This pattern is characteristic of most species; however, in pink and chum salmon that spawn in coastal streams, the fry are washed directly into the ocean. *(Adapted with permission from Ueda, H.,* General and Comparative Endocrinology, *170, 222–232, 2011.)*

FIGURE 7-13 **Plasma hormone levels during smoltification of coho salmon.** Prolactin, growth hormone, thyroxine, and cortisol all peak during smoltification. Insulin peaks in the parr and declines as smoltification gets under way. *(Adapted with permission from Dickhoff, W. et al.,* Journal of Experimental Zoology, *256 (Suppl. S4), 90–97, 1990. © John Wiley & Sons.)*

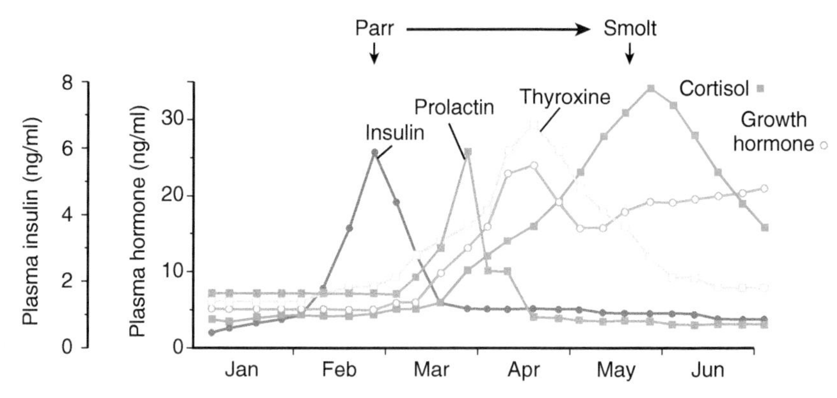

cocoon of dried mud when moistened may involve a neuroendocrine mechanism associated with the thyroid axis. One hypothesis, based upon limited experimental data, suggests that increasing humidity activates the HPT axis and stimulates thyroid hormone secretion. Thyroid hormones in turn would increase the sensitivity of olfactory centers that evoke normal feeding behavior as well as other behaviors associated with wakening. Additional studies of thyroid function in these fishes are needed to substantiate this interesting hypothesis, which is similar to the role of thyroid hormones in the awakening of mammals from hibernation (see Chapter 6).

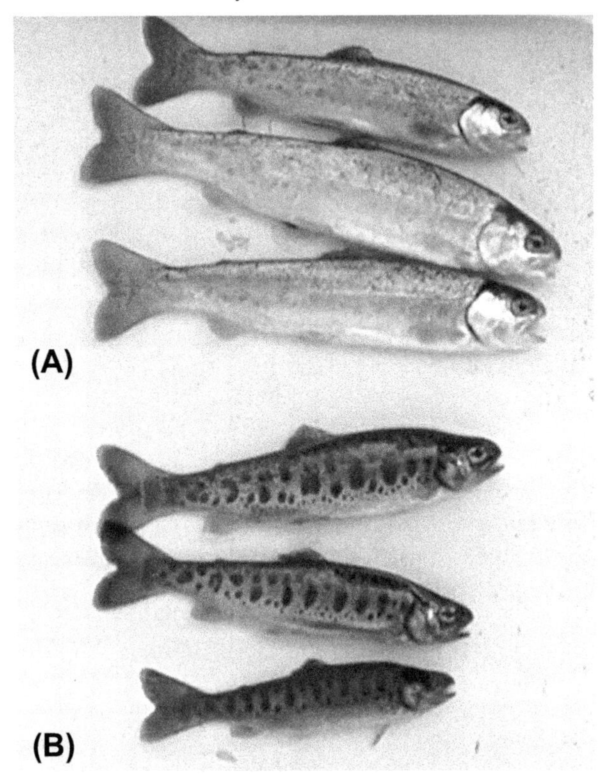

FIGURE 7-14 **Smolting in steelhead trout (*Oncorhyncus mykiss*).** Thyroid hormone stimulates deposition of guanine in the scales. (A) Smolts. (B) Parr.

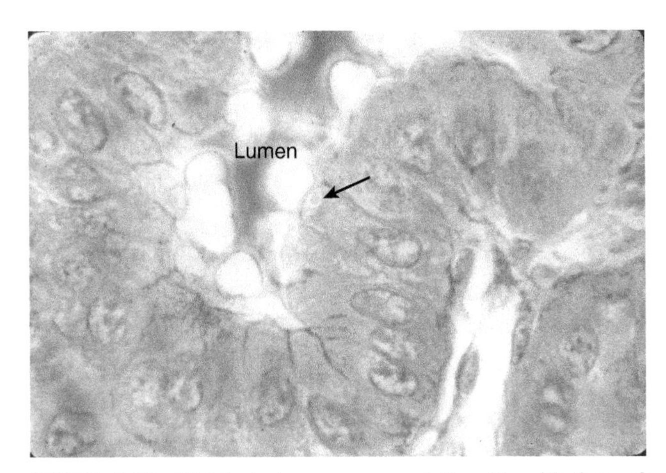

FIGURE 7-15 **Histological appearance of thyroid epithelium of a migrating adult sockeye salmon (*Oncorhynchus nerka*).** Note that the epithelium appears goitrous in that it is highly columnar, indicative of intense TSH stimulation. Little colloid is associated with the lumen, suggesting depletion. Arrow indicates colloid droplet endosome in the apical end of the follicle cell. Would you consider this a hyperthyroid or hypothyroid condition? Why?

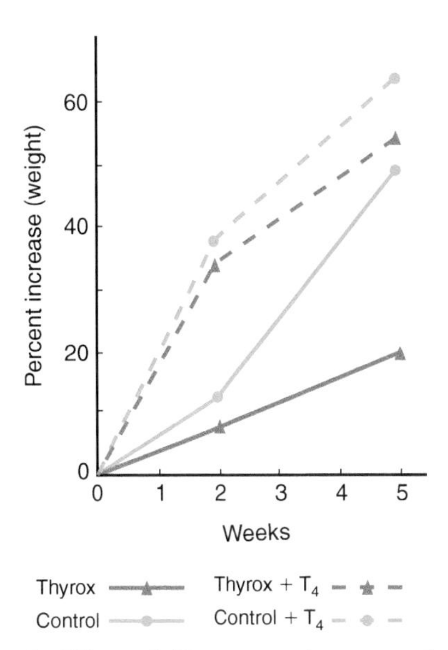

FIGURE 7-16 **Effect of T_4 treatment on growth of radiothyroidectomized steelhead trout.** *(Adapted with permission from Norris, D.O.,* Transactions of the American Fisheries Society, *97, 204–206, 1969.)*

F. Thyroid Functions in Amphibians

Thyroid hormones influence many processes in both anurans and urodeles, including reproduction, metamorphosis, metabolism, growth, and molting. The most dramatic and best studied events among amphibians are the direct actions of thyroid hormones on metamorphosis from an aquatic larva to a terrestrial or semiterrestrial form.

1. Thyroid and Reproduction in Amphibians

Experimental reduction of thyroid function, such as thyroidectomy or administration of goitrogens, accelerates gonadal development in several anurans. In one urodele, *Ambystoma tigrinum*, circulating levels of T_4 are inversely correlated with seasonal gonadal development, although such correlations do not substantiate cause–effect relationships. Some conflicting results with respect to thyroid hormones and seasonal maturation have been reported for some adult anurans, but the general relationship of thyroid hormones to reproduction in amphibians appears to be an antagonistic one. All of the data reported to date are circumstantial, however, and an absolute antagonism by endogenous thyroid hormones in natural populations is yet to be demonstrated.

2. Thyroid and Oxygen Consumption in Amphibians

The relationship between thyroid hormones and oxygen consumption is not clear. Some studies have shown a positive correlation between thyroid state and oxygen consumption, whereas most studies show no effects.

Oxygen consumption is not elevated during either spontaneous metamorphosis caused by elevated endogenous thyroid hormones or during induced metamorphosis caused by exogenous thyroid hormones. Liver slices prepared from T_4-treated adult frogs, *Rana pipiens*, exhibit significantly greater oxygen consumption *in vitro* than appropriate controls at 25°C. Oxygen consumption of treated slices is the same as for controls when observed at 15°C. These data suggest that the respiratory response of amphibian tissues to thyroid hormones may be temperature-dependent and that the response occurs only at higher temperatures. Most studies reporting no effect for thyroid hormones on oxygen consumption were performed below 20°C. These data suggest that thyroid hormones may enhance metabolic responses when environmental temperatures reach higher levels to provide more energy for maintenance and activity.

3. Thyroid and Metamorphosis in Amphibians

Thyroid hormones induce metamorphosis, a marked biochemical, physiological, morphological, and behavioral transformation from an aquatic larva to a terrestrial or semiterrestrial form. Metamorphosis occurs in the life history of most amphibians, although the developmental sequence and duration of the various stages may be modified extensively in those species that incubate their eggs on land or are live-bearing. Some salamander species never undergo metamorphosis and become sexually mature while remaining aquatic and retaining a larval body form. The prolongation of larval life or the retention of larval characteristics in sexually mature animals is termed **paedomorphosis**. If paedomorphosis occurs due to delayed development of non-reproductive (somatic) tissues but normal development of reproductive tissues it is called **neoteny**. If it is due to precocious development of the HPG axis, it is called **progenesis**.

Amphibian metamorphosis has been studied at all levels of the HPT axis and provides an excellent illustration of hormonal interactions of thyroid hormones with other hormones (e.g., PRL, corticosteroids) in a variety of tissues. Furthermore, metamorphosis is closely linked to the interaction of complex environmental factors that trigger the endocrine-related events responsible for this phenomenon.

Metamorphosis in amphibians involves regulation of specific genes, often initiated by an external factor, that orchestrate a programmed sequence of biochemical (see Table 7-7), morphological (e.g., tail and gill resorption; see Figure 7-17 and Table 7-8), physiological (e.g., nitrogen excretion; see Figure 7-18), and behavioral events. William Etkin proposed a model to explain the stimulatory role of the thyroid hormones on the HPT axis that he described as a "positive feedback" resulting in the acceleration of thyroid activity resulting in metamorphosis. At that time it was believed that TRH was elevated and resulted in increased TSH secretion. Although little support has been generated for this effect during the past 50 years, recent evidence suggests that thyroid hormones cause a decrease in the sensitivity of the HPT axis to thyroid hormone negative feedback that increases the level of HPT activity and induces metamorphosis. Hence, Etkin's hypothesis of "positive feedback" is actually a reduction of "negative feedback," leading to the same result.

As discussed in Chapter 5, we now know that CRH, and not TRH, is the endogenous hypothalamic stimulator of the thyroid axis in larval amphibians. Metamorphosis can be induced with CRH but not with TRH treatment (Figure 7-19). CRH causes release of TSH as well as **corticotropin (ACTH)** from the pituitary, resulting in an increase in corticosterone secretion (Figure 7-20). Consequently, activation of the stress response may be an important trigger for metamorphosis in some species under certain conditions (e.g., drying up of a pond; high density of larvae). Secretion of TSH is accelerated by increasing thyroid hormones via reduced sensitivity to thyroid hormone feedback, thereby increasing TSH secretion and resulting in metamorphosis. Increased TRH release may be responsible for the observed surge of PRL that also accompanies metamorphosis.

An important mechanism for induction of metamorphosis involves changes in deiodinase activities. In premetamorphic larvae, the presence of D3 deiodinase keeps T_3 levels low through the conversion of T_4 preferentially to rT_3 and hastens deiodination of T_3. During the activation of metamorphosis, there is a measurable increase in D2 deiodinase that enhances formation of T_3 from T_4 as well as a decrease in D3 deiodinase activity. Patterns of deiodinating activity in liver and target tissues (Figure 7-21) affect the availability of active T_3 and facilitate metamorphosis.

Thyroid hormones influence gene activity in target cells and initiate metamorphic changes (Table 7-7). For example, in the hindlimbs, intestine, and tadpole tail fins of anurans, thyroid hormones activate numerous genes while a few others are downregulated, resulting in rapid hindlimb growth, morphological and enzymatic changes in the digestive tract, and tail resorption, respectively, all characteristic of the later stages of metamorphosis. Many of these genes are called direct- or **early-response genes**, and they are activated directly by thyroid hormones, whereas **late-response genes** (e.g., urea cycle genes in liver, keratin genes in skin) are not activated until about 2 days after application of thyroid hormones. This delay implies that synthesis of other molecules, including transcription factors, is dependent on protein synthesis activated by the original stimulus by thyroid hormones. Marked increases in NIS and changes in tissue thyroid receptor levels also occur during metamorphosis (Figure 7-22). Changes in both

BOX 7A Endocrine Disruption

The presence of the anti-thyroid compound perchlorate in drinking water and certain foods has raised concerns about human health, especially with respect to potential effects on development of the mammalian fetal nervous system (see Chapter 6). Perchlorate contamination of natural waters also is a concern for thyroid function in wildlife; for example, the presence of perchlorate in natural waters of east and central Texas is associated with environmental contamination from U.S. military bases. Chorus frogs collected from perchlorate contaminated sites exhibit hypertrophied thyroid epithelia as compared to thyroids of frogs from reference sites (Box Figure 7A-1). Furthermore, frogs exposed to high but ecologically relevant ammonium perchlorate (14 mg/L) in the laboratory exhibit thyroid follicular hypertrophy and gland enlargement (goiter) compared to controls (Box Figure 7A-2), supporting the conclusion that environmental perchlorate may depress thyroid physiology and have detrimental effects on natural wildlife populations.

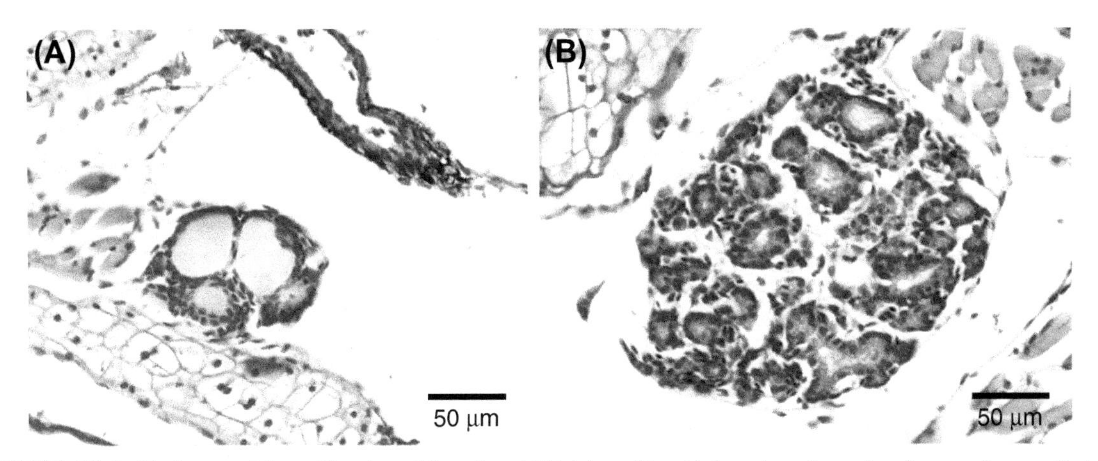

BOX FIGURE 7A-1 **Environmental perchlorate and frog thyroid histology.** Thyroid tissue in a chorus frog from a reference site in Texas appears at the left. On the right is a hyperstimulated thyroid gland from a perchlorate-contaminated site, presumably a result of increased thyrotropin secretion. *(Photomicrographs courtesy of Dr. James A. Carr, Texas Tech University.)*

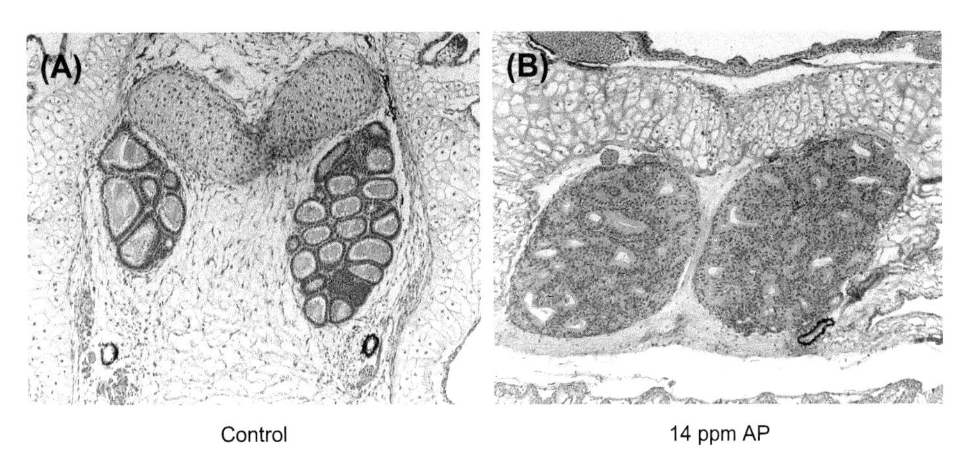

BOX FIGURE 7A-2 **Effect of perchlorate exposure in the laboratory.** The paired thyroid glands of a control frog can be seen at the left in marked contrast to the goitrous thyroids in the frog exposed to 14 ppm of ammonium perchlorate. *(Photomicrograph courtesy of Dr. James A. Carr, Texas Tech University.)*

α- and β-type thyroid receptors occur in different amphibian tissues. For example, both TRα and TRβ levels increase in tail tissue of three species of spadefoot toads during tail resorption.

Programmed cell death or **apoptosis** is another component of tail regression that is influenced by thyroid hormones. Induction of **ubiquitin**, a biochemical marker for apoptosis, is caused by T_4 treatment of isolated tadpole tail tips.

TABLE 7-7 Genes Affected by Thyroid Hormones during Amphibian Metamorphosis

Gene	Tissue	Regulation
Early-response genes		
Tail 1 (zinc finger region SPI)	Tail	Up
TRβ (thyroid hormone receptor)	Tail and intestine	Up
Tail 8/9 (bZIP[P] of E4BP4)	Tail and intestine	Up
Tail 14 (stromelysin 3)	Tail and intestine	Up
Tail 15 (type I 5′-deiodinase)	Tail	Up
Late-response genes		
Carbamyl-phosphate synthetase I	Liver	Up
Argininosuccinate synthetase	Liver	Up
Arginase	Liver	Up
Albumin	Liver	Up
Myosin heavy chain	Limb	Up
Keratin	Epidermis	Up
Trypsin	Pancreas	Down
Intestinal fatty acid-binding protein	Intestine	Down

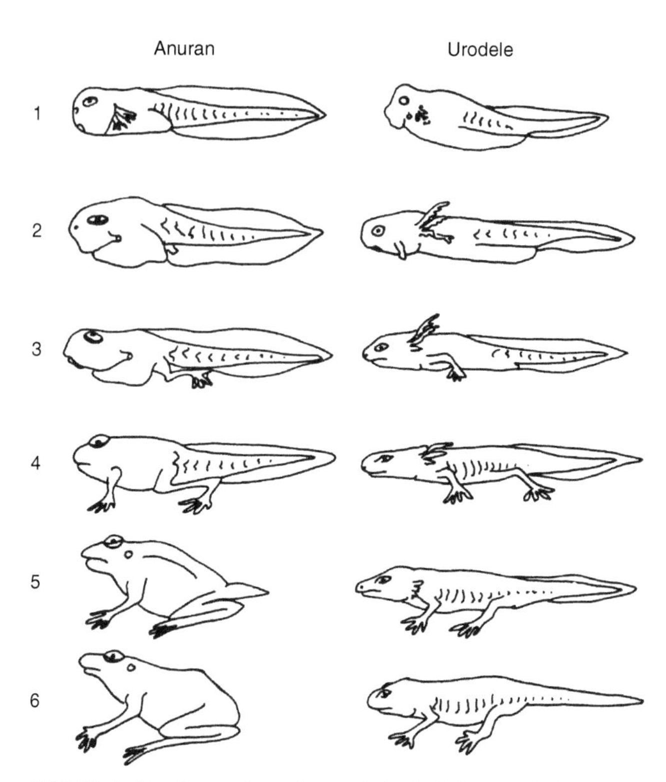

FIGURE 7-17 **Comparison of morphological changes during metamorphosis of an anuran (left) and a urodele (right) amphibian.** Urodeles quickly reach stage 4 and remain in this stage with external gills most of their larval lives. Anurans may spend a few weeks or up to 2 years at stage 2 before limbs emerge. Resorption of the anuran tail (metamorphic climax) may be delayed for some time after emergence of the hind limbs but occurs rapidly once the forelimbs emerge. Note that when urodeles undergo metamorphosis, the external gills are resorbed as is the tail fin (stages 5 and 6).

Environmental factors such as photoperiod may operate through catecholaminergic mechanisms in the brain to influence neurohormone release from hypothalamic nuclei (Figure 7-23). The pineal gland may play an important role in mediating the observed stimulatory actions of light on thyroid function and metamorphosis. In part, this is based on inference from observations that long day photoperiods stimulate metamorphosis in both salamanders and frogs. Numerous studies by Mary Wright have demonstrated an inhibition of thyroid secretion and thyroid hormone action by melatonin in frog tadpoles.

Thyroid hormones are also involved in a second amphibian metamorphic event, the so-called **water drive** associated with reproduction in newts that is induced by PRL. In fact, induction of water-drive behavior is used as a bioassay for PRL (see Appendix D). Small quantities of thyroid hormones facilitate the preinduced water drive and associated morphological changes in the integument. Larger amounts of thyroid hormones inhibit water drive, and thyroid hormones appear to be responsible for land-drive behavior of recently metamorphosed tiger salamanders.

TABLE 7-8 Life History of *Rana Pipiens* at 23°C

Period	Events	Duration
Embryonic	Embryogenesis	8 days
Premetamorphic	Growth	5–6 weeks
Prometamorphic	Accelerated growth of hind limbs; skin changes occur	3 weeks
Metamorphic climax	Forelimbs emerge; tail resorbs; gills resorb; head and gut reconstruction; etc.	1 week
Juvenile	Growth	?
Adult	Reproductive maturation and breeding	?

4. Thyroid and Growth of Amphibians

The involvement of thyroid hormones in the growth of amphibians, like the relationship to reproduction, deviates

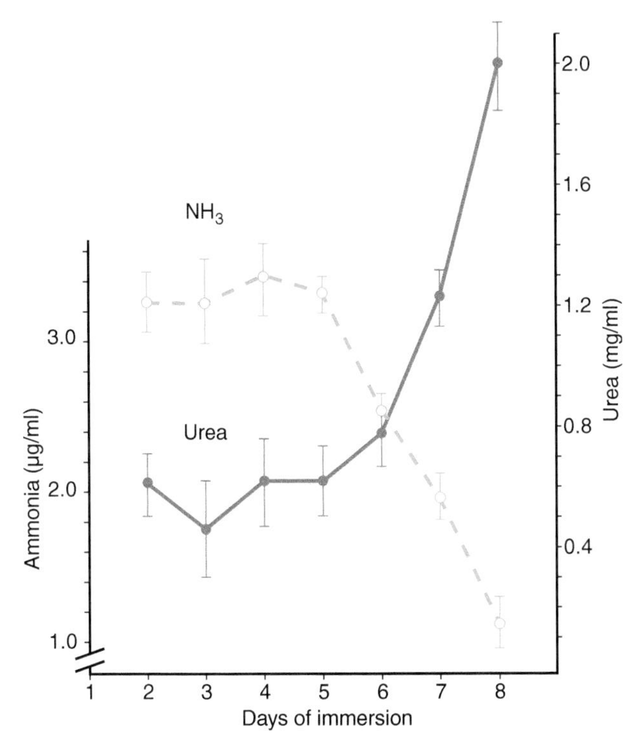

FIGURE 7-18 **Urea–ammonia excretion during anuran metamorphosis.** Aquatic animals excrete ammonia as their principal nitrogenous wastes, whereas terrestrial amphibians produce urea. Immersion of tadpoles in water containing thyroxine induces a switch from ammonia excretion to urea excretion similar to that observed during normal metamorphosis. A similar pattern to the ammonia change is seen for other metamorphic changes such as regression of tail whereas growth of hind limbs exhibits a urea-like pattern.

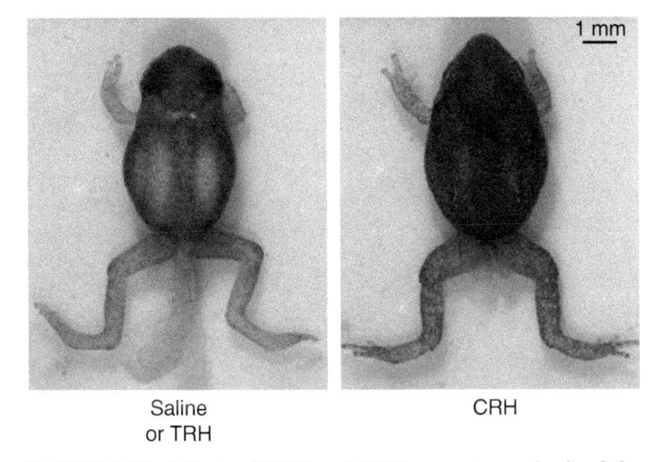

FIGURE 7-19 **Effects of TRH and CRH on metamorphosis of the coqui (*Eutherodactylus coqui*).** Neither saline nor TRH affected the rate of metamorphosis, but CRH treatment accelerated metamorphosis. Note tail regression and changes in body and head shape. *(Adapted with permission from Kulkarni, S.S. et al.,* General and Comparative Endocrinology, *169, 225–230, 2010.)*

from the general vertebrate pattern. Growth in natural populations is arrested during metamorphosis. The surge of thyroid activity that initiates metamorphosis in larval amphibians apparently arrests growth for a time. Normal growth resumes in juvenile animals after metamorphosis is complete. Treatment of anuran tadpoles with goitrogens or mammalian PRL accelerates larval growth and prevents metamorphosis. The complicated metamorphic process itself involves many drastic physiological changes and tissue rearrangements that may be responsible for arrested growth. These observations associated with larval growth do not eliminate participation of low levels of thyroid hormones in growth of the subadult or postmetamorphic juvenile forms. Participation of thyroid hormones in the growth of metamorphosed amphibians has not been reported, however.

5. Molting and Other Skin Effects of Thyroid Hormones in Amphibians

Molting or shedding of skin (**ecdysis**) in larval and adult urodeles is under direct stimulatory control by thyroid hormones, and frequent molting accompanies and follows metamorphosis in these animals. Molting in adult anurans, however, does not appear to be influenced by thyroid hormones. Here, molting is stimulated by corticosteroids or indirectly by ACTH operating through its action of increasing corticosteroid secretion. The reasons for this marked difference in hormonal control of molting between urodeles and anurans are not known.

Thyroid hormones cause a number of skin changes related to successful invasion of the terrestrial environment. Both a thickening of the epidermis and keratinization are induced by thyroid hormones in urodeles and anurans. In the red-spotted newts (*Notophthalmus viridescens*), thyroid hormones antagonize the effects of PRL on the skin of the juvenile eft prior to its return to water. PRL reduces the keratinization and enhances mucus production that helps return the skin to a smooth, moist condition characteristic of aquatic newts.

The large and conspicuous **Leydig skin cells**, which are characteristic of urodele larvae, disappear during metamorphosis, presumably a consequence of thyroid hormone actions (Figure 7-24). The role of these cells in the larvae is unknown. Histochemical observations suggest a contribution of their contents to formation of a keratinized layer in the skin that probably aids in reducing desiccation.

G. Thyroid Functions in Reptiles

There are relatively few studies on thyroid function in reptiles. Changes in thyroid state in reptiles are correlated with reproduction, environmental temperature, and activity level, although cause–effect relationships have not been

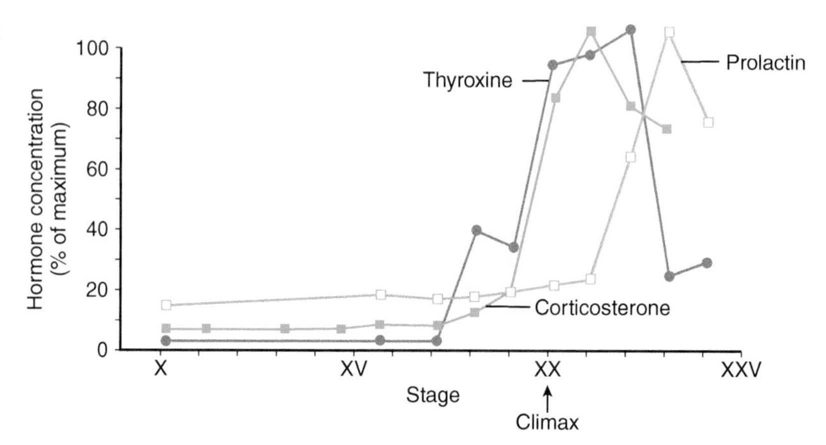

FIGURE 7-20 **Hormonal changes during bullfrog metamorphosis.** Thyroxine, prolactin, and corticosterone all peak at metamorphic climax. These changes are characteristic of both anurans and urodeles. Compare to the pattern of hormone secretion in smoltification of salmonid fishes (Figure 7-13). *(Adapted with permission from Dickhoff, W. et al.,* Journal of Experimental Zoology, *256 (Suppl. S4), 90–97, 1990. © John Wiley & Sons.)*

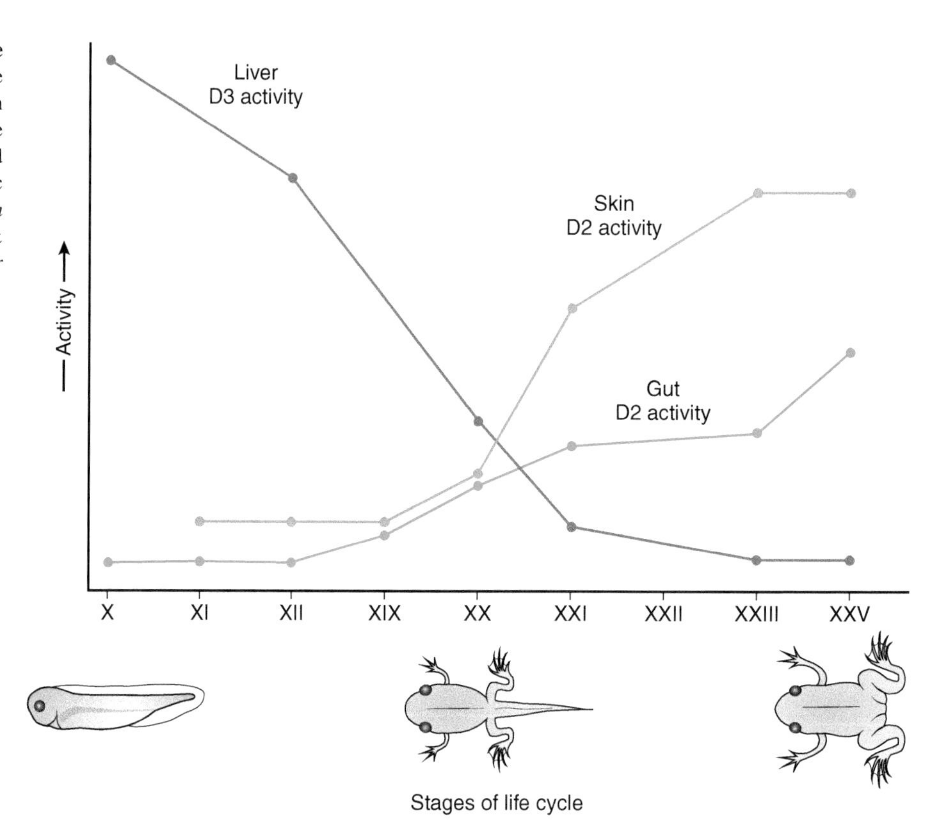

FIGURE 7-21 **Changes in deiodinase activities during metamorphosis.** The reduction of D3 deiodinase activity in liver and increases in D2 deiodinase activity in skin and gut leads to increased conversion of T_4 to T_3 and metamorphic climax. *(Adapted with permission from Galton, V.A.,* Trends in Endocrinology & Metabolism, *3, 96–100, 1992. © Elsevier Science, Inc.)*

established. Exogenous thyroid hormones also stimulate oxygen consumption, growth, and molting under certain conditions. Iodinated tyrosines (MIT, DIT) are present in the blood of some snakes, such as striped racer (*Elaphe taeniura*) and cobra (*Naja naja*). Thyrotropin stimulates *in vitro* release of MIT, DIT, and T_4 from chunks of thyroid tissue isolated from a turtle (*Geochemys reevsii*), a lizard (*Gekko gecko*), and a snake (*E. rachiata*). These studies suggest that thyroid glands of these reptiles may lack any deiodinase to convert MIT and DIT to inorganic iodide and tyrosine. However, studies have demonstrated that all deiodinase types are present in other tissues of reptiles and so the thyroid gland of reptiles appears to be unique in this regard.

1. Thyroid and Reproduction in Reptiles

Seasonal changes in thyroid function have been correlated positively with a number of reproductive events in lizards, snakes, and turtles. An active thyroid, usually assessed histologically, is associated with spermatogenesis, ovulation, and mating in a number of lizards. Similar

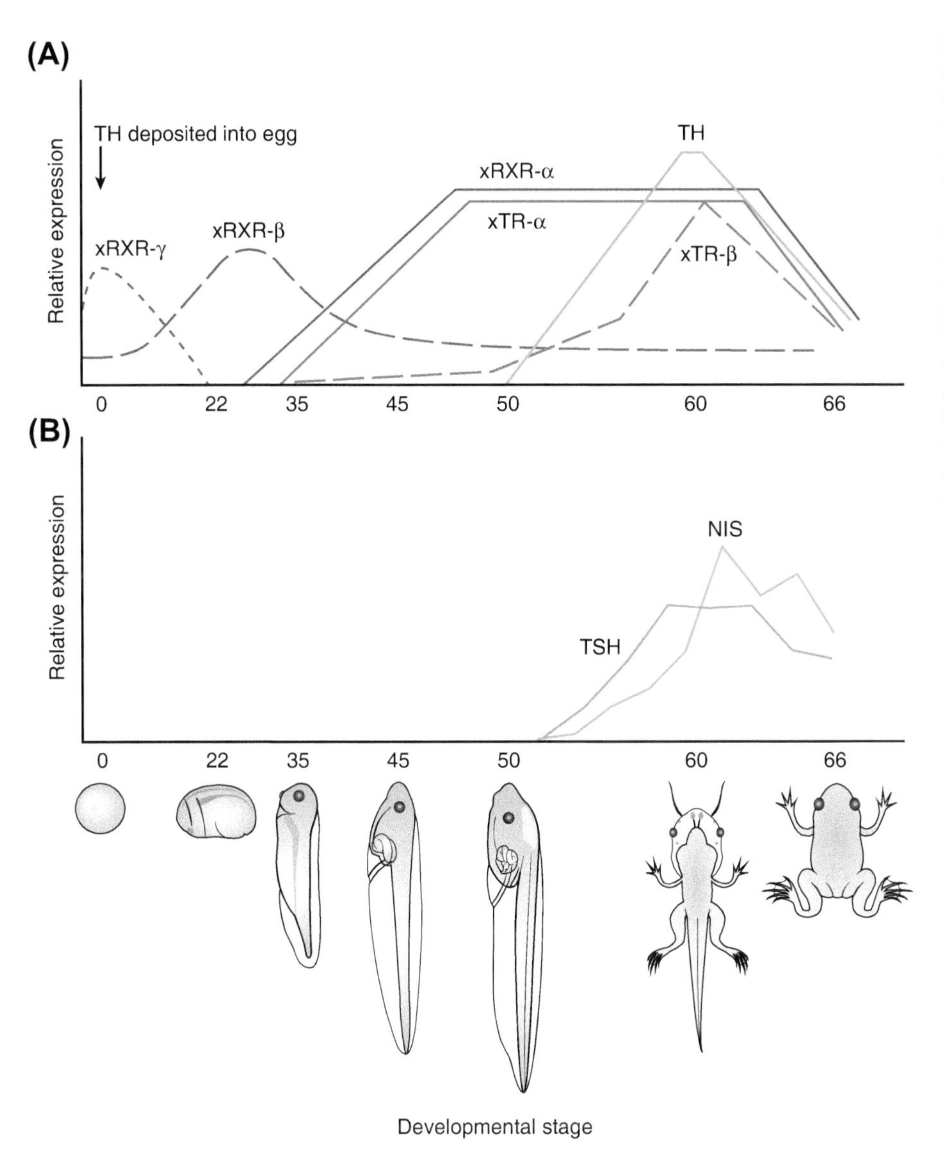

FIGURE 7-22 **Expression of thyroid hormone receptors (TRα and TRβ) as well as RXRs during early embryonic development and metamorphosis.** (A) TRs are expressed very early in embryonic development, even before the thyroid gland differentiates, possibly to respond to TH deposited into the egg by the mother. Note the increase in tadpole synthesis of thyroid hormones (TH). (B) Na^{+}-I symporter (NIS) and thyrotropin (TSH) during metamorphic climax. *(Adapted with permission from Furlow, J.D. and Neff, E.S.,* Trends in Endocrinology & Metabolism, ***17****, 38–45, 2006; Korte, J.J. et al.,* Gen. Comp. Endocrinol., ***171****, 319–325, 2011; Opitz, R. et al.,* J. Endocrinol., *190, 157–170, 2006.)*

observations have been reported in snakes and in at least one turtle. Among live-bearing reptiles, there is no evidence for increased thyroid function during gestation. However, thyroidectomy of pregnant lizards (*Lacerta vivipara*) six weeks prior to term causes premature discharge of most eggs, and those eggs that are retained fail to hatch.

2. Environmental Temperature and Reptilian Thyroid

In general, thyroid function in reptiles varies proportionally with changes in temperature; however, behaviors such as basking in lizards make it difficult to estimate body temperature when assessing thyroid state. Thyroid activity in temperate lizards and alligators is highest during the warmer seasons. Lowering the environmental temperature artificially during the summer months causes reduction in thyroid activity, and lizards maintained in the laboratory at high temperatures (35°C), exhibit greater thyroid activity than when maintained at 15°C. Thyroid activity in snakes is greater during warm periods and lowest during hibernation. The greatest level of thyroid activity, however, correlates with reproductive events in some lizards and snakes and not with temperature. Similar data have been reported for turtles. An exception to the relationship of thyroid activity and temperature described above is the situation found in several species of lizard inhabiting warmer climates. These lizards do not exhibit depressed thyroid function at lower temperatures. In fact, thyroids of such lizards tend to be more active during cool periods. There is a positive correlation between thyroid state and physical activity in

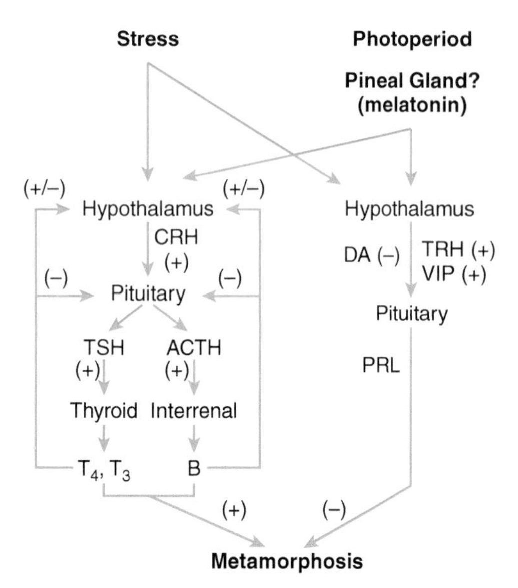

FIGURE 7-23 **Environmental factors affecting metamorphosis in amphibians.** Stress (e.g., pond drying, high population densities, salinity changes) and long photoperiods can stimulate metamorphosis in populations of amphibians through effects on hypothalamic regulation of pituitary secretion and release of T_4 and increased conversion of T_4 to T_3 as well as elevations in corticosterone (B) from the interrenal (adrenal). *(Adapted with permission from Denver, R.J.,* Comparative Biochemistry and Physiology C, *119, 219–228, 1998.)*

lizards and snakes. Increased humidity and activity are correlated with increased thyroid function in the lizard *Agama agama savattieri*, suggesting that a more complex relationship exists among thyroid state, physical activity level of the animal, and environmental factors.

3. Thyroid and Oxygen Consumption in Reptiles

The relationship between thyroid state and oxygen consumption is temperature dependent. Thyroid hormones, TSH, and thyroidectomy have little effect on lizards maintained at 20°C, but a positive relationship appears between oxygen consumption and thyroid state at 30°C. Heart, brain, liver, muscle, and lung tissues from lizards incubated *in vitro* at 30°C respond to T_4 with increased consumption of oxygen. Homogenates of both liver and skeletal muscles from T_4-treated snakes (*Natrix piscator*) exhibit higher oxygen consumption at 30°C than do tissues from untreated snakes. Thyroidectomy causes a decrease in oxygen consumption in these animals that is restored to normal by treatment with T_4. Metabolic rate and cytochrome oxidase activity also are stimulated by T_4 at 25°C. These observations suggest that homeothermy in birds and mammals (as well as homeothermy proposed for certain large, extinct reptiles) is an adaptation to maximize some basic features of oxidative metabolism and its stimulation by thyroid hormone.

4. Thyroid and Molting in Reptiles

Shedding of the skin is stimulated in lizards by thyroid hormones and is retarded by thyroidectomy. Implants of thyroid tissue into muscles of thyroidectomized *Lacerta* spp. restore molting that has been interrupted by thyroidectomy. Apparently either mammalian TSH or PRL is capable of restoring molting in some hypophysectomized lizards, and PRL enhances the effect of T_4 on intact animals. Such a relationship between PRL and the thyroid axis in lizards is unlike the role for PRL in premetamorphic amphibians but fits with the synergistic roles of these hormones observed during late stages of metamorphosis in both urodele and anuran amphibians. In marked contrast to lizards, thyroidectomy in snakes increases molting frequency, and cessation of molting follows administration of thyroid hormones. No explanation for this contradiction between snakes and lizards has been offered. This situation in snakes is different from that found in amphibians in which thyroid hormones are stimulators of molting in urodeles and corticosteroids are involved in anurans.

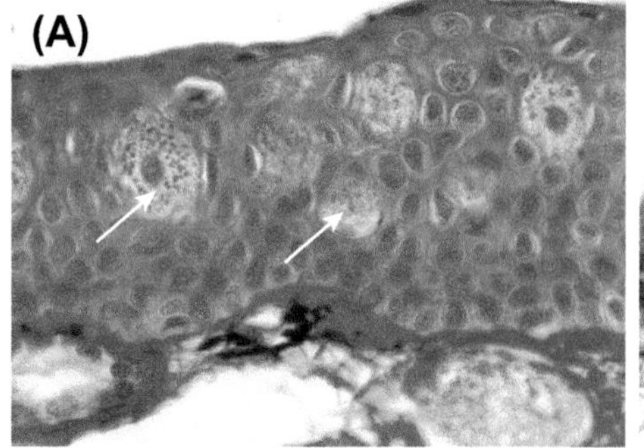

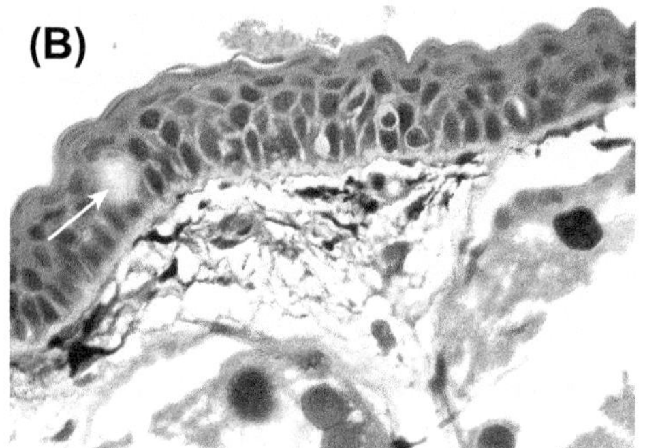

FIGURE 7-24 **Skin changes associated with metamorphosis in a urodele amphibian.** (A) Histological section through skin of larval tiger salamander (*Ambystoma tigrinum*) showing prominent Leydig cells (arrows). (B) During metamorphosis, the Leydig cells degenerate and release their secretions that are thought to contribute to the relative impermesbility of the skin of a terrestrial adult. Arrow indicates mucous cell (slightly lower magnification).

5. Thyroid and Growth of Reptiles

Although detailed studies of the relationship of thyroid state to growth are lacking for reptiles, a variety of studies employing embryonic, juvenile, and adult reptiles suggest a relationship between thyroid hormones and growth that is similar to the mammalian pattern. Thyroidectomy of the lizard *Sceloporus undulatus* retards growth. Other studies in reptiles are needed to substantiate this relationship.

H. Thyroid Functions in Birds

The structural and functional features of avian thyroid glands are very similar to those of mammals. Thyroid hormones have been reported to affect reproduction, growth, metabolism, temperature regulation, molting, and various behaviors. Most studies on avian species have concentrated heavily on certain domestic birds (chicken, duck, pigeon, Japanese quail), and few data are available for wild bird species. Recent development of fecal thyroid hormone assays for wild birds may rectify this situation. With respect to most of thyroid based events, the results are similar for wild and domestic species studied.

1. Development of Thyroid Function in Birds

The thyroid axis develops early in some birds, being functional in the chicken by day 13 of development. The embryonic chick thyroid actually responds to TRH and TSH administration on the sixth day of incubation, indicating that, although the thyroid gland's secretory capacity is all ready to go, the endogenous activity of the hypothalamic–pituitary connection is still immature. Plasma levels of T_4 increase steadily after day 13 to maximal levels by day 20. In contrast, levels of T_3 remain low until day 19 when there is a surge of T_3 production during pipping and hatching. This surge corresponds to a rise in D1 deiodinase activity in the liver and a marked decrease in D3 deiodinase similar to events described in tail tissues for D2 and D3 during amphibian metamorphosis. Thus, conversion of T_4 to T_3 increases as degradation of T_3 decreases, causing a surge of T_3 availability. This relationship continues for a time after hatching. Hypophysectomy of hatchlings, however, causes an increase in D3 deiodinase activity which can be prevented by treatment with GH. A similar effect of GH is seen in adult chickens as well, suggesting an additional side to the traditional synergistic actions between thyroid hormones and GH in growth (see Chapters 4 and 5). In contrast, D1 deiodinase is unaffected by GH. This dramatic change in thyroid activity has been documented in **precocial birds** (chicken and Japanese quail) but not in **altricial birds** (such as ring doves). Altricial birds, unlike precocial species, are totally dependent on their parents after hatching. In ring doves, thyroid and deiodinase activities change much more gradually as the birds develop greater independence from their parents.

2. Thyroid and Reproduction in Birds

Domestic birds require thyroid hormones for normal gonadal development. For example, T_4 stimulates testicular growth, whereas thyroidectomy or goitrogen administration impairs testicular function or induces gonadal regression. T_4 can stimulate testicular growth out of season in some wild birds as well; however, exogenous T_4 under some circumstances may exert suppressive actions on gonads. In several wild species, thyroidectomy prior to the time for normal gonadal growth results in precocious gonadal growth and maintenance at maximal condition. Treatment with T_4 results in gonadal regression in thyroidectomized birds at any time of year. In other wild species, thyroidectomy simply prolongs the active gonadal phase and shortens the time during which the gonads are regressed. Obviously, one cannot generalize about the relationships between thyroid hormones and reproductive events in birds since each species may prove to be a special case.

3. Thyroid, Thermogenesis, and Oxygen Consumption in Birds

The role for thyroid hormones in cold adaptation in birds can be considered a relatively new functional acquisition for thyroid hormones (see Figure 7-25) that may have appeared among the reptilian ancestors of both birds and mammals. **Thermogenesis** (heat production) in birds as in mammals is closely linked to oxygen consumption. It is not known if thyroid hormones uncouple oxidative phosphorylation or simply increase metabolic rate in birds. Some wild birds in temperate regions exhibit heightened thyroid activity in late autumn and early winter that may be related to thermogenesis. Thyroidectomy of adult birds depresses their ability to produce heat, and treatment with thyroid hormones increases oxygen consumption.

4. Thyroid, Carbohydrate Metabolism, and Growth in Birds

Thyroid hormones reduce glycogen stores in liver, increase free fatty acid levels and induce mild hyperglycemia. Thyroidectomy causes a decrease in blood glucose and liver fatty acid levels but an increase in blood cholesterol levels. These effects by thyroid hormones may not be of primary importance to overall metabolism in birds since a number of other hormones, such as glucagon and epinephrine, are more important than thyroid hormones with respect to controlling carbohydrate and lipid metabolism in birds (see Chapters 12 and 13). However, thyroid hormones may produce a permissive effect with respect to the actions of these other hormones, possibly through an

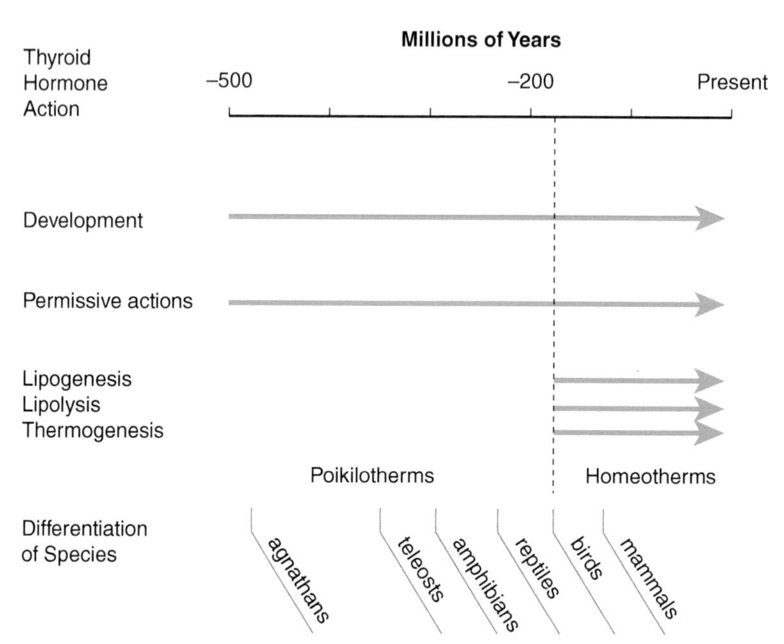

FIGURE 7-25 **Distribution of developmental and thermogenic action of thyroid hormones in vertetrates.** Perhaps developmental and/or permissive actions were the initial roles for thyroid hormones. Thermogenesis appears with the acquisition of homeothermy possibly in the reptilian ancestors of birds and mammals. *(Adapted with permission from Oppenheimer, J.H. et al., in "Molecular Endocrinology: Basic Concepts and Clinical Correlations" (B.D. Weintraub, Ed.), Raven Press, New York, 1995, pp. 249–268.)*

TABLE 7-9 **Plasma levels of T_3 and T_4 in Migrating and Post-Migrating Canadian Goose, *Branta canadensis interior***

Sampling Time	Plasma T_4 (ng/mL)	Plasma T_3 (ng/mL)	T_3/T_4
Spring pre-migration	11.8	3.5	0.28
Spring post-migration	17.0	2.0	0.10
Fall pre-migration	13.3	1.0	0.05
Fall post-migration	11.8	2.0	0.18

Note: Values estimated from Figure 27 in George, J.C. and John, T.M., *Cytobios*, 61, 97–115, 1990.

effect on adenylyl cyclase levels in the target tissues. Thyroid hormone effects on carbohydrate metabolism also may be related to their thermogenic action. There is evidence that thyroid hormones act cooperatively with GH in birds. Thyroidectomy causes depression or retardation of growth in birds, and seasonal changes in circulating GH and T_4 are correlated. Unlike the action of thyroid hormones in mammals that release GH, in birds GH is reduced by elevated thyroid hormones.

5. Thyroid and Molting in Birds

Thyroid hormones produce stimulatory effects on the skin and feathers that are usually associated with the molting process. It is generally accepted, however, that the HPG axis provides the factors that directly regulate the molting process in birds (see Chapter 11). Unlike the condition for urodele amphibians and lizards, the role for thyroid hormones in the molting process of birds may be permissive rather than causative.

6. Thyroid and Migration in Birds

Migratory birds possess active thyroid glands as compared to non-migrating or postmigrating individuals of the same species, suggesting some role in the migratory process. Although some of the older literature indicates that thyroid hormones may directly influence migratory behavior, the nature of this influence has not been confirmed. Thyroid hormones may alter metabolic patterns associated with energy requirements during migration and/or "tune" the nervous system to increase its sensitivity to environmental or endocrine cues. In the Canada goose (*Branta canadensis interior*), thyroid histology is activated and plasma T_4 levels are increased in association with spring premigratory restlessness or **Zugunruhe**, but the plasma T_3/T_4 ratio decreases dramatically (Table 7-9). However, in relation to the fall migration, there is a marked increase in T_3 in postmigrants, increasing the T_3/T_4 ratio. This latter observation may reflect differences in thermal conditions and/or the role of thyroid hormones in thermogenesis during the cooler fall.

III. SUMMARY

The thyroid axis of non-mammalian vertebrates shows many parallels to mammals with respect to thyroid structure and to the synthesis and metabolism of thyroid hormones as well as to their actions. T_3 appears to be the more active form of thyroid hormones in all vertebrates. Thyroid hormones appear to interact with a variety of other endocrine-regulated systems where they play permissive or possibly synergistic roles, especially in development, growth, and reproduction. The system of deiodinases seems to be responsible for allowing surges in T_3 to occur at critical times. Direct actions of thyroid hormones on development occur in all vertebrates, but their participation in lipid metabolism and thermogenesis is correlated with homeothermy. Some notable phylogenetic differences stand out with respect to hypothalamic regulation in teleost fishes and amphibians. TRH appears to exert a negative control in many fishes over TSH release, and in amphibians evidence is mounting to identify the CRH peptide as the endogenous TSH-releasing hormone. Considerable work remains to be done on thyroid systems in non-mammalian vertebrates, especially in elasmobranchs, a greater variety of teleosts, and reptiles. Once this has been accomplished, a clearer pattern of vertebrate thyroid function should appear.

STUDY QUESTIONS

1. Compare the details of the physiology of the HPT axis among the major vertebrate groups with that of mammals. Be sure to include the manner of hypothalamic control, TSH actions on thyroid gland activity, and the importance of deiodinases.
2. Compare and contrast the regulation and process of metamorphosis in various vertebrates.
3. Compare and contrast the actions of thyroid hormones among the major vertebrate groups with that of mammals.

SUGGESTED READING

Books

McDiarmid, R.W., Altig, R., 1999. Tadpoles: The Biology of Anuran Larvae. University of Chicago Press, Chicago, IL.

Shi, Y.-B., 2000. Amphibian Metamorphosis: From Morphology to Molecular Biology. John Wiley & Sons, New York.

General

Crockford, S.J., 2009. Evolutionary roots of iodine and thyroid hormones in cell–cell signaling. Integrative and Comparative Biology 49, 155–166.

Norris, D.O., 1999. Thyroid hormones in subavian vertebrates. In: Knobil, E., Neill, J.D. (Eds.), Encyclopedia of Reproduction, vol. 4. Academic Press, New York, pp. 807–812.

Oppenheimer, J.H., Schwartz, H.L., Strait, K.A., 1995. An integrated view of thyroid hormone actions *in vivo*. In: Weintraub, B.D. (Ed.), Molecular Endocrinology: Basic Concepts and Clinical Correlations. Raven Press, New York, pp. 249–268.

Sutija, M., Joss, J.M.P., 2006. Thyroid hormone deiodinases revisited: insights from lungfish: a review. Journal of Comparative Physiology B 176, 87–92.

Protochordates

Denver, R.J., 2008. Chordate metamorphosis: ancient control by iodothyronines. Current Biology 18 (3), R567–R569.

Holland, L.Z., Albalat, R., Azumi, K., Benito-Gutierrez, E., Blow, M.J., Bronner-Fraser, M., et al., 2008. The amphioxus genome illuminates vertebrate origins and cephalochordate biology. Genome Research 18, 1100–1111.

Ogasawara, M., 2000. Overlapping expression of amphioxus homologs of the thyroid transcription factor-1 gene and thyroid peroxidase gene in the endostyle: insight into evolution of the thyroid gland. Development Genes and Evolution 210, 231–242.

Paris, M., Brunet, F., Markov, G.V., Schubert, M., Laudet, V., 2008. The amphioxus genome enlightens the evolution of the thyroid hormone signaling pathway. Development Genes and Evolution 218, 667–680.

Patricolo, E., Cammarata, M., D'Agati, P., 2001. Presence of thyroid hormones in ascidian larvae and their involvement in metamorphosis. Journal of Experimental Zoology 290, 426–430.

Cyclostomes

Kluge, B., Renault, N., Rohr, K.B., 2005. Anatomical and molecular reinvestigation of lamprey endostyle development provides new insight into thyroid gland evolution. Development Genes and Evolution 215, 32–40.

Bony Fishes: Non-teleosts

Plohman, J.C., Dick, T.A., Eales, J.G., 2002. Thyroid of lake sturgeon, *Acipenser fulvescens*. I. Hormone levels in blood and tissues. General and Comparative Endocrinology 125, 47–55.

Bony Fishes: Teleosts

Björnsson, B.T., Stefansson, S.O., McCormick, S.D., 2011. Environmental endocrinology of salmon smoltification. General and Comparative Endocrinology 170, 290–298.

Eales, J.G., 2006. Modes of action and physiological effects of thyroid hormones in fish. In: Reinecke, M., Zaccone, G., Kapoor, B.G. (Eds.), Fish Endocrinology, vol. 2. Science Publishers, Enfield, NH, pp. 767–808.

Habibi, H.R., Nelson, E.R., Allan, E.R.O., 2012. New insights into thyroid hormone function and modulation of reproduction in goldfish. General and Comparative Endocrinology 175, 19–26.

Johnson, K.M., Lema, S.C., 2011. Tissue-specific thyroid hormone regulation of gene transcripts encoding iodothyronine deiodinases and thyroid hormone receptors in striped parrotfish (*Scarus iseri*). General and Comparative Endocrinology 172, 505–517.

Larsen, D.A., Swanson, P., Dickhoff, W.W., 2011. The pituitary–thyroid axis during the parr–smolt transformation of coho salmon, *Oncorhynchus kisutch*: quantification of TSH β mRNA, TSH, and thyroid hormones. General and Comparative Endocrinology 171, 367–372.

MacKenzie, D.S., Jones, R.A., Miller, T.C., 2009. Thyrotropin in teleost fish. General and Comparative Endocrinology 161, 83–89.

Morgado, I., Santos, C.R.A., Jacinto, R., Power, D.M., 2007. Regulation of transthyretin by thyroid hormones in fish. General and Comparative Endocrinology 152, 189–197.

Nelson, E.R., Habibi, H.R., 2009. Thyroid receptor subtypes: structure and function in fish. General and Comparative Endocrinology 161, 90–96.

Power, D.M., Llewellyn, L., Faustino, M., Nowell, M.A., Björnsson, B.Th., Einarsdottir, I.E., Canario, A.V.N., Sweeny, G.E., 2001. Thyroid hormones in growth and development of fish. Comparative Biochemistry and Physiology C 130, 447–459.

Subhash Peter, M.C., 2011. The role of thyroid hormones in stress response of fish. General and Comparative Endocrinology 172, 198–210.

Taillebois, L., Keith, P., Valade, P., Torres, P., Baloche, S., Dufour, S., Rousseau, K., 2011. Involvement of thyroid hormones in the control of larval metamorphosis in *Sicyopterus lagocephalus* (Teleostei: Gobioidei) at the time of river recruitment. General and Comparative Endocrinology 173, 281–288.

Ueda, H., 2011. Physiological mechanism of homing migration in Pacific salmon from behavioral to molecular biological approaches. General and Comparative Endocrinology 170, 222–232.

Amphibians

Bonett, R.M., Hoopfe, E.D., Denver, R.J., 2010. Molecular mechanisms of corticosteroid synergy with thyroid hormone during tadpole metamorphosis. General and Comparative Endocrinology 168, 209–219.

Carr, D.L., Carr, J.A., Willis, R.E., Pressley, T.A., 2008. A perchlorate sensitive iodide transporter in frogs. General and Comparative Endocrinology 156, 9–14.

Carr, J.A., Patiño, R., 2011. The hypothalamus–pituitary–thyroid axis in teleosts and amphibians: endocrine disruption and its consequences to natural populations. General and Comparative Endocrinology 170, 299–312.

Denver, R.J., 1998. The molecular basis of thyroid hormone-dependent central nervous system remodeling during amphibian metamorphosis. Comparative Biochemistry and Physiology C 119, 219–228.

Duarte-Guterman, P., Langlois, V.S., Pauli, B.D., Trudeau, V.L., 2010. Expression and T_3 regulation of thyroid hormone- and sex steroid-related genes during *Silurana (Xenopus) tropicalis* early development. General and Comparative Endocrinology 166, 428–435.

Furlow, J.D., Neff, E.S., 2006. A developmental switch induced by thyroid hormone: *Xenopus laevis* metamorphosis. Trends in Endocrinology & Metabolism 17, 38–45.

Hollar, A.R., Choi, J., Grimm, A.T., Buchholz, D.R., 2011. Higher thyroid hormone receptor expression correlates with short larval periods in spadefoot toads and increases metamorphic rate. General and Comparative Endocrinology 173, 190–198.

Korte, J.J., Sternberg, R.M., Serrano, J.A., Thoemke, K.R., Moen, S.M., Lillegard, K.E., Hornung, M.W., Tietge, J.E., Degitz, S.J., 2011. Thyroid-stimulating hormone (TSH): measurement of intracellular, secreted, and circulating hormone in *Xenopus laevis* and *Xenopus tropicalis*. General and Comparative Endocrinology 171, 319–325.

Kulkarni, S.S., Singamsetty, S., Buchholz, D.R., 2010. Corticotropin-releasing factor regulates the development in the direct developing frog *Eleutherodactylus coqui*. General and Comparative Endocrinology 169, 225–230.

Okada, R., Kobayashi, T., Yamamoto, K., Nakakura, T., Tanaka, S., Vaudry, H., Kikuyama, S., 2009. Neuroendocrine regulation of thyroid-stimulating hormone secretion in amphibians. Annals of the New York Academy of Sciences 1163, 262–270.

Okada, R., Miller, M.F., Yamamoto, K., De Groef, B., Denver, R.J., Kikuyama, S., 2007. Involvement of the corticotropin-releasing factor (CRF) type 2 receptor in CRF-induced thyrotropin release by the amphibian pituitary gland. General and Comparative Endocrinology 150, 437–444.

Opitz, R., Trubiroha, A., Lorenz, C., Lutz, I., Hartmann, S., Blank, T., Braunbeck, T., Kloas, W., 2006. Expression of sodium-iodide symporter mRNA in the thyroid gland of *Xenopus laevis* tadpoles: developmental expression, effects of antithyroidal compounds, and regulation by TSH. Journal of Endocrinology 190, 157–170.

Opitz, R., Kloas, W., 2010. Developmental regulation of gene expression in the thyroid gland of *Xenopus laevis* tadpoles. General and Comparative Endocrinology 168, 199–208.

Page, R.B., Monaghan, J.R., Walker, J.A., Voss, S.R., 2009. A model of transcriptional and morphological changes during thyroid hormone-induced metamorphosis of the axolotl. General and Comparative Endocrinology 162, 219–232.

Paris, M., Hillenweck, A., Bertrand, S., Delous, G., Escriva, H., Zalko, D., Cravedi, J.-P., Laudet, V., 2008. Active metabolism of thyroid hormone during metamorphosis of amphioxus. Integrative and Comparative Biology 50, 63–74.

Sachs, L.M., Damjanovski, S., Jones, P.L., Li, Q., Amano, T., Ueda, S., Shi, Y.-B., Ishizuya-Oka, A., 2000. Dual functions of thyroid hormone receptors during *Xenopus* development. Comparative Biochemistry and Physiology B 126, 199–211.

Safi, R., Vlaeminck-Guillem, V., Duffraisse, M., Seugnet, I., Plateroti, M., Margotat, A., et al., 2006. Pedomorphosis revisited: thyroid hormone receptors are functional in *Necturus maculosus*. Evolution & Development 8, 284–292.

Sternberg, R.M., Thoemke, K.R., Korte, J.J., Moen, S.M., Olson, J.M., Korte, L., Tietge, J.E., Degitz Jr., S.J., 2011. Control of pituitary thyroid-stimulating hormone synthesis and secretion by thyroid hormones during *Xenopus* metamorphosis. General and Comparative Endocrinology 173, 428–437.

Reptiles

Boggs, A.S.P., Hamlin, H.J., Lowers, R.H., Guillette Jr., L.J., 2011. Seasonal variation in plasma thyroid hormone concentrations in coastal versus inland populations of juvenile American alligators (*Alligator mississippiensis*): Influence of plasma iodide concentrations. General and Comparative Endocrinology 174, 362–369.

Hulbert, A.J., 1985. A comparative study of thyroid function in reptiles and mammals. In: Follett, B.K., Ishii, S., Chandola, A. (Eds.), The Endocrine System and the Environment. Springer-Verlag, Berlin, pp. 105–115.

Villalobos, P., Orozco, A., Valverde-R, C., 2010. Molecular cloning and characterization of a type 3 iodothyronine deiodinase in the pine snake *Pituophis deppei*. General and Comparative Endocrinology 169, 167–173.

Birds

Chen, Y., Sible, J.C., McNabb, F.M.A., 2008. Effects of maternal exposure to ammonium perchlorate on thyroid function and the expression of thyroid-responsive genes in Japanese quail embryos. General and Comparative Endocrinology 159, 196–207.

De Groef, B., Geris, K.L., Vandenborne, K., Darras, V.M., Kühn, E.R., 2005. CRH control of thyroid function in the chicken. In: Dawson, A., Sharp, P.J. (Eds.), Functional Avian Endocrinology. Narosa Publishing House, New Dehli, pp. 414–426.

May, J.D., 1989. The role of the thyroid in avian species. Critical Reviews in Poultry Biology 2, 171–186.

Nabb, F.A.M., 2006. Avian thyroid development and adaptive plasticity. General and Comparative Endocrinology 147, 93–101.

Sharp, P.J., Klandorf, H., 1985. Environmental and physiological factors controlling thyroid function in galliformes. In: Follett, B.K., Ishii, S., Chandola, A. (Eds.), The Endocrine System and the Environment. Springer-Verlag, Berlin, pp. 175–188.

Van der Geyten, S., Verhoelst, C.H.J., Reyns, G.E., Kühn, E.R., Decuypere, E., Darras, V.M., 2005. Differential regulation of intracellular thyroid hormone availability in developing chicken brain and peripheral tissues. In: Dawson, A., Sharp, P.J. (Eds.), Functional Avian Endocrinology. Narosa Publishing House, New Dehli, pp. 427–436.

Wasser, S.K., Azkarate, J.C., Booth, R.K., Hayward, L., Hunt, K., Ayres, K., Vynne, C., Gobush, K., Canales-Espinosa, D., Rodríguez-Luna, E., 2010. Non-invasive measurement of thyroid hormone in feces of a diverse array of avian and mammalian species. General and Comparative Endocrinology 168, 1–7.

Chapter 8

The Mammalian Adrenal Glands: Cortical and Chromaffin Cells

The responses an animal makes to adverse or **stressful stimuli**, called **stressors**, leads to a **physiological response, stress,** that includes the release of hormones from the adrenal glands and their subsequent effects to enable the animal to cope with the stressor. These responses are controlled largely through the **hypothalamus-pituitary-adrenal (HPA) axis** and secretion of **corticotropin-releasing hormone (CRH)** and **corticotropin (ACTH)**. The adrenal hormones induce changes in metabolism and/or ionic regulation that work to combat physiological and psychological factors and eventually to eliminate or at least neutralize the stressor. Knowledge of this stress system contributes to our understanding of how animals adapt physiologically to physical and psychological traumas. In addition, these adrenal hormones play other important roles that are described in this chapter.

Mammals typically possess two adrenal glands, one located superior to each kidney (*ad-renal* or *supra-renal*; Figure 8-1). This anatomical arrangement is responsible for their present name, adrenals or adrenal glands, and for an alternative name in humans, the suprarenal glands.

Each adrenal gland actually consists internally of three almost separate endocrine glands. The outer portion or **adrenal cortex** represents three glandular regions and is composed largely of lipid-containing, steroidogenic **adrenocortical cells**. The adrenal cortex surrounds a fourth endocrine region consisting of an inner mass of **chromaffin cells** called the **adrenal medulla**. Chromaffin cells are so named because they contain intracellular granules containing catecholamines that can be stained by certain chromium compounds (see ahead).

The adrenocortical cells are derived from the coelomic epithelium in the pronephric region of the embryo adjacent to the genital ridge that gives rise to the gonads. These cells produce steroid hormones including **glucocorticoids**, **mineralocorticoids**, and weak androgens such as **dehydroepiandrosterone, DHEA** (see Chapter 3, Figures 3-22 to 3-25). Two regions secrete glucocortiocoids and DHEA, respectively, under the direct stimulatory influence of ACTH from the pituitary gland (see Chapter 4), although DHEA secretion also may be stimulated by LH and LH-like hormones. Secretion of mineralocoticoids by the third region is stimulated by the **renin-angiotensin system** (see ahead). Glucocorticoids are named for their influences on glucose metabolism and mineralocorticoids for their effects on Na^+ and K^+ balance. The major glucocorticoids synthesized by mammals are **cortisol**, **corticosterone**, and to some extent **11-deoxycortisol**. The major mineralocorticoids are **aldosterone** and **deoxycorticosterone**. Representative blood levels for glucocorticoids and mineralocorticoids in mammals are

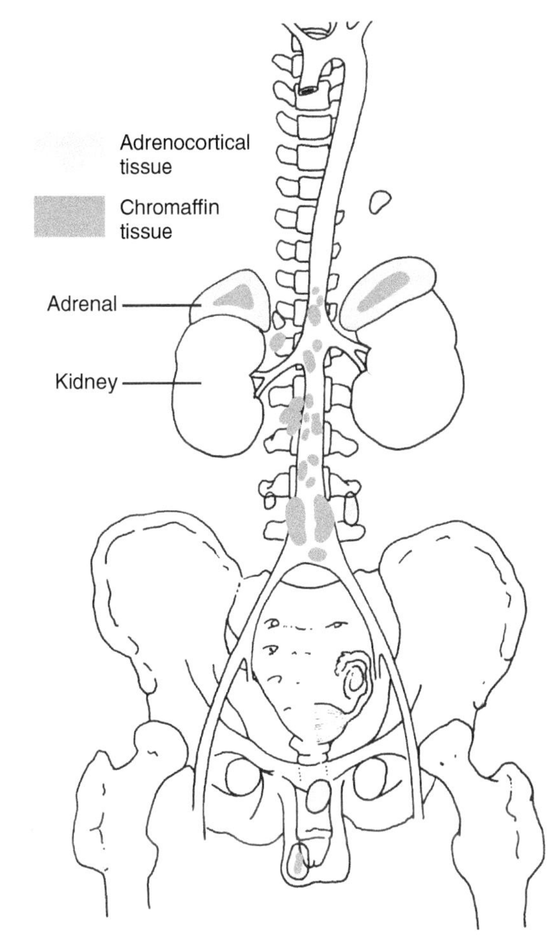

FIGURE 8-1 **Location of adrenal tissue in the human.** Note in addition to the expected location of adrenocortical and chromaffin cells in the adrenal gland heterotopic locations where either tissue may be found in either sex. *Adapted with permission from Bethune, J.E. The Adrenal Cortex: A Scope Monograph. Upjohn Co., 1974.)*

Vertebrate Endocrinology. http://dx.doi.org/10.1016/B978-0-12-394815-1.00008-2

TABLE 8-1 Circulating Levels of Corticoids in Selected Mammalian Species (Prototheria, Metatheria, and Eutheria)

Species	Mean Steroid Levels ± SD in Peripheral Blood				
	Aldosterone (ng/dL)	Corticosterone (μg/dL)	Cortisol (μg/dL)	Deoxycorticosterone (ng/dL)	11-Deoxy-cortisol (μg/dL)
Prototheria					
Echidna[a]	1.5	0.35	0.18	–	–
Echidna[b]	–	0.14 ± 0.07	0.07 ± 0.03	–	–
Echidna + ACTH[b]	–	1.06 ± 0.56	0.42 ± 0.23	–	–
Metatheria					
Black-tailed wallaby	8.2 ± 4.3	0.12 ± 0.02	1.1 ± 0.2	3.3 ± 0.6	0.13 ± 0.08
Common wombat	0.9 ± 1.4	0.06 ± 0.02	0.04 ± 0.04	2.0 ± 1.3	0.14 ± 0.08
Dingo	6.7 ± 4.3	0.04 ± 0.21	1.90 ± 1.90	23.6 ± 23.7	0.19 ± 0.09
Koala	1.6 ± 2.2	0.20 ± 0.05	*[c]	6.0 ± 8.3	0.06 ± 0.11
Eutheria					
Sheep	2.1 ± 1.7	0.09 ± 0.04	0.52 ± 0.50	2.5 ± 2.5	0.05 ± 0.03
Dog	2.1 ± 3.6	0.20 ± 0.13	0.85 ± 0.39	11.3 ± 6.0	0.07 ± 0.05
Fox	13.9 ± 3.2	0.59 ± 0.17	2.30 ± 1.2	37.4 ± 22.7	0.17 ± 0.03
BALB/cfC3H mice[d]	–	12.00 ± 2.16	–	–	–
Sand rat[e]	2.8 ± 13.3	0.20 ± 0.8	10.7 ± 24.9	–	–
Human, male	12.4	0.42	14.4	6.6	0.05
Harbor seals[f]			0.494 ± 0.153		

[a]*Oddie, G.J. et al., Journal of Endocrinology,* ***69****, 341–348, 1976.*
[b]*Sernia, C. and McDonald, I.R., Journal of Endocrinology,* ***72****, 41–52, 1977.*
[c]**, Undetectable.*
[d]*Hawkins, E.F. et al., Journal of Experimental Zoology,* ***194****, 479–484, 1975.*
[e]*Amirat, Z. et al., General and Comparative Endocrinology,* ***40****, 36–43, 1980.*
[f]*Oki, C. and Atkinson, S., General and Comparative Endocrinology,* ***136****, 289–297, 2004.*

provided in Table 8-1. Relative glucocorticoid and mineralocorticoid activities of some common natural and synthetic corticosteroids can be found in Table 8-2.

The chromaffin cells originate from the neural crest, and the medulla functions essentially like a modified sympathetic ganglion. The adrenal medulla is under direct neural control (preganglionic cholinergic sympathetic neurons originating in the thoracic spinal cord) and releases **norepinephrine** and/or **epinephrine** into the blood. These secretions are usually termed "hormones," but, considering the embryonic origin of the medulla, they could also be called "neurohormones."

At first glance, there seems to be no functional significance for the close anatomical relationship of the adrenocortical and chromaffin cells. Although there is participation of both systems with respect to adaptations to stressors, the factors controlling release of their secretions differ in significant ways and their biological actions at the receptor level do not overlap. The anatomical closeness of the cortex and medulla in mammals, as well as relationships of their homologous tissues in non-mammals, may simply be a function of the physical closeness of their embryological sites of origin, although the progressive evolution of this arrangement is obvious when non-mammals are examined (see Chapter 9). CRH, ACTH, and glucocorticoids do have effects on catecholamine synthesis by chromaffin cells, and these systems are not entirely independent.

I. THE MAMMALIAN ADRENAL CORTEX

The adrenal cortex of adult mammals may be subdivided by means of histological criteria into three well-defined

TABLE 8-2 Activities of Corticosteroids

Steroid	Relative Activity in Assay				
	MR[a]	Na+[b]	GLY[c]	GR[d]	AI[e]
Aldosterone	1	1	0.15	1	–
DOC[f]	0.8	0.03	0.02	1	–
Corticosterone	0.2	0.004	0.36	2	0.3
Cortisol	0.1	0.001	1	1	1
18-OH-DOC[f]	0.015	0.004	–	0.02	–
Dexamethasone	0.05	0.001	17–250	10	25–169

[a]MR, mineralocorticoid; MR receptor assay is based on competition for [^{3}H]aldosterone-binding sites in rat kidney.
[b]Na^+ bioassay, normally based on change in urinary Na^+/K^+ in the adrenalectomized rat.
[c]GLY, glycogen; GLY bioassay is based on glycogen deposition in the adrenalectomized rat liver.
[d]GR, glucocorticoid receptor; GR assay is based on competition with [^{3}H]dexamethasone-binding sites in rat kidney.
[e]AI, anti-inflammatory; AI bioassay is based on anti-inflammatory activity.
[f]DOC, deoxycorticosterone; 18-OH-DOC, 18-hydroxy-DOC.

regions: **zona glomerulosa, zona fasciculata**, and **zona reticularis** (Figure 8-2). These regions are arranged as concentric shells surrounding the adrenal medulla. In addition, there are inner zones described between the outer zones and the medulla in some mammals. These regions may perform unique functions and may be transitory.

A. Zonation of the Adrenal Cortex

The cells of the outermost region of the adrenal cortex, the zona glomerulosa, are smaller and more rounded and contain less lipid than those of the more central zona fasciculata. The zona glomerulosa is responsible for synthesis of aldosterone as well as some other corticosteroids. There are few cytological changes in the zona glomerulosa following hypophysectomy or administration of ACTH, suggesting that the secretion of aldosterone is independent of pituitary control. Although ACTH is not necessary for the synthesis and release of aldosterone, the responsiveness of the glomerulosal cells to agents that normally elicit these events is reduced in hypophysectomized mammals and is enhanced with ACTH treatment. Consequently, ACTH does have a permissive effect on cells of the zona glomerulosa.

The zona fasciculata is the largest zone in the adrenal cortex. It is located between the zona glomerulosa and the innermost zona reticularis and is histologically distinct from both. It consists of polyhedral (many-sided) cells that are sources of the glucocorticoids. The cells of the zona fasciculata are arranged in narrow columns or cords surrounding blood sinusoids that allow the cells to be bathed directly with blood (i.e., there is no tissue-blood barrier). The proportion of cortisol and corticosterone secreted differs markedly, from secretion of primarily cortisol (human), through mixtures of both (cat, deer), to primarily corticosterone (rat). Thickness of the zona fasciculata is most sensitive to circulating levels of ACTH. It exhibits hypertrophy and hyperplasia in response to prolonged elevation of ACTH secretion caused by stress or treatment with the drug **metyrapone**, which blocks the 11-hydroxylation step necessary for glucocorticoid synthesis (see Chapter 3) and hence elevates ACTH in the blood. Unlike the zona glomerulosa, the zona fasciculata atrophies markedly following hypophysectomy or prolonged glucocorticoid therapy and hypertrophies as a result of prolonged ACTH therapy.

The zona reticularis typically borders the adrenal medulla, and it contains numerous thin, extracellular reticular fibers (hence its name). It is a primary source of adrenal androgens but some glucocorticoids may be synthesized here as well. The zona reticularis also hypertrophies in response to ACTH and atrophies following hypophysectomy, but not so dramatically as the zona fasciculata. Gonadotropins (**luteinizing hormone [LH]** or **human chorionic gonadotropin [hCG]**) can also stimulate secretion of adrenal androgens.

It should be noted that the "typical" anatomical pattern described here within the adrenal cortex and the anatomical relationship of cortex to medulla varies considerably within mammals as a group. Furthermore, ectopic nodules of functional cortical tissue are not uncommon (Figure 8-1), and this accessory adrenocortical tissue may become a source for corticosteroids following surgical adrenalectomy.

B. Additional Zonation

Several unique adrenocortical zones are known only for certain species, but the adrenals of most mammalian species have not been examined in detail. These special zones may be conspicuous only at certain times in the life of an animal or in only one sex.

1. The Fetal Zone

In primates, a very conspicuous zone occupies the bulk of the adrenal gland prior to birth. This region is called the **fetal zone** and is responsible for the relatively large size of the adrenal at birth (Figure 8-3). In humans, the neonate adrenal may be as large as the adrenal gland of a 10- to 13-year-old. During gestation, the fetal zone, which is found between the cortex and the medulla, synthesizes and releases relatively large quantities of DHEA and lesser amounts of its sulfated derivative, **DHEA-S**. These adrenal androgens serve as precursors for the synthesis of estrogens by the placenta (see Chapter 10). Failure of the fetal zone to produce adequate amounts of DHEA/DHEA-S results in premature termination of gestation. Formerly, it was believed that CG from the

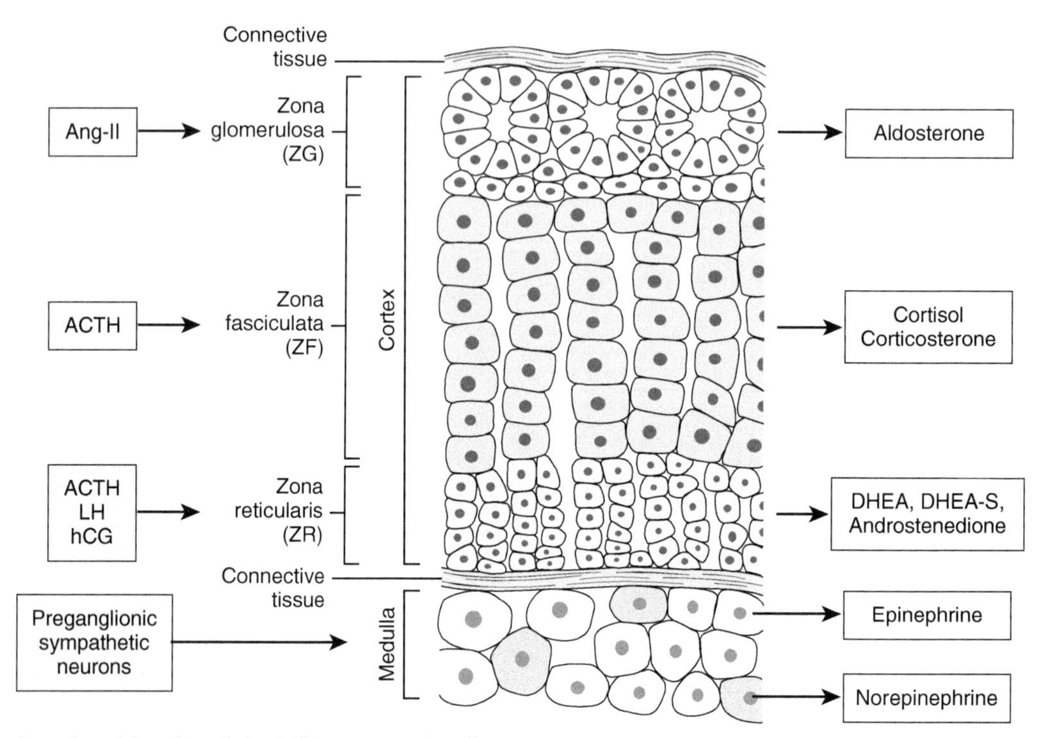

FIGURE 8-2 **Zonation of the adrenal gland.** The cortex consists of an outermost layer of connective tissue (CT); the zona glomerulosa (ZG), which produces aldosterone; the zona fasciculata (ZF), which secretes most of the glucocorticoids (cortisol and/or corticosterone); and the inner zona reticularis (ZR), which specializes in adrenal androgen production (DHEA, DHEA-S, androstenedione). The adrenal medulla is separated from the cortex by another layer of CT and consists primarily of chromaffin cells that secrete epinephrine or norepinephrine. Numerous other secretions have been associated with the adrenal medulla, including dopamine and endogenous opiates (EOPs).

placenta (see Chapters 4 and 10) was responsible for stimulating fetal adrenal androgen production necessary to maintain pregnancy, but recent studies suggest that adrenal androgen secretion is strongly influenced by other placental hormones, too. The fetal adrenal and placenta also play important roles in the birth process (see Chapter 10). Following birth, the fetal zone ceases to function in humans and degenerates rapidly. The fetal zone typically disappears completely by one year of age. The zona reticularis begins to synthesize DHEA/DHEA-S at about age 5 or 6, and DHEA synthesis accelerates during puberty. Maximal DHEA production is achieved around age 20, after which its production slowly declines. It has been suggested that DHEA may be an anti-tumor substance and/or a precursor for synthesis of other androgens or estrogens, especially in postmenopausal women.

2. The Mouse X-Zone

The cortex of the mouse adrenal contains a unique **X-zone** located between the zona reticularis and the medulla. The X-zone appears to be unrelated to the fetal zone of primates, although it appears in the same anatomical location. This zone degenerates in males at puberty and in females of most mouse strains during the first pregnancy. Degeneration in the males is correlated with production of androgens by the testes. The function of this X-zone is not known.

3. The "Special Zone"

In at least one marsupial, the brush-tailed possum (*Trichosurus vulpecula*), there is a large inner special zone that appears only in the adult female. Its function has not been elucidated, and it is not known whether it is comparable to either the primate fetal zone or the X-zone of mice.

II. BIOSYNTHESIS AND TRANSPORT OF CORTICOSTEROIDS

Glucocorticoids and mineralocorticoids differ not only in their actions but also in their synthesis and transport in the blood. As outlined in Chapter 3, cholesterol is transported to the inner mitochondrial membrane by the StAR protein where corticosteroids are derived from conversion of **cholesterol** to **pregnenolone** in the mitochondria by the action of the **side-chain cleavage enzyme $P450_{scc}$**. Pregnenolone is converted into progesterone and then further modified via the Δ^5 pathway (see Chapter 3, Figure 3-24). Cholesterol is readily obtained from circulating **low-density lipoprotein droplets (LDLs)** (see Chapter 12) or from stored deposits in the adrenal cells. Synthesis of cholesterol from acetyl groups also can occur in adrenal cells as described in Chapter 3.

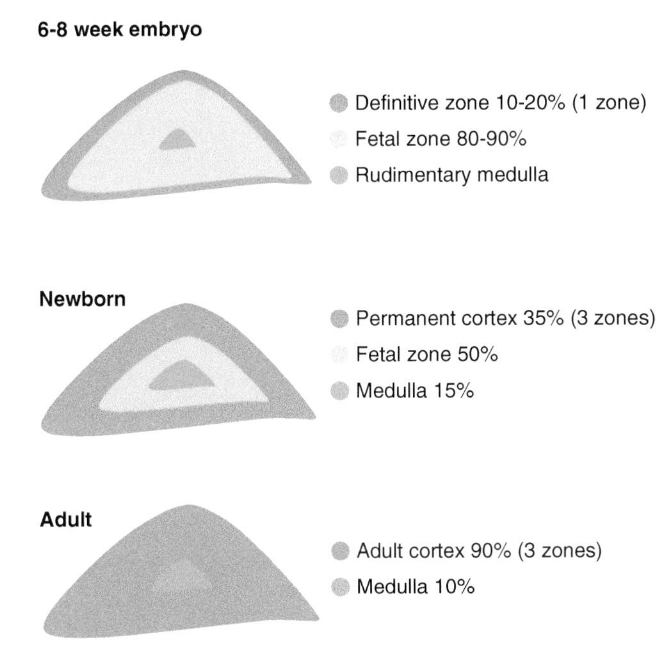

FIGURE 8-3 **Comparison of fetal zone to remainder of adrenal gland in humans.** Prior to birth, the bulk of the adrenal consists of the androgen-secreting fetal zone that regresses rapidly after birth. Whereas the fetal zone is gone by one year of age, the medulla and cortex continue to grow until puberty. In the actual adrenal gland, the fetal zone appears between the zona reticularis and the medulla but is shown above the cortex here for effect. *Adapted with permission from Tsakiri, S.P., Chrousos, G.P., and Margioris, A.N., Molecular development of the hypothalamic-pituitary-adrenal (HPA) axis. In E,A. Eugster and O.H. Pescovitz, eds. "Contemporary Endocrinology: Developmental Endocrinology" From Research to Clinical Practice, Humana Press, Inc., pp 359-380.*

A. Synthesis of Corticosteroids

At the membranes of the smooth endoplasmic reticulum, pregnenolone can be metabolized along either the **17α-hydoxysteroid pathway** or the **17-deoxysteroid pathway**. The former pathway leads to **11-deoxycortisol** and is accomplished by the actions of the enzyme **21-hydroxylase**. 11-Deoxycortisol may be secreted or may reenter the mitochondria where the **11β-hydroxylase enzyme** ($P450_{11\beta1}$) converts it to cortisol. Conversion of pregnenolone to progesterone in the zona fasciculata or glomerulosa, via the 17-deoxysteroid pathway yields **11-deoxycorticosterone** which is converted to corticosterone by employing the same enzymes as did the17α-hydroxysteroid pathway. Cortisol and/or corticosterone are the major end products of corticosteroidogenesis in the zona fasciculata and to a lesser extent in the zona reticularis.

In the zona glomerulosa, the 17-deoxysteroid pathway is favored, and corticosterone is further modified to **18-hydroxycorticosterone** and then to aldosterone. Another mitochondrial enzyme called **aldosterone synthase** ($P450_{11\beta2}$ or $P450_{.aldo}$) is responsible for aldosterone synthesis in rodents and humans. Aldosterone and 18-hydroxycorticosterone have mineralocorticoid activity, but aldosterone is more potent and is typically the dominant secretory product of the zona glomerulosa *in vivo*. However, deoxycorticosterone may be the major secretory product in some species (see Table 8-1).

As mentioned previously, adrenal androgens are synthesized primarily in the zona reticularis or by the fetal zone of primates. Both the Δ^4 and Δ^5 pathways may be utilized (see Chapter 3), but the Δ^5 pathway is more common, resulting in DHEA which is then sulfated to form DHEA-S. The primate fetal zone makes little or no **androstenedione,** and this weak Δ^5 androgen is more commonly found in the adult adrenal. DHEA also is a weak androgen, and the sulfated form does not penetrate readily into potential target cells, reducing the threat of masculinization to a female fetus or the mother. DHEA-S is converted by the placenta into **16α-hydroxyDHEA** and then by the **aromatase** enzyme ($P450_{aro}$) to estriol. Androstenedione is converted by $P450_{aro}$ to estrone, which can be processed further to estradiol or estriol. These conversions of fetal androgens are essential for maintaining pregnancy. Because the placenta cannot synthesize androgens, it must depend entirely on the fetal adrenal to provide the androgen precursor for estrogen synthesis.

B. Release of Corticosteroids

Circulating corticosteroids reflect synthesis rates, as little hormone is stored in the adrenal and corticosteroids are released as they are made. In adult humans, daily episodic release of cortisol is maximal between 6 a.m. and 9 a.m., with lower values occurring in the afternoon and minimal levels at night (see Figure 2-3). This daily rhythm can be disrupted by eating lunch that causes increased cortisol secretion but is not affected by eating dinner. In nocturnal animals, such as the rat, the rhythmic pattern of secretion is reversed.

Aldosterone secretion shows a similar rhythm to glucocorticoids in humans with peak levels observed between 6 a.m. and 9 a.m. The drug **propranolol**, a β-adrenergic blocker, abolishes the aldosterone rhythm in rats, suggesting that catecholamines from sympathetic neurons are responsible for the rhythm, although elevated ACTH may influence the responsiveness of the zona glomulerosa.

In humans, as is generally true for other mammals, adrenal androgen synthesis accelerates at the onset of puberty (**adrenarche**) in both males and females (Figure 8-4) and continues at this level until the early 20s when it begins to decline. Adrenal function in general declines after age 40 to 50. This age-related decline in adrenal production has been termed **adrenopause** (Figure 8-5).

Because of the episodic nature of corticosteroid release and the strong diurnal aspect of secretion, measurements to determine appropriate levels of corticosteroids are

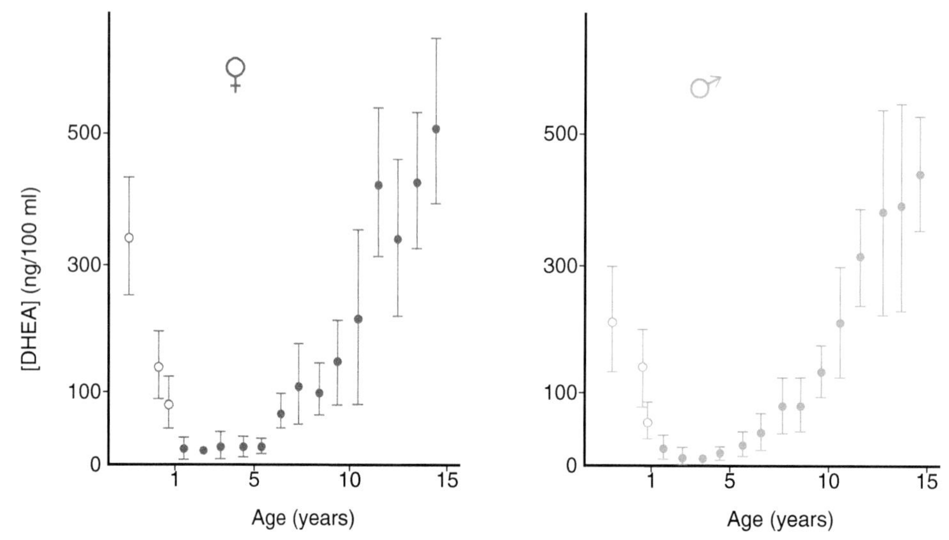

FIGURE 8-4 **Adrenarche in humans.** After birth, adrenal DHEA (open circles) declines as the fetal zone regresses. Although the fetal zone has regressed completely by age one, production of the adrenal androgen DHEA continues in the zona reticularis (solid circles). DHEA secretion begins to increase at about age five and peaks at puberty. *(Adapted with permission from de Peretti, E. and Forest, M.G.,* Journal of Clinical Endocrinology and Metabolism, ***43****, 982–991, 1976.)*

complicated by the need for repeated sampling of blood or saliva. (Saliva contains corticosteroids in proportion to their concentrations in the blood, and their measurement can be used as an index of blood levels.) Furthermore, sampling itself can evoke corticosteroid secretion, causing an overestimate of secretory activity. Analysis of steroid metabolites in 24-hour urine samples can provide an index of integrated secretion over that time period. Two non-invasive methods for assessing corticoid secretion have been developed recently using fecal and hair samples. Collecting fecal droppings in the field has proven to be a very effective way to monitor wild individuals or populations without invoking a sampling stress (Figure 8-6). Similarly, shed hair from wild primates can be collected in the field and analyzed for cortisol and provide evidence of recent cortisol secretion.

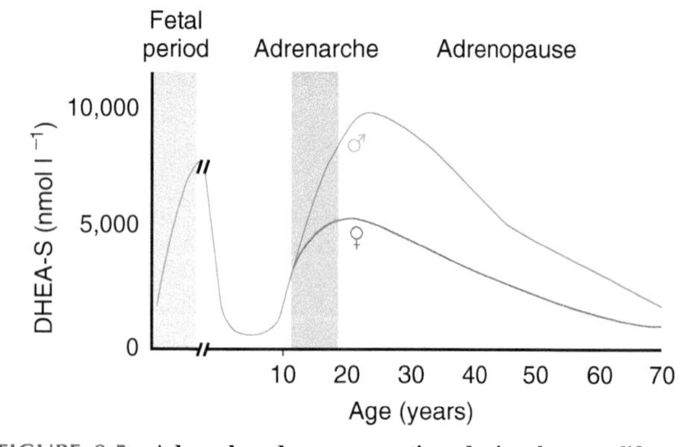

FIGURE 8-5 **Adrenal androgen secretion during human lifespan.** Other adrenal androgens follow the pattern shown here for DHEA-S. Note the marked decline associated with regression of the fetal zone after birth followed by a sharp increase at puberty (adrenarche) and a steady decline after mid-20s (adrenopause). *(Adapted with permission from Rainey, W.E. et al.,* Trends in Endocrinology and Metabolism, ***13****, 234–239, 2002.)*

C. Transport of Corticosteroids in the Blood

Probably, more than 90% of the glucocorticoids are bound to a specific plasma transport protein called **corticosteroid binding globulin (CBG)** or **transcortin**. Similar to TBGs (Chapter 3), CBGs are small proteins that belong to a larger class of serine protease inhibitors (*SERPINA6*). Some of the circulating glucocorticoid also binds nonspecifically to albumin, as described for thyroid hormones (see Chapter 6). Although there is a bias for assuming that free hormone is most important for entering target cells and producing effects, there also is evidence to support a role for interaction of occupied CBG with target cell membranes that facilitates dissociation of glucocorticoids from CBG and their entrance into the target cell. Aldosterone does not bind significantly to CBG or other plasma proteins; consequently, aldosterone is present primarily in the free state at much lower total concentrations in the blood than are the glucocorticoids. Aldosterone is cleared from the blood much more rapidly because of the lack of binding proteins.

D. Metabolism of Corticosteroids

The liver is the major site for metabolizing corticosteroids (see Chapter 3), although other tissues, such as the kidneys and intestines, also can perform this task. Most corticosteroids are conjugated with sulfates or glucuronides to

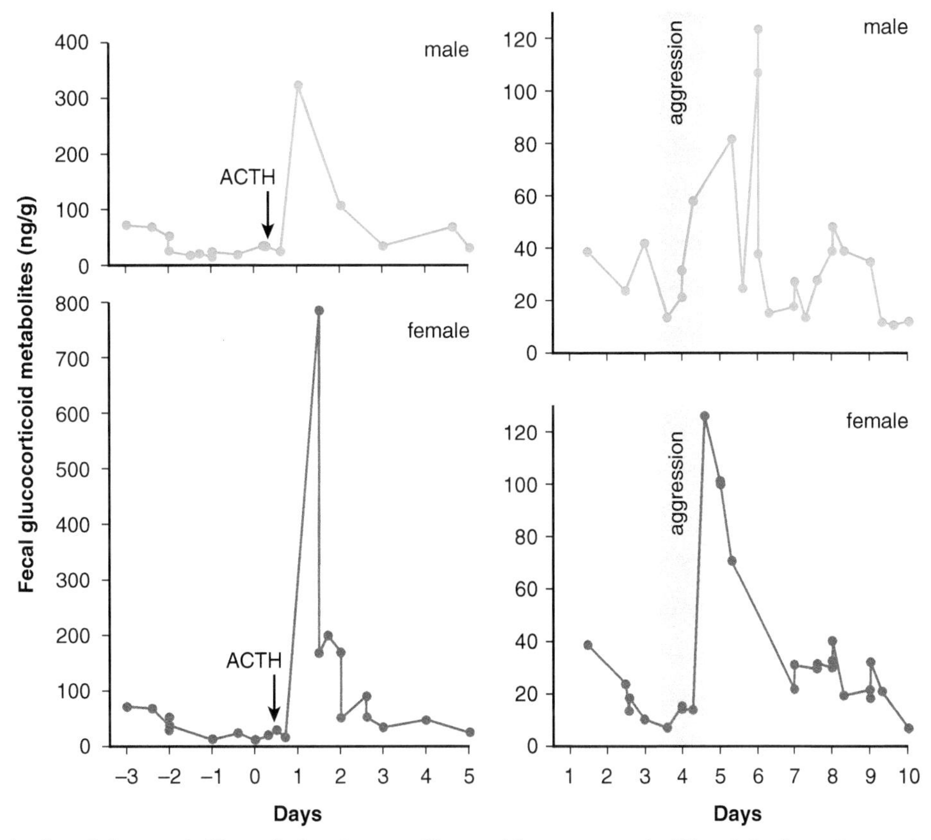

FIGURE 8-6 **Monitoring of glucocorticoid metabolites in spotted hyenas (*Crocuta crocuta*).** Although fecal samples contain a mixture of glucocorticoid metabolites, extraction of fecal samples and measurement of fecal glucocorticoid metabolites by immunoassay can serve as a noninvasive means for monitoring adrenal activity in wild animal species such as hyenas from which it would be difficult or harmful to collect a blood sample. Changes in fecal glucocorticoid content reflect the adrenal response to ACTH (left) and aggression (right) in both male and female hyenas. *(Adapted with permission from Goymann, W. et al.,* General and Comparative Endocrinology, ***114****, 340–348, 1999.)*

increase their solubility in water, and they are readily excreted via the urine, bile, or feces. Cortisol, aldosterone, and most adrenal androgens are conjugated as glucuronides, but DHEA is excreted primarily as DHEA-S. In humans, the majority of corticosteroid metabolites are excreted via the urine, whereas in the rat the biliary and fecal routes are more popular. One way in which corticosteroids are metabolized is by reduction of the A-ring and addition of hydroxyl groups at C_3 or C_{20} followed by side-chain cleavage to yield C_{19} steroids. Hydroxylation by cytochrome P450 enzymes also occurs at C_6 and C_{16}. In humans, up to a third of the cortisol may be oxidized to corotic acids (see Figure 3-30).

III. SECRETION AND ACTIONS OF GLUCOCORTICOIDS

Secretion of glucocorticoids from the zona fasciculata and zona reticularis is directly controlled by the HPA axis involving CRH and ACTH (see Chapter 4). Circulating ACTH levels are depressed by elevated levels of glucocorticoids and are increased following adrenalectomy, establishing the existence of direct negative feedback of corticosteroids on ACTH release. **Arginine vasopressin (AVP)** potentiates the action of CRH on the corticotropic cells of the pars distalis during responses to chronic stress (see Chapter 4). Thus, stressors can bring about and maintain a marked increase in glucocorticoid levels. The ability of traumatic stimuli to override the normal negative feedback mechanism verifies the importance of multiple control mechanisms governing ACTH release.

A. Actions of Glucocorticoids

Glucocorticoids produce marked effects on energy metabolism at physiological doses as a result of changes in transport of materials into cells and induction of new enzyme syntheses. Their major effect is to supplant and conserve the energy normally derived from circulating glucose by (1) inhibiting glucose utilization by peripheral tissues (especially skeletal muscle), (2) stimulating entry of amino acids into liver cells and their conversion to glucose and storage as glycogen, and (3) enhancing the mobilization of fat stores in non-neural tissues. By blocking glucose utilization, glucocorticoids spare blood glucose so that the brain has a preferential source of glucose. As described in Chapter 3, glucocorticoids can bind to both intracellular **glucocorticoid receptors (GR = type II GR)** or to intracellular **mineralocorticoid receptors (MR = type I GR)**. However, glucocorticoids usually are inactivated by enzymatic conversion to cortisone in the cytosol of cells involved with mineral balance. Cortisone does not bind to the MR; however, some of the glucocorticoid feedback on brain neurons is accomplished through MR receptors (see Chapter 4). Membrane GR in the vertebrate brain (see ahead) are involved in local glucocorticoid feedback on CRH neurons and may be associated with behavior. The molecular structure of the membrane glucocorticoid receptor is unknown, but recent studies suggest that it may be a processed form of the intracellular receptor that translocates to the plasma membrane.

B. Mechanism of Genomic Glucocorticoid Cellular Action

Glucocorticoid effects on target cells are described in Chapter 3, and only a brief summary is provided here. The initial requirements for glucocorticoid action on liver cells are binding to GR, translocation of occupied GR into the nucleus, and eventual stimulation of nuclear RNA synthesis (both messenger RNA and ribosomal synthesis). Approximately 2 to 4 hours following application of glucocorticoids, there is an increase in new enzymes that brings about the changes in cellular metabolism characteristic of glucocorticoid action.

1. Glucocorticoids and Metabolism

In liver cells, these new enzymes include those associated with the conversion of fatty acids or amino acids into glucose and the polymerization of glucose to form glycogen (see Appendix H). Amino acid transport into liver cells also is stimulated. In contrast, glucocorticoids inhibit the uptake of amino acids and the metabolism of glucose in peripheral tissues such as skin, skeletal muscle, and adipose cells. This action makes glucose more readily available to the brain during stress. In the brain, glucocorticoids may act classically through intracellular receptors or may bind to plasma membrane receptors and bring about rapid effects in target cells. These membrane receptors may operate through second-messenger systems and retrograde synaptic signaling (see ahead) to produce their rapid effects. Elevated glucocorticoids or excessive doses of glucocorticoids inhibit protein synthesis in certain tissues (heart and diaphragm) and accelerate protein catabolism in others (especially skeletal muscle, bone, and lymphoid tissue). Protein catabolism provides amino acids for glucose synthesis and contributes to hyperglycemia. The metabolic conversion of amino acids into glucose is termed **gluconeogenesis**, the production of glucose from a non-carbohydrate precursor—in this case, amino acids. Glucocorticoids promote gluconeogenesis in the liver in a number of ways including stimulating liver cells to produce the gluconeogenic enzyme **phosphoenolpyruvate carboxykinase (PEPCK)** and to produce the enzymes required to strip the NH_2 group from amino acids (so-called deamination or transamination enzymes) prior to their incorporation into the gluconeogenic pathway. Glucocorticoids also enhance the actions of **growth hormone (GH)** on lipolysis in adipose tissue. This causes release of fatty acids and glycerol into the circulation and their conversion by liver cells into glucose, another form of gluconeogenesis.

2. Glucocorticoids and Reproduction

The onset of puberty is correlated with activation of adrenal secretion (adrenarche). Animals with reduced adrenocortical function may exhibit delayed puberty and/or seasonal breeding. Glucocorticoid secretion is always elevated during the breeding season and is considered to be essential for successful reproduction. Interestingly, dominant males exhibit a more moderate elevation of glucocorticoid levels than do subordinate males. In male marsupial mice (e.g., *Antechinus* spp.), high cortisol is associated with aggressive breeding that results in the death of all males by the end of the breeding season. The females exhibit lower levels of cortisol than the males and survive to rear the young. Although stress levels of glucocorticoids typically reduce breeding success, stress does not prevent breeding in male *Antechinus* even though it contributes to their death.

3. Glucocorticoids and Immunity

It is common to observe an increased incidence of disease in stressed animals. In fact, the immune response system is inhibited by the HPA axis at several levels; for example, elevated levels of glucocorticoids inhibit the inflammatory response as well as production of antibodies. Furthermore, CRH not only stimulates ACTH release but also causes the release of somatostatin (SST) and dopamine, which reduce secretion of GH and prolactin (PRL), respectively. Either GH or PRL can restore immune function that is normally depressed following hypophysectomy. Secretion of GH and PRL occurs at night when ACTH and glucocorticoid levels normally are lowest, and renewal of immune function occurs. Cytokines such as interleukins and tumor necrosis factor may enter the brain through special structures called **circumventricular organs** such as the **organum vasculosum of the lamina terminalis (OVLT)** of the preoptic area or the area postrema in the brainstem. Circumventricular organs contain fenestrated capillaries that allow access of many bloodborne materials directly into the brain. Other evidence suggests that cytokines also can enter at certain sites via carrier-mediated transport. Regardless of how they enter, cytokines such as interleukins and interferons released during increased immune activity can activate CRH release directly from the hypothalamus or indirectly via ascending brainstem pathways that ultimately causes elevated glucocorticoids and suppression of the immune response. Recent studies of cytokine effects on brain function suggest that at least some of their effects are mediated via sensory fibers of the vagus nerve that terminate in the nucleus of the solitary tract and stimulate the activity of noradrenergic neurons (see Figure 4-2), which innervate CRH-producing neurons in the **paraventricular nucleus (PVN)**. In rats, vagotomy blocks several responses classically attributed to cytokine effects on the brain including fever, hypothalamic depletion of norepinephrine, and elevated plasma corticosterone. The actions of interleukins on the blood–brain barrier also may allow access of other regulatory factors and even leukocytes into the nervous system. Not all actions of glucocorticoids are suppressive of immune function. Certain aspects of cellular immunity are enhanced by glucocorticoids at physiological levels, even though chronic elevations may be suppressive.

4. Adrenal Hormones and Stress

In 1929, the physiologist Walter Cannon coined the term "fight or flight" to describe how an animal's body responds to an emergency life situation by releasing more epinephrine into the blood stream. Over the decades our knowledge of stress as a physiological response has grown to include the importance of the HPA axis and glucocorticoid hormones. The role of the adrenal cortex in stress was first demonstrated in rats by Hans Selye in the 1930s and incorporated into a general theory of adaptation to stressors termed the **general adaptation syndrome**. Selye noted that, regardless of the nature of the stressor (cold, heat, noxious chemicals), the physiological response resulted in the same three effects: gastric ulcers, enlarged adrenals, and shrunken immune tissues. Thus, initially stress was defined as an all-encompassing term to include the responses to all stimuli (stressors) that are harmful or potentially harmful to the organism.

According to Selye, there are three stages of adaptation by an organism to stressors as part of the general adaptation syndrome. The first phase of Selye's general adaptation syndrome is the **alarm reaction**, which includes a generalized increase in sympathetic stimulation involving the adrenal medulla followed closely by increased secretion of glucocorticoids. Because of the activation of glucocorticoid secretion, the alarm reaction is not identical to the "emergency" or "fight-or-flight" response of Cannon that emphasizes only epinephrine release.

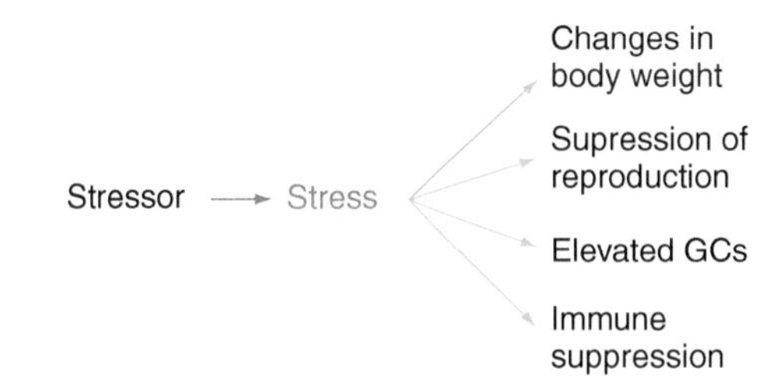

FIGURE 8-7 **Evidence of chronic stress.** See discussion in text.

The next phase of adaptation is termed the **stage of resistance** and is characterized by prolonged increased secretion of glucocorticoids. During this stage, the organism adapts to the continued presence of the stressor. The resistance phase is frequently marked by enlargement of the adrenal glands, primarily due to hypertrophy of the zona fasciculata and, to a lesser extent, the zona reticularis in response to elevated ACTH. Finally, under continuation of extremely stressful conditions, the ability of the organism to function normally is impaired. Chronic stress and prolonged glucocorticoid elevation are characterized by changes in body weight, loss of reproductive activity, and immune suppression (Figure 8-7, and see discussion on allostatic load ahead).

The continuous presence of the stressor causes the organism to enter the final **stage of exhaustion** that leads to death. Although Selye seemed to focus on an exhaustion of the HPA axis as the cause of death, it is the general debilitation of the body (e.g., responses to prolonged hyperglycemia, increased neurodenegeration, immune suppression) that leads to death.

Although Selye's findings suggested that all stressors evoke an identical stress response, it is clear that all stressors do not affect the animal in precisely the same manner. In fact, there appear to be at least two generalized pathways by which information about stressors reach CRH neurons in the PVN: (1) a pathway conveying information about anticipatory or psychological stressors and (2) a pathway conveying information about reactive or physiological stressors (Figure 8-8). **Anticipatory stressors** include bereavement, fear, restraint, and unpredictability, while **physiological stressors** include such reactive stimuli as inflammation, noxious chemical exposure, hemorrhage, or low blood sugar (hypoglycemia). Information about anticipatory stressors reaches the PVN from forebrain and limbic brain structures such as the amygdala and bed nucleus of the stria terminalis. Information regarding reactive stressors is conveyed primarily by noradrenergic/glucagon-like peptide 1 (GLP-1) producing neurons in the nucleus of the solitary tract in the brainstem (Figure 8-8; also see Figure 4-2).

The relative roles of the adrenal cortex and the adrenal medulla are not fixed but can vary. For example, humans

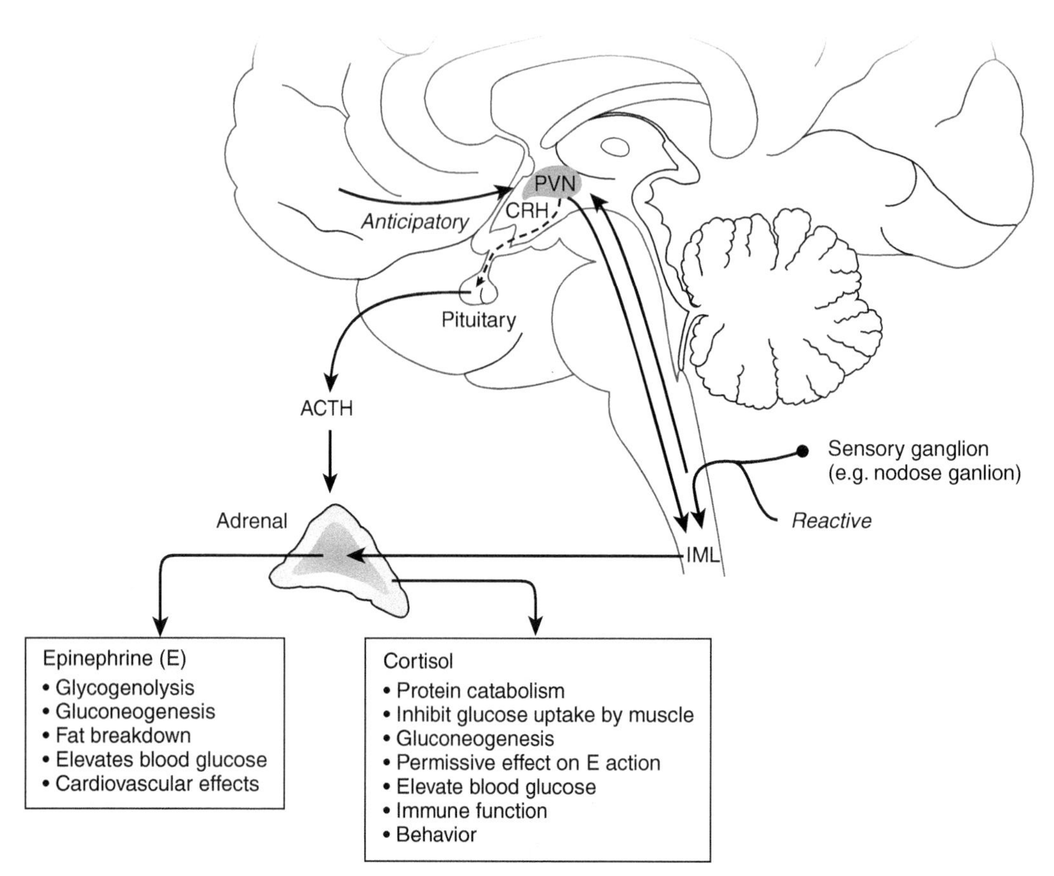

FIGURE 8-8 **Pathways in the response to anticipatory and reactive physiological stressors.** Signals regarding anticipatory stressors arise in forebrain and limbic brain structures that project to the PVN. Reactive stressors are perceived subconsciously via sensory nerves entering the brainstem or bloodborne factors such as cytokines that can enter the brainstem via circumventricular organs. Noradrenergic neurons located in the nucleus of the solitary tract (see Figure 4-2) convey information about reactive stressors from the brainstem to the PVN. Preganglionic sympathetic neurons in the intermediolateral (IML) cell column receive signals about both kinds of stressors from both forebrain and brainstem sites and activate epinephrine release from the adrenal medulla.

were examined performing the same task under self-paced conditions or externally paced conditions that challenged their ability to keep up. Plasma epinephrine levels were similar in both groups, suggesting an equal intensity of effort, but cortisol levels were much higher in the externally challenged subjects. Social position or status, especially in species exhibiting a clear dominance hierarchy, can act as a potent stressor. For example, following establishment of a new laboratory colony, mice becoming subordinate were found to have higher levels of plasma corticosterone, whereas mice emerging as dominants had higher levels of catecholamines as evidenced by increased tyrosine hydroxylase activity (see Chapter 3).

Stress in experimental animals is important when attempting to interpret data on corticosteroid levels or nutrient levels (amino acids, glucose, etc.). Laboratory conditions alone can influence the adrenal axis. Stress also can influence the levels of other hormones. For example, the order in which groups of rats were removed from a common holding facility and killed at a remote site influenced the mean levels of PRL that were measured for each group. Because it is not possible to undertake experimental work on animals free from stress, detailed knowledge of how animals were maintained and all procedures involved with an experiment is essential when interpreting results.

While it is clear that *chronic* stress and allostatic load (see ahead) can have deleterious effects on health, activation of the HPA axis in the short term during stress is adaptive, as the metabolic, behavioral, and cardiovascular changes that result will help an organism cope with the stressor. Just consider for a moment that the ability of the HPA axis to respond to stressors is evolutionarily ancient, dating back more than 400 million years (Chapter 9). The fact that the HPA axis and sympathetic nervous system response to stressors has remained relatively unchanged for such a long period of time is a testament to the highly adaptive value of the stress response.

5. Permissive Actions of Glucocorticoids

It has been suggested that, in addition to their roles during stress and participation in immune suppression, glucocorticoids may act as "permissive agents" as described earlier for thyroid hormones (see Chapter 6). Through changes in membrane permeabilities to important metabolites and by stimulating the synthesis of new enzymes or receptors, glucocorticoids may provide the appropriate cellular environment in which other hormones operate.

6. Pharmacological Actions of Glucocorticoids

The glucocorticoids possibly are better known for their pharmacological actions and therapeutic effects than for their biological actions. The tremendous potential for glucocorticoids in the treatment of the rare hypoadrenocorticism described by Thomas Addison in 1855 (Addison's disease) was not recognized for many years (see Chapter 4, Box 4C). However, the discovery of the **anti-inflammatory effects** of glucocorticoids and their use for treatment of rheumatoid arthritis spurred a tremendous explosion in therapeutic applications of glucocorticoids. The debilitating symptoms of rheumatoid arthritis are the consequence of inflammation associated with an autoimmune response in which the patient produces antibody against his/her own connective tissue. Glucocorticoid therapy alleviates painful inflammation occurring as a result of the immune reaction but does nothing to correct the causative factors.

A critical mechanism for glucocorticoid inhibition of inflammation is the direct interaction of occupied GRs with the **nuclear factor-kappaB (NF-κB)** and **activator protein-1 (AP-1)**, both transcription factors that play pivotal roles in inflammation at the level of gene transcription (Figure 8-9). Glucocorticoids can inhibit most

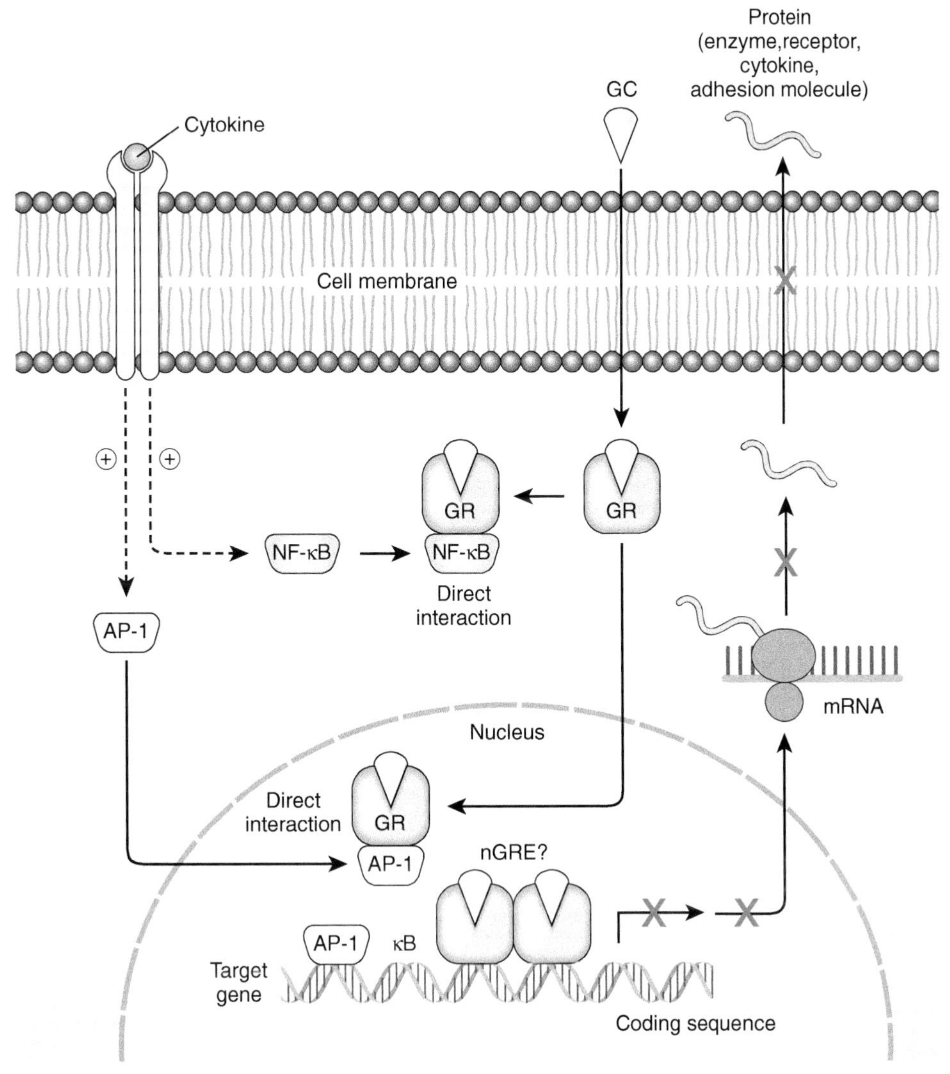

FIGURE 8-9 Glucocorticoid receptors (GR) bound to glucocorticoids may directly interact with AP-1 and NF-κB to cause mutual repression inside inflammatory cells. See text for explanation of abbreviations. *(Adapted with permission from Barnes, P.J.,* Clinical Science (London), ***94****, 557–572, 1998.)*

NF-κB- and AP-1-driven gene expression. Interestingly, this repression may be mutual because direct interaction of the glucocorticoid receptor with AP-1 or NF-κB may also suppress glucocorticoid signaling. This mutual repression may underlie certain glucocorticoid-insensitive inflammation disorders such as glucocorticoid-resistant asthma. Increased basal levels of cellular AP-1 in glucocorticoid-resistant forms of asthma may prevent the GR from interacting with its ligand.

Glucocorticoids also reduce inflammation by interfering with immunoglobulin E (IgE)-induced secretion of histamine from mast cells or with its actions in mediating the inflammatory response, which includes local hyperemia and resultant edema. One postulated mechanism for this interference is the glucocorticoid-induced inhibition of the **kallikreins**, enzymes that catalyze formation of **kinins** from a plasma precursor protein. Kinins induce inflammation by causing the release of histamine normally observed following the combination of antigen and antibody. Another suggestion for glucocorticoid antiinflammatory activity stems from observations of their effects on lysosomes. Glucocorticoids stabilize lysosomal membranes, thereby reducing release of hydrolytic enzymes following cell injury and hence reducing the spread of the inflammatory reaction. Inhibition by glucocorticoids of the cyclooxygenase enzyme necessary for prostaglandin synthesis reduces prostaglandin induction of inflammation. Glucocorticoids also inhibit the synthesis of leukotrienes and cytokine agents (e.g., interleukins) that mediate inflammation and cell-mediated immunity. Recent evidence also points to a role for glucocorticoids in directly modulating the activity of cellular kinases, such as **mitogen-activated protein kinases (MAPKs)** and **cyclin-dependent kinases (CDKs)**, and phosphatases, such as **dual-specificity phosphatases (DUSPs)**, enzymes that play a role in the inflammation pathway.

Massive doses of synthetic glucocorticoids are used to suppress the entire immune response so as to allow tissue and organ transplantations. Although normal therapeutic anti-inflammatory doses of glucocorticoids do not interfere with normal antigen–antibody interactions, very high doses depress new antibody synthesis. Immune suppression prevents the host from producing antibodies against the foreign transplants. However, continuous treatment of the transplant recipient is necessary for retention of the transplanted tissue and can lead to deleterious effects as well as susceptibility to diseases (see above).

7. Glucocorticoid Feedback on the HPA Axis

Glucocorticoid inhibition of the HPA axis is critical for limiting activation of the HPA axis and excessive exposure of tissues to the potent actions of glucocorticoids. Glucocorticoid feedback mechanisms are complex and involve both fast and slow feedback pathways that can be divided into three categories: (1) fast non-genomic feedback; (2) genomic or non-genomic feedback on RNA stability in neurons directly innervating CRH neurons; and (3) genomic feedback on forebrain structures not directly innervating CRH-producing cells. Fast feedback occurs locally within the vicinity of CRF neurons in the PVN and is mediated by non-genomic mechanisms initiated by glucocorticoids binding to a membrane GR, the structure of which is presently unknown (Figure 8-10). Glucocorticoid binding to the membrane receptor initiates the synthesis of **endocannabinoids** from lipid precursors in CRH neurons. The endocannabinoids are secreted from CRH neurons and then act by travelling backwards (retrograde) to inhibit excitatory glutamate-producing neurons that are responsible for activating CRH neurons (Figure 8-10).

In the second type of feedback, glucocorticoids directly inhibit other excitatory neuronal pathways that make direct synapses with CRH neurons in the PVN; for example, brainstem pathways arising in the nucleus of the solitary tract innervate CRH neurons in the PVN and are critical for stimulating CRH secretion during stress. Glucocorticoids inhibit the expression of a prohormone encoding GLP-1 in brainstem pathways and thereby short circuit the route by which sensory stimuli reach CRH neurons. Interestingly, this effect of glucocorticoids is not mediated by changes in gene transcription as described in Chapter 3 but through destabilization of RNA pools destined for translation. Finally, in the third type of feedback, glucocorticoids act through GR-mediated genomic pathways in the prefrontal cortex, hippocampus, and bed nucleus of the stria terminalis to engage inhibitory GABAergic pathways that innervate the PVN and inhibit CRH neurons (Figure 8-10).

IV. ALLOSTASIS AND ALLOSTATIC LOAD

How do scientists or clinicians know if an individual or a population of individuals are stressed? Moreover, can we determine how the accumulation of stress over a lifetime affects health? While it is convenient to think of stress as any perturbation that disrupts homeostasis, how do we account for the subtle, day-to-day adjustments our bodies make to changes in the environment? These "minor" adjustments to the curve balls that life throws our way may seem trivial on their own, but over the course of a lifetime may accumulate to affect our health. In the late 1980s, Sterling and Eyer proposed the concept of **allostasis** to encompass the subtle changes an organism makes to adjust in the short term to predictable and unpredictable changes in the environment. The HPA axis and sympathetic nervous system are both important mechanisms for "maintaining stability through

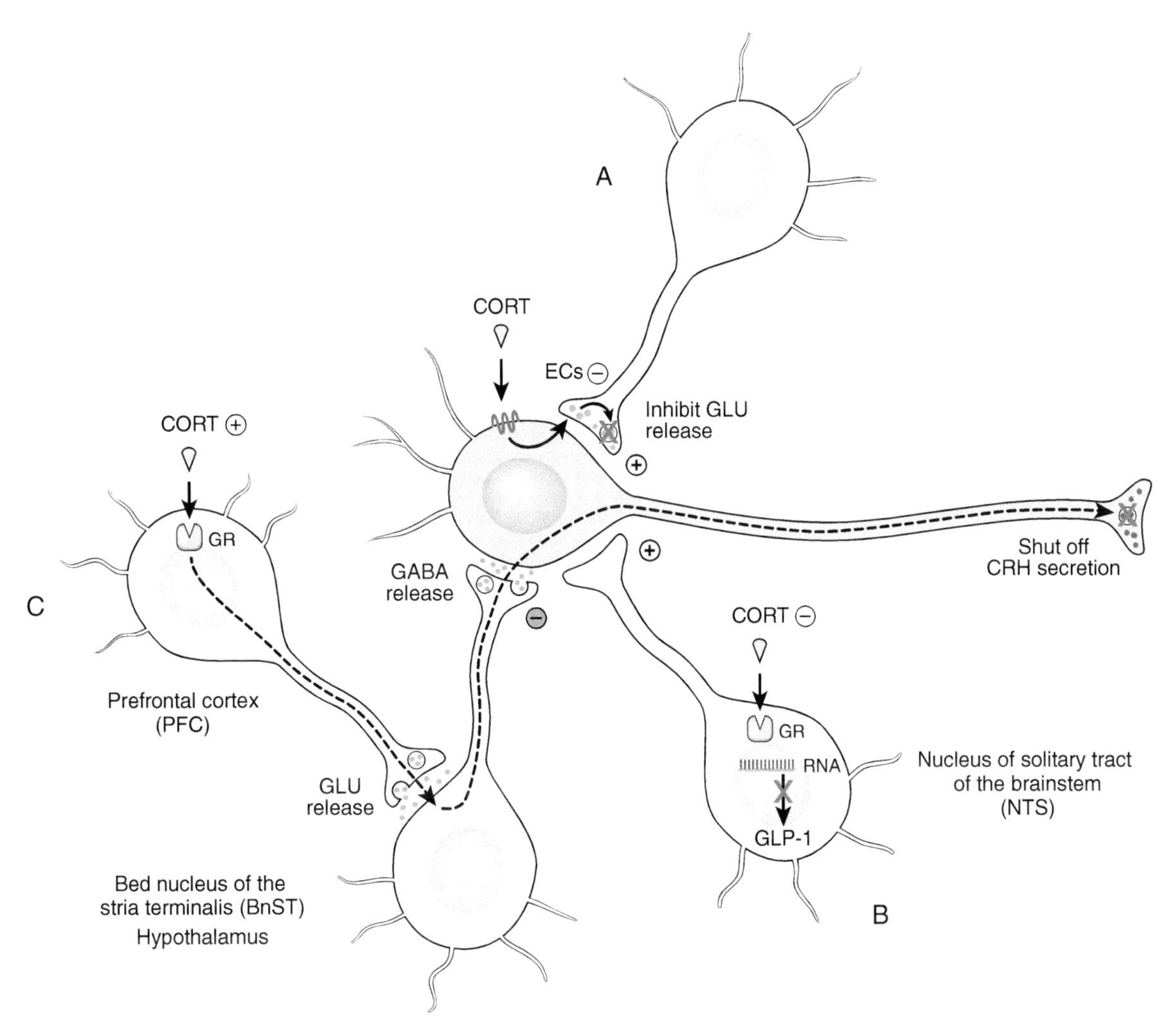

FIGURE 8-10 **Pathways for glucocorticoid feedback on CRH neurons.** (A) Glucocorticoids (CORT) act on membrane receptors to initiate endocannabinoid (EC) synthesis. ECs travel backwards (retrograde) through the synapse to inhibit glutamate (GLU) release. (B) CORT reduces the stability of RNA encoding the prohormone for glucagon-like peptide 1 neurons in the nucleus of the solitary tract (NTS) in the brainstem, thereby short-circuiting ascending signals from sensory areas of the brainstem. (C) CORT acts through the glucocorticoid receptor to activate glutaminergic neurons in the prefrontal cortex (PFC) that travel to the bed nucleus of the stria terminalis (BnST) and hypothalamus to excite inhibitory GABAergic neurons that shut off CRH secretion. GLP-1, glucagon-like peptide 1. *(Adapted with permission from Myers, B. et al.,* Cellular and Molecular Neurobiology, ***32****, 683–694, 2012.)*

change" as the name allostasis implies (*allo*, from the Greek word for change or divergence; *stasis* from the Greek word for stability). **Allostatic load** is defined as the cumulative *cost* to an organism of the changes required to maintain allostasis. To understand this cost, remember that glucocorticoids and epinephrine have important roles in elevating blood sugar and mobilizing lipids. These changes, while adaptive in the short term, can have deleterious effects on health over the long-term. Thus, when allostatic load reaches a point where it can lead to chronic disease, the condition of **allostatic overload** is reached. From a practical point of view allostatic load is useful for determining the effects of multiple stressor effects on human populations. Greater allostatic load has been linked to lower socioeconomic status and to greater risk of cardiovascular and metabolic disease and death. In humans, allostatic load is estimated by examining a number of metabolic (e.g., blood glucose, hip-to-waist ratio), cardiovascular (e.g., blood pressure), and endocrine (e.g., salivary cortisol, salivary α-amylase as a marker of sympathetic nervous system activity) end points (Table 8-3).

V. ALDOSTERONE: THE PRINCIPAL MAMMALIAN "MINERALOCORTICOID"

The zona glomerulosa secretes aldosterone independently of direct pituitary control, although, as mentioned earlier, ACTH appears to play a permissive role in maintaining the responsiveness of these cells to other controlling factors. The major action of aldosterone is maintenance of the normal sodium and potassium balance in body fluids, and

TABLE 8-3 Parameters Commonly Used to Measure Allostatic Load in Human Studies

Parameter	Description	Use in Assessing Allostatic Load	Cutoff[a]
Waist to hip ratio	Ratio of the circumference of the waist to that of the hips	Marker of long-term adipose deposition thougth to be produced by elevated glucocorticoid exposure	0.95
Body mass index	Ration of body mass to height	Marker of stress-induced weight gain	$\geq$ 30 kg/m^2
Glycosylated hemoglobin (Hemoglobin A1C)	Form of hemoglobin with carbohydrate attached	Marker of prolonged elevations in blood glucose	$\geq$ 0.06%
Blood pressure	Arterial blood pressure measured using a sphygmomanometer	Marker of cardiovascular health; higher levels indicate greater risk	Systolic $\geq$ 140 mmHg; Diastolic $\geq$ 90 mmHg
Blood total cholesterol	LDL cholesterol, HDL cholesterol, and VLDL cholesterol	Marker of cardiovascular health; higher levels indicate greater risk	$\geq$ 4.6 mmol/L
Blood HDL cholesterol	Cholesterol bound to HDLs	Marker of cardiovascular health; higher levels indicate less risk	$\leq$ 1.73 mmol/L
Blood triglyceride	Most common circulating lipid	Marker of cardiovascular health; higher levels indicate greater risk	$\geq$ 1.45 mmol/L
Blood cortisol	Primary glucocorticoid in humans	Marker of HPA axis activity; higher levels indicate higher allostatic load	$\geq$ 662.5 nmol/L
C-reactive protein	Secreted from liver into blood to enhance phagocytosis	Marker of inflammation	$\geq$ 6 mg/L
Salivary cortisol	Primary glucocorticoid in humans	Non-invasive marker of HPA axis activity; higher levels indicate higher allostatic load	$\geq$ 20 nmol/L[b]
Salivary alpha-amylase	Salivary enzyme	Non-invasive marker of sympathetic nervous system activity; higher levels indicate higher allostatic load	Usually compared to baseline or pretest level in a challenge test such as the Trier Social Stress Test (TSST)

dL, deciliter, 100 mL; HDL, high density lipoprotein; LDL, low density lipoprotein; VLDL, very low density lipoprotein.
[a]*Most parameters based upon 75% percentile for expected clinical range (Juster, R-P., et al., 2011. A clinical allostatic load index is associated with burnout symptoms and hypocortisolemic profiles in healthy workers. Psychoneuroendocrinology 36, 797–805). Measurements outside normal values would be considered to contribute to allostatic load. For example, outside normal values would be given a score of '1' otherwise a '0' is assigned. Scores for all parameters are added up to measure allostatic load. Salivary and blood cortisol will vary depending upon time of day, see text for more details.*
[b]*Adapted from Aardal, E. and Holm, A.C., European Journal of Clinical Chemistry and Clinical Biochemistry, **33**, 927–932, 1995.*

its secondary action is to regulate extracellular fluid volume. Aldosterone stimulates sodium reabsorption into the blood and potassium secretion into the urine by kidney nephrons. The mechanism controlling secretion of aldosterone involves a most complex and seemingly circuitous series of events involving the lungs, liver, and kidneys: the **renin–angiotensin system**.

A. The Renin–Angiotensin System and Aldosterone Secretion

Renin is an enzyme produced in the kidney by the **juxtaglomerular body**, a modified group of smooth muscle cells located in the afferent arteriole carrying blood to the glomerulus (Figure 8-11). This enzyme is a glycoprotein (MW $\approx$ 40,000 Da) possibly secreted as a larger, inactive form or **prorenin** (MW = 63,000 Da). Conversion of prorenin to renin may be accomplished by the activity of kidney kallikrein. Development of renin activity from prorenin also occurs following mild acidification of the plasma. Renin has a plasma biological half-life of about 15 minutes and is rapidly degraded. The juxtaglomerular body is intimately associated with a modified region of the distal convoluted portion of the nephron known as the **macula densa**. Together these two structures comprise the **juxtaglomerular apparatus**.

BOX 8A Whitehall Studies

The theory that chronic elevation of glucocorticoids and sympathetic nervous system activity can lead to cardiovascular and metabolic disease is supported by a number of studies in humans. The most comprehensive of these studies, collectively called the "Whitehall Studies," has tracked the health status of nearly 30,000 British civil servant workers at various ranks for nearly 50 years beginning in the late 1960s. The Whitehall Studies have reported that employees at lower occupational ranks have a greater risk of developing a number of diseases, including cardiovascular disease, obesity, and cancer, despite the fact that access to health care is equivalent within all six ranks of British civil service.

Although there may be many factors contributing to psychosocial stress in individuals with lower socioeconomic status (SES) (i.e., lower occupational grade and lower wealth), studies of Whitehall participants have shown an inverse link between SES and cortisol levels; that is, workers with a lower pay grade have increase adrenal cortisol secretion (Box Figure 8A-1). Elevated cortisol in lower SES individuals also has been linked to cardiovascular disease and obesity. Interestingly, both male and female participants in the Whitehall Studies exhibit higher salivary cortisol on working days then on weekends, presumably due to anticipation of workplace stress (Box Figure 8A-2).

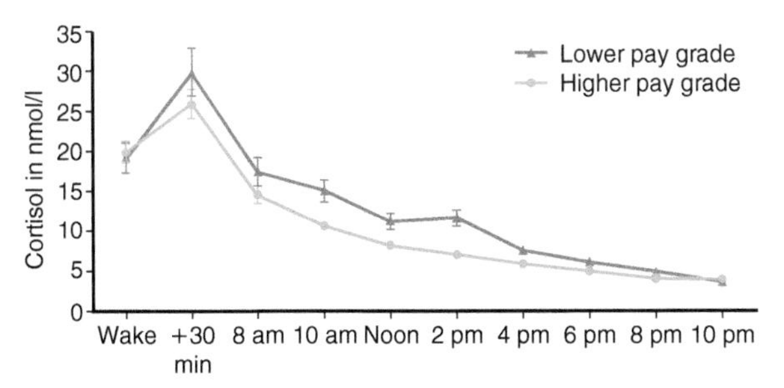

BOX FIGURE 8A-1 **Salivary cortisol during the day in workers from the Whitehall II study with lower and higher socioeconomic status.** Cortisol levels show the characteristic higher level at waking with a decline during the workday. Men with a lower pay grade showed elevated cortisol throughout the day compared to men of a higher pay grade after smoking and alcohol consumption were factored out. Salivary cortisol levels were no different between the groups after work. *(Adapted with permission from Steptoe, A. et al.,* Psychosomatic Medicine, ***65****, 461–470, 2003.)*

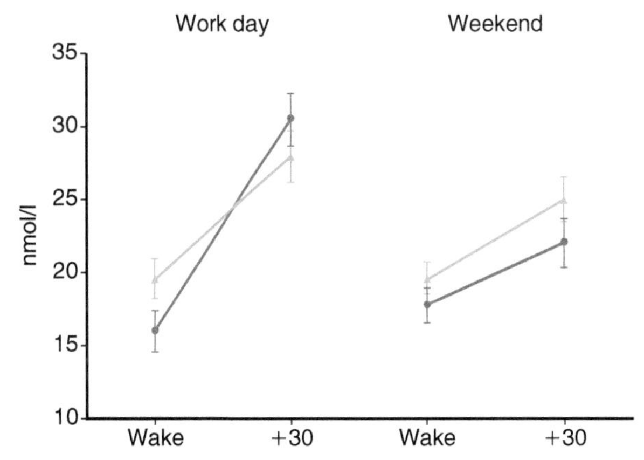

BOX FIGURE 8A-2 Salivary cortisol is lower on weekends than on work days in male (blue) and female (pink) British civil servants examined from the Whitehall II study. Participants were from higher, intermediate, and lower occupational grades. *(Adapted with permission from Kunz-Ebrecht, S.R. et al.,* Psychoneuroendocrinology, ***29****, 516–528, 2004.)*

Release of renin is controlled by blood volume as reflected in blood pressure within the renal arterioles and/or by sodium concentration in the glomerular filtrate as it enters the proximal (convoluted) tubule of the nephron. Intrarenal blood pressure is monitored by stretch receptors in the juxtaglomerular body. Renin is released in response to a decrease in this pressure. Sodium concentration in the tubular lumen is monitored by cells of the macula densa, and low sodium levels somehow trigger communication between macula densa and juxtaglomerular cells, resulting in renin release. Changes in either or both parameters influence renin secretion.

Once renin enters the blood, it comes in contact with a plasma protein termed **renin substrate (= angiotensinogen)** that was synthesized by the liver. Mammalian renin substrates are large glycoproteins (MW = 58,000 to 110,000) that behave like α_1- globulins, α_3-globulins, or albumin in humans, herbivores, and rodents, respectively. In venous blood returning to the heart from the kidneys, renin causes the enzymatic release of a decapeptide known as **angiotensin-I (Ang-I)** from renin substrate. Although Ang-I can facilitate release of norepinephrine from the adrenal medulla and can produce direct and indirect pressor effects on the cardiovascular system, it is generally believed to be of little physiological importance due to the pattern of blood flow and the rapid enzymatic degradation of Ang-I. Venous blood containing Ang-I travels first to the the heart and then directly to the lungs, where **angiotensin converting enzyme (ACE)**, located in the endothelial cells of the lung capillaries, hydrolyzes most of the Ang-I to an octapeptide, **angiotensin-II (Ang-II)**. **Captopril** and **lisinopril** are two of a number of potent synthetic inhibitors of ACE that block formation of Ang-II, and these drugs are used clinically to treat hypertension.

The actions of the Ang-II are summarized in Figure 8-12. Blood containing mostly Ang-II returns from the lungs to the heart and then is pumped into the general circulation, where it produces multiple short-term effects. Ang-II only has a biological half-life of about 2 minutes. When Ang-II reaches the adrenal cortex, it stimulates synthesis and release of

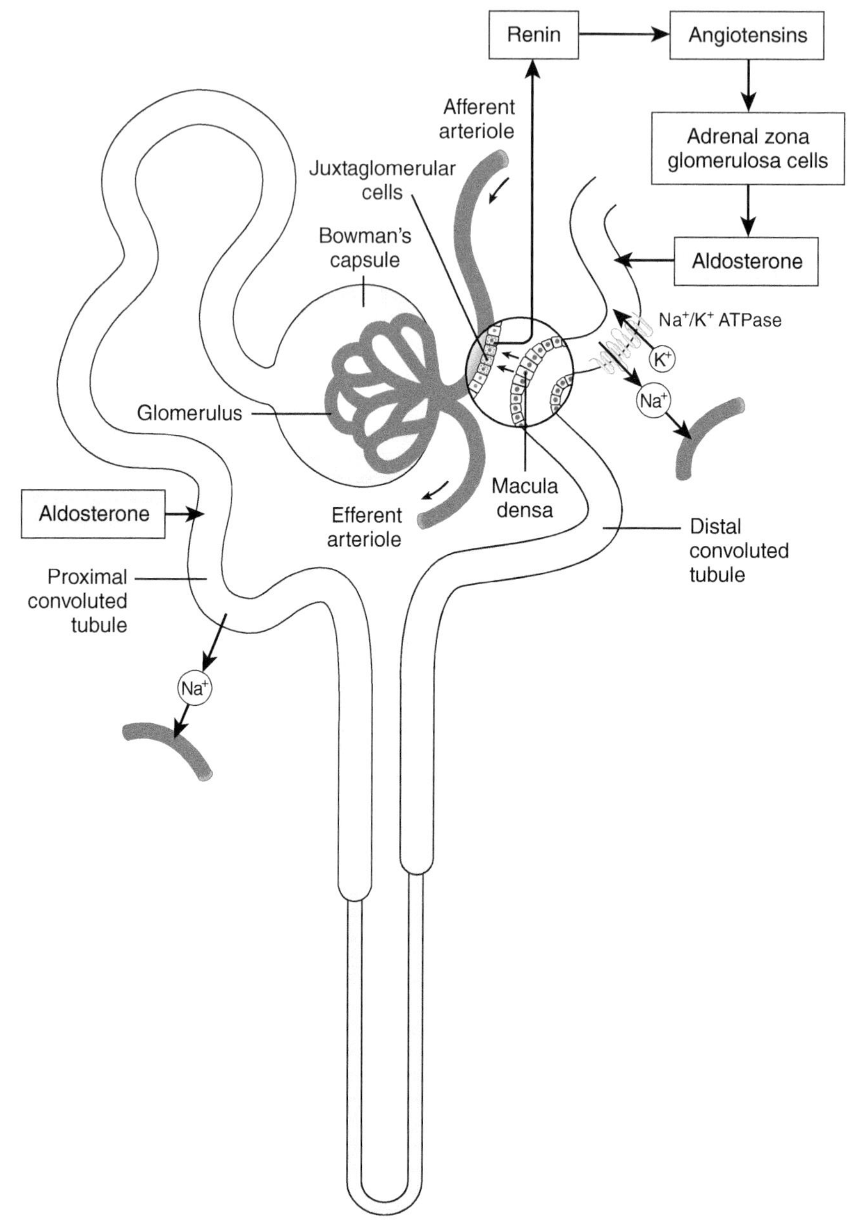

FIGURE 8-11 **Schematic representation of the juxtaglomerular apparatus.** The red structures represent a portion of the vasculature of the nephron including the afferent arteriole bringing blood to the glomerulus (a capillary bed) housed in Bowman's capsule of the nephron and exiting via the efferent arteriole prior to entering the peritubular capillaries (not shown). The juxtaglomerular apparatus consists of the renin-secreting juxtaglomerular cells embedded in the walls of the afferent arteriole and the macula densa, which consists of modified cells in the distal convoluted portion of the nephron.

aldosterone from cells of the zona glomerulosa. Aldosterone release appears to be a result of activation of **inositol trisphosphate (IP_3)** and **diacylglycerol (DAG)** as second messengers and their effects on calcium channels in zona glomerulosa cells (see Chapter 3). Aldosterone travels through the blood to the kidney and stimulates increased reabsorption of sodium and some increased excretion of potassium. This increased sodium reabsorption aids in water retention, reduces urine volume, and helps to restore normal fluid volume.

In addition to stimulating aldosterone release, Ang-II is a potent vasoconstricting agent (about 40 times more potent than norepinephrine) and quickly helps restore blood pressure to normal by causing contraction of vascular smooth muscle, resulting in a decrease in arteriole diameter. Ang-II also induces hypertrophy of cardiac myocytes as well as mitosis in mesangial cells. This hypertrophic response in cardiac cells is mediated by transcription and synthesis of **transforming growth factor β (TGFβ)**. Prolonged elevation of Ang-II can induce significant cardiac hypertrophy

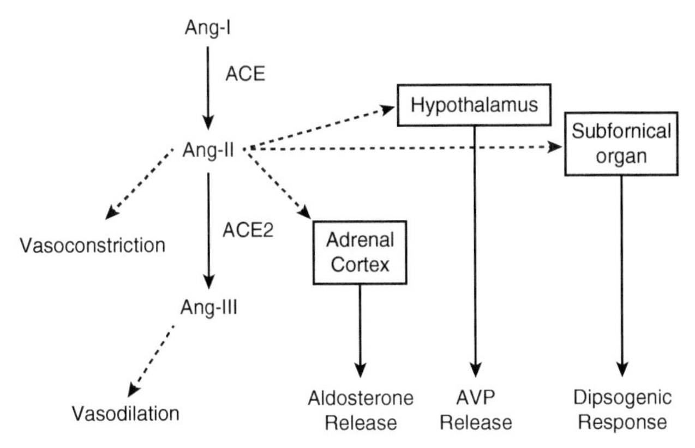

FIGURE 8-12 **Generalized actions of angiotensins.** The major endocrine actions of angiotensins are included but not its vasopressor action. In humans, the major angiotensin is angiotensin-II (Ang-II), and angiotensin-III (Ang-III) has a local depressor action (see text).

and contribute to chronic hypertension. Furthermore, Ang-II causes release of norepinephrine from atrial sympathetic nerve endings, causing increased heart rate and/or contraction strength, contributing further to increased blood pressure.

Ang-II has two important central nervous system effects that also contribute to an elevation in blood pressure. Through actions on the subfornical organ in the brain, Ang-II stimulates drinking (dipsogenic response), and absorption of ingested water from the gut contributes to increased blood volume and pressure. Furthermore, Ang-II causes release of AVP through actions on the hypothalamus. These two effects contribute to the resumption of normal fluid balance.

The levels of an enzyme, **aminopeptidase a**, are increased by actions of Ang-II. This enzyme degrades Ang-II. In addition, a second form of ACE called **ACE2** is present in numerous tissues (especially kidney, heart, and arteriole smooth muscle) and converts Ang-II into a hetapeptide called **angiotensin-III (Ang-III)** or angiotensin 1–7 (Figure 8-14). Ang-III locally counteracts the effects of Ang-II by causing vasodilation. In the heart, Ang-III is also antihypertrophic. Although Ang-III can bind to the Ang-II receptor, it is much less potent than Ang-II on aldosterone release and on vasoconstriction. However, the vasodilation and hypertrophy effects of Ang-III are mediated via a unique Ang-III receptor. Some conditions of hypertension are related to the absence of sufficient ACE2 expression. The regulation of blood pressure by the renin–angiotensin system is summarized in Figures 8-13 and 8-14).

B. Intracellular Renin–Angiotensin Systems

Since the discovery of the renin–angiotensin system involving the cooperative efforts of several organs to regulate blood pressure and sodium/potassium balance, complete renin–angiotensin mechanisms have been demonstrated within the brain, pituitary, gonads, and the adrenal cortex itself. The presence of a complete renin–angiotensin system within the adrenal cortex suggests the possibility of an adrenal paracrine regulatory system for controlling basal secretion of aldosterone. Evidence suggests a more active brain renin–angiotensin system may be present in hypertensive rats. In the pituitary, Ang-II may stimulate proliferation of tropic hormone-secreting cells. There is evidence to support a role for Ang-II in ovulation. In the cow ovary, the number of Ang-II receptors increases with size of the follicle. The amount of prorenin in seminal fluid is proportional to the sperm count, but neither the origin nor the significance of seminal prorenin is known.

C. Additional Factors Controlling Aldosterone Secretion

Aldosterone secretion is influenced by several factors in addition to the renin–angiotensin system. For example, elevated blood potassium levels can stimulate aldosterone release directly. Release of aldosterone can be inhibited by **natriuretic peptides** secreted by the heart during chronic hypertension.

1. Potassium

High levels of potassium in extracellular fluids directly stimulate aldosterone secretion from cells of the zona glomerulosa which, in turn, promotes renal potassium loss. Potassium may increase the sensitivity of cells in the zona glomerulosa to Ang-II or may directly cause aldosterone release. Either K^+ or Ang-II can induce expression of StAR protein and further enhance aldosterone synthesis and secretion. In contrast, extracellular sodium variations do not directly influence aldosterone secretion unless unusually large variations are produced.

2. Natriuretic Peptides

Two similar peptides are known to inhibit release of aldosterone through direct actions on cells of the zona glomerulosa. The first natriuretic peptide discovered was believed to be secreted by the atria of the heart and was named **atrial natriuretic peptide (ANP)** for its ability to increase sodium excretion via the urine and its presumed origin. A second peptide was discovered in the brain and became known as **brain natriuretic peptide (BNP)**; however, both ANP and BNP are found in the heart (including in the ventricles). A third form **(CNP)** has been found in the arcuate nucleus PVN of the hypothalamus and with gonadotropes in the pituitary.

In humans, ANP is the predominant circulating form and consists of 28 amino acids (Figure 8-15). BNP occurs

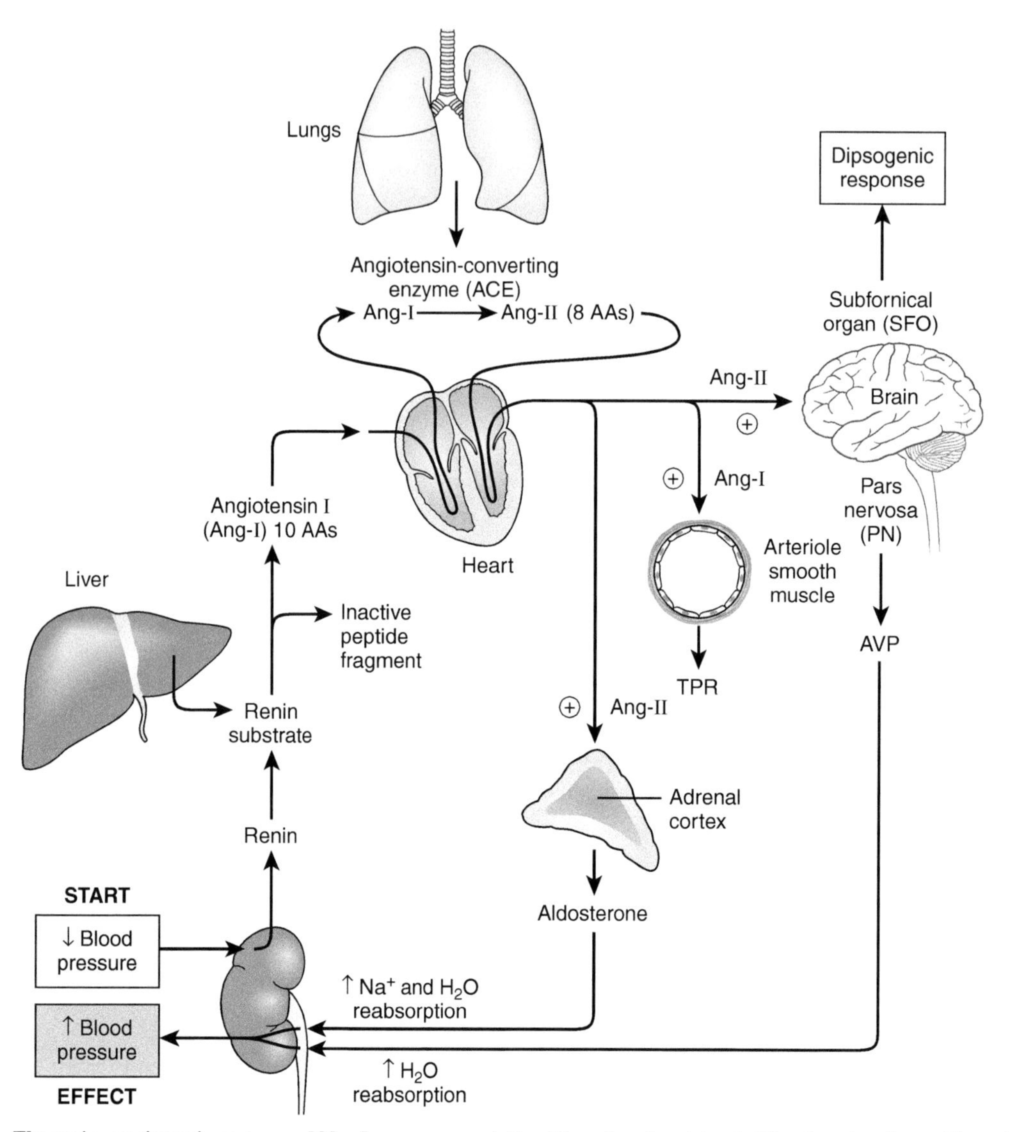

FIGURE 8-13 **The renin–angiotensin system and blood pressure regulation.** The roles of angiotensin III and natriuretic peptides on blood pressure are not included. See text for explanation. TPR, total peripheral resistance.

in two forms of 26 and 32 amino acids, both of which are structurally similar to ANP. Peripheral **guanylyl cyclase-A (GC-A)** receptors bind ANP and BNP equally well. CNP has a unique receptor, the **guanylyl cyclase-B (GC-B)** receptor, that has high specificity for CNP. All of these peptides activate guanylyl cyclase and produce cGMP as a second messenger. CNP may be the oldest member of this group of related peptides as it is phylogenetically the most conservative of the three natriuretic peptides (Figure 8-15). GC-B receptors occur on hypothalamic GnRH neurons as well as on gonadotropes, and CNP is believed to be a paracrine inhibitor of gonadotropin release. CNP inhibits ACTH release and stimulates GH secretion.

Chronically elevated blood volume or blood pressure stretches the atria and causes the release of ANP, which not only inhibits aldosterone release but also inhibits vasopressin release and directly promotes sodium excretion and water loss in the kidney. Release of ANP also is stimulated directly by elevated plasma sodium. ANP also lowers renin production in the kidney. All of these effects will reduce blood volume and blood pressure.

3. Neurotransmitters

The adrenal cortex is well supplied with neurons employing a variety of neurotransmitters. Serotonin, norepinephrine, acetylcholine, **vasoactive intestinal peptide (VIP)**, AVP, and prostaglandins are found in the adrenal cortex and can stimulate aldosterone release. SST may be produced locally and is known to inhibit Ang-II-induced release of aldosterone.

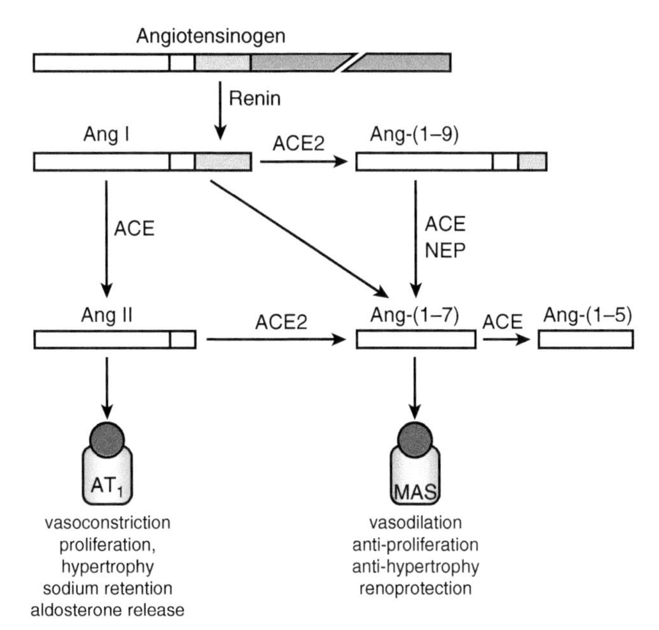

FIGURE 8-14 **Role of ACE2.** See text for details. AT_1, angiotensin type 1 receptor; MAS, MAS protooncogene receptor. *(Reprinted with permission from Simões, E. et al.,* Pediatric Nephrology, *27(10), 1835–1845, 2011.)*

D. Mechanism of Aldosterone Action

Aldosterone stimulates sodium reabsorption and potassium excretion in the distal (convoluted) tubule of the kidney and possibly enhances some sodium reabsorption in the proximal portion as well as in the intestinal mucosa, salivary glands, and sweat glands. Changes in ion transport are caused by aldosterone regulating the expression of various ion transporters in the epithelium including the Na^+/K^+-ATPase, the Na^+/H^+ exchanger, the **renal outer medullary potassium channel (ROMK)**, and the **epithelial sodium channel (ENaC)**. Aldosterone produces the typical steroidal pattern of action on target cells following binding to cytoplasmic MRs including increased synthesis of nuclear RNAs encoding transporter subunits and transcription factors that go on to influence the expression of regulatory proteins such as the kinase enzyme **SGK1 (Serine/threonine-protein kinase 1)** in the epithelial cells. Most actions of aldosterone on renal epithelium can be blocked by MR antagonists such as **spironolactone**. However. some actions of aldosterone cannot be blocked using traditional MR antagonists which has led some to suggest that there also is a membrane MR that may mediate rapid actions of aldosterone on renal epithelial cells.

VI. ENDOCRINE DISRUPTION OF ADRENAL CORTEX FUNCTION

Compared to what is known about the impact of environmental EDCs on reproduction and the thyroid axis, much less is known about EDCs impacting the adrenal gland. However, because the adrenal cortex receives a tremendous amount of blood flow, has the capacity to accumulate lipid based compounds, and produces a number of CYP family enzymes capable of activating toxins, it is considered to be highly vulnerable to disruption by EDCs. Due to the high degree of lipid transport that occurs in the adrenal cortex, this portion of the adrenal gland has been shown to accumulate lipophilic contaminants such as DDT and PCB metabolites. Many compounds have been shown to increase or decrease adrenal steroidogenesis by altering CYP family enzymes, including **mitotane** (*o,p*′-DDD, a derivative of the pesticide DDD), PCBs, and the pesticide atrazine, which increases the expression of aromatase (*CYP19*) in human adrenocorticocarcinoma H295R cells. Interestingly, the adrenal-suppressing effects of some of these compounds have been harnessed for therapeutic use in treating Cushing's disease (mitotane, etomidate).

Licorice, not usually at the top of the list when one thinks of EDCs, can lead to a false form of elevated aldosterone secretion (pseudohyperaldosteronism) when consumed in excessive amounts. Glycyrrhizinic acid, the adrenal active component of licorice, inhibits the **11β-hydroxysteroid dehydrogenase 2 isoform (HSD11B2**; see Chapter 3), thereby elevating cortisol levels. Cortisol, in excess because it is not being inactivated by HSD11B2, binds to mineralocorticoid receptors, leading to increased blood pressure, sodium retention, and potassium excretion; in other words, presenting a situation that on the surface resembles elevated levels of aldosterone.

BOX 8B Etomidate, Sedation, and Adrenal Suppression

Etomidate is a short-acting sedative that has been used since the early 1970s for anesthesia. Etomidate also is a potent inhibiter of 11β-hydroxylase, a major enzyme involved in corticosteroid biosynthesis. When given intravenously, etomidate is thus a potent suppressor of adrenal steroid synthesis, lowering cortisol levels in patients for 72 hours after one injection. In the early 1980s, etomidate began to be used as a replacement for benzodiazepine-type sedatives in intensive care units. Within a year after its more widespread use, a team of British physicians published a study demonstrating a dramatic increase in mortality associated with etomidate use. In a cohort of the most severely injured patients, every one that was administered etomidate died shortly thereafter. These results sparked a tremendous amount of controversy about the continued use of etomidate. Further work showed that the cause of the high mortality was adrenal insufficiency. The mortality associated with etomidate use in critically ill patients serves to point out the consequences of removing glucocorticoids from the blood stream at a time when their action is most critically needed.

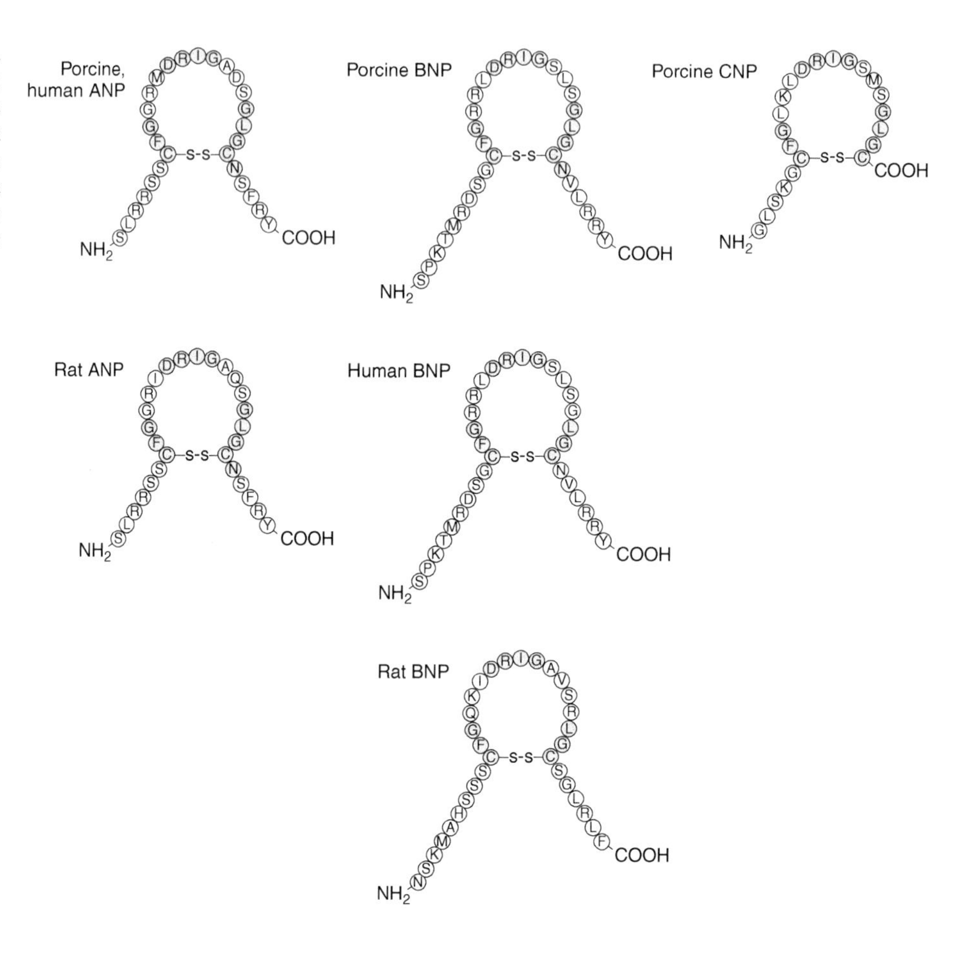

FIGURE 8-15 **Natriuretic peptides.** Three forms have been identified: ANP, BNP, and CNP. Shared amino acids are highlighted in purple. Amino acid designations can be found in Appendix C. *(Adapted with permission from Samson, W.K.,* Trends in Endocrinology and Metabolism, *3, 86–90, 1992. © Elsevier, Inc.)*

VII. CLINICAL ASPECTS OF THE ADRENAL AXIS

Glucocorticoid pathologies generally are distinct from those for aldosterone. However, high levels of glucocorticoids can bind to and activate MR receptors and mimic aldosterone effects on Na^+ and K^+ balance as well as on blood pressure. Primary and secondary disorders as well as hypersecretion and hyposecretion pathologies have been described.

A. Glucocorticoid Hypersecretion

The major disorders of glucocorticoid hypersecretion or **hyperadrenocorticism** are **Cushing's disease** and **Cushing's syndrome**. The latter term encompasses any adrenal disorder accompanied by elevated glucocorticoids, whereas the former refers to the specific condition of hyperadrenocorticism caused by hypersecretion of ACTH by the pituitary.

1. Cushing's Disease

This condition was described by Cushing in 1932 and is the most common cause of Cushing's syndrome. Hypersecretion of ACTH by a basophilic pituitary adenoma is the cause of this condition. Such adenomas are not sensitive to feedback by glucocorticoids. In this secondary adrenal disorder, the adrenal cortices are hypertrophied and plasma cortisol levels also are elevated (hypercortisolism). Hyperpigmentation (excessive darkening of the skin) may occur due to the α-MSH-like actions of chronically elevated ACTH (see Chapter 4). Excessive cortisol levels have adverse effects on the metabolism of many tissues, including the brain, muscles, skin, vascular tissue, kidney, liver, and skeleton. Hypophysectomy alleviates the symptoms (see Table 8-4) but necessitates extensive replacement therapies because all of the tropic hormones are removed by this procedure, whereas adrenalectomy requires only corticosteroid therapy. Cushing's disease is three to four times more common in women than in men.

2. Cushing's Syndrome

The symptoms of this disorder are due to production of an excess of glucocorticoids, and the cause usually is due to excessive secretion of ACTH by pituitary cells (i.e., Cushing's disease) but may be due to elevated cortisol from adrenal adenomas or adrenal carcinomas or due to ACTH

TABLE 8-4 Some Symptoms Commonly Presented with Cushing's Disorders

Symptom	Percentage of Cases
Obesity or weight gain	80
Thin skin	80
Hypertension	75
Purple skin striae	65
Hirsutism	65
Amenorrhea (women only)	60
Abnormal glucose tolerance test	55
Acne	45
Osteoporosis	40
Hyperpigmentation	20

secreted by non-pituitary tumors (e.g., bronchial carcinoid tumor, pancreatic carcinoid tumor, or medullary thyroid carcinoma). Such corticosteroid-secreting tumors suppress CRH and ACTH and bring about atrophy of normal adrenal cortical cells; consequently, hyperpigmentation only occurs in the pituitary-dependent form of hyperadrenocorticism, Cushing's disease. Cushing's syndrome may occur in a small percentage of diabetic patients as well. Excessive glucocorticoid secretion can be associated with obesity-related diabetes and escape diagnosis as an adrenal disorder.

B. Glucocorticoid Hyposecretion

The best known examples of **hypoadrenocorticism** are the primary disorders called **Addison's disease** and **congenital adrenal hyperplasia (CAH)**. Secondary hypoadrenocorticism is not so common.

1. Addison's Disease

This disease is characterized by a shortage or absence of cortisol resulting in hypersecretion of ACTH. Hyperpigmentation occurs in most cases due to elevated ACTH and its ability to bind to α-MSH receptors on melanocytes. A person with Addison's disease is usually hypoglycemic due to lack of cortisol and a reduced capacity for gluconeogenesis. Concomitant glucocorticoid therapy typically employs sufficient glucocorticoids that also stimulates MR receptors. Absence of aldosterone causes other symptoms, including muscle weakness, water loss, hypotension, and salt craving. Although Addison's disease once was commonly associated with tuberculosis infections, today the most common cause of Addison's disease is bilateral atrophy of the adrenals resulting from an autoimmune attack. It also may occur as a result of drug-induced or congenital deficiencies in steroidogenetic enzymes or as a complication in AIDS.

BOX 8C Thomas Addison and the Simultaneous Discovery of Pernicious Anemia and Adrenal Insufficiency

In 1849, Thomas Addison gave a talk to the South London Medical Society about patients of his that exhibited weakness and a general lack of energy that persisted until death. Addison astutely recognized a distinction in the general condition of the patients who died from idiopathic anemia, as it was referred to in those days. Virtually all of the patients who died of anemia exhibited altered adrenal size upon postmortem examination; some exhibited adrenal atrophy, and some exhibited adrenal hypertrophy. However, some patients dying of anemia had darkened skin while others had skin that remained pale and pallored until death. The discovery of adrenal abnormalities in patients exhibiting what we now refer to as "pernicious anemia" is considered one of the most fortuitous coincidences of the modern medical era. What Addison had accomplished was the simultaneous discovery of two unrelated disorders: pernicious anemia and adrenal insufficiency. In his 1855 monograph entitled "On the Constitutional and Local Effects of Disease of the Supra-Renal Capsules," Addison dedicated a part of the introduction to redefining idiopathic anemia and described in detail how important hyperpigmentation was to distinguishing the adrenal disorder (Box Figure 8C-1). Of course, we now know that pernicious anemia results from a lack of vitamin B12 absorption and is unrelated to adrenal insufficiency.

BOX FIGURE 8C-1 The marked skin discoloration associated with Addison's disease is evident in James Wooten, Addison's first patient. *(Reproduced with permission from Graner, J.L.,* Canadian Medical Association Journal, ***133****, 855–880, 1985.)*

2. *Congenital Adrenal Hyperplasia (CAH)*

CAH is a genetic defect in one or more genes coding for steroidogenic enzymes involved in corticosteroid synthesis. It can cause reduced glucocorticoid synthesis and hence excessive ACTH secretion due to the absence of normal feedback. Hypersecretion of ACTH, in turn, causes hypertrophy and hyperplasia of the adrenal cortex, principally the zona fasciculata and zona reticularis. Depending upon what enzymes are affected, there usually is an increase in adrenal androgen production caused by the elevated levels of ACTH.

The most common form of CAH is caused by the absence of the enzyme 21-hydroxylase ($P450_{c21}$), which blocks synthesis of cortisol, corticosterone, and aldosterone. This disorder is responsible for 90 to 95% of the cases of CAH, with the remaining cases being attributed mainly to the absence of enzymes such as 11β-hydroxylase ($P450_{c11}$) and $P450_{scc}$. Failure to convert progesterone to deoxycorticosterone in the absence of a functional C_{21}-hydroxylase causes a buildup of 17α-hydroxypregnenolone and 17α-hydroxyprogesterone, which are consequently converted into DHEA (Δ^5 pathway) and androstenedione (Δ^4 pathway). These weak androgens may be converted by peripheral tissues into testosterone, causing virilization. Newborn males with CAH may not be readily recognized but a newborn female with CAH may be diagnosed as an ambiguous sex or as a male depending on the extent of masculinization of the external genitalia. Early diagnosis is important because both sexes suffering from CAH tend to exhibit rapid somatic growth associated with early cessation of long bone growth and attainment of short stature. The penis and clitoris also show continuous growth. Aldosterone synthesis is impaired in about two-thirds of these cases due to the unavailability of suitable precursors resulting in loss of sodium and elevation of plasma potassium. Hence, these subjects also are said to suffer from salt-losing or **salt-wasting disease** because they cannot retain sodium ions.

There is a much milder form called **cryptic CAH** in which excessive androgen production does not occur until puberty. Subjects with cryptic CAH caused by C_{21}-hydroxylase deficiency can be identified by a simple test that shows elevated 17α-hydroxyprogesterone levels in the blood following a 60-minute challenge with exogenous ACTH.

3. *Secondary Hypoadrenocorticism*

Hypothalamic or pituitary lesions that block production of CRH or ACTH can produce the symptoms of hypoadrenocorticism. People with secondary hypoadrenocorticism lack the hyperpigmentation that usually accompanies Addison's disease but have the other symptoms. This condition can develop as a result of hypophysectomy, autoimmune disorders, viral illnesses, prolonged morphine administration, and other causes. The most common origin of this disorder is a consequence of prolonged therapy with cortisol or related steroids (e.g., dexamethasone) chronically employed as anti-inflammatory agents or immune suppressants that block the HPA axis through feedback effects.

C. Disorders of Aldosterone Secretion

Hyperaldosteronism is characterized by low blood potassium, high blood sodium, and muscle weakness. Elevated plasma sodium levels bring about water retention and may produce hypertension. Hyperaldosteronism may result from an adenoma or carcinoma in the zona glomerulosa that autonomously secretes excessive amounts of aldosterone. These symptoms also develop when ACTH or ACTH-like peptides are elevated chronically. Treatment with aldosterone inhibitors can reduce these symptoms until the source of the excessive aldosterone is removed. Hyperaldosteronism also may be associated with reduced insulin sensitivity in insulin target tissues caused by aldosterone-induced reductions in insulin receptor synthesis and affinity (Figure 8-16). Insulin release in response to insulin receptor downregulation causes more aldosterone secretion via direct action on the zona glomerulosa (Figure 8-16). Loss of the zona glomerulosa caused by adrenalectomy or Addison's disease, congenital (e.g., CAH) or drug-induced depression of aldosterone production, or defects in the renin–angiotensin system can induce **hypoaldosteronism**. Potassium excretion is reduced, sodium is lost in the urine, and water retention is impaired. Imbalances in sodium/potassium ratios can alter muscle and nerve function. These conditions can be alleviated by treatment with aldosterone therapy.

D. Adrenal Excesses in Androgen Production

Adrenal androgen production may be elevated in several hyperadrenocorticoid conditions such as Cushing's syndrome or with hypoadrenocorticism as in the case of CAH. Adrenal tumors may secrete excessive quantities of adrenal androgens. Symptoms of excessive production of adrenal androgens in adult women include hirsutism (excessive body hair production), acne, seborrhea (dandruff), irregular menses, reduced fertility, lowered voice pitch, atrophy of the breasts, possible thinning of hair and recession of the scalp in the temporal region, clitoral enlargement, and hypertrophy of skeletal muscles. These same symptoms may be present in males but are not as noticeable since many of the symptoms overlap with male features. Young adult women that exhibit anorexia nervosa (a disorder characterized by greatly reduced food intake and an elevated adrenal axis) may develop facial hair

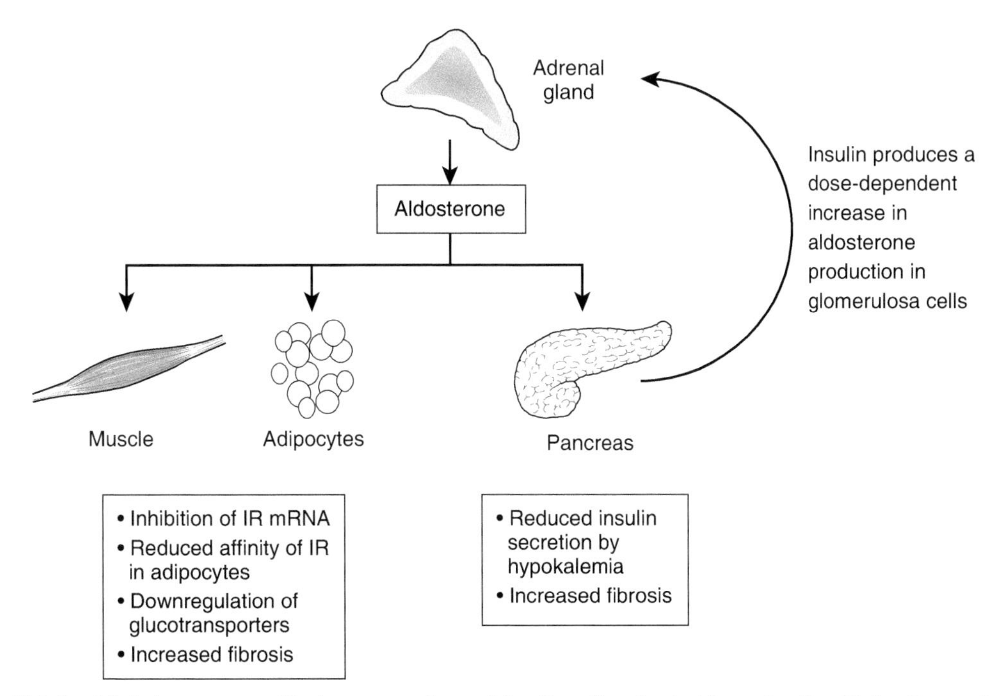

FIGURE 8-16 **Relationship between excess aldosterone secretion and insulin action.** See text for explanation. IR, insulin receptor. *(Adapted from Giacchetti, G. et al.,* Trends in Endocrinology and Metabolism, ***16****, 120–126, 2005.)*

(hirsutism) as a result of ACTH-induced secretion of adrenal androgens.

E. Side Effects of Corticosteroid Therapy

Corticosteroids typically are administered in high doses to achieve their therapeutic effects. This is especially true of glucocorticoid use for treating inflammations that vary in their intensity from skin rashes to muscle injuries or arthritis. Adverse side effects on metabolism and/or osmotic and ionic balance that occur as a consequence of prolonged glucocorticoid therapy are predictable from a simple knowledge of glucocorticoid actions.

1. Adverse Effects of Glucocorticoid Therapy

The beneficial therapeutic effects of glucocorticoids are manifest only when applied at doses two to three times physiological levels. Consequently, a number of adverse side effects occur with prolonged administration of glucocorticoids, including mild diabetes mellitus as a consequence of a prolonged hyperglycemic antagonism induced by glucocorticoids (see Chapter 12). Subjects also experience muscle weakness due to extensive protein catabolism (an effect that does not occur with physiological doses), osteoporosis due to destruction of bone substance, reduced activity of the immunological response system, and mental depression. Glucocorticoids also produce mineralocorticoid-like effects on Na^+ and K^+ balance due to high levels that operate through saturation of the MR and/or of the mechanisms for intracellular inactivation of cortisol and corticosterone (see Chapter 3).

2. Adverse Effects of Aldosterone Therapy

Excessive doses of mineralocorticoids cause sodium retention and consequent accumulation of fluids (edema), as evidenced by rapid weight gain following their administration. However, an "escape" phenomenon due to unknown factors usually occurs, and the condition is alleviated before serious complications arise and irreparable damage has occurred. In cases of cardiac failure, the escape mechanism also fails. Aldosterone therapy results in decreased blood potassium and increased urine potassium. Excessive aldosterone can cause severe potassium losses that can induce muscle cramps and muscle weakness. These events occur primarily because of adverse effects on cell membrane characteristics and resultant alterations of normal muscle cell physiology. Because pharmacological doses of glucocorticoids result in increased binding to the MR, excess glucocorticoids can mimic the effects of excess aldosterone.

VIII. THE MAMMALIAN ADRENAL MEDULLA

The medullary portion of the mammalian adrenal consists of sympathetic preganglionic neuronal endings (cholinergic) and modified cells derived from neural crest and homologous to postganglionic sympathetic neurons (adrenergic). In other words, the adrenal medulla is a modified sympathetic ganglion that secretes either norepinephrine or epinephrine directly into the blood. Both epinephrine and norepinephrine (as well as small quantities of dopamine) can be extracted from the adrenal medulla, but the ratio in most adult mammals strongly favors epinephrine (Table 8-5). The proportion of norepinephrine to epinephrine may vary throughout life, however. Fetal and neonatal adrenals secrete predominantly norepinephrine followed by a gradual increase for most species in the proportion of epinephrine so that eventually epinephrine dominates in adults. Whales are an apparent exception in that the adult whale adrenal consists of about 83% norepinephrine.

Treatment of adrenal medullary cells with potassium dichromate or chromic acid results in formation of a yellowish or brown oxidation product, the **chromaffin reaction**. Cells that exhibit a positive chromaffin reaction are termed *chromaffin cells*. The catecholamine-secreting cells of the adrenal medulla show a positive chromaffin reaction, but so do other catecholamine-secreting cells in the body (for example, in the brain, intestinal epithelium, and skin). Cells containing the tryptophan derivative serotonin also exhibit a positive chromaffin reaction; however, the term "chromaffin cell" is usually applied only to catecholamine-secreting cells of the adrenal medulla.

Norepinephrine-secreting cells can be distinguished from epinephrine-secreting cells by the formaldehyde treatment devised by Hillarp and Falck. Formaldehyde combines chemically with norepinephrine storage granules, and the resulting complex will fluoresce. Today, we can readily separate epinephrine- and norepinephrine-secreting cells using immunohistochemical techniques to localize either the specific catecholamine or the enzyme responsible for synthesis of epinephrine (see below).

TABLE 8-5 Percentage of Catecholamines in Adult Adrenal Medulla Represented by Norepinephrine

Vertebrate Group/Species	Percent NE
Whale	83[a]
Ungulates	15–50[a]
Carnivores	27–60[a]
Rodents	2–50[a]
Lagomorphs	0–12[a]
Rabbit (*Oryctolagus cuniculus*)	8–13[b]
Primates	0–20[a]

[a]*Data from Gorbman, A. and Bern, H.A., "A Textbook of Comparative Endocrinology," Wiley, New York, 1962.*
[b]*Data from Coupland, R.E., Journal of Endocrinology,* **9**, *194–203, 1953.*

A. Synthesis and Metabolism of Adrenal Catecholamines

Two distinct cellular types in the mammalian adrenal medulla are related to production of norepinephrine and epinephrine, respectively. Both epinephrine and norepinephrine are synthesized from the amino acid tyrosine and employ the same biochemical pathway (see Chapter 3, Figure 3-1). However, only the epinephrine-secreting cell possesses the critical enzyme **phenylethanolamine *N*-methyltransferase (PNMT)** necessary for converting norepinephrine to epinephrine through addition of a methyl group donated by *S*-adenosylmethionine. Norepinephrine and epinephrine may be found circulating free in the plasma or as conjugates with sulfate or glucuronide. Most of the circulating epinephrine is bound to plasma proteins, especially albumin. Norepinephrine binds to plasma proteins to a much lesser degree than does epinephrine. Circulating catecholamines have a short biological half-life and are rapidly excreted via the urine in either free or conjugated forms. The biological half-life for epinephrine is about 5 minutes. The most common metabolic pathway for inactivation of catecholamines involves liver **monoamine oxidase (MAO)** followed by aldehyde oxidase to produce inactive metabolites that appear in the urine. The most common urinary metabolites of chromaffin cell cartecholamines are **3-methoxy-4-hydroxymandelic acid**, **metanephrine**, and **normetanephrine**.

B. Regulation of Catecholamine Secretion

Primary control of secretion by the adrenal medulla is by the sympathetic nervous system from cells originating in a part of the thoracic spinal cord called the **intermediolateral cell column (IML)**. Environmental stimuli operate through the sympathetic system under emergency conditions or stress. In addition, some control of medullary secretion is exerted directly by ACTH and by glucocorticoids.

1. The Central Nervous Pathway

Stimulation of the cholinergic sympathetic fibers innervating the medulla causes local release of **acetylcholine (ACh)**, which in turn stimulates release of norepinephrine, epinephrine, or both from the chromaffin cells. This action of ACh involves Ca^{2+} uptake by the chromaffin cells. These cholinergic sympathetic neurons arise from the IML. The IML in turn receives innervation from a number of different brain areas, including the hypothalamus, but the most important brain area for regulating sympathetic outflow through the adrenal medulla is a population of neurons in the **rostral ventrolateral medulla (RVLM)** of the brain stem. The RVLM is a critical area for integrating adrenal medulla and sympathetic nervous system output during stress. Norepinephrine and epinephrine may be released separately under differing physiological conditions, and these two catecholamines have independent physiological roles in homeostasis.

2. Environmental Factors and Catecholamine Release

As noted earlier, Cannon first formulated the fight-or-flight hypothesis involving secretions of the adrenal medulla and activity of other portions of the sympathetic nervous system in response to a potential threat. The emergency responses include increased heart rate, vasodilation of arterioles in skeletal muscle, general venoconstriction, relaxation of bronchiolar muscles, pupillary dilatation, piloerection (elevation of hair), and mobilization of liver glycogen and free fatty acids. All of these responses contribute to increased efficiency of operation so the organism can best respond to whatever emergency has arisen. This type of response to short-term stress may be distinguished from the response to chronic stress associated with the glucocorticoids and the general adaptation syndrome of Selye described earlier. However, the emergency reaction of Cannon may be thought of as being similar to the adrenal medulla component in the alarm reaction of the Seyle hypothesis.

The major physiological actions of adrenal medullary hormones are their effects on metabolism in response to emotional stress (anxiety, apprehension), physical stress (injury, exercise), or what has been distinguished as physiological stress (temperature, pH, oxygen availability, hypotension, and hypoglycemia). The actions of adrenal catecholamines on cardiovascular events other than the acceleration of heart rate (which is due to metabolic effects of epinephrine on cardiac muscle) are probably secondary to the effects of norepinephrine released from postganglionic sympathetic fibers or, in the case of the skeletal muscle arterioles, the release of ACh from postganglionic fibers. This secondary role for adrenal catecholamines is further supported by the relatively low percentages of norepinephrine in the adrenals of most adult mammals. Sympathetic control mechanisms are not well developed in fetal and neonatal mammals, and this observation could be linked to the high proportion of norepinephrine in their adrenals.

Epinephrine stimulates hydrolysis of liver glycogen to glucose by a cAMP-dependent activation of phosphorylase-a (see Chapter 3) and production of lactate from muscle glycogen stores. Circulating norepinephrine produces a similar effect on liver glycogen, but muscle glycogen stores are not affected by norepinephrine. This would explain the ineffectiveness of exogenous norepinephrine as a cardioacceleratory drug, whereas epinephrine is very potent. The mobilization of lipids and release of free fatty acids from adipose tissue is largely under neural sympathetic control and is not influenced appreciably by adrenal catecholamines.

Emotional and severe physical stress increase circulatory levels of catecholamines via outflow from the sympathetic premotor cells in the IML (see Figure 8-8). Information regarding stressors may reach the IML directly from descending neuronal pathways in the hypothalamus and forebrain or via the RVLM. Response to emotional stress such as written or oral examinations involves an increase only in epinephrine, whereas the adrenal response to anticipation involves primarily norepinephrine. Exercise causes an increase in norepinephrine levels, presumably from both adrenal and neural sources. Epinephrine secretion is not influenced by moderate exercise, but it is markedly increased during long-distance running. Several physiological factors such as cold and heat stress, alkalosis or acidosis, and hypotension do not appear to be primary stimulators of adrenal medulla hormone secretion but rather via direct neural actions. However, responses to asphyxia or anoxia and to hypoglycemia are major factors influencing epinephrine release from the adrenal medulla in adult mammals. Asphyxia causes an increase in epinephrine release, probably through direct actions of oxygen deprivation on the nervous system. In fetal or neonatal animals, asphyxia directly evokes catecholamine release from the adrenal.

Insulin-induced hypoglycemia results in cardiac acceleration through increased epinephrine release. Hypoglycemia induces epinephrine release primarily through direct effects on glucose-sensitive centers in the hypothalamus that then communicate with brainstem (RVLM) and spinal cord areas directly controlling the adrenal medulla. Epinephrine also retards the insulin-induced decrease in blood sugar through its antagonistic actions on liver glycogen. The release of insulin can be inhibited by epinephrine.

3. ACTH and Glucocorticoids on Catecholamine Secretion

Development of a close anatomical association between adrenocortical and chromaffin tissues during vertebrate evolution has suggested a concomitant development of a functional relationship as well. Extensive studies by Wurtman and his coworkers have demonstrated that ACTH exerts a stimulatory effect on epinephrine secretion through the action of ACTH on circulating glucocorticoid levels. Hypophysectomy reduces adrenal epinephrine levels, and treatment with either ACTH or glucocorticoids restores adrenal levels of epinephrine to normal. Furthermore, glucocorticoids increase the activity of adrenal medullary PNMT, the enzyme responsible for conversion (methylation) of norepinephrine to epinephrine. Some studies indicate that ACTH may have a direct action on the medulla as well. Levels of both **tyrosine hydroxylase** (rate-limiting enzyme for catecholamine synthesis) and **dopamine-β-hydroxylase** but not PNMT are increased by ACTH treatment. These observations indicate that chronic stress may influence epinephrine secretion not only during the alarm reaction but also in the later stages of the response. It has been reported that epinephrine can cause release of ACTH through actions at either the hypothalamic or adenohypophysial level, but the physiological significance of these observations is not clear.

C. Mechanism of Action for Adrenal Catecholamines

The presence of specific receptors for epinephrine was first postulated in 1906 by Sir Henry Dale, who showed that ergot alkaloids (drugs such as ergocornine and ergocryptine) blocked some of the actions of epinephrine. Later studies suggested there are two kinds of adrenergic receptors in target cells that are capable of binding adrenal catecholamines: α- and β-receptors. These receptors also respond to a number of epinephrine-like drugs that have been termed *sympathomimetic drugs* because they mimic actions of sympathetic catecholamines (see Table 8-6; also see Table 3-2). Two common sympathomimetic drugs are **isoproterenol** and **phenylephrine**. Norepinephrine binds mainly to α-receptors, whereas epinephrine binds to both. When both α- and β-receptors are present on a target cell that binds epinephrine, the α-effect predominates unless epinephrine is administered with an α-blocking agent (for example, **phentolamine**).

Detailed studies of the mechanism of adrenal catecholamine actions on target cells have concentrated on the effects of epinephrine in cardiac muscle and liver cells. In fact, it was studies in cardiac cells that led Earl Sutherland and his coworkers to the discovery of the second-messenger role for cyclic adenosine 3′,5′-monophosphate (cAMP) and an eventual Nobel Prize in Physiology or Medicine (see Chapter 3). Epinephrine stimulates the breakdown of glycogen to glucose in both liver and muscle cells by first stimulating an increase in intracellular cAMP. The glucose released from liver glycogen tends to enter the general circulation, whereas the glucose liberated from muscle glycogen is utilized for rapid ATP synthesis and production of lactate. For a further explanation of this difference, see Chapter 12.

TABLE 8-6 Some Catecholamine Agonists and Antagonists and the Receptor Type to Which They Preferentially Bind

Adrenal Catecholamine	Receptor Type α_1	α_2	β_1	β_2
Agonists				
Clonidine		x		
Isoproterenol			x	x
Phenylephrine	x			
Ritodrine				x
Antagonists				
Butoxamine				x
Metoprolol			x	
Propranolol			x	x
Phentolamine	+	x		
Yohimbine	x	+		

Epinephrine binds to all types, but best to β-receptors. Norepinephrine binds best to α-receptors, but will bind β_1-receptors (although not as well as epinephrine).
x, binding; +, weak binding compared to other receptor types.

D. Clinical Aspects of the Adrenal Medulla

Most disorders of the adrenal medulla arise from neoplasms or "new growth" of chromaffin cells. **Pheochromocytomas** are the most common type of tumor occurring in chromaffin cells, but as a disease are very rare, with a frequency of about two in a million people. These tumors are usually found in middle-aged individuals and produce episodic releases of epinephrine and norepinephrine. While this uncontrolled release of catecholamines may lead to high blood pressure, pheochromocytomas overall are responsible for only a very small (0.1 to 0.2%) percentage of hypertension cases. Many pheochromocytomas go undiagnosed but can now be detected with genetic screening, given the familial basis for this disease. Three genes associated with risk of pheochromocytoma are *VHL* (von Hippel-Lindau disease), *NF1* (neurofibromatosis type 1), and *RET* (multiple endocrine neoplasia type 2). Patients diagnosed with pheochromocytomas or other catecholamine producing tumors of the adrenals are usually prescribed α- and β-adrenergic blockers until such time as the tumor can be surgically removed. In very rare cases individuals may be diagnosed with dopamine

BOX 8D Jokichi Takamine and the Discovery of Adrenaline

There are many small discoveries that eventually led to the purification and use of adrenaline at the beginning of the 20th century. The most important contribution was the crystallization of pure adrenalin by **Jokichi Takamine** in 1900 (Box Figure 8D-1). By 1900, Otto von Furth and John Jacob Abel had isolated pharmacologically active compounds named *suprarenin* from pig adrenal glands (chemical structure, $C_5H_9NO_2$) and epinephrine from sheep adrenal glands (chemical structure, $C_{17}H_{15}NO_4$), respectively. Both claimed that their discoveries represented the biologically active principle produced by the adrenal medulla. Amazingly, the chemical structures of both suprarenin and "epinephrine" varied significantly, and both differed considerably from the pure adrenaline crystallized by Takamine ($C_9H_{13}NO_3$). Of course, the proof is in the pudding (or in this case the vasoconstricting properties of adrenaline), and when it was demonstrated that adrenaline had a biological potency more than 2000 times greater than suprarenin or Abel's epinephrine compound, much of the debate was settled. Takamine went on to patent adrenaline and licensed the patent to Parke, Davis & Company, which marketed adrenaline for therapeutic use worldwide. In an ironic twist, the compound that Takamine isolated and named adrenaline is now internationally known as epinephrine except in the United Kingdom, where the British name for this hormone remains unchanged.

BOX FIGURE 8D-1 Jokichi Takamine was a Japanese scientist who moved to the United States and was the first to purify adrenaline in the early 1900s.

β-hydroxylase deficiency, a heritable disorder that results in virtually no catecholamines being produced in the adrenal medulla or in sympathetic nerves.

IX. SUMMARY

The mammalian adrenal gland consists of an outer region (cortex) of adrenocortical cells (steroidogenic) and an inner region (medulla) of chromaffin cells (adrenergic). The cortex consists of a zona glomerulosa that secretes primarily the mineralocorticoid aldosterone and two inner zones, zona fasciculata and zona reticularis. The zona fasciculata secretes primarily glucocorticoids (typically cortisol or corticosterone). Adrenal androgens (DHEA and DHEA-S) are produced primarily in the zona reticularis. The medulla contains two chromaffin cellular types that secrete the two catecholamine hormones norepinephrine and epinephrine, respectively.

The synthesis and release of aldosterone is controlled by the renin-angiotensin system. Renin is released from the juxtaglomerular apparatus in the kidney in response to reduction in sodium levels of extracellular fluids or reduction in blood pressure. Renin acts on a protein substrate (renin substrate or angiotensinogen) in the blood to release Ang-I. Angiotensin-converting enzyme in lung capillaries transforms Ang-I to Ang-II, which in turn stimulates aldosterone release. Ang-II may be converted to Ang-III by ACE2. Ang-III may function locally as a vasodilator. ACTH plays only a permissive role in aldosterone secretion. Aldosterone regulates sodium levels by increasing sodium reabsorption by the kidney. This response of the cells of certain regions in the nephron to aldosterone involves the synthesis of a specific protein that is responsible for sodium reabsorption. Aldosterone plays a secondary role in regulating volume of the extracellular fluids through its action on sodium reabsorption. It also regulates potassium excretion. Ang-II may stimulate hypertension, drinking, and AVP release to aid in fluid volume regulation. The renin–angiotensin system may have evolved to regulate blood pressure and secondarily acquired control of aldosterone secretion. Natriuretic peptides (ANP and BNP) oppose the actions of the renin–angiotensin system and aldosterone on blood pressure. Pharmacological levels of cortisol or corticosterone also can bind effectively to the MR and mimic aldosterone actions.

Secretion of cortisol or corticosterone is under direct pituitary control through ACTH, and these hormones produce their effects after binding to intracellular or membrane GRs. The main physiological actions of these glucocorticoids are related to their effects on transport of materials into cells and the induction of new cellular enzymes. Glucocorticoids inhibit glucose utilization by peripheral tissues, stimulate amino acid uptake and conversion to glucose and storage as glycogen, and stimulate mobilization of fat stores. They are also antiinflammatory and may produce immunosuppression. Cytokines from immune responses may also activate adrenal activity. Another important contribution may be the permissive action whereby glucocorticoids create an intracellular environment favorable to the actions of many other hormones. Glucocorticoids certainly are important in the adaptive mechanisms whereby an organism combats chronic stressors (general adaptation syndrome of Selye). Glucocorticoids also are important mediators of the physiological adjustments that lead to allostatic load. Some important pharmacological actions of glucocorticoids include anti-inflammatory properties, a diabetogenic action, and excessive catabolism of body proteins.

Release of epinephrine and norepinephrine in mammals is directed by cholinergic preganglionic sympathetic neurons in the IML cell column of the thoracic spinal cord. Glucocorticoids and ACTH may also influence the ability of the medulla to secrete catecholamines by stimulating the synthesis of key enzymes. Epinephrine is primarily a metabolic hormone; for example, it stimulates hydrolysis of glycogen in liver and skeletal muscle to provide glucose for combating hypoglycemia (liver) or for use as an immediate energy source (in skeletal muscle). Epinephrine is responsible for the fight or flight response described by Cannon and also is involved in the alarm reaction of the general adaptation syndrome of Selye. Administration of epinephrine causes increased glucose metabolism and ATP availability within cardiac cells, leading to cardiac acceleration. Epinephrine binds to either α- or β-receptors in target cell membranes. Norepinephrine binds significantly only to α-receptors, and its major physiological action is venoconstriction. Circulating norepinephrine is not a cardiac stimulator since cardiac muscle cells possess predominantly β-receptors on their exposed surfaces. In most adult mammals, circulating norepinephrine appears to be of secondary importance to the sympathetic postganglionic neurons for control of vascular tone.

STUDY QUESTIONS

1. What is the origin of the name for *chromaffin* tissue?
2. Discuss the embryonic origins of adrenocortical and adrenal medullary tissue.
3. Identify at least two ways in which adrenal medullary cells resemble neurons.
4. Discuss the structural basis for distinguishing cells within the three layers of the adrenal cortex. How do the phenotypes of cells in the three layers of the adrenal cortex differ? How does steroid biosynthesis differ in the layers of the adrenal cortex?
5. Hypophysectomy, a procedure in which the pituitary gland is surgically removed, leads to atrophy of the adrenal cortex. Which of the three layers of the adrenal cortex would be most affected by this treatment?
6. What is the major steroid produced by the fetal adrenal gland?
7. Discuss the role of corticosteroid-binding globulin in corticosteroid transport in blood.
8. Explain how glucocorticoid secretion varies during a 24-hour period.
9. Describe the differences in epinephrine and glucocorticoid action on energy metabolism.
10. Distinguish between the discoveries of Walter Cannon and Hans Selye related to coping with stressors.
11. Discuss differences between the concepts of homeostasis and allostasis.
12. Identify the criteria that are used clinically to diagnose allostatic load.
13. What is socioeconomic status and how is it related to allostatic load?
14. Identify three modes of glucocorticoid feedback on CRH neurons.
15. Explain the roles of renin, angiotensin converting enzymes, angiotensinogen, Ang-I, and Ang-II in the regulation of aldosterone secretion.
16. Identify two factors other than the renin-angiotensin system that are involved in controlling aldosterone secretion.
17. Explain how aldosterone affects salt and water balance. What type of receptor does aldosterone act upon? Is this receptor selective for mineralocorticoids and not glucocorticoids? Identify target tissues for aldosterone action.
18. What types of evidence have been used to argue that membrane receptors exist for both glucocorticoids and mineralocorticoids?
19. Explain the differences among Cushing's disease, Cushing's syndrome, and Addison's disease.
20. Outline the biosynthetic steps in the synthesis of epinephrine.
21. What types of stimuli lead to increased adrenomedullary catecholamine secretion and where do these stimuli originate?

SUGGESTED READING

Books

Brown, M.R., Koob, G.F., Rivier, C., 1990. Stress: Neurobiology and Neuroendocrinology. Marcel Dekker, New York.

Margioris, A.N., Chrousos, G.P. (Eds.), 2010. Adrenal Disorders. Humana Press, Totowa, NJ.

Mulrow, P.J., 1986. The Adrenal Gland. Elsevier, New York.

Vinson, G.P., Whitehouse, B., Hinson, J., 1993. The Adrenal Gland. Prentice Hall, Upper Saddle River, NJ.

General Articles

Barnes, P.J., 1998. Anti-inflammatory actions of glucocorticoids: molecular mechanisms. Clinical Science (London) 94, 557–572.

Beck, I.M., Vanden Berghe, W., Vermeulen, L., Yamamoto, K.R., Haegeman, G., De Bosscher, K., 2009. Crosstalk in inflammation: the interplay of glucocorticoid receptor-based mechanisms and kinases and phosphatases. Endocrine Reviews 30, 830–882.

Bhargava, A., Pearce, D., 2004. Mechanisms of mineralocorticoid action: determinants of receptor specificity and actions of regulated gene products. Trends in Endocrinology and Metabolism 15, 147–153.

Campbell, B., 2011. Adrenarche in comparative perspective. American Journal of Human Biology 23, 44–52.

Carr, J.A., Summers, C.H., 2002. Is stress more than a disease? A comparative look at the adaptiveness of stress. Integrative and Comparative Biology 42, 505–507.

Dhabhar, F.S., 2009. Enhancing versus suppressive effects of stress on immune function: implications for immunoprotection and immunopathology. Neuroimmunomodulation 16, 300–317.

Dallman, M.F., 2010. Stress-induced obesity and the emotional nervous system. Trends in Endocrinology and Metabolism 21, 159–165.

de Kloet, E.R., Karst, H., Joëls, M., 2008. Corticosteroid hormones in the central stress response: quick-and-slow. Frontiers in Neuroendocrinology 29, 268–272.

de Peretti, E., Forest, M.G., 1976. Unconjugated dehydroepiandrosterone plasma levels in normal subjects from birth to adolescence in human: the use of a sensitive radioimmunoassay. Journal of Clinical Endocrinology and Metabolism 43, 982–991.

Funder, J.W., 1993. Aldosterone action. Annual Review of Physiology 55, 115–130.

Fung, M.M., George, C.F., 1996. Naming of drugs: pass the epinephrine, please. British Medical Journal 312, 1315–1316.

Greenberg, N., Carr, J.A., Summers, C.H., 2002. Causes and consequences of stress. Integrative and Comparative Biology 42, 508–516.

Herman, J.P., Figueiredo, H., Mueller, N.K., Ulrich-Lai, Y., Ostrander, M.M., Choi, D.C., Cullinan, W.E., 2003. Central mechanisms of stress integration: hierarchical circuitry controlling hypothalamo–pituitary–adrenocortical responsiveness. Frontiers in Neuroendocrinology 24, 151–180.

Hinson, J.P., Raven, P.W., 2006. Effects of endocrine-disrupting chemicals on adrenal function. Best Practice & Research Clinical Endocrinology & Metabolism 20, 111–120.

Juster, R.P., McEwen, B.S., Lupien, S.J., 2010. Allostatic load biomarkers of chronic stress and impact on health and cognition. Neuroscience & Biobehavioral Reviews 35, 2–16.

Kunz-Ebrecht, S.R., Kirschbaum, C., Marmot, M., Steptoe, A., 2004. Differences in cortisol awakening response on work days and weekends in women and men from the Whitehall II cohort. Psychoneuroendocrinology 29, 516–528.

Kvetnansky, R., Sabban, E.L., Palkovits, M., 2009. Catecholaminergic systems in stress: structural and molecular genetic approaches. Physiological Reviews 89, 535–606.

Labrie, F., Luu-The, V., Labrie, C., Simard, J., 2001. DHEA and its transformation into androgens and estrogens in peripheral target tissues: intracrinology. Frontiers in Neuroendocrinology 22, 185–212.

Myers, B., McKlveen, J.M., Herman, J.P., 2012. Neural regulation of the stress response: the many faces of feedback. Cellular and Molecular Neurobiology 32, 683–694.

Pacák, K., Palkovits, M., 2001. Stressor specificity of central neuroendocrine responses: implications for stress-related disorders. Endocrine Reviews 22, 502–548.

Rainey, W.E., Carr, B.R., Sasano, H., Suzuki, T., Mason, J.I., 2002. Dissecting human adrenal androgen production. Trends in Endocrinology and Metabolism 13, 234–239.

Samson, W.K., 1992. Natriuretic peptides: a family of hormones. Trends in Endocrinology and Metabolism 3, 86–90.

Seeman, T.E., McEwen, B.S., Rowe, J.W., Singer, B.H., 2001. Allostatic load as a marker of cumulative biological risk: MacArthur studies of successful aging. Proceedings of the National Academy of Sciences U.S.A. 98, 4770–4775.

Steptoe, A., Kunz-Ebrecht, S., Owen, N., Feldman, P.J., Willemsen, G., Kirschbaum, C., Marmot, M., 2003. Socioeconomic status and stress-related biological responses over the working day. Psychosomatic Medicine 65, 461–470.

Ulrich-Lai, Y.M., Herman, J.P., 2009. Neural regulation of endocrine and autonomic stress responses. Nature Reviews Neuroscience 10, 397–409.

Yamashima, T., 2003. Jokichi Takamine (1854–1922), the Samurai chemist, and his work on adrenalin. Journal of Medical Biography 11, 95–102.

Renin–Angiotensin System

Burrell, L.M., Johnston, C.I., Tikellis, C., Cooper, M.E., 2004. ACE2, a new regulator of the renin-angiotensin system. Trends in Endocrinology and Metabolism 15, 166–169.

Kumar, R., Thomas, C.M., Yong, G.C., Chen, W., Baker, K.M., 2012. The intracrine renin-angiotensin system. Clinical Science (London) 123, 273–284.

Spät, A., Hunyady, L., 2004. Control of aldosterone secretion: a model for convergence in cellular signaling pathways. Physiological Reviews 84, 489–539.

Clinical Articles

Bratland, E., Husebye, E.S., 2011. Cellular immunity and immunopathology in autoimmune Addison's disease. Molecular and Cellular Endocrinology 336, 180–190.

Giacchetti, G., Sechi, L.A., Rilli, S., Carey, R.M., 2005. The renin-angiotensin-aldosterone system, glucose metabolism and diabetes. Trends in Endocrinology and Metabolism 16, 120–126.

Goymann, W., Möstl, E., Van't Hof, T., East, M.L., Hofer, H., 1999. Noninvasive fecal monitoring of glucocorticoids in spotted hyenas. *Crocuta crocuta*. General and Comparative Endocrinology 114, 340–348.

Graner, J.L., 1985. Addison, pernicious anemia and adrenal insufficiency. Canadian Medical Association Journal 133, 855–880.

Rossi, G.P., Sechi, L.A., Giacchetti, G., Ronconi, V., Strazzullo, P., Funder, J.W., 2008. Primary aldosteronism: cardiovascular, renal and metabolic implications. Trends in Endocrinology and Metabolism 19, 88–90.

Simões, E., Silva, A.C., Flynn, J.T., 2011. The renin-angiotensin-aldosterone system in 2011: role in hypertension and chronic kidney disease. Pediatric Nephrology 27 (10), 1835–1845.

Husebye, E., Løvås, K., 2009. Pathogenesis of primary adrenal insufficiency. Best Practice & Research Clinical Endocrinology & Metabolism 23, 147–157.

Lacroix, A., Baldacchino, V., Bourdeau, I., Hamet, P., Tremblay, J., 2004. Cushing's syndrome variants secondary to aberrant hormone receptors. Trends in Endocrinology and Metabolism 15, 375–382.

Magiakou, M.-A., Chrousos, G.P., 1995. Diagnosis and treatment of Cushing's disease. In: Imura, H. (Ed.), The Pituitary Gland, second ed.). Raven Press, San Diego, CA, pp. 491–508.

Nimkarn, S., New, M.I., 2008. Steroid 11beta-hydroxylase deficiency congenital adrenal hyperplasia. Trends in Endocrinology and Metabolism 19, 96–99.

Sandeep, T.C., Walker, B.R., 2001. Pathophysiology of modulation of local glucocorticoid levels by 11beta-hydroxysteroid dehydrogenases. Trends in Endocrinology and Metabolism 12, 446–453.

Stewart, P.M., Wallace, A.M., Valentino, R., Burt, D., Shackleton, C.H., Edwards, C.R., 1987. Mineralocorticoid activity of liquorice: 11-beta-hydroxysteroid dehydrogenase deficiency comes of age. Lancet 2, 821–824.

Viveros, O.H., O'Connor, D.T., 2008. Diseases of the adrenal medulla. Acta Physiologica 192, 325–335.

Chapter 9

Comparative Aspects of Vertebrate Adrenals

Physiologists naturally think of "stress" and "stressors" in the same breath with "adrenals." Such a mindset has caused us to overlook the important daily and seasonal roles that adrenal steroids perform and focus on the responses to life-threatening events. Stressors may be physical or chemical factors or simply a result of perceptions by the organism. Working independently or together, these stressors may evoke **primary changes** within the endocrine system at the level of the pituitary, **secondary changes** on physiological parameters due to elevated corticosteroids, and/or **tertiary changes** at the organismal level caused by corticosteroids; for example, behavior, growth, or immune status (see Figure 9-1).

Adrenal hormones operate at three physiological levels as outlined in Figure 9-2. Adrenocorticosteroids have basal physiological roles on ion and water balance as well as on metabolism (physiological state A). As reflected in consistent daily and seasonal increases in glucocorticoid levels, the adrenals are involved in regulation of what have been termed **predictable stressors** such as daily increases in activity and feeding or seasonally recurring events such as migration or breeding (physiological state B). Finally, **unpredictable stressors** such as predation, disease, or weather may require additional surges of glucocorticoids and catecholamines from the adrenals to aid in the survival of the animal (physiological state C). As described in Chapter 8, **allostasis** is defined as the process whereby the appropriate level of adrenal responses is determined by the combination of physiological states present at any given time. Consequently, the level of adrenal activity fluctuates

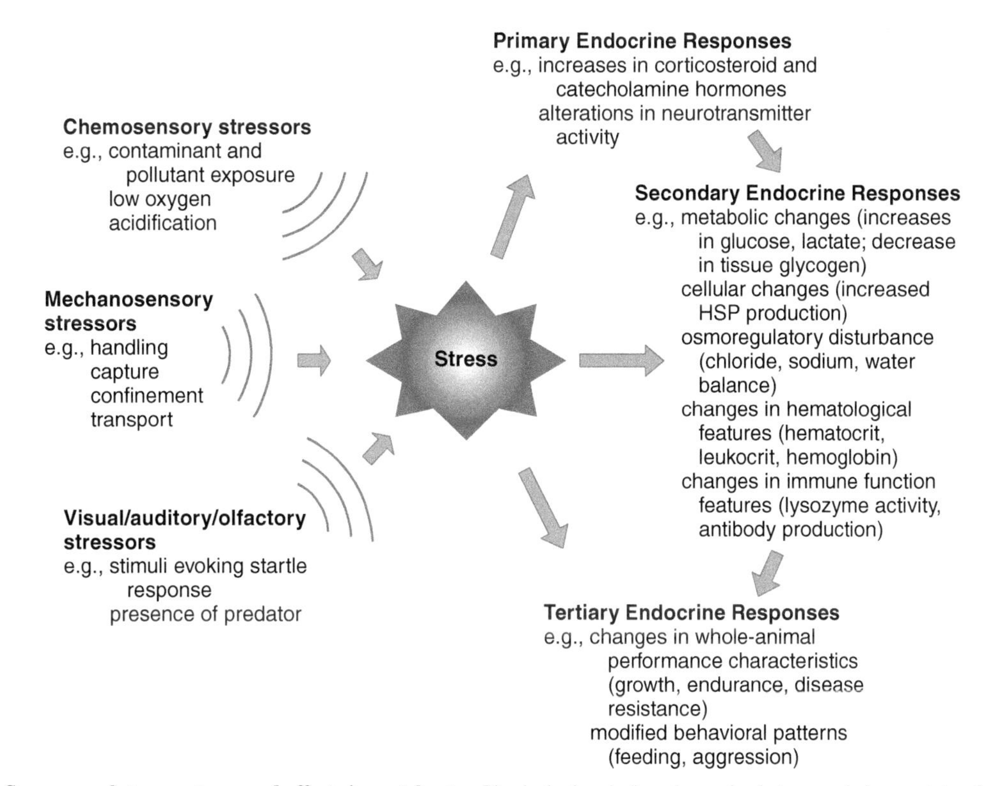

FIGURE 9-1 **Summary of stressor types and effects in vertebrates.** Physical, chemical, and perceived stressors induce a state of stress that in turn brings about changes at the brain and pituitary (primary), physiological changes (secondary), and changes at the level of the organism (tertiary). *(Adapted with permission from Barton, B.A.,* Integrative and Comparative Biology, *42, 517–525, 2002.)*

Vertebrate Endocrinology. http://dx.doi.org/10.1016/B978-0-12-394815-1.00009-4

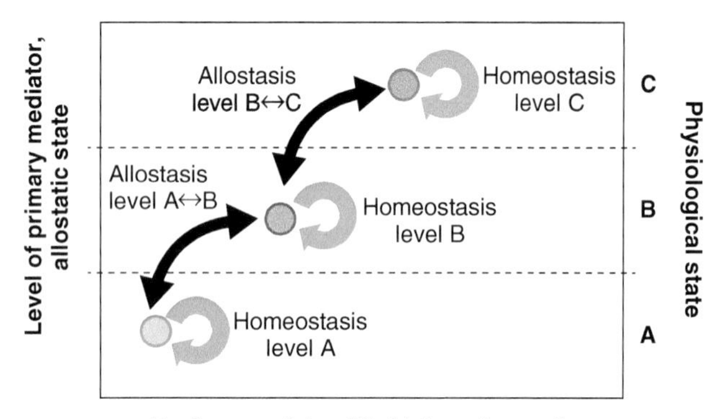

FIGURE 9-2 **Environmental or life history demands: allostasis and homeostasis.** Physiological state A represents basic physiological and behavioral processes necessary for simple existence. Homeostatic mechanisms (level A, arrow) operate around specific set points to maintain balance (blue circle). Predictable or manageable demands of the environment or life history events such as reproduction activate allostatic mechanisms that raise the physiological state to B and homeostatic level B (arrow) involving different set points and new maintenance conditions (purple circle). In the face of unpredictable and potentially life-threatening events, additional allostatic mechanisms drive the physiological state to C, and hence new homeostatic mechanisms (level C, arrow) are required to maintain survival conditions (red circle). *(Adapted with permission from Landys, M.M. et al.,* General and Comparative Endocrinology, *148, 132–149, 2006.)*

predictably in the face of predictable stressors and episodically with respect to unpredictable stressors (Figure 9-3). In this chapter, we examine the roles of glucocorticoids, mineralocorticoids, and catecholamines at the three physiological states described above.

The **hypothalamus–pituitary–adrenal (HPA)** axis is described in Chapters 4, 5, and 8 and the chromaffin cells are described in Chapter 8. Briefly, **corticotropin-releasing hormone (CRH)** produced in the hypothalamus causes secretion of **corticotropin (ACTH)** from the adenohypophysis which in turn stimulates glucocorticoid secretion (mainly **cortisol** and **corticosterone**) from the adrenocortical cells. **Arginine vasopressin (AVP)** in mammals and **arginine vasotocin (AVT)** in non-mammals may participate in the stimulation of ACTH release. Secretion of epinephrine and norepinephrine from the

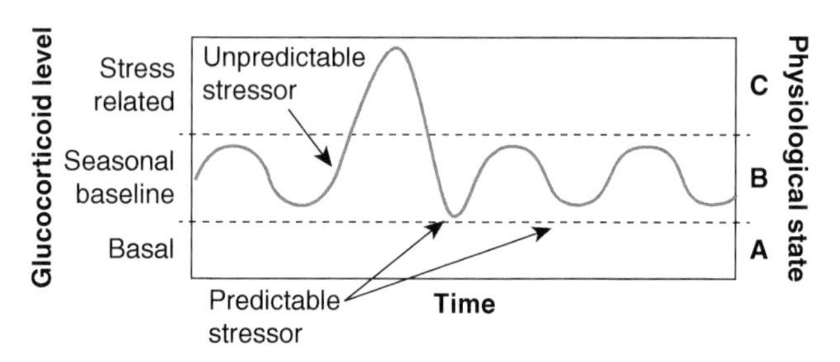

FIGURE 9-3 **Predictable and unpredictable stressors and effects on glucocorticoid levels.** *(Adapted with permission from Landys, M.M. et al.,* General and Comparative Endocrinology, *148, 132–149, 2006.)*

chromaffin cells is regulated from the hypothalamus by descending projections to preganglionic sympathetic nerves arising in the spinal cord.

I. COMPARATIVE ASPECTS OF ADRENOCORTICAL TISSUE

Cytologically, the adrenocortical cells of non-mammals resemble the steroidogenic cells of the mammalian zona fasciculata. Adrenocortical cellular types of cyclostomes, teleosts, and non-mammalian tetrapods possess well-developed smooth endoplasmic reticula, mitochondria with tubular cristae, and numerous osmiophilic (lipoidal) inclusions. Following stimulation with ACTH, pituitary extracts, or appropriate environmental stimuli, adrenocortical cells exhibit increased basophilia, increased activity of Δ^5,3β-hydroxysteroid dehydrogenase (3β-HSD), and decreased lipid content. Nuclear and cellular enlargement and hyperplasia occurs as a consequence of chronically elevated ACTH or treatment with the drug metyrapone (SU4885), which blocks the 11β-hydroxylation step catalyzed by $P450_{11\beta}$ which prevents glucocorticoid synthesis and causes elevation of plasma ACTH. As described for the cells in the zona fasciculata of mammalian adrenals, atrophy of non-mammalian adrenocortical cells follows hypophysectomy. The discovery that **melanocortin receptor accessory protein 1 (MRAP1)** is required for **melanocortin-2 receptor (MC2R**; see Chapter 4) translocation and activation of adrenal steroidogenesis in teleosts indicates that MC2R regulation of corticosteroid secretion is conserved in vertebrates.

Zonation of adrenocortical cells is suggested cytologically in some anurans, reptiles, and birds, and two separate cellular types have been claimed for the bullfrog (*Lithobates (Rana) catesbeianus*) and for birds. However, little work has been done to establish firmly the existence of more than one functional type of adrenocortical cell in fishes and most other non-mammals.

Adrenocortical cells of fishes differ most from the general mammalian pattern of corticosteroidogenesis with respect to some of the hormones produced. However, the general sequences for corticosteroidogenesis are similar in all the vertebrates with respect to precursor–product relationships, with many of the same enzymes being involved (Chapter 3). Among non-mammalian tetrapods, the nature of corticosteroid secretion is nearly identical, and the pattern of secretion is very similar to that described for cells of the mammalian zona fasciculata.

Daily rhythms and seasonal variations in corticosteroid secretory patterns occur in non-mammals as well as in mammals. Peak seasonal adrenocortical activity is roughly correlated to periods of reproductive activity, although a cause-and-effect relationship cannot be categorically applied. Some studies suggest that "stress" may be the critical factor and that stressors associated with reproduction may be only one of the components involved in stimulating adrenocortical function (albeit a major one). The responses of nonmammals to stressors such as surgery, forced exercise, and handling are very much like those described for mammals.

A. Agnathan Fishes: Cyclostomes

In lampreys, presumed adrenocortical cells have been identified as islands of cells above the pronephric funnels in the kidney as well as in the walls of the large dorsal blood vessels (postcardinal veins) in this same region. However, neither *in vitro* studies employing radioactively labeled steroidal precursors (pregnenolone, progesterone, or both) nor immunocytochemistry have demonstrated the ability of these presumed adrenocortical cells in lampreys or hagfishes to produce corticosteroids. Early studies reported that cortisol, cortisone, corticosterone, and 11-deoxycortisol were present in hagfish and lamprey plasma, but a more recent study in sea lampreys (*Petromyzon marinus*) found only 11-deoxycortisol and 11-deoxycorticosterone in the circulation and no evidence for cortisol or corticosterone. These chemicals are precursors for the synthesis of cortisol and/or corticosterone in gnathostomes. A generalized **corticosteroid receptor (CR)** has been described in the sea lamprey gill that has a strong affinity for 11-deoxycortisol.

B. Chondrichthyean Fishes

Some elasmobranchs and holocephalans have one large unpaired adrenal gland. Because this gland consists exclusively of adrenocortical cells and is located between the posterior ends of the kidneys, it is truly interrenal in position. In others, the adrenocortical tissue may occur as paired strands along the medial border of the posterior kidney. In addition, small islands or islets of adrenocortical cells may be found on the surface of the kidneys extending anteriorly. Chromaffin tissue occurs as small masses along the medial border of each kidney.

In 1934, Grollman and coworkers used extracts of the interrenal glands from three skates (genus *Raja*) to maintain adrenalectomized rats, demonstrating the presence of corticosteroid activity in these glands. Soon afterwards, elasmobranchs were shown to produce a unique corticosteroid, **1α-hydroxycorticosterone (1α-OHB)** as well as a small amount of 11-deoxycorticosterone. The enzyme necessary for synthesizing this unique steroid, **1α-hydroxylase**, is found only in the elasmobranch interrenal gland and nowhere else. Even holocephalans, such as the ratfish *Hydrolagus colliei*, lack 1α-hydroxylase and secrete primarily cortisol. However, the absence of a specific antibody for 1α-OHB and reliance on corticosterone ELISA procedures for detection of 1α-OHB have hampered research efforts. Nevertheless, numerous studies employing corticosterone antibodies show that shark (*Scyliorhinus canicula*) and ray (*Dasyatis sabina*) interrenals incubated *in vitro* appear to synthesize primarily 1α-OHB with lesser amounts of 11-deoxycorticosterone. Plasma levels of 1α-OHB have been reported for several elasmobranchs (see Table 9-1).

TABLE 9-1 Plasma Levels of 1α-Hydroxycorticosterone (1α-OHB) in Wild Caught Elasmobranchs Held in Captivity for Varying Periods of Time

Species	1αOHB (ng/ml)
Raja clavata (thornback ray)	<0.8-139
Raja radiata (thorny or starry skate)	2-43
Scyliorhinus caniculus (common dogfish)	1-7.8

Based on Pankhurst, N.W., *General and Comparative Endocrinology*, 170, 265–275, 2011.

C. Ray-Finned Osteichthyean Fishes: Actinopterygians

The anatomical arrangement of adrenocortical cells in the actinopterygian fishes differs markedly from that described for all other fish groups and could be described anatomically as "intrarenal." In the sturgeons and polypterine fishes (chondrosteans) and in the ganoid fishes (holosteans *Amia* and *Lepisosteus*), the adrenocortical cells are scattered in small clumps throughout the kidney. The identification of these cells is hampered in the ganoids by the presence in the kidney of large numbers of **corpuscles of Stannius**, which although not steroidogenic do resemble cytologically the adrenocortical cells. The corpuscles of Stannius may play a role in calcium metabolism in bony fishes, including teleosts (see Chapter 14). The teleost adrenocortical cells are embedded in the most anterior portion of the kidney, known as the **head kidney** (Figure 9-4). Frequently, these cells are associated with the dorsal posterior cardinal veins as described for cyclostomes. The head kidney has lost its renal function and consists mostly of lymphoid tissue, nonfunctional pronephric tubules, and small islands of adrenocortical cells (Figure 9-5). In a few species, all of the adrenocortical cells surround the posterior cardinal veins, and none is associated with kidney elements. Because of the diffuse nature of the adrenocortical tissue in teleosts, it is not possible to remove these cells surgically, and one must resort to the use of selective inhibitors of corticosteroid synthesis such as metyrapone, which blocks the synthesis of corticosteroids (see Chapter 8).

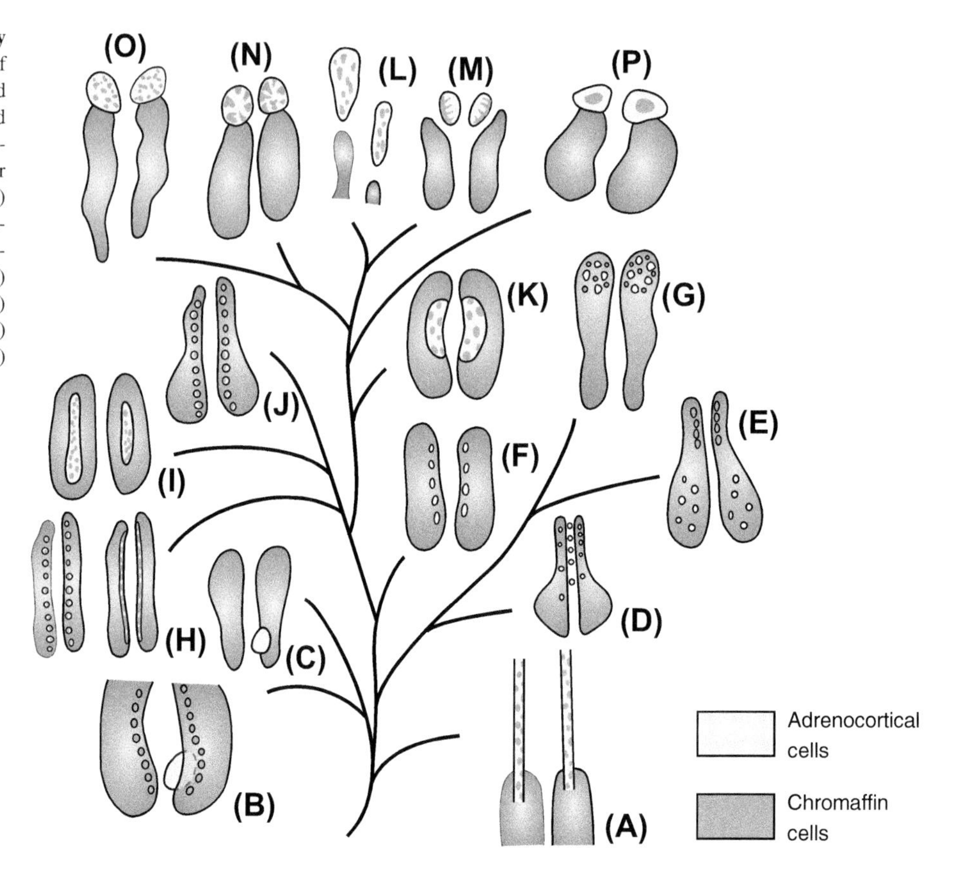

FIGURE 9-4 **Comparative anatomy of adrenal tissues.** Comparision of distribution of chromaffin (black) and adrenocortical tissue (clear) associated with the kidneys (stippled) in the vertebrates. Chromaffin tissue is not shown for the ratfish (C) and sturgeon (D). (A) cyclostomes; (B) selachians; (C) holocephalans; (D) chondrosteans; (E) holosteans; (F) dipnoans; (G) teleosts; (H) gymnophionans; (I) anurans; (J) urodeles; (K) chelonians; (L) snakes; (M) lizards; (N) crocodilians; (O) birds; (P) mammals.

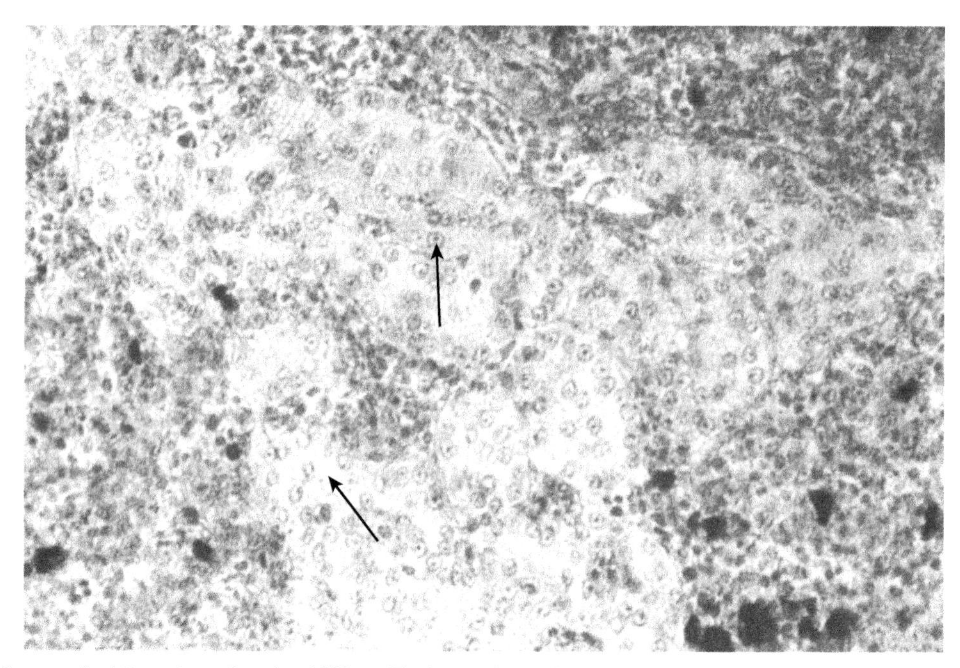

FIGURE 9-5 **Adrenocortical tissue in a teleost head kidney.** The larger, clear cells are adrenocortical (interrenal) cells located in clumps among the much smaller and darker lymphoid cells of the head kidney of a brown trout (*Salmo trutta*).

The principal circulating steroid in the chondrostean, holostean, and teleost fishes is cortisol, (Table 9-2) with corticosterone, aldosterone, and some others present in much lower quantities. Bony fishes lack the 1α-hydroxylase of elasmobranchs and consequently do not synthesize 1α-OHB. *In vitro* studies with teleost adrenocortical cells indicate that they convert pregnenolone preferentially to cortisol. When progesterone is supplied as a precursor *in vitro* the principal product is corticosterone, but *in vivo* cortisol primarily is produced.

The HPA axis of teleosts differs from other vertebrates in that hypothalamic neurons directly innervate the pars distalis of the pituitary (see Chapter 5). Teleosts respond rapidly to ACTH and to acute or chronic stress with markedly elevated corticosteroids within 5 to 10 minutes (Table 9-2). Treatment with metyrapone elevates ACTH secretion, causing adrenocortical hypertrophy and hyperplasia. Chronically stimulated adrenocortical cells also exhibit enlarged nuclei. The adrenocortical cells respond to treatment with ACTH by secreting more cortisol, and both ACTH and cortisol levels are elevated under conditions of stress. Release of ACTH is directed through innervation by neurons that release CRH to the corticotropes residing in the pars distalis. Immunoreactive CRH cells in the preoptic nucleus (PON) of the common white sucker also contain AVT but those in the **nucleus lateralis tuberis (NLT)** that send fibers into the adenohypophysis contain only CRH. Studies in the goldfish suggest that the PON regulates release of ACTH, whereas the NLT controls synthesis of ACTH. The CRH-like activity in the NLT may be due to either urotensin I or a urotensin-like peptide.

D. Lobe-Finned Osteichthyean Fishes: Sarcopterygians

The lungfishes (Dipnoi) have been of special interest to comparative endocrinologists seeking to understand the evolution of corticosteroids because they represent close relatives to both the fish and tetrapod lines. In dipnoan fishes, the adrenocortical cells are found as small cords located between renal and perirenal tissues adjacent to branches of the postcardinal veins. Adrenocortical cells from estivating *Protopterus* synthesize corticosterone *in vitro* from progesterone. However, only cortisol was identified in the plasma of the aquatic phase, suggesting that a tetrapod-like secretion (corticosterone) occurs during its moist, air-breathing phase and a teleost-like secretion (cortisol) during its aquatic phase. Although it is tempting to speculate on the evolutionary significance of these data, it would be premature to do so without further investigation. Aldosterone, cortisol, corticosterone, and a trace of 11-deoxycortisol circulate in the blood of the

TABLE 9-2 Effects of Stressors on Plasma Cortisol Levels in Bony Fishes

Species	Treatment	Unstressed (ng/ml)	Stressed (ng/ml)
Chondrosteans			
Scaphirhynchus albus (pallid sturgeon)	Handling stress, 30-sec	2.3	3
Polyodon spathula (spoonbill)	Handling stress, 30-sec	2.2	11
Teleosts			
Salmo trutta (brown trout)	Mild confinement, 1 hr	20	75
	Severe confinement, 1 hr	–	150
Salmo salar (Atlantic salmon)	Mild confinement, 30 min	–	200 (50–600)
Salvelinus namaycush (lake trout)	Acute handling and examined 1 hr later	20	270
Oncorhynchus mykiss (rainbow trout)	Acute handling and examined 1 hr later	2–25	300
Perca flavescens (yellow perch)	Acute exposure to air	8	90
Sander vitreus (walleye)	Acute exposure to air	11	229
Anoplopoma fimbria (Sablefish)	Chronic confinement	8	170
Gadus morhua (cod)	Chronic chasing	3	27
Rutilus rutilus (roach)	Chronic confinement	1.4	600

Based on numerous published accounts.

predominently aquatic South American lungfish (*Lepidosiren paradoxa*). Recent studies show that the obligate aquatic Australian lungfish (*Neoceratodus fosteri*) secretes only corticosterone and 11-deoxycorticosterone. It would be interesting to know which corticosteroids are secreted by adrenocortical cells of the crossopterygian coelacanths *Latimeria chalumnae* and *L. menadoensis*, which represent the only other living branch of sacropterygians.

E. Amphibians

The adrenocortical cells of amphibians are extrarenal and extremely variable with respect to their location. The anatomical pattern of anurans generally differs markedly from that of apodans and urodeles. However, the specific adrenal secretions correlate more to habitat than to anatomy or phylogeny. Adrenal corticosteroid secretion is stimulated by ACTH, which in turn is controlled by CRH. Evidence suggests that CRH also causes release of thyrotropin (TSH) from the adenohypophysis of larval amphibians (see Chapter 7).

1. Anatomical Features of Amphibian Adrenocortical Tissue

In anurans, adrenocortical tissue is found as irregular nodules organized loosely into a pair of interrenal glands on the ventral surface of the kidneys (Figure 9-4). In most anurans, some chromaffin cells are associated with the interrenal glands, and in one anuran, *Rana hexadactyla*, there are more chromaffin cells than adrenocortical cells in the interrenal glands. In addition to the adrenocortical cells and chromaffin cells, a third cellular type, the summer or **Stilling cell**, has been found in ranid frogs (Figure 9-6). This Stilling cell appears in summer and regresses in winter frogs. It is an eosinophilic cell and resembles a tissue mast cell (histamine-producing cell). The functional significance of the Stilling cell is unknown. Adrenocortical cells of both apodans and urodeles occur in scattered islands on the ventral surface of the kidney (Figure 9-4). This anatomical arrangement in part explains the virtual absence of studies employing apodan or urodele adrenocortical cells *in vitro*. Curiously, in one anuran, *Xenopus laevis*, the adrenocortical tissue also is organized as small islets on the ventral surface of the kidney. Each of these adrenocortical islets contains two or three chromaffin cells as well.

2. Amphibian Adrenocortical Secretions

The major corticosteroids synthesized by adult anuran adrenocortical tissue are aldosterone and corticosterone, and both hormones have been identified in adult amphibian plasma (Table 9-3). In addition, *in vitro* syntheses result in production of a large quantity of 18-hydroxycorticosterone, which is a precursor for the synthesis of aldosterone (see Chapter 3). The ratio of aldosterone to 18-hydroxycorticosterone to corticosterone *in vitro* is 6:3:1;

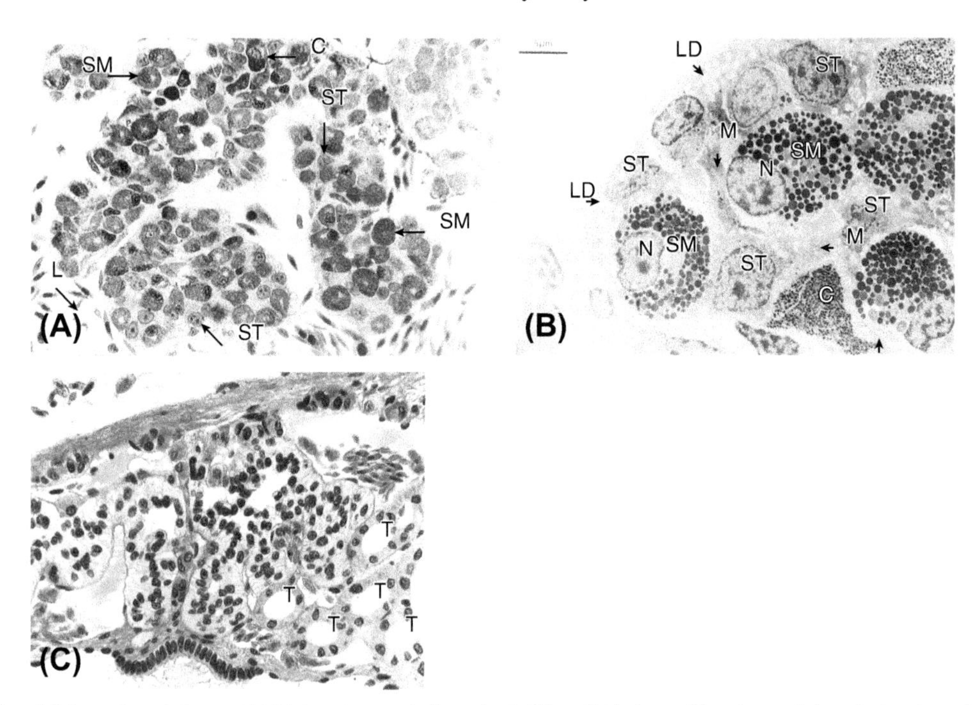

FIGURE 9-6 **Amphibian adrenal tissue.** (A) Light micrograph from the bullfrog (*Lithobates (Rana) catesbeianus*) showing adrenocortical (ST), chromaffin (C), and summer or Stilling cells (SM). (B) Electron micrograph from the bullfrog. The cells with the large secretory granules are Stilling cells (SM). The steroidogenic cells contain lipid droplets (LD) and mitochondria with tubular cristae. N, nucleus. (C) Adrenocortical cells in a salamander are located close to kidney tubules (T). *(Parts A and B reprinted with permission from Matsumoto, A. and Ishii, S., "Atlas of Endocrine Organs," Springer-Verlag, Berlin, 1989. Part C reprinted with permission from Vinson, G.P. et al., "The Adrenal Cortex." Prentice-Hall, Englewood Cliffs, NJ, 1993.)*

TABLE 9-3 Basal Plasma Corticosterone and Stress in Selected Amphibians

Species		Basal Levels (ng/ml)	Stressed Levels (ng/ml)
Lithobates (Rana) catesbeianus (American bullfrog)	Adults	2-16	180
Rhinella (*Bufo*) *marina* (Cane toad)[a]	Male	X	?
Plethodon shermani[b] (red-legged salamander)	Male	12.5	22.8
Desmognathus orchophaeus[b] Allegheny Mountain dusky salamander	Male	22.2	33.8
Cryptobranchus alleganiensis (hellbender)	Paedomorphic male	0.3	0.86
	Paedomorphic female	0.21	0.47
	Juvenile	0.33	1.10

[a]*Based on urinary levels that are elevated following captivity/handling stress; response is reduced in breeding males.*
[b]*values provided by Dr. Sarah Woodley*

however, corticosterone is the predominent circulating corticosteroid, and the high levels of corticosterone (an aldosterone precursor) and aldosterone may simply be an artifact of *in vitro* conditions. Ovarian production of significant quantities of 11-deoxycorticosterone has been reported, and this may be an important source for circulating corticosteroids in sexually mature females.

Although corticosterone is the dominant corticosteroid reported for terrestrial amphibians, cortisol appears to be the major corticosteroid in metamorphosing ranid tadpoles, in the permanently aquatic frog *X. laevis*, and in some permanently aquatic urodeles such as the mudpuppy (*Necturus maculosus*). Similarly, aquatic-phase, adult red-spotted newts (*Notophthalmus viridescens*) produce substantial amounts of cortisol, whereas the terrestrial efts secrete primarily corticosterone. There appears to be a transition in amphibians from secretion of cortisol to corticosterone that takes place during or immediately following metamorphosis of the larvae to a terrestrial adult (see Chapter 7). These observations would support the hypothesis that cortisol is important for maintaining sodium balance in freshwater amphibians, as reported for fishes, and that corticosterone becomes more important following metamorphosis to a terrestrial-phase amphibian. Corticosterone is also the major corticosteroid in reptiles, birds, and many mammals, although some adult mammals rely on cortisol (e.g., humans; see Chapter 8).

3. Direct Regulation of Adrenocortical Cells by AVT in Amphibians

AVT directly stimulates secretion of corticosterone (*X.*) and aldosterone (*X. laevis*, *L. catesbeianus*) by adrenocortical cells. Recall that in teleosts and mammals, AVT has a positive influence on the HPA axis but does so by increasing ACTH release. It is not known if this is a general condition among amphibians.

4. Other Regulators of Amphibian Adrenocortical Cells

There appear to be many local regulators of amphibian adrenocortical tissue. Nerve fibers innervating amphibian interrenal tissue have been shown to produce **substance P** and **neurotensin**, and newly identified polypeptides called **endozepines** (endogenous regulators of the benzodiazepine receptor) have been identified in amphibian chromaffin and Stilling cells. Substance P, **pituitary adenylate cyclase-activating polypeptide (PACAP)**, neurotensin, endozepines, and **endothelin** all increase corticosterone secretion from amphibian interrenal tissue *in vitro*.

F. Reptiles

Chelonians, crocodilians, and most snakes have paired suprarenally positioned adrenal glands similar to mammals. There is a variable degree of intermingling of chromaffin cell cords within a mass of adrenocortical cells that may show evidence of zonation (Figure 9-7). In lizards and some snakes, the adrenocortical cells are partially encapsulated by chromaffin cells, resulting in an anatomical "cortex" homologous to the mammalian medulla. Some chromaffin cells also are found within the central mass of adrenocortical cells. The chromaffin cells of *Sphenodon* (the primitive rhynchocephalian reptile) surround the dorsal aspect of the gland and form islets within the mass of adrenocortical cells. For the first time in vertebrates, the interrenals of reptiles have attained their own vascular

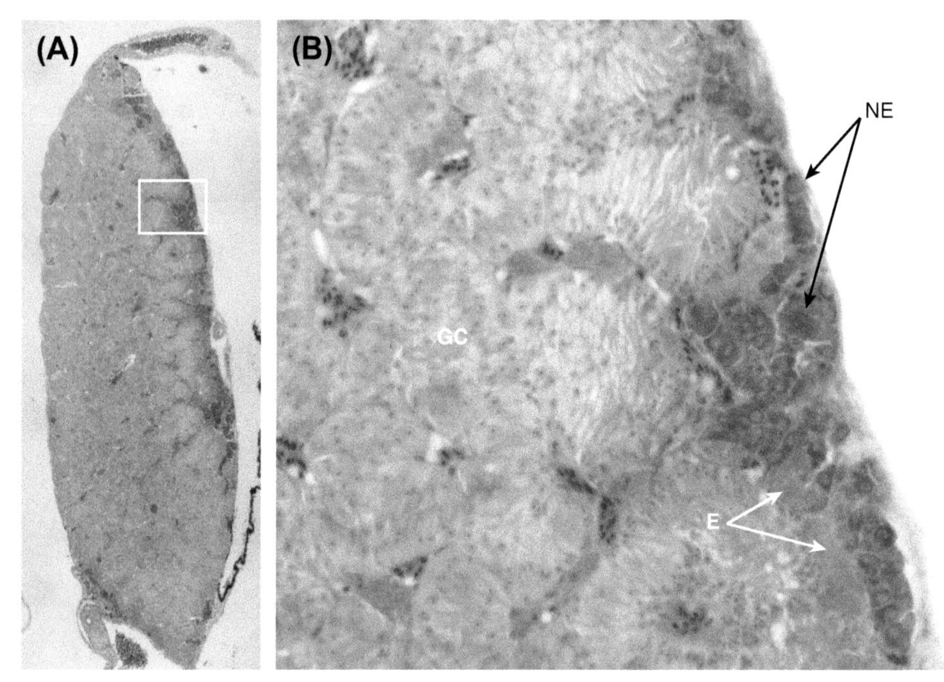

FIGURE 9-7 **Adrenal gland of the lizard *Podacris sicula*.** (A) Entire adrenal gland. (B) Section from A enlarged to show chromaffin cells (dark green = norepinephrine (NE)-secreting cells; light green = epinephrine (E)-secreting cells) and steroidogenic (GC) cells (light purple cells). Some chromaffin cells are distributed among the steroidogenic cells in addition to being along the surface. Giemsa stain. Photomicrographs courtesy of Salvatore Valiante.

supply and venous drainage, no longer relying on the kidney and a renal portal system for distribution of their secretory products. The metanephric kidney appears first in reptiles and has its own blood supply.

Although very few species have been studied, all of those examined under *in vitro* conditions (including turtles, lizards, snakes, and the American alligator) synthesize aldosterone and corticosterone as the major corticosteroids as well as 18-hydroxycorticosterone. Plasma levels of corticosterone are provided in Table 9-4. Corticosterone synthesis predominates *in vitro*, with the amounts of 18-hydroxycorticosterone exceeding the levels of aldosterone, suggesting a problem with *in vitro* conversion of 18-hydroxycorticosterone to aldosterone.

Adrenocortical cells from turtles (*Chrysemys picta*) secrete corticosterone *in vitro* following the addition of mammalian ACTH or crude extracts of avian, chelonian, or anuran pituitaries to the culture medium. Adrenocorticoid synthesis in the cobra *Naja naja* also is stimulated by mammalian ACTH. Corticosterone levels *in vivo* and *in vitro* are elevated following administration of ACTH to *Caimen crocodilus* or *C. sclerops*. ACTH also elevates corticosterone levels in lizards but does not increase aldosterone levels. Hypophysectomy causes a reduction in the adrenal weight of reptiles, whereas treatment with metyrapone elevates ACTH levels and causes adrenal hypertrophy and hyperplasia, indicating that a normal HPA feedback system is in place in reptiles.

One extra-adrenalocortical source of corticosteroids has been reported for reptiles. As demonstrated in fishes and amphibians, isolated ovaries from the desert night lizard (*Xantusia vigilis*) synthesize 11-deoxycorticosterone, but the importance of this observation to either corticosteroid physiology or to reproductive function in reptiles is uncertain.

G. Birds

The adrenal glands of birds are organized in the same manner as described for turtles, crocodilians, and most snakes. The relative quantities of chromaffin with respect to adrenocortical cells varies, however. There appears to be some zonation of cell types on histochemical and cytological bases similar to that seen in mammals but not clearly defined. The major corticosteroids synthesized by adrenocortical cells taken from domestic species and incubated *in vitro* are predictably corticosterone, aldosterone and 18-hydroxycorticosterone. This sequence of steroidogenesis (corticosterone to 18-hydroxycorticosterone to aldosterone) is characteristic of birds as well as anuran amphibians and reptiles and is essentially the same pattern found in cells of the mammalian zona glomerulosa. However, only corticosterone (Table 9-5) and aldosterone have been found in avian plasma. The absence of 18-hydroxycorticosterone in plasma further supports the conclusion that it is only a precursor for aldosterone synthesis, and its accumulation *in vitro* is an artifact. In addition to corticosterone and aldosterone, 11-deoxycorticosterone has been reported in the herring gull.

Although mammalian ACTH stimulates corticosterone secretion in chickens, hypophysectomy does not stop

TABLE 9-4 Effect of Stressors on Basal Plasma Corticosterone in Selected Reptiles. Based on Numerous Published Accounts

Chelonia		Basal Levels (ng/ml)	Stressed Levels (ng/ml)
Chelonia mydas (green sea turtle)	Non-breeding	0.3	15
	Breeding	0.3	7
Lepidochelys olivacea (olive ridley turtle)	Basking	0.11	6.5
Caretta caretta (logger head turtle)		0.2	
Crocodilians			
Crocodylus johnstoni (Australian crocodile)		4	≈40
Squamata			
Hyplodactylus maculatus (common gecko)	Female	≤6.4	
Sceloporus undulatus (fence lizard)	Male, breeding	6.8	12.1
	Male, post-breeding	1.7	52.5
Amblyrhynchus cristatus (marine iguana)	Remote population	≈5	≈12
	Tourist-exposed	≈5	≈7
Boiga irregularis (brown tree snake)		4.4	49.3

Based on numerous published accounts.

TABLE 9-5 Basal Plasma Corticosterone and Stress in Selected Birds

Species	Basal Levels (ng/ml)	Stressed Levels (ng/ml)
Sternus vulgaris (starling)	≈10	≈40
Zonotrichia leucophrys gambeli (White crowned sparrow)	≈15	≈27

corticosteroidogenesis. Thus, there may be a basal level of secretion that does not require ACTH, It also has been proposed that **melanophore-stimulating hormone (α-MSH)** produced and released by the hypothalamus maintains corticosteroidogenesis in hypophysectomized chickens. Apparently, bird adrenals also secrete corticosteroids in response to PRL, GH, serotonin, or parathyroid hormone. The response to ACTH is also enhanced by exposure to **parathyroid-hormone-related peptide (PTHrP)**, although the physiological significance of this observation is not clear.

II. PHYSIOLOGICAL ROLES FOR CORTICOSTEROIDS IN NON-MAMMALIAN VERTEBRATES

Many of the studies of corticosteroid function in non-mammals have focused on their effects on salt transport, particularly sodium—that is, mineralocorticoid activity. Aldosterone, cortisol, or corticosterone may possess strong mineralocorticoid activity when administered to non-mammals, a response also observed in mammals to high levels of cortisol or corticosterone (see Chapter 8). The roles of 11-deoxycortisol (cyclostomes) and cortisol (elasmobranchs, teleosts) on the Na^+ balance of fishes are well known.

The effects of corticosteroids on metabolic activities have not been studied extensively but considerable effort has been made to examine the response of glucocorticoid secretion to various natural and artificial stressors. In this regard, one must distinguish between predictable events (reproduction, migration, feeding, etc.) and unpredictable events (e.g., storms, droughts, predators, reduction in food supply) that can result in elevated corticosteroids. Whereas predictable events may result in moderately elevated corticosteroids, typically there is no disruption of normal activities. Unpredictable events result in a greater secretion of corticosteroids and may even shut down some activities such as reproduction. Such severe stressor effects often are observed in captive populations, but these responses may be abnormal in the sense that placing the animal in the laboratory has severely limited the repertoire of activities in which it might engage to reduce or eliminate the impact of a stressor.

Although chronic high elevations of glucocorticoids are associated with neurodegeneration in mammals, there is a positive association of moderately elevated glucocorticoids with improved memory recall. Young salmon imprint on the chemical composition of their natal streams and lakes; this imprint is recalled and directs them to their natal site. Olfactory centers in the salmon brain express GRs. It has been hypothesized that the elevated cortisol levels observed in salmon during their spawning migration facilitates memory recall of their natal stream. Furthermore, treatment of immature salmon with cortisol causes GRs in these olfactory centers to translocate from the cytoplasm to the nucleus. Elevated corticosterone in food-catching birds has been proposed to facilitate recall of the location of stored food, thereby increasing winter survival.

A. Corticosteroids and Agnathan Fishes: Cyclostomes

The blood of myxinoids (hagfishes) is isosmotic to seawater, but there are some minor differences in concentrations of specific ions. Therefore, although osmotic balance per se is no problem, the differential distribution of certain ions must be maintained actively. Injections of aldosterone or deoxycorticosterone acetate alter electrolyte composition of the body fluids with respect to sodium ions, but cortisone has no effect. Lampreys are either freshwater organisms or migrate between fresh water and the sea. While lampreys are in fresh water, their body fluids are hyperosmotic to their surroundings, and they produce a large quantity of dilute urine. Sea lampreys (*Petromyzon marinus*) do not secrete corticosteroids when in seawater, but when they are in fresh water corticosteroids can be identified in the circulation. Although aldosterone treatment can reduce renal and extrarenal sodium losses from lampreys held in fresh water, 11-deoxycortisol binds to the lamprey corticosteroid receptor in the gill and appears to be the natural regulator of Na^+ transport by gills. Treatment of adult lampreys in fresh water with physiological levels of 11-deoxycortisol increases gill Na^+/K^+-ATPase, thus establishing a mineralocorticoid role for 11-deoxycortisol. Stressing these adults causes a significant elevation of 11-deoxycortisol, supporting a glucocorticoid-like action in response to stressors. Evidence suggests that corticosteroids can reduce reproductive functions in lampreys. Chronic elevation of 11-deoxycortisol or 11-deoxycorticosterone in adult lampreys reduces circulating levels of androgens and estradiol, suggesting inhibition of gonadal functions at some level in the HPG axis.

B. Corticosteroids and Chondrichthyean Fishes

In chondrichthyian fishes, the oldest living jawed vertebrates (gnathostomes), both a mineralocorticoid receptor (MR) and a glucocorticoid receptor (GR) are present. It is assumed that a single corticosteroid receptor (such as occurs in living cyclostomes) was duplicated when the entire vertebrate genome duplicated following divergence of cyclostomes before formation of gnathostomes. Stressors result in elevation of blood glucose and 1α-OHB, establishing a glucocorticoid type of role on metabolism in elasmobranchs. Marine elasmobranchs have plasma that is hyperosmotic to seawater and hence do not have the osmoregulatory problems exhibited by most marine fishes. This hyperosmotic condition is achieved by maintaining high circulating levels of urea and trimethylamine oxide. Additionally, 1α-OHB binds to cells in the gills, rectal gland, and nephrons of the kidney and stimulates Na^+ retention. The elasmobranch rectal gland is especially important in Na^+ secretion, whereas special cells of the gills perform this function in teleosts. Exogenous angiotensin stimulates release of 1α-OHB and suggests a mammalian-like regulatory pattern for corticosteroid secretion and a mineralocorticoid action.

C. Corticosteroids and Bony Fishes

Among the bony fishes, corticosteroids have been investigated with respect to function primarily among the teleosts, where cortisol functions as both a mineralocorticoid and a glucocorticoid. Chondrostean fishes apparently respond like teleosts.

1. Mineralocorticoid Effects of Cortisol in Teleosts

In general, cortisol stimulates sodium transport across gills (both influx and efflux) via activation of

Na^+/K^+-ATPase as well as across the mucosa of the gut and in the kidney of freshwater fishes. Recent studies show that aldosterone and 11-deoxycorticosterone are without effect. In the spring, plasma corticosteroids rise markedly in juvenile salmonid fishes prior to their migration from fresh water to the ocean or while adapting to seawater. Seawater-adapted eels (*Anguilla* spp.) exhibit a marked turnover of Na^+ (50 to 60%/hour), but ion flux is very low in freshwater-adapted eels (<1%/hour). Eels in seawater are faced with an influx of Na^+ that they must eliminate, whereas freshwater eels must conserve body Na^+ that readily can be lost to their Na^+-poor surroundings. Cortisol treatment increases the activity of Na^+/K^+-ATPase in gills, gut epithelial cells, and kidneys. Circulating levels of cortisol are similar in freshwater-adapted and seawater-adapted eels, suggesting that cortisol is not important. However, other studies show that elevated cortisol occurs during the initial process of adapting to fresh water and later drops back to a basal level. Neither cortisol nor ACTH is completely effective in maintaining normal Na^+ balance in hypophysectomized freshwater eels, and PRL, another osmoregulatory hormone in fishes, may be important as well.

2. Glucocorticosteroid Roles in Teleosts

Cortisol also has been reported in several species to maintain gluconeogenesis and increase levels of the gluconeogeneic enzyme **phosphoenol pyruvate carboxykinase (PEPCK)** to provide a balance between lipid, carbohydrate, and protein metabolism. Feeding of fishes results in elevated cortisol as does starvation, implying an important role for cortisol in the management of energy stores.

Stressors produce marked elevations of both glucose and corticosteroids in newly hatched, juvenile, and adult teleost fishes (Table 9-2), although increases seen in chondrosteans (sturgeon, paddlefish) are rather moderate. Capture and confinement of wild trout cause a marked elevation in cortisol over time, and bringing wild trout into the laboratory can induce sustained increases in levels of plasma cortisol. Severe stresses can induce extremely high levels of cortisol in some species.

Chronic exposure to a stressor may produce adaptations that result in what appear to be normal levels of plasma cortisol. For example, adult brown trout (*Salmo trutta*) in streams polluted with elevated levels of cadmium and zinc exhibit plasma cortisol levels similar to upstream populations that are not exposed. Subjecting the two populations to a 24-hour confinement stressor reveals an impaired ability to maintain a normal stress response in the metal-exposed fish with respect to both ACTH and cortisol (see Box 9A).

Cortisol is elevated in migrating juvenile and adult Pacific salmon (genus *Oncorhynchus*) and may reflect a predictable response of the HPA axis to chronic stress with no acclimation as described earlier for brown trout. Adult salmonids exhibit extensive adrenocortical cell hyperplasia and elevated corticosteroid levels during their spawning migration from the ocean to fresh water. This increase in corticosteroid levels is accompanied by accelerated catabolism of protein, suggesting a major role in protein metabolism similar to that in mammals (see Chapter 8).

The salmon brain contains numerous nuclei that express GRs (Figure 9-8). Chronically elevated cortisol in reproducing Pacific salmon is correlated with accelerated neurodegeneration in these same brain nuclei followed by extensive deposition of β-amyloid peptide (Figure 9-9). Other pathological changes occur in spawning salmon similar to those described for Cushing's disorders in humans (see Chapter 9) and what appears to be adrenal exhaustion after spawning that contributes to the death of the spawned fish (Table 9-6).

3. Corticosteroids and Reproduction in Teleosts

As seen for mammals (Chapter 8), corticosteroids may have either positive or negative impacts on reproduction in fishes. Cortisol or 11-deoxycorticosterone are elevated during spawning in a number of species, but it is not clear whether this is associated with stress or is part of the normal endocrine regulation of reproduction. These observations may reflect normal, predictable roles of the HPA axis. Some studies show direct participation of corticosteroids in reproduction. For example, cortisol peaks along with estradiol at ovulation in the killifish (*Fundulus heteroclitus*), and secretion of ovarian 11-deoxycorticosterone in the Indian catfish (*Heteropneustes fossilis*) may be responsible for induction of ovulation. Conversely, chronic elevation of cortisol is associated with reductions in androgen and estrogen levels in tilapia and salmonids.

D. Corticosteroids and Amphibians

The HPA axis is important in stress and metabolism, ionic–osmotic regulation, reproduction, and development in amphibians. Recognition of stress and its implications for reproduction and survival is especially important for the many endangered species of amphibians. Given the rate of amphibian population declines worldwide, it is critical that we understand the roles of the HPA axis in both metamorphosis and reproduction of amphibians and how anthropogenic changes in the environment affect fitness in amphibians. Functioning of the HPA axis is dependent upon hypothalamic CRH and pituitary ACTH. Predictably, hypophysectomy induces atrophy of adrenocortical tissue due to the absence of ACTH.

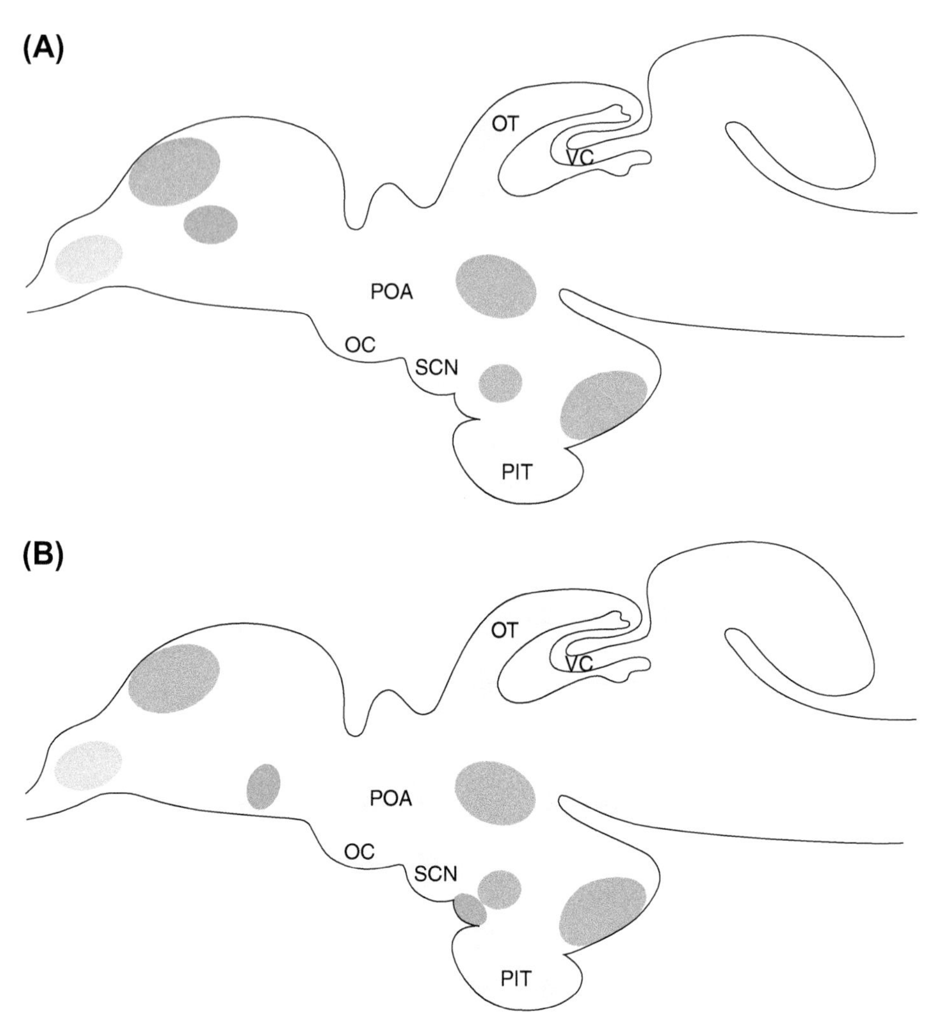

FIGURE 9-8 **Glucocorticoid receptor distribution compared to areas of neurodegeneration and amyloid deposition in the spawning salmon brain.** (A) Distribution of glucocorticoid receptors in brain of Kokanee salmon. (B) Sites of neurodegeneration and deposition of β-amyloid in the spawning Kokanee salmon brain. The OT, VC, POA, and SCN exhibit both GR and amyloid as do additional sites not named here (purple areas); green indicates a common olfactory region exhibiting both GR and amyloid whereas the orange areas are unrelated sites positive for GR (in A) and amyloid only in (B), respectively. Abbreviations: OC, optic chiasm; OT, optic tectum; PIT, pituitary; POA, preoptic area; SCN, suprachiasmatic nucleus; VC, valvula cerebelli. *(Part A adapted with permission from Carruth, L.L. et al.,* General and Comparative Endocrinology, *117, 66–76, 2000. Part B adapted with permission from Maldonado, T.A. et al.,* Brain Research, *858, 237–251, 2000. © Elsevier Science, Inc.)*

1. Corticosteroids, Stress, and Metabolism in Amphibians

Stressors cause elevation in plasma corticosteroids in amphibians. Placing freshly collected bullfrogs in sacks for up to 24 hours doubles their corticosterone levels. Stress can also alter secretion of other hormones. The secretions of PRL, GTHs, and androgen all decrease in some species of stressed amphibians following capture. The failure of certain species to breed in captivity and the stimulus for some larvae to undergo metamorphosis (see Chapter 7) following capture and laboratory confinement may be consequences of stress. Other species, however, breed readily in captivity and should be examined more closely in this regard.

2. Mineralocorticoid Activity in Amphibians

Exposure of ranid frogs to seawater results in atrophy of the adrenocortical tissue of amphibians presumably because of reduced requirements for corticosteroids and HPA activity. Similarly, adrenalectomized frogs show a decrease in plasma sodium and an increase in plasma potassium as observed in mammals. Winter ranid frogs usually survive adrenalectomy, but summer frogs die soon afterward; perhaps the death of these summer frogs is related to the Stilling cells. Aldosterone seems to be the important salt-regulating hormone (mineralocorticoid) acting on the skin and urinary bladder, increasing sodium influx and retention, respectively. The action of aldosterone on the toad urinary bladder involves synthesis of new protein and would appear to be analogous to the production of aldosterone-induced protein in the mammalian distal convoluted tubule (see Chapter 9). ACTH elevates aldosterone levels in *Rana esculenta*, implying a direct action of ACTH on aldosterone secretion, a condition different from that described for mammals and other vertebrates. Aldosterone treatment reverses the depression of plasma sodium caused by treatment of tiger salamanders with **aminoglutethamide** (an inhibitor of the $P450_{scc}$ enzyme necessary for synthesis of pregnenolone from cholesterol; see Chapter 3). Larger doses of corticosterone

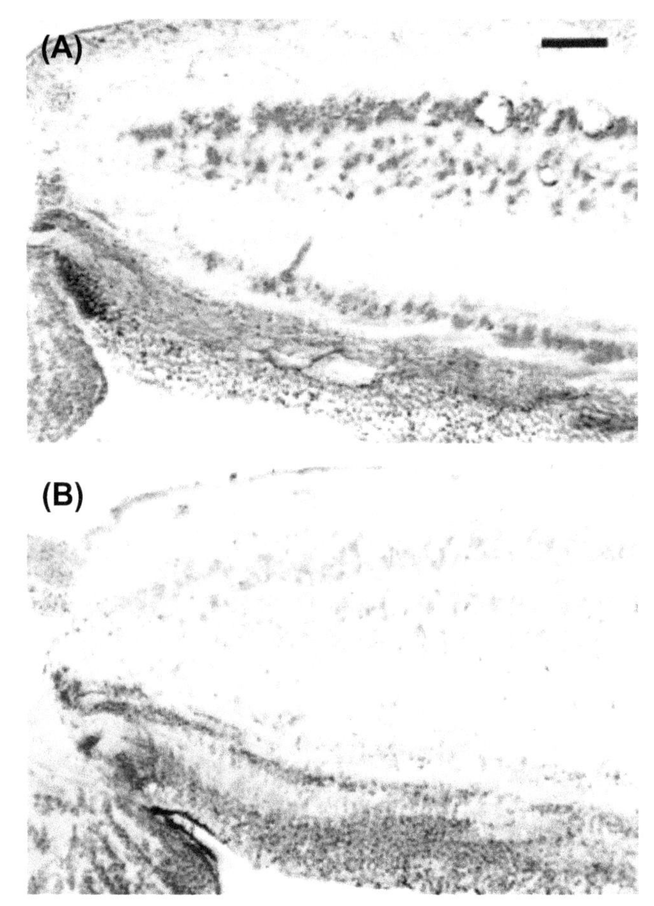

FIGURE 9-9 **Amyloid deposition in the salmon brain.** Brains of spawning Pacific salmon are characterized by extensive neurodegeneration and the deposition of immunoreactive β–amyloid as shown in the optic tectum (A) compared to an adjacent section of the same brain that was pretreated with immune serum (B). Staining in the optic tectum of a pre-spawning animal would resemble the immune control. *(Reprinted with permission from Maldonado, T.A. et al.,* Brain Research, *858, 237–251, 2000. © Elsevier Science, Inc.)*

were needed than aldosterone to bring plasma sodium back to normal, supporting a physiological role for aldosterone principally as a mineralocorticoid in urodele amphibians. Administration of either aldosterone or corticosterone induces marked hyperglycemia as well as some increase in liver and muscle glycogen levels, but probably only corticosterone is a physiological glucocorticoid.

3. Corticosteroids and Reproduction in Amphibians

Seasonal rhythms of corticosterone levels have been reported in plasma samples from several unstressed anurans and at least one urodele showing in general that corticosteroids are greatest during breeding. Although high blood levels of corticosterone can inhibit reproduction (see ahead), and there is generally a reciprocal relationship between plasma testosterone and corticosterone in males, amphibian reproductive behavior is energetically costly (especially male calling) and thus it is not surprising that corticosterone is elevated during reproduction in order to help meet energy demands. In at least one amphibian, the male toad *Rhinella* (*Bufo*) *arenarum*, corticosterone is converted by the testicular enzyme **5α-reductase** to **5α-dihydroxycorticosterone (5α-DHB)** that binds to a testicular GR. 5α-reductase also converts testosterone to **5α-dihydrotestosterone (5α-DHT)**, a more potent androgen in amphibians (see Chapter 11). Thus, corticosterone serves as a competitor of testosterone and lowers production of 5α-DHT. In liver, 5α-DHB can activate the gluconeogenic enzyme PEPCK and increase glucose production. The importance of these observations to reproduction has not been elucidated but serves to indicate there is still much to be learned.

As in other vertebrates, stress can inhibit reproductive behavior. Extensive studies by Frank Moore and associates

TABLE 9-6 **Comparison of Changes Associated with Spawning in Salmon to Cushing's Syndrome and Hormone-Induced Hyperadrenocorticism in Mammals**

Tissue	Spawning Salmon	Cushing's Syndrome	Experimental Hyperadrenocorticism
Adrenal	Hyperplasia and degeneration	Hyperplasia and tumors	Hyperplasia
Pituitary	Degeneration	Hyaline change in basophils	Minimal degeneration
Spleen	Depletion of lymphocytes, fibrosis	No reported change	Depletion of lymphocytes
Thymus	Involution, depletion of thymocytes	Involution, occasional tumor	Depletion of thymocytes
Liver	Degeneration	Occasional fatty degeneration	No change

(*Continued*)

TABLE 9-6 Comparison of Changes Associated with Spawning in Salmon to Cushing's Syndrome and Hormone-Induced Hyperadrenocorticism in Mammals—cont'd

Tissue	Spawning Salmon	Cushing's Syndrome	Experimental Hyperadrenocorticism
Kidney	Degeneration	Degeneration	Degeneration
Pancreas	Hypertrophy of islets	Hypertrophy of islets; variable	Hypertrophy of islets
Stomach	Atrophy and generation	Occasional ulcers	Atrophy of epithelium and occasional ulcers
Thyroid	Atrophy and generation	Atrophy of follicular epithelium	Atrophy of follicular epithelium
Gonads	Degeneration of testes	Atrophy	Degeneration
Muscle	Degeneration of masseter	Atrophy	No change
Cardiovascular system	Degeneration and beginning arteriosclerosis	Arteriosclerosis	Arteriosclerosis
Skin		Hypertrophy	Atrophy

[a]Adapted from Robertson, O.H., and Wexler, B.C. (1960). Histological changes in the organs and tissues of migrating and spawning Pacific salmon (*genus Oncorhynchus*). *Endocri.* 66, 222–239.

at Oregon State University of male rough-skinned newts (*Taricha granulosa*) have implicated CRH and corticosterone in the inhibition of clasping of females, an essential reproductive behavior. Corticosterone, operating through neural membrane receptors in male newts, blocks clasping behavior that is stimulated by AVT (see Chapter 11); however, in the male cane toad, *Rhinella* (*Bufo*) *marina*, where corticosterone levels are very high, there is no inhibitory effect on testosterone secretion as seen in cyclostomes and teleosts.

4. The HPA Axis and Development in Amphibians

As described in Chapter 5, CRH not only activates the HPA axis but also regulates thyrotropin release. This may be particularly important in amphibian larvae inhabiting ephemeral ponds. As ponds dry up and water levels decrease, CRH is elevated, leading to accelerated metamorphosis. Both thyroid hormones and corticosteroids have positive roles in amphibian metamorphosis as described in Chapter 7.

E. Corticosteroids and Reptiles

Most of our understanding of the reptilian HPA axis comes from studies of stress as it relates to reproductive behaviors and reproductive success. Much of this work has been done with lizards because of the ease of observing their behaviors both in the field and in the laboratory. Much less is known for the role of corticosteroids in metabolism and ionic regulation.

1. Corticosteroids and the Stress Response in Reptiles

The reptilian HPA axis functions similar to that of mammals in response to a great variety of stressors. The seasonal pattern of corticosterone secretion of turtles, snakes, and lizards generally shows a peak in corticosterone levels during the breeding season although some species show no breeding increase and a few exhibit a decrease with breeding. Regardless of the pattern of secretion, the response to stress is blunted during breeding as seen in other vertebrates. Studies on the male eastern fence lizard (*Sceloporus undulatus*) suggest that this change is due to alterations at the level of the hypothalamus and/or pituitary, possibly in response to elevated testosterone. In comparing dominant to subordinate male *Anolis carolinensis*, corticosterone plasma levels were greater in the subordinate animal.

2. Adrenal Hormones and Metabolism in Reptiles

In the desert iguana *Dipsosaurus dorsalis*, it appears that corticosterone does not play much of a role in lactate utilization by skeletal muscles during recovery from exercise. Epinephrine, however, produces a marked stimulation of lactate incorporation into glycogen under similar conditions. Both hormones are elevated in the blood after five minutes of exhaustive exercise in the laboratory. Following aggressive bouts in the American chameleon *Anolis carolinensis*, winners have much higher levels of epinephrine and norepinephrine than do losers, possibly relating to this metabolic action.

3. Corticosteroids and Ionic Regulation in Reptiles

The roles of corticosteroids in ionic regulation by reptiles have not been studied extensively. Like the situation in the non-mammalian vertebrate groups discussed previously, there does not seem to be any clear distinction between mineralocorticoids and glucocorticoids, with both aldosterone and corticosterone producing similar effects. Reptiles have nasal (orbital) salt-excreting glands similar to those found in aquatic birds. Corticosterone increases Na^+ secretion by these glands, and aldosterone decreases Na^+ excretion while stimulating Na^+ reabsorption in the kidneys and urinary bladder. Injections of concentrated sodium chloride solutions (salt-loading) depress plasma aldosterone levels in several species of lizard. Similar observations are reported for the tortoise *Testudo hermanni*. Salt-loaded lizards also have elevated levels of AVT that is consistent with its antidiuretic role (see Chapter 5).

F. Corticosteroids and Birds

1. Corticosteroids and Stress in Birds

Stress activates the HPA axis in birds in the same manner as for other vertebrates; however, this effect is blunted under laboratory conditions as compared to free-ranging birds, indicating an accommodation to captivity conditions (Figure 9-10). The stress response appears very early after hatching. In 10- to 14-day-old chicken embryos, there is a surge of catecholamines (dopamine and epinephrine) in the face of short-term stress. This includes release of epinephrine from the adrenal medulla. Activation of the endocrine stress mechanism also influences other endocrine glands. Profound effects of stress are noted in adult birds, too; for example, chasing (the stressor) male zebra finches for 15 minutes results in a significant depression in plasma androgens measured 2 hours later. Isolation of males in small cages for 12 hours virtually obliterates testosterone in the circulation.

2. Corticosteroids in Development and Reproduction of Birds

Corticosteroids generally peak seasonally with breeding as seen in other vertebrates, and the sensitivity to stressors is depressed, although different species may not conform to this generalized pattern. Corticosterone levels do not change in giant petrels (*Maconectes* spp.) throughout the reproductive season or during molting, although there is a possible small increase in corticosterone in the males of the Southern giant petrel species at the onset of molting. One role for elevated corticosterone has been proposed for females. The female deposits corticosterone in the egg during its production and, although the role of corticosterone in early development is not known, the amount

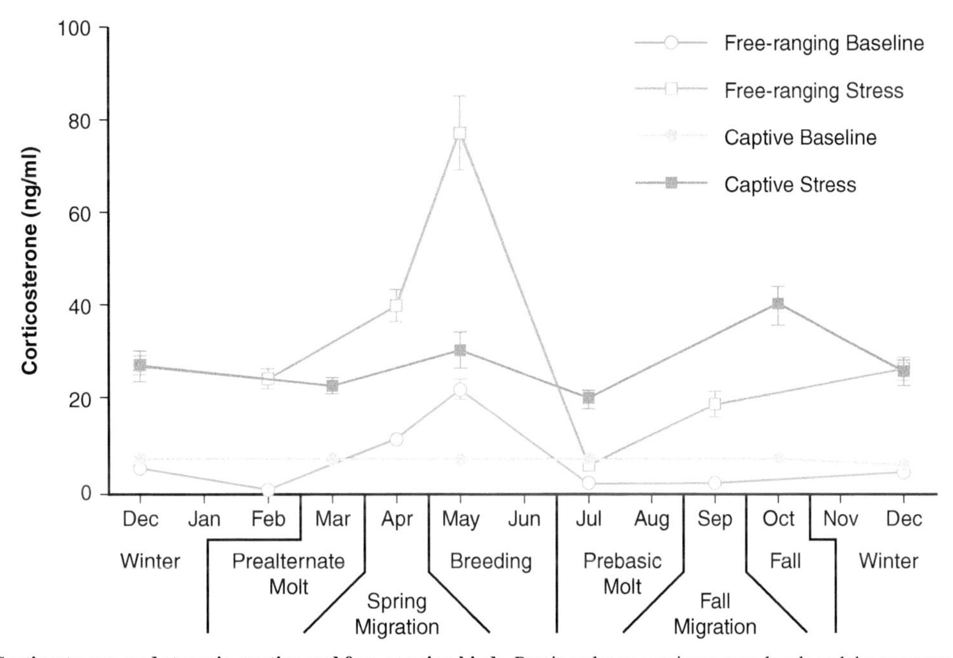

FIGURE 9-10 **Corticosterone and stress in captive and free-ranging birds.** Resting plasma corticosterone levels and the response to stress in captive birds are blunted as compared to free-ranging birds. Note also the seasonal differences in the sensitivity to a stressor. *(Reprinted with permission from Romero, L.M. and Wingfield, J.,* Comparative Biochemical and Physiology B, *12, 13–20, 1999.)*

deposited in the egg may influence the time of hatching. Levels of corticosterone tend to be higher in earlier eggs than later eggs.

3. Corticosteroids and Ionic Balance in Birds

Corticosteroids cause the nasal salt-excreting glands of certain aquatic birds to secrete a hypertonic NaCl solution. This salt secretion is enhanced by treatment with ACTH or corticosterone. Furthermore, corticosterone uptake by nasal salt glands is followed by an increase in salt excretion. Adrenalectomized ducks cannot excrete a salt load, but corticosteroid therapy restores this salt-excreting ability. Metyrapone treatment blocks nasal gland secretion. Aldosterone also stimulates nasal salt excretion, unlike the situation in reptiles where aldosterone diminishes excretion of Na^+ as it normally does in kidney. However, due to aldosterone's powerful action on Na^+ reabsorption in the kidney, there is a net positive effect on Na^+ retention in spite of what occurs locally at the salt glands. Both aldosterone and corticosterone promote Na^+ retention in the lower intestine (called the *copradeum*). Increases in environmental salinity are correlated with increases in Na^+/K^+-ATPase in salt gland cells, and this event is possibly related to increased protein synthesis caused by corticosteroids. Marine birds have larger adrenal glands than do freshwater or terrestrial birds, supporting the constant role for corticosterone in nasal salt gland regulation. Predictably, birds inhabiting brackish water have intermediate-sized adrenals.

4. Corticosteroids and Metabolism in Birds

In addition to being a mineralocorticoid, elevated corticosterone causes metabolic effects in birds, including weight loss, which probably is a reflection of increased protein catabolism as reported in mammals. In addition, chronic treatment with corticosterone elevates blood glucose and liver glycogen, probably through gluconeogenesis.

5. Corticosteroids and Migration in Birds

Circadian rhythms for corticosteroids have been described in some birds, usually peaking early in the morning and exhibiting lowest levels at dusk. The time of the daily corticosteroid peak is believed to determine the behavioral response following injection of PRL into migratory white-crowned sparrows. These observations have led to hypotheses concerning the control of migratory behavior by the phase relationships between rhythms of corticosterone and PRL secretion. North American migratory birds exposed to a day–night cycle with a long photophase release PRL, which induces premigratory fattening. Seasonal PRL and corticosteroid rhythms differ markedly throughout the year, and fattening can be altered experimentally simply by changing the timing of the corticosteroid peak with respect to the pattern in timing of the daily PRL peak.

III. RENIN–ANGIOTENSIN SYSTEM IN NON-MAMMALS

The mammalian renin–angiotensin system discussed in Chapter 8 is briefly summarized here. The enzyme **renin** is secreted into the blood by the kidney in response to lowered blood pressure or lowered blood Na^+. Renin hydrolyzes **renin substrate** to release the decapeptide **angiotensin-I (Ang-I)**. In the lung capillaries, Ang-I is converted by **angiotensin-converting enzyme (ACE)** into the octapeptide **angiotensin-II (Ang-II)**. Ang-II is a stimulator of aldosterone release from the adrenal cortex (zona glomerulosa), and in turn it stimulates Na^+ and water reabsorption that increases blood pressure. Ang-II also stimulates contraction of arteriole smooth muscle, secretion of AVP, and drinking behavior that all contribute to an elevation of blood pressure. It has been hypothesized that the renin–angiotensin system evolved as a mechanism for regulating blood pressure. Control over mineralocorticoid synthesis and regulation of sodium/potassium balance presumably was acquired later. All of the elements of the renin–angiotensin system are also present in diverse mammalian cell types, where they perform intracrine roles and regulate cell metabolism; for example, the entire system is found within the mammalian brain. In some cases, the intracrine renin-angiotensin system has also been described in non-mammalian vertebrates.

A. The Renin–Angiotensin System in Fishes

Angiotensinogens (angiotensinogen = renin substrate) are part of the **family A serine protease inhibitors (serpins)** and are expressed in lampreys. Administration of lamprey angiotensin-II has a vasodepressor effect in adult Japanese lamprey. No evidence of either renin or a juxtaglomerular apparatus has been found in cyclostomes. This kidney specialization may not have evolved until later in the chondrichthyean and bony fishes or in their immediate common ancestor. However, the system apparently evolved subsequent to the appearance of jawed vertebrates.

1. The Renin–Angiotensin System in Chondrichthyean Fishes

Although it has been reported repeatedly that the renin–angiotensin system is absent in chondrichthyean fishes and that attempts to demonstrate the presence of renin have been unsuccessful, careful cytological and histological studies have shown otherwise. Examination of the mesonephric kidneys of four selachian species (two sharks, one ray, one skate) reveals that definite modified smooth muscle cells containing renin-like secretion granules are associated with the afferent arterioles and form juxtaglomerular-like structures in the kidneys. Furthermore, a distinct macula densa-like modification of the distal tubule where it passes between the afferent and efferent arterioles is present.

Elasmobranch Ang-I (and hence Ang-II) has a unique sequence of amino acids not seen in other vertebrate groups (Table 9-7). **Angiotensin-converting enzyme-like action (ACELA)** is present in gills and spleen as well as in lesser amounts in brain and kidney of *Scyliorhinus canicula.* Ang-II stimulates production of 1α-OHB by shark interrenals *in vitro*. Although ACTH stimulated steroidogenesis in the Atlantic stingray (*Dasyatis sabina*), this action was not altered by application of human Ang-II.

Receptors for Ang-II occur in gills, kidney, and rectal glands of elasmobranchs as well as in blood vessels, indicating that Ang-II is probably important in osmoregulation and blood pressure regulation. Dogfish (*Squalus acanthias*) exhibit increased drinking behavior with increasing salinity, and drinking can be stimulated by Ang-II. Additional physiological studies are still needed, but it seems likely that a mammalian-like renin–angiotensin system involved with osmotic and blood pressure regulation is present in these primitive fishes and appeared before the evolution of the metanephric kidney.

2. The Renin–Angiotensin System of Actinopterygian Fishes

Renin activity is present in all of the spiny-rayed fish groups, and juxtaglomerular-like cells have been identified in the mesonephric kidneys of several teleosts (Figure 9-11). Histological and cytological identification of renal cells exhibiting renin granules has not been verified in any of the non-teleost actinopterygian fishes, however. No macula densa-like structure has been reported for any bony fish. Ang–II stimulates secretion of cortisol, which is the principal salt-regulating corticosteroid in these fishes, with actions at both the gills and kidney. Aldosterone is not an important mineralocorticoid in teleosts. The renin–angiotensin system appears to regulate osmotic and ionic adaptation in teleosts. Freshwater-adapted eels exhibit greater renin activity when in seawater. Furthermore, there are gradual changes in plasma renin activity during adaptation of eels to either fresh water or seawater. Administration of either ACTH or renin causes elevation of circulating cortisol levels in eels. Treatment of freshwater-adapted eels with captopril, an inhibitor of ACE, reduces the level of cortisol following a seawater challenge.

TABLE 9-7 Amino Acid (AA) Sequences for Vertebrate Angiotensin-I

Position	1	2	3	4	5	6	7	8	9	10
Mammals	D	R	V	Y	I	H	P	F	H	L
Birds					V				S	
Reptiles	D or N				V				H or Y	
Teleosts	D or N				V				H,Y, or G	
Elasmobranchs	N		P						Q	

Note: Blanks mean the amino acid is the same as in mammalian Ang-I. See Appendix C for explanation of amino acid symbols.
Source: Based on Hazon, N. et al., *Journal of Experimental Zoology*, 284, 526–534, 1999.

3. The Renin–Angiotensin System in Sarcopterygian Fishes

Renin activity and the presence of renal cells containing renin granules have been observed in the coelacanth and in two genera of lungfishes. Infusion of isosmotic sodium chloride solution into the lungfish *Neoceratodus fosteri* causes reduction in plasma renin activity, suggesting that the involvement of a renin mechanism in osmotic adaptation Ang–II causes secretion of corticosteroids in *N. fosteri*.

B. The Amphibian Renin–Angiotensin System

Renin activity is present in amphibian renal tissue, although the secretory renin-containing granules differ morphologically from those of mammals. Several studies report no macula densa in amphibians (Figure 9-11), but a macula densa-like structure has been reported for one toad. This report is of considerable interest since a macula densa appears to be absent in reptiles. Salt-depleted frogs exhibit elevated kidney levels of renin, suggesting the presence of both mechanisms for regulating aldosterone production, although plasma renin levels do not appear to differ between distilled-water-adapted toads and saline-adapted toads. Ang–II stimulates both corticosterone and aldosterone synthesis *in vitro* by adrenocortical fragments from *Rana ridibunda*, although aldosterone is assumed to be the important mineralocorticoid of amphibians.

C. The Reptilian Renin–Angiotensin System

Renin activity has been reported for kidneys of turtles, lizards, and snakes, but no macula densa has been described (Figure 9-11). This is curious considering the possibility of a macula-like structure in amphibians and evidence for a distinct macula in elasmobranchs, birds, and mammals. Ang–II stimulates corticosterone release in both turtles and lizards. No data are available on other reptilian groups. Ang–II stimulates corticosterone and aldosterone secretion in lizards but only corticosterone secretion in turtles. However, sodium depletion apparently does not affect renin

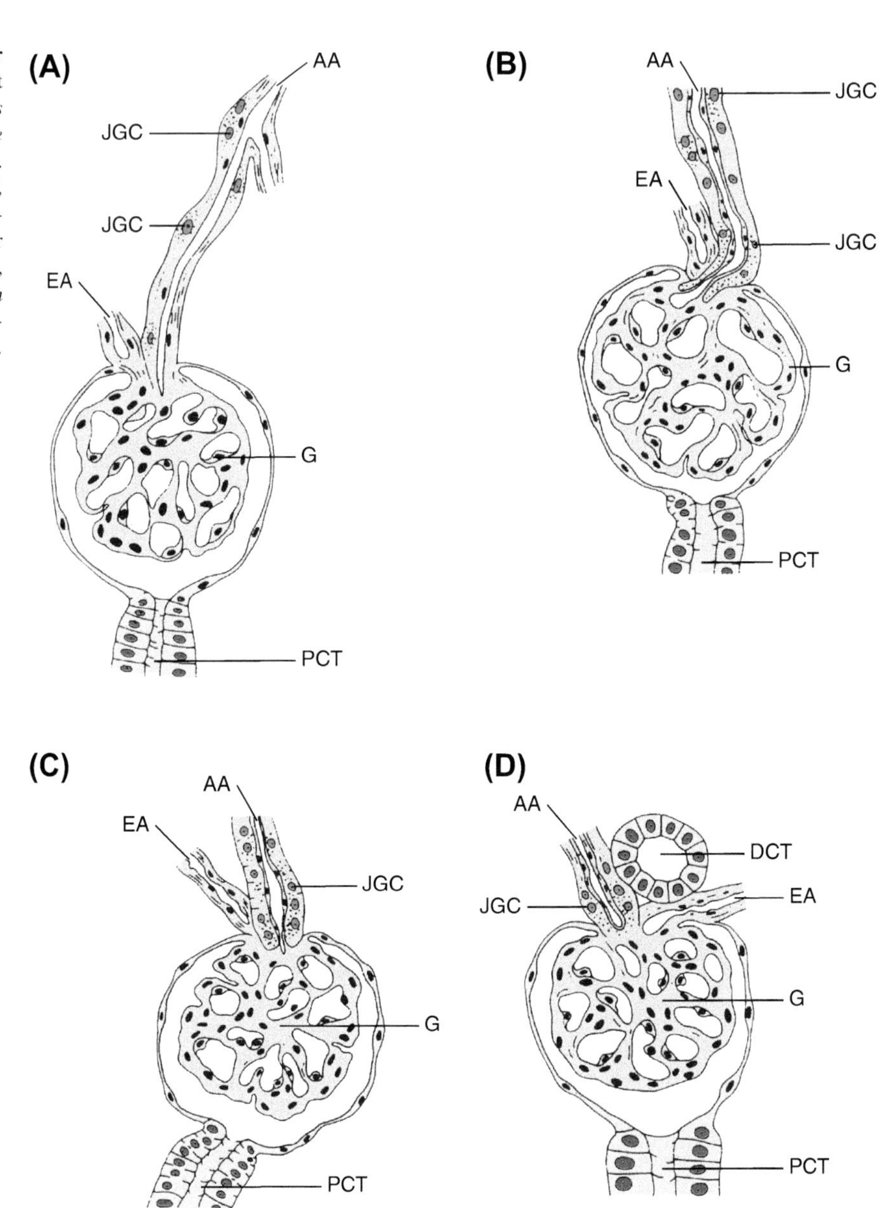

FIGURE 9-11 **The juxtaglomerular apparatus in non-mammals.** (A) Teleost *(Carassius auratus)*. (B) Bullfrog (*Lithobates (Rana) catesbeianus*). (C) Snake *(Elpahe quadrivirgata)*. (D) Domestic chicken. Abbreviations: AA, afferent arteriole; DCT, distal convoluted tubule; EA, efferent arteriole; G, glomerulus; JGC, juxtaglomerular cells; PCT, proximal convoluted tubule; MD, macula densa. *(Adapted with permission from Matsumoto, A. and Ishii, S., "Atlas of Endocrine Organs," Springer-Verlag, Berlin, 1989.)*

activity in turtles. Whether or not there is a role for the renin–angiotensin system *in vivo* is not clear.

D. The Avian Renin–Angiotensin System

Renin activity is present in birds, and both the juxtaglomerular apparatus and a macula densa have been described. Cells of the avian macula densa are similar to those of mammals, with only some minor differences. This implies that the macula densa developed in reptiles but is inexplicably absent from extant reptiles. Renin activity and angiotensin levels in blood plasma of ducks and pigeons increase following hemorrhage, establishing a physiological role for the renin–angiotensin system that is similar to mammals. Furthermore, elevated dietary Na^+ causes a reduction in circulating aldosterone without affecting corticosterone levels. Conversely, a reduction in dietary Na^+ raises aldosterone levels. Ang–II, stimulation of aldosterone secretion is reduced by administration of **atrial natriuretic peptide (ANP)**; however, in turkeys, ANP stimulates aldosterone release as it does in teleosts.

IV. NATRIURETIC PEPTIDES (NPs) IN NON-MAMMALIAN VERTEBRATES

Natriuretic peptides (NPs) decrease blood pressure by increasing urinary output (resulting in decreased blood volume) and increased Na^+ excretion (reducing osmotic concentration and decreasing water reabsorption). In mammals, three naturiuretic peptides (NPs) are produced (Chapter 8). Atrial natriuretic peptide was the first one isolated from the atria of the heart. The second NP was first isolated from the brain and originally was named **brain natriuretic peptide (BNP)**. However, it was soon shown that both ANP and BNP are produced in the heart and constitute the bulk of the NPs in the circulation. A third or **C-type NP (CNP)** was discovered primarily in the brain, where it probably functions in a paracrine/autocrine fashion. Two forms of CNP are present in mammals (CNP-3, CNP-4). With the exception of ANP, the appearance of NPs in a variety of tissues has prompted a name change, and BNP is now known as the **B-type NP**. NPs have been discovered in all non-mammalian groups, and an even greater variety of NPs occurs among the fishes. The presence of NPs in cyclostomes indicates that this system evolved early among the chordates. Molecular studies also indicate a common gene origin for NPs, with genes found in a number of vertebrate venoms.

Two classes of NP receptors have been described in teleosts that parallel those found in mammals. One class is a kinase-like transmembrane receptor linked to guanylyl cyclase that causes formation of cGMP and the second is a GPCR type of membrane receptor that may inhibit cAMP formation. Four forms have been isolated from the Japanese eel (*Anguilla japonica*). Two are of the cGMP-activating class (**NPR-A** and **NPR-B**) and two are of the GPCR class (**NPR-C** and **NPR-D**). The latter form, NPR-D, has only been reported from this fish. The greatest affinity of NPR-A, -C, and -D is equal for ANP, whereas CNP is the major ligand for NPR-B.

A. NPs in Fishes

Early studies reported that extracts prepared from atria and ventricles of the hagfish (*Myxine glutinosa*) produce relaxation of the precontracted rabbit aorta, a standard bioassay for the hypotensive action of ANP. In addition, ANP binding sites were identified in gills and kidney of this species, suggesting an osmoregulatory role for ANP-like peptides. The CNP gene is considered to be the ancestral form of NP genes and has been identified in three genera of lampreys and three genera of hagfishes. Neither ANP nor BNP has been found in cyclostomes.

Chondrichthyean fishes also have a single form of NP but this one most closely resembles mammalian CNP (Figure 9-12). Most of the NP activity resides in the atria of the heart. Radiolabeled ANP binds to the secondary

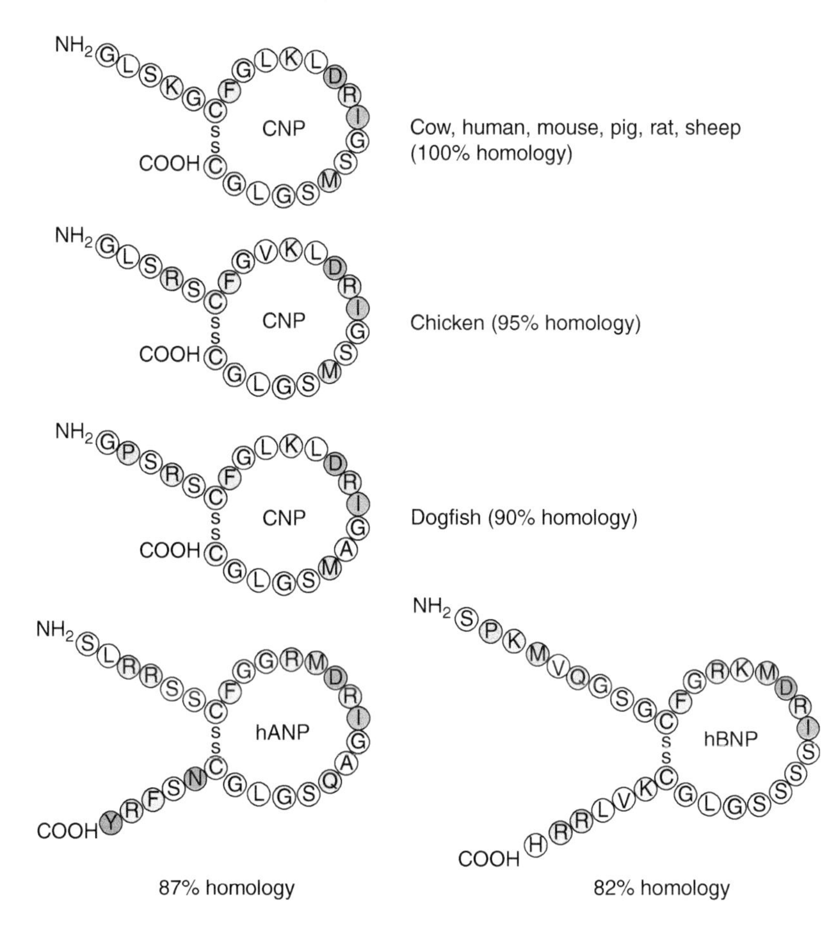

FIGURE 9-12 **Natriuretic peptides in non-mammals.** Comparison of dogfish shark and chicken CNPs to some mammalian CNPs as well as human (h) ANP and BNP. *(Adapted with permission from Fowkes, R.C. and McArdle, C.A.,* Trends in Endocrinology and Metabolism, *11, 333–338, 2000. © Elsevier Science, Inc.)*

lamellae of gill filaments and to kidney glomeruli in the shark *Scyliorhinus canicula*. hANP causes release of 1α-OHB from elasmobranch adrenals that is not consistent with ANP's inhibitory action on corticoid release described in tetrapod vertebrates. In the eryhaline Atlantic skate (*Dasyatis sabina*), adapted to seawater, there is a high concentration of NPR-B in the kidney and in the rectal gland. NPR-B concentration is also high in the rectal gland of freshwater-adapted *D. sabina*.

Chondrosteans and teleosts produce a unique NP called **ventricular NP (VNP)** in addition to ANP, BNP, and four CNPs (CNP-1–4). CNP–1 and CNP-2 are found in the brain and CNP-3 and CNP-4 are found in the heart, but only CNP-4 is expressed in the gills and kidney. Most of the bioassayable NP resides in the atria. Binding of ANP to secondary lamellae of gills occurs in both chondrosteans and teleosts, and it is clear that the gill is a major target for NPs in seawater-adapted fish whereas CNP is more important in freshwater-adapted fish. Data on osmoregulatory effects by NP on the kidney are contradictory. Furthermore, it is difficult to study this in seawater-adapted fish that produce little urine. Data on effects of NPs in stenohaline fish are scarce, indeed. Several studies have shown that NPs stimulate cortisol secretion, which in turn stimulates gill Na^+/K^+–ATPase to excrete Na^+ and Cl^- through the gills. This is similar to observations in elasmobranchs.

Few data are available on NPs in sarcopterygian fishes. ANP is present in myocardial tissue of the African lungfish (*Protopterus aethiopicus*) and may play a role in controlling corticosteroid release.

B. NPs in Tetrapod Vertebrates

ANP, BNP, and CNP are secreted by amphibians. Amphibian NPs are potent vasodilators that cause diuresis and natriuresis. Like mammals, NPs inhibit the actions of ACTH or Ang-II stimulation of corticosterone secretion. However, in *Xenopus laevis*, ANP blocked aldosterone release but did not affect corticosterone secretion. Effects of NPs on the amphibian urinary bladder have yielded conflicting results, and their role if any on this important osmoregulatory site requires clarification.

In reptiles, cDNA for ANP has been found only in turtles but not in any other group, whereas BNP is present in all reptilian groups. A CNP precursor protein has been demonstrated in the venom gland of snakes. ANP-like binding sites have been described for turtle kidney, brain, gastrointestinal tract, adrenal gland, and epididymis. Rat ANP blocks basal and Ang–II-induced secretion of aldosterone and corticosterone in cultured adrenocortical cells from the *S. undulatus*. In general, although the reptiles are the most poorly studied group with respect to NPs, they appear to have similar response to NPs as described for birds and amphibians.

Receptors have been described for birds that bind ANP, although the ANP gene is absent as is also seen for all reptiles except turtles (Figure 9-13). Birds do produce BNP and CNP. Mammalian ANP has been shown to lower blood pressure in some bird species. Nasal salt glands have NP receptors, and NPs have been shown to stimulate their secretion of salt. BNP also blocks Ang–II-stimulated aldosterone secretion in birds with the exception of the domestic turkey, where it stimulates aldosterone secretion.

V. EVOLUTION OF CHROMAFFIN TISSUE AND ADRENAL MEDULLARY HORMONES

Morphologically, there has been a general evolutionary trend within the vertebrates to develop a closer anatomical relationship between chromaffin tissues homologous to the

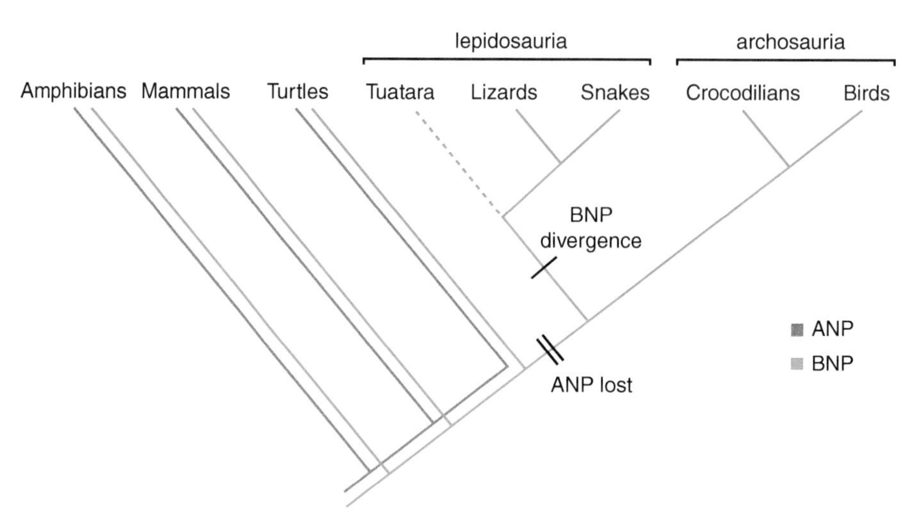

FIGURE 9-13 **Distribution of natriuretic peptides among tetrapod vertebrates.** This figure stresses the absence of ANP in most reptiles and in all birds suggesting that the gene was lost prior to the evolution of the modern Lepidosauria (squamates and the tuatara) and the archosaurs (crocodilians and birds). Bony and chondrichthyean fishes have the ANP gene, although it is absent from cyclostomes. *(Adapted with permission from Trajanovska, S. and Donald, J.A.,* General and Comparative Endocrinology, *156, 339–346, 2008.)*

mammalian adrenal medullary cells and the adrenocortical cells. Chromaffin tissue in agnathans is found in association with the posterior cardinal veins as are the separate clusters of presumed adrenocortical cells. In the cartilaginous fishes, the chromaffin tissue is intrarenal and is more widely separated from the "interrenal" adrenocortical cells than in any other vertebrate group. Although some chromaffin cells have been described in association with the cardinal veins of lungfishes, chromaffin cells are found mostly in the heart. These cardiac chromaffin cells are claimed to represent the homologue of the adrenal medulla in lungfishes. There is considerable anatomical variation among the adrenals of teleosts, but no species are known to exhibit cardiac chromaffin tissue. The chromaffin cells may form clumps entirely separate from the adrenocortical cell clusters in the head kidney or may be intermingled with adrenocortical cells. In some species, both conditions may be found (i.e., separate and mixed clusters of cell types). Clusters of adrenocortical cells (islets) in amphibians typically contain only a few chromaffin cells, and the latter usually are separated from adrenocortical cells. The anatomical distribution of chromaffin cells in reptiles is varied. The adrenal glands of squamates and the tuatara *Sphenodon punctatus* have peripherally located norepinephrine-secreting chromaffin cells and central islets of epinephrine-secreting cells intermingled among the adrenocortical cells. The avian adrenal glands consist of such a mixture of cortical and chromaffin cells that no distinction of "cortex" and "medulla" is possible.

Early studies on the proportion of adrenal catecholamines in non-mammals suggest that the mammalian fetal pattern of high norepinephrine production with increasing production of epinephrine following birth (see Chapter 8) is an example of the old principle that ontogeny recapitulates phylogeny. Analysis of extractable catecholamines from the shark *Squalus acanthias* (which has entirely separate chromaffin tissue), the frog *Rana temporaria* (which exhibits some mixing of chromaffin and adrenocortical cells), and the rabbit *Oryctolagus cuniculus* suggests an evolutionary change from reliance on norepinephrine to the progressive reliance on epinephrine. Histochemical and chromatographic procedures indicate that norepinephrine predominates in some bird species, whereas others produce mainly epinephrine. Still other groups show only a slight preference for producing one catecholamine over the other. Although norepinephrine predominates in extracts of adrenals prepared from chickens, turkeys, and pigeons, the major circulating catecholamine in these species is epinephrine. These data point out the dangers of extrapolating too broadly from gland content to secretory activities, especially when substrate–product relationships exist.

Few comparative studies of adrenal catecholamines on carbohydrate metabolism have been reported. The hyperglycemic action of epinephrine seems to be a primitive action in non-mammalian vertebrates, although the mechanisms involved are not clear. In chinook salmon, for example, both epinephrine and norepinephrine stimulate glycogen utilization in the liver and release of glucose. Incorporation of lactate into glycogen also is stimulated by epinephrine in carp liver, but this effect is overshadowed by a general breakdown of glycogen that occurs concomitantly. Epinephrine produces glycogen breakdown in the liver of all birds examined. Norepinephrine, on the other hand, produces hypoglycemia in the ratfish and in 6- to 9-week-old Japanese quail. The bases for these unusual observations on norepinephrine action are unknown.

Studies of catecholamine effects on lipid metabolism are less common than studies of catecholamines on carbohydrate metabolism in lower vertebrates, although numerous studies of effects by other hormones have been reported (e.g., GH, PRL, thyroxine). One of the more thorough studies shows that while acting through β-adrenergic receptors, norepinephrine stimulates, in a dose-dependent manner, a triglyceride lipase that causes lipolysis in liver slices from coho salmon. Epinephrine has no lipolytic action in this system although it has been shown to have lipolytic effects in other fishes. More work in this area clearly is needed.

VI. ENDOCRINE DISRUPTION OF CORTICOSTEROID FUNCTIONS IN NON-MAMMALIAN VERTEBRATES

Endocrine disruption the HPA axis has not been well studied in any vertebrate group compared to observations of disruption in the HPT and HPG axes. However, the presence of pharmaceutical corticosteroids such as dexamethasone and a variety of receptor antagonists and enzyme inhibitors, as well as natural corticosteroids present in wastewater effluents and the heavy metals from mining activities (see Box 9A) that find their way into the aquatic environment, should be of concern. The use of biosolids from waste treatment plants as enrichments for agricultural fields may be of concern for amphibians that breed in ponds supplied by runoff from the treated fields.

VII. SUMMARY

Non-mammalian vertebrates do not exhibit the anatomical cortex–medullary relationship like mammals for the homologous adrenocortical cells and chromaffin cells. However, there is a general trend for a closer anatomical relationship between these tissues from being entirely separate in ancient fishes with direct associations with the kidneys to the evolution of discrete, separate glands in reptiles and birds.

The synthesis of corticosteroids is the same as in mammals with the exception of elasmobranchs that produce a unique corticosteroid, 1α-OHB. Teleosts rely on cortisol as the principal mineralocorticoid and glucocorticoid. Amphibians, reptiles, and birds secrete corticosterone and aldosterone as their glucocorticoid and mineralocorticoid, respectively. Some larval and aquatic amphibians, however, rely on cortisol. The role in non-mammals for cortisol and corticosterone in gluconeogenesis, the stress response, and aging are similar to those of mammals.

Aldosterone secretion is regulated by a renin–angiotensin system in non-mammalian tetrapods. Blood pressure regulation involves the renin–angiotensin system in non-mammals similar to that of mammals and natriuretic peptides antagonize Ang-II actions. Chromaffin cells participate in emergency reactions and stress responses as described for mammals. In non-mammals, chromaffin cells generally secrete more norepinephrine than epinephrine. Clearly, cells of non-mammals have the capacity to convert NE to E,

BOX 9A Some Cryptic Aspects of the HPA Axis and the Stress Response System

Elevation of plasma cortisol or corticosterone is commonly used as a measure of how much "stress" an animal is experiencing. However, the HPA axis is able to adapt to chronic conditions such that measurement of a single parameter may not provide an accurate assessment. Consider this case of fish populations in nature.

Brown trout (*Salmo trutta*) tend to spend most of their lives living in a small section of a river and return to their home location after spawning; consequently, brown trout captured at a specific location can be found at that site year after year. Basal plasma cortisol levels measured immediately after capture of adult brown trout living their entire lives in a cadmium-polluted region of the Eagle River near Vail, Colorado, were not different from values determined for brown trout living in an upstream cadmium-free location. From these data, one might conclude that cadmium was not a stressor for these fish. However, an immunocytochemical and histological examination of the brain and adrenocortical (interrenal) tissue revealed that there were significantly more CRH-immunoreactive cells in the hypothalamus of these cadmium-exposed fish and that there was a proliferation of adrenocortical cells in the head kidney. Furthermore, the adrenocortical cells exhibited enlarged nuclei, which is evidence of increased stimulation by ACTH.

When both populations of brown trout were subjected to a 24-hour confinement stress test, marked differences were seen. Although fish from both populations achieved a comparable level of plasma cortisol after 3 hours of confinement (Box Figure 9A-1), the cadmium-exposed trout were slower to respond. Furthermore, the cadmium-exposed trout could not sustain the response at that higher level for 24 hours. Examination of plasma ACTH levels indicates that the cadmium-exposed trout required twice the level of ACTH to achieve the 3-hour level of plasma cortisol, and when the fish could no longer maintain that level of ACTH secretion the levels of cortisol decreased (see Box Figure 9A-2).

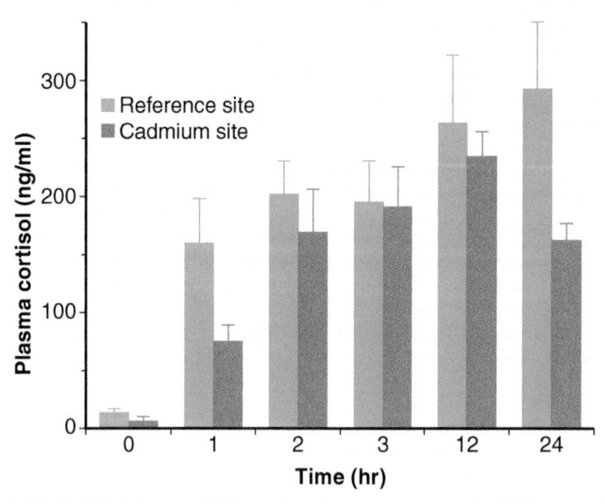

BOX FIGURE 9A-1 **Effects of cadmium exposure on HPA axis of brown trout.** Fish living their entire life in cadmium-contaminated water exhibit a delayed response in cortisol secretion to confinement stress but catch up to trout from an uncontaminated reference site by 3 hours; however, after 12 hours of confinement, cadmium-exposed trout can no longer maintain a high level of cortisol secretion. *(Data used with permission from Norris, D.O. et al.,* General and Comparative Endocrinology, *113, 1–8, 1999.)*

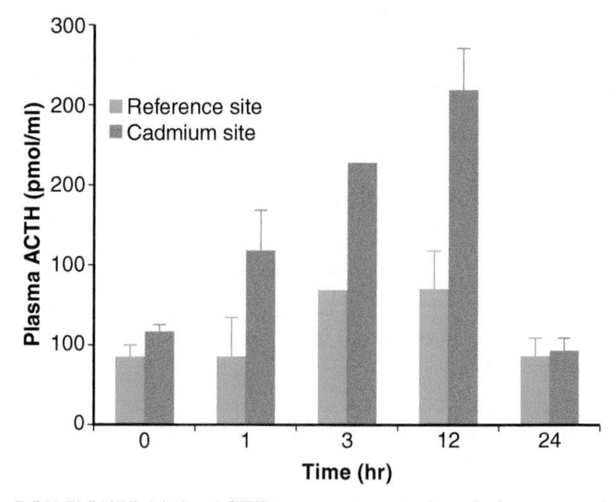

BOX FIGURE 9A-2 **ACTH response to stress in cadmium-exposed brown trout and reference-site brown trout.** A significantly greater amount of ACTH secretion is necessary for trout chronically exposed to cadmium to maintain normal cortisol secretion in response to confinement stress. After 12 hours, these fish can no longer maintain ACTH secretion and cortisol secretion also decreases. *(Data used with permission from Norris, D.O. et al.,* General and Comparative Endocrinology, *113, 1–8, 1999.)*

indicating that this biosynthetic step is phylogenetically ancient.

STUDY QUESTIONS

1. Compare the arrangement of adrenocortical and chromaffin tissues in fishes, amphibians, reptiles, and birds with that of mammals.
2. How have the roles of cortisol and corticosterone changed among the various major groups of vertebrates? What features are more or less common to all?
3. Compare the stress response of each non-mammal group to what you learned about mammals.
4. How is reproduction affected by the HPA axis?
5. Describe the endocrine regulation of ionic–osmoregulation in vertebrates, including the relationships between kidney functions and cardiovascular physiology.

SUGGESTED READING

Books

Chester-Jones, I., Henderson, I.W., 1976–1980. General, Comparative, and Clinical Endocrinology of Adrenal Cortex. Academic Press, New York.

Delrio, G., Brachet, J., 1984. Steroids and Their Mechanism of Action in Nonmammalian Vertebrates. Raven, San Diego, CA.

Pickering, A.D., 1981. Stress and Fish. Academic Press, New York.

Vinson, G.P., Whitehouse, B., Hinson, J., 1993. The Adrenal Cortex. Prentice-Hall, Englewood Cliffs, NJ.

General Articles

Chan, D.K.O., Wong, C.K.C., 1996. The adrenal cortex and the renin-angiotensin system of lower vertebrates. In: Vinson, G.P., Anderson, D.C. (Eds.), Adrenal Glands, Vascular System and Hypertension. Journal of Endocrinology, pp. 23–35. Bradley Stoke, Bristol, U.K.

Chester-Jones, I., 1987. Structure of the adrenal and interrenal glands. In: Chester Jones, I., Ingleton, P.M., Phillips, J.G. (Eds.), Fundamentals of Comparative Vertebrate Endocrinology. Plenum, New York, pp. 95–121.

Chester-Jones, I., Phillips, J.G., 1986. The adrenal and interrenal glands. In: Pang, P.K.T., Schreibman, M.P. (Eds.), Vertebrate Endocrinology. Fundamentals and Biomedical Implications, vol. I. Academic Press, San Diego, CA, pp. 319–349.

de Kloet, E.R., 2010. From vasotocin to stress and cognition. European Journal of Pharmacology 626, 18–26.

Denver, R.J., 2009. Structural and functional evolution of vertebrate neuroendocrine stress systems. Annals of the New York Academy of Sciences 1163, 1–16.

Fowkes, R.C., McArdle, C.A., 2000. C-type natriuretic peptide: an important neuroendocrine regulator? Trends in Endocrinology and Metabolism 11, 333–338.

Greenberg, N., Wingfield, J.C., 1987. Stress and reproduction: reciprocal relationships. In: Norris, D.O., Jones, R.E. (Eds.), Reproduction in Fishes, Amphibians, and Reptiles. Plenum, New York, pp. 461–503.

Hazon, N., Tierney, M.L., Takei, Y., 1999. Renin–angiotensin system in elasmobranch fish: a review. Journal of Experimental Zoology 284, 526–534.

Henderson, I.W., Kime, D.E., 1987. The adrenal cortical steroids. In: Pang, P.K.T., Schreibman, M.P. (Eds.), Vertebrate Endocrinology: Fundamental and Biomedical Implications, vol. 2. Academic Press, San Diego, CA, pp. 121–142.

Landys, M.M., Ramenofsky, M., Wingfield, J.C., 2006. Actions of glucocorticoids at a seasonal baseline as compared to stress-related levels in the regulation of periodic life processes. General and Comparative Endocrinology 148, 132–149.

Romero, L.M., 2002. Seasonal changes in plasma glucocorticoid concentrations in free-living vertebrates. General and Comparative Endocrinology 128, 1–24 (Actually only deals with tetrapods.).

Fishes

Auperin, B., Geslin, M., 2008. Plasma cortisol response to stress in juvenile rainbow trout is influenced by their life history during early development and by egg cortisol content. General and Comparative Endocrinology 158, 234–239.

Barton, B.A., 2002. Stress in fishes: a diversity of responses with particular reference to changes in circulating corticosteroids. Integrative and Comparative Biology 42, 517–525.

Brelin, D., Petersson, E., Winberg, S., 2005. Divergent stress coping styles in juvenile brown trout (*Salmo trutta*). Annals of the New York Academy of Sciences 1040, 239–245.

Close, D.A., Yuna, S.-S., McCormick, S.D., Wildbilla, A.J., Li, W., 2010. 11-Deoxycortisol is a corticosteroid hormone in the lamprey. Proceedings of the national Academy of Sciences (USA) 107, 13942–13947.

Donald, J.A., Toop, T., 2006. The natriuretic peptide system of fishes: structure, evolution, and function. In: Reinecke, M., Zaccone, G., Kapoor, B.G. (Eds.), Fish Endocrinology. Oxford and Science Publishers, Enfield, NH, pp. 443–486.

Flik, G., Klaren, P.H.M., Van den Burg, E.H., Metz, J.R., Huising, M.O., 2006. CRF and stress in fish. General and Comparative Endocrinology 146, 36–44.

Guerriero, G., Ciarcia, G., 2006. Stress biomarkers and reproduction in fish. In: Reinecke, M., Zaccone, G., Kapoor, B.G. (Eds.), Fish Endocrinology. Oxford and Science Publishers, Enfield, NH, pp. 665–692.

Holloway, A.C., Reddy, P.K., Sheridan, M.A., Leatherland, J.F., 1994. Diurnal rhythms of plasma growth hormone, somatostatin, thyroid hormones, cortisol and glucose concentrations in rainbow trout, *Oncorhynchus mykiss*, during progressive food deprivation. Biological Rhythm Research 25, 415–432.

Hontela, A., Rasmussen, J.B., Audet, C., Chevalier, G., 1992. Impaired cortisol stress response in fish from environments polluted by PAHs, PCBs, and mercury. Archives of Environmental Contamination and Toxicology 22, 278–283.

Iger, Y., Balm, P.H.M., Wendelaar Bonga, S.E., 1994. Cellular responses of the skin and changes in plasma cortisol levels of trout (*Oncorhynchus mykiss*) exposed to acidified water. Cell and Tissue Research 278, 535–542.

Kawakoshi, A., Hyodo, S., Nozaki, M., Takei, Y., 2006. Communications in genomics and proteomics identification of a natriuretic peptide (NP) in cyclostomes (lamprey and hagfish): CNP-4 is the ancestral gene of the NP family. General and Comparative Endocrinology 148, 41–47.

Matsuda, K., Uchiyama, M., Konnoa, N., Kurosawaa, M., Kaiyab, H., Miyazatob, M., 2010. Molecular cloning and characterization of V2-type receptor in two ray-finned fish, gray bichir, *Polypterus senegalus*, and medaka *Oryzias latipes*. Peptides 31, 1273−1279.

McCormick, S.D., 1995. Hormonal control of gill Na^+K^+-ATPase and chloride cell function. In: Wood, C.M., Suttle-Worth, T.J. (Eds.), Fish Physiology. Cellular and Molecular Approaches to Fish Ionic Regulation, vol. 14. Academic Press, San Diego, CA, pp. 285−315.

Mosconi, G., Cardinaletti, G., Carotti, M., Palermo, F., Soverchia, L., Polzonetti-Magni, A.M., 2006. Neuroendocrine mechanisms regulating stress response in cultured teleost species. In: Reinecke, M., Zaccone, G., Kapoor, B.G. (Eds.), Fish Endocrinology. Oxford and Science Publishers, Enfield, NH, pp. 693−720.

Norris, D.O., 2000. Endocrine disruptors of the stress axis in natural populations: how can we tell? American Zoologist 40, 393−401.

Norris, D.O., Hobbs, S.L., 2006. The HPA axis and functions of corticosteroids in fishes. In: Reinecke, M., Zaccone, G., Kapoor, B.G. (Eds.), Fish Endocrinology. Oxford and Science Publishers, Enfield, NH, pp. 721−765.

Pankhurst, N.W., 2011. The endocrinology of stress in fish: an environmental perspective. General and Comparative Endocrinology 170, 265−275.

Schreck, C.B., 2010. Stress and fish reproduction: the roles of allostasis and hormesis. General and Comparative Endocrinology 165, 549−556.

Toop, T., Donald, J.A., 2004. Comparative aspects of natriuretic peptide physiology in non-mammalian vertebrates: a review. Journal of Comparative Physiology B 174, 189−204.

Wedemeyer, G.A., Barton, B.A., McLeay, D.J., 1990. Stress and acclimation. In: Schreck, C.B., Moyle, P.B. (Eds.), Methods in Fish Biology. American Fisheries Society, Bethesda, MD, pp. 451−489.

Wunderink, Y.S., Engels, F., Halm, S., Yúfera, M., Martínez-Rodríguez, G., Flik, G., Klaren, P.H.M., Mancera, J.M., 2011. Chronic and acute stress responses in Senegalese sole (*Solea senegalensis*): the involvement of cortisol, CRH and CRH-BP. General and Comparative Endocrinology 171, 203−210.

Amphibians

Donald, J.A., Trajanovska, S., 2006. A perspective on the role of natriuretic peptides in amphibian osmoregulation. General and Comparative Endocrinology 147, 47−53.

Guardabassi, A., Muccioli, G., Andreoletti, G.E., Pattono, P., Usai, P., 1991. Prolactin and interrenal hormone balance in *Xenopus laevis* adult specimens adapted to brackish water. Atti della Accademia delle Scienze di Torino 125, 55−69.

Hopkins, W.A., DuRant, S.E., 2011. Innate immunity and stress physiology of eastern hellbenders (*Cryptobranchus alleganiensis*) from two stream reaches with differing habitat quality. General and Comparative Endocrinology 174, 107−115.

Iwamuro, S., Hayashi, H., Yamashita, M., Kikuyama, S., 1991. Arginine vasotocin (AVT) and AVT-related peptide are major aldosterone-releasing factors in the bullfrog intermediate lobe. General and Comparative Endocrinology 84, 412−418.

LaForgia, V., Capaldo, A., 1992. The interrenal gland of *Triturus cristatus* after insulin administration during the annual cycle. Journal of Morphology 211, 87−93.

Lihrmann, I., Netchitailo, P., Feuilloley, M., Cantin, M., Delarue, C., Leboulenger, F., De Lean, A., Vaudry, H., 1988. Effect of atrial natriuretic factor on corticosteroid production by perifused frog interrenal slices. General and Comparative Endocrinology 71, 55−62.

Moore, I.T., Jessop, T.S., 2003. Stress, reproduction and adrenocortical modulation in amphibians and reptiles. Hormones and Behavior 43, 39−47.

Schubert, S.N., Wack, C.L., Houck, L.D., Feldhoff, P.W., Feldhoff, R.C., Woodley, S.K., 2009. Exposure to pheromones increases plasma corticosterone concentrations in a terrestrial salamander. General and Comparative Endocrinology 161, 271−275.

Tesone, A.J., Regueira, E., Canosa, L.F., Ceballos, N.R., 2012. 5α-Reductase, an enzyme regulating glucocorticoid action in the testis of *Rhinella arenarum* (Amphibia: Anura). General and Comparative Endocrinology 176, 500−506.

Wack, C.L., DuRant, S.E., Hopkins, W.A., Lovern, M.B., Feldhoff, R.C., Woodley, S.K., 2012. Elevated plasma corticosterone increases metabolic rate in a terrestrial salamander. Comparative Biochemistry and Physiology A 161, 153−158.

Wack, C.L., Lovern, M.B., Woodley, S.K., 2010. Transdermal delivery of corticosterone in terrestrial amphibians. General and Comparative Endocrinology 169, 269−275.

Zerani, M., Gobetti, A., 1991. Effects of β-endorphin and naloxone on corticosterone and cortisol release in the newt (*Triturus carnifex*): studies *in vivo* and in vitro. Journal of Endocrinology 131, 295−302.

Reptiles

Carsia, R.V., John-Alder, H.B., 2006. Natriuretic peptides are negative modulators of adrenocortical cell function of the eastern fence lizard (*Sceloporus undulatus*). General and Comparative Endocrinology 145 1457−161.

Dauphin-Villement, C., Xavier, F., 1987. Nychthemeral variations of plasma corticosteroids in captive female *Lacerta vivipara* Jacquin: influence of stress and reproductive state. General and Comparative Endocrinology 67, 292−302.

Gabe, M., 1970. The adrenal. In: Gans, C. (Ed.), Biology of the Reptilia. Morphology C, vol. 3. Academic Press, New York, pp. 263−318.

Gleeson, T.T., 1993. Plasma catecholamine and corticosterone and their *in vitro* effects on lizard skeletal muscle lactate metabolism. American Journal of Physiology 265, R632−R639.

Klukowski, M., 2011. Effects of breeding season, testosterone and ACTH on the corticosterone response of free-ranging male fence lizards (*Sceloporus undulatus*). General and Comparative Endocrinology 173, 295−302.

Mahapatra, M.S., Mahata, S.K., Maiti, B.R., 1987. Influence of age on diurnal rhythms of adrenal norepinephrine, epinephrine, and corticosterone levels in soft-shelled turtles (*Lyssemys punctata punctata*). General and Comparative Endocrinology 67, 279−281.

Palacios, M.G., Sparkman, A.M., Bronikowski, A.M., 2012. Corticosterone and pace of life in two life-history ecotypes of the garter snake *Thamnophis elegans*. General and Comparative Endocrinology 175, 443−448.

Reinhart, G.A., Zehr, J.E., 1994. Atrial natriuretic factor in the freshwater turtle *Pseudemys scripta*: a partial characterization. General and Comparative Endocrinology 96, 259−269.

Summers, C.H., 2002. Social interactions over time: implications for stress responsiveness. Integrative and Comparative Biology 42, 591–599.

Summers, C.H., Watt, M.J., Ling, T.L., Forster, G.L., Carpenter, R.E., Korzan, W.J., Lukkes, J.L., Overli, O., 2005. Glucocorticoid interaction with aggression in non-mammalian vertebrates: reciprocal action. European Journal of Pharmacology 526, 21–35.

Tokarz, R.R., Summers, C.H., 2011. Stress and reproduction in reptiles. In: Norris, D.O., Lopez, K.H. (Eds.), Hormones and Reproduction of Vertebrates. Reptiles, vol. 3. Academic Press, San Diego, CA, pp. 169–213.

Trajanovska, S., Donald, J.A., 2008. Molecular cloning of natriuretic peptides from the heart of reptiles: loss of ANP in diapsid reptiles and birds. General and Comparative Endocrinology 156, 339–346.

Birds

Crossin, G.T., Dawson, A., Phillips, R.A., Trathan, P.N., Gorman, K.B., Adlard, S., Williams, T.D., 2012. Seasonal patterns of prolactin and corticosterone secretion in an Antarctic seabird that moults during reproduction. General and Comparative Endocrinology 175, 74–81.

Gray, D.A., Schutz, H., Gerstberger, R., 1991. Interaction of atrial natriuretic factor and osmoregulatory hormone in the Peking duck. General and Comparative Endocrinology 81, 246–255.

Kocsis, J.F., McIlroy, P.J., Carsia, R.V., 1995. Atrial natriuretic peptide stimulates aldosterone production by turkey (*Meleagris gallopavo*) adrenal steroidogenic cells. General and Comparative Endocrinology 99, 364–372.

Trajanovska, S., Donald, J.A., 2008. Molecular cloning of natriuretic peptides from the heart of reptiles: loss of ANP in diapsid reptiles and birds. General and Comparative Endocrinology 156, 339–346.

Wingfield, J.C., 2005. Modulation of the adrenocortical response to acute stress in breeding birds. In: Dawson, A., Sharp, P.J. (Eds.), Functional Avian Endocrinology. Narosa Publishing House, New Delhi, pp. 225–256.

Chapter 10

The Endocrinology of Mammalian Reproduction

Reproduction is the process that perpetuates a species through evolutionary time. It includes the process of sex determination and sexual differentiation (conversion of the indifferent gonads into testes or ovaries), embryonic development and birth, sexual maturation or puberty, development of gametes, physiological and behavioral aspects of mating, fusion of gametes, and development of the resulting zygote. In addition, a period of complex parental care is intercalated between birth and sexual maturation and possibly extends longer, such as in the case of humans. Every step in this complicated reproductive process is controlled directly or is modified by bioregulators secreted within the body or by pheromones from other members of the species.

The "reproductive system" includes the complex **hypothalamus–pituitary–gonad (HPG)** axis as well as the targets of steroid hormones secreted by the gonads. Environmental factors (chemical, visual, photic, thermal, and tactile stimuli) operating through effects on neural and endocrine factors frequently determine the timing of many reproductive events. Regulators of other endocrine axes, such as the thyroid (HPT) and adrenal (HPA) axes, have important effects on reproductive events, too (see Chapters 6 and 8).

Reproductive mechanisms are of central importance to survival of a species and hence are under intense evolutionary selection. Consequently, the reproductive system has been highly responsive to selective forces throughout the long evolutionary history of vertebrates. Because the same selective forces in the environment act upon all animal species, it should not come as a surprise that similar mechanisms have evolved in diverse vertebrates to achieve reproductive success in the face of similar pressures. Progressive "improvements" in the endocrine mechanisms regulating reproduction may not show progressive "development" from fishes to mammals, but instead we see specific adaptations to solve common environmental problems appearing in diverse groups of vertebrates. A case in point would be the achievement of viviparity in all but two extant vertebrate classes, the jawless fishes (Agnatha) and birds (Aves), as specific adaptations that, when coupled with varying degrees of parental care, result in a greater percentage survival of a small number of offspring. Viviparity represents only one solution, however, to similar selective pressures that confront all species. In birds, extensive and complicated parental behavior serves a similar evolutionary role as viviparity does in mammals to ensure reproductive success. In spite of the problems of environmental adaptations that tend to confuse evolutionary relationships, there remain numerous conservative features in the regulatory mechanisms of reproductive biology, and it is these that are emphasized in this chapter and in Chapter 11, where reproduction of non-mammalian vertebrates is discussed.

I. GENERAL FEATURES OF MAMMALIAN REPRODUCTION

Mammals can be separated into three, distinct taxonomic groups: Prototheria (monotremes), Metatheria (marsupials), and Eutheria (placentals). All possess **mammary glands**, specialized skin glands that are employed in secretion of milk to feed their young. The egg-laying monotremes comprise the most ancient group of mammals of which only a handful of species are extant. The marsupials or pouched mammals (**marsupium** = pouch) are confined mostly to Australia with a few species in North, Central, and South America. The marsupials are evolutionarily intermediate between the prototherians and the placental mammals, and there are about 230 extant species of marsupials. Their survival in Australia is due largely to the late arrival of eutherian mammals to that continent, whereas competition with eutherians has severely limited them in North and South America. The placental mammals are the dominant group of living mammals in number of species, distribution over the Earth, and abundance. The **placenta** is a specialized structure that develops through interactions of zygote-derived extra-embryonic tissues and some maternal uterine tissues. It provides nutritional, respiratory, excretory, and endocrine support for the offspring developing within the uterus. Although a placenta is present in marsupials, it is very short-lived and the marsupial fetus soon exits the uterus via the vagina and finds its way into the pouch. Most development for this

Vertebrate Endocrinology. http://dx.doi.org/10.1016/B978-0-12-394815-1.00010-0

exteriorized fetus takes place within the pouch rather than *in utero*, with the mammary glands providing the nutrition for continued development.

Reproduction is influenced by environmental variables that are perceived and integrated by the **central nervous system**. Communication between the nervous system and the reproductive system is achieved through the HPG axis, which coordinates specific gonadal events through regulation by circulating **gonadotropins (GTHs)**. This gonadal axis is modified by other systems, especially the HPT axis (Chapter 6) and the HPA axis (Chapter 8). Factors influencing GTH release and hence gonadal functions were discussed in Chapter 4 and are summarized only briefly here (see also Table 10-1).

Reproductive events in mammals are controlled through the release of **luteinizing hormone (LH)** and **follicle-stimulating hormone (FSH)** from the adenohypophysis under the control of a single hypothalamic hormone, **gonadotropin-releasing hormone (GnRH)**. Pulsatile release is an innate feature of GnRH neurons. The **tonic center**, located in the hypothalamus of both males and females, maintains a relatively constant pattern of pulsatile release of GnRH and produces rather static circulating levels of both LH and FSH. The **surge center** found only in the brains of females is responsible for the midcycle **LH surge** observed in mature females in response to elevated estrogen. In some species, these neural centers are distinctly separated in females (e.g., the rat), and in others there is no obvious anatomical separation (e.g., human). Although the pulsatile rhythm of GnRH secretion is ultimately due to calcium cycling within GnRH neurons and autocrine feedback on GnRH neurons by GnRH (see Chapter 4), oscillations in GnRH secretion can be stimulated by catecholaminergic neurons such as **norepinephrine (NE)** and inhibited by **endogenous opioid peptides (EOPs)**. Positive feedback of estrogens on GnRH release from the surge center in females is mediated by **GABA (γ-amino butyric acid)** neurons. GnRH secretion also is modulated by **kisspeptin** acting via **GPR54** receptors located on GnRH neurons (see Chapter 4), whereas **gonadotropin inhibitory hormone (GnIH)** acts directly on gonadotropes to inhibit FSH and LH secretion (Chapter 4).

Gonadotropins stimulate gamete maturation in males and females as well as steroidogenesis and release of **estrogens**, **androgens**, and **progestogens** (= progestins) into the general circulation. **Gametogenesis** (**oogenesis** in females and **spermatogenesis** in males) is controlled primarily by FSH, whereas LH is mainly responsible for controlling androgen synthesis as well as release of gametes in both sexes. Estrogen synthesis is influenced by FSH in both males and females. In some species, **prolactin (PRL)** may play a role in regulating ovarian steroidogenesis. The details of these events are discussed later. GTHs may be responsible for synthesis by the gonads of a variety of paracrine or autocrine factors that play roles in steroidogenesis or gametogenesis.

Induction of ovulation and formation of corpora lutea from the remnants of the ovulated follicle in the ovaries are due to LH. Surges in both LH and FSH occur in response to elevated GnRH prior to ovulation, but the magnitude of the LH surge greatly exceeds the FSH surge. LH release is enhanced selectively by the neuropeptide **galanin**, a peptide that is co-released with GnRH just prior to the midcycle LH surge. Galanin has no effect on FSH release. Under the influence of FSH, the ovaries secrete a peptide called **inhibin** that selectively blocks FSH release from the pituitary and contributes to the reduced FSH surge. In **polyestrous** species (poly = many; i.e., a very short diestrous phase), the importance of the FSH surge may be related to initiation of follicle development for the next cycle.

The gonadal steroids, secreted as a result of the action of GTHs on special cells in the ovaries and testes, control differentiation and maintenance of many **primary sexual characters** (such as the uterus) and **secondary sexual characters** (such as muscle development and beard growth in men). These gonadal actions were recognized hundreds of years ago by the Chinese, who used gonadal (and placental) preparations routinely to treat conditions ranging from impotence in men to the inability of a woman to bear sons. The hypothalamic centers regulating GnRH release are sensitive to circulating steroids that generally produce negative feedback on GnRH release, the exception to this pattern being the positive feedback effect by estrogens on release of GnRH from the surge center and subsequent stimulation of LH release from the pituitary.

Several other hormones are involved in mammalian reproduction in addition to those of the HPG axis. The HPT and HPA axes as well as PRL have already been mentioned. In addition, nonapeptides from the pars nervosa influence reproductive events including courtship, birth, and parental behavior. The placenta of eutherian mammals assumes an endocrine role in pregnant females, producing steroids (primarily estrogens and progesterone) and polypeptide hormones (e.g., **chorionic gonadotropins, CG; chorionic somatomammotropin, CS; corticotropin-releasing hormone, CRH;** GnRH; PRL). In addition, the endocrine glands of the fetus may influence reproductive events—for example, contribution of the adrenal cortex to steroidogenesis by the placenta. The **pineal gland** may be a modulator of photoperiod and a source of antigonadotropic factors that influence gonadal function and prevent early onset of puberty (see Chapter 4). Finally, there are numerous reports of chemical agents termed **pheromones** that are produced by one sex to influence reproductive physiology, behavior, or both in the opposite sex.

TABLE 10-1 Summary of Generalized Hormone Actions in Mammalian Reproduction

	Action in:	
Hormone	**Females**	**Males**
Kisspeptin	Enhances GnRH secretion	Enhances GnRH secretion
GnIH	Inhibits LH/FSH secretion	Inhibits LH/FSH secretion
GnRH	Stimulates FSH and LH secretion	Stimulates FSH and LH secretion
FSH	Initiates follicle growth; conversion of androgen to estrogen; synthesis of inhibin, $P450_{aro}$	Initiates spermatogenesis; secretic of androgen-binding protein, STP, and inhibin by Sertoli cell conversion of androgen to estrogen by Sertoli cell
LH	Androgen synthesis; ovulation; formation of corpus luteum from granulosa; secretion of progesterone initiated in corpus luteum	Androgen secretion by interstitial cell (Leydig)
Prolactin	Synthesis of milk	Stimulates certain sex accessory structures (with androgen)
Oxytocin	Contraction of uterine smooth muscle; menstrual sloughing; birth; orgasm; milk ejection from mammary	Ejaculation of sperm; orgasm
Androgens	Precursors for estrogen synthesis; stimulates sexual behavior	Complete FSH-initiated spermatogenesis; stimulate prostate gland, other sex accessory structures; stimulate secondary sexual characters, such as beard growth in man
Estrogens	Stimulate proliferation of endometrium; induces LH surge; sensitize uterus to oxytocin; negative feedback on pituitary release; may be primate luteolytic factor (estrone); may induce PRL surge; maintain pregnancy; involved in birth	Converted from androgens; induces male hypothalamus; stimulates sexual behavior
Progesterone	Maintains secretory phase of uterus; inhibits release of gonadotropins from adenohypophysis; maintains pregnancy	Facilitates/inhibits sexual behaviors depending on physiological context; influences brain development
Prostaglandins	Causes corpus luteum to degenerate at end of luteal phase in some animals: may be involved in birth initiation (induction of labor)	Ejaculation
Relaxin	Softens pelvic ligaments and cervix; possible role in lactation	Enhances sperm motility (secreted by prostate)
Placental CRH	Stimulates fetal HPA axis	Stimulates fetal HPA axis
Chorionic gonadotropin	Stimulates corpus luteum to produce progesterone	Not present in males
Chorionic somatomammotropin	Stimulates mammary to synthesize milk during late pregnancy; growth hormone-like (somatotropin) actions on metabolism	Not present in males
Inhibin (Sertoli cell factor, folliculostatin)	Inhibits FSH secretion from pituitary	Inhibits FSH secretion from pituitary

A. Embryogenesis of Gonads and Their Accessory Ducts

Primary sexual characters include the vagina, uterus, and oviducts of the female and the penis, vasa deferentia, seminal vesicles, and prostate gland of the male. Secondary sexual characters are often dependent on gonadal hormones and usually enhance mating success but are not necessarily required for physically mating and producing offspring.

1. The Gonads

The paired gonadal primordia arise from the intermediate mesoderm of the mammalian embryo as a genital ridge on either side of the midline in close association with the transitory mesonephric kidney of the embryo. Numerous derivatives of the mesonephric kidney and its duct system are retained as functional portions of the adult reproductive system, although the bulk of the mesonephric kidney degenerates. A gonadal primordium consists of an outer **cortex** derived from peritoneum and an inner **medulla** (Figures 10-1 and 10-2). Germ cells do not arise within the gonadal primordium itself but migrate from their site of origin in the yolk sac endoderm to either cortex (female) or medulla (male) depending upon the genetic sex (Figure 10-2). The basic pattern of germ cell migration is evolutionarily conserved from fruit flies to humans and requires a complex interplay between (1) guidance signals and extracellular matrix attachment proteins that ensure directed migration of the germs cells to the genital ridge mesoderm, and (2) a host of chemical signals involved in alignment of the germ cells within the gonad and coalescence of the developing gonad. Some of the genes involved in regulating primordial germ cell differentiation, migration, and meiosis are listed in Table 10-2.

Initially, the medullary component in males and females differentiates into **primary sex cords**. Differentiation of the primary sex cords into **seminiferous cords** and regression of the cortex result in a testis. Each testis consists of seminiferous tubules derived from the primary sex cords. The germ cells migrate into the seminiferous tubules, give rise to spermatogonia, and eventually produce sperm. The **Sertoli** or **sustentacular cells** support sperm development. Steroidogenic **interstitial cells** or **Leydig cells** are located between the seminiferous tubules. These interstitial cells arise from medullary tissue surrounding the primary sex cords and become sources of androgens.

In females, the primary sex cords degenerate, and **secondary sex cords** differentiate from the cortical region. These secondary sex cords become the definitive ovary. In the ovary, the germ cells give rise to **oogonia**, which soon enter meiosis to form **primary oocytes**. The ovaries contain **follicles** that consist of one or more layers of **follicular cells** surrounding a primary oocyte.

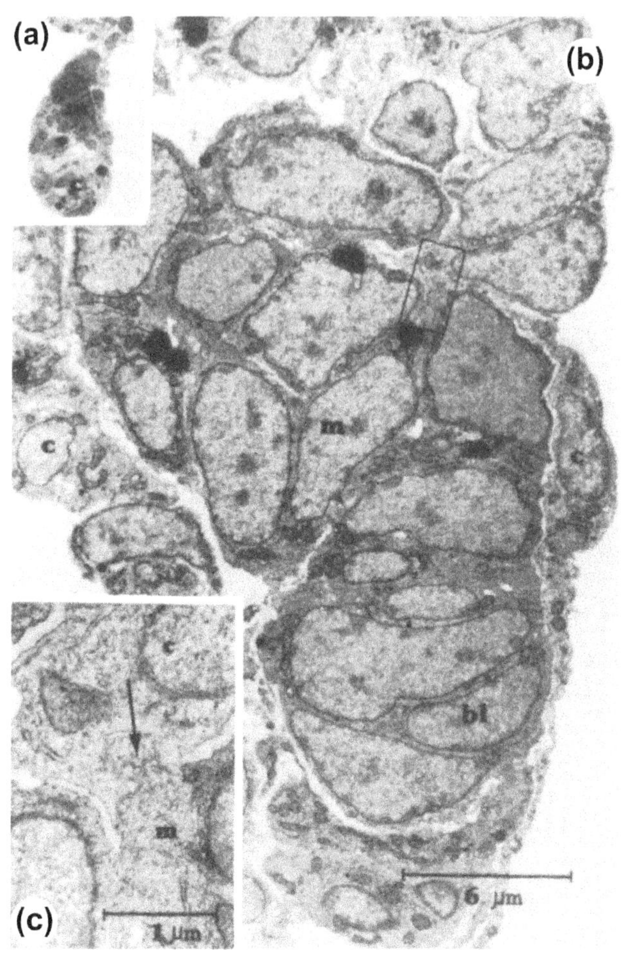

FIGURE 10-1 **Undifferentiated gonad.** Section of gonad from 25-mm tadpole of *Rana pipiens* showing cortical (c) and medullary (m) cells separated by a basal lamina (bl = basement membrane). (a) total gonad (upper left); (b) enlargement; (c) further enlargement showing contact between cortical and medullary cells (arrow). *(Reprinted with permission from Merchant-Larios, M., in "The Vertebrate Ovary" (R.E. Jones, Ed.), Plenum, New York, 1978, pp. 47–81.)*

2. Accessory Ducts

In males, the central portion of each differentiating testis forms a network of tubules, known as the **rete testis**, that do not contain seminiferous elements. The rete testis forms a connection between the seminiferous tubules and a surviving portion of the primitive mesonephric kidney duct called the **wolffian duct**, which, under the influence of testosterone, differentiates into the **vas deferens** and conducts sperm from the testis to the urethra. Most of the mesonephric kidney in mammals degenerates, with the exception of some of the anterior mesonephric kidney tubules (see Box 10A). In the presence of testosterone, this tissue together with a portion of the wolffian duct forms two glandular structures, the **epididymis** and the **seminal vesicle** (Figures 10-2 and 10-3).

A second pair of longitudinal ducts develops in the embryo from the mesial wall of each wolffian duct and lie parallel to them. These structures are known as the **ducts**. In genetic females, the müllerian ducts develop into the oviducts, uterus and the upper part of the vagina (Figure 10-3), usually fusing together to form a common vagina and, in some species, a single uterus as well. The wolffian ducts degenerate in female mammals. In males, it is the müllerian ducts that are suppressed in favor of wolffian duct development.

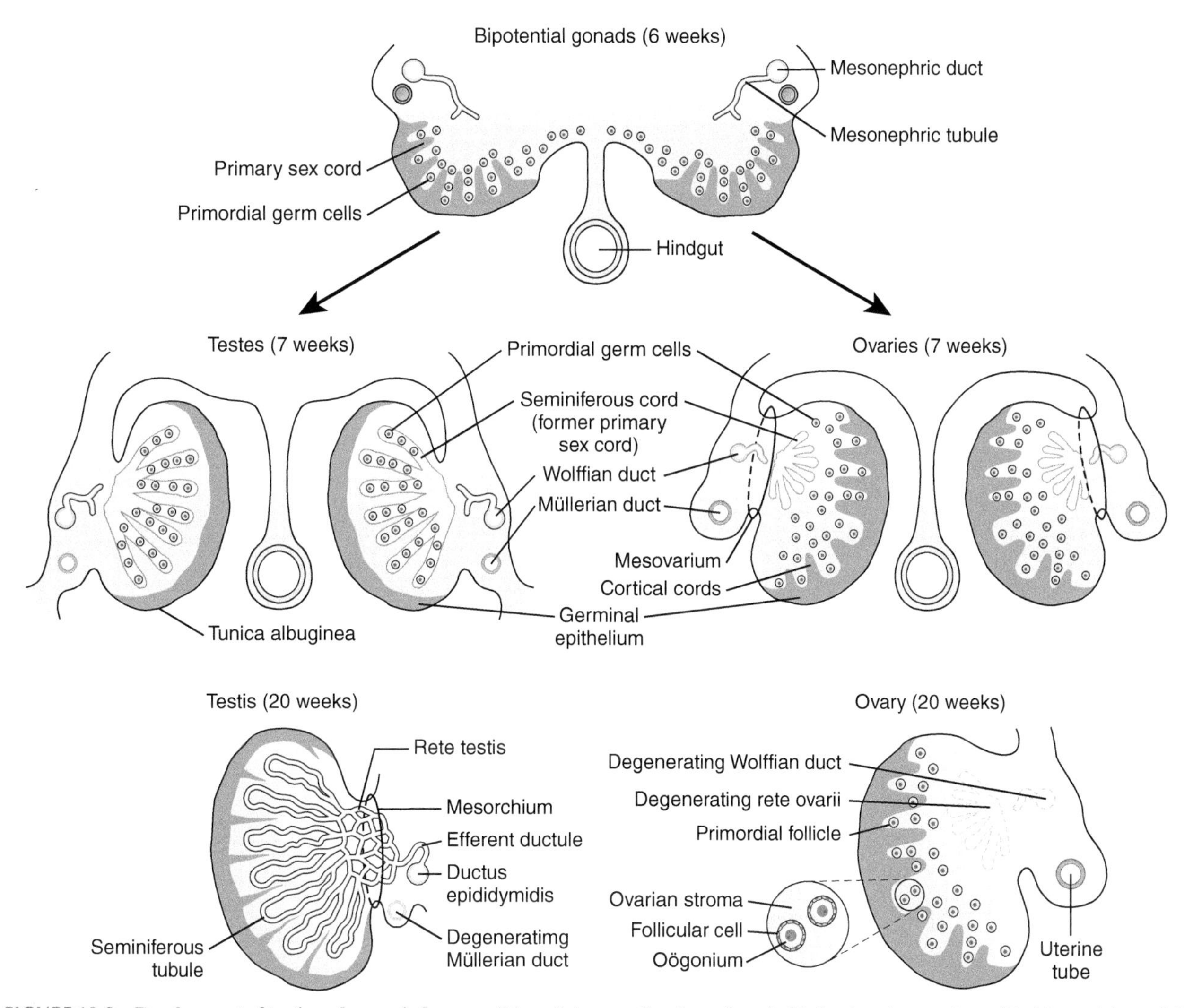

FIGURE 10-2 **Development of testis and ovary in humans.** Primordial germ cells migrate from the hindgut into the mesoderm of the bipotential gonad. In the male, the cortical tissue (orange) degenerates and the medullary tissue develops into the testis cords, which give rise to the seminiferous tubules including the Sertoli cells. Mesonephric tubules give rise to the intratesticular ducts such as the rete testis and the efferent ducts and vas deferens. In the female, the medullary cords degenerate, and the cortical cords (orange) give rise to an ovary. Some mesonephric elements remain in the female as well. The vasa deferentia are retained in amphibians but eventually they degenerate in reptiles, birds, and mammals in which the ureters develop to drain the metanephric kidneys (not found in anamniotes). *(Adapted with permission from Paxton, M., "Endocrinology Biological and Medical Perspectives," William C. Brown, Dubuque, IA, 1986.)*

Müllerian-inhibiting substance (MIS) was first proposed by Alfred Jost in the 1940s to explain the inhibitory effect of the testes on development of müllerian ducts in rabbit embryos. It also has been called the **anti-Müllerian hormone**, or AMH. AMH is a dimeric glycoprotein encoded by the *amh* gene that acts via a membrane serine/threonine kinase type-II receptor located in the gonads and in connective tissue near the müllerian ducts. Implantation of a testis into a female embryo results in sufficient AMH secretion to prevent development of the müllerian ducts. AMH not only blocks müllerian duct development but also is capable of inhibiting growth of tumors from ovaries and müllerian duct derivatives. It appears that AMH acts cooperatively with testosterone in producing these effects on the müllerian ducts. The ovary also makes AMH, but the müllerian ducts are protected by local estradiol secreted by the ovary.

Maleness in eutherian mammals is dependent upon secretion of androgens from the testis. In the absence of androgens or androgen receptors the male animal (genotype XY) will develop a female phenotype. Similarly, exposure of developing males to estrogens will result in female phenotype development to a degree proportional to the amount of estrogen and the timing of the exposure (see Table 10-3). Conversely, treatment of newborn females with androgens destroys the cyclical secretory pattern of the HPG axis and replaces it with a noncyclical or tonic pattern like that of males (see Box 10B). Becoming a male mammal, then, involves overcoming the basic tendency for mammalian embryos to develop as females. A gene

TABLE 10-2 Some Genes Involved in Primordial Germ Cell (PGC) Induction, Specification, Migration, and Meiosis

Gene	Name	Role
bmp 2/4/8	Bone morphogenic protein	Induction and competence of PGCs
prdm1 and *prdm14/blimp1* (mouse)	PR domain zinc finger protein 1	Required for PGC specification
pou5f1	POU domain class 5, transcription factor 1	PGC marker and specification
vasa; ddx4	DEAD box family of ATP-dependent RNA helicases	PGC marker and specification
nanos3	Nanos homolog 3	Migration, entry into mesoderm
dnd1	Dead end homolog 1	Migration, entry into mesoderm
kit	Mast/stem cell growth factor receptor (SCFR); proto-oncogene c-kit	Migration, entry into mesoderm
dazl	Deleted in azoospermia-like	Meiosis competency

BOX 10A Vertebrate Kidney Evolution

The first kidney in vertebrates was the **pronephros**. It appears in vertebrates only as a transitory structure during early development, and only remnants of the pronephros remain as the anteriormost part of the fish kidney that is known as the head kidney. This structure consists largely of lymphoid tissue as well as adrenocortical cells. The duct that drained the pronephros to the cloaca is retained. It is called the **pronephric** or **archinephric duct**. Posterior to the pronephros develops a second kidney, the **mesonephros**, which co-opts the pronephric duct as its conduit to drain urine to the cloaca. Developmentally, this **mesonephric duct** is called the wolffian duct in the embryo. The mesonephros becomes the definitive kidney of fishes and amphibians, where it is often designated as the **opistonephric kidney**. The wolffian duct is retained in both male and female fishes as a kidney duct and may also be used as a sperm duct in males of elasmobranchs as well as in amphibians. In amniote vertebrates, a third kidney that develops posterior to the opistonephros is called the **metanephric kidney**. A new urinary duct, the **ureter**, develops to connect the metanephric kidney to the urogenital sinus. The wolffian duct is retained as the epididymis and the vas deferens in males. A portion of the wolffian duct also gives rise to the seminal vesicles that retain a connection to the vas deferens. In addition, some of the mesonephric kidney tubules form the rete testis, which connects the seminiferous tubules of the testes to the epididymis. In female amniotes, the wolffian duct degenerates. Some mesonephric tubules are retained in females and become associated with the ovaries. In elasmobranchs, amphibians, and amniotes, a pair of müllerian ducts develops adjacent to the wolffian ducts. In females, these ducts give rise to the oviducts and uteri but usually degenerate in males. The utricle of the prostate gland in male mammals actually is a müllerian remnant. It is the stimulation of this female remnant by estrogens that is responsible for most prostate cancer.

seemingly responsible for male sex determination called ***sry*** (sex-determining region of Y chromosome) has been localized on the short arm of the Y chromosome that is characteristic of genetic males. In mice, the *sry* gene is activated in gonads of genetic males before they begin to differentiate into testes. Insertion of the *sry* gene into XX mice followed by its activation leads to formation of male-specific structures and regression of female ducts. The activated gonad secretes AMH, which causes regression of the müllerian ducts. The *sry* gene produces a factor called **testis determining factor (TDF)** (Figure 10-3) that activates the *amh* gene. Androgens secreted by the transformed gonad cause male-like differentiation of the external genitalia and the wolffian ducts as well as changes in the hypothalamus to suppress development of the surge center. This establishes the tonic secretory pattern for GnRH and GTHs that characterizes males. Studies with estrogen receptor knockout (ERKO) mice verify that defeminization of the male brain requires conversion of androgens to estradiol. Genetically male ERKO mice will exhibit female behavior, whereas wild-type males do not.

Although the female has been called the default sex in mammals, becoming a female is not just the absence of androgens. For example, studies have shown specific genes are required to be expressed in order for the ovary to form and that estrogens are necessary for development of the female difference in the corpus callosum of the brain.

Androgens and estrogens may alter basic traits through what are termed **organizational effects**. Stimulation of the development of male genitalia by androgens is an example of an organizational effect. Organizational effects are permanent and cannot be reversed later by exposure to other gonadal steroids. In contrast, **activational effects** can be induced by

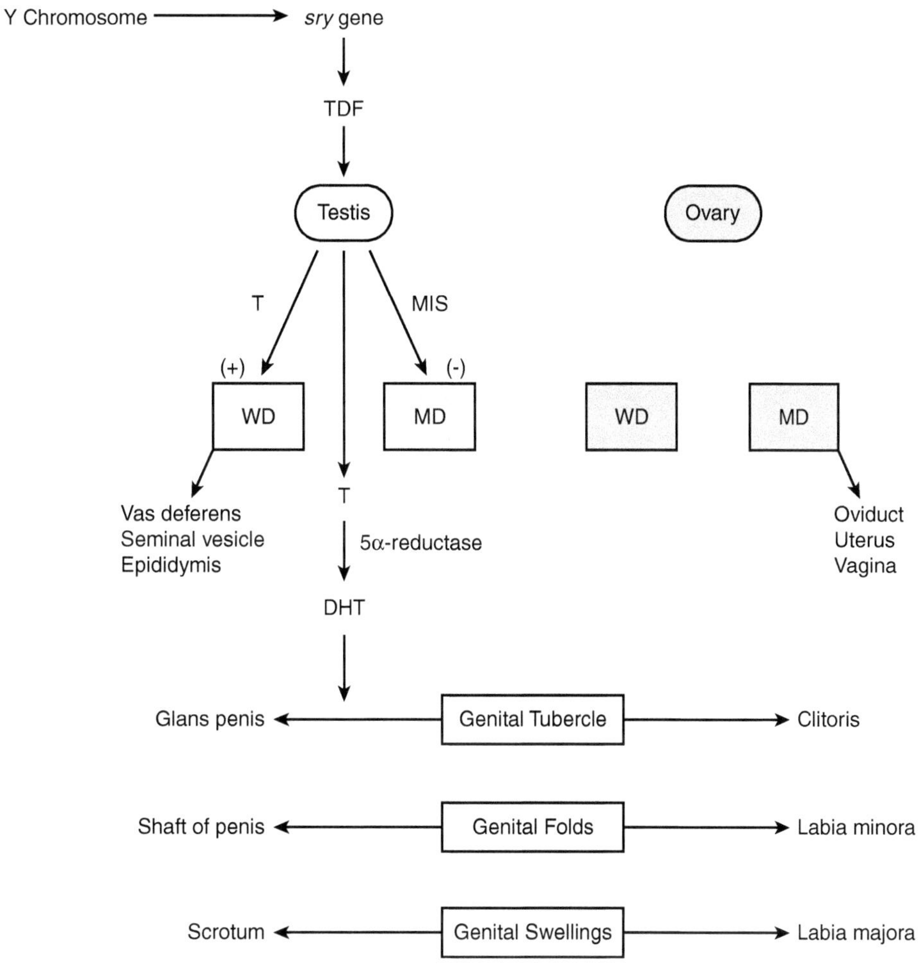

FIGURE 10-3 **Patterns of development for ducts and genitalia.** In males, the testis secretes testosterone (T), which stimulates differentiation of wolffian ducts, and müllerian-inhibitory substance (MIS), which causes regression of müllerian ducts. Dihydrotestosterone (DHT) is either produced by the testes or converted from T in the genital tubercle, genital folds, or genital swellings, causing them to differentiate in the male direction. Estradiol from the ovary prevents MIS (also secreted by ovary) from causing müllerian duct regression, and the absence of sufficient androgens determines the fate of the other structures. Abbreviations: *sry*, sex-determining region of the Y chromosome; TDF, testis determining factor.

TABLE 10-3 **Critical Periods for Sexual Differentiation of the Brain in Mammals**

Species	Gestation Period (days)	Critical Period (days)
Hamster	16	16–21
Laboratory rat	21–22	18–28
Laboratory mouse	18–22	20
Guinea pig	68	30–35
Human	270	84–126

gonadal steroids—for example, by inducing a specific behavior in adults. The type of behavior induced depends on the steroid applied, not on the genetic sex of the individual.

II. REPRODUCTION IN MONOTREMES AND MARSUPIALS

A. Monotremes

The monotremes have retained the reptilian feature of laying eggs but have mammary glands for feeding the young after they hatch. Unlike the eggs of reptiles and birds, monotreme eggs contain little yolk, and embryonic nutrition is supplied through uterine secretions that pass through the porous leathery shell as the egg passes through the female genital tract. Upon hatching, the new monotreme appears as a tiny, fetus-like creature with only a few well-developed features that enable it to attach itself to its mother's mammary gland and obtain nourishment. The mammary glands of monotremes lack external teats; consequently, milk is secreted onto a special common area, the **areola**. The newly hatched monotreme must attach itself to its mother with its forelimbs and suck or lick milk from the areola. Development continues and the offspring, because of its primitive nature, is considered an exteriorized fetus until it has completed development.

1. Monotreme Reproductive Patterns

Seasonally breeding monotremes are difficult to study because they are largely nocturnal animals and are carefully protected in nature. Furthermore, they do not breed readily in captivity. The duckbill platypus lays its eggs in a nest, but the female spiny echidna places her freshly laid eggs in a brood pouch that develops seasonally on her ventral surface. Formation of the pouch is probably dependent upon estrogens, although no experimental data are available to confirm this. The eggs develop and hatch within the pouch, and mom remains in her burrow while they develop. The brood pouch regresses after the breeding season. Echidnas enter a period of torpor and inactivity prior to the breeding season when they arouse to mate. Attempts to measure

BOX 10B Gene Regulation of Gonadal Development

Gonadal phenotype in mammals ultimately is determined by the presence or absence of the Y chromosome and *sry* gene. However, even before the molecular switches governing testis formation are called into play there are cellular events that must occur for the bipotential gonad precursor to form from mesoderm. To simplify the seemingly complex process underlying sexual fate it is helpful to divide sexual development into the two major events that bracket gonad formation: **sexual determination**, the mechanisms involved in testis or ovary formation, and **sexual differentiation**, which are the events requiring normal gonadal hormone secretion resulting in the overall male or female phenotype (see Figure 10-3). In order for normal gonadal development to occur, many genes are believed to be involved in formation of the bipotential gonad, the testis, and the ovary. Identification of the genes regulating gonadal development is important not only for understanding subsequent problems with sexual differentiation but also for understanding the full spectrum of **disorders of sexual development** (**DSDs**) in which gonadal phenotype is atypical of the normal male or female structure. Understanding how these genes and their protein products interact and are modulated by endocrine-disrupting chemicals obviously will lead to a better understanding of the genetic and epigenetic basis for directing the development of the ovary or testis. A simplified scheme for gene regulation of gonadal development is shown in Box Figure 10B-1. Links between defects in these genes and known DSDs are summarized later in this chapter.

The fact that the absence of *sry* expression leads to ovary development may lead one to conclude that ovary development is the default *bauplan*. We now know this to be inaccurate, as both testis and ovary formation requires a cascade of gene expression. *dax-1*, a gene encoding an orphan member of the nuclear hormone receptor family, is a gene that was initially shown to be critical in ovary formation (Box Figure 10-B1) and at the same time function in XY cells. Its role in males as initially proposed is to antagonize *sry* function (Box Figure 10-B1). More recent evidence suggests a more complicated role for *dax-1* in gonad formation, with *dax-1* expression during critical windows of development (7 weeks gestation in humans). If *dax-1* expression exceeds or falls below normal levels during the critical period, problems in testis formation may occur (Box Figure 10B-2).

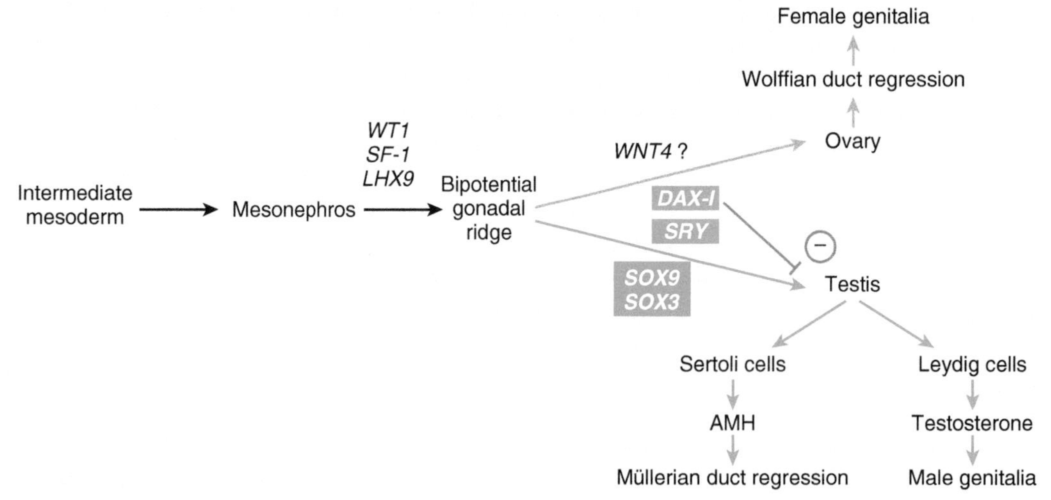

BOX FIGURE 10B-1 **Genes regulating gonadal development in humans.** Abbreviations: AMH, anti-müllerian hormone; *DAX-1*, dosage-sensitive sex reversal, adrenal hypoplasia critical region, on chromosome X, gene 1; *LHX9*, LIM homeobox 9; *SF-1*, steroidogenic factor 1; *SOX3*, sex-determining region Y box 3; *SOX9*, sex-determining region Y box 9; *SRY*, sex-determining region of Y chromosome; *WNT4*, wingless-type MMTV integration site family, member 4; *WT1*, Wilms tumor 1. *(Compiled from many sources.)*

BOX 10B Gene Regulation of Gonadal Development (Continued)

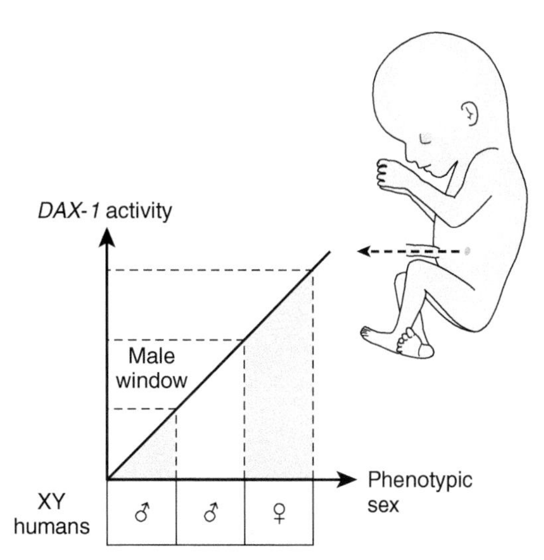

BOX FIGURE 10B-2 **Testes formation requires *DAX-1* activity during a critical period of development.** The *SRY* gene initiates testes development by the seventh week of pregnancy in humans. Full development of the testes requires a set level of *DAX-1* activity during this period (blue); *DAX-1* activity that is too high or too low will result in abnormal testes development in XY individuals. *(Reprinted with permission from Ludbrook, L.M. and Harley, V.R.,* Trends in Endocrinology & Metabolism, ***15****,116–121, 2004.)*

circulating estrogens by radioimmunoassay in platypus and echidnas have been unsuccessful as plasma levels are below the detectable level (<1 ng/mL). Examination of steroid metabolite levels in feces employed with chromatographic and mass spectrophotometry methods (see Chapter 2) would provide the sensitive noninvasive approach needed for these species. Measurable progesterone levels in female echidnas rise after mating as the pouch undergoes development and peak just prior to the appearance of an egg in the pouch (Figure 10-4). When males are in breeding groups with females, they exhibit elevated testosterone levels (Figure 10-4).

B. Marsupials

The marsupial placenta is rather primitive and apparently has no major endocrine function when compared to the eutherian placenta. The period of pregnancy or gestation is very short in marsupials (Table 10-4; Figure 10-5), and the young marsupial is born in an extremely immature condition. For example, among the macropodid marsupials (kangaroos, wallabies), the **joey** must find its way essentially unaided to the mother's pouch, where it permanently attaches to the nipple of a mammary gland. The attached joey, like the newly hatched monotreme, continues its development as an exteriorized fetus. After a long period of pouch development (about 200 days in the red kangaroo), the young marsupial disengages itself from the teat and ventures outside of the pouch, returning first at regular and later at irregular intervals for milk.

1. Marsupial Reproductive Patterns

Proestrus in females is characterized by follicular enlargement, estrogen-dependent uterine proliferation and enlargement of elements of the vaginal complex. Peak uterine and vaginal development coincides with estrus and copulation. Ovulation occurs spontaneously one to several days after the onset of estrus, and postovulatory follicles transform into corpora lutea that maintain a short secretory uterine condition. Progesterone continues the secretory uterine phase in castrates and is undoubtedly the hormone responsible for maintaining gestation as it is in eutherian mammals (Figure 10-5). In marsupials, estradiol is the main circulating estrogen and in males testosterone is the main circulating androgen (Table 10-5).

The luteal phase is the same in mated and unmated females, and no "pregnancy-recognition signal" is necessary. Pregnancy is very short (Table 10-4) and does not affect ovarian function. Both pregnant and unmated females return to proestrus at about the same time in most species. Equivalent mammary gland development occurs during post-estrus in both pregnant and non-pregnant females, and newborn foster young will develop normally if attached to virgin or non-lactating females at the equivalent post-estrous state in relation to the time of parturition or

BOX 10C Changes in Sexual Differentiation Caused by Exposure to Gonadal Steroids

Exposure of developing mammals to external (exogenous) sources of either androgens or estrogens can alter the sexual phenotype regardless of the genetic sex. The most elegant demonstration of the subtle effects of exposure to exogenous steroids was that of Frederick vom Saal, who observed that the position of the mouse embryo *in utero* could determine anatomical, physiological, and behavioral traits in the offspring (Box Figure 10-C1). Thus, a genetic female that developed between two males could be influenced by male hormones. When examined as newborns or adults, such females exhibited male traits (see Box Figure 10-C1A). Similarly, a male developing between two females will later exhibit some degree of feminization (see Box Figure 10-C1B). In cattle, when male and female twins share a common blood supply, the female will be masculinized to such a degree that it will be born as an intersex incapable of reproduction (called a *freemartin*). Similarly, recent studies of human dizygotic twins of opposite sexes provide evidence for masculinization of the hearing apparatus and behaviors in the female presumably as a result of *in utero* exposure to androgens.

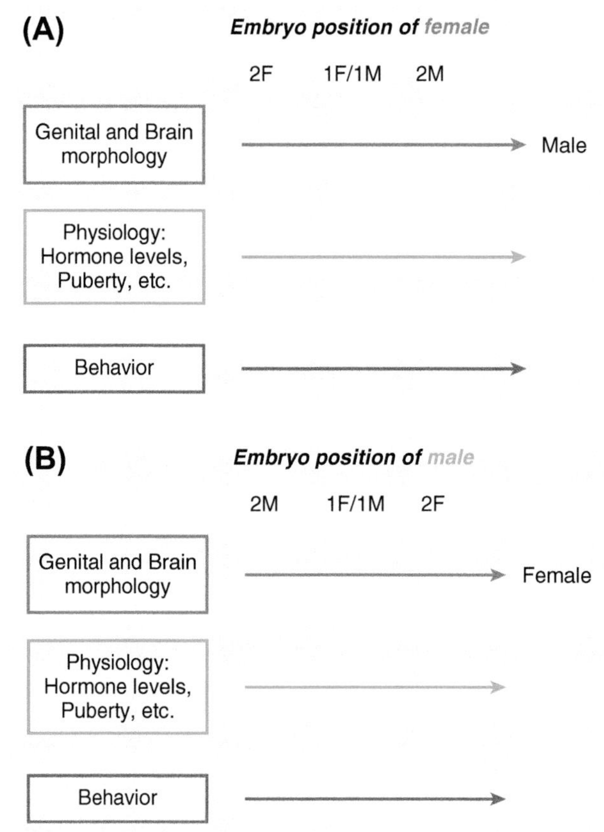

BOX FIGURE 10-C1 **The impact of position of embryos in utero on morphology, physiology and behavior.** A female that develops between 2 females vs. between 1 female and 1 male vs. between 2 males shows a progressive tendency toward maleness as a newborn and an adult (Figure 10C-1A). Similarly, a male developing between 2 males vs. between 1 male and 1 female vs. between 2 females shows a progressive tendency in femaleness (Figure 10C-1B). *(After the work of Dr. Fredrick vom Saal and associates.)*

cycle cessation. Circulating progesterone and urinary pregnanediol levels are similar in pregnant, unmated, and luteal-phase females. That there are no endocrine differences between the pregnant and non-pregnant post-estrous phase supports the conclusion that the marsupial placenta is not an endocrine organ.

Birth in wallabies involves an increase in **relaxin**, an insulin-like hormone discovered in eutherian mammals, and a reduction in progesterone. Relaxin is thought to soften the cervix of the uterus to ease the birth process as it does in eutherians (see ahead). Increases in prostaglandins and upregulation of **mesotocin (MST)** and **oxytocin (OXY)** receptors in the uterus aid in birth. Uterine contractions are probably initiated by MST and/or OXY. Fetal glucocorticoids are also elevated at the end of gestation and may be involved in birth events as described for eutherian mammals (see ahead).

Lactation is somewhat unique in marsupials. Marsupials can simultaneously produce two kinds of milk. As in eutherian mammals, the first milk differs markedly in

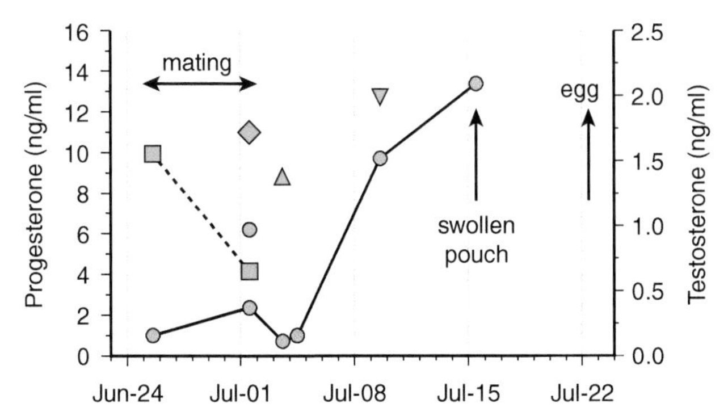

FIGURE 10-4 **Progesterone and egg laying in the echidna.** Progesterone levels (green circles) rise following mating and prior to the appearance of an egg. Testosterone levels in five attendant males are indicated by the blue symbols. *(Adapted with permission from Nicol, S. et al.,* General and Comparative Endocrinology, ***144****, 204–210, 2005.)*

composition from that produced later during lactation. However, macropodid marsupials, which may have both a newborn joey and one that has already detached itself from a teat, will produce early and late milk simultaneously in the respective glands. The developmental state of a particular mammary gland would seem to be independent of endocrine conditions and strongly influenced by external conditions—that is, the joey. Furthermore, some marsupials (i.e., kangaroos and wallabies) exhibit a **lactating diapause** that inhibits implantation of a developing embryo (see Box 10F).

A number of rodent-like marsupials have a reproductive cycle in which all of the males die after a single breeding. For example, in the brown antechinus (*Antechinus stuartii*), annual reproduction is controlled strictly by photoperiod. Mating is restricted to the same two-week period each year, with the males exhibiting extreme aggression to one another and mating vigorously and frequently with females. This mating frenzy is accompanied by reduced levels of **corticosteroid binding globulin (CBG)** (see Chapter 8) and elevated cortisol secretion in the males that leads to a spike in free or available cortisol and subsequent neural, gastrointestinal, and kidney degeneration and death. Interestingly, castration of the males prevents the reduction of CBG and increases survival. The females generally survive breeding and rear their young with no help from the departed males. In years of good environmental conditions, the females may live to breed a second year.

A major developmental difference between eutherians and marsupials has had a profound influence on reproductive patterns in the latter group. marsupials exhibit a primitive reptilian pattern of wolffian and müllerian duct origins. Instead of developing medially to the kidneys and ureters as in eutherians, these ducts develop laterally. Consequently, it is not possible for left and right Müllerian ducts of females to fuse in the midline without placing considerable strain on the ureters. It is believed that the short gestation period of marsupials (Figure 10-5) is

TABLE 10-4 Comparison of Length of Estrous Cycle, Gestation Period, and Ratio of Body Weight of Neonate to Body Weight of Mother in Metatherian and Eutherian Mammals

Species	Length of Estrous Cycle[a](days)	Length of Gestation Period(days)	Neonate:Mother Body Weight Ratio
Eutheria			
Rat	4–5	22	—
Sheep	16	148	1:14
Metatheria			
Virginia opossum	29	12	1:8,300
Long-nosed bandicoot	26	12	1:4,250
Brush possum	26	17	1:7,250
Dama wallaby	30	29	1:10,000
Swamp wallaby	31	37	—
Red kangaroo	35	33	1:33,400
Western gray kangaroo	35	30	—

[a] *The estrous cycles are similar to the gestation period in metatherians and may not require a pregnancy-recognition mechanism.*
Adapted with permission from Sharman, G.B., in "Reproduction in Mammals, Vol. 6" (C.R. Austin and R.V. Short, Eds.), Cambridge University Press, Cambridge, U.K., 1976, pp. 32–70.

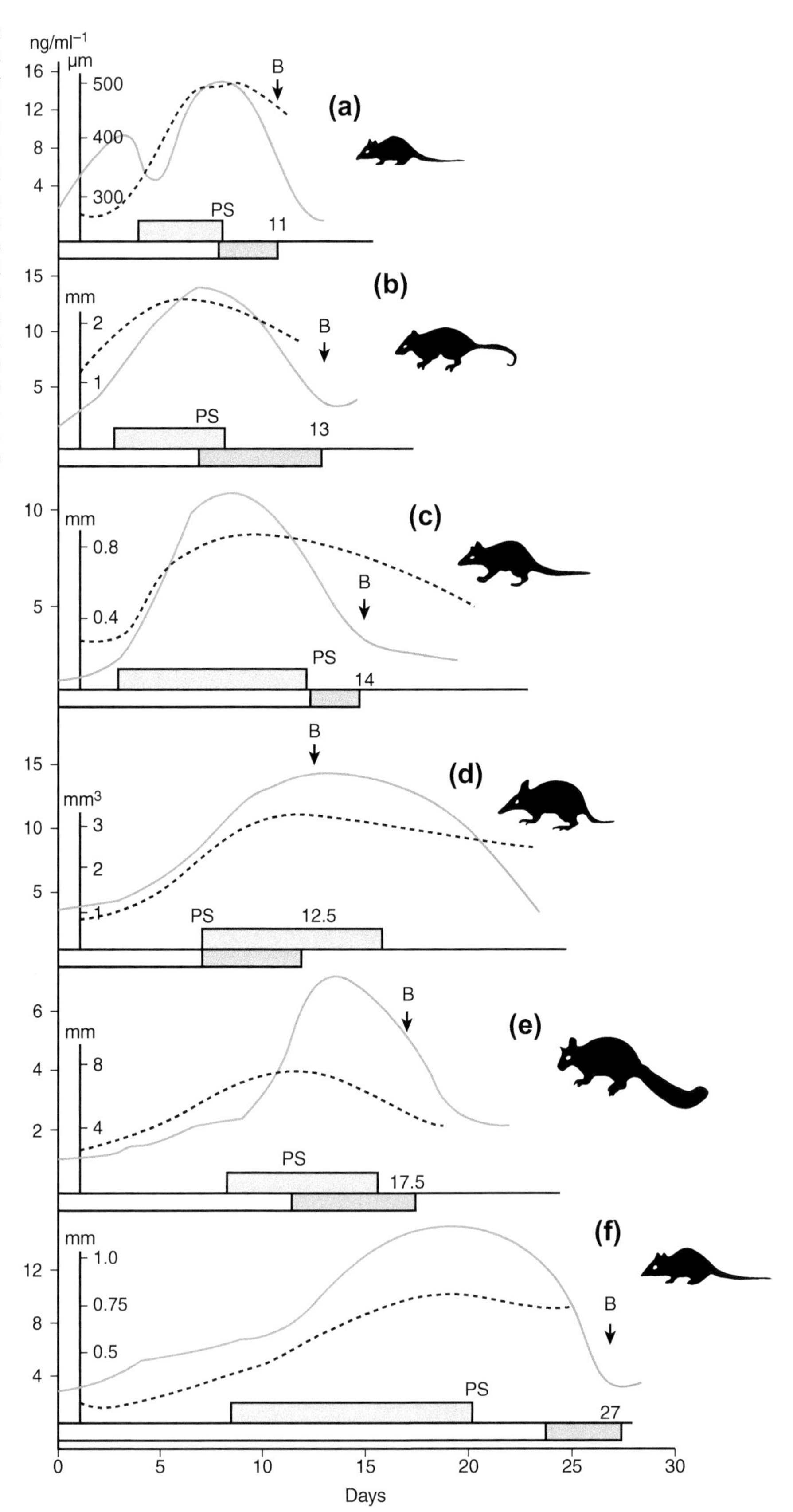

FIGURE 10-5 **Reproductive cycles in six marsupial species show the relatively short gestation period before birth (B).** Pink bar represents the luteal phase, and the dashed line represents the size of the corpus luteum with circulating progesterone levels shown in the green line. Pregnancy is shown by the bar at the bottom where the open bar represents the attached phase and the shaded bar represents the unattached phase. PS, time to primitive streak stage of development. (a) Stripe-faced dunnart (*Sminthopsis macroura*), (b) Virginia opossum (*Didelphis virginiana*), (c) eastern quoll (*Dasyurus viverrinus*), (d) northern brown bandicoot (*Isoodon macrourus*), (e) brushtail possum (*Trichosurus vulpecula*), (f) brown antechinus (*Antechinus stuartii*). *(Adapted with permission from Bradshaw, F.J. and Bradshaw, D.,* General and Comparative Endocrinology, ***170****, 18–40, 2011.)*

TABLE 10-5 Steroid Levels in the Tammar Wallaby (*Macropus eugenii*)

Tammar Wallaby	Reproductive Status	E_2 (pg/mL)	T (ng/mL)
Female	Day 5 of estrous cycle	15	N.D[a]
	Rest of cycle	5–10	N.D[a]
Male	Non-breeding	N.D	1–3
	Breeding	N.D	6

[a]N.D.,non-detect.

a consequence of separate uteri and vaginas and limited space for uterine hypertrophy. A special birth canal must be formed so that parturition can occur, and in some species it forms anew each season. This development of separate vaginas has influenced evolution of the male reproductive system as well. Males of some species have a bifid penis, with left and right prongs apparently being inserted into the separate vaginas of the female during copulation.

III. REPRODUCTION IN EUTHERIAN MAMMALS

Reproductive events such as spermatogenesis and oogenesis and their hormonal control are similar in monotremes and marsupials to events described for eutherian mammals. Consequently, most of the events and associated terms described ahead for eutherians probably apply to monotremes and marsupials, although the latter groups have not been studied to the extent of eutherians. Only marked differences will be notes in the following accounts.

Eutherian mammals employ the placenta not only as an endocrine organ to maintain gestation but also as a replacement for the mammary glands to supply early nutrition to the fetus. Consequently, birth (**parturition**) is delayed considerably, and the newborn or **neonate** of most placental mammals is at a comparable stage of development to the young joey when it first ventures out of the pouch. Like the juvenile marsupial, the placental neonate relies at first on the mammary gland as the exclusive source of nourishment but gradually abandons it for other foods.

Three distinct reproductive patterns occur in sexually mature eutherians: one typical for all males and two among females. Males of some domesticated species and humans are characterized by continuous secretion of GTHs and occasionally continuous spermatogenesis, and these males are capable of siring offspring at any time of year. Most eutherian species, however, exhibit seasonal episodes of spermatogenesis, sexual activity, or both such as displayed by most non-mammalian vertebrates (see Chapter 11).

Although considered a continuous breeder, humans show some dramatic seasonal patterns of breeding activity correlated with photoperiod and latitude (based on the past 30 to 50 years of birth records). In Europe and Japan, there are more births in the early spring (March–April) with the amplitude of the birth peak increasing in a south-to-north gradient. In South Africa, Australia, India, and New Zealand, the birth peak is associated with the months of September through November. The amplitude of the birth peak decreases from north to south, which is opposite to the latitudinal change observed in the Northern Hemisphere. However, in North America, the greatest number of births for both white and non-white people (many having their origins as transplants from Europe and Africa, respectively, in the past 300 years) occurs from August to October, with a decrease in amplitude from warmer to colder latitudes as occurs in countries of the Southern Hemisphere of the Old World. It has been proposed that North American patterns are a response to environmental temperature superimposed over the Old World patterns that correlate more strongly with photoperiod. These variations with latitude are less obvious for the Northern Hemisphere in the Old World and are opposite to that observed in the New World and the Southern Hemisphere of the Old World. In spite of these differences, it is clear that humans show seasonal tendencies in reproductive activity, resembling those of other eutherians.

Females exhibit cyclic patterns of GTH secretion that can be traced to a basic rhythmicity probably residing within the hypothalamus. There are two types of female cycles: the ovarian-based estrous cycle and the uterine-based menstrual cycle. The **estrous cycle** is typical for mammals (except possibly humans) and consists of a series of precisely regulated endocrine events repeated in each cycle. **Proestrus** is characterized by hormonal changes that bring about follicular development and ovulation. **Estrus** is a short period when the female is receptive to the male and during which mating can occur. It immediately follows proestrus and coincides with ovulation. Estrus represents the time when fertilization is most likely to lead to pregnancy and successful birth of offspring. The interim between estrus and the onset of hormonal changes characteristic of proestrus in cases in which pregnancy did not result is termed **diestrus** and usually is associated with ovarian inactivity and uterine regression. Carnivores and some other mammals may be classified as **monestrous**. If mating does not occur in a monestrous species or if mating occurs but fertilization and implantation are unsuccessful, the female will not return to estrus until the next breeding season (mono = one). Hence, these species have a long diestrous phase. Many mammalian species are polyestrous,

BOX 10D Serftoli Cells and Organ Transplant

The testes are considered an "immune privileged" organ because they produce a number of chemical signals that protect cells of the spermatogenetic lineage from being recognized as being foreign by cells of the immune system. Sertoli cells modulate the activity of local white blood cells by secreting cytocrines that act in a paracrine fashion to influence resident macrophage cells in the testis (Box Figure 10D-1). The ability of Sertoli cells to modulate immune cells holds promise for organ transplant technology. For example, Sertoil cells protect pancreatic islet implants from being attacked by the hosts immune system. Co-implantation of Sertoli cells with pancreatic islets represents a promising treatment for type 1 diabetes.

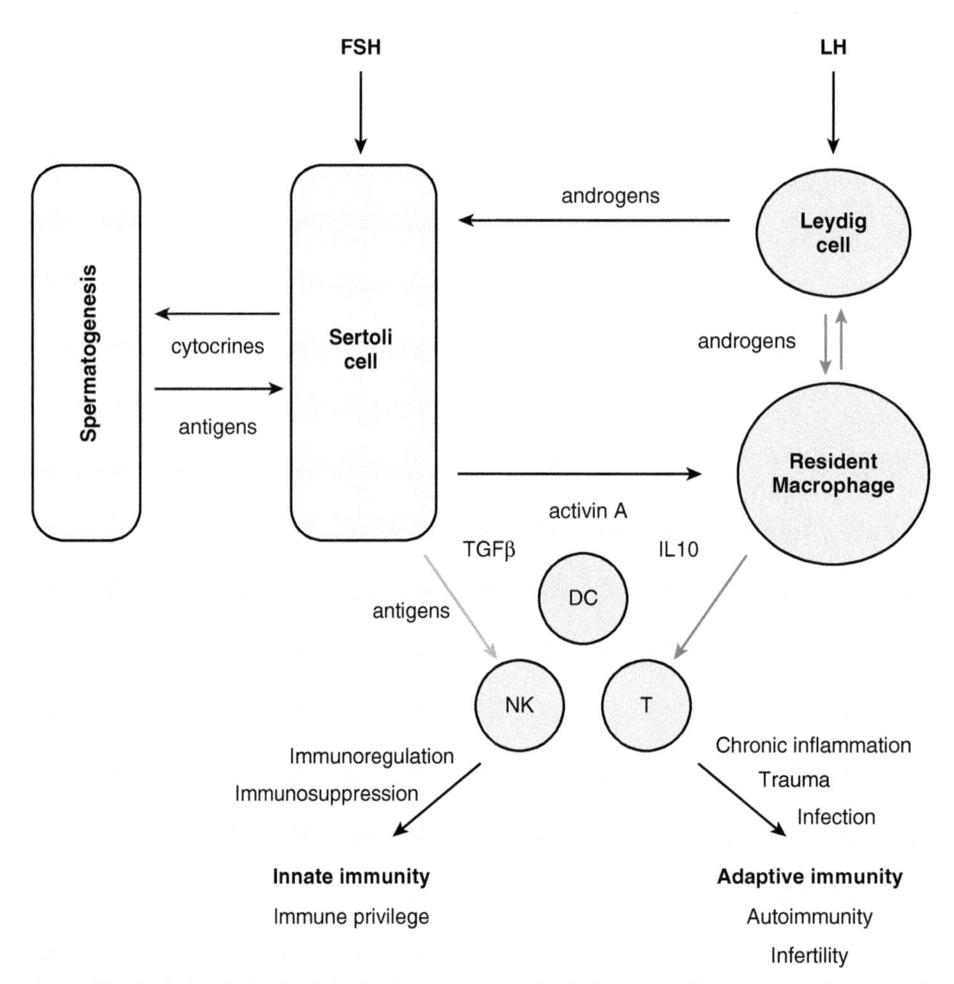

BOX FIGURE 10D-1 **Chemical signals involved in the immune network of the testes.** The immune cell cohort of the testis includes macrophages, natural killer lymphocytes (NKs), T lymphocytes (T), and dendritic cells (DCs) that aid in antigen presentation to lymphocytes and macrophages. In the process of phagocytosing residual cellular material from developing sperm, Sertoli cells generate potential antigens to be recognized by the immune cells. In order to prevent the immune system from attacking developing sperm cells, adaptive immunity pathways mediated via T lymphocytes are suppressed, whereas innate immunity via NK cells is enhanced through signaling from transforming growth factor β (TGFβ), activin, and interleukin-10 (IL-10). Failure to keep adaptive immunity pathways in check will lead to T cells attacking spermatogenetic cells and infertility. *(Adapted with permission from Meinhardt, A. and Hedger, M.P.,* Molecular and Cellular Endocrinology, ***335,*** *60–68, 2011.)*

however, and will return immediately to proestrus if mating does not occur or if mating is unsuccessful. Eventually, the female will enter diestrus. In some mammals (e.g., certain rodents), mating without successful fertilization and implantation may result in a short period of simulated or false pregnancy, termed **pseudopregnancy**, after which the female re-enters proestrus without experiencing diestrus.

Some eutherian mammals exhibit a sequence of uterine events, known as a **menstrual cycle**, that is characterized by one phase in which rapid sloughing of the uterine epithelium (**menses**) occurs if fertilization does not lead to pregnancy rather than the slow regression seen in diestrus. Often, these animals do not experience diestrus and may breed continuously, whereas others are seasonal breeders. Menstrual cycles are found in humans, monkeys, gibbons, the slender loris, marmosets, and several shrews and bats. The sloughing of the uterine epithelial lining often results in a vaginal discharge (**menstruation**) of uterine epithelial cells and trapped blood. The blood is trapped due to constriction of special spiral-shaped arteries that supply most of the blood flow to the uterine epithelium. The lack of blood flow to the outer portion of the epithelium results in cell death and hastens sloughing. A few of these mammals with menstrual cycles are known to exhibit **covert menstruation** where the sloughed tissues are resorbed and there is no uterine discharge. The onset of menses marks the end of one cycle and the beginning of the next. Although reproductive endocrinologists consistently focus upon the cyclic nature of estrous and menstrual cycles and their restarting with the failure of pregnancy to occur, it is important to remember that the normal sequel to ovulation is pregnancy. In nature, it is probably unusual for a female to enter estrus and not become pregnant.

Most primates exhibit estrous behavior to some degree at about the time of ovulation, including species characterized as having menstrual cycles, although a well-defined estrus is not observed in the human female. Exhibition of menstrual or estrous cycles should not be considered as alternative strategies, but that, with the possible exception of humans, the menstrual cycle is one variation within the estrous cycle.

Although there have been numerous hypotheses developed as to why menstrual cycles evolved, most are discounted because they cannot be generalized to all mammals exhibiting menses. Some, for example, relate menstruation to a placental type common to menstruating mammals, yet other species with this same type of placental anatomy do not have menstrual cycles. If it is simply a byproduct of endometrial function, then it is a costly one in terms of nutrient losses. One intriguing hypothesis suggests that menstruation evolved as an adaptation in mammals to neutralize and eliminate pathogens introduced during copulation and/or by sperm deposited in the vagina. However, this does not explain the scarcity of menstrual cycles among eutherian species. Perhaps the repeating menstrual cycle is simply another adaptation in favorable climates that allows for continuous breeding activity to ensure pregnancy will occur at some point but so that not all females in the population are pregnant at the same time.

Uterine bleeding and discharge occur at other times in the estrous cycles of some mammals; for example, the cow, domestic dog, and coyote discharge blood prior to ovulation and the onset of estrous behavior. This discharge is estrogen induced and does not involve degeneration of the uterine lining. Periovulatory bleeding also occurs either overtly or covertly in several primates, including humans, white-throated monkeys, vervet monkeys, and possibly in the cottontop tamarin.

A. Puberty

The achievement by the gonads of their full hormonal and gametogenetic capacity for reproduction is termed **puberty**. Remarkably, much remains to be learned regarding the mechanisms regulating the timing of puberty. Ultimately, puberty is the result of an increase in GnRH secretion (Figure 10-6), but the fact that GnRH is produced by one out of every 100 millions neurons in the human brain presents some technical difficulties in determining the precise mechanisms involved in the timing of GnRH secretion. Puberty may be gradual or rather sudden, depending on the species, and may be accompanied by a variety of morphological changes as well. In mammalian groups that have been the best studied such as rats, monkeys, and humans, puberty is associated with a marked increase in GnRH release from the hypothalamus resulting in elevated GTH secretion. In children, the frequency of these GTH pulses increases twofold at night and later during the day as puberty progresses. The amplitude of the GTH pulses also increases as the sensitivity of pituitary gonadotropes to GnRH is enhanced. Experimental studies suggest that the most critical factor in the induction of puberty is the increase in the frequency of GnRH pulses.

These changes in the HPG axis are paralleled in other endocrine systems. Nocturnal release of PRL also is elevated along with FSH and LH, although its role in puberty is uncertain. Adrenal androgens increase in boys and girls prior to the onset of puberty and continue to rise during puberty. This process is called **adrenarche** (see Chapter 8). These events are independent of the changes in the gonadal axis but are important contributions to puberty; for example, adrenal androgens stimulate the prepubertal growth spurt and the appearance of axillary and pubic hair.

There are several hypotheses concerning the mechanism(s) for the normal onset of puberty (Figure 10-6). The **gonadostat hypothesis** suggests that decreased feedback sensitivity to gonadal steroids develops in the hypothalamus and brings about increased release of GnRH. Accelerated GTH release from the pituitary activates gonadal steroidogenesis and gametogenesis or **gonadarche**. Gonadal responses also may involve receptor synthesis; for example, receptors for FSH are present in early ovarian follicles but not LH receptors. FSH stimulates production of LH receptors and increases the levels of **aromatase** (**$P450_{aro}$**). Now, the ovary can respond to both GTHs. The **missing link hypothesis** implies that some factor is missing and that it is the brain that is functionally incompetent prior to puberty. Data obtained from experimental and clinical observations that support these hypotheses also

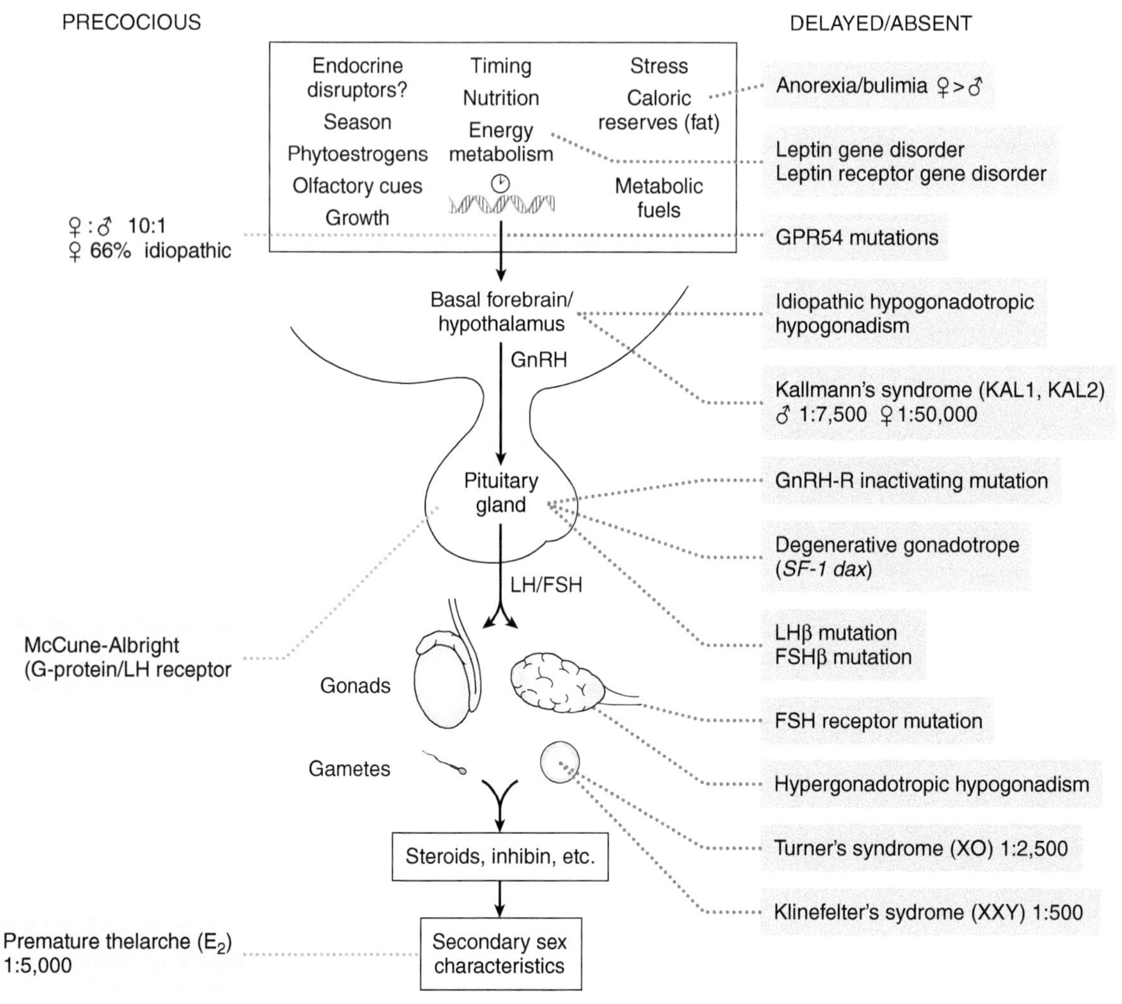

FIGURE 10-6 **Factors involved in the timing of puberty.** Puberty involves an increase in GnRH secretion from the hypothalamus. Many genetic and environmental factors come into play in determining the timing for the onset of GnRH secretion. Factors leading to precocious puberty are shown on the left. Factors that lead to delayed or absent puberty are shown on the right. Central nervous system mechanisms that may affect the onset of GnRH secretion are shown in the purple shaded box. *(Adapted with permission from Ebling, F.J.,* Reproduction, ***129****, 675–683, 2005.)*

support the **active inhibition hypothesis** that currently is in favor. This hypothesis states that puberty occurs because of a progressive decrease in physiological inhibition. The well-known actions of the pineal gland and melatonin on inhibiting reproductive function (see Chapter 4), and the observed reduction in melatonin secretion with precocial puberty offer strong support to the inhibitory hypothesis (see ahead). Early puberty events in humans also have been linked to exposures to endocrine-disrupting chemicals.

A separate hypothesis has been formulated to explain puberty and **menarche**, the onset of menstruation in girls. One fact supporting a separate mechanism in girls is that the likelihood of precocious puberty is ten times greater in girls than boys (Figure 10-6). This **lipostat hypothesis** or **critical weight hypothesis** implies that attainment of puberty is in part a function of fat storage in the body; consequently, girls who have more body fat reach menarche sooner than leaner girls. Furthermore, chronic strenuous exercise reduces body fat and may prevent young girls from reaching menarche just as it can block menstruation in women. Once the excessively lean girls are placed on a less severe regimen so that body fat increases, menarche usually is achieved. This mechanism probably is not related to puberty per se but rather may be an evolutionary mechanism to ensure that there is evidence of sufficient environmental resources (as evidenced in body fat) to support the energy demands of pregnancy and lactation for successful rearing of young. Recent studies of the peptide **leptin** that is secreted by adipose cells may influence GnRH release in the brain. Higher levels of leptin occur when lipid stores are greater, and leptin may provide a signal to the brain of the extent of fat storage in addition to suppressing appetite (Figure 10-6; also see Chapter 12). Excess kisspeptin secretion also may lead to precocious puberty. It is worth mentioning that **McCune–Albright syndrome** also leads to precocious puberty and menstruation in girls (Figure 10-6), although the mechanism is linked to mutations in the *GNAS1* gene that encodes the α subunit of the G protein Gs. This condition also leads to gigantism and discoloration of the skin.

There are many potential causes for delayed or absent puberty (Figure 10-6). Mice lacking the GPR54 (kisspeptin) receptor fail to go through puberty, and many loss-of-function mutations in the *GPR54* gene have been described in patients with delayed or absent pubertal development (Figure 10-6). **Kallmann's syndrome** represents a wide spectrum of disorders associated with reduced GnRH secretion. Delayed puberty may manifest itself in individuals with eating disorders that lead to reduced caloric intake (Figure 10-6).

IV. ENDOCRINE REGULATION IN EUTHERIAN MALES

In some species, mature males are capable of copulating with a female whenever she is receptive. Secretion of GnRH and hence of GTHs is more or less continuous in these males but with daily and often seasonal fluctuations occurring in circulating levels of some GTHs. Daily secretory patterns for GTHs show considerable variation among different species. Hourly fluctuations of LH have been reported in bulls, and these variations in LH are correlated with following increases in circulating testosterone (Figure 10-7). In human males, FSH shows no cyclic variation in blood levels, although LH and testosterone exhibit obvious daily patterns with peak levels occurring during early morning hours and minimum values reported for the afternoon. Most wild mammals exhibit distinct seasonal breeding, and active spermatogenesis may be restricted to only a few months of the year or less.

A. Spermatogenesis

Each testis develops primarily from the medullary portion of an embryonic gonadal blastema as described previously (see Figure 10-2). Differentiation of the medullary portion with concomitant regression of the cortical components (progenitor of the ovary) appears to be controlled by local embryonic androgen secretion activated by the *sry* gene. The medullary region differentiates into seminiferous tubules and interspersed masses of interstitial cells. These interstitial cells are located between the seminiferous tubules and synthesize and release androgens into the general circulation.

The seminiferous tubules consist of large Sertoli cells, germ cells, spermatogonia, and cells derived from the spermatogonia (Figures 10-8, 10-9, and 10-10). Each tubule is surrounded by a thin layer of connective tissue. Under the influence of androgens, **peritubular myoid cells** develop in this connective tissue layer during puberty. They surround and provide support for the seminiferous tubules and are believed to be responsible for contractile activity of the tubules that propels sperm to the

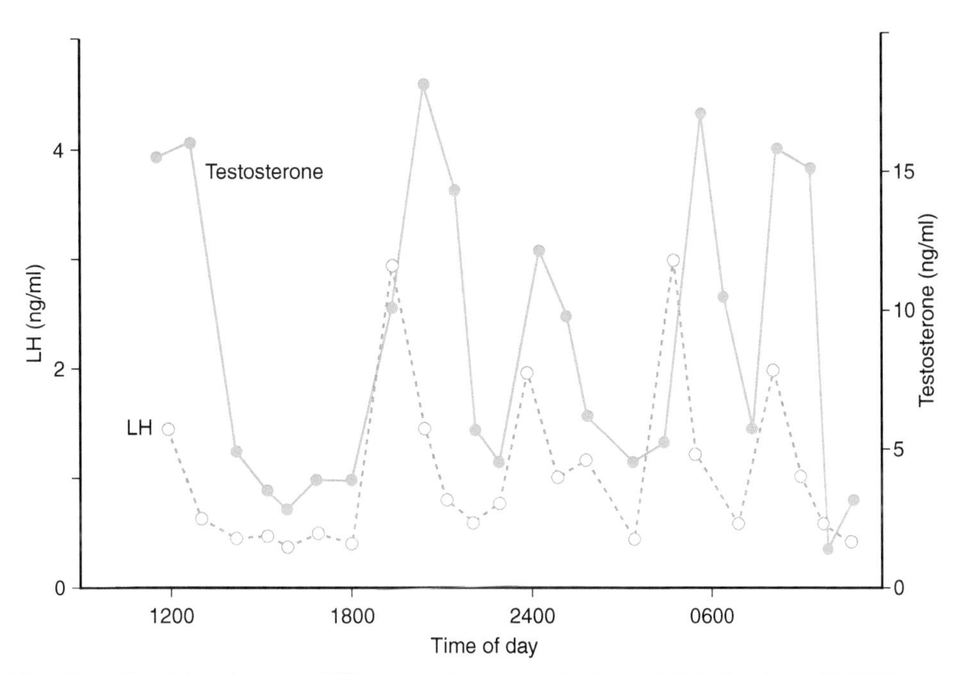

FIGURE 10-7 **Male pattern of luteinizing hormone (LH) and testosterone secretion in a bull.** Pulsatile release of GnRH (not shown) would precede each LH peak that precedes each testosterone peak. (*Adapted with permission from Short R.V.,* Reproduction in Mammals, Book 3, *Cambridge University Press, 1972.*)

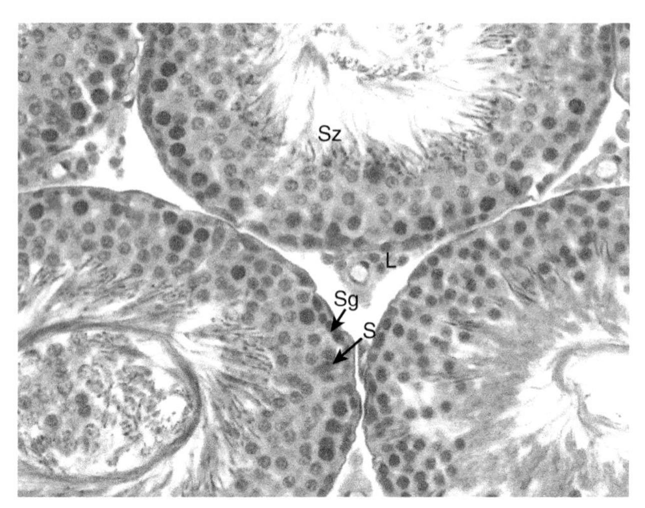

FIGURE 10-8 **Spermatogenesis in rat testis.** Section of rat testis showing edges of adjacent seminiferous tubules. Note the position of the Leydig cells adjacent to a capillary. Abbreviations: L, Leydig cells; S, nuclei of Sertoli cells; Sg, spermatogonia; Sz, sperm.

epididymis. The Sertoli cell has an extensive cytoplasm extending from the outer edge to the lumen of the tubule. The angular nucleus of the Sertoli cell is located at the outer edge of the tubule. Sertoli cells form tight junctions with adjacent Sertoli cells and together with the peritubular myoid cells they secrete the various collagen and laminin proteins that form the basement membrane. This basement membrane along with their **tight junctions** together form the blood/testis barrier, quite literally a cellular fence that isolates the seminiferous tubules into a 'backyard" (basal) compartment containing Sertoli cells and spermatogonia and a "frontyard" (adluminal) compartment containing the spermatogenetic cells (Figure 10-10). As a result of the blood testis barrier, nutrients and leukocytes cannot move between the Sertoli cells into the seminiferous tubules, thereby protecting developing spermatogenetic cells from exposure to cells and antibodies of the immune system. All chemicals must pass through the Sertoli cells in order to reach the spermatogenetic cells (Figure 10-10).

Germ cells are present along the outer margins of the seminiferous tubules and differentiate into **spermatogonia**. Spermatogonial cells proliferate mitotically under the influence of FSH to produce more spermatogonia. Eventually some of these spermatogonia will undergo differentiation characterized by nuclear enlargement and will become **primary spermatocytes** that are capable of entering spermatogenesis (Figure 10-11). Testicular androgens are somehow necessary for initiation of meiosis in primary spermatocytes that undergo the first meiotic division to give rise to two smaller **secondary spermatocytes**. These latter cells are infrequently observed in histological preparations because, once formed, they quickly enter the second meiotic division to yield four haploid **spermatids** which are transformed to sperm (spermatozoa) by concentrating the chromatin material into the sperm head and by elimination of the majority of the cytoplasm. The process of transformation of spermatids to sperm is termed **spermiogenesis**.

Spermatogenesis is a temperature-sensitive process, and high temperatures such as found within the body cavity of terrestrial eutherians can impair normal spermatogenesis and produce temporary sterility. Consequently, at some time prior to the attainment of sexual maturity or prior to the annual breeding season, the testes descend into the scrotum, where spermatogenesis can proceed at a slightly lower temperature. The failure of the testes to descend, a condition known as **cryptorchidism** (*crypto-*, hidden;

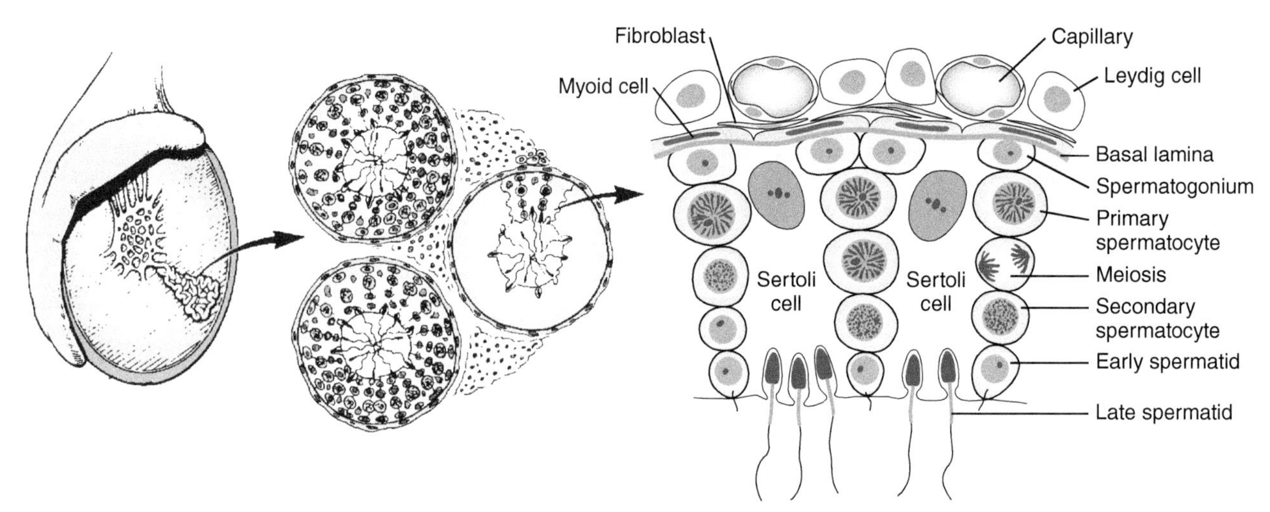

FIGURE 10-9 **Organization of mammalian testis.** Detail at right shows events of spermatogenesis and spermiogenesis in relation to Sertoli cells. The endocrine roles of interstitial (Leydig) cells, peritubular myoid cells, and Sertoli cells are described in the text. *(Adapted with permission from from Skinner, M.K.,* Endocrine Reviews, ***12****, 45–77, 1991; Junqueira, L.C. and Carneiro, J., "Basic Histology," 11th ed., McGraw-Hill, New York, 2005.)*

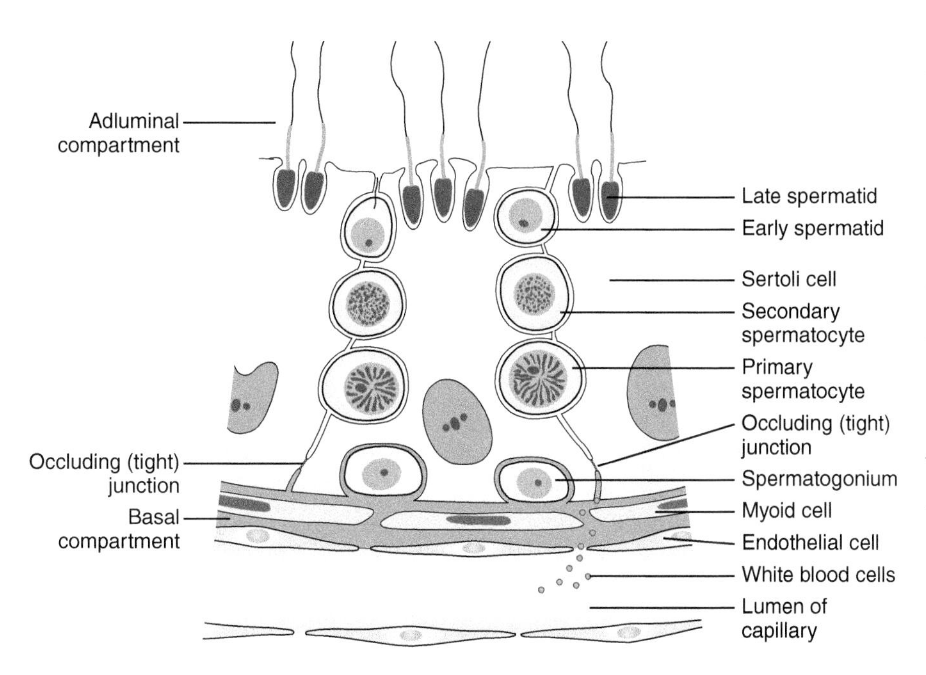

FIGURE 10-10 **Occluding (tight) junctions between adjacent Sertoli cells form the blood testis barrier.** The presence of the occluding junctions establishes an adluminal compartment isolating developing spermatocytes, spermatids, and spermatozoa from materials or white blood cells, leaving the capillaries in the basal compartment. Substances or leukocytes may diffuse or move by diapedesis through capillaries in the interstitium but cannot move past the occluding junctions linking the Sertoli cells. Spermatogonia lie basal to the occluding junctions in the basal compartment. *(Adapted with permission from Junqueira, L.C. and Carneiro, J., "Basic Histology," 11th ed., McGraw-Hill, New York, 2005.)*

orchi, testis), may cause irreparable damage to the seminiferous epithelium in most species. Some mammals lack a scrotum (e.g., elephants, whales, seals), and the testes are permanently located within the abdominal cavity. Male elephants, however, are capable of producing viable sperm and copulating with a female at any time of year. In such species, either spermatogenesis does not exhibit the same temperature sensitivity characteristic for scrotal species or these animals possess other mechanisms to reduce testicular temperature.

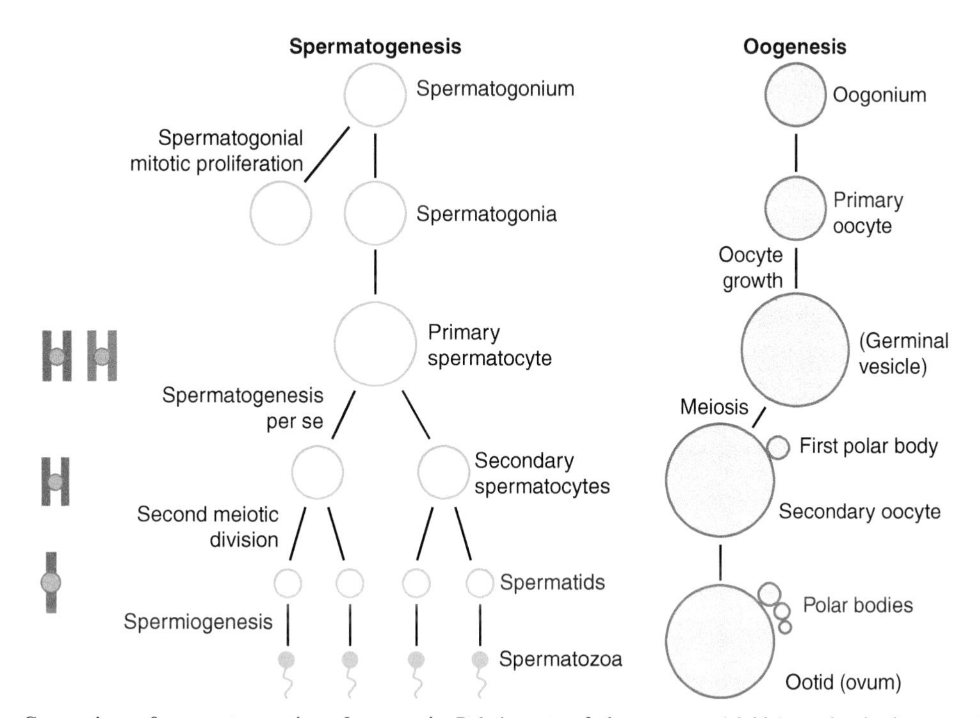

FIGURE 10-11 **Comparison of spermatogenesis and oogenesis.** Relative sets of chromosomes (ploidy) number is shown on the left. A major difference between males and females is the mitotic proliferation of gonial cells after birth in males, whereas the germ cells have all progressed to the primary oocyte stage in females and no new oocytes appear after birth. The primary spermatocyte undergoes meiosis to produce four spermatids, whereas meiotic division of the primary oocyte with unequal distribution of cytoplasm produces only one oocyte plus up to three polar bodies. The first polar body (a secondary oocyte with very little cytoplasm) often does not undergo the second division.

A given histological section of a seminiferous tubule may show varying numbers of spermatogonia, primary spermatocytes, possibly a few secondary spermatocytes, spermatids, and sperm in sequence from the outer margin (gonia) to the lumen (sperm). The tails of the sperm extend into the lumen, and the heads of the sperm typically are still surrounded by highly folded margins of Sertoli cells (Figure 10-9).

Millions of maturing sperm may be sloughed off into the lumina of the seminiferous tubules each day. This process is termed **spermiation** and is stimulated by LH. These sperm pass through the tubules that eventually coalesce into larger ducts of the rete testis that eventually join the epididymis associated with each testis. Vast numbers of maturing sperm are stored in the epididymis. Under the influence of androgens and PRL, the epididymis secretes materials into its lumen where the sperm are being held. Included in this secretion are protein-bound **sialic acids** (sialomucoproteins), **glycerylphosphoryl-choline**, and **carnitine**. These particular substances are involved directly in maturing and maintaining sperm in viable condition until ejaculation. Androgens and **androgen-binding protein (ABP)** produced by Sertoli cells in the seminiferous tubules are released along with sperm and travel to the epididymis. Androgen molecules freed from ABP in the lumen or androgen-ABP complexes or possibly both are absorbed by the epididymal cells. These androgens stimulate epididymal cells to secrete materials involved in maintenance of the sperm.

Contraction of the smooth muscles of the epididymis and vas deferens causes ejection of sperm (ejaculation). The mature sperm leave the epididymis, enter the vas deferens, travel to the urethra, traverse the length of the penis via the urethra, and are deposited in the female's vagina during coitus. **Seminal vesicles** and the **prostate gland** add their fluid secretions to the sperm and epididymal secretions to form a watery mixture of sperm and various organic and inorganic substances known as **semen**. The **bulbourethral gland** or **Cowper's gland**, which is homologous to the Bartholin's gland in females, produces a pre-ejaculate that cleanses and lubricates the urethra prior to the arrival of the semen. The entire ejaculatory event may be induced by the release of OXY from the pars nervosa in response to a neural reflex initiated by mechanical stimulation of the penis.

B. Endocrine Regulation of Testicular Functions

GTHs and a number of paracrine factors including testosterone (Table 10-6) control production of sperm. Many of the details of the endocrine regulation of these processes are not clear, but a generalized picture is emerging. Relatively separate roles have been defined for LH and FSH in males, although FSH and testosterone work cooperatively in some cases. Spermatogenesis is initiated indirectly by FSH through mitotic proliferation of spermatogonia and formation of primary spermatocytes. Spermatogonia lack FSH receptors, and the mechanism of spermatogonial activation is mediated by paracrine factors from the Sertoli cells that do have FSH receptors. One paracrine factor identified as having an important role in proliferation of spermatogonia is **glial cell line-derived neurotrophic factor (GDNF)**. The intermediary role of the Sertoli cell is supported by observations that FSH stimulates mitosis in Sertoli cells whose number in the adult testis is directly proportional to sperm abundance. Testosterone may initiate meiotic divisions of primary spermatocytes that differentiate from spermatogonia, resulting eventually in formation of spermatids. The production of ABP, cytoskeleton proteins (**actin, vinculin**), and $P450_{aro}$ by the Sertoli cell also is stimulated by FSH.

Testosterone is the major circulating androgen in mammals, although other androgens such as **androstenedione** or **5-α-dihydrotestosterone (5α-DHT)** may circulate in significant amounts (Table 10-7). Some 5α-DHT also is produced and stimulates red blood cell production in bone marrow and contributes to the higher hematocrit found in males as compared to females. Prior to attainment of puberty in bulls, androstenedione is the principal circulating androgen, but it is gradually replaced by testosterone at puberty. Testosterone within the testis, however, seems to be the most important androgen influencing sperm production.

Testosterone has several important paracrine effects on spermatogenesis and spermiogenesis. Testosterone concentrations within the testis are generally orders of magnitude greater than circulating testosterone levels (Table 10-8) and, for reasons still not understood, much higher than needed in order to maintain spermatogenesis. Perhaps testosterone is in part a precursor for testicular estrogen synthesis; for example, testicular estrogens reach dramatic levels in the stallion. Androgens are required for spermatogenesis to proceed normally, in part through the activation of a multitude of genes, one of which produces **gonadotropin-regulated testicular helicase (GRTH/DDX25)** in Leydig cells and germ cells. GRTH is a 56-kDa protein that is transported into the nucleus, where it binds to nuclear RNAs as a component of messenger ribonucleoprotein (mRNP) complexes. It regulates the activity and turnover of certain mRNAs involved in meiosis. GRTH-knockout mice are incapable of producing functional sperm. Androgen actions probably are indirect, as testosterone receptors appear to be absent or occur in very low numbers on germ cells. However, estrogen receptors (ERs) are present on germ

TABLE 10-6 **Possible Local Actions for Gonadal Secretions**

Factor	Source	Proposed Action
Activin	Granulosa cells Sertoli and interstitial cells	Unknown Specific receptors shown on germ cells
Epidermal growth factor (EGF)	Thecal/interstitial cells of ovary	Stimulates granulosa cell proliferation
Estradiol	Interstitial cells in male epididymis	Blocks androgen synthesis Stimulates fluid resorption
Fibroblast growth factor (FGF)	Granulosa cell	Causes epithelial proliferation in early follicular development and conversion of thecal cells to ovarian interstitial cells after ovulation
	Testicular germ cells	Binds to receptors on Sertoli cell; function unknown
Gonadotropin-releasing hormone (GnRH)	Ovary	Working through IP_3 second messenger, GnRH alters steroidogenesis by granulosa cells in certain follicular stages; may signal atresia
	Sertoli cell	Alters androgen synthesis by interstitial cells; increases local permeability of capillaries
Growth differentiation factor-9 (GDF-9)	Ovary	Maintains FSH receptors on oocyte
	Oocyte	Follicle growth and maturation (folliculogenesis)
Growth hormone-releasing hormone (GHRH)	Corpora lutea, oocyte	Promotes follicular development and ovulation
	Germ cells of tests	Stimulates sertoli cells to make stem cell factor
Insulin-like growth factor (IGF-I)	Granulosa cells	Increases number of LDL receptors on granulosa cells; stimulates cholesterol and inhibin synthesis
Interleukin 1(IL-1)	Sertoli cells	Decreases steroidogenesis in interstitial cells
PmodS protein	Peritubular myoid cells	Non-mitogenic factor that regulates differentiation and function of Sertoli cells
Interleukin 6 (IL-6)	Ovarian T cells	Suppresses response of granulosa cells to FSH,i.e., decreased progesterone synthesis; induces apoptosis and atresia of granulosa cells
KIT ligand	Oocyte	KIT ligand is a cytocrine that acts on a tyrosine kinase receptor (KIT) to promote folliculogenesis
Nerve growth factor (NGF)	Ovarian cells	Stimulates follicular formation and organization as well as differentiation of ovarian interstitial cells
Testosterone	Interstitial cell	Regulates functions of Seroli cells; stimulates meiosis in primary spermatocytes
Transforming growth factor (TGF-α)	Thecal cells	Facilitates proliferation of thecal and granulosa cells but slows their GTH-induced differentiation
	Sertoli and peritubular myoid cells	Causes EGF-like growth stimulation in interstitial cells; decreases steroidogenesis

cells that also possess aromatase activity. Furthermore, at least in rodents, the local metabolism of DHT to an androgen metabolite **5α-androstane-3β, 17β-diol (3β-diol)** (see Chapter 3; Table 10-8) has been reported in brain, prostate, and testes. 3β-Diol binds and activates ERs. Furthermore, aromatase knockout male mice exhibit disruption of spermatogenesis, supporting a local role for estrogen receptor-binding regulators.

TABLE 10-7 Plasma Levels of Reproductive Steroids in Mammals

Species	Testosterone (ng/mL)	Dihydrotestosterone (ng/mL)	Estradiol (pg/mL)	Progesterone (ng/mL)
Mustela ermina (stoat)				
Male (annual range)	4.5–26			
Ursus americanus (black bear)				
Female				
Nonpregnant			35–71	17.8
Pregnant			32,000–35,000	29.40
Elaphus maximus (Asian elephant)				
Male (annual range)	0.7–45			
Female				
Not pregnant (peak)			26	0.153–0.195
Pregnant (peak)			26	4
Macaca fuscata (Japanese monkey)				
Male (annual range)	0.2–19.8			
Female				
Follicular phase (peak)			150	2
Luteal phase (peak)			250	5.3
Homo sapiens				
Male (mean)	7.9	0.4	50	
Female				
Follicular phase (range)	0.6	0.3	60–600	0.3–1.5
Luteal phase (range)	0.6	0.3	200	3–20
Pregnant (range)			5,500–30,000	45–210

TABLE 10-8 Intratesticular Concentrations of Various Steroids in Humans

Steroid	Intratesticular Concentration(ng/mL)	Range
Testosterone[a]	572±102	103–1,085
DHT	13.4±1.8	4.9–26
3 β-diol	12.3±1.9	3.6–26
E_2	15.7±2.3	8–29

[a]*Plasma concentrations averaged 3.8 ng/mL.*
Abbreviations: 3 β-diol, 5α-androstane-3α, 17β-diol; DHT, dihydrotestosterone; E_2, estradiol.
Data from Jarow, J.P. and Zirkin, B.R., *Annals of the New York Academy of Sciences*, 1061, 208–220, 2005.

The attachment of Sertoli cells to spermatids involves cytoskeletal actin and vinculin interactions with the spermatids, as well as indirect effects of testosterone. Peritubular myoid cells (Figure 10-9) are stimulated by testosterone to release two proteins, **PModSA** and **PModSB**, that cause Sertoli cells to secrete additional paracrine regulators that may alter spermiogenesis (Table 10-6).

The Leydig cells of the testis also synthesize and release small quantities of estrogens. Locally, estradiol can block androgen synthesis by interstitial cells and can influence the responsiveness of these cells to GTHs. Estradiol also binds to receptors in the epididymis, where it regulates resorption of excess testicular fluid that was used to conduct sperm to the epididymis. The ratio of testosterone to estradiol in the general circulation may alter the ratios of FSH and LH being released from the pituitary through negative feedback. Finally, conversion of androgens to estrogen occurs in certain brain cells, and circulating estrogens themselves may influence male sexual behavior.

C. Actions and Metabolism of Androgens in Males

Circulating androgens influence development and maintenance of several glands and related structures associated with the male genital tract, such as the prostate gland and seminal vesicles, and induce development of certain secondary sexual characters such as growth of the beard and development of skeletal muscles in men. Androgens also exert a negative feedback effect upon the secretion of GTHs primarily through actions at the level of the hypothalamus (see Chapter 4).

The action of testosterone in some of its target cells involves its cytosolic conversion to DHT by the enzyme **5α-reductase**. DHT has a greater affinity for the androgen receptor than does testosterone and hence is more potent than testosterone. Timely development of prostate and bulbourethral glands, the penis, and scrotum are dependent upon conversion of testosterone to DHT by target cells in these structures. This conversion of testosterone to the stronger androgen DHT is necessary due to the low circulating level of testosterone at this time that is insufficient to activate these tissues. The testes generally do not synthesize DHT until puberty. Neurons in some brain areas express 5α-reductase, and some effects of testosterone on behavior may involve prior conversion to DHT.

Many androgenic responses, however, are not mediated by DHT, and this conversion is not necessary for testosterone to produce these effects. For example, development of wolffian duct derivatives (the epididymis, vas deferens, and seminal vesicles) is accomplished by a local level of testosterone that is sufficient to activate receptors without prior conversion to DHT. This effect may occur because local testosterone levels from the developing testes are sufficiently high to effectively activate androgen receptors in these structures. In the tammar wallaby, a marsupial, testosterone is converted to DHT in the wolffian duct, but this has not been shown to occur in a eutherian mammal.

In some target tissues, androgens have been shown to undergo conversion to estrogens through aromatization of the A-ring and removal of the C_{19} carbon atom (see Chapter 3). Conversion of androgens to estrogens also occurs in both Sertoli and Leydig cells. This process also occurs in the central nervous system, where aromatization may be essential to some androgen actions. Induction of some male behaviors in castrates requires aromatization and cannot be induced by essentially non-aromatizable androgens such as DHT, whereas others may be induced by either aromatizable androgens or by DHT. The Sertoli cells, under the influence of FSH, secrete two forms of inhibin that selectively block FSH release from the adenohypophysis. Inhibin activity also has been found in rete testis fluid, seminal plasma, testicular extracts, and ejaculate, suggesting local actions (see Table 10-6). Two forms of inhibin have been isolated: **inhibin A** and **inhibin B** (Figure 10-12). These molecules are glycoprotein heterodimers of 31 to 35 kDa that possess a common α-subunit combined with one of two β-subunits (A or B). Inhibins are believed to be the major factor responsible for negative feedback in the selective regulation of FSH release in both males and females. In addition, a β-subunit heterodimer called **activin** has been isolated from gonads. Activin is composed of two β-subunits (Figure 10-12) and is a potent releaser of FSH from the pituitary gland in laboratory experiments, although its physiological role is still undetermined. Activins bind to serine−threonine kinase type I and type II receptors on the plasma membrane (see Chapter 3), leading to phosphorylation of downstream signaling proteins (so-called **R-SMADs** for **receptor-regulated SMADs**, short for an isoform of the humorously named "mothers against decapentaplegic") and changes in gene transcription (Figure 10-12). Inhibins interfere with activin receptor interation by binding to activin type II receptors and blocking recruitment of type I receptors (Figure 10-12). Activin may be a local regulator, as activin levels within the testes are highly modulated during development (Figure 10-13) and activin receptors occur on spermatogenetic cells in the testis.

V. ENDOCRINE REGULATION IN EUTHERIAN FEMALES

Regardless of whether or not an animal exhibits a menstrual cycle in addition to an estrous cycle, there are numerous distinctive features that characterize reproduction in all female mammals. In addition to the roles of GTHs and gonadal steroids described below, numerous other regulators (including GnRH, inhibins, AMH, prostaglandins, and growth factors) are synthesized in the ovaries and are suspected of playing important paracrine roles in ovarian events. A partial listing and some demonstrated autocrine and paracrine ovarian regulators are provided in Table 10-6. One of those, **growth differentiation factor 9 (GDF-9)**, is produced by the oocyte and is required for proliferation of follicular cells and subsequent ovarian follicle development (Figure 10-14). Other candidate ovarian growth factors include **bone morphogenetic proteins 6 and 15 (BMP-6, 15)**, **fibroblast growth factor 8 (FGF-8)**, and **transforming growth factor β2 (TGF-β2)**. Developing follicles also produce AMH, which is responsible for recruitment of new follicles. Ovarian and uterine cyclical events during reproduction are similar. The ovarian events are discussed first followed by the uterine events that are linked to the ovarian cycle.

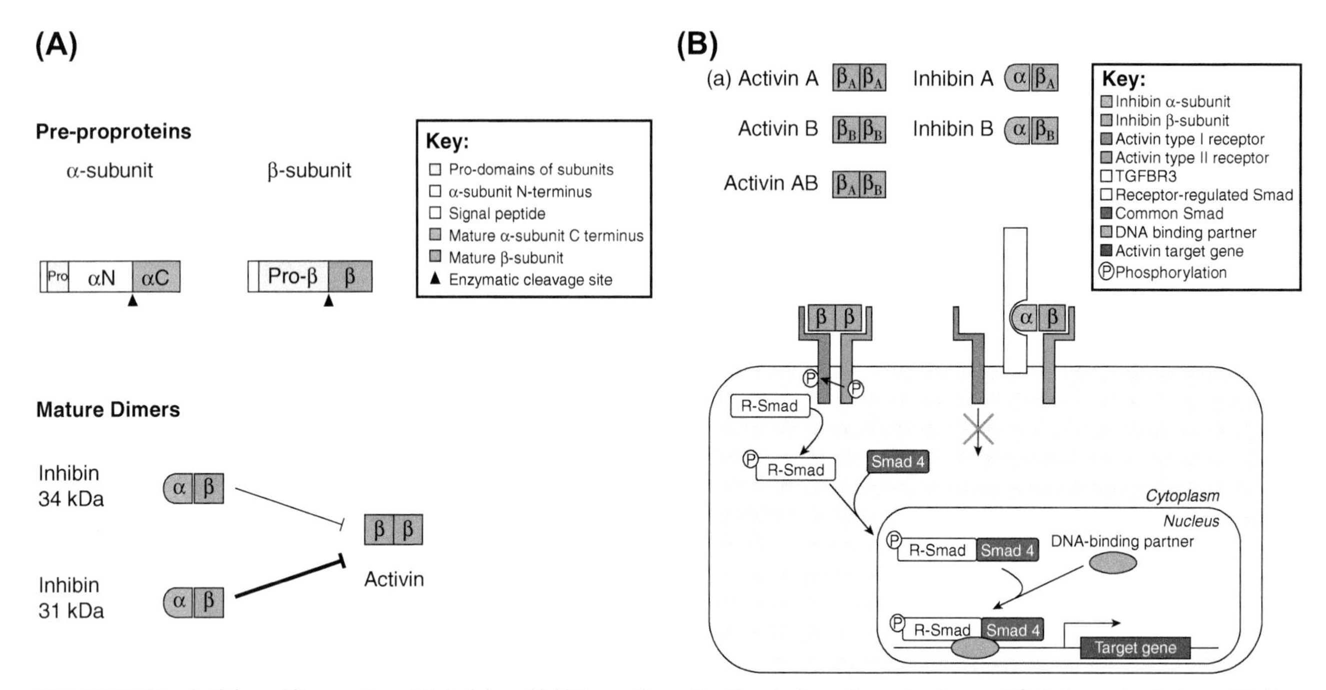

FIGURE 10-12 **Activin and its receptors.** (A) Activin and inhibins are formed by dimerization of two subunits, α and β, that are extensively modified by glycosylation and proteolysis after translation from either the α or β subunit mRNA. Activin is formed from two β subunits. Activins and inhibins act as natural inhibitors of each other's activity. (B) Activins bind to serine–threonine kinase type I and type II receptors on the plasma membrane. Binding to the type II receptor (green) leads to phosphorylation of the type I receptor (purple) and subsequent phosphorylation of downstream signaling proteins (R-SMADs). Inhibins interfere with activin receptor interaction by binding to activin type II receptors as well as TGFBR3 thereby blocking recruitment of type I receptors. Abbreviations: R-SMAD, receptor regulated mothers against decapentaplegic homolog 1; SMAD-4, SMAD family member 4; TGFBR3, transforming growth factor, beta receptor III. *(Adapted with permission from Stenvers, K.L. and Findlay, J.K.,* Trends in Endocrinology & Metabolism, ***21**, 174–180, 2010.)*

For sexually mature mammals in nature, fertilization and pregnancy are normal events, and coitus occurs frequently during estrus. The character of the ovarian cycle, with its rapid resumption in polyestrous species or following a short menses as in primates if fertilization and successful implantation do not occur, enhances the chances for successful reproduction. Either rapid reentry into estrus or rapid appearance of one or more new ova or both can occur, and a second opportunity to produce offspring is made possible during that season. In some cases, polyestrous animals (especially rodents) produce more than one litter in a single breeding season.

A. The Ovarian Cycle

The basis for the cyclical nature of female reproductive events resides in the hypothalamus and is a genetically determined female characteristic. The ovary undergoes cyclical development in response to GTHs. The duration of the ovarian cycle is characteristic for each species. During proestrous, the growth of one or more follicle occurs in the ovaries. This portion of the ovarian cycle is the **follicular phase**. The follicular phase results in development of one or more mature follicles, each containing one oocyte. Following ovulation, which ends the follicular phase, the remains of a ruptured follicle are transformed into a **corpus luteum**. Thus, ovulation also marks the onset of the **luteal phase** of the ovarian cycle. The luteal phase may last from a few days to weeks, depending upon the species. Some species, like humans, may begin another follicular phase during the latter portion of the luteal phase, whereas others may enter an inactive period (diestrus) that may last until the next breeding season when a new follicular phase is initiated.

1. The Follicular Phase of the Ovarian Cycle

During the follicular phase of the ovarian cycle (Figure 10-15), the tonic hypothalamic center releases small quantities of GnRH into the portal circulation and relatively low but rather constant circulating levels of FSH, LH, or both are maintained. Prior to puberty, which is characterized by increased GTH levels, the ovary contains **primordial follicles** consisting of **primary oocytes** invested with an additional layer of flattened follicle cells derived from the germinal epithelium that surrounds the ovary. In most mammals, it is assumed that there are no oogonia in the ovary because all of them entered meiosis and became primary oocytes prior to or shortly after birth. However, recent studies in mice have demonstrated new follicle development occurs after birth, and production of new oocytes has been described among certain primates as well.

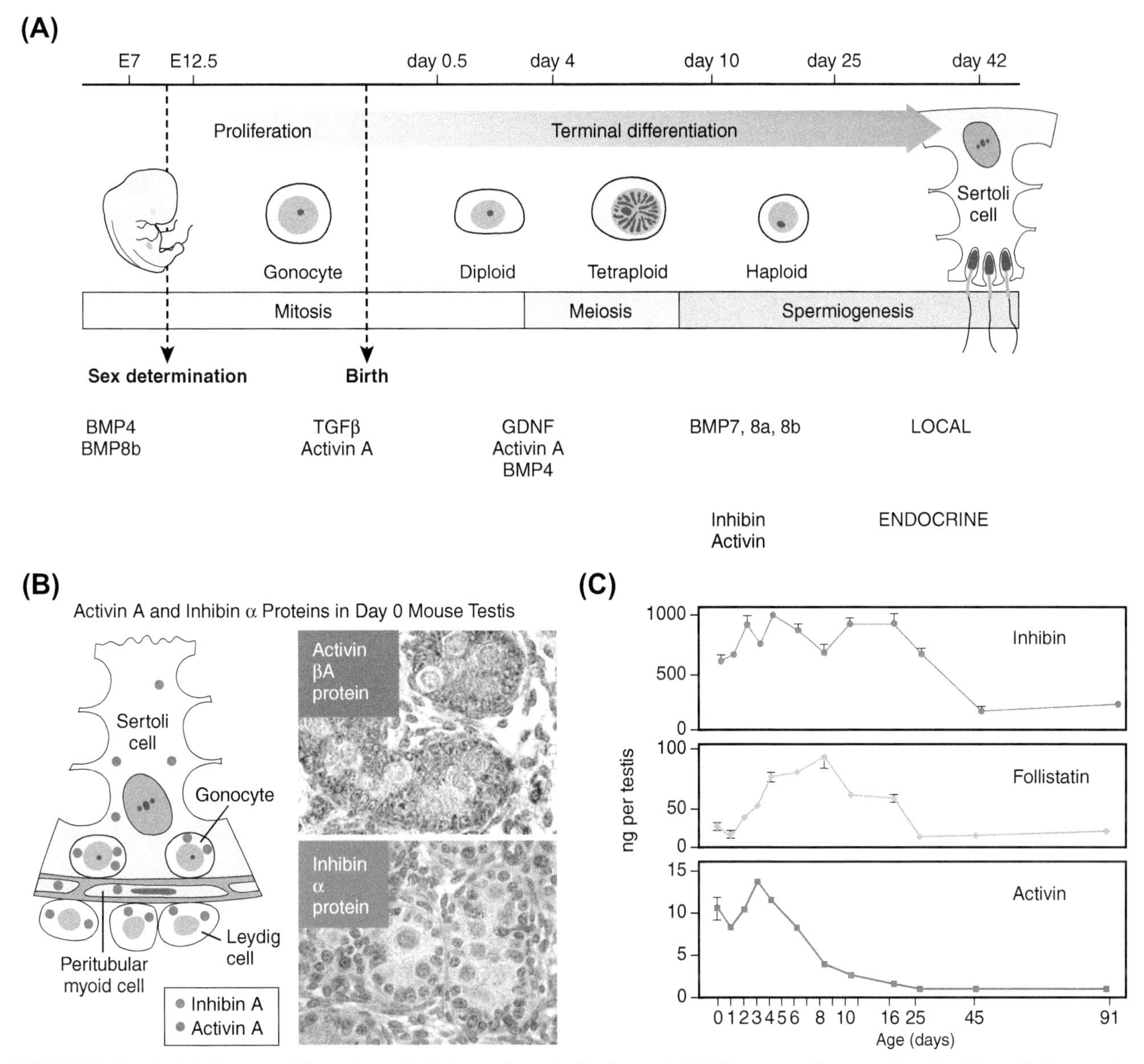

FIGURE 10-13 **Activin is secreted from the testis during embryonic development.** (A) Time course for expression of activin and other putative growth factors in testis during mouse development. Abbreviations: BMP, bone morphogenic protein; GDNF, glial cell-derived neurotrophic factor; TGFβ, transforming growth factor β. (B) Both inhibin A and activin A are highly expressed in the testis at birth. (C) Patterns in activin, inhibin, and follistatin production in mouse testis during postnatal development. See text for explanation of functions of these regulators. *(Adapted with permission from Barakat, B. et al.,* Molecular and Cellular Endocrinology, ***359****, 66–77, 2012.)*

The arrival of FSH at the ovary causes local release of AMH that stimulates a number of primordial follicles to begin to enlarge and differentiate into **primary follicles**. The follicle cells surrounding the growing oocyte develop into **granulosa cells**, which are in contact with the oocyte. The granulosa cells secrete a basement membrane along their outermost surfaces. A second layer of **thecal cells** derived from the ovarian stroma surrounds the granulosa outside of the basement membrane. Thecal cells further differentiate into inner and outer layers: the endocrine **theca interna** and the connective tissue-like **theca externa**. The rich supply of capillaries in the thecal layer do not cross the basement membrane and penetrate the granulosa layer. The presence of FSH receptors on the granulosa cells and their ability to proliferate mitotically is initiated by the oocyte through production of GDF-9 (Figure 10-14). In *Gdf9*-knockout mice, the ovary develops normally but follicle formation is blocked by the failure of follicle cells to respond to FSH.

As the follicle grows, the granulosa cells secrete the **liquor folliculi** or **antral fluid** that is primarily an ultrafiltrate of blood plasma. Increasing production of antral fluid

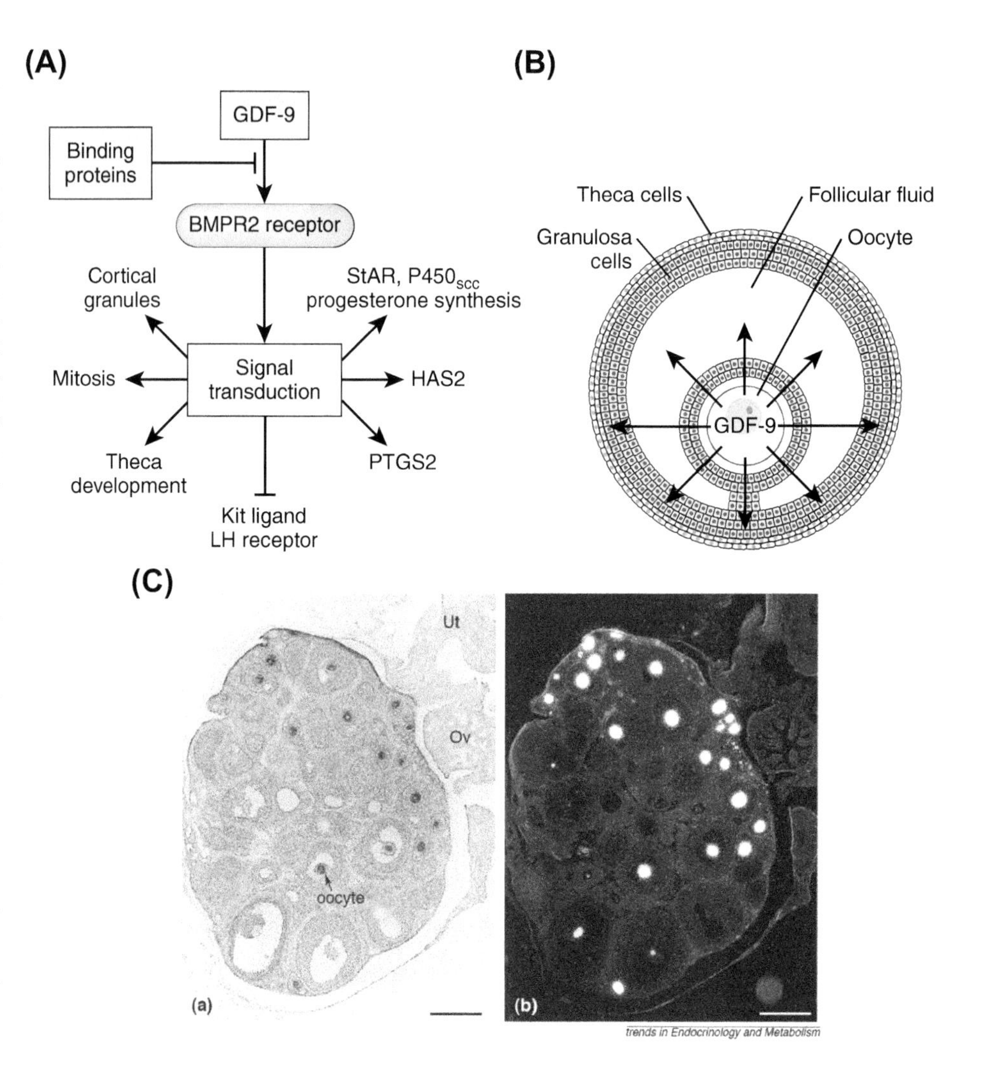

FIGURE 10-14 **Growth differentiation factor-9 (GDF-9) is required for folliculogenesis in the mammalian ovary.** (A) GDF-9 is produced by the oocyte and binds to surface membrane BMPR2 on follicle cells to initiate a wide variety of functions important for follicle growth including cholesterol transport and steroidogenesis (StAR, $P450_{scc}$), follicle growth (HAS2), prostaglandin synthesis (PTGS2), follicle recruitment (Kit ligand), cell division, theca growth, and the formation of cortical granules. (B) Gradient theory explaining the role of GDF-9 and other morphogens from the oocyte. (Adapted with permission from Erickson, G.F. and Shimasaki, S., *Trends in Endocrinology & Metabolism*, **11**, 193–198, 2000.) (C) Location of GDF-9 within oocytes. Mouse ovary showing follicle histology (a) and autoradiography showing location GDF-9 within oocytes only (b). Abbreviations: BMPR2, bone morphogenetic protein type II receptor; GDF-9, growth differentiation factor-9; HAS2, hyaluronan synthase 2; $P450_{scc}$, side-chain cleaving enzyme; StAR, steroid acute regulatory protein; Ut, uterus; Ov, oviduct; PTGS2, prostaglandin-endoperoxide synthase 2. *(Reprinted with permission from Erickson, G.F. and Shimasaki, S.,* Trends in Endocrinology & Metabolism, ***11**, 193–198, 2000. © Elsevier Science, Ltd.)*

results in formation and progressive enlargement of a fluid-filled cavity within the follicle, the **antrum**. The follicle is now called a **secondary follicle** or an **antral follicle**.

Under the influence of LH and FSH as well as a variety of paracrine factors (see Table 10-6), growing ovarian follicles synthesize and release estrogens, predominantly estradiol (= 17β-estradiol), into the general circulation. The synthesis of estrogens in the ovary appears to be a cooperative effort between cells of the theca interna and the granulosa (Figure 10-16). LH stimulates the thecal cells to produce androgens (principally androstenedione) that are aromatized by the granulosa cells to form estradiol. Conversion of androgens to estradiol by the granulosa cells is stimulated by FSH. In turn, FSH increases $P450_{aro}$ levels in these cells. In addition, FSH causes the granulosa cells to produce inhibins, which feed back on the pituitary to selectively inhibit FSH release as was described earlier for males. Inhibin also inhibits $P450_{aro}$ activity locally in granulosa cells, whereas the related peptide activin increases $P450_{aro}$ activity.

The final stage of follicle growth is the **tertiary** or **mature follicle** (also called a *graafian follicle*). This follicle has reached maximal size and often is characterized by a single large antrum surrounded by a relatively thin layer of granulosa cells with the oocyte relegated to and surrounded by a small mass of granulosa cells, the **cumulus oophorus**. The final meiotic maturation of the oocyte apparently is influenced by paracrine factors from cells of the cumulus oophorus following ovulation. The mature follicle is located just beneath the surface of the ovary.

Most (99%) of the follicles that begin to grow during a given ovarian cycle will exhibit apoptosis and degenerate, a process called **atresia** (Figure 10-17). These degenerating follicles are called **corpora atretica**. Atresia can occur at any stage of follicle development. Some of the steroidogenic cells from these atretic follicles will remain active and contribute to what has been called the **interstitial gland** of the ovary. Androstenedione produced in the interstitial gland by LH stimulation supplements thecal cell contributions for synthesis of estradiol by the granulosa cells.

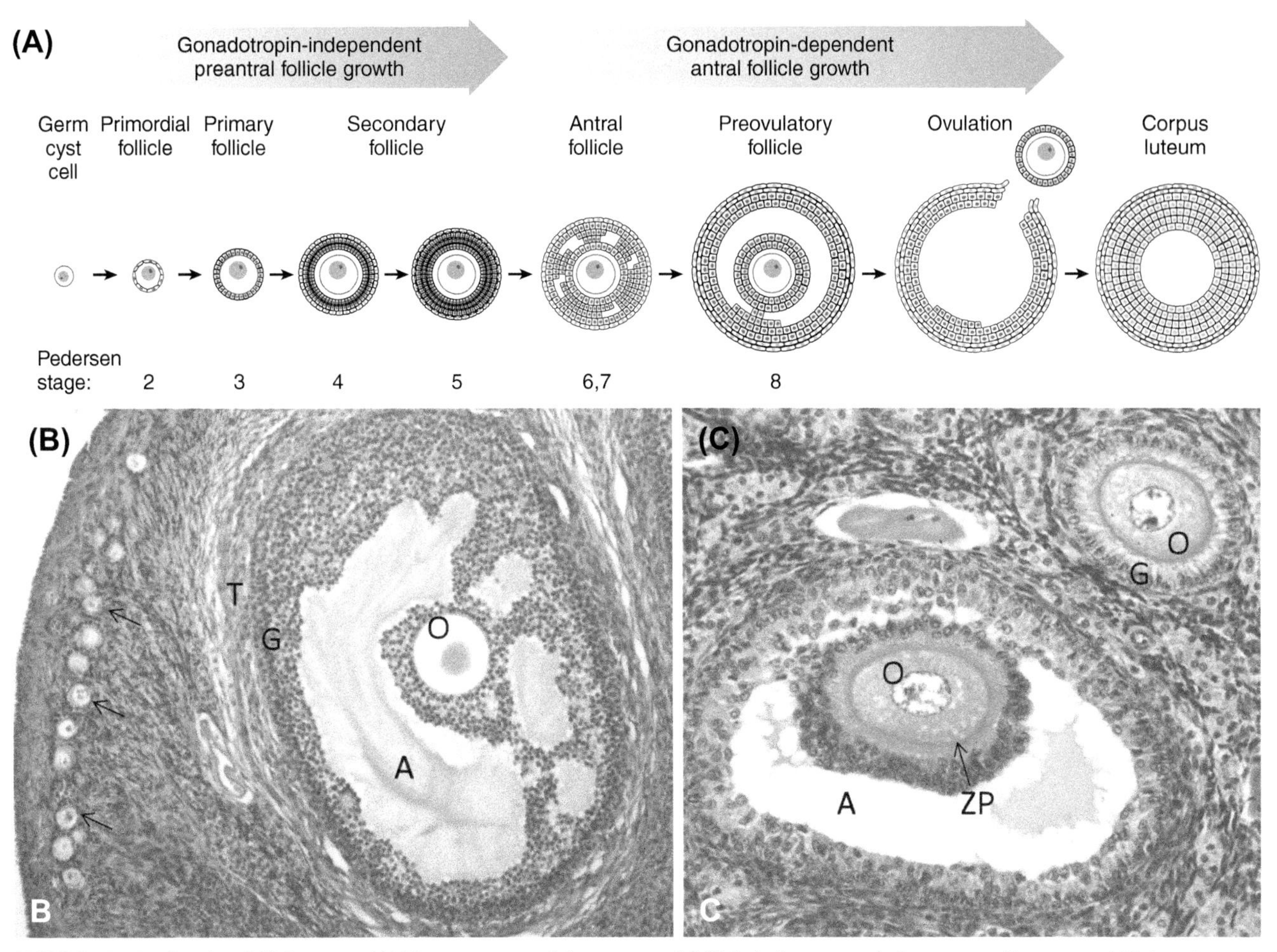

FIGURE 10-15 **Ovarian follicle stages.** (A) The appearance of the oocyte and follicle during oogenesis from germ cell to mature follicle and corpus luteum. (B, C) Sections of rat ovary showing stages of follicular development. (B) Mature or Graffian follicle with a large antrum (A) and clearly defined theca (T) and granulosa (G) cell layers. Primordial and primary follicles are indicated by arrows. (C) A secondary follicle and early antral follicle with a clearly defined zona pellucida (ZP). O, oocyte.

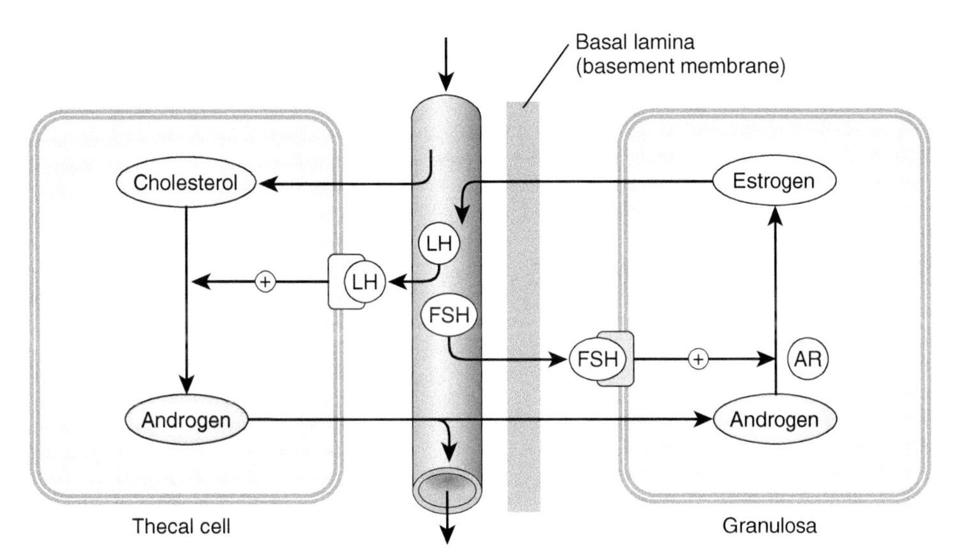

FIGURE 10-16 **Two cell model for steroidogenesis.** Binding of LH to receptors found only on thecal cells (or ovarian interstitial cells) stimulates androgen synthesis, most of which diffuses through the basal lamina (basement membrane) to the granulosa cell. FSH stimulates aromatase (AR) production, which transforms androgens into estrogens. A similar two-cell system is present in the testis but with both FSH and LH receptors associated with the interstitial cells (Leydig cells) and FSH receptors on the Sertoli cells. Androgens reach the Sertoli cell by diffusion through the basal lamina surrounding the seminiferous tubule.

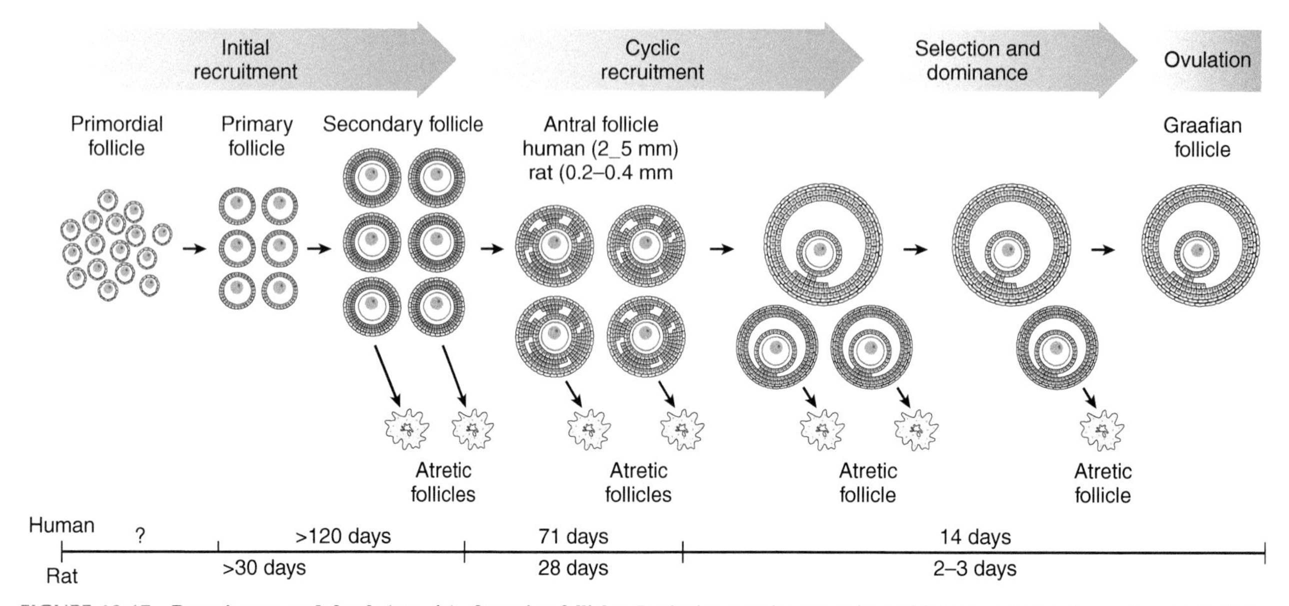

FIGURE 10-17 **Recruitment and death (atresia) of ovarian follicles.** Beginning at puberty a cohort of follicles selected from primordial follicles begins to grow. The majority of these follicles undergo programmed cell death, and one dominant follicle is selected for ovulation during each ovarian cycle from puberty to menopause. *(Adapted with permission from Norris, D.O. and Lopez, K.H., in "Hormones and Reproduction of Vertebrates. Vol. 5. Mammals" (D.O. Norris and K.H. Lopez, Eds.), Academic Press, San Diego, CA, 2011, pp. 59–72.)*

2. Ovulation

The process of ovulation involves the rupture of the mature follicle and release of the oocyte from the ovary into the body cavity. This event marks the end of the follicular phase and the beginning of the luteal phase in the ovary and is correlated with estrus. Ovulation occurs as a result of the progressive increase in circulating estradiol that occurs with the growth of the follicles. Increased estradiol also is responsible for estrous behavior in the female and the enhanced attractiveness of the female to the male at this time. A maximal or critical estrogen level in the blood in most cases activates the surge center in the hypothalamus, which releases a large pulse of GnRH as described in Chapter 4. The pulse of GnRH released results in the LH surge (Figure 10-18) that causes ovulation of one or more follicles within a matter of hours (usually 12 to 24 hours, regardless of the species). The LH surge results in a remarkable series of genetic switches being turned off (FSH gene expression program) and on (genes required for oocyte meiosis; expansion of the **cumulus cell oocyte complex**, or **COC**; and corpus luteum formation). The number of follicles that reach maturity and ovulate is species specific, varying from a norm of one in women to a dozen or more in the sow. The determining factors appear to be the amount of GTH available, and increased numbers of mature follicles are produced following supplementation with exogenous GTHs. The physical mechanism by which LH causes the mature follicle to rupture and release the mature oocyte is not understood completely but involves both an increase in growth of the COC and the production

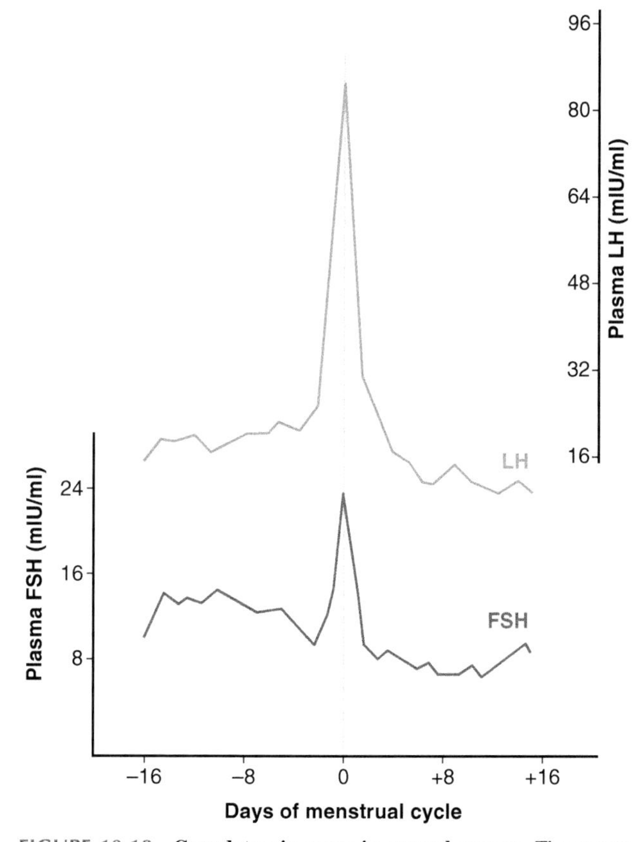

FIGURE 10-18 **Gonadotropin surge in normal women.** The greater release of LH is probably due to the presence of galanin released with GnRH at this time as well as negative feedback effects of inhibins on FSH release at the pituitary level. *Adapted with permission from McCann, S.M. (1974). Regulation of secretion of follicle-stimulating hormone and luteinizing hormone. Handbook of Physiology, Sec. 7, Endocrinology 4, 489–518.*

of **matrix metalloprotease (MMP)** enzymes that digest the collagen and elastic fiber components of the extracellular matrix (Box 10E).

In some species, meiosis in the oocyte that began early in life is not completed until after fertilization. Prior to fertilization, the ovulated cell is an arrested oocyte. If meiosis were completed prior to ovulation, this cell would be termed an ovum. The situation in mammals apparently varies from ovulation of oocytes to ova, but in the following discussions the ovulated cell in every case will be referred to as an ovum to simplify terminology.

Some mammals ovulate following coitus and are termed **induced ovulators**. Several carnivores (e.g., ferret, mink, raccoon, cat), rodents (e.g., *Microtus californicus*), lagomorphs (e.g., cottontail and domestic rabbits), at least one bat (lump-nosed bat), and several insectivores (e.g., hedgehog, common shrew) are confirmed induced ovulators. In these species, coitus is immediately followed by an LH surge that induces ovulation (Figure 10-19). Some other species are suspected to be induced ovulators, including the elephant seal, nutria, and long-nosed kangaroo rat (a marsupial). Most mammals are believed to be **spontaneous ovulators** in that the LH surge and ovulation are independent of coitus; however, even some spontaneous ovulators can be induced to ovulate following copulation under special conditions. Evidence from humans suggests that ovulation may be induced in rape cases especially if the female is very young.

3. The Luteal Phase of the Ovarian Cycle

Ovulation marks the onset of the **luteal phase** of the ovarian cycle as well as the end of the follicular phase. In addition to causing ovulation, the LH surge induces granulosa cells as well as some theca interna cells to differentiate into the corpus luteum (Box Figure 10E-1). This process, known as **luteinization**, results in the corpus luteum, which functions as an endocrine gland, secreting both estrogens and progesterone into the general circulation. One corpus luteum will form from each ovulated follicle. In addition, other developing follicles may undergo premature luteinization and function as **accessory corpora lutea** during pregnancy. The corpus luteum begins secreting large quantities of progesterone, along with lesser amounts of estradiol as well as other estrogens and progestogens. Circulating progesterone and estrogens inhibit both the tonic and cyclic hypothalamic GnRH centers during the luteal phase so that additional follicular development is arrested and a second ovulatory episode is prevented. All developing follicles that do not ovulate undergo atresia or form accessory corpora lutea. Some of the follicular cells from the atretic follicles will persist as part of the ovarian interstitial gland that is responsive to LH and synthesizes androstenedione that can be used as a substrate by the corpus luteum to form estradiol.

Depending on the species, regulation of corpora lutea function may require LH or be independent of LH once it has formed. In sheep, PRL together with LH apparently stimulates steroid secretion by the corpus luteum; however, only PRL is necessary to maintain the activity of the rat corpus luteum. Preovulatory estradiol can produce a surge of PRL release in several species and might be related to corpora lutea function. These actions of PRL on the corpus luteum were the basis for the older name of luteotropic hormone for this molecule; however, PRL has no role in corpus luteum functions in primates and most other mammals, and the older name should not be used.

The corpus luteum secretes steroids for only a relatively short period in many species (5 to 8 days in humans) after which it begins to degenerate if mating and fertilization were not successful. As the corpus luteum undergoes degeneration, steroidogenesis declines, and the uterus enters a regressive phase. In some species, the corpus luteum is relatively long-lived, especially in carnivorous species like the dog. Corpora lutea in the bitch are active for about 63 days after ovulation, which is equivalent to the normal gestation period, regardless of whether fertilization and pregnancy occurred. If unmated, the bitch will not reenter estrus until the next breeding season.

The predetermined life span for the functional corpus luteum has provided one of the most intriguing mysteries of the ovarian cycle. Apparently, the corpus luteum sows the seeds of its own destruction (Figure 10-20). In female rats, mice, hamsters, rabbits, guinea pigs, and ewes, breakdown of the corpus lutem (luteolysis) requires the production of the prostaglandin **$PGF_{2\alpha}$** from the uterine lining. Production of luteolytic pulses of $PGF_{2\alpha}$ requires the coordinated action of estradiol, OXY, and progesterone on their respective receptors in uterine epithelial cells (Figure 10-20). First, estradiol secretion from mature follicles increases the expression of **progesterone receptors (PRs)**, **OXY receptors (OXTRs)**, and **estrogen receptors (ERs)** in uterine epithelial cells. Progesterone action on the PR causes uterine cells to build up phospholipid in order to generate arachidonic acid, the precursor for prostaglandin synthesis. (see Chapter 3). Progesterone action then causes a downregulation or block of ERs and OXTRs during this period of phospholipid buildup followed by an upregulation or release of ERs and OXTRs due to suppression of PR. Pulses of OXY secreted from the posterior pituitary and corpus luteum act on uterine OXTRs to stimulate $PGF_{2\alpha}$ synthesis and luteolysis. The exact mechanism of luteolytic activity caused by $PGF_{2\alpha}$ is not clear, although it may relate to reducing blood flow to the corpus luteum via an interaction locally with **angiotensin II** and **endothelin 1**. In primates, the destruction of the corpus luteum toward the end of the luteal phase is not influenced by the uterus but appears to be

BOX 10E Mechanism of Ovulation

Ovulation occurs as a consequence of a series of events (Box Figure 10-E1) that take place in the follicular wall at an avascular site called the **stigma**. LH action on its receptor during ovulation triggers not only the **protein kinase A (PKA)** signaling pathway (see Chapter 3) but also the **rat sarcoma signaling (RAS)** cascade. Stimulation of the PKA pathway by LH results in the production of a series of epidermal growth factor-like signaling molecules, including **amphiregulin (AREG)**, **betacellulin** and **epiregulin (EREG)**. These **epidermal growth factor (EGF)**-like messengers act on their respective receptors on granulosa cells and cells in the cumulus oophorous to stimulate RAS and **extracellular signal-regulated kinases (ERK1/2)** that have several important effects. First, activation of ERK1/2 turns off FSH signaling and FSH-induced gene expression. Second, ERK1/2 activation increases the transcription of a number of genes involved in (1) **follicle rupture**, including chemokines (IL-6, for example), protease enzymes, and **prostaglandin-endoperoxide synthase 2 (PTGS2)**; (2) **luteinization**, including genes regulating steroidogenesis; and (3) **COC complex formation**, including **hyaluronan synthase 2 (HAS2)**. HAS2 plays an important role just prior to ovulation in making a hyaluronan-rich matrix that envelops the oocyte. The activity of a number of protease enzymes acts to weaken the follicular wall. Enzymatic breakdown products bring about an inflammatory response and the release of **prostaglandin (PGE_2)** that causes local blood vessels to constrict, leading to local ischemia and cell death. These events weaken the follicular wall further. Pressure builds up in the antral cavity and causes the follicular wall to rupture at its weakest point, the stigma. The ovum and surrounding cells of the cumulus oophorus (now called the **corona radiata**) are expelled along with the antral fluid. It is speculated that the smooth-muscle-like cells located in the follicular wall may be responsible for producing the increased antral pressure that triggers ovulation. Another hypothesis suggests that the increased pressure results from water influx into the antral fluid.

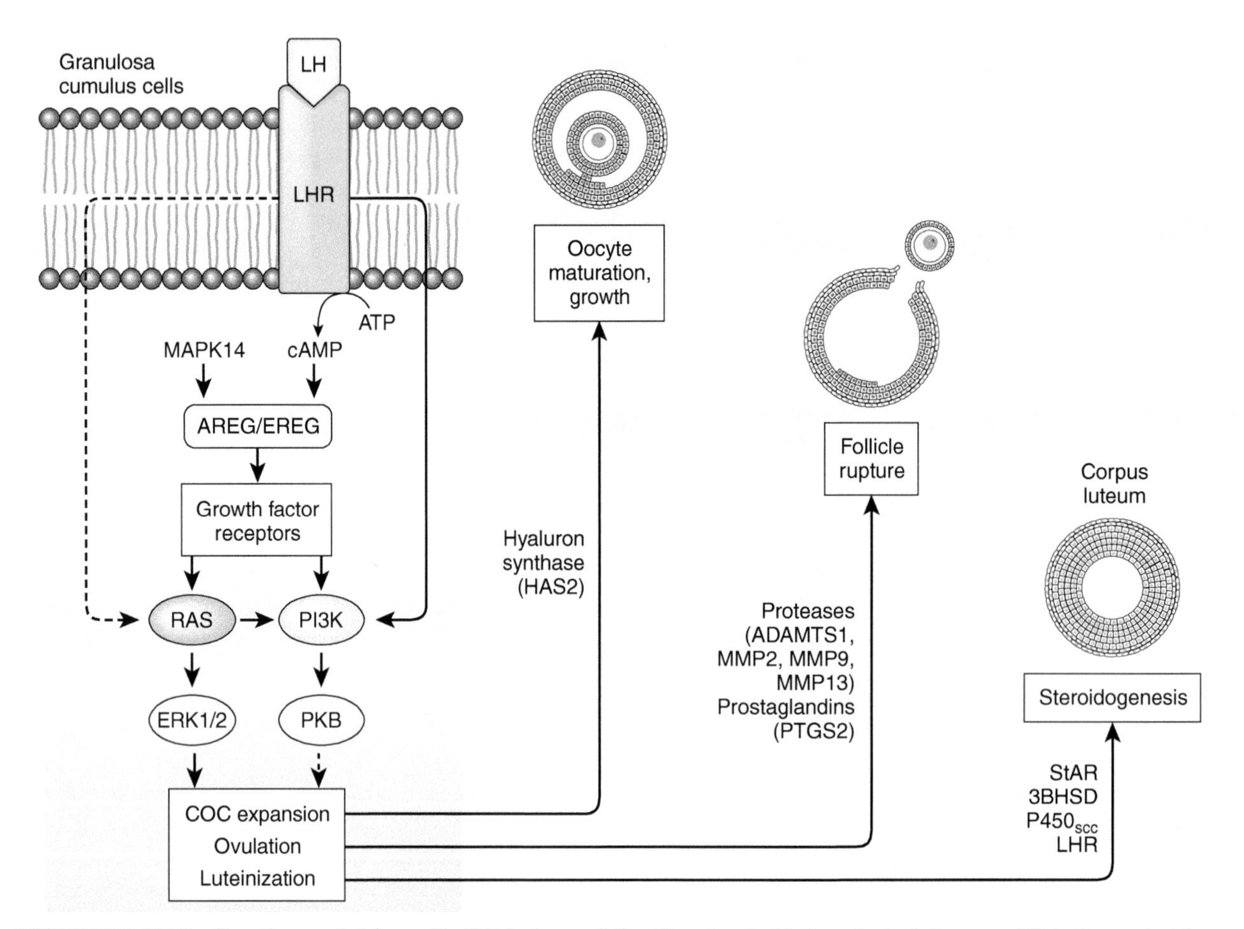

BOX FIGURE 10E-1 **Signaling events triggered by LH during ovulation.** How does the binding of a single hormone, LH, to its receptor trigger such a diverse array of cellular changes? In part because the hormone's signal is **amplified** by activation of multiple signaling pathways within the granulosa cells. The LH receptor triggers changes in the cAMP, RAS, and PI3K/AKT signaling pathways that lead to the production of proteins involved in oocyte maturation and cell growth, follicle rupture, and steroidogenesis required for ovulation and corpus luteum formation. 3βHSD, 3β-hydroxysteroid dehydrogenase; ADAMTS1; A disintegrin and metalloprotease (ADAM) with thrombospondin type 1 motif, 1; AKT, protein kinase B; AREG, amphiregulin; COC, cumulus cell oocyte complex; EREG, epiregulin; ERK, extracellular-signal-regulated kinases; LHR, LH receptor; MMP2, matrix metalloproteinase-2, 72 kDa type IV collagenase; MMP9, matrix metalloproteinase-9, 92 kDa type IV collagenase; MMP13, matrix metalloproteinase-13, collagenase 3; $P450_{SCC}$, P450 side-chain cleaving enzyme; PI3K, phosphatidylinositol 3-kinases; PKB, protein kinase B; RAS, rat sarcoma; StAR, Steroid acute regulatory protein. *(Adapted in part from Fan, H.Y., Liu, Z., Mullany, L.K., Richards, J.S., Consequences of RAS and MAPK activation in the ovary: the good, the bad and the ugly.* Molecular and Cellular Endocrinology, ***356(1–2)****, 74–79, 2012).*

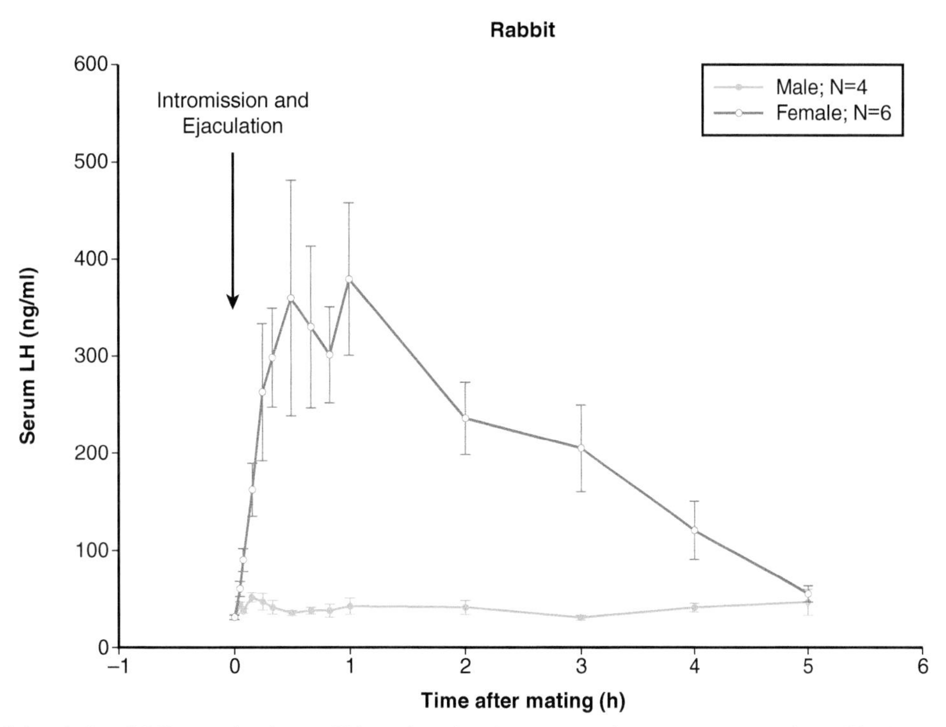

FIGURE 10-19 **Coitus induced LH secretion in a rabbit.** *(Adapted with permission from Jones, E.F. et al.,* Fertilization and Sterilization*, 27, 848–852, 1976.)*

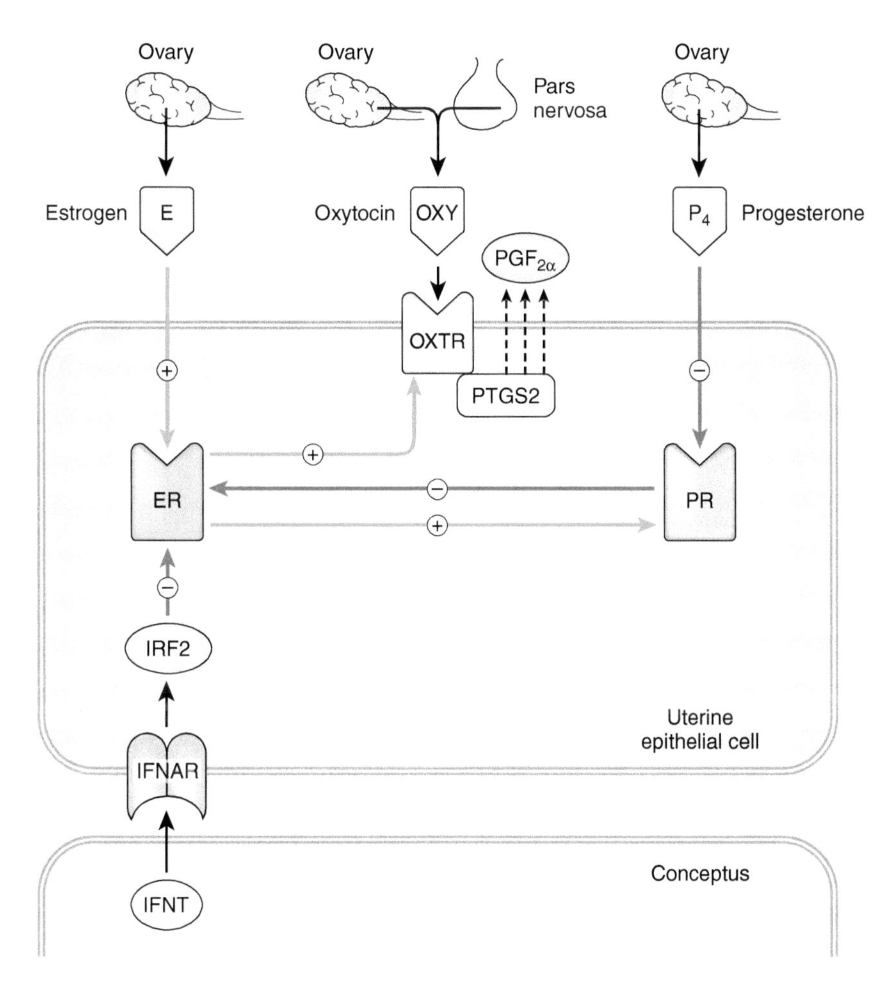

FIGURE 10-20 **Mechanism of luteolysis.** Estradiol action on uterine epithelial cells leads to increased expression of estrogen receptor alpha (ER), progesterone receptor (PR), and oxytocin receptor (OXTR). Increased PR availability leads to an increase in arachidonic acid, the substrate for prostaglandin PGF_2 synthesis. Increased progesterone secretion downregulates PR, leading to even greater expression of ER and OXTR. Oxytocin secretion from the pars nervosa leads to pulses of PGF_2 secretion, causing luteolysis. Signals (interferon tau, IFNT) from the developing embryo and associated tissues (conceptus) block this pathway, prolonging the life of the corpus luteum. Abbreviations: IFNAR, type 1 interferon receptor; IRF2, interferon regulatory factor 2; PTGS2, prostaglandin-endoperoxide synthase 2. *(Adapted from Bazer, F.W. and Spencer, T.E., in "Hormones and Reproduction of Vertebrates. Vol. 5. Mammals" (D.O. Norris and K.H. Lopez, Eds.), Academic Press, San Diego, CA, 2011, pp. 73–94.)*

caused locally by a luteolytic factor (e.g., estrone), produced by the corpus luteum itself. Once fertilization has taken place, there are several mechanisms by which corpus luteum degeneration may be prevented and the life of the corpus luteum prolonged (see ahead).

Degeneration of corpora lutea frees the hypothalamic GnRH centers from the inhibitory influence of estrogens and progesterone, resulting in a moderate increase in circulating GTHs and consequent renewal of follicular development. In fact, increased FSH release occurs in many species during the later stages of the luteal phase so that follicular growth may resume even before regressive uterine events become obvious.

The importance of the corpus luteum in maintaining pregnancy varies considerably as does the role of pituitary hormones in stimulating corpus luteum function. For example, in rats, PRL is necessary for maintaining the first half of gestation through actions on the corpus luteum. However, in pigs, the corpora lutea secrete progesterone to maintain the uterine secretory phase during the early portion of the gestation period without aid of any pituitary hormones. In ewes, both LH and PRL are necessary for maintaining corpus luteum function during the first third of pregnancy, but maintenance of pregnancy actually resides in the ability of the conceptus to neutralize the uterine luteolytic factor $PGF_{2\alpha}$. Estrogens of placental origin apparently are responsible for prolonging the life span of corpora lutea in rabbits as well as promoting progesterone synthesis. If the estrogen-secreting placental cells are damaged (for example, by x-rays), pregnancy is abruptly terminated.

B. The Uterine Cycle

The uterine cycle can be separated into a **proliferative phase** corresponding to the ovarian follicular phase, a **secretory phase** corresponding roughly to the ovarian luteal phase, and a **post-luteal phase** (Figure 10-21). The proliferative phase is separated from the secretory phase by the occurrence of ovulation. The end of the luteal phase marks the entrance into the post-luteal phase, which is a quiescent period of slow regression of the uterine epithelium in most mammals and is termed the **menses** in mammals that exhibit menstrual cycles. The wall of the uterus consists of an outermost connective tissue covering, a thick intermediate layer of smooth muscle called the **myometrium**, and an innermost epithelium, the **endometrium**, that contacts the uterine lumen. The endometrium can be further separated into an outer **basal layer** that proliferates during each uterine cycle to produce an inner **functional layer** that later regresses.

1. The Proliferative Phase of the Uterine Cycle

Estradiol produced during the follicular phase by the ovary stimulates differentiation and proliferation of the endometrium in preparation for implantation of the **blastocyst**, an embryonic stage formed from the first series of cellular divisions following fertilization. The blastocyst consists of an outer extraembryonic layer of cells, the **trophoblast**, which will form the fetal component of the placenta, and an **inner cell mass**, which will become the embryo proper. The proliferative phase of the uterine cycle is characterized by hyperplasia of the basal layer of the endometrium to form the functional layer in response to estrogens secreted by the growing ovarian follicles. In addition, there is a marked increase in the vasculature (hyperemia) of the functional layer as a result of estradiol stimulation. In higher primates, this hyperemic response includes the development of special **spiral arteries** that play an important role in the menstrual cycle of these mammals. There may be two independent targets for estradiol in the uterus. One target is the basal epithelial cell that responds to estrogens with new protein synthesis and mitosis. This results in an increase in the functional layer as well as in the development of tubular uterine exocrine glands. A second target for estrogens is the uterine eosinophil, which is a white blood cell that has infiltrated the uterine lining. These eosinophils possess specific receptors for estradiol and appear to be responsible for the rapid uptake of water and release of histamine that causes local hyperemia.

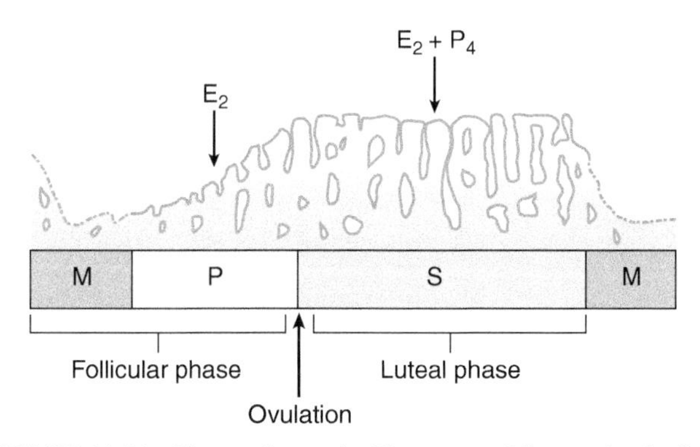

FIGURE 10-21 **The uterine cycle.** The menses (M) occupies the first 5 days of the cycle. The endometrium is stimulated by estradiol (E_2) during the proliferative stage (P). The follicular phase in the ovary corresponds to M + P. Following ovulation the corpus luteum secretes E_2 and progesterone (P_4) during the luteal phase which maintain the vascularity of the endometrium as well as secretion by exocrine glands during the secretory phase (S). Following the death of the corpus luteum, the uterine lining degenerates and the uterus reenters menses.

2. The Secretory Phase of the Uterine Cycle

Progesterone and estradiol secreted by corpora lutea in the ovary are the hormones controlling this phase, which is characterized by secretion of the exocrine glands in the functional layer of the endometrium. During the secretory phase, estradiol maintains the proliferated uterine endometrium and increased hyperemia that were initiated during the proliferative phase. Progesterone stimulates the uterine glands to secrete a fluid called **uterine milk** or **embryotroph**. Uterine milk is believed to be a source of nourishment for unimplanted blastocysts. The endometrium of both eutherian and marsupial mammals produces uterine milk. Progesterone and estradiol also maintain the highly vascularized state of the uterus necessary for implantation and early development of the embryo. Uterine muscle becomes desensitized by progesterone, reducing the chance that rhythmic contractions of uterine smooth muscle might dislodge an implanting or recently implanted blastocyst.

3. The Post-Luteal Phase of the Uterine Cycle

If fertilization is successful and implantation occurs, the secretory phase will continue throughout pregnancy. Should implantation not occur, the corpus luteum of many eutherian mammals will rapidly degenerate, resulting in a marked decrease in circulating levels of progesterone and estrogens. This decrease in circulating ovarian steroids causes regressive changes in the endometrium following steroid withdrawal. The endometrium becomes less vascular, and secretion by the uterine glands is reduced. Thus, the uterus becomes less capable of supporting implantation of a blastocyst. In monestrous species such as carnivores, the uterus would enter the quiescent post-luteal phase called diestrus during which the functional layer would be slowly resorbed. In polyestrous species, however, the female may quickly re-enter proestrus, resulting in the resumption of endocrine secretions that would prevent uterine regression.

Instead of a quiescent post-luteal phase, animals with a menstrual cycle exhibit a rapid regression and actual sloughing of the outer portion of the endometrium during the menses if implantation is not successful. In higher primates, the spiral blood vessels constrict, preventing flow of blood to the functional layer of the endometrium and causing extensive cell death. The degenerating tissue and trapped blood are sloughed into the uterine lumen, where they are resorbed or discharged as menstrual flow. Following the menses, considerable rebuilding of the endometrium (another proliferative phase) must occur during the next follicular ovarian phase to prepare for implantation of blastocysts resulting from the next ovulation.

C. The Pregnancy Cycle

Estrus usually occurs just prior to ovulation and normally leads to mating. The recently ovulated ovum still surrounded by some of the granulosa cells, the **corona radiata**, enters the open upper end of the fluid-filled oviduct and is propelled toward the uterus by the action of cilia lining the oviduct and the muscles of the oviduct wall. Contractility of the oviductal smooth muscle is controlled by adrenergic nerves, steroid hormones, nitric oxide, and prostaglandins. Sperm deposited in the vagina by the copulating male during estrus are transported at least in part by peristalsis through the uterus and ascend into an oviduct in which recently ovulated ova are descending. Fertilization leading to successful implantation typically occurs in the upper third of the oviduct. Cleavage begins soon after fertilization, and the **zygote** or fertilized egg rapidly becomes a minute, multicellular blastocyst. The trophoblast produces enzymes that enable the blastocyst to implant (i.e., erode the highly vascularized, secretory uterine endometrium and settle in for development). Following implantation, the outer layer of the blastocyst, the trophoplast, will give rise to the extraembryonic membrane called the **chorion**. Implantation marks the beginning of gestation or pregnancy. In some species, the blastocyst may not implant immediately into the endometrium but may remain in the uterine lumen for a period of time before implanting. This **delayed implantation** (see Box 10F) allows species with a short developmental period to prolong the time before birth will take place after mating. The gestation period is specific for each species and may be as short as 12 days in the opossum (a marsupial) or as long as 22 months in an elephant (eutherian).

In carnivores, such as the domestic dog, the corpora lutea normally function throughout gestation since the length of the normal luteal phase is equal to the gestation period (see before). In others, the corpora lutea would degenerate much earlier with respect to the time required for gestation if fertilization and implantation were not successful. A central question puzzled reproductive physiologists for many years: How did mammals “know” they were pregnant and how did they prolong corpora luteal function and prevent premature regression or sloughing of the endometrium? It turns out there are several mechanisms.

The signal for prolongation of corpora luteal function in some species is the synthesis of an LH-like **chorionic gonadotropin** (CG) by the blastocyst even before implantation. Placental GTHs are structurally very similar to pituitary GTHs and generally produce LH-like effects (see Chapter 4). Their synthesis and release, however, are not influenced in a negative way by steroids in the manner of the steroidal feedback on pituitary GTHs. The trophoblast of the developing human blastocyst begins to secrete hCG prior to implantation. hCG appears in maternal blood within a few days of fertilization and soon after appears in sufficient quantities in urine to be detected with antibody-based pregnancy kits. Later, the trophoblast will contribute to the placenta following implantation and will continue to secrete hCG throughout pregnancy.

In the mare, only fertilized ova ever reach the uterus, implying some sort of early chemical recognition that

BOX 10F Delayed Implantation

Several eutherian mammals, such as mink, bats, and skunks, have evolved a mechanism known as **delayed implantation** whereby development of the blastocyst is arrested and the unimplanted blastocyst remains in the oviduct or uterus for an extended period prior to implantation. Among some eutherian mammals, delayed implantation appears to be an adaptation allowing copulation to occur at a particular time that is especially advantageous to the parent while ensuring that the young are born at the most favorable time for their survival. Neither the basis for causing the blastocyst to remain in a healthy, arrested state nor the stimulus to bring about implantation is known.

Macropodid marsupials have developed a form of delayed implantation called **embryonic diapause**. Embryonic diapause has been reported for 14 macropodid species but does not occur in at least one species, the western gray kangaroo. One major difference from delayed implantation occurring in eutherian mammals is the condition of the resting blastocyst. The macropodid blastocyst consists of about 70 to 100 cells of a uniform type termed **protoderm**. The macropodid blastocyst is surrounded by a shell membrane and an albumen layer. It has not yet differentiated into embryonic (inner cell mass) and extraembryonic (trophoblast) regions like that of the eutherians.

Presence of a joey suckling on a teat presumably evokes release of OXY from the pars nervosa. Oxytocin is believed to arrest corpus luteum functions while allowing lactation to occur. Removal of the suckling joey will allow the resting blastocyst to implant. Ovariectomy following ovulation induces diapause, but if ovariectomy is performed during diapause there is no effect on the duration of diapause. Progesterone administered to either intact or ovariectomized females stimulates cessation of diapause and reinstates blastocyst development. Estrogen is also effective, but continued embryonic development is not as successful as following progesterone treatment.

In the red kangaroo, embryonic diapause may be an adaptation to renew pregnancy immediately following the death of the joey living in the pouch. The gestation period for the red kangaroo is 33 days. After birth the newborn must find its way to the pouch virtually unaided. When it reaches the pouch, the joey attaches itself permanently to a teat and continues development as an exteriorized fetus. Soon after parturition, the mother kangaroo enters estrus again and mates. The presence of one joey in the pouch inhibits implantation of the new blastocyst resulting from the second mating. The new blastocyst remains in a suspended state of development for up to 200 days, at which time the first joey normally disengages itself from the teat and ventures into the outside world as a juvenile kangaroo. The newly liberated kangaroo will return at intervals to the teat to which it was formerly attached for nourishment. Meanwhile, the detachment of the first joey from the teat either releases an inhibition to implantation or provides an endocrine stimulus for implantation of the waiting blastocyst.

In about 4 weeks, the gestation period terminates in birth of the second joey. The new joey enters the pouch and attaches to a teat. The mother kangaroo again enters estrus and mates, and another blastocyst enters embryonic diapause. Thus, a female red kangaroo may have a young juvenile that requires occasional nourishment, a joey attached to a teat, and a blastocyst "waiting in the wings." Secretion of PRL stimulates milk secretion and prevents progesterone synthesis by the corpus luteum that is necessary to sustain an implantation. During extensive periods of drought, the older joey could be denied milk and allowed to die. The implantation of the waiting blastocyst soon provides another joey whose demands upon the mother's nutritional reserves and water supply would be very small in comparison to the demands of the larger joey. Next, the mother would reenter estrus, produce another diapausing blastocyst, and be ready to continue reproducing should conditions improve.

fertilization has occurred. The equine chorionic gonadotropin is called **pregnant mare serum gonadotropin (PMSG)** and appears in large amounts in the urine of pregnant horses.

LH-like GTHs can prolong the life of the corpus luteum that continues to secrete progesterone and estradiol, thus maintaining the secretory phase of the uterus. Secretion of ovarian steroids by the corpus luteum inhibits hypothalamic centers controlling pituitary GTH release so that follicular development and subsequent ovulation are blocked in pregnant animals. Suppression of subsequent follicular development and ovulation prevents having several embryos in the uterus at different stages of development. This would create the problem of expulsion of the younger embryos and fetuses during parturition of the oldest one(s).

While the HPG system is more or less shut down during pregnancy, the placenta begins to function as a composite HPG in the female. Thus, we observe secretion of hypothalamic peptides (e.g., GnRH, TRH, CRH), tropic-like hormones (e.g., ACTH, CG), and gonadal steroids (estrogens, progesterone). During the last third of pregnancy, another pituitary-like hormone, **chorionic somatomammotropin (CS)**, is secreted by the placenta in a number of species (primates, mice, rats, voles, guinea pigs, sheep, chinchillas, and hamsters but not bitches or rabbits). This placental hormone has both growth hormone (GH)-like and PRL-like activities. Antibodies to CS will cross-react with both GH and PRL in at least some of these species. The major roles for CS appear to be effects on metabolism (GH-like) and stimulation of the mammary gland to begin milk synthesis during the later stages of pregnancy. In humans, hCS formerly was called **human placental lactogen (hPL)**. The human placenta also secretes PRL that is identical to pituitary PRL. Placental

PRL accumulates in the amniotic fluid during pregnancy where it is thought to regulate volume and ionic composition of amniotic fluid. Levels of amniotic PRL are not affected by drugs that block maternal pituitary PRL release or even by hypophysectomy of the mother.

D. Birth (Parturition)

The birth process requires coordinated hormonal changes that culminate in the expulsion of the fetus and the associated placenta. Birth can be related to levels of estrogens, progesterone, OXY, prostaglandins, relaxin, corticosteroids, and CRH. The pattern varies in different mammals as to what hormones are involved and the patterns of their secretions. In many mammalian species, the end of gestation is marked by a dramatic decrease in circulating progesterone levels (Figure 10-22). However, in humans and guinea pigs, progesterone levels do not decline before birth and parturition (Figure 10-23).

In sheep, there is a marked reduction in circulating progesterone levels just prior to birth that, presumably, sensitizes the uterus to OXY. The contractions of the uterus initiated by OXY result in expulsion of the fetus as well as of the **afterbirth** (the detached placenta). Experimental studies in sheep show that the fetal adrenal axis plays an essential role in the initiation of the birth process (Figure 10-24). Factors that interfere with adrenal function at any level retard the normal onset of labor, and premature birth can be induced by addition of ACTH or corticosteroids. In sheep, the timing of parturition appears to depend on maturation of the fetal hypothalamus and CRH (Figure 10-24).

In humans, there is no drop in progesterone to trigger birth, although there is a relative increase in estrogen production as compared with progesterone (see Figure 10-23). Not only is the fetal adrenal essential for maintaining pregnancy, but it also is involved in the events associated with birth (Figure 10-25). Recent studies demonstrate that CRH from the placenta acts upon the fetal adrenal to produce DHEA, which is converted to estriol and estradiol, thereby altering the blood ratio of estrogens to progesterone (Figure 10-25). Estrogens increase synthesis

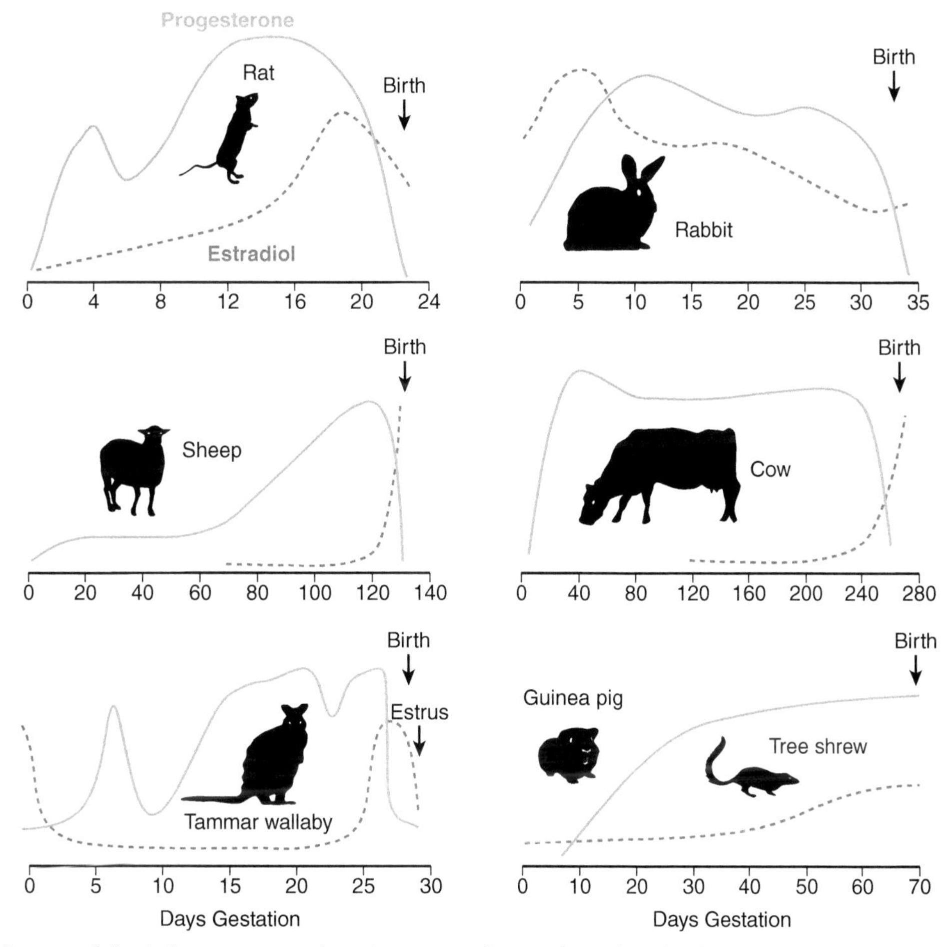

FIGURE 10-22 **Patterns of circulating progesterone in various mammalian species.** *(Adapted with permission from Young, I.R. et al., in "Hormones and Reproduction of Vertebrates. Vol. 5. Mammals" (D.O. Norris and K.H. Lopez, Eds.), Academic Press, San Diego, CA, 2011, pp. 95–116.)*

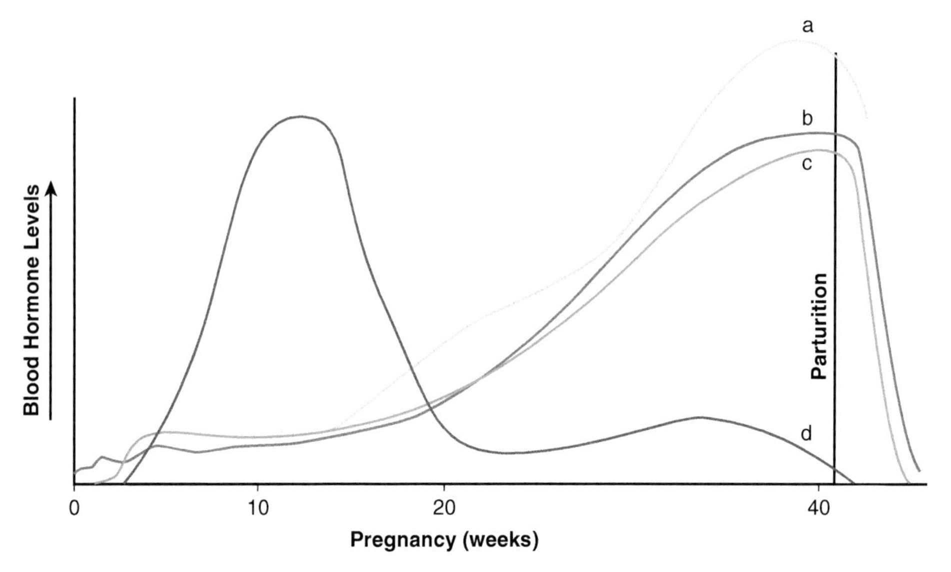

FIGURE 10-23 **Pattern of hormone secretion during human pregnancy.** Note that maternal progesterone levels do not decrease until detachment of the placenta. (a) hCS; (b) estrogens; (c) progesterone; (d) hCG. *(Adapted with permission from Bolander, F.F.* Molecular Endocrinology, *Academic Press, 1989.)*

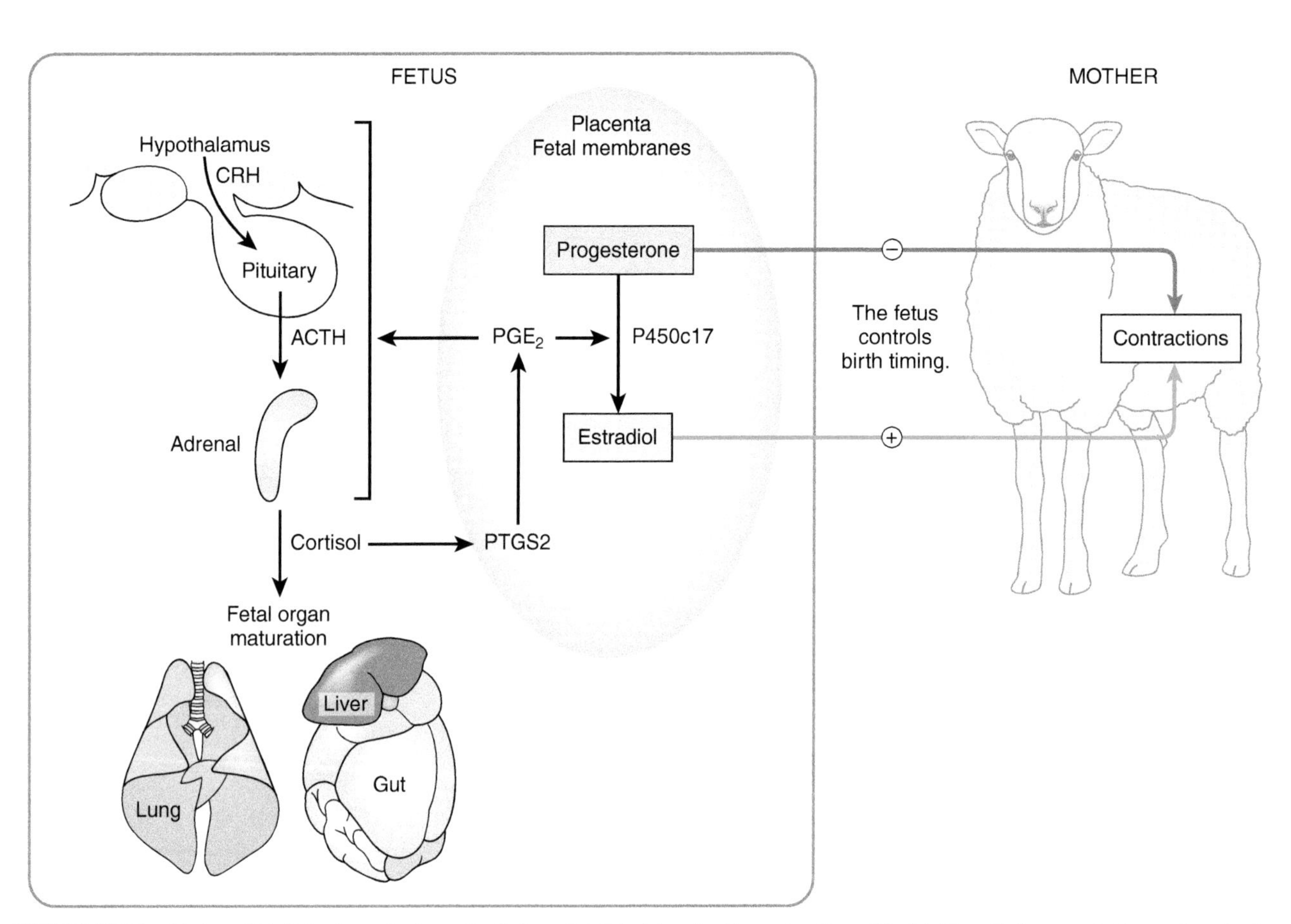

FIGURE 10-24 **Timing of birth in sheep.** In sheep, the timing of birth is determined by maturation of the fetal hypothalamus–pituitary–adrenal axis. Abbreviations: P450c17, 17α-hydroxylase/17,20 lyase; PGE_2, prostaglandin E_2; PTGS2, prostaglandin synthase 2. *(Adapted with permission from Young, I.R. et al., in "Hormones and Reproduction of Vertebrates. Vol. 5. Mammals" (D.O. Norris and K.H. Lopez, Eds.), Academic Press, San Diego, CA, 2011, pp. 95–116.)*

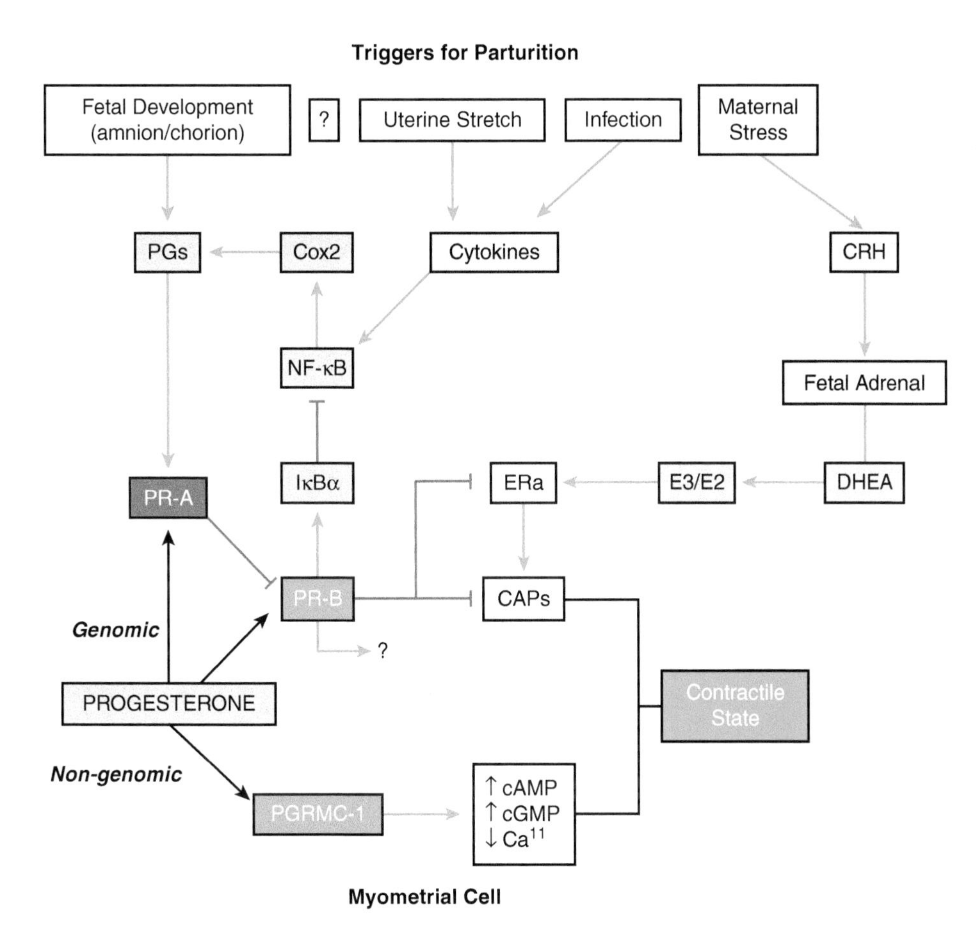

FIGURE 10-25 **Timing of birth in humans.** Contraction of the uterus is stimulated by placental CRH stimulating estrogen synthesis by the placenta and by blocking the PR-B form of the progesterone receptor, which promotes uterine relaxation. Both cytoplasmic (PR-A, PR-B) and surface membrane receptors for progesterone are involved in the control of uterine muscle contraction. The blockade of PR-B receptors is initiated by NF-κB (nuclear factor kappa-light-chain-enhancer of activated B cells), a transcriptional regulatory protein found in virtually all cells. NF-κB promotes the synthesis of PR-A, which represses the expression of PR-B, thereby allowing for stronger uterine contractions. Abbreviations: CAPs, contraction-associated proteins; COX2, cyclooxygenase isoform 2 (also known as prostaglandin-endoperoxide synthase 2, PGS2); DHEA, dehydroepiandrosterone; E_3, estriol; ER, estrogen receptor; Iκβα, nuclear factor of kappa light polypeptide gene enhancer in B-cells inhibitor, alpha; P_4, progesterone; PG, prostaglandin; PGRMC-1, progesterone receptor membrane component 1. *(Adapted with permission from Young, I.R. et al., in "Hormones and Reproduction of Vertebrates. Vol. 5. Mammals" (D.O. Norris and K.H. Lopez, Eds.), Academic Press, San Diego, CA, 2011, pp. 95–116.)*

of prostaglandins and of OXY receptors in the uterine myometrium, facilitating contractions. Cortisol also increases CRH production by the placenta at this time, setting up a positive feedback on the fetal adrenal via placental CRH as well as increasing the effects of CRH on the myometrium. Women with higher levels of CRH early in pregnancy are more likely to exhibit higher levels later on and give birth prematurely (Figure 10-26). Cortisol from the fetus near term also induces the lungs to begin production of surfactants that reduce surface tension on the lung surface and will be essential for the switch to breathing air that occurs after birth.

Mechanical stimulation of the vagina, cervix, or uterus can release OXY in humans and induce a fetal ejection reflex. Administration of prostaglandins also can induce uterine contractions, and OXY may stimulate prostaglandin synthesis in the uterus. Synthetic OXY is normally used to induce labor in women and frequently is given to reduce postpartum bleeding following detachment of the placenta. OXY is preferred over prostaglandins for clinical uses even though both are involved in normal births because administration of prostaglandins tends to produce strong contractile effects on non-reproductive smooth muscle as well (i.e., gastrointestinal smooth muscle).

Relaxin causes relaxation and softening of estrogen-primed pelvic ligaments, allowing the pelvis to stretch and expand (relax) during birth. This expansion allows the relatively large head of the eutherian fetus to pass through the pelvis during parturition. Relaxin reaches peak levels prior to birth and rapidly disappears from the maternal circulation afterward. Spontaneous motility of the uterus may be inhibited by relaxin in some mammals, thereby reducing the risk of premature birth. Relaxin working with estrogens, progesterone, and prostaglandins actually can alter the structural collagen of the uterine cervix, increasing its distensibility at parturition. There are data supporting an action of relaxin in combination with steroids and PRL on the mammary gland and the onset of lactation following birth.

The corpus luteum is the major source of relaxin in species where the corpus luteum is retained throughout gestation (pig, rat, carnivores). Relaxin is produced by the human corpus luteum during early gestation and to some extent by the placenta. Only a little relaxin is found in placentas of sheep, rats, cows, and rabbits, but in horses the placenta is the major source of relaxin. In humans, the

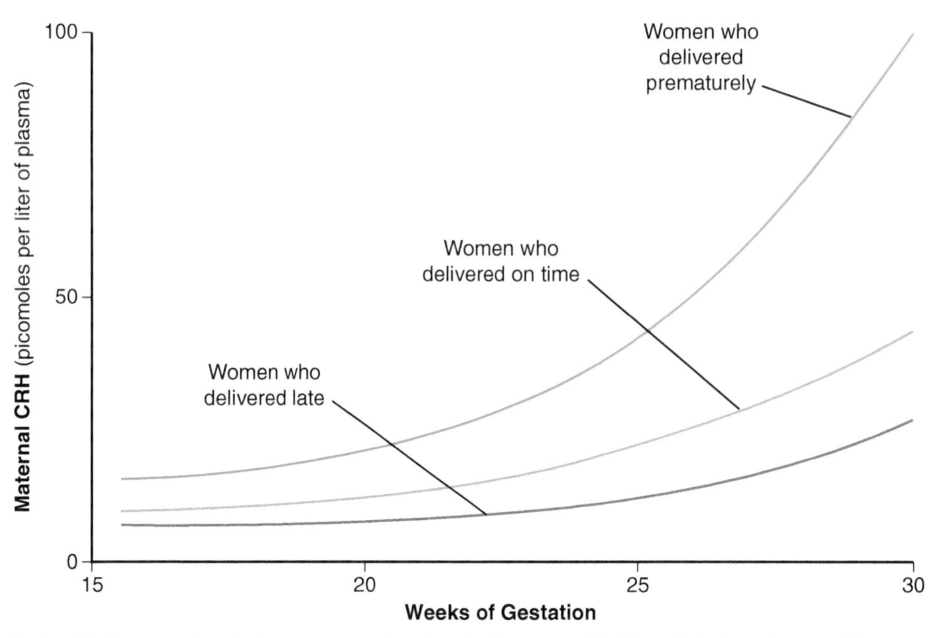

FIGURE 10-26 **Relationship between circulating corticotropin-releasing hormone (CRH) and timing of parturition.** As early as 15 weeks of pregnancy, it is possible to predict which women are at risk for premature birth by monitoring maternal blood levels of CRH. Low CRH delays birth, whereas high CRH levels indicate premature delivery is very likely. *(Adapted with permission from Young, I.R. et al., in "Hormones and Reproduction of Vertebrates. Vol. 5. Mammals" (D.O. Norris and K.H. Lopez, Eds.), Academic Press, San Diego, CA, 2011, pp. 95–116.)*

ovarian interstitial cells continue to be the major site for relaxin synthesis during pregnancy even after death of the corpus luteum.

Relaxin is chemically similar to insulin and insulin-like growth factors (IGFs) and consists of two short A-chains (22 to 24 amino acids) and a longer B-chain (26 to 35 amino acids) joined together by disulfide bonds (see Chapter 3, Figure 3-4). The positioning of the disulfide bonds is the same as for insulin and the IGFs, although there are many differences in amino acid sequences. It has been suggested that the relaxin gene arose by duplication from the insulin gene. Subsequent to this duplication, there has been considerable divergence in the relaxin genes among mammals as evidenced by considerable variation in amino acid sequences of mammalian relaxins. Relaxin acts on two G-protein-coupled receptors called the **relaxin/insulin-like family peptide 1 and 2 receptors (RXFP1 and 2)** to carry out its actions on the reproductive tract. Both receptors stimulate the cAMP/PKA signal transduction pathway (Figure 10-27). There also is evidence that relaxin/RFXP interaction can transactivate receptor tyrosine kinase and that relaxin can interact directly with glucocorticoid receptors (Figure 10-27). Relaxin receptors are part of a family of G-protein-coupled receptors called the **leucine-rich repeat-containing G-protein-coupled receptors (LGRs)**. Other family members include the receptors for the glycoprotein hormones FSH, LH, and TSH. Relaxin receptors were first identified as members 7 and 8 (LGR-7, LGR-8) of this receptor family.

E. Lactation

The development of mammary glands, their synthesis of milk, and the ejection of milk to the suckling offspring are all regulated by hormones. Mammary glands in eutherian mammals usually occur as paired structures, from 2 to 18, and may be located on the thorax (human, elephant, bat), along the entire ventral thorax and abdomen (sow, rabbit), in the inguinal region (horse, ruminants), along the abdomen (whale), or even dorsally (nutria, a South American rodent). These glands are apparently modified sweat glands, glands that are also unique to mammals. The internal structure is rather uniform and includes supporting stromal cells and a glandular epithelium that is organized into clusters of minute, sac-like structures called **alveoli**. It is this glandular epithelium that is responsible for the synthesis of milk. The alveoli are continuous with ducts and various duct-derived enlargements for storing milk. In addition, there are modified epithelial cells that contain muscle-like myofilaments parallel to the long axis of the cells. These cells are termed **myoepithelial cells** and are capable of contracting and causing ejection of milk from the alveoli into the duct system and out of the gland in the region of the nipple.

Information obtained from the mouse and rat indicates that differentiation of mammary glands from ectoderm involves specific induction by a particular underlying mesenchyme. These glands in both mother and fetus normally undergo hyperplasia and hypertrophy with the aid of estrogens during the last third of the gestational period. The

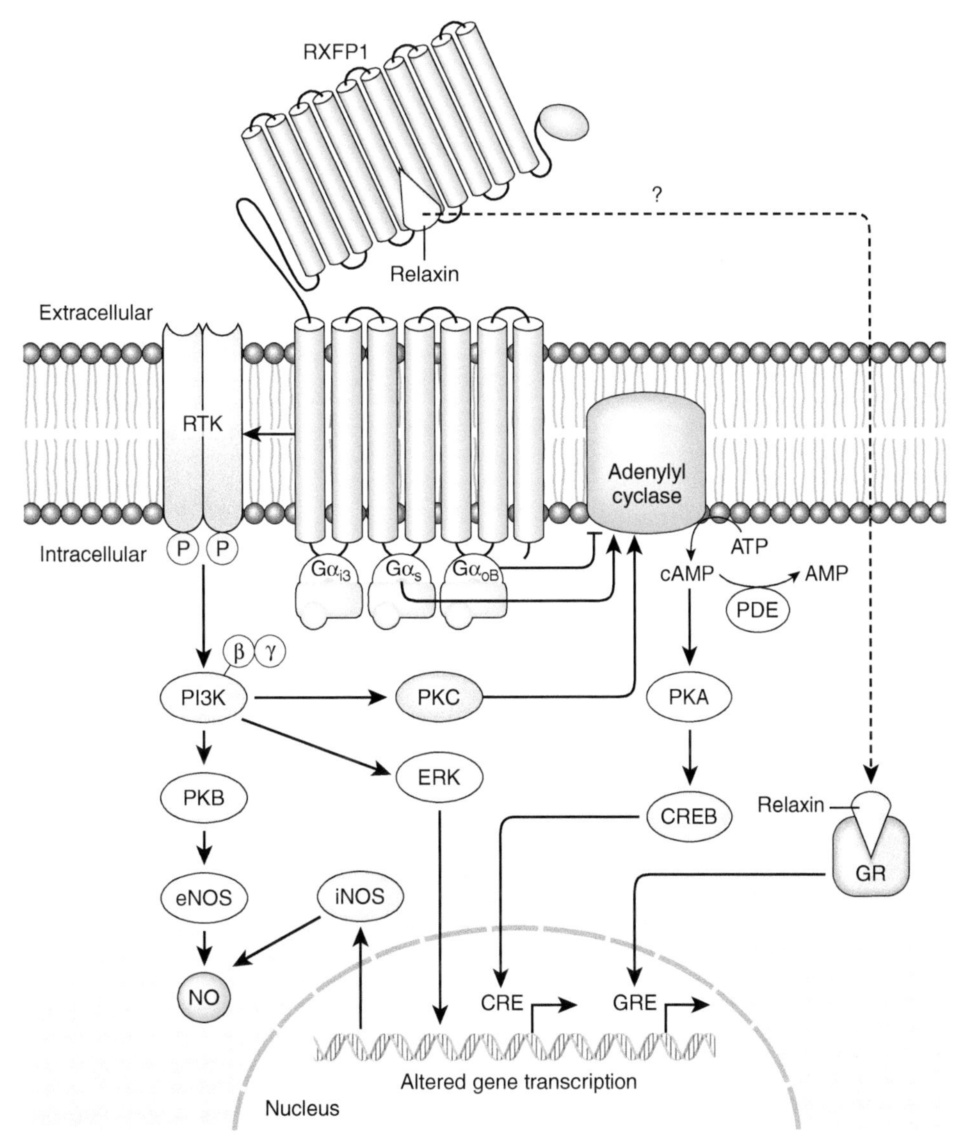

FIGURE 10-27 **Relaxin signal transduction and control of transcription.** Relaxin binds to RFXP1 receptors to stimulate cAMP production, triggering the PKA pathway and leading to phosphorylation and activation of cAMP response element-binding (CREB) protein which acts as a transcription factor. Relaxin bound to RFXP1 may transactivate receptor tyrosine kinase (RTK) pathways. Relaxin also may interact directly with the glucocorticoid receptor to modulate transcription, although the mechanism is unknown. Abbreviations: cAMP, cyclic AMP; CRE, cAMP response element; eNOS, endothelial nitric oxide synthase; ERK, extracellular signal-regulated kinase; GR, glucocorticoid receptor; GRE, glucocorticoid response element; iNOS, inducible nitric oxide synthase; NO, nitric oxide; PDE, phosphodiesterase; PKA, protein kinase A; PKB, protein kinase B; PKC, protein kinase C; PI3K, phosphoinositide 3-kinase; RXFP1, relaxin family peptide receptor 1. *(Adapted with permission from Du, X.-J. et al.,* Nature Reviews Cardiology, *7, 48–58, 2010.)*

placenta is the source of these estrogens. Androgens partially suppress mammary gland development and are presumably responsible for the lack of stimulation seen in the male fetus.

Postnatal mammary development involves hormones from the pituitary, ovaries, and adrenal cortex, at least in mice and rats. Growth of mammary ducts requires estrogens, GH, and corticosterone working in concert. However, expansion of the alveoli (called lobuloalveolar growth) is dependent upon the direct interactions of estrogens, progesterone, PRL, GH, relaxin, and corticosteroids.

Lactation can be separated into two basic processes or phases under separate endocrine control mechanisms. The first phase is milk secretion or **lactogenesis**. This process primarily is controlled by pituitary PRL (or placental CS), growth factors, and glucocorticoids. In primates, lactogenesis also is stimulated by GH. Lactogenesis involves synthesis of milk fat, milk protein, and milk sugar, typically lactose. The synthesis of lactose ultimately depends upon

TABLE 10-9 Some Bioregulators Found in Milk[a]

Regulator Type	Examples	Regulator Type	Examples
Adenohypophysial hormones	PRL	Gastrointestinal peptides	VIP
	GH		CCK
	TSH		Gastrin
	FSH		GIP
	LH		Substance P
	ACTH		Neurotensin
Growth factors	IGF-I	Steroid hormones	Estradiol
	NGF		Progesterone
	EGF		Testosterone
	TGF-α		Corticosterone
	PDGF		Vitamin D
Neurohormones	TRH	Other regulators	Prostaglandins (PGE, $PGF_2\alpha$)
	GnRH		cAMP
	SS		Delta sleep-inducing peptide
	GHRH		Relaxin
	Oxytocin		Thyroid hormones (T_3, T_4)
			Calcitonin
			Parathyroid hormone

[a]*See Appendix A for abbreviations.*

protein synthesis; that is, the enzyme responsible for lactose synthesis, lactose synthetase, must be induced. Lactose synthetase is composed of two protein units, one of which is lactalbumin, which also is found in milk. Lactose, fat, and milk protein (largely casein) are secreted into the lumen of the alveolus. Water and numerous water-soluble substances enter the lumen by osmosis and result in a watery liquid known as milk. Many hormones are present in milk, including hypothalamic peptides, pituitary hormones, growth factors, steroids, gastrointestinal peptides, and others (see Table 10-9). In addition, the mammary route may conduct lipid-soluble pollutants such as PCBs and pesticides accumulated by the mother to the offspring.

The composition of milk produced by the mammary gland associated with suckling the young is very different at birth from what it will be shortly thereafter. This first milk, known as **colostrum**, is characterized by having a greater concentration of protein and less carbohydrate than does later milk. Colostrum contains antibodies and other substances that serve to protect the neonate against allergies and diseases while its own immune response system is developing.

The second phase of lactation is **milk ejection**, a simple reflex mechanism controlled by OXY from the pars nervosa. Mechanical stimulation of the nipple (suckling) evokes release of OXY from the pars nervosa via a spinohypothalamic neuronal pathway. Release of PRL also occurs when milk is ejected and stimulates further milk synthesis. OXY stimulates contraction of myoepithelial cells which causes milk to be ejected from the alveoli into the ducts and storage channels of the mammary gland. Suckling by the young animal strips this milk from the gland by expressing it between the tongue and hard palate.

The milk ejection neurohormonal reflex exhibits classical conditioning responses as evidenced by the stimulation of milk flow in the cow by sight and sounds of the milking parlor or in women by the cries of their hungry infant. This reflex can be influenced by other neural or chemical inputs to the hypothalamus; for example, stress or physical discomfort can inhibit ejection of milk in the presence of the stimulus that would normally elicit release of OXY.

F. Menopause

In nature, few animals live beyond their peak of reproductive activity due to predation, disease, or other environmentally related phenomenon. In contrast, life after reproductive age is a common occurrence in human females. Whereas men may produce viable sperm most of their lives, the ovary becomes refractory to GTHs, usually during the mid to late 40s. This transitional stage is called **menopause**. Cycles of these women become irregular and eventually they cease to ovulate and menstruate. This is accompanied by a marked reduction in circulating levels of gonadal steroids as well as of adrenal androgens and by an elevation in GTH levels. The transition from **premenopausal** (actively reproductive) to **postmenopausal** (nonreproductive) usually is gradual over several years and may be accompanied by additional symptoms, including vaginal atrophy, hot flashes or flushes, reduced libido, and accelerated bone resorption leading to calcium deficiency syndromes such as osteopenia and osteoporosis (see Chapter 14). Many studies have shown that heart disease and other cardiovascular disorders increase exponentially in postmenopausal women and deaths due to cardiac disease are several-fold greater than for uterine and breast cancer combined.

Estrogen replacement therapy, usually in combination with a progestogen, alleviates many of the symptoms of menopausal and postmenopausal women. Estrogen therapies also have been associated with improvement in cognitive skills due to their stimulatory actions on neural development and learning tasks. When taken with calcium supplements and a regimen of weight-bearing exercise, steroid therapy also can prevent bone resorption. Studies report that estrogens reduce the risk of heart disease by as much as 50% in postmenopausal women as well as slow skeletal calcium losses. Apparently, estrogens or estrogens plus progestogens elevate high-density lipoproteins (HDLs) (see Chapter 12) which are associated with reduced cardiovascular risk. Potential benefits of estrogen therapy need to be considered in the light of other evidence linking estrogen replacement therapy to breast cancer. The decision for a woman to elect estrogen therapy involves many complicating factors that must be weighed. For example, is osteoporosis, heart disease, or breast cancer a serious problem in her family? How severe are the symptoms of menopause and/or osteoporosis and how do they affect her family life, her job? Exposure to other estrogenic chemicals through food, water, cosmetics, and other sources (e.g., phytoestrogens, bisphenol A, nonylphenols, ethinylestradiol, phthalates, certain pesticides) should also be considered as they do contribute to the total estrogen exposure. Careful scientific studies have verified that mixtures of estrogenic chemicals at levels unlikely for each to produce estrogenic effects are additive when they all work through the same mechanism (e.g., binding to and activating the estrogen receptor).

Men also experience reproductive decline with age, although this "male menopause" or **andropause** occurs more gradually and is not so evident as female menopause. Testosterone levels begin to decline after about age 30 and can lead to clinical signs in the 50s and 60s. Symptoms of andropause include reduced libido (sex drive), depression, loss of skeletal muscle mass, increased body fat, declines in cognitive ability, and osteoporosis. Metabolic changes may be responsible for the correlation of low testosterone with increased risk for cardiovascular disease. Although testosterone therapies are readily available, the known relationships between excess androgens and cancer induction should be considered before embarking on either a preventative or restorative course.

VI. REPRODUCTIVE CYCLES IN SELECTED EUTHERIAN FEMALES

In this section, four reproductive cycles are presented as being representative of eutherian mammals: four-day cycling rats, ewes, women, and elephants. These four examples emphasize both the features described previously that are characteristic of eutherian mammals and some of the differences seen among different species. The cycles of these species are among the best known, but not necessarily representative of all mammals. Ewes and rats are polyestrous species with distinct periods of estrus, whereas women have no seasonal estrous behavior and exhibit a menstrual cycle. Reproductive cycle length varies from 4 or 5 days in rats to 16 days in sheep, 28 days in humans, and 16 weeks in elephants. Cows and pigs have cycles that are essentially like the ewe cycle although they differ somewhat in timing of the various events. Both rats and women are continuous breeders, but ewes, like elephants, are distinctly seasonal breeders. Rats have a short gestation period lasting only 22 days, whereas elephants at the other extreme have a 22-month gestation period. None of these species exhibits delayed implantation, and all are believed to be spontaneous ovulators except under special conditions for the rat and possibly the human.

A. The Four-Day Cycling Rat

The laboratory rat cycle (Figure 10-28) is separable into **proestrus** (1 day), **estrus** (1 day), and **diestrus** (2 days in the four-day rat and 3 days in the five-day rat) and is cued closely to environmental events. Typically, several follicles develop and ovulate during each cycle, resulting in multiple corpora lutea in the postovulatory ovary.

GTHs from the pituitary stimulate ovarian follicle development and steroidogenesis. On the morning of the day prior to estrus (that is, during proestrus), the levels of

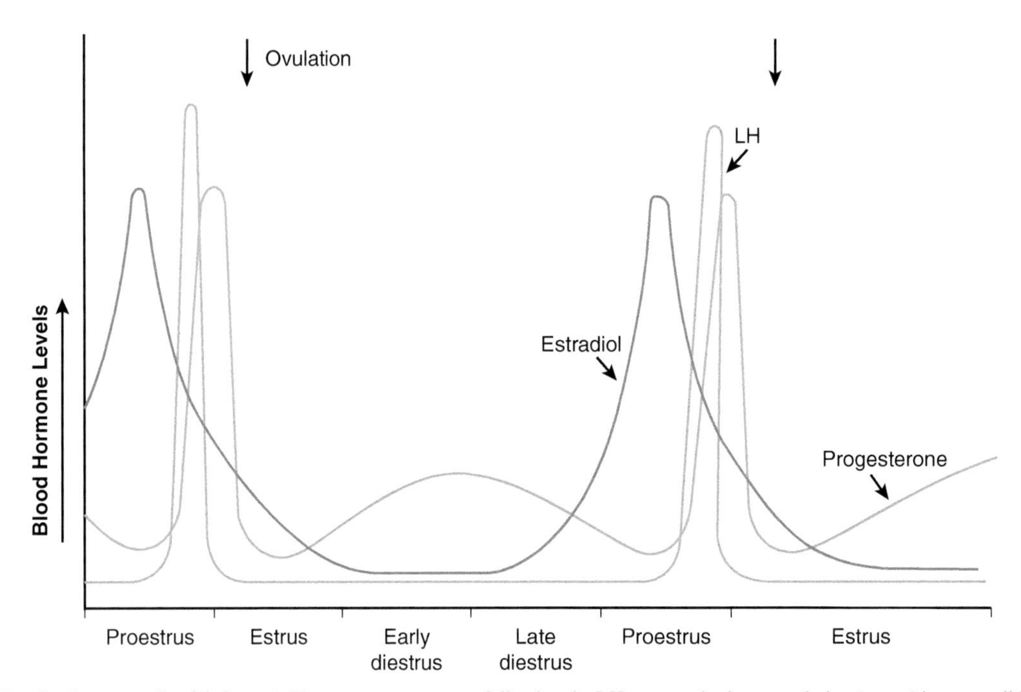

FIGURE 10-28 **Ovulatory cycle of 4-day rat.** The progesterone surge following the LH surge and prior to ovulation (arrow) is responsible for increased receptivity in the female for the male. There is a secondary, slow increase and decline in progesterone secretion from the corpus luteum that occurs during diestrous in the unmated female. *(Adapted with permission from Short R.V.,* Reproduction in Mammals, Book 3, *Cambridge University Press, 1972.)*

estrogen in the plasma reach a peak that stimulates an LH surge accompanied by a small surge in FSH. The GTH surge occurs on the afternoon of proestrus and is followed rapidly by a marked surge of progesterone from several short-lived corpora lutea. Ovulation occurs a few hours after midnight on the day of estrus. Several follicles usually mature simultaneously, and multiple ovulations commonly occur. Estrus lasts about 9 to 15 hours, during which time the female is highly receptive to the male. Ovulation occurs during estrus. Cornified cells, which were produced by the actions of estrogens during proestrus, appear in the superficial layers of the vagina, and their presence in vaginal smears characterizes estrus (Figure 10-29). In fact, this action was used as a bioassay for many years to define whether any compound had estrogenic activity (see Appendix D). Today, we have substituted more molecular techniques to assess estrogen activity (see Chapter 2 and ahead in Section VII of this chapter).

The third and fourth day of the cycle are termed **diestrus I** and **diestrus II**. Vaginal smears prepared during diestrus are characterized by the absence of cornified cells and a pre-dominance of leukocytes in the smear. There is limited secretory function by the corpora lutea, as evidenced by a slight increase in plasma progesterone. Much of this progesterone is produced by ovarian interstitial cells, which constitute what has been called a permanent corpus luteum in the rat ovary. This interstitial tissue responds to LH by secreting both progesterone and **20α-dihydroprogesterone**.

Many researchers recognize a short transitional period between estrus and diestrus in rats termed **metestrus**. The female in metestrus is no longer receptive to the male, but some cornified cells still appear in smears prepared from the vaginal mucosa.

Mating stimulates the GnRH surge followed by ovulation and consequently the corpora lutea form and begin to secrete significant amounts of progestogens. This increased production of progestogens will inhibit hypothalamus–pituitary function and delay the onset of the next cycle. If fertilization and implantation do not result from mating, the rat will not return immediately to proestrus but will delay resumption of proestrus for a few days. This period of copulation-induced **pseudopregnancy** often occurs in laboratory rodents and sometimes in other domestic mammals. The condition of pseudopregnancy is often accompanied by PRL-like effects on lactation and behavior presumably due to pituitary release of PRL.

If implantation does occur, the corpora lutea continue to secrete progestogens under the influence of placental CG and begin to secrete relaxin. The rat placenta also produces CS, which contributes to stimulation of mammary gland development and lactogenesis prior to birth.

Pheromones play central roles in mating and successful pregnancy in rodents. Most of the scientific studies have been done with mice rather than rats or wild rodents but may apply to many rodent species. Crowded female mice become anaestrus when no males are present (called the Lee–Boot effect). However, simply the odor from a male

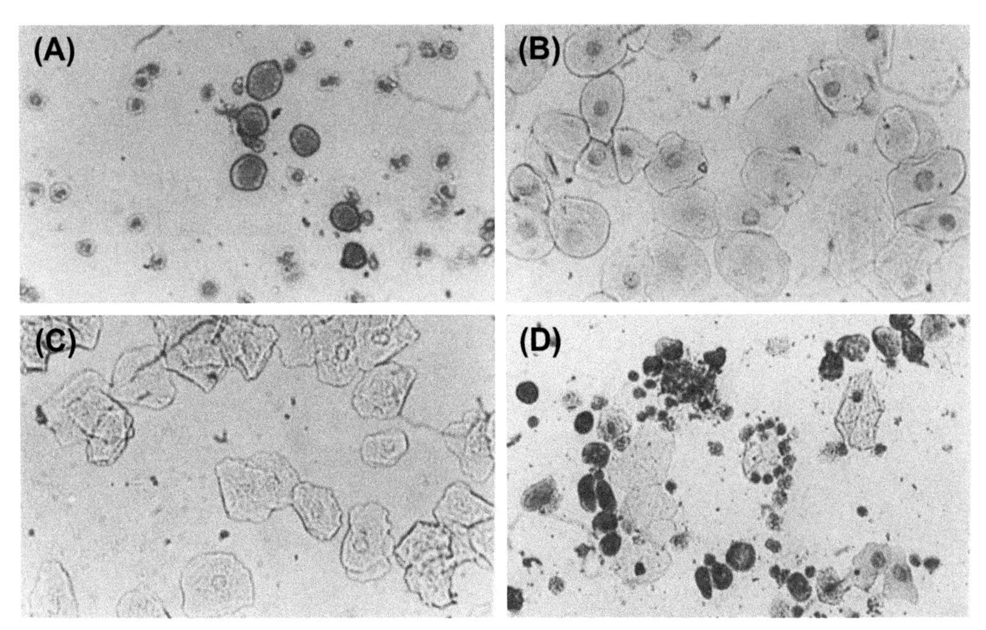

FIGURE 10-29 **Vaginal smears from rats during estrous cycle.** (A) *Diestrus:* absence of cornified (keratinized) cells and presence of small leukocytes. (B) *Proestrus:* many live epithelial cells with smooth margins; leukocytes absent. (C) *Estrus:* large cornified cells with irregular margins. (D) *Metestrus:* leukocytes have infiltrated among the cornified cells; this stage is transitional between estrus and diestrus. (*Adapted with permission from Short R.V.,* Reproduction in Mammals, Book 3, *Cambridge University Press, 1972.)*

mouse can cause them to synchronously ovulate and enter estrus (Whitten effect). The endocrinological basis for these effects is suggested by observations that pheromones from female mice suppress pituitary release of FSH, whereas male pheromone stimulates GTH release that is followed in normal sequence by an LH surge and ovulation. A newly mated female mouse may abort if placed with a "strange" male (not the previous mate), and the likelihood of spontaneous abortion increases with the genetic dissimilarity of the strange male to the male with whom she originally was mated (this is called the Bruce effect). If offspring result, they are always from the "strange" male.

The sexual pheromones involved in the Lee–Boot and Bruce effects are probably modified steroids (steroid metabolites) and are transmitted via the urine of the male to the olfactory apparatus of the female. Male mouse urine induces and accelerates estrous cycles of females (Whitten effect), and the effect is most pronounced on Lee–Boot groups of females. The time of vaginal closing in females is also influenced by male urine. Anosmic females (animals whose nostrils have been blocked or whose olfactory bulbs have been removed surgically) do not respond to male urine.

Pregnant and lactating rats produce pheromones that influence other females. Odors from a lactating female with pups lengthens estrous cycles of non-pregnant females. Thus a socially dominant, lactating female may suppress fertility of other females until she is again in estrus herself.

Males also may be influenced by female pheromones. Pairing of a previously paired male mouse with a strange female results in elevation of plasma testosterone in the male, indicating that endocrine responses of both males and females may be influenced through bisexual encounters.

B. The Ewe

Sheep estrous cycles occur seasonally, and the duration of one complete cycle is 16 days. The ewe may return to proestrus at least once if fertilization does not occur (Figure 10-30). Reproductive cycles can be blocked in ewes by **genistein**, a phytoestrogen found in certain clovers (genistein is also present in some plant products consumed by humans such as soy). The structure of genistein is provided in Chapter 3, Figure 3-20. Grazing of sheep in pastures containing this clover can have contraceptive effects.

During the follicular phase (= proestrus), there is a marked increase in estrogen and androgen levels in ewes, a peak being reached about 24 hours after the onset of proestrus. About 12 hours later, a surge of plasma LH occurs caused by the action of estradiol on the cyclic hypothalamic neurosecretory center. A high level of circulating androstenedione actually may be responsible for inducing estrous behavior by being a substrate for estradiol synthesis. Usually, a single ovulation follows the LH surge by about 24 hours, and a corpus luteum forms from the ruptured follicle under the influence of LH. Low levels of LH following ovulation and the estrogen-induced surge of PRL stimulate the corpus luteum to secrete progesterone. Under the influence of progesterone, the

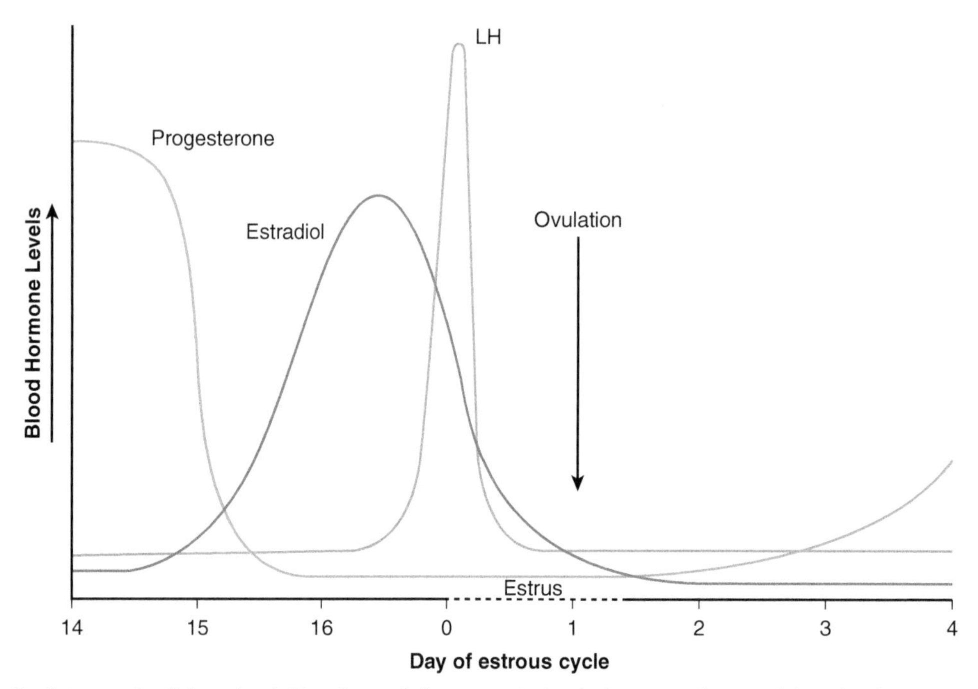

FIGURE 10-30 **Ovulatory cycle of sheep (ewe).** Note that ovulation occurs during the later part of estrus. *(Adapted with permission from Short R.V.,* Reproduction in Mammals, Book 3, *Cambridge University Press, 1972.)*

uterine endometrium synthesizes a luteolytic prostaglandin ($PGF_{2\alpha}$) that causes degeneration of the corpus luteum and resumption of proestrus. It appears that LH alone can cause follicle growth, ovulation, and luteinization, although several studies have shown that FSH can stimulate follicle growth, too; however, both postovulatory LH and PRL are necessary to induce progesterone synthesis by the sheep corpus luteum.

Fertilization followed by implantation delays degeneration of the corpus luteum. Although oCG is produced, the conceptus somehow neutralizes the PGFs synthesized under the influence of progesterone so that the corpus luteum may continue to secrete progesterone until the placenta is capable of producing sufficient steroids to maintain gestation. The placenta also secretes oCS throughout gestation. Birth seems to be triggered by a marked reduction in progesterone production near term (birth) (see Figure 10-24) and an increase in the activity of the HPA axis in the fetus.

C. Women

The human female exhibits continuous reproductive cycling with a mean non-pregnancy cycle length of 28 days for most reproductively active women (Figure 10-31). Timing of reproductive events is related to the uterine menstrual cycle. The rhesus monkey also has a menstrual cycle of 28 days, and there are many parallels in the menstrual cycles of these two primates. Thus, studies of the rhesus monkey have provided valuable insight into factors

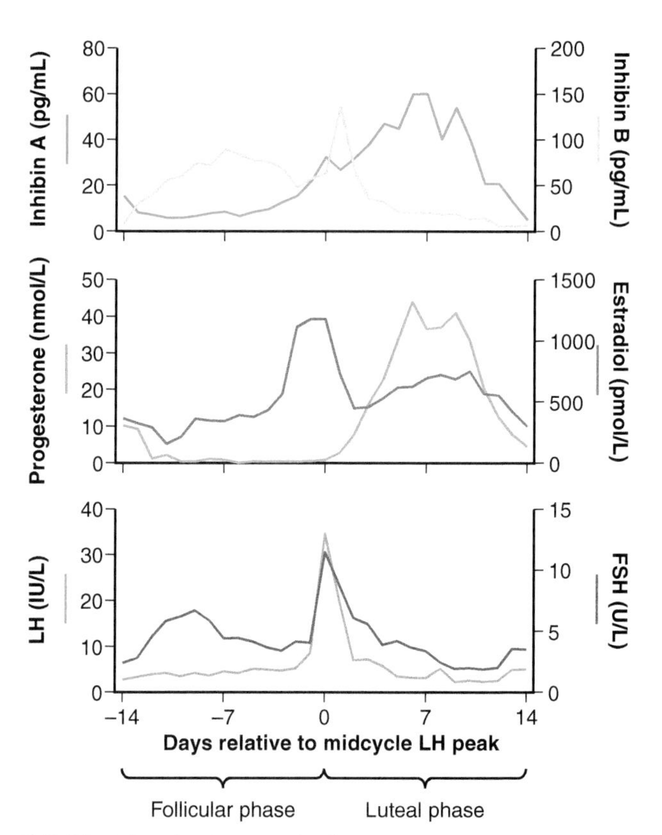

FIGURE 10-31 **Ovulatory cycle of woman.** Changes in blood levels of inhibin A and B (top), progesterone and estradiol (middle), and LH and FSH (bottom). See Appendix E for explanation of LH and FSH units. *(Adapted with permission from Burger, H.G. et al.,* Recent Progress in Hormone Research, ***57****, 257–275, 2002.)*

regulating the human menstrual cycle. Although this menstrual cycle is sometimes referred to as a lunar cycle because its periodicity is equivalent to approximately one lunar month, the human menstrual cycle is not correlated with any particular phase of the lunar month and should not be termed lunar. It could be that at one time the menstrual cycle was correlated more closely with moon phases but it has become highly modified by numerous environmental and internal factors.

The duration of menstrual cycles of women can vary from as short as 14 days to as long as 360 days, depending upon both endocrine and psychological factors. Furthermore, considerable variation can occur in cycle length in a given woman at different times in her life history. In general, short cycles and irregular cycles are associated with the onset of puberty and with the end of the reproductive life prior to menopause when the ovaries become refractory to pituitary GTHs and both estrogen synthesis and ovulation cease. The reduction or absence of circulating estrogens and progesterone results in elevated levels of circulating GTHs in menopausal and postmenopausal women through the normal negative feedback mechanism.

By convention, the menstrual cycle begins at the onset of the menses, which occupies the first 5 days of the cycle (Figure 10-31). However, as soon as the corpus luteum of a previous cycle begins to regress prior to the onset of menses, there is a depression in plasma gonadal steroids and a resultant moderate elevation in plasma FSH that initiates growth of new follicles. Typically, only one follicle (dominant follicle) in one of the ovaries will reach maturity in a given cycle, with ovulation often occurring in the alternate ovary during the following cycle (Figure 10-17). The largest follicle in the other ovary will ovulate the next month; hence, the true ovarian cycle is at least twice as long as the menstrual cycle (56 days, and some estimates suggest as long as 98 days). At least two periods of follicular growth are correlated with the proliferative phase of two successive uterine cycles.

The thecal cells of the growing follicles in both ovaries begin to secrete estrogens, progestogens, and inhibin B (Figure 10-31). Secretion of estrogens and progestogens peaks on about day 14 of the normal menstrual cycle. Steroidogenesis is regulated by LH and FSH operating on the thecal and granulosal cells, respectively. The peak in follicular phase estrogen level stimulates accelerated pulsatile GnRH release from the cyclic center, causing a large surge of LH accompanied by a lesser surge of FSH. Galanin released with GnRH enhances the LH response, whereas inhibin B secreted by the ovary under the stimulation of FSH actually reduces the release of FSH in response to the GnRH surge. Ovulation of typically only the largest follicle follows the LH surge by about 24 hours, and the LH-stimulated granulosa cells and some thecal cells from the ruptured follicle undergo luteinization to form a corpus luteum. The resulting corpus luteum is independent of pituitary hormones and begins production of progesterone and estradiol and inhibin A. The hypothalamic centers governing LH and FSH release are inhibited by the high levels of these circulating steroids and inhibins so that neither follicular growth nor ovulation can occur during the early portion of the luteal phase of the cycle. However, the corpus luteum functions for only a few days if the ovum is not fertilized. Unlike the case for sheep, the human uterus plays no active role in causing degeneration of the corpus luteum. Surgical removal of the uterus (hysterectomy) does not affect the duration of the luteal phase. Production of estrone by the corpus luteum of the rhesus monkey increases markedly prior to the onset of luteal degeneration. A similar mechanism employing estrone as a luteolytic factor may be operating in women. As the production of steroids by the corpus luteum decreases, the pituitary is released from steroid-induced inhibition, and a new cycle of follicular growth begins during the latter portion of the luteal phase. The outer portion of the uterine lining begins to slough off following the decline in progesterone levels, and the menses begins.

The limiting factor for follicle growth appears to be the availablility of GTH. Treatment with fertility drugs that enhance GnRH secretion and hence GTH levels often can cause development and ovulation of multiple follicles resulting in multiple births. Hence, women may transition from being non-reproductive to producing litters.

Should fertilization occur at the appropriate time during the menstrual cycle, the trophoblast of the blastocyst begins to secrete hCG prior to implantation, and the production of hCG continues at an accelerated rate during early pregnancy (Figure 10-30). Under the influence of hCG, the corpus luteum continues to secrete both progesterone and relaxin for about 60 days, after which it degenerates even in the presence of exogenous hCG. However, the normal fetal adrenal−placenta unit by this time produces sufficient estrogens and progesterone to continue the inhibition of pituitary GTH release and to maintain the secretory and hyperemic condition of the uterus. The placenta can synthesize progesterone but lacks the necessary enzymes to synthesize androgens. The fetal adrenal provides weak androgens (androstenedione, DHEA, DHEA-S) so that the placenta using $P450_{aro}$ can convert them into estrogens. Failure of the fetal adrenal to produce adequate androgens before the corpus luteum degenerates invariably results in a miscarriage at about 2 to 3 months following implantation.

At one time, a synthetic estrogen **diethylstilbestrol (DES)**, discovered in 1938, was prescribed for thousands of women who were unable to sustain sufficient estrogen levels to maintain pregnancy. Many women were treated with DES as a precautionary therapy even without a history of miscarriage. Although this treatment was claimed to be effective in preventing miscarriages, it has not been proven, and DES was found to increase the incidence of cancer in both the treated women and their offspring. For these reasons, this practice was discontinued in the early 1970s. Similar observations had been made in animals years before, but the medical community was slow to recognize the danger and the Federal Drug Administration (FDA) approved its use. There are recent published reports of similar problems in the grandchildren of the DES-treated women.

Pheromones have been documented as important agents for coordinating events of sexual reproduction among many mammals as discussed earlier for rodents, and may even play important roles in human reproduction as inferred from studies of monkeys. Female rhesus monkeys produce a mixture of fatty acids of low molecular weight. The compounds appear in vaginal secretions and stimulate sexual interest of males as well as mounting behavior and ejaculation. The major volatile agents are fatty acids, including acetic acid, butanoic acid, propanoic acid, methylbutanoic acid, and methylpropanoic acid. Synthetic mixtures of these fatty acids in appropriate ratios stimulate male interest in females. Estrogens stimulate fatty acid secretions, and progesterone is inhibitory, observations that correlate well with levels of fatty acids observed in vaginal secretions throughout the menstrual cycle. Human vaginal discharges exhibit a similar cyclical variation in fatty acid composition (Figure 10-32), although human females produce a much greater percentage of acetic acid than do rhesus monkeys. The administration of oral contraceptives to these females effectively obliterates the preovulatory increase in volatile fatty acids, suggesting either an inhibition caused by high levels of estrogen or that the specific pheromonal agents are GTH dependent (recall that negative feedback by the contraceptive steroids blocks GTH release and hence prevents follicle development and ovulation).

Although studies of pheromones in humans are complicated by a number of psychological and social considerations, some evidence exists for production of pheromones and their roles in reproduction. A "dormitory effect" of menstrual synchrony has been described by McClintock for all-female living groups. Even though it is generally accepted that the olfactory sense in humans is limited as compared to most mammals, several studies have shown definite sensitive olfactory discriminations, including sexually based differences in the abilities to perceive certain odors. Trained perfumers can apparently distinguish between different skin and hair types, and some psychiatrists claim to be able to smell schizophrenics because of abnormal production and elimination of *trans*-3-methylhexanoic acid. The ability to detect some odors is sex dependent, such as the greater sensitivity of women to "boar taint" associated with spoiled pork. Some odors (e.g., licorice, lavender, doughnuts, pumpkin pie) are

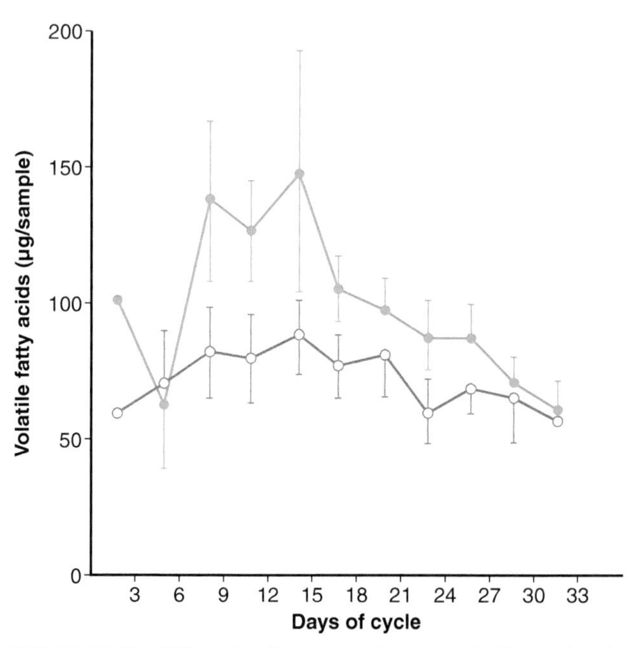

FIGURE 10-32 **Effect of oral contraceptives on vaginal secretions in humans.** Lipid composition of human vaginal secretions collected at 3-day intervals during the menstrual cycle (blue). Treatment with oral contraceptives (red) reduced the normal midcycle rise observed in the volatile fatty acid content of vaginal secretions in 47 women (blue line). *(Data originally reported in Michael, R.P. et al.,* Science, ***186****, 1217–1219, 1974.)*

claimed to induce sexual arousal. Clearly, controlled experimental studies are needed to establish what roles pheromones have in controlling human behavior and reproductive functions.

D. Elephants

Asian (*Elephas maximus*) and African (*Loxodonta africana*) elephants have similar seasonal reproductive cycles with a 4- to 5-year birth interval in females due to the relatively long gestation period and extended lactation. The following account merges information from both species. Females are most likely to mate with older males that exhibit **musth**, a period of increased activity, increased association with females, and elevated aggressiveness toward other males. A female may signal her receptiveness to a male by calling or through release of pheromones. In the Asian elephant, a forthcoming ovulation is advertised by the excretion of **(Z)-7-dodecen-1-yl acetate** in urine that arouses interest by males. When a male is in musth, he secretes more fluid from his temporal glands and dribbles strong urine about that is attractive to females during the follicular phase of their ovarian cycles.

The ovarian cycle appears to be 16 weeks in duration with ovulation and a one-week period of estrus occurring near the middle of the cycle (Figure 10-33). There appear

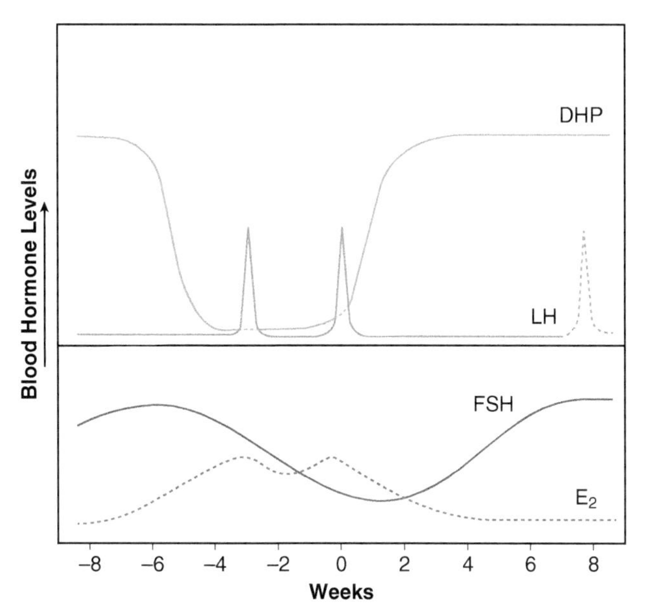

FIGURE 10-33 **Ovulatory cycle of the elephant.** Two LH surges are observed but it is believed that only the second results in ovulation and corpora lutea formation as evidenced by the levels of progestogens (DHP). *(Adapted with permission from Hodges, J.K.,* Animal Reproduction Science, *53, 3–18, 1998. © Elsevier Science, Inc.)*

to be be two successive LH surges about 2 to 3 weeks apart and both are related to elevated estradiol levels. The first surge is believed to stimulate development of accessory corpora lutea and the second LH surge stimulates ovulation. The single follicle that usually ovulates following the second LH surge forms a corpus luteum that together with the other corpora lutea are responsible for the observed rise in circulating progestins characterizing the luteal phase. Typically, only one follicle probably ovulates as twins are rare among elephants and usually only a single offspring is produced following 22 months of gestation. Progestogen levels rise during the luteal phase and remain elevated in the pregnant elephant until about 30 weeks prior to birth, reaching prepregnancy levels about 2 to 5 days before parturition. Relaxin appears in the blood about 20 weeks after the onset of pregnancy and also shows a decline with the progestogens; however, relaxin rises again during the last 8 weeks of pregnancy. Nothing is known about the factors responsible for initiation of birth in elephants.

Elephants are unique in producing very low amounts of progesterone but instead produce large amounts of the progestogens **5α-dihydroxyprogesterone (DHP)** and **5-pregnan-3-ol-20-one**. DHP has the strongest affinity for the progesterone receptor and may be the physiologically important progestogen in elephants. Following birth of the offspring, the female elephant enters a prolonged period of lactation during which GTH release is suppressed and the ovary remains quiescent.

VII. ENDOCRINE DISRUPTORS AND MAMMALIAN REPRODUCTION

In July 2012, the U.S. Food and Drug Administration banned the use of a plasticizer named **bisphenol A** (BPA) in baby bottles. Concerns regarding the health effects of BPA arose from several lines of evidence. First, over 6 billion pounds of BPA are produced worldwide every year and used for manufacuring the plastic lining of metal cans, the polycarbonate plastic in many household items and drinking bottles, and in dental sealant products. BPA has been measured in human tissues at concentrations higher than those found in the environment and has been measured in umbilical cord serum and amnionic fluid. Decades of research indicate that BPA is an estradiol mimic on estrogen receptors, and new data support a role for nongenomic actions of BPA by acting on membrane estrogen receptors. While BPA in many ways is a model for understanding how environmental estrogen mimics can adversely impact reproduction, there are many endocrine-disrupting chemicals (EDCs) that have been proposed to impact reproduction ranging from natural estrogen mimics such as genistein, a phytoestrogen found in soy, to industrial contaminants such as PCBs that may disrupt both androgen and estrogen actions. The structures of some selected EDCs are shown in Figure 10-34.

A. Estrogen Receptor (ER) Agonists and Antagonists

A seemingly endless number of manmade and natural chemicals have the ability to bind to ERs and produce estrogen-like effects or inhibit ER functions by acting as an antagonist. Considerable effort has been directed at developing so-called **high throughput assays** to rapidly screen hundreds and thousands of chemicals for ER action. **Molecular docking** and **three-dimensional quantitative structure–activity relationships (3D-QSARs)** also have been developed to perform so-called *in silico* or desktop identification of a chemical's potential interaction with ERs. Whereas high-throughput assays are certainly needed to test the sheer number of contaminants released into the environment and to reduce the number of animals required for testing, whole animal studies are also needed, especially given the number of chemicals such as dioxins and PCBs that may indirectly interfere with ERs (and in fact androgen receptor) signaling by activating the **aryl hydrocarbon receptor (Ah receptor)**. The U.S. Environmental Protection Agency Endocrine Disruptor Screening Program (EDSP), tasked with reviewing pesticides and environmental contaminants for endocrine disruption, currently employs five assays for ER action in its initial (Tier 1) battery of tests (Table 10-10). Two of these assays are high-throughput cell-culture-based assays (ER binding assay,

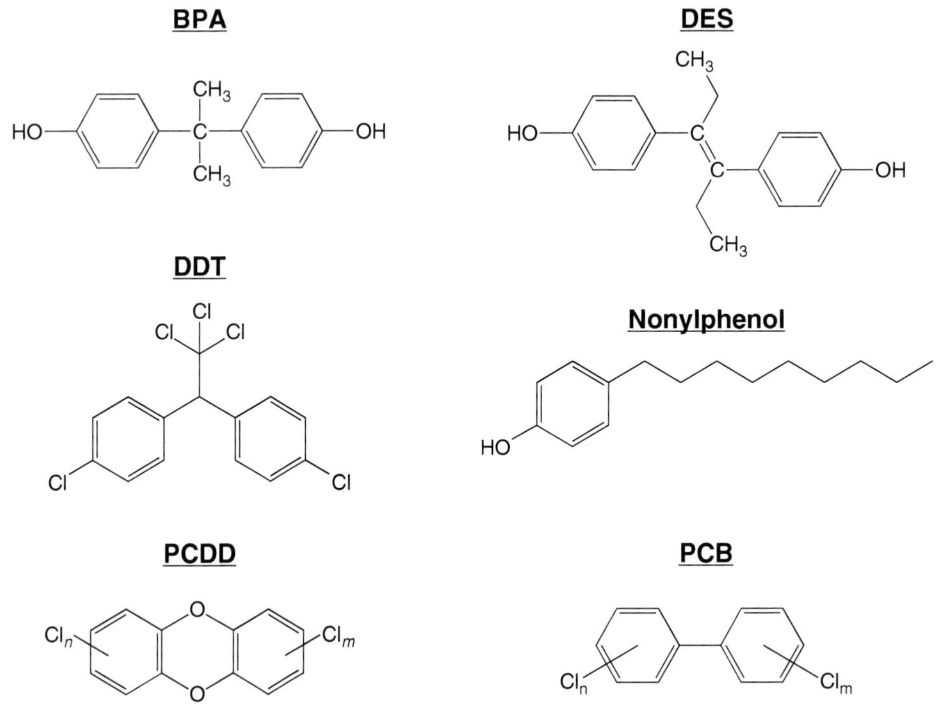

FIGURE 10-34 **Chemical structures of some reproductive endocrine disruptors.** Abbreviations: BPA, bisphenol A; DDT, dichlorodiphenyltrichloroethane; DES, diethylstibestrol; PCB, polychlorinated biphenyl; PCDDs, polychlorinated dibenzodioxins.

TABLE 10-10 Screening Assays Approved for the U.S. Environmental Protection Agency Endocrine Disruptor Screening Program

	Mode of Action						
	Receptor Binding				Steroidogenesis		
Assay	**ER agonist**	**Anti-ER**	**AR Agonist**	**Anti-AR**	**Estradiol**	**Androgen**	**HPG Axis**
Cell based							
ER binding	X	X					
ERα transcptional activity	X						
AR binding			X	X			
Steroidogenesis					X	X	
Aromatase					X		
Whole animal							
Uterotrophic	X						
Hershberger			X	X		X	
Pubertal male			X	X		X	X
Pubertal female	X	X			X		X
Fish reproduction	X	X	X	X	X	X	

ER, estrogen receptor; AR, androgen receptor.
Data from U.S. EPA Endocrine Disruptor Screening Program (EDSP), Harlan Laboratories, Indianapolis, IN.

ER transcriptional activation assay) and three are whole animal assays (uterine growth in rats, female puberty in rats, and fish reproduction). Chemicals known to interfere with reproduction by binding to ERs include pesticides (aldrin, DDT, dieldrin), phthalates, dioxins, nonylphenols, and PCBs. There is even evidence that cadmium, a heavy metal contaminating water supplies as a result of mining activity, can bind to ERs. Undoubtedly, as the U.S. EPA proceeds with the enormous task of reviewing the thousands of chemicals released into the environment for endocrine activity the list of chemicals known to bind to ERs will grow.

B. Androgen Receptor (AR) Agonists and Antagonists

As is the case with ER disruptors, there are both cell culture and whole animal assays for testing the ability of contaminants to activate or inhibit ARs including the **Hershberger assay**, which measures the weight of secondary sex organs such as the seminal vesicles and prostate gland which are androgen sensitive. In the Hershberger assay, rats are castrated and potential AR agonist effects are tested by administering the contaminants and examining the effect on organ weight. Potential AR antagonist effects are tested by administering the contaminants along with testosterone. Contaminants known to act on the AR include pesticides (*p,p′*-DDE, a DDT metabolite; fenitrothion; lindane; linuron; permethrin; and vinclozolin), phthalates, benzopyrene, and PCBs, all of which are AR antagonists. Very few data exist on contaminants that may activate the AR.

C. EDCs That Impact Steroidogenesis and Aromatase Activity

Steroidogenesis can be tested using a number of cell lines or primary culture with gonadal tissue, but the most widely used assay utilizes a human adrenocortical cell line called **H295R**. Activity of aromatase, the enzyme converting testosterone to estradiol, is generally measured for EDSP purposes in cells engineered to express the aromatase (*cyp19*) gene, although any tissue expressing the aromatase enzyme could be used in theory. Flavones, a type of phytoestrogen, are potent inhibitors of aromatase activity, possibly by competing with the natural substrates for the aromatase enzyme. Atrazine, a pesticide that is widely used in the United States as a pre-emergent herbicide on corn and sugarcane, increases aromatase activity in cell lines and has been implicated in altering steroidogenesis in the male pubertal development assay at high doses, although the mechanism may involve indirect action on GnRH neurons in the hypothalamus. BPA and the anti-AR pesticide vinclozolin also are suspected of disrupting aromatase activity. Many fungicides alter aromatase activity because of the similarity of this enzyme to the CYP proteins targeted by chemically engineered fungicides. Fadrozol, an aromatase inhibitor widely used by researchers and as a therapy for breast cancer, belongs to a class of chemicals called **azoles** that includes many fungicides. **Organotins** are organic metal hybrid chemicals known to interfere with aromatase activity. Tributyltin (TBT), widely used as an antifungal and wood preservative and as an anti-fouling agent in marine vessels, is a potent aromatase inhibitor. The organochlorine pesticide methoxychlor interferes with the expression of a wide number of genes involved in steroidogenesis.

D. EDCs with Epigenetic Effects on Reproduction

In 2005, Michael Skinner's laboratory at Washington State University reported in the journal *Science* that exposing pregnant rats to vinclozolin and methoxychlor led to aberrant spermatogenesis in the initial offspring and all other generations after that. This was the first published evidence of EDCs acting to disrupt reproduction through epigenetic mechanisms (i.e., altering gene expression and phenotype in offspring without altering the maternal DNA sequence). As we discussed in Chapter 1, EDCs may affect offspring phenotype without altering maternal DNA by altering DNA methylation and histone proteins. Epigenetic effects may also explain why short-term exposure of neonates to EDCs may have life-long consequences for reproduction. A diverse array of EDCs have been reported to cause epigenetic changes. BPA treatment of neonatal rats alters DNA methylation in the testis, possibly explaining the adverse effects of BPA exposure on spermatogenesis and fertility. Fetal exposure to methoxychlor also leads to hypermethylation of DNA in the ovaries of female offspring, leading to alterations in the expression of ERβ but not ERα genes. The epigenetic effects of methoxychlor extend to effects on the preoptic region of the hypothalamus which controls gonadotropin secretion. Embryonic rats exposed to methoxychlor had altered methylation of the ERα gene that led to earlier reproductive aging and altered estrous cycles. Genistein, a phyto-estrogen from soy, and phthalates, used as plasticizers, has also been implicated in epigenetic disruption of reproduction. Some epigenetic effects of EDCs are listed in Table 10-11.

VIII. MAJOR HUMAN ENDOCRINE DISORDERS RELATED TO REPRODUCTION

Reproductive disorders have been studied extensively in order to prevent their occurrence as well as to correct

TABLE 10-11 Examples of Endocrine-Disrupting Chemicals (EDCs) That Disrupt Reproduction Through Epigenetic Mechanisms

EDC	Exposure Period	Gene Targeted	Tissue	Epigenetic Effect	Phenotype	Reference
Bisphenyl SA	Neonate	*Esr1, Esr2* (estrogen receptors 1 and 2; estrogen receptors α and β)	Testis		Altered spermatogenesis, fertility	1
Diethylstilbestrol	Neonate	*FOS*	Uterus	Demethylation	More tumors	2
Methoxychlor	Embryonic	*Esr1*	Hypothalamus	Hypermethylation of *Esr1* promotor	Early reproductive aging; altered estrous cyclicity in females	3
Methoxychlor	Embryonic	*Esr2*	Ovary	Hypermethylation of *Esr2* promotor	Increase in the number of preantral and early antral follicles and a reduced number of corpora lutea	4

[1] *Doshi, T. et al.*, Toxicology, ***289***, *74–82, 2011.*
[2] *Li, S. et al.*, Molecular Carcinogenesis, ***38***, *78–84, 2003.*
[3] *Gore, A.C. et al.*, Molecular Endocrinology, ***25***, *2157–2168, 2011.*
[4] *Zama, A.M. and Uzumcu, M.*, Endrocinology, ***150***, *4681–4691, 2009.*

defects and increase reproductive capacities of both men and women. Some of the more common reproductive disorders are described here. Additional disorders related to reproduction include congenital adrenal hyperplasia (see Chapter 8), osteoporosis (see Chapter 14) and a host of **disorders of sex development (DSDs)** (see Box 10B), or **genetic disorders of gonadal development**, which are summarized in Table 10-12. The discussion of major endocrine disorders is separated into three major categories. The first two categories consider factors that influence the timing of puberty. The third deals with major genetically based disorders many of which can result in ambiguous sexual determination.

A. Precocious Puberty

Puberty is a delayed period of development focusing upon activation of the HPG axis and the functional integrity of sex accessory structures that may lead to successful sexual reproduction. **Precocity** is defined as the appearance of any one indicator of puberty at an age earlier than 2.5 to 3 standard deviations below the mean age at which the indicator normally appears in that population (see Table 10-13). The sequence and mean age for appearance of these indicators should be considered only as a guide. Implied precocity may not be evidence of an endocrine disorder, and some variation in the sequence of these events is normal. Major deviations in a number of indicators may signal precocial endocrine activity of a pathological nature (Figure 10-6). **Isosexual precocity** involves early appearance of the genetically determined sex. It is termed **heterosexual precocity** if male features develop precocially in a female or if female features appear precocially in a male.

It should be noted that there has been a shortening of the mean prepubertal period in the last several decades resulting in earlier puberty. Evidence from studies of endocrine disruption (see Section VII) through accidental exposures to estrogenic compounds such as phthalates suggest environmental causes for this accelerated timing of puberty. Exposure to artificial lighting that extends daylight and perhaps to television and computer monitors also has been hypothesized to influence pineal function (see ahead) and contribute generally to precocity.

1. Precocity with Normal Endocrinology

Idiopathic (of unknown cause) precocity may be familial. Sexual development and body growth appear normal but are accelerated. Reproduction may be possible at an early age. For example, the youngest mother on record was 5 years 8 months of age at delivery.

2. Precocity and Pineal Tumors

Pineal tumors are not common but occur most frequently in young males. Precocious sexual maturation occurs in about one third of these cases. Pineal tumors may impair release of melatonin from the pineal and allow sexual maturation to occur prematurely (see Chapter 4). In contrast, some pineal

TABLE 10-12 Some Genes Involved in Diseases of Sexual Development (DSDs) in Humans

Gene	Protein	DSD Phenotype
Bipotential gonad formation		
NR5A1	Steroidogenic factor 1 (SF-1)	Loss leads to XY gonad failing to form properly (gonadal dysgenesis)
M33/CBX2	Chromobox protein homolog 2, required for *SRY* expression	Loss results in male-to-female sex reversal, XY develops ovary, uterus, full female phenotype due to lack of *SRY* expression and function
WT1	Wilms tumor protein, transcription factor	Loss results in Denys-Drash syndrome, gonadal dysgenesis, pseudo-hermaphroditism
Testes formation		
DMRT1	Doublesex and mab-3 related transcription factor 1	Loss leads to XY gonadal dysgenesis, failure of Sertoli cells to develop
SOX9	*SRY* (sex-determining region Y)-box 9 protein, transcription factor	Loss leads to XY ovary and female phenotype; mutations upstream of *SOX9* leads to female-to-male sex reversal
SRY	Sex-determining region of Y chromosome encodes testis-determining factor (TDF), transcription factor	Loss leads to male-to-female sex reversal, XY develops ovary
Ovary formation		
FOX12	Forkhead transcription factor 12	Loss leads to premature ovary failure (POF)
RSPO1	R-spondin-1, possibly ligand for surface membrane protein	Loss leads to female-to-male sex reversal, ovotestis
WNT4	Wingless-type MMTV integration site family, member 4, ligand surface membrane receptor	Loss leads to müllerian duct agenesis

From Eggers, S., Sinclair, A. (2012). Mammalian sex determination—insights from humans and mice. Chromosome Res. **20,** 215–238.

tumors have been related to delayed puberty in a few cases. These tumors possibly secrete more melatonin than does a normal gland and hence delay puberty.

3. Precocity from Ectopic Gonadotropins or Gonadal Steroids

Rarely, pituitary tumors secrete excessive amounts of gonadotropins, and sometimes a non-pituitary tumor secretes chorionic gonadotropin, causing premature gonadal maturation. Likewise, certain ovarian or testicular tumors can produce sufficient steroids to cause external evidence of puberty although the gonads themselves are still quiescent with respect to gamete production. Heterosexual precocity can occur from feminizing tumors in testes or androgen-secreting tumors of the female adrenals or ovaries. (See also the discussion of compensatory adrenal hyperplasia in Chapter 8.)

4. Precocity from Endocrine Disrupting Chemicals

Exposure of animals to compounds such as phthalates and BPA associated with plastics have been shown to accelerate puberty in laboratory animals. Young girls exposed to phthalates exhibit early and sometimes excessive breast development (thelache).

B. Delayed Puberty

Numerous examples of delayed puberty are known. The causes and characteristics of delayed puberty are unique for males and females. However, mice lacking the G-protein-coupled receptor **GPR54** fail to undergo puberty (Figure 10-6). The natural ligand for GPR54 is kisspeptin, a peptide of 54 amino acids that is produced by the *KISS-1* gene. Kisspeptin was originally named metastin for its ability to suppress the metastatic potential of melanoma

TABLE 10-13 Mean Age for Normal Attainment of Certain Indicators of Puberty in Humans

	Mean Age (in years ± SD)
Female	
Budding of breasts	11.2 ± 1.1
Sparse pubic hair	11.7 ± 1.2
Peak vertical growth rate	12.1 ± 1.0
Menarche[a]	
United Kingdom	13.5 ± 1.0
United States	12.9 ± 1.2
Male	
Enlargement of testes and scrotum	11.6 ± 1.1
Lengthening of penis	12.8 ± 1.0
Sparse pubic hair	13.4 ± 2.2
Peak vertical growth rate	14.1 ± 0.9
Adult genital size and shape	14.9 ± 1.1

[a]*Ninety-five percent reach menarche between ages 11 and 15.*

and breast carcinoma cells. Knockout mice for GPR54 (the so-called Harry Potter strain) fail to undergo puberty. (See Chapter 4 for a review of kisspeptin's actions on the HPG axis.) Recent clinical studies have observed mutant GPR54 genes in patients with idiopathic hypogonadotropic hypogonadism.

1. Causes of Delayed Puberty in Males

Prevalence of undescended testes or cryptorchidism is common at birth (10%) but is reduced to only 1% of males by one year of age. Only about 0.3% of adult males exhibit cryptorchidism, and the case of only one undescended testis is much more common than the bilateral condition. Testicular descent normally is initiated by local effects of DHT. Because of higher temperatures experienced by an undescended testis, the spermatogenetic tissue degenerates at about the time when spermatogenesis would normally begin (at about age 10). Androgen production usually is normal but may be reduced in some cases. The external testis of a unilateral cryptorchid develops normally, and these males usually are fertile. There are several other causes for **hypogonadism** in males including insufficient levels of LH and FSH due to hypothalamic or pituitary dysfunction. In some very rare cases, the interstitial cells may be unresponsive to GTHs. Genetic causes for hypogonadism and gonadal dysgenesis are summarized in Table 10-12. Abuse of anabolic steroids for performance-enhancing effects can cause hypogonadism by reducing FSH and LH secretion via negative feedback.

2. Causes of Delayed Puberty in Females

Primary amenorrhea is the failure for menarche to occur at the normal time (Table 10-13). This condition can be related to many different causes, including disorders of the hypothalamus, pituitary, and ovaries. Poor nutrition, stress, or rigorous athletic training programs can delay puberty through inhibitory actions on the HPG axis. For example, levels of LH and FSH are depressed in women suffering from **anorexia nervosa**, a disorder in which food intake is greatly reduced. Simple weight loss can depress FSH levels for a time but does not inhibit LH secretion.

Secondary amenorrhea occurs after menarche and can result from many endocrine disorders including thyrotoxicosis, drug therapy, and premature menopause as well as from a variety of hypothalamic, pituitary, and gonadal disorders. Two of the more common gonadal disorders associated with secondary amenorrhea are **polycystic ovarian syndrome (PCOS)** and **luteinization of atretic follicles (LAF)**.

Approximately 10% of women are affected by PCOS. The name for PCOS is derived from the general thickening and simultaneous luteinization of several ovarian follicles resulting in formation of numerous cysts in the ovaries. These cysts develop from thecal cells, and there is a corresponding decrease of granulosa cells in these follicles. Current research indicates that excessive AMH production decreases the sensitivity of ovarian follicles to FSH but at the same time increases the recruitment of new follicles (Figure 10-35). Progesterone and estrogen production are diminished and gonadotropin secretion consequently is elevated. This resultant elevation of gonadotropins stimulates excessive production of androgenic steroids (DHEA, androstenedione) that cannot be aromatized to estrogens in the absence of granulosa cells. These androgens are responsible for the masculinization that accompanies PCOS. Uterine abnormalities and infertility result as well as obesity, hirsuitism, and occasional balding from the androgens. An association has been recognized between diabetes mellitus and the tendency to develop PCOS and it appears to be linked to a persistent metabolic disturbance.

The LAF syndrome results from premature luteinization of ovarian follicles prior to formation of the cumulus oophorus. GTH levels are elevated, but the masculinization described for PCOS usually does not happen. Numerous small ovarian cysts may be present, but these are easily distinguished from the large cysts that characterize PCOS.

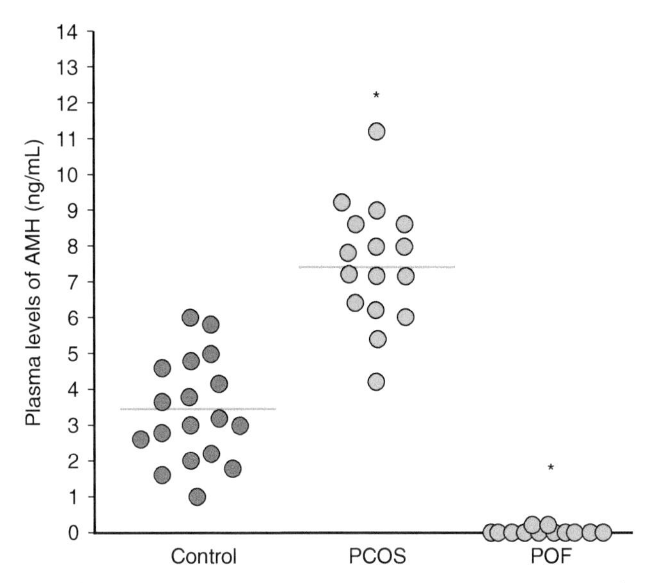

FIGURE 10-35 **Anti-müllerian hormone (AMH) and polycystic ovarian syndrome (PCOS).** AMH is responsible for stimulated growth of new follicles. Elevated levels of AMH cause additional follicle recruitment but prevents follicles from reaching the antral stage. The arrested follicles produce large amounts of the androgens dehydroepiandrosterone and androstenedione, causing the associated masculinization of the woman. In premature ovarian failure (POF) no AMH is produced and masculinization is rare. *(Adapted with permission from Broekmans, F.J. et al.,* Trends in Endocrinology & Metabolism, ***19****, 340–347, 2008.)*

C. Hereditary Disorders

Mutations in key genes regulating gonadal development and chromosomal rearrangements are common causes for deviations in sexual determination and/or expression (Table 10-12). A common disorder is the consequences of meiotic **nondisjunction** associated with the sex chromosomes; that is, the paired sex chromosomes fail to separate during the first meiotic division, resulting in gametes with either two or no sex chromosomes. Zygotes produced by such gametes would have only one sex chromosome or would have three sex chromosomes. Thus, one can obtain XO and XXY or XXX individuals following fertilization of these zygotes. YY zygotes are not viable.

Occasionally, people are born with a combination of ovarian and testicular tissue and sometimes are termed **hermaphrodites**. These are rarely functional in mammals but are often functional among non-mammalian vertebrates (see Chapter 11). This term comes from Greek mythology, combining the names of Hermes, a god who sometimes played tricks on lovers, and Aphrodite, the goddess of love. Most hermaphrodites possess an ovotestis on one or both sides of the body and are infertile or exhibit reduced fertility. These conditions are commonly termed **intersexes**. Rarely, **gynandromorphs** are discovered; these are individuals with an ovary and attendant müllerian derivatives on one side and a testis with its wolffian duct derivatives on the other. **Pseudohermaphrodites** have gonads of the genetic sex but externally resemble the opposite sex.

1. Klinefelter's Syndrome (XXY)

Persons with **Klinefelter's syndrome** are born with an abnormal number of sex chromosomes due to nondisjunction: 47,XXY (total number of chromosomes, sex chromosomes). This familial disorder occurs at fertilization and is present in about 0.2 to 0.3% of males. Klinefelter's syndrome can exist without obvious somatic abnormalities, although these persons are infertile and exhibit differing degrees of mental retardation. Similar syndromes have been described with additional sex chromosomes (48,XXYY, 48,XXXY, etc.). Severity of the symptoms increases with the number of X chromosomes present.

2. Turner's Syndrome (XO)

Another disorder arising at fertilization is **Turner's syndrome**, in which there is a loss of one sex chromosome so that the resulting genotype is 45,XO (one X chromosome and no Y or no second X chromosome). Sometimes this condition occurs when a twin is found to exhibit Klinefelter's syndrome. Individuals with Turner's syndrome have a female phenotype but are infertile. They also exhibit a number of anatomical defects as well as cardiovascular and kidney disorders. These patients do exhibit H-Y antigen that is characteristic of males but at a lower concentration, suggesting some activation of an H-Y antigen gene on the remaining X chromosome. Thus, the presence of H-Y antigen in blood cells is not indicative of the presence of a Y chromosome.

3. Galactorrhea

Secretion of a lactescent (milky) fluid from the breasts of either sex is called **galactorrhea**. It is usually caused by excessive secretion of PRL. Breast enlargement is not a prerequisite for its appearance. Galactorrhea frequently occurs in severe hypothyroidism characterized by elevated circulating levels of TSH and TRH. Prolactin release may be evoked by the high TRH levels.

4. Complete Androgen Insensitivity Syndrome (CAIS)

A person with CAIS is a genetic male (46,XY) that lacks androgen receptors. The testes are normal and secrete testosterone but do not descend from the body cavity. The testes may be removed, generally after puberty, to prevent testicular cancer in the undescended testes. Müllerian duct derivatives are absent because the embryonic testes also secreted AMH. However, the external appearance is that of a woman because of the congenital absence of androgen receptors in the tissues. This is an example of male pseudohermaphroditism.

5. 5α-Reductase Deficiency

A most unusual example of pseudohermaphroditism is the apparent shift of sex at puberty. This condition was first reported from several small villages of the Dominican Republic. This disorder is the result of a genetic deficiency for the ability to synthesize the enzyme 5α-reductase. Males with this defect are born with undescended testes that synthesize testosterone like normal testes. However, these males lacking 5α-reductase cannot convert testosterone to DHT, and early development of male external genitalia does not take place. Although testosterone also can bind to the receptors in these tissues, DHT has a much greater affinity for them. Levels of testosterone in normal prepubertal males are too low to activate these receptors sufficiently, hence the need for 5α-reductase to convert testosterone to the more effective androgen, DHT. Apparently, in men with 5α-reductase deficiency, there is sufficient testosterone to stimulate other androgen-dependent structures that develop normally (such as the vasa deferentia and epididymi) but the normally DHT-dependent structures such as the prostate gland, penis, and scrotum do not develop. These males understandably are raised as girls until these latter structures appear at puberty when testicular testosterone levels are elevated suitably to stimulate the appropriate tissues. The marked increase in circulating testosterone at puberty allows the penis and other structures to enlarge and causes facial hair to appear. Although these men usually change gender roles after puberty, they are infertile.

IX. SUMMARY

The reproductive system includes the HPG axis and sex accessory structures. Primary control resides in pulsatile production of GnRH, which controls pituitary production of FSH and LH, which in turn cause gamete formation, gonadal steroid secretion, and ultimately the regulation of sexual characters and reproductive behaviors. FSH is involved primarily with gamete production whereas LH is responsible for initiating steroid secretion and release of mature gametes (ovulation and spermiation). FSH and testosterone work cooperatively to produce mature sperm. Androgens, progestogens, and estrogens stimulate other primary reproductive structures and secondary sex characters as well as mating behavior. Numerous paracrine regulators stimulated by GTHs or gonadal steroids contribute to reproductive events. Pheromones may play important roles in reproductive behavior and in the timing and success of reproduction.

In mammals, the male sex must be determined by the *sry* gene, which is responsible for production of AMH and androgens by the testis. AMH causes regression of some potential female structures (e.g., oviducts, uterus), and androgen production allows development of male sex accessory structures and reprograms hypothalamic reproductive centers in the brain to the male secretory pattern. Brain neurosteroids may also play a role in development of male brain differences. The default sex in mammals is female although a number of transcription factors and nuclear receptor proteins are required for ovary development and estrogens are essential for completely normal female development.

Mammalian reproductive cycles show great variations among the major taxonomic groups. Monotremes are egg-laying mammals that nourish their hatchlings with milk from their mammary glands. Marsupials allow their young to develop in a pouch after a short gestation period involving a nonendocrine placenta, relying on the mammary glands to support continued development of an exteriorized fetus. Eutherian mammals have a prolonged period of intrauterine fetal development supported by an endocrine placenta. Mammary glands are used for extended nutritional support after birth in all mammals.

Most female mammals exhibit an ovarian-based estrous cycle characterized by a period of enhanced receptivity of the female to the male called estrus. In some species (e.g., primates, including humans and the rhesus monkey), a special phase of the uterine cycle occurs, the menses, which involves sloughing and discharge of a portion of the endometrium and trapped blood. Because of this special uterine cycle, their reproductive cycles are typically called menstrual cycles and most may (e.g., rhesus monkey) or may not (human) exhibit a distinct period of estrus. The proliferative phase of the uterine cycle corresponds closely to the follicular phase of the ovarian cycle, and the secretory phase of the uterine cycle that follows ovulation coincides exactly with the luteral phase of the ovarian cycle. The uterine secretory phase is followed by a quiescent phase called diestrus in most mammals or by the menses in species having a uterine menstrual cycle. The menses corresponds to the first few days of the next follicular phase in the ovary.

Species may exhibit one (monestrous), two (diestrous), or many reproductive cycles (polyestrous) during the breeding season should fertilization or pregnancy not occur. In eutherians, the cycle can be separated into a follicular phase during which one or more ova develop in follicles and a luteal phase that prepares the uterus for implantation of the blastocyst. The luteal phase is named for one or more corpora lutea that develop from ruptured follicles and possibly some atretic follicles that continue to secrete estrogens and progestogens. These phases are clearly separated by ovulation.

During pregnancy, the eutherian placenta in cooperation with the fetal adrenal functions as an endocrine gland to maintain pregnancy, initiate birth, and prepare the mammary glands for postnatal functions. The corpus luteum performs various roles in pregnancy depending on the species. The placenta also functions in gaseous, nutrient, and metabolic waste exchanges. Numerous modifications of the placenta have evolved in different eutherian groups. OXY, prostaglandins, fetal adrenal steroids, and placental CRH, CG, CS, etc. play important roles in pregnancy and in the birth process. Postnatal functions of the mammary gland are controlled by PRL and OXY from the pituitary gland.

Gonadal secretory activities involve two special cell types responsive to FSH and LH, respectively. Ovarian granulosa cells and testicular Leydig cells are responsive primarily to LH and synthesize androgens. Ovarian thecal cells and testicular Sertoli cells as well as Leydig cells respond to FSH with conversion of androgens into estrogens ($P450_{aro}$ activity). FSH also stimulates Sertoli cells to synthesize inhibin, activin, and other local bioregulatory factors. Gonadal cells make many other local regulators (e.g., GnRH, IGFs, AMH, OXY, GDF-9, prostaglandins) in response to GTHs or independently, and these regulators may have autocrine and/or paracrine actions that control local events in the gonads.

STUDY QUESTIONS

1. Identify key steps in germ cell migration to the undifferentiated gonad. How would an endocrinologist identify primordial germ cells before they have reached the undifferentiated gonad?
2. What embryonic tissue type gives rise to the undifferentiated gonad?
3. Outline the major steps involved in differentiation of the bipotential gonad into a testes and ovary.
4. What information is coded in the mammalian Y chromosome that leads to the formation of a testis from a bipotential gonad?
5. Identify genes involved in determining whether the bipotential gonad differentiates into a testis or an ovary.
6. Identify how the differentiated gonad controls secondary sexual development.
7. As a researcher you are interested in performing tests to determine whether androgen receptors are involved in sexual libido. You have the choice of administering either testosterone or dihydrotestosterone to the test animals. Which androgen would you choose to unequivocally demonstrate an action at androgen receptors and why?
8. Explain general differences in reproduction between monotremes and eutherian mammals.
9. Outline the major steps involved in the process of spermatogenesis.
10. Identify which cells of the spermatogenetic lineage possess $4N$ DNA and $2N$ chromosomes.
11. Which cells of the spermatogenetic lineage are haploid?
12. Identify factors that can lead to precocious or delayed puberty.
13. Explain the role of the Sertoli cell tight junctions in forming the blood−testis barrier.
14. Which parts of the seminiferous tubules reside in the adluminal and basal tissue compartments and why is this segregation important?
15. Describe the regulation of testosterone synthesis by Leydig cells.
16. Identify secretory products of the Sertoli cells.
17. Differentiate between the roles of FSH and LH in testes regulation.
18. Describe how inhibin and activin peptides are synthesized and their mode of action.
19. Describe how Sertoli and Leydig cells modulate the activity of macrophage, T lymphocyte, and natural killer cells in the testes. Describe why the testes are considered to be an immune privileged organ.
20. Describe the roles of AMH and growth differentiation factor-9 in folliculogenesis.
21. Identify the major features of oogenesis.
22. How does the recruitment of new germ cells differ in the ovary and testis?
23. What is follicular atresia and how is it essential for normal reproduction?
24. Outline the major steps involved in corpus luteum lysis in non-primate mammals. Identify specifically the roles of the pituitary gland and uterus in this process.
25. Discuss factors involved in the timing of birth in mammals.
26. Describe the "two cell, two hormone" theory of estrogen synthesis in the ovarian follicle and compare to gonadal steroidogenesis in males.
27. Describe how the single event of LH binding to its receptor on ovarian follicle cells can lead to the transcription of multiple genes involved in such diverse processes as steroidogenesis and follicle rupture during ovulation.
28. Identify modes of action for endocrine disruptors targeting the reproductive system.
29. Identify assays used to test chemicals that are released into the environment for potential endocrine-disrupting activity.
30. Compare the reproductive cycles of the rat, woman, ewe, and elephant.

SUGGESTED READING

Books

Baxter, J.D., Melmed, S., New, M.I., 2002. Genetics in Endocrinology. Lippincott Williams & Wilkins, Philadelphia, PA.

Campbell, K.L., Woods, J.W., 1994. Human Reproductive Ecology: Interactions of Environment, Fertility, and Behavior. New York Academy of Sciences, New York, 431 pp.

Johnson, M., 2007. Essential Reproduction, sixth ed. Wiley-Blackwell, New York.

Jones, R.E., Lopez, K.H., 2006. Human Reproductive Biology, third ed. Academic Press, Burlington, MA.

Leung, P.C.K., Adashi, E.Y., 2004. The Ovary, second ed. Academic Press, Burlington, MA.

Neill, J.D., 2006. Knobil and Neill's Physiology of Reproduction, third ed. Elsevier, San Diego, CA.

Norris, D.O., Lopez, K.H. (Eds.), 2011. Hormones and Reproduction of Vertebrates. Mammals, vol. 5. Academic Press, San Diego, CA.

Tyndale-Biscoe, H., Renfree, M., 1987. Reproductive Physiology of Marsupials. Cambridge University Press, Cambridge, U.K.

General Articles

Development

Barsoum, I., Yao, H.H., 2006. The road to maleness: from testis to Wolffian duct. Trends in Endocrinology & Metabolism 17, 223–228.

Glickman, S.E., Cunha, G.R., Drea, C.M., Conley, A.J., Place, N.J., 2006. Mammalian sexual differentiation: lessons from the spotted hyena. Trends in Endocrinology & Metabolism 17, 349–356.

Hiort, O., Holterhus, P.M., 2000. The molecular basis of male sexual differentiation. European Journal of Endocrinology 142, 101–110.

Lim, H.N., Hawkins, J.R., 1998. Genetic control of gonadal differentiation. Bailliére's Clinical Endocrinology and Metabolism 12, 1–16.

Ludbrook, L.M., Harley, V.R., 2004. Sex determination: a 'window' of DAX1 activity. Trends in Endocrinology & Metabolism 15, 116–121.

Eggers, S., Sinclair, A., 2012. Mammalian sex determination—insights from humans and mice. Chromosome Research 20, 215–238.

Robinson, J., 2006. Prenatal programming of the female reproductive neuroendocrine system by androgens. Reproduction 132, 539–547.

Sim, H., Argentaro, A., Harley, V.R., 2008. Boys, girls and shuttling of SRY and SOX9. Trends in Endocrinology & Metabolism 19, 213–222.

Starz-Gaiano, M., Lehmann, R., 2001. Moving towards the next generation. Mechanisms of Development 105, 5–18.

Endocrine Disruptors

Anway, M.D., Skinner, M.K., 2006. Epigenetic transgenerational actions of endocrine disruptors. Endocrinology 147, S43–S49.

Doshi, T., Mehta, S.S., Dighe, V., Balasinor, N., Vanage, G., 2011. Hypermethylation of estrogen receptor promoter region in adult testis of rats exposed neonatally to bisphenol A. Toxicology 289, 74–82.

Gore, A.C., Walker, D.M., Zama, A.M., Armenti, A.E., Uzumcu, M., 2011. Early life exposure to endocrine-disrupting chemicals causes lifelong molecular reprogramming of the hypothalamus and premature reproductive aging. Molecular Endocrinology 25, 2157–2168.

Li, S., Hansman, R., Newbold, R., Davis, B., McLachlan, J.A., Barrett, J.C., 2003. Neonatal diethylstilbestrol exposure induces persistent elevation of c-fos expression and hypomethylation in its exon-4 in mouse uterus. Molecular Carcinogenesis 38, 78–84.

Skinner, M.K., Manikkam, M., Guerrero-Bosagna, C., 2011. Epigenetic transgenerational actions of endocrine disruptors. Reproduction and Toxicology 31, 337–343.

Welshons, W.V., Nagel, S.C., vom Saal, F.S., 2006. Large effects from small exposures. III. Endocrine mechanisms mediating effects of bisphenol A at levels of human exposure. Endocrinology 147, S56–S69.

Viñas, R., Jeng, Y.-J., Watson, C.S., 2012. Non-genomic effects of xenoestrogen mixtures. International Journal of Environmental Research and Public Health 9, 2694–2714.

Zama, A.M., Uzumcu, M., 2009. Fetal and neonatal exposure to the endocrine disruptor methoxychlor causes epigenetic alterations in adult ovarian genes. Endocrinology 150, 4681–4691.

Gonadotropins

Pak, T.R., Chung, W.C.J., 2011. Neuroendocrine control of gonadotropins in mammals. In: Norris, D.O., Lopez, K.H. (Eds.), Hormones and Reproduction of Vertebrates. Mammals, vol. 5. Academic Press, San Diego, CA, pp. 25–43.

Menon, K.M., Menon, B., 2012. Structure, function and regulation of gonadotropin receptors: a perspective. Molecular and Cellular Endocrinology 356, 88–97.

Lactation

Bernt, K.M., Walker, W.A., 1999. Human milk as a carrier of biochemical messages. Acta Paediatrica 88 (Suppl.), 27–41.

Brooks, C.L., 2012. Molecular mechanisms of prolactin and its receptor. Endocrine Reviews 33 (4), 504–525.

Kleinberg, D.L., Barcellos-Hoff, M.H., 2011. The pivotal role of insulin-like growth factor I in normal mammary development. Endocrinology & Metabolism Clinics of North America, 40,461–471.

Newman, J., 1995. How breast milk protects newborns. Scientific American 273 (6), 76–79.

Monotremes and Marsupials

Bradshaw, F.J., Bradshaw, D., 2011. Progesterone and reproduction in marsupials: a review. General and Comparative Endocrinology 170, 18–40.

Griffiths, M., 1984. Mammal: monotremes. In: Lamming, G.E. (Ed.), Marshall's Physiology of Reproduction. Reproductive Cycles of Vertebrates, Vol. 1. Churchill Livingston, London, pp. 351–385.

Nicol, S., Anderson, N.A., Jones, S.M., 2005. Seasonal variations in reproductive hormones in free-ranging echidnas (*Tachyglossus aculeatus*): interaction between reproduction and hibernation. General and Comparative Endocrinology 144, 204–210.

Shaw, G., Fenelon, J., Sichlau, M., Auchus, R.J., Wilson, J.D., Renfree, M.B., 2006. Role of the alternate pathway of dihydrotestosterone formation in virilization of the Wolffian ducts of the tammar wallaby. Macropus eugenii. Endocrinology 147, 2368–2673.

Tyndale-Briscoe, C.H., 1984. Mammals: marsupials. In: Lamming, G.E. (Ed.), Marshall's Physiology of Reproduction. Reproductive Cycles of Vertebrates, vol. 1. Churchill Livingston, London, pp. 386–454.

Ovary

Broekmans, F.J., Visser, J.A., Laven, J.S., Broer, S.L., Themmen, A.P., Fauser, B.C., 2008. Anti-Müllerian hormone and ovarian dysfunction. Trends in Endocrinology & Metabolism 19, 340–347.

Burger, H.G., Dudley, E.C., Robertson, D.M., Dennerstein, L., 2002. Hormonal changes in the menopause transition. Recent Progress in Hormone Research 57, 257–275.

Erickson, G.F., Shimasaki, S., 2000. The role of the oocyte in folliculogenesis. Trends in Endocrinology & Metabolism 11, 193–198.

Fan, H.Y., Liu, Z., Mullany, L.K., Richards, J.S., 2012. Consequences of RAS and MAPK activation in the ovary: the good, the bad and the ugly. Molecular and Cellular Endocrinology 356, 74–79.

Kwintkiewicz, J., Giudice, L.C., 2009. The interplay of insulin-like growth factors, gonadotropins, and endocrine disruptors in ovarian follicular development and function. Seminars in Reproductive Medicine 27, 43–51.

Edson, M.A., Nagaraja, A.K., Matzuk, M.M., 2009. The mammalian ovary from genesis to revelation. Endocrine Reviews 30, 624–712.

Matsuda, F., Inoue, N., Manabe, N., Ohkura, S., 2012. Follicular growth and atresia in mammalian ovaries: regulation by survival and death of granulosa cells. Journal of Reproduction and Development 58, 44–50.

Norris, D.O., Lopez, K.H., 2011. The endocrinology of the mammalian ovary. In: Norris, D.O., Lopez, K.H. (Eds.), Hormones and Reproduction of Vertebrates. Mammals, vol. 5. Academic Press, San Diego, CA, pp. 59–72.

Richards, J.S., Pangas, S.A., 2010. The ovary: basic biology and clinical implications. Journal of Clinical Investigation 120, 963–972.

Richards, J.S., Pangas, S.A., 2010. New insights into ovarian function. Handbook of Experimental Pharmacology 1 (198), 3–27.

Robker, R.L., Wu, L.L., Yang, X., 2011. Inflammatory pathways linking obesity and ovarian dysfunction. Journal of Reproductive Immunology 88, 142–148.

Pregnancy

Bealer, S.L., Armstrong, W.E., Crowley, W.R., 2010. Oxytocin release in magnocellular nuclei: neurochemical mediators and functional significance during gestation. American Journal of Physiology—Regulatory, Integrative and Comparative Physiology 299, R452–R458.

Bazer, F.W., Spencer, T.E., 2011. Hormones and pregnancy in eutherian mammals. In: Norris, D.O., Lopez, K.H. (Eds.), Hormones and Reproduction of Vertebrates. Mammals, vol. 5. Academic Press, San Diego, CA, pp. 73–94.

Newbern, D., Freemark, M., 2011. Placental hormones and the control of maternal metabolism and fetal growth. Current Opinion in Endocrinology, Diabetes and Obesity 18, 409–416.

Young, I.R., Renfree, M.B., Mesiano, S., Shaw, G., Jenkin, G., Smith, R., 2011. The comparative physiology of parturition in mammals: Hormones and parturition in mammals. In: Norris, D.O., Lopez, K.H. (Eds.), Hormones and Reproduction of Vertebrates. Mammals, vol. 5. Academic Press, San Diego, CA, pp. 95–116.

Weetman, A.P., 2012. Thyroid disease in pregnancy in 2011: thyroid function—effects on mother and baby unraveled. Nature Reviews Endocrinology 8, 69–70.

Puberty

Ebling, F.J., 2005. The neuroendocrine timing of puberty. Reproduction 129, 675–683.

Elias, C.F., 2012. Leptin action in pubertal development: recent advances and unanswered questions. Trends in Endocrinology & Metabolism 23, 9–15.

Testis

Akingbemi, B.T., 2005. Estrogen regulation of testicular function. Reproductive Biology and Endocrinology 3, 51.

Barakat, B., Itman, C., Mendis, S.H., Loveland, K.L., 2012. Activins and inhibins in mammalian testis development: new models, new insights. Molecular and Cellular Endocrinology 359, 66–77.

Delbès, G., Levacher, C., Habert, R., 2006. Estrogen effects on fetal and neonatal testicular development. Reproduction 132, 527–538.

Dores, C., Alpaugh, W., Dobrinski, I., 2012. From *in vitro* culture to *in vivo* models to study testis development and spermatogenesis. Cell and Tissue Research 349, 691–702.

Dufau, M.L., Tsai-Morris, C.H., 2007. Gonadotropin-regulated testicular helicase (GRTH/DDX25): an essential regulator of spermatogenesis. Trends in Endocrinology & Metabolism 18, 314–320.

Jarow, J.P., Zirkin, B.R., 2005. The androgen microenvironment of the human testis and hormonal control of spermatogenesis. Annals of the New York Academy of Sciences 1061, 208–220.

Meinhardt, A., Hedger, M.P., 2011. Immunological, paracrine and endocrine aspects of testicular immune privilege. Molecular and Cellular Endocrinology 335, 60–68.

Josso, N., Picard, J.Y., Rey, R., di Clemente, N., 2006. Testicular anti-Müllerian hormone: history, genetics, regulation and clinical applications. Pediatric Endocrinology Reviews 3, 347–358.

Plant, T.M., 2010. Undifferentiated primate spermatogonia and their endocrine control. Trends in Endocrinology & Metabolism 21, 488–495.

Sofikitis, N., Giotitsas, N., Tsounapi, P., Baltogiannis, D., Giannakis, D., Pardalidis, N., 2008. Hormonal regulation of spermatogenesis and spermiogenesis. Journal of Steroid Biochemistry and Molecular Biology 109, 323–330.

Walker, W.H., Cheng, J., 2005. FSH and testosterone signaling in Sertoli cells. Reproduction 130, 15–28.

Zirkin, B.R., Brown, T.R., Jarow, J.P., Wright, W.W., 2011. Endocrine and paracrine regulation of spermatogenesis. In: Norris, D.O., Lopez, K.H. (Eds.), Hormones and Reproduction of Vertebrates. Mammals, Vol. 5. Academic Press, San Diego, CA, pp. 45–57.

Miscellaneous

Ball, G.F., Balthazart, J., 2006. Androgen metabolism and the activation of male sexual behavior: it's more complicated than you think! Hormones and Behavior 49, 1–3.

Ivell, R., Einspanier, A., 2002. Relaxin peptides are new global players. Trends in Endocrinology & Metabolism 13, 343–348.

Mihm, M., Gangooly, S., Muttukrishna, S., 2011. The normal menstrual cycle in women. Animal Reproduction Science 124, 229–236.

Stenvers, K.L., Findlay, J.K., 2010. Inhibins: from reproductive hormones to tumor suppressors. Trends in Endocrinology & Metabolism 21, 174–180.

Clinical Articles

Bertelloni, S., Dati, E., Baroncelli, G.I., Hiort, O., 2011. Hormonal management of complete androgen insensitivity syndrome from adolescence onward. Hormone Research in Paediatrics 76, 428–433.

Chlebowski, R.T., Anderson, G.L., 2012. Changing concepts: menopausal hormone therapy and breast cancer. Journal of the National Cancer Institute 104, 517–527.

Cole, L.A., 2012. The hCG assay or pregnancy test. Clinical Chemistry and Laboratory Medicine 50, 617–630.

Conrad, K.P., 2011. Emerging role of relaxin in the maternal adaptations to normal pregnancy: implications for preeclampsia. Seminars in Nephrology 31, 15–32.

Du, X.-J., Bathgate, R.A., Samuel, C.S., Dart, A.M., Summers, R.J., 2010. Cardiovascular effects of relaxin: from basic science to clinical therapy. Nature Reviews Cardiology 7, 48–58.

Guttilla, I.K., Adams, B.D., White, B.A., 2012. ERα, microRNAs, and the epithelial–mesenchymal transition in breast cancer. Trends in Endocrinology & Metabolism 23, 73–82.

Houk, C.P., Lee, P.A., 2012. Update on disorders of sex development. Current Opinion in Endocrinology, Diabetes, and Obesity 19, 28–32.

Nimkarn, S., New, M.I., 2009. Prenatal diagnosis and treatment of congenital adrenal hyperplasia due to 21-hydroxylase deficiency. Molecular and Cellular Endocrinology 300, 192–196.

Skakkebaek, N.E., Rajpert-De Meyts, E., Main, K.M., 2001. Testicular dysgenesis syndrome: an increasingly common developmental disorder with environmental aspects. Human Reproduction 16, 972–978.

Chapter 11

Comparative Aspects of Vertebrate Reproduction

Survival of all vertebrate species depends on successful reproduction and production of successful new generations of offspring. It should not be surprising that reproduction involves an interaction of environmental and endogenous factors to coordinate events within the **hypothalamus–pituitary–gonad (HPG) axis** as well as the regulation of complicated behavioral events involving individuals and often entire populations. Nutritional state is closely tied with reproduction, and chemical signals from adipose tissue appear to be important signals that modulate hypothalamic functions. Reproduction also requires close cooperation of the adrenal **(HPA)** and thyroid (**HPT**) axes, as well as many other hormones that influence reproduction and metabolism in a myriad of ways.

An understanding of reproductive patterns and their hormonal control in vertebrates is central to our concerns about environmental quality and the future of aquatic and terrestrial ecosystems that are affected adversely by human activities. Levels of environmental contamination previously considered "safe" because they were not immediately toxic are now being seen to influence reproductive efforts through more subtle mechanisms than the dramatic thinning of bird eggshells by the pesticide DDT described some decades ago. Furthermore, documented changes in human populations dealing with sex ratios, timing of puberty, declines in reproductive potentials in human males, and dramatic increases in reproductive cancers provide even more incentive to understand endocrine-related reproductive mechanisms. It is imperative that biologists learn more about the endocrine-regulated reproductive mechanisms that are most prone to disturbance, how these disturbances occur, and what remedies might be applied. In this respect, we need more information about the roles of natural environmental influences and nutrition on reproduction as well as the influences of environmental contaminants on reproduction and metabolism.

Because of the great diversity among vertebrates and the important role of natural selection on reproductive phenomena, it is even more difficult to generalize about reproductive patterns in non-mammalian vertebrates than it was for mammals in Chapter 10. Considerable diversity may be observed even within a relatively small taxonomic group. The descriptions of reproductive anatomy and physiology in this chapter are based largely on the mammalian terminology provided in Chapter 10. Those terms defined previously in Chapter 10 will not be redefined here. Similarly, the details of the organization and operation of the HPG axes of vertebrates are provided in Chapters 4 and 5.

I. SOME GENERAL FEATURES OF VERTEBRATE REPRODUCTION

The HPG axis regulates the reproductive success of all vertebrates. It is influenced by a variety of internal factors (i.e., hormones and other bioregulators) and external cues such as temperature, photoperiod, and pheromones. Attainment of sexual maturity (puberty) occurs at a time characteristic for each species and is followed by a series of reproductive cycles closely attuned to certain environmental factors. Bony fishes represent the largest, most diverse, and one of the oldest groups of living vertebrates. Among these fishes, the teleosts illustrate the full range of evolutionary reproductive strategies known for vertebrates, and numerous excellent models for the study of natural reproduction and its disruption by human activities can be found among them. For example, sexual maturity may be achieved during the first year of life (many teleosts), after more than 15 years of juvenile existence (e.g., Atlantic eel, sturgeon) or at some intermediate period. Some animals are **semelparous** and breed only once after attaining sexual maturity and die soon afterward (e.g., Pacific salmon, *Oncorhynchus* spp.), whereas most species are **iteroparous** and exhibit two or more reproductive cycles. Some of these may produce successive broods in a given year or season or may exhibit only one or two cycles per year. A few species may breed as one sex, then change to the opposite sex (termed a *sex reversal*), and breed again. Males may exhibit an **associated reproductive pattern** (also called **prenuptial**) in which gonadal steroids are highest during mating or a **dissociated reproductive pattern** (also called **postnuptial**) where mating occurs when androgens are reduced (Figure 11-1). Natural environmental factors, such as temperature and photoperiod and the presence of suitable breeding or nesting sites, influence the central nervous system and the HPG axis and regulate gonadal maturation and secretion of sex hormones. Steroid hormones, pituitary hormones, or both determine the development of

Vertebrate Endocrinology. http://dx.doi.org/10.1016/B978-0-12-394815-1.00011-2

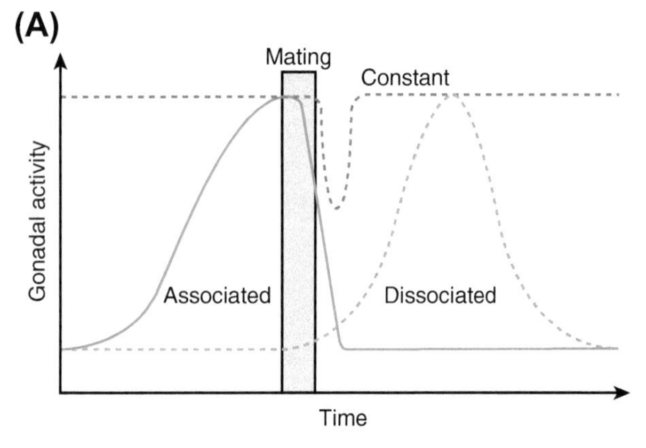

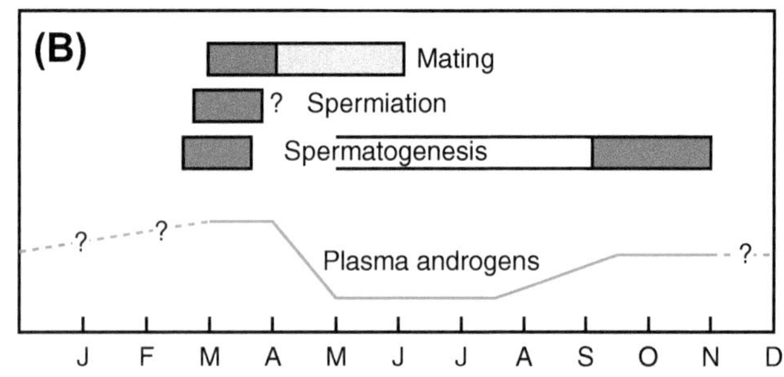

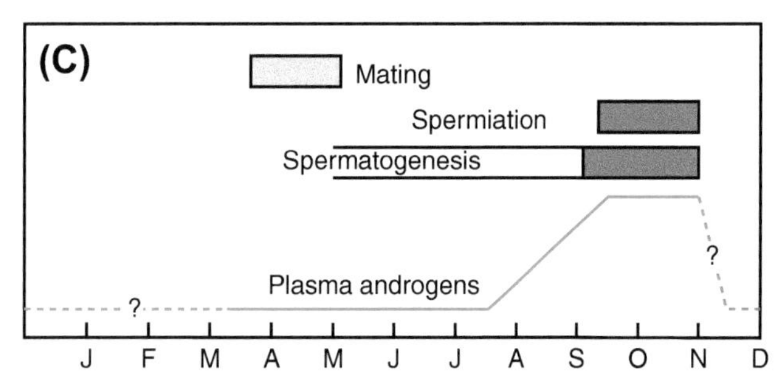

FIGURE 11-1 **Associated and dissociated reproductive patterns.** Mating occurs at the peak of gonadal activity in species exhibiting the associated pattern, whereas mating occurs when gonadal activity is low in the dissociated pattern. *(Part A is adapted with permission from Whittier, J.M. and Crews, D., in "Hormones and Reproduction in Fishes, Amphibians, and Reptiles" (D.O. Norris and R.E. Jones, Eds.), Plenum Press, New York, 1987, pp. 385–410. Parts B and C are adapted with permission from Houck, L.D. and Woodley, S.K., in "Amphibian Biology. Vol. 2. Social Behaviour" (H. Heatwole, Ed.), Surrey Beatty & Sons, Chipping Norton, New South Wales, Australia, 1994, pp. 677–703.)*

TABLE 11-1 Patterns of Reproduction Described as Ovoviviparity

Pattern	Description
1	Internal fertilization; partial development of eggs within female reproductive tract; eggs at oviposition contain visible embryos
2	Restricted to anamniotes; site of metamorphosis is central to recognization; young at birth are premetamorphic larvae (amphibians)
3	Nutrients all supplied by yolk and not by placenta; oviductal secretions, or sibling embryos (including yolk)
4	A trace of egg shell appears whereas in viviparous species, no egg shell is shown
5	Includes anurans that brood eggs in vocal sacs, stomachs, dorsal skin pouches, etc.

Adapted with permission from Blackburn, D.G., *Herpetological Journal*, **4**, 65–72, 1994.

various sex-dependent characters and influence courtship, breeding, and parental behaviors. Pheromones may play critical roles in coordination of male and female physiology and behavior leading to successful reproduction. Many species are known to migrate between feeding and breeding sites.

Like mammals, chondrichthyean and bony fishes, amphibians, and reptiles may be either **viviparous** or **oviparous** whereas cyclostomes and birds are all oviparous. Females of oviparous species lay eggs but viviparous species give birth to live young. Oviparous species all lay eggs with protective coverings from which a larval or juvenile form later will hatch regardless of the state of development at oviposition. The term **ovoviviparity** has been applied somewhat inconsistently in vertebrates but will not be used here as suggested by Blackburn (1994) (see Table 11-1). For simplicity, use of the term "viviparous" here will indicate live-bearing species regardless of whether there is a placental relationship or not; hence, "viviparous" will include retention of eggs in the body of the parent prior to hatching so that free-living young are released into the environment.

Fertilization of eggs after they leave the female's body (**external fertilization**) is a common practice among fishes and amphibians. However, a prerequisite for viviparity requires a technique for transferring sperm from the male to the female prior to release of eggs (**internal fertilization**). Some viviparous anurans (for example, *Nectophrynoides*) transfer sperm through cloacal apposition (as do the oviparous birds) or what has been termed the "cloacal kiss." Aquatic fishes and urodele amphibians, which practice internal fertilization, rely on **spermatophores** for the transfer of sperm. The spermatophore consists of a bundle of sperm that are aggregated and enclosed in a gelatinous substance that will not rapidly dissolve in water. The spermatophore allows the male to directly or indirectly transfer sperm to the female without excessive dilution of the semen. Spermatophore transfer often is facilitated by a sex accessory structure such as a modified fin in fishes or by a copulatory organ. Elasmobranchs, viviparous teleosts, gymnophionid (= caecilians, apodans) amphibians, two anuran species (*Ascaphus*), and many reptiles possess **intromittent organs** that allow direct transfer of sperm or spermatophores from male to female. In birds, sperm transfer is accomplished by direct apposition of the male's cloaca with that of the female. In urodele amphibians, a terrestrial or aquatic male typically deposits his spermatophores on the ground or on the bottom of a pond, respectively, and through a complicated behavioral ritual induces the female to pick one up with her cloaca, thus accomplishing indirect transfer of sperm to the female. Frequently, the female receiving a spermatophore has a special storage site, the **spermatotheca**, which is capable in some species of storing viable sperm for months. The spermatotheca has special mechanisms to disperse and nourish sperm so that they can perform their destined functions at a later time. Eggs can be fertilized and laid at a later time even when males are absent.

A. Gonad Features in Non-Mammals

There is a major difference in the structure of testes in anamniotes and amniote vertebrates. Whereas testes of mammals, birds, and reptiles (amniotes) exhibit a tubular pattern of seminiferous elements with interspersed clumps of interstitial cells, the testes of anamniotes (fishes and amphibians) exhibit a cystic organization. In cyclostomes and elasmobranchs, the testes consist of isolated cellular **cysts** in which spermatogenesis occurs, whereas in bony fishes and amphibians the testes consist of lobes or **lobules**, each of which is composed of large cellular cysts in which spermatogenesis is synchronous. Each testicular cyst is derived from a single **germ cell** that is surrounded by a **Sertoli (sustentacular) cell** (Figure 11-2). Mitotic proliferation of the germ cell occurs, and all of the cells within a cyst (and usually all the cysts within a lobule) will be in the same stage of spermatogenesis. The more posterior lobules may be in a more advanced stage of spermatogenesis in repeating breeders than are the more anterior lobules. Spermiation in anamniotes is usually followed by complete evacuation of sperm from each mature cyst and degeneration of the associated Sertoli cell. In iteroparous species, differentiation of lobules containing new cell nests occurs anteriorly from connective tissue elements and residual germ cells in the connective tissue covering of the testis, the **tunica albuginea**. In some fishes, however, all lobules develop and discharge sperm more or less simultaneously, and if breeding recurs there must be extensive regeneration of new cysts and spermatogonial nests prior to the next breeding season. In urodele amphibians, different lobules mature each breeding season and spent lobules do not regenerate. Although the gonads of anurans have been described by some as consisting of seminiferous tubules, they actually exhibit a pattern of cystic spermatogenesis as seen in other anamniotes.

Cystic spermatogenesis, at least in fishes, is considered to be more efficient than tubular spermatogenesis as exemplified by mammals. Generally, there is less apoptosis of germ cells (about 30 to 40% in fishes), whereas 60 to 80% of all germ cells undergo apoptosis in rats and mice. Furthermore, more sperm are produced in fish per Sertoli cell (e.g., about 100 in guppy, tilapia, or zebrafish) but only 8 to 10 per Sertoli cell in rats and mice. It is hypothesized that a second type of stem cell gives rise to new Sertoli cells throughout the reproductive life of fishes. Sertoli cells in fishes are also extremely efficient at phagocytosis of apoptotic germ cells and at destroying unused sperm remaining in the testes after spawning.

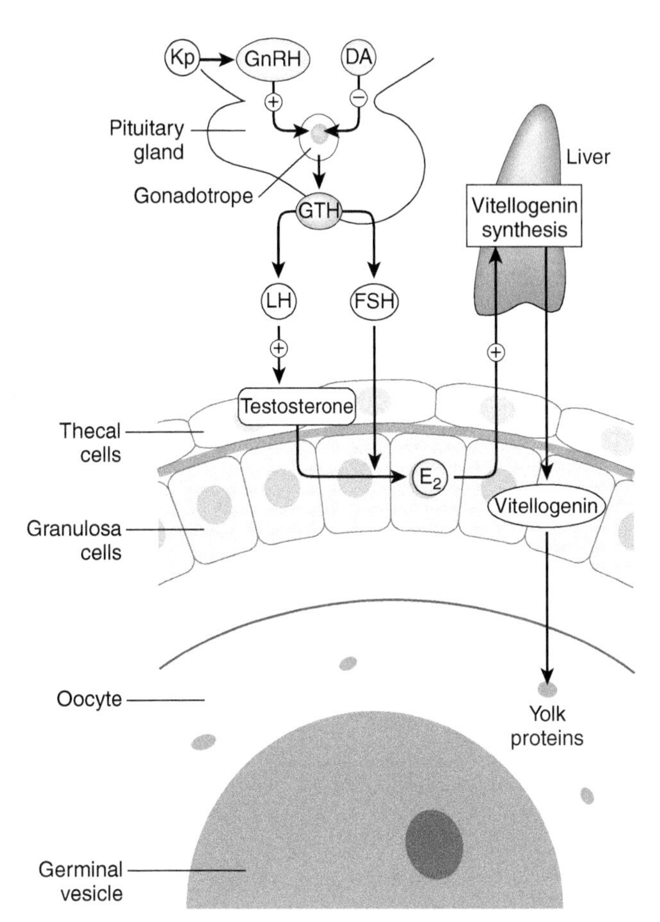

FIGURE 11-2 **Vitellogenin (Vtg) synthesis and incorporation into oocytes in teleosts.** Gonadotropin (GTH) secretion is stimulated by gonadotropin-releasing hormone (GnRH) produced in the hypothalamus of the brain and inhibited by dopamine. Luteinizing hormone (LH) stimulates production of testosterone by thecal cells. Follicle-stimulating hormone (FSH) stimulates conversion of testosterone in granulosa cells to estradiol (E_2) that is secreted into the blood. Estradiol travels to the liver, where it stimulates synthesis of the phosprotein Vtg, which returns via the blood to the ovary. Vtg in turn is incorporated into oocytes and converted under the influence of FSH into yolk proteins. The germinal vesicle is the nucleus of the oocyte. This process occurs in all vertebrates that produce yolky eggs. Kp, kisspeptin. *(Adapted with permission from Connaughton, M.A. and Aida, K., in "Encyclopedia of Reproduction, Vol. 2" (E. Knobil and J.D. Neill, Eds.), Elsevier, Amsterdam, 1999, pp. 193–204.)*

It was formerly believed on the basis of light microscope observations that interstitial tissue (i.e., **Leydig cells**) was lacking in many anamniotes, and the synthesis of androgenic hormones was thought to occur in **lobule boundary cells** associated with the testicular lobule walls. However, the cells formerly identified as lobule boundary cells are actually Sertoli cells. In most cases, Leydig cells are present between cysts in anamniotes, in the periphery of the testes, or in a few cases adjacent to the testes. In the amniote testis, Sertoli cells are associated with spermatogonia in seminiferous tubules and the Leydig cells develop between the tubules in the interstitial regions. Furthermore, the amniote Sertoli cell is involved with the entire range of spermatogenetic stages in levels or layers from the spermatogonia at the outside of the tubule to spermatids and sperm bordering the tubule's lumen (see Chapter 10, Figure 10-9). Furthermore, the amniote Sertoli cell does not degenerate after releasing sperm as occurs in anamniotes. However, in birds and reptiles as well as in some mammals, the Sertoli cell regresses considerably after the breeding season when GTH levels are low.

Ovarian structures and events occurring in the gonads of non-mammalian vertebrates are similar to those described for mammals. Ovarian follicles consist of a surrounding layer of **granulosa cells** that in turn are surrounded by **thecal cells**. Oocyte development is regulated by pituitary **gonadotropins (GTHs)**. The process of yolk protein formation is called **vitellogenesis** (Figure 11-3). Synthesis of lipoprotein yolk precursors or **vitellogenins** by the liver is stimulated by estrogens. When released into the blood, these vitellogenins bind calcium ions and result in an elevation of total blood calcium in females undergoing vitellogenesis. Thus, marked increases in blood calcium can be used as an indicator of vitellogenesis and are a reflection of circulating estrogen levels. Furthermore, vitellogenins are phosphoproteins, and an increase in plasma phosphoproteins can be monitored to provide information on reproductive status. However, sensitive immunoassays (e.g., ELISA, see Chapter 2) are available to measure plasma vitellogenin levels, and measurement of plasma calcium levels or phosphoprotein levels to approximate vitellogenin levels are no longer necessary. Incorporation of vitellogenins by growing oocytes and their conversion to yolk proteins are controlled by estrogens synthesized in the ovary.

An additional steroidogenic tissue, the **interstitial gland**, may develop in the ovaries of non-mammalian gnathostomes similar to that described for mammals in Chapter 10. Interstitial glands develop mainly from cells derived from atretic previtellogenic follicles. It has been suggested that much of the androgen and estrogen synthesized in females during reproductive cycles is from the interstitial gland of the ovary and not from the ovarian follicles.

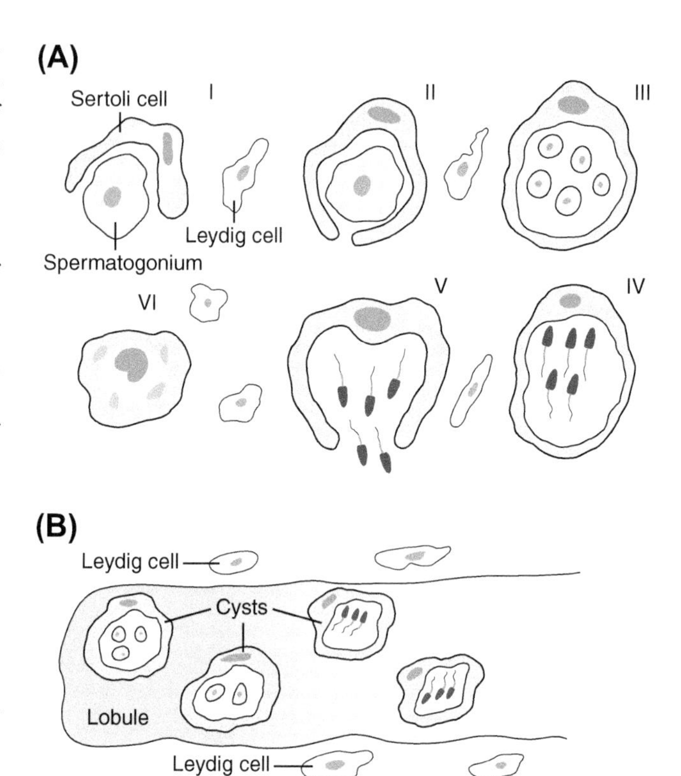

FIGURE 11-3 **Phylogenetic organization of the testis.** (A) Development of cysts in agnathans and elasmobranches where Sertoli cells envelop a spermatogonium cell to form a cyst. Leydig cells occur in the connective tissue surrounding the cycts. (B) In teleosts and amphibians, the cysts develop in lobules and Leydig cells are located between lobules. *(Adapted with permission from Pudney, J., in "Encyclopedia of Reproduction, Vol. 2" (E. Knobil and J.D. Neill, Eds.), Elsevier, Amsterdam, 1999, pp. 1008–1020. © Elsevier Science, Inc.)*

B. Reproductive Ducts in Non-Mammals

In mammals (Chapter 10), sperm are conducted from the epididymis associated with each testis via a **vas deferens** that is derived from the primitive pronephric (archinephric duct) and embryonically is referred to as a **wolffian duct**. The paired **müllerian ducts** develop adjacent to or possibly from the wolffian ducts and give rise to the oviducts and uterus of females when present. The upper end of the müllerian duct, and hence the entire oviduct, is open to the peritoneal cavity.

You will recall that the müllerian ducts degenerate in male mammals due to production of **anti-müllerian hormone (AMH)**, but in females but they may persist in some non-mammals (e.g., male amphibians). In elasmobranch fishes, the müllerian ducts definitely develop from the pronephric duct. The müllerian ducts give rise to oviducts in primitive bony fishes but they do not develop in teleosts. Instead, some teleosts have a short duct of uncertain homologies that transports ova, but most teleosts have an oviduct associated with each ovary that develops from a fold of peritoneum. In still others, a temporary opening develops in the body wall to allow the extrusions of gametes.

The sperm ducts of teleost fishes like the oviducts are derived from the coelomic walls and are not homologous to the vasa deferentia of other vertebrates that develop from the wolffian ducts. Amphibians, reptiles, and birds all retain the müllerian duct in females, and some male amphibians retain a rudimentary müllerian duct. The wolffian ducts function both as urinary ducts and sperm ducts in male fishes and amphibians. They are also retained in all fish and amphibian females as a urinary duct to drain the mesonephric kidney. In amniotes, the wolffian ducts degenerate in all females since new ducts, the ureters, develop to drain the metanephric kidney.

C. Endocrine Features in Non-Mammals

The endocrine factors in non-mammalian vertebrates are similar to and in many cases identical to those already described for mammals. The reader is reminded, however, that relatively few vertebrates have been examined with respect to endocrine factors and their involvement in reproduction; for example, about 50 of the 4600 mammals and even fewer of the more than 27,000 teleosts have been studied thoroughly. The same is true of amphibians, reptiles, and birds. So, there is much to be learned about the diversity of reproductive patterns in vertebrates.

It is clear in every case that the control of reproduction resides in the hypothalamus that controls pituitary and ultimately gonadal functions. One or more **gonadotropin-releasing hormones (GnRHs)** have been identified in all vertebrate groups (see Chapter 5) and GnRH-1s are responsible for release of GTHs: **follicle-stimulating hormone (FSH)** and **luteinizing hormone (LH)**.

There appear to be two distinct GTHs in most non-mammals that are FSH- and LH-like in their actions. Their release generally is under stimulatory hypothalamic control. Follicular development in females and spermatogonial mitoses in males are stimulated by FSH, with meiotic events in males being influenced locally by androgens. Spermiation and ovulation are generally controlled by LH-like GTHs. Teleost fishes have two distinct GTHs now termed FSH and LH (formerly GTH-I and GTH-II, respectively; see Chapter 5). Amphibians, birds, and most reptiles have separate FSHs and LHs. In contrast to fishes and the other tetrapod vertebrates, reproduction in squamate reptiles requires only an FSH-like GTH, and mammalian LH is ineffective in these reptiles (see Chapter 5).

The major circulating estrogen in non-mammals is **estradiol**, and **testosterone** or a structurally similar androgen such as **5α-dihydrotestosterone (DHT)** or **11-ketotestosterone (11-KT)** is characteristic for males. **Progesterone** or a closely related steroid is commonly produced by non-mammalian ovaries as well as by testes. In males, androgens are typically synthesized by Leydig cells, although Sertoli cells may be important sources in some cases. Ovarian production of androgens is typically carried out by thecal cells, with conversion to estrogens in the granulosa cells as described for mammals in Chapter 10.

Reproductive steroid hormone concentrations in blood plasma generally are greater than in mammals possibly due to high circulating levels of **steroid hormone-binding globulin (SHBG)**. Furthermore, relative levels of androgens and estrogens are not always correlated with a particular sex. For example, female fishes may exhibit levels of androgens at certain times that exceed estrogen levels. Similarly, males may secrete significant amounts of estrogens at particular times. The actions of gonadal steroids, including negative feedback effects on the HPG axis, are similar in non-mammals to those described for mammals. In addition to their feedback effects, gonadal steroids regulate gonaduct differentiation and function, differentiation and maintenance of sex accessory structures, and induction of certain behaviors.

As in mammals, **prolactin (PRL)** exhibits some specialized functions in certain species that are closely linked to reproductive events. The specific involvements of PRL will be discussed in some of the accounts that follow.

The relationships between thyroid hormones and reproductive events in non-mammals were discussed in Chapter 7, and actions of corticosteroids on reproduction were described in Chapter 9. Briefly, thyroid hormones appear to enhance the onset of gametogenesis, especially in males. It is only in amphibians and certain avian species that a negative correlation has been reported between thyroid activity and the onset of sexual maturation. Generally, corticosteroids enhance reproduction and typically are elevated during breeding. Stressful stimuli, however, can activate the hypothalamus–pituitary–adrenal (HPA) axis to a level that results in a reduction in or complete cessation of reproductive activities.

D. Sex Determination in Vertebrates

There are several mechanisms of sex determination in vertebrates (see Figure 11-4). The determination of sex in many vertebrates is associated with distinct differences in one pair of chromosomes known as **sex chromosomes**. The remaining pairs are referred to as *autosomes*. In the **homogametic sex**, both sex chromosomes are identical; hence, all gametes produced by members of that sex are the same with respect to chromosome morphology (i.e., they are homogametic). The opposite sex has unlike sex chromosomes that separate during meiosis, producing two kinds of gametes with respect to the sex chromosomes, hence the

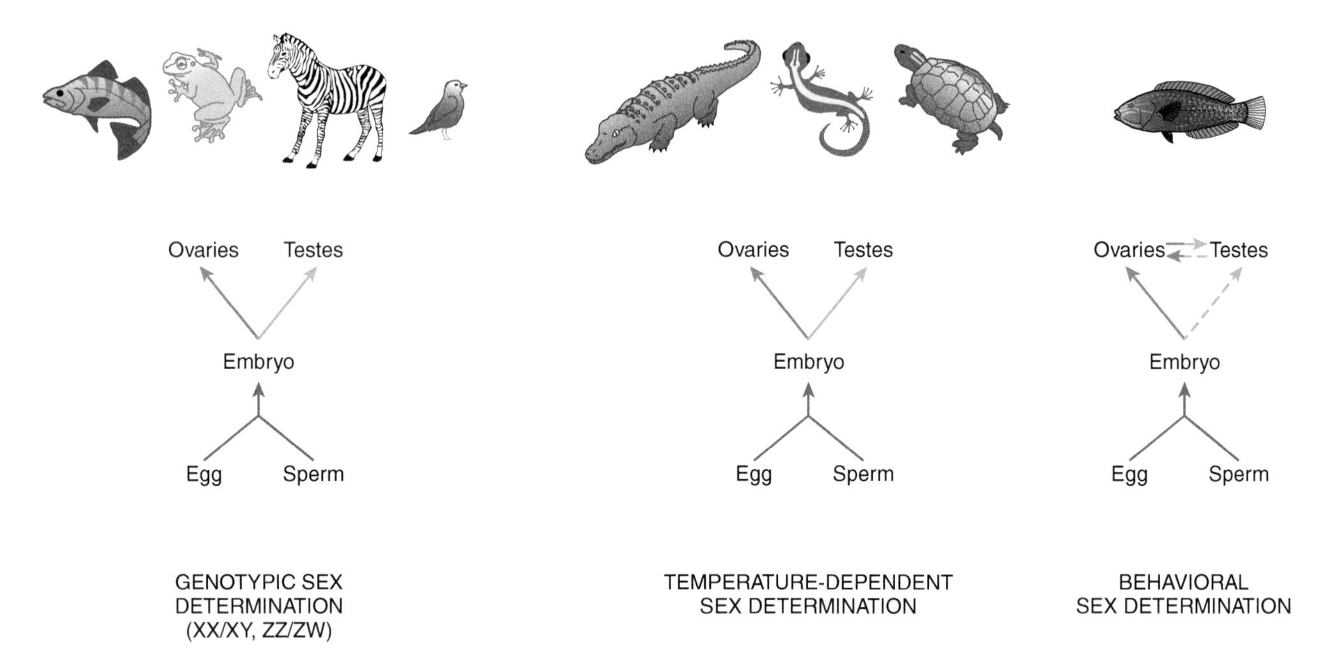

FIGURE 11-4 **Mechanisms of sex determination.** Genotypic sex determination is found among most fishes, amphibians, mammals, and birds. Temperature-dependent sex determination occurs in crocodilians and turtles and some squamate reptiles and amphibians; however, most squamates appear to have genotypic sex determination. Behavioral sex determination involves visual perceptions driving sex changes in adult individuals such as in certain wrasse species that exhibit transformation of mature females into mature males. See text for discussion.

heterogametic sex. When the female is homogametic, as in mammals, most anuran amphibians, some reptiles, and some teleosts, the common sex chromosome type is termed an X chromosome and her diploid sexual genotype is XX. The male is heterogametic, with one X chromosome and one Y chromosome (genotype = XY). When the male is the homogametic sex, as in birds, some reptiles, most urodele amphibians, and numerous teleosts, the male is designated as ZZ and the heterogametic female is ZW. Typically, the homogametic sex does not require gonadal steroids for early differentiation (i.e., it is the default sex), but gonadal steroids must be present for the heterogametic sex to overcome the default sex.

The *sry* gene on the Y chromosome of mammals and *sry*-like genes in some other vertebrates are responsible for initiating heterogametic sexual differentiation (Box 11A). When sex is determined by unique genes located on one chromosome, it is termed **genotypic sex determination (GSD)**, regardless of whether or not there are distinct sex chromosomes. Sex determination in most non-mammals appears to involve GSD even where chromosomal dimorphism is absent. However, specific sex-determining genes have not been identified in non-mammals with the exception of two teleost species identified by Yoshi Nagahama and his colleagues in Japan, although differential sequences of gene activation have been described for males and females of many non-mammalian groups (see Box 11A). Although in numerous instances, genes unique to a given sex have been identified in some non-mammals, they are downstream of unidentified events that determine if and when they become active.

Many vertebrates do not have distinct sex chromosomes and some of these exhibit environmental mechanisms of sex determination. Although the lack of a morphologically distinct pair of sex chromosomes does not mean there is no genetic basis to sex determination, at least two mechanisms depend on environmental temperature or behavior of conspecifics. **Temperature-dependent sex determination (TSD)** has been demonstrated in all crocodilians, many turtles, some lizards and snakes, some amphibians, and some fishes. TSD is correlated with early nest or water temperatures (Table 11-2). Incubation at a high temperature produces all one sex whereas at a lower temperature all the offspring are the other sex. These

TABLE 11-2 Temperature-Dependent Sex Determination in Vertebrates

Group	Male-producing temperatures (°C)	Female-producing temperatures (°C)
Crocodilians	> 34	< 30
Turtles	23–27	30–33
Lizards	29–33	24–29
Teleosts	17–25	11–19

BOX 11A Genes Involved in Sex Determination of Non-Mammalian Vertebrates

The pathway of gene activation responsible for sexual differentiation in mammals is provided in Box Figure 11A-1. The *sry* (sex-determining region Y) gene located on the Y-chromosome in mammals is required for testis formation, but the *sry* gene has not been found outside of the placental mammals. Interestingly, the downstream genes or their homologues are widely distributed among vertebrates. Steroidogenic factor-1 (SF-1), an orphan receptor of the steroid receptor superfamily, is a transcription factor, and the *sf-1* gene is thought to be an important activator of *sry* in mammals.

Among teleosts with XX/XY sex determination, a unique male sex-determining gene was discovered in the medaka (*Orizias latipes*), the *dmy/drmt1bY* gene (DM domain of the Y chromosome/doublesex- and mab-3-related transcription factor 1). This medaka gene has not been found in any other teleost except for another species of *Orizias*. However, another gene called the gonadal-soma-derived factor (*gsdf*) is activated in other XX/XY teleosts early in the sex-determining process. This gene appears to be unique to fishes. In males, expression of *gsdf* increases in what will become Sertoli cells. The gene *gsdf* activates the *drmt1* gene coincident with testicular differentiation, starting a cascade of gene activation, including *sox9a2* (Box Figure 11A-2). In mammals, the *sox9* (SRY-related high-mobility group box 9) gene regulates differentiation of Sertoli cells during development of a testis. A different cascade has been identified in females involving expression of *foxl2* (forkhead box L2) and *cyp19a1a* that encodes $P450_{aro}$ (Box Figure 11A-2). In teleosts that exhibit TSD, the expression of *cyp19a1a* is diminished in males. Conversely, *cyp19a1a* expression is increased in females (Box Figure 11A-2).

Non-mammalian tetrapods employ genes for sex determination that are similar to those found in both fishes and mammals. The gene *dax1* (dosage-sensitive sex reversal, adrenal hypoplasia critical region, on chromosome X, gene 1) is expressed early in the gonads of both sexes of amphibians and reptiles. In male amphibians, *dmrt1* and *sox9* are expressed later in interstitial cells and Sertoli cells within the testes. In female amphibians, *foxl2* is responsible for activation of the aromatase gene. Among reptiles exhibiting TSD, administration of testosterone and 5α-reductase inhibitors to turtles results in a large proportion of females at male-producing temperatures whereas inhibition of aromatase at female-producing temperatures yields mostly males. Although there are few gene studies done in reptiles, the *sox9* gene appears to be activated after the *amh* gene (anti-müllerian hormone), whereas in mammals the sequence is reversed. In ZZ/ZW birds, a Z-linked *dmrt1* gene activates the *sox9* gene for testis development. Both *foxl2* and the aromatase gene are involved with ovarian differentiation in birds, but the activating gene has not been identified.

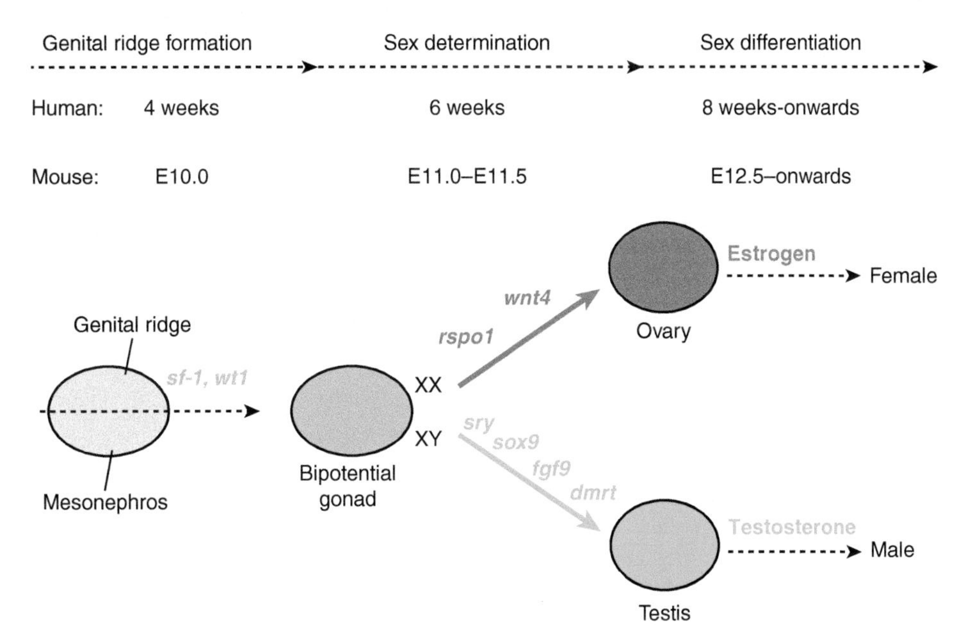

BOX FIGURE 11A-1 **Mammalian genes and gonadal development and differentiation.** In XY male mice, activation of some genes inhibits activity of genes in the XX pathway. Similarly, in a female mouse, activation of certain genes prevents activation of male genes downstream of *sry*. E10.0, etc. represent day of embryonic development in the mouse. Developmental times for humans are shown for comparison. The pathway for humans and other mammals is thought to be essentially like the mouse. Genes are indicated in italics. *(Adapted with permission from Sim, H. et al.,* Trends in Endocrinology & Metabolism, ***19****, 213–222.)*

BOX 11A Genes Involved in Sex Determination of Non-Mammalian Vertebrates—cont'd

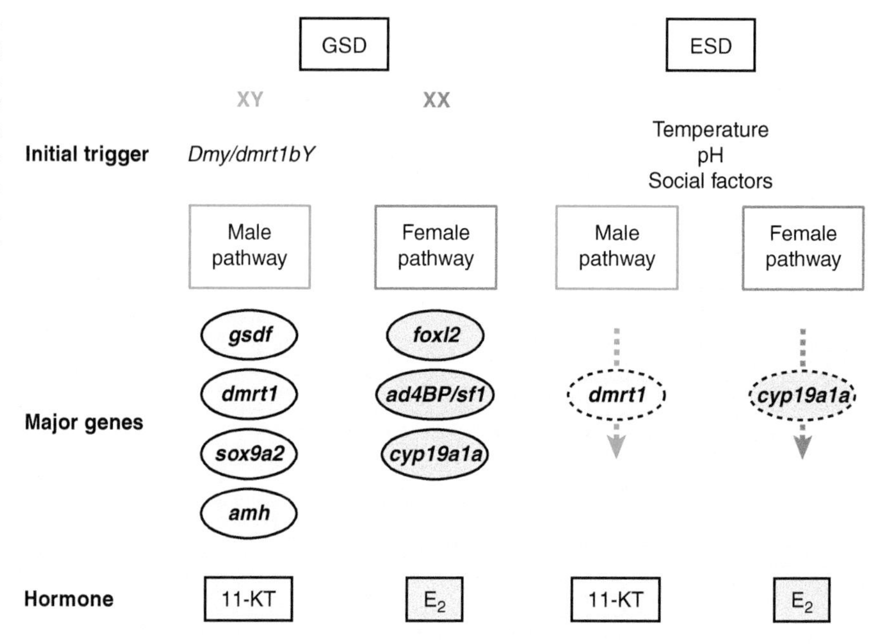

BOX FIGURE 11A-2 **Gonadal differentiation pathway in the medaka, a teleost fish.** GSD, genetic sex determination; ESD, environmental sex determination. See text for gene explanations. *(Adapted with permission from Paul-Prasanth, B. et al., in "Hormones and Reproduction of Vertebrates. Vol. 1. Fishes" (D.O. Norris and K.H. Lopez, Eds.), Academic Press, San Diego, CA, 2011, pp. 1–14.)*

temperature effects at the extremes are "all-or-none" but intermediate temperatures produce equal ratios of males and females (50:50) rather than intersexes. Hormones do not play an initiating role in TSD, although the levels of certain enzymes, such as **aromatase ($P450_{aro}$)** and **Δ^5-3β-hydroxysteroid dehydrogenase (3β-HSD)** are greatest in adrenocortical tissue and mesonephric kidney during sexual differentiation. Brain $P450_{aro}$ may also be involved in brain changes associated with sexual differentiation.

Another type of environmental sex determination is based on social situations (i.e., behavior) that can initiate changes in gonadal sex. In sequential hermaphrodites, sex change may be induced by behavioral events, and such species are said to exhibit **behavioral sex determination (BSD).** Most examples of sex change (often termed *sex reversal*) occur in adult coral reef fishes. **Protogyny** (female to male sex change) is the more common and appears to be triggered by environmental cues. Male to female sex change is called **protandry**. Some sex-changing species are termed **diandric** in that two types of males are found. One male is a genetically determined or **primary phase male**, and the **secondary** or **terminal phase male** results from sex change of an adult female. In many cases of protogyny, sex change is initiated by loss or removal of a dominant male in the social grouping, and the largest ranking female takes his place as the dominant fish. Often there are immediate behavioral changes (e.g., increased aggression) in the dominant female that prevent sex reversal in subordinate females. Changes in skin coloration and transformation of the ovaries to testes usually follow the behavioral changes.

The bluehead wrasse (*Thalassoma bifasciatum*) is a diandric, protogynous sex-changing teleost living in the Caribbean. Spontaneous reversal is accompanied by a twofold increase in the number of GnRH-immunoreactive cells in the preoptic area of the brain. A similar increase in GnRH cells can be induced with implants of the androgen 11-KT into females. Treatment of females with **human chorionic gonadotropin (hCG)** induces gonadal sex change within 1 to 6 weeks. How GnRH and GTHs produce a sex change in females previously employing these same hormones for female reproduction is unclear. Possibly the answer lies in changes in brain function that are initiated by environmental cues.

A similar pattern of sex change occurs in a closely related diandric species, the saddleback wrasse (*Thalassoma duperrey*) living on coral reefs in Hawaii. A very different mechanism for triggering sex change operates in *T. duperrey*, a promiscuously mating fish that does not live in male-dominated groups like the bluehead wrasse. Long-term studies of these fish in underwater cages have determined that the cue for sex change is visual rather than chemical or tactile. The largest female somehow perceives a change in the ratio of larger fish (usually males) to smaller fish (usually females) and undergoes sex reversal when sufficient males are not present. Experimental studies suggest a role for biogenic amines in the events that influence GnRH and GTH release. Isolation of a female *T. duperrey* from the population results in immediate changes in catecholaminergic neuronal activity leading to sex change that can be duplicated with appropriate pharmacological treatments.

1. Steroid Hormones and Sex Determination

Regardless of the mode of sexual differentiation (GSD, TSD, or BSD), gonadal steroids and neurosteroids appear to have important roles. Androgens and estrogens from the gonads traveling in the blood or locally produced in the brain often determine the pattern of HPG function early in development that persists in adults. Furthermore, exposure to exogenous steroids can interfere with this process and cause reversal of one sex to the other. Typically, there is

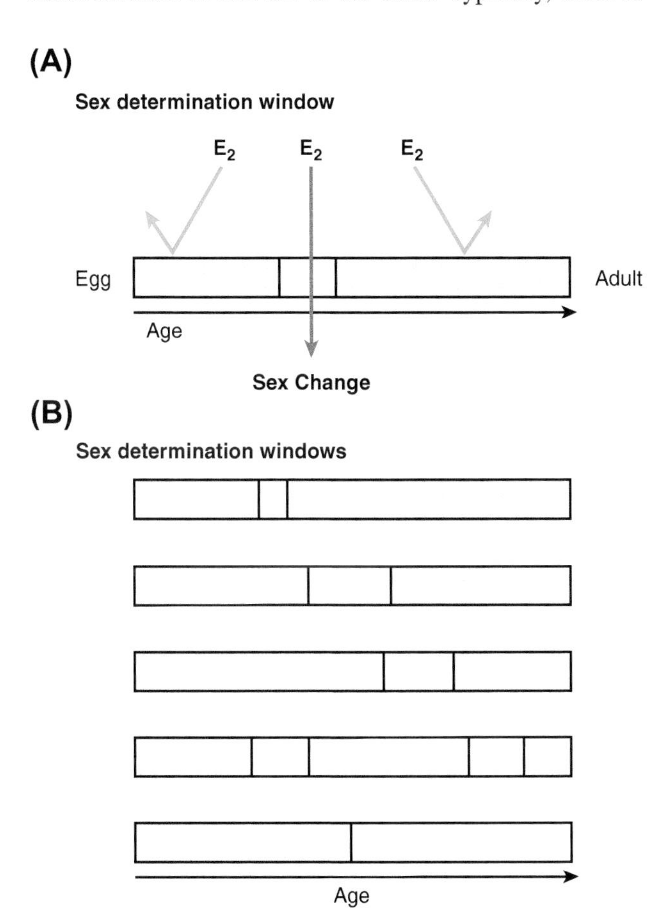

FIGURE 11-5 **Sex determination windows in non-mammals.** (A) Estrogen (E_2) can alter the genetic sex only if it appears during the short time or window (clear area) when sex determination is normally occurring in the embryo or larva. Phenotypic sex is an example of a trait that is determined by an organizational action of steroids. Exogenous estrogen often alters the genetic sex from male to female if it appears at this time. (B) Illustration that such windows (clear areas) can exist at different times in the life history of animals and can be variable in their duration. Thus, adults of some species may undergo sex reversal even after functioning as a male or as a female.

a window of sensitivity in development for this process (Figure 11-5). When during development this window occurs and its width (how long the window is open) vary with the species. Treatment of larval fish, for example, with reproductive steroids during the sensitive window can produce populations of all one sex that may be advantageous in fish farming where one sex may grow more rapidly than the other or where sterile fish spend energy producing muscle and not gametes. Estrogen treatment can reverse the effects of a male-producing temperature in turtles and augment the actions of a female-producing temperature, but androgens usually cannot override effects of a female-producing temperature. In turtle eggs incubated at a neutral temperature (one that should produce a 50:50 ratio), the sex ratio is sensitive to either androgens or estrogens. However, addition of both androgens and estrogens simultaneously at the neutral temperature will produce intersexes.

One of the major concerns of exposure of developing animals in nature to environmental estrogens relates to the appearance of an **endocrine-disrupting chemical (EDC)** during this window of sensitivity. Depending on dose and time of exposure, we may see complete or partial (intersexes) sex changes.

Paradoxical effects of steroid exposure are frequently observed, especially in fishes and amphibians. For example, exposure of a juvenile female catfish to testosterone causes precocial but otherwise normal sexual development, whereas exposure to a large dose of testosterone may feminize larval male fish.

This early sexual differentiation based on gonadal steroids or temperature resulting in a permanent designation of sex or of a sexual characteristic is known as an **organizational effect**. Sexual differentiation of the HPG axis, for example, is an example of an early organizational event in most vertebrates. In the situation described above for the coral reef fishes, sex and related structures and behaviors are not organized at an early age but show that sex determination and differentiation of the gonads are due to an **activational effect** that occurs later in life.

II. REPRODUCTION IN AGNATHAN FISHES: CYCLOSTOMES

It appears that the cyclostome gonad arises entirely from the embryonic cortex, whether it is destined to be a testis or an ovary. This singular embryonic origin may account for the common observations of what are believed to be hermaphroditic gonads among the hagfishes. In contrast, the single ovary of hagfishes is due to the failure of one primordium to develop. Some data on circulating levels of steroids for selected species of cyclostomes are provided in Table 11-3. Gonaducts that are characteristic of most vertebrates are absent in cyclostomes, and the gametes are

TABLE 11-3 Circulating Steroid Levels in Fishes (ng/mL)

Class and species	Testosterone[a]	Estradiol	Progestogen
Agnatha			
Petromyzon marinus (sea lamprey)			
Male, prespermiating	275	0.4–1.4	
with mature sperm	216	0.6–1.0	
Female, preovulatory	156		
Female, ovulating		10–13	
	0.02		
Chondrichthyes			
Torpedo marmorata (male)	15.6–35		
Raja radiata			
Male	28–102		
Female	0.2–6		
Raja eglanteria			
Male	42.7	0.022	0.150
Female	14	2.48	0.042
Scyliorhinus canicula (male)	2-6		
Osteichthyes			
Oncorhynchus nerka			
Male	17		
Female	78		
Salmo trutta			
Male	2–33		
Female	20–77		
Oncorhynchus mykiss			
Prespawning female	52–235	24–48	8–15[b]
Spawning female	65–84	2–3	354–416[b]
Spent female	2–5	1–2	8–19[b]

[a] *15α-hydroxylated testosterone in cyclostomes.*
[b] *17,20β-dihydroxy-4-pregnen-3-one.*

shed into the coelom from which they exit via abdominal pores.

A. Lampreys

Lampreys (Petromyzontidae) are characterized by having no breeding cycle, since all individuals die after a single spawning (i.e., semelparous). The indifferent gonad develops very slowly and appears female-like in both sexes. In males, the posterior portion develops into testicular tissue and oocytes may persist for a time but eventually undergo a form of degeneration called **atresia**. In females, the anterior gonad becomes the functional ovary and the posterior region degenerates. Sex accessory structures include enlarged fins, fusion of the two dorsal fins, and cloacal swelling. Steroidogenesis is unique in lampreys, with the dominant steroids being 15α-hydroxylated compounds. The physiological roles of these compounds is not clear as specific receptors have not been identified for either **15α-hydroxyprogesterone** or **15α-hydroxytestosterone**. Testosterone appears to be a precursor for both 15α-hydroxytestosterone and estradiol.

1. Male Lampreys

Male lampreys exhibit a single median lobular testis with a cystic pattern of spermatogenesis. Testicular development is slow, beginning at or after metamorphosis of the ammocoete to the adult form. The testes may contain only primary spermatocytes at the time of migration to the breeding grounds. These spermatocytes are transformed rapidly near the time of spawning into sperm masses. When the cysts have completed formation of sperm, they simultaneously rupture and release sperm into the body cavity.

Typical interstitial cell masses can be identified cytologically between the lobules in testes of migrating lampreys. These cells accumulate cholesterol-positive lipids and have become densely lipoidal by spawning time. Cytologically, interstitial cells appear to be steroidogenic and exhibit maximal 3β-HSD activity in February and March prior to the time of spawning. Plasma testosterone levels are extremely low and do not change with administration of lamprey GnRH nor does gonadal tissue respond *in vitro* with testosterone secretion. However, 15α-testosterone levels in male sea lampreys exhibit a dose–response relationship to administered lamprey GnRH.

2. Female Lampreys

Because of fusion of the paired primordia early in development, the adult female lamprey has a single ovary. Oogenesis has been carefully examined in the parasitic sea lamprey *Petromyzon marinus* and in the river lamprey *Lamperta fluviatilis*. In *P. marinus*, oogonia proliferate mitotically in the larvae to form primary oocytes, but by the time of metamorphosis of the larva to the juvenile, there are no oogonia remaining in the ovary. The primary ovarian follicles become more vascularized at this time. During the prolonged parasitic phase of body growth (about 10 to 20 months), the oocytes continue to enlarge slowly. The single follicular cell layer becomes thinner and less vascularized as spawning approaches, and the oocyte enters a period of rapid enlargement to reach the preovulatory condition. The mature follicles rupture immediately before spawning, and the eggs enter the coelom. Follicular atresia occurs throughout the history of ovarian development, and many oocytes undergo atresia, establishing this basic pattern early in the phylogeny of vertebrates. Phagocytes derived from the follicular cells ingest the yolk, and the follicle layers and surrounding stroma collapse into the area formerly occupied by the oocyte.

Oocyte growth in *L. fluviatilis* accelerates markedly just prior to spawning. The granulosa contacts the oocyte only at the vegetal pole and reaches maximal development about one month prior to spawning. The thecal cells are greatly reduced and cover the granulosa layer and the animal pole. The theca interna consists of a single layer of cells in which there is a marked increase in smooth endoplasmic reticulum and mitochondrial differentiation during vitellogenesis. These cells show maximal cytological activity prior to the time of most intensive vitellogenesis, following which they undergo progressive regression until the time of ovulation. The theca externa consists of fibroblasts, collagen fibers, and capillaries. 3β-HSD activity is apparently confined to the thecal cells, where peak activity is observed about one month prior to the appearance of secondary sex characters and the acceleration of follicular development.

Vitellogenesis appears to be an estrogen-dependent event in lampreys involving cooperative action of the liver, which produces proteins that are secreted into the blood and are sequestered by the ovary to be incorporated in the developing oocyte. Estrogens stimulate liver hypertrophy and elevate plasma protein-bound calcium, suggesting the presence of a mechanism such as the one that has been documented so carefully in other non-mammals (see ahead).

In the free-living lampreys that do not feed after metamorphosis, the ovarian events occur over a much shorter time and are consequently more dramatic. In brook lampreys, the immediate post-metamorphic period is marked by the onset of both vitellogenesis and massive atresia. As many as 70% of the follicles present at metamorphosis may become atretic, and phagocytosis of the yolk may provide an essential nutritional source for growth of the remaining oocytes to maturity.

3. Endocrine Function in Lampreys

The importance of gonadal steroids to development of sex accessory structures has been demonstrated through classical experiments involving hypophysectomy, gonadectomy, and appropriate hormone therapy to either hypophysectomized or castrate animals. However, lampreys secrete primarily 15α-hydroxylated steroids that are not secreted by any other vertebrates. Lamprey GnRH or pituitary GTHs from various vertebrates stimulate gonadal hormone secretions, which in turn stimulate formation of secondary sex characters. Spermiation also can be induced with lamprey GnRH. The gonads of hypophysectomized *L. fluviatilis* are less developed than those of sham-operated controls, indicating a reliance on the HPG axis.

B. Hagfishes

There is a single gonad in adult hagfishes similar to that described in lampreys. In *Myxine glutinosa*, about 41% of the gonads examined are hermaphroditic, about 58% are all female, and less than 1% are male; however, there is to date no evidence of whether hagfishes are functional hermaphrodites. All gonads appear female after hatching, and sexual differentiation of the gonads does not occur until the hagfishes reach about 20 cm in length. Both "preovulatory corpora lutea," referred to here as **corpora atretica** or **atretic follicles**, and "postovulatory corpora lutea," referred to here as **corpora lutea**, have been described in mature females, and in *M. glutinosa* they can convert pregnenolone to progesterone. Although male hagfishes appear to lack identifiable interstitial cells, they do secrete testosterone.

Hagfishes (Myxinoidea) apparently breed more than once, and several species exhibit seasonal migrations and reproduction. The reproductive biology of deepwater myxinoids is poorly known. They typically live on muddy ocean bottoms at depths between 100 and 300 meters in northern or arctic waters. Several hagfishes are seasonal breeders and some exhibit continuous reproduction. *Eptatretus* species, which migrate periodically into the shallow coastal waters of Japan, show a definite seasonal cycle of gonadal activity. The brain content of GnRH of the Atlantic hagfish (*M. glutinosa*) exhibits a seasonal cycle, and peaks of GnRH activity in females are followed by peaks in plasma estradiol and progesterone.

Hypophysectomy of *Eptatretus stouti* results in testicular degeneration in males, indicating evidence of gonadotropic control. However, hypophysectomy of mature females has no effect on either ovarian structure or circulating steroid hormone levels. Vitellogenesis is stimulated by estradiol treatment of *E. stouti*.

III. REPRODUCTION IN CHONDRICHTHYEAN FISHES

Chondrichthyean fishes are oviparous (holocephalans, skates, and some sharks) or viviparous (all rays, 70% of sharks). The prevailing view is that viviparity evolved independently 9 to 10 times, although a more recent suggestion considers the smaller bodied oviparous species as being derived from viviparous species. Viviparity is present in ten families of sharks and four families of rays. Many viviparous sharks and batoid rays are "continuous breeders"; for example, the spotted dogfish (*Scyliorhinus canicula*) is sexually active throughout the year and may have embryos in different stages of development *in utero* at the same time. In contrast, many rays and some viviparous sharks are "seasonal breeders," with entire populations synchronized to environmental factors. Some viviparous and oviparous species are intermittent or "punctuated breeders," meaning that reproductively active periods may be interspersed with one or more non-breeding years.

Fertilization is internal regardless of whether they are viviparous or oviparous, and special intromittent organs called **claspers** are present in mature males. Holocephalans have not been studied extensively, and the reproductive biology of the more common elasmobranchs is better known. The elasmobranch reproductive system is illustrated in Figure 11-6, and plasma levels of gonadal steroids for selected species are provided in Table 11-3.

A. Male Elasmobranchs

Male elasmobranchs have paired testes and paired **epigonal organs**. Efferent ducts connect each testis to a genital duct (ductus deferens) and a sperm storage organ (epididymis). Claspers are associated with ventral (pelvic) fins located near the cloaca and are used for transferring sperm to the female genital tract. In sharks, claspers are not used to grasp the female during mating as the name implies, but rather males grasp the females with their teeth during

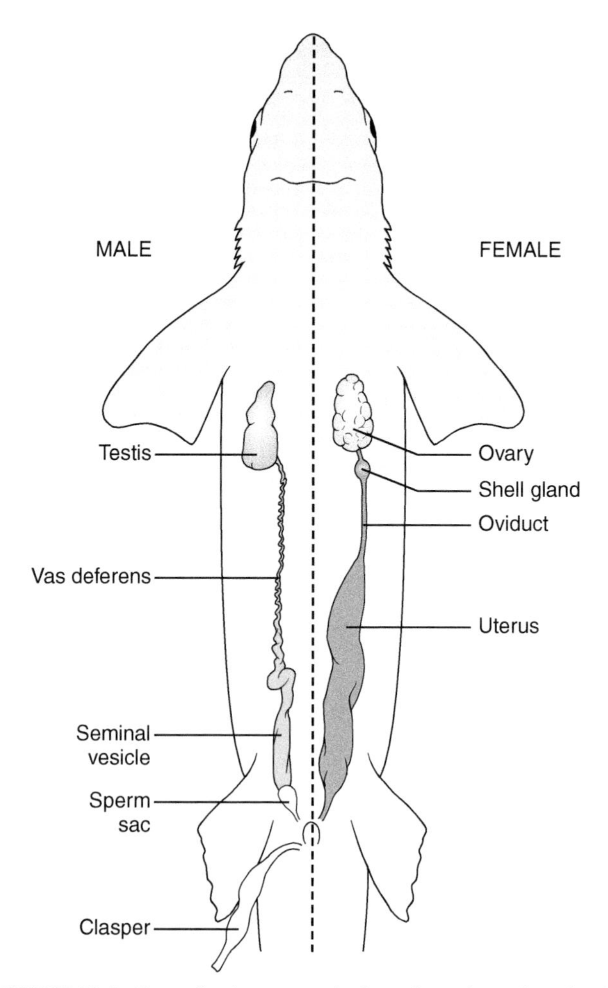

FIGURE 11-6 **Reproductive systems in elasmobranchs.** *(Adapted with permission from Maruska, K.P. and Gelsleichter, J., in "Hormones and Reproduction of Vertebrates. Vol. 1. Fishes" (D.O. Norris and K.H. Lopez, Eds.), Academic Press, San Diego, CA, 2011, pp. 209–237.)*

copulation. Recent bite marks on a female shark are used as evidence of recent copulation.

Spermatogenesis in paired testes is of the cystic type. Sertoli cells are present and possess 3β-HSD and $P450_{aro}$ activity and are considered to be the primary source of androgens and estrogens. Sertoli cells become densely lipoidal and cholesterol positive following spermiation and are eventually resorbed. Sertoli cells for the next cycle differentiate from connective tissue cells (fibroblasts or possible stem cells) in the wall of the testis. Nests of spermatogonia proliferate from germ cells in the same regions, and they are responsible for producing sperm utilized during the next breeding period. In a few species, Leydig cells have been described between cysts although they may not be readily discernible throughout the testicular cycle. They are not considered an important source of androgens, however.

Spermatophores produced by elasmobranchs are the result of secretory activities of male accessory ducts. After spermiation, sperm pass through vasa efferentia and enter the coiled tubules of the **Leydig gland**, which is derived from the anterior portion of the mesonephric kidney. It is not steroidogenic and should not be confused with the Leydig cells. Sperm and secretions of the Leydig gland pass on to an expanded region of the vas deferens known as the **ampulla epididymis**. Here the sperm are consolidated and receive additional secretory material to form complex spermatophores typical for each species.

1. Endocrine Factors in Male Elasmobranchs

The importance of the ventral lobe of the elasmobranch pituitary as the source of GTH controlling spermatogonial proliferation (mitosis) has been demonstrated in the spotted dogfish. Degenerative changes in the testes appear 6 weeks after removal of only the ventral lobe, and 22 months later the testes contain only spermatogonia and mature sperm, indicating that removal of the ventral lobe blocks further differentiation of spermatogonia to spermatocytes, whereas all spermatocytes present at the time of surgery are able to complete meiosis and spermiogenesis. In addition to testosterone, 11-KT and DHT have been reported in many but not all elasmobranchs. Androgen levels are elevated when mature spermatocysts are present in the testes, during development of sex accessory structures, and during semen transport. Circulating estradiol and progesterone also have been reported in males. Estradiol levels are lower in male *Raja eglanteria* than in females, but progesterone levels are greater in males. Corticosterone levels in seasonally breeding elasmobranchs are elevated during testis growth, spermatogenesis, and mating.

A **relaxin** molecule, structurally more similar to mammalian insulins than to mammalian relaxin, is produced by the testis of the shark *Sypyrna tiburo* and is elevated in the circulation during late spermatogenesis and at copulation. The level of relaxin in semen is 1000× greater than that found in the blood. The precise role of relaxin in males needs to be determined.

B. Female Elasmobranchs

The elasmobranch ovary is covered by germinal epithelium and may contain a cavity derived from large lymph spaces within the stroma. Elasmobranch follicles are similar to those of mammals in possessing several distinct layers of cells. The connective tissue near a nest of oogonia will differentiate into the theca. As each follicle begins to develop, some epithelial cells from the germinal epithelium undergo hypertrophy and hyperplasia to become the granulosa. In some species, the granulosa may consist of only a single layer of cells. These cells are responsible for transfer of vitellogenin during follicle growth as well as for yolk resorption should a given follicle become atretic. Granulosa cells also are thought to be the source of testosterone and estrogens since they exhibit more 3β-HSD activity than do thecal cells. Most estrogen synthesis occurs in mature follicles that have a well-developed granulosa. During follicular development, a theca interna and theca externa can be discerned; however, both layers largely consist of connective tissue elements, and only a small amount of 3β-HSD activity has been observed in the theca interna cells. The granulosa cells are the major source of steroidogenic cells in both atretic follicles and corpora lutea, although thecal cells may contribute as well. The connective tissue layers surrounding these structures are derived from the theca.

Elasmobranch females have well-developed müllerian ducts that give rise to the oviducts as well as to the uterus of viviparous species. Oviducts have been examined in oviparous species that secrete horny shells to protect the eggs laid in the ocean as well as in viviparous species, and they possess a number of specialized features. **Oviductal** or **nidamental glands** occur in the anterior oviduct and secrete albumen and mucus. The oviductal glands of oviparous species are often differentiated into an anterior albumin-secreting area and a posterior shell-secreting region. An intermediate mucus-secreting zone may be found in some species. The "shells" secreted by the oviduct are rigid in oviparous species but are pliable in viviparous species. In some species, the anterior oviduct may function as a spermatotheca.

The lower portion of the oviduct is expanded into a uterus where eggs are held prior to spawning or in which young develop in viviparous species. Villus-like structures may develop in the uterine portion of the oviducts of certain viviparous females, and they provide nourishment for their young. Even in aplacental viviparous species, the uterus often supplies nutrients and oxygen and removes metabolic wastes during gestation. In *Squalus acanthias*, contraction of the myometrium flushes fresh seawater into the uterus, brings in fresh electrolytes, and removes wastes. Such mechanisms may operate in other species as well, although this has not been documented. In some aplacental species, unfertilized eggs may be consumed later by developing young. In still others, the first hatching young may cannibalize less developed uterine inhabitants.

1. Endocrine Factors in Female Elasmobranchs

Removal of the ventral lobe of the adenohypophysis blocks oviposition in female *S. canicula*, and all follicles containing oocytes larger than 4 mm in diameter undergo atresia, demonstrating dependence on GTHs. Progesterone can inhibit vitellogenesis in the spiny dogfish *S. acanthias* and prevents reinitiation of oocyte growth during early gestation. Estradiol stimulates vitellogenin synthesis in the liver and growth of the oviduct.

In the oviparous skate, *Raja erinacea*, estradiol and testosterone levels increase during follicular growth but progesterone shows a peak just prior to ovulation. Testosterone rises in females of some species at ovulation or during oviposition by oviparous species. Elevated testosterone has been reported in *S. tiburo* females at mating and during retention of sperm.

After ovulation, there is a decline in steroid secretion until oviposition (Figure 11-7). However, estradiol peaks prior to ovulation in the egg-retaining viviparous *S. acanthias* and progesterone remains elevated during the early part of gestation (Figure 11-8). Corpora lutea of several species possess 3β-HSD activity, and corpora lutea persist during gestation in *S. acanthias*. The corpora lutea from pregnant *S. acanthias* produce twice as much progesterone *in vitro* as do those from nonpregnant females, which possess only atretic follicles and no corpora lutea, supporting an endocrine role for atretic follicles in this species. Viviparous rays, *S. tiburo* and *Dasyatis sabina*, exhibit patterns of estradiol and progesterone similar to those of *S. acanthias* (Figure 11-9).

The ovaries of viviparous sharks produce relaxin with similar biological actions when tested for its uterine-relaxing activity in the guinea pig. A similar molecule called **raylaxin** has been isolated from rays and skates. Relaxin prevents premature uterine contractions in sharks and enlargement of the cervical region prior to pupping or egg laying. Relaxin appears early in vertebrate evolution and performs a similar function in sharks as in mammals.

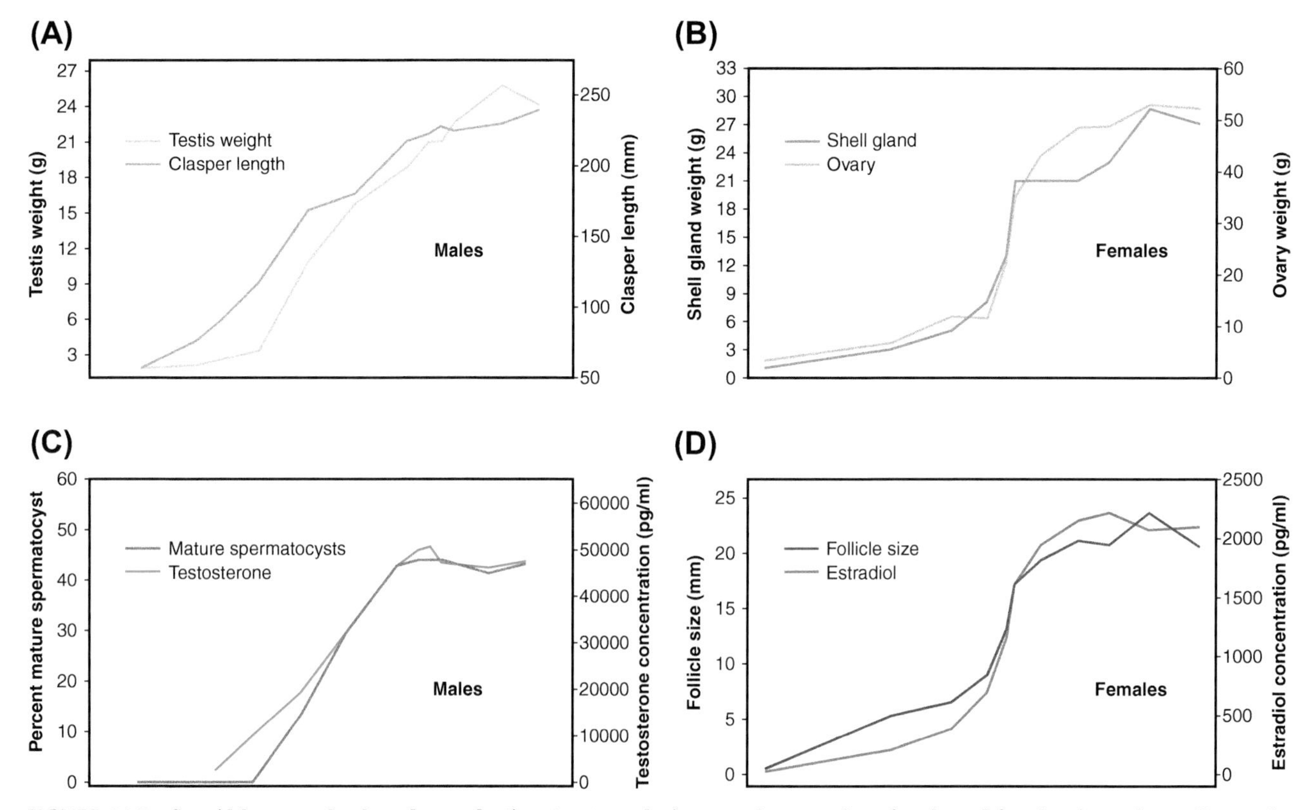

FIGURE 11-7 **Steroid hormone levels and reproductive structures during sexual maturation of male and female winter skates (*Leucoraja ocellata*).** *(Adapted with permission from Maruska, K.P. and Gelsleichter, J., in "Hormones and Reproduction of Vertebrates. Vol. 1. Fishes" (D.O. Norris and K.H. Lopez, Eds.), Academic Press, San Diego, CA, 2011, pp. 209–237. Figure originally appeared in Sulikowski, J.A. et al.,* Environmental Biology of Fishes, *72, 429–441, 2005.)*

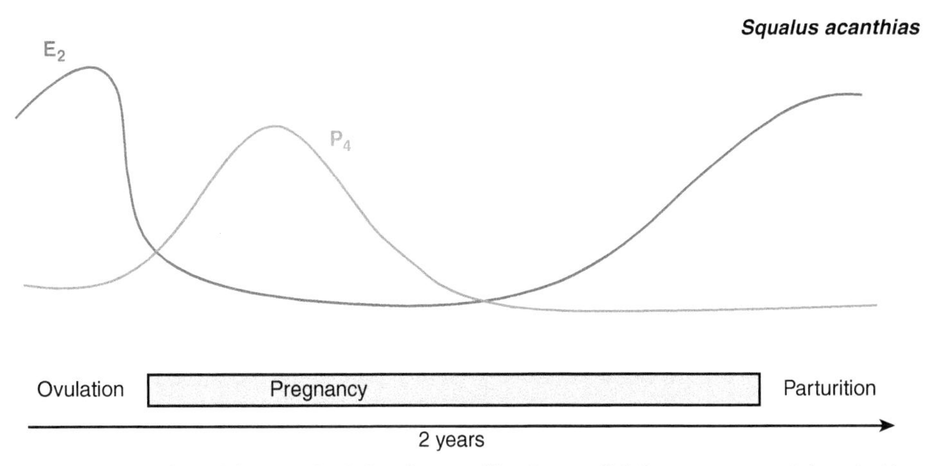

FIGURE 11-8 **Reproductive cycle of the viviparous shark *Squalus acanthias*.** E_2, estradiol; P_4, progesterone. *(Adapted with permission from Koob, T.J. and Callard, I.P.,* Journal of Experimental Zoology, ***284****, 557–574, 1999.)*

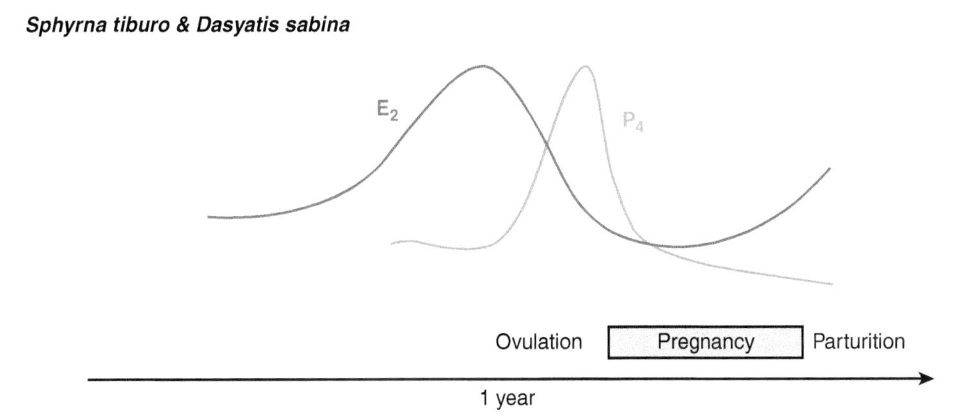

FIGURE 11-9 **Reproductive cycles of viviparous skates, *Sphyrna tiburo, Dasyatis sabina*.** E_2, estradiol; P_4, progesterone. *(Adapted with permission from Koob, T.J. and Callard, I.P.,* Journal of Experimental Zoology, ***284****, 557–574, 1999.)*

However, this is the only group of non-mammalian vertebrates so far shown to produce a relaxin and to respond to mammalian relaxin.

IV. REPRODUCTION IN BONY FISHES

Among the bony fishes, the 27,000 species of teleosts exhibit almost every reproductive pattern and strategy known for vertebrates, including some that are unique to these fishes. Most of the account here is based on teleosts, but there are many similarities between teleosts and the other groups of bony fishes. Some features of these other bony fishes will be illustrated, also, but first some generalizations need to be made.

Like that of cyclostomes, the teleost gonad develops only from a cortical primordium. Bony fishes may be **gonochoristic** (separate sexes), **unisexual** (all female populations), or **ambisexual** (simultaneous or sequential hermaphrodites). **Hermaphroditism** implies that both functional sexes appear in the same body as opposed to **intersex**, which refers to the presence of features normally characteristic of the opposite sex (i.e., presence of a female character in a genetic male). The intersex condition is considered to be an anomaly. A few hermaphroditic species are simultaneous hermaphrodites but most are sequential hermaphrodites. Among the sequential hermaphrodites, there are numerous examples of protandry (function first as males and later transform to females) and protogyny (*gyno*, female). Most bony fishes are oviparous but viviparity has arisen in many forms as well.

Fertilization may be external or internal as in the viviparous teleosts and in the viviparous coelacanths, *Latimeria* spp. In some viviparous teleosts, the fertilized egg is known to develop within the ovary. Elaborate patterns of courtship, nest building, parental care, and other specific reproductive behaviors have been reported among diverse groups.

Breeding typically is cyclic with the exception of semelparous species. Some species are **synchronous spawners** such as in the semelparous Pacific salmon. Others are called **group-synchronous spawners** with most commonly two spawning episodes per season (e.g., Atlantic salmon, rainbow trout). Still other species are **asynchronous spawners** and spawn repeatedly all year (e.g., zebrafish, medaka, killifishes). Each species has a well-defined spawning period regulated by one or more environmental factors (seasonal changes in photoperiod, lunar cycles, temperature, etc.). Many iteroparous species spawn several times during a single breeding season (e.g., *Hyporhamphus melanochir*, an Australian garfish of the halfbeak family). In some viviparous species, such as the guppy *Poecilia reticulata*, a new batch of oocytes is released soon after birth of the young and another brood is raised. Other viviparous species require a longer "interbrood period" for oocyte maturation and vitellogenesis (*Mollienesia* and *Gambusia*). However, in *Quintana atrizoma*, oocyte development occurs during gestation so that a new batch of eggs can be fertilized as soon as the young are born.

The endogenous nature of seasonal or annual rhythms has been shown only in a few species; for example, the killifish (*Fundulus grandis and F. heteroclitus*) and the golden rabbit fish (*Siganus guttatus*), which spawn according to lunar cycles, exhibit free-running lunar-related reproductive cycles in the laboratory (see Box 11B). Seasonal reproductive cycles clearly are evident even in tropical species. The effects of artificial lengthening and decreasing of the photophase may accelerate spawning in spring and fall spawners, respectively. A classical demonstration of environmental phasing of reproduction has been demonstrated by transporting a poecilid, *Jenynsia lineata*, from South America to the Northern Hemisphere. In South America, this species normally spawns in January and February (summer). At the new pond location where photoperiod and related seasons were reversed, the fish switched to spawning in July and August. However, the possible importance of the temperature regimen, which was also switched, should not be overlooked. In some species, temperature is the critical factor in controlling recrudescence regardless of the light regimen imposed on the fish.

A. Male Bony Fishes

The major circulating androgens in teleosts are 11-KT and testosterone. Levels of androgens are correlated with testis size, which increases annually in breeding adults (Figures 11-10 and 11-11). Many species also show marked seasonal development in gonaducts, accessory glands, and secondary sexual characters that are presumed to be under androgenic control. For example, testosterone induces formation of **nuptial tubercles** on the heads of males in some species (Figure 11-12).

Spermatogenesis in the bony fishes is of the cystic type (Figures 11-3 and 11-13). The testes of more ancient groups (i.e., coelacanths, chondrosteans, salmonids, cyprinids) consist of anastomosing tubules, whereas the neoteleosts (e.g., killifishes, bass, cichlids, sunfishes, labrids) exhibit a lobular organization. Nests of spermatogonia proliferate from germ cells along the tubules or at the ends of the lobules located near the testicular surface. Gonadotropins (FSH and LH) stimulate production of a number of paracrine regulators that control various steps in the formation of sperm (Figure 11-14). Production of new spermatogonia for each breeding season is stimulated by FSH and estradiol acting through the Sertoli cell. Production of androgens by Leydig cells is stimulated by LH, and these androgens cause secretion of paracrine regulators by the Sertoli cells that regulate spermatogonial proliferation. Induction of meiosis and spermiogenesis is regulated by **17,20β-dihydroxy-4-pregnen-3-one (17,20β-DHP)** secreted from the Sertoli cell.

Fertilization in most bony fishes is external and gonadal development is synchronized in males and females to spawning. Fertilized eggs may be released directly into the water, attached to vegetation, or reared in nests. Some species are mouth-brooders, which retain the eggs during development that takes place in the parent's mouth. Viviparous teleosts, like their distant elasmobranch relatives, produce spermatophores, employing secretions by the male gonaducts. These males often have an androgen-dependent modified anal fin, called a **gonopodium**, to faciliate sperm transfer to the female.

B. Female Bony Fishes

The teleost ovary (Figure 11-15) has been studied in considerable detail with respect to gonadal differentiation, oogenesis, vitellogenesis, and ovulation, both in oviparous and viviparous species. The ovary of most teleosts is hollow, whereas solid ovaries have been reported in most lungfishes and chondrostean fishes. A few teleosts also have solid ovaries. Unlike the hollow ovary of elasmobranchs and amphibians, the teleost ovarian cavity is lined with germinal epithelium. Each hollow ovary is continuous, with an oviduct that is not homologous to the müllerian duct-derived oviducts of other vertebrates. Eggs are discharged from the ovary directly into the oviduct. In species with solid ovaries, the eggs are discharged into the body cavity from which they pass to the exterior via oviducts or directly through temporary openings in the body wall. Teleost GnRH-1 or LH induces ovulation in teleosts. Viviparous species acquire sperm from males and retain their eggs that are fertilized within the hollow ovary, where they develop prior to birth.

Basically, the teleost ovary consists of masses of follicles embedded in a rather sparse stroma. Development of mature eggs follows a set sequence of stages (Figure 11-16). Each follicle begins as a single-layered epithelium derived from the germinal epithelium and surrounding an oocyte. As the

BOX 11B Lunar Cycles and Teleost Reproduction

Rabbitfishes are widespread tropical reef fishes found in the Indo-Pacific, the Red Sea, and the eastern Mediterranean. Photoperiod and temperature provide relatively little variation as environmental clues, and many species of fishes spawn in synchrony with the lunar cycle. Reproduction of many estuarine fishes, such as members of the genus *Fundulus* are linked to lunar cycles. The rotation of the moon around the Earth requires approximately 28 days (called the *sidereal lunar month*, which is 27.32 days in length). Another lunar cycle of about 14 days refers to the interval between the moon's crossing of the equator. The semilunar or tidal cycle is also approximately two weeks long and is a consequence of the relationship of the gravitational influences of the moon, the Earth, and the sun. This cycle is accompanied by continual changes in water level and currents as well as by changes in weak electromagnetic fields.

The foxtail rabbitfish (*Siganus argenteus*), near Japan, shows peak gonadal development from May through July with spawning occurring during the full moon (Box Figure 11B-1). Variations in plasma hormone levels are provided in Box Figure 11B-2. Other rabbitfish species (e.g., the golden rabbitfish, *Siganus guttatus*) in the same area show similar peak cycles but the months of the peaks may vary in different latitudes. For example, *S. argenteus* in the Philippines spawns from February to September. The observation that the golden rabbitfish will spawn all year long in captivity if fed an adequate diet suggests that variations in food availability may influence the reproductive period of these fishes in nature.

Some tropical species in lakes (e.g., Lake Tanganyika in Africa) that show no tidal influences also spawn during a full moon. Captive *Fundulus* spp. and rabbitfishes also exhibit lunar periodicity in the absence of any tidal influence. These observations suggest that detection of variations in the weak electromagnetic field correlate to tides, as shown for beach crabs transported inland hundreds of miles from the seashore. After a period of adjustment, these crabs showed tidal-related behaviors as if the ocean had moved inland to Indiana.

In some freshwater and brackish water fishes, the reproductive cycles are tuned to the wet and dry seasons. Periodic flooding and drying cause marked changes in water availability and also influence salinity and chemical composition of the aquatic environment. During the dry season, the entire habitat of the South American annual killifish (*Austrofundulus limnaeus*) dries up and all adults die. However, the last eggs produced by these fish enter a state of diapause and survive in a desiccated state until the next rainy season. When the temporary ponds fill with water, the eggs hatch and the young grow rapidly, mature, reproduce repeatedly, and die when the ponds dry again.

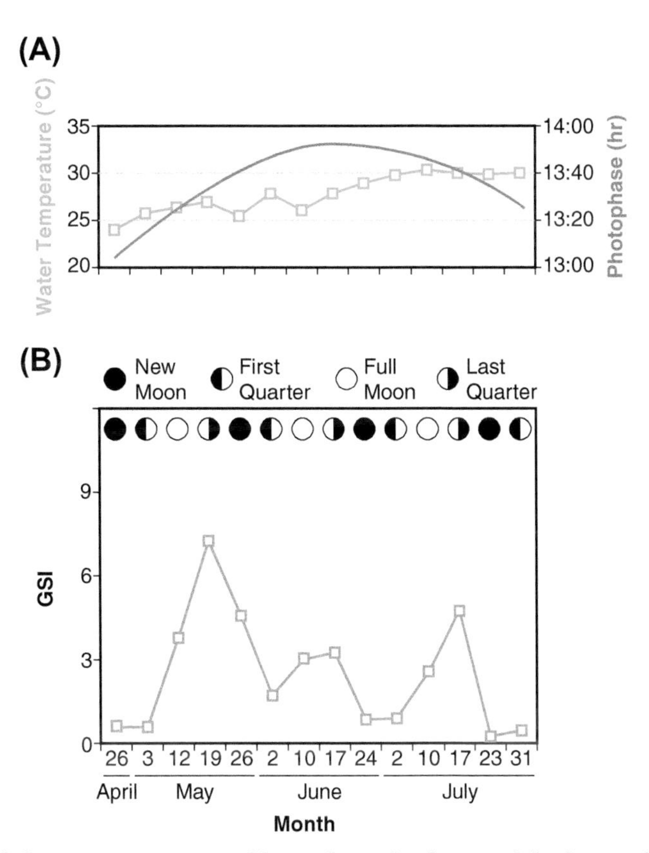

BOX FIGURE 11B-1 **Photoperiod, water temperature, and lunar phase related to gonad development in the foxtail rabbitfish (*Siganus argenteus*).** Note that the GSI (gonadosomatic index) decreases dramatically due to spawning under the full moon. *(Adapted with permission from Takemura, A. et al.,* Fish and Fisheries, *5, 317–328, 2004.)*

BOX 11B Lunar Cycles and Teleost Reproduction—cont'd

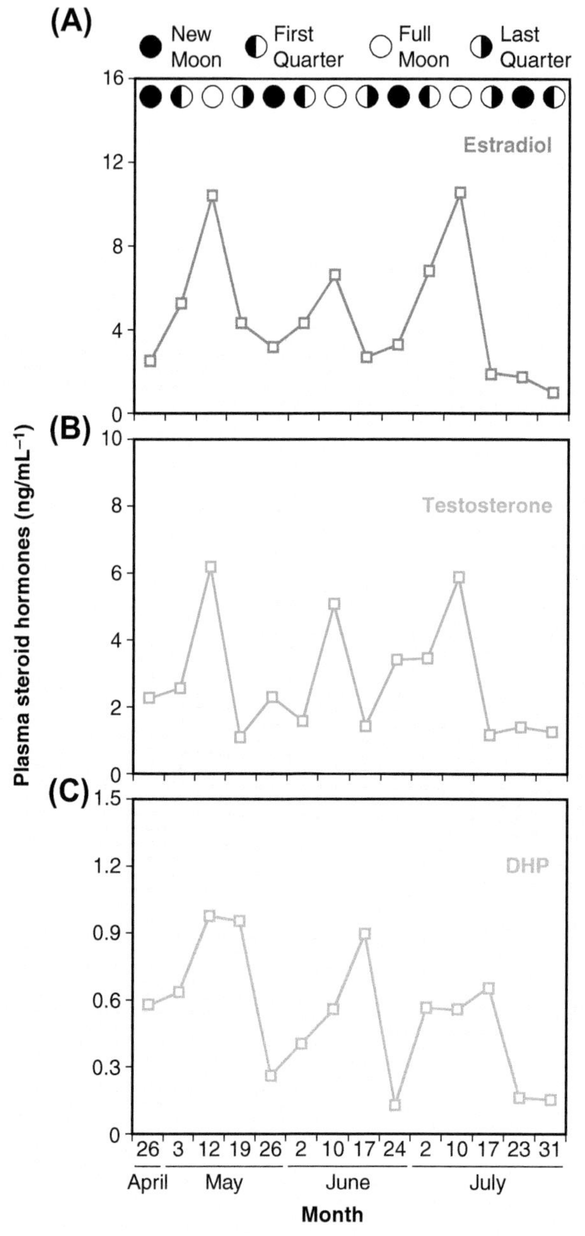

BOX FIGURE 11B-2 **Variations in plasma steroid levels in female foxtail rabbitfish (*Siganus argenteus*) with phases of the moon.** Note that all three steroids peak with the full moon. *(Adapted with permission from Takemura, A. et al.,* Fish and Fisheries, *5, 317–328, 2004.)*

follicle grows, these epithelial cells undergo hyperplasia and hypertrophy to form the granulosa. Connective elements in the stroma near the follicular nest will differentiate into a theca, which may further differentiate into a theca externa and a theca interna. The thecal and granulosa cells cooperate in the synthesis of androgens and estrogens, respectively, as described for mammals (Figure 11-17). The granulosa cells also are responsible for yolk deposition in the oocyte during follicular growth and resorption of yolk during atresia. In salmonids, receptors that will bind FSH are found on both thecal and granulosa cells, and FSH is responsible for synthesis of estrogens during follicular growth. The LH receptors occur only on granulosa cells, where they are associated with synthesis of 17,20β-DHP (sometimes called the **maturation promoting factor, MPF**) that is responsible for final oocyte maturation and ovulation in many species

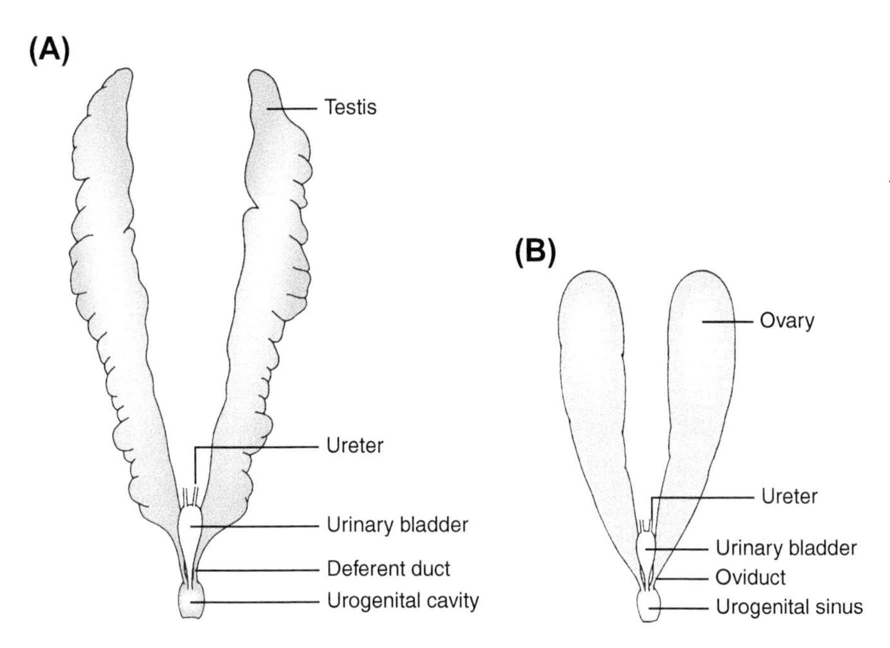

FIGURE 11-10 **Reproductive system of teleost fishes as exemplified by the carp *Cyprinus carpio*.** (A) Male. (B) Female. Note the absence of the elaborate system of ducts seen in chondrichthyeans. *(Adapted with permission from Matsumoto, A. and Ishii, S., "Atlas of Endocrine Organs: Vertebrates and Invertebrates," Springer-Verlag, Berlin, 1992.)*

(Figure 11-18). A related derivative of progesterone, **17,20β,21-trihydroxy-4-pregnen-3-one (17,20β,21-THP)**, is responsible for these events in some species. A corticosteroid, **deoxycorticosterone (DOC)**, is responsible for ovulation and final oocyte maturation in an Indian catfish, *Heteropneustes fossilis*.

Three developmental patterns for ovaries can be identified in teleosts. In the **synchronous ovary**, all oocytes are in the same stage of development. Species with a synchronous ovary are semelparous (for example, *Anguilla* spp. and most members of the genus *Oncorhynchus*). Species such as rainbow trout, white sucker, and flounder have a **group-synchronous ovary** with at least two populations of oocytes. These iteroparous species generally spawn once each year during a short breeding season. The last type is the **asynchronous ovary**, which has oocytes in all stages of development at all times during the breeding season. These species spawn frequently each year during a prolonged breeding season. Tropical species may be continuous breeders or may be tuned to wet and dry cycles.

Teleost ovarian tissue synthesizes testosterone, estrogens and DOC. These steroids have been identified in the peripheral plasma of females from several teleost species (Figures 11-19 and 11-20). The levels of testosterone often are greater in prespawning females than they are in males, suggesting a behavioral role for androgens in females as reported for some female mammals (e.g., hyenas, European moles of the genus *Talpa*).

All groups of bony fishes develop preovulatory, secretory corpora atretica as a result of atresia of developing follicles and develop short-lived corpora lutea following ovulation. However, a convincing endocrine function for corpora lutea is not yet established.

Vitellogenesis by the liver is stimulated by estradiol, and consequently total plasma calcium and phosphoprotein levels usually are elevated during oogenesis (Table 11-4). Calcium and vitellogenins are sequestered by the granulosa cells and are transferred to growing oocytes where they are transformed into yolk proteins (Figure 11-3).

C. Reproductive Behavior in Bony Fishes

Many aspects of reproductive behavior have been studied in teleosts, including migration to breeding sites, courtship, nest building, spawning, copulation, and parental care. Most of this work has concentrated on roles of testis, testosterone, and synthetic androgens in males. Male behaviors are regulated by 11-KT and hence do not appear in females that rarely produce 11-KT; however, elevated testosterone is correlated with aggressive behavior in the females of some species. Castration of males blocks breeding behavior and causes reversal to the non-breeding condition of androgen-dependent characters.

1. Breeding Behaviors in Teleosts

Male fishes often defend territory or females against other males or females through aggressive **agonistic (combative) behavior**. Androgens are responsible for agonistic behavior in some species. Additionally, dominant fish that hold territory tend to have higher androgen levels and lower cortisol levels than subordinate fish. Castrating male three-spined sticklebacks (*Gasterosteus aculeatus*) eliminates nest building one week later. Agonistic behavior remains high for three or four weeks if castration is performed less than one week within nest building. In some

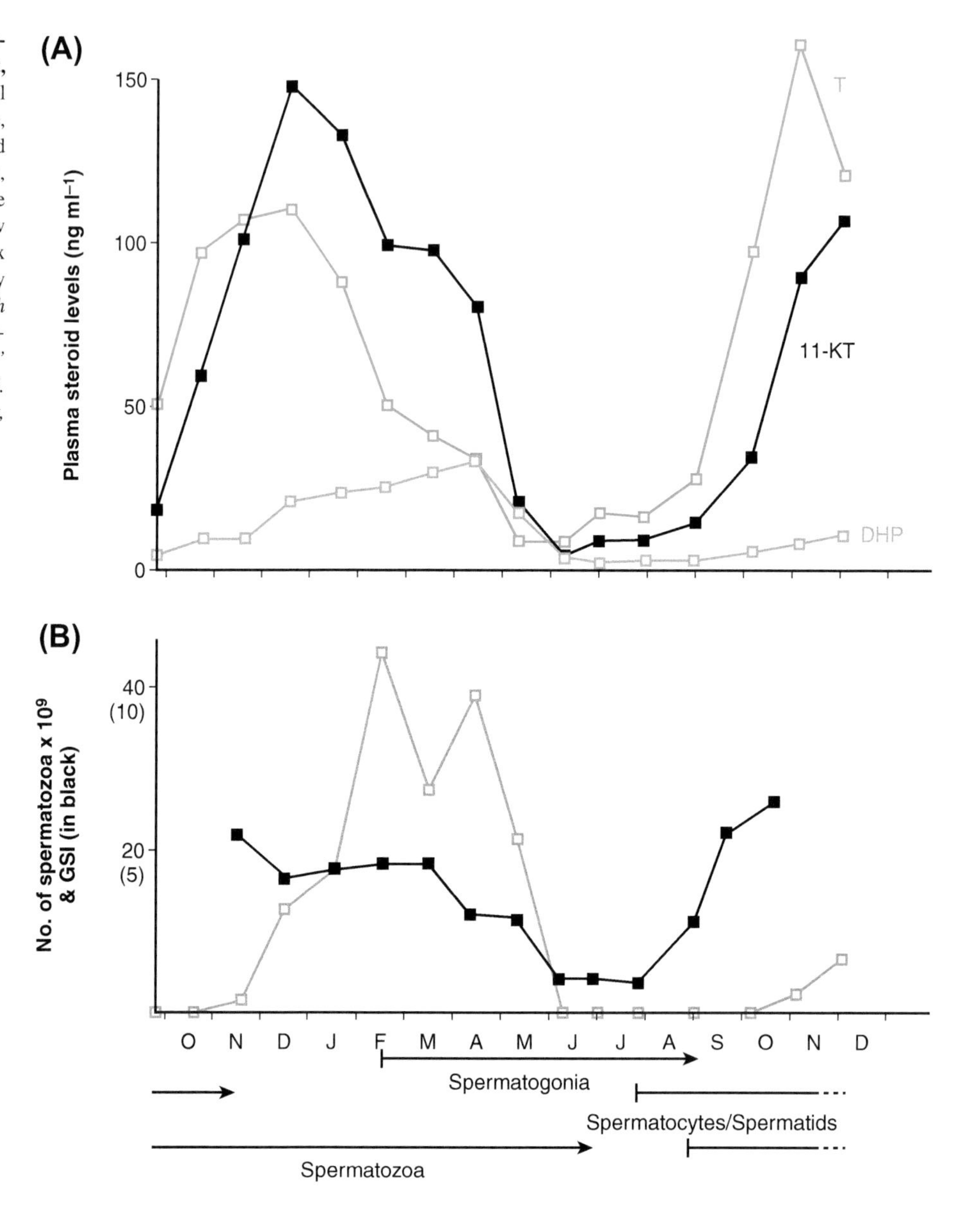

FIGURE 11-11 **Annual reproductive cycle of male rainbow trout, *Oncorhynchus mykiss*.** The top panel depicts plasma levels of testosterone, T; 11-ketotestosterone, 11-KT; and 17,20β-dihydroxy-4-pregnen-3-one, DHP. The lower panel shows the volume of sperm produced (yellow line) and the gonadosomatic index (GSI, gonad weight related to body weight; black line). *(Adapted with permission from Scott, A.P., in "Fundamentals of Comparative Endocrinology" (I. Chester-Jones, P.M. Ingelton, and J.G. Phillips, Eds.), Plenum Press, New York, 1987, pp. 223–256.)*

species, castration does not result in a decrease in agonistic behavior, and androgens may not be required to maintain the behavior once it has been induced in these fishes.

Males sometimes exhibit alternative reproductive tactics, and a population may consist of more than one male phenotype. For example, in some species there are "sneaker" males who live at the edges of a dominant male's territory awaiting a chance to dart in and fertilize some eggs of attendant females. Sneakers are often younger and smaller as compared to the territorial male that controls the nesting site and attracts the females. In some cases, the sneaker male may closely resemble a female and even exhibit female-like behaviors to fool the territorial male. Although androgen levels generally are greater in dominant male teleosts versus subordinate males, levels of testosterone, 11-KT and 17,20β-DHP in smaller sneaker coho salmon "jacks" did not differ from those of larger, hook-nosed males that defended the spawning nest.

The roles of estrogens and androgens in relation to the breeding behaviors of females have been studied less than the roles of androgens in males. Castration of females may result in complete abolition of all reproductive behaviors, loss of only some, or only a decrease in intensity. In one case, ovariectomized *G. aculeatus*, form *leiurus*, show more aggressive behavior than intact females, implying that ovarian steroids normally repress aggressive behavior in females. Estrogens have proven ineffective for inducing female behavior in females. Possible roles for androgens have not been studied, but the relatively high level of plasma androgens in prespawning females is suggestive of a behavioral role.

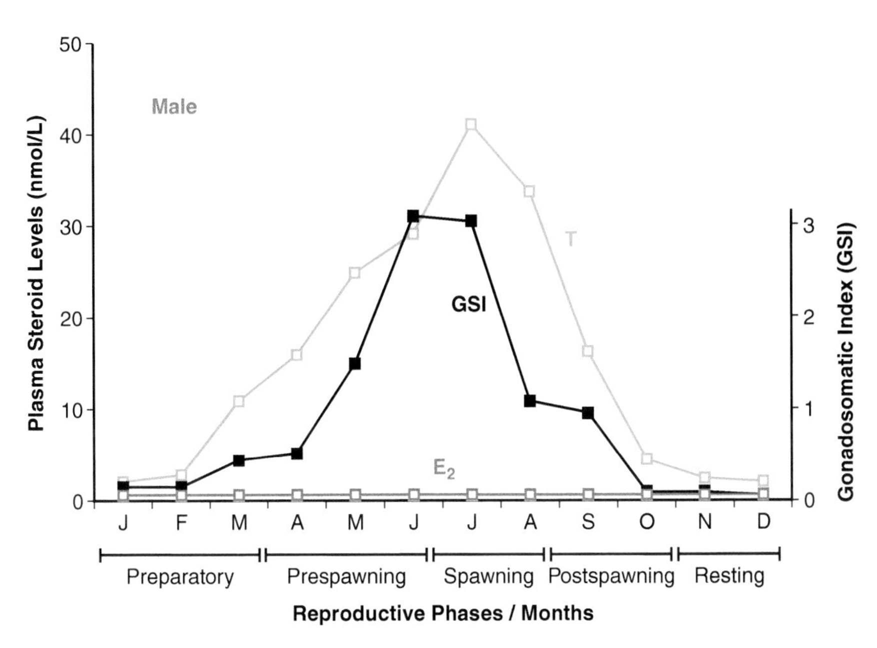

FIGURE 11-12 **Seasonal reproductive cycle in Indian major carp (*Labeo rohita*).** Testosterone (T) levels are highly correlated with growth of the testes (gonadosomatic index, GSI). Estradiol (E_2) levels are minimal in this species. *(Adapted with permission from Suresh, D.V.N.S. et al.,* General and Comparative Endocrinology, ***159****, 143–149, 2008.)*

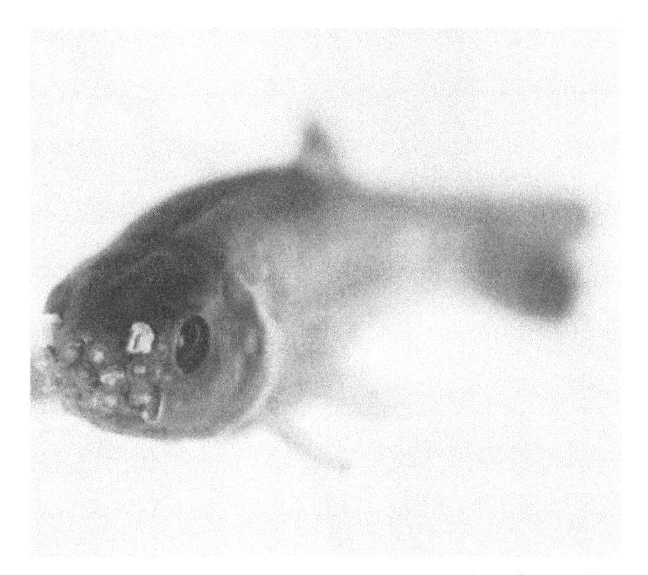

FIGURE 11-13 **Nuptial tubercles on the snout of a minnow.** These androgen-dependent secondary sexual characters are found on a number of cyprinid teleosts.

Recognition of conspecific (belonging to the same species) genders and/or breeding conditions may occur by behavioral displays, coloration patterns, special adornments, electrical signals, or chemical signals (pheromones). All of these may have endocrine mechanisms of production and/or determination by the receiver.

Spawning behavior in females appears to be under the control of GnRH and GTHs, although in *F. heteroclitus* neurohypophysial preparations or synthetic oxytocin induce reflexive spawning movements in hypophysectomized or castrated females (see Chapter 5). This spawning reflex is a behavior not dependent upon shedding of ova. Similar observations have been reported for a few additional species. Spawning behavior in female goldfish is stimulated by ovarian prostaglandins.

2. Parental Behaviors

Many teleosts exhibit extended care of eggs in a nest prior to hatching and may even protect young hatchlings from predators. Mouth-brooding species retain the fertilized eggs in their oral cavities, and young hatchlings may continue to seek refuge there when threatened. Relatively little is known about the endocrine regulation of these behaviors. Mammalian PRL influences certain aspects of parental behavior in fishes. Fanning behavior associated with aeration of the eggs can be stimulated in *Symphysodon aequifasciata* and *Pterophyllum scalars* by PRL treatment; however, similar treatment inhibits fanning behavior in sticklebacks. PRL stimulates secretion of mucus that is fed to young *S. aequifasciata*, but it is not clear if the behavior of feeding the young also is PRL dependent in these parent fish.

3. Pheromones and Behavior in Teleosts

Communication by pheromones is important in the reproduction of teleosts. Prostaglandins (e.g., $PGF_{2\alpha}$), and a variety of free and conjugated steroids function as pheromones. Following ovulation and prior to oviposition, females of numerous oviparous species emit pheromones that attract and arouse sexual activity in males. For example, female *Bathygobius soporator* secrete a pheromone from the ovary that elicits courtship behavior by

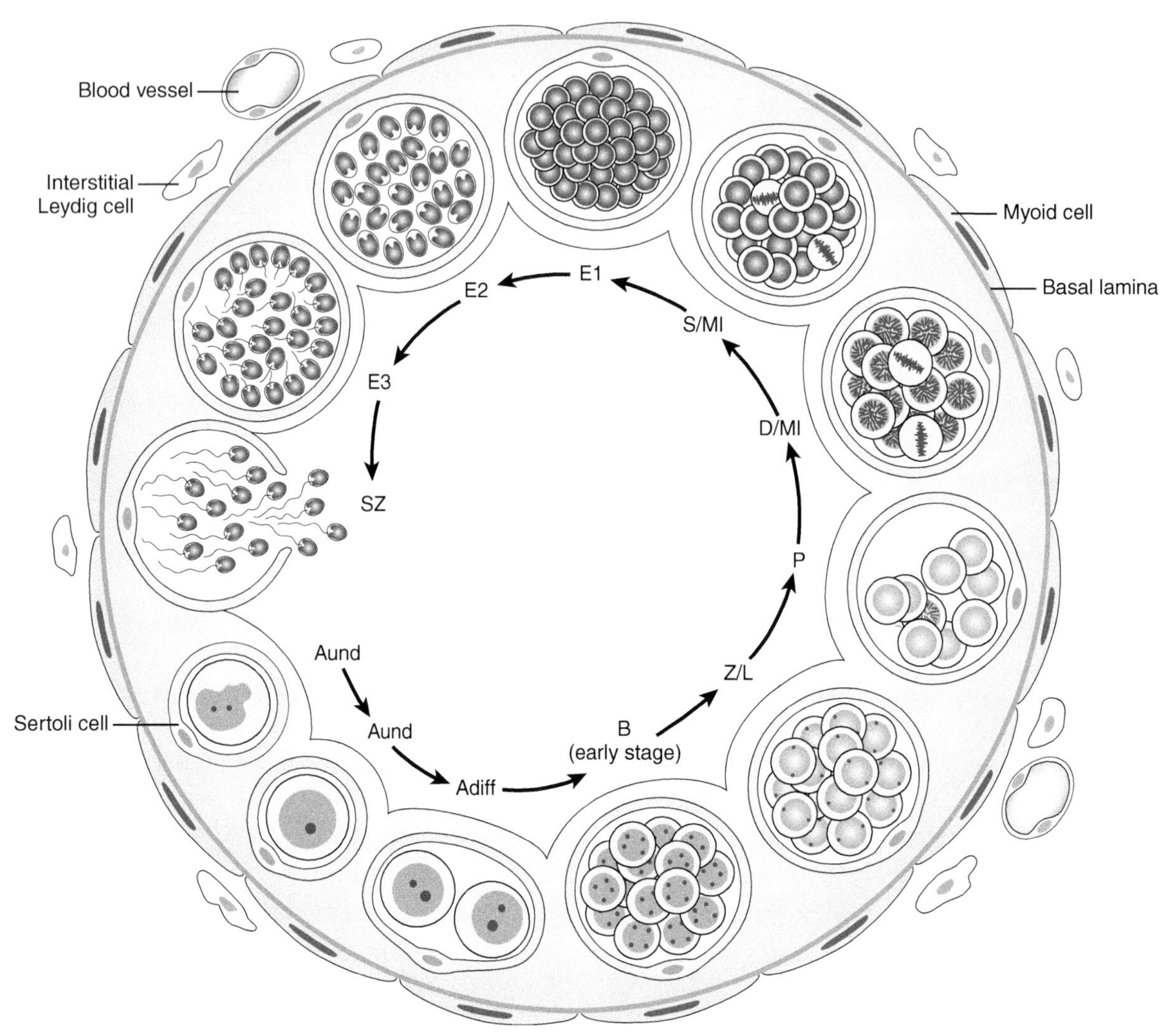

FIGURE 11-14 **Stages in teleost spermatogenesis.** Schematic depiction of spermatogenesis in a seminiferous tubule from zebrafish (*Danio rerio*, Cyprinidae, Cypriniformes) testes. Progression of spermatogenesis is depicted flowing from the lower left corner around to the upper left corner. Abbreviations: Adiff, type A differentiated spermatogonia; Aund, type A undifferentiated spermatogonia (potentially a stem cell); B (early—late), type B spermatogonia; D/MI, diplotene spermatocytes/metaphase I; E1, early spermatids; E2, intermediate spermatids; E3, final spermatids; S/MI, secondary spermatocytes/metaphase I; SZ, sperm; Z/L, leptotene/zygotene primary spermatocytes. *(Adapted with permission from Schulz, R.W. et al.,* General and Comparative Endocrinology, ***165**, 390—411, 2010.)*

intact males but not by anosmic males (treated to prevent olfactory detection). A priming pheromone released in the urine of female salmonids increases production of 17,20β-DHP in recipient males that in turn causes spermiation. The chemical nature of this pheromone is not known. Parental behavior is stimulated in *Heterochromis bimaculatus* by chemicals secreted by the young. Several species recognize their own young by using olfactory cues, and the offspring of some species use chemical recognition to identify their parents.

The goldfish (Figure 11-21) has become a model system for investigations of pheromonal communication in teleost reproduction primarily due to the pioneering discoveries by Norman Stacey at the University of Alberta and his collaborators. During the final stages of oocyte development, the ovaries synthesize a mixture of C_{21}-steroids including 17,20β-DHP and 17,20β,21-THP. These steroids not only induce final oocyte maturation but also act as pheromones in males to induce LH secretion, sperm release, and competence for spawning behavior. In addition, sulfated forms of these steroids are produced by goldfish ovaries and these too are potent stimulators of males. Recent evidence indicates that different regions of the male olfactory epithelium can detect 17,20β-DHP and distinguish selectively between 17,20β-DHP and its sulfated form, 17,20β-DHP-S.

In addition to releasing free and conjugated progestogens, goldfish ovaries also release considerable quantities of the androgen androstenedione that also plays a pheromonal role. Androstenedione release precedes release of

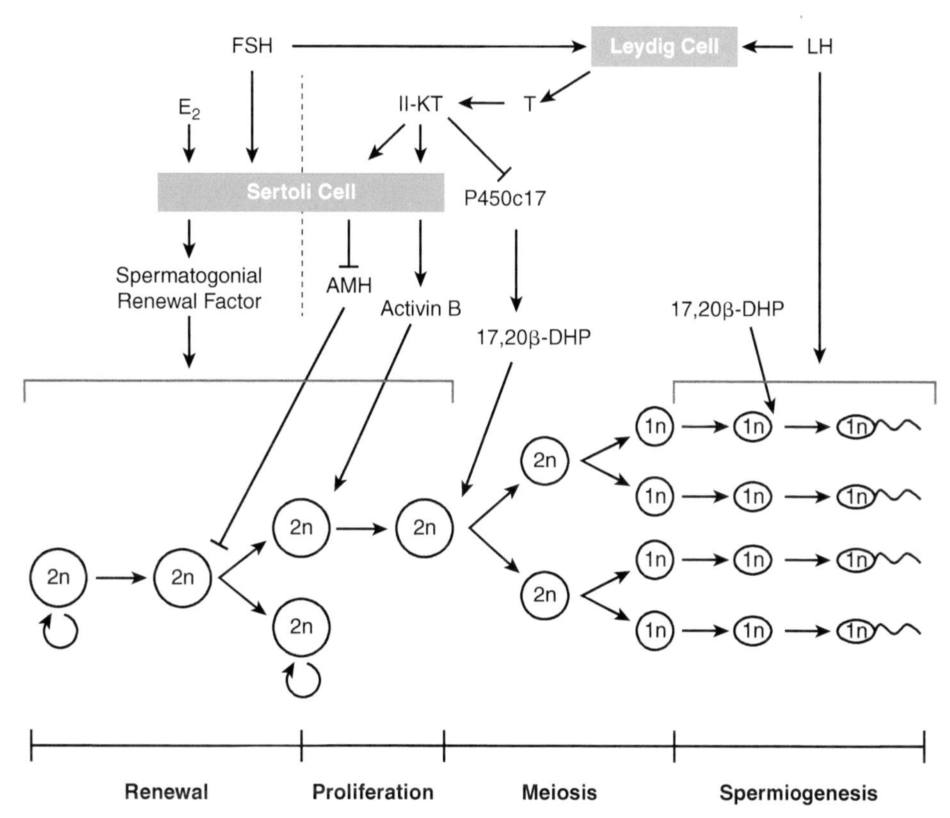

FIGURE 11-15 **Gonadotropin control of paracrine production in the teleost testis.** Note that some factors are inhibitory. Abbreviations: 2*n*, diploid number of chromosome; 1*n*, haploid number of chromosomes. See Appendix A for explanation of abbreviations. *(Adapted with permission from Knapp, R. and Carlisle, S.L., in "Hormones and Reproduction of Vertebrates. Vol. 1. Fishes" (D.O. Norris and K.H. Lopez, Eds.), Academic Press, San Diego, CA, 2011, pp. 43–63.)*

17,20β-DHP and inhibits the responsiveness of males to 17,20β-DHP. This mechanism may prevent premature gamete release in males.

4. Migratory Behavior and Reproduction in Teleosts

The implication of hormones in migratory behavior is largely circumstantial. The gonads and their secretions probably do not play a causative role, since gonadal maturation often occurs during migration. Thyroid hormones in salmon and other species have been claimed to be causative factors of reproductive migratory behavior, and increased thyroid activity coincides with migration as does increased corticosteroid secretion as seen also in seaward migration of juvenile salmon (see Chapter 7, Figure 7-13). It is possible that the increased activity of the thyroid and adrenal glands is correlated to "permissive" effects of these hormones related to metabolism and osmoregulation. Levels of mRNAs for GH, PRL, and **somatolactin (SL)** are elevated during the marine phase of the spawning migration. Levels of GH mRNA decrease when the fish enter the estuaries (brackish water), and both PRL and SL decrease after the fish enter freshwater. These changes also may be related to osmoregulation. These hormones may only enhance the physiological states favorable to migration, whereas the behavioral changes may be controlled through actions of environmental factors such as photoperiod and temperature operating on the nervous system or possibly by endogenous rhythmic neural cycles that are regulated by these environmental factors.

V. REPRODUCTION IN AMPHIBIANS

Many amphibians are characterized as terrestrial or semi-terrestrial adults with an aquatic larval form. Typically, eggs are laid in water where they hatch into a larval form. The importance of the aquatic intermediate larval stage is that it allows for an aquatic feeding stage where growth can be optimized without using energy stores or food resources of the terrestrial parents. During metamorphosis of these fish-like amphibian larvae to terrestrial-type tetrapods, the same problems are encountered and solved that confronted the evolution of terrestrial vertebrates from aquatic animals.

Although only a few of the almost 6000 known species of amphibians have been studied thoroughly, it is clear that there is a great diversity of reproductive patterns within this group of animals. Within the modern Amphibia, several trends in reproductive patterns are evident (Figure 11-22).

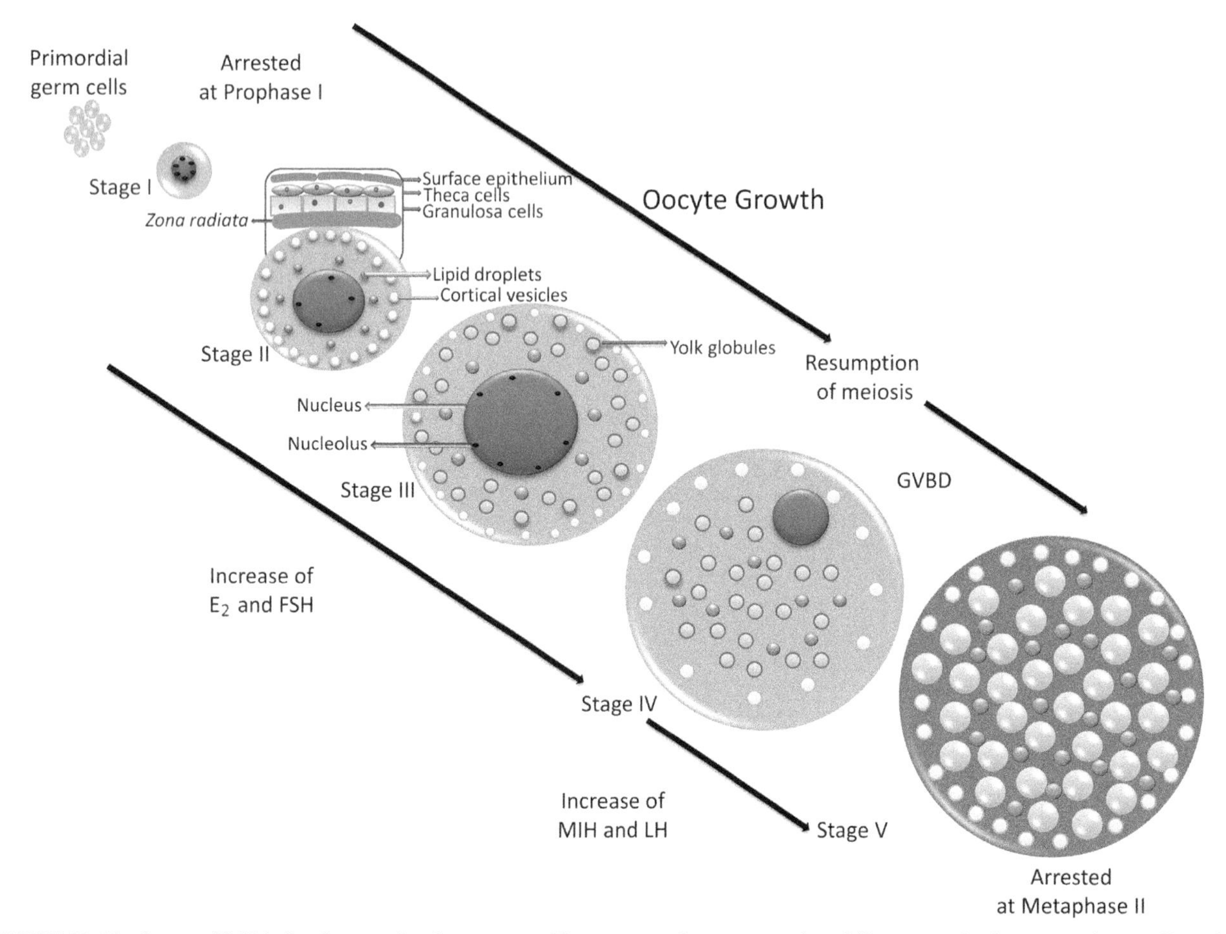

FIGURE 11-16 **Stages of follicle development in teleost oocytes.** The sequence of oocyte stages is as follows: stage I, primary growth; stage II, cortical alveoli growth period; stage III, early vitellogenic oocytes; stage IV, late vitellogenic phase; and stage V, mature/ovulated oocyte, full of yolk (with lipid and protein globules). Oocyte growth is controlled by 17β-estradiol (E_2) and follicle-stimulating hormone (FSH), whereas the resumption of meiosis is regulated by luteinizing hormone (LH) and maturation-inducing hormone (MIH). GVBD, germinal vesicle breakdown. *(Reprinted with permission from Urbatzka, R. et al., in "Hormones and Reproduction of Vertebrates. Vol. 1. Fishes" (D.O. Norris and K.H. Lopez, Eds.), Academic Press, San Diego, CA, 2011, pp. 65–82.)*

All three extant orders (anurans, urodeles, and gymnophionids) show a reduction in the use of the aquatic habitat with a tendency toward terrestrial development. This trend is accompanied by greater reliance on internal fertilization, a reduction in clutch size (number of eggs produced per breeding), and development of simple parental care of eggs and young. Mate selection and courtship patterns have become very elaborate in some species, often related to these complex patterns. Finally, oviparity apparently has given rise to viviparity independently on several occasions in each group of anurans, urodeles, and gymnophionids.

Sexes are separate and natural hermaphroditism is rare, but not undocumented, in adult amphibians. Some amphibians may exhibit varying degrees of juvenile hermaphroditisim during development as the gonads mature. A period of juvenile hermaphroditisim has been reported in *Bufo americanus*, *Hyla versicolor*, and *Rana sphenocephala* raised under laboratory conditions. Amphibians exhibiting a **semi-differentiated** type of gonadal development (*Rana curtipes*, for example) may develop ovaries first regardless of their genetic sex followed by further differentiation into testes in genetic males. Amphibians exhibit both XX/XY and ZZ/ZW forms of genetic sex determination, although the sex chromosomes are not morphologically distinct. The gonads of most species develop through an indifferent or bipotential stage and differentiate in the male or female direction early in development (Figure 11-23).

Conspicuous masses of adipose tissue called **fat bodies** are located adjacent to the gonads of amphibians (see ahead). These are structures unique to amphibians. In female anurans and urodeles, the size of the fat body is correlated inversely with gonadal weight, and it has been proposed that the lipoidal substances stored in the fat bodies are utilized for oocyte growth. However, the size of the fat bodies also may be correlated with food availability. Fat bodies of both male and female European newts (*Triturus cristatus*) can synthesize steroids and therefore may influence gonadal function, accessory sex structures, or

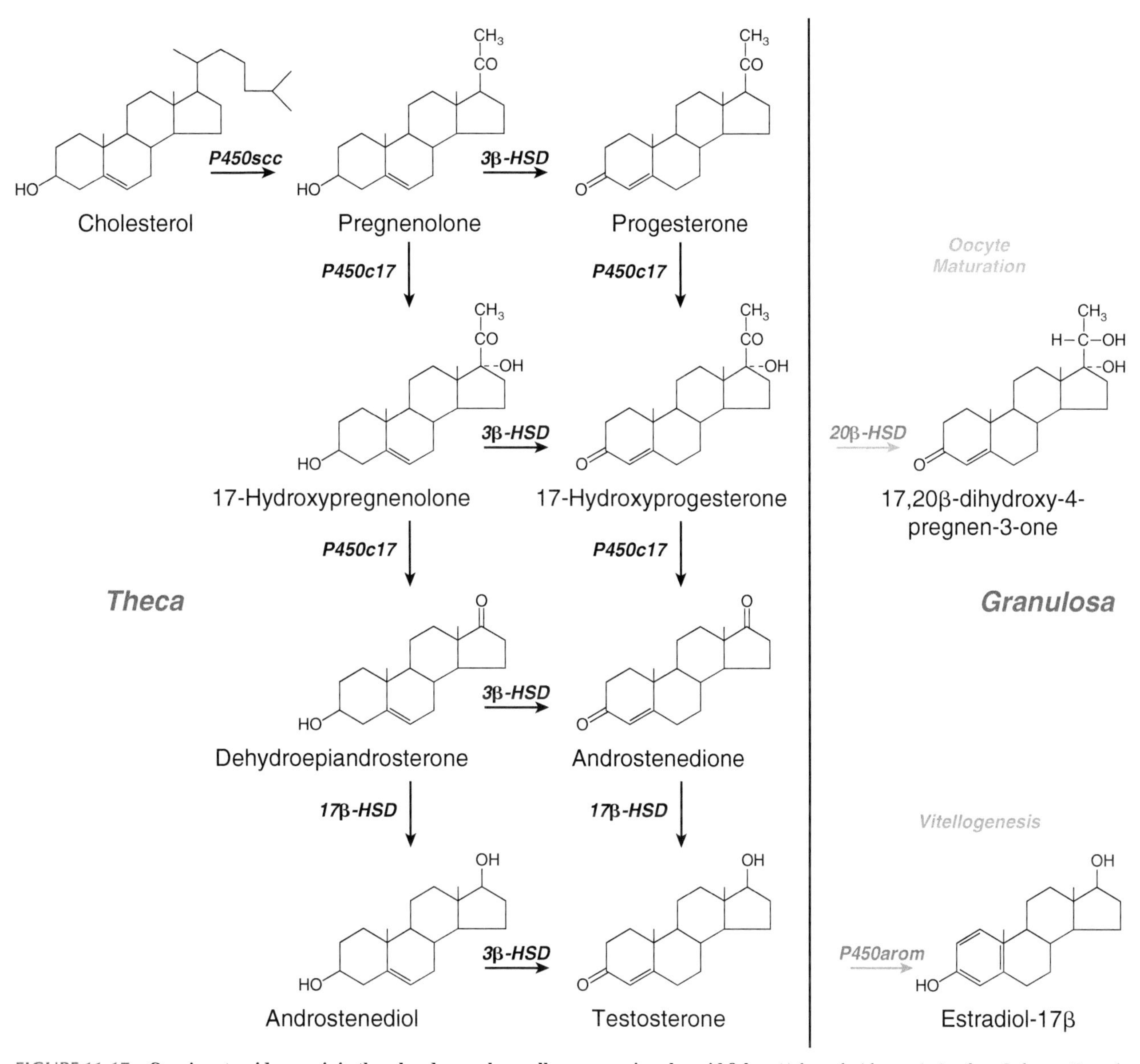

FIGURE 11-17 **Ovarian steroidogenesis in thecal and granulosa cells as occurs in salmonid fishes.** *(Adapted with permission from Lubzens, E. et al., General and Comparative Endocrinology, **165**, 367–389, 2010.)*

both. In the European frog *Rana esculenta*, the steroidogenic function of fat bodies appears to be regulated by pituitary GTHs.

The male gonaducts function as both urinary ducts and as sperm ducts or vasa deferentia. Enlargement and differentiation of the wolffian ducts to the functional male condition is caused by either testosterone or DHT but is antagonized by estrogens. In marked contrast, although normal ovarian differentiation of müllerian ducts can be stimulated by estrogens, both DHT and testosterone are effective at stimulating müllerian duct enlargement in tiger salamanders (Figure 11-24) although only testosterone is effective in frogs.

Reproduction in amphibians is controlled by the hypothalamus and pituitary, acting in response to environmental cues. Administration of GnRH along with a dopamine antagonist induces spawning in a variety of frogs, suggesting an inhibitory role for dopamine as seen in teleost fishes. Treatment with GTHs alone is also effective at inducing spawning and other reproductive events by acting directly on the gonads.

A. Oviparity in Amphibians

Most anuran amphibians studied are oviparous with external fertilization, although internal fertilization occurs in several species. Breeding in oviparous species is tied closely to a seasonal cycle involving photoperiod, temperature, availability of moisture, or a combination of

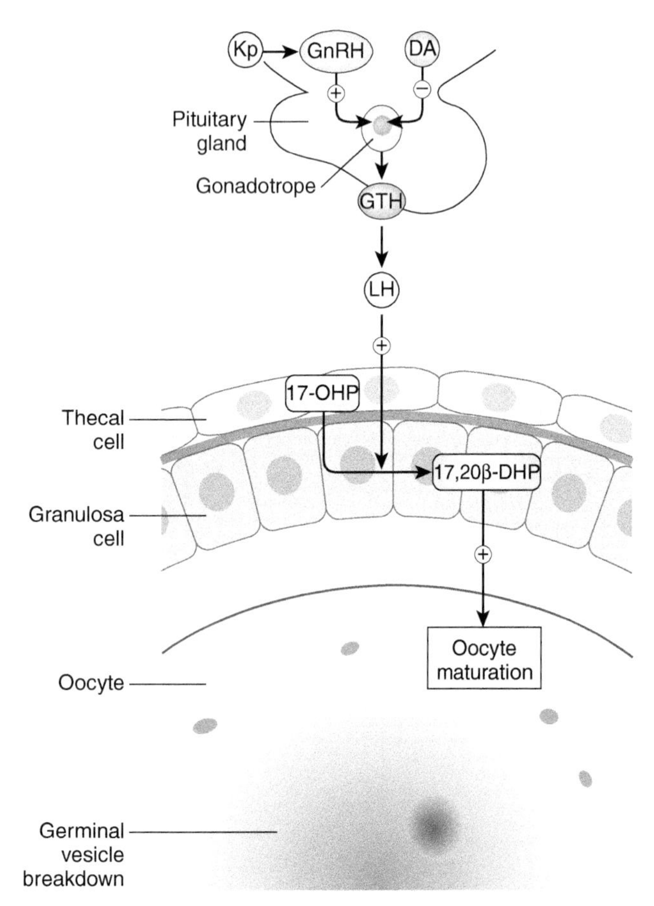

FIGURE 11-18 **Final oocyte maturation in teleosts.** LH stimulates production of 17α,20β-dihydroxyprogesterone (or a related progestogen) by follicle cells. This progestogen is also known as the maturation promoting factor (MPF). MPF causes a breakdown of the oocyte nucleus, the germinal vesicle. This process is called germinal vesicle breakdown (GVBD) and immediately precedes expulsion of the oocyte from the follicle. See text or Appendix A for explanation of abbreviations. *(Adapted with permission from Connaughton, M.A. and Aida, K., in "Encyclopedia of Reproduction, Vol. 2" (E. Knobil and J.D. Neill, Eds.), Elsevier, Amsterdam, 1999, pp. 193–204.)*

these, although a few species are continuous breeders (e.g., the Indian frogs *Rana tigrina* and *R. erytrea* and the South American toads *Bufo arenarum* and *B. paracmenis*). One predominant reproductive pattern is found in temperate oviparous anurans. Spermatogenesis and ovarian follicular development are completed in the fall, and the animals simply "hibernate" until suitable breeding conditions occur in the spring. Many oviparous species lay their eggs in temporary or permanent ponds with the eggs developing into free-swimming larvae. Tadpole larvae are the characteristic limbless, fish-like larval forms of anurans and differ markedly from the larvae of urodeles that possess external gills and four limbs at hatching. Anurans have internal gills like fishes and obtain their limbs later during metamorphosis. One anuran (*Ascaphus*) is known to lay its eggs in streams, and the tadpole larvae that result have special modifications to keep from being swept downstream. Some anuran and urodele species lay their eggs on land, usually in moist places such as under logs, in the axil of tree branches, or in temporary ponds held within the leaves of certain tropical plants. Terrestrial eggs require considerable parent care (see ahead). Terrestrial eggs that are heavily yolked often develop directly into miniature adults, and no aquatic larval stage exists except as a transitional state within the egg.

Oviparous urodeles exhibit several reproductive patterns. In some species (e.g., *T. cristatus*, *Notophthalmus viridescens*), the pattern is similar to that of oviparous anurans. In the hellbender *Cryptobranchus alleganiensis*, spermatogenesis occurs in July shortly before breeding in August and September. Other species such as the mudpuppy (*Necturus* spp.) transfer sperm to the females in the fall, and oviposition and fertilization occur the next spring when males are not present. A number of oviparous gymnophionids have been described, all of which lay terrestrial eggs. In *Ichthyophis*, the eggs are laid in a burrow near a stream, and each newly hatched larva must emerge from the burrow and find its way to the stream. Gymnophionids generally produce larger eggs than do the other amphibian groups, and clutch sizes are proportionally small.

B. Viviparity in Amphibians

Two European land salamanders, *Salamandra salamandra* and *S. atra*, give birth to live offspring that develop in the posterior portion of the oviducts. In *S. atra*, one young develops and undergoes metamorphosis in each oviduct during a 4-year gestation period. Gestation is shorter in *S. salamandra*, which gives birth to larval salamanders. The large size of these offspring indicates that considerable nutrient contributions are made by the mother during these prolonged gestation periods. Details of these contributions and the mechanisms for their transfer have not been reported.

Viviparity in anurans typically involves modifications that allow the eggs to develop into tadpoles on the body of the maternal animal. A fold of skin may develop that completely covers the eggs in a pouch such as that found in the so-called marsupial frogs, *Gastrotheca* spp., of South America. In others, such as the African frog *Pipa pipa*, each egg develops in its own dermal chamber that forms on the back of the parent. Oviductal incubation of eggs occurs in *Nectophrynoides* and *Eleutherodactylus*. One anuran incubates its young in its vocal sacs and at least one Australian species broods its young in its gut. Although this latter species was discovered very recently, it is now believed to be extinct. The South American tree frogs carry their eggs in a single mass on their back.

It is estimated that viviparity occurs in the majority of gymnophionid species. The contribution of maternal

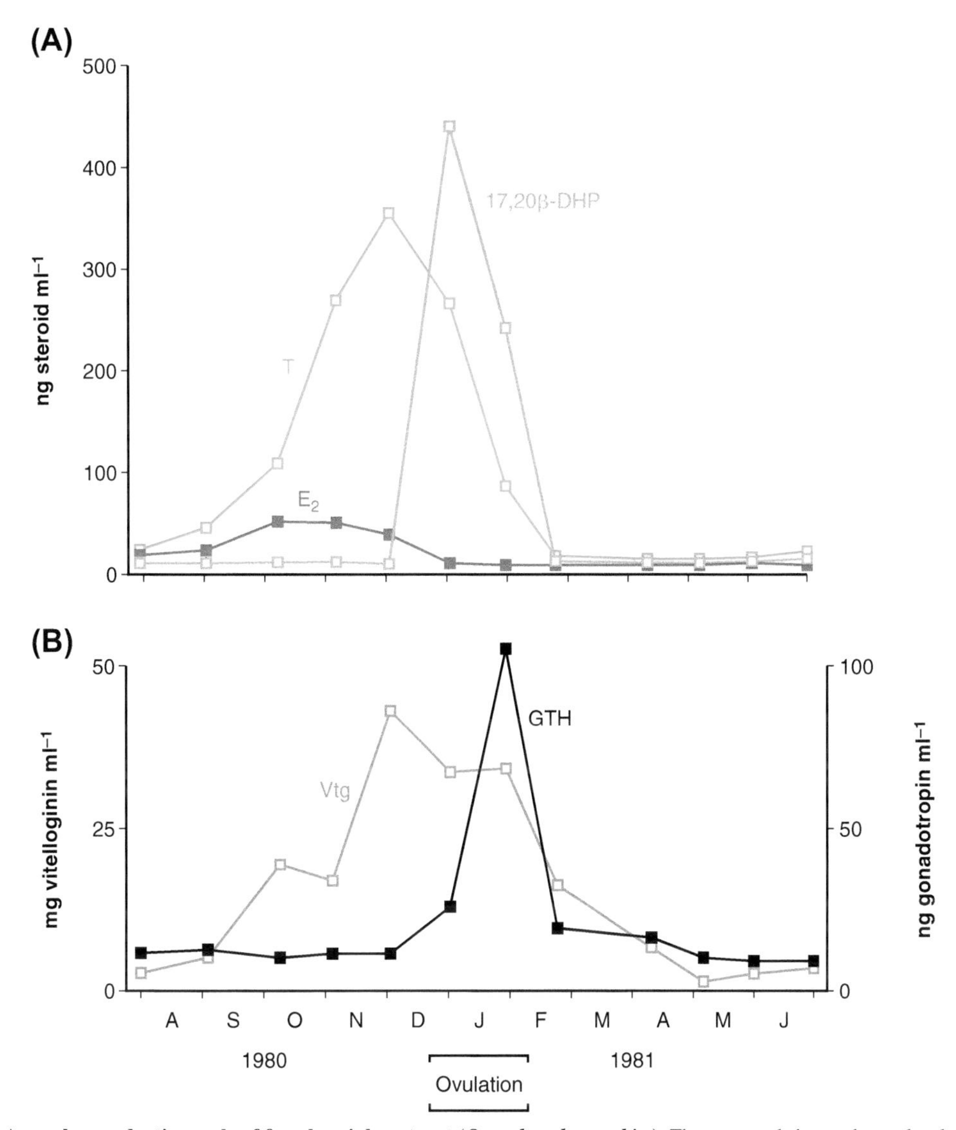

FIGURE 11-19 **Annual reproductive cycle of female rainbow trout (*Oncorhynchus mykiss*).** The top panel shows plasma levels of testosterone, T; estradiol, E_2; and 17,20β-dihydroxy-4-pregnen-3-one, DHP. Plasma levels of vitellogenin, Vtg and gonadotropin, GTH appear in the lower panel. *(Adapted with permission from Scott, A.P. and Sumpter, J.P.,* General and Comparative Endocrinology, ***52****, 79–85, 1983.)*

energy through oviductal secretion to support the developing young also is considerable. In *Typhlonectes*, one female may give birth to as many as nine larvae, each of which weighs about 40% of the mother's body weight at birth.

C. Reproduction in Male Amphibians

Male anurans and urodeles secrete both testosterone and **dihydrotestosterone (DHT)** in response to LH. Additionally, urodeles may produce some 11-KT. Circulating androgens, FSH, and LH vary seasonally, but estradiol levels are very low. Highest values for reproductive hormones occur during mating in some species (e.g., *Rana catesbeiana*) but not in others (*Rana esculenta*, *Ambystoma tigrinum*) where gametogenesis is dissociated from time of spawning and development of accessory structures is dependent on steroids (See Figure 11-1).

1. Male Urodeles

The reproductive system of a male urodele is provided in Figure 11-25. Spermatogenesis is of the cystic type in urodeles, and testicular structure and function are very similar to those of fishes. The urodele testis consists of one or more lobes, each containing several ampullae, which in turn are comprised of several germinal cysts (Figure 11-25). Germ cells associated with a germinal cyst divide mitotically to produce a cluster of secondary spermatogonia. These cells undergo synchronous differentiation to primary spermatocytes and enter meiosis. All

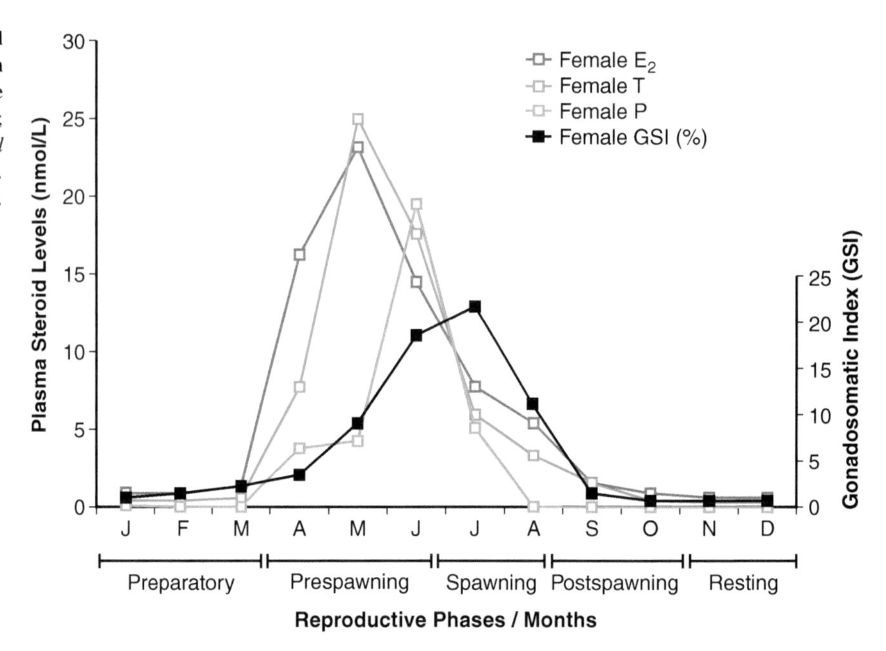

FIGURE 11-20 **Plasma steroid levels and gonadosomatic index (GSI) of female Indian major carp (*Labeo rohita*) during the five phases of the reproductive cycle.** E_2, estradiol; T, testosterone; P, progesterone. *(Adapted with permission from Suresh, D.V.N.S. et al.,* General and Comparative Endocrinology, ***159****, 143–149, 2008.)*

TABLE 11-4 Reproductive State and Serum Calcium Levels during the Spring in Steelhead Trout (*Oncorhynchus mykiss*) from a Natural Population

Sexual state	*N*	Mean body weight (g)	Mean serum calcium (mg/dL ± SE)
Immature males and females	12	38.4	11.6 ± 0.43
Sexually mature males prior to spawning	11	213.6	11.5 ± 0.74
Sexually mature females after ovulation but prior to spawning	9	204.9	15.5 ± 1.44[a]

[a]*Sexually mature female trout differ significantly ($p < 0.01$) from sexuantly mature males and immature trout.*

of the cysts within an ampulla develop synchronously, although it is typically only the more posterior ampullae that exhibit spermatogenesis prior to a given breeding season. The other ampullae represent the source of sperm for future breeding seasons. Sertoli cells develop from fibroblasts in the cyst walls while the spermatogonial divisions are taking place. As the ampullae mature, the posterior portion of the testis becomes swollen with sperm, whereas ampullae of the most anterior portion consist primarily of spermatogonia. The posterior portion of the testis becomes dense and whitish because of masses of sperm. After spermiation occurs the collapsed ampullae that have discharged their sperm into the male ducts are resorbed, and after breeding, spermatogenesis is initiated in the anterior portion of the testis. If spermiation occurs in the fall, spermatogenesis will not be resumed until the next summer. New ampullae differentiate from connective tissue elements and germ cells in the tunica albuginea.

Androgen levels exhibit distinct seasonal cycles in urodeles (Figure 11-26, Table 11-5). Seasonal patterns of androgen secretion have been reported for *Taricha granulosa*, *Cynops pyrrhogaster*, and *A. tigrinum*. Androgen levels appear to be low during breeding, and high levels are more closely associated with development and maintenance of the vas deferentia that store sperm until breeding. Androgens also stimulate development of cloacal glands in males that are associated with spermatophore production and possibly with production of pheromones used in breeding. Contraction of the vasa deferentia and discharge of sperm during mating and spermatophore production are caused by AVT.

Androgens also stimulate development of sex accessory structures called **nuptial pads** in the newt *Notophthalmus viridescens*, but maximal development is obtained by simultaneous treatment with PRL and androgen. Development of skin glands also is influenced by androgens. In urodeles such as *N. viridescens*, *T. cristatus*, and *A. tigrinum*,

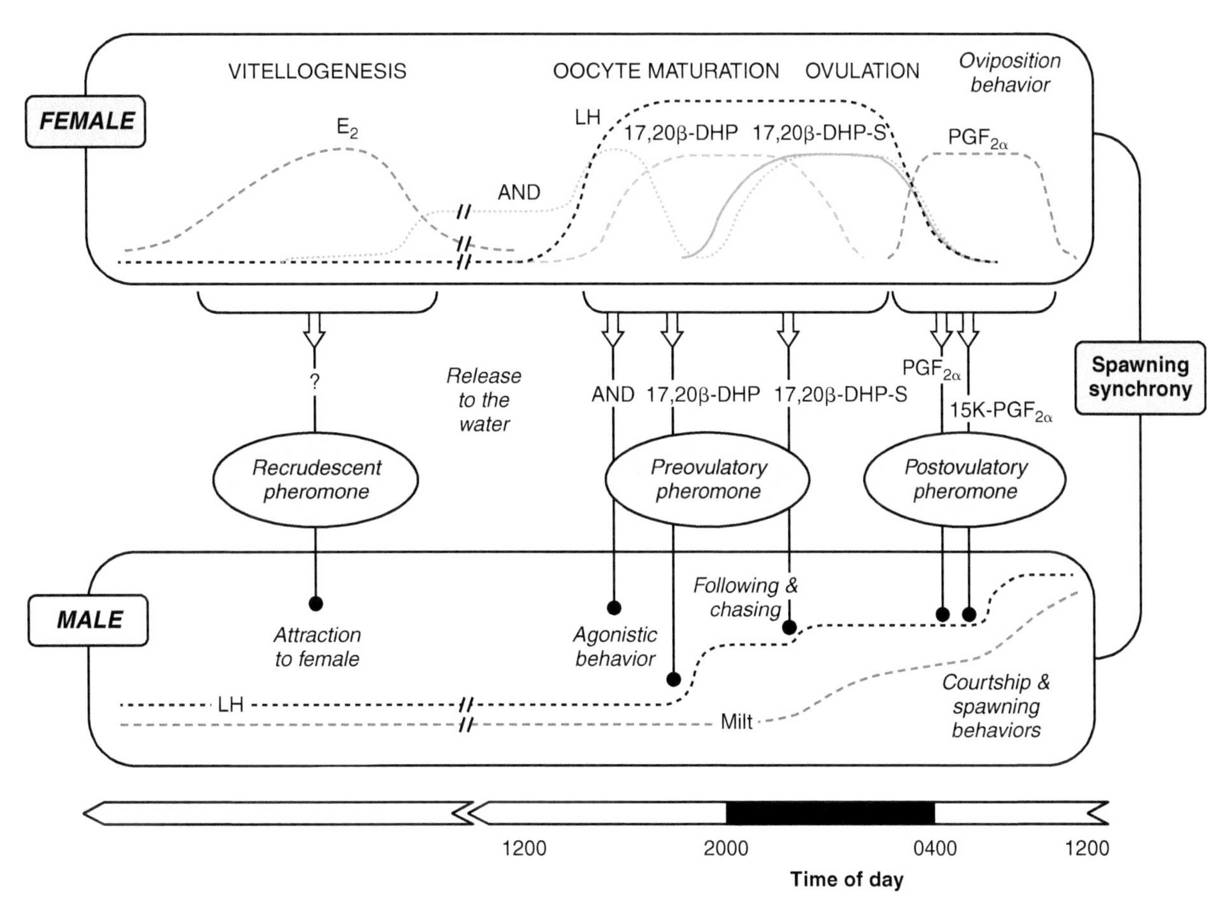

FIGURE 11-21 **Model of action for reproductive pheromones in the goldfish (*Carassius auratus*).** Luteinizing hormone (LH) is involved in production of the female pheromones and influences the responsivity of the male. Abbreviations: AND, androstenedione; $PGF_{2\alpha}$, prostaglandin; 15K-$PGF_{2\alpha}$, 15-keto metabolite of $PGF_{2\alpha}$, 17,20β-P, 4-pregnen-17,20β-dihydroxy-3-one; Milt = sperm. *(Adapted with permission from Stacey, N., in "Hormones and Reproduction of Vertebrates. Vol. 1. Fishes" (D.O. Norris and K.H. Lopez, Eds.), Academic Press, San Diego, CA, 2011, pp. 169–192.)*

PRL is known to influence the movement of land-phase animals to water for breeding and also induces heightening of the tail fin, which is a male secondary sex character in some species.

Internal fertilization in both aquatic and terrestrial urodeles occurs through transfer of a spermatophore from the male to the female. The spermatophore is produced through the actions of an array of specialized cloacal glands. The spermatophore consists of a glycoprotein matrix to which a packet of coagulated sperm is attached. The glycoprotein matrix acts as a base upon which the sperm packet rests. Following a complex courtship procedure, the female is induced to pick off the sperm packet using her cloacal lips. The sperm packet may then be stored in a specialized portion of the female's spermatotheca (Figure 11-27) until ovulation and fertilization occur.

2. Male Anurans

The reproductive systems of male frogs and toads are provided in (Figure 11-28). The anuran testis is often described as consisting of seminiferous tubules with a permanent germinal epithelium and conspicuous Leydig cells. However, spermatogenesis clearly is of the cystic pattern as described for all other anamniotes (Figures 11-3 and 11-29). The Leydig cells resemble mammalian Leydig cells ultrastructurally and possess 3β-HSD activity. The lipid cycle within the Leydig cells and the degree of 3β-HSD activity closely parallel the development of androgen-dependent sex accessory structures such as the enlarged thumb pads of ranid frogs. Leydig cells of post-spawning anurans exhibit considerable lipoidal accumulation but very low 3β-HSD activity, and thumb pads regress in male ranids at this time.

During winter months, Sertoli cells in testes of ranid frogs lack lipid, but these cells elongate and exhibit small lipoidal granules as the breeding season approaches. Sertoli cells of breeding animals have a well-developed smooth endoplasmic reticulum, and 3β-HSD activity is detectable. After spermiation the Sertoli cells detach from the tubule wall and degenerate. New cells for the next reproductive period differentiate from fibroblasts in the tubule walls.

Among male bufonid toads are found rudimentary ovaries or **Bidder's organs** (Figure 11-28) that develop

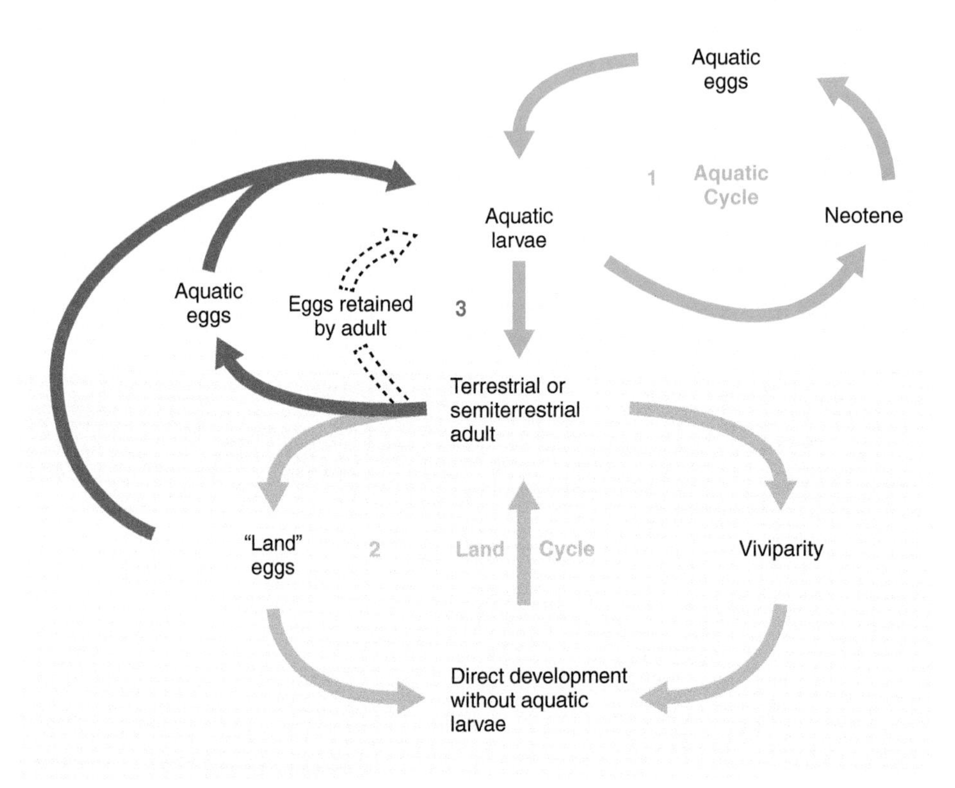

FIGURE 11-22 **Summary of amphibian life history patterns.** Three basic patterns are found in amphibians: (1) a totally aquatic cycle with sexually mature larvae (e.g., neotenes; see text), (2) a totally terrestrial or land cycle, (3) an aquatic—land pattern with terrestrial or semiterrestrial adults and aquatic larval stages. Within these patterns are some distinct variations such as viviparity versus laying an egg on land within the land cycle.

from cortical remnants of the embryonic genital ridge prior to normal gonadal differentiation. Histologically, the Bidder's organ consists of a compact mass of small oocytes. In males, the bidderian oocytes undergo a limited seasonal growth and degeneration cycle correlated with the testicular cycle (Figure 11-30). These bidderian oocytes never reach the vitellogenic stage in males, however. The presence of 3β-HSD activity suggests they are steroidogenic during this time. After castration of male bufonids, the Bidder's organ hypertrophies, presumably under the influence of increased GTHs, and forms a functional ovary. In castrated males, the rudimentary müllerian ducts may develop into functional oviducts, and such sex-reversed animals may breed as females. The presence of isolated oocytes in the testes of bufonids is a common condition in at least one species, *Bufo woodhousei*. Female bufonids also exhibit a Bidder's organ but it is assimilated into the ovary and bidderian oocytes are not distinguishable from ovarian oocytes.

3. Male Gymnophionids (Caecilians)

Male caecilians differ from urodeles and almost all anurans by possessing an elaborate intromittent organ, the **phallodeum**, associated with the posterior part of the cloaca. The pallodeum is employed for transferring sperm to the female reproductive tract; consequently, fertilization is internal in all caecilians. Another unique feature is the retention of only the most posterior portion of the müllerian ducts in males as the **müllerian glands**. These tubular apocrine glands are believed to produce seminal fluid and hence would be analogous to the prostate of mammals. The structure of the lobed gymnophionid testes is similar to that of urodeles, but the cell nests within an ampulla are not synchronized and do not exhibit the same stage of spermatogenesis.

D. Reproduction in Female Amphibians

1. Female Anurans and Urodeles

The reproductive systems of female urodeles and anurans are depicted in Figures 11-31 and 11-32, respectively. Amphibian ovaries are hollow, sac-like structures derived from the embryonic cortex and covered by germinal epithelium. A derivative of embryonic medullary tissues forms the inner lining of the ovary. Oogonia are present in the germinal epithelium, and they give rise to nests of oocytes. The follicular epithelium is derived from the germinal epithelium and consists of a single layer of granulosa cells throughout the maturation period. A very thin thecal layer does form around the follicle, but it is only obvious when viewed with the aid of the electron microscope (compare Figures 11-33 and 11-34).

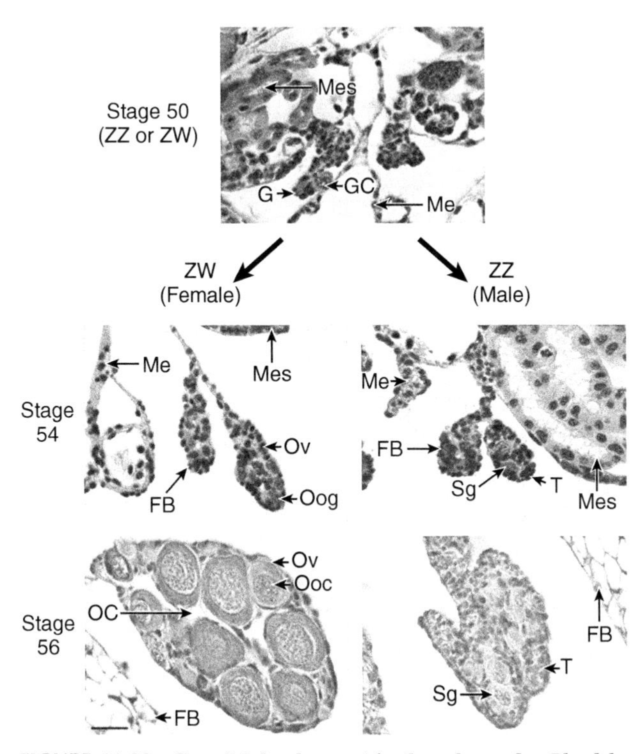

FIGURE 11-23 **Gonadal development in the salamander *Pleudeles wallti*.** Stage 50 larva has an indifferent or undifferentiated gonad. In ZW females, the beginnings of the ovary and oviduct can be seen. By Stage 56 the ovary is clearly differentiated. In the ZZ male, development is not so obvious as in the female. Fat body, FB; gonad, G; germ cell, GC; gut mesentery, Me; mesonephros, Mes; ovarian cavity, OC; oocytes, Ooc; oogonia, Oog; ovary, Ov; spermatogonia, Sg; testis, T. *(Reprinted with permission from Flament, S. et al., in "Hormones and Reproduction of Vertebrates. Vol. 2. Amphibians" (D.O. Norris and K.H. Lopez, Eds.), Academic Press, San Diego, CA, 2011, pp. 1–19.)*

Granulosa cells and thecal cells contain 3β-HSD activity, but there is no steroidogenic activity in the ovarian interstitial cells of the ovary. Both thecal and granulosa cells may be sources of circulating ovarian steroids, but cytological evidence favors the granulosa as the major source. GTHs stimulate these cells to synthesize estrogens.

At the end of a breeding season, the ovary typically contains young follicles that will become the next crop of mature oocytes, numerous cell nests that will become the young follicles of the next vitellogenic period, and primary germ cells that will give rise to new cell nests for future generations of oocytes. Progression from primary germ cells to mature oocytes may require three breeding seasons or more for completion.

Ovarian estrogens control development of sex accessory structures such as the hypertrophy of oviducts prior to sexual maturation and during each season prior to ovulation. The oviducts regress when estrogen synthesis declines after breeding. In the marsupial frog *Gastrotheca riobambae*, the development of the brood pouch that will be used to incubate the eggs is dependent upon estrogens secreted from preovulatory follicles.

Postnuptial ovaries frequently contain postovulatory corpora lutea, which are short-lived in oviparous species. Granulosa cells hypertrophy after ovulation and accumulate cholesterol-positive lipids. The follicle collapses and becomes a central mass of lipoidal cells surrounded by a fibrous capsule derived from the thecal layer. Postovulatory corpora lutea of *T. cristatus* and *R. esculenta* possess 3β-HSD activity and are sources for steroids; however, no functional endocrine role for postovulatory corpora lutea has been demonstrated in these oviparous species.

In viviparous amphibians such as the anuran *Nectophrynoides occidentalis*, postovulatory corpora lutea are capable of converting pregnenolone to progesterone and persist throughout gestation. Corpora lutea are required for about the first 25 to 30 days of the 100- to 125-day gestational period in the pouch of the marsupial frog *G. riobambae*, and ovariectomy after day 40 has no effect on gestation. The granulosa-lutein cells of the postovulatory corpora lutea in viviparous *S. salamandra* possess 3β-HSD activity and appear cytologically to be steroidogenic. Thirty or more corpora lutea persist in each ovary during the first 2 years of gestation and gradually decrease in both size and number over the next 2 years until the young are born. Development of new oocytes may be arrested by the presence of corpora lutea. In *N. occidentalis* as well as in oviparous *Taricha torosa*, the succeeding crop of follicles begins development only after degeneration of the postovulatory corpora lutea. These latter data suggest an inhibitory action of progesterone produced in the postovulatory corpora lutea on release of GTHs from the adenohypophysis.

Atresia occurs frequently during follicular development in amphibians. Granulosa cells are responsible for phagocytosis of yolk and formation of the corpora atretica as they are in fishes. However, these corpora atretica exhibit no 3β-HSD activity and hence are probably not steroidogenic.

The process of vitellogenesis in the liver and yolk deposition in oocytes of oviparous amphibians has been examined extensively. Estrogen levels are sufficient in females to stimulate vitellogenesis and elevate plasma vitellogenin (Figure 11-35). In *Xenopus laevis*, GTH stimulates micropinocytosis of vitellogenin by oocytes. Micropinocytotic vesicles of vitellogenin are hydrolyzed enzymatically in the yolk platelets to produce the yolk proteins **phosvitin** and **lipovitellin**. The yolk platelets containing these yolk proteins are utilized as an energy source during early embryogenesis. Estrogens will induce vitellogenin synthesis in both female and male livers when administered *in vivo*. Circulating vitellogenin binds free calcium ions, resulting in elevated total plasma calcium levels through release of calcium from storage sites.

Hypophysectomy results in atresia of all vitellogenic follicles in excess of about 0.4 mm in diameter, indicating the

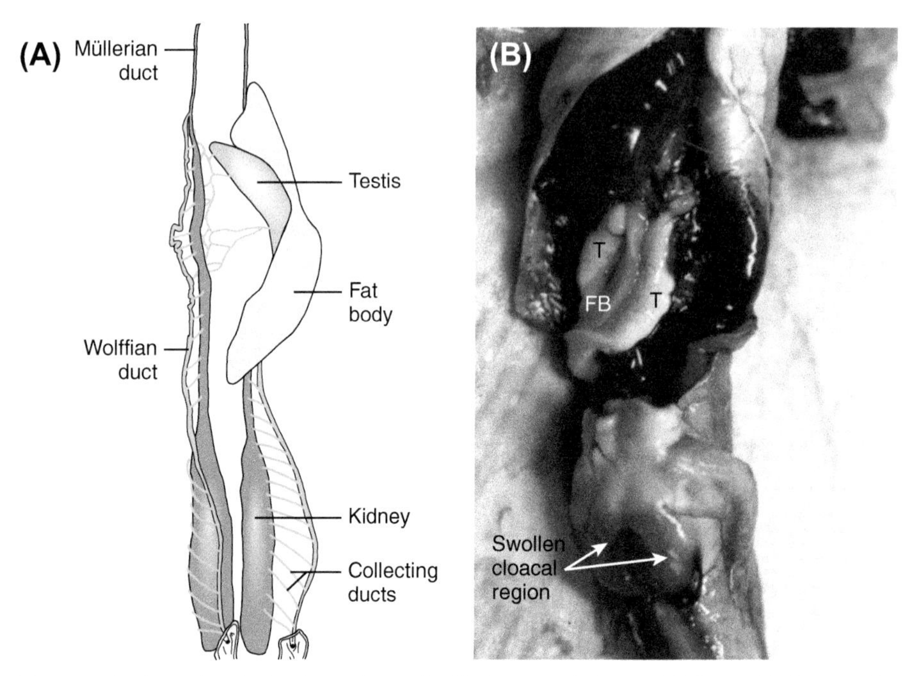

FIGURE 11-24 **Urodele reproductive system in the tiger salamander (*Ambystoma tigrinum*).** (A) Schematic diagram of a sexually mature metamorphosed male. The wolffian duct doubles as both vas deferens and ureter. Many urinary collecting ducts connect the lumbar portion of the kidney to the wolffian duct. One testis (T) and its corresponding fat body (FB) have been removed. Even in the adult, a remnant of the müllerian duct persists. (B) Dissected sexually mature male larvae (neotene) showing white testes swollen with sperm. Note the pair of prominent fat bodies. The swollen cloacal region is caused by development of multiple glands associated with production of the spermatophore and is both an indicator of sexual maturity and sex as the cloacal region of a mature female is only slightly swollen and lacks the darkly pigmented tubercles on the psoterior margin of the swellings. *(Part A adapted with permission from Rodgers, L.T. and Risley, P.L.,* Journal of Morphology, ***63**, 119–139, 1938. © John Wiley & Sons. Part B photograph provided by D.O. Norris.)*

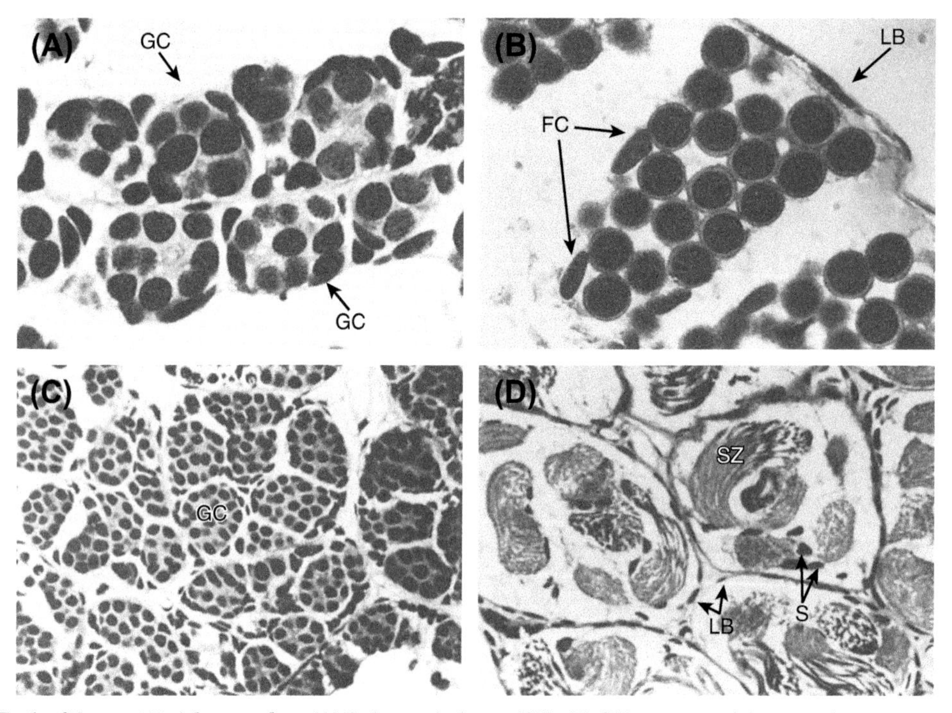

FIGURE 11-25 **Testis of the newt *Taricha granulosa*.** (A) Early germinal cysts (GC). (B) Older cyst containing secondary spermatogonia. Follicle cells (FC) have flattened nuclei. LB, lobule boundary. (C) Lower magnification showing several ampullae each containing six to eight cysts. (D) Enlargement of cysts from another region of the testis containing mature sperm (SZ) and prominent Sertoli cells (S) derived from follicle cells. *(Courtesy of Frank L. Moore, Oregon State University, Corvallis.)*

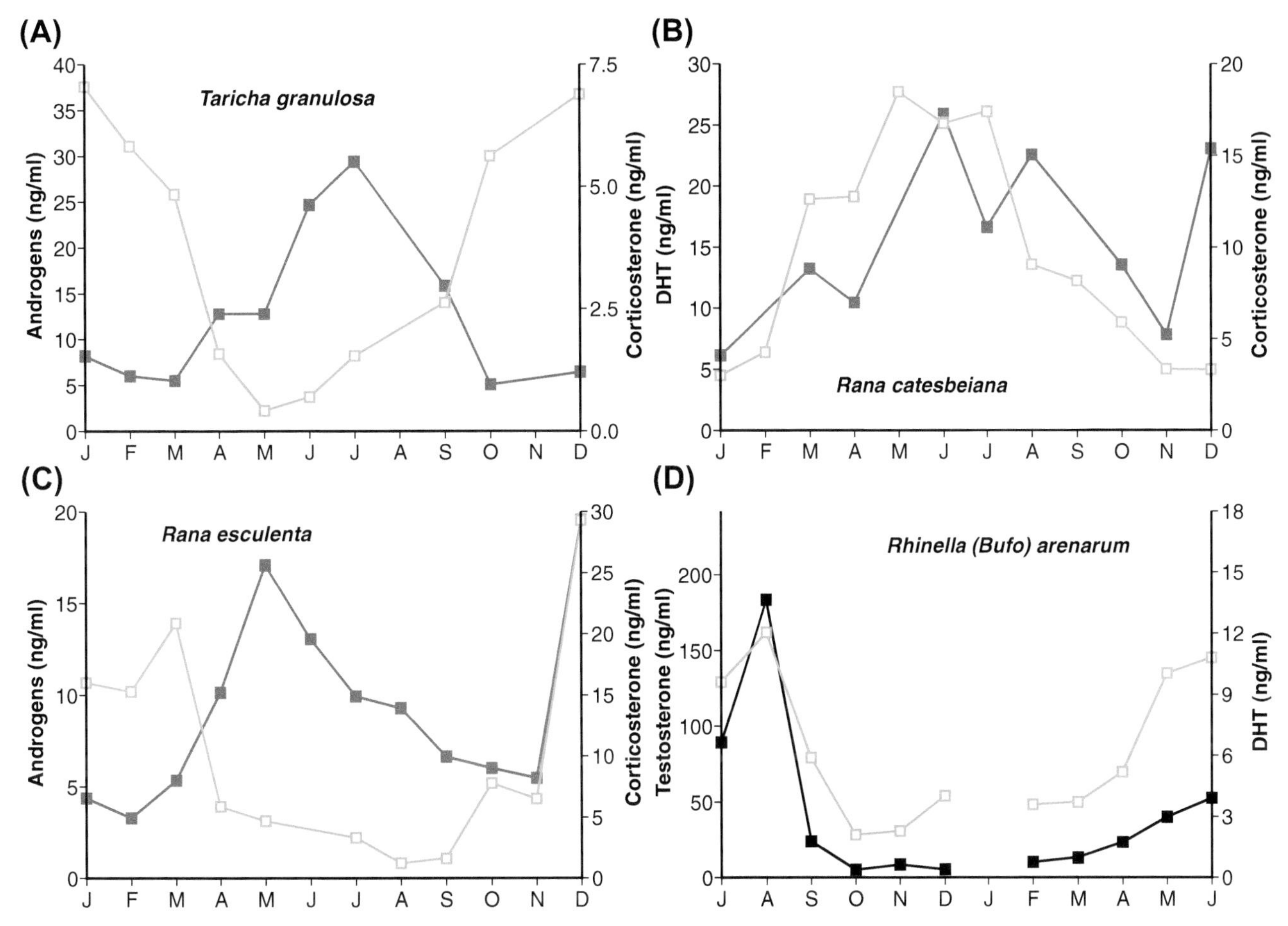

FIGURE 11-26 **Comparision of androgen and corticosterone levels during the reproductive cycles of selected amphibians.** Androgen levels appear to be negatively correlated to corticosterone levels in the newt *Taricha granulosa* (A) and the frog *Rana esculenta* (C) but dihydrotestosterone (DHT) is positively correlated with corticosterone in bullfrogs (*Rana catesbeiana*) (B). The toad *Rhinella* (*Bufo*) *arenarum* (D) shows a pattern of testosterone (blue) and dihydrotestosterone (DHT; black) secretion similar to androgen secretion in *T. granulosa* and of *R. esculenta*. *(Parts A to C adapted with permission from Moore, F.L. and Deviche, P., in "Processing of Environmental Information in Vertebrates" (M. Stetson, Ed.), Springer-Verlag, Berlin, 1988, pp. 19–45. Part D adapted with permission from Medina, M.F. et al.,* General and Comparative Endocrinology, ***136****, 143–151, 2004.)*

importance of endogenous pituitary GTHs in the process of vitellogenesis and maintenance of follicular growth. Mammalian FSH will augment the growth of vitellogenic follicles and prevent atresia following hypophysectomy. The failure of GTHs to stimulate formation and growth of previtellogenic follicles (i.e., follicles less than 0.4 in mm diameter) coupled with their apparent insensitivity to hypophysectomy has led to the suggestion that these processes are completely independent of pituitary control. However, experimental studies have not ruled out completely a role for GTHs in the development of previtellogenic follicles, and it is possible that low endogenous levels of amphibian GTHs are necessary for even the earliest events in gametogenesis.

Amphibian ovaries *in vitro* produce progesterone, estradiol, estrone, testosterone, DOC (like teleosts), androstenedione, and DHT. The high levels of circulating testosterone reported for some female amphibians may be related to a precursor role for peripheral aromatization to estrogens or to a behavioral role (Figure 11-36).

Ovulation is under control of an LH-like GTH and progesterone. *In vitro* studies of the ovary, initiated by Paul Wright about 60 years ago and subsequently elaborated by others, indicate that LH as well as a wide variety of steroids can induce oocyte maturation (completion of meiosis and **germinal vesicle breakdown, GVBD**) and ovulation *in vitro*. Progesterone is the most potent maturational steroid, and it is reasonable to presume that progesterone or a closely related steroid plays a key role in normal ovulatory events. The synthesis of progesterone is under the control of LH, and the action of progesterone is believed to be indirect, operating through stimulation of a "maturation-promoting factor". Recent studies with *Xenopus laevis* have shown that it may be the enzymatic conversion of progesterone into testosterone in the follicle that actually triggers final oocyte maturation and

TABLE 11-5 Seasonal Extremes in Androgen Levels (ng/mL) in Selected Male Amphibians

	Testosterone		5α-Dihydrotestosterone		Total androgens	
Species	High	Low	High	Low	High	Low
Urodeles						
Ambystoma tigrinum	32	0.4	7	0.3	39	0.7
Taricha granulosa					37	3
Triturus carnifex					28	2
Anurans						
Bufo japonicus	203	47	115	27	318	74
Bufo mauritanicus	183		52			
Rana blythi	4		1			
Rana catesbeiana					5	
Xenopus laevis	39	19			35	3

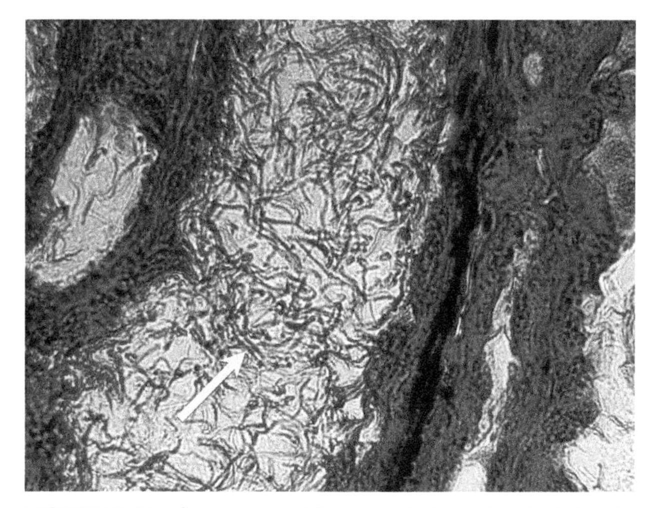

FIGURE 11-27 **Sperm (arrow) in spermatotheca of neotenic female tiger salamander.** *(Reprinted with permission from Norris, D.O., in "Hormones and Reproduction of Vertebrates. Vol. 2. Amphibians" (D.O. Norris and K.H. Lopez, Eds.), Academic Press, San Diego, CA, 2011, pp. 187–202.)*

ovulation. PRL enhances the sensitivity of oocytes to GTH or progesterone (or testosterone) both *in vivo* and *in vitro*. This enhancement can be blocked by simultaneous *in vivo* treatment with thyroxine (T_4).

Ovarian maturation typically is completed in autumn, and ovulation is delayed over winter until favorable conditions occur in the spring. The endocrine basis for this diapause is not clear but may involve direct inhibition of ovulation by one or more pituitary hormones. Hypophysectomy of gravid anurans and at least one urodele results in ovulation and oviposition. Furthermore, hypophysectomy of gravid neotenic tiger salamanders increases their sensitivity to induced ovulation in response to a single injection of hCG. This "reflexive" ovulation could be due to inadvertent release of GTHs by the operation itself or to removal of an active inhibitory substance of pituitary origin or to both.

Growth of amphibian oviducts normally is stimulated by estrogens, and the size of the ovaries and the oviducts varies with season (Figure 11-37). Oviducts of some species may respond to treatment with either estrogens or androgens (Figure 11-38). The feminizing action of androgens is another example of a paradoxical effect. In mature females, contraction of oviducts is caused by AVT, which presumably is the hormonal stimulus in oviposition. Oviducts of breeding animals are more sensitive to AVT than are those of non-breeding adults. Progesterone induces responsiveness to AVT in immature oviducts of salamanders, but estrogens are not effective. Possibly the pre- or postovulatory follicle releases sufficient progesterone in response to LH to alter receptor levels for AVT in the muscles of the oviducts. Androgens and estrogens do not affect the sensitivity of oviducts to AVT although they stimulate growth of the oviducts.

2. Female Gymnophionids (Caecilians)

Fertilization in caecilians normally occurs in the upper portion of the oviduct following intromission by the male, and the fertilized eggs are either laid in burrows or are retained in the oviducts until the developing larvae

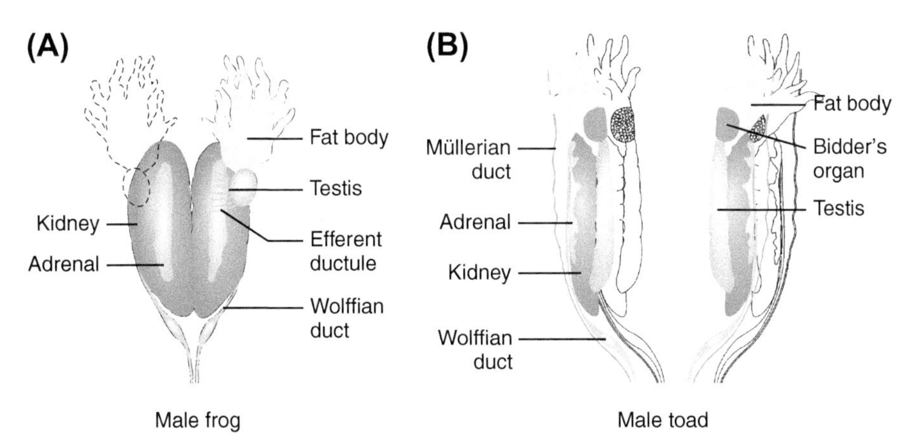

FIGURE 11-28 **Reproductive systems of male anurans.** (A) Bullfrog (*Rana catesbiana*). (B) Toad (*Bufo bufo*) with Bidder's organ, an ovarian remnant commonly found among species of *Bufo*. *(Adapted with permission from Matsumoto, A. and Ishii, S., "Atlas of Endocrine Organs: Vertebrates and Invertebrates," Springer-Verlag, Berlin, 1992.)*

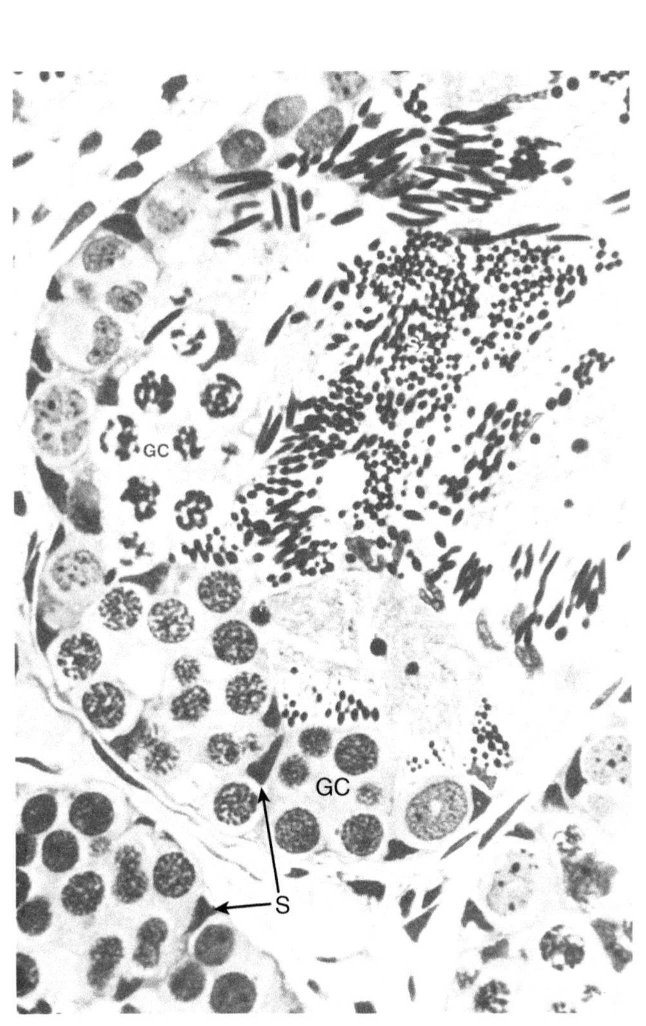

FIGURE 11-29 **Section through testis of the bullfrog (*Rana catesbeiana*).** Note how all cells in a germinal cyst (GC) are in the same stage of development. S, Sertoli cell nucleus. Compare to urodele testis in Figure 11-10. *(Courtesy of Charles H. Muller, University of Washington, Seattle.)*

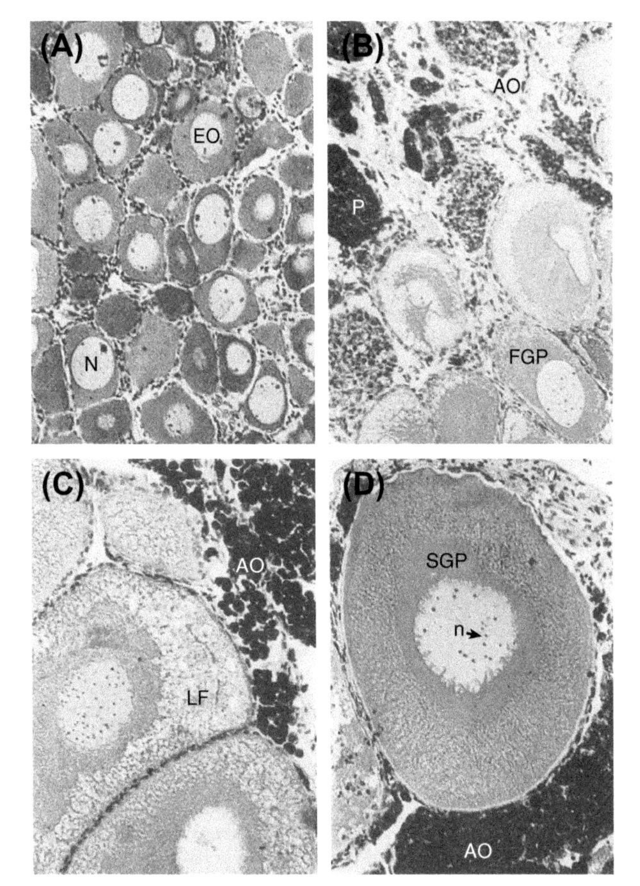

FIGURE 11-30 **Sections through Bidder's organ from male *Bufo woodhousii*.** These figures at the same magnification show the effects of castration and gonadotropin treatment on oocytes of Bidder's organ. (A) Sham-operated toad treated with saline injections for 26 days shows unstimulated follicles. EO, early previtellogenic follicle; N, nucleus. (B) Sham-operated toad treated with mammalian gonadotropins for 26 days shows moderately enlarged follicles. AO, atretic follicle; FGP, first growth phase follicle (previtellogenic); P, pigment granules. (C) Castrated toad treated with saline for 26 days shows oocyte and follicle growth into the late previtellogenic stage. Late first growth phase oocyte, LF. (D) Castrated toad treated with gonadotropins shows vitellogenic or second growth phase oocyte (SGP). *(Reprinted with permission from Pancak-Roessler, M.K. and Norris, D.O.,* Journal of Experimental Zoology, ***260****, 323–336, 1991. © John Wiley & Sons, Inc.)*

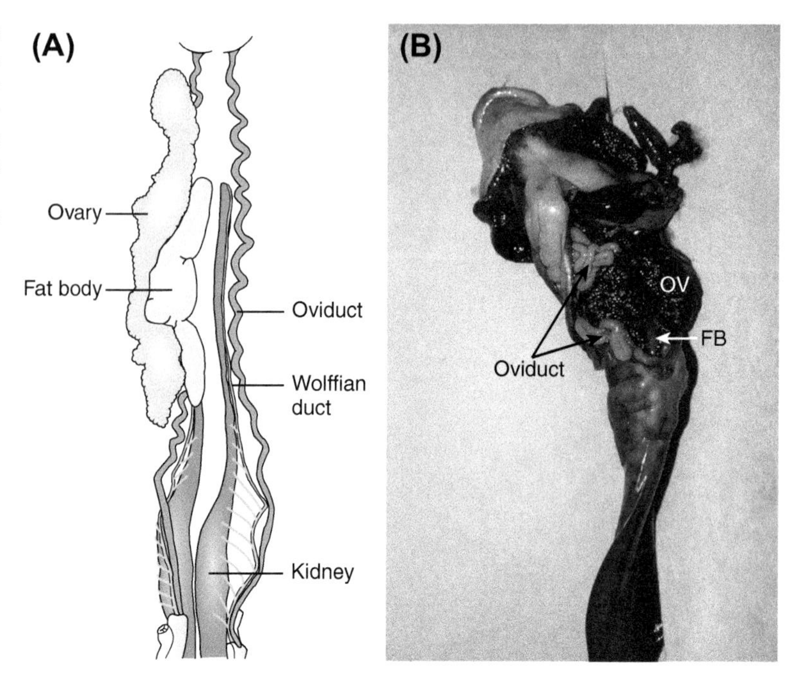

FIGURE 11-31 **Reproductive system of female urodele, the tiger salamander (*Ambystoma tigrinum*).** (A) Schematic diagram of a sexually mature metamorphosed female. (B) Dissection of an adult sexually mature female larva (neotene). Abbreviations: FB, fat body; OV, ovary. *(Part A adapted with permission from Rodgers, L.T. and Risley, P.L.,* Journal of Morphology, ***63****, 119–139, 1938. © John Wiley & Sons. Part B photograph provided by D.O. Norris.)*

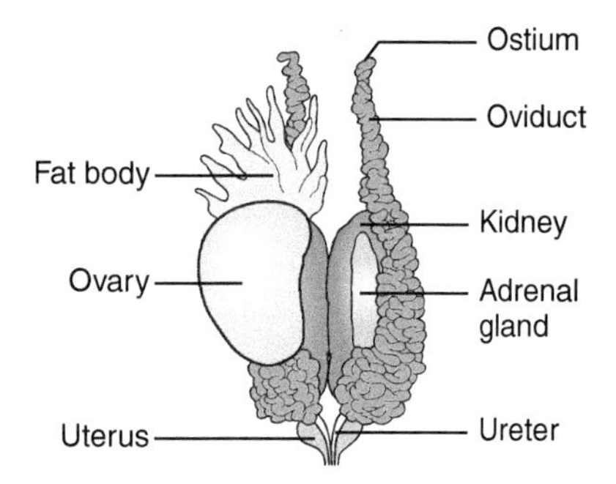

FIGURE 11-32 **Reproductive system of female anuran, the bullfrog (*Rana catesbeiana*).** Ovary has been removed from the right side to expose the kidney, adrenal, and the convoluted oviduct. *(Adapted with permission from Matsumoto, A. and Ishii, S., "Atlas of Endocrine Organs: Vertebrates and Invertebrates," Springer-Verlag, Berlin, 1992.)*

have completed metamorphosis. Ovarian development is very similar to that described for anurans and urodeles, but in caecilians the eggs tend to be larger and fewer. Postovulatory corpora lutea develop in the ovaries, and they appear to be important for maintaining oviductal secretion (even in oviparous species) and pregnancy. Oviductal secretions in viviparous species provide nutrition for the developing young, and these secretions may be controlled by hormones released from the corpora lutea. Relatively few caecilians have been examined, and it is not possible to describe the seasonality of breeding and ovarian cycles in caecilians. Consequently, the endocrine factors involved in reproductive events mainly must be inferred from studies on anurans and urodeles.

E. Reproductive Behavior in Amphibians

Numerous aspects of reproductive behavior have been described for amphibians, but little is known about the mechanisms of endocrine control for most species. Reproductive behavior includes migration, calling, courtship, clasping, spawning, and various kinds of parental care. Studies involving castration, hypophysectomy, and/or injections of pituitary hormones support the conclusion that testicular hormones are involved in calling, courtship, and clasping. In neotenic tiger salamanders, *A. tigrinum*, a species with dissociated mating, androgens are not high at the time of mating (see Figure 11-1). Although most species examined exhibit associated mating, attempts to stimulate reproductive behavior with androgens alone often have not been successful.

Even knowing plasma androgen levels may not tell the entire story. In the European crested newt (*T. carnifex*), inactive males have higher testosterone levels than courting males. Clasping of the female does not occur in these newts and courtship involves progressive stages: approaching, fanning, tail lashing, and spermatophore deposition. $P450_{aro}$ activity increases in the brain and gonad during courtship, and courting males have significantly higher levels of estrogen in brain and plasma during the approach stage of courtship.

Neuropeptides (AVT, GnRH, ACTH) trigger mating behavior in androgen-primed animals. For example, when a female frog that is not ready to spawn is clasped by a courting male, she croaks to signal her non-receptivity. A receptive female will not emit this release call. This

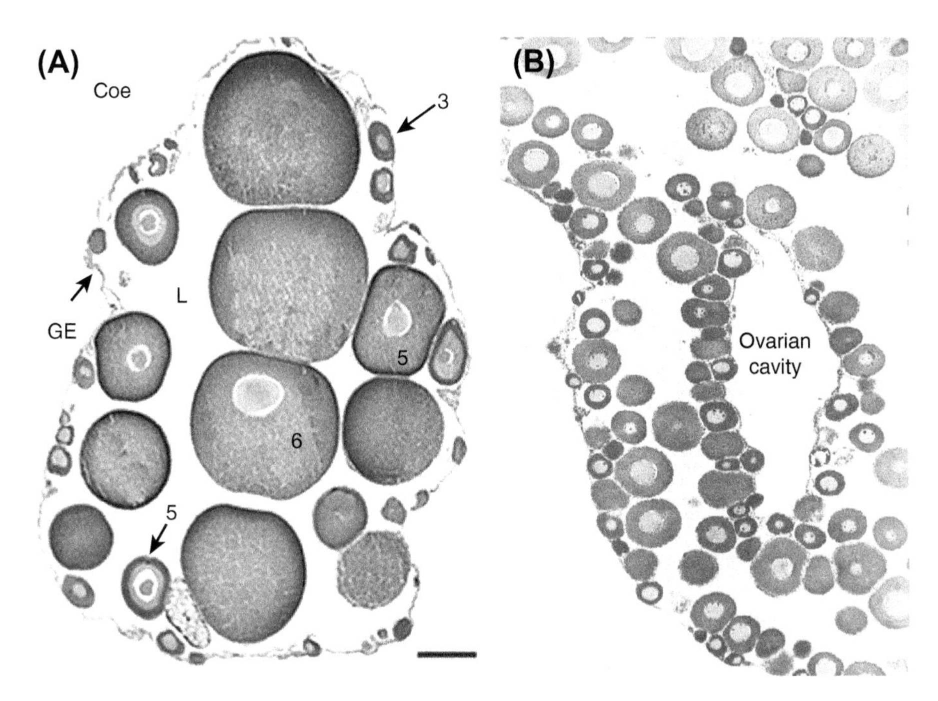

FIGURE 11-33 **Hollow amphibian ovaries.** (A) Ovary of *Ambystoma dumerlii* showing several stages of amphibian follicle growth according to Uribe (2011). Abbreviations: GE, germinal epithelium; L, ovarian lumen. (B) This ovary from the cane toad (*Bufo marinus*) is typical of the hollow amphibian ovary with follicles attached to the germinal epithelium. *(Part A reprinted with permission from Uribe, M.C.A., in "Hormones and Reproduction of Vertebrates. Vol. 2. Amphibians" (D.O. Norris and K.H. Lopez, Eds.), Academic Press, San Diego, CA, 2011, pp. 55–81. Part B courtesy of Charles H. Muller, University of Washington, Seattle.)*

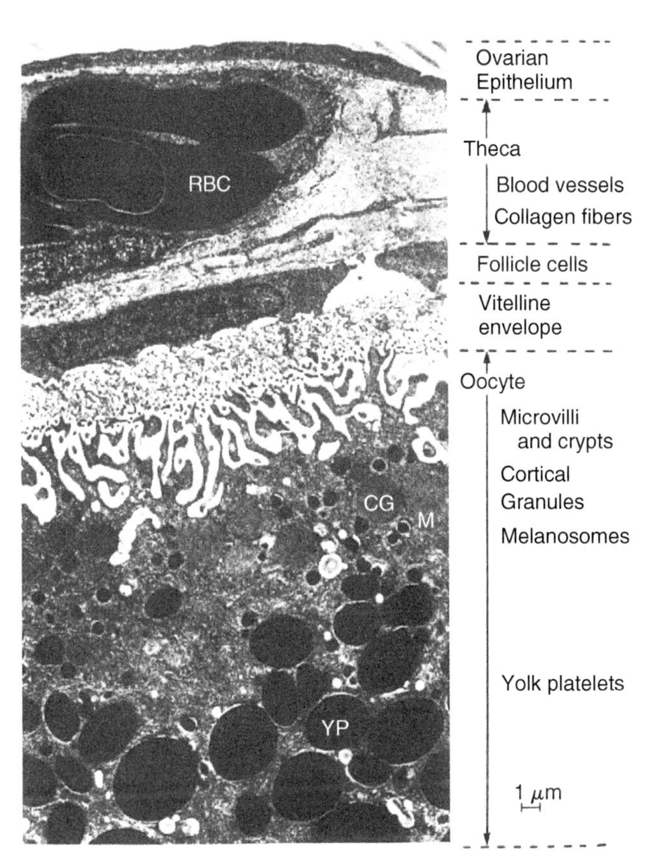

FIGURE 11-34 **Vitellogenic follicle of *Xenopus laevis*.** This transmission electron micrograph shows the vascular theca and its relationship to microvillous processes projecting from the surface of the developing oocyte. YP, yolk platelets. *(Reprinted with permission from Dumont, J.N. and Brummett, A.R.,* Journal of Morphology, ***155****, 73–97, 1978.)*

receptive female has accumulated water that will be used in ovulation and oviposition. Water retention is caused by AVT (see Chapter 5), and administration of AVT inhibits the release call, possibly through effects on the brain.

According to extensive studies by Frank Moore and associates at Oregon State University, reproductive behavior by male rough-skinned newts (*Taricha granulosa*) also is stimulated by AVT, although androgens and other factors are involved (Figure 11-39). During courtship, males in the breeding pond will attempt to clasp a female along the back. Attempts by several males to clasp the same female result in the formation of mating balls that persist for a time, but the unsuccessful males soon drop off leaving only one clasping the female. A series of behavioral and chemical interactions between the clasping pair eventually will lead to spermatophore transfer. Androgens play a priming role that enhances the sensitivity of the newts to AVT. Stressors or injection of corticosterone can rapidly repress clasping behavior within minutes. The finding that corticosterone rapidly inhibits reproductive behavior in newts has led to the hypothesis that corticosterone may interact with a membrane receptor, rather than a nuclear receptor, to modulate reproductive behavior. Clasping can be activated by cloacal stimulation and is controlled by neurons located in the rostral portion of the medulla. Furthermore, activity of these neurons is increased by AVT treatment but decreased by administration of corticosterone.

1. Pheromones in Amphibian Reproduction

Pheromones are used in courtship by many urodeles that possess **hedonic glands** or **cloacal glands**. These glands are

FIGURE 11-35 Estrogen secretion elevates plasma levels of vitellogenin in post-metamorphic *Xenopus tropicalis females.* Animals begin to secrete estrogens about 16 weeks after metamorphosis followed by rapid elevation of vitellogenin production, whereas vitellogenin production is minimal in males. *(Adapted with permission from Olmstead, A.W. et al.,* General and Comparative Endocrinology, ***160****, 117–123, 2009.)*

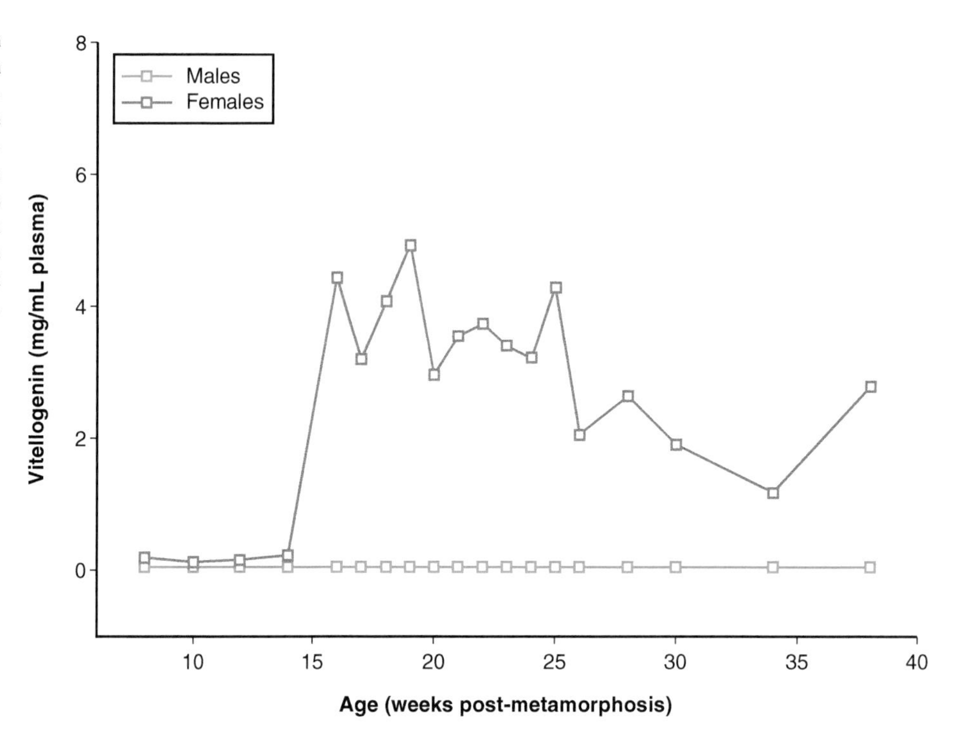

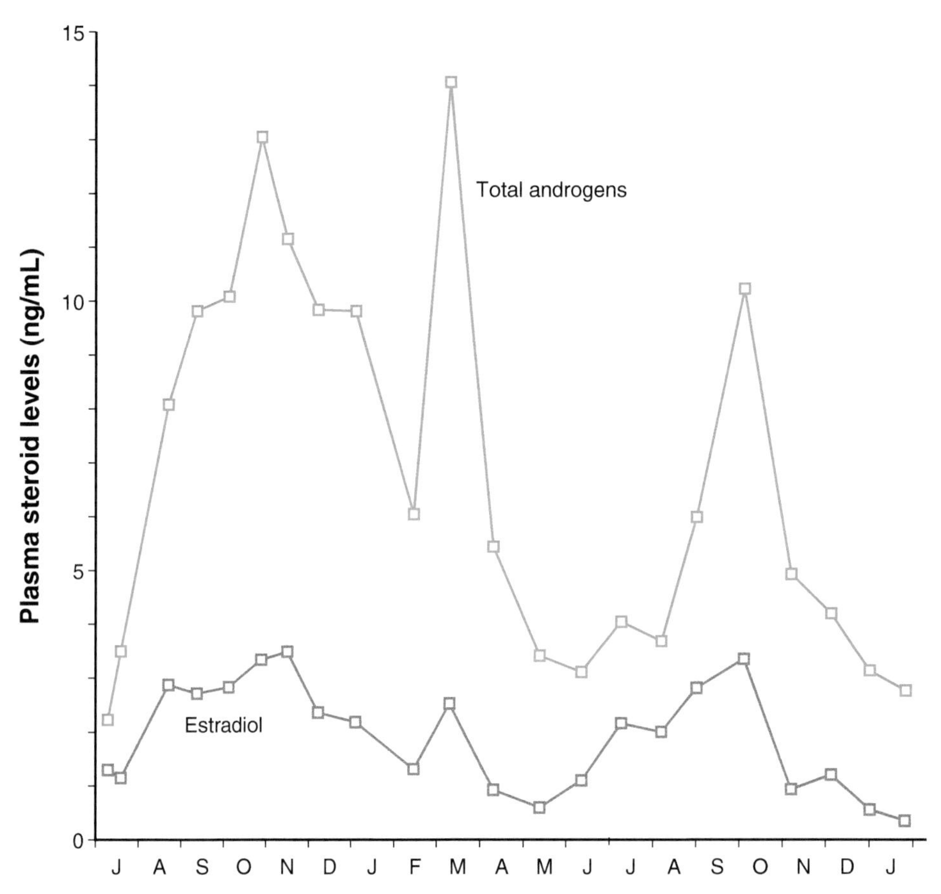

FIGURE 11-36 Seasonal variations in steroid levels in female *Pleurodeles waltl.* *(Adapted with permission from Garnier, D.H.,* General and Comparative Endocrinology, ***60****, 414–418, 1985.)*

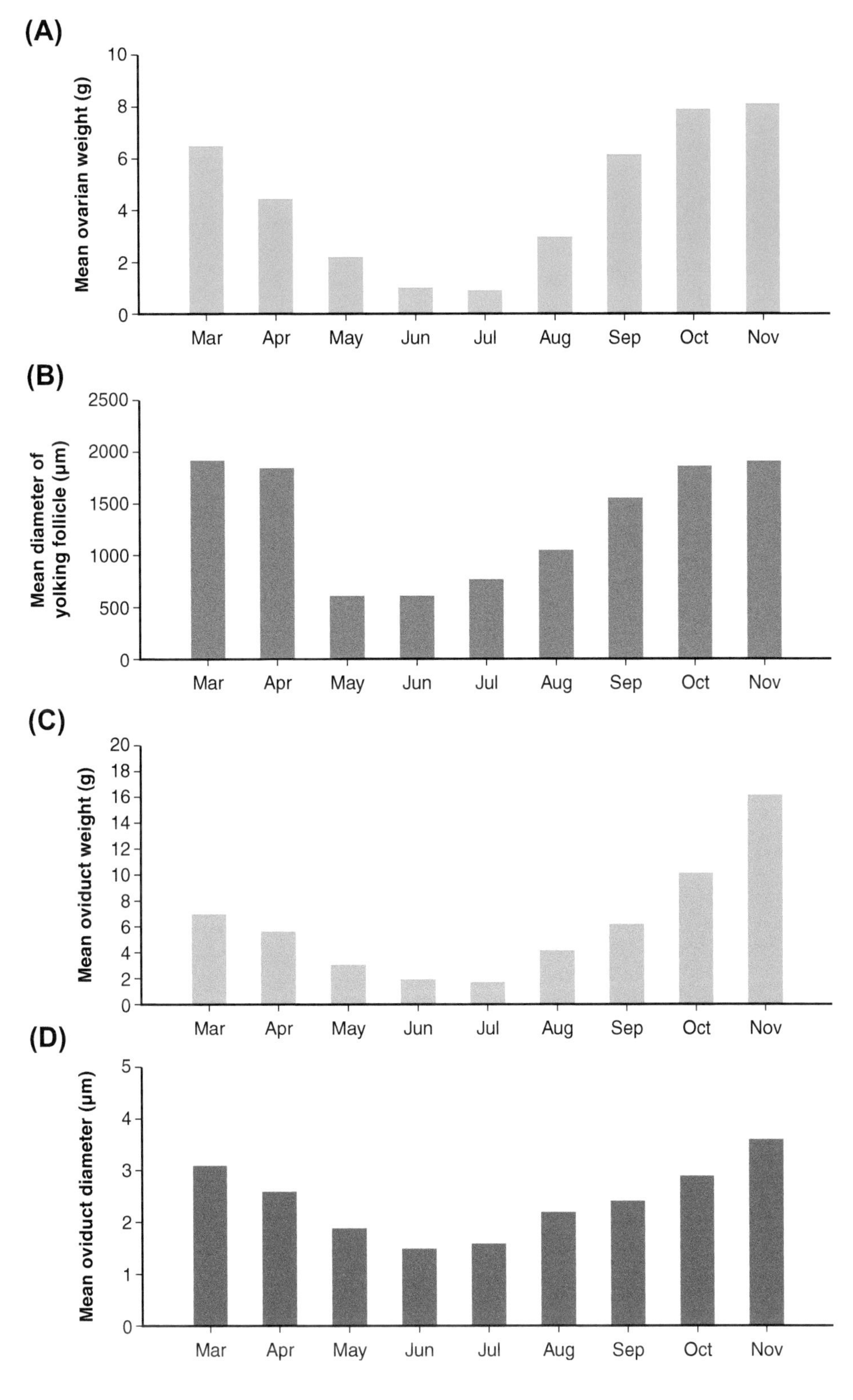

FIGURE 11-37 **Correlative changes in reproductive parameters of sexually mature female tiger salamander larvae (neotenes).** Oviposition occurs in March and April. Pond was frozen over from December through February. Mean ovarian weight, oviduct weight, follicle diameter, and oviduct diameter show identical patterns throughout the year. Numerous studies of estrogen levels in other species allow prediction of a similar pattern for estrogen levels in this species correlated to any one of these parameters. *(Adapted with permission from Norris, D.O., in "Hormones and Reproduction of Vertebrates. Vol. 2. Amphibians" (D.O. Norris, and K.H. Lopez, Eds.), Academic Press, San Diego, CA, 2011, pp. 187–202.)*

known to be sources of potent pheromones. Male plethodontid and desmognathid salamanders employ a tubular **mental (chin) gland** for stimulating courtship behavior in females. Ambystomatids and salamandrid salamanders rely on pheromonal secretions from one of their cloacal glands, the **abdominal gland**, to stimulate females. Female attractants released into a stream readily attract reproductively active male *T. granulosa*. Furthermore, males of this species

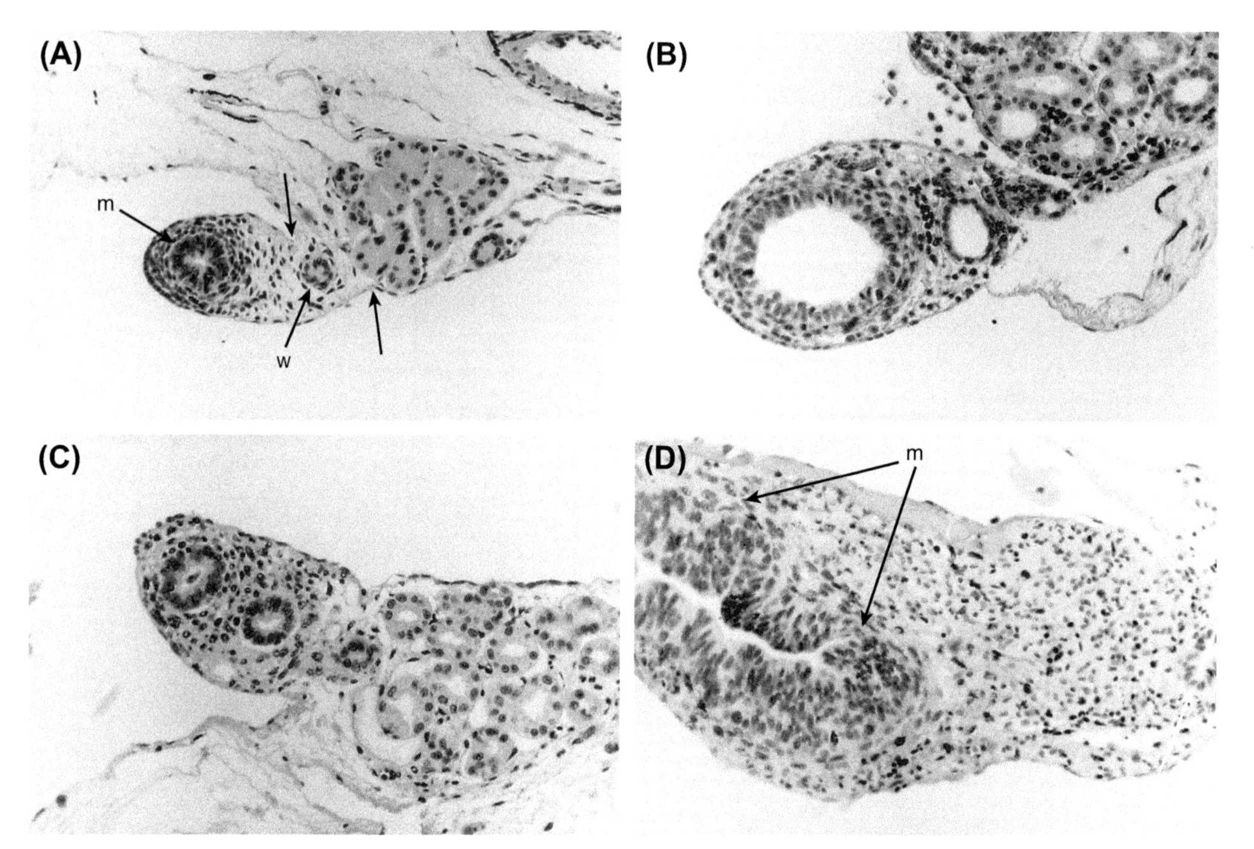

FIGURE 11-38 **Effects of steroid treatment on müllerian ducts of larval tiger salamanders.** (A) Saline-treated control. (B) Effect of 12.5 μg estradiol (E_2). (C) Effect of 12.5 μg dihydrotestosterone (DHT). (D) Synergistic effect of E_2 and DHT. Abbreviations: m, müllerian duct; w, wolffian duct. All photomicrographs were prepared at the same magnification. *(Reprinted with permission from Norris, D.O. et al.,* General and Comparative Endocrinology, ***106****, 348–355, 1997.)*

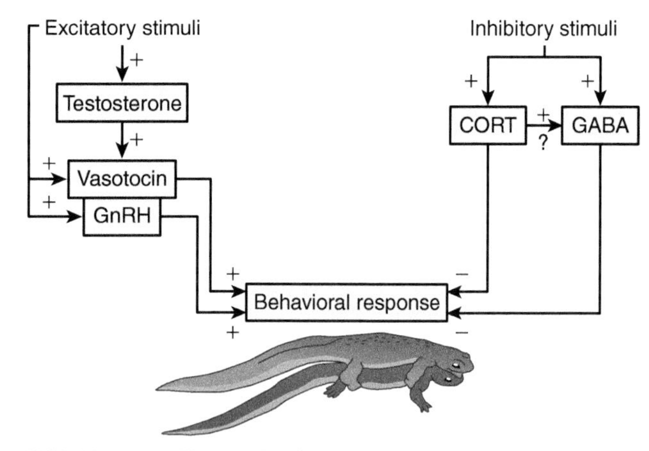

FIGURE 11-39 **Neuroendocrine control of clasping behavior in the rough-skinned newt.** Stress activates inhibitory pathways (CORT, corticosterone; GABA, γ-aminobutyric acid) that antagonize the actions of GnRH and arginine vasotocin activated by external sexual parameters. *(Adapted with permission from Moore, F.L. and Orchinik, M.,* Seminars in Neuroscience, ***3****, 489–496, 1991.)*

are attracted to females by airborne cues as well (see Table 11-6). Two peptide pheromones have been isolated from the abdominal cloacal glands of newts (*Cynops pyrrhogaster, C. ensicauda*) and a 22-kD protein was found in the mental gland of the salamander *Plethodon jordani*. These secretions are used to attract the female to a spermatophore. The pheromone from male *C. pyrrhogaster*, **sodefrin**, is released by the action of AVT on the abdominal gland, and it binds to receptors in the female's vomeronasal organ. The number of receptor cells for sodefrin in the female increases during the breeding season. The responsiveness of the vomeronasal epithelium to sodefrin is enhanced by exposure of the animals to PRL and estrogens.

In contrast, there is little evidence for chemical communication among anuran species, and mating is accomplished presumably by using auditory, tactile, and visual cues. However, another peptide with female attractant properties has been isolated from the skin glands of a male anuran, *Litoria splendida*.

2. Parental Care in Amphibians

Some parental care has evolved in certain amphibian species as a means of ensuring greater survival of young, mostly among the anurans. These parental behaviors are largely an alternative strategy for viviparity and help ensure survival to hatching or in some cases through metamorphosis. Female marsupial frogs (e.g., *G. riobambae*,

TABLE 11-6 Sex-Dependant Conspecific Odor Preferences in *Taricha granulosa*: Directed Locomotor Response of Newts to Newt and Non-Newt Odors

No. of tests	Sex of test newt	Sex of stimulus newt	Ratio of S/Ns[a]	Probability that choices were random
10	Male	Male	7/3	0.0570
10	Male	Male	10/0	0.0003
20	Male	Male/female	17/3	0.0004
10	Female	Male	4/6	0.625
10	Female	Female	6/4	0.625
20	Female	Male/female	10/10	0.987

[a]*Subject chose air from stimulus animal/air from non-stimulus source. Unpublished data of M. Schwartz, D. Duvall, and D.O. Norris.*

described earlier) develop an estrogen-dependent pouch on their backs into which eggs are stuffed and where they develop to hatching. Males of the midwife toad (*Alytes obstetricans*) wrap the strings of jelly containing the fertilized eggs and periodically take them to water to moisten them and finally to allow them to hatch as tadpoles. In terrestrial glass frogs of Central America (e.g., *Hyalinobatrachium fleischmanni*), eggs are attached under leaves; the male keeps them hydrated and defends them from predatory spiders (Figure 11-40A). In the gastric-brooding frog of Australia (*Rheobatrachus silus*), the female swallows the fertilized eggs and they remain in her stomach undigested until they hatch as tadpoles, which are then forcefully regurgitated alive. Unfortunately, this rare frog with its unusual brooding habit is now considered to be extinct.

Nest-building is a complex form of parental care that is exemplified by the tungara frog, which forms foam nests on the surface of ponds (Figure 11-40B), and the terrestrial foam-nest frog *Chiromantis xerampelina*. As many as 40 females of *C. xerampelina* may contribute to production of a huge, communal foam nest in which fertilized eggs are deposited. The females produce the foam by beating water with their hind limbs. Males are much smaller than females and they attract females by making soft clicking sounds. When a female arrives at a calling site, males drop from above onto her back. One male occupies the central position while up to seven peripheral males hang on. This central position is considered to be the most advantageous position for fertilizing eggs emanating from the female's cloaca. In experiments where the cloacae of males were sheathed to prevent sperm release, no eggs were fertilized. When only the central male was sheathed, about half of the eggs were still fertilized, but they were fertilized by the unsheathed peripheral males. In another study, 10 of 15 naturally breeding females were shown to produce clutches having 2 or more fathers. Thus, cooperation occurs not only in nest-building behaviors in this species but also in mating, which ensures a good mixing of genetic material in the following generation.

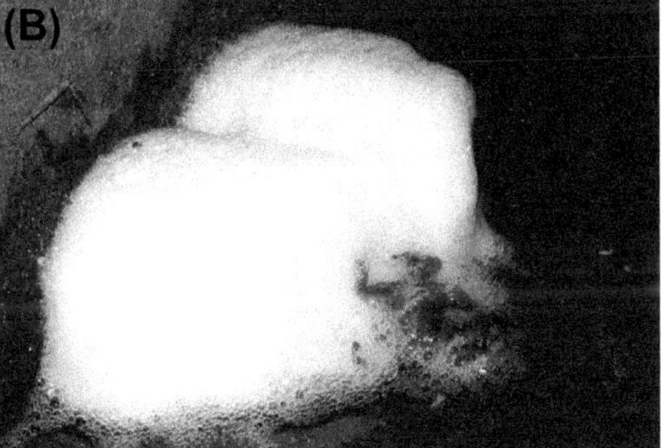

FIGURE 11-40 **Parental behaviors protecting young amphibians.** (A) A male glass frog, *Hyalinobatrachium fleischmanni*, moistens the developing eggs and protects them from predation by spiders (B) A foam nest produced on the pond surface by a breeding pair of Túngara frogs, *Engystomops (Physalaemus) pustulosus*, in which the fertilized eggs will develop protected from sunlight, dehydrations, predators, etc. *(Photographs courtesy of Jesse Delia.)*

F. Occurrence of Unisexual Species in Amphibia

The *Ambystoma jeffersonianum* complex consists of four salamander species found in the Great Lakes region of North America. Two of these species are normal bisexual, diploid ($n = 14$, $2n = 28$) species—*Ambystoma jeffersonianum* ("JEFF") and *Ambystoma laterale* ("LAT")—but two of these are unisexual, consisting of all triploid ($n = 14$, $3n = 42$) females: *Ambystoma platinium* ("PLAT") and *Ambystoma tremblayi* ("TREM"). The two $3n$ species are believed to have resulted from hybridization of the two $2n$ species where one of them produced diploid gametes by mitosis instead of meiosis (Figure 11-41). The $3n$ species are all female and produce $3n$ gametes via mitosis; however, these eggs will not develop unless penetrated by a sperm provided by one of the $2n$ species. At breeding time, the $3n$ females compete with the $2n$ females to get a male to provide sperm to activate their eggs, which then develop parthenogenetically into $3n$ females (the male gametes contribute no genetic material to the embryo). Interestingly, the LAT male prefers to mate with LAT females but his second preference is to mate with a TREM female who has two sets of LAT chromosomes and only one set of JEFF chromosomes. The LAT male's third choice is a PLAT $3n$ female with only one set of LAT chromosomes and two sets of JEFF chromosomes. Least favorite mating choice is a JEFF female. Similarly, The JEFF male shows the same hierarchy: JEFF female, then PLAT $3n$, then TREM $3n$, and finally LAT female.

VI. REPRODUCTION IN REPTILES

Living reptiles are members of diverse orders, and it is not surprising that considerable differences occur, making it difficult to generalize about reptilian reproduction. The squamate reptiles (lizards and snakes) possess features unique to their order, whereas the other orders may be more typical of primitive reptiles that gave rise to four extant reptilian groups (chelonians, crocodilians, squamates, and the tuatara) and also gave rise independently to birds and mammals. Hence, the study of reptilian reproduction is of special evolutionary interest. The reproductive system of a typical lizard is provided in Figure 11-42.

Most reptilian species are oviparous and exhibit well-defined annual reproductive cycles and breeding seasons. In addition, many examples of viviparity are known among snakes and lizards, and it is hypothesized that viviparity evolved 100 times in these groups. Only a few, heavily yolked eggs are produced by most reptilian species although clutch sizes may be relatively large in some turtles, crocodilians, and snakes. Fertilization is internal in all reptiles, and males have intromittent organs (often two hemipeni) for placing sperm into the cloaca of a female. Mating frequently follows complicated behaviors, including male–male territorial and aggressive interactions. Following mating, females of many species can store sperm in spermatothecae for months.

Crocodilians and turtles produce two distinct GTHs, but squamate reptiles rely on a single FSH-like hormone. The reptilian hypothalamus produces GnRH-1 that regulates

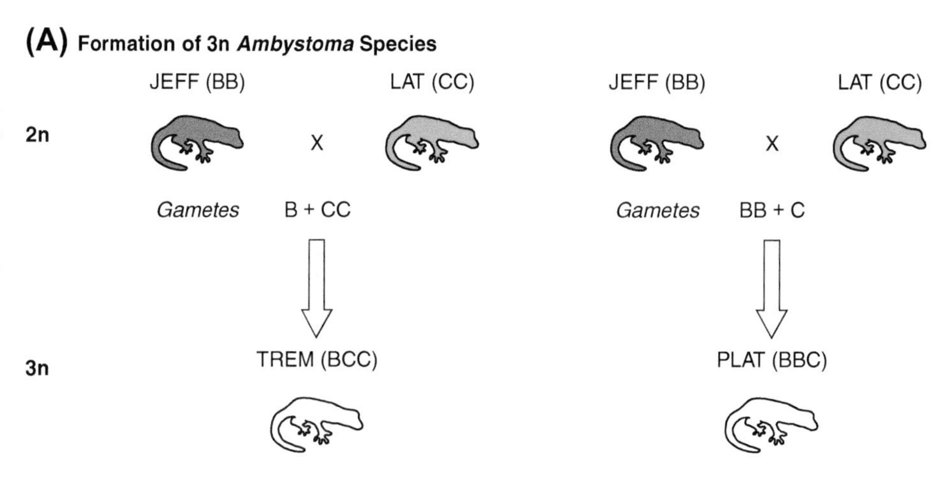

	First Choice	Second Choice	Third Choice	Fourth Choice
LAT Male (CC)	2n LAT female (CC)	3n TREM female (CCB)	3n PLAT female (CBB)	2n JEFF female (BB)
JEFF Male (BB)	2n JEFF female (BB)	3n PLAT female (BBC)	3n TREM female (BCC)	2n LAT female (CC)

FIGURE 11-41 **The *Ambystoma jeffersonianum* complex.** (A) Formation of the triploid involves hybridization between two diploid species where nondisjunction has occurred in either *A. laterale* (LAT) or *A. jeffersonian* (JEFF). The former results in formation of the $3n$ species *A. tremblayi* and the latter results in the $3n$ species *A. platinium*. (B) Mating preference for a diploid species is for females that have the closest genetic composition. Note that the chromosome sequences are changed in the triploid females to emphasize closeness to the male genotype.

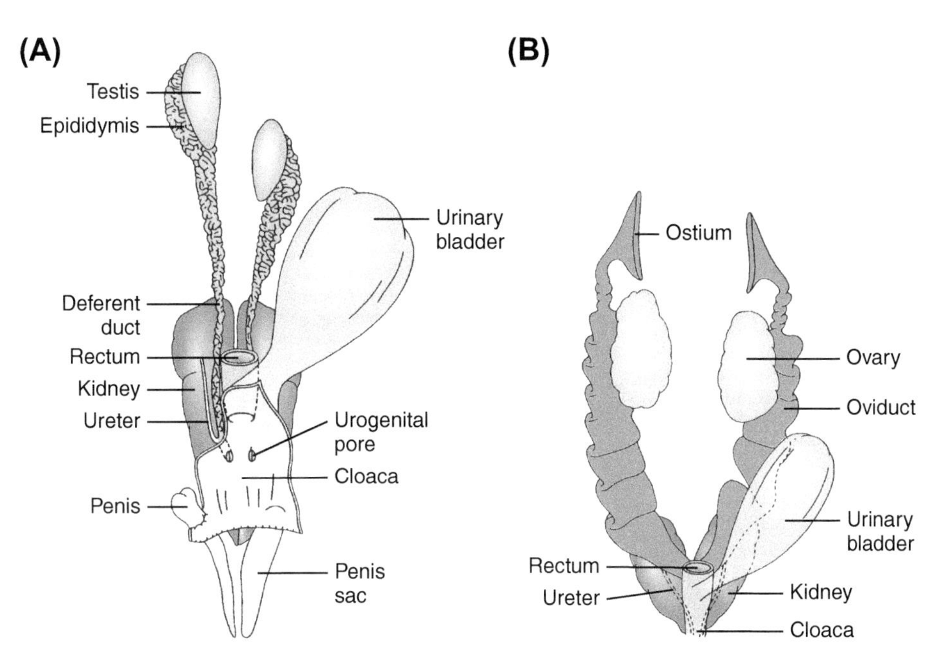

FIGURE 11-42 **Reproductive system of a lizard.** (A) Male. The top wall of the cloaca has been removed to show the duct openings and the two hemipeni. (B) Female. *(Adapted with permission from Matsumoto, A. and Ishii, S., "Atlas of Endocrine Organs: Vertebrates and Invertebrates," Springer-Verlag, Berlin, 1992.)*

GTH release in response to environmental stimuli such as photoperiod and temperature (see Chapter 5).

A. Sex Determination and Gonadal Development

Temperature-dependent sex determination (TSD) had been observed in all reptilian groups as well as in both XX/XY and ZZ/ZW types of sex determination. The same genes are involved in forming a testis in reptiles with TSD as were described for other vertebrates (see above) although there is some disagreement as to the exact activation sequence. Whether a higher temperature produces males or females varies considerably among different species (Table 11-20). Gonads typically go through an indifferent or bipotential stage prior to differentiation, and hormonal factors can influence which direction differentiation will follow. The development of sex accessory ducts (Figure 11-43) follows the same patterns as in other tetrapods. In males, the vas deferens develops from the wolffian ducts as described for amphibians. The vas deferens conducts no urine in reptiles, however, as the mesonephric kidney is completely replaced in reptiles by the evolution of the metanephric kidney with its own ureter connecting to the urinary bladder. As demonstrated in the American alligator, müllerian ducts degenerate in males prior to hatching as a consequence of AMH secreted by the testes.

B. Male Reptiles

The male reproductive system (Figure 11-42) is typical of amniotes, and the testes consist of convoluted seminiferous tubules (Figure 11-44A), each surrounded by a connective tissue sheath, the tunica propria. The entire testis is enclosed by a tunica albuginea. Spermatogenesis recurs soon after the breeding season and is completed in most species prior to the onset of winter. Pituitary GTHs stimulate spermatogenesis, and sperm may be stored in the vas deferens for up to several months prior to mating. Leydig cells (Figure 11-44A) and Sertoli cells are common, and they are both steroidogenic. After spermiation and testicular collapse, the Sertoli cells fill with cholesterol-positive lipid that is depleted under the influence of FSH at the time mitosis resumes in the spermatogonia. Leydig cells undergo cyclical changes associated with androgen secretion and sexual changes in androgen-dependent sex accessory structures. Representative plasma androgen levels are provided in Table 11-7. In species with associated reproductive patterns (typically crocodilians, squamates, and some turtles), there is typically a single peak of testosterone that coincides with spermatogenesis and development of the vasa deferentia. However, some species, such as the Northern Pacific rattlesnake *Crotalus oreganus*, may show two androgen peaks and breed in both the spring and fall of the year (Figure 11-45). In contrast, species exhibiting dissociated reproductive patterns (mostly turtles) also have two peaks of androgen secretion but neither is associated directly with spermatogenesis.

In sexually active squamates, a portion of the kidney tubules and possibly other portions of the excretory system develop into the **renal sexual segment** of the kidney (Figure 11-44B). Some suggest it may be homologous to the seminal vesicles of male mammals. The sexual segment undergoes hypertrophy under the influence of androgens

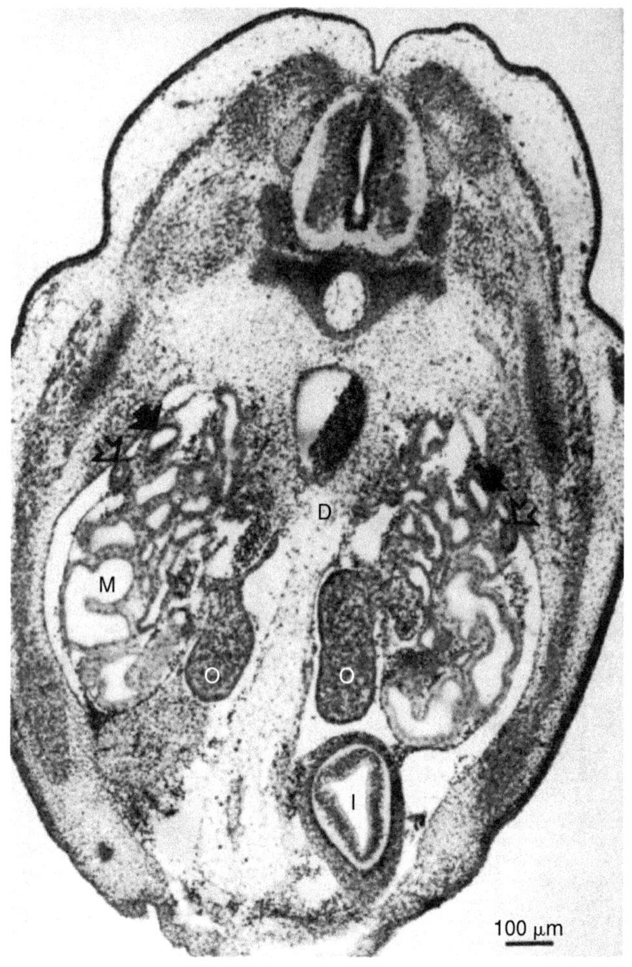

FIGURE 11-43 **Cross-section through an early embryo of the lizard *Sceloporus undulatus* prior to sexual differentiation.** Developing gonads (O), mesonephric kidney (M), müllerian ducts (open arrows), wolffian ducts (solid arrows), intestine (I), and dorsal mesentery (D). *(Reprinted with permission from Austin, H.,* General and Comparative Endocrinology, *72, 351–363, 1988.)*

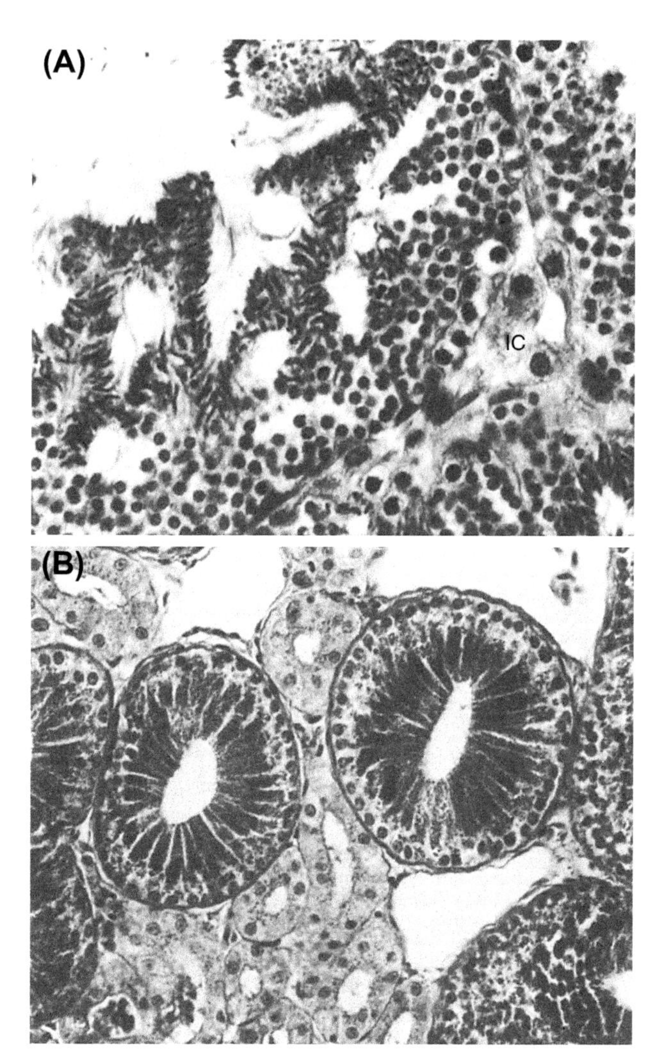

FIGURE 11-44 **Spermatogenesis and the sexual segment in scleoporine lizards .** (A) Spermatogenesis in the testis of *Scerloporus jarrovi*. Portions of three seminiferous tubules are separated by large Leydig cells (IC). (B) Section through kidney of *Sceloporus undulatus* showing small, lightly stained renal tubules and the large renal sexual segments modified for sperm storage. *(Photo courtesy of Dr. John Matter, Juanita College.)*

and appears to secrete materials that help maintain sperm stored in the vas deferens prior to ejaculation. The renal sexual segment is largest at the time of spermiation. Apparently, in some species it may contribute secretions to the formation of the **copulatory plug** produced to block the female cloaca after copulation and prevent entry of sperm from another male.

C. Female Reptiles

Reptiles have paired hollow ovaries with little stromal tissue. The reproductive system of a female lizard is provided in Figure 11-42. Ovarian cycles as exemplified by steroid hormone secretion for the American alligator and the North Pacific rattlesnake are provided in Figures 11-46 and 11-47, respectively. Oogonia are present in the mature ovary as described for anamniotes and give rise to primary oocytes and follicle growth throughout reproductive life. The developing oocyte becomes invested with granulosa cells derived from the germinal epithelium. The granulosa cells are separated from the surrounding thecal cells by a connective tissue layer, the **membrana propria**. In reptiles as in mammals, the theca is differentiated into an inner, glandular **theca interna** surrounded by a fibrous **theca externa**. The cells of the granulosa in most species are considered the primary source of follicular estrogen during ovarian recrudescence, although histochemical evidence in skinks (genus *Lamproholis*) suggests that only the thecal cells are steroidogenic in those lizards. Follicle cells exhibit changes in cholesterol-positive lipid inclusions and 3β-HSD activity paralleling estrogen-dependent oviductal growth and changes in other sex accessory structures. Changes in

TABLE 11-7 Plasma Steroid Levels (ng/mL) in Reptiles

Species	Testosterone	Estradiol	Progesterone
Turtles			
Chrysemys picta			
Male	15–40		
Preovulatory female	3.2–5.7	0.79–1.37	1.2–1.5
Preovulatory female	0.2	UND[a]	0.3–0.5
Stenotherus ordonuttus			
Female	0.025–1.5	0.5–5	0.7–4
Male	10–75		
Lizards			
Uromastix hardwicki			
Preovulatory female	0.37	0.18	1.6
Gravid female	1.57	0.46	13.41
Iguana iguana			
Male	0.1	0.079	
Female	0.003	0.27	
Snakes			
Natrix fasciata (female)	0.05–1.065	0.010–0.54	0.090–1.445
Naja naja			
Male	0.6–2.3		
Female	0.03–0.7	10–310	1.4–25
Nerodia sipedon (male)	2–21		

[a]*UND, undetectable.*

the gonadotropes of the adenohypophysis parallel follicular changes indicating an important role for gonadotropin. As the oocyte enlarges, it begins to project into the ovarian cavity and out from the surface of the ovary.

The squamate granulosa contains a unique flask-shaped cell type, the **pyriform cell** (Figure 11-48) that is in direct contact with the developing oocyte. These cells apparently are involved with early steps in oocyte development as they either degenerate or transform into typical granulosa cells soon after the onset of vitellogenesis. Thus, the multilayered granulosa becomes a single layer of cells in mature follicles.

As ovulation approaches, the granulosa cells as well as some thecal cells accumulate cholesterol-positive lipids and, following ovulation, proliferate and luteinize to form corpora lutea. These corpora lutea are well vascularized, exhibit 3β-HSD activity, and synthesize progesterone. They persist throughout egg laying in oviparous species or during gestation in most viviparous forms. Corpora lutea of viviparous species synthesize greater amounts of progesterone than do those of oviparous species and persist much longer. Plasma progesterone levels are greatest following ovulation and are maintained at elevated levels throughout gestation in most viviparous lizards and snakes. Preovulatory peaks of progesterone are found in oviparous turtles, crocodilians, and lizards (Table 11-8).

Only a few of the follicles that begin development reach maturity at a given time; the majority undergo atresia. Follicular atresia and formation of corpora atretica is a common occurrence in reptilian ovaries as in other vertebrates. The importance of corpora atretica is unknown, but the absence of 3β-HSD activity in corpora atretica makes it unlikely they would have an endocrine function. However, steroidogenic cells from atretic follicles may give rise to an "interstitial gland" that is believed to be a major source of ovarian estrogens.

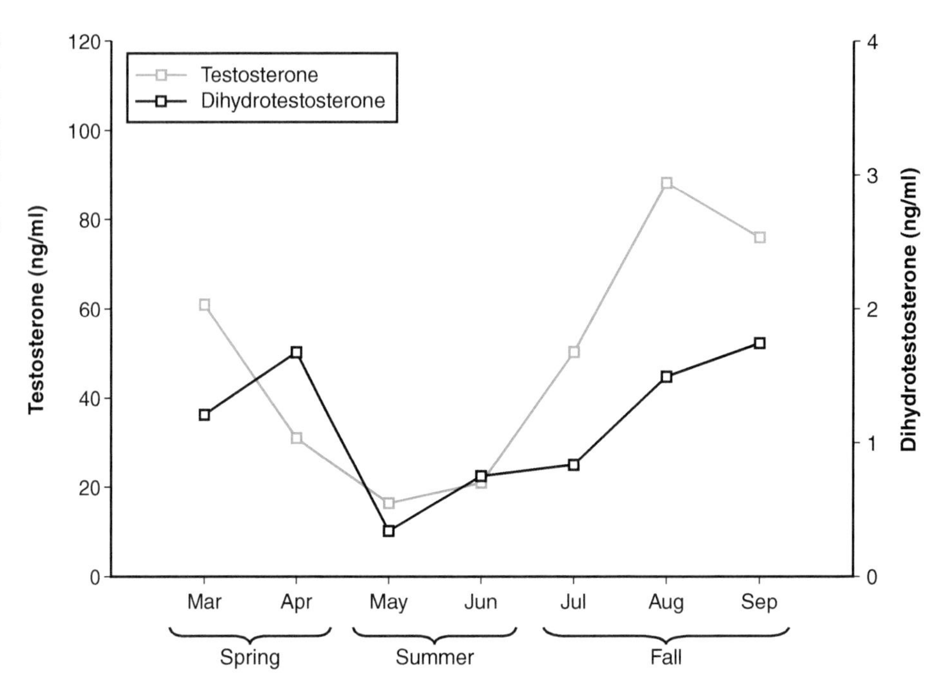

FIGURE 11-45 **Seasonal androgen levels in male northern Pacific rattlesnakes (*Crotalus oreganus*).** Testosterone is the predominant androgen. Males may mate with females either in the spring or fall when androgen levels are highest. *(Adapted with permission from Lind, C.M. et al.,* General and Comparative Endocrinology, ***166****, 590–599, 2010.)*

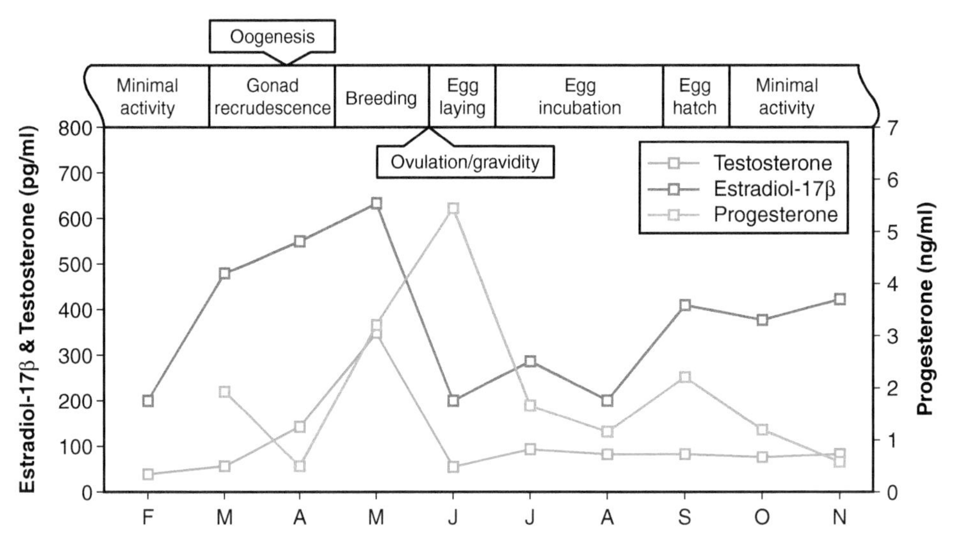

FIGURE 11-46 **Seasonal reproductive cycle for female *Alligator mississippiensis*.** *(Adapted with permission from Milnes, M.R., in "Hormones and Reproduction of Vertebrates. Vol. 3. Reptiles" (D.O. Norris and K.H. Lopez, Eds.), Academic Press, San Diego, CA, 2011, pp. 305–319.)*

Ovaries of reptiles show different patterns of follicular maturation and ovulation. Some produce several eggs simultaneously from each ovary (most reptiles). Others may alternate production of a single egg from each ovary (anoline lizards). Differences in follicular atresia rather than in the number of oocytes beginning development may be responsible for these patterns. Still others (e.g., certain turtles) may produce most of their eggs in one ovary during one season and most from the other ovary the next season.

Exogenous GnRH produces direct actions on ovarian follicles that may reflect a paracrine role for ovarian GnRH. Treatment of lizards (*Podacris sicula sicula*) with salmon GnRH (sGnRH) increases secretion of prostaglandin ($PGF_{2\alpha}$) from mid- to late follicles and early corpora lutea and increases secretion of progesterone from follicles. Antagonists of GnRH produce opposite results. The physiological relevance of these observations is open to interpretation, but it does suggest that paracrine effects of GnRH are not limited to mammals.

Alternation of ovulation in the ovaries of the anoline lizard *Anolis carolinensis* has been investigated extensively. There is a definite alteration in catecholamine activity in the hypothalamus that mirrors ovarian alternation. These

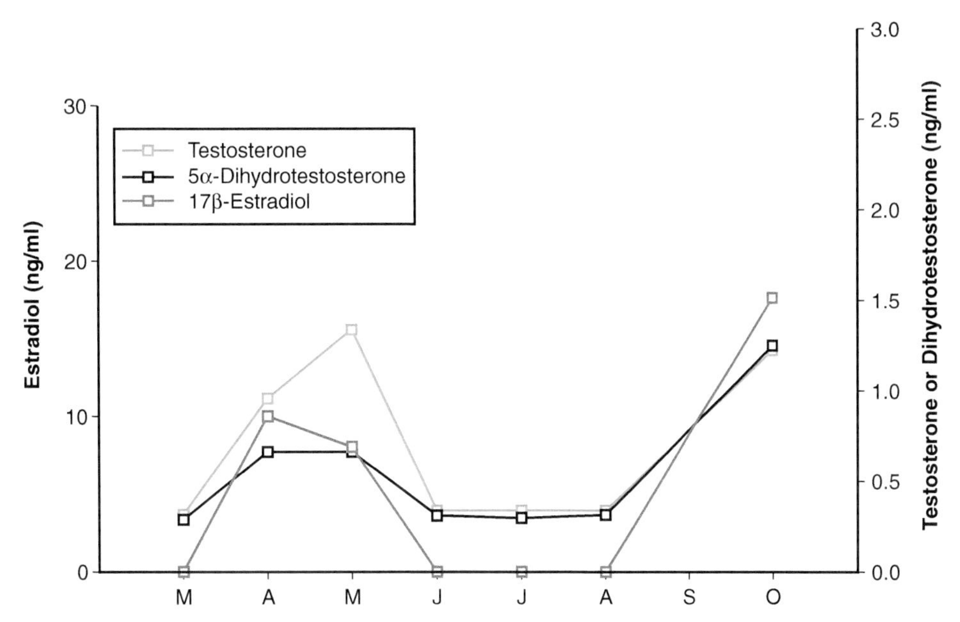

FIGURE 11-47 **Seasonal steroid levels in female northern Pacific rattlesnakes (*Crotalus oreganus*).** *(Adapted with permission from Lind, C.M. et al.,* General and Comparative Endocrinology, *166, 590–599, 2010.)*

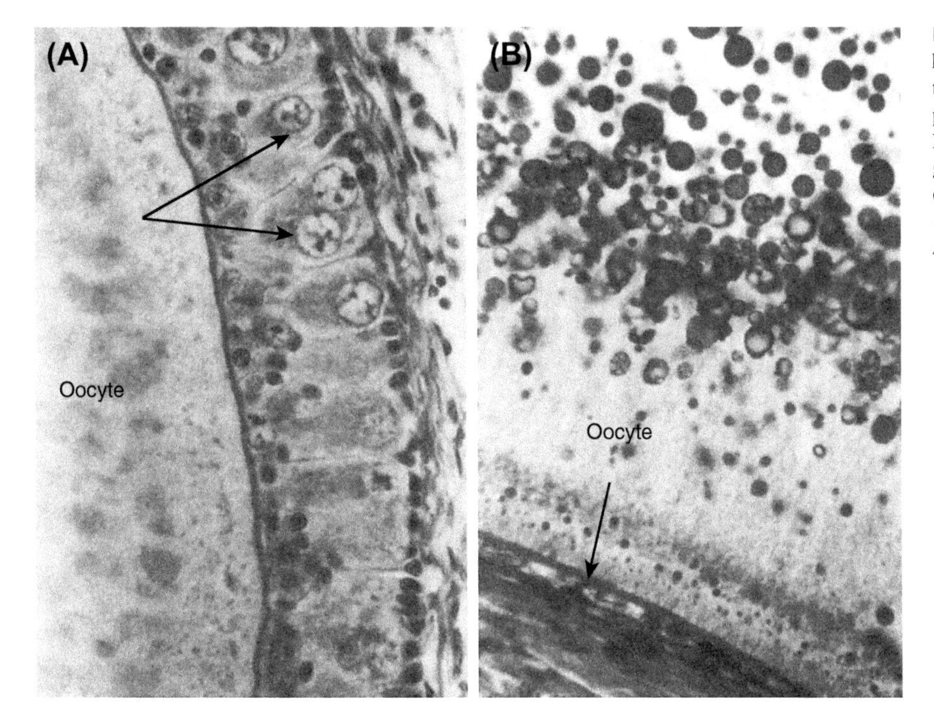

FIGURE 11-48 **Ovary of the iguanid lizard *Ctenosaura pectinata*.** (A) Section through a perivitellogenic follicle with large pyriform cells (arrows) in the granulosa layer. (B) Lower magnification of a vitellogenic follicle. Note the flattened granulosa (arrow). *(Courtesy of Dr. Mari Carmen Uribe, Facultad de Ciencias, UNAM, Mexico.)*

observations may be explained by sensory neural connections between the ovaries and brainstem areas projecting to the hypothalamus that are transmitting information responsible for regulating the alternating pattern of ovulation.

Reptilian oviducts are differentiated into several regions. Squamate oviducts have only three distinct regions: a thin-walled region, the **infundibulum**; a muscular **uterus** with many glands; and a **vagina** connecting the uterus to the cloaca. The uterine glands are responsible for first secreting a fibrous layer around the egg and a surrounding shell that is calcified in oviparous species. The oviducts of crocodilians and turtles, species that generally make much larger eggs than squamates, have two additional regions between the infundibulum and the uterus. The major region is the glandular **uterine tube** that secretes the egg white and may

TABLE 11-8 Circulating Progesterone Levels (ng/mL) during Reproductive Cycles of Turtles, Lizards, and Snakes

Species	Period of early follicle growth	Preovulatory stage	Early postovulatory	Mid-pregnancy	Late pregnancy
Turtles					
Chrysemys picta (oviparous)	0.2 ± 0.06	5.0 ± 1.02	0.5 ± 0.01	—	—
Chelonia mydas (oviparous)	0.2 ± 0.08	1.8 ± 0.13	0.7 ± 0.88	—	—
Lizards					
Sceloporus cyanogenys	0.7 ± 0.15	0.9 ± 0.38	3.3 ± 0.48	—	3.5 ± 0.34
Chamaelo pumillis	0.9	1.0 ± 0.71	5.0 ± 3.90	2.3 ± 0.34	—
Snakes					
Natrix taxispilota	0.4 ± 0.04	0.9 ± 0.08	1.9 ± 0.24	—	—
Nerodia sipedon	1.3 ± 0.19	3.9 ± 0.83	5.0 ± 1.41	6.9 ± 0.78	2.8 ± 0.44
Thamnophis elegans	—	—	1.7 ± 0.30	6.2 ± 1.00	—

Adapted with permission from Lance, V. and Callard, I.P., *General and Comparative Endocrinology*, **35**, 295–301, 1978.

constitute 40% of the oviduct in these oviparous species. A short isthmus connects the uterine tube to the uterus where the shell is secreted. This arrangement is very similar to the oviducts of birds.

Oviductal development apparently is under the influence of ovarian estrogens, and progesterone is without effect. In oviparous species, estrogens probably influence the secretion around the egg of albumin and shell from the anterior end of each oviduct. Estrogens also stimulate synthesis of vitellogenic proteins by the liver necessary for oocyte growth and cause increases in serum calcium of oviparous snakes, lizards, and turtles that contributes to shell formation (Table 11-9). Crocodilians produce yolk proteins that are biochemically similar to those of birds.

Oviposition or birth of live young is controlled by AVT, prostaglandins, and β-adrenergic innervation in turtles, lizards, and snakes. In the American chameleon (*A. carolinensis*), the sensitivity of the uterus to AVT is determined by the presence or absence of a corpus luteum in the adjacent ovary.

TABLE 11-9 Effect of Estradiol on Serum Calcium Levels of Ovariectomized Female Lizards (*Anolis carolinesis*)[a]

Treatment	*N*	Serum calcium (mg/dL ± SE)
Ovariectomized saline-injected females	5	14.8 ± 1.05
Reproductively active, sham-operated females	4	21.2 ± 3.47
Ovariectomized females injected with 1.0 μg of estradiol per day for 7 days	5	213.2 ± 9.88
Ovariectomized females injected with 10 μg of estradiol per day for 7 days	7	256.0 ± 22.13

[a]*Unpublished data of K. Faber and D. Norris.*

D. Environment, Behavior, and Reproduction in Reptiles

The role that physical and biological components of the environment play in sexual behavior and reproduction has been extensively studied in reptiles. Species living in temperate climates exhibit distinct seasonal patterns of hormonal secretion and reproductive events. Reproduction in tropical reptilian species varies from cyclic patterns to continuous breeding (see Figure 11-49). There is a strong tendency for an observed increase in the incidence of viviparity among species inhabiting colder climates (altitude or latitude), but it is not clear which is cause or consequence.

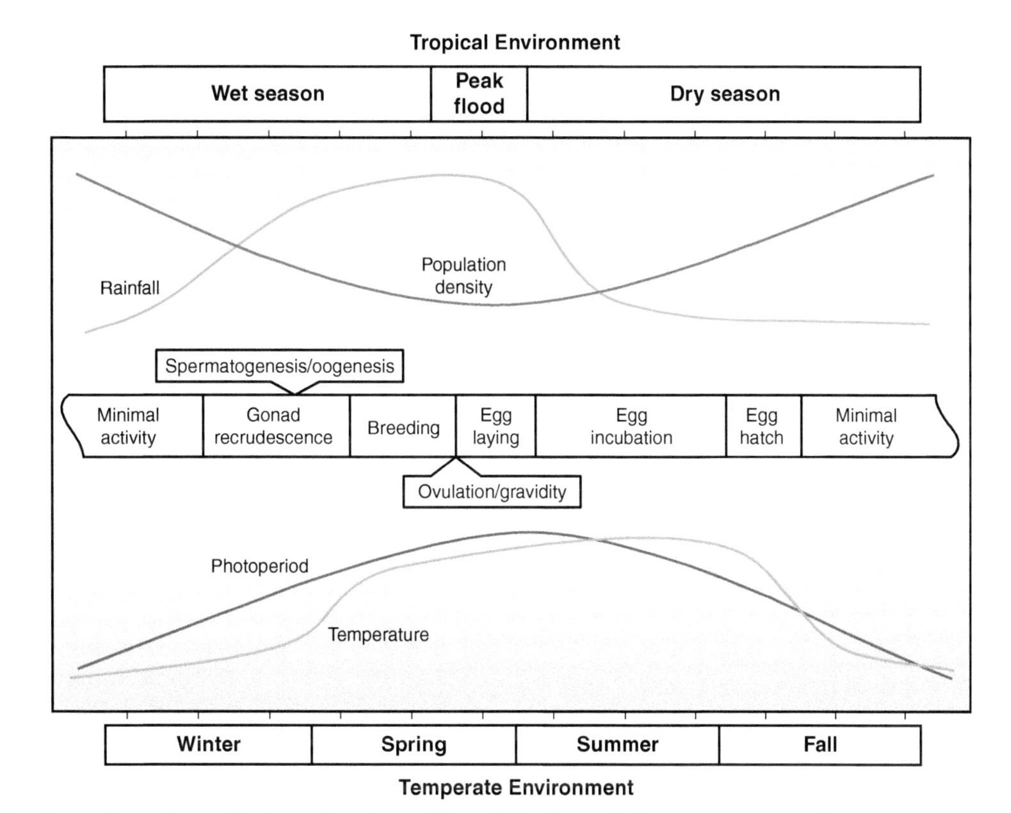

FIGURE 11-49 **Comparison of reproductive cycles of crocodilians living in tropical and temperate environments.** In tropical situations, population density increase during the dry season inhibits reproduction. Gonadal recrudescence occurs during the wet season. Temperate species are controlled by temperature and photoperiod. *(Adapted with permission from Milnes, M.R., in "Hormones and Reproduction of Vertebrates. Vol. 3. Reptiles" (D.O. Norris and K.H. Lopez, Eds.), Academic Press, San Diego, CA, 2011, pp. 305–319.)*

Among temperate lizards, temperature is the dominant environmental factor influencing reproduction. Photoperiod, humidity, and nutritional status play decisive roles in some species. Other groups of reptiles have not been studied as extensively as lizards.

Although visual cues are the primary mechanism employed in reptilian courtship, evidence for pheromonal communication can be inferred from some experimental studies in all major reptilian groups. Male lizards of several families (Scincidae, Lacertidae, Teidae, Gekkonidae) have androgen-dependent **femoral glands** located on the inner thighs as well as special cloacal glands that seem to play important roles in courtship and territorial behavior in association with breeding. Female garter snakes, *Thamnophis sirtalis parietalis*, produce pheromones that are exuded from the skin and elicit courtship by males. Cloacal secretions by brown tree snake (*Boiga irregularis*) females apparently prevent courtship by unwanted suitors. Inguinal and axillary glands of chelonians have been implicated in reproductive behavior, too. Crocodilians appear to use chemical communication in courtship, but few details are available.

When a female red-sided garter snake (*Thamnophis sirtalis*) emerges from its winter hibernaculum, she is immediately courted by a large number of males who had emerged previously. This behavior produces a mating ball of males, all attempting to copulate with a single female. The skin of the reproductive female produces methyl ketone that apparently in the presence of estrogens attracts the males and signals that she is ready to mate. As soon as one male successfully copulates with the female, she secretes another semiochemical that immediately turns off male mating behavior. Although injection of estrogens into adult males does not make them attractive to other males, some males with high testosterone and aromatase levels also are attractive to normal males, presumably because of the conversion of testosterone to estrogens. These "she-males" are more successful in achieving copulation with females than are normal males, presumably because the "she-males" confuse normal males who attempt to mate with them rather than with the true female, thus reducing their competition. Interestingly, males that are courting have lower corticosterone levels than males that are dispersing from the breeding area. Furthermore, the mating males show no stress response.

One model for studying reptilian behavior is exemplified by the work of David Crews and his collaborators at the University of Texas. In general, estrogen and progesterone are responsible in reptiles as in most vertebrates for stimulating female receptive and mating behaviors, whereas androgens, typically testosterone, control male behaviors. As previously mentioned, in a number of cases, androgens may be converted to estrogens by aromatase in order to produce behavioral effects. Among the species of whiptail lizards in the genus *Cnemidophorus*, about one-third are unisexual. These species consist only of females that are further unusual in that these females are all triploid (3*n*). Studies by Crews have employed one 3*n* species, the desert-grasslands whiptail (*C. uniparens*), which apparently evolved from hybridization of two 2*n* species, the rusty rump whiptail (*C. burti*) and the little striped whiptail (*C. inornatus*). Because *C. uniparens* has two sets of chromosomes derived from *C. inornatus* and only one from *C. burti*, Crews focused his behavioral studies on *C. uniparens* and *C. inornatus* (which exhibits normal sexual reproduction). In *C. uniparens*, mating does not occur with males of diploid species, but rather the 3*n* females alternate between expressing female (receptive) and male (mounting) mating behaviors (Figure 11-50). Although estradiol levels in the 3*n* females is five times less than in 2*n* females, this results in higher hypothalamic levels of estrogen receptor mRNA in the 3*n* brains and hence greater sensitivity to low plasma levels of estradiol. The postovulatory decrease in estrogen allows for the appearance of male-like mounting behavior or **pseudocopulation** although the exact chemical stimulus is not clear.

E. Parental Behavior in Reptiles

Reptiles classically have been characterized by a lack of parental behavior and, in some cases, even lack of recognition of their offspring. In recent years, however, studies have demonstrated that there is considerable investment in parental care even among oviparous species. Although members of the oldest extant group, the chelonians, typically abandon their nests once the eggs are laid, many squamates exhibit parental behavior. Crocodilian parents participate in the hatching process and in protecting the young. Evidence of nest building and parental care has been unearthed for some extinct dinosaurs, as well. Thus, complex parental care did not appear *de novo* in birds and mammals but already had evolved in early reptiles.

VII. REPRODUCTION IN BIRDS

Knowledge of the reproductive endocrinology of birds is based largely on the intense investigations of only a few domestic (e.g., chicken, Japanese quail, emu, turkey, ostrich) and wild migratory (e.g., white-crowned sparrow) species with limited information from numerous other species. The HPG axis of birds appears to be controlled by both the production of GnRH-1 in the PVN and by a GRIH neuropeptide from the same nucleus in the hypothalamus

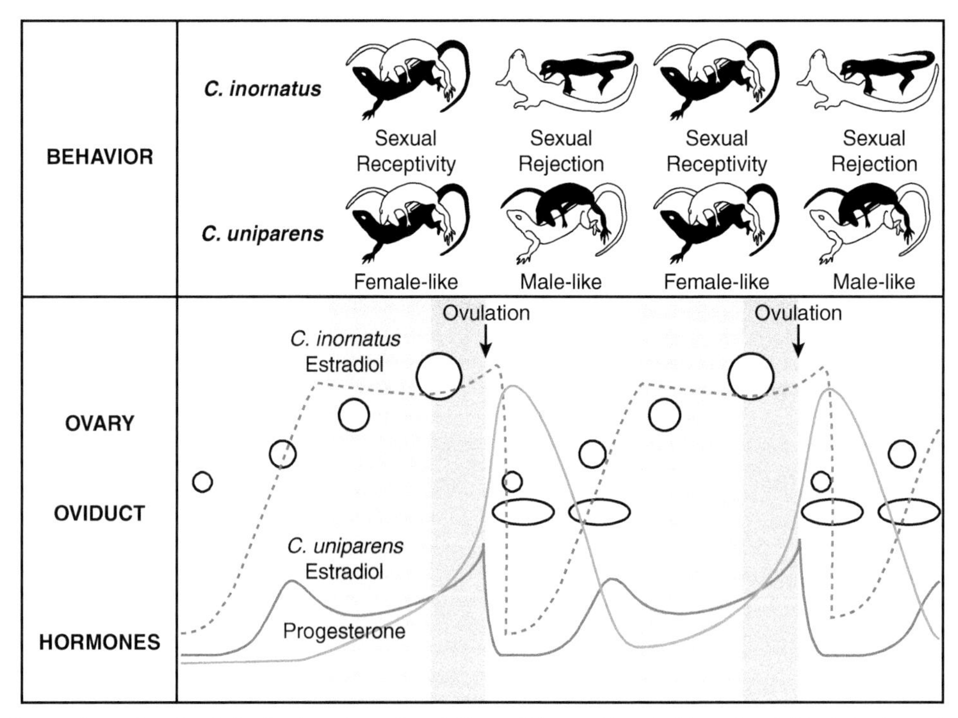

FIGURE 11-50 **Comparison of mating behavior in unisexual and bisexual lizards.** Female-like and male-like behavior in the all female desert-grasslands whiptail lizard (*Cnemidophorus uniparens*) is compared with female receptive behavior in a bisexual lizard, the little striped whiptail (*C. inornatus*). The differences in estradiol levels for the two species are illustrated (red lines). Female-like behavior in *C. uniparens* is elicited by a lower estrogen level and is followed by male-like copulatory behavior. The circles represent size of the ovarian follicles and the ovals indicate presence of eggs in the oviduct. *(Adapted with permission of the publisher from Young, L.J. and Crews, D.,* Trends in Endocrinology & Metabolism, ***6**, 317–323, 1995.)*

controlling secretion of LH and FSH from the pituitary (see Chapter 5). Secretion of GRIH appears to be regulated by melatonin from the pineal gland. These dual factors regulating reproduction in migratory birds are controlled by photoperiodic information.

Avian reproductive organs (Figure 11-51) reflect the anatomical adaptations to flight that characterize most bird systems. In females of most species, only the left ovary and its attendant oviduct develop, whereas the right-hand components remain in a rudimentary state. Should the left ovary be removed surgically or destroyed by disease, the right rudiment may develop, but it will usually form an ovotestis or a testis rather than another ovary. This left–right asymmetry is reflected in the male where the left testis usually is larger than the right although both are functional.

Avian gonads develop from a pair of undifferentiated primordia associated with the embryonic nephrotome. These primordia are invaded by primordial germ cells that migrate through blood vessels to reach the developing gonads, where they develop into the germinal epithelium. As in other vertebrates, the embryonic gonad goes through a bipotential state in which both cortical and medullary components are present. Differentiation of cortical tissue is necessary for ovarian development, and the medullary portion is suppressed. The reverse condition prevails in male birds. In contrast to mammals it is the male bird that is the homogametic sex (ZZ), and it is the female that has unlike sex chromosomes (ZW). Developing a female phenotype requires estrogens, and castration of a young female may result in development of male plumage.

Development or suppression of the müllerian and wolffian ducts eventually depends on the direction of gonadal development as it does in amphibians and reptiles. In females in which only the left half of the reproductive system usually develops, the left ovary receives the larger proportion of germ cells that migrate to the gonads. The mechanism behind this disproportionate distribution of germ cells is not known. Degeneration of the müllerian duct occurs only on the side of the smaller ovary and is induced by AMH produced in ovarian cells. Local secretion of estrogens by the larger ovary is believed responsible for preventing oviduct degeneration on that side.

All birds are oviparous but display significantly more parental behavior than any non-mammalian group of vertebrates. No placenta has evolved in birds although they are common in reptiles; however, studies of the **chorioallantoic membrane (CAM)** of birds indicate this membrane has the capacity to synthesize progesterone *in vitro* if given the appropriate precursor. This membrane gives rise to placentas in many viviparous reptiles and in mammals. Birds have evolved other mechanisms to provide extended care to their progeny (see ahead).

Birds are endothermic like mammals and use body heat to support development of the embryo within the egg much as the mammal does *in utero*; consequently, birds can breed successfully under conditions that are too cold for their reptilian relatives. The adaptation of long-distance flight coupled with high body temperatures allows utilization of polar and subpolar regions where winter conditions are too severe for survival but where the summer months provide adequate warmth and food to breed and rear young birds to a size sufficient for successful fall migration to warmer latitudes for winter.

Reproduction is decisively cyclic in adult birds and is closely attuned to environmental factors, especially temperature and photoperiod (Figure 11-52). Olfactory systems are minimally developed in most birds, and

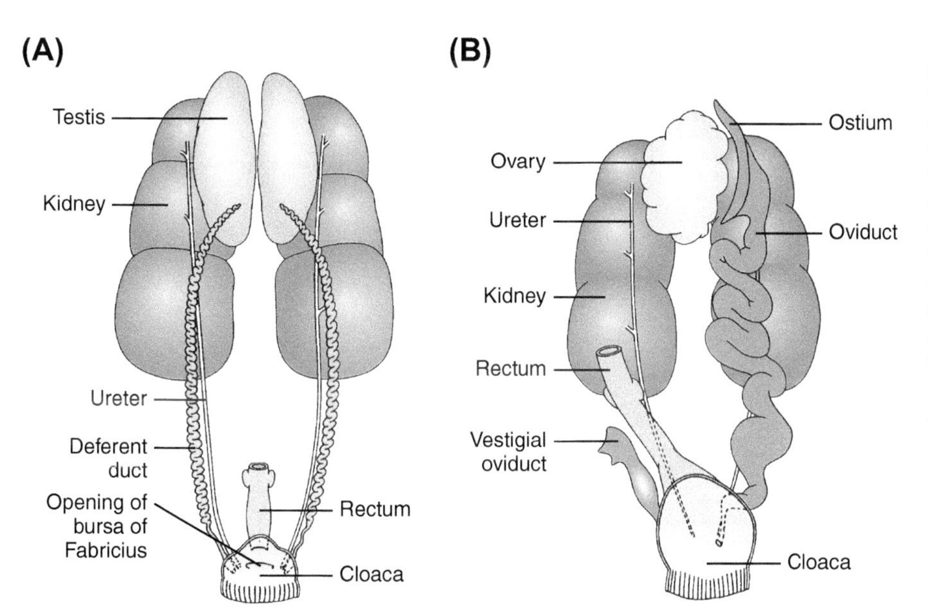

FIGURE 11-51 **Reproductive organs of the pigeon.** The top of the cloaca has been removed in both sexes to illustrate connection of the gonaducts. (A) Male. (B) Female. Note that the right ovary is absent in most birds (the left in others) as well as the corresponding oviduct. This regression is due to production of AMH in the embryo by the remaining ovary which secretes estradiol locally and protects the müllerian duct on that side from AMH. *(Adapted with permission from Matsumoto, A. and Ishii, S., "Atlas of Endocrine Organs: Vertebrates and Invertebrates," Springer-Verlag, Berlin, 1992.)*

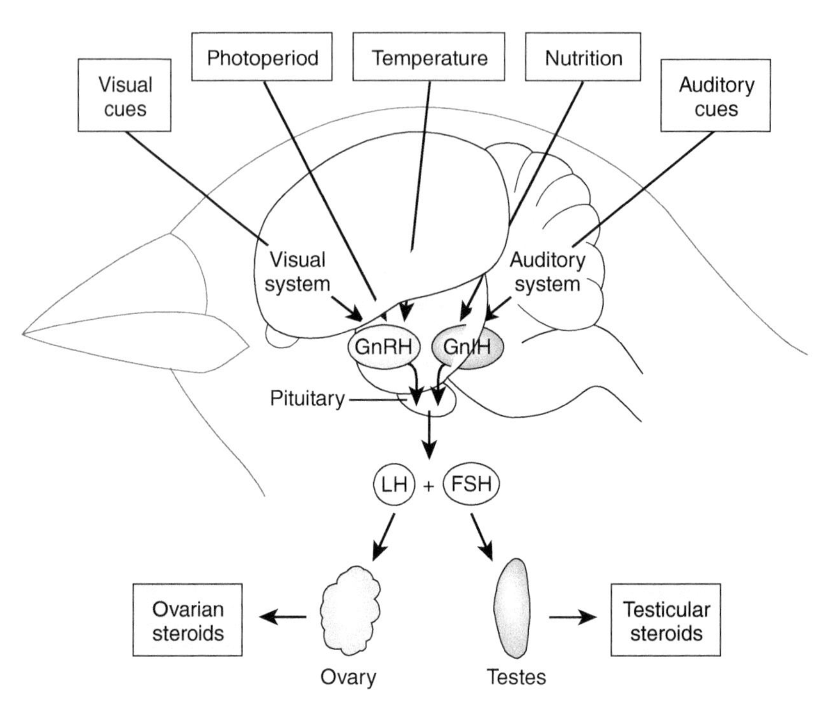

FIGURE 11-52 **Environmental and endocrine factors controlling reproduction in birds.** Environmental factors influence secretion of GnRH or GnIH from the hypothalamus that control release of gonadotropins (LH and FSH) from the adenohypophysis (Pit). The gonadotropins stimulate secretion of estrogens and other hormones by the ovary (OV) and androgens and other hormones by the testes (T). *(Adapted with permission from Ramenofsky, M., in "Hormones and Reproduction of Vertebrates. Vol. 4. Birds" (D.O. Norris and K.H. Lopez, Eds.), Academic Press, San Diego, CA, 2011, pp. 205–237.)*

pheromones do not appear to be important in regulating reproduction as compared to all other terrestrial vertebrates. Both migratory and non-migratory species exhibit seasonal cycles, with breeding occurring in the spring and sometimes continuing through much of the summer (Figure 11-53). On the other hand, species occupying arid regions may show irregular cycles cued to the availability of water, which may not occur with any seasonal regularity.

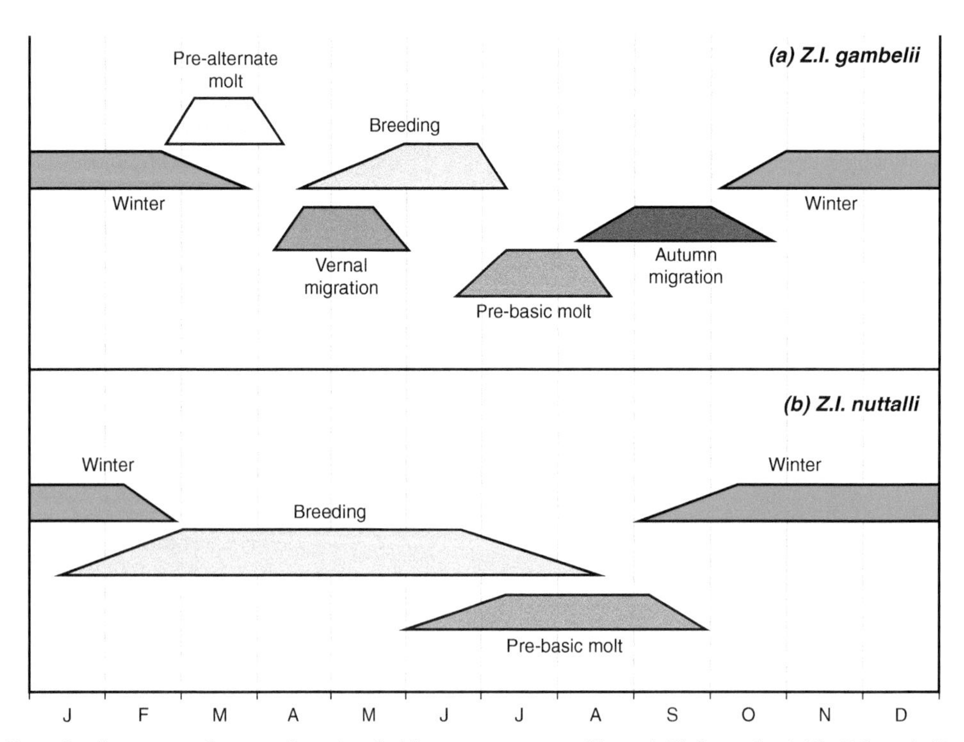

FIGURE 11-53 **Reproductive patterns for two subspecies of white-crown sparrows (*Zonotrichia leucophrys*).** The *Z. l. gambelii* subspecies migrates great distances to breed in northern habitats where winter survival is not a possibility. The sedentary subspecies *Z. l. nutalli* exhibits the pattern typical of temperate non-migratory species. Thickness of the lines reflects intensity of activity. *(Adapted with permission from Ramenofsky, M., in "Hormones and Reproduction of Vertebrates. Vol. 4. Birds" (D.O. Norris and K.H. Lopez, Eds.), Academic Press, San Diego, CA, 2011, pp. 205–237.)*

One American species, the rufous-collared sparrow (*Zonotrichia capensis*), has been reported to breed every 6 months regardless of rainfall as long as food is available.

Generally, ovaries and testes remain small in non-breeding birds and undergo tremendous hypertrophy in a very short time at the onset of the breeding season. This is especially advantageous for migratory species or species relying on particular stimuli for breeding where gonadal recrudescence can await arrival on the breeding grounds or appearance of suitable conditions such as abundant food.

A. Male Birds

Each testis consists of a mass of convoluted seminiferous tubules lined with a germinal epithelium and surrounded by connective tissue. Both developing germ cells and steroidogenic Sertoli cells as well as numerous fibroblasts can be seen in the germinal epithelium. As in other vertebrates the cytoplasm of the Sertoli cells completely envelops the germ cells. Typical steroidogenic interstitial cells occur between the seminiferous tubules as seen in reptiles and mammals.

In non-breeding birds, testes are very small, and histologically these quiescent testes appear to be composed largely of interstitial cells due to a marked post-breeding regression of spermatogenetic tissue. The onset of spermatogenesis (recrudescence) results in a rapid and marked increase in testicular size. Such rapid and extreme growth (to as much as 500 times the resting gonad weight in some species) results in considerable strain and damage to the tunica albuginea surrounding the testis, and it must be replaced each year during the postnuptial phase of the testicular cycle. Replacement is accomplished through differentiation of fibroblasts and formation of a new tunica directly beneath the damaged one. It is often possible to distinguish histologically between juvenile birds and postnuptial birds by the presence of two connective tissue capsules around the testis in the latter.

Testicular recrudescence may involve a single synchronous spermatogenetic event or separate spermatogenetic waves, depending on whether a given species produces successive clutches during a particular breeding season. In either event, following spermiation, sperm migrate to expanded distal ends of the vasa deferentia known as **seminal sacs** from which sperm will be ejaculated forcefully during mating.

1. The Avian Testicular Cycle

The annual testicular cycle of temperate birds has three more or less distinct phases: (1) regeneration or **preparatory phase**, (2) acceleration or **progressive phase**, and (3) **culmination phase**. Similar phases can be identified in all birds regardless of the seasonal nature of their reproductive cycles or what environmental factors control testicular events. The most common environmental factor influencing development of the avian testis is **photoperiod**. Placing quiescent, temperate birds, such as the ostrich (*Struthio camelus*) on long day photoperiods will typically stimulate testicular recrudescence, whereas maintenance of these birds on short-day photoperiods even into the normal breeding season represses anticipated testicular events. In contrast, some birds, such as the emu (*Dromaius novaehollandiae*), are short-day breeders with peak testicular function occurring during the winter months and quiescence being maintained under long-day photoperiods.

The preparatory phase of testicular development begins immediately after the reproductive period and is characterized by marked collapse of the testis. Animals in the preparatory phase are insensitive to effects of long photoperiod and are termed **photorefractory**. The end of the preparatory phase is heralded by restoration of **photosensitivity** that characterizes the following two phases. The endocrine basis for the photorefractory period in birds is not clear and more than one mechanism may be involved in different species. Some studies suggest that feedback of testosterone on the hypothalamus is responsible for induction of the photorefractory period and for lower levels of LH during the photorefractory period. Other investigations point to changes in hypothalamic sensitivity and/or steroid metabolism and not testosterone feedback. An intriguing hypothesis proposed by Peter Sharp involves the inhibitory role of PRL that determines GnRH secretion in both long-day and short-day breeders.

During the progressive phase, there is an increase in GTH secretion brought about by actions of lengthening photoperiod on the HPG axis. Increased circulating GTHs stimulate both spermatogenesis and androgen secretion by the interstitial cells. An increasingly intensive period of sexual activity (often including singing) occurs, and males of some species may begin exhibiting territorial behavior and mate selection. This effect of long photoperiod can be blocked by low temperatures.

The culmination phase coincides with the time of ovulation in females and includes the time of insemination. The male typically is ready for breeding before the female, and his testes will be bulging with sperm. Successful breeding involves a complex, hormonally dependent series of events involving precise male-female behavioral interactions.

2. Leydig Cells in Birds

A characteristic lipid cycle occurs in avian Leydig cells similar to that described for other vertebrates. There is accumulation of lipid in young birds followed by rapid depletion coincident with onset of the first breeding season and spermatogenesis. The Leydig cells of adult birds are small and sparsely lipoidal in winter, although they occupy a large proportion of the testis because of the regressed nature of the seminiferous tubules. There is a gradual accumulation of lipids, including cholesterol, throughout the progressive phase as well as an increase in 3β-HSD activity. At the time of maximal sexual display, there is rapid depletion of Leydig cell lipid. Cholesterol disappears completely, but 3β-HSD activity remains strong, indicating that lipid depletion is a consequence of rapid synthesis and secretion of androgens. The activity of 17α-hydroxylase is also high at this time (see Chapter 3 for its specific role in androgen synthesis). A massive disintegration of interstitial cells occurs during the preparatory phase, and new Leydig cells differentiate from fibroblasts.

3. Sertoli Cells in Birds

Cyclical changes in lipid content are characteristic of avian Sertoli cells that ultrastructurally resemble steroidogenic cells. Both 3β-HSD and 17β-HSD activities have been reported for these cells. They become densely lipoidal following the breeding season, and no detectable 3β-HSD activity remains. The stored lipid is depleted with the onset of the next period of spermatogenesis.

4. Endocrine Control of Testicular Function

The hypothalamus controls release of LH and FSH from the adenohypophysis through secretion of GnRH or GnIH (Figure 11-52). LH causes hyperplasia of Leydig cells, which become lipoidal and exhibit increased 3β-HSD activity. Androgens secreted by Leydig cells stimulate sex accessory structures and secondary sexual characters. Mitotic divisions of spermatogonia are stimulated by FSH, and local effects of androgens from Sertoli cells are responsible for stimulating meiosis. Once initiated, androgens can maintain spermatogenesis in hypophysectomized birds. The patterns of LH and androgen secretion are illustrated for ostrich (Figure 11-54) and king penguins (Figure 11-55). Testosterone levels for black kites are provided in Figure 11-56. Prolactin is present in the male pituitary and has been reported to inhibit FSH release and block spermatogenesis in some species. The formation of incubation patches (see ahead) on males of certain species

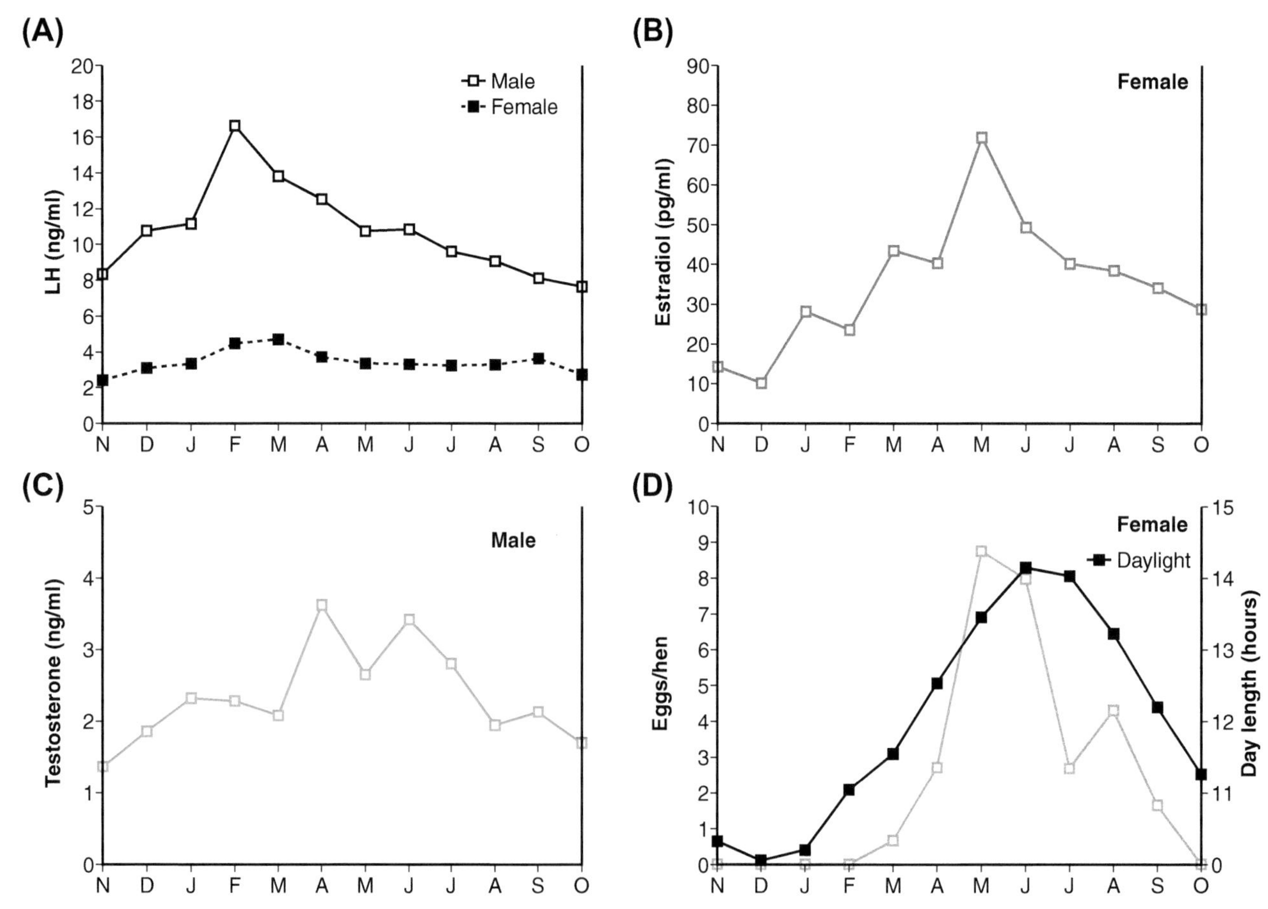

FIGURE 11-54 **Plasma LH, plasma estradiol, and egg production (light bars) in female South African ostriches are affected by photoperiod.** *(Adapted with permission from Degen, A.A. et al.,* General and Comparative Endocrinology, ***93****, 21–27, 1994.)*

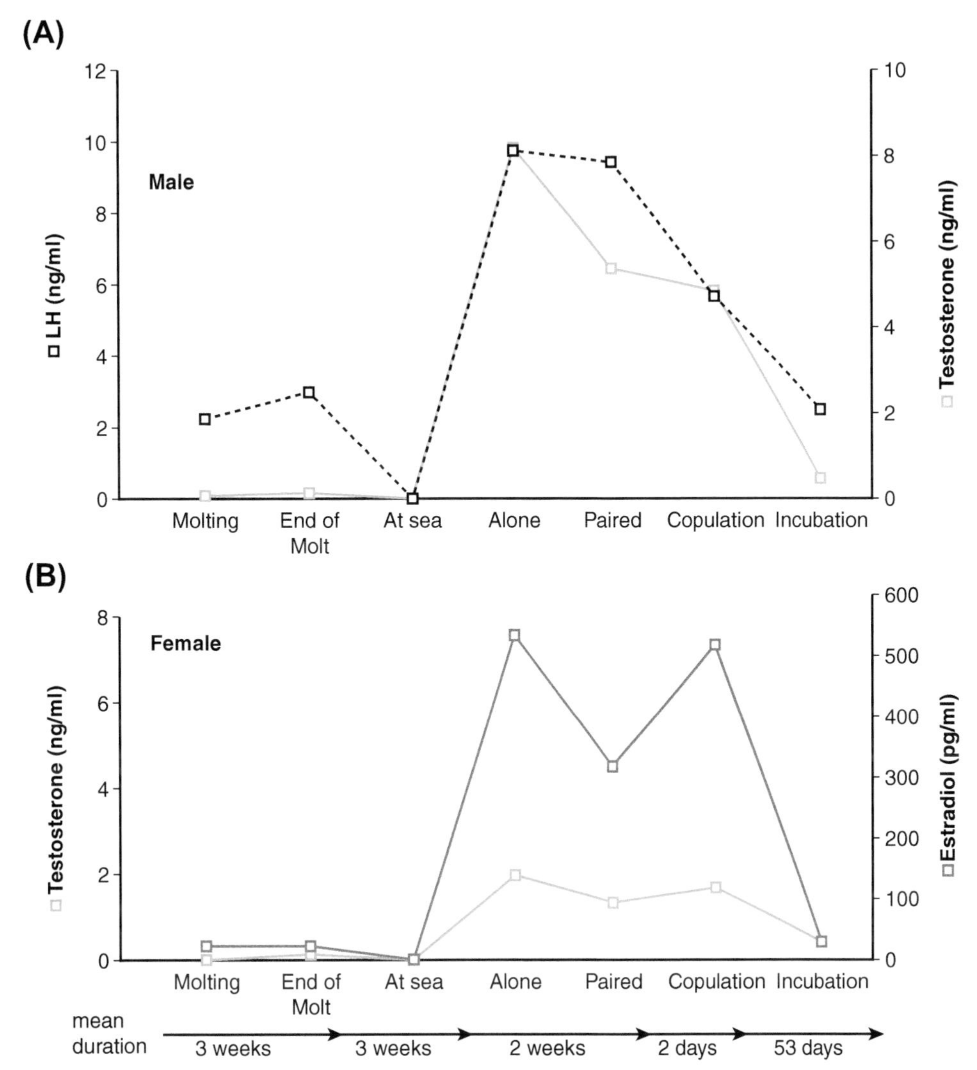

FIGURE 11-55 **Plasma hormone levels in male and female King penguins from molting to the onset of egg incubation.** (A) Plasma LH and testosterone levels in males. (B) Plasma, testosterone and estradiol in females. LH levels (not shown) paralleled that for the males in (A). *(Adapted with permission from Mauget, R. et al.,* General and Comparative Endocrinology, ***93****, 36–43, 1994.)*

is induced in part by PRL working cooperatively with testicular steroids.

5. Sex Accessory Structures in Male Birds

Wolffian ducts give rise to paired vasa deferentia, vasa efferentia, and the epididymides. These structures exhibit hypertrophy with the onset of sexual activity that is dependent on testicular androgens. Each testis is connected to vasa efferentia by small rete tubules in the tunica albuginea that become enlarged during the breeding season. The vasa efferentia show increased secretory activity during the breeding season and coalesce to form a long, coiled tube, the epididymis. Hypertrophy of the epididymis is accompanied by secretion of seminal fluid. Mature sperm leave the epididymis and enter the enlarged vas deferens for storage. The distal end of each vas deferens (seminal sac) fills with sperm. The posterior walls of the seminal sacs protrude into the cloaca as erectile papillae that facilitate transfer of sperm to the female. During copulation, the male's cloaca is everted and these erectile papillae are brought into contact with the vagina of the female. In some species, the cloaca actually is modified into a penis-like intromittent organ.

B. Female Birds

Assessment of ovarian function in domestic species is relatively complete and provides a basis for comparison with wild species. In many respects, ovarian function in birds is like that of their oviparous ancestors, the chelonians and especially the crocodilians. However, females of domestic species differ most importantly from wild birds

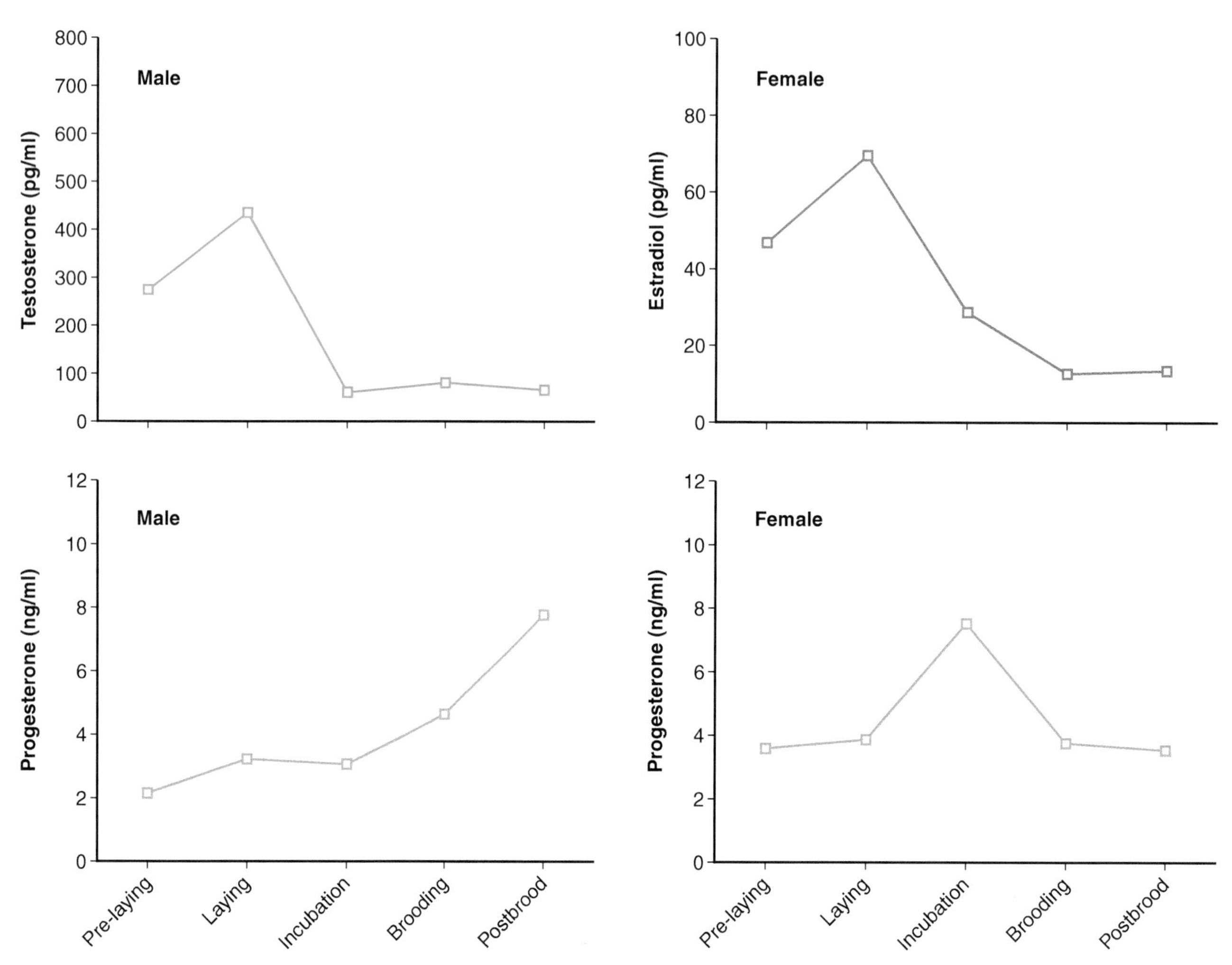

FIGURE 11-56 **Plasma steroid levels in male and female black kites (*Milvus migrans*).** *(Adapted with permission from Blas, J. et al.,* General and Comparative Endocrinology, ***168****, 22–28, 2010.)*

by being selected for continuous breeding. The following account is based largely on studies of the domestic hen.

Prior to hatching there is a proliferation of oogonia to produce thousands of primary oocytes that will serve the hen throughout her long and busy reproductive life. Whether new oocytes will be formed after hatching, as recently described for mammals, remains to be demonstrated. However, most of these oocytes will undergo atresia during early maturational stages.

A primary follicle consists of an oocyte surrounded by a layer of granulosa cells. As the follicle grows, thecal layers are added, and the follicle becomes highly vascularized. Both granulosa and thecal cells are steroidogenic, possess 3β-HSD activity and produce steroids as described for mammals. However, in birds, the entire steroidogenic process may involve three cells rather than two. For example, in the domestic turkey, progesterone is synthesized by the granulosa cell, whereas androgens are made in cells of the theca interna and converted to estrogens in the theca externa. This pattern has yet to be confirmed in wild species.

Estrogens secreted from the follicular cells cause the liver to produce large quantities of calcium-binding vitellogenin and triglycerides for the growing oocytes. This mechanism brings calcium to the oviducts where it is used to construct the shell. Vitellogenin is enzymatically hydrolyzed to produce phosvitin and lipovitellin, which are stored in the yolk of the egg. Incorporation of phosphoproteins and triglycerides into growing oocytes is stimulated by FSH. Avian follicles, like those of other nonmammals, do not form an antrum as described for mammalian follicles (Chapter 4).

Not only does the female place nutrients in the egg to support early development but she also provides a number of hormones that influence the offspring in many ways. For example, differences in androgen levels in eggs have been shown to influence behavior, growth morphology, immune functions, and survival. Studies show that a female can influence the levels of androgens deposited in different eggs or in different clutches of eggs, thus perhaps adjusting development to environmental conditions. Nestlings of the

Florida scrub jay (*Aphelocoma coerulescens*) that exhibit higher corticosterone levels of maternal origin are more fearful as adults. Furthermore, nestlings with lower corticosterone levels breed early than their siblings that have higher corticosterone levels as nestlings. Although the underlying mechanisms of controlling egg hormone levels are not understood, this is an intriguing area that deserves more attention.

The large developing follicles of birds bulge conspicuously from the surface of the ovary, giving it the appearance of a bunch of grapes. The largest follicles are highly vascularized except for a rough, avascular spot, the **stigma**, where the follicle will rupture at ovulation.

Atresia of developing follicles may occur at any time during follicular development. These atretic follicles can be easily recognized histologically by an influx of fibroblasts that phagocytize the yolk materials. Granulosa and thecal cells are lipoidal and contain cholesterol. There are many corpora atretica at all times in the ovary, but their endocrine importance is not known. However, some of the cells of the disintegrating corpora atretica may become stromal interstitial cells and secrete estrogens.

Birds are characterized by the absence of persistent corpora lutea following ovulation. This feature is correlated with the absence of viviparity among all of the more than 8000 known species of birds. Following ovulation, collapsed follicles consist largely of granulosa cells containing progesterone, abundant smooth endoplasmic reticulum, and considerable 3β-HSD activity. The only evidence for a functional role, however, is the observation that surgical removal of these ruptured follicles increases the time that the ovulated egg is retained in the oviduct.

1. Endocrine Control of Ovarian Functions in Birds

There is a close correlation between pituitary GTH content and ovarian function in both domestic and wild birds as evidenced by patterns of hormone secretion (see Figures 11-53, 11-54, and 11-55). Hypophysectomy results in ovarian regression and extensive follicular atresia, which can be prevented by GTH replacement therapy. Follicle development is stimulated by FSH, and FSH will maintain oviducts in hypophysectomized but not in ovariectomized birds. Secretion of estrogens is controlled by both LH and FSH. Although mammalian GTHs are not always as effective as avian GTHs, avian FSHs are very effective at stimulating follicle development in lizards, emphasizing the close similarity between reptilian and avian pituitary hormones.

As is the case for certain reptiles, growth of follicles and ovulation is a continual process throughout the breeding season. Ovarian function is regulated so that typically only one egg is discharged at a time. This condition is reminiscent of the human and the lizard *A. carolinensis*, in which only one ovum is discharged, and the ovaries alternate in providing the ovum. However, recall that birds have only one functional ovary with a hierarchy of graded follicle sizes. The endocrine basis for establishment and maintenance of this hierarchy is not known for reptiles, birds, or mammals, and it represents one of the major unanswered questions in reproductive biology.

Estrogens induce the synthesis of vitellogenin by the avian liver. Total serum calcium concomitantly increases, which is related to the binding of calcium by vitellogenin (Table 11-10). In addition to incorporation of vitellogenic proteins into the oocyte, circulating calcium is sequestered by the shell glands of the "uterus" (an expanded region of the oviduct; see ahead) for construction of the eggshell.

Pituitary LH is responsible for triggering ovulation of the fully mature follicle. Plasma LH peaks about 6 to 8 hours before ovulation in domestic hens as well as in Japanese quail, but the magnitude of the avian LH surge is considerably smaller than that observed in mammals. This lower surge of LH might be an adaptation to ensure only sufficient LH for ovulating the largest follicle.

Calcium availability may be a potent factor regulating reproduction in female birds. Production of shelled eggs in domestic species directs as much as 10% of the body calcium stores per day into eggs. If large amounts of calcium are not available in the diet, the reproductive axis of the egg factory is shut down before damage to the skeleton occurs. When sufficient calcium becomes available, the birds resume laying. Although continuous egg laying does not occur in wild birds as it does in domestic fowl, it is possible that calcium depletion in wild birds contributes to cessation of breeding and induction of the refractory period.

Another pituitary hormone, PRL, plays essential roles in reproduction, including the development of a specialized, de-feathered region in some species known as an **incubation patch**, which aids in incubating eggs (see ahead). Secretion of **crop milk** by the crop sac of some birds for use in feeding young hatchlings is induced by

TABLE 11-10 Effect of Mammalian Parathyroid Extract and the Influence of the Egg-Laying Cycle on Total Serum Calcium of Chicken

Subject	Control (mg/dL ± SE)	Treated with parathyroid extract (mg/dL ± SE)
Roosters	10.1 ± 0.2	19.5 ± 3
Nonlaying hens	13.4 ± 2	19.5 ± 4
Laying hens	29.8 ± 11*	47.7 ± 9[a]

[a]*Significantly different from roosters and non-laying hens.*
**From Assenmacher, I., in "Avian Biology, Vol. 3" (D.S. Farner and J.R. King, Eds.), Academic Press, New York, 1973, pp. 183–286.*

PRL. This feature of PRL has resulted in development of a most useful biological assay for tetrapod vertebrates called the **pigeon crop sac assay** (see Chapters 4 and 5 and Appendix D). Prolactin does not affect steroidogenesis in cultured chick granulosa cells and possibly has no effect on progesterone synthesis.

2. The Avian Oviduct

Estrogens secreted by the growing follicle stimulate hypertrophy of the oviduct and differentiation of secretory regions. Five differentiated regions can be identified in the mature avian oviduct which are like those of crocodilians (see above): **infundibulum**, **magnum**, **isthmus**, **shell gland**, and **vagina**. After ovulation, the ovum moves through the body cavity, enters the open end of the infundibulum, and is fertilized in the upper portion of the oviduct before the egg white protein **albumen** is added. The middle portion of the oviduct or magnum becomes highly glandular under the influence of estrogens, forming tubular glands and goblet cells. Estrogens stimulate synthesis of ovalbumen protein by these tubular glands, whereas progesterone stimulates the goblet cells to secrete the other major egg white protein, **avidin**. After accumulation of several coatings of albumen, the egg passes from the magnum to the muscular isthmus, where two shell membranes are applied. These membranes are composed of fibrous proteins cemented together with albumen. The shell consists largely of calcium salts supported by a fibrous protein matrix deposited on the outermost shell membrane by the shell gland or "uterus" (see Chapter 14 for the role of hormones in making Ca^{2+} available for eggshell formation). After the shell has been applied, contraction of a powerful sphincter muscle causes the egg to rotate in the muscular vagina and enter the cloaca pointed-end first. Movement of the egg into the cloaca as well as its extrusion into the nest (oviposition) is controlled by AVT and prostaglandins. An increase in plasma AVT together with a concomitant decrease in neurohypophysial AVT coincides with oviposition. Treatment with AVT can cause premature oviposition.

3. Incubation Patches

In many avian species, a ventral region of skin called the **apterium** becomes de-feathered, highly vascularized, and edematous just prior to or during egg laying to form the incubation or brood patch. In addition, the epidermis of this region may exhibit hyperplasia. When in contact with the eggs it provides an efficient transfer of warmth from the parent's incubation patch to the eggs. Incubation patches may form in females, males or both, depending upon the species and which sex is responsible for incubating eggs; however, possession of an incubation patch by a male is not requisite for exhibition of incubating behavior. Male house sparrows (*Passer domesticus*) have no incubation patch, yet exhibit incubating behavior. Conversely, male flycatchers (genus *Empidonax*) develop an incubation patch like that of the female but do not show incubating behavior.

Formation of incubation patches involves cooperative actions of both estrogens and PRL (see Figure 11-56). Estrogens stimulate vascularization of the patch region, and PRL stimulates de-feathering and epidermal hyperplasia. Typically, pituitary lactotropes transform at this time to a stimulated morphology known as "broody cells." Both PRL and estrogens are necessary for complete patch development. Furthermore, the response of the ventral skin epidermis to form a patch is both site specific and tissue specific. After transplant to the dorsal surface, ventral skin will still respond to PRL although not to estrogen. Vascularization of the ventral skin occurs only when it is in its normal location. Dorsal skin transplanted to the normal patch site will not respond to either estrogens or PRL.

C. Androgen-Dependent Secondary Sex Characters in Male and Females Birds

Androgens play important roles in both male and female birds. In a number of species, a change in bill color is associated with breeding. Such changes are induced in both sexes by androgens but not by estrogens or progesterone. However, there is at least one case in which bill color change occurs only in the female, and, in that case, the color change is induced by estrogens.

Plumage color changes also may be controlled by androgens. This can occur even in the case of phalarope birds where the females possess the more colorful or **nuptial plumage**. One cannot presume that androgens are responsible unless specific studies have been performed to verify this fact, because androgens are not always responsible for nuptial plumage. For example, development of nuptial plumage in castrated male weaver finches (*Euplectes orix*) has been used as the classical bioassay for LH (see Appendix D). In some instances, estrogens actively inhibit formation of nuptial dress, and castration of females will cause development of male plumage. Assumption of nuptial dress in males also can be blocked with estrogens, verifying that it is the absence of estrogens that allows male plumage to develop in both sexes and not the presence of androgens.

In some strains of domestic chickens, both sexes have female type plumage, and castration causes development of male plumage. Treatment of castrated males of these strains with testosterone causes a return to female-type plumage, but growth of the comb and wattle is stimulated (a normal male trait). If castrated males that developed male plumage are treated with DHT, the growth of the comb and wattle is stimulated but there is no reversion to female plumage. The explanation appears to be that, in these strains of chicken, the skin aromatizes testosterone to estrogens and stimulates female plumage. In the comb and wattle, 5α-reductase converts testosterone to DHT, causing comb and wattle development, whereas the plumage of DHT-treated, castrated males does not revert to the female type.

D. Reproductive Behaviors in Birds

Most wild avian species exhibit a precise sequence of endocrine-dependent behaviors such as migration, acquisition of territory, advertisement by song, attraction of mate, pairing, nest building, egg laying, incubating eggs, and rearing young birds. Some of these behaviors appear in domesticated species as well. The actual sequence of events and their endocrine bases are species specific and cannot easily be generalized. Successful breeding involves a complex interaction of male and female behaviors in precise sequences (that is, if male does A, then female does B, which stimulates male to do C, etc.) as well as presence of suitable environmental cues such as proper nesting material, availability of water, etc. Little experimental work has been done with wild birds, as it is difficult to get them to perform under laboratory conditions although several descriptive studies on hormone levels and behavior are available. Much of our knowledge of wild birds comes from the work of John Wingfield, Peter Sharp, Eberhard Gwinner, Thomas Hahn, Marilyn Rand, and their many associates.

Androgens appear to be responsible for territorial display and aggression in wild birds as it is in domesticated species. Aggressive behavior also can be stimulated by FSH but not by LH in males. Courtship appears to involve negative feedback of testosterone on FSH levels, which results in reduction in circulating androgens and allows for subsequent, less aggressive behaviors. In domestic ring doves, the initial aggressive behavior involves testosterone, and copulatory behavior coincides with reduced androgens and increased $P450_{aro}$ activity. Androgens antagonize incubation patch development, and a reduction in circulating testosterone may be necessary for patch development in males of certain species.

Bowing behavior in feral pigeons coincides with maximal androgen synthesis but decreases prior to egg laying, coincident with an increase in progesterone levels. Progesterone is a well-known stimulus for incubation behavior in laying pigeons, and removal of the post-ovulatory follicle from chickens blocks nesting behavior. In black kites, progesterone is elevated in females while they incubate eggs and are elevated in males as they take care of the hatchlings (Figure 11-55). Progesterone decreases at that time in the females that do not care for the hatchlings in this species.

E. Parental Behavior in Birds

Birds exhibit high levels of parental care of eggs and young to ensure survival of the relatively small number of offspring they produce as compared to fishes, amphibians, and reptiles. Generally, both sexes of birds are involved in parental care. Incubation is regulated largely by PRL levels (Figure 11-57) although PRL often interacts with other hormones such as described earlier for incubation patch development. In altricial birds, there is a long period of total dependence on the adults for both temperature

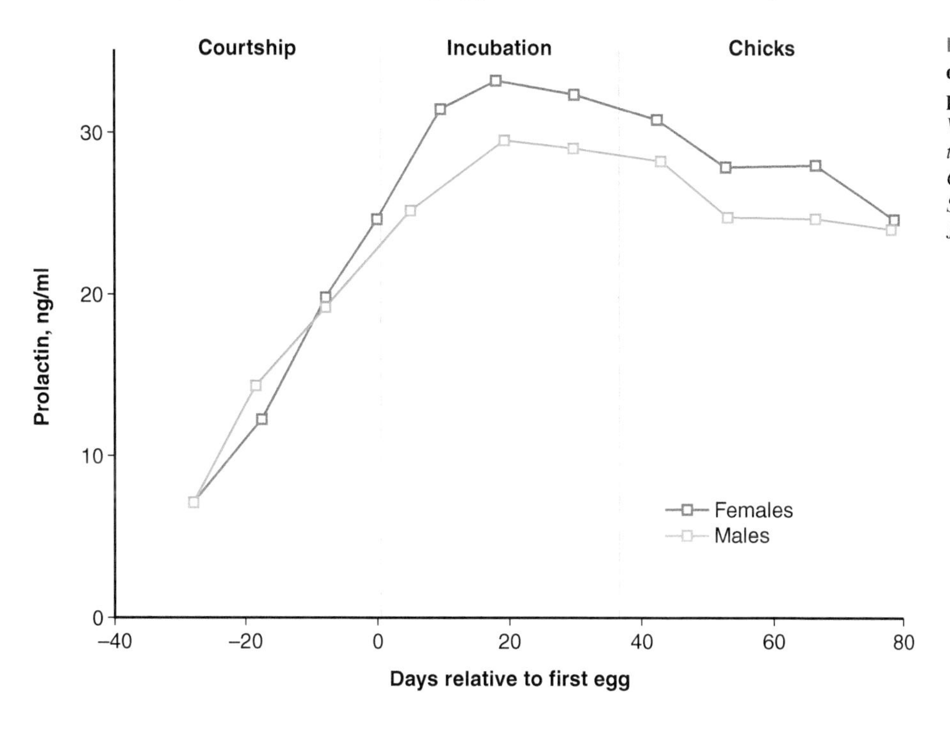

FIGURE 11-57 **Plasma prolactin levels during incubation by male and female penguins.** *(Adapted with permission from Vleck, C.M. et al., in "Proceedings of the 22nd International Ornithological Congress, Durban" (N.J. Adams and R.H. Slotow, Eds.), BirdLife South Africa, Johannesburg, 1999, pp. 1210–1223.)*

regulation and food requiring considerable effort by the parents to ensure their survival. Elevated androgen levels are not compatible with parental behaviors. Environmental stressors that result in elevation of corticosterone levels inhibit PRL secretion in adults and favor adult survival over that of the offspring, hence the need to reduce human interference in areas where wild birds are breeding.

VIII. ENDOCRINE DISRUPTION OF REPRODUCTION IN NON-MAMMALIAN VERTEBRATES

Endocrine disruption (see Chapter 1) of reproduction has been reported in all vertebrate groups, including mammals (see Chapter 10). The most detailed studies in non-mammals are those on teleost fishes and frogs. Because they are aquatic or semiaquatic they are more susceptible to chemicals that are dumped into the environment that readily find their way into lakes and rivers and eventually reach estuaries and the coastal areas of the continents. Summaries of the rapidly expanding literature on this topic are provided in the suggested readings at the end of this chapter.

In general, the effects observed in these groups are similar to those described in reptiles and birds. Fishes and amphibians tend to be most susceptible to laboratory exposures at environmentally relevant concentrations of natural estrogens, and estrogen mimics, including certain pesticides, **bisphenyl A (BPA)**, nonylphenols, and phytoestrogens such as genistein, as well as anti-androgens, cause induction of intersexes, feminization, and/or demasculinization of male fish, in addition to producing contraceptive-like effects in adult fishes of both sexes. Similar effects are seen in fish populations downstream of wastewater treatment plants (WWTPs), and these effects are directly a result of exposure to the wastewater effluent being added to the streams and rivers. One controlled Canadian study conducted in an experimental lake demonstrated that the contraceptive pharmaceutical ethinylestradiol at a concentration of less than 6 ng/L completely blocked reproduction and virtually eliminated a population of fathead minnows from the lake within 3 years. Intersex has been observed in populations of amphibians in proximity to suburban and agricultural landscapes, but identifying which anthropogenic chemicals may place amphibians most at risk for gonadal abnormalities, aside from cases where clear estrogenic contamination is evident, has been problematic due to the complex soup of chemicals to which animals may be exposed. Male alligators in a Florida lake exposed to metabolites of the pesticide DDT contaminating the lake exhibit feminization (e.g., reduced penis size) due to anti-androgenic actions. Females also showed numerous examples of abnormal follicles in their ovaries. Early exposure of female songbirds, specifically zebra finches (*Taeniopygia guttata*), to estrogens causes masculinization of the brain region necessary for production of song in males. Exposure of adult zebra finches to the insecticide methoxyclor at environmentally relevant doses decreases reproductive performance. In contrast, brains of developing male Japanese quail were feminized by estrogen exposure and they lacked normal male reproductive behavior.

Clearly, the contamination of the environment by chemicals that disrupt reproduction in wildlife are of great concern, as the human population depends on the aquatic and marine environments as major food sources. If concentrations of contraceptive-type chemicals continue to increase, the potential of widespread reproductive decline will markedly change the characteristics of Earth's ecosystem. It is also important to recognize that other classes of endocrine disruptors can alter thyroid and adrenal function as well as metabolism, all of which can influence reproduction. Additionally, neuroactive pharmacological agents (e.g., antidepressants affecting norepinephrine or serotonin activities) may also alter hypothalamic control of pituitary secretion, thus affecting reproduction.

IX. SUMMARY

Reproduction involves a precise integration of environmental factors (e.g., photoperiod, temperature, availability of nesting sites), physiological factors (e.g., nutritional state, general endocrine state with respect to thyroid hormones, adrenocortical functions), and specific endocrine secretions (e.g., FSH, LH, androgens, estrogens, progestogens, PRL). Reproductive patterns are finely tuned to environmental conditions in order to maximize evolutionary success, and this results in frequent observations of greater similarities in reproductive patterns between phylogenetically divergent species facing similar environmental problems than between closely related species living in diverse environments.

Environmental factors operate through the nervous system and specifically the hypothalamus to control the release of GTHs and in certain cases PRL. Prolactin molecules or PRL activity as well as FSH and LH molecules have been identified in all tetrapods. Fishes have a unique PRL-like hormone. In both fishes and tetrapods, FSH initiates spermatogenesis in males and follicular development in females. Local androgens secreted from testicular cells under the influence of FSH appear to be necessary for initiating reductional division (meiosis) of primary spermatocytes. Luteinizing hormone induces androgen synthesis by interstitial (Leydig) or lobule-boundary cells in males and estrogen synthesis and ovulation in females. Androgen synthesis in female mammals also may be stimulated in thecal cells or ovarian interstitial cells by LH. It is thought that thecal androgens are converted to estrogens by granulosa cells.

Follicular atresia associated with formation of corpora atretica is a common occurrence in female vertebrates. Atresia appears to be a mechanism for effectively reducing the biotic potential and placing reliance in production of a smaller number of offspring with better individual survival for evolutionary success. Corpora lutea form in many vertebrates primarily from granulosa cells of ruptured follicles, and corpora lutea usually synthesize progesterone, which is related to gestation or behavior in many viviparous species. Many examples of autocrine and paracrine regulation in the gonads are known.

Courtship, breeding, and parental behaviors appear to be controlled primarily by gonadal steroids although evidence is accumulating for participation of other hormones and pheromones. In addition, estrogens produce dramatic effects on vitellogenesis in non-mammalian liver and bring about a consequent disturbance in calcium metabolism. The basic oviparous mode of reproduction has become modified with respect to the development of viviparity in all non-mammalian classes except Aves and Agnatha.

Endocrine disruption of reproductive biology of non-mammals may contribute to environmental instability. It is imperative that we reduce levels of such chemicals in the environment.

STUDY QUESTIONS

1. Characterize the mechanisms for sex determination for bony fishes, amphibians, reptiles, and birds, indicating those aspects unique to non-mammals.
2. Describe the patterns of reproduction in non-mammalian vertebrates.
3. Compare and contrast hormonal components of the HPG axis of the major vertebrate groups discussed in this chapter with the mammalian HPG axis.
4. Describe the process of sperm formation and the roles hormones play in vertebrates.
5. Describe the process of oogenesis and the roles of hormones in viviparous and oviparous vertebrates.
6. Develop a scenario to explain the evolution of (a) sex-changing fishes, and (b) parthenogenesis.
7. How do environmental factors influence reproduction in teleosts, amphibians, reptiles, and birds?
8. What does the *Ambystoma jeffersonianum* complex in amphibian salamanders suggest to you about the type of sex determination operating in these species?
9. How would you explain the results of transplantation of ventral or dorsal skin described for birds that normally exhibit an incubation patch?
10. From an endocrine and evolutionary point of view, why are there no live-bearing birds?

SUGGESTED READING

Books

Dawson, A., Sharo, P.J., 2005. Functional Avian Endocrinology. Narosa Publishing House, New Delhi.

Duellman, W.E., Trueb, L., 1986. Biology of Amphibians. McGraw-Hill, New York.

Hamlett, W.C., 2005. Reproductive Biology and Phylogeny of Chondrichthyes: Sharks, Batoids and Chimaeras. Science Publishers, Enfield, NH.

Hoar, W.S., Randall, D.G., Donaldson, E.M., 1983. Fish Physiology, vol. IX. Reproduction, Part A. Endocrine Tissues and Hormones. Part B. Behavior and Fertility Control. Academic Press, New York.

Jamieson, B.G.M., 2003. Reproductive Biology and Phylogeny of Anura. Science Publishers, Enfield, NH.

Jones, R.E., 1978. The Vertebrate Ovary. Plenum Press, New York.

Lamming, G.E., 1984. Marshall's Physiology of Reproduction, vol. 1. Reproductive Cycles in Vertebrates. Churchill Livingston, Edinburgh.

Norris, D.O., Jones, R.E., 1987. Hormones and Reproduction in Fishes, Amphibians, and Reptiles. Plenum Press, New York.

Norris, D.O., Lopez, K.H. (Eds.), 2011. Hormones and Reproduction of Vertebrates, vol. 1. Fishes. Academic Press, San Diego, CA.

Norris, D.O., Lopez, K.H. (Eds.), 2011. Hormones and Reproduction of Vertebrates, vol. 2. Amphibians. Academic Press, San Diego, CA.

Norris, D.O., Lopez, K.H. (Eds.), 2011. Hormones and Reproduction of Vertebrates, vol. 3. Reptiles. Academic Press, San Diego, CA.

Norris, D.O., Lopez, K.H. (Eds.), 2011. Hormones and Reproduction of Vertebrates, vol. 4. Birds. Academic Press, San Diego, CA.

Rocha, M.J., Arukwe, A., Kapoor, B.G., 2008. Fish Reproduction. Science Publishers, Enfield NH.

Sever, D.M., 2003. Reproductive Biology and Phylogeny of Urodela. Science Publishers, Enfield, NH.

Sharp, P.J., 1993. Avian Endocrinology. The Society for Endocrinology, Bristol, U.K.

Van Tienhoven, A., 1983. Reproductive Physiology of Vertebrates, 2nd ed. Cornell University Press, Ithaca, NY.

Articles

General

Blackburn, D.G., 1994. Review: Discrepant usage of the term ovoviviparity in the herpetological literature. Herpetological Journal 4, 65–72.

Crews, D., 1993. The organizational concept and vertebrates without sex chromosomes. Brain, Behavior and Evolution 42, 202–214.

Jones, R.E., Baxter, D.C., 1991. Gestation, with emphasis on corpus luteum biology, placentation, and parturition. In: Pang, P.K.T., Schreibman, M. (Eds.), Vertebrate Endocrinology: Fundamentals and Biomedical Implications, vol. 4. Academic Press, San Diego, CA, pp. 205–302. Part A.

Lange, I.G., Hartel, A., Meyer, H.H.D., 2003. Evolution of estrogen functions in vertebrates. Journal of Steroid Biochemistry and Molecular Biology 1773, 1–8.

Polzonetti-Magni, A.M., Mosconi, G., Soverchia, L., Kikuyama, S., Carnevali, O., 2004. Multihormonal control of vitellogenesis in lower vertebrates. International Review of Cytology 239, 1–46.

Pudney, J., 1999. Leydig and Sertoli cells, nonmammalian. In: Knobil, E., Neill, J.D. (Eds.), Encyclopedia of Reproduction, vol. 2. Elsevier, Amsterdam, pp. 1008–1020.

Sim, H., Argentaro, A., Harley, V.R., 2008. Boys, girls and shuttling of SRY and SOX9. Trends in Endocrinology & Metabolism 19, 213–222.

Unniappan, S., 2010. Ghrelin: an emerging player in the regulation of reproduction in non-mammalian vertebrates. General and Comparative Endocrinology 167, 340–343.

Whittier, J.M., Crews, D., 1987. Seasonal reproduction: patterns and control. In: Norris, D.O., Jones, R.E. (Eds.), Hormones and Reproduction in Fishes, Amphibians, and Reptiles. Plenum Press, New York, pp. 385–409.

Young, L.J., Crews, D., 1995. Comparative neuroendocrinology of steroid receptor gene expression and regulation: relationship to physiology and behavior. Trends in Endocrinology & Metabolism 6, 317–323.

Sex Determination in Non-Mammalian Vertebrates

Angelopoulou, R., Lavranos, G., Manolakou, P., 2012. Sex determination strategies in 2012: towards a common regulatory model? Reproductive Biology and Endocrinology 10, 13.

Blázquez, M., Somoza, G.M., 2010. Fish with thermolabile sex determination (TSD) as models to study brain sex differentiation. General and Comparative Endocrinology 166, 470–477.

DeFalco, T., Capel, B., 2009. Gonad morphogenesis in vertebrates: divergent means to a convergent end. Annual Review of Cell Development and Biology 25, 457–482.

Devlin, R.H., Nagahama, Y., 2002. Sex determination and sex differentiation in fish: an overview of genetic, physiological, and environmental influences. Aquaculture 208, 191–364.

Flament, S., Chardard, D., Chensel, A., Dumond, H., 2011. Sex determination and sexual differentiation in amphibians. In: Norris, D.O., Lopez, K.H. (Eds.), Hormones and Reproduction of Vertebrates, vol. 2. Amphibians. Academic Press, San Diego, CA, pp. 1–19.

Göth, A., Booth, D.T., 2004. Temperature-dependent sex ratio in a bird. Biology Letters 1, 31–33.

Graves, J.A.M., Peichel, C.L., 2010. Are homologies in vertebrate sex determination due to shared ancestry or to limited options? Genome Biology 11, 205. http://genomebiology.com/2010/11/4/205.

Graves, J.A.M., Shetty, S., 2001. Sex from W to Z: evolution of vertebrate sex chromosomes and sex determining genes. Journal of Experimental Zoology 290, 440–462.

Manolakou, P., Lavranos, G., Angelopoulou, R., 2006. Molecular patterns of sex determination in the animal kingdom: a comparative study of the biology of reproduction. Reproductive Biology and Endocrinology 4, 59.

Morish, B.C., Sinclair, A.H., 2002. Vertebrate sex determination: many means to an end. Reproduction 124, 447–457.

Herpin, A., Schartl, M., 2011. *Dmrt1* genes at the crossroads: a widespread and central class of sexual development factors in fishes. FEBS Journal 278, 1010–1019.

Paul-Prasanth, B., Nakamura, M., Nagahama, Y., 2011. Sex determination in fishes. In: Norris, D.O., Lopez, K.H. (Eds.), "Hormones and Reproduction of Vertebrates, vol. 1. Fishes. Academic Press, San Diego, CA, pp. 1–14.

Warner, D.A., 2011. Sex determination in reptiles. In: Norris, D.O., Lopez, K.H. (Eds.), "Hormones and Reproduction of Vertebrates, vol. 3. Reptiles. Academic Press, San Diego, CA, pp. 1–38.

Yoshimoto, S., Ito, M., 2011. A ZZ/ZW-type sex determination in. *Xenopus laevis*. FEBS Journal 278, 1020–1026.

Fishes

Cyclostomes

Bryan, M.B., Young, B.A., Dlose, D.A., Semeyn, J., Robinson, T.C., Bayer, J., Li, W., 2006. Comparison of synthesis of 15α-hydroxylated steroids in males of four North American lamprey species. General and Comparative Endocrinology 146, 149–156.

Kavanaugh, S.I., Powell, M.L., Sower, S.A., 2005. Seasonal changes of gonadotropin-releasing hormone in the Atlantic hagfish. *Myxine glutinosa*. General and Comparative Endocrinology 140, 136–143.

Powell, M.L., Kavanaugh, S., Sower, S.A., 2006. Identification of a functional corpus luteum in the Atlantic hagfish. *Myxine glutinosa*. General and Comparative Endocrinology 148, 95–101.

Sower, S.A., Balz, E., Aquilina-Beck, A., Kavanaugh, S.I., 2011. Seasonal changes of brain GnRH-I, -II, and -III during the final reproductive period in adult male and female sea lamprey. General and Comparative Endocrinology 170, 276–282.

Sower, S.A., Kawauchi, H., 2011. Reproduction in agnathan fishes: lampreys and hagfishes. In: Norris, D.O., Lopez, K.H. (Eds.), Hormones and Reproduction of Vertebrates, vol. 1. Fishes. Academic Press, San Diego, CA, pp. 193–208.

Young, B.A., Bryan, M.B., Glenn, J.R., Yun, S.S., Scott, A.P., Li, W., 2007. Dose–response relationship of 15α-hydroxylated sex steroids to gonadotropin-releasing hormones and pituitary extract in male sea lampreys (*Petromyzon marinus*). General and Comparative Endocrinology 151, 108–115.

Chondrichthyes

Callard, G.V., 1988. Reproductive physiology. Part B. The male. In: Shuttleworth, T.J. (Ed.), Physiology of Elasmobranch Fishes. Springer-Verlag, Berlin, pp. 292–317.

Callard, I.P., Klosterman, L., 1988. Reproductive physiology. Part A. The female. In: Shuttleworth, T.J. (Ed.), Physiology of Elasmobranch Fishes. Springer-Verlag, Berlin, pp. 277–291.

Chapman, D.D., Shivji, M.S., Louis, E., Sommer, J., Fletcher, H., Prodo, P.A., 2007. Virgin birth in a hammerhead shark. Biology Letters 3, 425–427.

Feldheim, K.A., Chapman, D.D., Sweet, D., Fitzpatrick, S., Prodöhl, P.A., Shivji, M.S., Snowden, B., 2010. Shark virgin birth produces multiple, viable offspring. Journal of Heredity 101, 374–377.

Koob, T.J., Callard, I.P., 1999. Reproductive endocrinology of female elasmobranches: lessons from the little skate (*Raja erinacea*) and spiny dogfish (*Squalus acanthias*). Journal of Experimental Zoology 284, 557–574.

Maruska, K.P., Gelsleichter, J., 2011. Hormones and reproduction in chondrichthyan fishes. In: Norris, D.O., Lopez, K.H. (Eds.), Hormones and Reproduction of Vertebrates, vol. 1. Fishes. Academic Press, San Diego, CA, pp. 209–237.

Parsons, G.R., Grier, H.J., 1992. Seasonal changes in shark testicular structure and spermatogenesis. Journal of Experimental Zoology 261, 173–184.

Prisco, M., Valiante, S., Di Fiore, M.M., Raucci, F., Del Giudice, G., Romano, M., Laforgia, V., Limatola, E., Andreuccetti, P., 2008. Effect of 17β-estradiol and progesterone on vitellogenesis in the spotted ray *Torpedo marmorata* Risso 1810 (Elasmobranchii: Torpediniformes): studies on females and on estrogen-treated males. General and Comparative Endocrinology 157, 125–132.

Bony Fishes

Butts, I.A.E., Love, O.P., Farwell, M., Pitcher, T.E., 2012. Primary and secondary sexual characters in alternative reproductive tactics of Chinook salmon: associations with androgens and the maturation-inducing steroid. General and Comparative Endocrinology 175, 449–456.

Connaughton, M.A., Aida, K., 1999. Female reproductive system, fish. In: Knobil, E., Neill, J.D. (Eds.), Encyclopedia of Reproduction, vol. 2. Elsevier, Amsterdam, pp. 193–204.

Cottone, E., Camantico, E., Gustalla, A., Aramu, S., Polzonetti-Magni, A.M., Franzoni, M., 2005. Are the cannabinoids involved in bony fish reproduction? Annals of the New York Academy of Sciences 1040, 273–276.

Dickhoff, W.W., Yan, L., Plisetskaya, E.M., Sullivan, C.V., Swanson, P., Hara, A., Berrard, M.G., 1989. Relationship between metabolic and reproductive hormones in salmonid fish. Fish Physiology and Biochemistry 7, 147–155.

Dufour, S., Weltzien, F.-A., Sebert, M.-E., Le Belle, N., Vidal, B., Vernier, P., Pasqualini, C., 2005. Dopaminergic inhibition of reproduction in teleost fishes: ecophysiological and evolutionary implications. Annals of the New York Academy of Sciences 1040, 9–21.

Fuzzen, M.L.M., Bernier, N.J., van der Kraak, G., 2011. Stress and reproduction. In: Norris, D.O., Lopez, K.H. (Eds.), Hormones and Reproduction of Vertebrates, vol. 1. Fishes. Academic Press, San Diego, CA, pp. 103–117.

Gonçalves, D.M., Oliveira, R.F., 2011. Hormones and sexual behavior of teleost fishes. In: Norris, D.O., Lopez, K.H. (Eds.), Hormones and Reproduction of Vertebrates, vol. 1. Fishes. Academic Press, San Diego, CA, pp. 119–147.

Grier, H.J., Uribe, M.C., Parenti, L.R., 2007. Germinal epithelium, folliculogenesis, and postovulatory follicles in ovaries of rainbow trout, *Oncorhynchus mykiss* (Walbaum, 1792) (Teleostei, Protacanthopterygii, Salmoniformes). Journal of Morphology 268, 293–310.

Guiguen, Y., Fostier, A., Piferrer, F., Chang, C.-F., 2010. Ovarian aromatase and estrogens: a pivotal role for gonadal sex differentiation and sex change in fish. General and Comparative Endocrinology 165, 352–366.

Larson, E.T., Norris, D.O., Grau, E.G., Summers, C.H., 2003. Monoamines stimulate sex reversal in the saddleback wrasse. General and Comparative Endocrinology 130, 289–298.

Larson, E.T., 2011. Neuroendocrine regulation in sex-changing fishes. In: Norris, D.O., Lopez, K.H. (Eds.), Hormones and Reproduction of Vertebrates, vol. 1. Fishes. Academic Press, San Diego, CA, pp. 149–168.

Joss, J.M.P., 2011. Hormones and reproduction of sarcopterygian fishes. In: Norris, D.O., Lopez, K.H. (Eds.), Hormones and Reproduction of Vertebrates, vol. 1. Fishes. Academic Press, San Diego, CA, pp. 239–244.

Kah, O., Dufour, S., 2011. Conserved and divergent features of reproductive neuroendocrinology in teleost fishes. In: Norris, D.O., Lopez, K.H. (Eds.), Hormones and Reproduction of Vertebrates, vol. 1. Fishes. Academic Press, San Diego, CA, pp. 15–42.

Knapp, R., Carlisle, S.L., 2011. Testicular function and hormonal regulation in fishes. In: Norris, D.O., Lopez, K.H. (Eds.), Hormones and Reproduction of Vertebrates, vol. 1. Fishes. Academic Press, San Diego, CA, pp. 43–63.

Levy, G., Jackson, K., Degani, G., 2010. Association between pituitary adenylate cyclase-activating polypeptide and reproduction in the blue gourami. General and Comparative Endocrinology 166, 83–93.

Lubzens, E., Young, G., Bobe, J., Cerdà, J., 2010. Oogenesis in teleosts: how fish eggs are formed. General and Comparative Endocrinology 165, 367–389.

Nagahama, Y., Yoshikuni, M., Yamashita, M., Tanaka, M., 1994. Regulation of oocyte maturation in fish. In: Sherwood, N.M., Hew, C.L. (Eds.), CA. Fish Physiology. Molecular Endocrinology of Fish, vol. XIII. Academic Press, San Diego, pp. 393–439.

Parenti, L.R., Grier, H.J., 2004. Evolution and phylogeny of gonad morphology in bony fishes. Integrative and Comparative Biology 44, 333–348.

Raine, J.C., 2011. Thyroid hormones and reproduction in fishes. In: Norris, D.O., Lopez, K.H. (Eds.), Hormones and Reproduction of Vertebrates, vol. 1. Fishes. Academic Press, San Diego, CA, pp. 83–102.

Reinboth, R., 1999. Fish, mode of reproduction. In: Knobil, E., Neill, J.D. (Eds.), Encyclopedia of Reproduction, vol. 2. Elsevier, Amsterdam, pp. 365–372.

Ross, R.M., Hourigan, T.F., Lutnesky, M.M.F., Singh, I., 1990. Multiple spontaneous sex changes in social groups of a coral-reef fish. Copeia 1990, 427–433.

Schulz, R.W., de França, L.R., Jean-Jacques Lareyre, J.-J., LeGac, F., Chiarini-Garcia, H., Nobrega, R.H., Miura, T., 2010. Spermatogenesis in fish. General and Comparative Endocrinology 165, 390–411.

Scott, A.P., 1987. Reproductive endocrinology of fish. In: Chester-Jones, I., Ingelton, P.M., Phillips, J.G. (Eds.), Fundamentals of Comparative Endocrinology. Plenum Press, New York, pp. 223–256.

Scott, A.P., Sumpter, J.P., 1983. Seasonal variations in sex steroids and gonadotropin in females of autumn and winter spawning strains of rainbow trout (*Salmo gairdneri*). General and Comparative Endocrinology 52, 79–85.

Stacey, N., 2011. Hormonally derived sex pheromones in fishes. In: Norris, D.O., Lopez, K.H. (Eds.), Hormones and Reproduction of Vertebrates, vol. 1. Fishes. Academic Press, San Diego, CA, pp. 169–192.

Strüssmann, C.A., Nakamura, M., 2002. Morphology, endocrinology, and environmental modulation of gonadal sex differentiation in teleost fishes. Fish Physiology and Biochemistry 26, 13–29.

Suresh, D.V.N.S., Baile, V.V., Prasada Rao, P.D., 2008. Annual reproductive phase related profile of sex steroids and their carrier, SHBG, in the Indian major carp, *Labeo rohita*. General and Comparative Endocrinology 159, 143–149.

Takemura, A., Rahman, M.S., Nakamura, S., Young Ju Park, Y.J., Takano, K., 2004. Lunar cycles and reproductive activity in reef fishes with particular attention to rabbitfishes. Fish and Fisheries 5, 317–328.

Taranger, G.L., Carrillo, M., Schulz, R.W., Fontaine, P., Zanuy, S., Felip, A., Finn-Arne Weltzien, F.-A., Dufour, S., Karlsen, Ø., Norberg, B., Andersson, E., Hansen, T., 2010. Control of puberty in farmed fish. General and Comparative Endocrinology 165, 483–515.

Urbatzka, R., Rocha, M.J., Rocha, E., 2011. Regulation of ovarian development and function in teleosts. In: Norris, D.O., Lopez, K.H. (Eds.), Hormones and Reproduction of Vertebrates, vol. 1. Fishes. Academic Press, San Diego, CA, pp. 65–82.

Villeneuve, D.L., Larkin, P., Knoebl, I., Miracle, A.L., Kahl, M.D., Jensen, K.M., et al., 2007. A graphical systems model to facilitate hypothesis-driven ecotoxicogenomics research on the teleost brain–pituitary–gonadal axis. Environmental Science & Technology 41, 321–330.

Warner, R.R., Swearer, S.E., 1991. Social control of sex change in the bluehead wrasse, *Thalassoma bifasciatum* (Pices: Labridae). Biological Bulletin 181, 199–204.

Wourms, J.P., Lombardi, J., 1992. Reflections on the evolution of piscine viviparity. American Zoologist 32, 276–293.

Amphibians

Ben-Yehoshua, L.J., Lewellyn, A.L., Thomas, P., Maller, J.L., 2007. The role of *Xenopus* membrane progesterone receptor-β in mediating the effect of progesterone on oocyte maturation. Molecular Endocrinology 21, 664–673.

Carr, J.A., 2011. Stress and reproduction in amphibians. In: Norris, D.O., Lopez, K.H. (Eds.), Hormones and Reproduction of Vertebrates, vol. 2. Amphibians. Academic Press, San Diego, CA, pp. 99–116.

DiFiore, M.M., Chieffi-Baccari, G., Rastogi, R.K., Di Meglio, M., Pinelli, C., Iela, L., 2005. Hormonal regulation of secondary sex characters. In: Heatwole, H. (Ed.), Amphibian Biology, vol. 6. Endocrinology. Surrey Beatty & Sons, Chipping Norton, New South Wales, Australia, pp. 2228–2249.

Dumont, J.N., Brummett, A.R., 1978. Oogenesis in *Xenopus laevis* (Daudin). V. Relationships between developing oocytes and their investing follicular tissues. Journal of Morphology 155, 73–97.

Flament, S., Dumond, H., Chardard, D., Chesnel, A., 2009. Lifelong testicular differentiation in *Pleurodeles waltl* (Amphibia, Caudata). Reproductive Biology and Endocrinology 7, 21.

Garnier, D.H., 1985. Androgen and estrogen levels in the plasma of *Pleurodeles waltl* Michah. during the annual cycle. II. Female cycle. General and Comparative Endocrinology 60, 414–418.

Greven, H., 2011. Maternal adaptations to reproductive modes in amphibians. In: Norris, D.O., Lopez, K.H. (Eds.), Hormones and Reproduction of Vertebrates, vol. 2. Amphibians. Academic Press, San Diego, CA, pp. 117–141.

Houck, L.D., Woodley, S.K., 1994. Field studies of steroid hormones and male reproductive behaviour in amphibians. In: Heatwole, H. (Ed.), Amphibian Biology, vol. 2. Social Behaviour. Surrey Beatty & Sons, Chipping Norton, New South Wales, Australia, pp. 677–703.

Jorgensen, C.B., 1992. Growth and reproduction. In: Feder, M., Burggren, W.W. (Eds.), Environmental Physiology of the Amphibians. University of Chicago Press, Chicago, IL, pp. 439–466.

Kikuyama, S., Hasunuma, I., Toyoda, F., Haraguchi, S., Tsutsui, K., 2009. Hormone-mediated reproductive behavior in the red-bellied newt. Annals of the New York Academy of Sciences 1163, 179–186.

Kikuyama, S., Nakada, T., Toyoda, F., Iwata, T., Yamaoto, K., Conlon, J.M., 2005. Amphibian pheromones and endocrine control of their secretion. Annals of the New York Academy of Sciences 1040, 123–130.

Kikuyama, S., Yamamotoa, K., Iwataa, T., Toyodab, F., 2002. Peptide and protein pheromones in amphibians. Comparative Biochemistry and Physiology Part B 132, 69–74.

Kvarnryd, M., Grabic, R., Brandt, I., Berg, C., 2010. Early life progestin exposure causes arrested oocyte development, oviductal agenesis and sterility in adult *Xenopus tropicalis* frogs. Aquatic Toxicology 103, 18–24.

Lofts, B., 1984. Amphibians. In: Lamming, G.E. (Ed.), Marshall's Physiology of Reproduction, vol. 1. Reproductive Cycles in Vertebrates. Churchill Livingston, Edinburgh, pp. 127–205.

Lutz, L.B., Cole, L.M., Gupta, M.K., Kwist, K.W., Auchus, R.J., Hammes, S.R., 2001. Evidence that androgens are the primary steroids produced by *Xenopus laevis* ovaries and may signal through the classical androgen receptor to promote oocyte maturation. Proceedings of the National Academy of Sciences 98, 13728–13733.

Lynch, K.S., Crews, D., Ryan, M.J., Wilczynski, W., 2006. Hormonal state influences aspects of female mate choice in the Túngara Frog (*Physalaemus pustulosus*). Hormones and Behavior 49, 450–457.

Lynch, K.S., Wilczynski, W., 2006. Social regulation of plasma estradiol concentration in a female anuran. Hormones and Behavior 50, 101–106.

Medina, M.F., Ramos, I., Crespo, Gonzalez-Calvar, C.A., Fernandez, S.N., 2004. Changes in serum sex steroid levels throughout the reproductive cycle of *Bufo arenarum* females. General and Comparative Endocrinology 136, 143–151.

Moore, F.L., Orchinik, M., 1991. Multiple molecular actions for steroids in the regulation of reproductive behaviors. Seminars in Neuroscience 3, 489–496.

Norris, D.O., 2011. Hormones and reproductive patterns in urodele and gymnophionid amphibians. In: Norris, D.O., Lopez, K.H. (Eds.), Hormones and Reproduction of Vertebrates, vol. 2. Amphibians. Academic Press, San Diego, CA, pp. 187–202.

Olmstead, A.W., Korte, J.J., Woodis, K.K., Bennett, B.A., Ostazeski, S., Degitz, S.J., 2009. Reproductive maturation of the tropical clawed frog: *Xenopus tropicalis*. General and Comparative Endocrinology 160, 117–123.

Pierantoni, R., Cobellis, G., Meccariello, R., Palmiero, C., Fienga, G., Minucci, S., Fasano, S., 2002. The amphibian testis as model to study germ cell progression during spermatogenesis. Comparative Biochemistry and Physiology Part B 132, 131–139.

Polzonetti-Magni, A.M., 1999. Amphibian ovarian cycles. In: Knobil, E., Neill, J.D. (Eds.), Encyclopedia of Reproduction, vol. 1. Elsevier, Amsterdam, pp. 154–160.

Propper, C.R., 2011. Testicular structure and control of sperm development in amphibians. In: Norris, D.O., Lopez, K.H. (Eds.), Hormones and Reproduction of Vertebrates, vol. 2. Amphibians. Academic Press, San Diego, CA, pp. 39–54.

Rastogi, R.K., Iela, L., Di Meglio, M., Di Fiori, M.M., D'Aniello, B., Pinelli, C., Fiorentino, M., 2005. Hormonal regulation of reproductive cycles in amphibians. In: Heatwole, H. (Ed.), Amphibian Biology, vol. 6. Endocrinology. Surrey Beatty & Sons, Chipping Norton, New South Wales, Australia, pp. 2045–2178.

Rastogi, R.K., Pinelli, C., Polese, G., D'Aniello, B., Chieffi-Baccari, G., 2011. Hormones and reproductive cycles in anuran amphibians. In: Norris, D.O., Lopez, K.H. (Eds.), Hormones and Reproduction of Vertebrates, vol. 2. Amphibians. Academic Press, San Diego, CA, pp. 171–186.

Rose, J.D., Kinnaird, J.R., Moore, F.L., 1995. Neurophysiological effects of vasotocin and corticosterone on medullary neurons: implications for hormonal control of amphibian courtship behavior. Neuroendocrinology 62, 406–417.

Sasso-Cerri, E., De Faria, F.V.P., Freymüller, E., Miraglia, S.M., 2004. Testicular morphological changes during the seasonal reproductive cycle in the bullfrog *Rana catesbeiana*. Journal of Experimental Zoology 301A, 249–260.

Sena, J., Zhiming Liu, Z., 2008. Expression of cyclooxygenase genes and production of prostaglandins during ovulation in the ovarian follicles of *Xenopus laevis*. General and Comparative Endocrinology 157, 165–173.

Sever, D.M., Staub, N.L., 2011. Hormones, sex accessory structures, and secondary sexual characteristics in amphibians. In: Norris, D.O., Lopez, K.H. (Eds.), Hormones and Reproduction of Vertebrates, vol. 2. Amphibians. Academic Press, San Diego, CA, pp. 83–98.

Tsai, P.-S., 2011. Neuroendocrine control of reproduction in amphibians. In: Norris, D.O., Lopez, K.H. (Eds.), Hormones and Reproduction of Vertebrates, vol. 2. Amphibians. Academic Press, San Diego, CA, pp. 21–37.

Uribe, M.C.A., 2011. Hormones and the female reproductive system of amphibians. In: Norris, D.O., Lopez, K.H. (Eds.), Hormones and Reproduction of Vertebrates, vol. 2. Amphibians. Academic Press, San Diego, CA, pp. 55–81.

Wake, M.H., 1985. Oviduct structure and function in nonmammalian vertebrates. In: Duncker, H.-R., Fleischer, G. (Eds.), Functional Morphology in Vertebrates. Gustav Fischer Verlag, Stuttgart, pp. 427–435.

Woodley, S.K., 2011. Hormones and reproductive behavior in amphibians. In: Norris, D.O., Lopez, K.H. (Eds.), Hormones and Reproduction of Vertebrates, vol. 2. Amphibians. Academic Press, San Diego, CA, pp. 143–169.

Reptiles

Albergotti, L.C., Guillette Jr., L.J., 2011. Viviparity in reptiles: evolution and reproductive endocrinology. In: Norris, D.O., Lopez, K.H. (Eds.), Hormones and Reproduction of Vertebrates, vol. 3. Reptiles. Academic Press, San Diego, CA, pp. 247–275.

Al-Habsi, A.A., AlKindi, A.Y.A., Mahmoud, I.Y., Owens, D.W., Khan, T., Al-Abri, A., 2006. Plasma hormone levels in the green turtles *Chelonia mydas* during peak period of nesting at Ras Al-Hadd-Oman. Journal of Endocrinology 191, 9–14.

Blackburn, D.G., 1992. Convergent evolution of viviparity, matrotrophy, and specializations for fetal nutrition in reptiles and other vertebrates. American Zoologist 32, 313–321.

Blanvillain, G., Owens, D.W., Kuchling, G., 2011. Hormones and reproductive cycles in turtles. In: Norris, D.O., Lopez, K.H. (Eds.), Hormones and Reproduction of Vertebrates, vol. 3. Reptiles. Academic Press, San Diego, CA, pp. 277–303.

Cease, A.J., Lutterschmidt, D.I., Mason, R.T., 2007. Corticosterone and the transition from courtship behavior to dispersal in male red-sided garter snakes (*Thamnophis sirtalis parietalis*). General and Comparative Endocrinology 150, 124–131.

Cree, A., Guillette Jr., L.J., Cockrem, J.F., Brown, M.A., Chambers, G.K., 1990. Absence of daily cycles in plasma sex steroids in male and female tuatara (*Sphenodon punctatus*), and the effects of acute capture stress on females. General and Comparative Endocrinology 79, 103–113.

Crews, D., 1983. Alternative reproductive tactics in reptiles. Bioscience 33, 562–566.

Crews, D., 2005. Evolution of neuroendocrine mechanisms that regulate sexual behavior. Trends in Endocrinology & Metabolism 16, 354–361.

Duvall, D., Guillette Jr., L.J., Jones, R.E., 1982. Environmental control of reptilian reproductive cycles. In: Gans, C., Pough, H. (Eds.), Biology of the Reptilia, vol. 13. Academic Press, New York, pp. 201–231.

Gist, D.H., 2011. Hormones and the sex ducts and sex accessory structures of reptiles. In: Norris, D.O., Lopez, K.H. (Eds.), Hormones and Reproduction of Vertebrates, vol. 3. Reptiles. Academic Press, San Diego, CA, pp. 117–139.

Guillette Jr., L.J., 1990. Prostaglandins and reproduction in reptiles. In: Epple, A., Scanes, C.G., Stetson, M.H. (Eds.), Progress in Comparative Endocrinology. Wiley-Liss, New York, pp. 603–607.

Hews, D.K., Hara, E., Anderson, M.C., 2012. Sex and species differences in plasma testosterone and in counts of androgen receptor-positive cells in key brain regions of *Sceloporus* lizard species that differ in aggression. General and Comparative Endocrinology 176, 493–499.

Johnson, M.A., Wade, J., 2011. Neuroendocrinology of reptilian reproductive behavior. In: Norris, D.O., Lopez, K.H. (Eds.), Hormones and Reproduction of Vertebrates, vol. 3. Reptiles. Academic Press, San Diego, CA, pp. 39–61.

Jones, R.E., Propper, C.R., Rand, M.S., Austin, H.B., 1991. Loss of nesting behavior and the evolution of viviparity in reptiles. Ethology 88, 331–341.

Jones, S.M., 2011. Hormonal regulation of ovarian function in reptiles. In: Norris, D.O., Lopez, K.H. (Eds.), Hormones and Reproduction of Vertebrates, vol. 3. Reptiles. Academic Press, San Diego, CA, pp. 89–115.

Katsu, Y., Matsubara, K., Kohno, S., Matsuda, Y., Toriba, M., Oka, K., Guillette Jr., L.J., Ohta, Y., Iguchi, T., 2010. Molecular cloning, characterization, and chromosome mapping of reptilian estrogen receptors. Endocrinology 151, 5710–5720.

Kumar, S., Roy, B., Rai, U., 2011. Hormonal regulation of testicular functions in reptiles. In: Norris, D.O., Lopez, K.H. (Eds.), Hormones and Reproduction of Vertebrates, vol. 3. Reptiles. Academic Press, San Diego, CA, pp. 63–88.

Lance, V., Callard, I.P., 1978. *In vivo* responses of female snakes (*Natrix fasciata*) and female turtles (*Chrysemys picta*) to ovine gonadotropins (FSH and LH) as measured by plasma progesterone, testosterone, and estradiol levels. General and Comparative Endocrinology 35, 295–301.

Licht, P., 1984. Reptiles. In: Lamming, G.E. (Ed.), Marshall's Physiology of Reproduction, vol. 1. Reproductive Cycles in Vertebrates. Churchill Livingston, Edinburgh, pp. 206–282.

Lind, C.M., Husak, J.F., Eikenaar, C., Moore, I.T., Taylor, E.N., 2010. The relationship between plasma steroid hormone concentrations and the reproductive cycle in the Northern Pacific rattlesnake, *Crotalus oreganus*. General and Comparative Endocrinology 166, 590–599.

Lovern, M.B., 2011. Hormones and reproductive cycles in lizards. In: Norris, D.O., Lopez, K.H. (Eds.), Hormones and Reproduction of Vertebrates, vol. 3. Reptiles. Academic Press, San Diego, CA, pp. 321–353.

Martín, J., López, P., 2011. Pheromones and reproduction in reptiles. In: Norris, D.O., Lopez, K.H. (Eds.), Hormones and Reproduction of Vertebrates, vol. 3. Reptiles. Academic Press, San Diego, CA, pp. 141–167.

Mason, R.T., 1993. Chemical ecology of the red-sided garter snake, *Thamnophis sirtalis parietalis*. Brain, Behavior and Evolution 41, 261–268.

Milnes, M.R., 2011. Hormones and reproductive cycles of crocodilians. In: Norris, D.O., Lopez, K.H. (Eds.), Hormones and Reproduction of Vertebrates, vol. 3. Reptiles. Academic Press, San Diego, CA, pp. 305–319.

Moore, B.C., Forouhar, Kohno, S., Botteri, N.L., Hamlin, H.J., Guillette Jr., L.J., 2012. Gonadotropin-induced changes in oviducal mRNA expression levels of sex steroid hormone receptors and activin-related signaling factors in the alligator. General and Comparative Endocrinology 175, 251–258.

Sinervo, B., Miles, D.B., 2011. Hormones and behavior of reptiles. In: Norris, D.O., Lopez, K.H. (Eds.), Hormones and Reproduction of Vertebrates, vol. 3. Reptiles. Academic Press, San Diego, CA, pp. 215–246.

Taylor, E.N., DeNardo, D.F., 2011. Hormones and reproductive cycles in snakes. In: Norris, D.O., Lopez, K.H. (Eds.), Hormones and Reproduction of Vertebrates, vol. 3. Reptiles. Academic Press, San Diego, CA, pp. 355–373.

Torkarz, R.R., Summers, C.H., 2011. Stress and reproduction in reptiles. In: Norris, D.O., Lopez, K.H. (Eds.), Hormones and Reproduction of Vertebrates, vol. 3. Reptiles. Academic Press, San Diego, CA, pp. 169–213.

Valiante, S., Prisco, M., Sciarrillo, R., De Falco, M., Capaldo, A., Gay, F., Andreuccetti, P., Laforgia, V., Varano, L., 2008. Pituitary adenylate cyclase-activating polypeptide, vasoactive intestinal polypeptide and their receptors: distribution and involvement in the secretion of *Podarcis sicula* adrenal gland. Journal of Endocrinology 196, 291–303.

Birds

Adkins-Regan, E., 2005. Female mate choice. In: Dawson, A., Sharp, P.J. (Eds.), Functional Avian Endocrinology. Narosa Publishing House, New Dehli, pp. 341–350.

Albergotti, L.C., Hamlin, H.J., McCoy, M.W., Guillette Jr., L.J., 2009. Endocrine activity of extraembryonic membranes extends beyond placental amniotes. PLoS ONE 4 (5), e5452.

Ball, G.F., 1993. The neural integration of environmental information by seasonally breeding birds. American Zoologist 33, 185–199.

Balthazart, J., Ball, G.F., 1995. Sexual differentiation of brain and behavior in birds. Trends in Endocrinology & Metabolism 6, 21–29.

Balthazart, J., Taziaux, M., Holloway, K., Ball, G.F., Cornil, C.A., 2009. Behavioral effects of brain-derived estrogens in birds. Annals of the New York Academy of Sciences 1163, 31–48.

Blas, J., López, L., Tanferna, A., Sergio, F., Hiraldo, F., 2010. Reproductive endocrinology of wild, long-lived raptors. General and Comparative Endocrinology 168, 22–28.

Breuner, C.W., 2011. Stress and reproduction in birds. In: Norris, D.O., Lopez, K.H. (Eds.), Hormones and Reproduction of Vertebrates, vol. 4. Birds. Academic Press, San Diego, CA, pp. 129–151.

Degen, A.A., Weil, S., Rosenstrauch, A., Kam, M., Dawson, A., 1994. Seasonal plasma levels of luteinizing and steroid hormones in male and female domestic ostriches (*Struthio camelus*). General and Comparative Endocrinology 93, 21–27.

Deviche, P., 1995. Androgen regulation of avian premigratory hyperphagia and fattening: from ecophysiology to neuroendocrinology. American Zoologist 35, 234–245.

Deviche, P., Hurley, L.L., Fokidis, H.B., 2011. Avian testicular structure, function and regulation. In: Norris, D.O., Lopez, K.H. (Eds.), Hormones and Reproduction of Vertebrates, vol. 4. Birds. Academic Press, San Diego, CA, pp. 27–70.

Fivizzani, A.J., Colwell, M.A., Oring, L.W., 1986. Plasma steroid hormone levels in free living Wilson's phalaropes, *Phalaropus tricolor*. General and Comparative Endocrinology 62, 137–144.

Follett, B.K., 1984. Birds. In: Lamming, G.E. (Ed.), Marshall's Physiology of Reproduction, vol. 1. Reproductive Cycles in Vertebrates. Churchill Livingston, Edinburgh, pp. 283–350.

Johnson, A.L., 2011. Organization and functional dynamics of the avian ovary. In: Norris, D.O., Lopez, K.H. (Eds.), Hormones and Reproduction of Vertebrates, vol. 4. Birds. Academic Press, San Diego, CA, pp. 71–90.

Lofts, B., Murton, R.K., 1973. Reproduction in birds. In: Farner, D.S., King, J.R. (Eds.), Avian Biology, vol. 3. Academic Press, San Diego, CA, pp. 1–107.

Mauget, R., Jouventin, P., Lacroix, A., Ishii, S., 1994. Plasma LH and steroid hormones in King penguins (*Aptenodytes patagonicus*) during the onset of the breeding cycle. General and Comparative Endocrinology 93, 36–43.

Ramenofsky, M., 2011. Hormones in migration and reproductive cycles of birds. In: Norris, D.O., Lopez, K.H. (Eds.), Hormones and Reproduction of Vertebrates, vol. 4. Birds. Academic Press, San Diego, CA, pp. 205–237.

Riters, L.V., Alger, S.J., 2011. Hormonal regulation of avian courtship and mating behaviors. In: Norris, D.O., Lopez, K.H. (Eds.), Hormones and Reproduction of Vertebrates, vol. 4. Birds. Academic Press, San Diego, CA, pp. 153–180.

Sharp, P.J., 2005. Photoperiodic regulation of seasonal breeding in birds. Annals of the New York Academy of Sciences 1040, 189–199.

Staub, N., De Beer, M., 1997. The role of androgens in female vertebrates. General and Comparative Endocrinology 108, 1–24.

Schoech, S.J., Rense, M.A., Bridge, E.S., Boughton, R.K., Wilcoxen, T.E., 2009. Environment, glucocorticoids, and the timing of reproduction. General and Comparative Endocrinology 163, 201–207.

Ubuka, T., Bentley, G.E., 2011. Neuroendocrine control of reproduction in birds. In: Norris, D.O., Lopez, K.H. (Eds.), Hormones and Reproduction of Vertebrates, vol. 4. Birds. Academic Press, San Diego, CA, pp. 205–237.

Vleck, C.M., Vleck, D., 2011. Hormones and regulation of parental behavior in birds. In: Norris, D.O., Lopez, K.H. (Eds.), Hormones and Reproduction of Vertebrates, vol. 4. Birds. Academic Press, San Diego, CA, pp. 181–203.

Von Engelhardt, N., Groothuis, T.G.G., 2011. Maternal hormones in avian eggs. In: Norris, D.O., Lopez, K.H. (Eds.), Hormones and Reproduction of Vertebrates, vol. 4. Birds. Academic Press, San Diego, CA, pp. 91–127.

Wingfield, J.C., Ball, G.F., Dufty, A.M., Hegner, R.E., Ramenofsky, M., 1987. Testosterone and aggression in birds. American Scientist 75, 602–608.

Wingfield, J.C., O'Reilly, K.M., Astheimer, L.B., 1995. Modulation of the adrenocortical response to acute stress in arctic birds: a possible ecological basis. American Zoologist 35, 285–294.

Yoshida, N., Mita, K., Yamashita, M., 2000. Comparative study of the molecular mechanisms of oocyte maturation in amphibians. Comparative Biochemistry and Physiology Part B 126, 189–197.

Endocrine Disruption and Reproduction

Boggs, A.S.P., Botteri, N.L., Hamlin, H.J., Guillette Jr., L.J., 2011. Endocrine disruption of reproduction in reptiles. In: Norris, D.O., Lopez, K.H. (Eds.), Hormones and Reproduction of Vertebrates, vol. 3. Reptiles. Academic Press, San Diego, CA, pp. 373–396.

Norris, D.O., 2011. Endocrine disruption of reproduction in amphibians. In: Norris, D.O., Lopez, K.H. (Eds.), Hormones and Reproduction of Vertebrates, vol. 2. Reptiles. Academic Press, San Diego, CA, pp. 203–211.

Ottinger, M.A., Dean, K., McKernan, M., Quinn Jr., M.J., 2011. Endocrine disruption of reproduction in birds. In: Norris, D.O., Lopez, K.H. (Eds.), Hormones and Reproduction of Vertebrates, vol. 4. Birds. Academic Press, San Diego, CA, pp. 239–260.

Vajda, A.M., Norris, D.O., 2011. Endocrine-active chemicals (EACs) in fishes. In: Norris, D.O., Lopez, K.H. (Eds.), Hormones and Reproduction of Vertebrates, vol. 1. Fishes. Academic Press, San Diego, CA, pp. 245–264.

Chapter 12

Chemical Regulation of Feeding, Digestion and Metabolism

Survival of any vertebrate requires ingestion of a nutrient source, enzymatic digestion of its macromolecules, and absorption of the end products of digestion. Once absorbed, these substrates may be used as energy sources or for synthesis of various molecules, or they may be stored in some form for later utilization. The processes responsible for these events are closely regulated by the nervous system and by hormones arising from the gastrointestinal (GI) tract, the endocrine pancreas, and the hypothalamus-pituitary system. In this chapter, we examine hormones that influence feeding behavior, hormones that regulate the processes involved in digestion of food, and the biochemical pathways involved in metabolism of ingested nutrients in mammals. These processes for non-mammals are discussed in Chapter 13.

I. REGULATION OF FEEDING

Ingestion of food is regulated by a complex system of peptide hormones and neural factors that is incompletely understood at this time. Numerous bioregulators have been described that stimulate or inhibit appetite and feeding behavior. Bioregulators that stimulate appetite and feeding are termed **orexic agents** (*orexin*, appetite), whereas those that suppress appetite and feeding are termed **anorexic agents**. These factors are summarized later in Table 12-3. Most of these factors are known for other roles as well as being related to metabolism or other events. Some of these bioregulators are produced in the hypothalamus, whereas others come from the gastrointestinal tract or adipose tissue. A brief account of some of the better known factors is provided here. Other peptides from adipose cells appear to be involved and may soon be established as important regulators of feeding and obesity, the latter possibly being the major clinical concern of the 21st century (see ahead).

A. Peripheral Bioregulators Controlling Feeding

Animals must eat to maintain a positive energy balance, but there are costs and risks associated with overeating just as there are fitness consequences associated with not meeting energy demands. When engaged in feeding behavior, animals may be more vulnerable to predation, and hundreds of ecological studies indicate that there is a trade-off between the amount of time an animal spends feeding or avoiding predators, the so-called **foraging and predator avoidance trade-off**. Simply stated, most animals cannot always be vigilant for predators when feeding. Overfeeding also can reduce the ability of an animal to escape predators by making the animal less mobile. Because predator avoidance is a major evolutionary selection pressure, it is not surprising that feeding behavior is intimately linked to peripheral factors from the gastrointestinal tract and adipose tissue that inform the organism about energy status and food intake. Interestingly, the large majority of these peripheral factors are inhibitory (Table 12-1), suggesting that timely inhibition of feeding is evolutionarily adaptive. Factors that inhibit appetite are sometime referred to as *satiety signals* because they reduce the size of the meal that is eaten. Most of these peripheral bioregulators act within the central nervous system (CNS) to exert their effects on food intake (Figure 12-1).

1. Leptin and Other Anorexic Bioregulators of Feeding from Adipose Tissue

Researchers now recognize several bioregulators from adipose tissue that may affect feeding. Collectively, these adipose bioregulators are called **adipokines.** These are summarized in Table 12-2, although only those with a verified known role in feeding behavior will be discussed here. For decades scientists speculated that bioregulators from adipose tissue inform the brain about the quantity of lipid reserves, acting as a sort of **lipostat**. The first of these to be discovered was a large peptide (167 amino acids) called **leptin** (*leptos*, thin), which curbs appetite and increases metabolic rate. Leptin was first isolated from adipose tissue in mice as a result of studies of a mutant gene called the obese gene (*ob*). Mice homozygous for this mutant (*ob/ob*) are obese, are deficient in leptin, and have higher levels of endogenous **cannabinoids** (see ahead). Leptin is secreted in greater amounts when an animal puts on fat, reduces appetite, and causes the animal to lose weight.

Vertebrate Endocrinology. http://dx.doi.org/10.1016/B978-0-12-394815-1.00012-4

TABLE 12-1 Peripheral Peptides Regulating Appetite and Feeding in Mammals

Anorexic Peptides	Source	Orexic Peptides	Source
Amylin	Pancreas	Ghrelin	Stomach
Bombesin	Stomach		
Cholecystokinin	Small intestine		
Enterostatin	Small intestine		
Glucagon	Pancreas		
Glucagon-like Peptide 1	Small intestine		
Insulin	Pancreas		
Interleukin-6 (IL-6)	Adipose tissue		
Leptin	Adipose tissue		
Nesfatin-1	Adipose tissue		
Somatostatin	Pancreas, small intestine		
Tumor necrosis factor alpha (TNFα)	Adipose tissue		

Leptin binds to receptors in the hypothalamus and elsewhere in the brain that operate through a **JAK–STAT (Janus kinase-signal transducer and activator of transcription)** pathway (Figure 12-2). It is the long form of this receptor (**ObRb**) that is present in neurons within several brain areas within the hypothalamus (**arcuate nucleus, ARC; dorsomedial nucleus, DMN; ventromedial nucleus, VMN**) and outside the hypothalamus (**ventral tegmental area, VTA; nucleus of the solitary tract, NTS**) (Figure 12-1). Phosphorylation of ObRb by **Janus kinase 2 (JAK2)** leads to phosphorylation of **signal transducer and activator of transcription 3** (**STAT3**), a transcription factor responsible for altering the expression of certain genes involved in leptin signaling in neurons including a protein called **suppressor of cytokine signaling 3 (SOCS3)**, which acts in a feedback loop to inhibit ObRb (Figure 12-1). Leptin acting through the ObRb also induces **mitogen-activated protein kinases (MAPKs)**, which further alter transcription, eventually leading to the opening or closing of ion channels in the target neurons. A shorter form of the receptor (**ObRa**) may be involved in transporting leptin across the blood–brain barrier. Leptin action on ObRb receptors may partly block the orexic action of the peptide **ghrelin** in the brain (see ahead).

High leptin levels may lead to resistance to leptin action in the brain via a number of different mechanisms. During

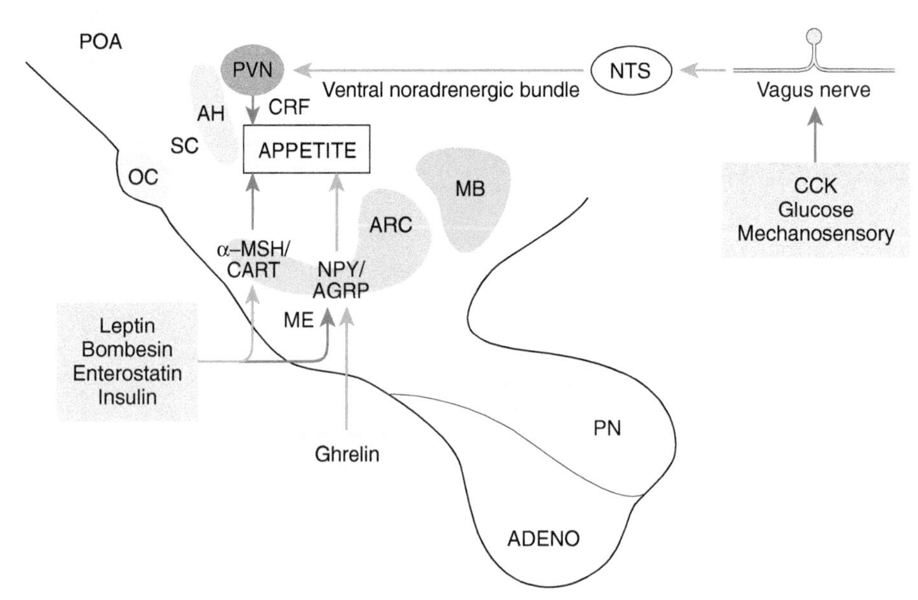

FIGURE 12-1 **Integration of peripheral bioregulators with hypothalamic feeding areas to regulate food intake.** Many bioregulators from adipose tissue (leptin), the pancreas (insulin), and the gastrointestinal tract (bombesin, enterostatin, ghrelin) modulate feeding by stimulating or inhibiting neurons in the arcuate nucleus of the hypothalamus that regulate appetite. Leptin and possibly other anorexigenic bioregulators (red) from the periphery turn on POMC neurons, producing alpha-melanocyte-stimulating hormone (α-MSH) and cocaine- and amphetamine-related transcript (CART) in the arcuate nucleus, and inhibit the neurons producing the orexigenic peptides neuropeptide Y (NPY) and agouti-related protein (AgRP). In addition, chemosensory (cholecystokinin, CCK; glucose) and mechanosensory (stretch) signals are transmitted via the vagus nerve to the nucleus of the solitary tract (NTS) in the brainstem, which conveys signals to the paraventricular nucleus (PVN) of the hypothalamus where CRH is secreted to inhibit appetite. See Chapter 4 for hypothalamus and pituitary nomenclature and Appendix A for other abbreviations.

TABLE 12-2 Adipokine Bioregulators in Mammals

Adipokine	Number of Amino Acids	Primary Function
Adiponectin	244	Stimulates fatty acid oxidation; insulin sensitizer
Adipsin	228	Stimulates insulin secretion; induces fatty acid and glucose uptake by adipocytes
Apelin	36,17,13	Promotes insulin sensitivity; induces glucose uptake by skeletal muscle
Interleukin-6	184	Inhibits feeding; induces cachexia
Leptin		Inhibits feeding; promotes fatty acid oxidation in muscle and adipose; insulin sensitizer
Resistin	94	Causes insulin resistance
TNFα	157	Inhibits feeding; induces cachexia
Visfatin; nicotinamide phosphoribosyltransferase (Nampt)	491	Regulation of cellular metabolism through nicotinamide (NAD) biosynthesis

pregnancy, blood levels of leptin in the expecting mother increase due to greater adipose tissue and leptin secretion from the placenta (Figure 12-3). By mid-pregnancy, expression of the gene encoding the leptin receptor in the ventromedial hypothalamus is reduced, thereby reducing the activation of satiety neurons in this brain area (Figure 12-3). Furthermore, expression of the leptin transporter in circumventricular organs is reduced and plasma binding protein for leptin is increased, reducing leptin entry into the CNS. All of these factors allow women to maintain their appetite despite elevated leptin during pregnancy, presumably due to increased **neuropeptide Y (NPY)** secretion in the ARC (Figure 12-3). In obese humans, leptin levels are about 5 times those of lean people,

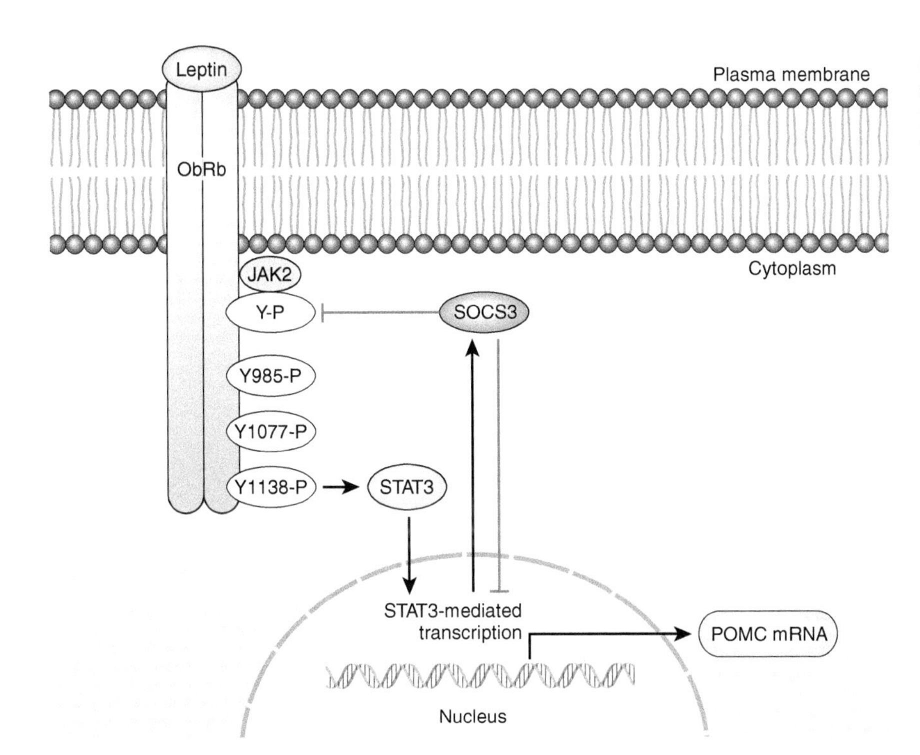

FIGURE 12-2 **Binding of leptin to its receptor leads to Janus kinase 2 (JAK2) tyrosine kinase phosphorylation of tyrosines (Y) in positions 985, 1077, and 1138.** Phosphorylation of Y1138 leads to activation and translocation of signal transducer and activator of transcription 3 (STAT3) to the nucleus, where it modulates transcription of genes involved in leptin action—for example, proopiomelanocortin (POMC)—as well as suppressor of cytokine signaling (SOCS3), which inhibits further leptin action. *(Adapted with permission from Ahima, R.S. and Lazar, M.A.,* Molecular Endocrinology, *22, 1023–1031, 2008.)*

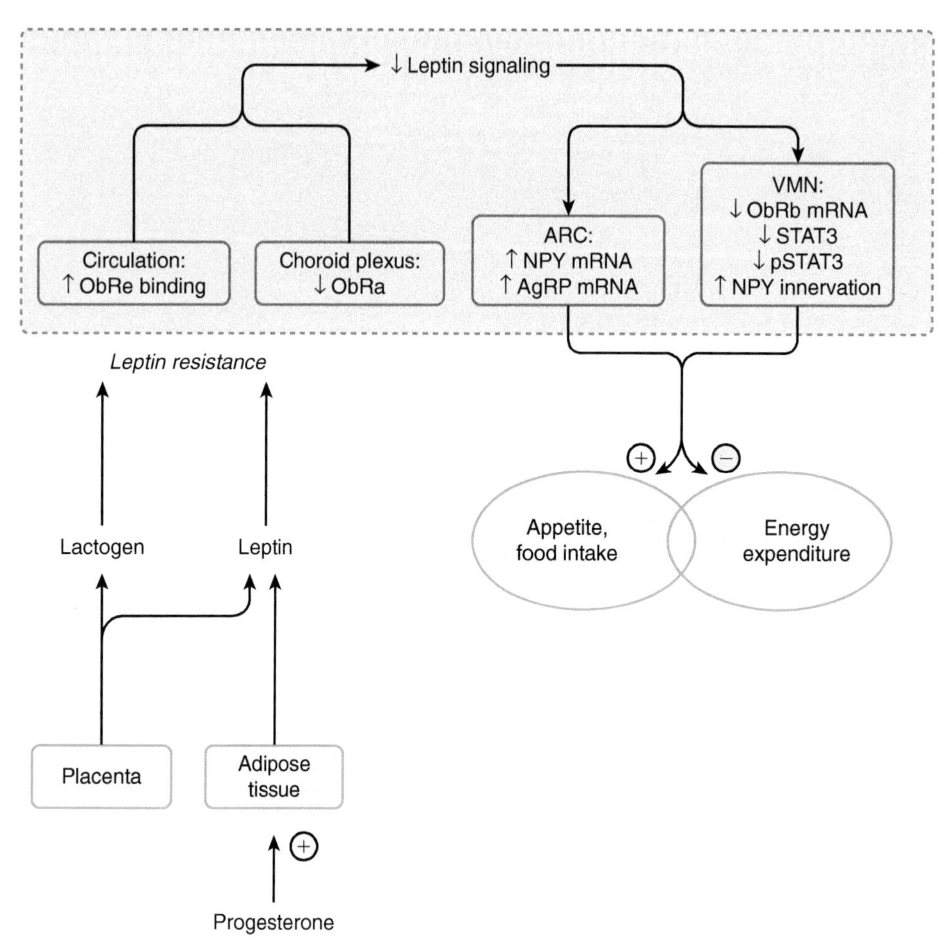

FIGURE 12-3 **Leptin resistance occurs at the two levels during pregnancy.** During pregnancy, leptin blood levels increase due to progesterone stimulation of adipose tissue, potentially harming both the health of both the mother and fetus by inhibiting appetite. However, circulating leptin becomes less available to target neurons in the hypothalamus due to an increase in circulating leptin binding protein (ObRe), which sequesters the circulating leptin, and reduced expression of the leptin transport protein (ObRa) in the choroid plexus, reducing leptin entry into the CNS. Reduced leptin signaling at the level of the arcuate (ARC) and ventromedial (VMN) nuclei of the hypothalamus leads to increased secretion of orexigenic peptides NPY and AgRP and increased appetite. Abbreviations: AgRP, agouti-related protein; NPY, neuropeptide Y; ObRa, short form of leptin receptor; ObRb, long form of leptin receptor; ObRe, plasma leptin binding protein; pSTAT3, phosphorylated signal transducer and activator of transcription 3; STAT3, signal transducer and activator of transcription 3. *(Adapted with permission from Brunton, P.J. and Russell, J.A.,* Nature Reviews Neuroscience, *9, 11–25, 2008.)*

suggesting that either this protein is not regulating fat levels in humans or that obese people do not respond to this signal in the same manner as do leaner people. Leptin also has been claimed to have affects on reproduction, and its anorexic action may delay puberty (see Chapter 10).

The discovery of leptin has led to a search for other adipokine bioregulators (Table 12-2). While several other adipokine bioregulators have been discovered, their role in feeding is not as well established as for leptin. **Adiponectin** is secreted in large amounts by white adipose tissue and circulates in the plasma complexed as trimers or higher molecular weight hexamers. **AdipoR1 and AdipoR2** have been identified as receptors for adiponectin and are found in key appetite-regulating areas of the brain, including the **proopiomelanocortin (POMC)/cocaine- and amphetamine-regulating transcript (CART)** and **NPY/agouti-related protein (AgRP)** neurons in the ARC and neurons in the brainstem. Both adiponectin receptors are seven-transmembrane domain receptors with an extracellular carboxy terminus and an intracellular amino terminus, which is just the opposite of every other known G-protein-coupled receptor (see Chapter 3). A potential role for adiponectin in appetite has yet to be firmly established, although adiponectin levels in blood are elevated during fasting and decreased in obesity, exactly opposite to the patterns observed for plasma leptin levels. In contrast, the adipokine **nucleobindin 2-encoded satiety- and fat-influencing protein (nesfatin)** appears to be a potent suppressor of appetite. Unlike leptin, nesfatin mRNA appears to be expressed in key appetite regulatory neurons of the ARC and NTS in the brainstem. The role of nesfatin as a hypothalamic regulator of appetite is well established and is discussed in more detail below. Two well-established cytocrines (see Chapter 3), **tumor**

necrosis factor alpha (TNFα) and **interleukin-6 (IL-6)**, are produced by adipose tissue, inhibit feeding, and have been linked to **cachexia**, a form of anorexia and weight loss that cannot be reversed by nutritional intervention. Cachexia accompanies such diseases as cancer or AIDS. **Visfatin**, an adipokine protein whose blood level closely correlates with white adipose tissue mass, was discovered in 2005. A central role for visfatin in feeding has not been established, although visfatin has been measured in the cerebrospinal fluid, suggesting that it may be transported across the blood–brain barrier into the brain.

2. Peripheral GI Peptides Inhibiting Feeding

In addition to mechanical signals that relay information about stretching from the stomach to brainstem areas regulating food intake, a number of peptides released from the gastrointestinal tract and pancreas inhibit appetite. The best studied of these is **cholecystokinin (CCK)**, a peptide that is secreted into the blood by duodenal cells in response to various nutrients present in lumen of the small intestine after a meal. Intravenous or direct administration of CCK into the intestinal lumen reduces meal size in a number of mammalian species, including humans. An endogenous role for intestinal CCK in regulating appetite is supported by data suggesting that administration of CCK-1 receptor antagonists before a meal leads to a larger meal size in rodents and humans.

Several other gastrointestinal peptides suppress appetite (Table 12-1). **Intestinal peptide YY (PYY)** is a gut peptide that also suppresses appetite. PYY inhibits release of NPY and stimulates **α-melanocyte-stimulating hormone (α-MSH)** release through actions at the ARC (see ahead). Furthermore, **pancreatic polypeptide (PP)** from the endocrine pancreas appears to reduce appetite, but its physiological role in not yet confirmed. **Oxyntomodulin**, a byproduct of posttranslational processing of the glucagon preprohormone in intestinal and brain tissue, also suppresses appetite. One of the more interesting peptides in this grouping is **glucagon-like peptide 1 (GLP-1)**, an intestinal hormone that reduces appetite and food intake. GLP-1 also enhances pancreatic secretion of insulin (see ahead).

3. Ghrelin

Ghrelin, a 28-amino acid peptide (Figure 12-4), is the only peripheral orexic bioregulator known. It was first discovered as an endogenous ligand for the **growth hormone (GH) secretagogue receptor (GHS-1a)** that can result in release of GH from the pituitary. Ghrelin is secreted by cells in the stomach epithelium and stimulates feeding (Table 12-1). Plasma levels of ghrelin increase prior to mealtimes as well as at night (Figure 12-5) and may be a hunger signal for the brain. Ghrelin-secreting neurons also are present within the brain and appear to be involved

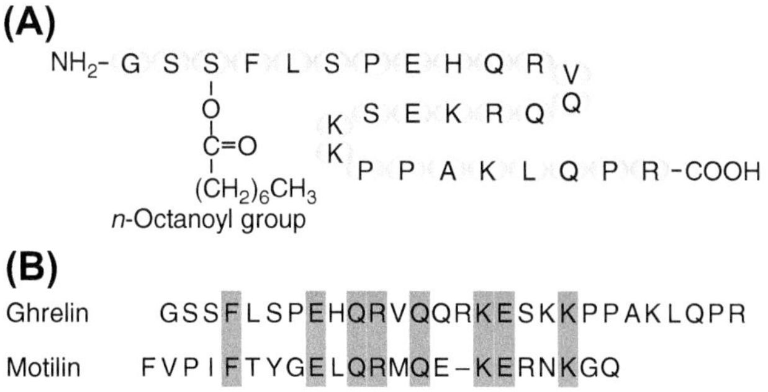

(A)
NH_2- G S S F L S P E H Q R V Q Q R K E S K K P P A K L Q P R -COOH
O–C=O–$(CH_2)_6CH_3$
n-Octanoyl group

(B)
Ghrelin GSSFLSPEHQRVQQRKESKKPPAKLQPR
Motilin FVPIFTYGELQRMQE–KERNKGQ

FIGURE 12-4 **Structure of ghrelin.** See Appendix C for an explanation of amino acid abbreviations. *(Adapted with permission from Korbonits, M. et al.,* Frontiers in Neuroendocrinology, *25, 27–68, 2004. © Elsevier Science, Inc.)*

in control of feeding. It has been associated with reduced gastric emptying that probably contributes to a sense of satiety. Some reports claim that **galanin**, the same peptide that stimulates LH release from the pituitary (Chapter 4), triggers a craving for fatty foods. Levels of galanin rise in the general circulation prior to lunch and dinner. Galanin is also thought to cause a weight increase in adolescent girls at puberty. This action fits with its known role in reproduction.

B. Central Bioregulators Controlling Feeding

Early brain lesion studies identified the ventromedial region of the hypothalamus as the "satiety center" and the lateral hypothalamus as the "feeding center." Subsequent studies on the neural circuits involved and the location of receptors for regulatory factors that influence feeding have implicated a number of hypothalamic centers, including the ARC, VMN, DMN, **lateral hypothalamus (LatH)**, and **paraventricular nucleus (PVN)**. These factors are summarized in Table 12-3.

1. Inhibition of Feeding

POMC neurons (see Chapter 4) in the brain are intimately associated with feeding in a complex manner. β-Endorphin released from POMC neurons is clearly orexic and may be involved in the increase in meal size that is observed with gentle or mild stressor exposure as well as hedonic or pleasurable eating, also called binge-eating. α-MSH, on the other hand, is a potent anorexic peptide. In simple terms, one would think that the activity of both peptides would just cancel each other out, as both are released simultaneously from POMC neurons in equimolar amounts; however, intensive research over the last 18 years has revealed that POMC neurons are a dominant anorexic influence in the hypothalamus. Why the effects of endorphin release from these neurons are masked is unclear but it may be due to

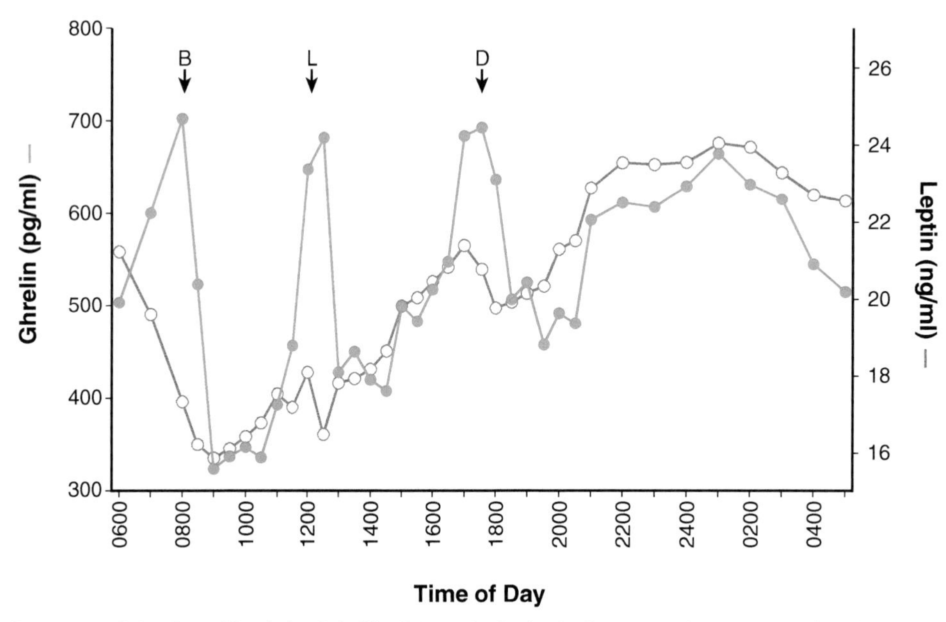

FIGURE 12-5 **Daily pattern of ghrelin and leptin levels in blood.** Note the rise in ghrelin occurs prior to each meal (B, breakfast; L, lunch; D, dinner), whereas there is a steady rise in appetite-suppressing leptin during the day. *(Reprinted with permission from Cummings, D.E. et al.,* Diabetes, ***50****, 1714–1719, 2001.)*

TABLE 12-3 **Brain Peptides Regulating Appetite and Feeding in Mammals**

Peptide	Number of Amino Acids	Source	Primary Function
Agouti-related peptide (AgRP)	132	Neurons in ARC	Orexic
CCK-8	8	Neurons in PVN	Anorexic
Cocaine and amphetamine-regulated transcript (CART)	40,47	Neurons in ARC, NTS, PVN	Anorexic
Corticotropin-releasing hormone (CRH)	41	Neurons in PVN	Anorexic
β-Endorphin	31	Neurons in ARC; neurons in NTS	Orexic
Galanin	29	Neurons in ARC, PVN	Orexic
Melanin-concentrating hormone (MCH)	19	Neurons in lateral hypothalamus	Orexic
α-Melanocyte-stimulating hormone (α-MSH)	13	Neurons in ARC; neurons in NTS	Anorexic
Neuropeptide Y	36	Neurons in ARC	Orexic
Orexin A	33	Neurons in lateral hypothalamus	Orexic
Orexin B	28	Neurons in lateral hypothalamus	Orexic
Urocortin-I (Ucn-I)	40	Neurons in Edinger-Westphal nucleus	Anorexic
Urocortin-II (Ucn-II)	38	Neurons in ARC and PVN	Anorexic
Urocortin-III (Ucn-III)	38	Neurons in rostral perifornical hypothalamic area	Anorexic

other anorexic bioregulators released from POMC neurons in addition to α-MSH, including the anorexic peptide CART. Interest in the role of POMC neurons in appetite regulation hit the mainstream when it was discovered that these neurons are a hotbed for leptin receptors. Exactly how leptin reaches ARC neurons is still a matter of debate, although there appear to be leptin transporters in the median eminence and arcuate nucleus that shuttle leptin from the blood circulation into the extravascular spaces in the ARC. α-MSH binds to **MC3R** and **MC4R melanocortin receptors** in the hypothalamus (see Chapter 4). Knockout mice lacking MC4R receptors or mice possessing a mutant MC4R gene exhibit obesity. In humans, mutations in the POMC gene lead not only to early onset obesity but changes in hair color due to insufficent melanin production as well as adrenal insufficiency (Figure 12-6). AgRP, which is an inverse agonist of the MC4R and is coexpressed in NPY neurons in the ARC, has orexigenic effects due to its blockade of MC4R receptors in the CNS. **Neurotensin (NT)** neurons also appear to be leptin targets and can stimulate neurons that release **corticotropin-releasing hormone (CRH)**, a potent anorexic peptide in the PVN (Figure 12-1).

Some additional peptides have been found to have anorexic actions (Table 12-3). A peptide originally named **prolactin-releasing peptide (PrRP)** has been reported to be a more affective appetite suppressant, especially in fishes, where it also has prolactin-releasing action. **Urocortin-II (Ucn-II)**, a hypothalamic peptide that binds to the **CRHR2 receptor** (see Chapter 4) has been shown to be a powerful appetite suppressant. CCK_8, an endogenous and potent agonist of the CCK-1 receptor (CCK1R), is secreted by hypothalamic neurons and decreases food intake (appetite suppression) in rats. **Nesfatin-1**, which is proteolytically cleaved from the protein **nucleobindin 2** (Figure 12-7), is co-localized within α-MSH/CART neurons in the ARC and is a potent inhibitor of food intake. Nesfatin-1 acts through an undiscovered G-protein-coupled receptor that leads to changes in K^+ and Ca^{2+} flux within target neurons.

2. *Stimulation of Feeding*

An important hypothalamic orexic agent in mammals is NPY. Repeated or continuous administration of NPY produces hyperphagia (excessive eating), weight gain, and obesity. Blocking NPY neuronal activity reduces hyperphagia. AgRP is colocalized in NPY neurons, and overexpression of AgRP also can lead to obesity. Overexpression of the hypothalamic peptide called **melanin-concentrating hormone (MCH)** (see Chapter 4) in the LatH or administration of MCH also elevates feeding. Furthermore, MCH knockout mice are hypophagic (reduced feeding) and are excessively lean. Studies of the lipid-based **endogenous endocannabinoids**, such as **anandamide**, suggest a central role for these compounds and their receptors, especially the **CB_1 receptor**. Discovery of these bioregulators was a result of investigations of the mechanisms of action for the active ingredient of marijuana, **9-tetrahydrocannabinol (THC)**, in the late 1960s and early 1970s. Endogenous endocannabinoids are potent stimulators of food intake and appear to be important in the development of obesity. They also have stimulatory peripheral effects on lipogenesis in liver and adipose cells. Knockout mice for the CB_1 receptor ($CB_1^{-/-}$) are very lean and are resistant to dietary-induced obesity.

Orexins are peptides that stimulate feeding. There are two forms of orexin that are derived from the same prohormone. Human **orexin A** consists of 33 amino acids, and **orexin B** consists of 28 amino acids (Figure 12-8). Orexins A and B were also named **hypocretin-1** and **hypocretin-2** by another laboratory that isolated them at about the same time. In addition to stimulation of appetite, orexins have been found to have multiple actions on hormone release, gastric secretion, metabolic rate, and other behaviors (Figure 12-9). **Galanin-like peptide (GALP)**, a second member of the galanin peptide family, is produced within ARC neurons and is even a more potent orexigenic peptide than galanin when administered into the brain.

II. REGULATION OF DIGESTION

The gastrointestinal tract secretes a variety of peptides that control digestion. Many of these peptides also have important roles in the nervous system as well as in the HP system (e.g., **vasoactive intestinal peptide, VIP**). You may recall that the study of endocrinology began with identification by Bayliss and Starling of the first hormone, **secretin**, produced by the duodenal mucosa. A few years later, Edkins proposed that **gastrin** from the antral stomach controlled secretion of acid. CCK was soon proposed to stimulate gall bladder contraction and release of bile into the gut, and this was followed by evidence of **pancreozymin** that stimulates enzyme secretion by the pancreas. However, it was many years later that these gastrointestinal hormones were finally isolated and characterized chemically due to the diffuse nature of the secretory cells responsible for their production. These endocrine cells do not form distinct glands but occur as isolated cells or small groups of cells distributed along the GI epithelium (Figure 12-10). A summary of these GI regulators is provided in Table 12-4.

A. The Human Digestive System

In addition to the secretions of the GI epithelium, digestion is aided by three essential exocrine glands: (1) salivary glands, (2) liver, and (3) exocrine pancreas. Several pairs of salivary glands secrete a solution called **saliva**, containing salts and the starch-digesting enzyme salivary amylase. The

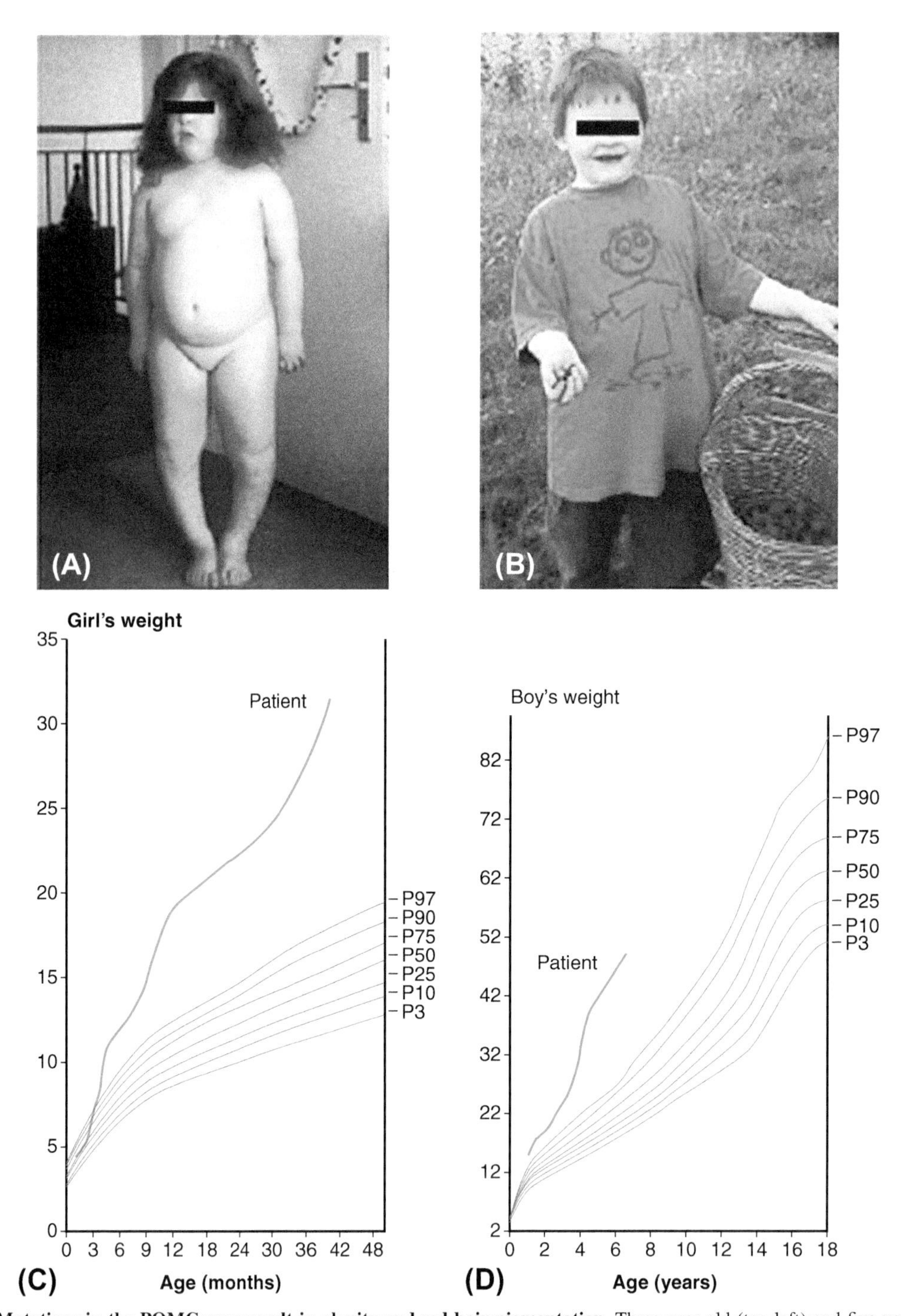

FIGURE 12-6 **Mutations in the POMC gene result in obesity and red hair pigmentation.** Three-year-old (top left) and five-year-old (top right) children with mutations in their proopiomelanocortin (POMC) genes. Notice the red pigmentation of the hair in both children due to melanocyte-stimulating hormone insufficiency. Growth charts for each patient are shown in the lower panels, showing the early age onset of obesity. *(Reprinted with permission from Krude, H. et al.,* Nature Genetics, ***19****, 155–157, 1998.)*

liver and pancreas have exocrine functions, too, and they secrete enzymes and other substances directly into the small intestine via ducts.

The first detailed and systematic knowledge of human digestive processes came in the early 1800s from observations of Dr. William Beaumont on his patient Alexis St. Martin, a French Canadian who, while visiting Fort Mackinac, Michigan, was accidentally shot in the chest from close range with a shotgun. St. Martin had two fractured ribs, his lungs were lacerated, and his stomach was perforated. Beaumont assumed that St. Martin would not live the night, but he miraculously survived. The wound in

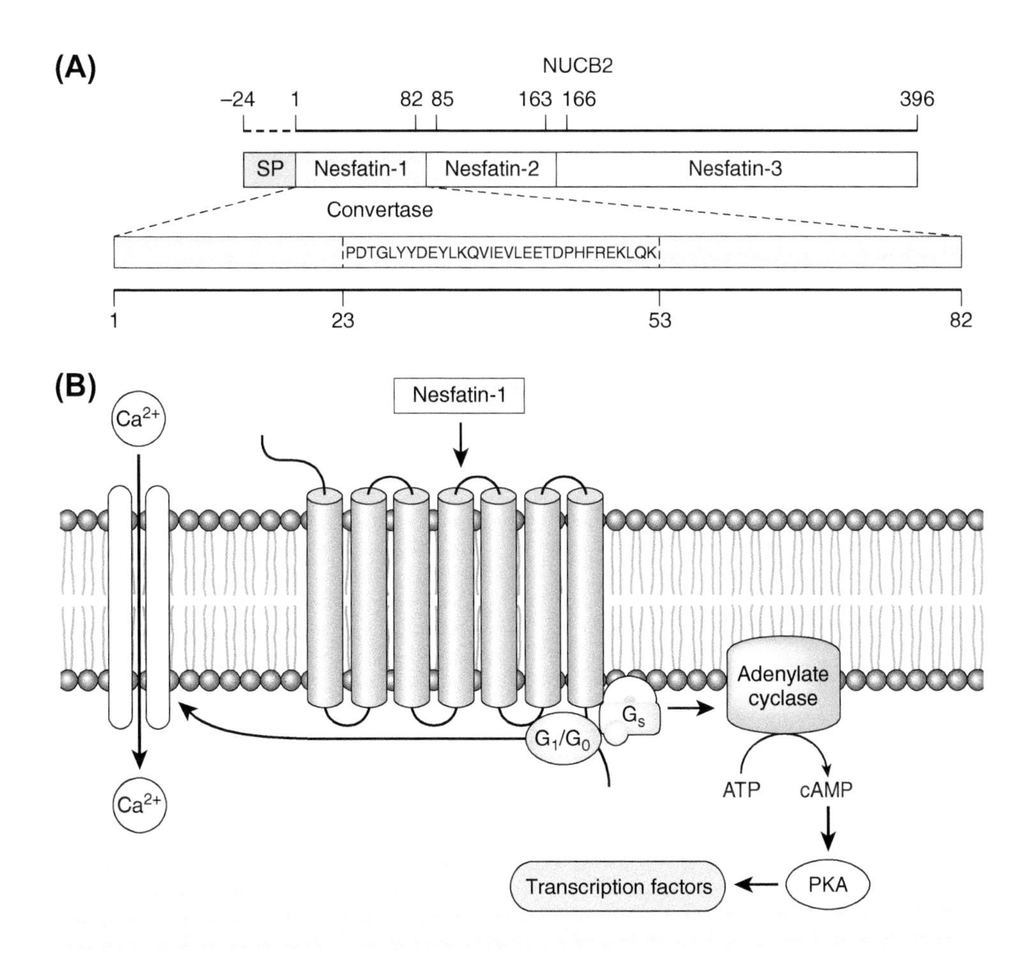

FIGURE 12-7 **The anorexic peptide Nesfatin-1 is cleaved from the protein nucleobindin 2 and acts upon an as-yet unidentified G-protein-coupled receptor.** The effective sequence of Nesfatin-1 on appetite is a 30 amino acid segment from the middle of the peptide (A). Nesfatin-1 acts on its receptor to open calcium channels via G_i/G_o G proteins and to increase cAMP and transcription in target neurons through a protein kinase A pathway (B). See Appendix C for explanation of amino acid abbreviations. *(Adapted with permission from Palasz, A. et al.,* Neuropeptides, ***46****, 105–112, 2012.)*

St. Martin's stomach, however, never healed completely, resulting in a permanent opening to the outside (what we now call a **gastric fistula**) through which Beaumont was able to observe the progression of gastric digestion under varying conditions over a period of years and also through which he could remove samples of gastric contents for closer analysis. The accidental production of a gastric fistula in St. Martin provided the inspiration for a variety of surgical techniques in other animals, including production of gastric fistulas and gastric or intestinal pouches (isolated pouches no longer connected with the lumen of the gut). Transplantation of denervated pouches or pieces of digestive tract or pancreas to sites under the skin where revascularization can occur has enabled subsequent investigators to separate endocrine and extrinsic nervous regulatory mechanisms. Such isolated organs still contain functional elements of the enteric division of the autonomic nervous system, however, and may exhibit endogenous nervous regulation. Development of crossed circulatory systems between experimental animals confirmed the transfer of chemical factors (later to be called *hormones*) through the blood to target tissues. *In vitro* studies of pancreatic slices or mucosal tissues from various regions of the gut were used to identify the various GI regulatory peptides as well as the factors regulating their release. Table 12-4 summarizes the actions of some GI peptides in mammals as well as their cellular sources.

B. Embryonic Origin of Gastrointestinal Endocrine Cells

Although the use of immunological and fluorescent techniques has enabled investigators to identify the actual cellular sources for many of the GI peptides, there is still considerable controversy with respect to the embryonic origin or origins of these cells in mammals. Pearse proposed that all of these GI cellular types as well as calcitonin-secreting C-cells of the thyroid gland, parathyroid chief cells, endocrine cells of the pancreas, all

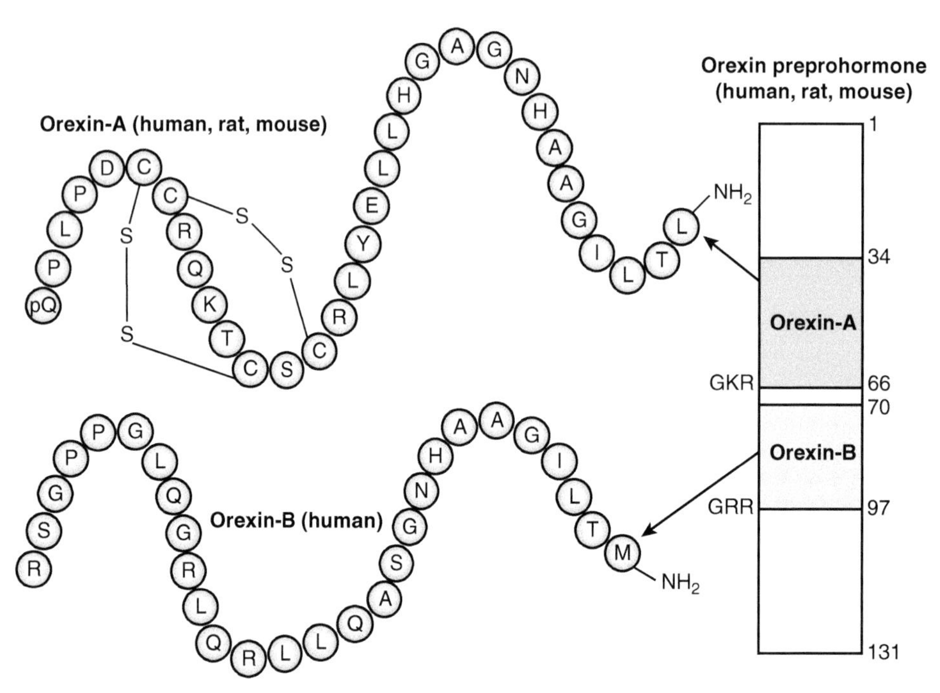

FIGURE 12-8 **Structure of Orexins.** Orexin A (= hypocretin-1) and orexin B (= hypocretin-2) are derived from the same prohormone. See Appendix C for an explanation of amino acid abbreviations. *(Adapted with permission from Samson, W.K. and Resch, Z.T.,* Trends in Endocrinology & Metabolism, ***11****, 257–262, 2000. © Elsevier Science, Inc.)*

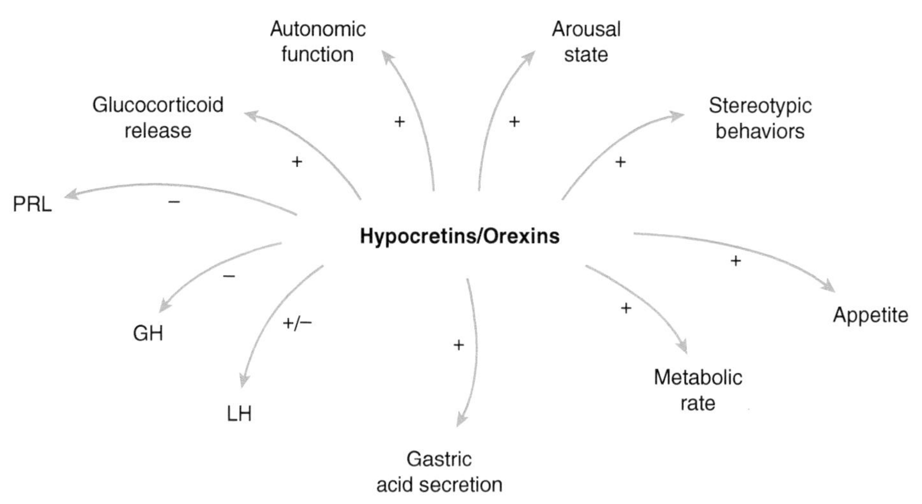

FIGURE 12-9 **Orexin actions.** *(Adapted with permission from Samson, W.K. and Resch, Z.T.,* Trends in Endocrinology & Metabolism, ***11****, 257–262, 2000. © Elsevier Science, Inc.)*

melanin-containing cells, adenohypophysial cells, and chromaffin cells of the adrenal medulla belong to the **APUD (amine content and amine precursor uptake and decarboxylation)** cellular series (see Chapter 4). These APUD cells are derivatives of neural crest or other neural ectoderm cells, suggesting that GI endocrine cells in the mucosa are of neural ectodermal origin rather than of endodermal origin and that they have migrated into the gastric or intestinal mucosa during early development. The neural origin for many of these endocrine cells has been confirmed, although the majority of these GI endocrine cells do not come from neural crest cells per se as originally postulated by Pearse. The occurrence of many of the GI peptides in brain tissue and peripheral neurons argues strongly for the neural origin for these GI hormone-secreting cells. A single evolutionary origin for GI endocrine cells is supported further by observations that immunoreactive gastrin, CCK, and glucagon appear to be

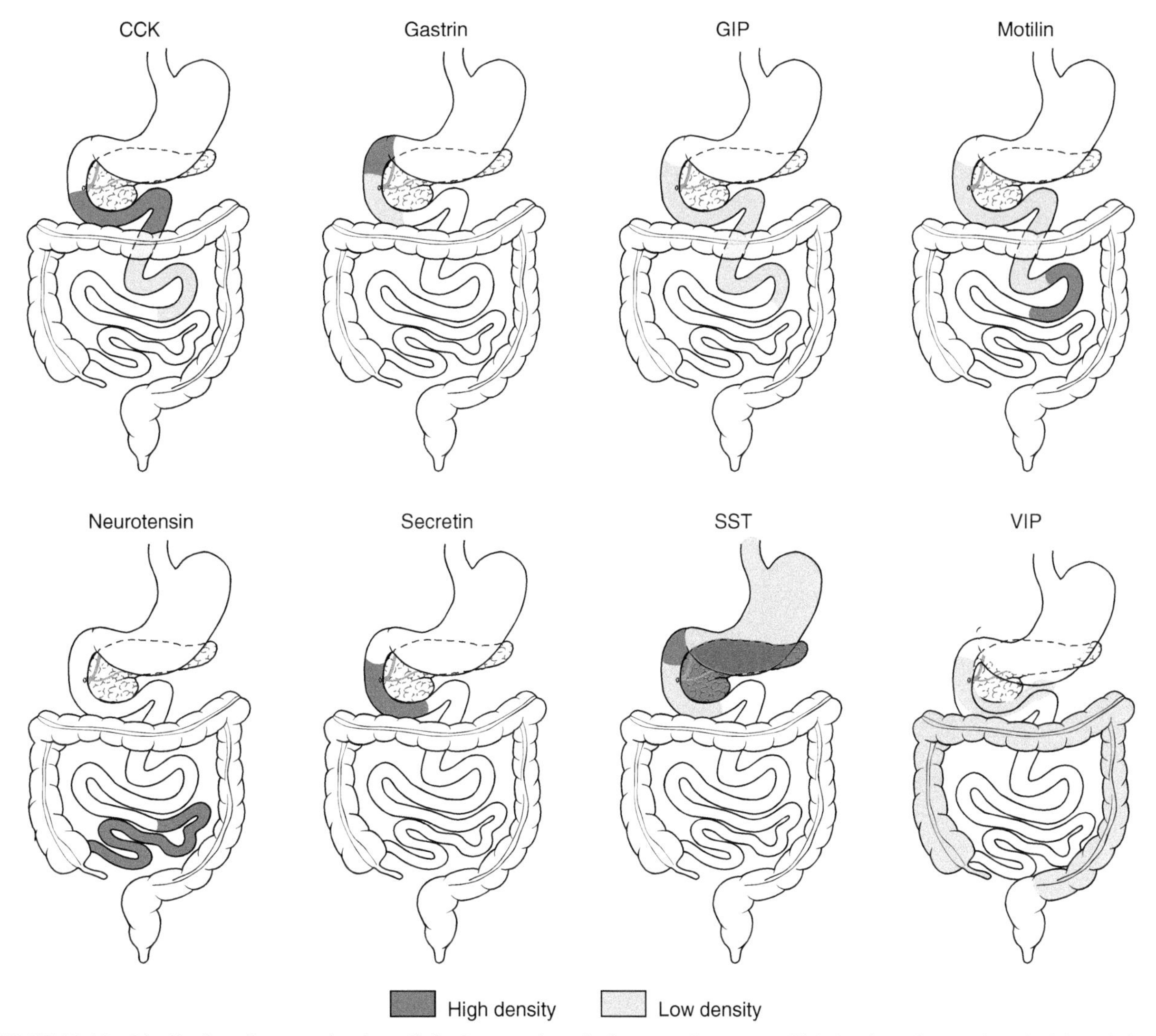

FIGURE 12-10 **Distribution of some endocrine cells in the gastrointestinal tract and pancreas.** High-density regions are in red; pink shaded areas indicate lower density. See text for an explanation of abbreviations. (Based on numerous sources.)

colocalized in a single intestinal cellular type in an invertebrate chordate (amphioxus) and in the cyclostome fishes.

C. Hormonal and Neural Regulation of Gastric Digestion

Neural control of gastric secretion occurs at two levels. The **cephalic phase** involves stimulation of secretion via parasympathetic discharges elicited by the same stimuli that cause salivation—that is, sight, smell, taste, thought, or presence of food. In the **gastric phase** of secretory control, the presence of food in the stomach elicits secretion through vagovagal reflexes and/or through the gastrin mechanism. In addition, neural factors can stimulate secretion of a hormone produced in the gastric epithelium that stimulates certain aspects of gastric secretion. It has not been possible to determine which of these mechanisms is the more important in controlling gastric secretion; probably all of these mechanisms operate in the normal digestive process.

1. The Gastrin Theory and Acid Secretion

Edkins in 1905 showed that extracts prepared from the most posterior portion of the stomach, the antrum, stimulated acid secretion by the gastric glands, and he suggested the name of *gastrin* for the active substance in these extracts. He found no gastrin activity in extracts prepared from other portions of the stomach. Edkin's **gastrin hypothesis** temporarily lost credibility with the discovery of **histamine**, a potent stimulator of gastric acid secretion, and the demonstration of histamine in extracts of the gastric mucosa. It was almost 30 years before it was shown that histamine-free extracts from the mucosa of the antral

TABLE 12-4 Gastrointestinal Peptides Regulating Digestion in Mammals

Peptide	Number of Amino Acids	Cellular Source	Primary Function
Calcitonin gene-related peptide (CGRP)	37	Gastric and intestinal neurons	May inhibit gastric secretion
Cholecystokinin (CCK)	8,33,39,58	Intestinal I cell	Stimulates pancreatic enzyme secretion, gall bladder contraction, and bile release
Gastrin	17,34	Antral G cell	Stimulates gastric acid secretion
Glucose-dependent insulinotropic peptide (GIP)	43	Intestinal K cell	Stimulates insulin secretion in the presence of glucose
Gastrin-releasing peptide (GRP)	27	Gastric and intestinal neurons	Stimulates gastrin release
Glucagon-like peptide-1 (GLP-1)	36	Intestinal L cells (EG cells)	Stimulates insulin release and inhibits glucagon release; anorexic
Motilin	22	Intestinal M cells	Causes intestinal contractions
Neurotensin	13	Intestinal N cells and neurons	May inhibit gastric acid secretion and motility
Peptide YY	36	Intestinal L cell	Inhibits pancreatic secretion; anorexic
Peptide histidine isoleucine (PH)	27	Intestinal neurons	Released with VIP and may have same actions
Secretin	27	Intestinal S cell	Stimulates pancreatic water and bicarbonate secretion
Somatostatin	14,28	Gastric D cell	Inhibits gastric acid secretion
Vasoactive intestinal peptide (VIP)	28	Neurons	Relaxes arteriole smooth muscle and increases blood flow to intestines

portion of the stomach possessed the ability to stimulate the acid-secreting **parietal cells** of the gastric glands. Nevertheless, it was not until gastrin was finally isolated almost 60 years later and characterized chemically by Gregory and Tracey that the term "gastrin theory" finally was discarded, and the hormone gastrin was confirmed.

There are two peptide forms of gastrin, each composed of 17 amino acids: **gastrin I** and a sulfated form, **gastrin II**, which has a sulfate group attached at position 6 near the *C*-terminal end. A larger form of gastrin called **big gastrin** also has been found in the circulation. Big gastrin consists of gastrin I or II plus a different 17-amino-acid peptide component. Only about 5% of the circulating gastrin occurs as big gastrin, however. The preprohormone for gastrin is composed of 104 amino acids and is cleaved several times to release big and little gastrins. Most of the biological activity of these gastrins resides in the four carboxy-terminal amino acids consisting of Trp—Met—Asp—Phe—NH_2. Several peptides that possess this terminal sequence (see Figure 12-11) have been shown to stimulate acid secretion. A synthetic pentapeptide (**pentagastrin**) incorporating the terminal tetrapeptide sequence is frequently used for experimental studies.

The control mechanism for acid secretion combines the observations that parasympathetic stimulation, **acetylcholine (ACh)**, histamine, gastrin, and some other peptides all cause acid secretion (see Figure 12-12). In contrast, atropine (a muscarinic cholinergic receptor antagonist), procaine (an anesthetic), sympathetic stimulation, and certain antihistamines tend to reduce acid secretion under some experimental conditions. Cephalic stimulation through the parasympathetic system (ACh) can release gastrin from the **G cell** in the mucosa of the antral portion of the stomach. Gastrin travels via the blood through the systemic circulation to the body of the stomach, where it stimulates the release of histamine from **enterochromaffin-like (ECL) cells** situated in the mucosa. Gastrin also activates the synthesis of **histidine decarboxylase**, the enzyme responsible for

Peptide	Residues				
Human gastrin II	17	EEPWL	EEEEE	A Y*	G W M D F-NH_2
Caerulein	10		QQD	Y* T	G W M D F-NH_2
CCK_8	8		D	Y* M	G W M D F-NH_2

FIGURE 12-11 **Peptides that stimulate gastric acid secretion.** Caerulein is a peptide isolated from frog skin that has the common amino-terminal pentapeptide and hence similar activity to gastrin II. The same sequence occurs in all CCKs, although only CCK_8 is shown. Y* indicates that a sulfate group is attached to that tyrosine in each of these peptides. See Appendix C for an explanation of the letters coding for individual amino acids. *(Adapted with permission from Walsh, J.H. and Dockray, G.J., "Gut Peptides," Raven, New York, 1994.)*

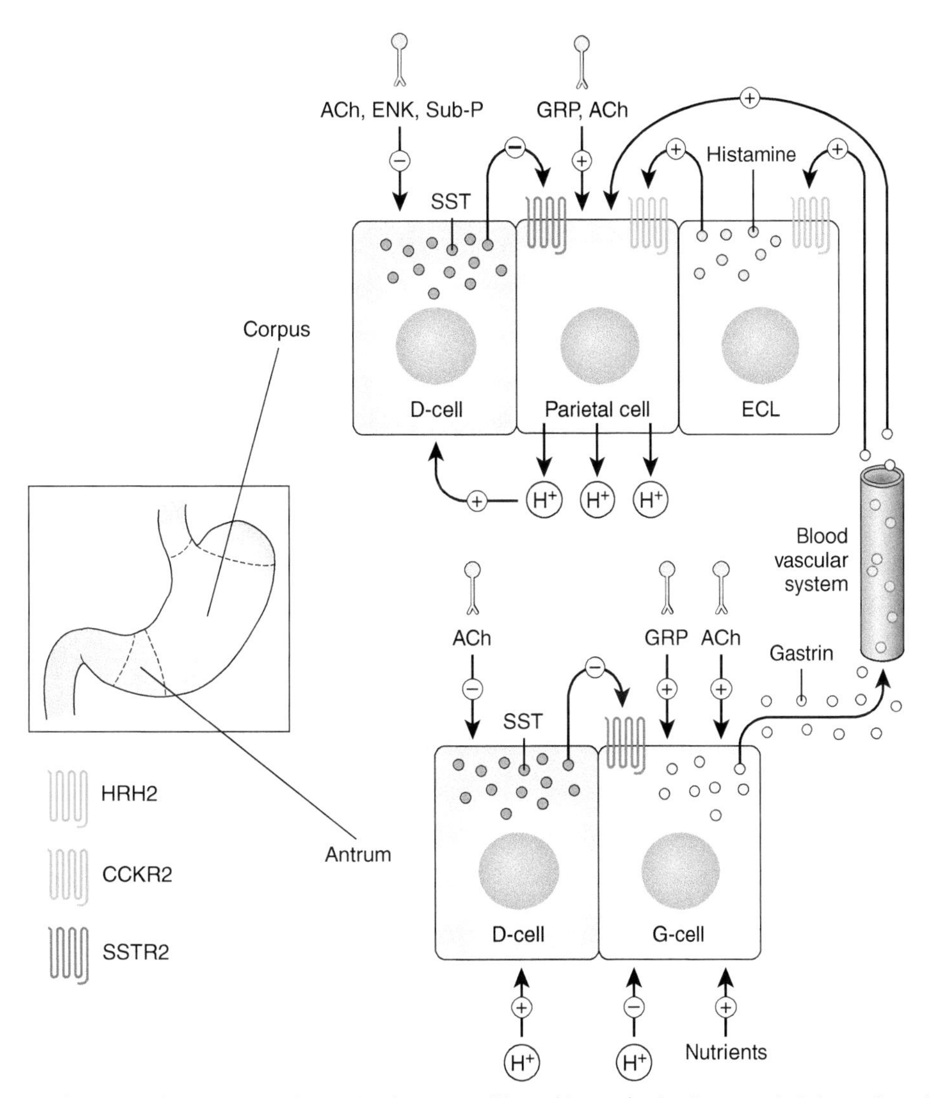

FIGURE 12-12 **Neurocrine, endocrine, and paracrine mechanisms controlling acid secretion by the stomach.** Release of gastrin from G cells in the antral stomach can be induced by nutrients in the gut lumen or by the neuropeptide gastrin-releasing peptide (GRP) or by acetylcholine (ACh). Gastrin travels through the blood and by acting on CCKR2 receptors directly stimulates the parietal cell in the corpus to secrete H^+ into the stomach lumen. Gastrin also stimulates histamine secretion from enterochromaffin-like (ECL) cells. Histamine in turn acts on histamine H2 receptors (HRH2) to stimulate HCl secretion from parietal cells. The parietal cell is also stimulated by ACh and GRP. D cells in the antrum and corpus secrete somatostatin (SST), which blocks gastrin release and parietal cell secretion, respectively, by acting on somatostatin subtype 2 receptors (SSTR2). D cells may be stimulated by H^+ and inhibited by neurotransmitters. H^+ in the stomach lumen also inhibits the G cell directly. ENK, enkephalins; Sub-P, substance P.

histamine synthesis. Histamine in turn stimulates release of HCl from the parietal cells. Gastrin also stimulates parietal cells directly without employing histamine as an intermediate. Vagal stimulation or application of ACh may stimulate the parietal cells directly to secrete HCl. Gastrin acts on the **CCK receptor 2 (CCKR2)** on parietal cells and ECL cells to stimulate acid and histamine secretion, respectively.

2. Somatostatin

During active digestion the pH of the human stomach may be between 1 and 2. Low pH in the antral portion of the stomach (especially near the pyloric sphincter) reduces gastrin release (feedback), probably acting through local paracrine release of **somatostatin (SST)** by **D cells** in the antral stomach. SST acts on the **somatostatin receptor subtype 2 (SSTR2)** to directly inhibit gastrin secretion from G cells and to directly inhibit acid secretion from parietal cells. In the 1930s, the term **enterogastrone** was coined to designate humoral inhibitors of intestinal origin that reduced gastric secretion and/or motility. Several peptides secreted by the small intestine may be candidates as enterogastrones either by evoking SST release or through more direct inhibitory actions (Figure 12-13).

3. Gastrin-Releasing Peptide

A novel peptide isolated from the mammalian stomach and intestine stimulates gastrin release and hence was named **gastrin-releasing peptide (GRP)**. This peptide of 27 amino acids bears a remarkable structural similarity to the 14-amino-acid peptide **bombesin (BBS)** that previously was isolated from the skin of frogs in the genus *Bombina.* In the stomach, immunoreactive GRP appears exclusively in post-ganglionic parasympathetic neurons, but the axonal tips of these neurons do not contact the G cells directly, implying that this neurocrine behaves in a paracrine fashion, whereas intestinal GRP acts in the manner of a true hormone. In addition to its action on gastrin release, GRP stimulates the secretion of pancreatic enzymes, contraction of gastric, intestinal, and gallbladder smooth muscle, and release of several gastrointestinal and pancreatic hormones. GRP also produces mitogenic effects resulting in hyperplasia of pancreatic, intestinal, and other tissues; thus, GRP may play multiple roles in modulating GI physiology. Administration of GRP or BBS into the ventricles of the brain causes a dramatic cessation of gastric acid secretion regardless of the stimulus used to evoke gastric secretion. This effect appears to be mediated via sympathetic nerves and suggests another level for the involvement of this peptide in gastric function.

4. Secretion of Pepsinogen

The major gastric enzyme in adult humans and carnivorous vertebrates is the protease **pepsin**, which is secreted by the **chief cells** of the gastric mucosa in an inactive form called **pepsinogen**. Conversion of inactive pepsinogen to pepsin is accomplished by the presence of an excess of H^+ supplied by the parietal cells. The optimum pH for vertebrate pepsins lies between 1 and 2, the normal pH range observed in the stomach following stimulation of acid secretion. Pepsin is inactive above a pH of about 4.5. The presence of

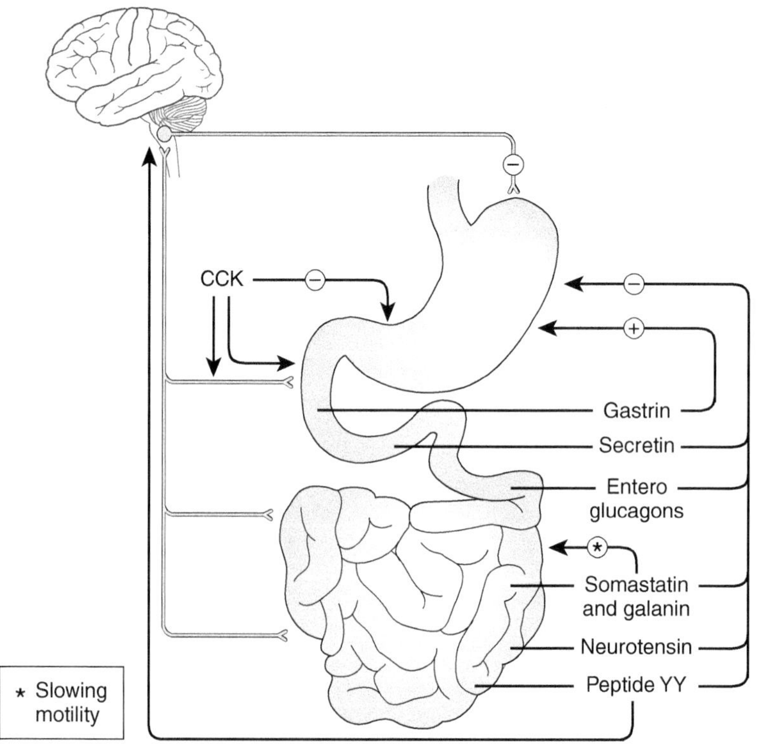

FIGURE 12-13 **Enterogastrones.** Several peptides released from the small intestine inhibit acid secretion and slow processing of food in the stomach. *(Adapted with permission from Lloyd, K.C.K. and Walsh, J.H., in "Gut Peptides" (J.H. Walsh and G.J. Dockray, Eds.), Raven, New York, 1994, pp. 147–173.)*

acid on the surface of the gastric mucosa may activate a cholinergic reflex that evokes pepsinogen release. Parasympathetic stimulation via the vagus nerve causes release of pepsinogen, and hormonal control of pepsinogen secretion may be absent. Gastrin causes release of pepsinogen only when applied in doses large enough to inhibit acid secretion by the parietal cells, implying that gastrin is not the normal factor causing pepsinogen release from the chief cells. Several other GI peptides can invoke pepsinogen release but do so only when applied in pharmacological doses. However, one duodenal peptide, **motilin** (see ahead), has been implicated in regulating pepsinogen secretion at physiological levels. Inhibition of pepsinogen release is caused by SST or by sympathetic stimulation.

D. Regulation of Intestinal Digestion

The duodenal mucosa contains many secretory cells, including cells responsible for secretion of several digestive enzymes as well as a variety of hormone-secreting and mucus-secreting cells. The intestinal mucosa is organized into thousands of tiny finger-like projections called *villi*. The presence of villi greatly increases the total surface area of the small intestine for releasing digestive secretions and for absorption of digestive products. The number of villi and secreting cells in the mucosa of the small intestine decreases progressively from the duodenum to the jejunum and ileum.

Three stages or phases of intestinal regulation involving neural and endocrine mechanisms can be identified. There appears to be a distinct **cephalic phase of intestinal regulation** mediated via vagal stimulation that influences pancreatic secretion. A **gastric phase of intestinal regulation** involves vagal and vagovagal stimulation of duodenal gastrin release that appears to influence pancreatic secretion and increases HCl. Finally, the **intestinal phase** relies primarily on release of intestinal peptides stimulated directly by the composition of the intestinal contents and is divided into an early (acid stimulatory) phase and a late (acid inhibitory) phase. During the early intestinal phase, the pH of the chyme entering the duodenum is above 4.5 and duodenal gastrin cells release the peptide into the circulation to increase stomach acid production. During the late intestinal phase the pH of the chyme entering the duodenum is 4 or below, and duodenal peptides are secreted in part to inhibit further stomach acid production.

1. Secretin

The presence of acidic chyme (pH less than 4) from the stomach directly stimulates the **S cell** in the duodenal mucosa to release the peptide secretin into the blood. Although H^+ are the primary stimulus, secretin release also is stimulated by bile salts, fatty acids, sodium oleate, and several herbal extracts. Secretin stimulates the pancreas to secrete basic juice (rich in HCO_3^-) and helps to neutralize the acidity of the chyme that has entered the small intestine. Although secretin levels in the blood do not increase following ingestion of a meal, the action of secretin on the exocrine pancreas is potentiated by another intestinal peptide hormone, CCK, that also increases in the blood following ingestion of a meal (see ahead).

Originally it was believed that secretin also was responsible for stimulating secretion of the digestive enzymes normally present in the pancreatic juice, including the proteases chymotrypsin and trypsin, pancreatic lipase, pancreatic amylase, and nucleases for DNA and RNA. After 40 years of controversy following the demonstration of secretin, it was confirmed finally by Harper and Raper (1943) that purified secretin stimulates secretion of pancreatic fluid that is rich in sodium bicarbonate but poor in digestive enzymes. Secretion of the digestive enzymes was attributable to a second duodenal peptide that was found to contaminate some secretin preparations. Because zymogen granules represent vesicles of stored enzyme within the acinar (exocrine) pancreatic cell, the peptide that caused extrusion of zymogen granules from pancreatic acinar cells initially was called *pancreozymin* (pancreas–zymogen). It was postulated that the release of pancreozymin into the blood in response to the presence of peptides and amino acids in the chyme is due to direct actions of these molecules on pancreozymin-producing cells, similar to the action of H^+ on the S cell. Sometimes the term **secretagogue** is applied to substances present in food, substances secreted from the mucosa into the gut lumen, or products of digestion that induce gastric or intestinal secretions. Pancreozymin later turned out to be another intestinal peptide hormone previously named for a different function (see ahead).

Secretin consists of 27 amino acids and chemically is related to several other peptides of the **PACAP** (**pituitary adenylate cyclase-activating polypeptide**) family (see Chapter 4), several of which are involved in digestion. Secretin has been isolated from several mammalian species and the primary sequence appears to be conserved (Figure 12-14), although mammalian secretins differ markedly from avian secretin. Receptors for secretin are G-protein linked, and secretin apparently operates through production of a cAMP second messenger to stimulate pancreatic HCO_3^- secretion (see Chapter 3).

2. Cholecystokinin (CCK)

In 1928, Ivy and Goldberg proposed that fat in the chyme or some of the products from fat digestion were a secretagogue for release of an intestinal peptide hormone that they named CCK (*chole*, bile; *kystis*, bladder; *kinein*, move). CCK travels via the blood to the gallbladder, where it stimulates contraction of smooth muscles comprising the walls of the gallbladder. At the same time, it causes

	1	11	21
Pig, cow	HSDGTFTSEL	SRLRDSARLQ	RLLQGLV
Dog	HSDGTFTSEL	SRLRESARLQ	RLLQGLV
Rat	HSDGTFTSEL	SRLQDSARLQ	RLLQGLV
Human	HSDGTFTSEL	SRLREGARLQ	RLLQGLV
Chicken	HSDGLFTSEY	SKMRGNAQVQ	KFIQNLM

FIGURE 12-14 **Comparison of secretins from mammals and the chicken.** Mammalian secretins are very conservative whereas more than half of the amino acids are different in the bird. See Appendix C for an explanation of the letters coding for individual amino acids. *(Adapted with permission from Leiter, A.B. et al., in "Gut Peptides" (J.H. Walsh and G.J. Dockray, Eds.), Raven, New York, 1994, pp. 147–173.)*

relaxation of the sphincter of Oddi, a muscle that controls the exit of bile from the gallbladder. As a result, bile is expelled from the gallbladder, enters the bile duct, and is transported to the duodenum. Bile is a viscous, complex mixture consisting largely of bile salts (derived from excess cholesterol; see Chapter 3) and bile pigments. Bile salts are powerful emulsifiers (i.e., surfactants). Bile pigments are breakdown products of hemoglobins, and they give bile and feces their characteristic colorations. Many other substances produced by the liver are present in bile, including metabolites of steroid hormones, thyroid hormone metabolites, and inorganic iodide from deiodination of thyroid hormones. When bile enters the duodenum, the bile salts emulsify globules of fat, causing them to be dispersed as small fat droplets within the aqueous digestive fluids. The emulsification of fats or **triacylglycerides (TAGs)** allows for marked reduction in the volume-to-surface ratio of the fat droplets, facilitating hydrolytic attack by pancreatic lipase to release **monoacylglycerides (MAGs)** and **nonesterified fatty acids (NEFAs)**, also called **free fatty acids (FFAs)**, for absorption by intestinal mucosal cells.

Isolation and purification of the first GI peptides was accomplished by Mutt and Jorpes in Sweden more than half a century after the discovery of secretin by Bayliss and Starling. Mutt and Jorpes discovered that the biological functions previously ascribed to pancreozymin and CCK resided in the same peptide consisting of 33 amino acids. Consequently, the rather cumbersome name **pancreozymin–cholecystokinin (PZCCK)** was proposed to designate the single peptide that for many years had been described in the literature under separate names. Because CCK was described first, Grossman (1970) proposed that we continue to use CCK for both actions and that we drop the PZ. In general, this proposal has been adopted. The **I cell** in the intestinal mucosa has been identified as the synthetic source for CCK.

Several CCK peptides have been identified, including molecules composed of 8 amino acids (CCK_8), 33 amino acids (CCK_{33}), 39 amino acids (CCK_{39}), and 58 amino acids (CCK_{58}). Recent studies indicate that the larger forms predominate and all forms are cleaved from the same gene product. In the CNS, CCK_8 is the common form present. CCK activity is very low in the general circulation and it is difficult to obtain quantitative information on its various potential forms. However, CCK_8, CCK_{39}, and CCK_{58} are measureable in blood plasma of the dog, but it is not known if they are secreted in these forms or whether the shorter forms are produced by peptidase activity in the blood following secretion of the larger forms.

The separate functional roles of secretin and CCK have been demonstrated elegantly *in vitro* using slices of exocrine pancreas. Physiological levels of CCK do not evoke release of basic pancreatic juice except in the presence of secretin, implying a permissive role for CCK, enhancing the action of secretin. As mentioned above, purified CCK, but not secretin, causes extrusion of zymogen granules from the acinar cells of the exocrine pancreas. ACh also causes extrusion of zymogen granules, suggesting direct parasympathetic (cephalic) control of enzyme release from the exocrine pancreas. Parasympathetic influence via the vagus nerve and CCK probably operates via separate mechanisms, as atropine blocks the effects of ACh but not of CCK on enzyme release. Furthermore, CCK does not require neural factors for its action.

Two G-protein-linked receptors have been characterized for CCK: CCKR1 and the CCKR2. One type predominates in target cells of the digestive or alimentary tract (CCK1R receptor) and preferentially binds larger or sulfated forms of CCK. The second receptor type (CCK2R) occurs in the brain and has highest affinity for CCK_8, although this isoform also is responsible for mediating gastrin actions in the stomach (see Figure 12-12). Once occupied, CCK receptors stimulate production of second messengers, IP_3, and DAG. The former causes an increase in cytosolic Ca^{2+}, and the latter activates phosphokinase C. There also is evidence suggesting that CCK may stimulate some cAMP production, although the physiological relevance of this is unclear as cAMP is not involved in any of the known biological actions of CCK. Fats, proteins, and amino acids in the gastric effluent entering the duodenum are the most potent, direct secretagogues for CCK release. Conversely, active proteases such as trypsin in the duodenum inhibit CCK secretion (negative feedback). The presence of bile elements also may reduce CCK release. Basal levels of CCK are very low (about 1 pM/L of plasma) but increase following eating to between 5 and 8 pM in about 10 to 45 minutes followed by a slow decline until the stomach is empty.

In addition to releasing enzymes in response to CCK, the rat exocrine pancreas also secretes a **pancreatic secretory trypsin inhibitor (PSTI)-61** (61 amino acids) that stimulates CCK release that in turn increases pancreatic secretion (positive feedback). The intestine also secretes a similar peptide that causes CCK release during early phases of digestion. During periods of fasting, this peptide is rapidly destroyed by the low levels of trypsin present in the duodenal lumen. However, when food enters from the stomach, the trypsin is further diluted so that the intestinal peptide is not degraded so rapidly which in turn elevates CCK release.

In rats and dogs, BBS and GRP are also potent releasers of CCK and hence pancreatic enzyme secretion; however, direct actions of BBS and GRP have been demonstrated on pancreatic acinar cells, as well. Furthermore, acinar cells are innervated by GRP-secreting neurons, providing another possible mechanism for enzyme release.

In several species, CCK slows the rate of gastric evacuation (emptying) through relaxation of gastric smooth muscle and contraction of the pyloric sphincter, thus slowing peristalsis and preventing passage of material from the stomach. These contradictory actions of CCK on gastric smooth muscle and the pyloric sphincter are similar to the effects observed on gallbladder smooth muscle (contraction) and the sphincter of Oddi (relaxation). CCK receptor antagonists accelerate gastric emptying in some human studies, suggesting a similar enterogastrone-like role for CCK. CCK inhibits gastric acid secretion by causing gastric D cells to secrete SST (Figure 12-12). Intestinal blood flow also may be altered by CCK. Certain neurons supplying blood vessels in the gut can release CCK, which causes relaxation of smooth muscle in arterioles and increases intestinal blood flow.

E. Intestinal Regulation of Gastric Secretion

In 1930, Kosaka and Lim proposed the existence of an intestinal peptide that they called *enterogastrone*. This hypothetical hormone was released in response to the arrival of fat from the stomach and was believed to inhibit gastric motility (peristalsis) and reduce acid secretion by the gastric parietal cells. These actions would slow the entrance of fat into the small intestine and allow more time for proper processing of fat already present in the duodenum. There may be numerous enterogastrones, and several peptides with enterogastrone-like activity have been isolated from the small intestine (Figure 12-13). The possible action of CCK on gastric secretion and peristalsis was discussed earlier. Three additional intestinal peptides with enterogastrone-like actions are described below.

1. Gastric-Inhibitory Peptide or Glucose-Dependent Insulinotropic Peptide

A peptide first called **gastric-inhibitory peptide (GIP)** was isolated from the intestine and shown to inhibit gastric function. GIP is produced by the **K cell** in the duodenal mucosa. Chemically, GIP is similar in amino acid sequence to secretin but is considerably larger (42 amino acids). Introduction of fat or glucose into the duodenum evokes GIP release. Ingestion of a meal results in a five- to sixfold elevation in plasma GIP, which remains elevated for about 6 hours. GIP has been extracted from several mammals, and its structure is rather conservative (Figure 12-15). GIP blocks both gastrin-stimulated secretion of acid and enzyme by the stomach but is effective only in the experimentally denervated stomach, suggesting this may not be a physiological role for GIP. As with CCK, it may produce its inhibitory effects by stimulating release of SST, which in turn directly inhibits acid secretion by the parietal cell. In humans, large (pharmacological) doses of GIP are required to obtain even a weak effect on gastric acid secretion, implying that this is not a major role for GIP. Other studies suggest that the physiological role for GIP is to stimulate insulin release from the endocrine pancreas to facilitate absorption and distribution of glucose and amino acids. A dose-dependent release of insulin is caused by physiologically relevant amounts of GIP both *in vivo* and *in vitro*. Because it is believed that this action represents the true role for GIP, the peptide has been renamed **glucose-dependent**

	1	11	21	31	41
Porcine GIP	YAEGTFISDY	SIAMDKIRQQ	DFVNWLLAQK	GKKSDWKHNI	TQ
Human GIP	YAEGTFISDY	SIAMDKIHQQ	DFVNWLLAQK	GKKNDWKHNI	TQ
Bovine GIP	YAEGTFISDY	SIAMDKIRQQ	DFVNWLLAQK	GKKSDWIHNI	TQ
Rodent GIP	YAEGTFISDY	SIAMDKIRQQ	DFVNWLLAQK	GKKNDWKHNL	TQ

FIGURE 12-15 **Comparison of mammalian GIP structures.** These peptides are highly conservative and there have been few substitutions. See Appendix C for an explanation of the letters coding for individual amino acids. *(Adapted with permission from Pederson, R.A., in "Gut Peptides" (J.H. Walsh and G.J. Dockray, Eds.), Raven, New York, 1994, pp. 217–259.)*

insulinotropic peptide, providing a more descriptive name while retaining the original acronym, GIP.

2. Peptide YY

Another potential enterogastrone may be PYY, which was first isolated from the pig intestine. It also is called *peptide tyrosine-tyrosine* and has both a *C*-terminal and an *N*-terminal tyrosine, for which the abbreviation is "Y." The peptide consists of 36 amino acids and is chemically very similar to NPY and pancreatic polypeptide (see below). PYY inhibits secretion by the exocrine pancreas and reduces gastric acid secretion. Release of PYY into the circulation occurs following the introduction of fat into the duodenum, and physiologically relevant levels of PYY produce a dose-dependent inhibition of gastric acid secretion.

3. Calcitonin Gene-Related Peptide

Calcitonin is a calcium-regulating hormone produced in mammals by the thyroid C cells (see Chapters 6 and 14). **Calcitonin gene-related peptide (CGRP)** is formed in neurons by an alternative posttranslational processing of the calcitonin gene product that results in this peptide of 37 amino acids. CGRP has been isolated from human, rat, rabbit, bovine, and porcine sources (Figure 12-16) and is known to inhibit gastric acid secretion. Two forms have been isolated from humans and rats, each produced by a separate gene. A related peptide is **amylin** (also consisting of 37 amino acids), which has about 40 to 50% homology to CGRP (Figure 12-16). Amylin, however, is co-secreted with insulin when pancreatic B cells are exposed to glucose.

Unlike most of the GI regulators, the distribution of CGRP-immunoreactive axons differs greatly among the species examined to date. For example, CGRP is virtually absent from gastric neurons in humans and pigs but is abundant in the stomachs of rat, hamster, mouse, mole, and ferret. Intermediate amounts of CGRP occur in stomachs of cats, dogs, and guinea pigs.

Most of the gastric CGRP-containing neurons are primary (sensory) afferent neurons. The cell bodies of these sensory neurons are located in the nodose ganglia (vagus nerve sensory neurons), the inferior ganglia of the glossopharyngeal nerve, or in the dorsal root ganglia. CGRP neurons occur throughout the mammalian digestive tract, but endocrine cells secreting CGRP have been found only in the human small intestine and the rat pancreas. CGRP-containing motor neurons from the brain stem have been described in the esophagus of monkeys and cats, and some CGRP-containing motor neurons as well as some CGRP-enteric neurons are seen in the stomach and intestine.

CGRP is released from the gastric mucosa of rats by the neurotoxin **capsaicin** (the active component in hot peppers) or by an increase in acidity of the duodenal contents to pH 6. Capsaicin is a potent agonist for a class of calcium channels called the **transient receptor potential vanilloid subtype 1 (TRPV1)**. TRPV1s are pain receptors recognizing a wide variety of stimuli and appear to be the a major receptor mediating the effects of acid and temperature on CGRP release from enteric neurons. CGRP probably stimulates SST release within the gastric mucosa that in turn inhibits gastric acid secretion.

4. Motilin

A small peptide (22 amino acids) that stimulates motility in the gastrointestinal tract and possibly the secretion of pepsinogen by the stomach has been isolated from the duodenum. This peptide is called **motilin** for its stimulatory action on smooth muscle that hastens the passage of material through the small intestine. Motilin is structurally unlike the other GI peptides (Figure 12-17). In most

CGRPS	1	11	21	31	37
Human CGRP-I	A C D T A T C V T H	R L A G L L S R S G	G V V K N N F V P T	N V G S K A	F-NH_2
Human CGRP-II	A C N T A T C V T H	R L A G L L S R S G	G M V K S N F V P T	N V G S K A	F-NH_2
Rat CGRP-I	S C N T A T C V T H	R L A G L L S R S G	G V V K D N F V P T	N V G S E A	F-NH_2
Rat CGRP-II	S C N T A T C V T H	R L A G L L S R S G	G V V K D N F V P T	N V G S K A	F-NH_2
Rabbit CGRP	G C N T A T C V T H	R L A G L L S R S G	G M V K S N F V P T	N V G S E A	F-NH_2
Bovine CGRP	S C N T A T C V T H	R L A G L L S R S G	G V V K S N F V P T	N V G S E A	F-NH_2
Porcine CGRP	S C N T A T C V T H	R L A G L L S R S G	G M V K S N F V P T	D V G S E A	F-NH_2
Chicken CGRP	A C N T A T C V T H	R L A D F L S R S G	G V G K N N F V P T	N V G S K A	F-NH_2
Amylins					
Human amylin	K C N T A T C A T Q	R L A N F L V H S S	N N F G A I L S S T	N V G S N T	Y-NH_2
Rat, mouse, amylin	K C N T A T C A T Q	R L A N F L V D S S	N N F G P V L P P T	N V G S N T	Y-NH_2
Guinea-pig amylin	K C N T A T C A T Q	R L T N F L V D S S	H N F G A A L L P T	D V G S N T	Y-NH_2
Cat amylin	K C N T A T C A T Q	R L A N F L I D S S	N N F G A I L S P T	N V G S N T	Y-NH_2

FIGURE 12-16 **Amino acid sequences of CGRPs and amylins.** Mammalian CGRPs are compared to chicken CGRP. Although there is considerable conservatism among the CGRPs and among the amylins, there are numerous differences between these groups. Conserved sequences underlined in red. See Appendix C for an explanation of the letters coding for individual amino acids. *(Data from Holzer, P., in "Gut Peptides" (J.H. Walsh and G.J. Dockray, Eds.), Raven, New York, 1994, pp. 493–523.)*

	1		11		21
Canine	FVPIF	THSEL	QKIRE	KERNK	GQ
Porcine	FVPIF	TYGEL	QRMQE	KERNK	GQ

FIGURE 12-17 **Amino acid sequences of canine and porcine motilins.** See Appendix C for an explanation of the letters coding for individual amino acids. *(Data from Poitras, P., in "Gut Peptides" (J.H. Walsh and G.J. Dockray, Eds.), Raven, New York, 1994, pp. 261–304.)*

mammals, alkaline conditions in the duodenum cause release of motilin from the **M cells** of the small intestine; however, in humans motilin apparently is released by acidic solutions. In the fasting human, an endogenous rhythmic secretion of motilin activates a wave of muscular contraction called the **migrating motor complex (MMC)**. This MMC begins at the antral end of the stomach every 75 to 90 minutes and moves along the intestine to sweep materials into the colon. This mechanism also prevents movement of bacteria from the large intestine into the small intestine, especially when food is absent.

5. Vasoactive Intestinal Peptide (VIP)

Relaxation of vascular smooth muscle increases blood flow to the viscera. This relaxation is caused by another peptide first isolated from the intestinal mucosa, VIP. Increased blood flow into the small intestine may enhance absorption of digestion products into the blood. In addition, VIP relaxes gastric smooth muscle and slows gastric processing. VIP consists of 28 amino acids and is structurally similar to both secretin and GIP (see Chapter 13). The immunoreactive VIP found in the brain is structurally identical to intestinal VIP. In the intestine, VIP is produced typically in neurons, but a pyramidal-shaped **H cell** found in the small intestine and in the colon of some species also may secrete VIP. Posttranslational processing of the preprohormone produces not only VIP but also another similar peptide consisting of 27 amino acids with an *N*-terminal histidine and a *C*-terminal isoleucine reflected in its name: **peptide histidine isoleucine (PHI)**. In humans, this VIP-like peptide exhibits methionine in place of isoleucine and is termed **peptide histidine methionine (PHM)**. Some of the actions previously attributed to VIP may be due to an interaction of PHI and VIP on the target cells. Sufficiently high doses of VIP have been shown to produce weak secretin-like effects on the pancreas as well as to induce a hyperglycemic response. These actions of VIP on the exocrine pancreas and blood sugar levels may be pharmacologic as a consequence of the similarity of amino acid sequence when applied at pharmacological doses.

A major role for VIP is that of an inhibitory neurotransmitter produced by intestinal nerves. It inhibits vascular smooth muscle but stimulates glandular epithelia. It is the inhibitory action on vascular smooth muscle that increases blood flow into intestinal tissues. Another proposed action for VIP is on the secretion of Brunner's glands in the duodenal mucosa. In response to the entrance of acidic chyme into the duodenum, Brunner's glands secrete a viscous, alkaline mucus that is believed to help in neutralizing gastric acid and protecting the duodenal mucosa.

6. Enteroglucagons

The **L cells** (also called **EG cells**) in the small intestinal mucosa produce several peptides that collectively can be called the **enteroglucagons**. These peptides are structurally and functionally like pancreatic glucagons (see ahead). Enteroglucagons are extractable in a large form named **glicentin** (69 amino acids). GLP-I and GLP-II are also cleaved from the same preprohormone as glicentin (Figure 12-18). A truncated (shortened) form of GLP-1 (**tGLP-1**) has been reported, as well. In addition to their possible role in feeding, enteroglucagons may stimulate insulin secretion while inhibiting pancreatic secretion of glucagon following ingestion of a meal. Furthermore, enteroglucagons may inhibit gastric acid secretion and delay gastric emptying. The physiological role for enteroglucagons and their relationship to pancreatic glucagon and glucose metabolism are considered ahead.

7. Other Gastrointestinal Peptides

SST, which acts as a paracrine inhibitor produced by **D cells** in the stomach, also is produced by neurons and D cells located throughout the small intestine. Here, it seems to function in paracrine fashion by inhibiting release of virtually all other GI peptides. SST has been demonstrated in the arterial circulation of dogs following a meal and may act as a hormone to inhibit release of gastrin, insulin, glucagon, and pancreatic polypeptide (PP) (see below). Two forms of SST have been isolated. The first is identical to the hypothalamic tetradecapeptide neurohormone (SST_{14}). The second consists of SST_{14} plus an additional 14 amino acids (SST_{28}). NT, a tridecapeptide, is another neurotransmitter first identified in the CNS that is found in the intestine. In addition to localization in neurons, NT also is found in epithelial **N cells** that directly contact the lumen of the small intestine. Although it has been shown to produce many effects, including acting as an enterogastrone, no clear role has been established for intestinal NT.

Substance P (onadecapeptide) was the first peptide to be identified in both the intestine and brain (in 1931). It is not clear whether substance P and NT function only as neurotransmitters or if they have paracrine functions in the

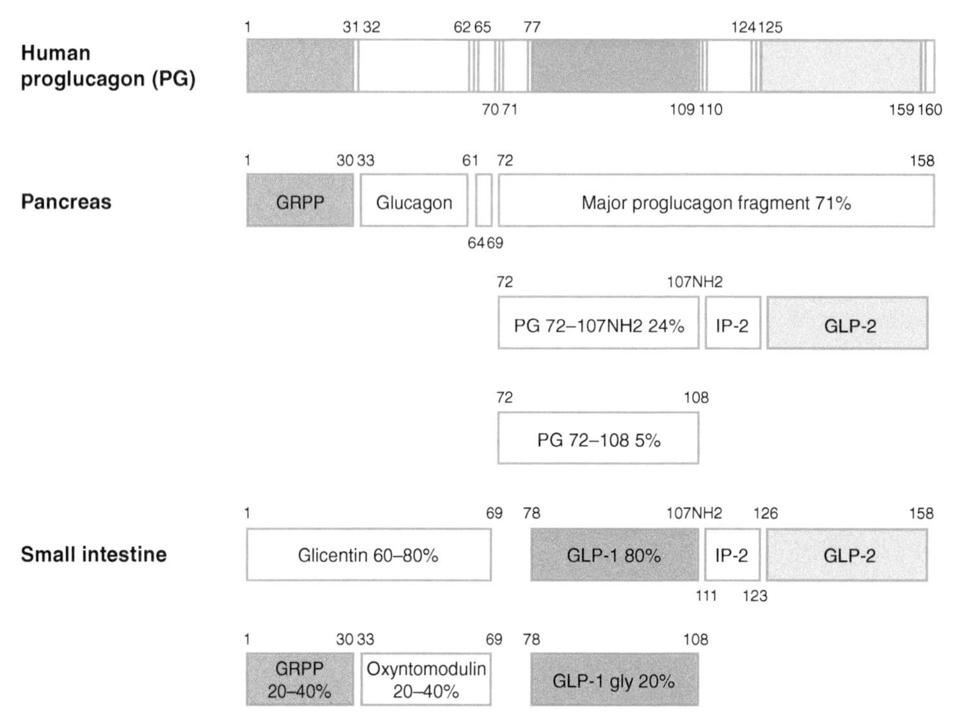

FIGURE 12-18 **Products from proglucagon in pancreas and small intestine.** Cleavage of the prohormone in the pancreas produces glucagons and glucagon-related polypeptide (GRPP). Further cleavage of the remaining proglucagon major fragment yields one of the glucagon-related peptides (GLP-2). In the small intestine, proglucagon yields glicentin as well as GLP-1 and GLP-2. Glicentin may be cleaved further to yield GRPP and oxyntomodulin that includes the glucagon sequence. IP-2, intervening peptide 2; PG, proglucagon. *(Adapted with permission from Holst, J.J.,* Trends in Endocrinology & Metabolism, ***10****, 229–235, 1999. © Elsevier Science, Inc.)*

intestine. Some of the substance P-containing neurons may be primary afferent (sensory), nociceptive (pain) neurons. **Dynorphin**, a heptadecapeptide opioid in the CNS (see Chapter 4), is also present in the intestine. It is produced by neurons and probably acts as a local inhibitor of the neurotransmitter substance P.

Galanin consists of 29 or 30 amino acids (Figure 12-19) and is concentrated in the duodenum. It is capable of inhibiting gastric acid secretion and can block secretion of several other GI regulators including NT, enteroglucagons, SST, PYY, and pancreatic hormones. Galanin may be a modulator of GI functions, including motility, secretion, and blood flow. Hypothalamic galanin also is a releaser of luteinizing hormone (LH) from the pituitary (see Chapter 4) and functions as a hypothalamic orexic factor (see above).

F. Complex Interactions of Gastrointestinal Peptides

Many studies have been published that involve observation of the effects of administering combinations of GI peptides as well as the influences of one peptide on the release of another. Studies of this type indicate considerable overlap in the actions of the various peptides (for example, glucagon-like activity in the secretin peptide), although pharmacological doses usually are necessary to produce these effects. The observation that many of these peptides double as neurotransmitters makes it even more difficult to interpret studies employing pharmacological doses. At the present time, it is difficult to sort out interactions due to structural similarities, use of pharmacological doses, or

	1	6	11
Common sequence (1–15)	G W T L N	S A G Y L	L G P H A -
Variable sequence (16–30)	16	21	26
Pig (29)	I D N H D	S F H D K	Y G L A -NH_2
Sheep (29)	I D N H D	S F H D K	H G L A -NH_2
Cow (29)	L D S H D	S F Q D K	H G L A -NH_2
Rat (29)	I D N H D	S F S D K	H G L T -NH_2
Chicken (29)	V D N H D	S F N D K	H G L T -NH_2
Human (30)	V G N H D	S F S D K	N G L T S-NH_2

FIGURE 12-19 **Comparison of mammalian galanins with chicken galanin.** All have the same 15 *C*-terminal amino acids with only moderate differences among the *N*-terminal sequence. See Appendix C for an explanation of the letters coding for individual amino acids. *(Data from Rokaeus, A., in "Gut Peptides" (J.H. Walsh and G.J. Dockray, Eds.), Raven, New York, 1994, pp. 425–552.)*

both from those interactions that might represent true synergisms, functional overlaps, or inhibitions.

III. THE MAMMALIAN PANCREAS

The mammalian pancreas is a mixed gland of exocrine and endocrine components that play essential roles in digestion and metabolism, respectively. The digestive role of the pancreas is accomplished by the exocrine portion that produces digestive enzymes and secretes an alkaline pancreatic fluid. The endocrine portion of the pancreas secretes hormones that regulate carbohydrate, lipid, and protein metabolism. These basic metabolic processes are reviewed in Appendix H.

The exocrine or acinar pancreas secretes critical digestive enzymes into the pancreatic duct whereby they reach the lumen of the small intestine. (An acinus is similar in structure to an endocrine follicle except that the lumen of the acinus makes contact with a duct through which the secretory products of the epithelium [acinar cells] are transported.) This pancreatic juice also contains bicarbonate ions that buffer the acidic material entering the small intestine from the stomach. As described above, these pancreatic activities are regulated by the hormones secretin and CCK from the intestines.

The endocrine pancreas or the **islets of Langerhans** consists of small masses or islands of endocrine cells scattered among the acinar tissue (Figure 12-20). The pancreatic islets secrete at least four regulators: **insulin, glucagon,** SST, and an additional peptide known as PP. Insulin is primarily a hypoglycemic agent that lowers blood glucose, whereas glucagon is a hyperglycemic hormone. Glucagon and insulin may produce opposing effects on lipid metabolism as well, with glucagon promoting lipolysis. The major role of pancreatic SST may be as a paracrine agent released by neural stimulation that inhibits local release of the other pancreatic peptides. The role for PP is not well established, but its increase in the circulation following ingestion of a meal suggests a role in postabsorptive metabolism.

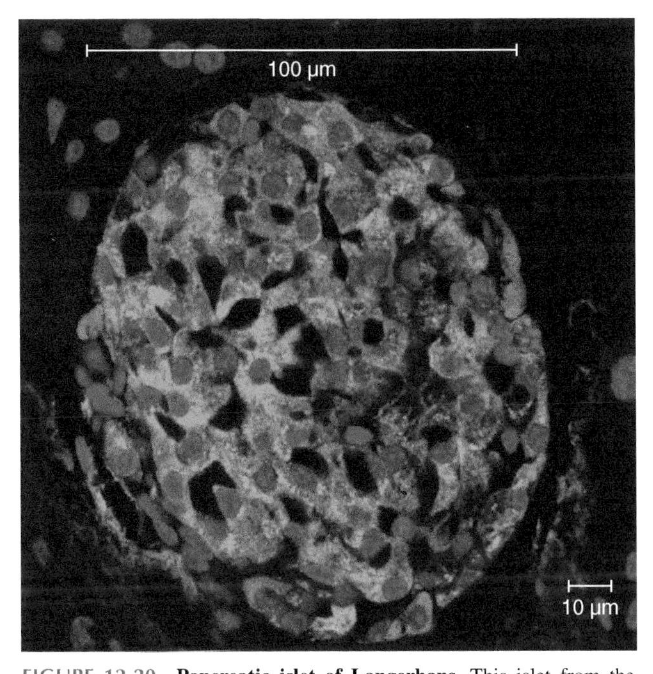

FIGURE 12-20 **Pancreatic islet of Langerhans.** This islet from the mouse pancreas exhibits insulin-producing B cells in green and glucagon-producing D cells in red. *(Photograph by Christin Süß, Jakob Suckale, and Michele Solimena, Solimena Lab, University of Technology, Dresden, Germany.)*

The endocrine pancreas has been linked with blood glucose regulation since the first observations of the clinical syndrome **diabetes mellitus**. This disease was first described by the Greek physician Aretaeus in about 1500 BC as a condition where "flesh and bones run together" and are siphoned into the urine. The term "diabetes" comes from the Greek word *diabainein* (*dia*, through; *bainein*, to go). Indian physicians reported the sweet taste of the diabetic's urine in the 6th century, but it was not until the 18th century that the word "mellitus" referring to the presence of glucose in the urine was added to the name. Diabetes mellitus was always fatal and even today shortens the lives and decreases the quality of life for afflicted people. It is rapidly increasing in developed countries and as a consequence is a leading contributor to death in the United States. We now recognize two major forms of diabetes mellitus; one is related to insufficient production of insulin, and the second involves an unresponsiveness of target cells to insulin (i.e., insulin resistance; see ahead).

Diabetes mellitus generally is characterized by the production of large quantities of glucose-containing urine. Normally little or no glucose should appear in the urine, and its presence contributes osmotically to the larger urine volume produced by the untreated diabetic. The occurrence of glucose in the urine results from excessively high circulating levels of glucose as a result of the failure by the endocrine pancreas to secrete adequate amounts of insulin and/or the inability of cells to react to insulin. Reduction in the ability to secrete insulin or to respond to insulin creates a number of metabolic disturbances resulting in glucosuria, including excessive lipolysis and gluconeogenesis of NEFAs. Consequently, blood levels of glucose become elevated, resulting in more glucose in the glomerular filtrate produced in the kidney. The oxidation of NEFAs for gluconeogenesis increases production of ketone bodies by the liver, yielding **ketonemia**. Under normal conditions, the proximal convoluted portion of each nephron transports all glucose that appears in the glomerular filtrate back into the blood. However, should blood sugar levels be sufficiently high, so much glucose enters the glomerular filtrate that the carrier-mediated transport mechanism in the nephron is saturated, and all of the glucose cannot be returned to the blood. The glucose remaining in the lumen of the nephron upsets the osmotic balance and less water can be reabsorbed from the filtrate, resulting in production of a greater volume of urine. **Ketonuria** also contributes to osmotic water losses. Because of the increased water loss associated with glucosuria and ketonuria, this syndrome achieved the descriptive title of "the pissing evil" during the Middle Ages. Several endocrine and metabolic alterations occur in an attempt to compensate for these nutrient losses.

Other consequences of the inadequacy of insulin production contribute to metabolic disturbances and additional water losses. Low insulin levels allow not only increased utilization of fats but also amino acids through gluconeogenesis to replace the lost glucose. Proteins are hydrolyzed to amino acids that are in turn transaminated or deaminated to produce carbohydrates plus ammonia. Most of the highly toxic ammonia is converted to less toxic urea in the liver and excreted through the urine. Elevated urea in the urine also contributes to increased urine production and additional water losses. The loss of glucose into the urine and increased production of nitrogenous wastes and ketone bodies all contribute to a process of desiccation or dehydration. Should the diabetic person enter a coma due to ketonemia-caused acidosis, he or she would be unable to compensate for water losses through drinking. Resultant dehydration may cause a marked decrease in blood volume and blood pressure that can lead to cardiac failure and death if proper treatment does not occur.

TABLE 12-5 Appearance of Pancreatic Cell Types in Mammalian Fetuses[a]

Cell Type	Human	Rat
A cells (glucagon)	8 weeks	12 days
B cells (insulin)	8 weeks	17 days
D cells (SS)	8 weeks	15 days
PP cells	10 weeks	-

[a] *As determined by the presence of immunoreactive material (humans) and cell ultrastructure (rat).*

A. Development of the Mammalian Pancreas

The mammalian pancreas, like that of other vertebrates, develops from the endodermal lining of the primitive or embryonic gut. A dorsal bud from the embryonic intestine fuses with one or two ventral buds to form the definitive pancreas. Usually only the dorsal connection to the intestine is retained as the exocrine pancreatic duct. The exocrine tissue in the developing pancreas can be identified by the formation of small ductules. These ductules coalesce into larger ducts until they form a large pancreatic duct connecting the exocrine pancreas with the lumen of the small intestine. Small buds develop from the ductules and become the islets of Langerhans. In addition, some of the endocrine cells may be scattered singly or in groups of only a few cells throughout various regions of the pancreas. Immunoreactivity of the various peptides occurs in humans at about 8 to 10 weeks of development, marking the first synthesis of pancreatic hormones (Table 12-5).

Although the islets originate as outgrowths from the pancreatic ductules, the actual embryonic origin for the endocrine cells is not clear. It has been suggested that they arise from mesoderm or from neural crest cells of neuroectodermal origin or from endoderm of the gastrointestinal tract. However, experiments have shown that the islet cells are not derived from neural crest (see discussion of APUD cells in Chapter 4), at least in birds, and it is probable that mammalian islet cells are not derived from either neural crest or other neuroectoderm. Transplantation of quail neural crest cells containing a unique cytological marker into chick embryos confirms their incorporation into the avian pancreas, but they do not give rise to the pancreatic endocrine cells. Instead, they differentiate into parasympathetic ganglia that, as we shall see, are important for regulating islet cell activity. Until the origin of its secretory cells is confirmed, the endocrine pancreas can be considered to be of endodermal origin.

Several functional schemes have been proposed for the origins of the endocrine pancreatic cells that are divorced from their possible germ layer origins. These endocrine cells may represent modified gastrointestinal mucosal cells that initially synthesized "inducers." Exocrine cells in teleosts that are considered to be homologous to avian and mammalian islet cells often possess a concentration of protein granules (zymogen mantle), presumably induced by secretions of the adjacent pancreatic endocrine cells. Another suggestion is that these cells originally secreted hydrolytic enzymes related to digestion but that these "enzymes" lost their catalytic properties and acquired hormonal functions. In the fetal guinea pig pancreas, immunoreactive glucagon, PP, insulin, and SST can be demonstrated in cells of the pancreatic tubules about 10 to 15 days before they appear in the islets, supporting the notion that they are derived from exocrine enzyme-secreting cells. These modified mucosal cells presumably lost their contact with the gut mucosa and specialized as centers for internal secretion. The similarity of peptides produced in the islet cells and those of the gastrointestinal tract described earlier supports a common origin for these pancreatic and intestinal cells.

B. Cellular Types in Pancreatic Islets

At least five different cellular types have been identified in the mammalian endocrine pancreas: B, A, D, PP, and **amphophils**. Although first identified by their staining characteristics, they now are identified by their immunoreactivity to antibodies prepared against the specific pancreatic peptides. Immunocytochemistry reveals that several peptides may be localized in one cell type, and the pattern of colocalization may differs markedly in different animals.

1. B Cells

The first cellular type identified in the pancreatic islets was the **B cell** (= β-cell in an alternative scheme for naming pancreatic cellular types) that stains with **aldehyde fuchsin (AF+)** and **pseudoisocyanin (PIC+)**. The latter staining procedure applied following an oxidation step has been claimed to be specific for insulin granules, but it stains other intracellular structures in non-pancreatic tissues that do not contain insulin (for example, some neurosecretory granules). Immunocytochemical techniques have verified that insulin is produced and stored in the B cell as granules of about 300 nm in diameter. Several drugs, such as **alloxan** and **streptozotocin**, selectively destroy the B cells and impair the ability of the pancreas to secrete insulin. The affected B cells undergo a process of **hydropic degeneration** that is characterized by clumping of nuclear chromosomal material (formation of pyknotic nuclei) and eventual cell death. This technique of chemically induced degeneration is often used in experimental animals to selectively remove the insulin-secreting cells without altering the ability of the islets to secrete other pancreatic peptides.

2. A Cells

Cells of the pancreatic islets that are acidophilic and argyrophilic (affinity for silver- staining techniques) are called **A cells** (= α or α_2 cells). They do not stain with AF or PIC procedures. The A cell is the source of the second major pancreatic hormone, glucagon, which is shown by immunocytochemistry to be stored in secretory granules of about 235 nm diameter. In some species, the secretory granules of A cells may contain other peptides in addition to glucagon (see Table 12-6). The round secretory granules in the cytoplasm of the A cells are morphologically distinct from the angular insulin granules of the B cells, making these cells easy to distinguish with the aid of the transmission electron microscope. This difference in granule shape is not found in all mammals, however. Treatment with cobalt selectively impairs the ability of A cells to secrete glucagon, but such treatment does not lead to destructive degeneration such as alloxan produces in the B cells.

3. PP Cells

PP has been localized by immunocytochemical techniques in 125-nm granules of cells found at the periphery of the

TABLE 12-6 Some Peptides Co-localized with Pancreatic Hormones in Mammals[a]

Cell Type and Hormone	IGF-II	Gastrin	CRH[b]	GIP[b]	PP[b]	Endorphin	CCK[b]
A cell (glucagon)			Cat	Human	Rat	Rat	Rat
			Mouse	Rat	Human		Human
			Monkey	Pig			
			Rat	Dog			
			Human	Cat			
				Guinea pig			
B cell (insulin)	Mammal						
D cell (somatostatin)		Human	Human				
			Rat				
			Guinea pig				
PP cell (pancreatic polypeptide)							

[a]*Co-localization was determined by immunocytochemistry.*
[b]*IGF, insulin-like growth factor; CRH, corticotropin-releasing hormone; GIP, glucose-dependent insulinotrophic peptide; PP, pancreatic polypeptide; CCK, cholecystokinin.*

endocrine islets as well as in cells scattered throughout the exocrine pancreas. These **PP cells** are distinct cytologically and immunologically from other cell types. The distribution of PP-secreting cells (= F cells) on the periphery of the islets varies greatly among different species, and it is difficult to generalize for all mammals.

4. D Cells

Immunoreactive SST is localized in the cytoplasmic granules of **D cells** (also called α_1 cells or δ cells). Although D cells contain cytoplasmic granules similar in size to those of A cells, they are cytochemically distinguished from A cells by applying the toluidine blue staining procedure and from both A cells and B cells by their staining with PIC following methylation but not following oxidation. Granule size (230 nm) is very similar to that of A cells.

5. Amphophilic Cells

Amphophils have been demonstrated in the islets of many mammalian species as well as in sharks, teleosts, amphibians, and reptiles, but no conclusions have been generated regarding their functional roles. They may represent either ungranulated and/or differentiating or granule-depleted degenerating forms of any of the four cellular types described above.

C. Hormones of the Mammalian Endocrine Pancreas

The two major pancreatic hormones are the hypoglycemic, antilipolytic, lipogenic hormone insulin and the hyperglycemic, lipolytic hormone glucagon. In addition, the pancreatic endocrine cells produce somatostatin and PP, but their roles in the bioregulation of metabolism are not as dramatic as are those for insulin and glucagon.

1. Insulin: The Hypoglycemic Hormone

Von Mering and Minkowski in 1889 first observed the correlation between the pancreas and blood sugar regulation when they induced diabetes mellitus in dogs following pancreatectomy. The suspected hypoglycemic factor of the pancreas was later named "insuline" (Latin *insula*, island) to emphasize its origin from the islets of Langerhans previously identified from histological observations. Purified insulin, however, was not isolated until a physician, Frederick Banting, and a graduate student in physiology and biochemistry, Charles Best, teamed up at the University of Toronto in the summer of 1921. Much of their success was due to the collaborative efforts of the biochemist J.B. Collip, who purified insulin, and the project's director, J.J.R. Macleod.

The presence of proteolytic enzymes in the exocrine pancreas had thwarted earlier attempts to extract insulin from a whole pancreas. Because it had been shown that tying off the pancreatic duct caused degeneration of the acinar pancreatic tissue but did not affect the islet tissue, Banting suggested they employ this technique to avoid contamination of their extracts with digestive enzymes. The hypoglycemic factor they isolated was first named isletin, but they later changed its name to conform to the name proposed earlier by von Mering and Minkowski. The importance of the successful isolation of insulin and its use to alleviate the fatal symptoms of diabetes mellitus led to the awarding of the Nobel Prize in Physiology and Medicine in 1923 to Banting and Macleod. Best and Collip were not officially recognized in the award, although the recipients acknowledged them for their contributions. This pioneering research with experimental animals also is recognized for having saved millions of human lives.

Insulin may be bioassayed *in vitro* for its ability to stimulate uptake of glucose from the medium into cells. Muscle cells from the rat diaphragm are often employed in this bioassay. Radioimmunoassay (RIA) or ELISAs now are routinely employed to measure circulating insulin levels, and many commercial RIA kits are available for performing these analyses. Insulin RIAs are highly specific, and some can detect the difference between porcine and human insulins that vary in only one amino acid substitution.

The standard measure of the ability of the pancreas to secrete insulin is the **glucose tolerance test**. Following a period of fasting, a glucose load (excessive amount of glucose) is administered orally or intravenously, and its rate of disappearance (clearance) from the blood is measured. The clearance rate for a given glucose load is directly proportional to the secretion of insulin and hence reflects the ability of the pancreas to respond to hyperglycemia with elevated insulin secretion (Table 12-7). Another common indicator of normal insulin secretion is to measure plasma glucose following a 12-hour fast.

2. Chemistry and Synthesis of Mammalian Insulins

Insulin is a very conservative and phylogenetically old molecule, occurring throughout the animal kingdom and even among unicellular organisms (see Chapter 13). The insulin molecule is a small protein composed of two different polypeptide chains linked together by disulfide bridges. It is structurally very similar to the **insulin-like growth factors (IGFs)** and **relaxin** described earlier, although each has its own receptor type (see Chapter 3). The insulin prohormone is a single polypeptide chain in which three regions are recognized: A, B, and C. The C-peptide is cleaved posttranslationally. Secretion granules

TABLE 12-7 Comparison of Human Responses to a Standard Glucose Tolerance Test[a]

	Venous Glucose Level (mg/dL)			
Subject status	**Fasting**	**30 min**	**60 min**	**120 min**
Normal person	<100	<160	<160	<100
Probable diabetic	<100 or 100–120	130–159	160–180	110–120
Diabetic	<100 or >100	>150	>160	>120

[a]An abnormal glucose tolerance test can also occur as a result of factors other than diabetes mellitus, including improper feeding prior to test, malnutrition, obesity, infection, fever, hyperthyroidism or hypothyroidism, acromegaly, kidney disease, and islet cell tumors.

of B cells contain crystals of insulin and C-peptide fragments as well as some uncleaved proinsulin. The insulin fragment forms a complex with available zinc ions to produce a crystalline hexamer within the secretory granule. It is this crystalline structure that imparts the unique, angular structure to insulin secretion granules in the B cell.

3. Regulation of Insulin Release

Hyperglycemia can stimulate release of insulin through a direct action of glucose on the B cell. Glucose is transported into B cells by the **glucose transporter 2 (GLUT2)** (Figure 12-21). Oxidation of glucose by the B cell leads to production of ATP, which closes ATP-sensitive K^+ channels in the plasma membrane, thereby preventing K^+ efflux and depolarizing the B cell. Depolarization results in the opening of voltage-sensitive Ca^{2+} channels in the plasma membrane, allowing Ca^{2+} to flood into the B cell, thereby causing exocytosis of insulin-containing secretory granules (Figure 12-21). Decreases in circulating glucose cause a reduction in the secretion of insulin, and it appears that the basic regulatory control mechanism is directed through changes in blood levels of glucose. Any agent capable of elevating circulating glucose levels normally will evoke insulin release. Amino acids, especially arginine, also can stimulate insulin release, although the mechanism is not well-understood.

There is evidence to support a role for certain GI peptides in stimulating insulin secretion. These so-called **incretins** are most likely peptides that have already been characterized for their action in GI function. GIP is a leading incretin candidate. Elevated glucose appears in human hepatic portal blood within 2 minutes after ingesting 75 g of glucose, and elevated GIP and insulin follows within 5 minutes. Insulin, GIP, and glucose all reach maximal levels within 30 minutes and then decline to normal within 3 hours. Following ingestion of a "standard" mixed meal, maximal levels of circulating GIP are achieved within 45 minutes and remain elevated for 6 hours. A similar elevation of circulating glucose induced through direct intravenous injection of glucose evokes a smaller insulin response than did the meal. Although GIP release has been demonstrated following ingestion of long-chain but not short-chain fatty acids, the augmentation of insulin release by GIP apparently requires a simultaneous increase in glucose levels. Release of insulin by GIP only occurs when accompanied by hyperglycemia. This seems to relegate the insulinotropic role of GIP to that of a permissive effect, but a physiologically important action nevertheless.

GLP-1 also is a candidate incretin based upon its role in stimulating insulin secretion. The actions of incretins in treating type 2 diabetes can be enhanced by co-administration of a new class of drugs called **dipeptidyl peptidase-4 inhibitors (DPP-4 inhibitors)**, which prevent incretins from proteolytic digestion.

In recent years, the importance of inhibitory nervous regulation of insulin secretion has been demonstrated. Non-myelinated autonomic axonal fibers are present in the islets. Similarly, acetylcholine or activation of parasympathetic nerves stimulates insulin release, whereas the anticholinergic drug atropine or norepinephrine from sympathetic neurons block insulin release. Epinephrine, which is a hyperglycemic agent affecting liver, muscle, and adipose tissue, also directly blocks the release of insulin. Thus, circulating epinephrine can potentiate its own hyperglycemic action by blocking the normal response of the B cell to increased blood glucose resulting from epinephrine's action on adipose and liver cells.

SST inhibits release of insulin by a direct action on the B cell, and it may play a paracrine role in regulating insulin release locally in the pancreas. It is not known what regulates SST release from the pancreatic D cells, but studies show that both acetylcholine (parasympathetic) and α-adrenergic (sympathetic) agents can block SST release whereas VIP and β-adrenergic agents stimulate its release.

The neuropeptide galanin has been identified in pancreatic islet neurons of dogs and humans, sometimes colocalized with other peptides such as NPY. Release of insulin from the B cell by glucose or amino acids is blocked directly by

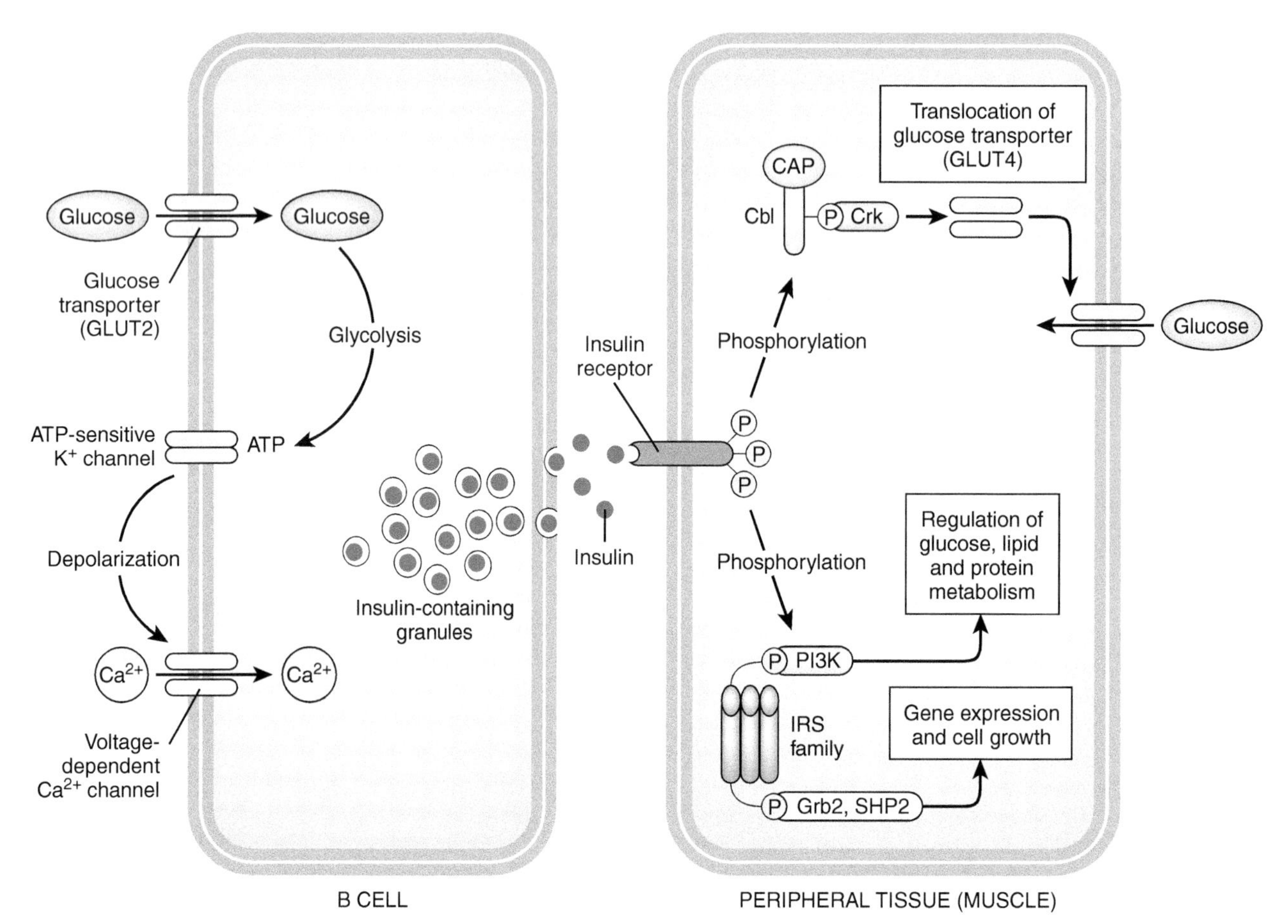

FIGURE 12-21 **Regulation of insulin secretion from the B-cell and insulin regulation of growth, glucose entry, and metabolism in target (fat, muscle) cells.** Glucose is transported into B cells by GLUT2. The B cell oxidizes glucose, resulting in ATP, which closes ATP-sensitive K^+ channels in the plasma membrane, thereby preventing K^+ efflux and depolarizing the B cell. Depolarization causes voltage-sensitive Ca^{2+} channels to open in the plasma membrane, allowing Ca^{2+} to enter the B cell and causing exocytosis of insulin-containing secretory granules. Insulin binds to its receptor, phosphorylating insulin receptor substrate proteins, which in turn dock with many intracellular signaling molecules to induce cell growth (Grb2 and SHP2) and metabolism (PI3K). Phosphorylation of insulin receptors also leads to phosphorylation of Cbl—CAP proteins, causing the translocation of GLUT4 proteins to the plasma membrane and glucose entry into the target cell. Abbreviations: CAP, Cbl-associated protein; Cbl, casitas B-lineage lymphoma (Cbl) protein; Crk, v-Crk sarcoma virus CT10 oncogene homolog; Grb2, growth factor receptor-bound protein 2; PI3K, phosphatidylinositol 3-kinase; SHP2, protein tyrosine phosphatase, non-receptor type 11. *(Adapted with permission from Notkins, A.L.,* Journal of Biological Chemistry, *277, 43545–43548, 2002.)*

galanin. Various stimulators of insulin release operate via different cellular mechanisms (e.g., blocking K^+ channels, activating IP_3/DAG-mediated Ca^{2+} influx, inhibition of adenylyl cyclase) all of which can be blocked by galanin. The physiological importance of this inhibitory mechanism and its relation to inhibition of insulin release by SST is not clear.

A peptide of 49 amino acids was isolated first from the porcine pancreas and named **pancreastatin (Pst)** for its ability to block glucose-induced release of insulin. Pst may be a product of a large acidic protein, **chromogranin A**, a member of a family of similar proteins that also is associated with the secretory granules of catecholamine-secreting cells of the adrenal medulla (see Chapter 8), some gastrointestinal endocrine cells, and both catecholaminergic and peptidergic neurons. Although its physiological role is uncertain, a phylogenetic survey reveals Pst-immunoreactivity in the gastric mucosa and pancreatic islets of all vertebrates examined including pig, human, chicken, Japanese quail, lizard, frog, shark, ratfish, and hagfish. No Pst-immunoreactivity was observed in either the rat or in cephalochordates or urochordates.

4. Actions of Insulin on Target Tissues

After insulin binds to its membrane receptor, there is an increase in the transport of glucose, amino acids, fatty acids, nucleotides, and various ions into the target cell. Within a few minutes, enhancement of anabolic pathways and a decrease in catabolic pathways occur. Through delayed effects on nuclear transcription and protein synthesis, some cellular growth is stimulated. This growth-promoting action is due to an interaction between insulin and GH-dependent, circulating IGFs which operate through separate receptors but through similar intracellular pathways.

Insulin brings about hypoglycemia by stimulating increased uptake of glucose through activation of the **glucose transporter 4 (GLUT4)** protein, which is responsible for facilitated diffusion of glucose from the blood into insulin target cells (Figure 12-21). Insulin facilitates transport of glucose into muscle and fat cells as well as transport of amino acids into muscle cells. Activity of hexokinase also is enhanced by insulin, stimulating glucose oxidation and favoring lipogenesis (adipose tissue) and protein synthesis (muscle). Insulin also prevents glycogen breakdown (glycogenolysis) and enhances glycogen synthesis following uptake of glucose in muscle and liver. This is accomplished through simultaneous inhibition of the cAMP-dependent enzyme **phosphorylase-a** and stimulation of the enzyme **glycogen synthetase**.

Glucose oxidation increases the intracellular pools of precursors for fat synthesis—that is, glycerol, acetyl-CoA, and fatty acids. In addition to indirectly enhancing esterification by stimulating acylglycerol synthetases, insulin inhibits **triacylglycerol lipase** and prevents lipolysis in adipose tissue. This reduces release of NEFAs and monoacylglycerol into the blood and hence reduces hepatic fatty acid oxidation and ketogenesis. These actions of insulin on adipose tissue and lipid metabolism are obvious in carnivores, but there is little effect of insulin on either glucose metabolism or lipogenesis in herbivorous mammals. Consequently, carnivorous mammals are more sensitive to pancreatectomy than are herbivores.

The increase in intracellular glucose-6-phosphate and amino acids in both liver and muscle cells promotes protein, lipid, and glycogen synthesis. These are important actions of insulin immediately following ingestion of a meal and the entry of large quantities of digestive products such as glucose and amino acids into the blood.

5. Mechanism of Action for Insulin in Fat and Muscle Cells

The many actions of insulin on its target cells have prompted multiple schemes to explain these diverse actions. At last, a unified picture is beginning to emerge that eventually may explain all of insulin's actions. Insulin binds to a tyrosine kinase transmembrane receptor that occurs as a dimer in the cell membrane (see Chapters 3 and 13). The occupied receptor undergoes autophosphorylation and then acts as a tyrosine kinase to phosphorylate various substrates, including **insulin receptor substrate (IRS) proteins** (Figure 12-22). IRS protein docking with downstream adaptor proteins is incredibly complex, as evidenced by the number of functional domains in the IRS protein (Figure 12-22). The phosphorylated IRS proteins work through independent signaling pathways to promote cell proliferation and metabolism (Figure 12-21). Phosphorylation of the insulin receptor also leads to phosphorylation of **Casitas B-lineage lymphoma (Cbl) protein** and its associated protein, **CAP (Cbl-associated protein)**, resulting in translocation of GLUT4 proteins to the plasma membrane and entry of glucose into the cell (Figure 12-21).

6. Oral Insulins

The term "**oral insulin**" refers to the synthetic hypoglycemic drugs known as sulfonylureas and related compounds (Figure 12-23). Because they are not proteins and thereby not subject to enzymatic degradation in the gut, the sulfonylureas may be taken orally and can relieve some of the symptoms of diabetes mellitus. Oral insulins thus have considerable therapeutic value in certain types of insulin-deficiency disorders. The effectiveness of these drugs may be related to their abilities in displacing insulin from nonspecific binding sites and non-effective linkages in plasma, connective tissues, and islet cells to provide marginally effective quantities of insulin for binding to effective (target) sites. If too high a dose is employed, the oral insulin also may block the effective sites as well, and no hypoglycemic effect is observed. The oral insulins are useful only in disorders related to insulin insufficiency and would not be useful in totally pancreatectomized animals or in humans who cannot produce functional insulin.

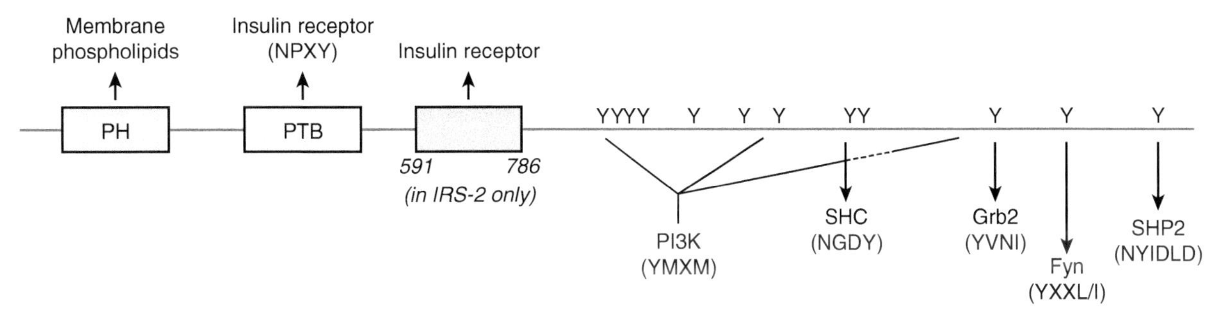

FIGURE 12-22 **Mouse insulin receptor substrate (IRS) protein.** The amino terminus contains a pleckstrin homology (PH) domain and a phosphotyrosine-binding (PTB) domain. Other tyrosine phosphorylation residues in the mid-region and carboxy terminus of the protein allow for interaction with other signaling proteins in the cascade including phosphatidylinositol 3-kinase (PI3K), Src homology 2 domain-containing protein (SHC), growth factor receptor-bound protein 2 (Grb2), proto-oncogene tyrosine-protein kinase (Fyn) and protein-tyrosine phosphatase SHP2. The amino acid motifs in the IRS protein are indicated in parentheses, where x stands for any amino acid. See Appendix C for explanation of amino acid abbreviations. *(Adapted with permission from Thirone, A.C.P. et al.,* Trends in Endocrinology & Metabolism, ***17****, 72–78, 2006.)*

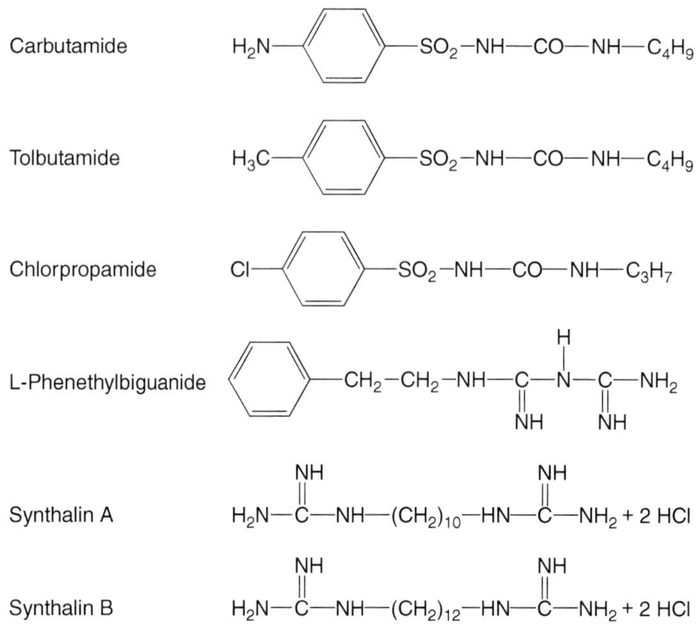

Carbutamide $H_2N-C_6H_4-SO_2-NH-CO-NH-C_4H_9$

Tolbutamide $H_3C-C_6H_4-SO_2-NH-CO-NH-C_4H_9$

Chlorpropamide $Cl-C_6H_4-SO_2-NH-CO-NH-C_3H_7$

L-Phenethylbiguanide $C_6H_5-CH_2-CH_2-NH-C(=NH)-N(H)-C(=NH)-NH_2$

Synthalin A $H_2N-C(=NH)-NH-(CH_2)_{10}-HN-C(=NH)-NH_2 + 2\ HCl$

Synthalin B $H_2N-C(=NH)-NH-(CH_2)_{12}-HN-C(=NH)-NH_2 + 2\ HCl$

FIGURE 12-23 **Oral insulins.** These synthetic molecules cannot substitute for insulin but enhance the activity of insulin already present. These drugs are useful only in the treatment of type 2 diabetes.

Furthermore, oral insulins are effective only if diet is carefully controlled.

B. Glucagon: The Hyperglycemic Hormone

The discovery of glucagon was not marked by great publicity or special prizes, as was that of insulin, and this second pancreatic hormone has always been overshadowed by the clinical importance of insulin. Glucagon is only one of several hyperglycemic factors, and more attention has been given to epinephrine as hyperglycemic hormones. Glucagon is more important in fasting carnivores and in herbivores than it appears to be in fed humans or carnivores. It also is more important than insulin in birds and lizards for regulating metabolism.

1. Chemistry of Glucagon

Glucagon is a straight-chain polypeptide hormone consisting of 29 amino acid residues and with a molecular weight of 3485 Da. It is a member of the PACAP family of peptides (see Chapter 4). Thus, glucagon is chemically similar to secretin and CCK. In addition to the peptide glucagon, proglucagon also contains the two glucagon-like peptides GLP-1 and GLP-2, discussed above. Glucagon is a conservative peptide among mammals that has been identified immunocytochemically in all vertebrates as well as in a number of invertebrates. Non-mammalian glucagons are structurally conservative, too. Another extrapancreatic glucagon-like hyperglycemic factor has been characterized in extracts of submaxillary salivary glands obtained from rats, mice, rabbits, and humans. Release of this salivary glucagon is influenced *in vitro* in the same manner as pancreatic glucagon. SST, however, does not block release of salivary glucagon.

2. Actions of Glucagon on Target Tissues

Glucagon promotes glycogenolysis in liver cells, its primary target with respect to raising circulating glucose levels. This effect appears to be mediated through stimulation of adenylyl cyclase and production of intracellular cAMP and activation of phosphorylase-a. Increased glycogenolysis accompanied by decreased intracellular oxidation of glucose directs the movement of glucose from liver cells into the blood. Lipolysis is stimulated by glucagon in fasting animals through activation of **hormone-sensitive lipase** (triacylglycerol lipase) in adipose tissues. The release of NEFAs into the blood further increases β-oxidation followed by ketogenesis and gluconeogenesis in liver. Glucagon also increases levels of PEPCK, the critical enzyme in gluconeogenesis, in liver cells (see Appendix H).

C. Pancreatic Polypeptide: A Third Pancreatic Hormone

Pancreatic polypeptide was discovered in the chicken pancreas and since has been purified from the usual mammalian sources (ovine, bovine, canine, porcine, rat, and human pancreatic tissue) as well as from wild equines (Przewalski's horse, zebra), rhino, tapir, several birds, and the alligator. The average molecular weight for PP is 4200, and all peptides isolated consist of 36 amino acid residues (see Chapter 13, Figure 13-7). Intestinal and pancreatic PPs are identical. Structurally, PP is similar to NPY and PYY. The physiological role of PP is unknown, although there are several reasons for giving PP hormonal status. In humans, ingestion of a protein meal such as ground beef causes an increase in plasma levels of PP from pre-ingestion levels of 57 pg/mL plasma to 229 pg/mL, 400 pg/mL, and 580 pg/mL, respectively, in 5, 10, and 240 minutes. Infusion or ingestion of glucose does not evoke this sort of increase in hPP, and it appears that the rapid increase in plasma hPP following ingestion of protein is at least in part a neural response mediated via the vagus nerve. Release of hPP can be blocked by SST, although it is not known whether SST is involved in the normal regulation of PP release. A second phase of elevated PP occurs later and may last for several hours. The initial phase of PP secretion can be induced with experimental gastric distention, but the later response occurs only following protein- or fat-rich meals. The biphasic release of aPP in chickens apparently is independent of protein, fat, or carbohydrate content of the diet. Bovine PP is reported to be a potent inhibitor of pancreatic exocrine secretion, but it does not seem to influence carbohydrate (glucose) metabolism. Resting levels of plasma PP in humans increase with age from about 50 pg/mL in 25-year-olds to more than 200 pg/mL at age 70. This increase is paralleled by a progressive increase in the number of PP cells in the islets.

D. Somatostatin: Paracrine Regulator of Pancreatic Secretion

The D cell is believed to be the source of pancreatic SST that is chemically identical with hypothalamic SST. In mammals, both SST_{14} and SST_{28} have been identified in the pancreas. Levels of SST in pancreatic extracts of rats are equivalent to those of hypothalamic extracts. Pancreatic SST is believed to locally inhibit the release of insulin, glucagon, and PP in a paracrine fashion. Regulation of pancreatic SST release is not understood. The observed inhibitory actions of acetylcholine and norepinephrine mentioned earlier possibly mimic parasympathetic and sympathetic innervation, respectively.

IV. HORMONES REGULATING MAMMALIAN METABOLISM

Metabolism is affected by many hormones that directly or indirectly affect key enzymes in glycolysis, gluconeogenesis, glycogen synthesis or degradation, lipogenesis, lipolysis, and the synthesis and utilization of proteins (Table 12-8; see also Appendix H). These hormones include insulin and glucagon from the endocrine pancreas as well as epinephrine from the adrenal medulla (Chapter 8), glucocorticoids from the adrenal cortex (Chapter 8), GH from the adenohypophysis, and IGFs from liver (Chapter 4), thyroid hormones (Chapter 6), and gonadal steroids (Chapter 10). A hormone may be classified according to its general actions as being **glycolytic** (favoring glycolysis), **glycogenolytic** (favoring glycogen hydrolysis to glucose–phosphate), **gluconeogenic** (favoring conversion of lipids and/or proteins or their intermediates into glucose production), **lipogenic** (favoring synthesis of TAGs or steroids), **lipolytic** (favoring hydrolysis of TAGs to NEFAs and MAGs), protein anabolic (favoring protein synthesis), or **protein catabolic** (favoring breakdown of proteins to amino acids). A hormone may affect directly or indirectly many metabolic processes and several of the above descriptors may be applied to the same hormone (e.g., glucocorticoid effects) (Table 12-9). These metabolic actions may be due to effects on enzyme activity or effects on transport at the cell membrane and hence affect the availability of substrates for various pathways. Although a detailed analysis of this topic

TABLE 12-8 Generalized Actions of Hormones with Respect ot Protein, Carbohydrate, and Lipid Metabolism

Hormone	Protein Metabolism	Carbohydrate Metabolism	Lipid Metabolism
Insulin	Anabolic	Hypoglycemic, glycogenic, antiglycogenolytic	Lipogenic, antilipolytic
Epinephrine		Glycogenolytic, hyperglycemic, gluconeogenic	Lipolytic
Glucagon		Hyperglycemic, glycogenolytic	
Growth hormone	Anabolic	Hyperglycemic, gluconeogenic	Lipolytic
Glucocorticoids	Antianabolic	Hyperglycemic, gluconeogenic, inhibits peripheral utilization	
Thyroxine	Anabolic or catabolic Anabolic	Hyperglycemic, glycogenolytic	Lipolytic
Androgens	Anabolic		

TABLE 12-9 Metabolic Actions of Glucocorticoids in Different Mammalian Tissues[a]

Process	Liver	Adipose	Heart, Diaphragm	Skeletal Muscle
Glucose uptake		−	−	
Glycogenesis	+		+	−
Amino acid uptake	+		+	−
Gluconeogenesis	+		+	
Protein synthesis			−	
Protein catabolism				+
Lipogenesis	+			
Lipolysis		+P		

[a]Note that the responses of each tissue to the same hormone are specific to the tissue. +, stimulatory; −, inhibitory; P, permissive effect.

is beyond the scope of this account, some hormonal interactions in mammals are briefly described in the following sections.

A. Endocrine Regulation Following Feeding

In carnivores and omnivores, the most important hormones following ingestion of a meal and following its absorption into blood are insulin and GH as well as GIP. This time period is referred to as **postprandial** and is characterized by elevated blood levels of glucose, amino acids, and lipids in proportion to the composition of the meal. Release of insulin is evoked directly by an increase in circulating glucose in the blood supplying the pancreatic islets. In addition, there is evidence of early enhancement of the glucose-stimulated release of insulin caused by release of GIP from the small intestine in response to the presence of glucose in the small intestine. It is not clear what the role of elevated circulating PP is in humans and dogs after a meal. No convincing evidence for a direct metabolic action of PP has been demonstrated, and this peptide may be limited to regulating digestive functions. Direct neural stimulation of insulin release also may occur following or during ingestion, but this does not appear to be a major regulatory pathway.

Insulin stimulates the uptake of glucose by muscle cells, liver cells, and adipocytes. In liver, insulin promotes glycogenesis and formation of **very-low-density lipoprotein (VLDL)** whereas in muscle cells it favors glycogenesis and glycolysis. These effects are mediated largely through reduction in the enzyme phosphorylase-a and an increase in the enzyme glycogen synthetase. In liver, there also is a reduction in glucose-6-phosphatase and glucokinase that also favors glycogenesis. In adipose cells, synthesis and storage of triacylglycerols as well as inhibition of lipolytic pathways are favored by the intracellular actions of insulin. Increased lipoprotein lipase activity also is favored in adipose tissue by insulin, which enhances incorporation of NEFAs and MAGs from VLDL droplets and **low-density lipoprotein (LDL) droplets**. In addition, insulin enhances transport of amino acids into muscle cells, thereby enhancing protein synthesis. In liver cells, insulin favors protein synthesis and reduces both deamination (transanimation) of amino acids and gluconeogenesis.

Release of GH may occur much later after a meal, and together with insulin it causes stimulation of protein synthesis in muscle cells. During sleep, insulin levels typically are low and GH release is maximal. GH stimulates additional protein synthesis as well as glycogenolysis and lipolysis.

B. Effects of Acute and Chronic Stressors on Metabolism

The physiological response to stressors is detailed in Chapter 9. Acute stressors, such as the "flight-or-fight" response and exercise, cause release of epinephrine from the adrenal medulla. Epinephrine brings about important changes in energy metabolism resulting in increased glycogenolysis and hyperglycemia (Figure 12-24). Glucagon levels also may increase, contributing to hyperglycemia. Chronic stress involves the gluconeogenic and hence hyperglycemic actions of the glucocorticoids (see Table 12-9). In cases of fasting or starvation, GH and glucagon as well as glucocorticoids play important metabolic roles.

1. Acute Stress and Metabolism

An early response to stress is the elevation in epinephrine release from the adrenal medulla. Epinephrine activates adenylyl cyclase in muscle cells, resulting in increased glycogenolysis (Figure 12-24). Because glucose-6-phosphate

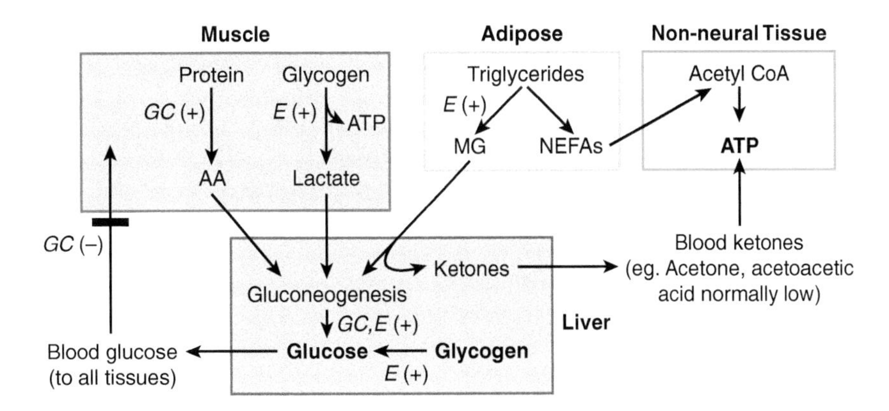

Epinephrine (E):
1. Promotes glycogenolysis.
2. Promotes gluconeogenesis in liver.
3. Promotes triglyceride breakdown in adipose tissue.
4. Elevates blood glucose.

Glucocorticoids (GC):
1. Promote protein catabolism.
2. Inhibit glucose uptake by muscle.
3. Promote gluconeogenesis in liver via stimulation of PEPCK.
4. Permissive effect on E action in adipose tissue.
5. Elevate blood glucose.

FIGURE 12-24 **Effects of epinephrine (E) and glucocorticoids (GC) on intermediary metabolism during stress. See text for explanation of abbreviations.**

resulting from glycogenolysis cannot be converted into free glucose and leave the muscle cell, it is used directly in glycolysis. This reduces the need for muscle cells to remove glucose from the blood and spares blood glucose for use by the nervous system during stressful times. Another cAMP pathway is stimulated by epinephrine in adipose cells and enhances the activity of hormone-sensitive lipase. This in turn increases the availability of NEFAs and glycerol for gluconeogenesis and a trend toward ketogenesis during stress. However, the lipolytic actions of epinephrine result primarily from sympathetic innervation of adipose tissue, and circulating epinephrine from the adrenal medulla does not play a major role.

2. Chronic Stress and Metabolism

If stressful conditions persist beyond the acute phase, the HPA axis is stimulated such that additional glucocorticoid is released. The most important role for glucocorticoids is to prevent utilization of amino acids for protein synthesis and favor conversion of amino acids to carbohydrate via gluconeogenesis. In liver, gluconeogenesis from amino acids provides glucose that can be added to the circulation, producing a subsequent hyperglycemia (Figure 12-24). This action also protects circulating glucose as an energy source for the nervous system by forcing muscle cells to employ amino acids for energy production and prevent use of circulating glucose by these cells.

During starvation, GH and glucagon have important roles in metabolism. Although glucagon normally stimulates glycogenolysis in liver, the absence of significant glycogen stores during starvation means that glucagon can have little direct effect on circulating glucose levels. However, GH and glucagon both stimulate lipolysis in adipose tissue and increase the levels of NEFAs for energy production as well as ketogenesis. Meanwhile, glucocorticoids continue to encourage muscle cells to catabolize protein stores and both muscle and liver cells to use the resulting amino acids as a major gluconeogenic source. However, gluconeogenesis from amino acids produces toxic nitrogenous wastes (ammonia, urea, uric acid) that must be eliminated by the kidney with attendant loss of water. In turn, muscle cells may utilize some of the circulating ketone bodies produced during the metabolism of NEFAs as an energy source. The rest of the ketone bodies are excreted in the urine along with the nitrogenous wastes from protein catabolism. As described earlier, addition of ketone bodies to the urine contributes to water loss and dehydration.

Chronic starvation can result in severe protein deficiencies as a consequence of the emphasis on gluconeogenesis from proteins to meet the glucose demands of the nervous system. Accompanying ketogenesis from the use of NEFAs can produce disturbances in acid–base balance as well. Calculations based on analyses of energy metabolism associated with acute starvation indicate that most humans could only survive about 21 days of total starvation before they would die of protein deficiencies. However, as early as 3 days after the onset of total starvation, the brain undergoes a remarkable decrease in its metabolic requirement for glucose and begins to utilize ketone bodies produced from lipid metabolism to provide the bulk of its energy needs. This change not only enhances lipolysis as a source of ketone bodies for brain metabolism and for gluconeogenesis in the liver but also spares the need for amino acids as an energy source and reduces the rate at which proteins are catabolized. Consequently, the anticipated protein deficiencies do not occur until much later; however, once the body's lipid stores are exhausted, protein remains as the only source to sustain life.

What causes this shift in brain metabolism during total starvation is unknown at this time, but it only occurs under conditions of total starvation. If there is any caloric intake, the switch to ketone utilization is prevented. Consequently, people experiencing semi-starvation (where caloric intake is less than required to balance energy expenditures) commonly suffer from protein deficiencies sooner because they employ protein stores to make up the energy deficit in preference to lipid stores. Humans who undertake self-regulated semi-starvation (commonly called "dieting") run the risk of developing protein deficiencies and reducing body weight more in terms of lean muscle mass than of fat. The losses of water involved with additional excretion of nitrogenous wastes accounts for most of the initial weight losses associated with "dieting" unless balanced carefully by increased water intake. Furthermore, it appears that protein stores can only be protected by simultaneously stimulating protein synthesis. Periodic bouts of strenuous exercise result in enhanced protein synthesis during non-exercise periods that is directed in part by episodes of GH release.

C. Protein Anabolic Hormones

Alterations in nitrogen retention and excretion reflect episodes of protein synthesis and protein catabolism for gluconeogenesis, respectively. Many hormones have been shown to enhance nitrogen retention through actions on protein synthesis. These **protein anabolic** agents include GH, androgens, estrogens, thyroid hormones, and insulin. We have already considered these actions of GH and insulin above. Because thyroid hormones are protein anabolic in thyroidectomized animals but not in hypophysectomized animals (lacking GH as well as the other tropic hormones), it is safe to conclude that the thyroid effect on protein synthesis is brought about by its enhancement of GH (see Chapters 4 and 6).

The earliest suggestion of a protein anabolic role for androgens was the claim for their rejuvenating power by Brown-Sequard in 1889 that he reportedly discovered after treating himself with extracts prepared from animal testes. In the 1930s, Charles D. Kochakian and his coworkers began 40 years of research that clearly established the unique role of androgens as **protein anabolic steroids**. Although estrogens were shown to stimulate protein synthesis in a few tissues (e.g., uterus, mammary gland, skin, skeleton), only the **androgenic–anabolic steroids** had such a profound effect on skeletal muscles that they determined an overall nitrogen retention by the body. Progesterone is protein anabolic on uterus but overall causes increased nitrogen excretion.

The greater muscle masses of male mammals as compared to female mammals long has been recognized and is attributed to the actions of androgens. In non-mammalian vertebrates, androgens have been shown to have sex-specific effects on certain muscles in males. Many analogs of the naturally occurring androgens have been prepared, but it has not been possible to separate their androgenic action from their protein anabolic effect. Androgenic–anabolic steroids do not influence transport of amino acids into muscles cells but apparently affect the levels of amino acid-activating enzymes employed in protein synthesis.

The potential effect of androgenic–anabolic steroids on muscle strength and endurance resulted inevitably in their use by humans for athletic training programs. Although their effect on muscle mass is easily demonstrated, the improvement of athletic performance is mostly anecdotal and is not well supported by controlled studies. Regardless, the established effects of androgenic–anabolic steroids on reduced reproductive performance (through negative feedback on the HPG axis), increased liver dysfunction (80% of subjects in one study), and correlations with hepatitis, liver failure, and fatal liver cancer should serve to warn people to avoid their use for unsubstantiated, short-term benefits on athletic performance.

V. CLINICAL ASPECTS OF PANCREATIC FUNCTION

A. Diabetes Mellitus

Diabetes mellitus is a complex, heterogeneous assemblage of disorders having the common feature of the appearance of glucosuria. Diabetes mellitus is always fatal if not treated, and even if treated serious complications may develop including blindness, kidney disease, and circulatory or circulation-based problems. Obesity is one of the major risk factors for development of diabetes mellitus. People with a **body mass index (BMI)** greater than 25 are considered to be overweight. Obesity is defined as having a BMI > 30. The average female in the United States is 5′4″ tall and should weigh 134 lbs to achieve a BMI = 25. For the average 5′9″ U.S. male, the appropriate weight would be 156 lb. Approximately 78 million adult

Americans over the age of 20 are obese (Figure 12-25). In addition, another 9 million can be categorized as extremely obese (BMI > 40). As of 2009, the U.S. Centers for Disease Control and Prevention estimated that one-third of all Americans exhibit **metabolic syndrome**, or "syndrome X," which is characterized by insulin resistance, obesity, hypertension, and elevated indicators of cardiovascular disease (high blood levels of triglycerides and cholesterol).

In the early 1990s, obesity levels exceeded 15% in populations of only four states. By 2003, only 15 states exhibited rates as low as 15 to 19% obesity, whereas 31 states had obesity rates of 20 to 24%, and the populations of four states exceeded 25% obesity (Figure 12-25). Of increasing concern is the rapid increase in obesity among children and adolescents. The effects of obesity are especially reflected in the incidence of diabetes mellitus. Current figures indicate that approximately 6% of the U.S. population exhibits diabetes mellitus, with another 14% exhibiting metabolic syndrome (dangerously elevated blood glucose levels). Moreover, cardio–renal–vascular complications resulting from diabetes mellitus are responsible for an additional 300,000 deaths annually, making diabetes a leading contributor to death in the United States.

Most cases of diabetes mellitus have been characterized as either **maturity onset diabetes** (ketosis-resistant, insulin-independent) or as **juvenile onset diabetes** (ketosis-prone, insulin-dependent). "Maturity" and "juvenile" are not entirely appropriate terms, as the former can occur occasionally in children and adolescents, whereas the latter can occur with low frequency in middle-aged or older adults. Consequently, these disorders were renamed first as **type 1 diabetes** and **type 2 diabetes** and sometimes are called **insulin-dependent diabetes mellitus (IDDM)** and **noninsulin-dependent diabetes mellitus (NIDDM)**, respectively. Several features of these two types of diabetes are compared in Table 12-10. There are some additional forms of diabetes now recognized, but they are rare.

Type 2 diabetes, accounting for approximately 90 to 95% of the reported cases, typically occurs in overweight people (80% of cases) who are over 40 years of age. Much of the increase in diabetes observed in the United States in recent years is due to an increasing proportion of elderly people in the population coupled with an increasing tendency toward obesity among children. Type 2 diabetes has a slow onset with no obvious symptoms, although evidence of metabolic syndrome (pre-diabetes) can be discovered by a simple fasting blood glucose test. Insulin levels in type 2 diabetes are usually normal or in excess of normal. Apparently, the target cells become insensitive to insulin, possibly due to a decrease in receptor populations or

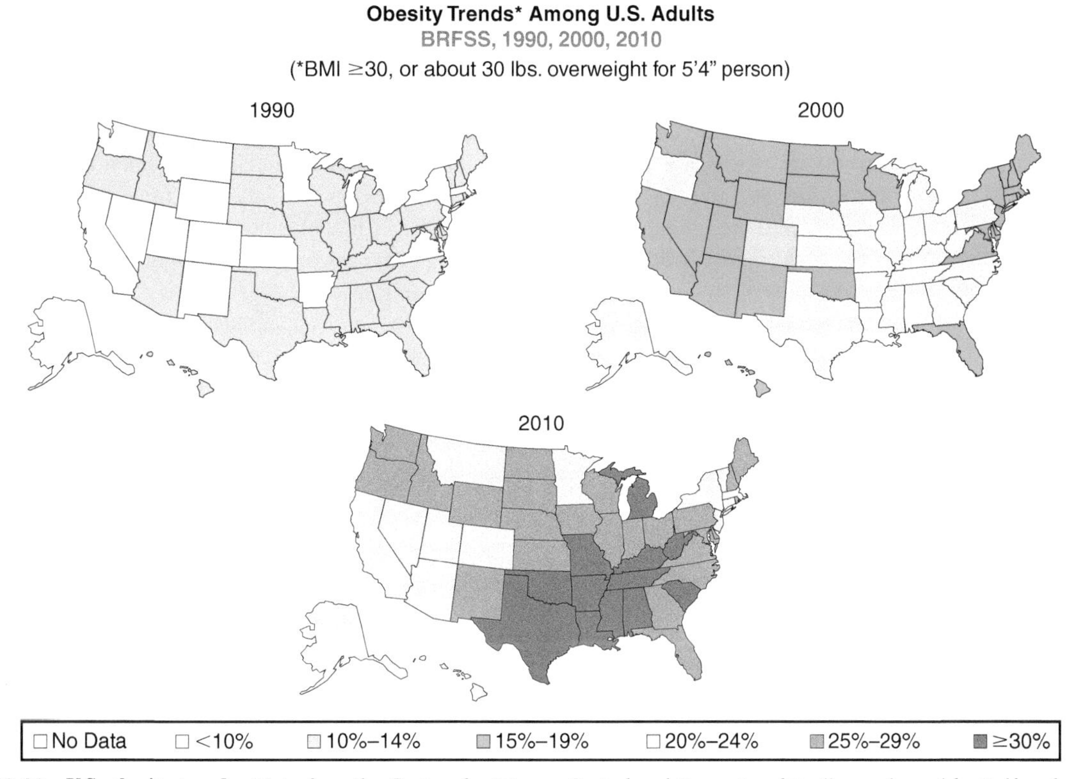

FIGURE 12-25 **U.S. obesity trends.** *(Data from the Centers for Disease Control and Prevention, http://www.cdc.gov/obesity/downloads/obesity_trends_2010.ppt.)*

TABLE 12-10 Comparison of Symptoms for Two Forms of Diabetes Mellitus

Feautures	Type 1 (IDDM)	Type2 (NIDDM)
Age at onset	Usually during childhood or puberty	Frequently over 45 years
Type of onset	Abrupt onset of overt symptoms	Usually gradual
Genetic basic	Frequently positive relationship	Commonly positive relationship
Nutritional status	Usually undernourished	Usually obesity present to some degree (about 80% of cases)
Symptoms	1. Polydipsia, Polyphagia, Polyuria 2. ketosis frequently present 3. Hepatomegaly rather common 4. Blood glucose fluctuates widely in response to small changes in insulin dose, exercies, and infection	1. May be none 2. Ketosis uncommon except under stress 3. Hepatomegaly uncommon 4. Blood sugar fluctuations less marked than for juvenile onset
Fasting blood sugar levels	Elevated	Often normal
Requirement for insulin	Necessary for all patients	Necessary for 20–30% of patients
Effectiveness of oral agents	Rarely effective	Effective when diet is controlled

to a deficiency in the mechanism of action of insulin after it is bound to the cell. Another possible cause might be an increase in non-specific binding of insulin to non-target cells that reduces insulin availability for specific receptors. Whatever the defect, dietary regulation and the removal of excess weight will bring these patients back into normal carbohydrate balance, and the regulation of carbohydrate metabolism will return to normal. In more severe cases, oral hypoglycemic agents (e.g., sulfonylureas) are used to reduce blood glucose levels along with rigid control of dietary intake.

There is evidence to support a familial or genetic component in type 2 diabetes. A series of studies on twins indicates that if one twin develops type 2 diabetes after age 50, the other twin develops type 2 diabetes in almost every case. Type 1 diabetes usually develops in persons under 20 years of age and is characterized by an abrupt onset of symptoms. A small percentage of people over 40 who contract diabetes mellitus also exhibit this form of the disease. Typically, the pancreatic B cells are reduced markedly (usually to less than 10% of normal), and insulin levels are very low or absent. Glycogen stores are depleted readily, lipolysis increases, and fatty acid metabolism is accelerated, resulting in excessive ketone body production. Accumulation of ketone bodies in the blood results in ketoacidosis and eventually leads to coma and death within 48 hours unless insulin is administered. Patients with type 1 diabetes have an absolute requirement for exogenous insulin for the rest of their lives. Because of this reliance on exogenous insulin and difficulties in administering the appropriate dose when needed, the diabetic must always be prepared for the onset of hypoglycemia following excessive insulin. Hypoglycemia can also induce coma and usually can be counteracted by consuming a high glucose source.

There is some evidence to suggest a role of infection as a cause of type 1 diabetes. The symptoms of type 1 diabetes appear abruptly, and there is a higher incidence of onset during the fall and winter months when the incidences of infectious diseases also increase. In certain genetic strains of mice, type 1 diabetes has been linked to a specific virus. There also appears to be a correlation between the occurrence of certain viral diseases (e.g., mumps, rubella *in utero*) and the later appearance of type 1 diabetes.

The major cause of type 1 diabetes, however, seems to be an autoimmune reaction that destroys the B cells. Although characterized by an abrupt onset of severe symptoms soon leading to death if untreated, type 1 diabetes actually has a long incubation period (sometimes years). Only after about 80% of the insulin-secreting capacity of the pancreas is destroyed do the classic symptoms appear, although death of B cells may be preceded by a decline in the so-called **first-phase insulin release (FPIR)** to a glucose challenge.

Identifying the precise mechanisms involved in inflammation and B-cell death in human type 1 diabetes has been difficult, and most of what we know about the causes of B-cell loss is based upon animal and *in vitro* cell culture models. The death of B cells in type 1 diabetes appears to be caused by an infiltration of white blood cells into the islets, and **cytotoxic T lymphocytes (CTLs)** have been implicated as the cell type responsible for initiating B-cell destruction. B-cell antigens that are recognized by CTLs have been identified in humans and include **IGRP265−273 (islet-specific glucose-6-phosphatase catalytic subunit-related protein)** and pro-insulin, the unprocessed biosynthetic precursor for insulin. Recognition of these antigens as "foreign" by CTLs leads to a the production of a complex array of signaling molecules from invading white blood cells including TNFα (see Box 12-A above) and interleukin-1β that trigger NF-κB and STAT-1 signaling pathways in B cells and programmed cell death (apoptosis).

Dietary restrictions are similar for treating both type 1 diabetes and type 2 diabetes patients although a reduction in total caloric intake also is necessary for overweight type 2 diabetes patients. Specific diets must be determined individually for every patient. Carbohydrates are still essential to the diabetic diet but need to be in the form of polysaccharides (e.g., starch). Mono- and disaccharides are usually avoided because of rapid uptake and resultant hyperglycemia soon after ingestion. The ingestion of simple sugars is less a problem for the type 2 diabetes patient as the pancreas can respond with some insulin release.

2. Extrapancreatic Tumor Hypoglycemia

This disorder sometimes appears in patients with cancer. Patients with **extrapancreatic tumor hypoglycemia (EPTH)** exhibit normal to low levels of insulin and normal IGF levels but have high levels of circulating proIGF-II. This proIGF-II is of tumor origin and can bind to the insulin receptor and produce insulin-like hypoglycemia.

VI. SUMMARY

Feeding is regulated by bioregulators secreted from adipose tissue and the GI tract as well as by local neuropeptides and endocannabinoids secreted by neurons in appetite-regulating areas of the NTS and hypothalamus. Peripheral factors affecting feeding are largely inhibitory (anorexic) and include leptin, which acts on cells in the hypothalamus, and CCK_8, which acts on vagal sensory neurons communicating with the brainstem. Only one peptide produced in the periphery, ghrelin, stimulates feeding (an orexic factor). CNS factors stimulating appetite include orexin A and B, NPY, AgRP, and endocannabinoids. Anorexic factors produced in the CNS include CCK_8, α-MSH, CART, CRH, Ucn-II, PYY, and PP. In addition, numerous adipokine proteins produced

BOX 12-A Adipokines, Insulin Resistance, and β-Cell Function

For many years it has been recognized that insulin acts to increase glucose uptake in adipose tissue, but it is now clear that signaling between the pancreas and adipose tissue is reciprocal. Chemical bioregulators from adipose tissue called adipokines (Table 12-2) have been implicated in regulating insulin secretion and sensitivity at the receptor. Leptin acts directly on pancreatic B cells, inhibiting insulin secretion at low doses and stimulating insulin secretion at higher doses. Pancreatic B cells express both the ObRb (long form) and ObRa (short form) leptin receptors. The complexity of the B cell response to leptin may reflect multiple signaling pathway activation and dual activation of both the long and short forms of the leptin receptor. In addition to the JAK/STAT pathway characteristic of the ObRb activation, leptin stimulates the MAPK/ERK pathway and the phosphatidyl inositol pathway in pancreatic B cells.

In 2001, researchers believed that the discovery of the adipokine **resistin** would signal a breakthrough in diabetes research, as resistin was shown to causes insulin resistance in rodent models. Unfortunately, resistin has no such role in causing insulin resistance in humans, possibly because human pancreatic islets also express resistin, and the mRNA for resistin is increased in human pancreatic islets during type 2 diabetes.

TNFα has been implicated in autoimmune destruction of B cells through the NF-κB signaling pathway in type 1 diabetes and is secreted from adipose cells as an adipokine (Table 12-2; also see Section I, above). TNFα levels in the blood do increase during obesity, although some researchers question whether blood levels reach high enough levels to cause B cell failure during type 2 diabetes.

Recent studies indicate that some adipokines stimulate B cell proliferation and activity. Visfatin (Table 12-2) stimulates B cell proliferation and reduces B cell apoptosis. Adiponectin also is implicated in maintaining B cell activity and survival. The role of adipokines in B cell function are summarized in Box Figure 12-A1.

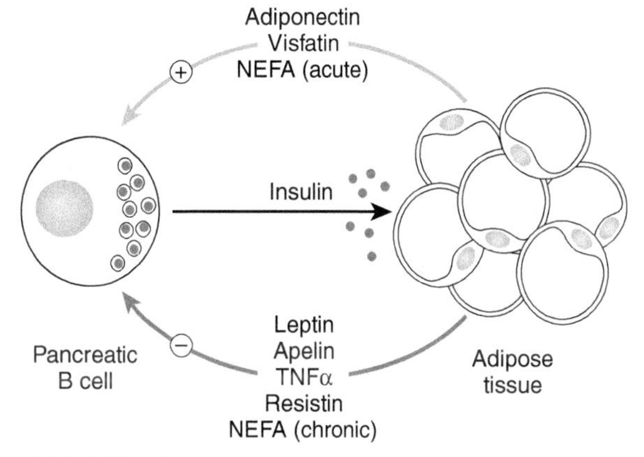

BOX FIGURE 12-A1 Adipokine regulation of pancreatic B cells. NEFA, non-essential fatty acids. *(Adapted with permission from Dunmore, S.J. and Brown, J.E.,* Journal of Endocrinology, *September 2, 2012 (Epub ahead of print) PMID: 22991412.)*

BOX 12-B Irisin: A New Muscle Hormone That Converts White Fat to Heat-Generating Brown Fat

In 2012, researchers reported the discovery of a hormone released by skeletal muscle cells that can affect fat metabolism. The hormone, **irisin** (after Iris, the Greek messenger goddess), was discovered by researchers studying the regulation of **uncoupling protein-1 (UCP-1)** in skeletal muscle cells. Irisin is a 111-amino acid peptide cleaved from a precursor protein in myocytes called **fibronectin type III domain containing 5 (FNDC5)**. Irisin is present in the circulation of humans and mice at concentrations proportional to skeletal muscle mass, and irisin levels increases within 30 minutes of the onset of exercise. Irisin acts on white adipose cells to increase the expression of UCP-1 and may mediate the beneficial effects of exercise on metabolism (Box Figure 12B-1). The primary amino acid sequence of irisin is amazingly conserved in zebrafish, chickens, mice, and humans, with 100% sequence identity between mice and humans (Box Figure 12B-2).

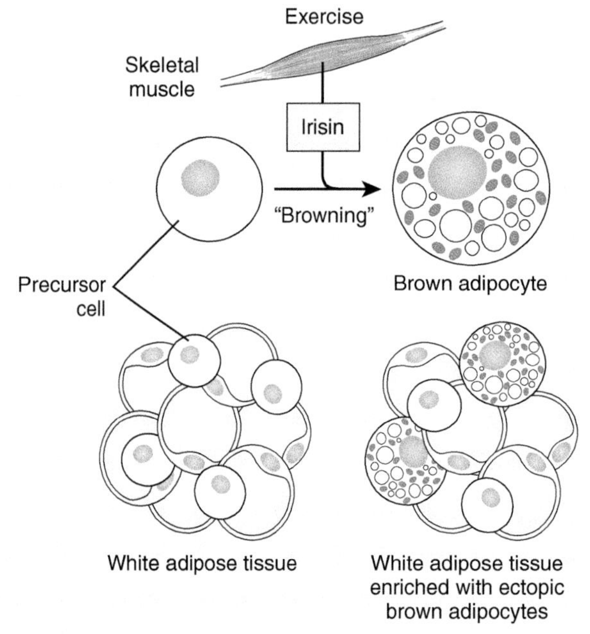

BOX FIGURE 12B-1 Irisin, a newly discovered hormone secreted by muscle tissue. Irisin recruits more heat-generating brown adipose tissue from white adipose tissue, thereby increasing metabolic rate. Irisin may be responsible in part for increasing metabolic rate during exercise. *(Adapted with permission from Villarroya, F.,* Cell Metabolism, ***15****, 277–278, 2012.)*

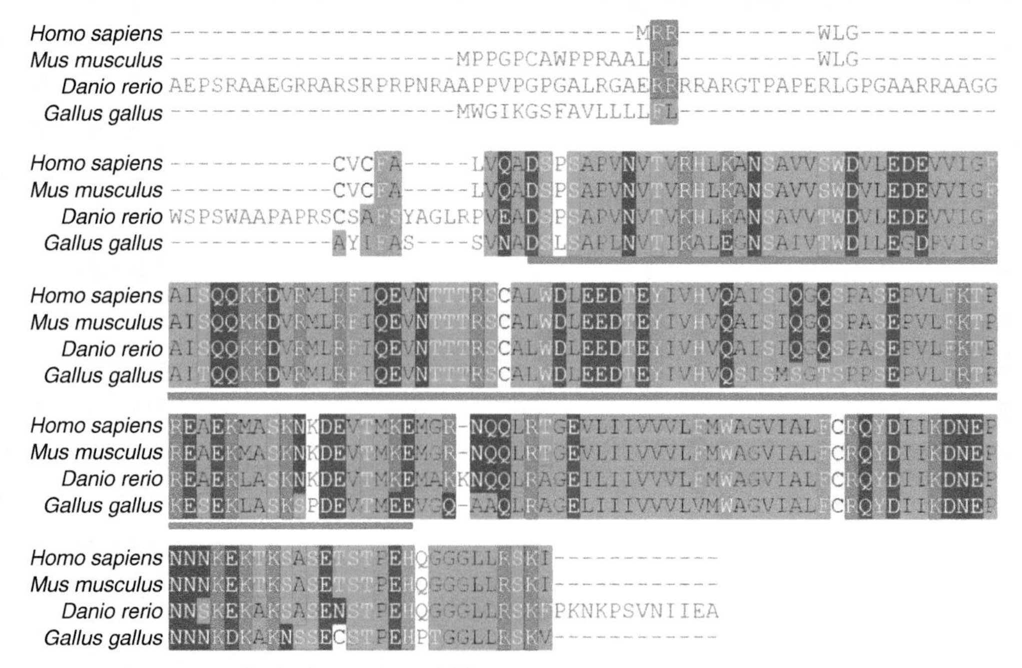

BOX FIGURE 12B-2 Sequence similarity in vertebrate irisins. Red line marks the irisin sequence. *(Reprinted with permission from Boström, P. et al.,* Nature, ***481****, 463–468, 2012.)*

BOX 12-C Regenerating B Cells Lost in Type 1 Diabetes

Advances in the ability to replace lost B cells in type 1 diabetes have been slow. Transplantation of islet grafts is fraught with problems in the donor rejecting the tissue graft. Interestingly, combination kidney/pancreas grafts in association with heavy immunosuppression have been somewhat successful, with approximately 70% of grafts remaining functional for 5 years or more. Transplantation of isolated grafts has a much lower success rate, about 10% after 5 years. The difference in success rate is not understood at present, and both transplantation methods suffer from a scarcity of pancreas/islet tissue donors. Manufacturing surrogate B cells to create so-called "insulin-producing cell factories" represents an immense technical challenge, as surrogate cells would need to manufacture insulin at a rate identical to that of normal B cells, about 10% of the cell's total protein synthesis, in order to be effective.

Considerable effort has been directed at expanding normal B cell populations from existing B cells or stem cells from adult or embryonic tissue. One promising technology for the production of insulin-producing B cells is through the growth and differentiation of adult B cells *ex vivo*. The problem with getting B cells to divide and replicate on their own is that they are considered to be a "terminal" cell type, with low or nonexistent rates of cell division in the normal pancreas. Scientists have had some luck in growing B cells by dedifferentiating B cells, allowing the dedifferentiated cells to divide and grow, and then redifferentiating the cells using chemical and molecular tools to achieve large populations of insulin-producing cells. To date, redifferentiated B cells have been maintained through up to 16 division cycles.

by adipose tissue (including adiponectin and apelin) have been implicated in appetite regulation.

The existence of three major GI hormones (gastrin, secretin, and CCK) postulated to be present in mammals at the beginning of this century has been established, and their chemical structures have been elucidated. Gastrin is produced by the G cell of the antral gastric mucosa in response to the presence of food. The parietal cell of the fundic and corpus portions of the stomach secrete HCl in response to gastrin or direct neural (vagal) stimulation. Histamine plays a role as an intermediate in the action of gastrin on the parietal cell. Parasympathetic stimulation stimulates acid secretion and also evokes secretion of pepsinogen from the chief cell in the gastric mucosa. SST produced in gastric D cells blocks gastric secretion. Release of gastrin also can be elicited by GRP.

Neural stimulation may be involved to a limited degree in the intestinal phases, but the endocrine factors predominate. Secretin is produced by the S cell of the duodenal mucosa in response to the presence of acidic chyme entering the small intestine from the stomach. The major hormonal action of secretin is to cause release of basic juices from the exocrine pancreas. The presence of peptides, amino acids, or fats in the chyme causes release of CCK from the I cell of the intestinal mucosa, which in turn stimulates secretion of pancreatic enzymes and release of bile from the gallbladder. This pancreatic action of CCK originally was postulated to be due to a hormone called pancreozymin that later was shown to be identical to CCK.

Four additional peptides have been isolated from the mucosa of the small intestine and established as GI hormones. GIP from the K cell stimulates release of insulin from the pancreas and may inhibit gastric activity under certain circumstances. Motilin from the M cell stimulates gastric motility and possibly secretion of pepsinogen but does not influence acid secretion. VIP is a neurotransmitter that stimulates blood flow to the viscera and is secreted along with PHI (PHM). Both VIP and PHI (PHM) are derived from the same preprohormone. Enteroglucagons (glicentin, GLP-I, GLP-II, tGLP-I) produced by the L cells have been identified in the intestinal mucosa and appear to stimulate insulin release but inhibit pancreatic glucagon secretion. Additional GI peptides that appear to have regulatory roles include NT, PYY, dynorphin, CRGP, and galanin. Several peptides have enterogastrone action (inhibit gastric function) including enteroglucagons, NT, CGRP, galanin, and GIP.

The endocrine pancreas secretes hormones that are directly involved in the regulation of metabolism. The predominant cellular types in the vertebrate endocrine pancreas are the A cell, B cell, and D cell. The A cell and B cell have been identified as the sources for glucagon and insulin, respectively. The D cell produces SST. A fourth cellular type located at the periphery of the mammalian pancreatic islet produces PP. It is not certain what the germ layer origins are for the pancreatic endocrine cells. They do not come from the neural crest in spite of their APUD characteristics, and it is not clear whether they are of endodermal or ectodermal origin.

Glucagon is one of several hyperglycemic regulatory factors in vertebrates, and it stimulates glycogenolysis and lipolysis in liver and adipose cells while inhibiting glucose oxidation. These actions of glucagon appear to be mediated through activation of adenylyl cyclase and formation of cAMP. Low blood sugar levels stimulate glucagon release, and SST blocks glucagon release. High levels of circulating NEFAs stimulate glucagon release in birds and herbivorous mammals. Glucagon may be the primary pancreatic regulator of carbohydrate metabolism in birds and lizards with insulin playing only a minor role. Glucagon seems to be hyperglycemic in amphibians and fishes. Immunoreactive glucagon also is present in several invertebrate phyla, including arthropods, mollusks, and the invertebrate chordates.

Insulin is the only naturally occurring hypoglycemic factor in vertebrates. It is present in many invertebrate species, as well, and may even function there as a hypoglycemic factor. The many actions of insulin include effects on nutrient transport in the plasma membrane, resulting in the increased uptake of glucose and amino acids into muscle cells and the uptake of glucose by adipose and muscle cells. Lipogenesis increases after binding of insulin to the plasmalemma of adipose cells because of the increased availability of substrates as a consequence of glucose uptake and oxidation as well as effects on enzymes that increase lipogenesis and inhibit lipolysis. Growth effects occur in cooperation with IGFs produced under the influence of GH.

High circulating glucose levels stimulate release of insulin from the B cells of the pancreatic islets. Both the uptake and the metabolism of glucose are actually responsible for causing insulin release. Neural factors (parasympathetic) and other hormones (for example, GIP) also may stimulate insulin release. SST, epinephrine, and gastrin, as well as low blood glucose levels, all inhibit insulin release.

Metabolism is controlled by the actions and interactions of many hormones under different nutritional conditions which maintain a healthy organism with the energy reserves necessary for maintenance of the organism and for reproduction. The major metabolic hormone is insulin, which is protein anabolic, lipogeneic, antilipolytic, glycogenic, antiglycogenolytic, and hypoglycemic. These actions are most important following ingestion of a meal and show lesser importance under postprandial conditions and during starvation. Insulin's actions are opposed primarily by six other hormones, including GH (protein anabolic, lipolytic, gluconeogenic, hyperglycemic), glucagon (lipolytic, glycogenolytic, hyperglycemic), epinephrine (lipolytic, glycogenolytic, gluconeogenic, hyperglycemic), glucocorticoids (protein catabolic, gluconeogenic, hyperglycemic), thyroid hormones (lipolytic, glycogenolytic, hyperglycemic), and anabolic−androgenic steroids (protein anabolic). Epinephrine directly inhibits insulin release. Glucocorticoids also prevent peripheral use of glucose, especially by muscle.

STUDY QUESTIONS

1. Identify peripheral bioregulators of appetite that are produced by adipose tissue and the GI tract.
2. Identify brain pathways used by peripheral bioregulators from adipose tissue and GI tract to affect appetite.
3. Distinguish between the roles of hypothalamic areas and brainstem areas in regulating feeding.
4. Identify orexic and anorexic peptides and where they are produced in the hypothalamus.
5. Describe how the leptin signal is transduced in hypothalamic neurons.
6. Outline the changes in leptin signaling that take place during pregnancy in humans.
7. What is the role of POMC peptides in the link between some forms of obesity and hair color?
8. Why is the endocrine system of the GI tract sometimes referred to as "diffuse"?
9. What is the relationship between APUD cells and endocrine cells of the GI tract?
10. Outline the pathways regulating gastric acid secretion during the cephalic, gastric, and intestinal phases of digestion.
11. Distinguish the different roles for GLUT2 and GLUT4 in insulin secretion and action.
12. Distinguish between the roles of catecholamine hormones and glucocorticoids in regulating metabolism during stress.
13. Identify some of the technical issues faced by scientists working to improve insulin secretion in patients with type I diabetes.

SUGGESTED READING

Books

Bliss, M., 2007. The Discovery of Insulin. University of Chicago Press, Chicago, IL.

Brown, D.R. (Ed.), 2012. "Gastrointestinal Regulatory Peptides" (Handbook of Experimental Pharmacology). Springer-Verlag, Berlin.

Greeley, G.R. (Ed.), 2010. Gastrointestinal Endocrinology. Humana Press, New York.

Islam, M.S. (Ed.), 2010. "The Islets of Langerhans" (Advances in Experimental Medicine and Biology). Springer-Verlag, Berlin.

Taché, Y., Goto, Y., Ohning, G., Yamada, T. (Eds.), 2002. Gut−Brain Peptides in the New Millennium: A Tribute to John Walsh by His Collaborators. Lippincott Williams & Wilkins, New York.

Articles

Regulation of Feeding

Ahima, R.S., Lazar, M.A., 2008. Adipokines and the peripheral and neural control of energy balance. Molecular Endocrinology 22, 1023−1031.

Bartness, T.J., Song, C.K., 2007. Brain-adipose tissue neural crosstalk. Physiology and Behavior 91, 343−351.

Brunton, P.J., Russell, J.A., 2008. The expectant brain: adapting for motherhood. Nature Reviews Neuroscience 9, 11−25.

Cottrell, E.C., Mercer, J.G., 2012. Leptin receptors. Handbook of Experimental Pharmacology 209, 3−21.

Cummings, D.E., Purnell, J.Q., Frayo, R.S., Schmidova, K., Wisse, B.E., Weigle, D.S., 2001. A preprandial rise in plasma ghrelin levels suggests a role in meal initiation in humans. Diabetes 50, 1714–1719.

Denver, R.J., Bonett, R.M., Boorse, G.C., 2011. Evolution of leptin structure and function. Neuroendocrinology 94, 21–38.

Engeli, S., 2012. Central and peripheral cannabinoid receptors as therapeutic targets in the control of food intake and body weight. Handbook of Experimental Pharmacology 209, 357–381.

Ferguson, A.V., Samson, W.K., 2003. The orexin/hypocretin system: a critical regulator of neuroendocrine and autonomic function. Frontiers in Neuroendocrinology 24, 141–150.

Holst, J.J., 1999. Glucagon-like peptide-1 (GLP-1): an intestinal hormone, signaling nutritional abundance, with an unusual therapeutic potential. Trends in Endocrinology & Metabolism 10, 229–235.

Korbonits, M., Goldstone, A.P., Gueorguiev, M., Grossman, A.B., 2004. Ghrelin: a hormone with multiple functions. Frontiers in Neuroendocrinology 25, 27–68.

Millington, G.W., 2007. The role of proopiomelanocortin (POMC) neurones in feeding behaviour. Nutrition & Metabolism (London) 4, 1–16.

Morris, D.L., Rui, L., 2009. Recent advances in understanding leptin signaling and leptin resistance. American Journal of Physiology: Endocrinology, and Metabolism 297, E1247–E1259.

Nixon, J.P., Kotz, C.M., Novak, C.M., Billington, C.J., Teske, J.A., 2012. Neuropeptides controlling energy balance: orexins and neuromedins. Handbook of Experimental Pharmacology 209, 77–109.

Näslund, E., Hellström, P.M., 2007. Appetite signaling: from gut peptides and enteric nerves to brain. Physiology & Behavior 92, 256–262.

Palasz, A., Krzystanek, M., Worthington, J., Czajkowska, B., Kostro, K., Wiaderkiewicz, R., Bajor, G., 2012. Nesfatin-1, a unique regulatory neuropeptide of the brain. Neuropeptides 46, 105–112.

Parker, J.A., Bloom, S.R., 2012. Hypothalamic neuropeptides and the regulation of appetite. Neuropharmacology 63, 18–30.

Sainsbury, A., Zhang, L., 2010. Role of the arcuate nucleus of the hypothalamus in regulation of body weight during energy deficit. Molecular and Cellular Endocrinology 316, 109–119.

Samson, W.K., Resch, Z.T., 2000. The hypocretin/orexin story. Trends in Endocrinology & Metabolism 11, 257–262.

Schulz, C., Paulus, K., Lehnert, H., 2010. Adipocyte-brain: crosstalk. Results and Problems in Cell Differentiation 52, 189–201.

Schellekens, H., Finger, B.C., Dinan, T.G., Cryan, J.F., 2012. Ghrelin signalling and obesity: at the interface of stress, mood and food reward. Pharmacology & Therapeutics 135, 316–326.

Trayhurn, P., Bing, C., 2006. Appetite and energy balance signals from adipocytes. Philosophical Transactions of the Royal Society B: Biological Sciences 361, 1237–1249.

Valette, M., Bellisle, F., Carette, C., Poitou, C., Dubern, B., Paradis, G., Hercberg, S., Muzard, L., Clément, K., Czernichow, S., 2012. Eating behaviour in obese patients with melanocortin-4 receptor mutations: a literature review. International Journal of Obesity. http://dx.doi.org/10.1038/ijo.2012.169 November 13 (Epub ahead of print).

Villarroya, F., 2012. Irisin, turning up the heat. Cell Metabolism 15, 277–278.

Regulation of Digestion

Chaudhri, O.B., Salem, V., Murphy, K.G., Bloom, S.R., 2008. Gastrointestinal satiety signals. Annual Review of Physiology 70, 239–255.

Dockray, G.J., 2012. Cholecystokinin. Current Opinion in Endocrinology, Diabetes, and Obesity 19, 8–12.

Low, M.J., 2004. Clinical endocrinology and metabolism. The somatostatin neuroendocrine system: physiology and clinical relevance in gastrointestinal and pancreatic disorders. Best Practice & Research Clinical Endocrinology & Metabolism 18, 607–622.

Simpson, K., Parker, J., Plumer, J., Bloom, S., 2012. CCK, PYY and PP: the control of energy balance. Handbook of Experimental Pharmacology 209, 209–230.

Regulation of Metabolism

Boller, S., Joblin, B.A., Xu, L., Item, F., Trüb, T., Boschetti, N., Spinas, G.A., Niessen, M., 2012. From signal transduction to signal interpretation: an alternative model for the molecular function of insulin receptor substrates. Archives of Physiology and Biochemistry 118, 148–155.

Boström, P., Wu, J., Jedrychowski, M.P., Korde, A., Ye, L., Lo, J.C., et al., 2012. A PGC1-α-dependent myokine that drives brown-fat-like development of white fat and thermogenesis. Nature 481, 463–468.

Castan-Laurell, I., Dray, C., Knauf, C., Kunduzova, O., Valet, P., 2012. Apelin, a promising target for type 2 diabetes treatment? Trends in Endocrinology & Metabolism 23, 234–241.

Dunmore, S.J., Brown, J.E., 2012. The role of adipokines in beta-cell failure of type 2 diabetes. Journal of Endocrinology September 2 (Epub ahead of print) PMID: 22991412.

Efrat, S., Russ, H.A., 2012. Making β cells from adult tissues. Trends in Endocrinology & Metabolism 23, 278–285.

Hansen, P.A., Corbett, J.A., 2005. Incretin hormones and insulin sensitivity. Trends in Endocrinology & Metabolism 16, 135–136.

Lightfoot, Y.L., Chen, J., Mathews, C.E., 2012. Immune-mediated β-cell death in type 1 diabetes: lessons from human β-cell lines. European Journal of Clinical Investigation 42, 1244–1251.

Notkins, A.L., 2002. Immunologic and genetic factors in type 1 diabetes. Journal of Biological Chemistry 277, 43545–43548.

Thirone, A.C.P., Huang, C., Klip, A., 2006. Tissue-specific roles of IRS proteins in insulin signaling and glucose transport. Trends in Endocrinology & Metabolism 17, 72–78.

Chapter 13

Comparative Aspects of Feeding, Digestion, and Metabolism

The endocrine regulation of feeding, digestion and metabolism in non-mammals has been poorly studied in comparison to mammals. Most of the factors were only discovered in mammals within the last decade or so, so it is not surprising that little is known for non-mammals with respect to feeding regulation. Most of the published research has been conducted with fishes, and relatively little is known about these processes in non-mammalian tetrapods. Chapter 12 provides a general description of these processes in mammals, and basic metabolic pathways are summarized in Appendix H.

I. HORMONES AND FEEDING IN NON-MAMMALS

Studies in fish have shown that several but not all mammalian orexigenic and anorexic factors are effective in fishes. A summary of known actions in fishes is provided in Figure 13-1. A few studies in amphibians and birds also have been done. Given that the hypothalamic and brainstem areas involved in regulating feeding in mammals are evolutionarily ancient, it is reasonable to suspect that the pathways for regulation of feeding are similar in mammals and non-mammals (see Chapter 12).

A. Orexigenic Factors in Teleost Fishes

A summary of known factors that stimulate feeding behavior in fishes and their sources is provided in Table 13-1. **Ghrelin** has been identified in the stomach and brain of several teleosts and in sharks. Both human and goldfish ghrelin stimulates feeding behavior in goldfish. Starvation increases production of mRNA for **pre-proghrelin** in goldfish. It is not known if any other peptides like ghrelin that are related to **growth hormone (GH)** secretion are involved in appetite regulation (e.g., **somatostatin, SST**; **GH-releasing hormone, GHRH**) but they are located in brain regions known to influence feeding. **Neuropeptide Y (NPY)** also is a potent stimulator of appetite as are pharmacological NPY agonists in teleosts. NPY increases in the brain prior to feeding and decreases after feeding. As in mammals, NPY appears to interact with a number of other factors, including **corticotropin-releasing hormone (CRH)**, **orexins**, and **galanin**. Galanin is distributed throughout several brain regions in goldfish and has been shown to stimulate food intake. Orexins synergize with galanin and also enhance NPY effects on feeding. Although changes have been observed in levels of **agouti-related peptide (AgRP)**, it is not clear if it has any regulatory role such as described for the related **agouti-signaling protein (ASP)** in mammals. Although **melanin-concentrating hormone (MCH)** has been characterized as an orexigen in mammals, it appears to be only a bioregulator of body color and an antagonist of **melanotropin (α-MSH)** in fishes. **Leptin** has been identified in fishes but it apparently has little if any effect on food intake.

B. Anorexogenic Factors in Teleost Fishes

A summary of anorexogenic factors affecting fishes is provided in Table 13-1. **Cholecystokinin (CCK_8)**, CRH, and the related **urotensin-I**, MCH, and **cocaine- and amphetamine-regulated transcript (CART)** all decrease feeding in goldfish, although CCK_8 stimulated feeding in rainbow trout. **Prolactin-releasing peptide (PrRP)** is an appetite suppressant in fishes in addition to its role in prolactin release (see Chapter 5). Many other factors have been shown to influence feeding although it is not clear if they play regulatory roles.

C. Feeding Regulation in Other Non-Mammals

Orexin A and **orexin B** have been identified in *Xenopus laevis*. Orexin immunoreactive neurons were present only in the **suprachiasmatic nucleus (SCN)** of the tree frog *Hyla cinera*, but immunoreactive (ir) orexin B was found in the SCN as well as in the anterior preoptic area and the magnocellular and ventral hypothalamus of *Rana ridibunda*. Leptin and leptin receptors have been characterized in the frog *X. laevis* and the newt *Triturus cristatus carnifex*. Interestingly, although the primary amino acid sequences of the amphibian and human leptin

Vertebrate Endocrinology. http://dx.doi.org/10.1016/B978-0-12-394815-1.00013-6

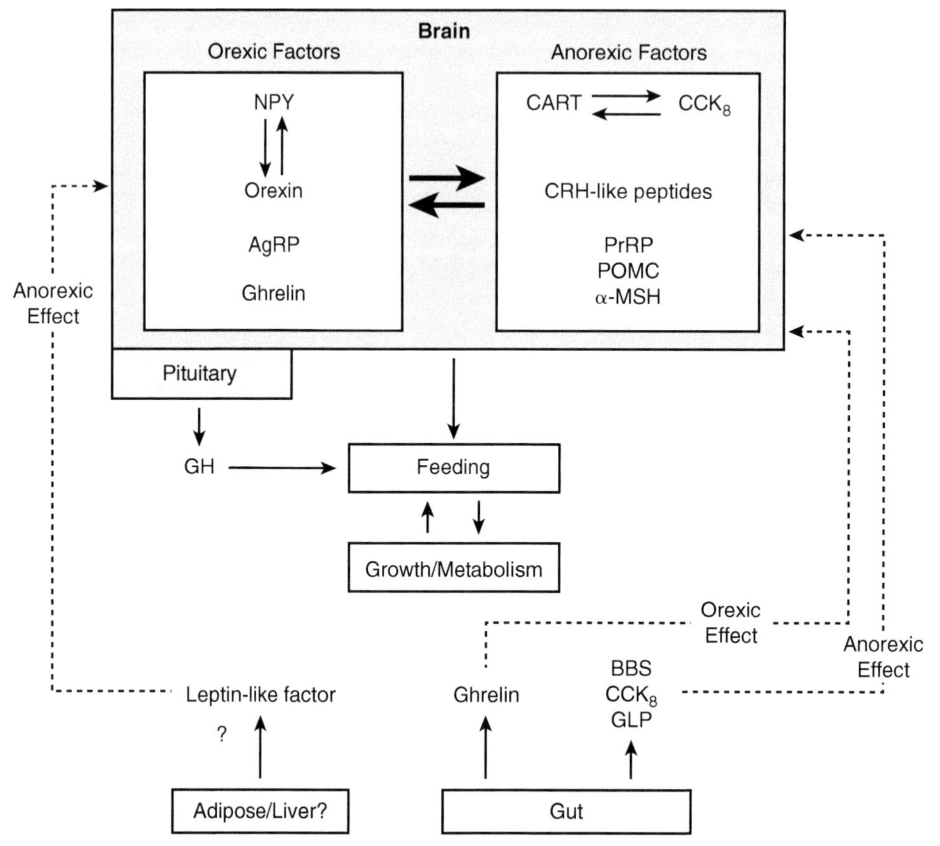

FIGURE 13-1 **Some neural and endocrine factors affecting feeding behavior in teleost fishes.** See text for explanation and Appendix A for abbreviations. *(Adapted with permission from Volkoff, H. et al.,* General and Comparative Endocrinology, ***142****, 3–19, 2005.)*

peptides share only about 35% sequence similarity, the tertiary structures of the peptides are virtually identical, and amphibian leptin is capable of activating the human **leptin receptors**. In addition, a version of a **26RFamide peptide** was isolated from the frog brain that can stimulate feeding in mice. CRH inhibits foraging and visually guided prey capture in Texas toads, *Bufo speciosus*.

Leptin also is present in the stomachs of a lizard and a snake. However, its role in feeding has not been reported. The leptin gene has been sequenced from chickens, and the chicken leptin peptide was shown to reduce food intake in chickens and great tits (*Parus major*). Circulating leptin is reduced by starvation in chicks. Administration of orexin to chicks did not stimulate feeding nor did fasting affect levels of orexin in either the lateral hypothalamus or the periventricular nucleus although there was a two-times increase in AgRP in the **infundibular nucleus (IFN = arcuate)**. An opioid-related peptide, **nociception/orphanin FQ**, decreases CART and increases AgRP, resulting in an orexigenic effect. β-Endorphin has been shown to increase feeding in white-crowned sparrows. Some factors affecting feeding in birds are provided in Table 13-2.

II. HORMONES AND DIGESTION IN NON-MAMMALS

The investigation into comparative regulation is hampered by the lack of basic understanding of the general physiology of digestion in non-mammalian species. For example, although extensive research in teleost fishes of commercial importance has been accomplished with respect to diets, growth, and feeding ecology, few experiments have been concerned with physiological control mechanisms. There are many unique situations with respect to digestive processes of non-mammals. For example, many fishes lack stomachs and may have other specialized structures (e.g., pyloric caeca) that imply major differences in control mechanisms. Also, digestion in fishes, amphibians, and reptiles may require days to accomplish what birds and mammals do in a few hours with elevated body temperatures.

Immunocytochemical data on gastrointestinal peptides in non-mammals indicates that many of the peptides described in mammals also are present in non-mammals. However, functional studies have lagged behind, and data on their functions are available for relatively few species.

TABLE 13-1 Some Factors Affecting Feeding in Fishes and Their Sources

Effect on Feeding	Hormone	Major Producing Site
Orexigenic	Apelin	GI tract/brain
	Ghrelin	GI tract
	Growth hormone (GH)	Pituitary
	Neuropeptide Y (NPY)	Brain/GI tract
	Orexins (OX)	Brain
Anorexigenic	Calcitonin gene-related peptide (CGRP)	Brain
	Cholecystokinin (CCK)	GI tract
	Cocaine- and amphetamine-regulated transcript (CART)	Brain
	Corticotropin-releasing hormone (CRH)	Brain
	Gastrin-releasing peptide (GRP)	GI tract
	Glucagon-like peptide-1 (GLP-1)	GI tract
	Leptin	Liver
	Melanin-concentrating hormone (MCH)	Brain
	α-Melanocyte-stimulating hormone (α-MSH)	Brain
	Prolactin-releasing peptide (PrRP)	Brain
	Urotensin I (UI)	Brain

TABLE 13-2 Some Factors Affecting Feeding in Birds

Orexigenic Factors	Anorexigenic Factors
AgRP	α-MSH
NPY	CART
Visfatin (chickens only)	CCK
Nociception/orphanin FQ	CRH
	PrRP
	Leptin

See Appendix A for explanation of abbreviations.

These observations suggest an ancient origin for these regulatory molecules. Using immunocytochemical techniques, a number of mammalian peptides have been demonstrated in invertebrate species as well. There appears to be a broad phylogenetic distribution for these peptides among invertebrate animals. Many of these peptides are similar to or identical with peptides located in vertebrate guts and vertebrate nervous systems, as well. For example, immunoreactive **gastrin/CCK-like peptides** have been extracted from the GI tract of many vertebrates (Figure 13-2) as well as a tunicate, two molluscan species, and several insects. However, only the octapeptide **cionin** extracted from the tunicate *Ciona intestinalis* appears to be a true gastrin/CCK peptide, establishing the origin for this peptide family at about 500 million years ago. Anatomically, there are several types of arrangements of pancreatic endocrine cells.

A. Digestive Hormones in Agnathan Fishes: Cyclostomes

Cyclostomes are stomachless fishes and therefore studies have emphasized the intestinal and pancreatic tissues. Cytological studies by Ostberg and coworkers on the intestine of the Atlantic hagfish (*Myxine glutinosa*) have revealed the presence of primitive open-type endocrine cells. These cells extend from the basal portion of the intestinal epithelium to border on the lumen of the gut. Hagfish intestinal endocrine cells do not possess APUD characteristics, although APUD-type cells have been reported in the pancreatic islets that differentiate into insulin-producing B cells. These intestinal endocrine cells in the Atlantic hagfish do not resemble zymogen cells, either, suggesting a separate origin for the endocrine and enzyme-secreting cells. Immunoreactive gastrin/CCK-like peptide, **glucagon-like peptides (GLP-1, GLP-2)**, SST, **substance P**, **glucose-insulinotropic peptide** (**GIP**, gastric inhibitory peptide), and **vasoactive intestinal peptide (VIP)** have all been reported in the hagfish intestine. These peptides are closely related to their mammalian counterparts (Figures 13-3 and 13-4). Gallbladder strips prepared from the Pacific hagfish (*Eptatretus stouti*), however, did not respond *in vitro* with contractions to porcine CCK (pCCK), although the neurotransmitter acetylcholine caused contractions. Secretion of intestinal lipase in this same species is stimulated by pCCK.

In contrast to the hagfishes, the intestinal epithelium of larval and adult lampreys (*Lampetra* spp.) contains APUD-type cells. Immunoreactive glucagon-like peptides, NPY, **peptide YY (PYY)**, SST_{14}, SST_{34}, and **pancreatic polypeptide (PP)** have been reported in the intestines of several lamprey species. **Secretin-like** and gastrin/CCK-like activities are present in extracts prepared from intestines of

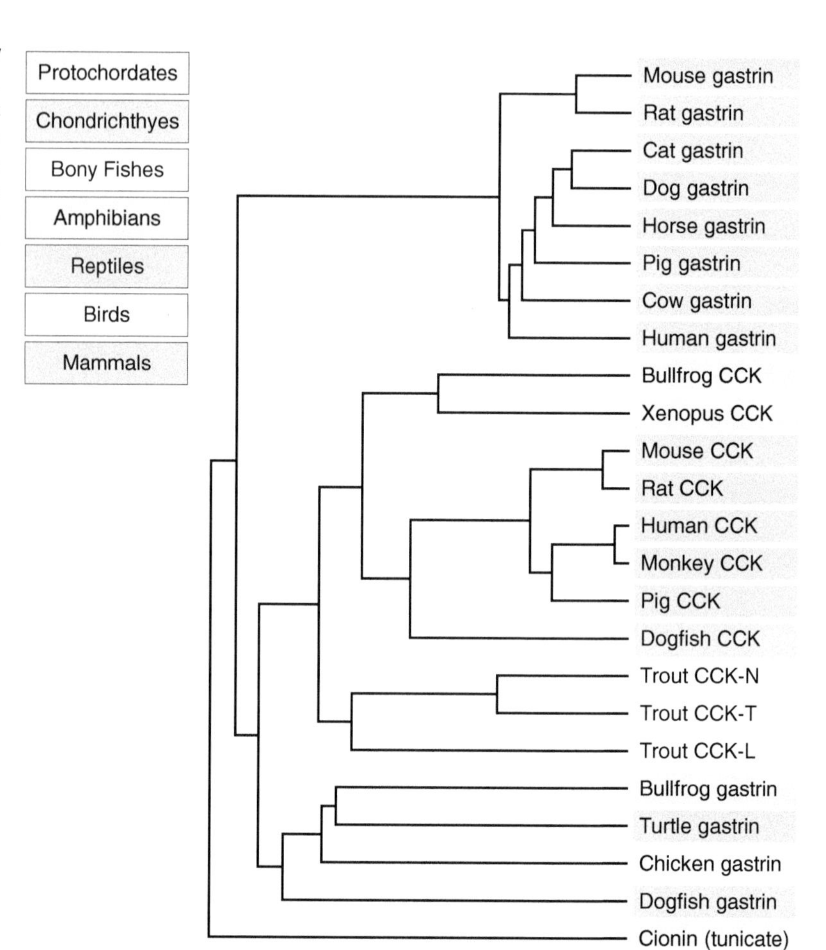

FIGURE 13-2 **Cladogram of the chordate CCK/gastrin family.** Based on preprohormone sequences. Three CCK variants are produced in rainbow trout with different substitutions at position 6 at the C-terminal end (represented by the single letters for the amino acids substituted for methionine; see Appendix C for names of amino acids). *(Adapted with permission from Johnsen, A.H.,* Frontiers in Neuroendocrinology, ***19****, 73–99, 1998.)*

river lampreys (*Lampetra fluviatilis*) and sea lampreys (*Petromyzon marinus*). Both secretin and gastrin/CCK-like activities in these extracts were assayed by monitoring pancreatic secretions in the anesthetized cat.

B. Digestive Hormones in Chondrichthyean Fishes

In their classical studies, Bayliss and Starling reported the presence of secretin-like activity in extracts prepared from dogfish shark and skate intestines when bioassayed in mammals. Intestinal extracts prepared from the holocephalan *Chimaera monstrosa* exhibit gastrin/CCK-like activity. Likewise, porcine CCK stimulates contractions in strips of gallbladder prepared from dogfish sharks, and the intensity of the response is proportional to the dose of CCK. Numerous factors known to affect digestion in mammals (e.g., gastrin, VIP, CCK, and **gastrin-releasing peptide, GRP**) are present in independent cells of the gastrointestinal tract of the dogfish shark (*Squalus acanthias*), in addition to hormones normally associated with cell types also found in the shark pancreas (i.e., insulin, glucagon, PP, GIP; see ahead). Separate gastrin-like and CCK-like peptides have been isolated from sharks, suggesting an early divergence of these genes prior to the separation of the cartilaginous and bony fishes (see Figure 13-2). Immunoreactive SST and GLPs have been reported in the gastric glands of five elasmobranch species (Figure 13-3), and VIP has been isolated from the dogfish (*Scyliorhinus canalicula*) (Figure 13-4).

C. Digestive Hormones in Bony Fishes: Teleosts

Only a few species of teleosts have been investigated. The first observations were those of Bayliss and Starling who reported that intestinal extracts prepared from salmon possessed secretin-like and possibly gastrin/CCK-like activities. Similar activities were reported for northern pike (*Esox lucius*) and Atlantic cod (*Gadus morhua*) using either avian or mammalian bioassays. Gastrin/CCK activity is present in the intestine of the Atlantic eel (*Anguilla anguilla*) and the pike. Isolated strips of gallbladder from Pacific salmon (*Oncorhynchus*) contract in the presence of

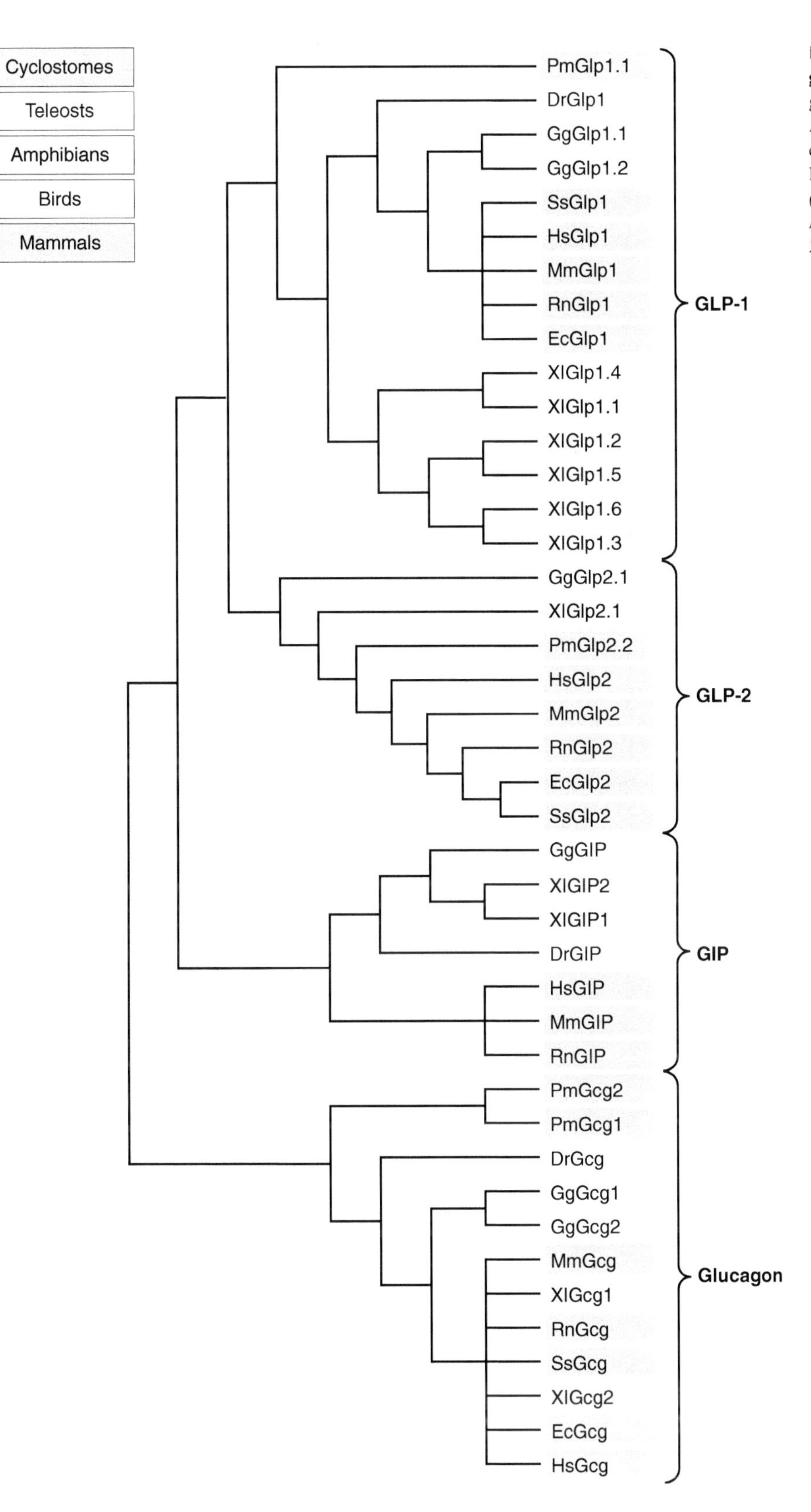

FIGURE 13-3 **Phylogeny of some glucagon, GLPs, and GIP genes.** Multiple genes are known to exist in some species. Abbreviations: Dr, zebrafish; Ec, horse; Gg, chicken; Hs, human; Mm, mouse; Pm, sea lamprey; Rn, rat; Ss, pig; Xl, *Xenopus*. *(Adapted with permission from Musson, M.C. et al.,* Regulatory Peptides, ***171****, 26–34, 2011.)*

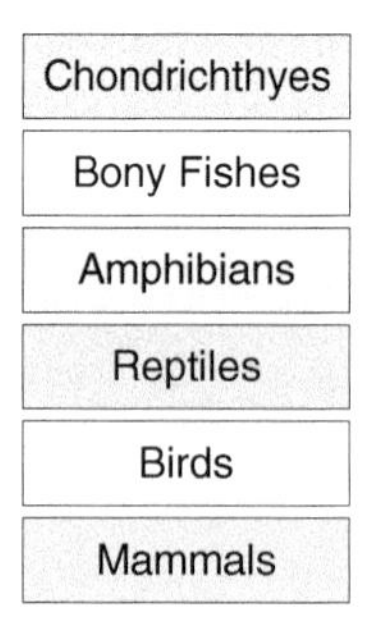

Pig	HSDDAV	FTDNY	TRLRL	QMAVK	KLYNS	ILN-NH_2
Guinea pig	HSDDAL	FTDTY	TRLRL	QMAMK	KLYNS	VLN-NH_2
Chicken	HSDDAV	FTDNY	SRFRL	QMAVK	KLYNS	VLT-NH_2
Alligator	HSDDAV	FTDNY	SRFRL	QMAVK	KLYNS	VLT-NH_2
Frog	HSDDAV	FTDNY	SRFRL	QMAVK	KLYNS	VLT-NH_2
Rainbow trout	HSDDAI	FTDNY	SRFRL	QMAVK	KLYNS	VLT-NH_2
Cod fish	HSDDAV	FTDNY	SRFRL	QMAAK	KLYNS	VLA-NH_2
Bowfin	HSDDAI	FTDNY	SRFRL	QMAVK	KLYNS	VLT-NH_2
Dogfish shark	HSDDAV	FTDNY	SRIRL	QMAVK	KLYNS	LLA-NH_2

FIGURE 13-4 **Comparison of amino acid sequences of some vertebrate VIPs.** Few substitutions have been made in the molecule through its long evolutionary history from fishes to mammals. *(Adapted with permission from Wang, Y. and Conlon, J.M.,* General and Comparative Endocrinology, *98, 94–101, 1995.)*

porcine CCK, indicating sensitivity of the salmon gallbladder to the mammalian peptide. Intestinal strips from codfish contract in response to either gastrin or CCK_8.

A gastrin–histamine type of mechanism is present in the stomach of the majority of teleosts examined (cyprinid and labrid fishes lack stomachs). Extracts from the gastric mucosa of sunfish (*Lepomis macrochirus*) stimulate acid secretion in bullfrogs, and large doses of histamine induce acid secretion in the European catfish *Silurus glanis*. Histamine-induced acid secretion in cod is blocked by certain antihistamines, supporting the existence of a mammalian-like regulatory system. Fish stomachs contain one cell type that secretes both pepsinogen and H^+ and probably secretes both products simultaneously in response to gastrin or histamine. Administration of CCK_8 slows gastric emptying in salmonids. Studies suggest one peptide that possesses both gastrin-like and CCK-like activity in teleosts (see Figure 13-2).

Ten immunoreactive cell types were identified in *Sparus auratus*, including stomach cells reacting to **neurotensin (NT)**, secretin, serotonin, substance P, and SST. Intestinal cells reacted to antibodies for gastrin/CCK, GLPs, PP, substance P, and the opioid **Met-enkephalin**. Similar examination of the intestine of a stomachless fish, *Barbus conchonius*, exhibited all but the substance P immunoreactive cell types. SST_{14} is present in the cod stomach, where it inhibits acid secretion. Both SST_{14} and SST_{28} have been found in the intestines of several teleosts but not in the mudsucker *Gillichthyes mirabilis*. VIP has been isolated and characterized chemically from the bowfin (a non teleost, bony fish) and two teleosts (Figure 13-4), but its physiological role is not understood.

D. Digestive Hormones Amphibians

Regulation of gastric mechanisms in amphibians has been studied almost exclusively in anurans, and immunoreactive cells for a variety of mammalian peptides have been demonstrated (Table 13-3). It appears that amphibians possess mechanisms very much like those of mammals, involving both neural and endocrine mechanisms. Distinct gastrin and CCK peptides have been isolated from amphibians (see Figure 13-2). Stomachs of intact frogs or isolated gastric mucosa prepared from frogs (including *Rana pipiens*, *R. catesbeiana*, *R. temporaria*, and *R. esculenta*) respond with acid secretion when subjected to ACh, histamine, synthetic pentagastrin, or crude gastrin preparations from non-mammals or mammals. Similarly, gastric mucosa isolated from a urodele, *Necturus*, secretes acid in response to a gastrin peptide. Treatment with atropine or surgical vagotomy reduces acid secretion in frogs as it does in mammals. Beyond the amphibians, CCK peptides are highly conserved but mammalian gastrin has diverged markedly from the gastrins of amphibians, reptiles, and birds. Prototherian mammals apparently do not produce either gastric acid or gastrin, and the distinct mammalian gastrin gene is demonstrable first among the marsupials.

The release of pepsinogen in *R. esculenta* can be induced by increasing parasympathetic activity. The peptides **caerulein** and **bombesin (BBS)**, originally isolated from frog skin, are associated with frog stomachs and may affect gastric secretion. Caerulin is known to stimulate acid secretion from stomach mucosa in a variety of vertebrates, and immunoreactive BBS has been reported in frog stomachs as well as in frog skin and brain. Chemical identification of GRP in the

TABLE 13-3 Distribution and Frequency of Some Immunoreactive Gastrointestinal Cells in the Fire-Bellied Toad (*Bombina orientalis*)

	Esophagus	Pylorus	Antrum	Duodenum	Ileum	Rectum
Somatostatin	±	++	++	++	+	±
CCK_8	−	−	++	+	±	−
Bombesin	−	++	+	−	−	−
Secretin	−	±	−	+	±	−
Glucagon		−	−	±	+	−
PP	−	−	±	+	+	+

CCK_8, cholecystokinin; PP, pancreatic polypeptide. Relative frequencies: +++, numerous; ++, moderate; +, a few; ±, rare; −, not detected.
Adapted with permission from Ku, S.-K. et al., *Journal of Veterinary Medical Science*, **62**, 589−594, 2000.

stomach of *R. ridibunda* suggests that GRP may be an endogenous peptide stimulator of gastrin release. Additionally, SST, CCK_8, glucagon, and PP have been demonstrated in the digestive tract of *R. rugosa.*

Bayliss and Starling reported that extracts from frog intestines would evoke pancreatic secretion in dogs, providing evidence for the presence of secretin-like or CCK-like factors or both. Frog gallbladders will contract in the presence of porcine CCK *in vitro*, supporting the possible existence of a CCK-like factor in amphibians as well as a role for CCK in regulation of gastric processes. VIP also has been isolated from a frog stomach (Figure 13-4).

E. Digestive Hormones in Reptiles

Bayliss and Starling reported that a factor or factors capable of causing pancreatic secretion in mammals is present in the intestine of a tortoise. Immunoreactive SST is present in the intestine of the lizard *Anolis carolinensis*, but **motilin** is not. In two chelonians, *Testudo graeca* and *Mauremys caspica*, immunoreactive BBS, NT, gastrin, GLPs, SST, PYY, and insulin were demonstrated in the intestinal mucosa, but no functional roles were indicated. The same study failed to find immunoreactive motilin, secretin, VIP, CCK, GIP, and opioid enkephalins; however, alligator VIP has been characterized chemically and found to be identical to chicken VIP (Figure 13-4). It is remarkable that so few studies have been reported for reptiles, suggesting that this is a group of vertebrates requiring thorough endocrine examination.

F. Digestive Hormones in Birds

Knowledge of the gastrointestinal physiology of domesticated birds is more extensive than for any other non--mammalian group because of the importance of gastrointestinal research to poultry science. However, little information on wild species is available. Gastrin can stimulate acid secretion in birds, and large dosages of CCK cause release of enzymes from the avian exocrine pancreas. Extracts prepared from chicken intestines are strong stimulants of pancreatic secretion when assayed in turkeys, but these extracts are only weak stimulants in mammals (cat, rat), suggesting that considerable molecular differences may exist in these avian factors. Glucagon and GIP appear to have no effects on pancreatic secretions in birds.

Other observations indicate there are some features in avian GI physiology that may be unique. For example, purified avian (chicken) or porcine secretin only weakly stimulates exocrine pancreatic secretion in turkeys, but purified mammalian VIP and **peptide histidine isoleucine (PHI)** are potent stimulators. VIP is a very conservative peptide in vertebrates, suggesting that its functions have been conserved through evolution. Chicken VIP has been isolated and differs structurally from porcine VIP at only four positions (Figure 13-4). Immunoreactive SST and motilin have been demonstrated in the intestines of Japanese quail and chickens, but nothing is known about their possible functions.

III. COMPARATIVE ASPECTS OF THE ENDOCRINE PANCREAS

All mammalian endocrine pancreas cell types have been described in non-mammalian vertebrates, including the cyclostomes, but their proportions very greatly from one major taxon to another (Table 13-4). However, it may be misleading to characterize classes of vertebrates as exhibiting a particular pattern (Table 13-5). The same terminology is applied to endocrine pancreatic cells of non-mammals as for mammals, and the same hormones are associated with each type (see Chapter 12 and ahead).

TABLE 13-4 General Patterns of Islet Cellular Type Distributions in Vertebrates with Respect to A and B Cells[a]

Category	Dominant Cytology	Representatives
I	Mostly B cells	Cyclostomes
II	More than 50% B cells	Teleosts, amphibians, mammals
III	Approximately 50% X cells[a]	Holocephalans
IV	More than 50% A cells	Most lizards, birds

[a]*X cells contain glucagon-like peptides, but their function is unknown.*
Adapted with permission from Epple, A. and Brinn, J.E., "The Comparative Physiology of the Pancreatic Islets," Springer-Verlag, Berlin, 1987.

A. Anatomical Features

Five anatomical arrangements for pancreatic islet systems can be distinguished by examination of the major vertebrate groupings. The **cyclostome type** is composed of aggregations of islet cellular types with no obvious relationship to the homologous intraintestinal equivalent of mammalian acinar (enzyme-secreting) cells. In hagfishes, pancreatic islets surround the base of the common bile duct (Figure 13-5A), and in lampreys they are located on the surfaces of the intestine (epi-intestinal), within the intestinal mucosa (intraintestinal; Figure 13-5B), and even within the liver (intrahepatic). Cyclostomes exhibit **B cells** (insulin-secreting) and **D cells** (SST-secreting) but lack **A cells** (glucagon-secreting) and **PP** cells. GLPs have been identified in the gastric mucosa. The cyclostome pattern of islet distribution is considered to be the most primitive pattern in vertebrates. Exocrine pancreatic elements are embedded in the intestinal lining of these primitive fishes.

The **primitive gnathostome type** is found in the elasmobranchs (Figure 13-6), holocephalans, and coelacanths. In these fishes, the islet tissue occurs as layers surrounding the smaller ducts of a compact pancreas. In addition, some scattered cells are located throughout the exocrine pancreas. B, A, D, and PP cells occur in sharks and holocephalans. Additionally, a separate cell type has been described in the pancreas that produces GIP.

The pancreas of lungfishes forms a unique arrangement referred to as the **lungfish type**. It is a compact, intraintestinal structure with a number of encapsulated islets similar to the teleosts (see below). Only B, A, and D cells have been identified in the Australian lungfish (*Neoceratodus fosteri*) islets but all four cell types are present in an African lungfish, *Protopterus aethiopicus*. In the latter species, all four cell types were also identified in the intestinal epithelium.

The diffuse **actinopterygian type** occurs in the other bony fishes where the exocrine pancreatic cells are scattered along the bile ducts and abdominal blood vessels and on the outer surfaces of the gastrointestinal tract, gallbladder, and liver. However, an intrahepatic pancreas has been observed in several species. The islet tissue often accumulates as clumps near the common bile duct but is not usually separated from the acinar tissues. Most teleosts possess

TABLE 13-5 Comparative Cytology of the Endocrine Pancreas with Respect to Relative Predominance of A, B, D, and PP Cells[a]

Class	A cells	B cells	D cells	PP cells
Agnathans				
Hagfish		++++	+	?
Chondrichthyeans				
Elasmobranchs	+++	++++	+	+
Osteichthyeans				
Teleosts	+++	++++	+	+
Amphibians				
Anurans and urodeles	+++	++++	+	+
Reptiles				
Saurians	++++	+	+	+
Crocodilians	+++	+++	+	?
Birds	+++++	+++	+	+
Mammals	++	+++++	+	+

[a]*The number of + marks in each column represents only an attempt to show relative abundance and should not be construed as precise ratios.*

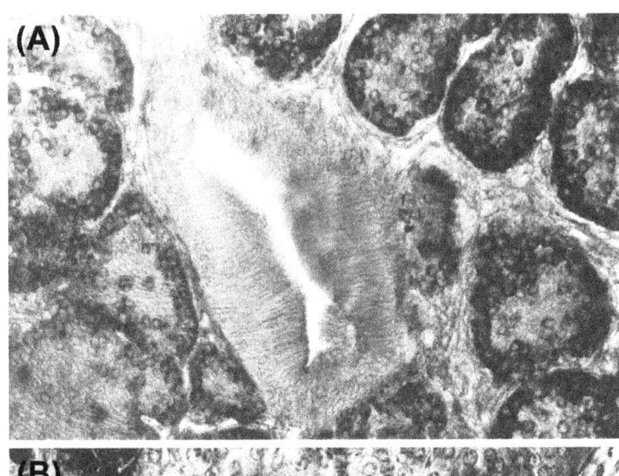

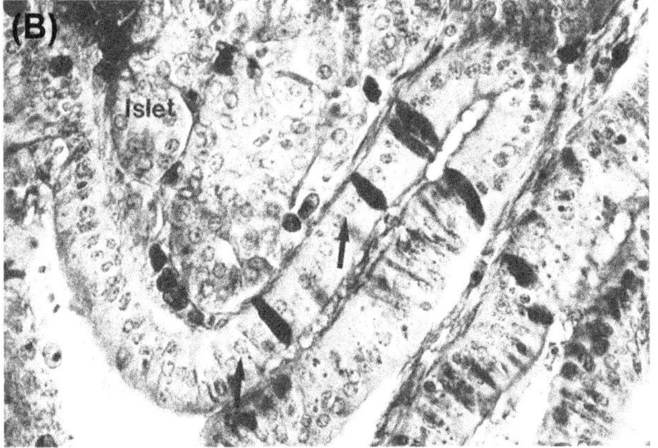

FIGURE 13-5 **Islet tissues of hagfish and lamprey.** (A) Follicles of islet tissue surrounding the lower bile duct from the hagfish *Eptatretus burgeri.* (B) Islet located in intestinal submucosa of the sea lamprey (*Petromyzon marinus*). *(Photographs courtesy of August Epple, Thomas Jefferson University, Philadelphia, PA.)*

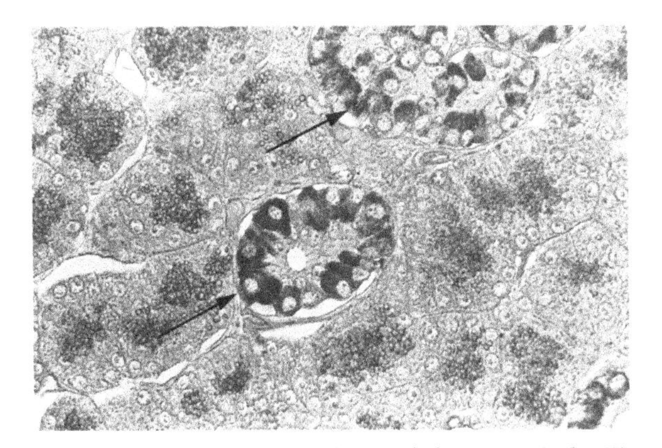

FIGURE 13-6 **Pancreas of dogfish *Scyliohrinus canicula.*** These primitive islets (arrows) consist of endocrine cells surrounding a small unstained ductile of the exocrine pancreas. *(Photograph courtesy of August Epple, Thomas Jefferson University, Philadelphia, PA.)*

isolated masses of pancreatic islet tissue called **Brockmann bodies**. A given species may have a number of Brockmann bodies, or the islet tissue may be concentrated largely in one "principal islet" as in the European eel (*A. anguilla*) or in bullheads (*Ictalurus* spp). Brockmann bodies of

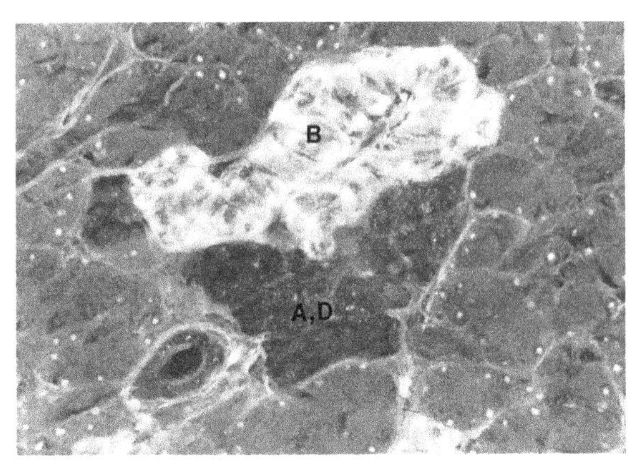

FIGURE 13-7 **B cells in a toad.** The presence of insulin is indicated by the light fluorescence. The dark area immediately surrounding the B cells consists mostly of A cells with a few D cells. *(Photograph courtesy of August Epple, Thomas Jefferson University, Philadelphia, PA.)*

actinopterygian fishes contain A, B, and D cells, but PP cells occur only in those Brockmann bodies located in the pyloric region. In some species (e.g., rainbow trout), glucagon, GLPs, and GIP occur in the same cell.

The **tetrapod type** of endocrine pancreas typically is a compact, extraintestinal structure containing scattered islets and occasionally some scattered endocrine cells. In the toad (*Bufo*) (Figure 13-7) and in birds (Figure 13-8), all four cell types known for mammals are present although their proportions often are very different from mammals. Additionally, an islet cell immunoreactive for secretin is common in the endocrine pancreas of the fire-bellied toad (*Bombina orientalis*). Islets may be concentrated in particular lobes of the pancreas, and islet types often are not evenly distributed among the lobes.

Immunoreactive **insulin-like growth factors (IGFs; IGF-I, IGF-II)** appear in the pancreatic islets or their homologues in protochordates, fishes, amphibians, reptiles, birds, and mammals. A phylogeny of **IGF-1 receptor gene (*IGF1r*)** is provided in Figure 13-9. In cephalochordates and urochordates, there appears to be a single gene that produces a common insulin/IGF molecule associated with the enteroendocrine cells. These observations support a common gene origin for both insulin and the IGFs. IGF-II is colocalized with insulin in all vertebrates, but IGF-I may be colocalized in different pancreatic cells in amphibians, reptiles, and birds (Table 13-6).

There are some generalizations we can make concerning the origin of the various components of the tetrapod pancreas and the relationship of their origin to the distribution of islets. The dorsal pancreatic bud gives rise to the tail and body of the pancreas (also known as the splenic portion), and the ventral buds give rise to the head or duodenal portion. Islets associated with the splenic lobe typically are larger and consist mainly of B, A, and D cells. The islets of the duodenal pancreas are smaller and may contain many PP cells.

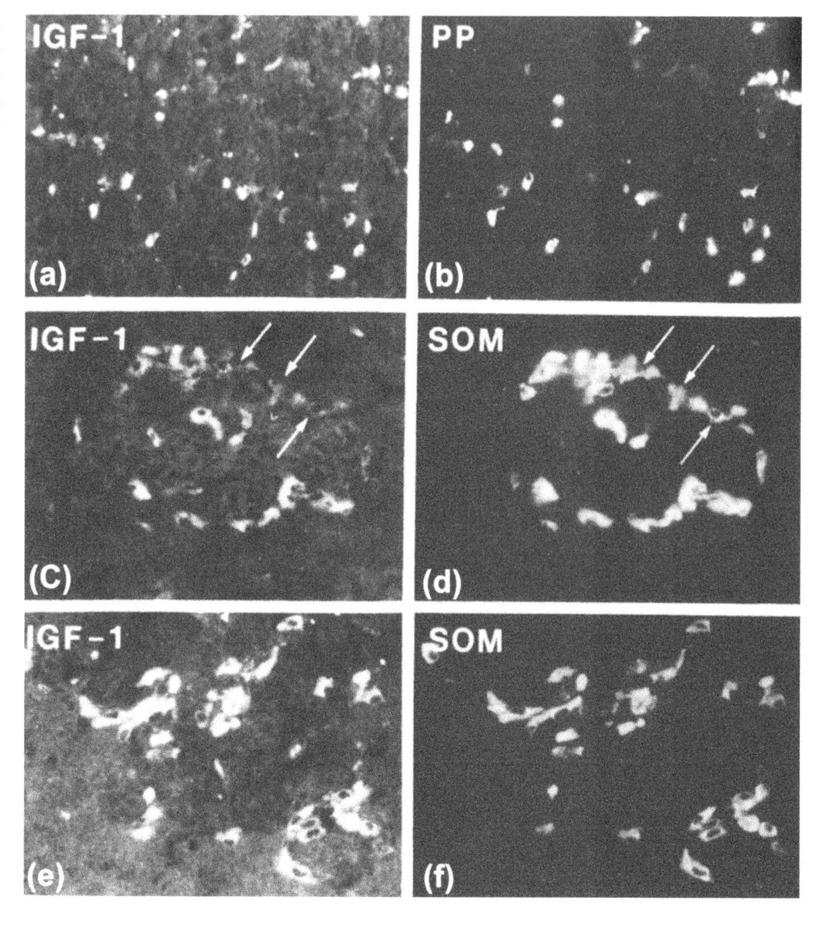

FIGURE 13-8 **Avian cell types in the pancreas.** (a–b) Colocalization of IGF1 and Pancreatic polypeptide (PP) in pancreas of Japanese quail. (c–d) Colocalization of IGF-1 and somatostatin (SOM) in D cells of Japanese quail. (e–f) Colocalization of IGF-I and SOM in D cells of the domestic chicken. *(Reprinted with permission from Reinecke, M. et al.,* General and Comparative Endocrinology, ***100,*** *385–396, 1995.)*

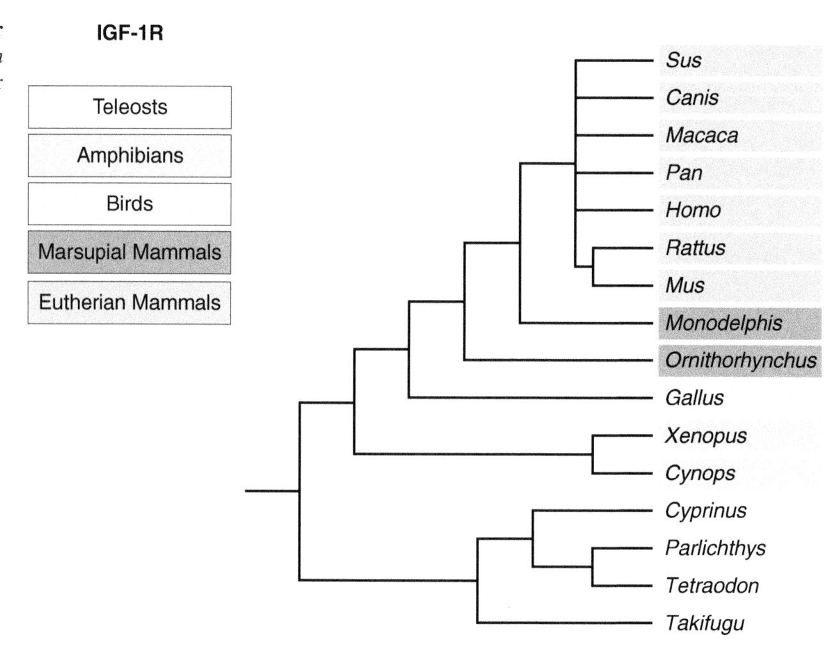

FIGURE 13-9 **Phylogeny of IGF-1 receptor genes in vertebrates.** *(Adapted with permission from Hernández-Sánchez, C. et al.,* Molecular Biology and Evolution, ***25,*** *1043–1053, 2008.)*

TABLE 13-6 Some Peptides Co-Localized in Selected Vertebrates with Pancreatic Hormones[a]

Cell Type and Hormone	IGF-I	IGF-II	CRH	PP
A cell (glucagon)	Frog		Catfish	Frog
			Toad	
			Lizard	
			Chicken	
B cell (insulin)		Fish		
		Frog		
		Lizard		
		Snake		
		Bird		
D cell (somatastatin)	Lizard			
	Snake			
	Bird			
PP cell (pancreatic polypeptide)	Frog			
	Lizard			

[a]*Co-localization was determined by immunocytochemistry.*
IGF, insulin-like growth factor; CRH, corticotropin-releasing hormone; GIP, glucose-dependent insulinotrophic peptide; PP, pancreatic polypeptide; CCK, cholecystokinin.

B. Pancreatic Hormones of Non-Mammals

All of the mammalian pancreatic hormones have been isolated from non-mammalian vertebrates including insulin, glucagon, GLPs, PP, and, in a few cases, GIP. Because the GIP gene is believed to be derived from the glucagon gene, its activation in some cells seems likely. SST is probably a paracrine factor in the non-mammalian vertebrate pancreas as described in mammals.

1. Insulin

Insulin is a conservative protein with few amino acid substitutions at the critical sites related to its structure and biological activity. Insulin **A chains** and **B chains** consist of 21 or 22 amino acids or 31 to 38 amino acids, respectively. The number of amino acids in the B chain is more variable when insulins from different vertebrate classes are compared. The **C-peptide** of **proinsulin** "tolerates" a greater number of substitutions, yet even portions of this peptide remain rather conservative—for example, the glycine-rich central core of the C-peptide. Single amino acid substitutions may have profound effects on the resultant molecule. One substitution in hagfish insulin is believed responsible for the observation that hagfish insulin does not bind Zn^{2+} and therefore does not form the typical hexamer crystals characteristic of most vertebrate insulins.

Examination of vertebrate **insulin receptor (*Ir*) genes** (Figure 13-10) reveals that most of the amino acid substitutions occur in the extracellular domain. The orphan **insulin-related receptor (*Irr*) gene** does not occur in teleosts but is present in all tetrapods examined (Figure 13-11).

B cells and insulin activity are present in the guts or digestive organs of protochordates and many non-chordate invertebrate animals (Table 13-7), and a number of actions for insulin are known in these animals. Mammalian insulin reduces blood sugar levels and promotes glycogen synthesis in a gastropod mollusk, *Strophocheilus oblongus*, supporting a physiological role for gastropod insulin. Furthermore, insulin-like molecules are found in distantly related taxonomic groups including protozoans, bacteria, and fungi. Insulin is therefore a widely distributed factor that appears to be capable of regulating carbohydrate metabolism in a great variety of organisms. The evolution of insulin must have involved a long and interesting process, most of which remains to be discovered.

2. Glucagon

Mammalian glucagons that have been characterized chemically exhibit no amino acid substitutions. Non-mammalian glucagons also are structurally conservative with the exception of teleosts. Turkey glucagon differs from mammalian glucagons only at position 28, where serine is substituted for asparagine. Chicken glucagon is identical to turkey glucagon, whereas duck glucagon has threonine at position 16 instead of serine. Glucagon extracted from teleosts is more variable. In contrast, glucagon isolated from the spiny dogfish shark (*S. acanthias*) and from *Torpedo marmorata* are structurally more like mammalian glucagon than like teleosts. Evolutionary relationships among vertebrate glucagons and GLPs are provided in Figure 13-3. Immunoreactive glucagon has been localized in the digestive glands of a crab, *Cancer pagurus*, and two gastropod mollusks, *Patella caerulae* and *Helix pomatia* (Table 13-7). A glucagon-like molecule is present in extracts of a number of invertebrate chordates, including both tunicates and cephalochordates.

3. Pancreatic Polypeptide (PP)

Evidence for PPs is found in chondrichthyeans, teleosts, amphibians, reptiles, and birds. A comparison of PPs extracted from amphibians, reptiles, and birds with those of several mammals is provided in Figure 13-12.

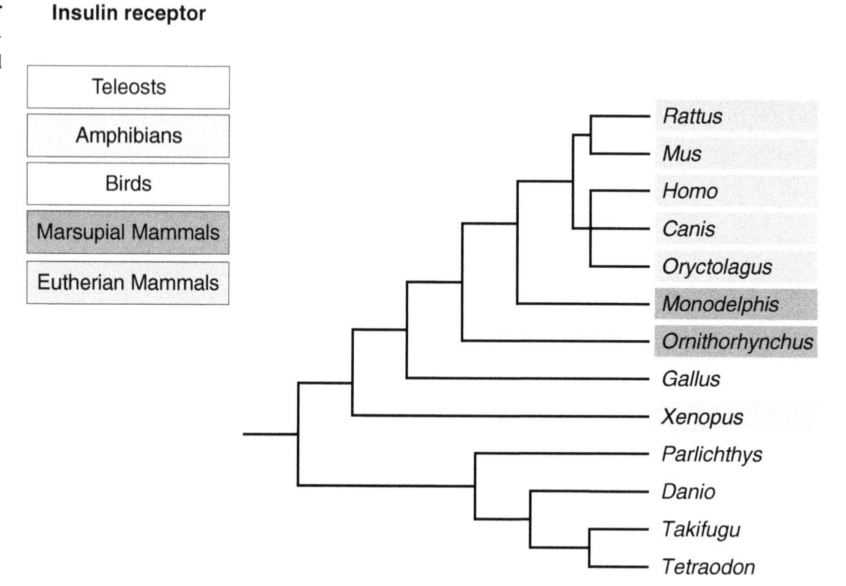

FIGURE 13-10 **Evolution of the insulin receptor gene (*Ir*).** *(Adapted with permission from Hernández-Sánchez, C. et al.,* Molecular Biology and Evolution, *25, 1043–1053, 2008.)*

C. Pancreatic Hormone Functions in Non-Mammals

1. Pancreatic Functions in Agnathan Fishes: Cyclostomes

Insulin has been purified from hagfishes and lampreys. The amino acid composition of hagfish and the Northern Hemisphere lampreys *L. fluviatilis* and *P. marinus* are more like mammalian insulins than like *Geotria australis* from the Southern Hemisphere, suggesting that *Geotria* diverged from the other lampreys very early in cyclostome evolution. Insulin injection causes hypoglycemia in lampreys, and a single injection can decrease plasma glucose levels significantly for several days.

Hagfish insulin behaves similarly to other insulins in mammalian assays. The biological activity of purified hagfish insulin in mammals, however, is only about 4 to 7% of that reported for mammalian insulins in the same bioassay, suggesting the evolution of greater specificity in receptors on target cells has occurred as well as some changes in the insulin molecule itself. When injected into hagfish, both mammalian and hagfish insulins cause hypoglycemia. Hagfish insulin stimulates both glycogen and protein synthesis in hagfish skeletal muscle but only stimulates protein synthesis in liver. Amino acids are known to stimulate insulin release in lampreys, providing additional support for insulin's role as a regulator of protein metabolism. Curiously, isletectomy of the hagfish *Myxine* does not cause hyperglycemia, and the response to a glucose load in isletectomized hagfish is normal. In contrast, total isletectomy of the lamprey *G. australis* results in a marked elevation in blood sugar within 24 hours. Hypophysectomy has no effect on islet cytology or blood sugar levels. There must be an explanation for the presence of insulin and the demonstration of its hypoglycemic action in the hagfish with the number of negative

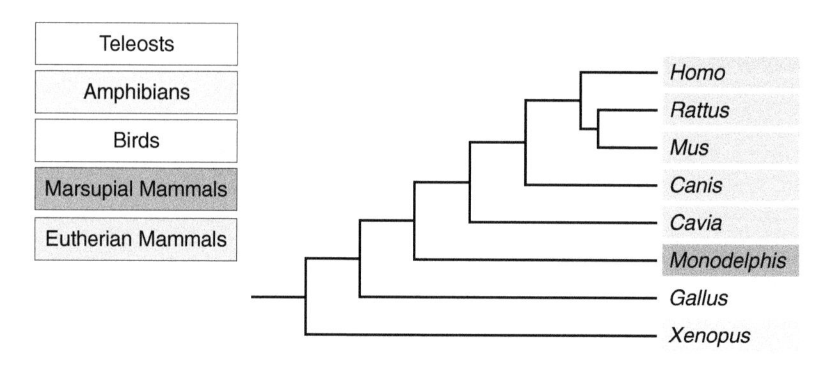

FIGURE 13-11 **Presence of insulin-related orphan receptor genes (*Irr*) in some tetrapods.** *(Adapted with permission from Hernández-Sánchez, C. et al.,* Molecular Biology and Evolution, *25, 1043–1053, 2008.)*

TABLE 13-7 Occurrence of Some Pancreatic Peptides in Invertebrate Animals

Group	Insulin	Glucagon	PP	SS
Gastropods (mollusks)	+	+		
Starfish (echinoderm)	+			
Crab (crustacean)		+		
Tunicate (urochordate)	+		+	+
Amphioxus (cephalochordate)	+	+		

observations just mentioned, but it will require more research to reconcile these contradictory data.

Some large SSTs have been isolated from lampreys and their amino acid sequences determined. As described for insulin, SST_{33} from *Geotria* is very different from the SST_{34} molecules that characterize the other lampreys that more closely resemble hagfish SST. Most of the variations occur among the first 18 amino acids, whereas little variation occurs in the rest of the molecules.

2. Pancreatic Functions in Chondrichthyean Fishes

Considerable variation in blood sugar levels occurs in chondrichthyean fishes; for example, values of 6 to 90 mg of glucose per 100 mL of blood have been reported for the dogfish shark. It appears that either blood sugar regulation is not very precise in these fishes or that blood sugar levels are very sensitive to capture and sampling procedures. Administration of mammalian insulin evokes hypoglycemia, and glucagon induces hyperglycemia in some species. Epinephrine also elevates blood glucose levels, and the effect of injected epinephrine occurs very rapidly. On a superficial basis, it appears that regulation of carbohydrate metabolism in these fishes is similar to the mammalian pattern.

There are many unusual features of carbohydrate metabolism in these fishes that require further study. Although liver glycogen levels in dogfish sharks conform with predicted responses to fasting and force feeding, the liver of holocephalans is either incapable of storing glycogen or is rapidly depleted of glycogen by the stress of capture and/or confinement. Injection of a glucose load does not alter tissue glycogen even though circulating glucose is increased. These observations may explain the inability of glucagon to increase blood glucose in the ratfish. Starvation forces sharks to rely on stored fats. Like mammals, circulating levels of ketone bodies are elevated in fasting sharks.

Protein and lipid metabolism with respect to the actions of pancreatic hormones have not been studied adequately in chondrichthyeans. Lipid metabolism in these fishes should be of special interest because hepatic fat content may be as high as 70% in the dogfish shark and 80% in the ratfish.

3. Pancreatic Function in Bony Fishes

Insulins prepared from several teleosts also are structurally similar to mammalian insulins. Like the chondrichthyean fishes, blood glucose levels of teleosts are very sensitive to handling and sampling stress, presumably due to the release and hyperglycemic action of epinephrine from chromaffin cells (see Chapter 9). Generally, both mammalian and teleost insulins produce hypoglycemia, and glucagon is hyperglycemic when tested in bony fishes. Injections of glucose or the amino acid leucine stimulate insulin secretion in the toadfish (*Opsanus tau*) and in the European silver eel (*A. anguilla*). Purified codfish insulin is a potent hypoglycemic agent in northern pike (*E. lucius*). Furthermore, mammalian insulin increases uptake and incorporation of labeled glucose and glycine into muscle lipids and protein in teleosts. Conversely, mammalian glucagon increases

	1	11	21	31
Human	APLEP VYPGD	NATPE QMAQY	AADLR RYINM	LTRPR Y-NH_2
Porcine/Canine	APLEP VYPGD	DATPE QMAQY	AAELR RYINM	LTRPR Y-NH_2
Tapir	APLEP VYPGD	NATPE QMAQY	AAELR RYINM	LTRPR Y-NH_2
Rhino	ASLEP VYPGD	NATPE EMAQY	AADLR RYINM	LTRPR Y-NH_2
Horse/Zebra	APMEP VYPGD	NATPE QMAQY	AAELR RYINM	LTRPR Y-NH_2
Bovine	APLEP EYPGD	NATPE QMAQY	AADLR RYINM	LTRPR Y-NH_2
Rat	APLEP MYPGD	YATHE QRAQY	ETQLR RYINT	LTRPR Y-NH_2
African bullfrog	ASPEP QHPGG	QATPE QLAQY	YSPLY QYITF	ITRPR F-NH_2
Alligator	TPLQP KYPGD	GAPVE DLIQF	YNDLQ QYLNV	VTRHR P-NH_2
Turkey	GPSQP TYPGD	DAPVE DLIRF	YNDLQ QYLNV	VTRHR Y-NH_2
Goose	GPSQP TYPGN	DAPVE DL?RF	YDNLQ QYRLV	VFRHR Y-NH_2

FIGURE 13-12 **Comparison of amino acid sequences for pancreatic polypeptide (PP).** There is considerable homology within a taxonomic group. Note the similarity between avian and alligator PPs that reflects their evolutionary closeness. The African bullfrog PP exhibits sequences similar to both mammals and the archosaurs. See Appendix C for explanation of the amino acid abbreviations.

blood glucose, for example, in channel catfish. Glycogenolysis in isolated hepatocytes (liver cells) of goldfish is stimulated by either glucagon or epinephrine.

4. Pancreatic Function in Amphibians

Mammalian insulin induces hypoglycemia in both anurans and urodeles, stimulates incorporation of amino acids into proteins, and lowers fatty acid levels. Following administration of a glucose load, blood sugar levels return to normal in about 24 hours. This response to a glucose load and the effects of pancreatectomy indicate an endogenous role for insulin. Mammalian glucagon causes hyperglycemia, although it is ineffective in some amphibians. Epinephrine is hyperglycemic in tadpoles and adults of *X. laevis*, and administration of epinephrine or norepinephrine causes reduction in liver and muscle glycogen.

Early research by Bernardo A. Houssay more than 50 years ago on the amphibian pancreas led to a widely employed procedure for research into mammalian diabetes mellitus. It was in the toad *Rhinella* (*Bufo*) *arenarum* that the antagonism of pituitary hormones and blood glucose of pancreatectomized or insulin-deficient animals was discovered. This discovery led to the widespread use of hypophysectomy to alleviate extreme diabetic conditions in experimental animals, and such doubly operated dogs, cats, and others became known commonly as **Houssay animals**. This action of hypophysectomy in the pancreatic-deficient animal is termed the **Houssay effect**.

5. Pancreatic Function in Reptiles

Reptiles exhibit mammalian-like responses to injected insulin and glucagon and respond typically to a glucose load, returning to preloading levels in about 48 hours. Lizards, however, are particularly insensitive to insulin, and large quantities of mammalian insulin are required to invoke hypoglycemic responses. It is virtually impossible to induce insulin shock even with massive doses to lizards irrespective of their diets (Table 13-8). However, pancreatectomy induces a hyperglycemic state in lizards as it does in snakes and alligators. These observations suggest that, in spite of the predominance of A cells in lizard islets and their demonstrated insensitivity to mammalian insulin, endogenous insulin may play an important metabolic role in lizards as well as in other reptiles.

Insulin has been purified from a rattlesnake, *Crotalis atrox*, and its amino acid composition was found to be considerably different from bovine insulin with some uncommon substitutions occurring in the B-chain. In contrast, alligator insulin is structurally very much like chicken insulin, and the alligator responds equally with hypoglycemia to bovine, alligator, or turkey insulin. Comparative analyses of reptilian insulins may shed some light on pancreatic regulation of carbohydrate metabolism in reptiles, especially the apparent insensitivity of many species to mammalian insulins. Curiously, cyclostome insulin is structurally more like mammalian insulins than is rattlesnake insulin, which points out the difficulties of reconstructing evolutionary events using molecular data obtained from one or a few species as being representative of a major taxonomic group and/or relying only on the structure of the biologically active portion of a large gene product.

Although glucagon has been isolated from the alligator, its role in carbohydrate metabolism in reptiles has been inferred largely from observations following mammalian glucagon treatment. Glucagon is hyperglycemic in squamate reptiles as in birds, and circulating levels of glucagon often are high. For example, fasting savannah monitor lizards (*Varanus exanthematicus*) have high glucagon levels whereas glucagon is lowest during times of food intake. Pancreatic polypeptide has been isolated from

TABLE 13-8 Fatal Doses of Insulin for Various Vertebrates

Animal	IU of Insulin/kg Body Weight	Animal	IU of Insulin/kg Body Weight
Nonmammals		Mammals	
Salamander	50	Rabbit	4–8
Lizard	>10,000[a]	Mouse	7–50
Canary	1000–4000	Rat	29–36
Pigeon	400–1200	Dog	16–300
Duck	50–500		

[a]*A dose of 10,000 IU of insulin had no effect on survival of lizards.*
Adapted with permission from Gorbman, A. and Bern, H.A., "A Textbook of Comparative Endocrinology," Wiley, New York, 1962.

alligators. The amino acid composition of alligator PP is more like that of birds than like mammals (Figure 13-12).

Of particular interest is the absence of studies in which lipid and protein metabolism of reptiles was examined. Insulin and other "metabolic" hormones have not been examined for their roles in lipid or protein metabolism. Until all aspects of the hormonal regulation of metabolism in reptiles are integrated with the observations of effects of insulin and glucagon, a complete picture for the roles of reptilian hormones in metabolism will not emerge.

5. Pancreatic Function in Birds

Hyperglycemia initially observed following pancreatectomy is now believed to have been a result of overlooking the splenic lobe (rich in A cells) during pancreatectomy. Total pancreatectomy results in hypoglycemia rather than hyperglycemia and suggests that glucagon plays a major role in avian blood sugar regulation. This suggestion is supported further by observations that a high glucose load is required to induce insulin release in the duck and by the relative insensitivity of avian tissues to injected mammalian insulin (Table 13-8). Elevated glucose levels depress glucagon release, whereas high circulating levels of **nonesterified fatty acids (NEFAs)** induce glucagon release.

Avian tissues respond with hyperglycemia, glycogenolysis, and lipolysis following administration of mammalian glucagon. The avian pancreas contains about five to ten times the extractable glucagon per gram of pancreas as compared to mammals, emphasizing its probable importance to metabolic regulation in birds. In one of the few studies performed in wild birds, mammalian glucagon stimulated both plasma levels of free fatty acids and glucose in penguins, whereas insulin caused a reduction in blood glucose. Additional studies with wild avian species representing diverse taxonomic groupings would be welcome in order to confirm or alter generalizations based largely upon observations of inbred domesticated species.

Insulin has been extracted from the avian pancreas, and its structure has been examined in a number of diverse species (Figure 13-13). Chicken insulin is more effective at inducing hypoglycemia in birds than it is in mammals, and it is more potent in birds than is mammalian insulin. Glucagon has been extracted and its primary structure determined for duck, turkey, and chicken hormones. These hormones are almost identical structurally to mammalian glucagon.

Like mammalian PP, avian PP (aPP) consists of 36 amino acid residues and has a molecular weight of approximately 4200 Da (Figure 13-12). Avian PP, unlike its mammalian counterpart, appears to have a strong stimulatory influence on gastric acid secretion in chickens. The physiological role for aPP, however, is uncertain at this time.

D. Non-Gastrointestinal/Pancreatic Hormones and Metabolism in Nonmammalian Vertebrates

In contrast to mammals, the endocrine regulation of metabolism has been neglected in vertebrate groups that have relatively little economic importance. The differences in body temperature, reliance on anaerobic metabolism, and great variety of activity levels make it difficult to generalize even within a reasonably defined taxonomic unit. A brief review of effects by corticosteroids, catecholamines, and thyroid hormones on metabolism in nonmammalian vertebrates is provided here.

1. Fishes

Lampreys exhibit hyperglycemia following treatment with either cortisol or epinephrine. Cortisol elevates liver glycogen, suggesting a gluconeogenic action. Hagfishes do not exhibit hyperglycemia to either cortisol or catecholamine treatment. Known insulin antagonists, such as glucagon, **thyroxine (T_4)**, **corticotropin (ACTH)**, and **prolactin (PRL)** are all ineffective in altering blood sugar levels in cyclostomes. Chondrichthyean fishes have been studied less than have the jawless fishes, and few generalizations are possible. Blood sugar levels are relatively low in selachians but can be elevated with treatment by either ACTH or glucocorticoids, implying similar metabolic actions to those described in mammals. Among teleosts, gluconeogenesis occurs mainly in the liver and kidney and is accelerated by high-protein diets or by starvation. Cortisol is the principal glucocorticoid secreted by the adrenocortical cells. In addition to its role as a mineralocorticoid, cortisol depletes lipid and protein stores and promotes gluconeogenesis as it does in mammals. High cortisol levels, for example, are associated with up to a 96% reduction in body fat and loss of more than half the body protein in spawning sockeye salmon. Catecholamines stimulate glycogenolysis, lipolysis, and gluconeogenesis. The possible role of thyroid hormones (see Chapter 7) is unclear at this time, but they do not appear to be important for regulating metabolism.

2. Amphibians

Corticosterone and, to a lesser extent, aldosterone are gluconeogenic and hyperglycemic in amphibians, but the primary metabolic role of these steroids as seen in fishes appears to be the regulation of ion and water balance. However, specific gluconeogenic enzymes are activated by corticosteroids as well as are urea cycle enzymes associated with elimination of nitrogenous wastes generated from amino acid degradation. Epinephrine is a potent stimulator of glycogenolysis in both liver and muscle and therefore is

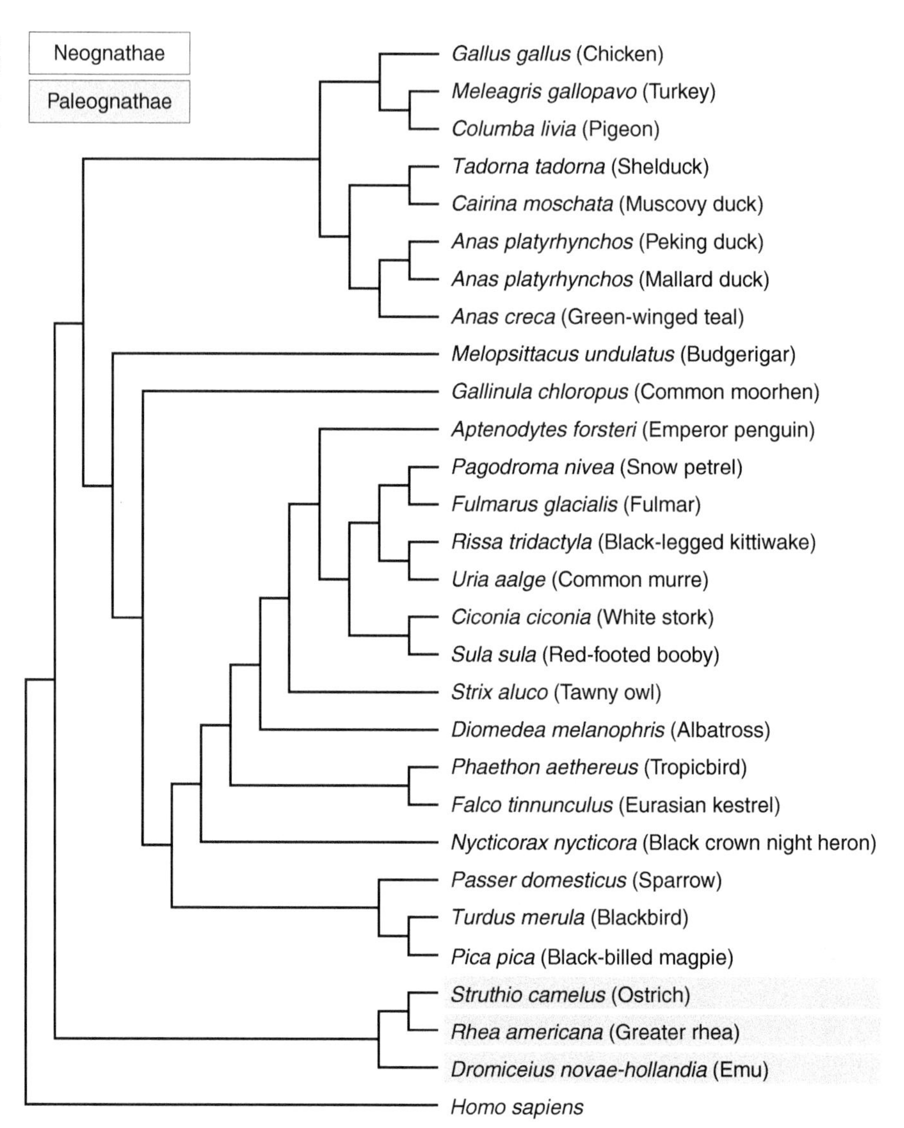

FIGURE 13-13 **Phylogenetic relations for avian preproinsulin gene.** *(Adapted with permission from Simon, J. et al.,* Molecular Phylogenetics and Evolution, ***30****, 755–766, 2004.)*

hyperglycemic. Thyroid hormones can increase oxidative metabolism, but apparently only at higher temperatures—for example, 25°C (see Chapter 7)—and any physiological metabolic role is highly speculative.

3. Reptiles

The metabolism of reptiles and the roles of corticosteroids and catecholamines in regulating reptilian metabolism are poorly studied. Although there is little evidence to establish a gluconeogenic action in reptiles, there is some indirect evidence. Corticosteroids are hyperglycemic and favor deposition of glycogen in lizards, turtles, alligators, and snakes. Epinephrine is hyperglycemic in representatives of all major reptilian groups and promotes glycogenolysis in liver and glycogen deposition in alligator muscle. As reported for amphibians, thyroid hormones stimulate oxygen consumption in lizards at warmer temperatures (30°C) but not at lower temperatures (see Chapter 7).

4. Birds

Domestic birds exhibit all of the glycolytic, glycogenic, and glycogenolytic pathways for carbohydrate metabolism described for mammals. Because of the dependence of developing birds on lipid-rich yolk, the normal glycolytic and oxidative pathways are not important until after hatching. Corticosterone is hyperglycemic and gluconeogenic primarily due to stimulation of protein catabolism. Corticosteroids also stimulate **triacylglycerol (TAG)** synthesis in the liver. These TAGs are transported to adipose tissue for storage, an important action during fattening prior to spring or fall migrations. This lipogenic action of corticosteroids is in marked contrast to the effects of glucocortioids in mammals and other vertebrates and may represent a special adaptation in long-distance migrators. Epinephrine and norepinephrine are potent glycogenolytic and lipolytic hormones in birds. Catecholamines cause hepatic and muscle glycogenolysis that can lead to hyperglycemia. Thyroid hormones affect oxygen consumption and therefore affect metabolism and heat production.

IV. SUMMARY

Regulation of feeding in non-mammals is similar to mammals although it has not been studied extensively in non-mammals other than teleost fishes. Orexigenic actions in fishes have been shown for ghrelin, NPY, galanin and orexins. CCK_8, CRH, urotensin-I, CART, and PrRP are anorexic. Orexins stimulate feeding in anurans and leptin is present in newts as well as in reptiles. Leptin has been isolated from birds but orexins are ineffective, at least in chickens.

Comparative studies of gastrointestinal hormones are limited primarily to demonstrations of the presence of immunoreactivity to antibodies prepared against various mammalian GI peptides. Some putative regulatory peptides have been isolated that are closely related to mammalian counterparts but their physiological roles are not clear. It appears that intestinal peptides with GI hormone activities occurred early in vertebrate evolution (class Agnatha), and some counterparts are in evidence among invertebrates, as well.

Five patterns of pancreatic islet morphology can be distinguished in vertebrates: (1) cyclostome type, (2) primitive gnathostome type, (3) actinopterygian type, (4) lungfish type, and (5) tetrapod type. It is assumed that the most primitive anatomical arrangement for the endocrine pancreatic tissue occurs in lampreys as the follicles of Langerhans are partially embedded in the gut lining. The primitive hagfishes, however, have a definite endocrine pancreas. Bony fishes tend to have discrete islets (Brockman bodies) but lack an acinar (exocrine) pancreas. Elasmobranchs, holocephalans, and *Latimeria* all exhibit the primitive gnathostome type of endocrine pancreas. Tetrapods have distinct pancreatic organs containing both acinar and islet tissues.

Insulin is highly conserved among the vertebrates, although insulin receptors appear to be more variable. The most primitive function for insulin may be its effect on amino acid incorporation into protein. Glucose administration does not invoke insulin release in lampreys, lizards, or birds, and insulin stimulates only protein synthesis in hagfish liver. These observations suggest limited involvement of insulin in carbohydrate metabolism of non-mammals and that its hypoglycemic actions may be largely due to effects on protein and/or lipid metabolism. However, the effects of insulin on lipid metabolism may be largely a mammalian occurrence.

Glucagon and the GLPs also are present and highly conserved throughout the vertebrates. In general, reptiles and birds appear to be more reliant on glucagon for regulation of carbohydrates than insulin.

STUDY QUESTIONS

1. Compare the regulation of feeding behavior in fishes with that of mammals (Chapter 12).
2. Based on our knowledge of fishes, birds, and mammals, what would you predict we would discover about regulation of feeding in reptiles?
3. Why do you think our understanding of gastrointestinal regulation is so limited?
4. What generalizations can you make about the evolution of insulin and insulin actions?
5. What generalizations can you make about the evolution of glucagon and glucagon-like peptides?

SUGGESTED READING

Adriaensen, D., Timmermans, J.-P., 2006. The endocrine pancreas of African lungfish: light and electron microscope immunocytochemistry and morphology. In: Reinecke, M., Zaccone, G., Kapoor, B.G. (Eds.), Fish Endocrinology, Vol. 1. Science Publishers, Boca Raton, FL, pp. 199–222.

Carr, J.A., Brown, C.L., Mansouri, R., Venkatesan, S., 2002. Neuropeptides and prey-catching behavior in toads. Reviews in Comparative Biochemistry and Physiology Part B: Biochemistry & Molecular Biology 132, 151–162.

Conlon, J.M., Patterson, S., Flatt, P.R., 2006. Major contributions of comparative endocrinology to the development and exploitation of the incretin concept. Journal of Experimental Zoology 305A, 781–786.

Crespi, E.J., Denver, R.J., 2006. Leptin (ob gene) of the South African clawed frog *Xenopus laevis*. Proceedings of the National Academy of Sciences USA 103, 10092–10097.

Epple, A., Brinn, J.E., 1987. The Comparative Physiology of the Pancreatic Islets. Springer-Verlag, Berlin.

Galas, L., Vaudry, H., Braun, B., Van den Pol, A.N., de Lecea, L., Sutcliffe, J.G., Chartrel, N., 2001. Immunohistochemical localization and biochemical characterization of hypocretin/orexin-related peptides in the central nervous system of the frog, *Rana ridibunda*. Journal of Comparative Neurology 429, 242–252.

Gorbman, A., Bern, H.A., 1962. A Textbook of Comparative Endocrinology. Wiley, New York.

Gutiérrez, J., Navarro, I., Planas, J.V., Montserrat, N., Rojas, P., Castillo, J., et al., 2006. Insulin and IGF receptors in fish. In: Reinecke, M., Zaccone, G., Kapoor, B.G. (Eds.), Fish Endocrinology, vol. 1. Science Publishers, Boca Raton, FL, pp. 131–166.

Hernández-Sánchez, C., Mansilla, A., de Pablo, F., Zardoya, R., 2008. Evolution of the insulin receptor family and receptor isoform expression in vertebrates. Molecular Biology and Evolution 25, 1043–1053.

Hoskins, L.J., Volkoff, H., 2012. The comparative endocrinology of feeding in fish: insights and challenges. General and Comparative Endocrinology 176, 327–335.

Johnsen, A.H., 1998. Phylogeny of the cholecystokinin/gastrin family. Frontiers in Neuroendocrinology 19, 73–99.

Kaiya, H., Sakata, I., Kojima, M., Hosoda, H., Sakai, T., Kangawa, K., 2004. Structural determination and histochemical localization of ghrelin in the red-eared slider turtle. *Trachemys scripta elegans*. General and Comparative Endocrinology 138, 50–70.

Kaiya, H., Sakata Yamamoto, K., Koda, A., Sakai, T., Kangawa, K., Kikuyama, S., 2006. Identification of immunoreactive plasma and stomach ghrelin and expression of stomach ghrelin mRNA in the bullfrog. *Rana catesbeiana*. General and Comparative Endocrinology 148, 236–244.

Kawauchi, H., 2006. Functions of melanin-concentrating hormone in fish. Journal of Experimental Zoology 305A, 751–760.

Ku, S.-K., Lee, H.-S., Koh, J.-K., Lee, J.-H., 2003. An immunohistochemical study on the neuropeptide-producing endocrine cells in the alimentary tract of the wrinkled frog, *Rana rugosa* (Ranidae). General and Comparative Endocrinology 131, 1–8.

Ku, S.-K., Lee, H.-S., Lee, J.-H., 2000. An immunohistochemical study of endocrine cells in the alimentary tract of the red-bellied frog, *Bombina orientalis*. Journal of Veterinary Medical Science 62, 589–594.

Musson, M.C., Jepeal, L.I., Finnerty, J.R., Wolfe, M.M., 2011. Evolutionary expression of glucose-dependent-insulinotropic polypeptide (GIP). Regulatory Peptides 171, 26–34.

Navarro, I., Capilla, E., Castillo, J., Albalat, A., Díaz, M., Gallardo, M.A., et al., 2006. Insulin metabolic effects in fish tissues. In: Reinecke, M., Zaccone, G., Kapoor, B.G. (Eds.), Fish Endocrinology, vol. 1. Science Publishers, Boca Raton, FL, pp. 15–48.

Nelson, L.E., Sheridan, M.A., 2006. Gastroenteropancreatic hormones and metabolism in fish. General and Comparative Endocrinology 148, 116–124.

Ng, S.Y., Lee, L.T.O., Chow, B.K.C., 2010. Insights into the evolution of proglucagon-derived peptides and receptors in fish and amphibians. Annals of the New York Academy of Sciences 1200, 15–32.

Reinecke, M., Broger, I., Brun, R., Zapf, J., Maake, C., 1995. Immunohistochemical localization of insulin-like growth factor I and II in the endocrine pancreas of birds, reptiles, and Amphibia. General and Comparative Endocrinology 100, 385–396.

Simon, J., Laurent, S., Grolleau, G., Thoraval, P., Soubieux, D., Rasschaert, D., 2004. Evolution of preproinsulin gene in birds. Molecular Phylogenetics and Evolution 30, 755–766.

Volkoff, H., Canosa, L.F., Unniappan, S., Cerdá-Reverter, J.M., Bernier, N.J., Kelly, S.P., Peter, R.E., 2005. Neuropeptides and the control of food intake in fish. General and Comparative Endocrinology 142, 3–19.

Wang, Y., Conlon, J.M., 1995. Purification and structural characteristics of vasoactive intestinal polypeptide from trout and bowfin. General and Comparative Endocrinology 98, 94–101.

Youson, J.H., Mahrouki, A.A., Amemiya, Y., Graham, L.C., Montpetit, C.J., Irwin, D.M., 2006. The fish endocrine pancreas: review, new data, and future research directions in ontogeny and phylogeny. General and Comparative Endocrinology 148, 105–115.

Chapter 14

Regulation of Calcium and Phosphate Homeostasis in Vertebrates

Physiologists often focus on the regulation of Na^+ and K^+ ion levels because of their importance in maintaining membrane potentials and their importance for neuronal and muscle function. However, calcium and phosphate also are extremely important for the regulation of many basic body functions, including the release of neurotransmitters and other bioregulators from cells as well as the mechanics of skeletal, cardiac, and smooth muscle contraction. They are involved in the structure of bones, scales, and teeth that serve as the major reservoirs for maintaining plasma and interstitial fluid levels of these ions in aquatic and terrestrial vertebrates. Development of a bony skeleton was essential for support of the vertebrate body prior to invasion of the land. In the aquatic environment, fishes have adequate access to external calcium and have evolved mechanisms to prevent excessive influx of calcium from their environment. In contrast, terrestrial vertebrates must rely on dietary sources of calcium to maintain a proper balance among skeletal stores, blood, and extracellular fluids; consequently, we find some interesting differences between fishes and other vertebrates in the manner of regulating calcium and phosphate balance.

I. IMPORTANCE OF CALCIUM AND PHOSPHATE

Although inorganic calcium and phosphate are not related closely with respect to most of their essential roles in vertebrate physiology, they are regulated by some of the same hormones; consequently, a discussion of the endocrine regulation of calcium homeostasis cannot be divorced entirely from the regulation of phosphate. A discussion of calcium and phosphate homeostasis in mammals will set the standard for making comparisons to other vertebrates.

A. Calcium Homeostasis

Calcium is an essential mineral for life. Calcium levels in extracellular fluids and in the cytosol of cells must be regulated precisely if normal body functions are to be maintained. In all vertebrates, blood calcium is tightly regulated and maintained between 9.0 and 10.5 mg/dL. In addition to its role in bone and tooth construction, calcium ions have other important roles to play (Table 14-1). They are responsible for excitation and contraction of muscle cells and are important in the induction of spontaneous excitations of cardiac pacemaker cells. Calcium ions are essential for exocytosis of secretion granules in neurons and glandular cells and serve as second messengers in many target cells (see Chapter 3). Certain key metabolic enzymes are activated by calcium ions that can also serve as cofactors for several blood-clotting proteins (factors VII, IX, and X).

The calcium level in adult mammalian blood plasma is maintained at about 10 mg/dL or approximately 2.5 m*M*. About half of this calcium is free in the plasma in ionic form (Ca^{2+}), and the remainder is bound to circulating proteins (40%) or occurs in other chemical complexes

TABLE 14-1 Some Physiological Roles for Calcium and Phosphate

Roles for calcium	Roles for phosphate
Structural component (with phosphate) of bones, teeth	Structural component (with calcium) of bones, teeth
Necessary for contraction of skeletal, smooth, and cardiac muscle	Activator of enzymes from inactive form
Maintenance of membrane potentials in some neurons	Buffer in blood and other body fluids
Unique role in membrane potentials of pacemaker cells	Essential component for energy storage in chemical bonds of ATP, creatine phosphate, etc.
Second messenger in hormonal and neurocrine mechanisms of action	Essentials component of nucleic acids
Role in exocytosis of secretory products from cells	Essential component of phospholipids in cell membranes
Necessary cofactor for certain enzymes	Involved in second messenger formation
Component of intercellular matrices (e.g., basement membrane)	Necessary for metabolism of glucose and other carbohydrates

Vertebrate Endocrinology. http://dx.doi.org/10.1016/B978-0-12-394815-1.00014-8

(10%). It is this free ionic calcium in the blood that is essential to so many important life processes because it can be exchanged readily with other extracellular fluids and cells. About 90% of the protein-bound calcium is linked to albumin, and this binding is pH sensitive. Acute acidosis decreases binding and elevates ionic plasma calcium. Acute alkalosis increases binding and reduces free plasma calcium.

After birth, calcium is obtained in the diet, absorbed through the small intestine, deposited in bones and teeth, or excreted via urine or feces. Urinary excretion of calcium is directly proportional to plasma levels, and little calcium is excreted unless plasma calcium levels exceed normal. Calcium deposited in bone serves as a reservoir to provide adequate plasma Ca^{2+} for minute-to-minute regulation of body needs and during acute or chronic periods of dietary deprivation. Human body calcium levels vary according to size and age (Table 14-2).

Blood calcium homeostasis is maintained by the cooperative actions of the bones and teeth and the intestine that together serve as internal and external sources of calcium (see Figure 14-1). The kidneys prevent loss of calcium to the urine or can allow excretion of excess calcium. The intestine also excretes calcium.

TABLE 14.2 Calcium Content of Human Body

Age (years)	Body weight (kg)	Calcium content (g)
1	10.6	100
5	19.1	219
10	33.3	396
15	55.0	806
20	67.0	1078

From Irving, J.T., "Calcium and Phosphorus Metabolism," Academic Press, New York, 1973.

B. Phosphate Homeostasis

Phosphate, like calcium, is an essential component of bones and teeth. Approximately 80% of the total body phosphate is sequestered in the skeleton as the calcium phosphate salt **hydroxyapatite** ($Ca_{10}(PO_4)_6(OH)_2$). In addition, many essential molecules contain phosphate, including structural phospholipids in cellular membranes, nucleic acids, nucleotides, and hexose phosphates. Furthermore, phosphate is indispensable for energy storage within cells in the form of ATP or creatine phosphate. Hydrolysis of ATP or guanosine triphosphate (GTP) to form cyclic adenosine 3′,5′-monophosphate (cAMP) or cyclic guanosine 3′,5′-monophosphate (cGMP), respectively, is necessary for mediating the actions of many hormones, for neural transmission, and for many other cellular processes. The actions of protein kinases determine the presence or absence of phosphate, which activates or inactivates key molecules in many biochemical pathways discussed within this textbook. Although phosphate ions play a minor role as buffers of hydrogen ions in most body fluids, they are the major buffer system in urine.

It is rare that phosphate becomes a limiting factor for an organism, and the usual phosphate disturbances in mammals are a result of excessive levels of phosphate. Inorganic phosphate (P_i) in mammalian plasma is generally about 3.1 mg/dL (about 1.0 m*M*). Most of this phosphate (about 80%) is in the form of HPO_4^{-2}, with almost 20% occurring as $H_2PO_4^{-1}$ and only a trace as PO_4^{-3}. Henceforth, the chemical formula HPO_4^{-2} will be used to represent all of the phosphate ions.

In normal plasma, about 90% of the P_i is free ionic (filterable) phosphate, and about 10% is bound to the

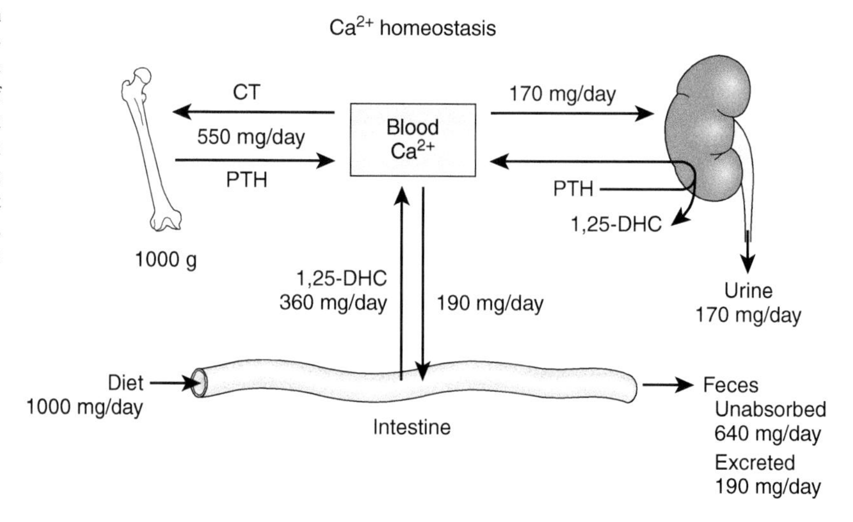

FIGURE 14-1 **Endocrine regulation of calcium homeostasis.** PTH and 1,25-DHC are hypercalcemic agents, and CT is the only hypocalcemic agent protecting the skeleton during periods of growth, pregnancy, or lactation. The skeleton represents the major reservoir of calcium ions in mammals. For humans, a daily intake of 1000 mg Ca^{2+}/day is necessary to offset what is lost normally through urine and feces. 1,25-DHC, 1,25-dihydroxycholecalciferol; CT, calcitonin; PTH, parathyroid hormone.

plasma proteins. In addition to P_i, plasma contains considerable amounts of lipid-bound phosphate and esterified phosphate so that total plasma phosphate is actually about 12.5 mg/dL. Phosphate values vary considerably with diet, age, and metabolic state, however, and it is difficult to provide a "normal" value without specifying the conditions under which this "normal" occurs.

C. Interrelationship of Ca^{2+} and HPO_4^{-2}

Calcium and phosphate are regulated in such a way that the product of the free plasma concentrations of Ca^{2+} and HPO_4^{-2} ($[Ca^{2+}][HPO_4^{-2}]$) equals some constant called *k*. This constant however, may change according to differing physiological states or pathological conditions. For example, *k* is greater in growing mammals than it is in adults. This relationship between Ca^{2+} and HPO_4^{-2} implies that if there is an increase in Ca^{2+} then a corresponding decrease in HPO_4^{-2} should follow. Likewise, an increase in HPO_4^{-2} should cause a decrease in Ca^{2+}. This generalization is useful to illustrate some of the relationships that exist between the regulatory mechanisms governing these ions. Minute-to-minute adjustments of Ca^{2+} and HPO_4^{-2} levels in extracellular fluids are accomplished primarily through a combination of bone destruction (resorption) or formation, absorption of dietary calcium by the small intestine, and renal excretion of phosphate.

II. BONE FORMATION AND RESORPTION IN MAMMALS

In bones and teeth, calcium phosphate occurs in the form of small submicroscopic crystals deposited upon an organic matrix composed primarily of **collagen fibers**. These crystals assume a uniform structural and complex chemical form known as **hydroxyapatite crystals**. Construction of bone through formation of calcium phosphate is not understood completely, but some of the major features are well accepted. Bone formation may involve **apatite formation** (deposition of new hydroxyapatite crystal) or simply **mineral accumulation** (the additional growth of existing crystals). Exchange of Ca^{2+}, HPO_4^{-2}, and water can occur between the surface of these crystals and the extracellular fluids. This exchange is inversely proportional to the size of the crystal. Thus, larger crystals contain considerable amounts of calcium phosphate that cannot engage in free exchange with the extracellular fluids. About 99% of bone calcium phosphate is found in these larger, nonexchangeable stable or **diffusion-locked crystals**.

The specific process of bone formation and growth also involves a number of local chemical factors that are responsible for collagen matrix formation, cartilage matrix deposition, and formation of hydroxyapatite crystals, as well as cellular replication and differentiation. Cells known as **osteoblasts** (literally, bone-forming cells) are responsible for bone formation and are targets for Ca^{2+} and HPO_4^{-2} regulation. (The homologous cell in teeth is termed an **odontoblast**.). The osteoblasts comprise the **endosteal membrane** that lines the cavities within bone, and they synthesize the collagen matrix upon which apatite formation occurs. Some osteoblasts get embedded in matrix and become **osteocytes** that are completely surrounded by bone except for minute channels through which the osteocytes communicate with one another. There seems to be little agreement on the role for osteocytes in calcium–phosphate metabolism, but they also may be important targets for hormonal regulation.

Resorption of bone may involve either removal of the collagen matrix and/or solubilization of hydroxyapatite crystals with consequent release of Ca^{2+} and HPO_4^{-2}, but both processes usually occur. Another bone cell, the **osteoclast** (literally, bone destroying), is primarily responsible for bone resorption. The osteoclast (Figures 14-2 and 14-3) is a large, multinucleate cell that is easy to distinguish from uninucleate osteoblasts or osteocytes. Osteoclasts arise from monocyte/macrophage precursors. These precursor cells interact with osteoblasts that induce them to differentiate into osteoclasts (Figure 14-4). Mature osteoclasts are characterized by possession of a number of unique biochemical features, such as the presence of the enzymes **tartrate-resistant acid phosphatase (TRAPase)**, **carbonic anhydrase**, and **cathepsin K**.

III. ENDOCRINE REGULATION OF CALCIUM AND PHOSPHATE HOMEOSTASIS IN MAMMALS

Three hormones regulate calcium and phosphate homeostasis in mammals. **Parathyroid hormone (PTH)** secreted by the **parathyroid glands** is a **hypercalcemic factor**; that is, its actions can cause an elevation in the level of plasma calcium. Its secretion is controlled primarily by a direct action of plasma Ca^{2+} levels on the parathyroid cells. **Calcitonin (CT)** is a **hypocalcemic factor** secreted by the **C cells** of the mammalian thyroid gland (see Chapter 6). Release of CT also is related primarily by changes in plasma Ca^{2+}. One major site of action for PTH is bone, where it may stimulate calcium release from bone (bone resorption), releasing both Ca^{2+} and HPO_4^{-2} into the circulation (see Figure 14-1). Parathyroid hormone also increases Ca^{2+} reabsorption by the nephron as well as the secretion of HPO_4^{-2} into the urine, resulting in a decrease in plasma HPO_4^{-2} levels and a concomitant increase in urinary HPO_4^{-2}. The effects of PTH on Ca^{2+} homeostasis are linked closely to the actions of the third regulatory hormone, **1,25-dihydroxycholecalciferol (1,25-DHC)**. Absorption of calcium from the gut lumen is stimulated by 1,25-DHC,

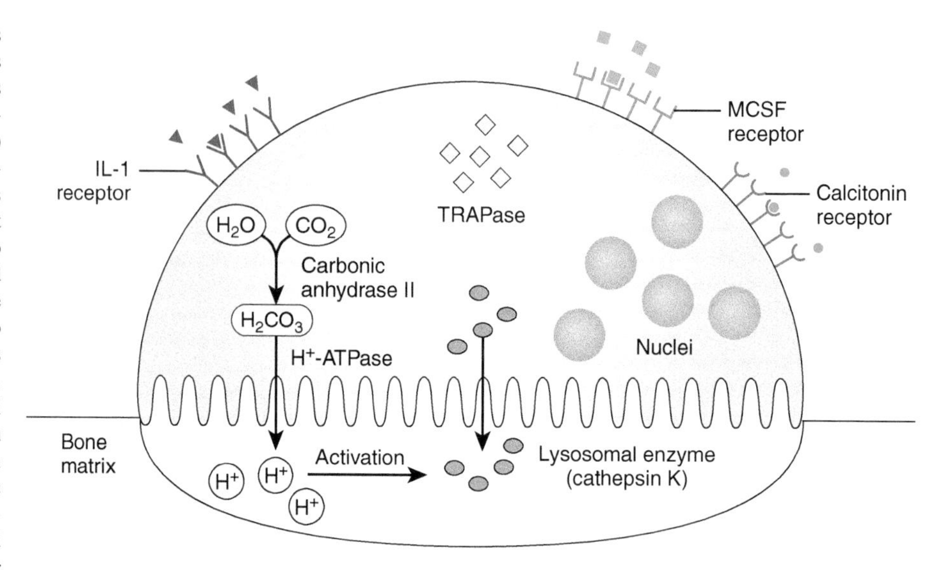

FIGURE 14-2 **An osteoclast.** This multinucleate bone-destroying cell is a target for calcitonin that inhibits its activity. Interleukin-1 (IL-1) and macrophage colony-stimulating factor (MCSF) can enhance osteoclast survival. Tartrate-resistant acid phosphatase (TRAPase) is considered a marker for osteoclast activity. Osteoclast activity and the ratio of osteoclasts to osteoblasts is increased by PTH, but indirectly (see text). Note the ruffled border of the cell adjacent to the resorption space (white area) that is involved in the active resorption of Ca^{2+}. Production of H^+ and the lysosomal enzyme cathepsin K results in dissolution of the bone matrix and release of calcium and phosphate ions. *(Adapted with permission from Martin, T.J. and Udagawa, N.,* Trends in Endocrinology and Metabolism, *9, 6–12, 1998. © Elsevier Science, Inc.)*

which also influences the actions of PTH on bone and kidney. Furthermore, as discussed later, PTH influences the production of 1,25-DHC by the kidney, and 1,25-DHC has other actions as well.

A. The Parathyroid Glands and PTH

Although the parathyroid glands had been observed previously, Sandstrom rediscovered and named them in 1880. In many species, the parathyroids, which develop like the thyroid from pharyngeal tissues (Figure 14-5), are embedded within the thyroid glands (e.g., mouse, cat, human). In other mammals, such as goats and rabbits, they are separate glands located near the thyroid. Some mammals have more than four separate parathyroid glands,

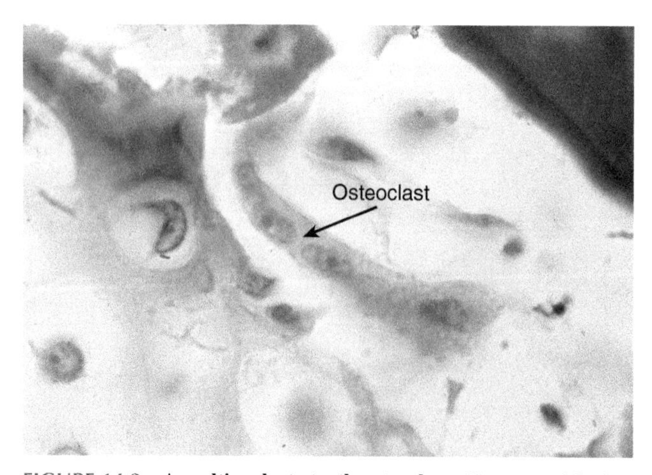

FIGURE 14.3 **A multinucleate turtle osteoclast.** *(Courtesy of the late Nancy B. Clark, University of Connecticut, Storrs.)*

and accessory parathyroid tissue is not uncommon. Thyroidectomy may result in lowered plasma Ca^{2+} levels in some species because of simultaneous removal of the embedded parathyroids. Decreasing plasma calcium by artificial means can increase PTH secretion several fold. This can be accomplished by infusion of calcium-chelating (calcium-binding) agents such as EDTA.

The parathyroid glands developmentally arise from pharyngeal endoderm; however, the parathyroid **chief cells** that secrete PTH arise from neuroectoderm, as shown first in the frog *Rana temporaria*. Chief cells are part of the **amine precursor uptake and decarboxylation (APUD)** series of peptide-secreting cells (see Chapter 4). Indirect evidence also supports a similar origin for PTH-secreting cells in birds and mammals. Chief cells are cuboidal with no unique cytological features other than the presence of granules containing immunoreactive PTH. They are the dominant cellular type and comprise about 99% of the parathyroid's cellular population in most species. Chief cells exhibit β-adrenergic receptors, suggesting that some neural influence on PTH release may exist. Glucocorticoid receptors also are present and when activated result in enhanced PTH release. Estrogens may diminish the response of the chief cells to lowered plasma calcium.

In a few species, such as deer, the parathyroids are composed exclusively of chief cells; however, in most species a second cell type called an **oxyphil** is usually present in the parathyroids. Oxyphils are eosinophilic cells rich in mitochondria (hence their name). The function of the oxyphil and the significance for its large number of mitochondria are unknown. Oxyphils represent only about 1% of the total parathyroid cells.

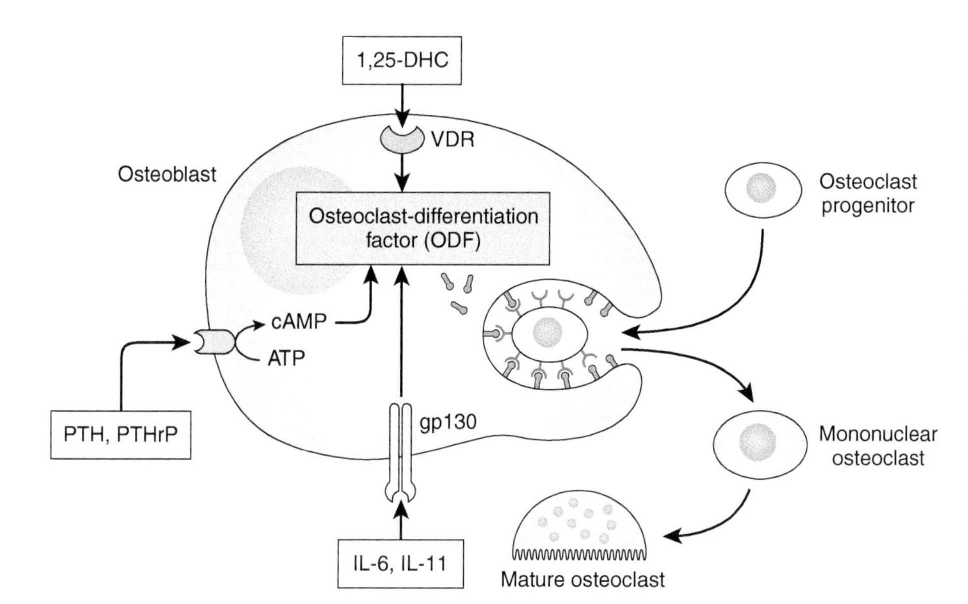

FIGURE 14-4 **An osteoblast.** Bone-forming osteoblasts are targets for PTH, PTHrP, and 1,25-DHC. Stimulated osteoblasts transform osteoclast progenitor cells into mature osteoclasts through direct surface contact. The osteoclast-differentiation factor (ODF) has not been identified. gp 130, glycoprotein 130; PTHrP, parathyroid hormone-related peptide; IL, Interleukin; VDR, Vitamin D receptor. *(Adapted with permission from Martin, T.J. and Udagawa, N.,* Trends in Endocrinology and Metabolism, *9, 6–12, 1998. © Elsevier Science, Inc.)*

1. Parathyroid Hormone

Mammalian PTH has been isolated and characterized from several species. It is a large polypeptide consisting of 84 amino acids (Figure 14-6). Apparently all of the biological activity resides in the first 34 amino acids. Synthesis of PTH occurs in two steps. **Preproparathyroid hormone (PreProPTH)**, consisting of 115 amino acid residues, is synthesized on the ribosomes of the rough endoplasmic reticulum (RER) but is rapidly cleaved at the NH_2-terminus to **proparathyroid hormone (ProPTH)** as it enters the cisterna of the endoplasmic reticulum. ProPTH consists of 90 amino acids and travels through the cisterna to the Golgi apparatus, where the remaining six NH_2-terminal residues of the prohormone are removed. The resulting PTH is then packaged into storage vesicles to await release.

Numerous fragments of parathyroid hormone appear in the circulation. At first it was assumed that they were formed from endopeptidase activity in the plasma, but we now know that these fragments are largely formed within the PTH cells. The biological half-life for PTH is about

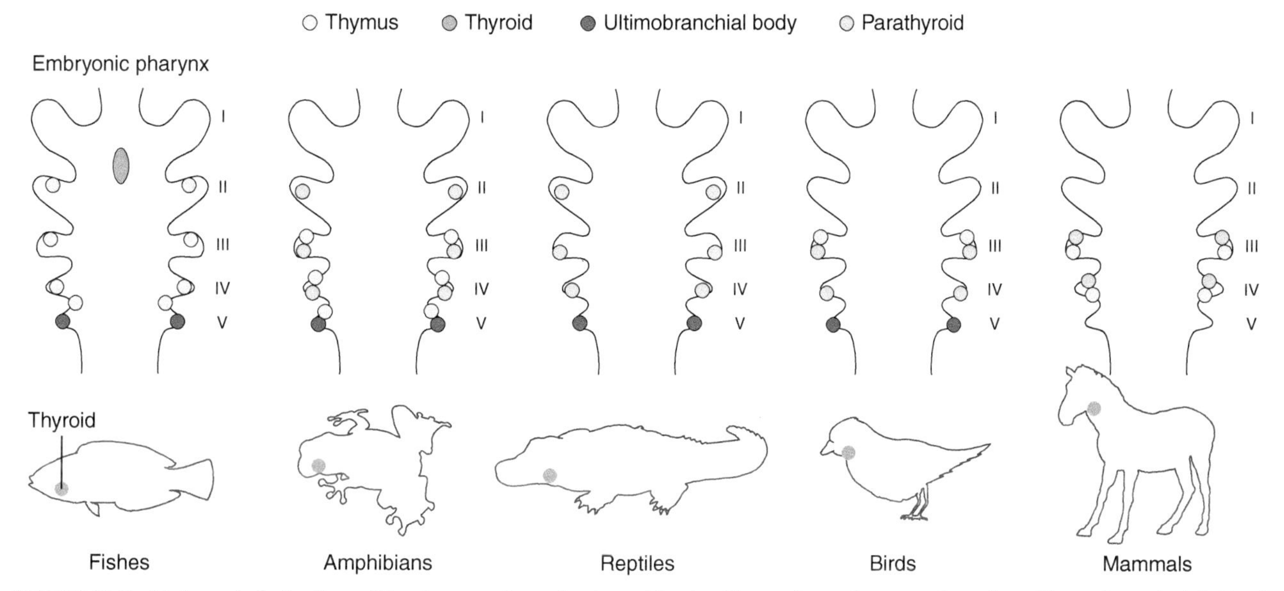

FIGURE 14-5 **Embryonic derivatives of the pharyngeal pouches in vertebrates.** The numbers refer to pouch number with anterior to the left. Pouch 1 remains only as the spiracle (selachians) or the eustachian tube. The thyroid actually develops from the pharynx floor between the first and second pouches. Parathyroids appear first in amphibians. In reptiles, the origin of the thymus (not shown) may be from pouches 2 and 3 (lizards), 3 and 4 (turtles), or 4 and 5 (snakes). In mammals, the origins of parathyroids and thymus are reversed. The ultimobranchial body (UB) is absent in mammals, and the calcitonin-secreting cells migrate to the thyroid instead of the UB.

	1	6	11	16	21	26	31
Bovine PTH	AVSEI	QFMHN	LGKHL	SSMER	VEWLR	KKLQD	VHNF–
Rat PTH	AVSEI	QLMHN	LGKHL	ASMER	VEWLR	KKLQD	VHNF–
Human PTH	SVSEI	QLMHN	LGKHL	NSMER	MQWLR	KKLQD	VHNF–
Human PTHrP	AVSEH	QLLHD	KGKSI	QDLRR	RFFLH	HHIAE	IHTA–

FIGURE 14-6 **Amino acid sequences of mammalian PTHs and PTHrP.** Only the first 34 amino acids are shown. Explanation of the amino acid abbreviations is provided in Appendix C.

20 minutes in cattle or rats. PTH is responsible for both deposition and resorption of bone as well as for reabsorption of Ca^{2+} and the excretion of excess HPO_4^{-2} via the urine. Constant PTH stimulation causes bone resorption and increases in blood Ca^{2+} and HPO_4^{-2}, whereas intermittent exposure to PTH produces bone growth. **Parathyroid hormone receptor 1 (PTHr1)** is the principle receptor for PTH in bone and kidney. PTHr1 is a GPCR type linked to production of cAMP and is found only on osteoblasts that in turn are responsible for both differentiation of osteoclasts and stimulation of their bone-resorbing actions (see Figure 14-4). A second receptor for PTH (**PTHr2**) has been identified, and the phylogeny of these genes is outlined in Figure 14-7. In the kidney, PTH stimulates Ca^{2+} reabsorption and HPO_4^{-2} excretion in the distal tubule of the nephrone. PTH also stimulates production of 1,25-DHC by cells of the proximal tubule (see ahead).

2. Regulation of PTH Secretion

Earlier (Chapter 3), we emphasized the role of increased Ca^{2+} in endocrine cell secretion of hormones through exocytosis. However, unlike most secretory cells, parathyroid cells release PTH when extracellular and intracellular Ca^{2+} levels are minimal. Release of PTH normally occurs at a high rate, but secretion of active hormone is decreased by elevated plasma Ca^{2+} levels primarily through increased degradation of PTH prior to secretion (see Figure 14-8). High levels of extracellular Ca^{2+} cause the PTH cell to secrete a higher proportion of inactive fragments and less bioactive hormone.

Although the precise mechanisms for regulating PTH secretion have not been worked out completely, several cellular events are established. Low intracellular Ca^{2+} is correlated with activated **protein kinase C (PKC)**, which is a known participant in cellular regulation (see Chapter 3). High extracellular Ca^{2+} induces formation of **inositol trisphosphate (IP_3)**, which releases intracellular stores of Ca^{2+} that activate neutral proteases called **calpains** and/or activate lysosomal hydrolytic enzymes. Degradation of intracellular PTH is accomplished by the actions of these enzymes. Furthermore, elevated calpains can cause downregulation of PKC that would prevent PTH secretion. Both Ca^{2+} and 1,25-DHC have been shown to reduce PTH

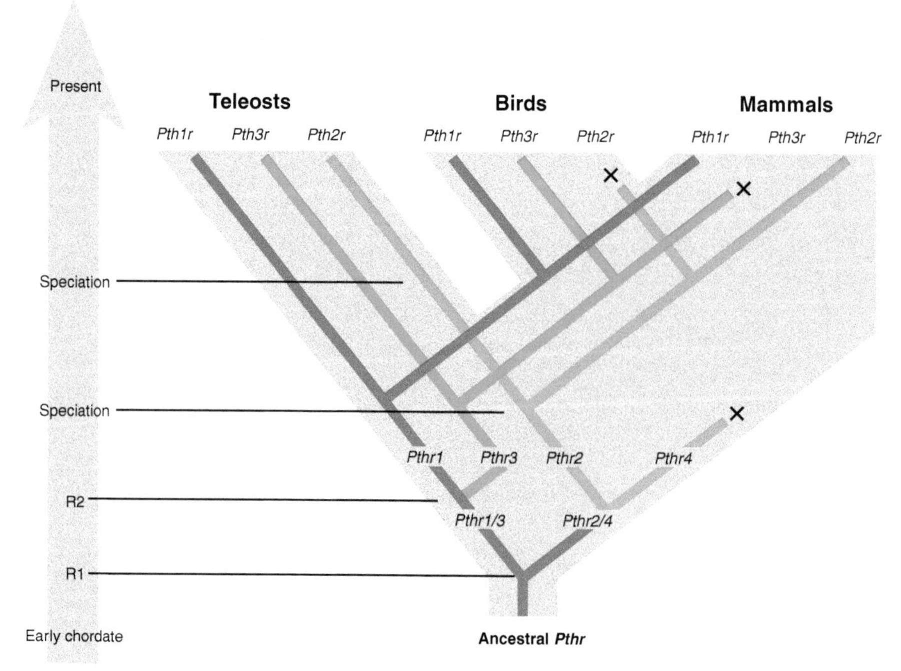

FIGURE 14-7 **A history of the parathyroid hormone receptor (*PTHr*) gene family.** The simplest explanation for this phylogenetic tree is that an ancestral *PTHr* gene gave rise to a precursor gene at the first vertebrate genome duplication (R1) to a *PTHr1/3* precursor gene that at the second vertebrate duplication (R2) gave rise to *PTHr1* (red line) and *PTH3r* (blue line). The second precursor gene resulting from R1 (*PTHr2/4*) gave rise at R2 to *PTHr2* (green line) and a fourth gene, *Pth4r* (yellow line), that somehow was lost. Initially, both fish and tetrapod lineages had three PTH receptor genes (*PTHr1*, *PTHr2*, and *PTHr3*), but later *PTHr3* was lost in mammals and *PTHr2* was lost in birds. *(Adapted with permission from Bhattacharya, P. et al.,* Journal of Endocrinology, ***211****, 187–200, 2011.)*

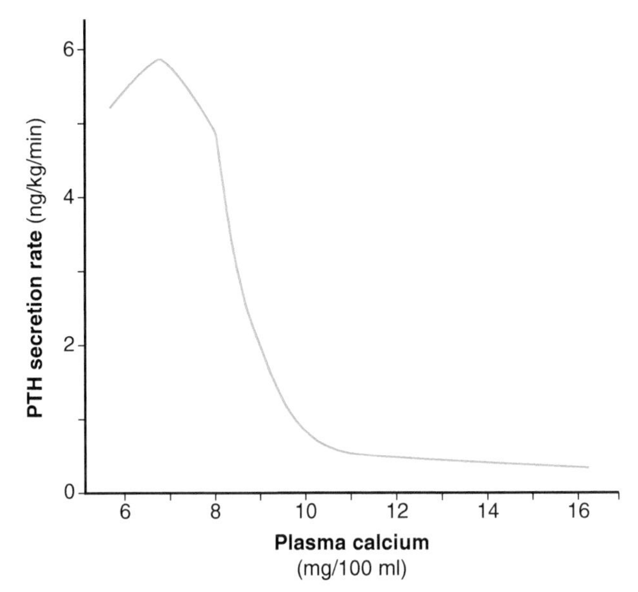

FIGURE 14-8 **Relationship of calf parathyroid secretion of PTH to experimental manipulation of plasma calcium levels.** Plasma Ca^{2+} levels above 8 mg/100 mL (about 1 n*M*) suppress PTH secretion whereas lower levels increase PTH secretion. *(Adapted with permission from Mayer, G.P. and Hurst, J.G.,* Endocrinology, ***102****, 1036–1042, 1978. © The Endocrine Society.)*

synthesis possibly through direct effects on transcription. Because PTH can elevate plasma levels of both Ca^{2+} and 1,25-DHC, their effects on PTH release can be considered feedback.

Extracellular Ca^{2+} may open calcium channels in the chief cell membrane by binding to a G-protein-linked membrane receptor system. There is also evidence to support a direct action of Ca^{2+} on membrane calcium channels. PTH release is not directly influenced by fluctuations in HPO_4^{-2} levels.

3. Parathyroidectomy Effects

In all cases, parathyroidectomy causes a reduction in plasma Ca^{2+} leading to detrimental muscular effects. As Ca^{2+} levels decrease, hyperexcitability of motor neurons and skeletal muscles occurs, resulting in twitches, spasms, and, in extreme cases, violent convulsions. This condition is known as **low-calcium-induced tetany**. If prolonged contractions (tetany) of the respiratory muscles occur, death due to asphyxiation may result. The severity of these neuromuscular effects however, differs markedly with respect to the species involved and/or to various physiological conditions. For example, exercise raises the body temperature and increases the breathing rate, causing a reduction of blood CO_2. Reduced blood CO_2 in turn alters blood pH (alkalosis) and retards ionization of Ca^{2+}. This further reduction in Ca^{2+} in a parathyroidectomized animal may precipitate a tetanic seizure. Animals on diets low in Ca^{2+} and high in HPO_4^{-2} exhibit tetany following parathyroidectomy more readily than do animals on normal diets. Hyperventilation reduces CO_2 levels, thus altering pH, and can induce tetany, especially when calcium levels are already low.

B. Parathyroid-Hormone-Related Peptide

A peptide with hypercalcemic activity in part related chemically to PTH was isolated from patients exhibiting **human hypercalcemia of malignancy (HHM)** as a result of cancer. Because of the *N*-terminal overlap with PTH (see Figure 14-6), it was named **parathyroid-hormone-related peptide (PTHrP)**. This peptide has proven to be a natural product of many cell types by a gene believed to have been derived from an ancestral gene that gave rise through duplication to both the *PTH* gene and the *PTHrP* gene (Figure 14-9). This latter gene is active in more than 20 tissues in humans, including the heart, arterial smooth muscle, lactating mammary gland, uterus, and amnion and parathyroid glands of the fetus. It also stimulates growth in several cell lines, including renal carcinoma, fibroblasts, osteoblasts, chondrocytes, and lymphocytes. PTHrP also binds to PTHr1. A third polypeptide, **tuberoinfundibular protein 39 (TIP39)**, has been discovered with similarities to PTH and PTHrP. TIP39 binds to PTHr2 that is located in the brain and is colocalized with **somatostatin (SST)** cells in the pancreatic islets. Human PTHr2 binds PTH but not PTHrP, whereas rat PTHr2 binds only TIP39. It is doubtful that TIP39 plays a role in the regulation of calcium–phosphate balance. Recently, an additional PTH-like gene that codes for a polypeptide called **PTH-L** has been reported in fishes and birds (Figure 14-10), although its physiological role is not clear.

During fetal development, PTHrP is synthesized by the amnion and accumulates in amniotic fluid. It is also a major product of the fetal parathyroid. PTHrP may modulate transfer of Ca^{2+} across the fetal–placental unit and may be responsible for the high Ca^{2+} levels observed in fetal as compared to maternal plasma.

Increased PTHrP gene activity appears in the rat lactating mammary gland within 24 hours after birth and remains elevated during suckling. Removal of the pups causes a drop within 1 to 2 hours, and PTHrP production ceases by 4 hours. Resumption of nursing reinstates lactating levels of PTHrP.

In humans, most but not all (12 of 19) breast-feeding women exhibited elevated plasma PTHrP (2 to 8 pmol/L), but PTHrP is absent from the plasma of bottle-feeding mothers (all of 16 examined). There are considerable amounts of PTHrP secreted into human and bovine milk, suggesting that it may be important for maintaining calcium homeostasis in newborns. Much lower levels of PTHrP are present in commercial, milk-based infant

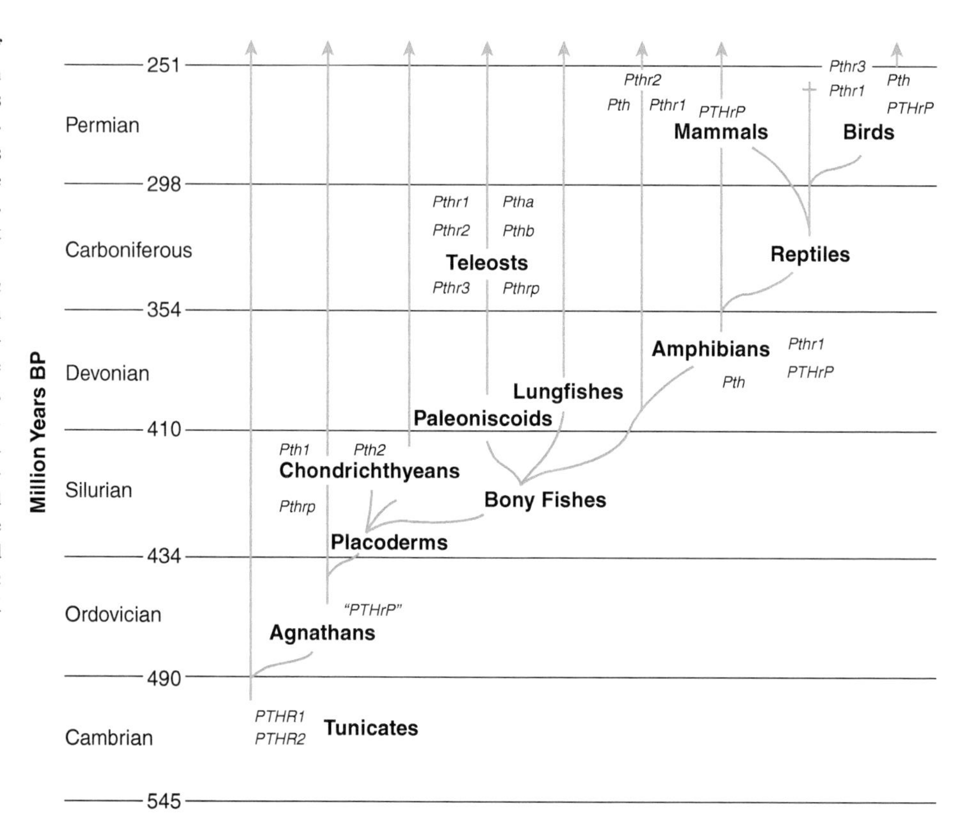

FIGURE 14-9 **A classification of chordates, including the tunicates, in relation to their phylogenetic origins and time in terms of palaeontological periods with some members of the parathyroid hormone gene family.** Parathyroid hormone genes have been identified in the elephant shark (*Callorhinchus milii*; *Pth1* and *Pth2*), the pufferfish (*Takifugu rubripes*; *Ptha* and *Pthb*), chicken, and mammal (*Pth*). Parathyroid hormone-related protein genes (*PTHrP*) have been identified in elephant shark, pufferfish, chicken, and mammals. *PTHr1* and *PTHr2* genes have been identified in an ascidian (*Ciona intestinalis*) and zebrafish (*Danio rerio*) and *PTHr1* in a frog (*Xenopus laevis*). The *PTHr3* gene has only been identified in zebrafish. *(Adapted with permission from Danks, J.A. et al.,* General and Comparative Endocrinology, ***170****, 79–91, 2011; Bhattacharya, P. et al.,* Journal of Endocrinology, ***211****, 187–200, 2011.)*

formulas, and PTHrP is absent from soy-based formulas (see Table 14-3).

PTHrP causes relaxation of both vascular (i.e., hypotensive) and nonvascular smooth muscle. During pregnancy, PTHrP may be an important relaxant of smooth muscle, and it decreases in the uterus prior to birth. PTHrP also relaxes smooth muscle of vascular tissue, urinary bladder, and the stomach, although a physiological role for PTHrP in these tissues is not established.

C. The C Cells and Calcitonin (CT)

The **C cells** (parafollicular) cells of the mammalian thyroid gland have been identified as the source of CT in most mammals; however, the actual origin of these cells is unclear. The most generally accepted origin is that proposed by Godwin, that these cells originated from the ultimobranchial body that develops from the sixth pharyngeal pouch (endoderm). These ultimobranchial cells become incorporated into the thyroid gland of mammals just as parathyroids frequently do. The ultimobranchial body remains as a distinct separate structure in other vertebrate groups. Like the chief cells of the parathyroids, the C cells of the mammalian thyroid exhibit APUD characteristics. Some elegant studies employing cellular chimeras of chicken and quail embryonic tissues have confirmed that the origin of the CT-secreting cells of the ultimobranchial body of birds is neural crest cells. These observations support an ultimobranchial origin for mammalian C cells. SST is co-localized in mammalian C cells that ultrastructurally resemble the SST-secreting D cells found in the intestinal lining. It has been suggested that SST released along with CT may play an autocrine or paracrine role, but what that role is has not been determined.

D. Calcitonin (CT)

Production of a potent mammalian hypocalcemic factor by cells of the parathyroid gland was first reported by Copp and associates, who named it calcitonin. Other investigators provided evidence that the source of this hypocalcemic factor in some mammals was the C cells of the thyroid gland and suggested the alternative name of **thyrocalcitonin** to reflect its origin. It was soon discovered that the ultimobranchial glands of sharks and chickens contained CT and that C cells of the thyroid were of ultimobranchial origin. Hence, thyrocalcitonin should be abandoned in favor of CT.

Mammalian CT is a small, single-chain peptide of 32 amino acids with a disulfide bond linking residues 1 and 7.

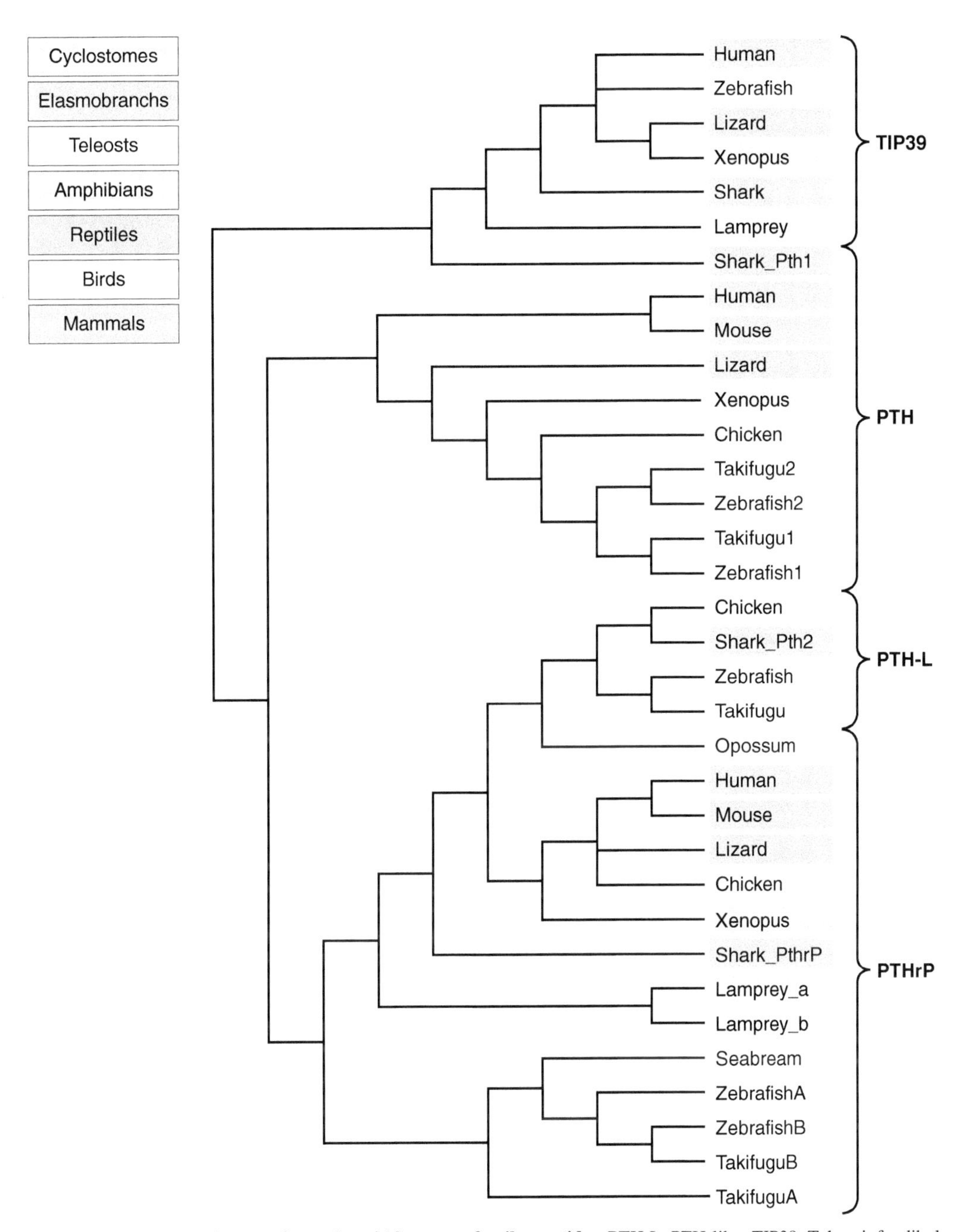

FIGURE 14-10 **Unrooted phylogenetic tree of parathyroid hormone family peptides.** PTH-L, PTH-like; TIP39, Tuberoinfundibular peptide of 39 residues. *(Adapted with permission from Pinheiro, P.L.C. et al.,* BMC Evolutionary Biology, ***12****, 110, 2012.)*

There are no active fragments of CT, and the entire molecule is necessary for biological activity. The amino acid sequence for some CT molecules is provided in Figure 14-11. Procalcitonin consists of 136 amino acids. Circulating levels of CT in humans are reported to vary from 5 to 100 pg/mL. CT has a short biological half-life (5 minutes) related to the presence of peptidases in blood and the fact that all fragments of CT are inactive.

The CT gene product may be processed through several mRNA pathways. This results in formation of several peptides known as **calcitonin gene-related peptides (CGRPs)**. CGRPs have no CT-like activity but may play other physiological roles (see Chapter 12).

E. 1,25-Dihydroxycholecalciferol (1,25-DHC)

Cholecalciferol or **vitamin D_3** is found throughout the animal and plant kingdoms, but the **vitamin D receptor (VDR)** is found only in vertebrates. It is essential for

TABLE 14.3 Approximate Levels of PTHrP in Milk and Infant Formulas[a]

Type of milk/formula	PTHrP (ng eq/mL)
Fresh milk	
Human	50
Bovine	96
Commercial milk	
Whole milk	81
Nonfat milk	118
Buttermilk, chocolate milk	5–7
Milk-based formula	
Six brand names	1–30
Soy-based formula	Undetectable

[a]*Using these contrived units, plasma PTHrP in an humoral hypercalcemia of malignancy (HHM) patient would be 0.08 ng eq/L. This means that PTHrP in breast milk is about 1000× more concentrated than in plasma and about 10,000× greater than normal plasma PTHrP.*
Data reprinted with permission from de Papp, A.E. and Stewort, A.F., *Trends in Endocrinology & Metabolism*, **4**, 181–187,1993.

calcium regulation, especially in terrestrial vertebrates, but has many additional functions as well.

1. Synthesis of 1,25-DHC

The common name "vitamin D" often is applied loosely to many of the intermediates between 7-cholesterol and the active form, 1,25-DHC. In the presence of ultraviolet light, 7-cholesterol is modified in the skin of terrestrial vertebrates to cholecalciferol which is no longer a steroid (see Figure 14-12), hence the designation of vitamin D as the "sunshine vitamin." Penetration of the skin by certain wavelengths of ultraviolet (UV) light is strongly dependent on the angle of incidence of sunlight (influenced by time of day, season, and latitude), as well as by cloud cover, air pollution, and the use of sun screens. Furthermore, window glass absorbs UV light effectively and reduces cholecalciferol synthesis. The brown skin pigment melanin absorbs UV light and decreases cholecalciferol synthesis. Ironically, prolonged exposure of skin to sunlight not only increases melanin synthesis but also causes cholesterol to be converted primarily to inert steroids known as **lumisterol** and **tachysterol** instead of to cholecalciferol. The lack of melanin pigment that evolved in northern Europeans may have served to increase UV penetration and hence enhanced cholecalciferol synthesis as an adaptation for living in northern latitudes.

Cholecalciferol is released from the skin into the blood from which it is removed by the liver, where the P450 mitochondrial enzyme **25-hydroxylase** (CYP2R1) converts it into **25-hydroxycholecalciferol (25-HC**; = **calcidiol**) by the addition of one –OH group. 25-HC is returned to the blood, where it is carried by binding proteins to the kidney, where the complex is filtered into the nephrons. Circulating levels of 25-HC reported for humans are between 7 and 42 ng/ml with an extended biological half-life of about 15 days. Next, cells in the proximal tubules of the kidney nephrons remove 25-HC

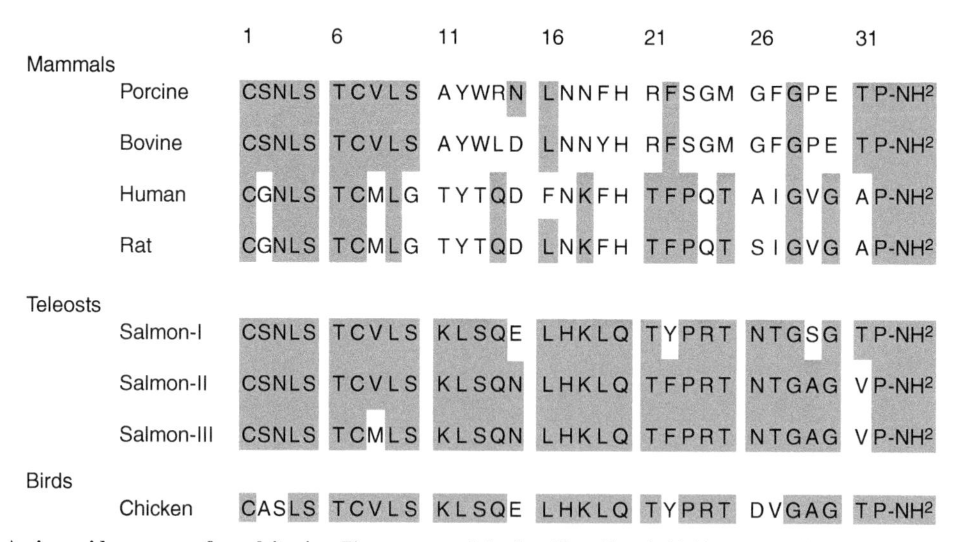

FIGURE 14-11 **Amino acid sequences for calcitonins.** The sequence of the first 10 residues is highly conserved when comparing teleosts, birds, and mammals. The presence of valine (V) at position 8 increases the biological activity of the molecule four to five times. Hence, salmon-I is more potent in humans than is human CT. Note also that human and rat CTs are more like each other than like bovine or porcine, which are both similar. Explanation of the amino acid abbreviations is provided in Appendix C. *(Adapted with permission from Matsumoto, A. and Ishii, S., "Atlas of Endocrine Organs," Springer-Verlag, Berlin, 1992.)*

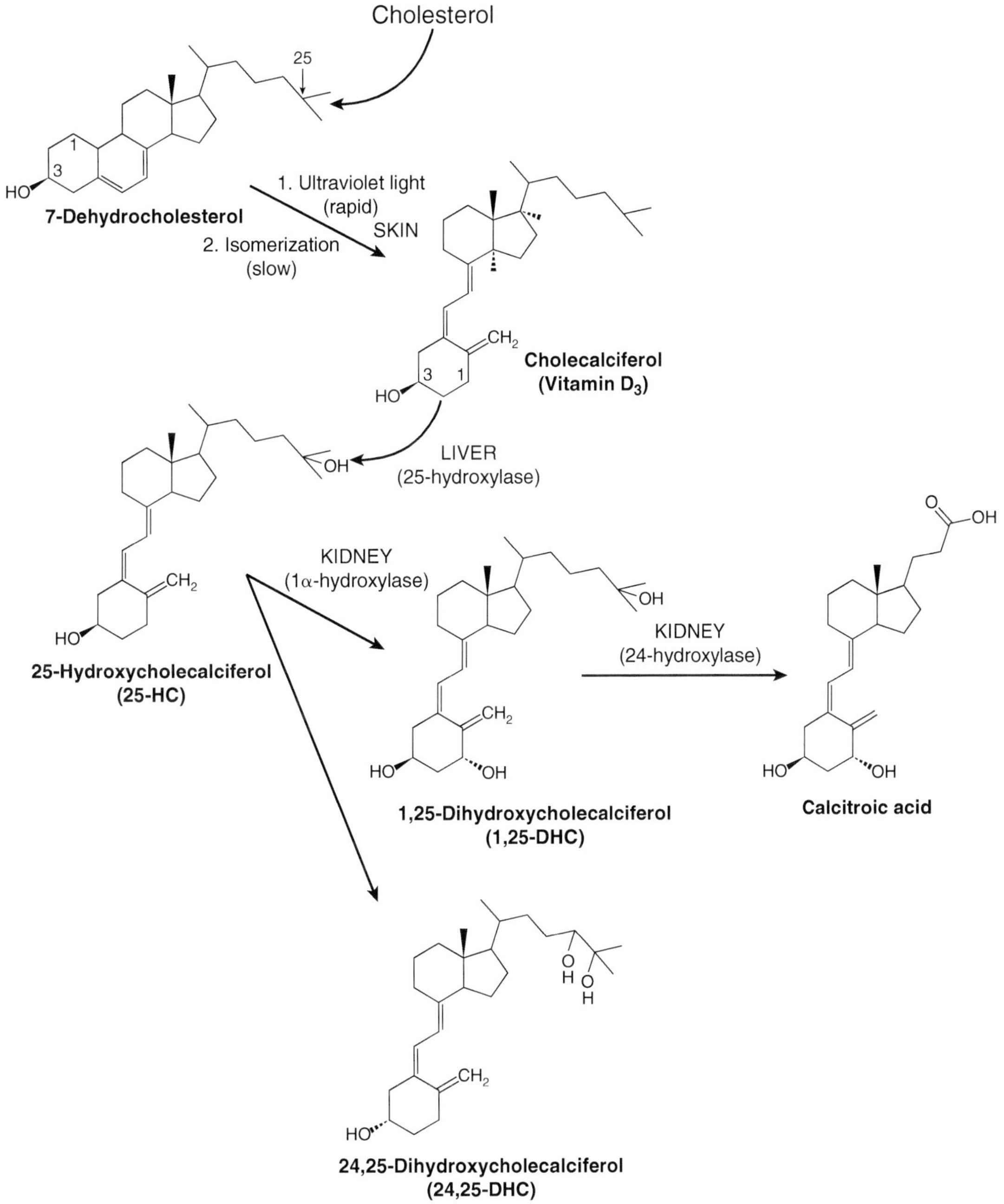

FIGURE 14-12 **Synthesis of 1,25-dihydroxycholecalciferol (1,25-DHC).** In terrestrial vertebrates, 7-dehydrocholesterol is changed sequentially by the skin, liver, and kidney. Although some 24,25-DHC is also produced, it exhibits very weak binding to the vitamin D receptor and is probably not very important. Note the rearrangement of the A ring as a consequence of opening up the B ring.

from the glomerular filtrate with the aid of **megalin**, a low-density lipoprotein receptor. Another P450 enzyme **1α-hydroxylase** (CYP27B1) adds an additional −OH to produce **1,25-dihydroxycholecalciferol (1,25-DHC;** = **calcitriol)** and releases it into the blood. The level of 1α-hydroxylase in the kidney is regulated by PTH acting through the PTHr1.

2. Actions of 1,25-DHC on Calcium Homeostasis

Blood levels of 20 to 50 pg/mL 1,25-DHC are reported for humans. These low levels are largely a product of its biological half-life of only 3 hours. Circulating 1,25-DHC enters certain cells of the intestinal mucosa, where it binds to VDRs. Following dimerization of VDR to the

retinoid X receptor (**RXR**), genes are activated that stimulate synthesis of special proteins responsible for calcium uptake from the intestinal contents and its transport into the blood. Differentiation of osteoblasts and osteoclasts increases and some of their functions in bone tissue also are stimulated by 1,25-DHC. Additionally, 1,25-DHC induces formation of the P450 enzyme **24-hydroxylase** (CYP24A1), which is responsible for degrading 1,25-DHC to its inactive metabolite, **calcitroic acid** (see Figure 14-12). Downregulation of 1α-hydroxylase also occurs as a consequence of elevated 1,25-DHC.

3. Other Actions of 1,25-DHC

In addition to its central role in calcium metabolism, 1,25-DHC has many other important actions in vertebrates (see Figure 14-13). 1,25-DHC is important in normal development of the nervous system, and 1,25-DHC deficiency has been linked to disorders such as autism, schizophrenia, and attention deficit hyperactivity disorder. Furthermore, deficiencies in 1,25-DHC altering immune function have been associated with multiple sclerosis, rheumatoid arthritis, inflammatory bowel disease, and type-1 diabetes. Stimulation of VDRs can inhibit cancer cell proliferation and decrease the metastatic potential of cancer cells related to breast, colon, and prostate cancer. Additionally, the renin gene in the kidney (see Chapter 8) has a VDR response element in its promoter region, and when occupied the renin gene is repressed.

During pregnancy, the placenta also will synthesize 1,25-DHC, which augments uptake of dietary calcium by the mother. This helps ensure an adequate supply of calcium

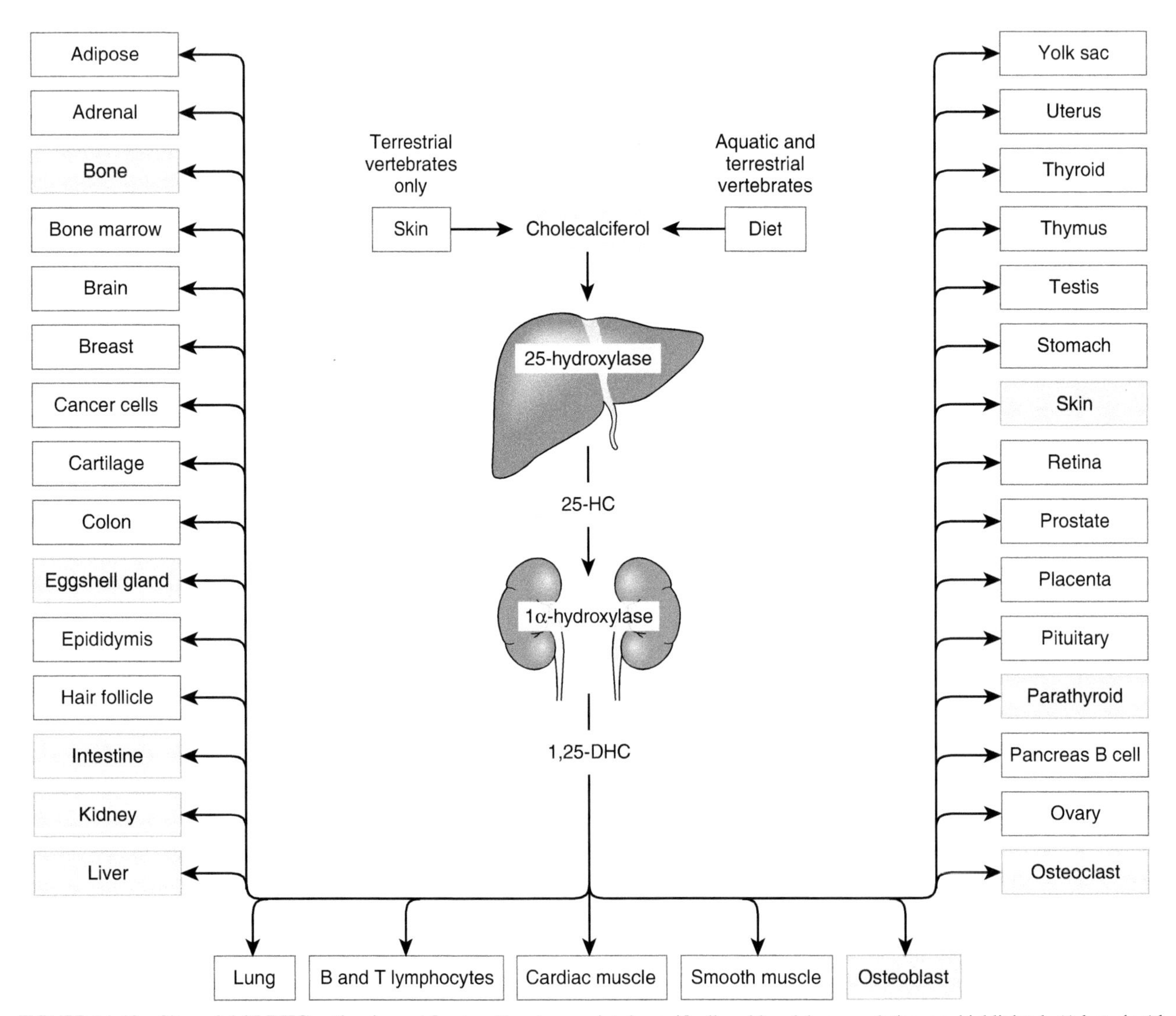

FIGURE 14-13 **Sites of 1,25-DHC action in vertebrates.** Targets associated specifically with calcium regulation are highlighted. *(Adapted with permission from LeBlanc, G.A. et al., "Detailed Review Paper: State of the Science on Novel* In Vitro *and* In Vivo *Screening and Testing Methods and Endpoints for Evaluating Endocrine Disruptors," prepared by RTI International for U.S. Environmental Protection Agency, Washington, D.C., 2012.)*

necessary for fetal growth. In cases of inadequate dietary calcium, the needs of the fetus will be met by resorption of maternal calcium stores (i.e., bones and teeth).

D. Interactions of PTH, CT, and 1,25-DHC

The major daily disturbance to calcium homeostasis is the influx of Ca^{2+} following ingestion of a meal. Periods of rapid growth cause an additional demand for dietary Ca^{2+} as do pregnancy and lactation. The small intestine, kidney, and bone are the primary sites where these regulatory hormones (PTH, CT, and 1,25-DHC) produce their actions during times of calcium excess or deficiency to maintain calcium homeostasis (see Figure 14-1).

1. Calcium and Phosphate Regulation in Bone

Minute-to-minute regulation of plasma Ca^{2+} and indirectly of plasma HPO_4^{-2} may reside in the activity of **osteocytes** embedded permanently in the bone matrix. PTH may cause **osteocytic osteolysis** (one form of bone resorption) through direct action on the osteocytes. This effect may be dependent on an interaction between PTH and 1,25-DHC. In contrast, major disturbances such as occur during growth, pregnancy, or lactation are associated with the interactions of PTH and CT on osteoclasts and osteoblasts.

PTH appears to stimulate osteoclast activity, and chronic elevation of PTH is correlated with an increase in the number of osteoclasts; however, these effects are indirect because the osteoclast has no receptors for PTH, although it is well endowed with CT receptors. In fact, it is the osteoblast that is the target for PTH. In addition to PTH, both PTHrP and 1,25-DHC also promote osteoclast formation. Differentiation of osteoclasts is then controlled by the physical interaction (cell-to-cell contact) between osteoblasts and osteoclast progenitor cells. Several paracrine regulators, such as **interleukin-6** and **interleukin-11**, are necessary for the differentiation and proliferation of osteoclasts (Figure 14-4). Hydrolytic lysosomal enzymes (e.g., cathepsin K) produced by osteoclasts destroy both mineral and bone matrix components of bone and are responsible for bone resorption and release of Ca^{2+} and HPO_4^{-2} to the plasma.

Another hypothesis concerning PTH action on bone resorption is focused on PTH effects on the bone-forming osteoblasts. According to this hypothesis, the osteoblasts possess a pumping mechanism for Ca^{2+} transport. PTH, through PTHr1 and subsequent activation of adenylyl cyclase and cAMP formation, increases the flux of Ca^{2+} into the osteoblast from the bone surface and out of the osteoblast on the blood side of the endosteal membrane. There is evidence to suggest that, present in intermittent pulses, PTH stimulates bone formation rather than bone resorption through an increase in osteoblast activity whereas constant infusion of PTH causes bone resorption.

Although CT may influence osteoblast functions to a limited extent, its major action appears to be a direct inhibition of the osteoclast (Figure 14-4). This role for CT is especially important during pregnancy, when CT protects the maternal skeleton while directing dietary Ca^{2+} directly to the fetus.

2. Regulation of Calcium Uptake in the Intestine

Calcium uptake by the intestinal mucosal cell is dependent upon a **calcium-binding protein** within these cells that is linked to a calcium-activated ATPase. Calcium is actively absorbed by this binding protein—ATPase complex at the mucosal surface that is in contact with the lumen of the gut. Once in the cell, Ca^{2+} is transported to the opposite (serosal) border of the cell where Ca^{2+} diffuses into the interstitial fluid and then into the blood capillaries. The negatively charged HPO_4^{-2} passively follow the movements of Ca^{2+}. Synthesis of the Ca^{2+}-binding protein and Ca^{2+}-activated ATPase is stimulated by 1,25-DHC.

No role has been hypothesized for CT in this intestinal mechanism for Ca^{2+} uptake. Following the influx of Ca^{2+} ions, however, CT may be released in response to the increase in plasma Ca^{2+} but it may also be released by gastrin, a gastrointestinal hormone released from the gastric mucosa during the early phase of digestion of a meal (see Chapter 12). This increased level of CT inhibits the action of bone osteoclasts and favors addition of the absorbed dietary Ca^{2+} to the bone matrix that can be stimulated by PTH.

CT becomes indispensable during pregnancy and lactation with respect to Ca^{2+} mobilization. Its role is apparently to protect the maternal skeleton from excessive destruction in meeting the calcium requirements of the fetus or neonate. The relative importance of CT in different species varies markedly according to the precise demands for maternal calcium.

3. Calcium and Phosphate Regulation in Kidney

The stimulation of calcium reabsorption by the kidney may be the most important physiological action of PTH, although some investigators believe the increased excretion of HPO_4^{-2} is equally or possibly more important. PTH also increases the enzymatic formation of 1,25-DHC in kidney and thus may enhance intestinal calcium absorption indirectly. Elevated 1,25-DHC feeds back and inhibits PTH secretion. Estrogens and **prolactin (PRL)** also enhance formation of 1,25-DHC, and these actions may be essential in pregnant and lactating mammals, respectively, for conserving Ca^{2+}. Although CT antagonizes the action of PTH on bone resorption, it apparently does not influence renal processes in normal animals when present at physiological levels.

TABLE 14-4 Hormones that Influence Calcium Metabolism in Mammals

Hormone	Action
Parathyroid hormone (PTH)	Stimulates bone resorption, renal calcium reabsorption, and synthesis of 1,25-DHC
Calcitonin (CT)	Antagonizes action of PTH on bone
1,25-Dihydroxycholecalciferol (1,25-DHC)	Facilitates calcium absorption from intestine
Growth hormone (GH)	Stimulates cartilage and bone growth
Thyroid hormones	Permissive effect on GH secretion and action
Insulin-like growth factors (IGFs)	Mediators of GH action on bone
Estrogens/androgens	Cause closure of epiphysial plate, blocking further long bone growth; protect adult skeleton from resorption; may reduce PTH release and prevent hypercalcemia
Glucocorticoids	High levels stimulate PTH release, causing increased bone resorption and resultant hypercalcemia

4. Other Hormones and Calcium–Phosphate Homeostasis

Estrogens, androgens, glucocorticoids, and thyroid hormones have direct and indirect effects on mineral homeostasis (see Table 14-4). In addition, GH has indirect effects through production of **insulin-like growth factors (IGFs)**. Although the effects of estrogens and androgens bring about cessation of long bone growth at puberty, these hormones also have important stimulatory effects on osteoblast activity, especially in adults. Estrogens not only protect the skeleton from resorption but also enhance the reabsorption of Ca^{2+} by the kidney and increase the production of 1,25-DHC that aids dietary uptake of Ca^{2+}. Androgens together with estrogens play a similar role in males and are responsible for the greater mass of the male skeleton. The decrease in gonadal steroids with menopause or adrenopause (see Chapter 10) has serious implications for calcium homeostasis in aging adults.

In addition to their ability to stimulate PTH secretion, chronic excesses of glucocorticoids, such as occur during prolonged stress, can reduce intestinal uptake and kidney reabsorption of Ca^{2+}. This could result in a significant calcium loss and could lead to low-calcium-induced tetany. Estrogens are hypothesized to alter the calcium setpoint in chief cells so that greater reduction in plasma Ca^{2+} is required to elicit PTH release. This action might explain, at least in part, the effects of estrogen withdrawal on loss of skeletal calcium in females during and after menopause. GH causes an increase in secretion of IGFs by the liver and stimulates IGF synthesis in bone. IGF-I stimulates bone proliferation as well as collagen synthesis by osteoblasts.

Hypothyroidism delays bone growth probably indirectly through its adverse effects on GH secretion and action. Delayed ossification of cartilage also is observed in hypothyroidism. In the hyperthyroid animal, excessive thyroid hormones augment bone resorption and can lead to weakening of the skeleton (see ahead).

IV. MAJOR CLINICAL DISORDERS ASSOCIATED WITH CALCIUM METABOLISM

In general, plasma Ca^{2+} cannot vary greatly or death will result; consequently, it is not easy to diagnose parathyroid abnormalities by examining plasma levels of Ca^{2+}. **Hypercalcemia** (elevation of plasma calcium) is not excessive but often is accompanied by a reduction in bone density and possibly increased Ca^{2+} in the urine. Similarly, **hypocalcemia** may show few overt symptoms.

A. Hypercalcemia

Excessive plasma Ca^{2+} (greater than 10 mg/dL) can result from a variety of causes. **Primary hypercalcemia** is characterized by chronically elevated levels of PTH. The most common cause is a single parathyroid adenoma (90% of cases). Carcinomas of the parathyroids are very rare and may account for less than 1% of primary hypercalcemic cases. Primary hypercalcemia is difficult to diagnose, since approximately one-third of these cases are without overt symptoms, and the remainder exhibit rather generalized and nonspecific symptoms such as weakness, nausea, and anorexia. Serum Ca^{2+} is moderately elevated (10.2 to 11.0 mg/dL) and is usually accompanied by lowered HPO_4^{-2} levels. Often, elevated serum levels of Ca^{2+} are not seen (i.e., there is no hypercalcemia per se) because the kidney compensates with increased calcium excretion or **hypercalcuria**. Among the many causative factors that can induce **secondary hypercalcemia** are carcinomas that spontaneously secrete PTHrP leading to HHM. Breast carcinomas also may produce vitamin D-like sterols that increase Ca^{2+} absorption. Chronic immobilization of an experimental animal or of humans can bring about extensive bone resorption and cause hypercalcemia.

B. Hypocalcemia

There are many different conditions that can lead to hypocalcemia. In most cases, there is a reduction in PTH secretion. This may be caused by abnormal development of the parathyroid glands, accidental damage, or removal by surgery. Hypomagnesemia impairs PTH secretion and indirectly can cause hypocalcemia. In some cases (e.g., pseudohypoparathyroidism), the target organs do not respond to PTH and in others an abnormal PTH is secreted that will not activate tissue receptors. Reductions in 1,25-DHC due to failure to convert 25-HC or to synthesize cholecalciferol also lead to hypocalcemia. Most cases are without serious overt symptoms. Tetany may be inducible under calcium stress but otherwise is absent.

C. Osteoporosis

This disease is characterized by decalcification and loss of bone matrix from the skeleton, resulting in shrinkage, distortion, and increased brittleness of the bones. Milder decalcification is termed **osteopenia** but may be an early signal of developing **osteoporosis**. More than 10 million Americans already have osteoporosis. Bones that are extensively decalcified become subject to easy fracture as a result of falls, blows, or lifting (stress fractures). Osteoporisis accounts for 70% of bone fractures in adults over 45 years of age. Approximately 1.3 million bone fractures that occur each year in the United States are attributed to osteoporosis. Risk factors for osteoporosis include age, sex, and race (see Table 14-5). Osteoporosis accompanies aging and is most common among postmenopausal women. It is four times more common in women than men largely as a consequence of women having smaller bone calcium reserves and the fact that men do not live as long as women. Maximal adult bone mass is achieved in women at about age 35 after which calcium losses exceed calcium gains. Beginning at menopause, there is a gradual reduction in estrogen levels that causes a disproportionate decrease in bone mass by allowing increased bone resorption to take place. Weight-bearing exercise and careful attention to dietary calcium and 1,25-DHC levels are important for postpubertal, premenopausal women to ensure maximal bone density prior to onset of menopause. Curiously, adult men exhibit estrogen levels comparable to those of postmenopausal women yet do not develop osteoporosis until much later in life than females. It is suggested that the higher androgen levels in males provide ready substrate for local conversion by aromatase (CYP19) to estrogens by bone cells that can in turn stimulate calcium deposition. However, elevated testosterone also is considered a factor in the shorter lifespan of men.

Type I osteoporosis may occur in women soon after the onset of menopause and is characterized by vertebral crush fractures or fracture of the arm just above the wrist. It is attributed to the marked reduction in estrogen levels accompanying menopause. **Type II osteoporosis** appears later in life and can affect both men and women. It is mainly a consequence of decreased ability to absorb sufficient dietary calcium with advancing age.

Estrogen replacement (women only), calcium supplements (usually fortified with extra cholecalciferol or a related compound to facilitate production of 1,25-DHC and hence increase intestinal uptake of calcium), and weight-bearing exercise are commonly prescribed. Frequently, a synthetic estrogen (e.g., ethinylestradiol) is given to women with a low dose of a synthetic progestogen (e.g., medroxyprogesterone acetate) that prevents uterine

TABLE 14-5 Risk Factors for Osteoporosis-Related Fracture

Risk Factors
Biologic Risk Factors
Low bone mass
Advancing age
Personal history of osteoporosis or fracture as an adult
Parental history of fracture
Family history of bone disease
Race (caucasian or Asian postmenopausal women)
Small, thin frame
Hypogonadism (premature estrogen deficiency in women <45 years of age; low testosterone levels in men)
Family history of bone disease
Weight loss of more than 1% per year in the elderly
Late onset of sexual development
Height loss or progressive spinal curvature
Lifestyle and Dietary Risk Factors
Low calcium rate
Vitamin D deficiency
High salt intake
Alcohol (≥3 drinks/day)
Inadequate physical activity
Low body mass index
Smoking
Weaker evidence:
Excess vitamin A intake
High caffeine intake
Immobilization

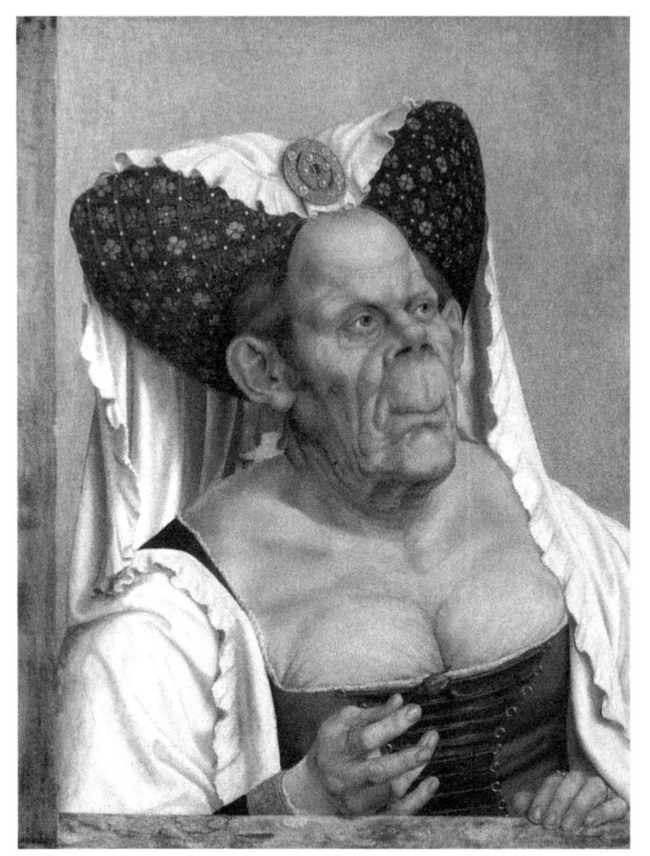

FIGURE 14-14 **"The Ugly Duchess."** This painting originally was thought to be done by Leonardo Da Vinci but is now attributed to the painter Quentin Matsys in 1513. The subject of the painting is Margaret, Countess of Tyrol, who is suspected to have had a form of Paget's disease that led to severely enlarged bones. The original painting resides in the National Gallery in London and is supposed to have been a source for some of the original illustrations in Lewis Carroll's "Alice's Adventures in Wonderland."

bleeding and reduces uterine hyperplasia and the risk of uterine cancer. When there is a family history of breast cancer, estrogen therapy may increase the risk of breast cancer. Even with the increased risk, it appears that the mortality rate for breast cancer is lower in women receiving estrogen. This apparent anomaly may be a result of earlier detection and hence higher cure rates in these women in spite of increased incidence. Although early studies suggested these therapeutic efforts only prevent or slow additional skeletal losses, recent studies show improvement in mineral density of bones even if estrogen replacement therapy is not begun until after age 70.

Hormonal manipulations are ineffective without close attention to diet and exercise. Emphasis should be placed on weight-bearing activities (e.g., walking). Swimming, while an excellent aerobic exercise, does not stimulate bone deposition. Weightlessness, like immobilization, accelerates bone resorption. Bone density in athletes is related directly to the type of exercise, with weight lifters having

TABLE 14-6 **Comparison of Relative Activity of Purified Human, Salmon, and Porcine Calcitonin as Determined by Standard Bioassay**

Source	Activity in MRC[a] units/mg hormone
Porcine	200
Human	120
Salmon	5000

[a] *MRC, Medical Research Council of England.*

TABLE 14-7 **Biological Half-Life for Purified Vertebrate Calcitonins When Incubated in Either Mammalian or Avian Blood Plasma**

Source	Half-life (min)
Porcine	2
Salmon	20
Chicken	90

the densest bone, swimmers the least dense bone, and runners intermediate. The racquet arms of men and women tennis players are 35% and 28% more dense than the other arm, respectively. Even mild activity for nursing home patients averaging 82 years of age prevented further bone loss and resulted in bone buildup over a 36-month period.

4. Paget's Disease

This disorder is caused by increased osteoclast activity resulting in accelerated bone resorption. It occurs in 3% of the population over age 40 and occurs with greatest frequency in North Americans of Western European or Mediterranean descent. About one-third of the people afflicted with Paget's disease do not exhibit any overt symptoms, but the bones become brittle as the disease progresses, leading to fractures. In some cases, abnormal remodeling of the bone can lead to severe enlargement of bones in the face and limbs (Figure 14-14). Serum levels of Ca^{2+} and HPO_4^{-2} usually are normal because the excess ions are excreted in the urine. Salmon CT has been used with some success to treat Paget's disease, because it is more effective than mammalian CT (Table 14-6), possibly due to its much longer biological half-life in mammals compared to mammalian CT (Table 14-7). However, therapeutic treatment is thwarted in part by downregulation of CT receptors on osteoclasts after a few days of treatment.

V. CALCIUM AND PHOSPHATE HOMEOSTASIS IN NON-MAMMALIAN VERTEBRATES

An attempt to discuss the evolutionary aspects of this problem is complicated by major differences, indeed a distinct dichotomy, between the fishes and the tetrapods. Some fishes lack bone (agnathans, chondrichthyeans), and most of the bony fishes possess acellular bone (no osteocytes) rather than cellular bone. Scales may provide a major physiological reserve of stored Ca^{2+} to bony fishes, and the surrounding waters (especially sea water) are an important source of Ca^{2+}. Because fishes readily accumulate Ca^{2+} from the surrounding water, regulation is focused on preventing influx or stimulating removal of excess ions whereas in terrestrial vertebrates the emphasis is on obtaining and maintaining adequate levels of Ca^{2+}.

Parathyroid glands are lacking in fishes but PTH and PTHrP genes have been described in chondrichthyeans and teleosts, suggesting that calcium regulation by these genes was a later acquisition (Figure 14-9). No PTH or PTHrP genes were found in lampreys although there is evidence for PTHrP mRNA and protein in lamprey skin. Bony fishes produce a pituitary hypercalcemic factor, **somatolactin (SL)**. PTHrP is synthesized in the pars intermedia of the teleost and amphibian pituitary and plays a role in calcium homeostasis. Additionally, teleosts have two sources of hypocalcemic factors: (1) the **corpuscles of Stannius**, embedded in the kidneys (Figure 14-15), which produce **stanniocalcin (STC**; = teleocalcin), and (2) the

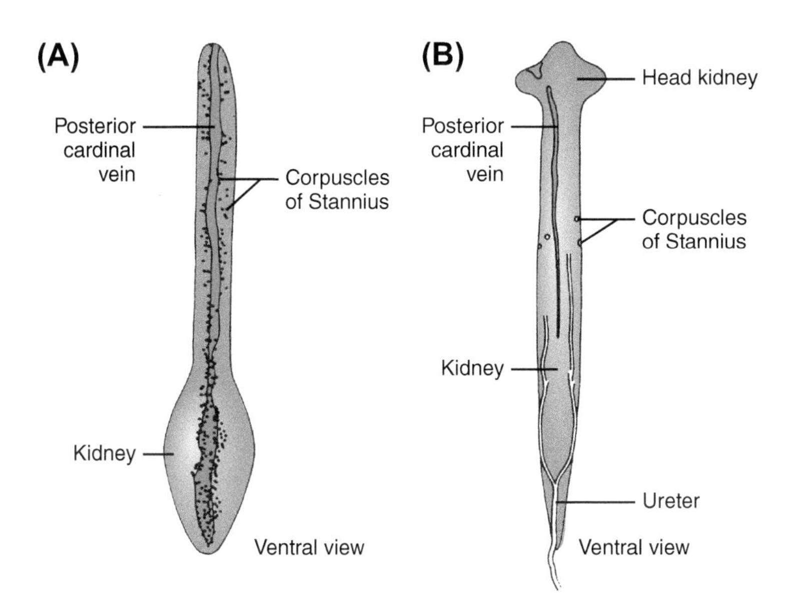

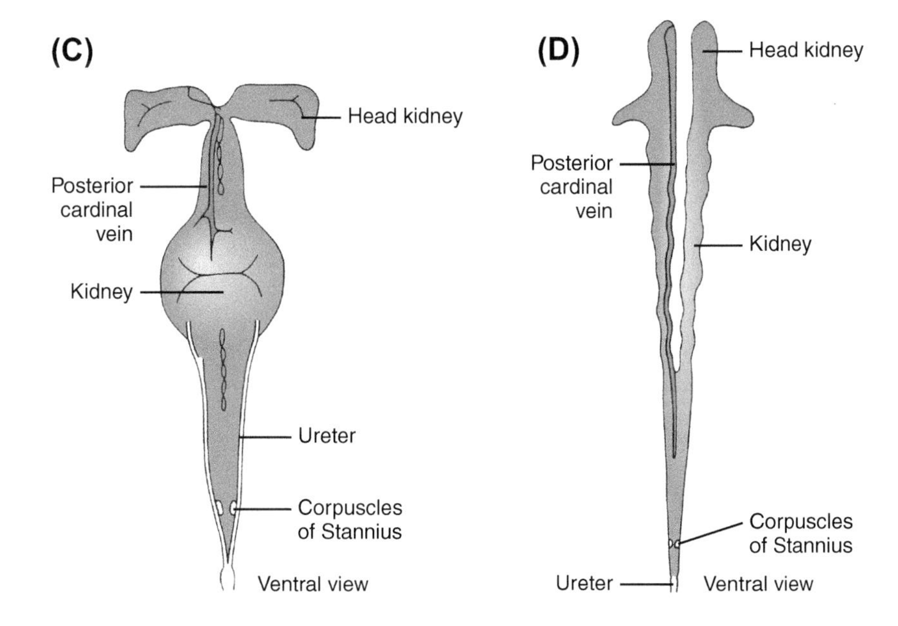

FIGURE 14-15 **Varying location of corpuscles of Stannius in selected bony fishes.** (A) The bowfin, *Amia calva*. (B) A salmonid, *Salvelinus pluvius*. (C) The Crucian carp, *Carassius carassius*. (D) The striped mullet, *Mugil cephalus*. Recall that the head kidney contains the adrenocortical cells in teleosts and is otherwise lympohid in nature. *(Adapted with permission from Matsumoto, A. and Ishii, S., "Atlas of Endocrine Organs," Springer-Verlag, Berlin, 1992.)*

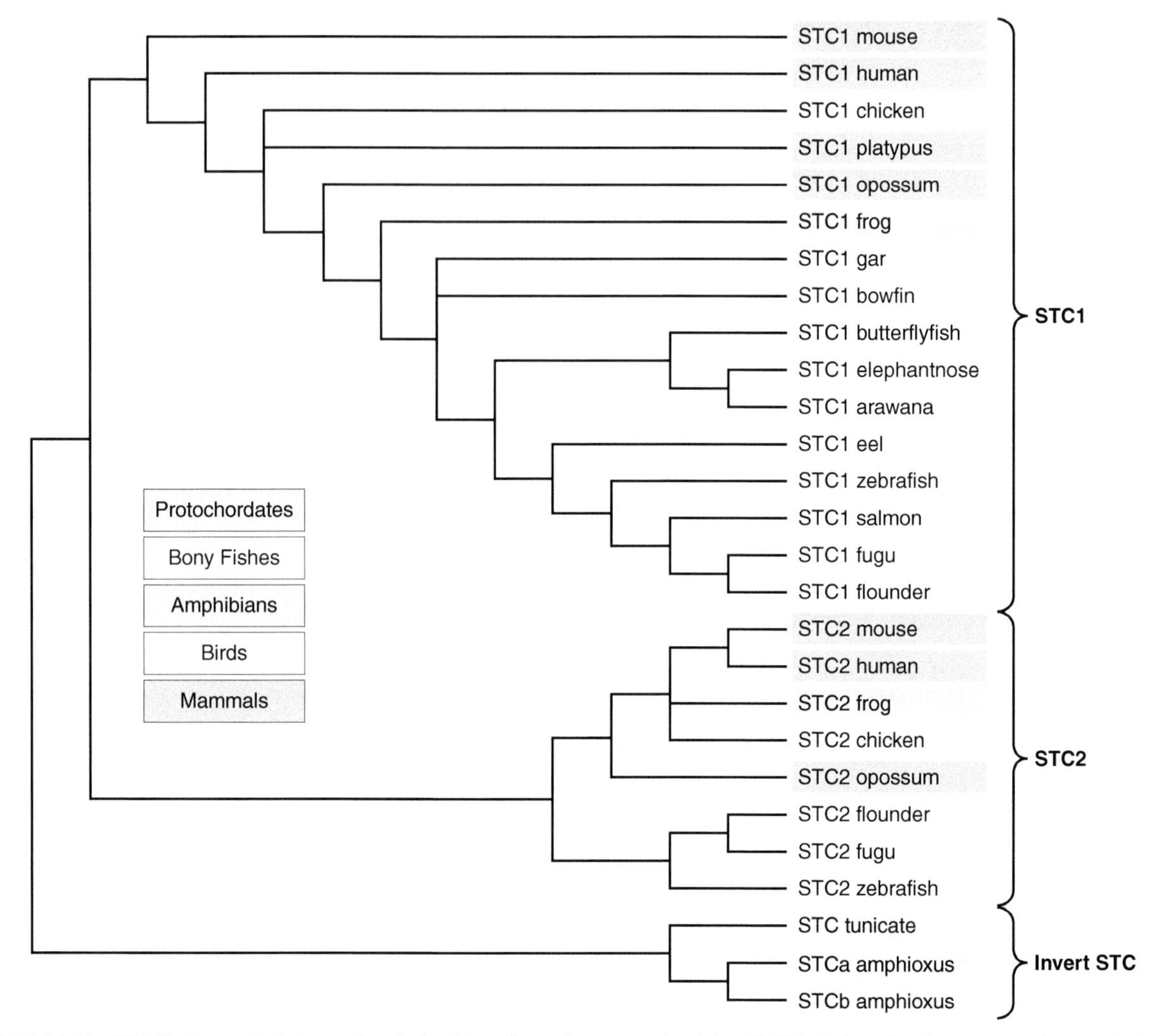

FIGURE 14-16 **Distribution and phylogenetic relationships of vertebrate stanniocalcins (STCs).** *(Adapted with permission from Roch, G.J. and Sherwood, N.M.,* Integrative and Comparative Biology, ***50****, 86–97, 2010.)*

ultimobranchial glands, which produce CT. STCs have been described in bony fishes, amphibians, birds, and mammals as well as STC orthologs in tunicates and amphioxus (Figure 14-16), but possibly perform functions other than calcium regulation in non-fishes. STCs probably are present in reptiles, but data are lacking.

Definitive parathyroid glands and PTH first appear fully differentiated in the amphibians. The comparative approach also is hampered by the lack of detailed information concerning calcium and phosphate regulation in, amphibians, and reptiles. The distribution of parathyroid and ultimobranchial glands in vertebrates is provided in Figures 14-17 and 14-18.

1,25-DHC appears to be important in regulating intestinal calcium uptake, especially in terrestrial vertebrates. VDRs are present in all vertebrates examined as well as in the tunicate *Ciona intestinalis* (Figure 14-19). Furthermore, the selectivity of VDRs for cholecalciferol metabolites is very similar from selected agnathans to mammals.

Estrogenic hormones elevate circulating calcium and phosphate levels indirectly in females of most non-mammalian species during the process of vitellogenesis associated with oocyte growth. Synthesis of calcium-binding, phosphoproteins (e.g., **vitellogenin**) by the liver is stimulated by estrogens. These proteins are released into the blood, through which they travel to the ovaries, where they are incorporated into growing oocytes. Vitellogenic proteins readily bind Ca^{2+}, indirectly decrease free plasma Ca^{2+} levels, and consequently stimulate release of more Ca^{2+} from reservoirs such as bone. The presence of vitellogenic proteins elevates total plasma levels of Ca^{2+}, although free Ca^{2+} remains about the same. The relationship between reproductive hormones and vitellogenesis is discussed in Chapter 11.

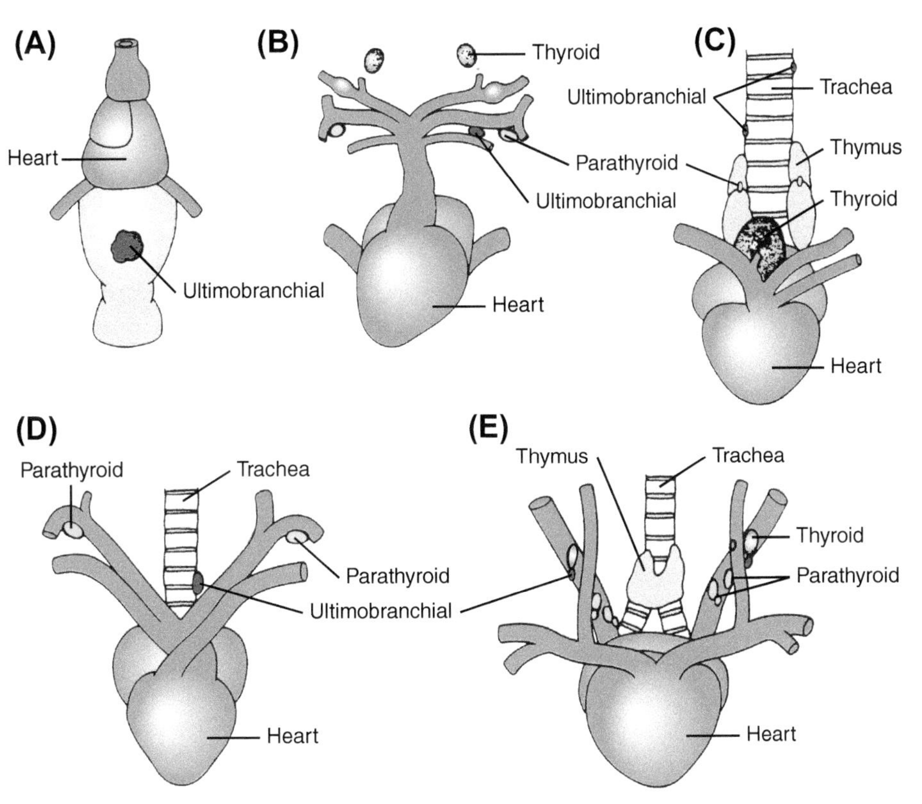

FIGURE 14-17 **Location of ultimobranchial tissue in nonmammalian vertebrates.** (A) Goldfish. (B) Newt. (C) Snake. (D) Lizard. (E) Bird. *(Adapted with permission from Matsumoto, A. and Ishii, S., "Atlas of Endocrine Organs," Springer-Verlag, Berlin, 1992.)*

A. Agnathan Fishes

Cyclostomes do not appear to have regulatory mechanisms specific for calcium and phosphate metabolism, although mammalian CT does decrease urinary excretion (Table 14-8). Whatever regulatory mechanisms are employed, they appear to be not as efficient as those found in other vertebrates as evidenced by circulating levels of calcium. Marine species exhibit an intermediate level of plasma Ca^{2+} between that of seawater and plasma of tetrapods, whereas the plasma Ca^{2+} level in freshwater lampreys is intermediate between that found in plasma of freshwater teleosts and freshwater. These observations raise some interesting questions about the roles of calcium and its regulation in the cartilaginous agnathans.

B. Chondrichthyean Fishes

Most elasmobranchs and other chondrichthyeans live in seawater, where calcium availability is not as serious a problem as it is for freshwater species. Although these fishes lack true bone tissue, calcium salts are added to their cartilaginous skeletons for additional strength, especially in larger species. Elasmobranch ultimobranchial glands contain a potent hypocalcemic factor when assayed in mammals, but extracts of shark ultimobranchial glands, salmon CT, and porcine CT are all ineffective in sharks. Curiously, estrogens do not produce any appreciable alteration in plasma calcium as they do in bony fishes and non-mammalian tetrapods. No experimental evidence has been reported for an adenohypophysial factor that would directly affect plasma calcium. Studies of sharks that are either freshwater residents or that penetrate long distances up major rivers, such as the Mississippi, might provide some interesting insight into calcium homeostasis in these fishes.

C. Bony Fishes: Teleosts

Calcium regulation appears to be under the control of a hypercalcemic adenohypophysial factor and hypocalcemic factors associated with the corpuscles of Stannius and the ultimobranchial bodies. Hypophysectomy of either *Anguilla* (which has cellular bone) or *Fundulus* (acellular bone) results in a decrease in serum calcium and induction of tetany. Teleost pituitaries also produce SL that can influence Ca^{2+} levels (see Chapter 5). In addition, PTHrP has been demonstrated in the brain, pituitary, and blood of several species. In the sea bream (*Sparus auratus*), two populations of pituitary cells are

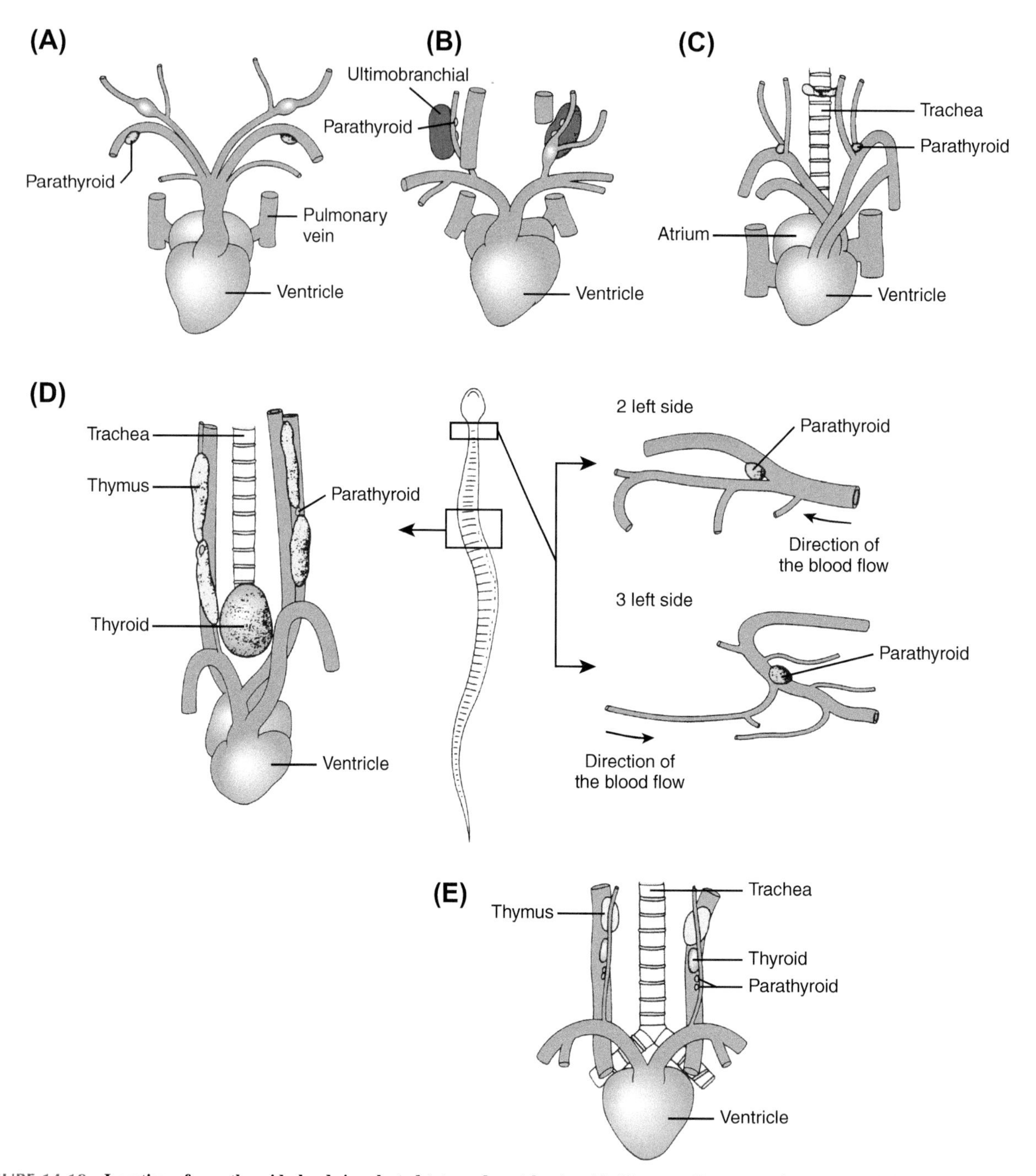

FIGURE 14-18 **Location of parathyroid glands in selected tetrapod vertebrates.** (A) The newt *Cynops pyrrhogaster*; urodele amphibians typically have one pair of parathyroids. (B) Frogs and other anurans typically have two pairs of parathyroids. (C) Lizards have one pair of parathyroids except for *Anolis* spp. that have two. (D) Snakes, such as *Agkistrodon halys*, have two pairs of parathyroids; the posterior pair is associated with the thymus gland. (E) Many birds such as the seagull (*Larus argentatus*), shown here, and domestic species have two pairs of parathyroids located near the thyroid although some (e.g., stork, quail) have one pair. *(Adapted with permission from Matsumoto, A. and Ishii, S., "Atlas of Endocrine Organs," Springer-Verlag, Berlin, 1992.)*

found that contain immunoreactive (ir) PTHrP. In the anterior group located in the rostral pars distalis, PTHrP is co-localized with β-thyrotropin (β-TSH). The second group of ir-PTHrP cells occurs at the borders of the proximal pars distalis and the pars intermedia. These cells do not react to antisera for β-TSH, CLIP, or ACTH. Although plasma levels of PTHrP in sea bream are ten times greater than in normal human plasma, no specific role for PTHrP is evident. The name **hypercalcin** has been applied to the pituitary hypercalcemic factor of teleosts, but this activity could be due to PTHrP, SL, and/or PRL (the latter shown to have some effect on plasma Ca^{2+} levels).

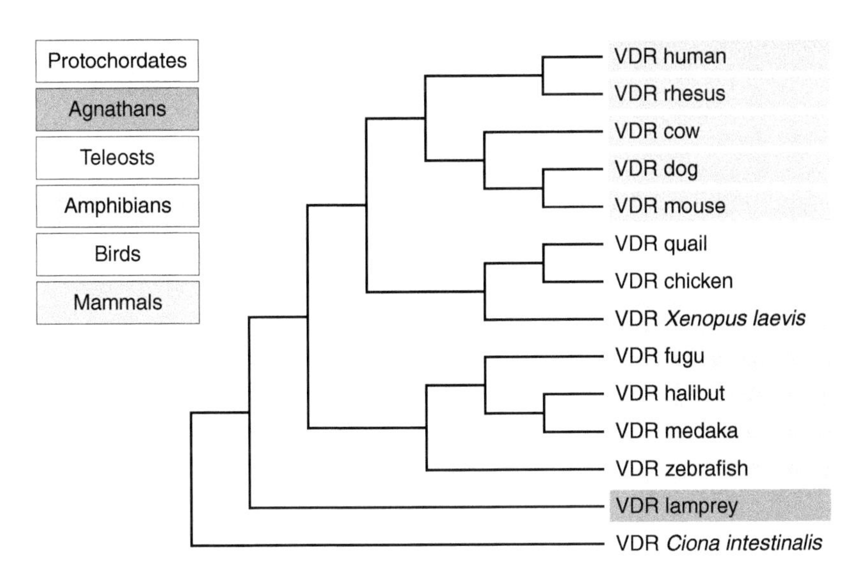

FIGURE 14-19 **Phylogenetic relationship of the vertebrate vitamin D receptor (VDR) for 1,25-dihydroxycholecalciferol (1,25-DHC).** *(Adapted with permission from Reschly, E.J. et al.,* BMC Evolutionary Biology, *7, 222, 2007.)*

The corpuscles of Stannius have been examined for their possible involvement in Ca^{2+} homeostasis. Little experimental work has been performed with species other than teleosts, and only a few teleosts have been examined thoroughly. Stanniectomy of the saltwater eel *Anguilla anguilla* results in an increase in serum calcium, and the administration of corpuscle extracts reduces calcium to normal levels. Stanniectomy of the closely related freshwater *Anguilla japonica* is followed by decreased urinary Ca^{2+} levels and increased urinary phosphate levels. It is possible that the corpuscles of Stannius are mainly active in reducing Ca^{2+} levels in fish adapted to high calcium environments such as seawater.

Two forms of STC have been identified in teleosts as well as in birds and mammals: **STC-1** and **STC-2** (Figure 14-15). Elevation of Ca^{2+} stimulates secretion of STC-1, which acts on gill chloride cells to slow the rate of Ca^{2+} uptake from the surrounding water. Additionally, STC-1 inhibits intestinal Ca^{2+} uptake and promotes phosphate

TABLE 14-8 **Effects of Calcitonin in Selected Non-Mammalian Vertebrates**

Species	CT source	Effect
Agnathan fishes		
Myxine glutinosa	Mammalian	Decreased urine flow and electrolyte content of urine
Bony fishes		
Periophthalmus schlosseri	Eel	Hypocalcemia
Carassius auratus	Goldfish	Hypocalcemia
Cyprinus carpio	Salmon	Hypocalcemia
Pacific salmon (*Oncorhynchus*)	Salmon	Decreased gill uptake of calcium
Amphibians		
Rana tigrina	Salmon	Transient hypocalcemia; increased calcium deposition in paravertebral calcium sacs
Ambystoma mexicanum	Eel	No effect on blood calcium; decreased calcium influx
Reptiles		
Dipsosaurus dorsalis	Salmon	No effect on kidney or basal salt gland handling of calcium

reabsorption by the kidney. Generally, STC-1 levels are higher in marine fishes than in freshwater fishes. STC-1 binds to a GPCR linked to a cAMP/PKA pathway but the actual mechanism of preventing Ca^{2+} influx is not known. In mammals, STC-1 and STC-2 do not appear to be involved in ion regulation and are widely distributed among many tissues. More research is needed to elucidate their specific roles in mammals.

Similar to STCs, treatment with eel, salmon, or goldfish CTs lowers calcium influx across the gill, and the gill may be a major site where CT regulates Ca^{2+} transport. Salmon CT is a powerful hypocalcemic agent when tested in avian or mammalian systems (Tables 14-6 and 14-8).

Some teleost kidneys are capable of hydroxylation of cholecalciferol to form 1,25-DHC. This hormone is apparently not necessary for stimulating Ca^{2+} uptake by the intestine, but large doses of 1,25-DHC do enhance Ca^{2+} uptake. Treatment with 1,25-DHC increases plasma calcium and phosphate in the freshwater catfish *Heteropneustes fossilis*. Calcium-binding proteins are present in the intestinal mucosa of teleosts (Table 14-9), and their synthesis may be controlled by 1,25-DHC as it is in mammals. In females, total plasma HPO_4^{-2} and Ca^{2+} increase markedly during oogenesis due to the actions of estrogens on liver vitellogenin synthesis and secretion.

TABLE 14-9 Calbindin Immunoreactivity in Vertebrate Tissues[a]

Species	Intestine	Kidney
Bony fishes		
Carassius auratus (goldfish)	+	
Salmo trutta (brown trout)	+	
Oncorhynchus mykiss (rainbow trout)		0
Amphibians		
Bufo bufo (toad)	+	0
Rana esculenta (frog)	+	0
Rana temporaria (frog)	+	
Xenopus laevis (frog)	+	
Triturus alpestris (newt)	+	
Reptiles		
Chrysemys (turtle)	+	
Psuedemys (turtle)		0
Gekko gecko (lizard)		0
Birds		
Japanese quail	+	
Chicken	+	+
Pigeon	+	
Mammals		
Rat	0	
Monkey	0	
Human	0	

[a]*All tetrapods examined have two forms of brain calbindin, but the fishes examined have only one. +, present; 0, absent; blank, no data.*

1. Thyroid State and Calcium Homeostasis in Teleosts

Serum calcium may be influenced by thyroid state, but the physiological importance of these observations has not been established. Serum calcium levels are decreased in juvenile steelhead trout (*Oncorhynchus mykiss*) that were radiothyroidectomized prior to complete resorption of the yolk sac (Table 14-10). It is not possible to distinguish between general effects of radiation that may have damaged some calcium-regulating mechanism and effects due to the absence of thyroid hormones, however. Growth of the skeleton is also abnormal following radiothyroidectomy, supporting an involvement of thyroid hormones at some level. Abnormal skeletal growth observed in these fish could be related to ineffective action of growth hormone in the absence of thyroid hormones (see Chapter 7).

D. Bony Fishes: Lungfishes

The lungfishes lack parathyroid glands and are relatively insensitive to tetrapod calcium-regulating hormones; however, there is evidence for the presence of PTRrP in the Australian lungfish (*Neoceratodus fosterii*). It is both surprising and somewhat disappointing that they do not exhibit some tetrapod-like feature. Parathyroid extracts,

TABLE 14-10 Serum Calcium Levels in Radiothyroidectomized Steelhead Trout (*Oncorhynchus mykiss*)

Treatment	Total serum calcium ± SEM (mg/100 mL)	Range (mg/100 mL)
Radiothyroidectomized as fingerlings	10	4–12
Intact controls	11	9–15
Radiothyroidectomized prior to yolk sac resorption	11.8 ± 1.01	
Intact controls	14.9 ± 0.51	

PTH, and salmon CT are all ineffective in altering Ca^{2+} levels in the South American lungfish *Lepidosiren paradoxa*. Surprisingly, however, CT and PTH are diuretic and antidiuretic, respectively, in these fishes. The physiological importance of these observations is not clear.

E. Amphibians

Studies of amphibians have shed little light on the origin of the parathyroid glands and the evolution of calcium regulation in tetrapods. Ultimobranchial and parathyroid glands are present in the apodans, anurans, and urodeles, and calcium regulation is similar to that observed in other tetrapods. Parathyroid glands are not present in some urodeles until after metamorphosis, and parathyroids never develop in some permanently neotenic aquatic species such as *Necturus*. These glands are more important in calcium balance of early terrestrial urodeles and especially of anurans. Ultimobranchial bodies are well developed and produce a hypocalcemic CT-like factor. As in fishes, a hypercalcemic factor (probably PRL) is present in pituitaries of urodele amphibians, and it is of greater importance for calcium regulation in the more aquatic species. Immunoreactivity to mammalian PTHrP has been reported in *Rana temporaria* at the time of emergence of metamorphosed animals from the water. PTHrP is synthesized in the pars intermedia of the pituitary but is transferred to the pars distalis from which it is released. Two PTH/PTHrP receptors have been detected in the genome of *Xenopus laevis* and have been shown in oocytes. However, the physiological role for PTHrP has not yet been established.

1. Amphibian Ultimobranchial Glands

The amphibian ultimobranchial glands develop as in all tetrapods from the fifth pharyngeal pouches. Most urodeles have only one ultimobranchial gland (usually on the left side). In apodans, anurans, and some urodeles (e.g., *Necturus*, *Amphiuma*), the ultimobranchial glands are paired. Immunoreactive CT is present in the ultimobranchial glands of *Rana temporaria* and *R. pipiens*. Cytologically, the ultimobranchial gland consists of one or more simple follicles composed of C cells which may appear in either a "dark" form (relatively electron dense) or in a less electron-dense "light" form, at least in anurans. In addition to the C cells, the ultimobranchial gland of the apodan *Chthonerpeton indistinctum* contains cholinergic and purinergic neuronal endings. Although the presence of sympathetic neurons has been demonstrated in the frog *R. pipiens*, most anurans investigated do not exhibit innervation of the ultimobranchial glands. The condition of the urodele ultimobranchial gland with respect to innervation has not been described. Salmon CT produces a transient decrease in plasma calcium and phosphate in anurans. Eel CT decreases calcium influx in aquatic axolotls (*Ambystoma mexicanum*) as it does in fishes.

2. Amphibian Parathyroid Glands

As in mammals, parathyroidectomy results in lowered calcium and usually leads to tetany and death. The parathyroid glands appear to develop from the third and fourth pharyngeal pouches as they do in mammals. Chief cells responsible for PTH secretion actually arise from neuroectodermal pharyngeal components rather than from endoderm. In apodan parathyroids, only chief cells are present, whereas two cellular types have been described for anurans. However, these appear to be two forms of chief cells that occur in light and dark phases as described above for the ultimobranchial bodies. Two distinct cellular types have been reported in parathyroids of the urodele *Cynops pyrrhogaster*. One of these cells is considered to be only a "supportive" cell, however.

3. Endolymphatic Sacs

In addition to cellular bone, the endolymphatic sacs located at the base of the skull and/or along the vertebral column, appear to be major targets for factors regulating calcium homeostasis. These sacs contain large amounts of calcium carbonate and may be important reservoirs of these ions, particularly during metamorphosis of larvae and subsequent ossification of bones. These structures also may provide bicarbonate ions for buffering the blood following the dissociation of calcium carbonate.

4. Amphibian Calcium and Phosphate Homeostasis

In general, the amphibians regulate plasma calcium and phosphate similarly to mammals. The parathyroids, pituitary gland, and ultimobranchial glands control calcium and phosphate metabolism, and the vitamin D complex seems to be related to at least some actions of PTH in amphibians. A calcium-binding protein occurs in the intestinal mucosa (Table 14-9), and this probably is a target for 1,25-DHC. Removal of the ultimobranchial glands in frogs generally causes an increase in osteoclast activity and a consequent increase in blood calcium levels. The ultimobranchial gland secretions also prevent removal of calcium from the endolymphatic sacs and block uptake of calcium through the gut. Parathyroidectomy causes a decrease in plasma Ca^{2+} in anurans and in the newt *Cynops pyrrhogaster*. However, no changes in serum Ca^{2+} occur following parathyroidectomy of immature giant salamanders (*Megalobatrachus davidianus*). Administration of bovine PTH to frogs increases plasma Ca^{2+} and decreases plasma phosphate, implying that the kidney may also be a target for PTH.

F. Reptiles

The regulation of calcium and phosphate in lizards has been well studied, and in recent years these investigations have been extended to include snakes and turtles. Ultimobranchial glands and parathyroids have been described, and regulation of calcium is essentially like that in other tetrapods. The major sites for endocrine regulation of calcium metabolism are the kidney, cellular bone, and the endolymphatic sacs, which in lizards, as in amphibians, are important reserves of calcium and carbonate ions. The presence of PTHrP and its possible roles in reptilian calcium metabolism have not been addressed.

1. Reptilian Ultimobranchial Glands

The ultimobranchial glands of reptiles are located near the thyroid and parathyroid glands. They are small glands consisting primarily of follicles. All the reptilian ultimobranchials are innervated, although the nature of these neuronal endings has not been examined extensively. In crocodilians, chelonians, and in some snakes, the ultimobranchial glands are paired. In lizards only the left gland persists. Seasonal changes in the cytology of ultimobranchial glands have been reported for a few species, but the relationship between these changes and physiological and environmental parameters has not been ascertained.

2. Reptilian Parathyroid Glands

Four reptilian parathyroids develop from the third and fourth pharyngeal pouches as described for the amphibians, and they resemble mammalian parathyroids cytologically. Adult lizards and crocodilians have only one pair of parathyroid glands, whereas snakes and turtles have four glands. In addition to cellular cords, the presence of follicles containing mucopolysaccharides is a common feature of reptilian parathyroids, although their functional significance is not known.

3. Calcium and Phosphate Homeostasis in Reptiles

Parathyroidectomy of lizards and snakes causes a marked decrease in plasma Ca^{2+} accompanied by tetany. However, there is little or no change in circulating Ca^{2+} in turtles following a similar operation, and tetany does not occur in turtles. This insensitivity of turtles to parathyroidectomy is apparently a consequence of the immense calcium reservoir represented by the shell. Treatment with mammalian PTH causes increased plasma Ca^{2+} and urinary phosphate as well as decreased urinary Ca^{2+} and plasma phosphate in both lizards and turtles.

The renal action of mammalian parathyroid extracts on phosphate excretion is marked in parathyroidectomized snakes (four species of *Natrix*), although calcium excretion is probably not affected significantly. Thus, it appears that reptiles possess basically the same regulatory control of calcium and phosphate homeostasis as exhibited by mammalian parathyroids.

The role of CT in reptiles is uncertain. Extracts of reptilian ultimobranchial glands produce hypocalcemia when injected into rats, and salmon CT has been shown to lower Ca^{2+} in the green iguana (*Iguana iguana*). As with fishes and amphibians, when reptilian ultimobranchial factors are tested in mammals the presence of hypocalcemic factors is noted, but their endogenous physiological roles are not known.

Snakes and lizards synthesize 25-HC and 1,25-DHC, and exposure of corn snakes (*Elaphe guttata*) or bearded dragons (*Pogona vitticeps*) to UV light increases circulating levels. At least one species, the panther chameleon (*Furcifer pardalis*) self-regulates its exposure to UV according to the availability of vitamin D_3 in its diet. A calcium-binding protein has been demonstrated in the intestinal mucosa of three species (Table 14-9) and is presumably the target for 1,25-DHC.

G. Birds

There are typically four parathyroid glands in birds, and separate ultimobranchial glands are present. Cytologically, the parathyroid glands resemble those of mammals, and the ultimobranchial cells are like the mammalian C cell. The pattern for regulation of calcium and phosphate homeostasis is typically mammalian with only minor differences. It should be noted, however, that few avian species have been investigated thoroughly, and most of the physiological studies have been conducted with domestic birds (e.g., chickens, pigeons, Japanese quail).

1. Avian Parathyroid Glands

Avian parathyroid glands are separate and usually distinct, except for a few species, such as the domestic chicken, in which some fusion of the separate parathyroids may occur. Cytologically the parathyroid glands contain only chief cells; no oxyphils have been reported.

2. Avian Ultimobranchial Glands

The ultimobranchial glands are usually separate structures, although some fusion with the thyroid gland occurs in the pigeon and turtledove. There are both light and dark cells in chicken ultimobranchials. The light cell is more abundant and resembles the mammalian C-cell cytologically, but both cell types secrete CT. A rich vagal innervation (parasympathetic) has been described for the chicken, but its importance has not been elucidated. Because removal of the ultimobranchials of chickens has no effect on plasma

BOX 14-A Bone As An Endocrine Organ: Osteocalcin

Osteocalcin is a 49-amino-acid peptide secreted by osteoblasts into bone matrix and the blood. During posttranslational processing, the glutamic acid residues on the osteocalcin prohormone are carboxylated (addition of a COOH group) by the enzyme gamma glutamyl carboxylase, which requires vitamin K as a cofactor. This leads to the formation of γ-carboxyglutamic acid (Gla) moieties on the prohormone; human osteocalcin contains three Gla residues per molecule. Carboxylated osteocalcin is then cleaved from its prohormone and secreted by osteoblasts where it plays a role in bone matrix building. Some uncarboxylated osteocalcin is also secreted and enters the bloodstream where it has been reported to act on pancreatic B cells to enhance insulin secretion and thus influence glucose metabolism.

Interestingly, the level of uncarboxylated osteocalcin in the blood can serve clinically as a marker of reduced bone growth, as elevated levels of the uncarboxylated form reflect less of the mature peptide being deposited into bone. Higher levels of uncarboxylated osteocalcin in blood may also reflect vitamin K deficiency, which has been linked clinically to bone fragility and risk of hip fracture. Normally only 10 to 40% of osteocalcin produced by osteoblasts enters the blood. Blood levels of osteocalcin are elevated by 10% in postmenopausal women with osteoporosis, and measurement of this peptide in blood may thus be a useful predictor of bone loss in postmenopausal women.

Ca^{2+}, bone metabolism, or egg laying, it does not appear that CT is essential for Ca^{2+} regulation.

3. *Calcium and Phosphate Homeostasis in Birds*

PTH, 1,25-DHC, and estrogens appear to be the most important hormones regulating Ca^{2+} metabolism, with CT playing only a minor role. Parathyroidectomy in birds usually causes marked hypocalcemia accompanied by tetanic seizures and death within 24 hours. This degree of sensitivity to parathyroidectomy is not manifest on this time scale in amphibians or reptiles and may be related to the much higher body temperature and metabolic rate of birds. Treatment with mammalian parathyroid extracts produces marked increases in plasma Ca^{2+}. Dietary deprivation of Ca^{2+} or vitamin D causes marked hypertrophy and hyperplasia of the parathyroid glands, whereas high Ca^{2+} diets result in regression of the parathyroids.

Formation of eggshells during egg laying is the most important use of calcium in adult birds. At sexual maturity, estrogens cause a shift from production of lamellar cortical bone to production of a unique nonstructural form called medullary bone. During the egg-laying cycle medullary bone is resorbed under the influence of PTH and 1,25-DHC to release Ca^{2+} for egg cell construction. Roughly 40% of the extracellular Ca^{2+} in birds is bound to serum albumin and the estrogen-dependent vitellogenin protein. Bone resorption occurs within minutes in response to PTH, whereas mammalian bone responds much more slowly (hours to days), which makes Ca^{2+} quickly available for eggshell construction. PTH also causes increased reabsorption of Ca^{2+} by the kidney and stimulates the kidney to secrete more 1,25-DHC. This is important because up to 70% of the Ca^{2+} needed for eggshell production must be absorbed from the diet. This absorption is dependent on 1,25-DHC.

Mammalian parathyroid extract stimulates renal excretion of phosphate in normal starlings but does not seem to alter renal treatment of calcium. The effect of mammalian parathyroid extract or PTH on avian bone is similar to that described for mammals. Avian PTH binds to a PTHr1 similar to the mammalian PTH receptor.

PTHrP has been isolated from chickens and is present in both embryos and adults. Numerous embryonic tissues express PTHrP, and it may be an important growth stimulator in bird embryos. Expression of the PTHrP gene in the isthmus and shell gland of the adult oviduct is related to entrance of the egg into the oviduct and calcification of the shell. PTHrP relaxes vascular smooth muscle and increases blood flow to the shell gland during calcification. It is not known whether it alters Ca^{2+} transport as well. Relaxation of oviductal smooth muscle caused by PTHrP allows the entrance and passage of the egg though the oviduct.

Calcitonins that are structurally similar to mammalian CT have been isolated from chickens and turkeys and are potent hypocalcemic agents in both birds and mammals. Unlike the condition for fishes, amphibians, and reptiles, the avian kidney specifically binds CT, suggesting a renal action for CT in birds. Dietary levels of calcium are directly related to ultimobranchial activities, and high levels of Ca^{2+} cause marked stimulatory changes and increased levels of plasma CT. An immunoreactive Ca^{2+}-binding exchanger protein similar to that of other vertebrates is present in the intestinal mucosa of birds (Table 14-9).

VI. SUMMARY

Precise regulation of calcium and phosphate homeostasis is necessary for many processes in vertebrates. In tetrapods, this regulation is accomplished primarily through secretion of hypercalcemic PTH from the parathyroid glands affecting bone and kidney function, 1,25-DHC affecting intestinal uptake of dietary Ca^{2+}, and to some extent by hypocalcemic CT produced by the C cells embedded between the follicles of the thyroid gland (mammals) or by the ultimobranchial glands (other tetrapods). CT primarily is antagonistic of PTH in bone. PTHrP may be an important regulator of embryonic development and growth. It also has important roles in placental function in mammals as well as for mammary gland development and lactation in relation to calcium homeostasis. Its presence in milk may help newborn mammals regulate calcium as well.

PTH increases plasma Ca^{2+} levels through direct and indirect action on bone, kidney, and intestine. Parathyroidectomy invariably brings about tetany and usually death as a consequence of the decrease in circulating Ca^{2+} following this operation. This condition can be alleviated through administration of PTH or Ca^{2+}. The actions of PTH on bone may be anabolic (intermittent secretion of PTH on bone formation mediated through the osteoblasts) or catabolic (bone reabsorption via the osteoclasts). Both actions involve binding of PTH to PTHr1 on osteoblasts that in turn influence osteoclast differentiation and functioning. In addition to the effect of PTH on bone, PTH stimulates renal excretion of phosphate ions and reabsorption of Ca^{2+} following binding to its receptor (PTHr1) in the kidney. Both the resorption of bone and the kidney action of PTH may be dependent on a derivative of vitamin D, 1,25-DHC.

The uptake of calcium through the intestinal wall requires 1,25-DHC, which stimulates production of a calcium exchanger protein responsible for the uptake of dietary Ca^{2+}. Synthesis of 1,25-DHC involves sequential intermediates produced in the skin (production of cholecalciferol or vitamin D_3), altered in the liver (conversion of cholecalciferol to 25-HC by 25-hydroxylase), and finally converted to 1,25-DHC by 1α-hydroxylase in the kidney. PTH stimulates production of 1α-hydroxylase, and estrogens and PRL also may affect production of 1,25-DHC.

Calcitonin has been isolated from mammalian parafollicular thyroid tissue and from ultimobranchial glands of various vertebrate groups. The major site of action for CT is bone, where it antagonizes the action of PTH on osteoclasts. CT is not very effective in regulating Ca^{2+} in fishes and may not be that important in birds as well.

The comparative aspects of calcium and phosphate regulation are complicated by the absence of parathyroid glands in fishes and the presence of the unique corpuscles of Stannius in teleosts, the source of STC. The pituitary of teleosts produces SL and PTHrP. Because PTHrP binds to PTHr1, it may have some effects on calcium homeostasis, but its main actions appear to be paracrine in nature. Cyclostomes and elasmobranchs have not been studied sufficiently and Ca^{2+}/HPO_4^{-2} regulation is not understood in these fishes. Immunoreactive PTHrP has been described in a lamprey but no evidence for a *Pth* gene, although they have been demonstrated in chondrichthyean fishes and teleosts despite the absence of distinct parathyroid glands. Among bony fishes, calcium regulation appears to be accomplished by a hypercalcemic pituitary factor (SL; PRL or hypercalcin) and a hypocalcemic factor (STC, hypocalcin) from the corpuscles of Stannius embedded in the kidneys of some bony fishes. Scales are important Ca^{2+} reservoirs in teleosts. The pituitary factors appear to be more important as Ca^{2+}regulators in freshwater fishes, and the corpuscles of Stannius appear to be more important for calcium regulation in marine fishes. STC works through the gill chloride cell, preventing Ca^{2+} influx. STCs are also present in other vertebrates but have not been shown to influence Ca^{2+} metabolism.

Mammalian or fish CTs can decrease blood Ca^{2+} in bony fishes, and their major target appears to be the gills, although CT does not seem to be a major regulator of Ca^{2+}. However, salmon CT is a more potent hypocalcemic factor in mammals than is mammalian CT, and this greater potency may be a result of its relative resistance to clearance from the circulation. Salmon CT is currently used for clinical treatment in some situations of human CT insufficiency.

Parathyroid glands are distinct in amphibians, reptiles, and birds, and the effects of PTH and parathyroidectomy are similar to those observed in mammals. The pituitary may have a hypercalcemic role in aquatic amphibians. Amphibians and some reptiles possess endolymphatic sacs, which may be important sites for calcium carbonate storage. CT has been demonstrated in amphibians, and extracts of their ultimobranchial glands have hypocalcemic activity when tested in birds or mammals. Treatment with CTs alters calcium and phosphate metabolism as in other vertebrates. Evidence for PTHrP is present for all vertebrate groups where it is widely distributed in many tissues, suggesting a paracrine role. PTHrP is important also in embryonic development and has roles in developmental Ca^{2+} regulation, at least in birds and mammals.

One feature that appears to be dominant throughout the tetrapods examined to date is the strong tendency for innervation of the ultimobranchial glands, although the nature of this innervation (e.g., cholinergic, adrenergic) and biological significance are not clear. A second consistent relationship appearing in all oviparous species is the process of vitellogenesis. Estrogens increase the production of vitellogenic phosphoproteins by the liver. These phosphoproteins bind Ca^{2+} when they are secreted into the blood, resulting in further movement of Ca^{2+} into the blood to maintain the proper ionic balance. Consequently, there is an elevation in total plasma calcium and phosphate associated with vitellogenesis in females that appears to be a mechanism whereby increased Ca^{2+} become available for production of egg shells (birds and reptiles) and/or for incorporation into the eggs (all oviparous species). Egg stores of Ca^{2+} are used for early developmental processes prior to hatching and feeding by the offspring. A similar mechanism operates in mammals whereby maternal dietary calcium is directed during development or during lactation to the offspring by actions of PTH with protection of the maternal skeleton afforded through the actions of CT. Estrogens also stimulate production of 1,25-DHC in both birds and mammals.

STUDY QUESTIONS

1. Describe the relationships between calcium and phosphate metabolism and why they are important minerals for survival. What major differences can be found among the various vertebrate groups?
2. List the major hormones that regulate calcium and phosphate metabolism in each major group of vertebrates and describe their mechanisms of action and general effects.

SUGGESTED READING

Books

Bilezikan, J., 2001. The Parathyroids. Academic Press, San Diego, CA.

Dacke, C.G., 1979. Calcium-regulating Hormones in Sub-mammalian Vertebrates. Academic Press, New York.

LeBlanc, G.A., Norris, D.O., Kloas, W., Kullman, S.W., Baldwin, W.S., Greally, J.M., 2012. Detailed Review Paper: State of the Science on Novel *In Vitro* and *In Vivo* Screening and Testing Methods and Endpoints for Evaluating Endocrine Disruptors. Prepared by RTI International for U.S. Environmental Protection Agency, Washington, D.C, 193 pp.

Licata, A.A., Lerma, E.V. (Eds.), 2012. Diseases of the Parathyroid Glands. Springer, New York.

General Articles

Abbink, W., Flik, G., 2007. Parathyroid hormone-related protein in fish. General and Comparative Endocrinology 152, 243–251.

Bhattacharya, P., Yan, Y.L., Postlethwait, J., Rubin, D.A., 2011. Evolution of the vertebrate *pth2 (tip39)* gene family and the regulation of PTH type 2 receptor (*pth2r*) and its endogenous ligand *pth2* by Hedgehog signaling in zebrafish development. Journal of Endocrinology 211, 187–200.

Bikle, D.D., 2010. Vitamin D: newly discovered actions require reconsideration of physiologic requirements. Trends in Endocrinology & Metabolism 21, 375–384.

Chirgwin, J.M., Guise, T.A., 2003. Cancer metastasis to bone. Science & Medicine 9, 140–151.

Danks, J.A., D'Souza, D.G., Gunn, H.J., Milley, K.M., Richardson, R.J., 2011. Evolution of the parathyroid hormone family and skeletal formation pathways. General and Comparative Endocrinology 170, 79–91.

Dempster, D.W., 2011. Osteoporosis and the burden of osteoporosis-related fractures. American Journal of Managed Care 11, S164–S169.

Lukacs, J.L., Booth, S., Kleerekoper, M., Ansbacher, R., Rock, C.L., Reame, N.E., 2006. Differential associations for menopause and age in measures of vitamin K, osteocalcin, and bone density: a cross-sectional exploratory study in healthy volunteers. Menopause 13, 799–808.

Martin, T.J., Udagawa, N., 1998. Hormonal regulation of osteoclast function. Trends in Endocrinology and Metabolism 9, 6–12.

Pinheiro, P.L.C., Cardoso, J.C.R., Gomes, A.S., Fuentes, J., Power, D.M., Canário, A.V.M., 2010. Gene structure, transcripts and calciotropic effects of the PTH family of peptides in *Xenopus* and chicken. BMC Evolutionary Biology 10, 373.

Pinheiro, P.L.C., Cardoso, J.C.R., Power, D.M., Canário, A.V.M., 2012. Functional characterization and evolution of PTH/PTHrP receptors: insights from the chicken. BMC Evolutionary Biology 12, 110.

Reschly, E.J., Dias Bainy, A.C., Mattos, J.J., Hagey, L.R., Bahary, N., Mada, S.R., Ou, J., Venkataramanan, R., Krasowski, M.D., 2007. Functional evolution of the vitamin D and pregnane X receptors. BMC Evolutionary Biology 7, 222.

Roch, G.J., Sherwood, N.M., 2010. Genomics reveal ancient forms of stanniocalcin in amphioxus and tunicate. Integrative and Comparative Biology 50, 86–97.

Talmage, R.V., Mobley, H.T., 2008. Calcium homeostasis: reassessment of the actions of parathyroid hormone. General and Comparative Endocrinology 156, 1–8.

Wagner, G.F., DiMattia, G.E., 2006. The stanniocalcin family of proteins. Journal of Experimental Zoology 305A, 769–780.

Appendix A

Abbreviations Part 1

Abbreviation	Definition
1,25-DHC	1,25-dihydroxycholecalciferol = calcitriol
11-KT	11-ketotestosterone
11β-HSD	11β-hydroxysteroid dehydrogenase
17,20β-DHP	17,20β-dihydroxy-4-pregnen-3-one
17,20β-P-S	17,20β-P-Sulfate
17,20β,21-P-S	17,20β,21-P-Sulfate
17,20β,21-THP	17,20β,21-trihydroxy-4-pregnen-3-one
17β-HSD	17β-hydroxysteroid dehydrogenase
1α-OHB	1α-hydroxycorticosterone
2-IHDA	2-iodohexadecanal
25-HC	25-hydroxycholecalciferol = calcidiol
3β-diol	5α-androstane-3β,17β-diol
3α-diol	5α-androstane-3α, 17β-diol
3β-HSD	Δ^5,3β-hydroxysteroid dehydrogenase
3D-QSAR	three-dimensional quantitative structure-activity relationship
$3T_1AM$	3-iodothyronamine
5-HIAA	5-hydroxyindoleacetic acid
5-HT	serotonin (5-hydroxytryptamine)
5α-DHB	5α-dihydroxycorticosterone
5α-DHP	5α-dihydroxyprogesterone
AA	arachidonic acid
AADC	aromatic L-amino acid decarboxylase
ABP	androgen binding protein
ACE	angiotensin converting enzyme
ACELA	angiotensin-converting enzyme-like action
ACh	acetylcholine
AChE	acetylcholinesterase
ACTH	corticotropin
Ad4BP	adrenal 4-binding protein = SF-1
ADHD	attention-deficit hyperactivity disorder
AdipoR	adiponectin receptor
AF	aldehyde fuchsin
AgRP	agouti-related protein
AHA	anterior hypothalamic area
ahR	arylhydrocarbon receptor
AMH	anti-müllerian hormone
Ang	angiotensin
ANP	atrial natriuretic peptide
AP-1	activator protein-1
APUD	amine content and amine precursor uptake and decarboxylation
AR	androgen receptor
ARC	arcuate nucleus

Abbreviation	Definition
AREG	amphiregulin
ASP	agouti-signaling protein
AspT	aspargtocin
AsvT	asvatocin
AVP	arginine vasopressin
AVPV	anteroventral periventricular area
AVT	arginine vasotocin
bark	β-adrenergic receptor kinase
BBS	bombesin
bFGF	basic fibroblastic growth factor
BMI	body mass index
BMP	bone morphogenetic protein
BMR	basal metabolic rate
BNP	natriuretic peptide B
BPA	bisphenyl A
BSD	behavioral sex determination
BTC	betacellulin
CAH	congenital adrenal hyperplasia
CAIS	complete androgen insensitivity syndrome
CAM	chorio-allantoic membrane
cAMP	cyclic adenosine monophosphate
CART	cocaine and amphetamine regulated transcript
CAT	computer axial tomography
CB1R	endocabinoid receptor
CBG	corticosteroid binding globulin = transcortin
Cbl	Casitas B-lineage Lymphoma
CC	chorionic corticotropin
CCK	cholecystokinin (= pancreozymin-cholecystokinin, PZCCK)
CCK_8	cholecytokinin (brain)
CCKR	cholecystokinin receptor
Cdk	cyclin-dependent kinase
cfGnRH	catfish GnRH
CG	chorionic gonadotropin
cGMP	guanosine monophosphate
cGnRH-I	chicken-I GnRH
cGnRH-II	chicken-II GnRH
CGRP	calcitonin gene-related peptide
ChIP	polymerase chain reaction
ClO_4^-	perchlorate ion
CLIP	corticotropin-like peptide
CNP	natriuretic peptide C
COC	cumulus cell oocyte complex
COMT	catechol-O-methyl transferase

Abbreviation	Definition
CR	corticosteroid receptor
CRE	cAMP response element
CREB	cAMP regulatory element binding protein
CRF	corticotropin-releasing factor = CRH
CRF-R1	CRH receptor
CRF-R2	urocortin receptor
CRH	corticotropin-releasing hormone
CS	chorionic somatomammotropin
CSF	colony-stimulating factor
CT	calcitonin
CTL	cytotoxic T lymphocyte
cyp19a1a	aromatase gene
D1	type I deiodinase
D2	type 2 deiodinase
D3	type 3 deiodinase
DAG	diacylglycerol
dax1	dosage-sensitive sex reversal, adrenal hypoplasia critical region, on chromosome X gene 1
DES	diethylstilbestrol
dfGnRH	dogfish (shark) GnRH
DHEA	dehydroepiandrosterone
DHEAS	DHEA sulfate
DHN	dorsal hypothalamic nucleus
DHT	dihydrotestosterone
DIT	3,5-diiodotyrosine
DMN	dorsomedial nucleus
dmy/drmt1bY	DM domain of the Y chromosome/ doublesex- and mab-3-related transcription factor 1 gene
DNA	deoxyribonuclei acid
DOC	deoxycorticosterone
DOPA	dihydroxyphenylalanine
DPP-4	dipeptidyl peptidase-4
DSD	disorder of sexual development
DSIP	delta sleep-inducing peptide
EAC	endocrine-active chemical
ECL	enterochromaffin-like cell
ECM	extracellular matrix
EDC	endocrine-disrupting chemical
EDSP	Endocrine Disruptor Screening Program
EE_2	ethynylestradiol
EGF	epidermal growth factor
ELISA	enzyme-linked immunoabsorbant assay
EOP	endogenous opioid peptide
EPTH	extrapancreatic tumor hypoglycemia
ER	estrogen receptor
EREG	epiregulin
ERK1/2	extracellular signal-regulated kinases
ERR	estrogen-related receptor
FFA	free fatty acid (see NEFA)
FGF	fibroblastic growth factor
FMRF	Phe-Met-Arg-Phe
FNDC5	fibronectin type III domain containing 5
foxl2	forkhead box L2 gene
FPIR	first phase insulin release
FSH	follicle-stimulating hormone
FSHR	FSH receptor
GABA	Gamma-amino butyric acid
GALP	Galanin-like peptide
GAP	GnRH-associated peptide
GDF-9	growth differentiation factor 9
GDP	guanosine diphosphate
GFP	green fluorescent protein
GH	growth hormone
GHR	GH receptor
GHRH	growth hormone-releasing hormone
GHS-1a	growth hormone secretogogue receptor
GLP	glucagon-like peptide
GLT	glumitocin
GLUT2	glucose transporter 2
GLUT4	glucose transporter 4
GnIH	gonadotropin inhibitory hormone
GnRH	gonadotropin-releasing hormone
GPA2	see GpHα
GPB5	see GpHβ
GPCR	G-protein coupled receptor
GpH	glycoprotein hormone
GpHα	α-subunit of pituitary glycoprotein hormones
GpHβ	β-subunit of pituitary glycoprotein hormones
GPR54	G-protein coupled receptor 54 = kiss1R
GR	glucocorticoid receptor
GR1	type 1 GR (= MR)
GR2	type 2 GR
GRB2	growth factor receptor-binding protein
GRH	growth hormone-releasing hormone
GRP	gastrin releasing peptide
GRTH/DDX25	gonadotropin-regulated testicular helicase
GSD	genotypic sex determination
gsdf	gonadal-soma-derived factor gene
GTH	gonadotropin
GTP	guanosine triphosphate
GVBD	germinal vesicle breakdown
HAS2	hyaluronan synthase 2
HAT	histone acetyltransferase
hCG	human chorionic gonadotropin
HDAC	histone deacetylase
HHM	human hypercalcemia of malignancy
HIOMT	hydroxy-O-methyltransferase
HPA	hypothalamus-pituitary-adrenal
HPG	hypothalamus-pituitary-gonad
HPH	hypothalamus-pituitary-hepatic
hPL	human placental lactogen
HPLC	high performance liquid chromatography
HPLC-MS	high performance liquid chromatography/ spectroscopy
HPT	hypothalamus-pituitary-thyroid
HRE	hormone response element
HSD11β2	11β-hydroxysteroid dehydrogenase 2
HSL	hormone-sensitive lipase
ICSH	interstitial cell-stimulating hormone
IDDM	insulin-dependent diabetes mellitus
IFN	infundibular nucleus
IGF	insulin-like growth factor
IGF1R	insulin-like growth factor-I receptor
IGF1r	IGF-1 receptor gene
IGRP265−273	islet-specific glucose-6-phosphatase catalytic subunit-related protein

Abbreviation	Definition
IL	interleukin
IML	intermediolateral cell column
IP_3	inositol trisphosphate
Ir	insulin receptor genes
Irr	insulin-related receptor gene
IRS	insulin receptor substrate
IST	isotocin
JAK	Janus kinase
JAK2	Janus kinase 2
kissR1	kisspeptin 1 receptor
KO	knock out
Kp	kisspeptin
LAF	luteinization of atretic follicles
LatH	lateral hypothalamus
LATS	long-acting thyroid-stimulator
LC-MS	liquid chromatography combined with mass spectroscopy
LDL	low-density lipoprotein droplet
lGnRHs	lamprey GnRHs
LH	luteinizing hormone
LHR	LH receptor
LHRH	luteinizing hormone-releasing hormone
LPH	lipotropin
LTH	luteotropic hormone
LTN	lateral tuberal nucleus
LVP	lysine vasopressin
MAG	monoacylglyceride
MAO	monoamine oxidase
MAPK	mitogen-activated protein kinase
MC1R	melanocortin-1 receptor
MC2R	melanocortin-2 receptor
MC3R	melanocortin-3 receptor
MC4R	melanocortin-4 receptor
MCH	melanin-concentrating hormone
MCT8	monocarboxylate transporter 8
mdGnRH	medaka GnRH
MPF	maturation promoting factor; see 17,20β-DHP
MG	menopausal gonadotropin
mGnRH	mammalian GnRH
MIH	maturation-inducing hormone
MIS	müllerian-inhibiting substance
MIT	3-monoiodotyrosine
MMC	migrating motor complex
MMP	matrix metalloprotease
mPON	medial preoptic nucleus
MR	mineralocorticoid receptor
MRAP1	melanocortin receptor accessory protein 1
MRIH	melanotropin release-inhibiting hormone
MRT	magnetic resonance tomography
MSH	melanotropin
MST	mesotocin
MTR	MST receptor
NAT	*N*-acetyltransferase
NEFA	non-esterified fatty acid (= free fatty acid)
NF-KB	nuclear factor-kappaB
NIDDM	noninsulin-dependent diabetes mellitus
NIL	neurointermediate lobe
NIS	Na-I symporter
NO	nitric oxide
NOS	nitric oxide synthetase
NP	natriuretic peptide
NPR	naturetic peptide receptor
NPY	neuropeptide Y
NS	neurosecretory
NT	neurotensin
NTS	nucleus of the solitary tract
ObRa	leptin receptor a
ObRb	leptin receptor b
OVLT	organum vasculosum lamina terminalis
OXTR	oxytocin receptor
OXY	oxytocin
P_4	progesterone
$P450_{.aldo}$	aldosterone synthase
$P450_{11\beta1}$	11β-hydroxylase enzyme
$P450_{11\beta2}$	aldosterone synthase
$P450_{aro}$	aromatase
$P450_{c17}$	17α-hydroxylase
PACAP	pituitary adenylate cyclase activating peptide
PAM	peptidylglycine alpha-amidating monooxygenase
PBDE	polybrominated diphenyl ether
PCB	polychlorinated biphenyl
PCOS	polycystic ovarian syndrome
PCR	polymerase chain reaction
PDGF	platelet-derived growth factor
PEPCK	phosphoenolpyruvate carboxykinase
PERIV	periventricular nucleus
PET	positron emission tomography
PG	prostaglandin
PGE_2	prostaglandin E_2
$PGF_{2\alpha}$	prostaglandin $F_{2\alpha}$
PGI_2	Prostacyclin
PhaT	phasvatocin
PHI	peptide histidine isoleucine
PHM	peptide histidine methionine
PIC	pseudoisocyanin
PIP_2	phosphatidylinositol bisphosphate
Pit-1	pituitary-specific transcription factor
PKA	protein kinase A
PKC	protein kinase C
PLC	phospholipase C
PMSG	pregnant mare serum gonadotropin
PNMT	phenylethanolamine *N*-methyltransferase
POA	preoptic area
POC	proopiocortin
POM	proopiomelanotropin
POMC	proopiomelanocortin
PON	preoptic nucleus
PP	pancreatic polypeptide
PPAR	peroxisome proliferator-activated receptor
PR	progesterone receptor
PreProPTH	Preproparathyroid hormone
PRH	Prolactin-releasing hormone
PRIH	prolactin release-inhibiting hormone
PRL	prolactin
ProPTH	proparathyroid hormone
PrRP	prolactin-releasing peptide

Abbreviation	Definition
Pst	pancreastatin
PSTI-61	pancreatic secretory trypsin inhibitor
PTGS2	prostaglandin-endoperoxide synthase 2
PTH	parathyroid hormone
PTH-L	parathyroid hormone-like
PTHR	parathyroid hormone receptor
PTHrP	parathyroid hormone-related peptide
PVN	paraventricular nucleus
PVP	phenypressin
PYY	peptide YY
PZ-CCK	pancreozymin-cholecystokinin (see CCK)
RAR	retinoic acid receptor
RAS	rat sarcoma signaling
RH	releasing hormone
RIA	radioimmunoassay
RIH	release-inhibiting hormone
RNA	ribonucleic acid
RNAi	RNA interference
rT3	reverse T_3
RVLM	rostral ventrolateral medulla
RXFP	relaxin/insulin-like family peptide receptor
RXR	retinoid X receptor
SAD	seasonal affective disorder
sbGnRH	sea bream GnRH
SCG	superior cervical ganglion
SCN	suprachiasmatic nucleus
SCN^-	Thiocyanate
SCO	subcommissural organ of Dendy
SDN	sexually dimorphic nucleus
SDS-PAGE	SDS-polyacrylamide gel electrophoresis
SERM	selective estrogen receptor modulator
serpin	serine protease inhibitors
SES	socioeconomic status
SF-1	steroidogenic factor 1
sGnRH	salmon GnRH
SHBG	sex hormone binding globulin
SHBG	steroid hormone-binding globulin
SIAD	syndrome of inappropriate antidiuresis
SL	somatolactin
SOCS3	suppressor of cytokine signaling
SON	supraoptic nucleus
SOS	son of sevenless
sox9	SRY-related high-mobility group box 9 gene
sry	sex-determining region Y gene
SST	somatostatin
SSTR2	somatostatin receptor subtype 2
StAR	steroidogenic acute regulating protein
STAT	signal transducer and activator of transcription
STC	stanniocalcin
SV	saccus vasculosus
T_2	3,3′-diiodothyronine
T_3	3,5,3′-triiodothyronine = triiodothyronine
T_4	3,5,3′,5′-tetraiodothyronine = thyroxine
TAAR1	trace amine associated receptor 1
TAG	triacylglyceride
TAM	thyronamine
TBA	transthyretin
TBPA	prealbumin
TBT	Tributyltin
Tc	technicium
TCAP	teneurin C-terminal-associated peptide
TCDD	2,3,7,8-tetrachlorodibenzo-*p*-dioxin
TcO_4^-	pertechnetate
TDF	testis determining factor
TETRAC	tetraiodothyroacetate
TBG	thyroid-binding globulin
tgb	thyroglobulin = TGB
TGF-β2	transforming growth factor β2
TGFβ	transforming growth factor β
tGLP-1	GLP-1 truncated form
THC	9-tetrahydrocannabinol
TIP39	tuberoinfundibular protein 39
TNFα	tumor necrosis factor alpha
T_0AM	thyronamine
TPO	thyroid peroxidase
TR	thyroid receptor
TRAP	T_3 receptor auxiliary protein
TRAPase	tartrate-resistant acid phosphatase
TRE	thyroid response element
TRH	thyrotropin-releasing hormone
TRIAC	triiodothyroacetate
TRPV1	transient receptor potential vanilloid subtype 1
TSD	Temperature-dependent sex determination
TSH	thyrotropin
ttf-1	thyroid transcription factor 1
TTR	transthyretin
TU	thiourea
Ucn	urocortin
UCP-1	uncoupling protein-1
ValT	valitocin
VDR	vitamin D receptor
VEGF	vascular endothelial cell growth factor
VHN	ventral hypothalamic nucleus
VIP	vasoactive intestinal peptide
VLDL	very low density lipoprotein droplet
VMN	ventromedial nucleus
VNO	vomeronasal organ
VNP	ventricular natriuretic peptide
VSCC	voltage-sensitive calcium channel
VT1aR	AVT receptor 1a
VT1bR	AVT receptor 1b
VT2R	AVT receptor 2
VT3R	Oxytocin gene-related receptor
VTA	ventral tegmental area
VTN	ventrotuberal nucleus
WWTP	wastewater treatment plants

Abbreviations Part 2

Term	Abbreviation
1,25-dihydroxycholecalciferol = calcitriol	1,25-DHC
11-ketotestosterone	11-KT
11β-hydroxylase enzyme	$P450_{11\beta1}$
11β-hydroxysteroid dehydrogenase	11β2-HSD
11β-hydroxysteroid dehydrogenase 2	$HSD_{11\beta2}$
17,20β-dihydroxy-4-pregnen-3-one	17,20β-DHP
17,20β-P-sulfate	17,20β-P-S
17,20β,21-P-sulfate	17,20β,21-P-S
17,20β,21-trihydroxy-4-pregnen-3-one	17,20β,21-THP
17α-hydroxylase	$P450_{c17}$
17β-hydroxysteroid dehydrogenase	17β-HSD
1α-hydroxycorticosterone	1α-OHB
2-iodohexadecanal	2-IHDA
2,3,7,8-tetrachlorodibenzo-*p*-dioxin	TCDD
25-hydroxycholecalciferol = calcidiol	25-HC
3-iodothyronamine	$3T_1AM$
3-monoiodotyrosine	MIT
3,3′-diiodothyronine	T_2
3,5-diiodotyrosine	DIT
3,5,3′-triiodothyronine = triiodothyronine	T_3
3,5,3′,5′-tetraiodothyronine = thyroxine	T_4
5-hydroxyindoleacetic acid	5-HIAA
5α-androstane-3β,17β-diol	3β-diol
5α-androstane-3α,17β-diol	3α-diol
5α-dihydroxycorticosterone	5α-DHB
5α-dihydroxyprogesterone	5α-DHP
9-tetrahydrocannabinol	THC
α-subunit of pituitary glycoprotein hormone	GpHα
acetylcholine	ACh
acetylcholinesterase	AChE
activator protein-1	AP-1
adiponectin receptor	AdipoR
adrenal 4-binding protein = SF-1	Ad4BP
agouti-related protein	AgRP
agouti signaling protein	ASP
aldehyde fuchsin	AF
aldosterone synthase	$P450_{aldo}$
amine content and amine precursor uptake and decarboxylation	APUD
amphiregulin	AREG
androgen binding protein	ABP
androgen receptor	AR
angiotensin	Ang
angiotensin converting enzyme	ACE
angiotensin-converting enzyme-like action	ACELA
anterior hypothalamic area	AHA
anteroventral periventricular area	AVPV
anti-müllerian hormone	AMH
arachidonic acid	AA
arcuate nucleus	ARC
arginine vasopressin	AVP
arginine vasotocin	AVT
aromatase A gene	*cyp19a1a*
aromatase inhibitor	AI
aromatic L-amino acid decarboxylase	AADC
aryl hydrocarbon receptor	ahR
aspargtocin	AspT
asvatocin	AsvT
atrial natriuretic peptide	ANP
attention-deficit hyperactivity disorder	ADHD
basal metabolic rate	BMR
basic fibroblastic growth factor	bFGF
behavioral sex determination	BSD
β-adrenergic receptor kinase	βARK
betacellulin	BTC
bisphenyl A	BPA
body mass index	BMI
bombesin	BBS
bone morphogenetic protein	BMP
calcitonin	CT
calcitonin gene-related peptide	CGRP
cAMP regulatory element binding protein	CREB
cAMP response element	CRE
Casitas B-lineage lymphoma	Cbl
catechol-*O*-methyl transferase	COMT
catfish GnRH	cfGnRH
chicken-I GnRH	cGnRH-I
chicken-II GnRH	cGnRH-II
cholecystokinin (= pancreozymin-cholecystokinin, PZCCK)	CCK
cholecystokinin receptor	CCKR
cholecytokinin (brain)	CCK_8
chorio-allantoic membrane	CAM
chorionic gonadotropin	CG
chorionic somatomammotropin	CS
cocaine- and amphetamine-regulated transcript	CART
colony-stimulating factor	CSF
complete androgen insensitivity syndrome	CAIS
computer axial tomography	CAT
congenital adrenal hyperplasia	CAH
corticosteroid binding globulin = transcortin	CBG
corticosteroid receptor	CR
corticotropin	ACTH
corticotropin-like peptide	CLIP
corticotropin-releasing factor = CRH	CRF
corticotropin-releasing hormone	CRH
CRH receptor type 1	CRH-R1
CRH receptor type 2	CRH-R2
C-type natriuretic peptide	CNP
cumulus cell oocyte complex	COC
cyclic adenosine monophosphate	cAMP
cyclin-dependent kinase	Cdk
cyclic guanosine monophosphate	cGMP
cytotoxic T lymphocyte	CTL

Term	Abbreviation
dehydroepiandrosterone	DHEA
deiodinase, type 1	D1
deiodinase, type 2	D2
deiodinase, type 3	D3
delta sleep-inducing peptide	DSIP
Δ^5,3β-hydroxysteroid dehydrogenase	3β-HSD
deoxycorticosterone	DOC
deoxyribonuclei acid	DNA
DHEA sulfate	DHEAS
diacylglycerol	DAG
diethylstilbestrol	DES
dihydrotestosterone	DHT
dihydroxyphenylalanine	DOPA
dipeptidyl peptidase-4	DPP-4
disorder of sexual development	DSD
doublesex- and mab-3-related transcription factor 1 gene	*drmt1*
DM domain of the Y-chromosome/ doublesex- and mab-3-related transcription factor 1 gene pair in medaka	*dmy/drmt1bY*
dogfish (shark) GnRH	dfGnRH
dorsal hypothalamic nucleus	DHN
dorsomedial nucleus	DMN
dosage-sensitive sex reversal, adrenal hypoplasia critical region, on chromosome X gene 1	*dax1*
endocabinoid receptor	CB1R
Endocrine Disruptor Screening Program	EDSP
endocrine-active chemical	EAC
endocrine-disrupting chemical	EDC
endogenous opioid peptide	EOP
enterochromaffin-like cell	ECL
enzyme-linked immunoabsorbant assay	ELISA
epidermal growth factor	EGF
epiregulin	EREG
estrogen receptor	ER
estrogen-related receptor	ERR
ethynylestradiol	EE_2
extracellular matrix	ECM
extracellular signal-regulated kinases	ERK1/2
extrapancreatic tumor hypoglycemia	EPTH
fibroblastic growth factor	FGF
fibronectin type III domain containing 5	FNDC5
first-phase insulin release	FPIR
follicle-stimulating hormone	FSH
forkhead box L2 gene	*foxl2*
free fatty acid (see NEFA)	FFA
FSH receptor	FSHR
G-protein-coupled receptor	GPCR
G-protein-coupled receptor 54 = kiss 1R	GPR54
galanin-like peptide	GALP
gamma-amino butyric acid	GABA
gastrin releasing peptide	GRP
genotypic sex determination	GSD
germinal vesicle breakdown	GVBD
GH receptor	GHR
GLP-1 truncated form	tGLP-1
glucagon-like peptide	GLP
glucocorticoid receptor	GR
glucocorticoid receptor, type 1 (= MR)	GR1
glucocorticoid receptor, type 2	GR2
glucose transporter 2	GLUT2
glucose transporter 4	GLUT4
glumitocin	GLT
glycoprotein hormone	GpH
glycoprotein hormone α-subunit	GpHα
glycoprotein hormone β-subunit	GpHβ
GnRH-associated peptide	GAP
gonadal soma-derived factor gene	*gsdf*
gonadotropin inhibitory hormone	GnIH
gonadotropin-regulated testicular helicase	GRTH/DDX25
gonadotropin-releasing hormone	GnRH
gonadotropin	GTH
green fluorescent protein	GFP
growth differentiation factor 9	GDF-9
growth factor receptor-binding protein	GRB2
growth hormone	GH
growth hormone secretogogue-1a	GHS-1a
growth hormone-releasing hormone	GHRH
guanosine diphosphate	GDP
guanosine triphosphate	GTP
high-performance liquid chromatography	HPLC
high-performance liquid chromatography/ mass spectroscopy	HPLC-MS
histone acetyltransferase	HAT
histone deacetylase	HDAC
hormone-response element	HRE
hormone-sensitive lipase	HSL
human chorionic gonadotropin	hCG
human hypercalcemia of malignancy	HHM
human placental lactogen	hPL
hyaluronan synthase 2	HAS2
hydroxy-*O*-methyltransferase	HIOMT
hypothalamus–pituitary–adrenal	HPA
hypothalamus–pituitary–gonad	HPG
hypothalamus–pituitary–hepatic	HPH
hypothalamus–pituitary–thyroid	HPT
IGF-1 receptor gene	*IGF1r*
infundibular nucleus	IFN
inositol trisphosphate	IP_3
insulin receptor gene	*Ir*
insulin receptor substrate	IRS
insulin-dependent diabetes mellitus	IDDM
insulin-like growth factor	IGF
insulin-like growth factor-I receptor	IGF1R
insulin-related receptor gene	*Irr*
interleukin	IL
intermediolateral cell column	IML
interstitial cell-stimulating hormone	ICSH
islet-specific glucose-6-phosphatase catalytic subunit-related protein	IGRP265–273
isotocin	IST
Janus kinase	JAK
Janus kinase 2	JAK2
kisspeptin	Kp
kisspeptin 1 receptor	kissR1
knock-out	KO
lamprey GnRH	lGnRH
lateral hypothalamus	LH

Term	Abbreviation
lateral tuberal nucleus	LTN
LH receptor	LHR
lipotropin	LPH
liquid chromatography combined with mass spectroscopy	LC-MS
long-acting thyroid-stimulator	LATS
low-density lipoprotein droplet	LDL
luteinization of atretic follicles	LAF
luteinizing hormone	LH
luteinizing hormone-releasing hormone	LHRH
luteotropic hormone	LTH
lysine vasopressin	LVP
magnetic resonance tomography	MRT
mammalian GnRH	mGnRH
matrix metalloprotease	MMP
maturation promoting factor; see 17,20β-dihydroxy-4-pregnen-3-one	MPF
maturation-inducing hormone	MIH
medaka GnRH	mdGnRH
medial preoptic nucleus	mPON
melanin-concentrating hormone	MCH
melanocortin receptor accessory protein 1	MRAP1
melanocortin-1 receptor	MC1R
melanocortin-2 receptor	MC2R
melanocortin-3 receptor	MC3R
melanocortin-4 receptor	MC4R
melanotropin	MSH
melanotropin release-inhibiting hormone	MRIH
menopausal gonadotropin	MG
mesotocin	MST
migrating motor complex	MMC
mineralocorticoid receptor	MR
mitogen-activated protein kinase	MAPK
monoacylglyceride	MAG
monoamine oxidase	MAO
monocarboxylate transporter 8	MCT8
MST receptor	MTR
müllerian-inhibiting substance	MIS
N-acetyltransferase	NAT
Na-I symporter	NIS
natriuretic peptide	NP
natriuretic peptide, B-type	BNP
natriuretic peptide, C-type	CNP
natriuretic peptide receptor	NPR
neurointermediate lobe	NIL
neuropeptide Y	NPY
neurosecretory	NS
neurotensin	NT
nitric oxide	NO
nitric oxide synthetase	NOS
non-esterified fatty acid (= free fatty acid)	NEFA
non-insulin-dependent diabetes mellitus	NIDDM
nuclear factor-kappaB	NF-κB
nucleus of the solitary tract	NTS
obese receptor a	ObRa
obese receptor b	ObRb
organum vasculosum lateralis terminae	OVLT
oxytocin	OXY
oxytocin gene-related receptor	VT3R

Term	Abbreviation
oxytocin receptor	OXTR
pancreastatin	Pst
pancreatic polypeptide	PP
pancreatic secretory trypsin inhibitor	PSTI-61
pancreozymin-cholecystokinin (see CCK)	PZ-CCK
parathyroid hormone	PTH
parathyroid hormone-like	PTH-L
parathyroid hormone receptor	PTHR
parathyroid hormone-related peptide	PTHrP
paraventricular nucleus	PVN
peptide histidine isoleucine	PHI
peptide histidine methionine	PHM
peptide YY	PYY
peptidylglycine alpha-amidating monooxygenase	PAM
perchlorate ion	ClO_4^-
periventricular nucleus	PERIV
peroxisome proliferator-activated receptor	PPAR
pertechnetate	TcO_4
phasvatocin	PhaT
Phe-Met-Arg-Phe	FMRF
phenylethanolamine *N*-methyltransferase	PNMT
phenypressin	PVP
phosphatidylinositol bisphosphate	PIP_2
phosphoenolpyruvate carboxykinase	PEPCK
phospholipase C	PLC
pituitary adenylate cyclase activating peptide	PACAP
pituitary-specific transcription factor	Pit-1
platelet-derived growth factor	PDGF
polybrominated diphenyl ether	PBDE
polychlorinated biphenyl	PCB
polycystic ovarian syndrome	PCOS
polymerase chain reaction	PCR
positron emission tomography	PET
prealbumin	TBPA
pregnant mare serum gonadotropin	PMSG
preoptic area	POA
preoptic nucleus	PON
preproparathyroid hormone	PreProPTH
pro-opiomelanocortin	POMC
progesterone	P_4
progesterone receptor	PR
prolactin	PRL
prolactin release-inhibiting hormone	PRIH
prolactin-releasing hormone	PRH
prolactin-releasing peptide	PrRP
proopiocortin	POC
proopiomelanocortin	POMC
proparathyroid hormone	ProPTH
prostacyclin	PGI_2
prostaglandin	PG
prostaglandin E_2	PGE_2
prostaglandin $F_{2\alpha}$	$PGF_{2\alpha}$
prostaglandin-endoperoxide synthase 2	PTGS2
protein kinase A	PKA
protein kinase C	PKC
pseudoisocyanin	PIC
radioimmunoassay	RIA

Term	Abbreviation
rat sarcoma signaling	RAS
relaxin/insulin-like family peptide receptor	RXFP
release-inhibiting hormone	RIH
releasing hormone	RH
retinoic acid receptor	RAR
retinoid X receptor	RXR
reverse T_3	rT3
ribonucleic acid	RNA
RNA interference	RNAi
rostral ventrolateral medulla	RVLM
saccus vasculosus	SV
salmon GnRH	sGnRH
sea bream GnRH	sbGnRH
seasonal affective disorder	SAD
selective estrogen receptor modulator	SERM
serine protease inhibitor	serpin
serotonin (5-hydroxytryptamine)	5-HT
sex-determining region Y gene	*sry*
sex hormone binding globulin	SHBG
sexually dimorphic nucleus	SDN
signal transducer and activator of transcription	STAT
socioeconomic status	SES
sodium dodecyl sulfate–polyacrylamide gel electrophoresis	SDS-PAGE
somatolactin	SL
somatostatin	SST
somatostatin receptor subtype 2	SSTR2
son of sevenless	SOS
SRY-related high-mobility group box 9 gene	*sox9*
stanniocalcin	STC
steroid hormone-binding globulin	SHBG
steroidogenic acute regulating protein	StAR
steroidogenic factor 1	SF-1
subcommissural organ of Dendy	SCO
superior cervical ganglion	SCG
suppressor of cytokine signaling	SOCS3
suprachiasmatic nucleus	SCN
supraoptic nucleus	SON
syndrome of inappropriate antidiuresis	SIAD
T_3 receptor auxiliary protein	TRAP
tartrate-resistant acid phosphatase	TRAPase
technicium	Tc

Term	Abbreviation
temperature-dependent sex determination	TSD
teneurin C-terminal-associated peptide	TCAP
testis determining factor	TDF
tetraiodothyroacetate	TETRAC
thiocyanate	SCN^-
thiourea	TU
three-dimensional quantitative structure–activity relationship	3D-QSAR
thyroglobulin	Tgb, TGB
thyroid-binding globulin	TBG
thyroid peroxidase	TPO
thyroid receptor	TR
thyroid response element	TRE
thyroid transcription factor 1	ttf-1
thyronamine	TAM
thyrotropin	TSH
thyrotropin-releasing hormone	TRH
thyroxine	T_4
trace amine associated receptor 1	TAAR1
transforming growth factor β	TGF-β
transforming growth factor β_2	TGF-β_2
transient receptor potential vanilloid subtype 1	TRPV1
transthyretin	TTR
triacylglyceride	TAG
tributyltin	TBT
triiodothyroacetate	TRIAC
triiodothyronine	T_3
tuberoinfundibular protein 39	TIP39
tumor necrosis factor alpha	TNF-α
uncoupling protein 1	UCP1
urocortin receptor 2	Ucn-R2
valitocin	ValT
vascular endothelial cell growth factor	VEGF
vasoactive intestinal peptide	VIP
ventral hypothalamic nucleus	VHN
ventral tegmental area	VTA
ventricular natriuretic peptide	VNP
ventromedial nucleus	VMN
very-low-density lipoprotein droplet	VLDL
vitamin D receptor	VDR
voltage-sensitive calcium channel	VSCC
vomeronasal organ	VNO
wastewater treatment plant	WWTP

Vertebrate Phylogeny and Evolution

A comparative study of vertebrate bioregulatory systems necessitates some knowledge of the major vertebrate groups, their evolutionary history, and their relationships to one another. Each group of vertebrates—each species, in fact—is a product of individualistic evolutionary change as well as a product of adaptations to similar environmental problems faced by unrelated groups. The comparative endocrinologist faces the task of sorting out similarities due to convergent evolution of structures and functions as opposed to similarities due to common ancestry.

The animals we classify as chordates are all members of the phylum Chordata that includes some invertebrate groups as well as the vertebrates that have evolved over millions of years. Some species are ancient and others are more recent. However, examination of an endocrine mechanism or the structure of a gene responsible for a given hormone in a phylogenetically old species may shed little light on its specific ancestry because of changes that are a divergence from the ancestral form.

All members of the phylum **Chordata** exhibit these basic features at some stage in their development: (1) **pharyngeal gill slits**; (2) **dorsal, hollow nervous system**; and (3) supportive endoskeletal element, the **notochord**, which lies beneath the dorsal nerve cord. Each of these features must be present at some stage in the life cycle to qualify for membership in the Chordata. In addition, most chordates possess a **postanal tail**, and fishes and amphibians have an aquatic **tadpole-like larva**. However, in some fishes and amphibians as well as in all reptiles, birds, and mammals, a free-living larval stage per se no longer exists, having been reduced to a transitory embryonic event. Earlier classifications of chordates were based largely on anatomical and developmental features, but modern schemes rely more heavily on a wide range of shared genetically analyzed characteristics of many kinds and attempts to distinguish among parallel, convergent, and divergent evolution.

In the more traditional approaches to classification (see Figure B-1), the phylum Chordata is subdivided into three subphyla: **Urochordata** (tailed chordates that are called *sea squirts*, *ascidians*, or *tunicates*), **Cephalochordata** (head chordates; i.e., amphioxus), and **Vertebrata** (vertebrates). These traditional subphyla are subdivided further into a variable number of Classes, Orders, Families, Genera, and species. However, in the cladistic approach to classification (i.e., phylogenetic systematics), taxonomic groups are designated by sharing derived characteristics such as the cranium and backbone that characterize all of the vertebrates, or the absence of these features, as is the case for urochordates, cephalochordates, and some jawless fishes. Although the cladistic approach also produces a hierarchy of relationships among groups, it results in different groupings (Figure B-2) that often transcend the earlier classifications that used categories such as classes and orders. This in turn necessitated coining new names for the newer groupings of organisms that are no longer connected to the historical groupings commonly used in the literature. In the following account, group names are used primarily from the older classification system without designating them as Subphyla, Classes, Orders, etc., because this approach is more consistent with much of the endocrine literature prior to the present decade.

I. THE INVERTEBRATE CHORDATES

A. Urochordates

The all-marine Urochordata is considered to be the most primitive chordate group. A free-swimming aquatic tadpole larva is characteristic. The larva possesses the three basic chordate features, but these are modified or eliminated when the larva undergoes drastic structural modifications during metamorphosis to become a sessile, aquatic adult that is attached firmly to some substrate. The dorsal, hollow nervous system degenerates to a single neural ganglion, the notochord is obliterated, and the animal secretes an exoskeleton or tunic that completely encases the adult (hence, the common name used for this group: tunicates).

The adult tunicate (sometimes called a sea squirt because of its habit of ejecting a fine stream of water when

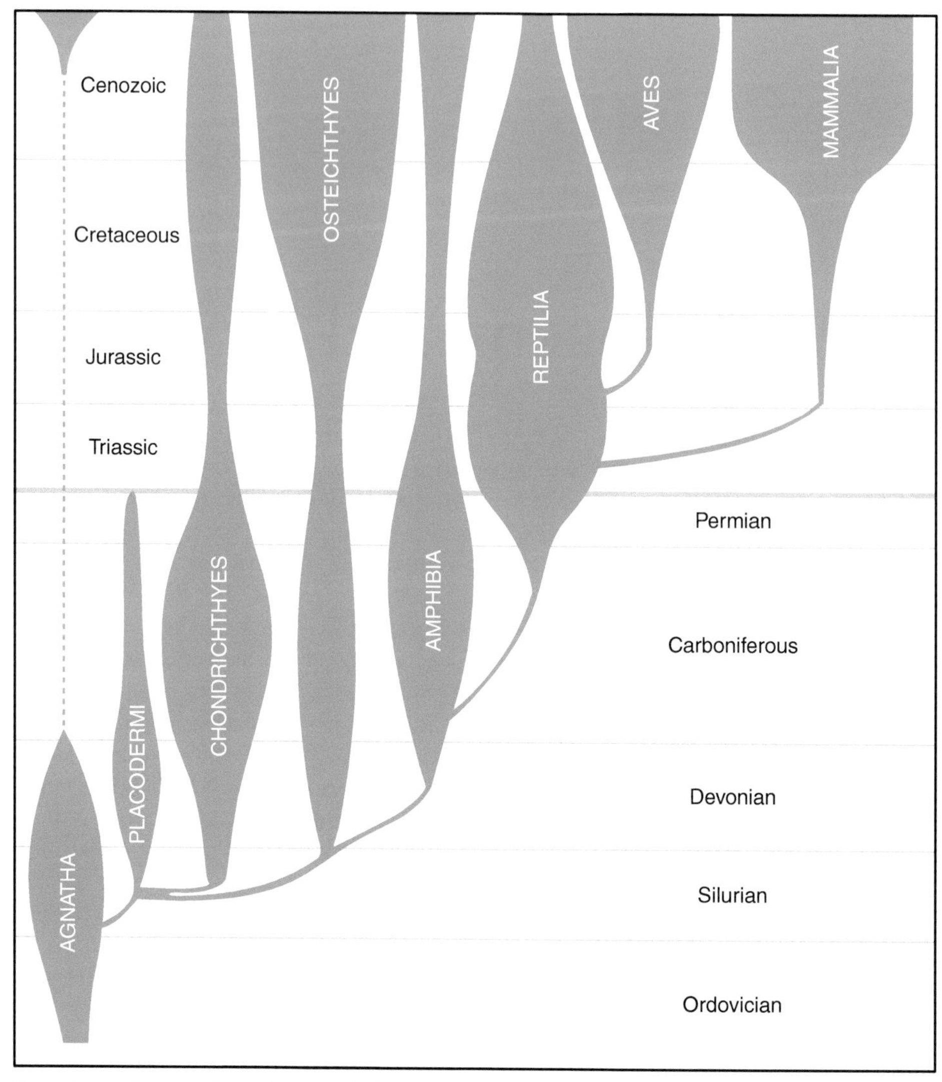

FIGURE B-1 **Traditional visualization of vertebrate evolution.** Relative abundance of some major vertebrate groups through geologic time. The width of each group representation is roughly proportional to the number of species known from the fossil record and those living today.

disturbed by a curious biologist) retains only one obvious chordate characteristic: a gill structure called a *branchial basket*. This apparatus is covered with cilia and has a mucus-secreting structure associated with it, the **endostyle**. Coordinated ciliary movements cause a current of water to flow into the branchial basket (via the mouth of the larva or the incurrent siphon of the adult) and out via the gill slits (the excurrent siphon of the adult). Mucus secreted by the endostyle traps minute food particles, and the mucus plus trapped food are moved by ciliary action into the gut, where both the mucus and trapped food particles are digested. This method for obtaining food is common among many invertebrates, including those groups believed to have given rise to the chordates. Organisms possessing such a feeding mechanism are called **ciliary mucus feeders**, or pharyngeal-filtration feeders.

B. Cephalochordates

The sessile urochordate or tunicate is considered to be an evolutionary dead end, but some ancestral form similar to the larval tunicate may have given rise to the marine Cephalochordata. Certain living tunicates (for example, *Oikopleura* spp.) never undergo metamorphosis to a sessile adult but remain free-living and attain sexual maturity while retaining the larval body form. Prolongation of larval life or retention of larval characteristics in sexually mature animals is termed **paedomorphosis**. If paedomorphosis is brought about by delayed development of nonreproductive (somatic) tissues, it is called **neoteny**. If it is thought to be a case of precocious gonadal development, it is called **progenesis**. Paedomorphosis has been evoked as an important mechanism in chordate evolution and there are numerous examples of it among extant vertebrates.

FIGURE B-2 **A cladogram emphasizing the phylogeny of vertebrates.**

The cephalochordates (such as the living amphioxus) have a body plan very similar to larval urochordates, including a branchial basket with a mucus-secreting endostyle and a persistent notochord throughout their life. Furthermore, cephalochordates anatomically resemble the larvae of lampreys (see Figure B-3).

Cephalochordate larvae and adults are ciliary-mucus feeders like the urochordates. Similarities among cephalochordates, urochordates, and larval lampreys support a common evolutionary origin for all three groups, but it is not certain how these groups are related to one another. The ancestral vertebrate may not have been a member of either invertebrate chordate group. Vertebrates have many features not found in either urochordates or cephalochordates—for example, specializations of the head, anterior nervous system, and pharyngeal breathing apparatus that are responsible for the active predaceous life of vertebrates.

II. THE VERTEBRATE CHORDATES

There are a number of major groups that comprise the vertebrates (Figures B-1 and B-2). Two groups of fishes are entirely extinct, whereas the other vertebrate groups have living members. In addition to possession of the three chordate characteristics, the vertebrates have spinal cords with protective **vertebrae** of cartilage or bony structures that surround and protect the spinal cord. Furthermore, there is a special protective case, the **cranium**, that surrounds the enlarged anterior portion of the nervous system, the brain. This latter feature is the basis for another name sometimes applied to vertebrates, the Craniata. Vertebrates are characterized by a duplication of the entire genome at some early time in their evolution. Among the bony fishes, there appears to have been a second complete duplication of the genome in the teleost fishes that explains the presence of duplicate forms of many chemical regulators and receptors.

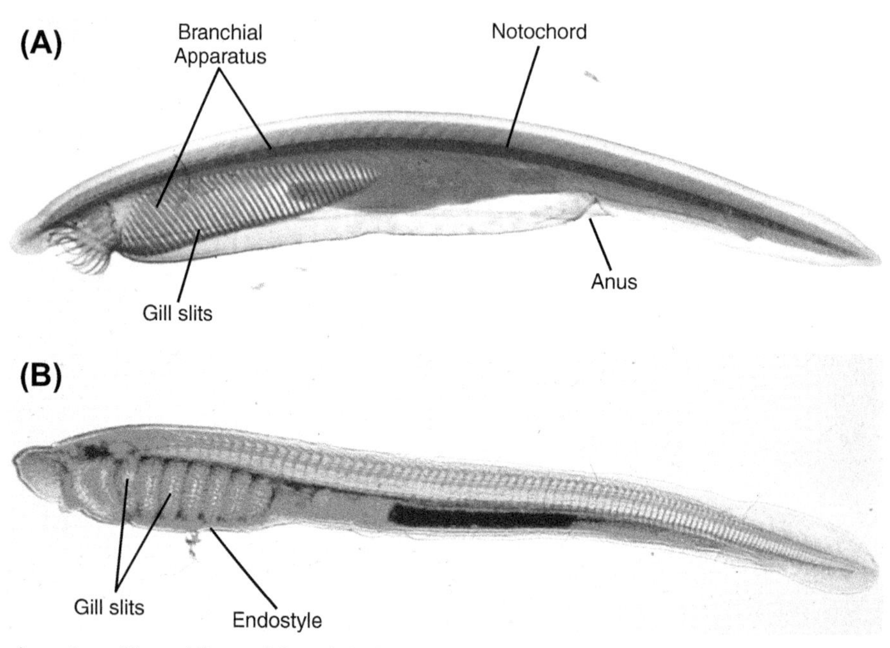

FIGURE B-3 **Comparison of amphioxus (A), an adult cephalochordate, with the ammocetes larva of a lamprey (B), an agnathan fish.** Note the prominent dark, dorsal notochord in amphioxus that appears clear in the ammocetes larva. An endostyle is present in both but is not evident in this photograph of amphioxus.

A. Agnathan Fishes

The Agnatha consists of ancient, jawless fishes (*a,* without; *gnathos,* jaws) believed to have evolved directly from cephalochordates or some cephalochordate-like ancestor. They traditionally are divided into two subgroups: the extinct ostracoderms and the extant cyclostomes (*kyklos,* round; *stoma,* mouth) consisting of the lampreys and hagfishes. The ostracoderms were all small fishes covered with bony plates (armor). They were limited to the ocean bottom, where they probably existed primarily as ciliary-mucus feeders. Although the ostracoderms were not sessile like the urochordates, most possibly lived a "sit-and-sift" existence close to the bottom sediments of the oceans; however, recent interpretations of some ostracoderm fossils suggest at least some of them may have been active swimmers and possibly predators. There are two taxonomic groups of living cyclostomes: Myxinoidea (marine hagfishes) and Petromyzontidae (essentially freshwater lampreys).

Many adult lampreys are parasitic on other vertebrate fishes, but the larvae are ciliary-mucus feeders. The ammocetes larva of lampreys has a branchial basket with an endostyle, and, in general, the body form looks much like the cephalochordate amphioxus (see Figure B-3). Although some biologists suggest that these structural similarities imply a close evolutionary relationship, others would argue against such an interpretation. When the ammocetes larva metamorphoses to the adult lamprey, the endostyle differentiates into the thyroid gland of the adult (see Chapter 7).

The Myxinoidea do not have a larval form and have many more primitive features than lampreys; for example, the hagfishes lack vertebrae and possibly should not be classified with the vertebrates. However, recent developmental studies have identified the presence of neural crests near the dorsal nervous system, a distinct vertebrate character. Although modern hagfishes may not be much like the first vertebrate, their apparent simplicity may be a degenerate condition rather than a representation of a primitive vertebrate state. Most cladistic analyses place the lampreys closer to the jawed vertebrates than to the hagfishes, indicating that hagfishes and lampreys diverged much earlier from a common ancestor. Lampreys are probably a better living example of a primitive vertebrate than are hagfishes.

B. Placoderm Fishes

The ostracoderm fishes were ancestors of the first jawed vertebrates, the placoderms, a heterogeneous collection of extinct, heavily armored fishes. Hinged jaws developed in the placoderm fishes from modifications of the first gill arch, and this same event can be observed early in embryonic development of all jawed vertebrates. Many zoologists consider development of jaws to be the most significant single event in vertebrate evolution. Certainly it was a significant advancement enabling the placoderm fishes to abandon the bottom-dwelling existence of their ancestors for a pelagic, predatory life style. The fossil record suggests, however, that some placoderms retained the ancestral filter-feeding habit. The placoderms were the dominant vertebrates of the Devonian period, but they declined and disappeared, leaving only a strong fossil record. During the Devonian, however, the placoderms gave rise to two important piscine taxa: the bony fishes (Osteichthyes) and the cartilaginous fishes (Chondrichthyes). These predatory descendants retained the jaws of their ancestors, but the bony armor was reduced to scales, allowing for an emphasis on speed and agility. The demise of the placoderms was probably due in no small part to the success of these more mobile predators.

C. Chondrichthyean Fishes

The chondrichthyeans (*chondros*, cartilage; *ichthy*, fish) have skeletons primarily composed of cartilage or calcified cartilage. Although often referred to as *the* "cartilaginous fishes," the agnathans also have a cartilaginous skeleton. True bone is not present in this group. Because cartilage forms prior to bone in the normal developmental sequence, some zoologists suggest that this group arose from the bony placoderms via neoteny. Included in the Chondrichthyes taxon are the sharks, rays, and skates (= **selachians** or **elasmobranchs**) and the ratfishes or chimaeras (**holocephalans**). The chondrichthyeans flourished for a time but then declined. Although in recent geological time they are increasing in abundance of species, they are believed to represent an evolutionary dead end and did not give rise to any other vertebrate group. The chondrichthyean fishes have not been as successful as the higher bony fishes (teleosts) in exploiting the aquatic environment (especially freshwater), and they represent a secondary fish fauna today.

D. Osteichthyes: The Bony Fishes

Osteichthyean fishes (*osteon*, bone) have excelled in exploitation of freshwater and marine habitats. These modern bony fishes may have evolved in fresh water and secondarily invaded the marine habitat. It appears that freshwater bony fishes also gave rise to the first terrestrial vertebrates, the amphibians, probably before they exploited the marine environment. The bony fishes can be readily separated into two major subgroupings: **Actinopterygii**, or ray-finned fishes, and the **Sarcopterygii**, or lobe-finned fishes. The Actinopterygii (spiny wings or fins) have distinct rays that support the fins, whereas the Sarcopterygii (fleshy fins) have lobed fins with internal skeletal supports that are homologous to the limb bones of tetrapods. Early in the evolution of the bony fishes (or possibly even in the placoderm group that gave rise to the bony fishes), a pouch developed ventrally off the gut anterior to the stomach, remained connected to the gut by a duct, and evolved into the **air bladder**. In the actinopterygian fishes, this air bladder was used as a flotation device or **swim bladder**. Among the sarcopterygians, the air bladder became modified as an accessory breathing device homologous to the lungs of terrestrial vertebrates.

E. Osteichthyes: The Actinopterygii

The four distinctive groups of actinopterygians include most of the living bony fishes: **Polypteri, Chondrostei, Holostei**, and **Teleostei**. The most ancient ray-finned fishes are the polypterans, of which the two living freshwater genera (*Polypterus* and *Calamoichthyes*) are found in Africa. The chondrosteans include sturgeons (e.g., *Acipenser* spp.) and spoonbills of Asia (*Psephrurus gladius*) and North America (*Polyodon spathula*). Some zoologists designate the polypterans as the most primitive chondrosteans. The chondrostean fishes all live in fresh water in the Northern Hemisphere. The living holosteans are restricted to North America, occurring today only in the Mississippi River drainage area. This group consists of the bowfin (*Amia calva*) and several species of gar (*Lepisosteus*). The teleost fishes (Teleostei) are the most advanced actinopterygians. They exhibit a tremendous adaptive radiation in fresh water and have secondarily invaded seawater where they are the most abundant and successful vertebrate group. The majority of all extant fish species are teleosts, with estimates from almost 29,000 described species to as many as 40,000 species. Sometimes the teleosts are divided into the so-called lower teleosts, exemplified by salmonid fishes (e.g., trout, salmon), and the higher teleosts, such as the centrarchids (e.g., largemouth bass, bluegill sunfish). The higher teleosts can be recognized by the strong tendency for the pelvic (ventral) paired fins to move anteriorly close to the pectoral (shoulder) paired fins.

F. Osteichthyes: Sarcopterygii

The ancestors of the first four-footed or tetrapod vertebrates were the sarcopterygian fishes. Two groups of sarcopterygian fishes have living representatives: two species of the **Crossopterygii** (fringe fins) and three genera of the **Dipnoi** (*dipnoos*, double breathing; i.e., in air and in water). Both of these groups represent offshoots from the main line of evolution within the Sarcopterygii.

Crossopterygian fishes were known only from their excellent fossil record until 1938 when the first living crossopterygian (*Latimeria chalumnae*), caught by a fisherman off the coast of Madagascar, attracted the attention of some scientists. Since that time a number of these bizarre, bluish giants have been captured and their anatomy, physiology, behavior, and ecology closely scrutinized by comparative zoologists. In 1998, a second species, *Latimeria menadoensis*, was discovered in Indonesia. Analysis of DNA from *L. menadoensis* confirmed its status as a separate species. *Latimeria*, like other crossopterygian fishes, has internal nares and a lung-like air bladder. It may reach 5 to 6 feet in length and is viviparous (live-bearing).

The Dipnoi consists of three genera of lungfishes (seven species) restricted to tropical regions on three continents: *Protopterus* in Africa, *Lepidosiren* in South America, and *Neoceratodus* in Australia. These fishes are gill breathers that use their lungs as accessory breathing structures. Only *Protopterus* survives breathing air alone. *Protopterus* can secrete a mucus-lined cocoon in which it resides and breathes air during periods of intense drought when its aquatic habitat may disappear altogether.

Recent molecular studies suggest that lungfishes are more closely related to the amphibians (hence to other terrestrial vertebrates) than are the crossopterygians. Furthermore, the single species of the freshwater Austalian lungfish (*Neoceratodus fosteri*) is considered the most primitive of the living lungfishes and therefore presumed to be closest to the ancestral vertebrates that first invaded the land. Studies by Jean Joss and her colleagues of the development and physiology of this species in an outdoor lungfish facility near Sydney suggest that this lungfish may be neotenic.

III. AMPHIBIA

There are three living groups of amphibians: (1) **Caudata** (= Urodela; both *uro* and *cauda* refer to "tail"), which includes salamanders and newts; (2) **Gymnophiona** (= Apoda; without feet), the limbless caecilians, a tropical group about which little endocrinology is known; and (3) **Anura** (= Salientia; without tail), the tail-less frogs and toads, which are the best-studied amphibians in spite of their obvious deviation from the basic tetrapod body plan. A primitive group of large-tailed amphibians (**labyrinthodonts**) most probably had their origin from close relatives of the lungfishes that had developed internal nares and lungs for air breathing and had supportive skeletal elements in their fleshy lobed fins. The labyrinthodonts also gave rise to the first reptiles as well as to the modern amphibians.

As their common name implies (*amphi*, both; *bios*, life), many amphibians lead double lives: one as an aquatic larva and the second as a terrestrial or semiterrestrial adult. This "typical" life history involves laying eggs in freshwater where they develop into tadpole-like larvae. The larvae remain in freshwater for a period of growth followed by a remarkable hormone-regulated metamorphosis involving drastic structural and biochemical alterations to attain the different adult body form and the physiology and behavior to survive in a desiccating environment (see Chapter 7). Although numerous species of amphibian have diminished their attachment to the aquatic environment through live bearing or laying eggs on land, a number of species have retained the larval body form (paedomorphosis) and become permanently aquatic.

IV. THE AMNIOTES

Traditional classification separates the reptiles, birds, and mammals into three distinct groupings, but modern phylogenetic analyses suggest that reptiles and birds represent a single clade and that mammals arose from one part of this clade. Indeed, the crocodilians resemble birds very closely in their anatomy, physiology, and biochemistry, and together these two living groups are often considered to constitute a unique grouping, the living archosaurs.

A. Reptiles

In a sense, it was a mistake for the amphibians to give rise to the reptiles, for the reptiles quickly replaced them as the dominant terrestrial vertebrates. Reptiles owed their success in exploiting the terrestrial environment in large part to the evolution of a unique "land egg," which enclosed the aquatic environment for embryonic development within a membrane called the **amnion**. This **amniote egg** could be laid on land where both egg and "larva" were safe from aquatic predators. Due to a leathery or calcified shell, the reptilian egg is much more resistant to desiccation than even the "terrestrial" eggs of amphibians. Because the reptiles were no longer tied to water for reproduction, there were fewer restrictions on their movements, and they moved into all sorts of terrestrial habitats, including deserts. The large amniote eggs of reptiles allow young to hatch at a size considerably greater than is possible from the smaller eggs of most oviparous fishes and amphibians.

Birds and mammals have retained many of the features of the reptilian egg, including the amnion and other membranes (**chorion, allantois**, and **yolk sac**). Bird eggs are very similar to reptilian eggs, and embryonic development also is very similar. In mammals, these membranes have been greatly modified, especially among the placental mammals. Reptiles, birds, and mammals are often referred to collectively as the amniote vertebrates, or **amniotes** (that is, they all possess an amnion), whereas fishes and amphibians are termed **anamniotes** (without an amnion as well as the other membranes).

Primitive amphibians gave rise to the **cotylosaurs**, or stem reptiles, which early in reptilian evolution diverged along several separate pathways. Only four of these pathways have living representatives. One pathway (**Anapsida**) separated early and gave rise to the heavily armored oviparous **chelonians**: turtles, tortoises, and terrapins. The chelonians are anatomically a conservative group, having changed little in appearance over several million years; however, it would be a mistake to assume that all of their physiology has remained equally conservative.

A second pathway from the cotylosaurs (**Lepidosaura**) produced two important groups, the **Squamata** (snakes and lizards) and the more ancient **Rhynchocephalia**, which has only one living species, the New Zealand tuatara (*Sphenodon punctatus*). The tuatara is of special interest because it is the only living representative of a very old reptilian group. Live bearing (**viviparity**) evolved numerous times within the squamates, with variations from simple retention and development of the eggs within the oviducts to development of a placenta.

The **Archosaura** represents a third evolutionary line from the cotylosaurs, of which only the crocodilians (crocodiles and alligators) have survived among the reptiles. The extinct dinosaurs were part of this evolutionary line. One group of reptilian archosaurs, the **thecodonts**, gave rise to modern birds (**Aves**).

The fourth reptilian group derived from the cotylosaurs, the now-extinct **Synapsidia**, gave rise to the mammals. This reptilian group separated early and gave rise to mammals even before the archosaurs gave rise to birds. Apparently, the ability to maintain a relatively constant body temperature developed in the synapsids (specifically in the **therapsid** group of synapsids) independently of its development in the thecodont reptiles, which gave rise to the birds.

B. Birds

Birds are characterized by having feathers, no teeth, a relatively constant body temperature (ranging from 37 to 43°C in different species), a four-chambered heart analogous but not homologous to that of mammals, and numerous structural modifications for flight. Their complicated mating rituals and intricate behavioral mechanisms associated with rearing of young have contributed greatly to their success. Although viviparity has developed in all other tetrapod classes (as well as in some fishes), all birds lay eggs. However, birds have been successful at exploiting the terrestrial habitat in such a way as to avoid undue competition with reptiles and mammals, and hence birds exhibit an impressive adaptive radiation. Lack of internal development is compensated in birds by extended parental care of eggs and young.

C. Mammals

The most distinctive and uniform features of mammals are the possession of hair and mammary glands, for which the group is named. Like birds, mammals maintain a high internal body temperature but, unlike birds, are almost exclusively live bearing. Mammals traditionally have been separated into three distinct subgroups: **Prototheria** (*protos*, first; *thereon*, animal), **Metatheria** (*meta*, middle), and **Eutheria** (*eu*, good).The **monotremes** are the most primitive group of mammals (Monotremata or Prototheria). This group includes the duckbill platypus (*Ornithorhynchus anatinus*) and the spiny anteaters (*Tachyglossus* spp.). All members of this group are oviparous and are found only in Australasia.

The Metatheria consists of the **marsupials**, or pouched mammals (*marsupium*, pouch). After a very short intrauterine period, newborn marsupials spend most of their early life in the pouch provided by their mother. The marsupials include the kangaroos, wallabies, koala, and many others species in Australia; the opossum of North America; and a few rodent-like marsupials still persisting in South America. The fossil record indicates that marsupials were just beginning their adaptive radiation when continental drift began to separate the continents. This explains in part their present skewed distribution, with the vast majority of extant species being found in Australia.

The Eutheria consist of the **placental mammals**. The insectivores represent the most primitive group of eutherian mammals from which 13 other groups have evolved. The eutherian mammals evolved in the Old World, entered North America from Asia, and eventually migrated southward and invaded South America.

Although marsupials once were common elements of the New World, they were replaced almost completely by the eutherians. The geographical isolation of Australia as a result of continental drift allowed the persistence of a great many marsupial species there. The primates are considered (by humans, of course) to be the most highly evolved group of mammals, a group showing the most advanced evolutionary adaptations, the greatest intelligence, and the highest ecological success. The latter remains to be seen.

Appendix C

Amino Acid Abbreviations

Amino acid	Old system	New system	Basic AA	Acidic AA
Alanine	Ala	A		
Arginine	Arg	R	+	
Asparagine	Asn	N	+	
Aspartate	Asp	D		+
Cysteine	Cys	C		
Glutamate	Glu	E		+
Glutamine	Gln	Q	+	
Glycine	Gly	G		
Histidine	His	H		
Isoleucine	Ile	I		
Leucine	Leu	L		
Lysine	Lys	K	+	
Methionine	Met	M		
Phenylalanine	Phe	F		
Proline	Pro	P		
Serine	Ser	S		
Threonine	The	T		
Tryptophan	Trp	W		
Tyrosine	Tyr	Y		
Valine	Val	V		

Appendix D

Bioassays

The bioassays described below represent the more common ones for the major peptide and protein hormones. Additionally, some bioassays are described for thyroid hormones and estrogens that are being used today to monitor **endocrine-disrupting chemicals (EDCs)**. Although largely of historical interest, these bioassays sometimes are useful when functional endpoints are required or when precise radioimmunoassay or biochemical identification is not possible.

I. GONADOTROPINS

Specific bioassays have been developed in a variety of vertebrates for distinguishing between **follicle-stimulating hormone (FSH)** and **luteinizing hormone (LH)** biological activities. A summary of some typical responses are provided in Tables D-1 and D-2.

A. Luteinizing Hormone (LH)

One of the first bioassays was the **spermiation test** observed following injection of pituitary extracts or purified LH into frogs or toads. At one time, this bioassay was widely employed by doctors to detect the presence of **chorionic gonadotropin (CG)** in urine of women and confirm early pregnancy. The structural and functional similarity of hCG and hLH to amphibian gonadotropins is the basis for this clinical test. A spermiation test, however, responds to highly purified FSH, too, and cannot be considered a specific bioassay for LH-like hormones. Pregnancy is now determined from urine samples with a simple and highly sensitive immunoassay involving an antibody produced specifically against hCG.

A popular early bioassay for LH was the **ovarian ascorbic acid depletion (OAAD) test**, which was based upon a quantitative reduction of ascorbic acid in rat ovarian tissue following administration of LH. The degree of depletion is proportional to the dose of LH. The OAAD bioassay is not affected by FSH. The relationship of ascorbic acid to hormone secretion is not understood.

The **pigment response** to LH by feathers of the African weaver finch *Euplectes franciscanus* was another highly specific bioassay for LH or CG, but it, too, is cumbersome. The **frog ovulatory response *in vitro*** to LH or CG (Tables D-1, D-2) has resulted in development of several similar bioassays. Administration of pituitary LH or progesterone stimulates ovulation in isolated fragments of anuran or urodele amphibian ovaries. Luteinizing hormone causes progesterone synthesis in follicular cells surrounding the oocyte, and it was believed that the local progesterone binding to oocyte receptors induces the ovulatory event. This *in vitro* ovarian bioassay also has been used for detection of progesterone; however, recent studies suggest that progesterone is further metabolized to testosterone and

TABLE D-1. Relative potencies of Follicle-Stimulating Hormone and Luteinizing Hormone Purified from Different Vertebrates and Tested in the *Anolis* and *Xenopus* Gonadotropin Bioassays[a]

Hormone	Relative potency	
	Anolis bioassay	*Xenopus* bioassay
Ovine FSH	100.0[b]	< 0.005
Ovine LH	0.04	2.0[c]
Snapping turtle FSH	3.0[b]	0.002
Snapping turtle LH	0.13	1.8[c]
Chicken FSH	4.3[b]	< 0.0015
Chicken LH	0.85	0.08[c]
Bullfrog FSH	7.0[b]	0.004
Bullfrog LH	1.2	0.33[c]

[a]Based on Licht, P., Papkoff, H., Farmer, S. W., Muller, C. H., Tsui, H. W., and Crews, D. (1977). Evolution of gonadotropin structure and function. Recent Prog. Horm. Res. 33, 169–248
[b]*Anolis* bioassay for FSH: maintenance of testis weight in hypophysectomized lizard.
[c]*Xenopus* bioassay for LH: in-vitro ovulation.

TABLE D-2. **Relative Effectiveness of Purified Ovine Gonadotropin (FSH and LH) in Some Gonadotropin Bioassays**

Bioassay	Minimal effective dose (μg/ml or μg/injections[a]) FSH	LH	Relative poteney
A. Testis weight maintenance in hypox lizard (*Anolis carolinensis*)	0.01	20.0	FSH >> LH
B. *In vitro* ovulation of frog ovary (*Xenopus laevis*)	> 200.0	0.5	LH >>> FSH
C. Spermiation response by frog (*Hyla regilla*)	0.5	1.0	FSH ≅ LH

[a]*Injection for bioassays A and C.*

the latter is responsible for final oocyte maturation. Although the system is somewhat responsive to a few other steroids, it is relatively unresponsive to FSH and other tropic hormones.

B. Follicle-Stimulating Hormone (FSH)

The major bioassay for FSH prior to development of RIAs was the **increase in testis or ovarian weight in hypophysectomized rats** following injection of pituitary extracts or purified preparations of tropic hormones. Development of any of the androgen-sensitive sex accessory structures in the test animals would be evidence of LH contamination. Another specific bioassay for FSH is the **maintenance of testis weight in male lizards** (*Anolis carolinensis*) following hypophysectomy (Tables D-1, D-2). The lizard testis regresses rapidly following hypophysectomy and normally is unresponsive to LH; hence, this bioassay is very specific for small quantities of highly purified FSH.

II. THYROTROPIN (TSH)

Stimulation of thyroid gland function can be quantified cytologically by means of the TSH dose-dependent stimulation of **epithelial cell height** of the thyroid follicles in hypophysectomized animals; however, this bioassay requires several days to obtain results of these tissue changes. A more rapid bioassay measures the amount of **radioiodide uptake** by thyroid follicles following administration of purified TSH or pituitary extracts. This bioassay can be performed *in vitro* as well as *in vivo*. Epithelial cell height and radioiodide accumulation indicate the degree of TSH stimulation and are also proportional to circulating TSH levels.

III. GROWTH HORMONE (GH)

The classical bioassay for GH is the histological measurement of the thickness of the **epiphysial cartilage** in the tibia of the hypophysectomized rat or mouse following treatment with pituitary extracts or purified molecules. Some typical results for GHs from various vertebrates are provided in Table D-3.

IV. INSULIN-LIKE GROWTH FACTORS (IGFS)

An *in vitro* bioassay for IGFs uses costal cartilage from pigs. The rate of incorporation of radioactive sulfate ions is measured and is directly proportional to the amount of IGF.

V. PROLACTIN (PRL)

Several specific bioassays for PRL have been reported utilizing animals representing four major vertebrate classes (teleosts, amphibians, birds and mammals). Not all vertebrate PRLs are effective in every assay as indicated in Tables D-3 and D-4. The bioassays are

TABLE D-3. **Potencies of Growth Hormone Determined in the Rat Tibia Assay[a]**

Species	Potency[b]
Sturgeon	0.40
Tilapia	0.05
Bullfrog	1.15
Leopard frog	0.36
Snapping turtle	0.24
Sea turtle	0.12
Ostrich	0.94
Duck	0.16
Marsupial	0.03
Mammal	1.00

[a]*From Chester-Jones, I., Ingleton, P. M., and Phillips, J. G. (1987). "Fundamentals of Comparative Vertebrate Endocrinology." Plenum Press, New York.*
[b]*Bovine GH standard = 1.0. Marsupial GHs were nonparallel.*

TABLE D-4. **Bioassayable Prolactin in Vertebrate Pituitaries**

Vertebrate group	*Gillichthyes*[a] xanthophore-yellowing response	Tilapia Na^+-retaining response	Red-eft water drive	Pigeon crop-sac assay	Mouse mammary *in vitro*
Agnatha	+	−	−	−	−
Chondrichthyes	+	−	+	−	−
Chondrostei	+	−	?	−	−
Holostei	+	−	?	−	−
Teleostei	+	+, −	+	−	−[b]
Dipnoi	+	−	+	+	+
Amphibia	+	−	+	+	+
Reptilia	+	?	+	+	+
Aves	+	?	+	+	+
Mammalia	+	+, −	+	+	+

[a]*This assay system may not be specific for prolactin (see text).*
[b]*Purified tilapia PRL stimulates casein synthesis in the rabbit mammary gland.*

1. **Sodium-retaining bioassay** in the cichlid teleost *Sarotherodon* (*Tilapia*) *mossambicus*
2. **Xanthophore-expanding bioassay** in the teleost *Gillichthyes mirabilis*
3. **Red-eft water drive bioassay** performed in the newt *Notophthalmus viridescens*
4. **Crop-sac bioassay** performed in the domestic pigeon
5. ***In vitro* mouse mammary gland bioassay**

A. Fish Bioassays for PRL

1. Sodium-Retaining Bioassay

This bioassay emphasizes the osmoregulatory actions of PRL in teleost fishes, and particularly its effect on sodium uptake across the gill and resultant alterations in plasma sodium levels. Only some pituitaries from teleosts, however, will produce a measurable response in this bioassay, and all other piscine and other non-mammalian preparations are inactive. Purified mammalian PRLs, curiously, work very well in this bioassay. This observation supports the hypothesis that the osmoregulatory portion of the piscine PRL molecule has been retained in the structure of mammalian PRL with the addition of new structural modifications that are responsible for crop-sac and mammary gland actions. This bioassay was first developed for fish PRL in hypophysectomized guppies (*Poecilia latipinna*). Later, it was modified to use intact, seawater-adapted tilapia. Adaptation of these fishes to seawater almost eliminates PRL from the circulation and alleviates the need for hypophysectomy.

2. Xanthophore-Expanding Bioassay

This bioassay is performed in the gobiid fish *Gillichthyes mirabilis* (longjawed mudsucker). It is an all-or-none bioassay for determination of the smallest dose of purified hormone or the greatest dilution of pituitary homogenate or extract that will cause local yellowing in 50% of the fish tested following injection of the test material beneath the preopercular skin. The bioassay does have the distinct advantage, however, of being extremely sensitive to piscine (including agnathan) and amphibian pituitary homogenates. For example, teleost pituitaries are more than 100,000 times more effective in the assay than is oPRL, which is used as a standard. As little as 1/100th of a tiger salamander pituitary gives a positive response in the xanthophore-expanding bioassay, whereas an entire salamander pituitary of the same size is required to produce a minimal response in the pigeon crop-sac bioassay.

B. Red Eft Water Drive Bioassay

Prolactin induces a second metamorphosis, or **water drive**, in newts that is characterized by migration of the juvenile terrestrial form, the eft, back to water where breeding will occur (Figure D-1). Water-drive behavior, which also occurs in salamanders, is accompanied by a series of physiological and morphological changes as well, but these are not related to the bioassay per se. Following injection of PRL, the adults leave the dry areas of their laboratory containers and enter the water where they remain for breeding. No other hormone has been found to induce

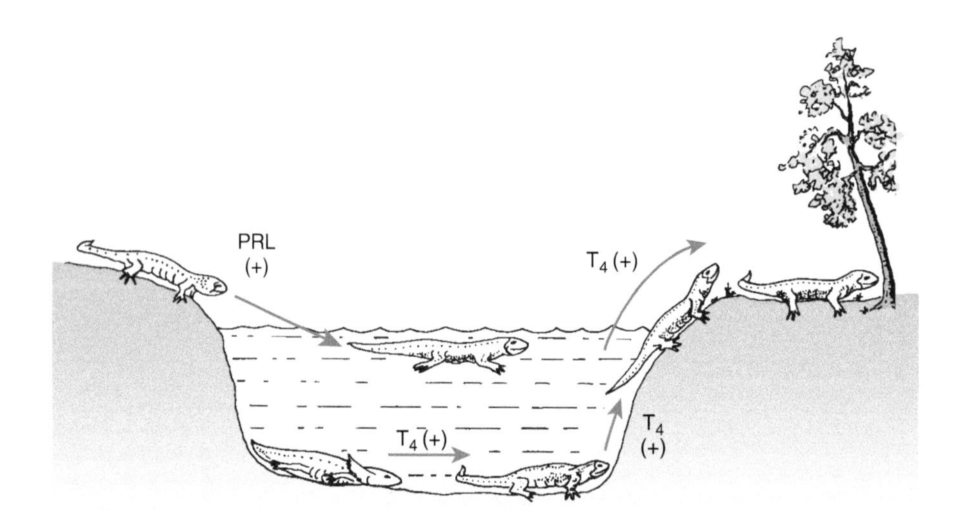

FIGURE D-1 **Water- and land-drives in a urodele amphibian.** Prolactin (PRL) is associated with increased locomotor activity and a preference for entering the water. Typically, this is related to reproduction and deposition of eggs in fresh water. Increased thyroid activity is associated with a preference for the terrestrial substrate and probably is responsible for emergence of newly metamorphosed animals from the water. T_4 = thyroxine.

water-drive behavior. This bioassay has the disadvantage of being an all-or-none response. Pituitary preparations for all classes of vertebrates except the class Agnatha possess water-drive-inducing activity (Table D-4). Treatment with mammalian PRL stimulates locomotor activity in tiger salamanders that may be a component of water-drive behavior. Curiously, terrestrial anuran amphibians that migrate to water for breeding do not exhibit elevated PRL levels. The failure for PRL to elicit water-drive behavior in anurans suggests that the PRL-based water drive is a phenomenon limited to caudate amphibians.

C. Pigeon Crop-Sac Bioassay for PRL

The pigeon crop-sac bioassay is a precise, quantitative, dose-related bioassay that yields a standard curve from which relative activities of unknown preparations can be assessed. The bioassay is performed on 6-week-old pigeons that have not reached sexual maturity. The pigeon crop-sac is a bilateral extension of the esophagus that in brooding birds produces a cytogenous secretion that is fed to the hatchlings. For this bioassay, the crop-sac first is "primed" by injection of the birds with a small amount of mPRL (usually ovine). Then, on one side of the crop, the test solution containing either a known or unknown amount of mPRL or pituitary extracts is applied by subcutaneous injection directly over the crop surface. On the opposite side, a control solution (usually the same saline vehicle used for the test solution) is similarly applied. After removal of the crop-sac several days later, a given area of the crop epithelium is scaped off around each control or test application site, dried, and weighed. The weight of the dried crop sac epithelial area is plotted against the known dose of mPRL used so that a standard curve is produced. The PRL content of unknowns can be identified by extrapolation from the standard curve. Positive results have been obtained in the crop-sac bioassay with lungfish pituitaries as well as pituitaries from all tetrapod groups, but most piscine PRLs will produce only an atypical response or no response at all. Although this bioassay is very consistent and sensitive, it has the major disadvantage of the time involved in performing it.

D. Mammary Gland *In Vitro* Bioassay for PRL

For this bioassay, mammary gland explants from pseudopregnant or midpregnant mice (or rabbits) are cultured in a precise medium that includes some other hormones (e.g., glucocorticoids, insulin). The addition of PRL or pituitary homogenates to the culture medium causes cytological changes that can be quantified according to a numerical index (Figure D-2). These cytological changes are correlated with the ability of PRL to stimulate milk synthesis. Like the crop-sac bioassay, mammotropic (lactogenic) activity is present in the pituitaries of all tetrapod species tested in the mouse mammary bioassay, but piscine pituitary preparations yield minimal responses, if any. Like the crop-sac bioassay, the mammary gland response is not a rapid assessment.

VI. ADRENOCORTICOTROPIN (ACTH)

The classical bioassay for ACTH has been the **adrenal ascorbic acid depletion test (AAAD)**, which is similar to the OAAD bioassay described for LH. The lowest effective dose of purified ACTH in this bioassay is 0.2 mU, or about 1.2×10^{-9} g (about 1 ng). The adrenal cortex secretes certain corticosteroids in response to ACTH stimulation. The AAAD test is rarely used today and has been replaced

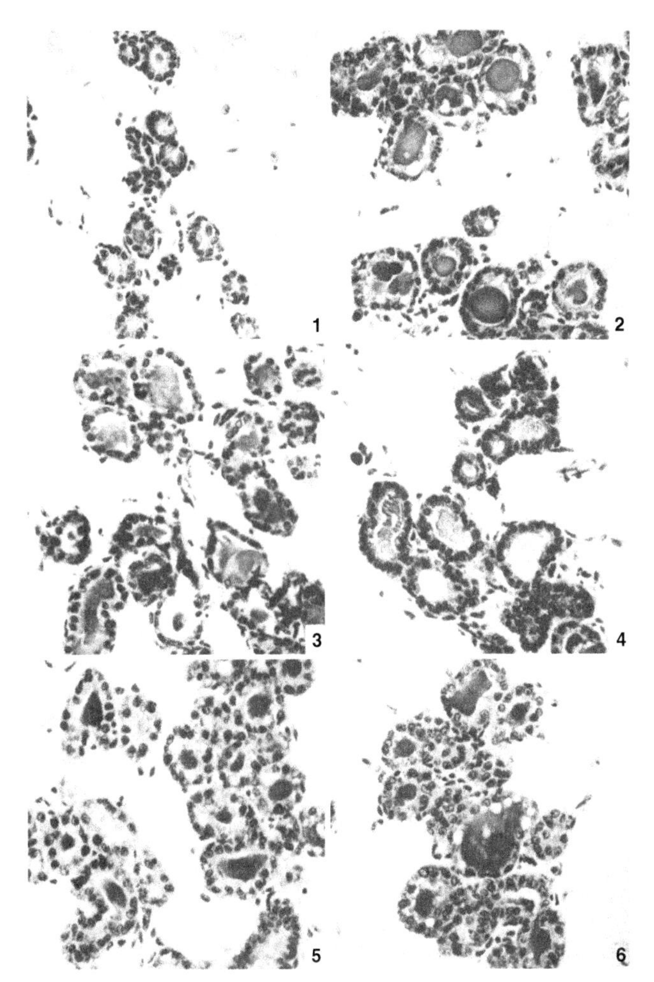

FIGURE D-2 **In-vitro mouse mammary gland bioassay for prolactin.** Secretion rating (SR) is determined using a visual scale of the cytological responses (lowest response =1; highest response = 5. (1) Control, = SR of 1. (2) SR= 5 is response to amphibian pituitary extract. (3) SR =4, response to reptile pituitary extract. (4) SR= 3 in response to a bird extract. (5) Guinea pig pituitary extract, SR = 4. (6) Purified ovine prolactin, SR = 5. *Reprinted with permission from Nicoll, C.S., Bern, H.A., Brown, D.* J. Endocrinol. *34, 343–354, 1966.*

by measurement of either circulating ACTH or corticosteroids with RIA.

VII. MELANOTROPIN (MSH)

The standard bioassays for MSH have been developed in amphibians and reptiles utilizing their ability to adapt to dark or light backgrounds. One bioassay involves measurement (using a reflectometer) of the quantity of light reflected from an isolated piece of skin removed from a light-adapted frog. A similar bioassay employs isolated pieces of skin from the light-background-adapted American chameleon (*A. carolinensis*). A rapid all-or-none response (darkening) occurs *in vitro* in the presence of MSH. The Hogben Index was a subjective analysis developed to measure the effects of MSH on the amphibian melanophore. It ranked the degree of melanosome dispersion within the melanophore on a scale of 1 to 5 (see Figure D-3).

VIII. LIPOTROPIN (LPH)

A standardized bioassay for LPH activity involves culturing mouse or rabbit epididymal fat pads (isolated from testes) and measuring the release of glycerol or free fatty acids or both into the culture medium. Another bioassay technique for LPH employs measurement of the inhibition of incorporation of ^{14}C-labeled acetate into lipid following the addition of LPH to the culture medium. This procedure is, in effect, a measurement of lipogenesis, which is inversely related to lipolysis. The value of the bioassay is questionable, as LPH does not seem to play a significant physiological role in lipid metabolism.

IX. NONAPEPTIDES OF PARS NERVOSA

A variety of bioassays have been developed in vertebrates to quantify the amount of oxytocin-like or vasopressin-like activity. Sample responses in each assay for various nonapeptides are found in Table D-5. These bioassays include

1. Antidiuretic bioassay for vasopressin (rat)
2. Pressor bioassay for vasopressin (rat)
3. Uterotonic bioassay for oxytocin (rat)
4. Depressor activity for oxytocin (chicken)
5. Milk ejection reflex (rabbit)

X. BIOASSAY OF GLUCAGON

Related to its role in fasting mammals, glucagon is often bioassayed in the fasting cat, in which it induces hyperglycemia. As in the case of insulin, however, rapid RIAs have been developed for measuring blood glucagon levels, and these are employed routinely.

XI. PARATHYROID HORMONE (PTH) BIOASSAY

The classical bioassay for PTH was an increase in plasma calcium following administration of parathyroid gland extracts or PTH to normal or parathyroidectomized dogs. Later bioassays employed parathyroidectomized rats maintained on a calcium-deficient diet.

XII. THYROID HORMONE BIOASSAYS

A. Land-Drive Bioassay

The antithesis of water drive is the return to land by newts and salamanders as well as by frogs and toads after breeding (see Figure D-1). This **land-drive** behavior appears to be a consequence of elevated thyroid hormones, and experimentally elevated thyroid hormones prevent breeding migrations of anurans to water.

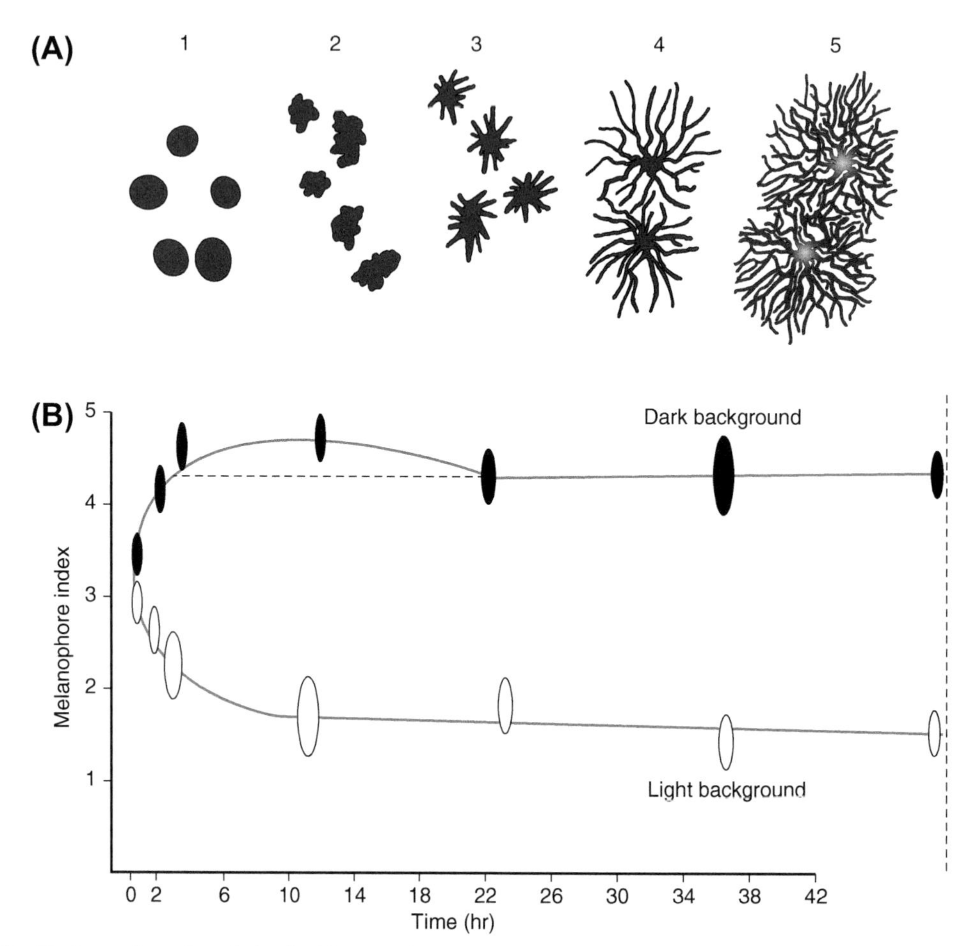

FIGURE D-3 **The Hogben melanophore index for bioassay of melanotropin.** (A) Five stages of melanophore dispersion from most concentrated (= 1) to most dispersed condition (= 5). *Adapted with permission from Hogben, L.T., Slome, D.* Proc. Roy. Soc. Lond. *108, 10–53, 1931.* (B) Changes in the melanophore index of anuran skin during adaptation to light or dark backgrounds. *Adapted with permission from Novales, R.R.* Handbook of Physiology, Section 7, Endocrinology, *Vol. 4, Pt. 2, 347–366, 1974.*

TABLE D-5. **Biological Activity of Neurohypophysial Peptides[a,b] in Various Bioassays**

	Activity units[c]				
Hormone	**Uterotonic (rat)**	**Depressor (chicken)**	**Milk ejection (rabbit)**	**Pressor (rat)**	**Antidiuretic (rat)**
Oxytocin	450	450	450	5	5
Mesotocin	291	502	330	6	1
Isotocin	145	310	290	0.06	0.18
Glumitocin	10	—	53	0.35	0.41
Valitocin	199	278	308	9	0.8
Aspargtocin	107	201	298	0.13	0.04
Arginine vasotocin	120	300	220	255	260
Arginine vasopressin	17	62	69	412	465
Lysine vasopressin	5	42	63	285	260

[a]*Based on data from Acher, R. (1974). Chemistry of the neurohypophysial hormones: An example of molecular evolution.* In *"Handbook of physiology, Sec. 7, Endocrinology, Vol. 4, Part 1," pp. 119–130. Williams & Wilkins, Baltimore, MD.*
[b]*For distribution of these hormones among the vertebrates, see Table 5–12.*
[c]*Units are expressed in international units per micromole of pure synthetic substance; 1 mg of synthetic oxytocin = 500 USP units.*

B. Xenopus Metamorphosis Bioassay

The induction of metamorphosis in *Xenopus* tadpoles (see Chapter 7) has proven to be an important bioassay for either thyroxine or triiodothyronine. Additionally, this bioassay has utility today for monitoring effects of potential thyroid EDCs.

XIII. ESTROGEN BIOASSAYS

A. Yeast Estrogen Screen (YES) Bioassay

The yeast estrogen screen (YES) bioassay employs a recombinant strain of the yeast *Saccharomyces cerevisiae* possessing the human estrogen receptor gene linked to a *lac Z* reporter gene. Although the YES bioassay is highly specific for natural and synthetic estrogenic chemicals, it does not always reflect the potency of a given estrogen when tested *in vivo*. The YES bioassay is very useful for screening the presence of estrogens in water samples collected in the environment, but the potency of a given suspected estrogen disruptor is better assayed using a whole animal system.

B. Use of Fish Bioassays for Estrogenic EDCs

Small fishes such as fathead minnows (*Pimephales promelas*) and zebrafish (*Danio rerio*) have proven very useful laboratory animals for bioassays. One useful bioassay is the induction of **vitellogenin (Vtg)**, a yolk protein precursor induced in male fishes following brief exposures to very low doses of estrogenic chemicals. Vtg can be readily measured by the ELISA method in a small quantity of blood plasma. Because male fathead minnows possess **external androgen-dependent characteristics** (e.g., nuptial tubercles, dorsal fat pad, pigmentation markings) (see Chapter 11) that are antagonized by estrogens, they are very useful in a non-invasive visual bioassay of the fish with regard to exposure to estrogenic chemicals. Simply adding known amounts of suspected estrogens to their aquarium water results in time- and dose-dependent effects. Furthermore, after removal of the estrogens, the fish soon return to normal.

Appendix E

Units for Measuring Hormones in Tissues

Prior to the isolation and characterization of many protein and peptide hormones, concentrations of hormones were expressed in arbitrary units achieved from bioassays based on how a particular partially purified extract behaved in the assay. Thus, for example, the National Institutes of Health (NIH) had a reference sample for follicle-stimulating hormone (FSH) called NIH-FSH-1. One milligram (mg) of that preparation produced a certain amount of increase in the rat testis bioassay (see Appendix D). All new preparations of FSH that were more potent (e.g., NIH-FSH-8) were related in potency to NIH-FSH-1 by having so many NIH units per mg (e.g., 50 NIH units/mg). Because it was unknown how many actual micrograms (μg) of pure FSH were in the preparation, people published their results using these preparations in NIH units per mg. The Medical Research Council of Canada had their own preparations, so we see data presented in MRC units.

Once highly purified preparations of these hormones became available, endocrinologists reported levels used for injections or present in blood in terms of actual amounts and concentrations, such as nanograms (ng) or μg (typically μg or mg/g of body weight of the injected animal or ng or μg/mL or dL of blood plasma). However, in order to compare hormones of very different molecular weights, hormone amounts may be expressed in terms of the number of molecules present, or moles per liter (e.g., n*M* or μ*M*). Consequently, it is often necessary to be able to convert n*M* to ng in order to compare data reported by investigators using different units. To make this conversion, one needs to know the molecular weight of the hormone. For example, estradiol (E_2) has a molecular weight of 272.4 daltons (Da). Thus, a mole of E_2 contains 272.4 g of pure E_2, so 1.0 μ*M* of E_2 would be the equivalent of 272.4 μg E_2/L or 272.4 ng/mL. Although a 1.0-μ*M* solution may seem to be a small amount, in endocrine systems it would be a huge amount considering that exposure of a male fish to 1 ng/L of E_2 can have feminizing effects and the apparent dissociation constant (concentration at which half of the binding sites are occupied; see Chapter 2) of E_2 for its receptors is generally in the low nanomolar concentration range, usually around 1 n*M*.

Appendix F

Vertebrate Tissue Types

I. THE ORIGIN OF VERTEBRATE TISSUES

During embryonic development the process of gastrulation defines the three **primary germ layers**: ectoderm, mesoderm, and endoderm. The **ectoderm** gives rise to the nervous system, including the neural crest and its derivatives, the epidermis, the lining of the oral cavity, and parts of certain sense organs. The **endoderm** gives rise to the mucosal lining of the gut and a number of derivatives of the gut, including the lungs, thyroid gland, liver, and pancreas. **Mesoderm** is the source of muscle, dermis, linings of the coelomic cavity (peritoneum, pleura, pericardium) and blood vessels (endothelium), and special organs such as the kidneys, adrenal cortex, and the gonads. The skeleton is formed from mesoderm as well as from neural crest (i.e., certain skull bones).

The primary germ layers give rise to four **primary tissues**: **epithelium**, **connective tissue**, **muscle**, and **nervous tissue**. Ectoderm gives rise to nervous tissue and certain epithelia. Endoderm gives rise to both covering epithelia and glandular epithelia. Mesoderm gives rise to special epithelia (mesothelia, endothelia), in addition to glandular epithelia, elements of the various connective tissues, and all muscle. The origin of primordial germ cells that eventually reside in the gonads in at least some vertebrates has been traced to endoderm, although it is possible that mesoderm may be involved in some species.

II. EPITHELIUM

An epithelium consists of closely associated cells organized into **sheets** that develop into coverings of outer and inner surfaces or are modified into glandular structures. Little intracellular material is found between the cells of epithelia, but they secrete a distinct basement membrane found adjacent to the basal portion of the cells. The basement membrane is secreted by cells of the epithelium. Epithelia also may occur as tubes (**ducts**, **cords**) or spheres (**follicles**, **acini**). Individual cells may differ markedly in shape, and some cells may exhibit specialized adornments such as **cilia** or **microvilli**, the latter forming a **brush border**. Brush borders greatly increase the total surface area of cells. Presence of microvilli is a clue to an epithelium's involvement in transport of molecules between the cells and extracellular fluids. Cilia may be responsible for producing currents to aid the flow of materials through a duct or other passageway as a result of coordinated rhythmic beating. Sometimes cilia are sensory structures linked to neurons.

Epithelia may consist of a single layer or sheet of cells (**simple**) or of several layers of one cell type (**stratified**). **Compound epithelia** consist of layers of more than one cell type. Some examples of sheet-like epithelia are

1. *Simple squamous epithelium*—Thin, flat cells organized in sheets such as the peritoneum
2. *Simple cuboidal epithelium*—Cube-shaped cells comprising the lining of certain ducts such as portions of the nephron
3. *Simple columnar epithelium*—Tall, rectangular cells that may also be found lining certain ducts as well as the intestinal lumina
4. *Stratified squamous epithelium*—Epidermis of skin

Glandular epithelia may appear as cords of cells as in the adrenal cortex, acini surrounding a duct as found in salivary glands or the exocrine pancreas, small clumps or **islets** as in the endocrine pancreas, or follicles, hollow balls of endocrine cells such as those comprising the thyroid gland.

III. CONNECTIVE TISSUE

The cells of connective tissue are generally separated from one another by extensive intracellular material (matrix) that they have secreted. Mesenchyme derived from mesoderm gives rise to four basic kinds of connective tissue: blood and lymph-forming tissues, connective tissue proper, cartilage, and bone.

A. Blood- and Lymph-Forming Tissues

The blood-forming elements give rise to circulating **erythrocytes** or **red blood cells (RBCs)** and **leukocytes** (leucocytes) or **white blood cells (WBCs)** in adult vertebrates. The RBCs of mammals are unique among the vertebrates in that the mature circulating cell lacks a nucleus. Most of the WBCs are produced in lymph nodes or the spleen of adult vertebrates. The WBCs are further subdivided into cells with granular cytoplasm: **granulocytes (eosinophils, basophils, neutrophils)** and **agranulocytes (lymphocytes, monocytes, plasma cells)**. Circulating eosinphils are identical to those associated with many organs such as the uterus and the lung. Basophils may be identical to **tissue mast cells** that secrete histamine. The extracellular matrix of blood is called **plasma**. Excess interstital fluid enters the lymphatic vessels where it is now called lymph. Lymph is conducted via these vessels and eventually is added to the blood vascular system via the vena cava.

B. Connective Tissue Proper

A large number of tissue types are lumped under the title of connective tissue proper, including loose fibrous connective tissue, dense fibrous connective tissue (e.g., tendons), elastic tissue, and reticular connective tissue. The fibrous components are proteins (collagen, elastin, reticular fibers) and are found in the different connective tissue types to various extents.

C. Adipose Tissue

Adipose (fat) tissue consists of cells (adipocytes) that contain large amounts of fat restricted to a central vacuole which confines the cytoplasm to a thin rim adjacent to the plasmalemma. Although fat storage and release when needed is an important role for adipose tissue, it is also responsible for production of a variety of bioregulators that may affect feeding, metabolism, and even reproduction. In rodents, there are two readily distinguishable types of adipose tissue termed white and brown. **White adipose tissue** shows regular variations in the amount of stored fat with nutritional state. **Brown fat** is more vascular than white adipose tissue, and the cells have numerous small vacuoles filled with fat whereas a single vacuole is present in the white adipose cell. It does not vary with nutritional state but has been correlated with hibernating behavior. In most mammals it is not possible to differentiate brown and white types of adipose tissue by appearance but some adipose may function like brown adipose tissue of rodents.

D. Cartilage

Cartilage cells or **chondrocytes** secrete a matrix consisting of a glycoprotein, **chondromucoid**, that contains the sulfonated polysaccharide chondroitin sulfate. The extensive matrix between the chondrocytes may have few fibrous components (**hyaline cartilage**) or may contain collagen (**fibrocartilage**) or elastin fibers (**elastic cartilage**). Cartilage may also be strengthened by addition of calcium salts. Cartilage is often a precursor for bone formation.

E. Bone

Bone is the strongest connective tissue and the densest. The extensive matrix of bone is comprised of crystalline calcium salts, primarily calcium phosphate, and very little water. Mammalian bone occurs in a uniformly dense, compact form (**compact bone**) and in a less dense, more easily modified form (**cancellous** or **spongy bone**). The bone-forming cells are known as **osteoblasts** and are rich in the enzyme phosphatase. Osteoblasts that have become embedded within the bone matrix are termed **osteocytes**. Giant multinucleated cells, the **osteoclasts**, produce hydrolytic enzymes and are responsible for bone destruction (resorption). The osteoblasts and osteoclasts are responsible for bone forming, resorbing, and reforming in accordance with physical stresses placed on bone and with physiological demands for calcium.

As a tissue, bone varies considerable from the mammalian patterns in many non-mammals. For example, bony fishes have acellular bone, and other structures such as scales provide readily available calcium. Endolymphatic sacs present at the base of the skull in amphibians are also an important alternative calcium source.

F. Muscle

Mesoderm gives rise to three basic muscle types: smooth, striated, and cardiac. **Smooth muscle** frequently is termed involuntary muscle, as it is under the control of the autonomic nervous system. It is associated with internal organs and is found in such places as the gut wall, blood vessels, various ducts, and wall of the uterus. Individual smooth muscle cells are small, uninuclear cells, and the contractile elements (myofibrils) are not highly organized within the cell. Smooth muscle is characterized by slow, rhythmic contractions. **Striated** or **skeletal muscle** is the so-called "voluntary" muscle tissue that is under conscious control, although it also may be influenced by the autonomic system. Striated muscle cells are large cells with myofibrils so highly organized as to produce regular bands or striations when the cells are viewed with the aid of a microscope. Movements of the skeleton are controlled by striated muscles that are attached to bones by dense fibrous connective tissue. In addition, a few sphincter muscles are also of this type and hence are under conscious control (e.g., the external urinary bladder sphincter). **Cardiac muscle** possesses properties of both skeletal muscle (striations due to highly organized myofibrils) and smooth muscle (rhythmic contractions that are innate properties of cardiac muscle cells). Instead of inserting on bones, the cardiac cells connect directly to one another through specialized tendinous attachments known as *intercalated disks*.

G. Nervous Tissue

Nervous tissue is specialized for integrative functions and conduction of information throughout the body. **Neurons** are specialized cells for conducting electrochemical neural impulses to coordinate body processes. Most neurons release chemical bioregulators known as *neurotransmitters* to control the activity of other neurons, muscle cells, and glands. **Neurosecretory neurons** produce neurohormones that are secreted into the blood vascular system and constitute a second type of bioregulatory control mechanism. The central nervous system also contains several types of **glial cells** (neuroglia) that perform many supportive functions for the neurons, including production of paracrine bioregulators. One important role of glial cells is the secretion of myelin that forms the white matter of the central nervous system and insulates neuronal axons from each other. The ependymal cells that line the brain ventricles are another nervous cell type. They control production of cerebral spinal fluid and movement of materials in and out of the central nervous system. The **Schwann cell**, a type of glial cell in the peripheral nervous system, secretes the myelin sheath characteristic of many peripheral neurons.

H. General Tissue Responses

In response to various stimuli a given tissue may exhibit regression (**atrophy**), proliferation (**growth**), or no morphologically observable response. Growth may be due simply an increase in cell size (**hypertrophy**) or an increase in cellular divisions with an actual increase in cell numbers (**hyperplasia**) or both. Tumors or neoplasms are abnormal proliferations of tissues (**neoplasia**) having no normal physiological function. Such growths may be classified as either **benign** (harmless) or **malignant** (very harmful or likely to cause death; i.e., a cancer). The term **adenoma** refers to a benign tumor of glandular origin that may or may not synthesize and release abnormal amounts of hormones. Connective tissue tumors are called **sarcomas**, whereas a lymphatic tumor is a **lymphoma**. Malignant growths of any epithelial tissue, including glandular epithelia, are termed **carcinomas**. Some adenomas or carcinomas also produce excessive quantities of hormones or hormone-like substances, as in the production of excessive amounts of growth hormone by pituitary adenomas or secretion of gastrin (a hormone that stimulates hydrochloric acid secretion in the stomach) by a pancreatic carcinoma.

Comparative Pituitary Cytology

TABLE G-1. Comparative Cytology of the Pars Distalis from Representatives of Different Vertebrate Groups

Hormone	Vertebrate group	Cellular	Alternate names	Ultrastructural determination of cytoplasenic granule size (nm)
TSH	Chondrichthyes: Selachians	Type 1 basophil	Type 1	90–120
	Osteichthyes			
	Holosteans	Type 1 basophil (amphophil)	δ-Basophil	—
	Teleosts	Type 1 basophil		400
	Amphibia: Anurans	Type 1 basophil		150–400
	Reptiles	Type 1 basophil		300–400 × 200–250
	Birds	Type 1 basophil	δ-Basophil	50, 100, 200
	Mammals	Type 1 basophil	β-Basophil	150
GTH	Chondrichthyes: Selachians	Type 2 basophil	Type V, VI	100–700
	Osteichthyes			
	Holosteans	Type 2 basophil	β- and γ-Basophils	60–160; 80–240 Polymorphous to 900
	Teleosts	Type 2 basophil		
	Amphibia: Anurans	Type 2 basophil		150–270; 600–800
	Reptiles	Type 2 basophil		120–200
	Birds	Type 2 basophil	β-Basophil: γ-Basophil	120–400
	Mammals	Type 2 basophil	δ-Basophil	200
ACTH	Chondrichthyes: Selachians	Type 3 basophil	Type II	140
	Osteichthyes			
	Holosteans	Type 3 basophil	Acidophil	110–250
	Teleosts			
	Amphibia: Anurans	Type 3 basophil	ε-Cell	100–200
	Reptiles	Type 3 basophil (amphophil)		—
	Birds	Type 3 basophil	ε-Cell	150–300
	Mammals	Type 3 basophil		200
PRL	Chondrichthyes: Selachians	Type 1 acidophil	Type IV	263
	Osteichthyes			
	Holosteans	Type 1 acidophil	η-Cell	—
	Teleosts	Type 1 acidophil		Polomorphic; 170–350
	Amphibia: Anurans	Type 1 acidophil		180–500

(Continued)

TABLE G-1. Comparative Cytology of the Pars Distalis from Representatives of Different Vertebrate Groups—cont'd

Hormone	Vertebrate group	Cellular	Alternate names	Ultrastructural determination of cytoplasenic granule size (nm)
	Reptiles	Type 1 acidophil		—
	Birds	Type 1 acidophil	η-Cell	Polomorphic;250–300
	Mammals	Type 1 acidophil	ε-Acidophil	Polomorphic; 600–900
GH	Chondrichthyes: Selachians	Type 2 acidophil	Type III	200
	Osteichthyes			
	Holosteans	Type 2 acidophil	α-Cell	–
	Teleosts	Type 2 acidophil		—
	Amphibia: Anurans	Type 2 acidophil		180–250
	Reptiles	Type 2 acidophil		310
	Birds	Type 2 acidophil	α-Acidophil	250–300
	Mammals	Type 2 acidophil	α-Acidophil	350

TABLE G-2. Some of the Dyes Used in Cytological Observation of Adenohypophysial Cells and Their Abbreviations

Dye or staining procedure	Abbreviation	Chemical specificity (if known)
Aldehyde fuchsin	AF	—
Alcian blue	AB	Disulfide bonds, glycosaminoglycans[a]
Periodic acid–Schiff	PAS	Glycoproteins, glycosaminoglycans[a]
Orange G	OG	—
Azocarmine	AZ	—
Lead hematoxylin	PbH	—
Iron hematoxylin	Feh	—

[a]= mucopolysaccharides

TABLE G-3. Some Light and Electron Microscopic Features of Cellular Type in the Mammalian pars Distails

Cellular type	Tropic hormone secreted	Stainability[a] for light microscope	*In situ* granule size (nm)
Thyrotrope (β-basophil)	TSH	PAS(+), AF(+)	150
Gonadotrope (δ-basophil)	FSH, LH	PAS(+), AF(−)	200
Corticotrope	ACTH	Weakly PAS(+), AF(+); maybe PbH(+)	200
Lactotrope (∊-acidophil)	PRL	Azocarmine(+)	600–900
Somatotrope (α-acidophil)	GH	OG(+)	Variable to 350

See Table G-2 for explanation of dye abbreviations.

Appendix H

Metabolic Pathways

The term *metabolism* represents the total of all enzymatic processes occurring within the cells of the body. Metabolism frequently is subdivided into specific pathways. One such subdivision is **intermediary metabolism**, which involves the metabolism of carbohydrates for energy and serves as the linkage between protein and lipid metabolism. Intermediary metabolism is especially important in the liver, which has the major responsibility for converting amino acids and fatty acids into carbohydrates for energy, especially during periods of fasting, and providing glucose to the circulatory system.

I. MAJOR ELEMENTS OF METABOLISM IN VERTEBRATES

This appendix provides an overview of the major features of carbohydrate, lipid, and protein metabolism and their interrelationships. In addition, differences among tissues in their metabolic capabilities are illustrated, setting the foundation for understanding how hormones regulate certain aspects of metabolism. Although the same metabolic enzymes and pathways are present in most vertebrates, the specific forms of these enzymes and their characteristics differ from group to group and from tissue to tissue within a group. The following account will address a generalized pattern of metabolism for all vertebrates.

II. INTERMEDIARY METABOLISM

The major pathway in intermediary metabolism involves the oxidation of glucose to produce adenosine triphosphate (ATP) to drive energy-requiring reactions in cells. This oxidative process is known as **cellular respiration**. About 70% of the glucose used in cellular respiration is oxidized through **glycolysis** (Figure H-1), with the remainder being oxidized via the pentose phosphate pathway or **pentose shunt** (Figure H-2). These two pathways operate under both anaerobic and aerobic conditions. The pentose shunt produces reduced (referring to a gain of an electron or electrons) **nicotinamide adenine dinucleotide phosphate (NADPH)**, which is necessary for the synthesis of fatty acids, for inactivation of steroids, and for detoxification of many drugs. Some intermediates produced in the pentose shunt can feed into the glycolytic pathway (e.g., glyceraldehyde 3-phosphate). Furthermore, some of the five-carbon sugars for which the pathway is named may be utilized for synthesis of ribose and deoxyribose used for making the nucleotides found in RNA and DNA, respectively.

A. Glycolysis

The initial steps of glucose oxidation all occur in the cytosol. There are three limiting steps in glycolysis, each of which is catalyzed by a unidirectional enzyme (i.e., the reaction is irreversible because of thermodynamic considerations). Free glucose enters a cell by facilitated transporters (GLUTs) and secondary active transporters (SGLTs). Once glucose enters a cell, it is immediately phosphorylated to glucose-6-phosphate in most tissues by a unidirectional enzyme called **hexokinase** (recall that a kinase is a phosphorylating enzyme). In liver, this enzyme often is called **glucokinase**. Once glucose has been phosphorylated, it can no longer be transported across the cell membrane and, in most tissues, is committed either to storage or to use for energy. However, in liver tissue (and in kidney to a limited extent), the enzyme **glucose-6-phophatase** can remove the phosphate from glucose-6-phosphate, creating free glucose that can leave the liver cell. Glucose-6-phosphatase hydrolytic activity is lacking in almost all other tissues, so the liver is the only tissue that can readily supply glucose to the rest of the body. In general, virtually all glucose in the blood is either absorbed following digestion of food or contributed by the metabolic activities of the liver.

Glucose-6-phosphate can be metabolized through glycolysis or the pentose shunt, or it can be converted to glucose-1-phosphate and polymerized with the aid of the enzyme **glycogen synthetase** (= glycogen synthase) to a storage form of polysaccharide called **glycogen**. This process is called **glycogenesis**. The enzyme **phosphorylase a** is responsible for hydrolyzing glycogen and conversion back to glucose-6-phosphate as needed (see Figure H-3). Following complete assimilation of a meal or during

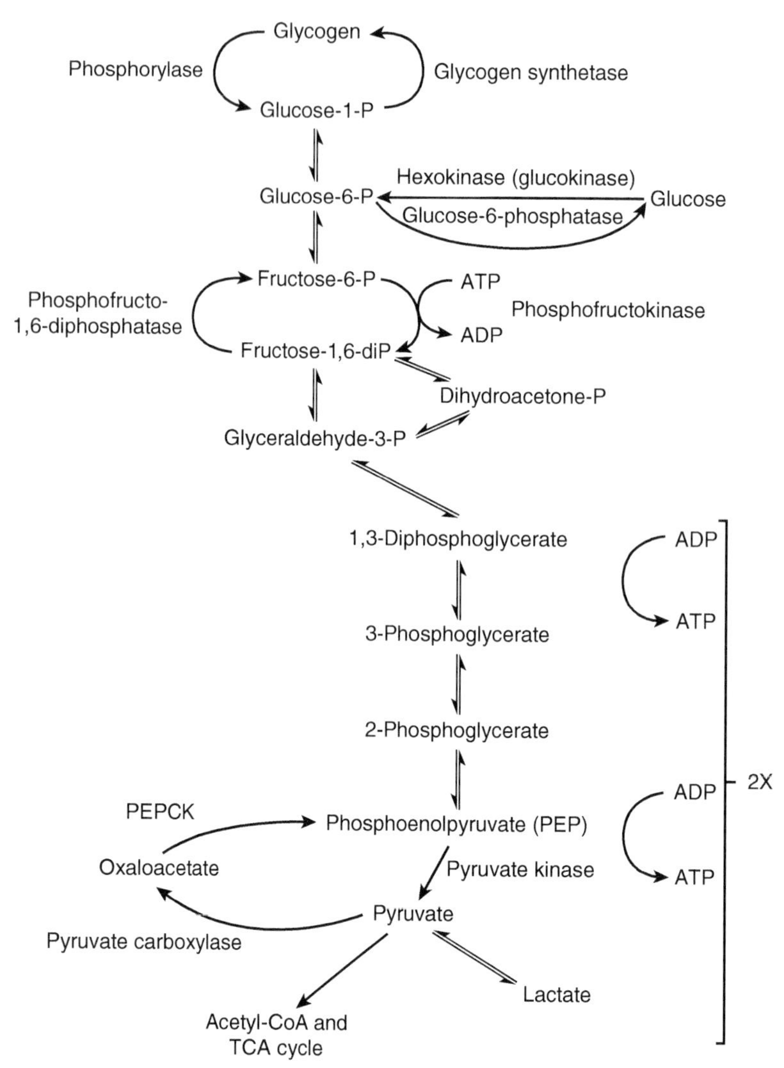

FIGURE H-1 **Glycolysis.** This cytosolic process is called the Emden-Mayerhoff pathway. It begins with glucose-1-phosphate (glucose-1-P) derived either from blood glucose or stored glycogen (Figure H-3) and ends with pyruvate (aerobic conditions) or lactate (anaerobic conditions). Lactate or pyruvate can be further oxidized under aerobic conditions via the TCA cycle (Figure H-4) or can serve as gluconeogenic substrates (Figure H-7) to resynthesize glucose. Only enzymes that perform regulatory roles in glycolysis are shown. PEPCK, phosphoenolpyruvate carboxykinase; TCA, Tricarboxylic Acid.

fasting, plasma glucose levels are maintained by the liver initially through conversion of glycogen to glucose-6-phosphate through a process called **glycogenolysis**. Glucose-6-phosphate can be dephosphorylated to glucose and exit the liver cell. The liver also produces glucose from amino acids, lactate, fatty acids, and glycerol (see gluconeogenesis ahead).

The second one-way conversion occurs early in glycolysis by the action of **phosphofructose kinase (PFK)** with ATP and fructose-6-phosphate to form fructose-1,6-diphosphate. This commits the cell to further oxidation by splitting that six-carbon sugar to two phosphorylated trioses (three carbons each) via the glycolytic pathway, yielding a small amount of ATP. In the third irreversible step, **pyruvate kinase (PK)** converts **phosphoenolpyruvate (PEP)** to **pyruvate**. In the presence of oxygen, PEP can be further oxidized to produce additional ATP or, in the absence of sufficient oxygen, converted to **lactate** (see Figure H-1).

B. Tricarboxylic Acid (TCA) Cycle and Electron Transport Chain

In the presence of oxygen and **coenzyme A**, pyruvate leaves the cytosol and enters a mitochondrion where it interacts with coenzyme A and is converted to **acetyl-coenzyme A (acetyl-CoA)** plus CO_2. Acetyl-CoA combines with the four-carbon compound called **oxaloacetate** to produce citric acid or **citrate** (Figure H-4). Through a series of reactions that generate two additional

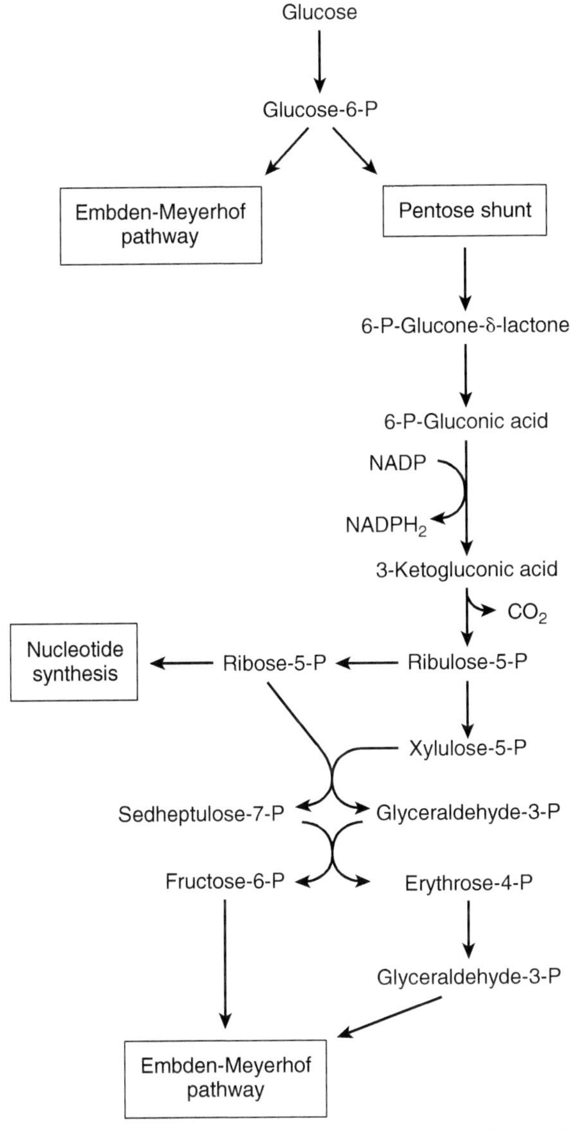

FIGURE H-2 **The pentose shunt.** This alternate pathway within glycolysis (Figure H-1) generates substrates for fatty acid and nucleotide synthesis.

CO_2 molecules and a number of important reduced compounds (NADPH, NADH, FADH), the citrate is oxidized back to oxaloacetate, which can combine with another acetyl-CoA generated from glucose. This series of cyclic events constitutes the **tricarboxylic acid (TCA) cycle**, so named because citrate has three carboxyl groups in its structure. It also is known as the **citric acid cycle**, as it begins with the combination of acetyl-coA and succinate to form citric acid, as well as the **Krebs cycle**, after the man who discovered it.

The reduced compounds generated by the TCA cycle are used to transfer electrons to a series or chain of electron-transfer molecules called **cytochromes**. This **electron transport chain** also is located in the mitochondria. A final electron transfer is made to oxygen that results in the generation of water. The energy released by the sequential transfers of electrons from higher to lower energy levels along the electron-transport chain is used to generate ATP. Thus, the complete oxidation of glucose to CO_2 and water requires glycolysis, the TCA cycle, the electron transport chain, and oxygen as the final electron acceptor in order to produce a large number of ATPs. This process is referred to as *aerobic cellular respiration*, or simply as **aerobic respiration**. In the absence of oxygen, the oxidative pathway stops in the cytosol at pyruvate with only a small amount of ATP produced. The excess pyruvate is converted by the enzyme **lactate dehydrogenase (LDH)** to lactate. This process is called **anaerobic respiration.** When oxygen is available, lactate can be converted back to pyruvate by LDH. In some animals, a portion of the pyruvate is converted irreversibly to ethanol and CO_2.

II. PROTEIN METABOLISM

Proteins are composed of more than 20 different amino acids. In animals that have a protein-rich diet, such as most predators, ingested proteins are hydrolyzed to amino acids that are transported across the epithelium of the small intestine. These amino acids can be used to synthesize proteins and the excess can be converted to carbohydrate for immediate energy use or to lipids for storage. Under conditions of starvation, hydrolysis of proteins located primarily in muscle cells provides amino acids for energy production in the muscle cells and amino acids for conversion to glucose by the liver (see gluconeogenesis ahead). Some organisms can synthesize all the amino acids required for protein synthesis, but many vertebrates cannot. Humans and some other vertebrates must rely on a dietary source of what have been called **essential amino acids**. In humans, nine amino acids are considered to be essential and must be obtained through the diet.

III. LIPID METABOLISM

Lipogenesis is the synthesis of lipids. The most important lipids are the **triacylglycerols** or **TAGs** (oils, fats), fatty acids, and steroids. TAGs are also known as triglycerides. Synthesis of a TAG begins with glycerol and one fatty acid and progresses stepwise to the TAG by the actions of a series of acylglycerol enzymes. This synthetic process is usually called **esterification**. The first step in esterification is formation of a **monoacylglycerol (MAG)** consisting of a molecule of glycerol esterified to a fatty acid (Figure H-5). This step is catalyzed by a **monoacylglycerol synthetase** and occurs more commonly in the liver. Addition of a second esterified fatty acid produces a **diacylglycerol (DAG)**, and a third addition produces a TAG. If the attached fatty acids have many double bonds (i.e.,

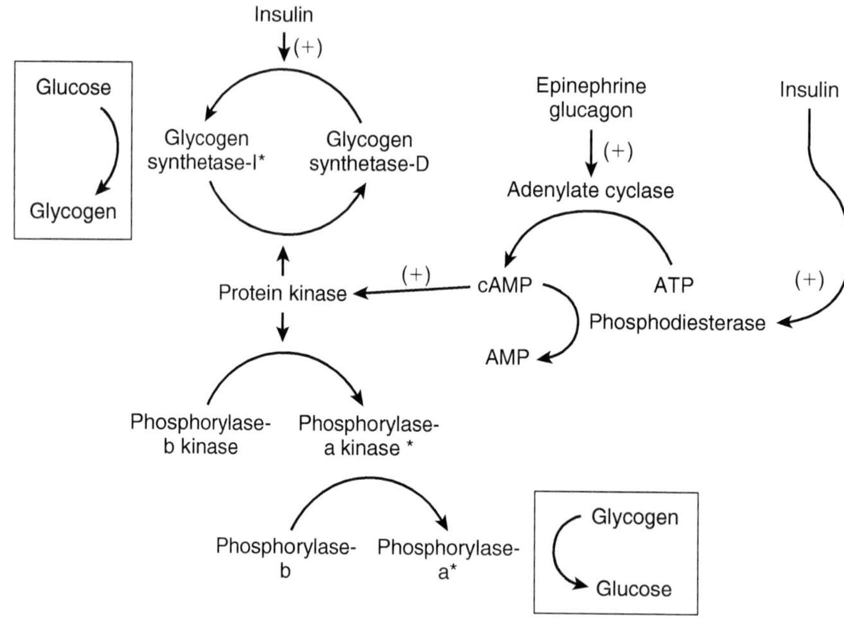

FIGURE H-3 **Glycogenesis and glycogenolysis.** Enzymes are converted from inactive to active forms by various hormones that alter the availability of glucose for metabolism. Insulin favors glucose storage as glycogen whereas epinephrine and glucagon favor glucose oxidation or release from the liver into the blood. The asterisk (*) denotes the active form.

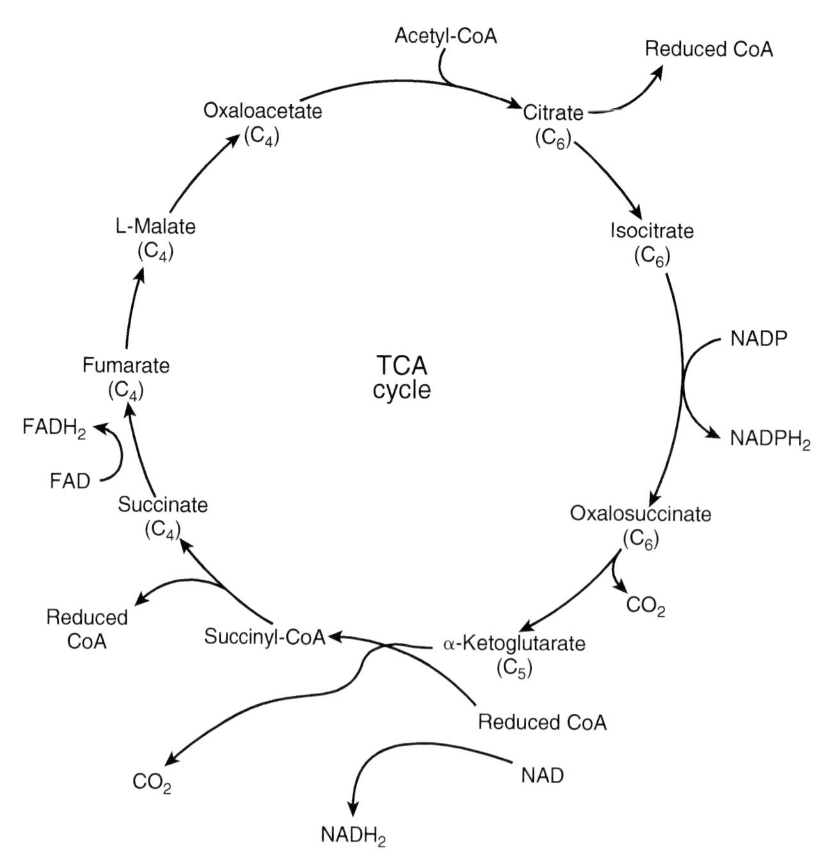

FIGURE H-4 **The tricarboxylic acid (TCA) cycle.** Pyruvate is completely oxidized in the TCA cycle to CO_2 that is released from the cell. Electrons (plus associated hydrogen ions) are transferred to electron-acceptors (FAD, NAD, NADP) that will be used by the electron-transport chain (cytochromes) to synthesize ATP and water. The TCA cycle and the cytochromes are confined within the mitochondria.

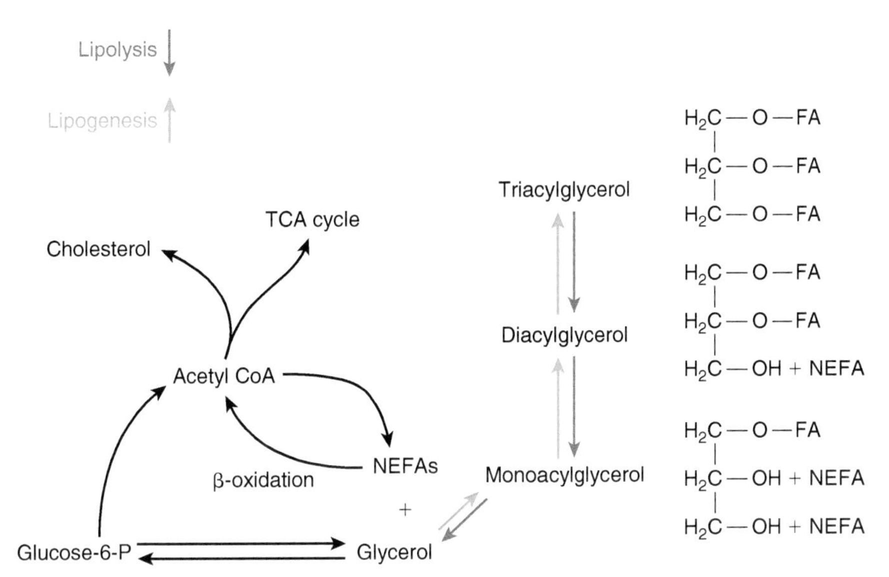

FIGURE H-5 **Triacylglycerols: esterification and lipolysis.** Fatty acids and glycerol produced from glucose may serve as the raw materials used to construct triacylglycerols or fats. Hydrolysis of fats yields non-esterified fatty acids (NEFAs) that can be utilized for energy through β-oxidation (Figure H-8) and monoacylglycerol. NEFAs also may be used as a gluconeogenic source for glucose production (see Figure H-7).

are unsaturated), the TAG is likely to appear as an oil at room temperature. If there are few double bonds in the esterified fatty acids, the compound is a solid at room temperature, and we call it a fat. We do not distinguish between these saturated and unsaturated TAGs in Chapters 12 and 13, although most of the natural compounds are saturated.

The liver sends TAGs to other tissues (mainly muscle and adipose tissues) for storage or utilization. Hydrolysis of TAGs in storage tissues is called **lipolysis** and occurs through the actions of **triacylglycerol lipase** and **diacylglycerol lipase**. Although lipolysis of stored TAGs is initiated primarily in muscle and adipose tissue, **monoacylglycerol lipase** is active only in the liver. Fatty acids are freed through lipolysis and become **non-esterified fatty acids** (**NEFAs** = free fatty acids). They can be released into the blood, where they bind reversibly to serum proteins and are transported to the liver.

NEFAs can be synthesized in the liver and other tissues from acetyl-CoA. However, acetyl-CoA generated within the mitochondria cannot cross the mitochondrial membrane and enter the cytosol where fatty acid synthesis occurs. Instead, citrate is transported into the cytosol and used to regenerate acetyl-CoA. The rate-limiting enzyme **acetyl-CoA carboxylase** initiates fatty acid synthesis from cytoplasmic acetyl-CoA. Glucose metabolism provides glycerol that is enzymatically coupled to a fatty acyl-CoA complex to form a MAG.

The steroid cholesterol may be absorbed following a meal or may be synthesized from cytoplasmic acetyl-CoA (see Chapter 3). This synthetic process involves activation of the rate-limiting enzyme **hydroxymethylglutaryl-CoA (HMG-CoA) reductase** and ends with the synthesis of cholesterol in the cytosol. Cholesterol and cholesterol esters (cholesterol linked by an ester bond to a fatty acid) may be extruded in lipoprotein droplets and transported through the blood to sites of cholesterol utilization or may be converted to bile salts and transported to the gall bladder for use in digestion (see Chapter 12). Cholesterol synthesis may be in direct competition with fatty acid synthesis for the cytoplasmic substrate acetyl-CoA, and the endocrine regulation of the rate-limiting enzymes for these pathways is critical (see Chapter 12). Cholesterol is necessary for building cell membranes (especially in dividing cells) and is the precursor for the synthesis of all of the steroid hormones as described in Chapter 3.

A. Lipoprotein Droplets and Lipid Transport

Following absorption of a meal, excess carbohydrates and amino acids are converted into TAGs by the liver. These TAGs plus cholesterol are surrounded by a phospholipid membrane in which special proteins called **apoproteins** are embedded. These lipoprotein droplets are extruded into the blood as **very-low-density lipoproteins (VLDLs)**. The apoproteins apparently act as "docking proteins" for attachment to receptors on the endothelial cell membranes of capillaries. These receptors are called **LDL receptors** and have a high affinity for **apoprotein E**. The endothelial cells express a TAG-digesting enzyme called **lipoprotein lipase**. This enzyme is most prevalent in the endothelial cells of capillaries in adipose tissue and to some extent in capillaries of skeletal muscle. Lipoprotein lipase converts

α-Ketoglutarate + Amino acid → Glutamate + α-Keto acid

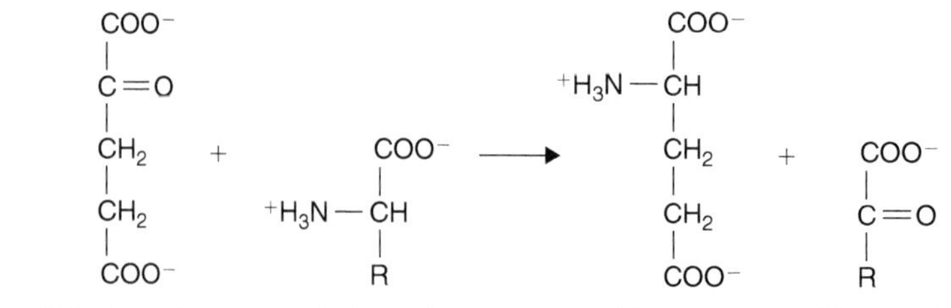

FIGURE H-6 **Transamination of amino acids.** This process makes amino acids available as gluconeogenic sources. The α-ketoglutarate is regenerated from glutamate by a separate reaction (not shown).

TAGs to MAGs and NEFAs, which diffuse into the adipose or muscle cells, where they are reconstituted to TAGs. The lipoprotein droplet that has lost some of its TAGs has a relatively higher concentration of protein and therefore is now denser. After the VLDLs have lost about 50% of their TAG content, they are classified as **intermediate-density lipoproteins (IDLs)**. When IDLs again reach the liver, they bind to LDL receptors and may be refilled with TAGs prior to making another journey to the storage sites as VLDLs. If the IDLs by chance remain in the circulation longer, they continue to lose TAGs to tissues. After they lose about 90% of their TAG content, they begin to lose the special protein apoprotein E and are renamed **low-density lipoproteins (LDLs)**. Furthermore, without apoprotein E, they bind poorly to the LDL receptors and remain longer in the circulation, meanwhile transferring their load of cholesterol to other cells, especially endothelial cells of arteries. However, **apoprotein B100** is still present on the surface of the LDL and can also bind to the LDL receptor but not as strongly as apoprotein E. This causes LDLs to be bound in the liver at a much slower rate than IDLs; hence, they have a much longer biological half-life in blood than do IDLs. Thus, LDLs tend to accumulate in the blood and continue to deliver cholesterol to arterial sites for deposition.

Another group of lipoprotein droplets called **high-density lipoproteins (HDLs)** represent a special group of droplets that are high in cholesterol as well as apoproteins. A special role for HDLs is to remove cholesterol from tissues. In addition, HDLs can transfer excess apoprotein E to LDLs, allowing them to be removed more rapidly from the blood and recycled by the liver thereby reducing their opportunity to interact with other cells. It is presumed that a higher ratio of HDLs to LDLs represents a shorter turnover time for LDLs and thus less opportunity for them to transfer cholesterol to arterial cells.

IV. GLUCONEOGENESIS

The synthesis of glucose from non-carbohydrate sources is termed **gluconeogenesis**, and this term is used today to encompass the conversion of amino acids, lactate, glycerol, or lipids into glucose. Because glucose and its hydrolysis products are used for ATP synthesis by all cells, gluconeogenesis is essential when the diet consists largely of protein and/or TAGs. Amino acids may simply be transaminated (deaminated) to produce intermediates of glycolysis or the TCA cycle (Figure H-6). For example, transamination of the amino acids alanine, serine, and glycine results in formation of pyruvate. Pyruvate also can be produced by conversion of lactate by mitochondria in liver or muscle cells in the presence of oxygen. It then can be a source for production of the TCA cycle intermediate oxaloacetate, resulting in ATP formation. Transamination of aspartate also yields oxaloacetate. These transaminations require α-ketoglutarate as a recipient for the amino group of the amino acid. Regeneration of α-ketoglutarate from glutamate can be accomplished by urea cycle enzymes or by direct deamination of the amino acids arginine, lysine, or histidine. Of the amino acids mentioned above, only lysine cannot be resynthesized in humans.

The rate-limiting enzyme for gluconeogenesis is **phosphoenolpyruvate carboxykinase (PEPCK)**, which converts oxaloacetate to PEP, which in turn can be used to synthesize glucose (Figure H-7). The ratio of PK to PEPCK can be used as an index to assess the gluconeogenic activity of a tissue under differing conditions.

TAGs contribute to energy metabolism and glucose synthesis in two ways. Hydrolysis of MAGs in the liver frees glycerol that can be used to resynthesize glucose-6-phosphate. NEFAs released from hydrolysis of TAGs are metabolized through special pathways (see ahead) to cytoplasmic acetyl-CoA and other metabolites that can be used for energy. Acetyl-CoA also can be used to synthesize cholesterol and bile salts.

V. FATTY ACID METABOLISM AND KETOGENESIS

NEFAs can be hydrolysed through a pathway known as **β-oxidation** in liver cells (Figure H-8). This process results in formation of cytoplasmic acetyl-CoA that can be transported into the mitochondria for ATP synthesis or to

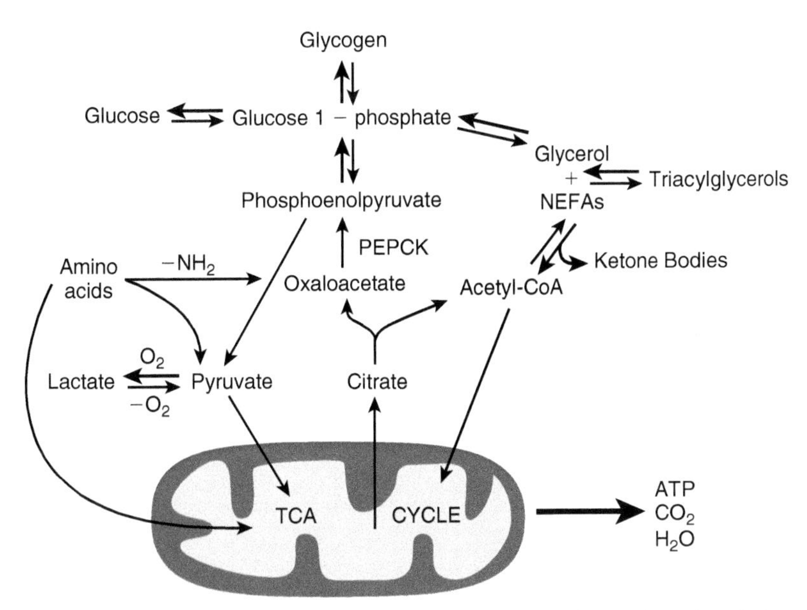

FIGURE H-7 **Gluconeogenesis.** This figure summarizes potential gluconeogenic pathways from amino acids, NEFAs, and lactate. Not all of these pathways are present in all cells. Some substrates must cycle through the mitochondria. Note the central position of the gluconeogenic enzyme PEPCK that is increased by hormones such as cortisol.

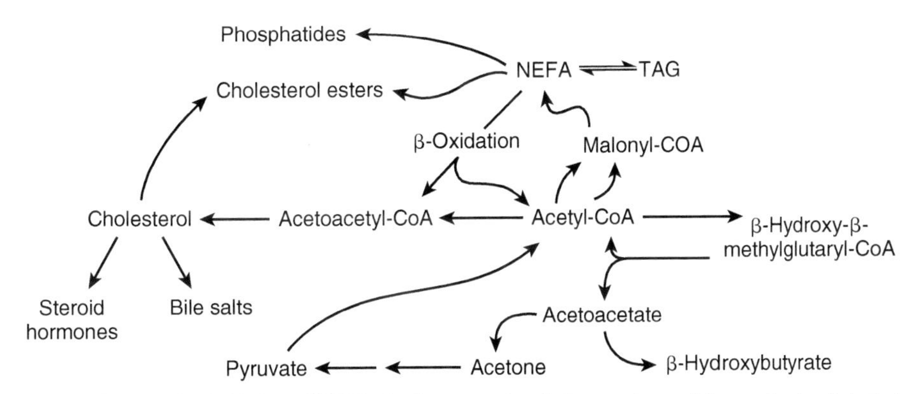

FIGURE H-8 **β-oxidation of fatty acids.** Oxidation of NEFAs leads to acetyl-coA that can be used for synthesis of cholesterol or that may enter the mitochondria for ATP production. Ketone bodies (acetone, acetoacetate, β-hydroxybutyrate) are byproducts of β-oxidation and usually are excreted in the urine. TAG, triacylglycerol.

form glucose through gluconeogenesis. However, certain byproducts of β-oxidation known as **ketone bodies** are produced whenever NEFAs are broken down. These ketone bodies are not metabolized in the liver but are released into the blood. The most common ketone bodies are **acetone, acetoacetate**, and **β-hydroxybutyrate**. Under certain circumstances (e.g., during starvation), other cells (e.g., brain cells) may utilize ketone bodies directly as energy sources. If the body is using a great deal of lipid for energy production, excessive liver ketogenesis results and can alter blood pH significantly. Ketone bodies are weakly acidic and their excessive production (**ketogenesis**) and subsequent elevation in the blood (**ketonemia**) leads to a lowering of blood pH or **ketosis** (acidosis). Thus, ketosis can reduce the ability of hemoglobin to deliver oxygen to the brain, leading to coma and eventually to death. Increased appearance of ketone bodies in the urine is termed **ketonuria**.

VI. SUMMARY

Overall metabolism is the sum of all anabolic and catabolic reactions occurring in the body. Intermediary metabolism consists of pathways involved in energy production from carbohydrates which are connected to pathways for protein and lipid metabolism such that excess carbohydrates can be converted to protein or lipid for storage or the latter can be used to supply carbohydrates for energy production via gluconeogenesis. The carbohydrate pathways include glycolyosis and the pentose shunt (anaerobic), the tricarboxcylic acid cycle, and the cytochrome system (which together with glycolysis and the pentose shunt constitute aerobic metabolism). Lipids (fatty acids, TAGs, cholesterol, phospholipids) are synthesized from carbohydrates and can be converted to carbohydrate via gluconeogenesis from glycerol and NEFAs (β-oxidation). Ketone bodies (ketogenesis) are formed during the β-oxidation of NEFAs. Amino acids derived from protein catabolism can be converted (by transamination) to carbohydrates and participate in gluconeogenesis. However, in many animals it is not possible to resynthesize all of the necessary amino acids from carbohydrate and some must be obtained through the diet (the essential amino acids).

Index

Note: Page numbers followed by "f" and "t" indicate figures and tables respectively

B

C

D

E

F

G

H

M

N

O

P

R

S

T

U

V